RESEARCH ON
PARTICLE
IMAGING
DETECTORS

World Scientific Series in 20th Century Physics

World Scientific Series in 20th Century Physics – Vol. 6

RESEARCH ON

PARTICLE IMAGING DETECTORS

Editor

Georges Charpak

CERN, Genevé, Switzerland

World Scientific
Singapore • New Jersey • London • Hong Kong

Published by

World Scientific Publishing Co. Pte. Ltd.

5 Toh Tuck Link, Singapore 596224

USA office: 27 Warren Street, Suite 401-402, Hackensack, NJ 07601

UK office: 57 Shelton Street, Covent Garden, London WC2H 9HE

British Library Cataloguing-in-Publication Data
A catalogue record for this book is available from the British Library.

The author and publisher would like to thank the following publishers of the various journals for their assistance and permission to reproduce the selected reprints found in this volume:

Italian Society of Physics (*Nuovo Cimento*); Organisation Européenne Pour La Recherche Nucléaire (CERN) (*CERN Reports*); Societe Francaise de Physique (*J. Phys. Radium, Rev. Phys. Appl., J. Phys.*); Society of Nuclear Medicine (*J. Nucl. Medicine*).

While every effort has been made to contact the publishers of the reprinted papers prior to publication, we have not been successful in some cases. Where we could not contact the publishers, we have acknowledged the source of the material. Proper credit will be accorded to these publishers in future editions of this work after permission is granted.

World Scientific Series in 20th Century Physics — Vol. 6
RESEARCH ON PARTICLE IMAGING DETECTORS

ISBN-13 978-981-02-1902-4
ISBN-10 981-02-1902-4
ISBN-13 978-981-02-1903-1 (pbk)
ISBN-10 981-02-1903-2 (pbk)

FOREWORD

I have always found it difficult to transform an idea about some new experimental way to investigate the marvels of the physical world into a reliable instrument.

I was fortunate to enter a field where the physicists were confronted with breathtaking interrogations on the infinitely small and infinitely large structures of our Universe, at the right time, when a clear road to progress was obvious because of new prospects for a substantial increase in the energy of particle accelerators.

Being an experimentalist, to my regret, I had no choice but to improve the instruments that were available in my laboratory. I was also fortunate to spend most of my professional life at CERN, the European Organization for Nuclear Research in Geneva, where I enjoyed perfect facilities and permanent contact with the best physicists in the world, experimentalists and theoreticians, who were competing in the field of high-energy physics.

I almost never started to study a detector without having to solve some problem for my own experiments or, very often, for the difficulties confronting the community in which I was working. In the selection of articles collected for this book, mainly describing detectors, I have included only a few on experiments in which I participated and which played a role in initiating some of my research on detectors. I have not hesitated to include some publications on detectors which did not take off, such as, for instance, an unrealistic high-pressure multitron air detector, imagined as a cheap megaton detector to address the fundamental problem of proton stability. This correlation between research on detectors and ambitious goals was an important element in my progress.

I have been attracted by some problems in medicine and biology, and have made attempts to build detectors for these applications. In this Foreward my aim is to give the reader a clue as to my apparently chaotic choice of fields of interest.

In 1948, I joined the laboratory of F. Joliot-Curie at the Collège de France, in Paris, being almost totally inexperienced in physics owing to serious disturbances in my life during the war.

For an experimentalist a big challenge was the relative backwardness of the equipment when compared with that in use in American or English laboratories which had undergone vigorous development during the war years, especially in the field of nuclear physics which benefited from the gigantic effort of the Manhattan Project. Geiger–Müller counters and cloud chambers were the instruments commonly used for the detection of particles.

Under the supervision of a slightly older physicist, F. Suzor, I started with an experiment relying on Geiger–Müller counters or proportional counters. Our first experiment was the study of the angular correlation between gamma-rays emitted in cascade by some nuclei, giving information on the spin of intermediate excited, nuclear states. Whilst this work introduced me to the capricious world of Geiger–Müller counters it, very soon, showed that we were wasting our time. The development of scintillation counters in the United States permitted our American colleagues to do the same experiments with an efficiency for gamma-rays about 100 times larger than that of gaseous detectors, and after a brilliant publication by M. Deutsch at MIT we left the subject.

We then switched to the construction of detectors, with quite novel features opening the path for us to intriguing physics problems.

F. Houtermans, a German physicist, had survived an extraordinary and dramatic story of successive emigration steps between Germany, the Soviet Union, and Switzerland; a physicist of deep originality he had developed large 2π wall-less proportional counters. We decided to build two such counters tangential to each other, with a radioactive source in the contact plane. Our idea was to study very low energy radiations coincident with the beta particle emitted in a radioactive decay. The instrument was unique in its capability of detecting ionizing radiation with a minimum detectable energy of a few tens of electronvolts, emitted very rarely in association with the beta electron.

This led us not only to an exciting and imaginary "big discovery" because of an artefact but also to the valuable measurement of some phenomena associated with the emission of low-energy radiation by radioactive atoms. It seems worthwhile to say a few words about the "big discovery". It consisted of the affirmation that the particles emitted by a nucleus undergoing a beta-decay were not electrons but were transformed into electrons after some time.

Our conclusion was based on the difference in the back-scattering properties on metallic supports, between the betas and the electrons. Both were emitted by radioactive sources, the electrons coming from excited long-lived isomeric states.

It appeared that our conclusions were derived from a systematic effect: the beta sources were really massless whilst the electron sources, being obtained from the evaporation of droplets of liquid, were concentrated in crystals with a thickness much larger, by orders of magnitude, than the average thickness calculated from the amount of material in the solution. It led us to a search for simple methods for making uniform deposits of radioactive sources and to a systematic study of the back-scattering properties of electrons by planar supports only to finally come to the conclusion that we were victims of an illusion. It is strange to think that a clean experiment on the scattering properties of betas and electrons would have led to a major discovery since the betas are totally longitudinally polarized because of parity violation in beta-decay, which was discovered six years later, and the scattering of electrons by nuclei is spin-dependent. But our experiment had no chance of observing such an effect.

The instrument was used for a systematic study of some phenomena giving rise to the emission of low-energy radiation by radioactive nuclei: let me quote the study of the double ionization of the K-shell of ^{55}Mn in the decay of ^{55}Fe by K-capture, the study of the shake-off electrons from the atomic shell of beta emitters, the study of some low-lying levels in nuclei after a radioactive decay, the study of the internal bremsstrahlung in beta emitters, the study of K-capture in ^{22}Na, which led me finally to the work of my thesis which encompassed most of these subjects.

Building the proportional counters and optimizing them for our experiments acquainted me with the theory and practice of gaseous detectors and especially proportional counters. I then had the good idea to replace the reading of the pulses produced at the wire of proportional counters by the detection of the light emitted by the avalanches. My work, undertaken with J. Renard, indeed showed the feasibility of this approach, but without a systematic study of the device we did not produce evidence of any advantage of the method.

Many years later, a Coimbra Group, led by A. Policarpo, made a big step forward while also starting with the detection of the light pulses from avalanches in proportional wire counters. They came to the conclusion, after a clean systematic study, that the interesting photons are emitted when no multiplication occurs, when the electrons accelerated in an electric field induce, in pure noble gases, inelastic radiative collisions without ionization. This improved the energy resolution of the detector by a large factor and produced a tool useful for applications in several fields of physics or astrophysics.

I did not pursue this research because I rightly thought I had found a far more interesting application of the light emitted by the electron avalanches in a gas. My interest in the light emitted by electron avalanches near the wire of a proportional counter had started when I watched the glowing of the avalanches around the wire of a Geiger–Müller counter built within a cylinder of low-resistivity glass painted externally with a conductive paint, the so-called Maze counter. By scratching away part of the paint the wire was visible. Above a certain rate the avalanches around a wire became visible in the darkness.

In 1956 I conceived a device where the avalanches would start from electrons liberated in the gas by an ionizing particle and would be amplified by an electric pulse, short enough to limit the avalanches to a very small length, and intense enough to give avalanches producing enough light to be visible, just like the wire of a Geiger counter at high counting rates.

An estimate of the size of the charged cloud liberated near the wire when it became visible and a realistic estimation of the maximum attainable size of an electron avalanche, which is about 10^8, led me to the right conclusion, i.e. that the avalanche would not be detectable with an ordinary optical system even with the most sensitive film.

I went to Hamburg to discuss this with Professor Raether, who was one of the highest authorities on gaseous electronics. In his laboratory, people were taking pictures of single avalanches with an image intensifier. I came to the conclusion that these avalanche chambers imperiously required such a device, which was not available commercially. While making long-range plans for the materialization of my idea, I built a preliminary device. It consisted of a high-voltage periodic source of very short duration made by charging up a capacitor bench and discharging it with a rotating device, a chamber made up of two plane electrodes, and the best optics and most sensitive films I could find.

I discovered that when I biased the electrodes with a small electric field, continuously eliminating the free electrons liberated in the gas by the intense alpha source deposited on one electrode, and applied intense pulses of the duration of a few nanoseconds, I could observe bright tracks, clearly following the trajectories of the particles.

This had never been observed. It is clear that a careful analysis of the results and a sizeable experimental effort would have led me to a substantial progress in instrumentation for particle physics. The door was open but I did not cross the threshold. Eight years later, after the development of the spark chambers by S. Fukui and S. Myamato, a Georgian physicist, G. E. Chickovany, overcame, with the same idea, the problems of light intensity. He produced limited streamers, i.e. strings of many avalanches along the applied electric fields, and later, with the progress of image intensifiers, beautiful avalanche chambers were built, in the US and in Europe, giving the best images of some of the most complex patterns in particle physics. My own work made no contribution to the subject. It was ignored. However, it played a major role for me since I presented it at the Padova Conference on Particle Physics in 1957, where I got acquainted with the latest problems in high-energy physics and where L. Lederman, after listening to my talk, proposed that I join a team he was setting up for an experiment at CERN in order to measure the anomaly factor in the magnetic moment of the muon.

I accepted with enthusiasm and came to CERN with the high-voltage radar modulator built for me to pursue my work on the sparks, but for three years this pulser was used to flip the spin of muons stopping in a graphite target imbedded in a magnetic pulsable coil.

It was, for me and most of the team, a period of intense learning. We had the privilege of working with physicists who had pioneered the nascent field of muon physics in the USA, such as L. Lederman, R. L. Garwin and V. Telegdi. We had the task of solving difficult experimental problems, for instance storing muons produced in the decay in flight of pions, in a magnet, for thousands of orbital revolutions, and measuring, as a function of the number of turns, the orientation of the muon spin direction.

The muons were coming from the decay of pions produced by the first accelerator built at CERN, a synchro-cyclotron accelerating protons to 600 MeV. They were fully longitudinally polarized, because of maximal parity violation in the weak forces, as demonstrated by some of our American co-workers; and, after each revolution in a magnet, their spin rotated, with respect to the direction of flight, by an amount equal to $2\pi\alpha$, where α is a small number, of the order of 10^{-3}, called the anomaly of the magnetic moment.

It was possible to measure the direction of the magnetic moment, collinear with the spin of the particle, but the accuracy of the final measurement was a function of the number of revolutions experienced by the muons. The game consisted in storing the muons in a magnet for a maximum number of turns, with a limitation coming from the finite lifetime of the muon, which is about 2.2 microseconds.

The stakes appeared to us to be very high. The muon is a particle with a mass of about 210 times that of the electron, showing no sign of a specific interaction not experienced by the electron which could have explained the considerable difference in masses. The measurement of the magnetic moment appeared as a sensitive probe for the search of this specific interaction. It had been measured with an accuracy of 10^{-9} for electrons and the agreement with the theoretical calculations, based on quantum electrodynamics, was at that level. Any disagreement with the predictions of quantum electrodynamics which we could have found in our measurement would have permitted us to imagine a new type of interaction for the muon. We brought the accuracy down to a limit of about 10^{-5} and the agreement with quantum electrodynamics was perfect, giving ground to the idea that Nature, for some mysterious reasons, was built with electrons of various masses. It was discovered later that a third generation of much heavier electrons exist, the tau (τ), with a mass of about 1.7 GeV, that each of the 3 "leptons" is associated with a specific neutrino, and also that this is the maximum number of lepton generations allowed by Nature, as beautifully demonstrated by the most recent experiments at CERN, with LEP (the Large Electron–Positron Storage Ring).

In the course of our measurements, we probed some properties of the muons, such as the upper limit of the electric dipole moment, the quenching of the polarization of muons stopping in materials in weak magnetic fields. But when our main goal, the high-accuracy measurement of the anomalous magnetic moment had been reached it was clear that a big new step in the accuracy could be achieved only by a major effort, storing muons of an energy well above that permitted by the synchro-cyclotron.

A fraction of our team, led by F. Farley, decided to design an experiment at the newly built CERN Proton Synchrotron, of 20 GeV, which marked an important step in the progress of CERN as a laboratory in the forefront of high-energy physics. Some very talented experimentalists, such as E. Picasso, joined the new team. As expected, they considerably improved the accuracy we had reached, confirming that the muons were indeed showing no evidence of a specific interaction different from that of electrons.

Feeling considerably strengthened by their experience in that experiment, several European members of the group, for example H. Sens and A. Zichichi, decided to split away and follow their own lines of inspiration while I decided to go back, for some time, to particle-detector research.

The spark chambers were flourishing in the experimental areas of accelerators. The bubble chambers had become a master tool in particle physics. With a sensitive volume of several cubic metres of liquid and an accuracy of the order of 100 microns in the determination of each visible bubble along a particle trajectory, they permitted the discovery of hundreds of new particles, which were indeed not elementary. This laid the basis for the discovery of the underlying symmetries which brought the theoreticians to the modern theory of Quantum Chromodynamics.

In the experimental areas around the accelerators we could for 20 years, at intervals of about 10 seconds, hear the deafening sound of the pistons of the bubble chambers: the pictures taken after each expansion were showing, with exquisite accuracy, the interactions of the beam of particles crossing the volume of the chambers imbedded in the field of large magnets curving the path of the trajectories, thus permitting the momentum of the particles to be obtained.

A maximum of 20 incoming charged particles was allowed in one spill, in order to avoid a jamming of the tracks on the film. The bubble chamber had no "memory". It was necessary to send the spill of particles into the chamber, after it had been made sensitive, for a time of the order of a millisecond, by the expansion of the fluid necessary to bring it momentarily to the superheated state.

With about 10 million pictures taken every year, the high-energy physics community around the world was reaching the maximum possible amount of photographs which could be reasonably treated by human examination of pictures. A drastic effort was being made to develop automatic scanning methods of the pictures in order to limit the amount of visual scanning to a minimum, but some questions in particle physics emerged which required the investigation of billions of interactions, or demanded masses of detectors of hundreds or thousands of tonnes, incompatible with bubble chambers.

The spark chambers developed by S. Fukui and S. Myamoto, with their perfect adequacy for particle-physics problems, as demonstrated by J. Cronin, entered impetuously into the experimental field of high-energy accelerators. They were blessed by the property of a built-in memory. It was possible to store the information of ionization electrons liberated in a gas by charged particles for about 1 microsecond. This permitted a decision on the application of a high-electric-field pulse producing the spark at the spot where the charged particles had crossed the gaps of successive chambers. Hundreds of thousands of particles per second were thus tolerable in a chamber, compared with 20 per burst in a bubble chamber. The overall picture of a complex interaction was very gross when compared to the one given by a bubble chamber, but a new class of experiments had become possible. Fast scintillation detectors, with resolution times of the order of 1 nanosecond, were used to select interactions imbedded in a large flow of uninteresting ones, and triggered the spark chambers, which for a microsecond were keeping the information left behind by the ionizing particles crossing the spark-chamber gaps. This led to major discoveries, such as the existence of a second type of neutrino associated with the muons. But this technique reached the same limit as the bubble chambers: the bottleneck of millions of pictures to analyse.

This is when I left the successfully completed measurement of the anomalous magnetic moment of the muon. I then thought of a method for the retrieval of the coordinates of a spark in a spark chamber, without having to take a picture.

A similar idea occurred simultaneously to several other researchers. They found solutions different from the one I reached, such as the sonic spark chamber or the wire spark chamber. This latter solution was the best adapted to high-energy physics.

I introduced the localization of the sparks by two methods: the building of electrodes with a delay-line structure permitting the spark position to be obtained by measuring the delay of the pulses generated by the sparks, and the measurement of the ratio or difference in the currents generated in spark-chamber electrodes at different points of an electrode. It is worth noting that these methods are at the heart of different devices introduced later, by myself and other groups, to retrieve the data from multiwire chambers used for the imaging of neutral radiations.

In order to study the properties of these automatic chambers I decided that it was appropriate to build a particle-physics experiment. It was the best way to get access to financing. The running-in of the Proton Synchrotron of 20 GeV had decreased the pressure for machine time at the Synchro-Cyclotron of 600 MeV. Under the influence of a nuclear-physics theoretician, M. Jean, from the Institut de Physique Nucléaire d'Orsay, I decided to study the (π^+2p) reactions in nuclei. This permitted me to enjoy the collaboration of a team of good experimentalists. We built an experiment with dozens of automatic gaps of spark chambers. We obtained results which seemed to interest the nuclear-physics community and gained considerable experience in the readout of electrodes by "current division". Several groups adopted our method in the equipment of the focal planes of spectrometers for the study of nuclear reactions.

I personally continued to investigate some intriguing problems in particle detector physics, enjoying the strength of the group which had been built up for the study of high-energy nuclear reactions.

With L. Dick and L. Feuvrais we showed that it is possible, using the same idea that permitted the localization of sparks in a delay-line spark chamber, to measure the position of the impact of a charged particle in a scintillator by exploiting the velocity of the scintillation light. This was made possible by the great expertise in the measurement of short times reached in L. Dick's group. This is now a routine method in many experiments dealing with large-size scintillators and is often used in order to correct for the time fluctuations introduced by the finite size of the scintillators.

At the same time, together with L. Massonnet and J. Favier, I tried to come back to my original idea of 1956 on the possibility of building visible avalanche chambers. With transparent electrodes, we observed beautiful tracks nearly parallel to the electrodes but with a high impedance discharge connecting the electrodes.

We started to work on triggering pulses short enough to permit the interruption of the streamers connecting the electrodes in order to obtain a three-dimensional picture of a track in one single gap. We reached our goal, but were preceded by a group of talented Georgian experimentalists led by G. E. Chickovany.

In large-gap chambers with a surface of the order of a square metre, he succeeded in obtaining interrupted streamers permitting beautiful pictures of some of the most complex events, such as high-energy electromagnetic showers. His work was completed by the work of the team of B. Dolgoshein, in Moscow, who improved the pulsing system which was at the heart of the success of the method. A large international effort was undertaken by groups attracted by the possibilities of these detectors to image the most complex events in high-energy interactions. Our work was again ignored, probably rightly so.

I then investigated several detectors which had no applications. I developed a method for localizing the photon impact on the photocathode of a photomultiplier. I also invented a bubble chamber free of the limitations of the normal ones: bubbles of gas were continuously produced in a liquid. The ionization produced in the bubbles by charged particles was used as the seed for a discharge produced by high-voltage pulses. The photography of the glowing bubbles was yielding the particle trajectory. It worked ideally with all the advantages of a spark chamber but, alas, as the density of bubbles grew in order to reach good resolution, the medium became opaque. I could not find practical conditions of pressure and temperature which would have permitted the same index of refraction for the liquid and the bubbles, and gave up because of the lack of practical interest.

Fortunately, I had found a more fruitful use for the energy I was deploying in my search for new detectors. Two methods of automatic readout of spark chambers had been adopted: the wire spark chambers with a ferrite core readout of every wire and the wire chambers with magnetostrictive readout of the wires carrying currents. Many experiments with successive spark-chamber gaps, of surfaces of tens of square metres, were studying reactions at rates of up to a few hundred per second. The memory of the chambers, which was reduced to a fraction of a microsecond, permitted the study of rare events and the accumulation of millions of events stored in the computers, which had started their exponential development. However, there was a limitation which could not be overcome: it is impossible to trigger spark chambers at rates above 1 kilohertz without again starting a spark at the location of old sparks. Several researchers thought it would be advantageous to operate the sparking gaps at low amplification in a proportional mode. This would give amplification factors a million times smaller than the one reached with a spark initiated by one single ionization electron, but the prospects opened up by the revolution in transistor amplifiers permitted the conception of an idea of a combination of moderate electron multiplication in the gas with a large quantity of low-cost, low-dimension amplifiers. This problem was superficially addressed by several groups. One mistake was the use of single-gap, wire spark chambers to study the limits of proportional amplification. Even with the best amplifier a sparking mode was reached at some places in the gap as soon as proportional amplification occurred in the volume of the gas, having a destructive effect on the amplifiers.

The physicists who came the closest to the solution used, correctly, planes of anode wires sandwiched between cathode planes, but, scared by the capacitive coupling between wires they used an electrostatic shielding between the wires. In the best case it consisted of a thick field wire between the anode wires thus limiting the accuracy. What was missing was the understanding of the pulse formation mechanism in a multiwire proportional chamber: the pulses on the anode wires are not due to the collection of the electrons of the avalanches, but to the motion of the positive ions inducing pulses of opposite polarity on the neighbouring wires.

With the experience I had gained in the construction of single-wire proportional counters I conceived a small chamber of 10×10 cm^2. It consisted of a plane of wires of 20 microns diameter sandwiched between two cathode planes. With the charge per unit length on each anode wire correctly computed, and with gases commonly used in single-wire counters, each wire had to work under the same conditions as in an ordinary single-wire counter. As a precaution, a guard strip separated the anode wire planes from the cathode planes at the edges in a way inspired by the field guard ring equipping single-wire proportional counters. When I tested the chamber with the 5.9 keV source of ^{55}Fe, I observed comfortable pulses on the wires. The avalanches collected on a wire generated positive pulses on the neighbouring wires, making it easy to identify the wire collecting the avalanche. The negative pulses were indeed due to the motion, away from the wire, of the positive ions moving in the intense field close to the wire. It was obvious to me that the collection of the electrons of the avalanches was producing an undetectable pulse under my experimental conditions.

The chamber delivered pulses proportional to the energy liberated in the gas. No sparking problem occurred and a high gain was possible, which was essential since the minimum ionizing particles crossing the chamber were losing an average of about 1 keV per cm gap and a threshold of detection of only a few ion pairs was necessary to obtain 100% efficiency. In other words, the multiwire chamber worked like the best old single-wire counter.

Testing with minimum ionizing particles immediately showed that the delay in the appearance of the pulses was related to the distance of the charged particle from the wire and that a new class of detectors, the "drift chambers", could be conceived where the position of the particle is determined by the time taken by the ionization electron to reach the amplification region close to the wire, at a distance of the order of the wire diameter.

The first measurements, in 1967, showed that counting rates close to 10^5 pulses per second could be reached on every wire with a resolution time of the order of a few tens of nanoseconds. Orders of magnitude were thus gained over the best imaging electronic detector, the spark chamber. Together with an enthusiastic community of experimental physicists, we decided to develop the detector and to bring it to the stage where large experiments corresponding to the needs of high-energy physics could be safely designed. Some of these developments were done with my team: some were done by groups working mostly at CERN and wanting to address ambitious problems with the new detectors. They tried, independently from us, to solve the problems connected with the transition from a 10×10 cm^2 counter to the many square metres required for the experiments.

In 1968 and 1969, I paid special attention to the drift chamber. I enjoyed the collaboration of D. Rahm and F. Steiner. We showed that accuracies better than 100 microns were easy to reach and that the combination of drift spaces and proportional bi-dimensional wire chambers opened up the prospect of a large-volume, three-dimensional chamber. But this powerful instrument was not adapted to the demands of the new accelerator built at CERN, the Intersecting Storage Rings (ISR). It is adapted to the conditions of the electron–positron colliders, and it is in this environment that D. Nygren constructed one of the most beautiful particle imaging detectors, the Time Projection Chamber, introducing original features which permitted it to be adopted universally.

During the same period, I tried to clarify the considerable potential of the proportional amplification approach in wire chambers: for high-rate data acquisition, for accurate low-cost measurement in large-surface detectors with drift chambers, for the retrieval of information from wire planes with analogic methods such as delay-line or current-division techniques. This was essential for simple neutral radiation imaging; and several authors improved the method, took out patents, and contributed to the use of wire chambers in several important applications.

In 1969, our former collaborator, J. Saudinos, started an ambitious project to equip the focal plane of particle spectrometers with electrons drifting over 20 cm before reaching the detecting wire. This required a small number of amplifiers and demonstrated clearly that the drift method could have been very useful before the transistor revolution, had it been discovered earlier.

We found that the detection of the pulses induced on segmented cathodes permitted a two-dimensional readout in single planes, which was essential for the imaging of neutral radiation, X-rays or neutrons, but also had interesting implications for relativistic charged particles.

I then divided my activities into two directions. At CERN, it was decided to build a large detector in a big magnet, the Split Field Magnet (SFM) conceived by J. Steinberger, at the intersection point of the newly built ISR at a beam energy of 20 GeV. About 50 000 wires were envisaged, with chambers requiring wires of a length larger than a metre. Many technical problems had to be solved: electrostatic stability of a structure made of honeycombs and fragile wires of 20 microns diameter, resistance to occasional sparks, maximum amplification permitting the use of the lowest cost amplifiers. The contributions of the high-energy physicists attached to the SFM project, which was led by A. Minten, have been of utmost importance.

We discovered new gas fillings, such as the one with four components, argon, Isobuthane, Methylal, freon-13, permitting considerable gains, in a saturation mode without entering into the Geiger–Müller or spark-mode operation. It was, for a long time, called "magic gas". We discovered later that the operation mode was totally original and ignored by the physicists who had developed the old single-wire proportional counters: it was the "limited streamer mode", where a series of avalanches develop in a string, starting from the anode wire, the string stopping at a great distance from the wire, short of reaching the cathode planes, which would lead to a spark, dangerous for the health of the wires or the amplifiers.

At the same time, a small group continued a systematic study of some possible important applications of the proportional chambers for particle physics, such as the particle separation by means of cascades of chambers which became the theme of the thesis of Z. Dimcovski. With F. Sauli, who joined our group after building a remarkable system of magnostrictive spark chambers, I started a study in several directions: operation of wire chambers in the Geiger–Müller mode; operation of fast parallel-plate counters for relativistic particles. With all the experience acquired by us in 1970 on the best quenching gases, we showed that parallel structures operating at atmospheric pressures could be operated with remarkable time resolutions of 5 ns. This did not threaten the wire chambers, which were much easier to operate in large surfaces. This experience was of value when, 15 years later, with F. Sauli we started a fruitful extension of these chambers with the multistep avalanche chambers.

We had succeeded in the construction of the large detector equipping the SFM. We had solved some practical problems that were essential for the successful, reliable operation of large chambers, such as finding gases showing no ageing problems in high-rate beams. Many other groups in the world undertook similar developments, often with original additions. C. Rubbia pioneered the field of large-surface drift chambers. When a group from Saclay wanted to copy his chambers for a gigantic neutrino detector, they asked me to collaborate with them to clarify some intriguing observations on anomalously large pulses. This study brought us to a deeper understanding of the so-called "magic gases". This was characteristic of the activity of our group and led me to give up any participation in a large experiment since I felt attracted by the many unsolved problems concerning gaseous detectors. I profited from the help of young physicists willing to invest their time in the study of new instruments and to join my group, which was stably established with F. Sauli and three precious technicians, R. Bouclier, G. Million and J.-C. Santiard, later reinforced by R. Benoit.

With A. Breskin, from 1971 to 1973, the study of discharge chambers with infinitely transparent electrodes was my last effort to operate chambers in sparking modes. With F. Sauli we built up an ongoing activity in the upgrading of wire chambers and drift chambers. We pushed to the limits the accurate measurement of the coordinates of an avalanche along the anode wires, and this has had a strong influence on the development of detecting systems in many laboratories all over the world. We clarified the geometrical shape of the avalanches or strings of avalanches around a wire and showed that it is possible to determine the azimuth of an avalanche around a wire.

This had consequences in some applications. For instance, it permits us to obtain, in drift chambers, the side of the development of an avalanche on an anode wire, the lifting of the so-called right–left ambiguity. It permitted us to transform a wire chamber into a detector with a continuous response in the direction orthogonal to the wire plane. F. Walenta, who in 1971 started publishing important work on drift chambers, independently came to the same conclusions at Brookhaven National Laboratory.

I started to be interested in the application of gaseous detectors to fields outside particle physics. The work of V. Perez-Mendez at Berkeley had given important results in the field of macromolecular structure studies by X-ray diffraction methods. After a discussion with R. L. Mössbauer, who expressed his interest in a gaseous detector but complained about the intrinsic defects of the wire chambers, I came to the conclusion that a spherical drift chamber, with a detecting space placed between a spherical entrance window and a concentric spherical grid for the exit of ionization electrons, was an ideal structure, free of the main defect of gaseous detectors, namely the parallax error from inclined tracks. It led to a detection system which is actively used with the synchrotron radiation facility at the Laboratoire d'Utilisation du Rayonnement Electromagnétique (LURE) at Orsay, and in many other places.

In 1974, in collaboration with a group of physicists from Denmark, we started to use high-accuracy drift chambers for the study of the channelling of high-energy charged particles at energies above 1 GeV. We observed beautiful channelling properties. We also developed high-accuracy drift chambers for a group at Fermilab, who started similar experiments based on a brilliant idea by a Soviet physicist, A. Tsiganov, of bending the crystal in order to use the high electric crystalline field to deflect high-energy particles. It was a spectacular success. Our Danish colleagues have since improved the method to the point where it is possible to bend particles of hundreds of GeV energy with the practical application of replacing heavy magnets by a simple bent crystal. Using the same drift chambers, this group has also discovered a whole range of new phenomena connected with particle channelling in crystals.

In 1973, V. Perez-Mendez had shown that with drift spaces filled with appropriate lead converters it was possible to substantially increase the efficiency of gaseous detectors for high-energy gamma-rays. In 1975, A. P. Jeavons, at CERN, was interested in developing high-efficiency detectors for the measurement of the angular correlation of the 511 keV photons emitted in the annihilation of β^+ particles stopping in solids. Beginning his work with our group, his systematic study of the properties of a solid channelling structure of heavy materials in a drift space led to an instrument which has been very successful in solid-state physics. The position resolution of about 1 mm coupled with an efficiency close to 10% was a unique feature in this field. The poor time resolution, owing to the slow drifting of electrons in the converting tubes, has excluded this instrument from being a candidate in the important field of Positron Emission Tomography, where the scintillation counter has so far been widely adopted, despite a much poorer position resolution.

In 1974, my group, together with J. Saudinos from Saclay, undertook the exploration of the human body with high-energy particle beams. With wire chambers giving the position of incoming and outgoing particles, we had the idea that it was possible to map the three-dimensional density distribution inside a living body without having to rotate the source or the body or the detector. The method presented several specific advantages over X-ray imaging because of the very different dependence of the nuclear cross sections as a function of the nature of the atoms. It was, for instance, very easy to separate the reaction on hydrogen from that on other nuclei. After obtaining good results on a human head and on some thick metal slabs, we interrupted the project since it was clear that no low-cost accelerators of a few hundred MeV were in view and that it was obviously more interesting to invest in other methods of investigation of the human body. But, with the advent of the development of accelerators for therapeutic goals in the treatment of some cancers, it may become useful to use a small fraction of the beam for the visualization of the treated body by our method.

In 1975, an article on the scintillating drift chamber illustrated the diversity of our research. It was co-signed with S. Majewski from the University of Warsaw, who for many years played an important role in several original developments for our group.

A thesis undertaken by G. Schultz in our group was an opportunity to clarify the phenomena connected with the exchange of charge between the positive ions liberated in the gaseous detectors in the complex mixtures commonly used. It justified several empirical findings.

While refining our understanding of the avalanche mechanisms in wire chambers until 1978, our group undertook a work which led us to develop gaseous detectors with electronic amplification but with radically different structures. It started from a discussion with L. Lederman, who complained that wire chambers had reached the limit in their capacity of handling particle flux from accelerators of growing intensity. The slow positive ions of an avalanche introduce a local dead time of about 10^{-4} seconds extended over the avalanche width. In the search for rare events the increasing intensity of background radiation paralysed the chamber. In 1978 we undertook an approach that proved very fruitful. The idea was to reach the final electron multiplication required in a gaseous detector in two successive steps, with an adjustable delay combined with a gating device introduced before the injection of the swarm of electrons produced in the first step.

This delay, produced by drifting the first swarm of electrons in a gas, permitted the taking of decisions with fast detectors and the triggering of a gating wire plane accepting the swarm of electrons if it was associated with a rare event identified by these detectors.

We started from the so-called hybrid spark chambers, developed in Brookhaven in 1970, where the group of J. Fischer, who had played an important role in the development of wire spark chambers, had tried to combine proportional amplification on a wire with a second delayed step, consisting of a spark chamber. This group wanted to use the considerable experience accumulated in the cheap readout of large-surface filmless spark chambers with the advantages of wire proportional chambers.

We had been intrigued by their observations, which could not be explained by their theory, of electrons being multiplied around a wire, then migrating away from the wire in a low-field drift space. It caused us some effort to understand why their interpretation was wrong; and S. Majewski, who had returned to work at the University of Warsaw, collaborated with us on the clarification of the process.

We discovered that the first multiplication step was obtained in the constant field between two parallel electrodes. The positive electrode is indeed made of wires or of an interwoven wire grid, but the amplification does not occur around the wires and it is easy to extract a constant fraction of the electrons of the avalanches to a drift space following this amplification gap. It led us to very much improve our mastering of the operation of detectors free of wire amplification. The amplification between parallel grids in successive steps permits the easy detection of single electrons liberated in a gas by any ionizing process, with much higher permissible counting rates than with wire chambers.

We were also led to investigating the coupling of secondary emission surfaces with gaseous electrons as another alternative for the construction of multistep detectors. The group of J. Saudinos at Saclay was having success in the detection of relativistic particles with thin, low-density layers of substances such as CsI deposited in a foam-type structure. They obtained a few electrons ejected from a thin layer for every incoming relativistic particle. Our idea was to accelerate these electrons, over a few centimetres, in a vacuum and then to transfer them to a gaseous detector separated by a thin window transparent to the accelerated electrons. Our first results were quite encouraging and I am convinced that this device can be useful for more applications, but our success with the multistep chambers led us to abandon this method for multistep detection.

Whilst our work on secondary emission layers had to be done in a vacuum, recent work by the group of A. Breskin, of the Weizmann Institute in Israel, has shown that the secondary emission layers are compatible with low-pressure gaseous detectors, which further extends their field of application.

In 1980 we showed that the multistep chambers offered an elegant solution for the imaging of betas emitted by surfaces such as the gels used in chromatography. Exploiting the fact that in a parallel gap the electrons close to the entrance cathodic surface experience maximum amplification, we found that it is possible to get rid of the effect plaguing the normal wire chambers that were tentatively used for this imaging, namely the large range in gases of the electrons emitted by most radioactive emitters. We reached an accuracy of 0.5 mm (full width at half maximum) for nuclides such as ^{3}H, ^{14}C, ^{35}S, ^{32}P. Two years later we applied the same principle to the imaging of X-rays in thick planar gaseous detectors free of parallax.

In 1970 we had the privilege of starting to co-operate with A. Policarpo of Coimbra University, who had, in Portugal, developed the proportional gas scintillation counter based on the emission of vacuum ultraviolet (VUV) photons by ionization electrons drifting in an electric field excluding electron multiplication. His experience on scintillating gaseous detectors led us to several developments based on the emission of VUV light by electrons drifting in pure gases and was also precious in the study of the phenomena we were trying to clarify. The multistep avalanche chamber is ideally suited for the detection and localization of single electrons liberated in a gas.

After the major work of J. Séguinot and T. Ypsilantis of CERN, which had shown that VUV photons are easy to detect, with a high efficiency, in multiwire chambers with an appropriate vapour filling, we coupled the multistep chambers with the scintillation chambers and obtained interesting results. We investigated some possible applications for gamma-ray imaging together with the group of H. Nguyen Ngoc at Orsay.

But our main effort, at CERN, was to develop our detectors in the direction of an efficient imaging of the Cherenkov light for the identification of relativistic particles, which was of prime importance for high-energy physics. It finally led us to a large collaborative effort with a group at Saclay and one at Fermilab in order to equip an experiment with a system of particle identification by the imaging of Cherenkov radiation. It was the first application of the multistep chambers in a high-energy physics problem and permitted the Cherenkov Ring Imaging method (RICH) to be applied to a high-rate experiment.

While studying instruments for the imaging of the Cherenkov light emitted by relativistic particles travelling in various radiators, it was clear to me that a considerable potential existed for many applications if we could extend the method to the many fields where the imaging of photons is of prime importance. This is why I became very interested when a young American physicist, working at Los Alamos, wrote to me about his projects.

He had found a prolific stock, in the US Army, of tetrakis(dimethylamine)-ethylene vapour (TMAE), which had the lowest ionization potential, 5.36 eV. In the liquid form it was easy to obtain in large quantities. The vapour pressure was easy to control at moderate temperatures and it had a considerable quantum efficiency in the VUV range. D. Anderson had the idea of using condensed layers of TMAE, since the ionization potential was lower than that of the vapour, and to exploit this property for the detection of photons from a heavy scintillator BaF_2. A French group, at Grenoble, had just shown that this scintillator could emit a light component at a wavelength of 220 nm, which could hardly be detected by the TMAE vapour but was perfectly suited to the liquid layer, which had a detection threshold of about 1 eV lower than the vapour.

I invited him to join our group at CERN. He came with his counter and found the support of a group which had some expertise in gaseous detectors. He discovered with us that in high-energy physics the small overlap between the emission spectrum of BaF_2 and the range of sensitivity of TMAE vapour was sufficient for the conception of a new style of fast electromagnetic calorimeters in the GeV range. He also discovered, while in our group, an intriguing new phenomenon, i.e. of the considerable increase of quantum efficiency of metallic photocathodes when they adsorb a few atomic layers of vapours such as TMAE.

Because of the potential importance of the new properties of these calorimeters for the future high-luminosity calorimeters foreseen for the year 2000, our group invested a great part of its energy in the construction of a prototype BaF_2 calorimeter. A high-energy physicist, P. Miné, led the experiment to a stage where most advantages and disadvantages of the method became clear. We also made some investigations on the applications of multistep chambers for the imaging of thermal neutrons.

In 1980 we started to collaborate with a laboratory, at the Ecole de Physique et Chimie, in Paris, on electrostatic imaging particles. Our work consisted of the combination of gas amplification and the capture of the charges on insulating surfaces read out by various original methods. We obtained a crop of intriguing results, which have never been exploited.

In 1982, with S. Majewski, we showed that thin wire chambers, of 2 mm thickness, could be built which have important applications for the construction of large-size high-energy calorimeters. This work resulted in an important instrument used in one of the very large detectors equipping an intersection of LEP.

However, for most of our group, it was clear that the coupling of gaseous detectors with photon-emitting radiators was a promising project. We were enriched by the coming of a Russian physicist, V. Peskov, who had worked in Moscow in the laboratory directed by P. Kapitza. He had been active in the characterization of hot plasma by detecting photons or X-rays and he had developed, in competition sometimes with us or other Western groups, many original gaseous detectors, totally unnoticed since the articles about them were mostly published in Russian.

For instance, only three months after J. Séguinot and T. Ypsilantis, but independently, Peskov began to use photosensitive wire chambers. He found new photosensitive vapours compatible with the chambers. But, as is clear from our work, the spreading of new techniques in detector physics is strongly coupled to the environment. Coming to CERN, Peskov found an opportunity to make important new contributions by bringing his experience to several groups.

He came with his own bottle of ethylferrolene, a liquid made of a substance with a low-ionization potential. It was not as suitable as TMAE, which was not available in Moscow, but much easier to handle. It reinforced our idea that it was absurd to rely on two or three vapours for such important applications and that it was worth making a systematic investigation based on a less empirical approach. So far this has failed. But, as can be seen from some articles, the investigations clarified some points and were finally a step forward. Some were done in collaboration with the pioneer group on gaseous photodetectors led by J. Séguinot and T. Ypsilantis.

An important step was made when it was shown that solid photocathodes, compatible with high-amplification gaseous detectors, could be used in a range of photon wavelengths where many applications could be foreseen. V. Peskov who had, in Moscow, shown that Cs and CuI photocathodes are compatible with large-gain gaseous detectors, observed when in our group that CsI photocathodes are compatible with gaseous amplification at large gain, without noticeable feedback problems.

The group of A. Breskin came independently to the same conclusion. But the real sensation, which started a widespread interest in this field, came from the observation, made in the group of J. Séguinot and T. Ypsilantis collaborating with V. Peskov, that a considerable quantum efficiency can be achieved by CsI, much higher than that observed in the first investigations. It took several years of work by competing groups to arrive at the conclusion that these high values were true. This opens the way to a new class of gaseous detectors, especially if further research can extend the sensitivity of the photocathodes towards the visible light and keep the compatibility with gas amplification.

While these studies were being developed by a kind of internal logic, each piece of progress inducing new questions which we tried to answer by new experiments, the CERN community was exerting specific pressure through questions arising from the evolution of high-energy physics.

For the future accelerators, there was a need for detectors with a high accuracy and granularity capable of analysing reactions where hundreds or thousands of particles are produced simultaneously, with resolution times in the nanosecond range and capable of resisting irradiation levels at which practically all known detectors die after a very short time. We joined the efforts of a large project, the Lepton Asymmetry Analyser (LAA), directed by A. Zichichi, and could investigate, thanks to their support and collaboration, several intriguing and promising questions: the use of BaF_2 scintillators coupled to gaseous detectors for fast electromagnetic pre-shower counters, the search for new scintillators, the search for high-rate, high-granularity imaging gaseous detectors.

With F. Sauli, I tried to address the last problem using tubes containing hundreds of single-wire cells, but a new counter, introduced by J. Oed, made of printed strips on insulators, captured the attention of the high-energy physics community. It has a counting-rate capacity exceeding that of wire chambers by, at least, one order of magnitude, and for the millions of channels foreseen for future accelerators the construction based on technologies used for the transistor industry seems to promise a higher reliability.

However, it does not seem that all problems have been solved. This is why I have been tempted, with other groups, to explore, mainly together with I. Giomataris, multiwire structures where the positive ions have to travel only a fraction of a millimetre, and we have then reduced the dead time due to the space charge of the slow-moving ions by at least one order of magnitude with respect to the most conventional wire chamber. I believe it is an easy way to adapt the wire chambers to the growing luminosity of accelerators.

Another demand in high-energy physics arises from several problems where the challenge is not coming from the characteristics of the new accelerators. The instability of the proton, predicted by modern theories, requires detectors of thousands of tonnes. Clever detectors, using the Cherenkov effect in water, have been built, up to about 10 000 tonnes in weight. The result on proton decay was negative but they had the unexpected privilege of observing the neutrino flux from the implosion of a supernova in 1987. It is clear that it is of the highest interest to improve this kind of detector.

Several other problems, such as the neutrino flux from the Sun and the measure of the neutrino mass, would greatly benefit from the existence of gigantic detectors with the best quality in pattern recognition, energy resolution, and time resolution. This is why I played with detectors which may seem of purely academic interest. The high-transparency, high-density spark chambers were inspired by this kind of research, as was as a gigantic air detector: dreaming of a megaton of compressed air, in a tank immersed in deep water, with millions of electronic cells measuring localized energy losses of a few MeV, led me to the study of a pulse ionization chamber in air. I discovered that it is pretty easy to detect, event by event, alpha particles of a few MeV in air and that a 1 GeV energy deposit of a decaying proton was not an impossible task to imagine, but I was still capable of appreciating that I would convince nobody to join me and finance such a project and that the economy made by replacing argon by air was totally negligible compared to the total cost of the instrument.

One of the most intriguing problems in our world is to find the nature of the missing mass of the Universe. Observations in astronomy point to the fact that the majority of the matter in the Universe is not made of visible galaxies. Many competing candidates are envisaged and some appeared to be detectable by gaseous detectors. However, it is clear that in the search for rare events the wire-chamber system suffered from a big disadvantage when compared to some of the detectors used at the beginning of nuclear physics or particle physics: cloud chambers or bubble chambers were capable of identifying a new phenomenon in one or a few pictures because of considerable redundancy in the information, just like a high-definition picture of a crowd sometimes permits the identification of one individual. Electronic detectors suffer from a cost-limited number of channels. This is the reason why I started to study the imaging of avalanches in gases by the emission of light by the excited atoms. After patient work, our group found out that, with intensified digital cameras and with avalanches in an electric uniform field, in some selected vapours it was possible to obtain beautiful images of ionizing events in a gaseous volume. M. Suzuki and W. Dominik played a leading role in this development.

An English group, independent from us, aiming at the photography of single Cherenkov photons in a complex nuclear reaction, obtained the same initial results but did not pursue their efforts.

The detector was developed in several ways: in Japan, M. Suzuki showed that it permits the building of a new class of neutron imaging cameras, and a group at CERN demonstrated that it is a valid approach for problems where an extremely high multiplicity of particles requires many channels of information. A young American physicist, N. Solomey, spent several years in our group. He invested his expertise in this project. Together with a group from Geneva University he demonstrated the quality of this approach for the study of reactions with a high multiplicity. He had earlier developed, in our group, a remarkable instrument: a multistep drift chamber, with electrons drifting at right angles to the avalanche multiplication, which may well have interesting applications. His stay in our group was also an opportunity to clarify the amplification mechanism in gaseous detectors of various structures. We discovered that it is pretty easy to design detectors where it is possible in a single gap to have gain in a uniform field combined with multiplication in the vicinity of the wires of the plane limiting the gap.

But optical imaging detectors introduced me to a new field. We found that it was a good way to image the starting point of a beta emitted by a radioactive element imbedded in compounds used routinely for many problems in biology; and with W. Dominik and N. Zaganidis we started a fruitful collaboration with biologists.

It may be that optical imaging still has more interesting applications in particle physics. It opens the way to continuously operating gaseous detectors, of a large volume or a very large surface, capable of visualizing complex events. One can imagine how to obtain the projection of an event and also the coordinate in the direction of drifting of the electrons: F. Sauli conceived a simple way for that by the time modulation of a second intensified camera.

It is impossible, being at CERN and active in detector research, not to respond to emerging problems. It is illustrated by our latest articles on the asymmetric high-rate wire chambers, on the "Hadron Blind" detector and on the "Optical Trigger for Beauty", started with L. Lederman and I. Giomataris. It reflects the considerable activity on new detectors which is observed in all the laboratories that have to be prepared for experiments with the new generation of accelerators expected near the end of this century. It is clear to me that, quite naturally, many improvements will be brought to all fields where the imaging of ionizing radiation is an important tool: biology, radiology, nuclear medicine. In high-energy physics laboratories where the technology is being developed, the existence of groups active in detector research can be a source of important achievements if experienced people have the freedom to also attack problems not directly connected with the main goal of their laboratories. Valuable solutions in many domains can result from the effort to build low-cost, high-rate, high-accuracy particle detectors required by the next generation of high-luminosity accelerators.

Georges Charpak

ACKNOWLEDGEMENTS

I am grateful to Anne Dirat who has, with competence and infinite patience, organized the editing of the book.

This book owes much to the skills of David Dallman who had to find very scattered articles, many of them forgotten by myself.

I thank J. Margaret Rabbinowitz for putting my English into good shape.

GC

CONTENTS

RESEARCH ON
PARTICLE
IMAGING
DETECTORS

1. LOW ENERGY NUCLEAR PHYSICS WITH
GASEOUS DETECTORS:
1950–1959

1. LOW ENERGY NUCLEAR PHYSICS WITH GASEOUS DETECTORS: 1950–1959

PHYSIQUE NUCLÉAIRE. — *Différence de comportement par diffusion des électrons et du rayonnement β.* Note (*) de MM. Francis Suzor et Georges Charpak, présentée par M. Frédéric Joliot.

Nous avons déjà exposé dans deux Notes (¹) le dispositif expérimental (*fig.* 1) nous permettant de mesurer la proportion de rayonnement diffusé dans tout l'espace situé d'un côté d'une feuille plane d'aluminium en fonction de l'épaisseur h de cette feuille (courbe A); la source sans matière est placée sur la feuille elle-même du côté où le rayonnement est mesuré. Pour un radioélément donné, nous pouvons en particulier tirer de la courbe A la proportion P de rayonnement diffusé en arrière dans l'angle solide 2 π par une feuille plane de 1 mg/cm² d'aluminium. La figure 2 représente P en fonction de l'énergie pour deux éléments émettant des électrons de conversion et trois éléments radioactifs β; pour les radioéléments β l'énergie portée en abscisse est le tiers de l'énergie limite du spectre correspondant environ au maximum d'intensité de ce spectre. Nous obtenons dans les deux cas (électrons, rayonnement β), deux courbes nettement différentes. Il serait nécessaire pour faire coïncider ces deux courbes de porter en abscisse pour les radioéléments β environ le 1/12 de l'énergie maximum du spectre. Autrement dit il faudrait que la montée

---

(¹) *Phys. Rev.,* 76, 1949, p. 933.

(*) Séance du 12 février 1951.

(¹) *Comptes rendus,* 231, 1950, p. 1471, et 232, 1951, p. 322.

rapide de A à son début pour les radioéléments β soit due de façon prépondérante à la partie très peu énergique du spectre. Afin d'élucider ce point nous avons disposé contre la feuille plane d'aluminium d'épaisseur variable h une autre feuille d'aluminium d'épaisseur constante 11,3 mg/mc², la source se

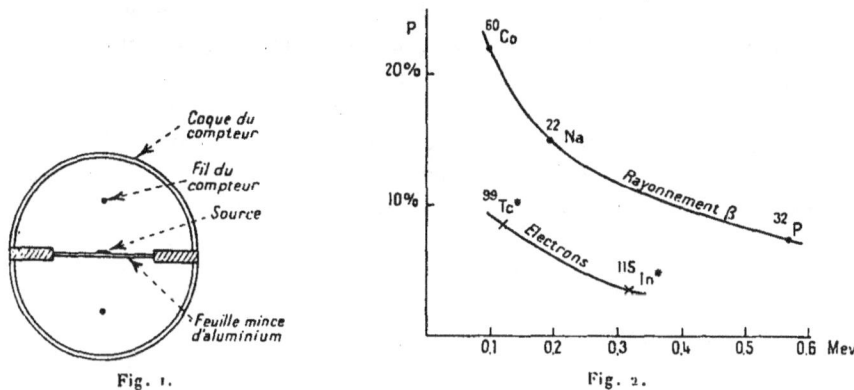

Fig. 1.　　　　　　　　　　　　Fig. 2.

trouvant en sandwich entre ces deux feuilles; nous obtenons ainsi la courbe A′ représentant, pour le rayonnement ayant traversé la feuille de 11,3 mg/cm², la proportion de celui-ci diffusée par la feuille d'épaisseur variable h. Un écran de 11,3 mg/mc² arrête tous les électrons d'énergie inférieure à 80 keV; nous avons de plus constaté que la proportion de rayonnement diffusé en arrière par une épaisseur saturante à travers une feuille de 11,3 mg/cm² était de $0 \pm 1\%$ dans le cas des électrons de 120 keV du ⁹⁹Tc*. La courbe A′ pour le ⁶⁰Co est représentée sur la figure 3, elle donne pour 1 mg/cm² P′ = 10 %.

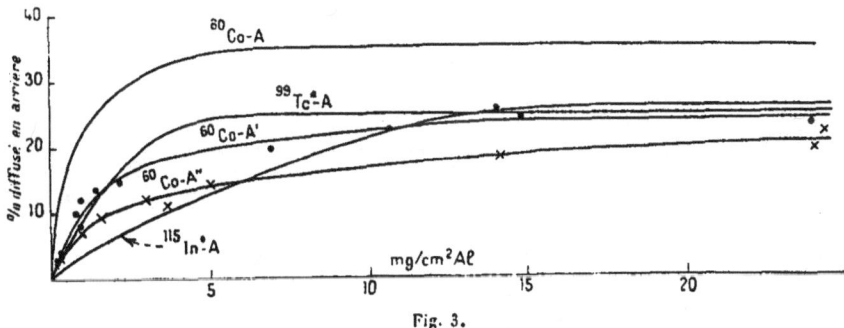

Fig. 3.

Ainsi la courbe « électrons » de la figure 2 est tout entière située nettement au-dessous de cette valeur 10 % obtenue avec la partie du spectre de ⁶⁰Co située, d'après ce que nous venons de dire entre 120 et 308 keV, énergie limite du spectre. Il en résulte qu'il n'est pas possible de faire appel aux électrons peu énergiques des spectres β pour faire coïncider les deux courbes de la figure 2, et que celles-ci représentent un argument en faveur de la non-identité des électrons et du rayonnement β que nous avons indiquée dans nos précédentes Notes ([1]). Ce raisonnement est vrai *a fortiori* si l'on remarque qu'il aurait fallu en réalité comparer la valeur $P' = 10\%$ pour le ^{60}Co non pas aux valeurs de P' portées sur la figure 2 pour les deux raies électroniques mais aux valeurs P' forcément inférieures ($P' = 0$) pour le rayonnement de 120 keV du ^{99}Tc^{m} que l'on aurait à travers 11,3 mg/cm^2 d'aluminium. Indiquons encore que la courbe A'' (*fig.* 3) à travers une feuille de 28,3 mg/cm^2 d'aluminium qui arrête les électrons jusqu'à 150 keV a donné pour le ^{60}Co, $P'' = 7,5\%$ (proportion de diffusion en arrière donnée par 1 mg/cm^2 d'aluminium).

([*]) Séance du 18 décembre 1950.

([1]) *Comptes rendus*, 230, 1950. p. 2279-2280.

([2]) *Comptes rendus*, 228, 1949, p. 1583-1584.

([3]) *La recherche du nuage radioactif*, Mémoire publié au S. D. I. T. de l'Aéronautique, Paris, n° 228, 1949.

([4]) *Comptes rendus*, 230, 1950. p. 1272-1274.

PHYSIQUE NUCLÉAIRE. — *Sur l'absorption et la diffusion en arrière des électrons et du rayonnement β*. Note (*) de MM. Georges Charpak et Francis Suzor, présentée par M. Frédéric Joliot.

Dans une première Note ([1]) nous avons exposé les faits expérimentaux qui nous ont amené à émettre l'hypothèse que le rayonnement pourrait être constitué de particules distinctes des électrons et se transformant en électrons dans des chocs inélastiques en traversant quelques milligrammes par centimètre carré de matière. Ces particules seraient instables et pourraient se désintégrer en électrons avec l'émission d'un ou plusieurs photons suivant une loi exponentielle en fonction du temps. L'énergie ainsi libérée en vol dans le vide sous forme de photons doit être suffisamment faible pour ne pas modifier sensiblement la direction de la vitesse, ce qui explique la possibilité de la focalisation des rayons dans les spectographes de petite et grande dimension. Ces particules en traversant la matière auraient une grande probabilité lors des chocs atomiques de se transformer en électrons. Une telle hypothèse, qui nous permet en grande partie d'interpréter nos expériences a l'avantage d'expliquer des observations faites par d'autres auteurs.

En particulier notre hypothèse explique la différence de la valeur de l'énergie moyenne des rayons du RaE mesurée par calorimétrie et calculée à l'aide des histogrammes obtenus à la chambre de Wilson. En effet l'énergie moyenne calorimétrique est environ ([2]) 15 % supérieure à celle déduite de l'histogramme. Dans le calorimètre la particule s'est transformée en électron avec émission de un ou de plusieurs photons de faible énergie qui sont absorbés. Dans l'appareil Wilson, après un parcours de l'ordre du centimètre, la majeure partie des β s'est transformée en électrons ; l'histogramme donnerait la valeur moyenne de l'énergie des électrons et non celle des β. Le même dispositif expérimental nous a permis de préciser certains faits. Nous avons refait avec une source beaucoup plus intense les expériences sur ⁹⁹Tc* donnant des électrons obtenus par conversion des γ de 141 keV. La courbe A ne présente pas l'aspect indiqué dans notre précédente Note, mais ceci ne modifie pas nos conclusions ; les premiers points relatifs à sa montée ont été très probablement faussés dans notre première expérience par la présence d'une impureté à vie courte. Les courbes B et S distinctes (17 % maximum) pour de faibles épaisseurs d'aluminium sont confondues pour des épaisseurs plus grandes.

(*) Séance du 15 janvier 1951.
([1]) *Comptes rendus*, 231, 1950, p. 1471.
([2]) C. D. Ellis, W. A. Wooster, *Proc. Roy. Soc.*, A, 117, 1928, p. 109; L. Meitner, W. Orthmann, *Zeits. Phys.*, 60, 1930, p. 1943; M. Lecoin, *Comptes rendus*, 224, 1947 p. 912.

Raisonnons sans tenir compte de notre hypothèse en assimilant le rayonnement β à des électrons. La théorie indique que pour des électrons la probabilité d'un choc avec déviation d'un angle donné est d'autant plus faible, que la vitesse de l'électron est plus grande; par conséquent le pourcentage p des électrons renvoyés en arrière par une feuille de $0,17$ mg/cm² d'aluminium, par exemple, ne pourra être qu'une fonction décroissante de l'énergie de ces

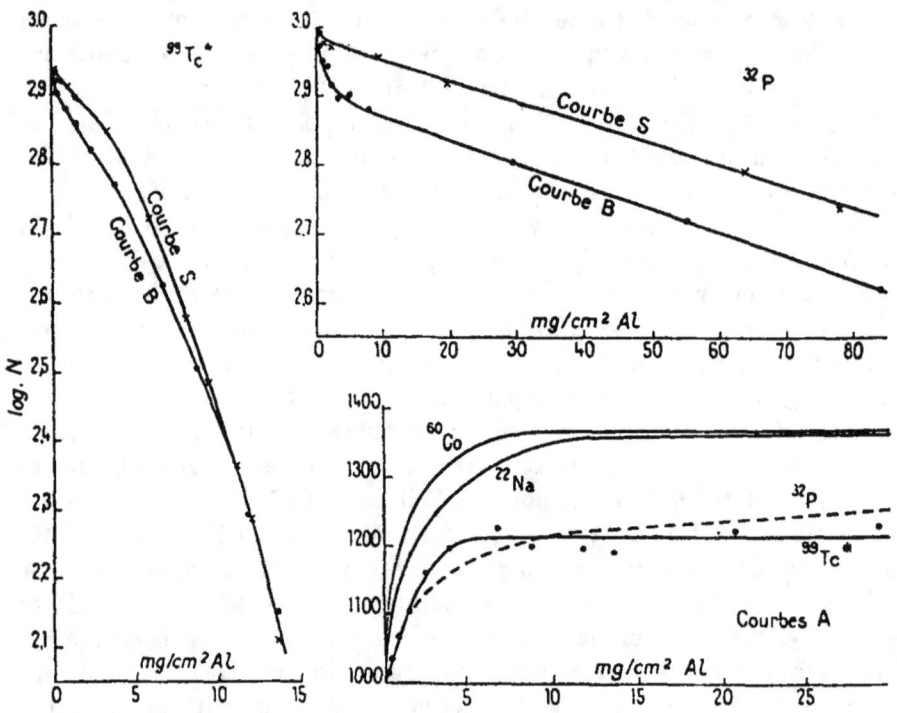

électrons. Nos expériences indiquent que $p = 2\%$ pour la raie électronique du ^{99}Tc* (120 keV), $p = 10\%$ pour le spectre du ^{60}Co dont l'énergie maximum est 308 keV et $p = 2\%$ pour le spectre du ^{32}P dont l'énergie maximum est $1,71$ MeV. Ceci impose que p soit nettement supérieur à 10% pour des énergies inférieures à 120 keV afin de trouver 10% en moyenne pour le spectre du ^{60}Co, et que p se maintienne à 2% pour des énergies allant de 120 keV jusqu'à $1,71$ MeV afin d'obtenir 2% pour le spectre du ^{32}P. Nos expériences permettent un raisonnement analogue pour une feuille d'aluminium de $0,6$ mg/cm². Ceci impose que le rayonnement du ^{60}Co renvoyé en arrière par une épaisseur saturante d'aluminium soit en très grande majorité formé d'électrons ayant une énergie inférieure à 120 keV, toute la partie du spectre comprise entre 120 et 308 keV étant relativement peu renvoyée en arrière. Nous sommes alors en présence d'une contradiction due à l'aspect des courbes B et S qui prouve que, après traversée d'une feuille de 12 mg/cm² d'aluminium, il passe dans le cas du ^{99}Tc* 20 % du rayonnement direct et o ± 1 % du rayonnement réfléchi et dans le cas du ^{60}Co, 15 % du rayonnement direct et 15 ± 4 % du rayonnement réfléchi.

PHYSIQUE NUCLÉAIRE. — *Sur la diffusion en arrière des électrons.*
Note de MM. Georges Charpak et Francis Suzor, présentée par
M. Frédéric Joliot.

Nous avons décrit dans plusieurs Notes ([1]) nos expériences relatives à la
diffusion en arrière du rayonnement β, dans un angle solide 2π, par la feuille
d'aluminium servant de support à la source radioactive. Nous avions étudié la
montée rapide à ses débuts de la courbe donnant la proportion de rayonnement
diffusé en arrière en fonction de l'épaisseur du support pour les radioéléments
β, ^{60}Co, ^{22}Na et ^{32}P; ayant obtenu ces mêmes courbes pour les émetteurs
électroniques ^{99}Tc° et ^{114}In° nous avions trouvé qu'elles étaient incompatibles
avec celles des radioéléments β en tenant compte de la forme connue de leurs
spectres énergétiques; c'est pourquoi nous avions dans ces Notes ([1]) émis
l'hypothèse que la particule β pourrait être un électron excité, dont la perte de
l'énergie d'excitation par chocs entraînerait une diffusion différente de celle
des électrons. Afin d'éliminer les erreurs qui peuvent s'introduire en calculant
l'influence des électrons très peu énergiques des spectres β ou ceux des électrons
Auger dans le cas des isomères, nous avons repris les expériences ([2]) de la
façon suivante.

La source était déposée sur une feuille d'aluminium d'épaisseur x et mise
en sandwich entre celle-ci et une feuille d'aluminium d'épaisseur h variable;
le rayonnement compté à travers x permettait d'obtenir la proportion diffusée
en arrière dans un angle solide 2π par la feuille d'épaisseur h en fonction de h.
Les électrons dont le parcours est inférieur à x n'interviennent pas et de plus
la mesure effective pour $h = 0$ est possible, ce qui évite l'extrapolation qui
était nécessaire dans les premières expériences. Nous appelons $P(x)$ l'ordonnée
de la courbe ainsi obtenue pour la valeur $h = 1$ mg:cm² d'aluminium; $P(x)$ est

---

([1]) *Comptes rendus*, 231, 1950, p. 1471: 232, 1951. p. 322 et 720.
([2]) *Journal de Physique* (sous presse).

donc la proportion (exprimée en pour-cent) de rayonnement diffusé en arrière par 1 mg:cm² à travers x mg:cm². Nous appelons de même $\sigma(x)$ la proportion de rayonnement diffusé en arrière à travers x mg:cm² par une épaisseur saturante d'aluminium. L'étude en fonction de x des nuclides ³⁵S, ⁶⁰Co et ³²P ainsi que la connaissance des spectres de ces nuclides et de l'absorption par x mg:cm² d'aluminium nous a permis d'obtenir sans grande ambiguïté les fonctions P(x) pour des raies β monocinétiques. Nous avons trouvé par exemple pour une raie β de 120 keV :

$$P(1) = 21,5 \pm 2\%, \qquad P(6) = 18 \pm 2\%, \qquad P(10) = 11 \pm 2\%.$$

Les premières expériences publiées que nous avions faites avec le ⁹⁹Tc* donnaient P(1) = 10 ± 2 % et $\sigma(6)$ = 15 % (les erreurs indiquées sont de caractère statistique). Utilisant une source de ⁹⁹Tc* dont la préparation avait été améliorée, nous avons obtenu les courbes de diffusion en arrière pour la raie électronique de 120 keV à travers $x = 1$ et $x = 6$ mg:cm² d'aluminium; ces nouvelles expériences nous ont donné

$$P(1) = 23 \pm 1\%, \qquad P(6) = 20,5 \pm 1\% \qquad \text{et} \qquad \sigma(6) = 29\%.$$

Un accord entre rayonnement β et électrons du même ordre que celui ainsi obtenu pour une énergie de 120 keV a été trouvé pour l'énergie de 320 keV en utilisant les électrons émis par le nuclide ¹¹³In*; il en résulte que l'hypothèse que nous avions avancée ne trouve aucune justification dans l'étude de la diffusion en arrière des électrons. L'écart de 13 à 14 % entre les deux expériences sur le ⁹⁹Tc* pour P(1) et $\sigma(6)$ s'explique par le fait que la première source n'était pas absolument sans matière et qu'une diffusion très importante se produisait dans la matière elle-même de la source faussant ainsi considérablement les résultats. Nous n'avons pas tenu compte dans les valeurs numériques précédemment indiquées du gaz se trouvant derrière la source (8 cm argon + 1 cm alcool); il faudrait le faire pour obtenir une valeur absolue de la proportion de rayonnement diffusé en arrière dans un angle solide 2π.

(¹) J. E. MOYAL, *Phil. Mag.*, octobre 1950, p. 1058-1077.

LETTRES A LA RÉDACTION

MÉTHODE SIMPLE DE PRÉPARATION DE SOURCES RADIOACTIVES UNIFORMES

Par G. CHARPAK et M. CHEMLA,

Laboratoire de Chimie Nucléaire,
Collège de France, Paris.

Les sources radioactives préparées par évaporation directe d'une solution sur un support sont constituées par un ensemble de cristaux isolés; de ce fait, la densité superficielle réelle est plusieurs centaines de fois plus élevée que celle calculée en supposant le dépôt uniforme [1]. Cet inconvénient peut être évité lorsque le radioélément est déposé par projection thermique dans le vide [2]. Cette méthode nécessite une installation importante et les rendements sont en général faibles.

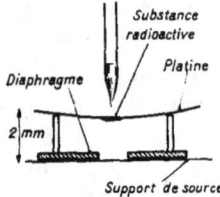

Fig. 1.

Dans ces expériences nous avons pu obtenir des sources minces et uniformes, par sublimation des radioéléments à l'air; pour cela, une goutte de solution active est évaporée à sec sur une feuille de platine de $1/20^e$ mm d'épaisseur; celle-ci est placée légèrement au-dessus du support, la face active tournée vers ce dernier (fig. 1); le platine est alors localement porté au rouge pendant une fraction de seconde par une petite flamme de chalumeau gaz-oxygène; dans ces conditions, un support de $10\,\mu g/cm^2$ de LC 600 ne subit aucun dommage et le sel radioactif se trouve déposé uniformément avec un rendement de l'ordre de 50 pour 100.

Cette méthode a été utilisée pour déposer des couches minces de chlorures de Na, K, Rb, Ag, Cs, sur des faces cristallines afin d'étudier la diffusion d'ions radioactifs dans les cristaux [3]. Elle a également permis d'obtenir des sources de ^{22}Na et ^{65}Zn, dans lesquelles des électrons Auger de $0,8$ keV subissaient une très faible autoabsorption [4].

Cependant, lorsque la source déposée sur le platine est insuffisamment séchée, des crépitements avec projection d'amas de matières peuvent se produire. Pous illustrer ces résultats, nous avons reproduit sur la figure 2 des autoradiographies de sources de ^{22}Na de même activité préparées respectivement par :

a. évaporation directe d'une goutte de solution sur aluminium;

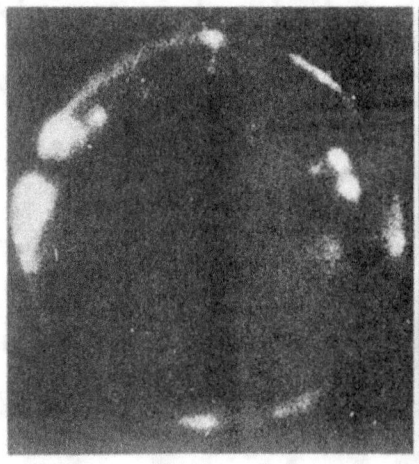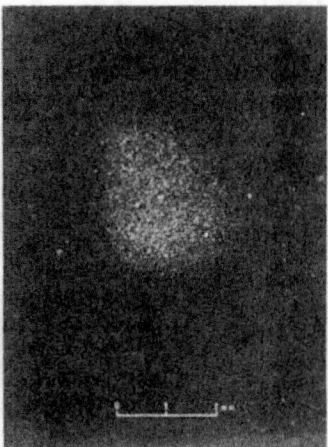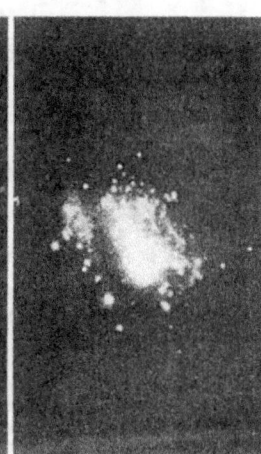

Fig. 2.

b. sublimation convenable par la méthode que nous décrivons;

c. sublimation d'une goutte insuffisamment séch'.

Cette méthode, très simple, n'est cependant pas générale; elle s'applique seulement aux sels volatils à l'air sans décomposition, c'est-à-dire la série des halogénures alcalins et les chlorures d'Ag, Tl, Pb, Bi, Hg; elle est encore applicable aux éléments dont les oxydes sont volatils, c'est le cas de Zn, Cd, As, Sb, etc. Enfin, dans les cas où le chauffage à l'air du chlorure conduit à un oxyde réfractaire (Fe, Sr, terres rares, etc.), cette méthode péut encore être utilisée à condition de travailler en atmosphère contrôlée (gaz inerte contenant de préférence de petites quantités de HCl ou Cl_2); La figure 3 représente un petit appareil de verre qui a été utilisé à cet effet.

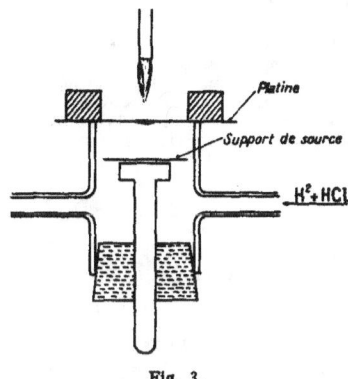

Fig. 3.

En résumé, le procédé de préparation de sources radioactives minces et uniformes que nous décrivons est simple; il nécessite l'emploi de composés chimiques convenables, en général halogénures; sa rapidité le rend applicable aux radioéléments de période courte.

Manuscrit reçu le 12 avril 1954.

[1] CHARPAK G. et SUZOR F. — *J. Physique Rad.*, 1954, **15**.
[2] FRAUENFELDER H. — *Helv. Phys. Acta*, 1950, **23**, 347.
[3] CHEMLA M. — *C. R. Acad. Sc.*, 1954, **238**, 82.
[4] CHARPAK G. — *J. Physique Rad.* (sous presse).

LE JOURNAL DE PHYSIQUE ET LE RADIUM.   TOME 16, JANVIER 1955, PAGE 62.

ÉTUDE DE LA CAPTURE ÉLECTRONIQUE DANS LA DÉSINTÉGRATION DU NUCLIDE 22Na

Par G. CHARPAK,

Laboratoire de Physique et Chimie nucléaires, Collège de France, Paris.

Sommaire. — L'étude de 22Na est faite avec un compteur Geiger 4π. On met en évidence l'émission d'un rayonnement de très basse énergie, indépendant des rayons β⁻, complètement absorbé dans un film de quelques microgrammes par centimètre carré d'aluminium ou de matière plastique LC 600 et attribué aux électrons Auger d'énergie maximum 0,85 keV, qui suivent la capture électronique. En raison du très faible parcours de ces électrons, nous sommes amené à discuter particulièrement une méthode simple de préparation de sources radioactives minces et uniformes. Nous obtenons la valeur du rapport

$$\frac{\text{Capture K}}{\text{Emission } \beta^+} = (6,5 \pm 0,9) \text{ pour 100.}$$

Introduction. — Le nuclide 22Na se désintègre suivant le schéma de la figure 1.

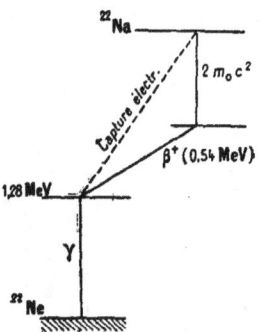

Fig. 1. — Schéma de désintégration du 22Na.

Divers auteurs ont tenté de mesurer le rapport

$$\frac{Pe}{P_+} = \frac{\text{Capture électronique}}{\text{émission } \beta^+}.$$

Ce problème, outre son intérêt théorique a une certaine importance pratique, car 22Na est souvent utilisé pour calibrer l'efficacité des compteurs à scintillations et il est important de savoir si le nombre des positons est égal à celui des quanta de 1,28 MeV.

Le tableau I donne le résultat de diverses mesures.

La nature de la transition du 22Na au 22Ne n'est pas connue avec certitude. En admettant que la transition soit permise ($\Delta J = 1$ ou 0, non), le calcul théorique donne $\frac{Pe}{P_+} = 0,11$ en excellent accord avec la mesure la plus précise de Miller et Sherr [5].

TABLEAU I.

$\frac{Pe}{P_+}$.	Auteurs.
0..............	Bothe (1945)
0 ±0,05....	Good W. M. et coll. (1946)
0,10 ±0,05....	Bouchez R. (1950)
0,04 ±0,03....	Major J. K. (1951)
0,07 ±0,02....	Hornyak W. F. et Coor T. (1953)
0,110 ±0,006...	Miller R. H. et Sherr R. (1954)

La méthode de ces auteurs repose sur la comparaison du nombre absolu de positons et de rayons γ de 1,28 MeV émis par une source de 22Na. Notre travail, entrepris avant la publication des résultats de Miller et Sherr, ne nous permet pas d'atteindre une aussi grande précision dans le cas particulier du 22Na. Cependant la méthode utilisée offre des possibilités d'emploi plus général.

Principe de la méthode. — La capture électronique dans le 22Na se traduit essentiellement, en raison du rendement de fluorescence voisin de zéro, par l'émission d'électrons Auger du néon dont l'énergie maximum est de 0,85 keV.

Le spectre β⁺ comporte certainement un nombre infime de positons de cette énergie à cause de la répulsion coulombienne avec le noyau.

Une méthode d'absorption peut donc en principe permettre la mise en évidence de la capture. La difficulté essentielle réside dans la détection des électrons et dans la préparation de sources suffisamment minces par rapport au parcours des électrons de 0,85 keV. Celui-ci est au maximum de quelques microgrammes par centimètre carré, soit quelques centaines de couches atomiques.

Mesures. — La source est préparée à partir d'une solution de chlorure de sodium dont l'activité spécifique est de 1 mC de ^{22}Na par milligramme de matière. Elle est déposée par la méthode de sublimation à l'air, décrite avec M. Chemla dans une Note récente [7], sur un support de source de LC 600, aluminisé de chaque côté et dont la masse superficielle totale est de 20 μg/cm^2.

Le support de source, tendu sur un trou de diamètre 1 cm constitue la surface de séparation de deux compteurs 2π. Le diamètre de la source est de 5 mm. Nous pouvons mesurer le nombre d'impulsions N_1 et N_2 dans chaque compteur, ainsi que les coïncidences N_c.

Les fils des deux compteurs peuvent également être connectés en parallèle. Le nombre d'impulsions alors compté, $N_{4\pi}$ est égal à $N_1 + N_2 - N_c$.

Afin d'obtenir une grande stabilité et une précision suffisante dans la mesure de $N_{4\pi}$ les impulsions du compteur sont envoyées dans un préamplificateur à temps mort T qui élimine pratiquement

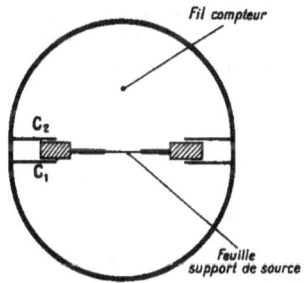

Fig. 2. — Coupe schématique du double compteur 2π. Le dispositif exact est décrit en détail par ailleurs [6].

tous les coups multiples. Avec $T = 5$ ms, la pente du compteur est inférieure à 0,5 pour 100 pour 100 V. Chaque mesure est affectée de la correction des pertes au moyen de la formule

$$N_{\text{réel}} = \frac{N \text{ compté}}{1 - T \times N \text{ compté}}.$$

T est connu avec une précision de 1 pour 100.

Nous appelons A_1 et A_2 les écrans situés entre la source et chaque compteur. Le tableau II résume les résultats obtenus dans diverses conditions :

TABLEAU II (^{22}Na.)

A_1 (μg/cm^2).	A_2 (μg/cm^2).	$N_{4\pi}$ (c/m).
20 LC 600	0.........	3030 ± 15
20 » + 170 Al	0.........	3035 ± 12
20 LC 600	170 Al......	2910 ± 5

On voit qu'il y a émission par la source d'un rayonnement peu pénétrant, ne pouvant pas tra-

verser le support LC 600 de 20 μg/cm^2 alors que les positons sont absorbés de façon négligeable par une feuille de 170 μg/cm^2 d'aluminium.

La source étudiée émet 120 rayons très peu pénétrants dans un des compteurs 2π.

Lorsque nous déposons sur la source, par projection thermique dans le vide une couche d'aluminium de 20 μg/cm^2, ce rayonnement est également totalement absorbé. Il est important de noter qu'il n'est pas associé au rayonnement β^+. En effet, l'absorption d'un rayon émis simultanément avec un positon qui n'est pas absorbé n'affecte pas $N_{4\pi}$.

Il est normal d'attribuer le rayonnement très « mou » émis par le ^{22}Na, indépendamment des positons, aux électrons Auger qui suivent la capture électronique. Notons que les rayons γ de 1,28 MeV, ne jouent qu'un rôle peu important en raison de leur faible efficacité (< 1 pour 100). Nous les négligerons dans la suite de la discussion.

Nos résultats expérimentaux conduisent à :

$$\frac{Pe}{P_+} = \frac{1}{\lambda_K} \frac{120}{2910} = \frac{1}{\lambda_K} (4,1 \pm 0,05) \text{ pour 100,}$$

λ_K est la probabilité pour qu'une désintégration par capture électronique produise une impulsion dans le compteur 2π situé du côté de la source. Théoriquement, λ_K peut être supérieur à $\frac{1}{2}$, soit en raison de la rétrodiffusion des électrons Auger, soit en raison des électrons secondaires produits dans le support de source par les électrons qui y pénètrent. Étant donné que $\lambda_K < 1$, nous pouvons affirmer que :

$$\frac{Pe}{P_+} > (4,1 \pm 0,5) \text{ pour 100.}$$

Étude de la source. Mesure de λ_K. — Nous avons essayé de mesurer λ_K et de vérifier du même coup la qualité de la source utilisée.

Nous tirerons profit du fait que l'énergie de liaison d'un électron L du cuivre (0,95 keV) est voisine de celle d'un électron K du néon (0,90 keV).

Nous préparons par la même méthode, sur un support d'aluminium de 170 μg/cm^2, une source de ^{65}Zn de même étendue que celle du ^{22}Na et d'activité spécifique voisine. Le support d'aluminium peut être placé contre des écrans de béryllium et d'or. Le tableau III résume nos résultats.

Les couches d'aluminium de quelques microgrammes par centimètre carré sont déposées sur la source par projection thermique dans le vide. Leur épaisseur est déterminée de deux façons :

1° On pèse la quantité d'aluminium qui a été évaporée et on admet que les atomes ont été projetés de façon isotrope;

2° On place le support de source, situé à 15 cm du filament sur une feuille d'aluminium de 100 cm^2 et l'on pèse avec une balance sensible au 1/100e de milligramme l'augmentation de poids de la feuille. On en déduit aisément la densité superficielle de la matière projetée sur la source.

TABLEAU III (65 Zn.)

A_1		A_2	N_1 (c/m).	N_2 (c/m).	$(N_c$ c/m).
170 μg/cm² Al + 40 mg/cm² Be	0	....	152	11 500	100 ± 4
»	»	3 μg/cm² Al	150	4 700	18 ± 3
»	»	7 » Al	150	4 220	13 ± 2
»	»	17 » Al	150	3 500	5.4 ± 1
170 μg/cm² Al + 40 mg/cm² Au	0	....	25	11 500	18 ± 2

actuellement bien déterminé (*fig.* 3), le compteur C_1 ne détecte pratiquement, à travers l'écran de beryllium, que les photons K du cuivre et les rayons γ d'énergie 1,114 MeV.

Si l'on fait varier l'épaisseur des écrans de beryllium à partir d'une épaisseur minimum de 20 mg/cm²,

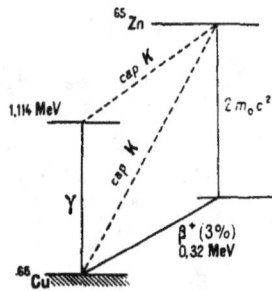

Fig. 3. — Schéma de désintégration du ⁶⁵Zn.

on constate que l'absorption du rayonnement, compte tenu de la géométrie particulière, a lieu suivant le coefficient d'absorption des photons K de 8 keV. Ceci montre que sur les 150 impulsions N_1 une fraction peu importante est produite par les rayons γ énergiques. On obtient une limite supérieure de celle-ci avec l'écran d'or de 40 mg/cm². Celui-ci absorbe, en effet, totalement les photons K tandis que l'efficacité du compteur aux rayons γ est sans doute augmentée. Nous pouvons ainsi conclure qu'approximativement 127 photons K sont comptés dans C_1 et provoquent 82 coïncidences.

Un photon K est lié le plus souvent à une vacance dans la couche L. Soit λ_L la probabilité pour qu'une vacance dans la couche L provoque l'émission d'un rayonnement qui déclenche le compteur C_2. On a

$$N_c = N_1 \times \lambda_L,$$

où N_c est le nombre de coïncidences entre C_1 et C_2 et N_1 le nombre de photons K comptés. Nos résultats expérimentaux nous permettent de conclure que $\lambda_L \sim 0,64$.

Les résultats des deux méthodes concordent à 50 pour 100 près.

D'après le schéma de désintégration du ⁶⁵Zn,

Or, les énergies de liaison des électrons K du néon et L du cuivre sont très voisines.

Dans les deux cas il y a émission, en majeure partie, après une vacance dans ces couches, d'électrons Auger d'énergies comparables.

Les sources ont été préparées par le même procédé et déposées sur l'aluminium.

L'activité spécifique du ⁶⁵Zn (3 mc/mg) est du même ordre que celle du ²²Na (1 mc/mg).

Si nous admettons que

$$\lambda_L (\text{zinc}) = \lambda_K (\text{sodium}),$$

ceci nous conduit à :

$$\frac{\mathrm{Pe}}{\mathrm{P_-}} = (6,5 \pm 0,9) \quad \text{pour 100.}$$

Discussion. — La chute très rapide du nombre des coïncidences, dans le cas du ⁶⁵Zn (tableau III), lorsque la source est couverte d'une couche de matière de quelques microgrammes par centimètre carré, prouve que la source est elle-même extrêmement mince et confirme la qualité de la méthode de préparation des sources minces uniformes par sublimation à l'air.

Nous avons, par le même procédé, recouvert la source de ²²Na, qui a servi aux mesures précédentes avec une nouvelle couche de ²²Na, trois fois plus intense. Aux erreurs statistiques près, nous observons le même nombre relatif de rayons de très basse énergie sortant de la source. Ceci montre que les électrons Auger ne subissent pas d'absorption notable dans la matière de la source.

Si nous confrontons notre résultat :

$$\frac{\mathrm{Pe}}{\mathrm{P_+}} = (6,5 \pm 0,9) \quad \text{pour 100}$$

avec celui de Miller et Sherr :

$$\frac{\mathrm{Pe}}{\mathrm{P_+}} = (11,0 \pm 0,6) \quad \text{pour 100}$$

et si nous admettons que la méthode de ces derniers ne comporte pas de cause d'erreur systématique supérieure à la précision statistique indiquée, nous voyons que nous commettons une assez grande erreur dans l'estimation de λ_K. Tout se passe comme si l'épaisseur de la source était telle que les électrons secondaires produits dans l'aluminium et la matière de la source par les électrons Auger K, ainsi que les rayonnements liés au réarrangement de la couche L du néon ne pouvaient pas être détectés par le compteur 2π situé du côté de la source.

Notre méthode pourrait donc comporter une incertitude d'un facteur voisin de deux quant au nombre de désintégrations se produisant par capture électronique. Elle s'applique toutefois dans des cas où le schéma de désintégration ne permet pas l'emploi de la méthode des coïncidences utilisé par Miller et Sherr.

Elle peut permettre une recherche systématique des désintégrations par capture électronique dans les noyaux impairs-impairs légers pour lesquels une telle éventualité est possible, par exemple le ³⁶Cl. Elle ne nécessite que de très faibles activités, ce qui facilite la réalisation de sources très minces.

Influence de la méthode de préparation des sources. — Nous avons repris les expériences relatives à ²²Na avec des sources préparées par séchage d'une goutte. Nous observons 2,5 fois moins d'électrons Auger lorsque la source est déposée sur un film de LC 600 aluminisé, et leur disparition totale lorsque la source est déposée sur une feuille d'aluminium battu de 170 $\mu g/cm^2$. Ceci montre que lorsque l'on prépare des sources de ²²Na par l'évaporation d'une goutte, même si la solution contient une quantité infime de matière, l'auto-absorption peut jouer un rôle notable et conduire à des erreurs voisines de 5 pour 100 dans l'estimation de l'intensité absolue des sources au moyen de la technique du compteur 4π Geiger.

Cette erreur peut être également importante dans le cas d'émetteurs β peu énergiques comme ³⁵S et ¹⁴C.

La méthode de sublimation à l'air que nous avons employée est aussi rapide que le séchage d'une goutte. Elle permet dans de nombreux cas d'éliminer ces difficultés tout en donnant d'excellents rendements.

Nous devons d'intéressantes discussions relatives à ce travail à MM. F. Suzor, P. Radvanyi et Mme Marty.

Nous remercions particulièrement M. Joliot d'avoir bien voulu suivre ce travail.

Manuscrit reçu le 12 avril 1954.

BIBLIOGRAPHIE.

[1] BOUCHEZ R. — *Physica*, 1952, **18**, 1171.

[2] GOOD W. M., PEASLEE D. et DEUTSCH M. — *Phys. Rev.*, 1946, **69**, 313.

[3] MAJOR J. K. — *Thèse*, Paris, 1951.

[4] HORNYAK W. F. et COOR T. — *Phys. Rev.*, 1953, **92**, 676.

[5] MILLER R. H. et SHERR R. — *Bull. Amer. Phys. Soc.*, 1953, **28**, 4; *Phys. Rev.*, 1954, **93**, 1076.

[6] CHARPAK G. et SUZOR F. — *J. Physique Rad.*, 1952, **13**, 1.

[7] CHARPAK G. et CHEMLA M. — *J. Physique Rad.*, 1954, **15**, 490.

5

LE JOURNAL DE PHYSIQUE ET LE RADIUM

PHYSIQUE APPLIQUÉE

SUPPLÉMENT AU N° 12.

TOME 19, DÉCEMBRE 1958, PAGE 167 A.

SPECTROMÉTRIE NUCLÉAIRE PAR COMPTEURS PROPORTIONNELS EN COINCIDENCE

Par G. CHARPAK et F. SUZOR,

Laboratoire de Synthèse Atomique, C. N. R. S.

Résumé. — Description d'un spectromètre à compteurs proportionnels spécialement adapté à l'étude des radiations de faible énergie (électrons ou photons au-dessous de 20 keV) émis par des atomes radioactifs avec une très faible probabilité. Deux compteurs proportionnels sont en contact, dans un plan qui contient le porte source. La source est en contact direct avec le gaz de l'un des compteurs. Les impulsions de ce compteur en coïncidence avec celles de l'autre sont analysées par un analyseur à canaux multiples. On décrit le montage électronique associé, en insistant sur le fait que les amplificateurs ont été rendus non saturables.

Abstract. — Description of a proportional counter spectrometer specially adapted to the study of low energy radiations (electrons or photons below 20 keV) emitted by radioactive atoms with very small probability. Two proportional counters of 18 cm diameter are in contact along a plane which contains the source holder. The source is in direct contact with the gas of one of the counters. Impulses of this counter in coincidence with impulses of the other are analyzed by a multichannel analyzer. The associated electronics in described with particular emphasis on the problems of non overloading qualities of the amplifiers.

Nous avons construit un appareil destiné à la mesure de l'énergie et de l'intensité d'un rayonnement photonique ou électronique de très basse énergie émis simultanément avec le rayonnement β afin d'étudier, par exemple, le freinage interne et l'autoionisation ainsi que certains schémas de désintégration radioactive contenant des transitions entre niveaux d'énergie très voisine.

La mesure de l'énergie est obtenue par l'utilisation de deux compteurs proportionnels. La durée importante et variable de la montée des impulsions ainsi fournies impose un grand temps de résolution lors de la détection en coïncidence de ces impulsions. C'est la raison pour laquelle nous avons placé les deux compteurs en contact l'un de l'autre suivant une génératrice horizontale, et la source dans le plan tangent vertical commun au milieu de cette génératrice ; chaque compteur est ainsi vu de la source sous un angle solide 2π, ce qui a pour effet de réduire à une valeur relative acceptable les coïncidences fortuites. Les compteurs sont de grandes dimensions pour augmenter l'efficacité lors de la détection de photons et pour que les électrons d'énergie la plus élevée possible soient totalement absorbés dans le compteur ; il en résulte aussi que des électrons d'énergie plus grande et au voisinage du minimum d'ionisation perdront, dans le compteur, une énergie suffisante pour pouvoir être distingués des électrons ou des photons de basse énergie que l'on se propose d'étudier. Indiquons, à titre d'exemple, que pour le rayonnement β du phosphore 32, 95 % des électrons perdent dans le compteur une énergie supérieure à 10 keV pour un remplissage de propane sous une pression voisine de la pression atmosphérique. On obtient, dans ces conditions, un nombre de coïnci-

dences fortuites relativement faible en mesurant les impulsions en coïncidence obtenues dans l'un des compteurs après discrimination à une valeur supérieure à une dizaine de keV environ (presque totalité du rayonnement β) et dans l'autre compteur après sélection dans une bande étroite explorant le domaine inférieur à 10 ou 20 keV où nous voulons étudier le rayonnement émis simultanément avec le rayonnement β. On peut ainsi mesurer, dans ce domaine d'énergie, des intensités de l'ordre de quelques 10^{-4} de celle du rayonnement β. Cet exemple typique illustre le domaine d'application de notre appareil. Une autre caractéristique importante réside dans l'existence d'un sas permettant l'entrée et la sortie rapide des sources sans perturbation du gaz remplissant le compteur ; il est ainsi possible de mettre la source directement en contact du gaz de l'un des compteurs et, à la condition d'employer des sources sans matière vaporisées sous vide [1], de pousser la limite inférieure de l'énergie des rayonnements étudiés jusqu'à quelques centaines d'électrons-volts.

I. Description de l'installation.

— Une enceinte cylindrique étanche en fonte à axe vertical de 50 cm de diamètre, 35 cm de hauteur et 2,5 cm d'épaisseur de paroi contient les deux compteurs accolés dont les axes sont horizontaux ; cette enceinte communique avec des dispositifs annexes permettant d'y faire le vide et de la remplir du gaz choisi ; aucun problème d'étanchéité ne se pose ainsi pour les compteurs eux-mêmes. Sur la figure 1, on voit, au-dessus de l'enceinte et suivant son axe, une barre qui, par translation verticale, permet l'entrée et la sortie de la source. Lorsque celle-ci est enlevée l'étanchéité de l'enceinte est obtenue par un clapet intérieur dont la commande extérieure se trouve sur la droite.

Une petite turbine actionnée de l'extérieur par

de champ en nickel [2] qui, portés à un potentiel intermédiaire exactement ajusté, ont pour effet de limiter le volume effectif du compteur à des plans perpendiculaires au fil et passant par l'extrémité de ces tubes. La forme et les dimensions exactes des

Fig. 2.

Fig. 1.

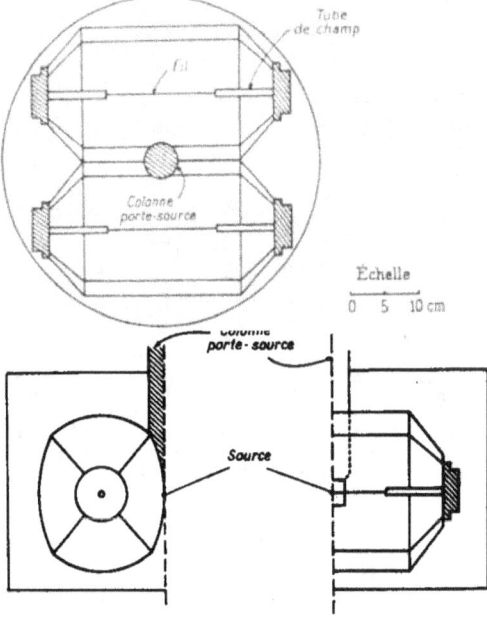

Fig. 3. — Vue en coupe des compteurs.

un volant magnétique établit d'une façon permanente un passage du gaz dans un four de purification contenant du calcium à une température réglable jusqu'à 400 ºC.

Les fils des compteurs qui sont en molybdène de 0,08 millimètre de diamètre, sont à un potentiel nul, les coques des deux compteurs étant portées à une même haute tension négative. La figure 2 représente les deux compteurs fixés par l'intermédiaire d'une plaque isolante en plexiglas sur la base de l'enceinte cylindrique. L'isolement électrique entre les coques des compteurs et l'enceinte est obtenu par des feuilles de plexiglas tapissant la paroi intérieure de celle-ci. Les coques des compteurs sont en cuivre ; ceux-ci sont munis de tubes

compteurs sont données sur la figure 3. La forme particulière et non circulaire de la section droite des coques des compteurs a été choisie pour obtenir un plus grand rayon de courbure au voisinage de la source et augmenter ainsi l'angle solide. Cette forme non circulaire est d'ailleurs sans influence notable sur la bonne marche du compteur. La forme conique des extrémités des compteurs n'a pas d'importance étant donné l'existence des tubes de champ et permet une utilisation maximum de la surface circulaire disponible dans l'enceinte étanche. A l'aide des cotes indiquées sur la figure 3, on peut calculer la longueur moyenne des rayons rectilignes issus du compteur ; on trouve 12,2 cm. Parmi tous les rayons issus dans l'angle solide 2π, 90 % ont un parcours supérieur à 8 cm, et, pour ceux-ci, la longueur moyenne est de 13,0 cm. Ces données sont importantes pour calculer l'efficacité des photons d'une énergie donnée dans le gaz remplissant le compteur, connaissant la nature et la pression de ce gaz.

Le porte-source lui-même est fixé d'une façon rigide à l'extrémité d'une colonne coulissante ; il est facilement démontable, permettant de mettre à volonté devant la source des écrans d'épaisseur et nature variables.

2. Appareillage électronique.

– L'appareillage électronique est destiné aux fonctions suivantes (fig. 4):

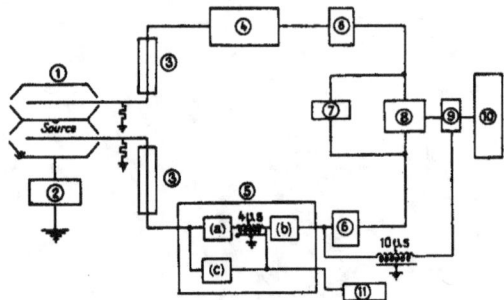

Fig. 4. — Diagramme des éléments de l'appareillage électronique.
1. Compteurs ; 2. Haute tension ; 3. Préamplificateurs ; 4. Amplificateur × 10 000 ; 5. Amplificateur avec blocage des saturées. (a) Étages préamplificateurs. (b) Étages finaux. (c) Porte à transistor ; 6. Discriminateur ; 7. Échelle ; 8. Sélecteur de coïncidences ; 9. Porte débloquée par les coïncidences ; 10. Sélecteur à 50 canaux ; 11. Comptage des impulsions saturées.

— alimentation haute tension des compteurs proportionnels ;
— amplification des impulsions dans chaque voie ;
— détection des coïncidences entre 2 voies ;
— comptage des impulsions et mesure, avec un sélecteur à 50 canaux, du spectre des impulsions de chaque compteur ou des impulsions en coïncidence.

Nous ne décrirons que les éléments particuliers à notre expérience. Le temps de déplacement d'un

électron libre dans le gaz du compteur, depuis le point où il a été libéré jusqu'à la zone voisine du fil où la multiplication commence est fonction de la distance au fil, de la nature et de la pression du gaz. Il atteint 5 microsecondes dans certains cas. Il en résulte que le temps de montée des impulsions produites par les rayons β est fonction de leur orientation dans le compteur et peut aller jusqu'à 5 microsecondes. Les intervalles entre l'instant où un électron de très faible parcours est libéré dans le gaz (par exemple par absorption d'un rayon X) et l'instant où l'impulsion se développe, fluctuent et peuvent également atteindre 5 microsecondes. Il est donc nécessaire, pour l'étude des impulsions en coïncidence dans les deux compteurs, d'utiliser un temps de résolution élevé. Celui-ci mesuré avec précision est voisin de 8 microsecondes dans nos expériences.

Grâce à la géométrie $2\pi \times 2\pi$, les sources utilisées peuvent être d'une intensité assez faible pour étudier les phénomènes rares avec précision, sans que les coïncidences fortuites soient gênantes.

Amplificateurs. — Ceux-ci devaient répondre aux conditions suivantes :
— gain élevé ($20 \times 10\,000$) ;
— bruit de fond d'entrée très bas ;
— grande stabilité ;
— insensibilité aux surcharges.

Il est fréquent, dans les expériences qui vont être décrites, de travailler dans les conditions suivantes :
— durée des impulsions : 15 microsecondes ;
— temps de montée : 1 à 5 microsecondes ;
— nombre d'impulsions saturées : 1 000 par seconde ;
— nombre d'impulsions dont le spectre est étudié : 20 par seconde.

Les impulsions saturées sont provoquées surtout par les rayons β absorbés dans le même compteur que celui où sont comptés les rayonnements de basse énergie. Elles correspondent à une perte d'énergie jusqu'à 200 fois plus forte que celle produite par les phénomènes étudiés. En raison de leur grande durée elles rendraient impossible l'observation des impulsions de faible amplitude sans les précautions suivantes : l'amplification se fait au moyen d'un préamplificateur de gain 20 et d'un amplificateur 200 kHz de gain 10 000. Les rebondissements sont éliminés aux divers étages à l'aide d'une contre-réaction par diodes qui permet de ne faire agir la contre-réaction que sur l'impulsion de signe indésirable [3]. Cette mesure est cependant insuffisante dans notre cas. L'amplificateur est en plus muni d'un dispositif [4] qui élimine dès les premiers étages les impulsions qui dépassent un certain niveau. Les impulsions sont amplifiées dans une voie parallèle à la voie proportionnelle. Lorsqu'elles dépassent un niveau prédéterminé,

12

elles débloquent, pendant 20 microsecondes, un transistor qui est placé en parallèle sur la résistance de fuite de grille d'une lampe amplificatrice située

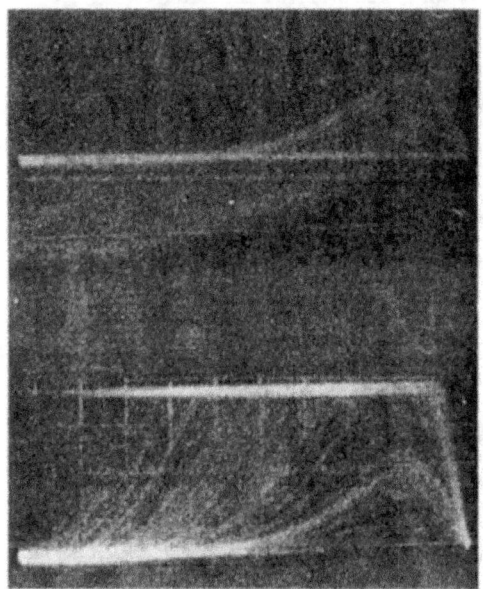

Fig. 5. — Impulsions produites dans le compteur par les rayons X du fer 55 en présence d'un grand nombre d'impulsions saturées fournies par les rayons γ du cobalt 60 (1 avec, 2 sans la porte à transistor — largeur d'un carreau = 2 micro-secondes).

dans les premiers étages de la voie proportionnelle. Les impulsions de la voie proportionnelle sont retardées de 4 microsecondes, à l'aide de ligne à

retard, avant d'arriver sur le transistor. On voit sur les clichés de la figure 5 l'effet de cette porte à transistor.

Ce dispositif limite l'effet des impulsions saturées ; il élimine celles-ci avant le sélecteur de coïncidences ce qui diminue de façon considérable les coïncidences fortuites.

Mesure du spectre. — Les impulsions sont analysées sur un sélecteur à 50 canaux du type Wilkinson, avec un temps mort de 1 milliseconde. Dans ses conditions normales de fonctionnement celui-ci permet d'analyser le spectre des impulsions de l'une ou l'autre voie. De plus il peut être bloqué en permanence ; le déblocage d'une durée de 20 microsecondes est commandé par les impulsions sortant du sélecteur de coïncidences. Ce dispositif permet l'analyse du spectre des impulsions en coïncidence. Les impulsions provenant des amplificateurs proportionnels sont retardées de 10 microsecondes avant d'entrer dans le sélecteur. On évite ainsi les pertes dues aux fluctuations du temps de montée des impulsions dans les compteurs.

Nous remercions M. Lanfrey et Mlle Clouet de leur collaboration dans la construction des compteurs et M. Pénège pour la mise au point de l'appareillage électronique. Nous devons une reconnaissance particulière au regretté Pr F. Joliot. Directeur du Laboratoire de Synthèse Atomique, pour l'intérêt constant qu'il a manifesté à ces expériences et pour les discussions dont il nous a fait bénéficier.

Ce travail a pu être réalisé grâce aux moyens mis à notre disposition par le Centre National de la Recherche Scientifique et à une subvention du Commissariat à l'Énergie Atomique.

Manuscrit reçu le 29 octobre 1958.

BIBLIOGRAPHIE

[1] MERINIS (J.), *J. Physique Rad.*, 1956, **17**, 308.
[2] COCKROFT (A. L.) et CURRAN (S. C.), *Rev. Scient. Inst.*, 1951, **22**, 37.

[3] MAGEE (F. I.), BELL (P. R.) et JORDAN (W.), *Rev. Scient. Inst.*, 1952, **23**, 30.
[4] PÉNÈGE (L.), *J. Physique Rad.*, 1958, **19**, 71 A.

LE JOURNAL DE PHYSIQUE ET LE RADIUM   TOME 20, JANVIER 1959, PAGE 25.

I. ÉTUDE DU RAYONNEMENT DE FREINAGE INTERNE,
DE L'AUTOIONISATION ET DES ÉLECTRONS ÉMIS SIMULTANÉMENT
AVEC LE RAYONNEMENT β DU PHOSPHORE 32

Par F. SUZOR et G. CHARPAK,

Laboratoire de Synthèse Atomique. C. N. R. S.

Résumé. — On étudie, par analyse des impulsions en coïncidences de deux compteurs proportionnels, les spectres de photons et d'électrons de basse énergie (1 à 25 keV) émis en même temps que les rayons β, par ^{32}P.

Le spectre de photons comprend la raie K de 2,3 keV provenant de l'autoionisation de l'atome résiduel d'intensité $5,5.10^{-4}$ par désintégration, et le spectre continu du rayonnement de freinage interne. Ce dernier concorde parfaitement avec les prévisions théoriques de 1 à 6 keV puis monte fortement au-dessus.

Le spectre électronique comprend la raie Auger de 2 keV et un spectre continu important, d'intensité supérieure à celle prévue par la théorie de l'auto-ionisation. Le désaccord devient considérable de 5 à 25 keV, où l'on trouve $6,6.10^{-3}$ électrons associés par désintégration.

Abstract. — By analysis of the coincident impulses in two proportional counters study is made of the low energy photon and electron spectrum (1 keV to 25 keV) emitted, simultaneously with beta rays, by ^{32}P.

The Photon spectrum comprises 2,3 keV K-line following autoionisation of the residual atom, of intensity $5,5. \times 10^{-4}$ per disintegration, and the internal bremsstrahlung continuous spectrum ; the latter agrees perfectly with the theoretical prediction from 1 keV to 6 keV and then it rises strongly above the thecretical values.

The electron spectrum comprises the 2 keV Auger line and an important continuous spectrum. The intensity of the latter is higher than predicted by autoionisation theory. The discrepancy becomes considerable from 5 keV to 25 keV where $6,6.10^{-3}$ associated electrons are found per disintegration.

I. Introduction. — Le spectrographe à double compteur proportionnel que nous avons construit [1] a été utilisé pour l'étude des photons et des électrons de basse énergie émis lors de la désintégration du phosphore 32 simultanément avec le rayonnement β. L'exposé de ces expériences montrera qu'il est possible d'éliminer les phénomènes parasites, et que l'on obtient entre 1 et 12 keV le spectre des photons et entre 1 et 25 keV celui des électrons avec une bonne précision, et seulement un ordre de grandeur du phénomène pour des énergies supérieures. Les résultats seront comparés avec les théories du freinage interne et de l'auto-ionisation, faisant apparaître, surtout dans le cas des électrons, un désaccord important.

La grande proximité des 2 compteurs est responsable de l'existence de phénomènes parasites qu'il est nécessaire d'éliminer ou tout au moins de réduire et d'évaluer avec précision. Les impulsions en coïncidence ainsi obtenues sont de deux sortes :

a) un rayon β enregistré dans un compteur projette dans le support de source, les écrans ou le gaz environnant un électron qui est enregistré dans l'autre compteur ;

b) un rayon β après un très court parcours de l'ordre de un ou plusieurs millimètres est rétrodiffusé par le gaz du compteur ou le bord de la rondelle sur laquelle est fixée la membrane supportant la source ou toute manière environnante et est ensuite enregistré dans l'autre compteur.

Le premier de ces phénomènes parasites a été systématiquement étudié [2] par l'addition d'écrans d'épaisseur variable à des distances variables de la source. Les résultats en accord avec les prévisions théoriques montrent que l'effet peut être considérablement diminué et rendu souvent négligeable à la condition d'employer des sources radioactives obtenues par vaporisation sur un support très mince en matière plastique et de ne pas employer des écrans en contact immédiat de la source mais à des distances d'au moins un demi-millimètre environ.

Le deuxième phénomène parasite dû à la rétrodiffusion du rayonnement β est beaucoup plus facilement contrôlable. On peut modifier la disposition géométrique de la matière entourant la source ; c'est ainsi qu'il est toujours préférable de disposer la membrane porte-source sur un support métallique percé d'un trou de petit diamètre. Nous avons choisi, dans la plupart des cas, comme support de source une membrane en matière plastique aluminisée d'une épaisseur de 40 microgrammes environ par centimètre carré, tendue sur une rondelle en cuivre ou en béryllium de 0,1 à 0,2 millimètre d'épaisseur avec un trou circulaire de 3 millimètres de diamètre. Le meilleur moyen pour contrôler et mesurer les coïncidences parasites de ce type consiste à diminuer la pression du gaz remplissant les compteurs. Pour toute rétrodiffusion sur une matière autre que le gaz, les conditions géométriques restant les mêmes, la grandeur de l'impulsion correspondant au parcours dans le premier compteur avant rétrodiffusion diminue proportionnellement à la pression du gaz ; pour la rétrodiffusion sur les molécules du gaz, il s'ajoute à cette même raison de diminution celle due à la diminution de la probabilité de diffusion.

II. Spectre des photons.

— Le domaine d'énergie des photons émis simultanément avec le rayonnement β que nous puissions étudier avec notre spectrographe est limité du côté des grandes énergies par l'efficacité du compteur et du côté des basses énergies par l'écran qu'il est nécessaire de mettre devant la source pour arrêter les électrons et diminuer les effets parasites. Cette limite inférieure est de 1 keV. Pour descendre plus bas, il faudrait laisser la source nue, sans écran, et on est alors considérablement gêné par les électrons émis simultanément avec le rayonnement β et les effets parasites précédemment décrits. Par contre pour l'étude d'une désintégration par capture K, ces raisons n'existent plus et rien ne s'oppose, la source étant nue dans l'un des compteurs, à la mesure de photons d'énergie inférieure à 1 keV en coïncidence avec des rayons X ou des électrons Auger.

La figure 1 représente la disposition des divers supports et écrans autour de la source dans les 2 cas choisis pour l'étude du spectre des photons. Ces 2 dispositifs ont l'avantage de rendre pratiquement nulles, dans le domaine d'énergie étudié, les coïncidences parasites par rétrodiffusion. Ils fournissent par contre des impulsions en coïnci-

dence dues à des électrons émis simultanément avec le rayonnement β et franchissant les écrans, ainsi que des coïncidences parasites dues à des électrons projetés. Nous ne tenterons pas de distinguer ces impulsions dues à des électrons de deux provenances différentes ; il nous suffira de remarquer qu'un changement de la nature ou de la pression du gaz remplissant les compteurs fournira toujours le même spectre d'impulsions dues aux électrons. En fait, nous obtiendrons le spectre des photons en comparant les résultats obtenus avec les trois remplissages suivants des compteurs :

a) argon : 60 cm de mercure + propagane 6 cm ;
b) propane : 57 cm de mercure ;
c) propane : 8 cm de mercure.

En négligeant l'efficacité des compteurs par effet paroi, il est facile de calculer l'efficacité en fonction de l'énergie pour chacun de ces remplissages, compte tenu des dispositifs A et B représentés sur la figure 1 et de la forme connue des compteurs ; ces courbes sont représentées sur la figure 2.

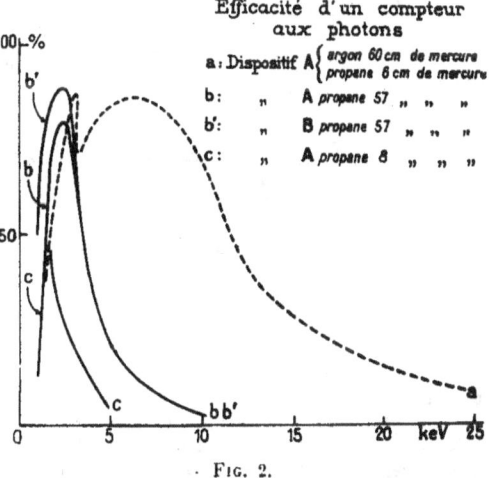

Efficacité d'un compteur aux photons

a : Dispositif A { argon 60 cm de mercure / propane 6 cm de mercure }
b : " A propane 57 " " "
b' : " B propane 57 " " "
c : " A propane 8 " " "

Fig. 2.

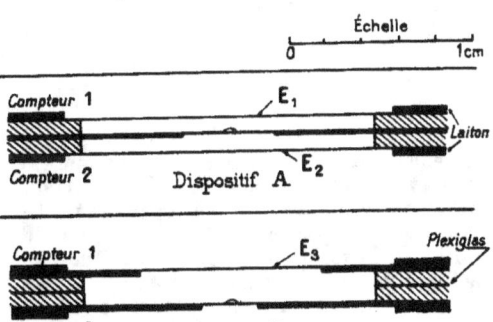

Échelle
0   1cm

Compteur 1 E_1
                              Laiton
Compteur 2 E_2 Dispositif A

Compteur 1 E_3 Plexiglas
Compteur 2   Dispositif B

Fig. 1.

Dispositif A : Les écrans E_1 et E_2 sont en béryllium de 3,3 mg/cm². La membrane mince supportant la source est tendue sur une plaque de béryllium de 0,2 mm d'épaisseur percée d'un trou de 5 mm de diamètre.
Dispositif B : L'écran E_3 est constitué par une membrane de matière plastique aluminisée de l'ordre de 40 microgrammes/cm² tendue sur une plaque de cuivre de 0,2 mm d'épaisseur percée d'un trou de 10 mm de diamètre. La membrane mince supportant la source est tendue sur une plaque de cuivre de 0,2 mm d'épaisseur percée d'un trou de 3 mm de diamètre.

L'étalonnage en énergie des compteurs est obtenu à l'aide des rayonnements X $(K\text{-}L)$ de l'aluminium $= 1,5$ keV, $(K\text{-}L)$ de l'argon $= 2,9$ keV, $(K\text{-}L)$ du manganèse $= 5,9$ keV, un mélange $(K\text{-}L)$ et $(K\text{-}M)$ du cuivre $= 8,5$ keV, $(L\text{-}M)$ du bismuth $= 13$ keV, $(L\text{-}N)$ du bismuth $= 16$ keV et la raie de 47 keV du RaD.

La figure 3 donne les résultats bruts en coïncidence après soustraction du mouvement propre et des coïncidences fortuites ; ils sont obtenus dans le sélecteur à 50 canaux avec le dispositif B et le remplissage b (propane : 57 cm de mercure) pour des gains de l'amplificateur de la voie 1 ayant les valeurs 2 500 et 5 000, et pour la voie 2 un gain de 580 et une discrimination ne laissant passer que les impulsions correspondant à une énergie supérieure à 14 keV.

La figure 4 représente les résultats complets dans les mêmes conditions, les abscisses étant graduées

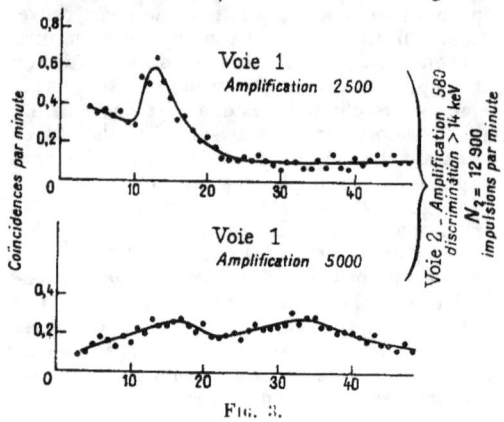

Fig. 3.

en keV et les ordonnées représentant le nombre d'impulsions comptées dans un intervalle de 1 keV. De plus, les courbes ont été ramenées à un taux de comptage de 10^4 dans le compteur 2 ($N_2 = 10\,000$ par unité de temps, pour une énergie supérieure à 14 keV). Ce qui nous intéresse est de connaître le spectre des impulsions dues au rayonnement enregistré dans le compteur 1 et associé aux 10 000 rayons β du compteur 2 ; il faut donc soustraire de la courbe de la figure 4 les coïncidences dues aux rayons β enregistrés dans le compteur 1 et associés au rayonnement d'énergie supérieure à 14 keV du compteur 2. Ces coïncidences représentées en pointillé sur la figure 4 sont calculées moyennant certaines hypothèses : on suppose que le rayonnement émis simultanément au rayonnement β est indépendant de l'énergie de ce dernier, ce qui donne pour la courbe pointillée le même spectre à une multiplication près des ordonnées ($n/10^4 . N_1$) que le spectre N_1 du rayonnement β enregistré dans le compteur 1 ; n est le nombre total d'impulsions d'énergie supérieure à 14 keV dans le compteur 2

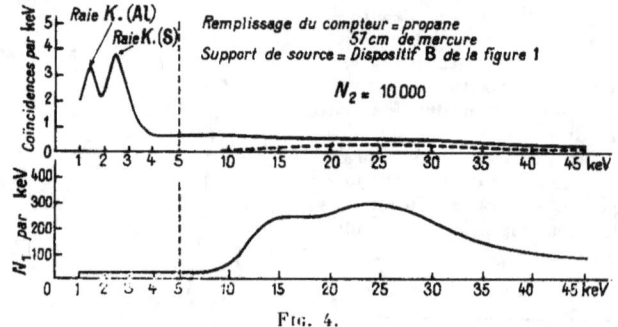

Fig. 4.

émis simultanément avec 10^4 rayons β enregistrés dans le compteur 1. Tenant compte des spectres intégrés de N_1 et des coïncidences pour des énergies supérieures à 14 keV, et connaissant les angles solides sous lesquels sont vus, de la source, chacun des compteurs, on en déduit pour n la valeur 9,2. On voit que la courbe pointillée ainsi obtenue et venant en déduction des coïncidences, représente une correction très faible pour des énergies inférieures à une dizaine de keV. Ce résultat est obtenu grâce aux grandes dimensions des compteurs rendant les valeurs de N_1 très petites pour des énergies inférieures à une dizaine de keV environ.

A côté des résultats représentés sur la figure 4, d'autres résultats semblables ont été obtenus avec chacun des remplissages a, b, et c du compteur ou avec chacun des dispositifs A et B (fig. 1). Les impulsions en coïncidence dues aux électrons étant indépendantes du remplissage, il est facile de les éliminer par différences entre les courbes obtenues

pour deux remplissages différents. Tenant compte des efficacités aux photons du compteur (fig. 2) nous avons obtenu le spectre des photons qui est représenté sur la figure 5. La courbe en trait plein représente le spectre des photons venant de la source, à l'exception de la bosse située autour de 9 keV qui est due à des photons d'énergie supérieure enregistrés dans le compteur par effet paroi (raie K du cuivre). Cette partie de la courbe a d'ailleurs été obtenue avec le dispositif A (fig. 1) où la plaque supportant la membrane porte-source est en béryllium. Avec le dispositif B où cette même plaque est en cuivre les résultats obtenus sont les mêmes, sauf autour de 9 keV où la bosse est considérablement augmentée, ce qui est dû à la diffusion des rayons β par les bords du trou et excitation de la couche K du cuivre. La comparaison de ces résultats avec plaque de cuivre et plaque de béryllium, ainsi que la prévision théorique permettent d'affirmer que la contribution due au freinage externe du rayonnement β est absolument négligeable. La courbe pointillée de la figure 5 se confond avec la courbe en trait plein à l'exception de 2 raies, l'une à 1,5 keV et l'autre à 2,9 keV. La première a été obtenue lorsque la membrane en matière plastique supportant la source est recouverte par vaporisation sous vide d'une couche d'aluminium d'une dizaine de microgrammes par centimètre carré pour la rendre conductrice. L'excitation de la couche K de l'aluminium par le rayonnement β est alors responsable de cette raie (celle-ci est visible sur les figures 3 et 4). La deuxième raie est obtenue avec le remplissage du compteur par de l'argon et est due à l'excitation par le rayonnement β de la couche K de ce gaz. Dans ces bandes d'énergie, la courbe en trait plein provient de mesures faites avec une source sans aluminium et un remplissage de propane.

En résumé, à l'exception de cette légère bosse autour de 9 keV, la courbe en trait plein de la

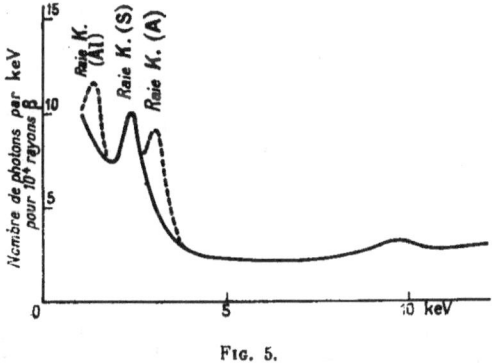

FIG. 5.

figure 5 fournit le spectre des photons émis simultanément dans un angle solide 4π avec 10 000 rayons β. Ce spectre se compose d'un fond continu d'intensité décroissante en fonction de l'énergie dû au rayonnement de freinage interne et d'une raie de 2,3 keV qui est la raie K de l'atome résiduel de soufre et qui est due au phénomène d'autoionisation. La séparation du fond continu et de la raie présente un certain arbitraire qui est considérablement réduit par l'obligation d'obtenir après séparation une raie symétrique ayant la largeur convenable imposée par le pouvoir de résolution connu du compteur. Le résultat, après séparation, est donné sur la figure 6 ; la surface de la raie correspond pour le phénomène d'autoionisation à l'émission de 5,5 photons K pour 10 000 rayons β. La figure 6 reproduit aussi en trait pointillé le résultat prévu pour le rayonnement de freinage interne par la théorie de Knipp et Uhlenbeck [3] dans notre dispositif expérimental. Cette courbe a

été calculée en tenant compte de l'intensité prévue par la théorie en fonction de l'énergie et aussi de la corrélation angulaire prévue par cette même théorie entre rayon β et photon de freinage. Cette corrélation angulaire présentant un maximum vers l'avant, il en résulte que, avec notre dispositif A

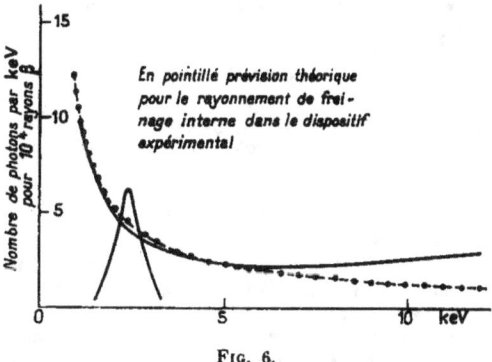

En pointillé prévision théorique pour le rayonnement de freinage interne dans le dispositif expérimental

FIG. 6.

(fig. 1), le nombre de photons enregistrés en coïncidence ne représente que 41 % de l'intensité que l'on obtiendrait avec le rayonnement isotrope. Avec le dispositif B, cette proportion est de 37 %. (Ces valeurs ont été obtenues par intégration graphique.)

On voit sur la figure 6 qu'il existe un accord excellent entre la théorie de freinage interne et notre résultat expérimental entre 1 et 5 keV. De 5 à 12 keV la courbe expérimentale s'élève progressivement et rapidement au-dessus de la courbe déduite de la théorie. Ce fait est à rapprocher des résultats obtenus par certains auteurs [4]. H. Langevin-Joliot étudiant le spectre des photons de freinage interne du phosphore 32 jusqu'à 15 keV a trouvé pour cette énergie un excès de 50 % par rapport à ce que prévoit la théorie. Nos expériences indiquent à 12 keV l'existence d'un nombre de photons presque 3 fois supérieur à la prévision théorique. Si l'on veut chercher un accord entre nos résultats et ceux de H. Langevin-Joliot, il faudrait admettre que l'excès de photons constatés dans les deux cas ne suit pas la même loi de corrélation angulaire que celle prévue par la théorie de freinage interne, mais est émis d'une façon pratiquement isotrope par rapport à la direction d'émission de la particule β. Indiquons pour terminer que nos résultats ne sont pas en accord avec les expériences de G. A. Renard [5] sur l'autoionisation et le rayonnement de freinage interne entre 1 et 5 keV du phosphore 32. Sauf erreur systématique dans une des expériences, le désaccord pourrait provenir d'une loi de corrélation angulaire photon-rayon β différente de la prévision théorique.

II. Spectre des électrons. — L'étude du spectre des électrons émis simultanément avec le rayonnement β lors de la désintégration du phosphore 32 a été effectuée avec le dispositif C comme porte-source (*fig.* 7) dans lequel aucun écran n'est inter-

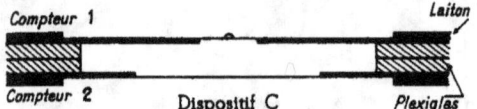

Fig. 7. — La membrane mince d'environ 40 microgrammes par cm² supportant la source est tendue sur une plaque de cuivre de 0,2 mm d'épaisseur percée d'un trou de 3 mm de diamètre.

posé entre la source et le gaz du compteur 1. Le dispositif C' est celui dans lequel un écran est placé directement contre la source du côté du compteur 1. Avec le dispositif C, nous enregistrerons dans le compteur 1 des photons, des électrons et les coïncidences parasites précédemment mentionnées. Les résultats que nous donnons dans la suite de cet article ont été obtenus après soustraction des coïncidences dues aux photons ; ces coïncidences ont été calculées en partant du spectre représenté sur la figure 6 et de l'efficacité connue du compteur (*fig.* 2).

Il existe pour des énergies faibles un spectre rapidement décroissant de coïncidences parasites dues à la rétrodiffusion du rayonnement β. L'importance de ce spectre avec un remplissage du compteur en argon rend impossible l'étude des électrons par ce procédé. Nous avons utilisé les remplissages b (propane : 57 cm de mercure) et c (propane : 8 cm de mercure). Avec le remplissage b ce spectre parasite est nul pour des énergies supérieures à 2 keV ; avec le remplissage c, il disparaît complètement pour des énergies supérieures à 1 keV, qui représente la limite inférieure de nos résultats. L'emploi du dispositif C' supprime également ce spectre parasite, avec le remplissage b, montrant ainsi qu'il est dû à la rétrodiffusion du rayonnement β sur les bords de la rondelle supportant la membrane sur laquelle est déposée la source, et non pas à une rétrodiffusion sur les molécules du gaz.

Le spectre d'électrons que reproduit la figure 9 se rapporte à 10 000 rayons β. Les coïncidences, d'ailleurs en petit nombre, dues aux rayons β enregistrés dans le compteur 1 ont été soustraites après calcul par une méthode analogue à celle décrite précédemment. Comme nous l'avons dit, les coïncidences dues aux photons, d'un ordre de grandeur inférieur à celles dues aux électrons, ont été également soustraites. Les coïncidences parasites dues à la rétrodiffusion n'apparaissent pas car le spectre a été obtenu avec le remplissage b pour une énergie supérieure à 2 keV et avec le remplissage c pour une

énergie supérieure à 1 keV. (Au-dessus de 2 keV, les deux résultats sont identiques, aux erreurs d'expérience près.) La figure 9 donne le spectre des électrons dans le dispositif C' après traversée d'un écran de 0,04 mg/cm² ou de 0,2 mg/cm² montrant que l'absorption est conforme à ce que l'on peut attendre.

S'il existe des coïncidences parasites dues à la projection d'électrons par le rayonnement β, ce phénomène ne peut se produire que dans la matière environnant la source (*fig.* 7). Cette matière est constituée par environ 2 à 3 millimètres d'épaisseur de gaz, par la membrane support de source et par la membrane faisant écran à la partie inférieure du dispositif C au contact du gaz du compteur 2. Ces deux membranes sont en matière plastique aluminisée d'une épaisseur totale de 0,04 mg/cm². Un calcul basé sur la probabilité de projection d'un électron et sur l'absorption de cet électron projeté nous a donné une valeur négligeable devant l'intensité du spectre de la figure 8. Ce point de vue est

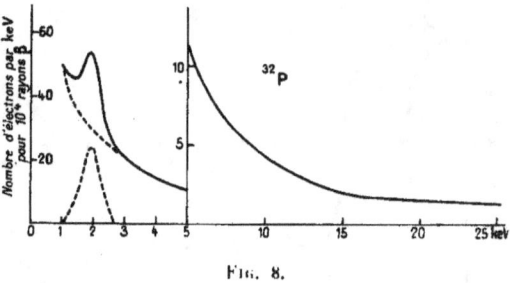

FIG. 8.

confirmé expérimentalement par l'étude analogue faite sur le soufre 35, que nous développons dans un autre article de ce même journal.

Sur la figure 8, nous avons reproduit en pointillé une décomposition du spectre des électrons donnant d'une part une raie de 2,0 keV et d'autre part, un fond continu décroissant avec l'énergie. Cette raie

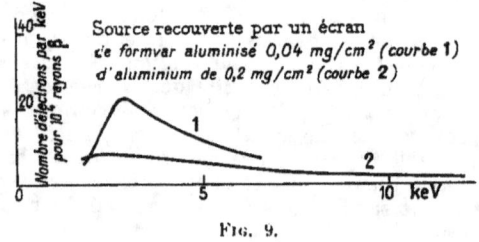

Source recouverte par un écran de formvar aluminisé 0,04 mg/cm² (courbe 1) d'aluminium de 0,2 mg/cm² (courbe 2)

FIG. 9.

est due aux électrons Auger consécutifs au réarrangement du cortège électronique de l'atome de soufre. Lorsqu'il y a autoionisation de l'atome, l'électron Auger est émis simultanément avec l'électron éjecté de la couche K, et cet électron Auger apparaît dans la raie de 2,0 keV seulement dans le cas où la direction d'émission de l'électron éjecté ne permet pas qu'il soit enregistré dans le même compteur. Par contre lorsqu'il y a autoexcitation de l'atome et que l'électron éjecté de la couche K est lié, l'électron Auger est toujours compté dans un cas sur deux, le compteur 1 étant vu de la source sous un angle solide 2π. La proportion inconnue des phénomènes d'autoionisation et d'autoexcitation ne permet pas de déduire de la surface de la raie d'électrons de 2,0 keV le nombre d'électrons Auger effectivement émis par la source. Il est important de remarquer que les électrons enregistrés dans notre expérience sont non seulement ceux émis dans un angle solide 2π, mais également ceux rétrodiffusés par la membrane supportant la source. Cette rétrodiffusion s'accompagne d'une dégradation de l'énergie ; il est pratiquement impossible de séparer la partie relativement faible du spectre qui correspond aux électrons rétrodiffusés ; indiquons seulement que la fraction d'électrons rétrodiffusés dans ce domaine d'énergie par une épaisseur saturante d'aluminium est de 15 % [7].

En résumé, nous trouvons des électrons monocinétiques de 2,0 keV et des électrons d'énergie variable dont le spectre est représenté approximativement par une loi en fonction inverse de l'énergie avec une intensité égale à 156.10^{-4} entre 1 et 25 keV.

La théorie de l'autoionisation [8] prévoit pour le phosphore 32 l'émission d'un spectre continu d'électrons que nous avons représenté en pointillé

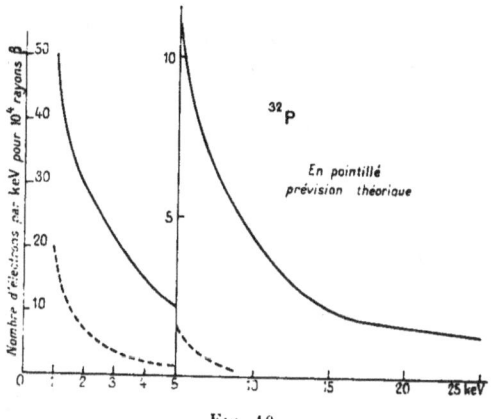

FIG. 10.

sur la figure 10, notre spectre expérimental, qui est le même que celui de la figure 8, y étant représenté en trait plein. On constate vers 1 et 2 keV environ 3 fois plus d'électrons que n'en prévoit la théorie ; à partir de 4 à 5 keV, le désaccord devient plus considérable, puisque l'intensité du spectre théorique pour des énergies supérieures à 5 keV est de $2,10^{-4}$ et que notre valeur expérimentale est de 66.10^{-4} entre 5 et 25 keV. Ce fait confirme et précise des expériences plus anciennes effectuées à l'aide de 2 compteurs Geiger-Muller fonctionnant en coïncidence [9].

Manuscrit reçu le 29 octobre 1958.

BIBLIOGRAPHIE

[1] CHARPAK (G.) et SUZOR (F.), *J. Physique Rad.*, 1958, **19**, 167 A.
[2] CHARPAK (G.), *Thèse*, Paris, 1955.
[3] β and γ ray spectroscopy, Siegbahn, Amsterdam. 1955, 649.
[4] LIDEN (K.) et STARFELT (N.), *Phys. Rev.*, 1955, **97**. 419. LANGEVIN-JOLIOT (H.), *Thèse*, Paris, 1956 : *Ann. Physique*, 1957, 2, 16.
[5] RENARD (G. A.), *J. Physique Rad.*, 1957, **18**, 681.
[6] MERINIS (J.), *J. Physique Rad.*, 1956. 17. 308.

[7] STERNGLASS (E. J.), *Research Report*, 1953, R. 94416-3-J Westinghouse.
[8] MIGDAL (A.), *J. Physique*, U. R. S. S., 1941, **4**, 449. FEINBERG (E. L.), *J. Physique*, U. R. S. S., 1941, **4**, 424. LEVINGER (J. S.), *Phys. Rev.*, 1953, **90**, 11. GRARD (F.), *Thèse*. Université libre de Bruxelles, 1958.
[9] SUZOR (F.) et CHARPAK (G.), *J. Physique Rad.*, 1954, **15**, 378. DUQUESNE (M.). *C. R. Acad. Sc.*, 1955, **241**, 195 et *Thèse*. Paris.

2. DETECTORS WITH OPTICAL IMAGING: 1956–1992

2. DETECTORS WITH OPTICAL IMAGING: 1956–1992

LE JOURNAL DE PHYSIQUE ET LE RADIUM — TOME 17, JUILLET 1956, PAGE 585.

ÉTUDE, AVEC UN PHOTOMULTIPLICATEUR,
DES IMPULSIONS LUMINEUSES PRODUITES DANS LE GAZ D'UN COMPTEUR PROPORTIONNEL

Par G. CHARPAK et G.-A. RENARD,
Laboratoire de Physique et Chimie Nucléaires du Collège de France.

Sommaire. — La lumière émise par les atomes excités pendant une décharge dans un compteur proportionnel a été étudiée au moyen d'un photomultiplicateur dont la photocathode a un diamètre de 25,4 mm. Les auteurs donnent des résultats préliminaires obtenus avec des mélanges argon-propane et krypton-propane comme gaz de remplissage. L'essai des compteurs avec des rayons X de 5,9 et 9,2 keV montre :

1º Que la lumière est émise par des états excités de vie moyenne voisine de 2.10^{-7} secondes.

2º Que l'intensité de la lumière est proportionnelle à l'énergie des rayons X quand le compteur travaille au régime proportionnel.

Dans les conditions expérimentales décrites, la limite inférieure de l'énergie des rayons X détectables est 1 keV. Elle peut être accrue par une meilleure géométrie.

Abstract. — The light emitted by the atoms excited during a discharge in a proportionnal counter has been studied with a photomultiplier with a 1 inch diameter photocathode. The authors report preliminary results with mixtures at argon-propane and krypton-propane as filling gases. Checking the counter with 5,9 keV and 9,2 keV X-rays shows :

1) That the light is emitted by excited states of half-life close to 2×10^{-7} seconds.

2) That the intensity of light is proportionnal to the energy of the X-rays when the counter is working in the proportionnal regim.

In the experimental conditions described, the lower limit for the energy of the X-rays detectable by this method is 1 keV. This can be improved in particular by better geometry.

Lorsqu'un rayonnement provoque une impulsion dans un compteur proportionnel, il y a émission de photons de longueurs d'onde diverses par les atomes excités lors de l'avalanche électronique, simultanément avec la collection des électrons sur le fil du compteur.

G. Charpak [1] a montré que ces impulsions lumineuses étaient d'intensité suffisante pour permettre, par cette méthode, la détection des rayons X de quelques kiloélectronvolts. A la suite de ces résultats une étude systématique a été entreprise dont nous exposons les résultats préliminaires.

Dispositif expérimental. — Dans la coque d'un compteur proportionnel, de longueur 25 cm, de diamètre 5 cm, du côté opposé à une fenêtre de 2/100 de millimètre d'aluminium, on a aménagé une ouverture de 3 cm devant laquelle vient se placer la photocathode d'un multiplicateur Du Mont 6 291 (figure 1). Une grille métallique, reliée à lacoque du compteur, est placée devant la photocathode pour éviter que l'ouverture ne perturbe le fonctionnement du compteur. La transparence de la grille est de 75 % environ. La haute tension positive est appliquée au fil du compteur. Deux ensembles d'amplification et de sélection d'amplitude permettent d'observer simultanément les impulsions électriques sur le fil du compteur et à l'anode du photomultiplicateur.

Nous avons travaillé dans les conditions suivantes : 90 % d'argon et 10 % de propane, ou 90 % de krypton et 10 % de propane, à des pressions totales variables entre 30 cm et 75 cm de

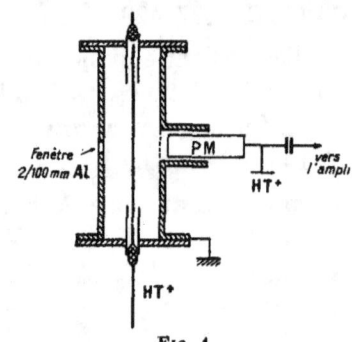

Fig. 1.

hauteur de mercure. Le gaz du compteur était excité avec des rayons X de 5,9 KeV émis par le ^{55}Fe et de 9,2 KeV émis par le ^{71}Ge.

Résultats expérimentaux. — Pour observer des impulsions dans le photomultiplicateur il faut fonctionner à la limite de la zone de proportionnalité du compteur. La figure 2 montre la raie de 5,9 KeV obtenue dans les conditions suivantes :

Remplissage argon-propane. Pression totale 75 cm.

Le rapport des amplitudes de la raie et du bruit de fond est de l'ordre de 10 du côté des impulsions lumineuses, alors qu'il est de 200 à 300 du côté des impulsions recueillies sur le fil. La limite de détection est donc de l'ordre de 1 KeV alors qu'elle est de l'ordre de l'énergie nécessaire à la formation d'une seule paire d'ions par la méthode normale. On peut toutefois gagner assez aisément un facteur 10 en améliorant l'angle solide, sous lequel le fil est vu par la photocathode. Notons que la lumière reçue par la photocathode provient de centres lumineux situés au voisinage du fil comme le prouve l'existence même d'une raie.

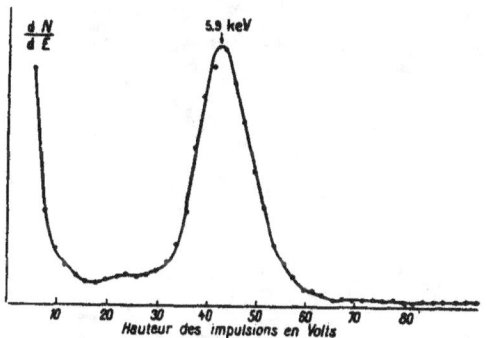

Fig. 2. — Spectres des impulsions obtenues avec le photomultiplicateur Du Mont 6291 avec des raies X de 5,9 keV.
Remplissage du compteur : propane 6,7 cm et argon 59,3 cm.
Tension compteur : 2 300 volts.

Nous avons examiné la forme des impulsions du photomultiplicateur (figure 3) à l'aide d'un oscilloscope et d'un amplificateur de largeur de bande 2 Mégacycles avec une résistance de charge de

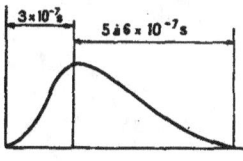

Fig. 3.

1 000 ohms à l'anode du photomultiplicateur. La montée de l'impulsion se produit en 0,3 microsecondes comme celle de l'impulsion recueillie sur le fil. La descente a lieu en 0,5 ou 0,6 microsecondes, ce qui montre que la vie moyenne des

atomes excités responsables de l'impulsion lumineuse est de l'ordre de quelques dixièmes de microsecondes. Ceci montre que l'émission de lumière n'est pas due aux atomes de krypton ionisés lors de l'avalanche électronique mais à des états excités d'énergie plus basse. Ce résultat est explicable. En effet, la vie moyenne des atomes ionisés est de l'ordre de 10^{-8} seconde, la désexcitation ayant lieu aussi bien par émission de radiations que par collision [2]. La propane absorbe la majeure partie de l'énergie d'excitation des atomes de krypton ionisés et la photocathode n'est pas sensible aux rayonnements situés dans l'ultraviolet lointain, émis par les atomes de krypton ionisés.

Notre résultat est à rapprocher de celui de Ward [3] qui a détecté avec un photomultiplicateur les impulsions lumineuses produites dans un gaz rare par des rayons α. Il observa que les impulsions lumineuses observées n'étaient pas dues aux atomes ionisés.

Par contre, dans une expérience analogue à celle de Ward, Grinfell, P. Boicourt et John E. Brolley [4] ont pu détecter les rayonnements ultraviolets produits en utilisant un film convertisseur de lumière, placé devant la photocathode du multiplicateur. Ce film convertit par fluorescence les rayonnements U.-V. lointains en rayonnements dont la longueur d'onde est dans le domaine de sensibilité de la photocathode. Ceci montre qu'il y a peut-être dans notre cas aussi une possibilité d'améliorer le rendement par cette methode.

Nous entreprenons des mesures analogues à celles que nous venons de décrire avec un compteur proportionnel à parois de quartz et grille interne, entouré de quatre P. M. Il est douteux que l'on puisse atteindre une résolution et un rapport signal sur bruit aussi bons que par la méthode conventionnelle. On peut cependant espérer obtenir des renseignements intéressants sur le fonctionnement du compteur.

Nos résultats préliminaires peuvent également suggérer certaines applications de la méthode décrite. Par exemple, avec un compteur proportionnel de grande longueur destiné à la détection de particules énergiques on pourrait, en « regardant » aux deux extrémités du compteur, avec un photomultiplicateur, les impulsions lumineuses, éliminer les particules qui n'ont pas parcouru toute la longueur du compteur. Ceci équivaudrait au dispositif de télescope à trois compteurs en coïncidences, en éliminant les inconvénients qui pourraient provenir du passage des particules dans deux compteurs supplémentaires.

BIBLIOGRAPHIE

[1] Charpak (G.), Travail non publié, 1955.
[2] Curran (S. C.) et Craggs (J. D.), Counting Tubes. Buttenwark Scientific Publications, London, 1949.
[3] Ward, Proc. Phys. Soc., London, septembre 1946, A 67, 841-846.
[4] Grinfell, Boicourt (P.) et Bolley (John E. Jr.), Rev. Sc. Instr., 1954, 25, 1218.

PRINCIPE ET ESSAIS PRÉLIMINAIRES
D'UN NOUVEAU DÉTECTEUR
PERMETTANT DE PHOTOGRAPHIER
LA TRAJECTOIRE DE PARTICULES IONISANTES
DANS UN GAZ

Par G. Charpak,

Laboratoire de Physique et Chimie Nucléaires,
Collège de France.

Considérons une particule ionisante qui traverse un gaz à la pression p, soumis à un champ électrique uniforme E, de durée T.

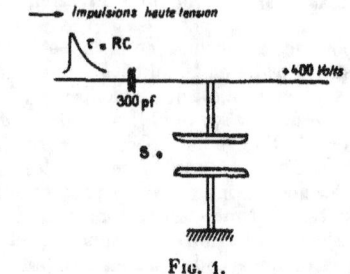

Fig. 1.

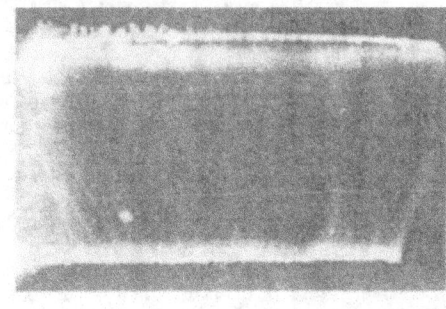
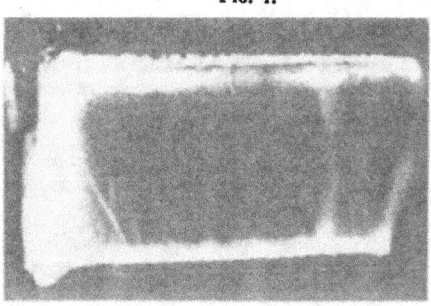
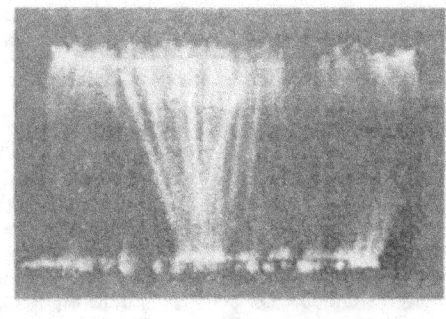
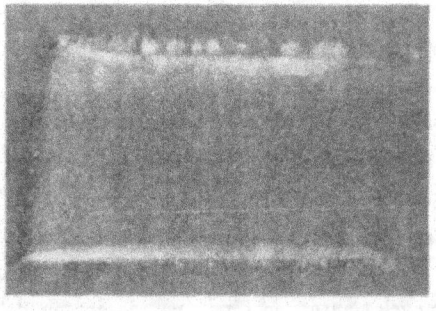
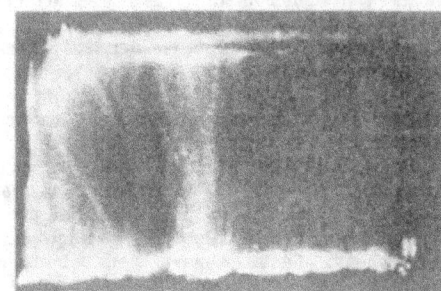
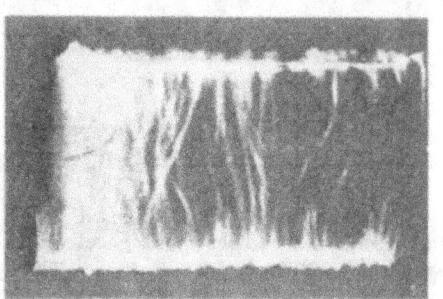

a

b

c

d

e

f

Fig. 2.

a) $V = 83\ 000$ volts, $\tau = 1{,}25 \times 10^{-9}$ s, 18 000 impulsions, source latérale.
b) $V = 83\ 000$ volts, $\tau = 1{,}25 \times 10^{-9}$ s, 6 000 impulsions, source latérale.
c) $V = 66\ 000$ volts, $\tau = 6{,}25 \times 10^{-9}$ s, 9 000 impulsions, source sur électrode inférieure.
d) $V = 66\ 000$ volts, $\tau = 6{,}25 \times 10^{-9}$ s, 6 000 impulsions, source sur électrode supérieure.
e) $V = 66\ 000$ volts, $\tau = 6{,}25 \times 10^{-9}$ s, 1 800 impulsions, source latérale et source sur électrode inférieure, simultanément.
f) $V = 66\ 000$ volts, $\tau = 6{,}25 \times 10^{-9}$ s. La pellicule de nickel, support de la source α, est retournée sur l'électrode inférieure.

Examinons s'il est possible de choisir la nature du gaz et les paramètres p, E et T, de sorte que chaque électron libéré le long de la trajectoire produise une avalanche de N électrons, le nombre N étant tel que les atomes excités dans l'avalanche émettent une lumière assez intense pour être photographiée, tout en restant confinés à un volume assez petit pour que l'ensemble des avalanches soit une image fidèle de la trajectoire.

D'après la théorie des étincelles, le nombre N est au maximum de l'ordre de 19^9 dans tous les gaz [1]. A partir de cette grandeur de l'avalanche il y a propagation de canaux prédisruptifs (« streamers ») vers l'anode et la cathode, avec des vitesses respectives comprises entre 2 à 4×10^7 cm/s et 4×10^7 à 2×10^8 cm/s. Lorsque ces canaux joignent les électrodes il y a passage de l'étincelle proprement dite (« return stroke ») à une vitesse de 10^{10} cm/s. [1], [3], [4].

La vitesse de formation de l'avalanche dépend du gaz et de la tension. Elle est, par exemple de 10^{-8} s pour une tension double de la tension de rupture statique, dans l'air [2]. En raison des grandes vitesses de propagation il est nécessaire, pour que la multiplication autour d'un électron soit localisée, d'employer des impulsions de tension très brèves (de l'ordre de 10^{-9} s) avec des gaz où la rapidité de formation de l'avalanche soit la plus brève possible.

Nous avons, pour cette raison, choisi de faire les essais avec un mélange gazeux exempt d'attachement électronique : 9 cm d'argon + 1 cm de propane. purifié dans un four à calcium. Le dispositif expérimental est représenté sur la figure 1.

Deux électrodes d'aluminium poli, de diamètre 3,5 cm, sont distantes de 2 cm, dans un tube de verre étanche dans lequel sont introduits les mélanges gazeux. La source radioactive est placée soit latéralement entre les électrodes, sur une tige isolante, soit placée sur l'une ou l'autre des électrodes.

L'objectif photographique, d'ouverture 1/2,8 est placé à 30 cm des électrodes. Les impulsions de tension sont produites en chargeant une capacité de 25 μμf à la haute tension et en la déchargeant dans une résistance R comprise entre 50 Ω et 250 Ω. (La constante de temps de l'impulsion (τ) est donc comprise entre $1,25 \times 10^{-9}$ et 6×10^{-9} s). La commutation a lieu par un moteur permettant d'effectuer l'opération 600 fois par minute. Les impulsions sont envoyées sur l'électrode à travers une capacité afin de pouvoir polariser l'électrode à 400 volts et balayer les électrodes et les ions produits par la source. L'expérience consiste à faire varier la tension depuis 50 kV jusqu'à 100 kV jusqu'à ce que la pellicule soit impressionnée et, à tension constante, à faire varier la constante de temps.

Résultats. — Avec une source de rayons β de 5 mC (^{35}S) nous n'observons rien d'autre que des étincelles filamentaires et irrégulières dont l'apparition est indépendante de la présence de la source.

Avec une source de rayons α (0,5 mc de Po^{210}), nous observons des trajectoires qui partent de la source (photos a, b, c, d, e). Les trajectoires émises par la source latérale sont toujours dirigées vers la cathode. Les trajectoires relient les deux électrodes. Ceci semble montrer qu'il y a dans ce cas intervention d'un processus d'amplification autre que celui que nous avons envisagé au début. Il y aurait non seulement une amplification du nombre des électrons de la trajectoire primitive par les avalanches mais également propagation de l'étincelle dans le canal préionisé ainsi créé (« return stroke ») ce qui multiplie le nombre initial d'électrons par un facteur voisin de 10^8. Ceci expliquerait qu'il soit nécessaire que la trajectoire atteigne la cathode pour être visible.

Pour obtenir la trajectoire, il est nécessaire de polariser l'électrode. Nous n'avons pas pu obtenir de trajectoires dans l'air pur et l'argon pur.

Lettre reçue le 17 juillet 1957.

BIBLIOGRAPHIE

[1] RAETHER (M.), *Z. Physik*, 1939, **112**, 464.
[2] RAETHER (M.), *Z. Physik*, 1941, **117**, 524.
[3] LÉONARD et LOEB (B.), *Handb. d. Physik*, 1956, **22**, 445.
[4] MEEK (J. M.) et CRAGGS (D.), Electrical breakdown of gases, Clarendon Press, Oxford, 1953.

Some Developments of the Discharge Chamber

G. CHARPAK*

CERN, Geneva, Switzerland

AND

L. MASSONET†

Faculté des Scienes de l'Université de Paris, Institut de Radium, Orsay, France

(Received 14 December 1962; and in final form, 11 March 1963)

A discharge chamber has been constructed with electrodes of transparent metallic mesh. By the addition of quenching agents, mainly I_2 vapor, to noble gases at atmospheric pressure, and the use of pulses of one or several microseconds duration, it is shown that a great number of discharges connecting the two electrodes can develop simultaneously, starting from the ionized path of a particle. Pulses in the microsecond range of 2 to 3 kV/cm are used, thus allowing the easy operation of large sensitive volumes. Tracks of good resolution are observed, with a variation of the luminosity as a function of the primary ionization. Repetition rates up to 10^4/sec are reached. Sparks following the trajectory of particles inclined at angles of 45° with respect to the electrodes are observed.

THE rapid development in the use of spark chambers, essentially in high energy physics, has to some extent left in the shade the different types of chambers also based on discharges in gases, but giving a more complete picture of the tracks, the so-called "track delineating chambers," as opposed to the standard "track sampling chambers".[1]

In the ordinary spark chamber also the spark may follow exactly the ionized path.[2,3] With gaps of the order of 1 cm, such an operation requires short pulses ($<10^{-7}$ sec) and small angles between the trajectory and the normal to the electrodes ($<20°$).[2] With large gaps (up to 10 cm), recent work at Dubna[4] has shown that this angle can be as big as 40°. Apart from being strongly anisotropic in their detection efficiency such chambers suffer from the same drawback as the sampling chambers: in general, no information on the primary ionization; relative inefficiency for simultaneous tracks.

The ideal delineating chamber is based on pure avalanche multiplication.[5,6] The conditions are to be such that each electron liberated along the track gives rise to an avalanche of limited size and sufficient intensity to produce an amount of light suitable for photography. A detailed analysis by Schneider[5] shows that in hydrogen, for instance, at a pressure of 10^2 cm Hg, pulses of 2×10^{-9} sec width and 130 kV/cm are necessary to obtain a sufficient amount of

light with a resolution of 1 mm. With oscillating pulses, however, the situation is more favorable[6] and a Milan group[7–9] has obtained satisfactory pictures of α tracks, with damped tracks of pulses of 0.4-μsec duration and 40-Mc frequency, with an amount of light still marginal for convenient photography.

The discharge chamber, invented by S. Fukui and S. Miyamoto,[3] offers some features of a perfect delineating chamber of very simple operation. The tracks are viewed through transparent glass electrodes. The conductive transparent coating is on the outside of the glass. A pulsed voltage is applied when a particle traverses the chamber, parallel to the electrodes. Discharges connecting the two electrodes start from seeds along the ionized trajectory. The important property exhibited by this chamber is that it is possible to have the energy of the discharge split into very many partial discharges. As emphasized by the authors, the difference from a spark chamber with metallic electrodes is that the electric field in the gap does not drop rapidly as a consequence of the large surface resistivity of the glass. The main limitation of these chambers is that they provide only a plane projection of the trajectory. The value of the applied fields, typically 6 to 10 kV/cm with neon–argon mixtures, limits the possible useful volume of such chambers. The recovery time is of the order $\frac{1}{10}$ of a second.

In this work we will show that it is possible to have also a great number of independent discharges between metallic transparent electrodes under conditions which change to a large extent the properties of the discharge chamber and extend its possible applications in particle physics.

---

* Visitor at CERN, on leave from Faculté des Sciences, Orsay and Centre National de la Récherche Scientifique, Paris.
† Present address: CERN, Geneva 23, Switzerland.
[1] A. Roberts, 1960 Int. Conf. on Instr. for High-Energy Physics, Berkeley, September 1960, and 1961 Argonne Symposium on Nuclear Spark Chambers, Rev. Sci. Instr. **32**, 480 (1961).
[2] G. Charpak, J. Phys. Radium **18**, 539 (1957) and Proceedings of the Int. Conf. on Mesons and recently discovered particles, Venice, 28 September 1957.
[3] S. Fukui and S. Miyamoto, Nuovo Cimento **11**, 13 (1959); J. Phys. Soc. Japan **16**, 2574 (1961).
[4] A. M. Govorov, V. I. Nikanorov, G. Peter, A. F. Pisarev, and Kh. Poze, Pribory i Tekhn. Eksperim. No. 6, 49–51 (November–December 1961).
[5] F. Schneider, CERN Internal report PS/int AR 59-6. The discharge chamber.

[6] S. Fukui, S. Hayakawa, T. Tsukishima, and H. Nukishima, Proc. Int. Conf. on Inst. for High-Energy Physics, Berkeley, September 1960, p. 267.
[7] F. T. Arecchi, G. Cavalleri, E. Gatti, and G. Redaelli, Energia Nucl. (Milan) **8**, 8, 539 (1961).
[8] C. Cavalleri, E. Gatti, M. Nasini, and G. Redaelli, Energia Nucl. (Milan) **8**, 12, 779 (1961).
[9] C. Cavalleri, E. Gatti, and G. Redaelli, Nuovo Cimento **25**, 1282 (1962).

664

THE SLOW DISCHARGE CHAMBER

In all types of chamber the applied voltages are much higher then the static breakdown voltages. This type of operation offers some advantages for obtaining short resolution times. The gases used are usually kept under conditions of low electron attachment. Because of the high mobility of the electrons, it is necessary to have very fast rise times in order to reach the breakdown potential before the electrons are swept away to the anode. The use of very overvolted gaps reduces the time lags between the application of the pulse and the breakdown to some tens of nanoseconds. Under such conditions, the production of discharges along several tracks is not easy. The fluctuations in the development of processes preceding the spark prevent a 100% efficiency in simultaneous tracks and, in order to reduce these fluctuations, one sometimes has to go to great overvoltages. Even so, in some gases, it is impossible to observe simultaneous tracks.

We observed that the operating conditions of the discharge and spark chambers can be changed and some novel features introduced by the following methods: (i) addition of a halogen vapor to the gas, (ii) use of pulses of long duration (0.5 to 10 μsec), (iii) use of fields not greatly exceeding the static breakdown potential. The main characteristics of the new chambers are (a) transparent electrodes of stainless steel mesh,[10] (b) no insulator between the electrodes, (c) great distance between the electrodes, (d) great possible repetition rates.

The feature (c) allows the introduction of targets inside the chamber. It leads, in some cases, to the complete knowledge of the trajectory position in space, despite the occurrence of discharge columns connecting the electrodes.

The applied voltage being of the same order of magnitude as the static breakdown voltage, it is possible to use pre-stressing dc voltage to reduce further the applied voltage pulses.

DESCRIPTION OF THE CHAMBERS

Two types of chambers have been constructed. Chamber A has 1-cm spacing between the electrodes. Chamber B has 6-cm spacing between the electrodes. The electrodes consist of a mesh of 0.05-mm wires with 83% transparency.[11]

The gas is admitted through an inlet at one side of the chamber and distributed through several holes. It flows out through the opposite side and flushes through a bubbler. Due to the irritating properties of the iodine vapors, it is necessary to exhaust it outdoors or to pass the gas through an absorber such as alcohol where it dissolves rapidly.

The electrode assembly is enclosed in a transparent

---

[10] Mesh electrodes have also been used in normal spark chambers. G. Culligan, D. Harting, and N. H. Lipman, CERN 61–25.
[11] This mesh is obtained from G. Michel Fils, 9 Fbg. de Colmar, Mulhouse (Haut Rhin, France).

Fig. 1. Discharge chamber with a useful gap of 30×30×6 cm. Mesh electrodes of 83% transparency. The top mirror allows viewing in the chamber interior through the top wall.

Lucite box allowing pictures to be taken simultaneously through one electrode and one side, with a mirror (Fig. 1).

ELECTRONICS

The pulses are produced by two pulse generators. G_1 is a hard tube radar modulator able to deliver pulses up to 25 kV. The final stage consists of two 4PR60A tetrodes in parallel. Its characteristics are: maximum current, 36 A; maximum voltage, 25 kV; maximum pulse width, a few microseconds [Fig. 2(a)], pulses of negative polarity; and maximum rate, 2000/sec. G_2 is an amplifier with a 4PR60A

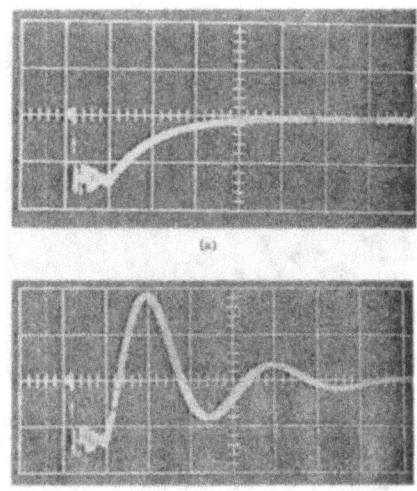

(a)

(b)

Fig. 2. Shape of the field pulses delivered by generator G_1. (a) Pulse shape, 1 horizontal division = 1 μsec. (b) Pulse with oscillating trailing edge.

in the final stage. It is used to amplify signals from a conventional laboratory pulse generator. We used it with the small chamber B to reach higher repetition rates (10^4/sec) and greater pulse width (up to 50 μsec).

The rise time of the pulses is larger than in the conventional spark chambers, triggered either by thyratrons or spark gaps. It is of the order of 20×10^{-9} sec. A decrease in the quality of the tracks is observed with increasing rise time.

The pulses are applied periodically when α sources are used.

The pulses are triggered from scintillation counter telescopes when cosmic rays or mesons from the CERN synchrocyclotron are studied. In this case, the over-all delay between the particle passage and the pulse rising is of the order of 0.25 μsec.

PROPERTIES OF THE CHAMBER

1. "Henogal" Filling

The best gas mixture we found available at CERN is the helium–neon mixture.[12]

Some tests were done by adding Br_2 and Cl_2 vapors. They show essentially the same features as I_2, which was used during the work described below. Tests with alcohol show also a quenching effect. But, in our conditions, the resolution was poorer, mainly for broad pulses.

The gas is saturated with I_2 by passing through a box full of I_2. At 20° the vapor pressure of I_2 is 0.20 mm.

The pictures of Figs. 3, 4, and 5 illustrate the results obtained under different conditions. They lead to the following conclusions:

(i) The voltage used is of the order of 2 kV/cm, and is a function of the pulse width and pulse shape. We have

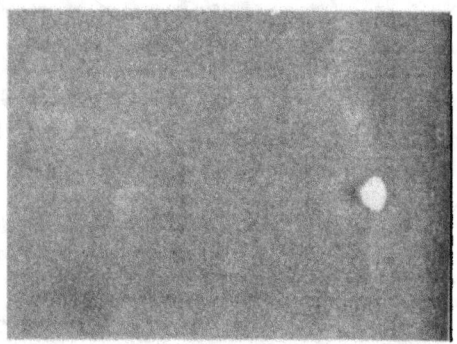

(a)

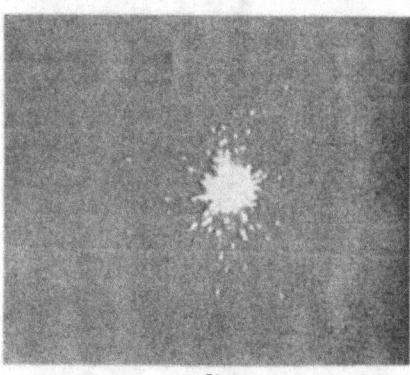

(b)

FIG. 3. Tracks in the slow discharge halogen chamber. Neon-helium filling (65%, 35%), atmospheric pressure, saturation by I_2 vapor. (a) Active deposit of mesothorium; α rays of 8.8 and 6.0 MeV; chamber B (6-cm gap); pulse shape (a) of Fig. 2; $E_{peak} = 12$ kV/6 cm; $f/4.7$; Polaroid ASA 3000. (b) α source emitting 10^6 α/sec in 2π, deposited on the anode of chamber A (1-cm gap); pulse shape (a) of Fig. 2; $E_{peak} = 2.5$ kV/cm; repetition rate 1000/sec; $f/1.4$; "Agfa Isopan Record" film. In each case (a) and (b), tracks of electrons emitted by the source are visible by eye and do not appear on these pictures.

(a)

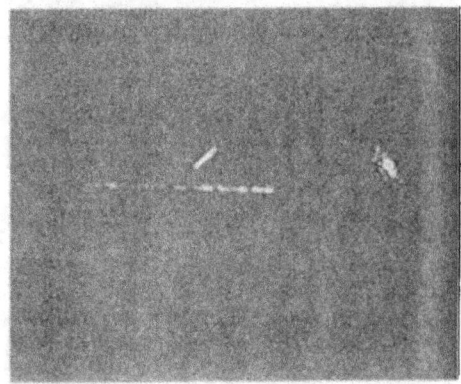

(b)

FIG. 4. Pion tracks obtained with "henogal" at atmospheric pressure and iodine vapor. (a) Chamber A (1-cm gap), 100-MeV pions, 2.5 kV/cm. (b) Chamber B (6-cm gap), 100-MeV pions. The projection on the two electrodes is visible.

---

[12] This mixture, called "henogal," contains 35% helium and 65% neon and is supplied by Air Liquide, Paris.

measured the threshold for γ-ray detection as a function of the voltages applied to chamber A (Fig. 6), the behavior being the same for any detected particles. We see a strong pulse-width dependence up to 2 μsec and then an asymptotic trend towards the static breakdown potential. However, it is necessary to be about 50 to 100% above threshold to have good tracks. The pulse shape also has an influence. By using the alternating pulse (b) of Fig. 2, the threshold is decreased by about 20% with respect to the pulse (a).

(ii) By using a voltage bias, the working potential is displaced by the same amount. We have worked with bias voltages up to ⅓ of the total potential. At higher voltages some difficulties arise, at moderate repetition rates, from the conductivity of the chamber.

(iii) Heavily ionizing particles produce a continuous track and the density of light increases at the end of the range.

(iv) We observe a strong asymmetry in the behavior of the cathode and anode. At low voltages, a bright red cathodic spot is visible while the columns are faint and bluish. This can be of use when the number of columns per cm is of interest, since with large gaps the columns may not conserve their individuality and could appear, on the picture, as a continuous track if the depth of the optics covers the whole gap.

(v) The discharge of the chamber is a slow process as is observed on the voltage pulse during the occurrence of a discharge. In an ordinary spark chamber the voltage drops sharply to zero, but here it drops only by a fraction of the total voltage, as is illustrated by the picture of Fig. 7. The breakdown occurs during the last part of the pulse. The discharge has a high intrinsic dynamic impedance and develops slowly. The speed of development is controllable by the value of the overvoltage. At higher values the

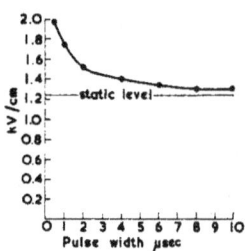

FIG. 6. Threshold for the detection of γ rays as a function of pulse width. Henogal filling.

voltage drops faster and closer to the ground. This behavior explains why this chamber does not require the additional decoupling of the discharges by insulating layers of glass or by highly resistive conducting layers. Halogens[13] have the property of a strong band absorption in the uv region. The low ionization potential of iodine molecules (9 eV) gives rise to free electrons by second-order collisions with the metastable states of the noble gas, thus making the amplification more efficient. An increase of the dynamic impedance of the discharge has also been observed, by J. Fischer and G. T. Zorn,[14] with alcohol added in a normal spark chamber.

(vi) The observed dead times are small compared to those of a spark chamber. We call dead time the minimum time separating two successive pulses without re-ignition of the discharge. With the small chamber A, we have applied up to 10 000 pulses/sec with an α source emitting 10^5 particles/sec into the chamber (pulses of 3 kV, type a of Fig. 2). The number of visible tracks increased roughly proportionally to the repetition rate. With the large chamber B we have reached 2000/sec, being limited by our power supplies.

(vii) Although this chamber suffers from the same drawback as the original ones, namely that only two coordinates of the seeds which originated the discharges are known, the large dimensions open the way for a determination of the

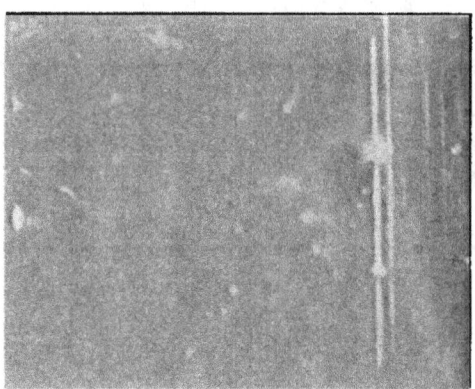

FIG. 5. Photoelectrons from an x-ray source. Chamber A (1-cm gap), He+I₂ filling. Photoelectrons from a Tm¹⁷⁰ source: 55- and 84-keV x rays. I₂ contributes importantly to the absorption of the x rays.

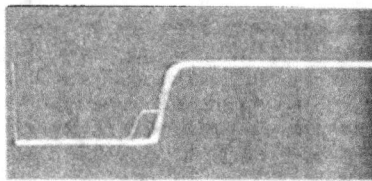

FIG. 7. Voltage pulse shape during the detection of a γ ray. The voltage drop is visible near the end of the pulse. One horizontal division = 1 μsec.

[13] S. Korff, in *Handbook of Physics*, edited by S. Flügge (Springer-Verlag, Berlin), Vol. XLV, p. 70.
[14] J. Fischer and G. T. Zorn, Rev. Sci. Instr. **32**, 499 (1961).

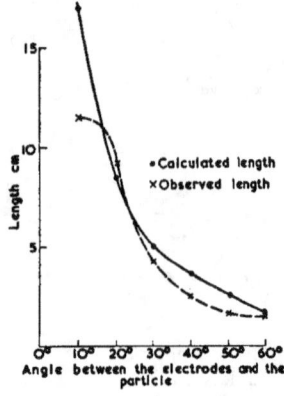

Fig. 8. Length of the track on the picture as a function of the angle of the trajectories with the electrodes. For angles above 30° the S shape of the slanting discharge reduces the apparent projection length. At 10° the track is limited by the sidewall of the chamber. The calculated length is obtained with the assumption of infinite lateral dimension of the chamber.

spatial position of the trajectory. For particles of a range much larger than the gas layer between the electrodes, the length of the track—as observed on the picture through one electrode—is proportional to $\cot \theta$, where θ is the angle between the trajectory and the electrodes. This has been checked by triggering the chamber on a narrow beam of mesons and rotating it by a variable angle around a vertical axis. Figure 8 shows the variation of the length measured on the picture compared to the expected behavior. For tracks of small inclination with respect to the normal, a systematic trend towards shorter projection appears. The simultaneous picture taken through the side gives the explanation to this behavior: the discharge follows the track but, close to the electrodes, the discharges are bent towards the electrodes. It should be mentioned that this slanting discharge is observed with pulses of several microseconds duration, in contrast to normal spark chamber operation with small gaps where this type of discharge disappears for pulses longer than 10^{-7} sec.

(viii) The memory properties were investigated with cosmic rays by varying the delay between the detection in the telescope and the application of the high voltage. The applied pulses were of exponential shape with a decay time of 1 μsec, and a peak voltage of 19 kV/6 cm. The effect of the clearing voltage was examined. We observed that with and without the clearing voltage of any polarity, the behavior was similar. The efficiency for detecting tracks drops to 50% at about 1.5 μsec, the quality of the tracks becoming poorer. No sharp cut in the efficiency is observed.

(ix) The sensitive depth of the chamber is close to 6 cm; S. Fukui and F. Miyamoto have observed a dead zone[2] at about 30% from the anode. By moving the chamber B parallel to the beam, with a telescope of 1-cm-thick counters defining the beam, tracks were observed in all positions inside the chamber except the few millimeters against the electrodes for the exploration of which the scintillation counters are too extended.

2. Helium Filling

With pure helium saturated with I_2 we observe a similar behavior. The differences are:

(i) The operating voltage is higher, compared to the preceding helium-neon mixture, by about 50%.

(ii) The appearance of the discharge is different. No cathodic spot of particular intensity is visible. With the small chamber A we could observe tracks even with pure helium without the use of I_2. However, the conditions were more critical, and the range of possible voltages smaller. This observation was made with a new chamber never contaminated by iodine. However, there might have been some traces of other organic quenching agents since the chamber is made of Lucite. With the "henogal" filling, operation without halogen was even more critical than in pure helium.

3. Argon Filling

With argon filling, the influence of iodine is very striking. We used commercial argon of the quality commonly used for welding. No track could be seen without I_2, and indeed S. Fukui and S. Miyamoto observed that, under their conditions, 38 kV/cm were necessary to get tracks.[3] By addition of I_2, the operating potential drops to a low value and at 3 kV/cm with pulses of 1 μsec, we observe correct tracks with α particles.

(i) The tracks in argon have a particular aspect. A bluish halo surrounds the tracks and may make it difficult to photograph.

(ii) Argon is reputed to be unsuitable for observation of simultaneous tracks. The fact that we can observe an α track with a continuous distribution of discharge columns along the tracks shows that the drawback can be overcome. This can be of interest since argon has a higher

Fig. 9. Cosmic-ray tracks; chamber B (6-cm gap); henogal filling; target of lead glass covered by a thin layer of polyethylene; pulses of 15 kV/6 cm, type (a) of Fig. 2.

efficiency than the usual neon-helium mixture and is much more common.

This effect of I_2 on the breakdown potential can be explained by the Penning effect. The ionization potential of I_2 is 9.0 V, while the energies of the metastable states of noble gases are Ar(11.53, 11.72) Ne(16.62, 16.72) He(20.96, 20.62). Thus I_2 takes energies of excitation of noble gas molecules and converts it to ionization of iodine.

4. Addition of Targets inside the Chamber

The possible great depth of the chamber allows the introduction of targets inside the sensitive volume. Polyethylene targets do not perturb the operation, while lead glass, for instance, shows a tendency to produce a widespread glow in the chamber. Figure 9 shows an event obtained with a cosmic ray reaction in a lead glass target covered by a thin CH_2 layer. It illustrates the capability of this chamber to give a picture of complex events involving the simultaneous detection of several particles of different ionization.

ACKNOWLEDGMENTS

This work has been performed with the excellent technical help of R. Bouclier. We are grateful to J. Favier from the Faculté des Sciences d'Orsay for his participation at several stages of this work. Part of this work was accomplished with the financial help of the Centre National de la Recherche Scientifique.

NUCLEAR INSTRUMENTS AND METHODS **20** (1963) 482–486; NORTH-HOLLAND PUBLISHING CO.

A TRIGGERED LIQUID TRACK CHAMBER WITH SHORT RESOLVING TIME

G. CHARPAK*

CERN, Geneva

1. Introduction

The principle of a new type of track detector is described in a recent note[1]: bubbles of gas are produced in oil by injection through a porous surface** at the bottom of the oil container, which has transparent walls. An electric field can be applied to the liquid by means of plane electrodes, one of which is transparent (conductive glass or

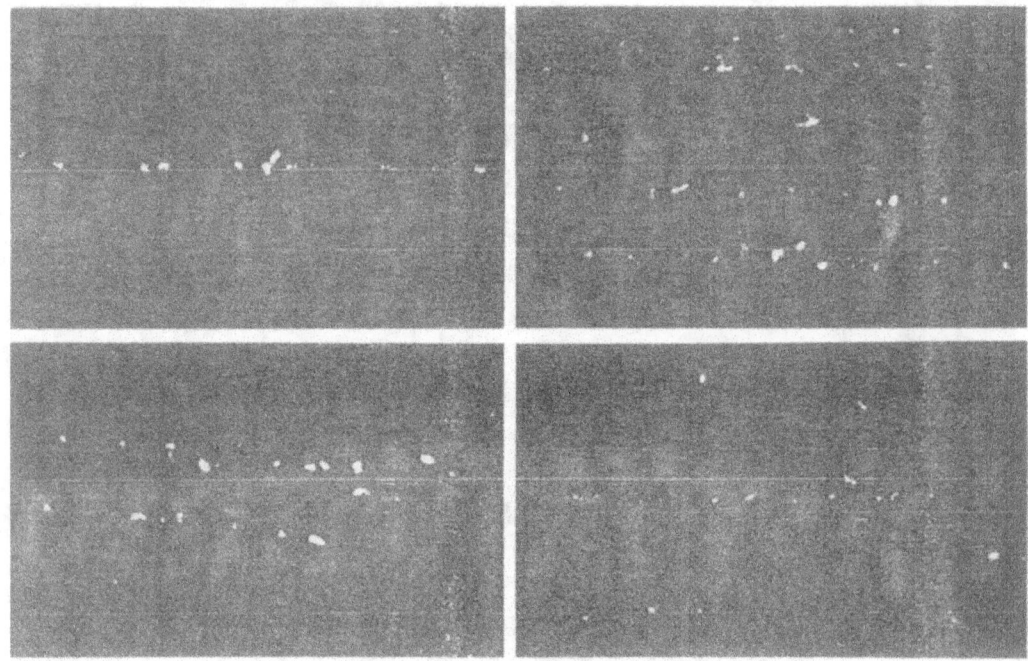

Fig. 1. Tracks of muons in paraffin oil, bubbles of neon. Chamber dimensions: width 210 cm, height 250 cm, depth 0.6 cm. Aperture $f/2$, isopan record film.

metallic grid). When a charged particle, detected by scintillation counters, traverses the liquid between the electrodes, the voltage is applied to the liquid. Those bubbles which have been ionized by the particle give rise to a discharge limited to the bubble so that all the track is made visible.

With neon bubbles of about 2 mm in paraffin

* On leave from Centre National de la Recherche Scientifique, Paris.

** Porous ceramics and porous metallic sheets have been obtained from Electro-physique, 40 rue du Cherche-Midi, Paris.

[1] G. Charpak, Compt. Rend. **254** (1962) 3155.

482

oil one observes bright tracks, visible at several metres from the chamber and easy to photograph (fig. 1 shows typical results). The operation of the chamber has some features in common with that of an ordinary spark chamber[2-5]) and of the neon tubes hodoscope[6]).

The applied field ranges from a few kV/cm to 30 kV/cm. The brightness of the bubbles is, for a given liquid, mainly a function of the field strength. By making tests with pulses with exponential decay times ranging from 0.5 μs to 30 μs we observe first an increase in light output and then, from a value of 2 μs, very little dependence of the pulse width. However, with an oscillating pulse (of frequency 1 Mc/s) we observe an increased intensity. A saturation of the light produced in each bubble is probably reached when the space charge effects cancel the external field.

2. Steady Field or Slowly Varying Field Operation

Application of a steady field in the liquid or of a 50 c/s alternating field shows also some interesting features.

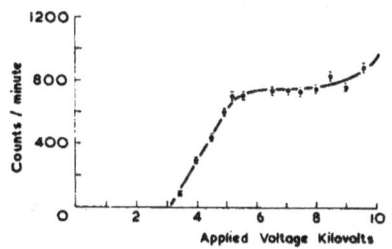

Fig. 2. Counting rate versus d.c. voltage. Co⁶⁰ source, 10 MΩ resistor in series with the chamber.

With bubbles of helium, argon and neon, the chamber behaves like a spark counter with self-quenching produced by the oil. Fig. 2 shows the

²) T. E. Cranshaw and J. F. De Beer, Nuovo Cimento 5 (1957) 1107.

³) G. Charpak, J. Phys. et le radium, 18 (1957) 539 and Proc. of the Intern. Conf. on Mesons, Venice, 1957.

⁴) S. Fukui and S. Miyamoto, Nuovo Cimento 11 (1959), 113.

⁵) O. C. Allkofer, E. Bagge, P. G. Henning and L. Schmieder, Atomkernenergie 2 (1957) 88.

⁶) M. Conversi and A. Gozzini, Nuovo Cimento 2 (1955) 189.

plateau obtained with helium bubbles and a γ-ray source, the pulses being counted on a scaler. The length of the plateau decreases strongly with increasing intensity.

The spark, started in the bubbles under these conditions, breaks through the liquid. The effect of bubbles on the minimum breakdown voltage in liquids is well known[7]). It is still remarkable that this can occur at such a low field as 3.5 kV/cm in transformer oil which is supposed to stand very high voltages. It shows how the occurrence of a bubble in any high voltage device making use of oil as an insulating medium can lead to unexpected breakdown when the bubble is ionized by a cosmic ray or any source of radiation, even at voltages 100 times smaller than the supposed minimum breakdown potential. The efficiency of this liquid spark counter is low for γ-rays (between 10^{-3} and 10^{-4}). For charged particles, however, the efficiency is much higher and one may think of various applications for this type of counter. Fig. 3 shows a picture obtained in a beam with d.c. voltage. When looking in the dark in the presence of a horizontal beam with a steady field or slowly varying field, one observes, beside the bright sparks, a smaller luminosity produced by the charged particles traversing the bubbles. To get rid of the sparks we cover one electrode with a thin layer of insulating material (mylar or araldite). One then observes that with the 50 c/s alternating field the bubbles keep their luminosity, moving up in the liquid, thus making it difficult to observe in a clear way the tracks produced by the beam, which are still visible. Another trouble is that the liquid starts foaming under the influence of the alternating field becoming rapidly opaque; this happens only when a layer of insulator covers one of the electrodes. Thus, it is the triggered mode of operation that seems the most promising for detection of the trajectory of charged particles.

3. Pulsed Field Operation

The source of pulsed field is a thyratron. The delay between the passage of the particle and the rising of the high-voltage pulse is 300 ns. An adjust-

⁷) A review article on this subject is found in "Progress in Dielectrics", vol. 1 (1959) p. 99, by T. J. Lewis.

X. HIGH MAGNETIC FIELDS

Fig. 3. Horizontal beam of pions. Bubbles of helium with applied d.c. field of
5 kV/cm. Picture taken in daylight using $f/20$, $\frac{1}{10}$ sec. exposure, polaroid film ASA 3000.

able clearing field can be applied to the electrodes. The relative efficiency for the occurrence of a track, as measured visually, is a function of this field

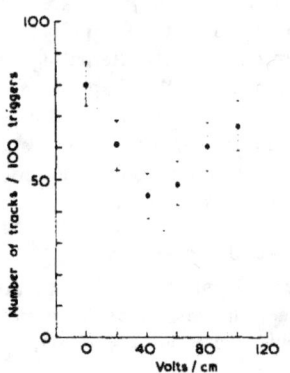

Fig. 4. Relative efficiency versus clearing field.

(fig. 4), and exhibits a behaviour similar to the one observed in spark chambers, showing that the memory properties of the chamber are similar to those of a spark chamber. An essential difference, however, is that since the condenser made of the chamber and of the capacities in parallel is not discharged by a track, it is possible to adjust the sensitive time by adjusting the width of the applied pulse. For instance, by triggering the chamber by a counter telescope detecting pions entering and stopping in the chamber, the electron decay from the muon emitted by the stopped pion is observed with pulses of 4 μs duration (fig. 5).

An attempt has also been made to use as liquid a scintillator showing a low vapour pressure at normal temperature and not attacking our lucite containers*. The tracks that can be observed visually are much fainter than with oil, due to the

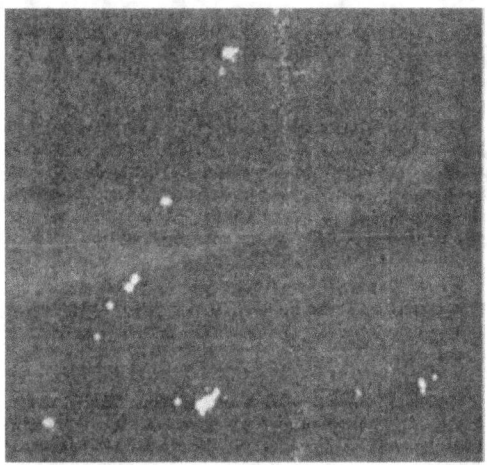

Fig. 5. Pion stopping in the liquid. The decay electron from the muon is visible. Pulse of 4 μs duration.

quenching action of the vapours of this liquid. However, this possibility of a self-triggered chamber seems worth mentioning.

The main weakness of the chamber is now the small depth of the sensitive volume. While all other dimensions can be made arbitrarily big, the depth is limited by the necessity of having a medium trans-

---

* I am indebted to Messrs. H. Faissner, F. Ferrero, A. Ghani and M. Reinharz from CERN who have developed it, for the supply of this scintillator, called decalin.

parent enough to let the light from the luminous bubbles escape with a minimum of scattering. From that point of view, liquids with a low index of refraction like liquid hydrogen would be most favourable. If the light passing through a bubble is fully lost, the maximum number of illuminated bubbles per cm, for a track parallel to the electrode at a depth δ, is $l = xe^{-\delta z}$, where $x = 0.75 \, \nu/\rho$, ν is the fraction of the volume occupied by the bubbles and ρ is the radius of the bubbles. Also in order to have 100 % efficiency for bubbles of small size, operation at pressures higher than the normal one would be preferable. It will also be better to trigger the chamber with a delay much shorter than the one used so far. The influence of all these factors is now under investigation.

Acknowledgements

It is a pleasure to acknowledge the collaboration and contributions of Mr. Massonnet at different stages of this work. The skilful technical assistance of Mr. Bouclier has allowed rapid construction of a great variety of experimental devices.

I am indebted to Dr. Meunier for interesting suggestions concerning the optical problems connected with this chamber. I am grateful to Professor Bernardini for his continuous and stimulating interest in this work, and many colleagues at CERN for fruitful discussions.

DISCUSSION

KOLBRAK: What is the magnitude of the electric field you use to trigger the chamber?

CHARPAK: The intensity of the light goes as the intensity of the electric field. If you just look at the tracks, you start seeing them at 4—5 kV. These pictures which I showed were taken with 15 kV pulses.

I investigated the influence of the duration of the pulse. You have to use very short pulses, much below a μs, because after a microsecond it saturates. In fact long pulses of say 100 μsec will produce a plasma in the bubbles. This plasma will reduce the field to a very low value which will stop the phenomenon. Looking with a photomultiplier into these bubbles, we discover that it takes some tens of nanoseconds for the light to reach a maximum.

One way to increase the light is to use a high frequency field, but even at 1 Mc/s I observed an increase in light. The space charge effects that stop the developing of the avalanche work for you in that case.

KOLBRAK: What is your electrode spacing for these 15 kV?

CHARPAK: About 1 cm.

TENG: Is the space resolution determined by the size of the bubble, or can you see the position of the spark in the bubble?

CHARPAK: You cannot see it in the bubble. If you increase the size of the bubble, the whole bubble becomes luminous. There is a glow discharge in the bubble. What would be very interesting is to try liquids which have a very low index of refraction, such as helium. Then you will get much less trouble due to the diffusion of light and, maybe, you can also increase the depth.

Using d.c. fields, you observe a rather amusing phenomenon. At 5 kV the chamber works just like a self-quenched Geiger counter. The quenching is done by the liquid and it is the simplest counter you can imagine. You can have a very large pulse and you do not need to amplify. The chamber can have any dimension you want. Instead of a d.c. voltage you can use an alternating voltage of 5 kV with an ordinary transformer. It works as well. I have detected 500 counts per sec with this chamber. My power

X. HIGH MAGNETIC FIELDS

supply did not allow me to go further. I have taken several pictures in the daylight with an *f*-stop of about 20. One sees bubbles which seem aligned. The resolution is poor. If one takes pictures in the dark one observes luminous sparks in between the bright spots.

It is important to avoid spurious sparks because their light is bright enough to obscure the tracks. In order to avoid spurious sparks in pulsed operation, I put an insulator of about 1 mm of lucite against one of the electrodes.

SHAW: You said the resolving-time depends on the characteristics of the pulse you apply to the electrodes. The memory effect you describe would suggest to me that there may be some other factors.

CHARPAK: I said that this is a difference between this chamber and a spark chamber. In a spark chamber the resolving-time does not depend on the pulse characteristics. The memory-time seems to be the same as in a spark chamber. As for the dead-time, I am now investigating it. The question is, how long does it take to neutralize the amplified ionization which is a factor 10^{11} higher than the initial ionization. As the liquid is insulating it is not clear whether it can be very fast. The fact that I can work with 300 counts per second means that it does not take a very long time to neutralize the ionization.

JONES: I wondered if you or anyone had tried a gadget that has been discussed many times, of using liquid xenon for example in an a.c. or d.c. pulsed spark chamber. One knows that the light output of xenon increases in an electric field, but I do not know of anyone who has tried to photograph tracks in this way.

CHARPAK: I know at least 5 people who have tried without success to apply voltage pulses to a liquid scintillator after a particle has passed through. In my case you see tracks because you have bubbles.

O'NEILL: I might be confusing this with something else, but I know that some years ago somebody in England made a small liquid xenon parallel plate chamber. He may have seen tracks or he may have seen the breakdown when the particle had passed through. I do not remember well, but I think they had trouble with polarization.

HEYMAN: Could you tell us something about the time characteristics of the pulses that you get out of the chamber when you operate it under d.c. conditions as a "Geiger counter"?

CHARPAK: I had a resistance of 1 MΩ or 10 MΩ, depending on the power supply, in series with the high voltage and I looked at the voltage pulse with an oscilloscope. The observed rise-time of pulses was as high as the rise time of the scope. I do not know the delay or the jitter, but the rise-time was very fast.

HUTCHINSON: Could you tell us something about the porous plate that you used; in particular whether it is critical?

CHARPAK: It is not critical. I used plates which are common in chemical industries. They can be made of various materials. The one I used was of ceramic.

NUCLEAR INSTRUMENTS AND METHODS 100 (1972) 157–164; © NORTH-HOLLAND PUBLISHING CO.

CERN
SERVICE D'INFORMATION
SCIENTIFIQUE

SOME PROPERTIES OF STACKS OF DISCHARGE CHAMBERS WITH INFINITELY TRANSPARENT ELECTRODES

G. CHARPAK, A. BRESKIN and F. PIUZ

CERN, Geneva, Switzerland

Received 11 November 1971

It is shown that it is possible to build projection discharge chambers with electrodes infinitely transparent to the light emitted by the discharges. This is achieved mainly by having the wires of the electrodes perfectly aligned either in parallel arrays, or in arrays converging to the objective of the optics. Chambers with almost isotropic properties and many attractive features can thus be built.

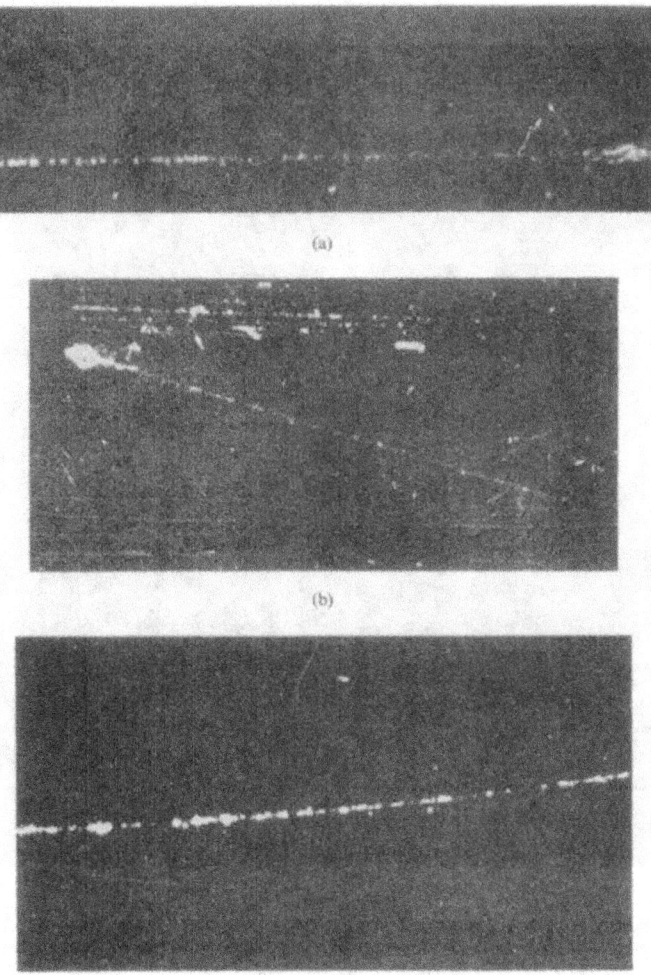

(a)

(b)

(c)

157

1. Introduction

In the evolution of visual detectors based on gaseous discharges, the streamer chamber[1,2] and the avalanche chamber[3,4] have represented the ultimate steps in development. With a perfect control of the duration and the intensity of the field pulses, it has become possible to control a discharge starting from a single electron, to any stage of its development, the single

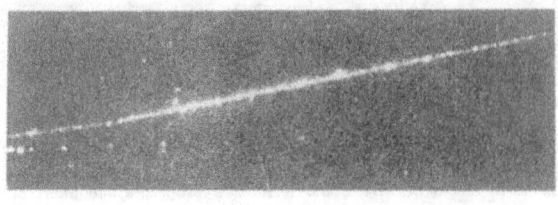

(d)

(e)

Fig. 1. Examples of tracks in projection chambers. a) Gap of 1 cm, helium–neon + iodine, pulses of 2.5 kV/cm. b) Gap of 5 cm, helium–neon + iodine, pulses of 3 kV/cm. Reaction in the wall of the chamber. c) Gap of 5 cm, helium–neon + iodine, pulses of 3 kV/cm. 600 MeV protons in a magnetic field. d) Gap of 5 cm, helium + iodine, pulses of 5 kV/cm. Scattering of a proton in the gas. e) Gap of 5 cm, helium–neon filling, pulses of 3 kV/cm. Cosmic-ray event. Tracks of about 20 cm length. From refs. 10 and 11 (1962).

avalanche or the streamer of any desired length. The elimination of intermediate, closely spaced electrodes has resulted in an almost isotropic device, with the minimum amount of matter. A heavy price has had to be paid for this process; it is the serious loss of light intensity as compared to spark chambers and the loss of simplicity of the high-voltage pulsers, since pulses of hundreds of kilovolts are required in most of the streamer chambers, with very strict requirements on the rise-time and the duration.

In the course of the detector development which preceded the streamer chambers, an intermediate step has been reached by many authors[5−11] [first of all by Fukui and Myamoto[5,6]] which bears a strong resemblance to the streamer chambers – the two-dimensional discharge chambers. Tracks parallel to transparent electrodes were made visible by a series of discharges along the tracks.

Very good quality tracks were obtained under such

conditions, but only with a two-dimensional delineation of the tracks. Using appropriate quenching agents, in 1962 a group at CERN[10,11] obtained tracks that were very similar to those of streamer chambers, as can be seen from fig. 1. But these tracks were obtained under much easier conditions than those in streamer chambers and in most projection chambers; fields of only 3 kV/cm, long pulses, and much greater brightness of the tracks.

It is striking to see that in many experiments with streamer chambers one has to let the streamers grow to a length of 1 cm in order to obtain a sufficient amount of light. It is not a serious drawback in many cases, since the coordinates that are relevant to momentum measurements are usually those in the plane parallel to the electrodes. This is the case if the magnetic field is parallel to the electric field.

2. Projection chambers with infinitely transparent electrodes

If one could build electrodes with an infinite transparency it would be more advantageous to use a stack of projection chambers, with a gap width equal to the length of a streamer in a streamer chamber.

A development of the discharge in steps as illustrated in fig. 2b would lead to a very small loss of information about the depth of the track, when compared to a streamer of the same length as shown in fig. 2a.

Moreover, in large-size projection chambers only the rare tracks exactly parallel to one electrode have an uncertainty of position equal to the gap width. As soon as tracks cross several gaps, the accuracy is considerably improved. With a target external to the chamber, as is often the case, it is possible to orient the chamber in such a way that no track is ever parallel to the electrode.

The advantage would be a considerable decrease of the applied voltage in proportion to the number of steps, and eventually an increase of the luminosity of the track, since it is possible to control at will the discharge in a short gap from the streamer mode to the straight spark mode.

We propose several solutions that are in many respects equivalent to a chamber of type (2b) with infinitely transparent electrodes.

1. One imperfect solution consists of using very thin wires to build an electrode with a high transparency. We have checked that with 20 μm wires spaced by 2 mm, electrodes with 99% transparency can be built and can operate well in a discharge mode giving bright tracks with no harm to the wires. Even a stack of 50 such chambers

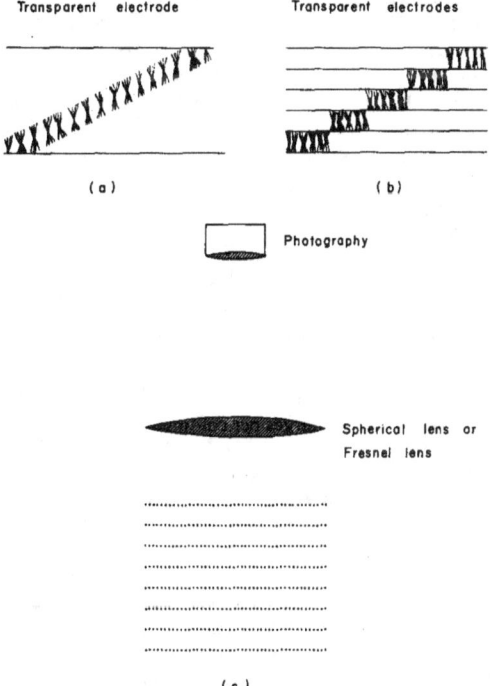

Transparent electrode   Transparent electrodes

(a)   (b)

Photography

Spherical lens or Fresnel lens

(c)

Fig. 2. Some types of discharge chambers. a) The streamer chamber. b) The multistack projection chamber. c) The spark or projection chamber with a regular structure. Either the chamber is infinitely transparent from a finite point (convergence to the camera objective), or the chamber is transparent at infinity.

Fig. 3. Chamber with a periodic wire structure. Ten gaps of 1 cm with 1 mm wire spacing. Chamber illuminated from behind. Beta-particle tracks. Camera at a finite distance. Aperture $f/5.6$. Polaroid type 107.

will have no more than 40% flight absorption. Two other solutions present the advantage of an infinite transparency for any number of gaps.

2. The electrodes are made of wires aligned with the objective of the lens placed some distance away. This leads to a construction where the wire spacing is variable with the depth of the chamber.

3. The wires are perfectly aligned according to a periodic structure (fig. 3c). By using a spherical lens it is possible to look at any depth without absorption by intermediate electrodes. This is also true for several privileged axes.

By taking pictures along two of these directions one obtains a stereo view giving the depth of the track. This is quite similar to the situation of a streamer chamber with two cameras where the possibility of taking pictures at any angle is of no use whatsoever.

One evident difficulty is that sparks occurring at a great depth will suffer from a greater collimation than the sparks occurring near the surface of the chamber. However, if a discharge régime can be found where the necessary aperture is so small that the width of the lens used is of the same order of magnitude as the wire spacing s, then the difference of luminosity between sparks at different depths can become acceptable.

Our first attempts have shown us that it can be a potentially powerful method leading to large detectors, almost isotropic, working at very low voltages of the order of a few kilovolts with an average amount of matter controllable at will. We are going to present our first results in this article.

3. Properties of the tracks in the projection chambers

A small prototype of 10×10 cm has been built for us at CERN by R. Benoit. It consists of 10 gaps of 1 cm, with electrodes made of wires of 50 μm of stainless steel, spaced every millimetre. The wires are perfectly aligned, and the alignment is conserved when the chamber is rotated by angles in steps corresponding to the wire distance and gap, in our case about $\theta = 100$ mrad.

We observed that such a chamber works, as expected, as a projection chamber. It is possible with pure heno-gal (helium–neon mixture) to have good tracks with 3 kV pulses of a duration of about 1 μsec applied to the chamber. This voltage could have been reduced by adding I_2, as in ref. 10. Because of the corrosive character of this component, we avoided its use and instead tried to obtain régimes giving an even higher luminosity. We found henogal bubbling through methylal* to be satisfactory, although we think that most gas mixtures used in spark chambers are suitable. At 6 kV we obtain bright discharges with no spurious edge effects, which were existing in our chamber when working at high voltages in pure henogal in the spark mode.

Fig. 3 shows a view of the chamber, uniformly illuminated from behind, with no lens. We see a darkening of the image at positions where the density of wires is maximum. The intensity of the sparks is not modulated since most of the tracks cross the chamber in the middle, and the transparency of the electrode, even with no alignement, is still very good.

Fig. 4a shows a picture of the chamber with a lens collecting the light from a direction perpendicular to an electrode, and an example of a track. Figs. 4b and 4c show tracks with the chamber oriented at different angles. The following observations could be made. The number of discharges per centimetre is of the order of 2, as in most streamer chambers. The tracks are as bright as in most spark chambers. Fig. 5 shows tracks observed at variable apertures with sensitive films It is possible to take pictures with apertures of $f/8$. This is a very important feature for this method, as had already been mentioned. If one would be forced to work with a large aperture, the additional collimation of the beam introduced in one dimension by the wire arrays would considerably reduce the amount of light collected from discharges that would lie at a great depth in the chamber.

One may wonder where the spark position is with respect to the wires. The sparks seem to be centred on a

* Methylal: $(CH_2O)_2CH_3$, 2% in the helium–neon.

wire, as can be seen from fig. 6. With thin wires we thus have a collimation width of two times the wire spacing for the collection of light. In the régime where we operated, the spark has an interesting structure: a bright cathodic spot on one wire, as already observed by Charpak et al.[10] in the henogal–halogen mixture, and a fainter discharge connecting this spot to several anode wires. The spot is probably due to the emission of light by the region of the positive ion sheath surrounding the cathode wire. A side view of discharges illustrating these properties is shown in fig. 7.

We verified that the tracks are visible with an ordinary vidicon camera, which can be of the greatest interest for digitalization.

4. Discussion

We have seen that it is possible to construct a stack of chambers using electrodes made of well-aligned wires with gaps of 1 cm. We have easily found an operational régime where the brightness of the tracks is such that small-aperture optics can be used. This permits the necessary depth of focus and, to a large extent, suppresses the loss of light for deep tracks. There is no reason to be limited to a 1 cm gap. We think that it

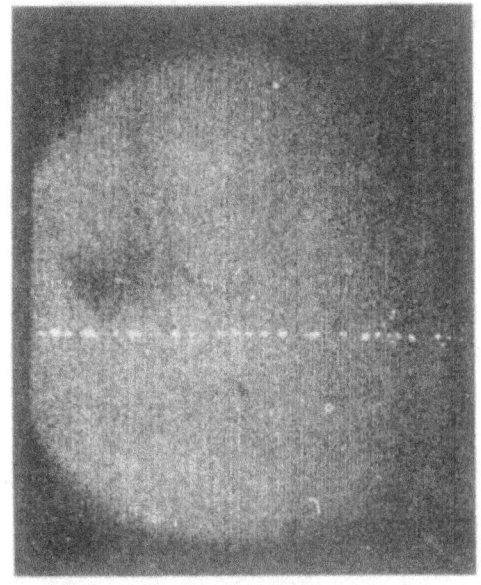

(b)

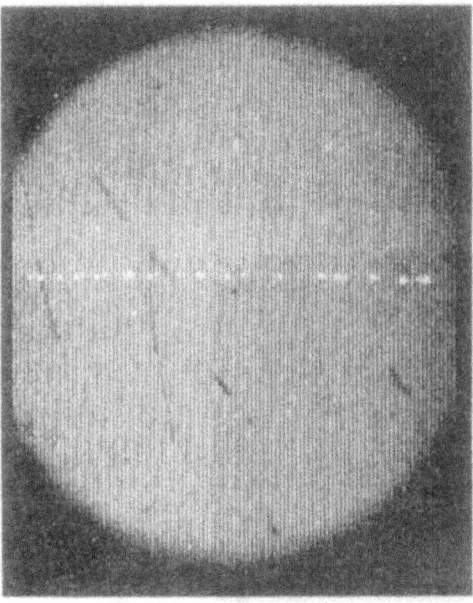

(a)

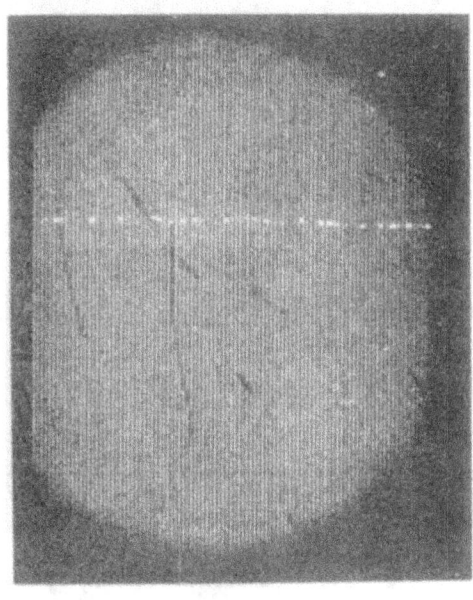

(c)

Fig. 4. Chamber with a periodic wire structure. Ten gaps of 1 cm with 1 mm wire spacing. Chamber illuminated from behind. Beta-particle tracks. Camera at infinite distance (lens). Aperture $f/5.6$. Polaroid type 107. a) Axis of camera perpendicular to the planes. b) Chamber turned by 50 mrad. c) Chamber turned by 100 mrad.

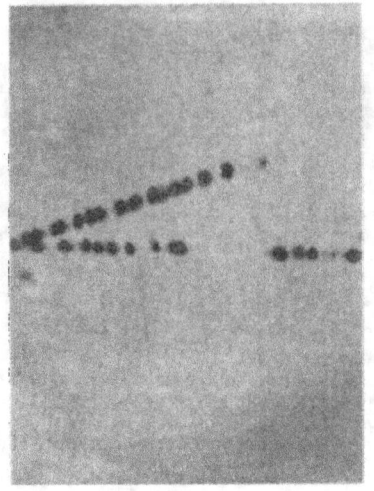

(a)

Fig. 5. Examples of tracks. 5.5 kV pulses, 1 cm **gap,** henogal-methylal. Kodak film 2475. a–c) Aperture $f/1.2$. d) Aperture $f/5.6$. e) Aperture $f/8$.

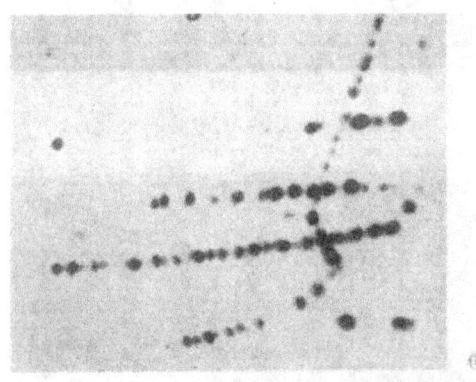

(b)

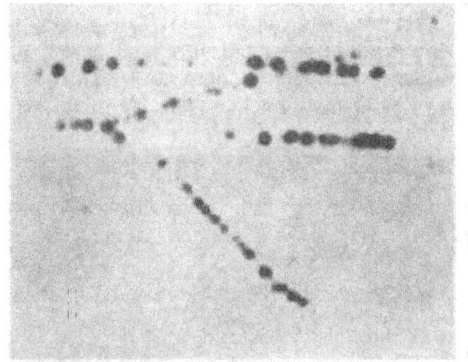

(c)

(d)

(e)

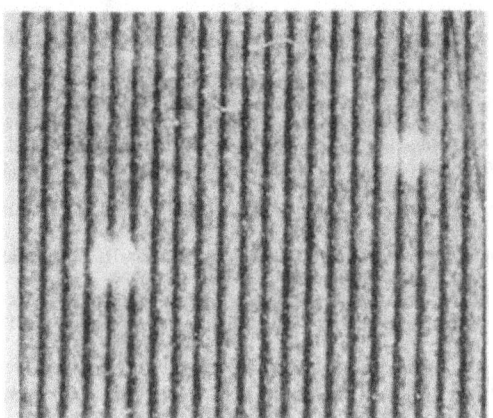

Fig. 6. Position of the light source. The camera is focused on the front wires. One sees that the spots of light emission are centered on the wires.

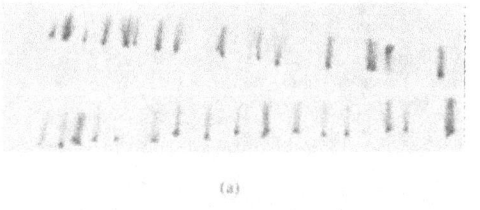

(a)

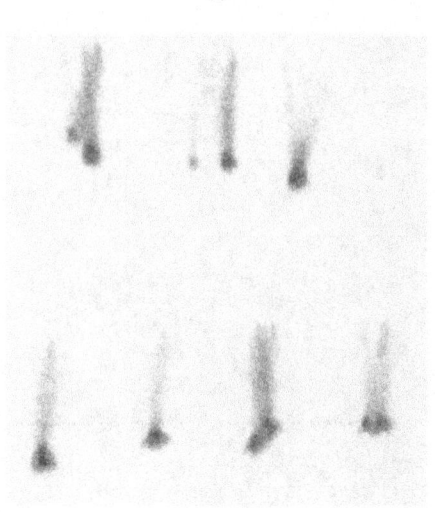

(b)

should be possible to work at smaller gap distances even at atmospheric pressure. The accuracy of such a chamber should then be very similar to the one of a good streamer chamber. However, it would operate at voltages of a few kilovolts. The optical problem is not a formidable one.

The distortion introduced by the optical systems is compensated by the built-in fiducial marks constituted by the arrays of well-aligned wires, which can be illuminated at will. One objection may be the amount of matter. It is clearly greater than with a streamer chamber. However, with aluminium wires of 50 μm placed every 1 mm, with a gap of 3 mm, the average density is still of the same order as the pure gas density, which is still much better than in any spark chamber.

An interesting direction that the development can take is to increase the average amount of matter at will. With tungsten wires of 1 mm, spaced by 3 mm and 3 mm gap, one obtains an average density of 2 g/cm^3 which gives an average radiation length of 2.5 cm. If the wires are thicker than the sparks, the possibility to take pictures at the inclined angles corresponding to the transparent directions of the array is an essential feature.

One could thus build very attractive chambers with, for instance, a partial volume of low density for charged particles and a partial volume of high density for the detection of γ-rays with a high efficiency.

In developing this technique we have in mind that it

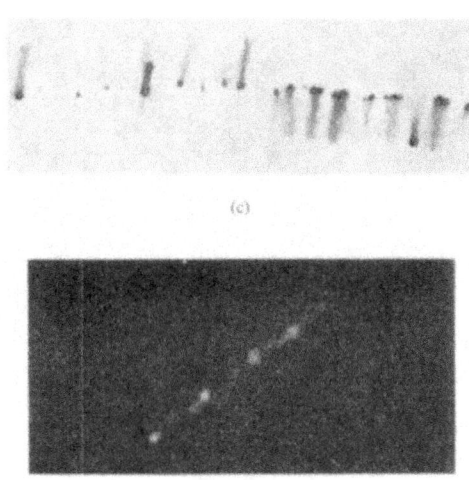

(c)

(d)

Fig. 7. Structure of a discharge. 5.5 kV pulses, 1 cm gap, henogal–methylal. Views of some tracks taken at a large angle. a) Discharges from trajectories not crossing a gap. b) Enlarged view of some discharges showing the bright cathodic spot and the faint discharges going to several anode wires. c) Discharges from a trajectory crossing the cathode. d) Discharges from a trajectory orthogonal to the electrodes and crossing seven gaps. The bright spots on the four crossed cathodes are visible.

may prove difficult to use multiwire proportional chambers with large systems of spark and streamer chambers, while with such "crystal chambers" the problem is considerably simpler because of the very low operating voltage.

We are grateful to Mr. R. Benoit for his technical help and the construction of the prototypes used in this work. One of us wishes to thank Dr. R. Meunier for useful and stimulating discussions concerning the optics of this detector.

References

1) G. E. Chicovani, V. Mikailov and V. N. Roynischvili, Phys. Letters 6 (1963) 254.
2) B. H. Dolgoshein, B. V. Rodionov and B. I. Luchkov, Nucl. Instr. and Meth. 29 (1964) 270.
3) F. Schneider, The discharge chambers, CERN Internal Report PS/Int. AR J9-6 (1959).
4) F. Schneider, A spark chamber of three-dimensional resolution, Proc. Informal Meeting *Filmless spark chamber techniques and associated computer use*, CERN 64-30 (1964) p. 351.
5) S. Fukui and S. Myamoto, Nuovo Cimento 11 (1959) 113.
6) S. Fukui and S. Myamoto, J. Phys. Soc. Japan 16, no. 12 (1961) 2574.
7) S. Fukui and B. Zacharov, Nucl. Instr. and Meth. 23 (1963) 24.
8) A. A. Borisov, B. A. Dolgoshein, B. I. Luchov, L. V. Reshetin and V. I. Ushakov, Instr. Exp. Tech. 1 (1962) 48.
9) A. I. Alikhanian, T. L. Asatiani, E. M. Matevosian, A. A. Nazaryan and R. O. Sharkhatunian, Phys. Letters 7 (1963) 272.
10) G. Charpak and L. Massonnet, Rev. Sci. Instr. 34 (1963) 664.
11) G. Charpak, L. Massonnet and J. Favier, Progr. Nucl. Tech. Instr. 1 (1964) 323.

NUCLEAR INSTRUMENTS AND METHODS 107 (1973) 361–363; © NORTH-HOLLAND PUBLISHING CO.

SOME PROGRESS IN PROJECTION CHAMBERS

A. BRESKIN, G. CHARPAK and K. GEISSLER

CERN, Geneva, Switzerland

Received 9 October 1972

With a simple method of construction each wire of a "crystal" projection chamber is provided with its own storage capacity. Excellent efficiency for multitracks is obtained. The accuracy is shown to be independent of the orientation with respect to the wire and to be better than 150 μm.

The density of discharges is 3/cm for a 5 mm gap, and 5/cm for a 3 mm gap.

1. Introduction

In order to investigate the feasibility of large-size crystal projection chambers and to have a clear idea as to how their properties compare with those of streamer chambers, we have started a small program of research on which we wish to report.

Let us first recall the principle of these chambers[1]: they consist of stacks of gaps with electrodes made of wires, perfectly aligned in such a way that the electrodes are infinitely transparent along privileged directions. The operating conditions are close to those of small-gap spark chambers. The pictures are taken through the transparent electrodes. Tracks similar to those of streamer chambers are observed, with the following main advantages:

– operation at very low voltage (typically 5 kV for a 5 mm electrode spacing and 2.5 kV for 3 mm spacing);

– greater luminosity of the tracks.

The following investigations were performed:

1) Determination of the range of variation of the chamber parameters: wire diameter, wire spacing, electrode spacing.

It is clear that if the field around a wire is high enough for initiating spontaneous discharge along the wire, no amplification can occur in the region of uniform field. A typical example of the constraint imposed on the wire spacing as a function of the wire diameter is shown in fig. 1 for gaps of 5 mm and 10 mm. A detailed study has been published elsewhere[2].

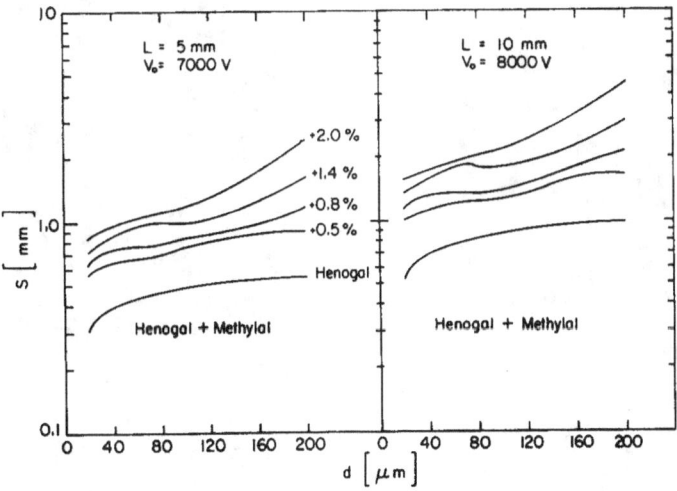

Fig. 1. Critical wire spacing. – Neon-helium mixture (65-35), with variable methylal addition [(OCH₃)₂CH₂] – Critical wire spacing for different wire diameters.

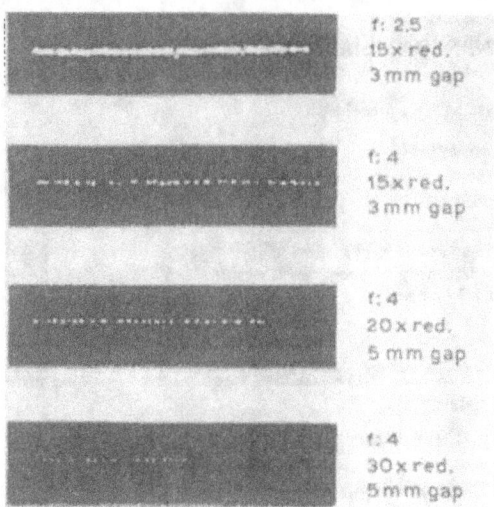

Fig. 2. Typical trajectories in track chambers with a storage capacity for each wire. Tracks perpendicular to the wires. Track length 10 cm; neon-helium mixture +0.8% methylal; Kodak film SO 265.

2) Construction of chambers with a high efficiency for multitracks.

2. Construction of the chambers

For the tracks parallel to a plane, hundreds of sparks have to occur simultaneously, and it seemed to us important to provide optimum conditions for an independent development of the discharges.

Several methods with automatic wire spark chambers have already proved to be successful, but in general they do not fit the multigap structure or the transparency of the electrodes.

One method tried at CERN* seemed very well suited. Each wire of a chamber is connected individually to a storage capacity.

We found this idea easy to apply, by using capacitors made of printed circuits on a thin kapton layer. On one side we have a continuous layer to which the dc high voltage is applied. On the other side, we have printed

* CERN-Trieste High-Energy Group.

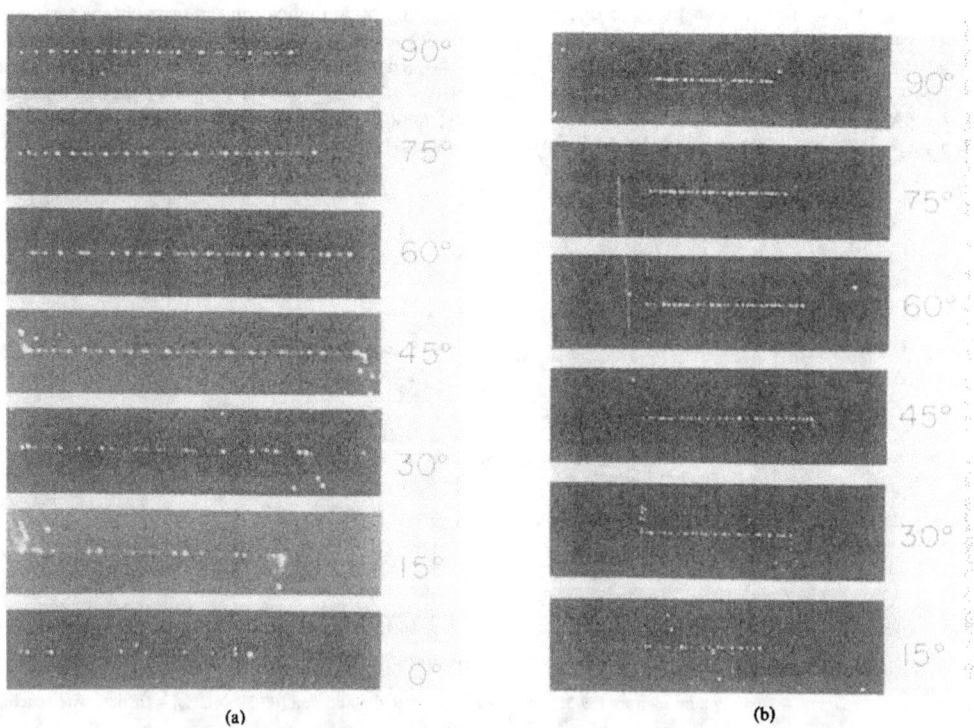

(a)   (b)

Fig. 3. Tracks at different angles. Same conditions as fig. 2. The tracks are oriented at variable angles with respect to the wires. 0° = wire direction. (a) 5 mm gap; 5 kV; reduction × 10; f: 4. (b) 3 mm gap; 2.4 kV; reduction × 15; f: 4.

circuits to which the wires are attached. By switching to ground the high-voltage plane we transmit to each wire the energy stored by its capacity to this plane. We easily obtain capacities of the order of 30 pF, which prove well suited for this application. It would be very easy to go to values of the order of 100 pF if it would prove necessary.

Resistances in parallel on each wire fix the time constant to about 0.8 μs.

The operation has been studied in a chamber having the following characteristics:

width:	100×100 mm^2;
double gap:	5 mm or 3 mm;
wire diameter:	100 μm;
wire spacing:	1 mm;
kapton thickness:	0.13 mm;
capacity per wire:	30 pF.

The tracks obtained with such a chamber had perfect regularity. Fig. 2 shows some typical tracks obtained with high-energy particles, using a gas mixture of helium-neon, with 0.8% of methylal. The operating conditions were: with a 3 mm gap, f: 2.5 (fig. 2a), f: 4 (fig. 2b), pulses of 2.4 kV; with a 5 mm gap, f: 4, pulses of 5 kV (fig. 2c, d).

3. Accuracy in the trajectory measurements

We have undertaken a measurement of the accuracies as a function of the angle with respect to the direction of the wires, using particle beams of 300 MeV/c and 1000 MeV/c. Fig. 3 shows the traces obtained at different angles. We see that when the tracks are exactly parallel to a wire, there is a reduction of brightness owing to energy sharing. With a small gap and wires 1 mm apart this corresponds to a very small solid angle.

The measurements were performed with manual methods. Only a straight line fit has been applied to the string of discharges.

With the chambers of 5 mm gap and 3 mm gap we observe an accuracy of about $\sigma = 0.15$ mm independent of the angle of the track with respect to the wire, as can be seen in fig. 4.

Two sources of error may account for the size of the deviations from the straight lines. One is the operator's setting error. It is characteristic that the same error is observed even when the sparks are on one single wire. The diffusion of the electrons during the delay between the passage of the particle and the application of the voltage may also account for a fraction of the error. In this respect the situation is more favourable than for a streamer chamber since the delay can be much smaller

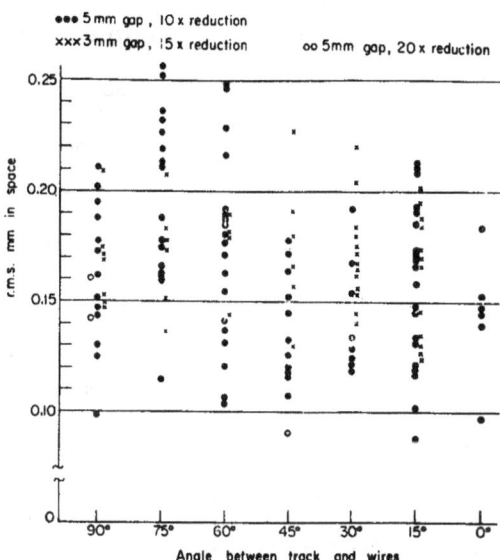

Fig. 4. Accuracy in trajectory measurements. Straight lines are fitted to the tracks – manual measurements. r.m.s. as a function of the angle between the track and the wire. 0° = wire direction. Neon-helium mixture + 0.8% methylal.

for a small pulse than for the huge streamer chamber pulses.

At this stage of our investigation we may say that the crystal chambers compare favourably with the streamer chambers. The increased amount of light permits the use of smaller apertures ($f/4$ to $f/5.6$) and, for a given depth of field, smaller demagnifications ($\times 20$) than is usual with streamer chambers.

4. Conclusion

These preliminary tests seem to us an encouragement to further investigate the properties of these chambers.

In our program we have the construction of a large-size chamber with the wires equipped with their individual storage capacity. We also want to investigate more closely the conditions of operation in gaseous hydrogen.

We wish to thank Messrs A. Scharding and G. Brandon for their collaboration and Dr J. Meyer for many stimulating discussions.

References

[1] G. Charpak, A. Breskin and F. Piuz, Nucl. Instr. and Meth. 100 (1972) 157.
[2] A. Breskin, CERN NP Int. Report 72-20 (1972).

NUCLEAR INSTRUMENTS AND METHODS 108 (1973) 427–429; © NORTH-HOLLAND PUBLISHING CO.

A GLASS TRACK CHAMBER

A. BRESKIN and G. CHARPAK

CERN, Geneva, Switzerland

Received 8 January 1973

A hodoscope of glass tubes filled with neon + helium has been constructed using tubes of a diameter of the order of 1 mm. Good operation is observed with electric fields nearly parallel to the tubes. A simple detector with high spatial resolution can thus be constructed.

1. Introduction

In a recent article[1]), chambers with a periodic-electrode structure were proposed which allowed pictures to be taken of discharges occurring deep in layers of matter.

The electrodes of these projection spark chambers are made of wires. The wires of the successive electrodes are perfectly aligned, in such a way that for priviledged angles the chamber is infinitely transparent.

This principle proved to be successful[2,3]), and tracks similar to those of a streamer chamber were observed.

One may wonder whether such a method is not more general and should not be used in liquid and solid media.

We have studied two arrangements of a solid medium, which gave very different results despite their resemblance.

First, we have tried a medium made of capillar glass tubes (see fig. 1), with the following characteristics :
- internal diameter: 0.7 mm,
- external diameter: 1.2 mm,
- length: 20 mm.

The stack of tubes of 60×70 mm^2, which was built to be a film support in a fast camera and is accurately machined to have flat surfaces, is sandwiched between two electrodes made of transparent metallic meshes. The tubes are open at their ends and a standard spark chamber gas mixture (70% Ne, 30% He) is flushed through the chamber.

As we shall see, we encountered difficulties in obtaining good tracks with such a device.

We then tried another arrangement which was made of small glass tubes with the following characteristics:

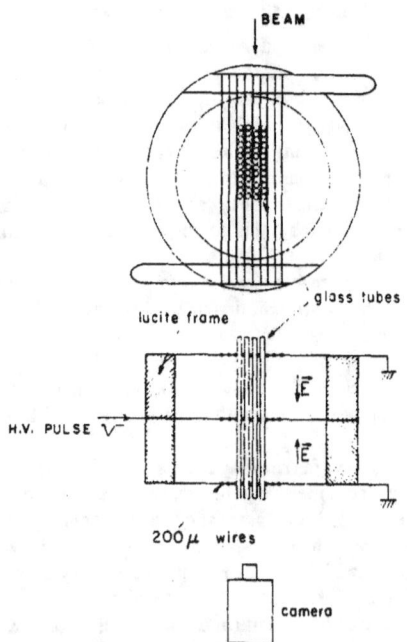

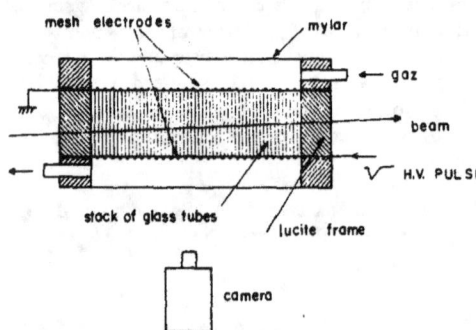

Fig. 1. Track sensitive device made with a stack of tubes glued together.

Fig. 2. Hodoscope chamber made with 50 closed glass tubes stacked between the wires of a double-gap wire spark chamber.

427

– internal diameter: 1 mm,
– external diameter: 2 mm,
– length: 50 mm.

The tubes are closed and filled with Henogal (70% Ne, 30% He) at atmospheric pressure. They are stacked between the wires of a double-gap (2 × 2 cm²) wire spark chamber in such a way that the electric field is longitudinal to the wires (see fig. 2).

2. Results

The first chamber was tested in a 300 MeV/c beam

of the CERN Synchrocyclotron. We applied to the chamber 11 kV pulses of approximately 1 μs length. Unfortunately it gave a very poor performance. Whilst perfect tracks are obtained in the chamber without the glass tubes, the operation changes considerably with the glass in. Very bright sparks occur. It seems that the development of the sparks is very unbalanced and that the fastest one eats up the available energy.

Fig. 3 shows two examples of tracks. One observes the multiple scattering in the high-density material.

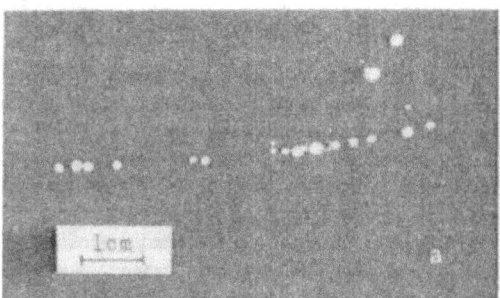 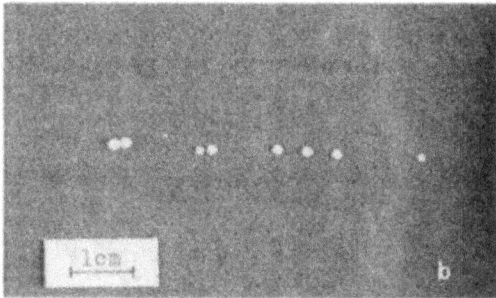

Fig. 3. Examples of tracks in the first chamber. One can observe the scattering of particles of 300 MeV/c in the high density material. Pictures taken with 90 mm lenses, f:8, on a Kodak S0265 film.

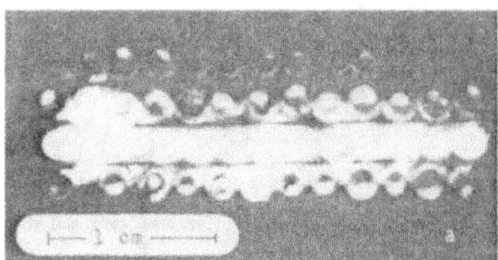 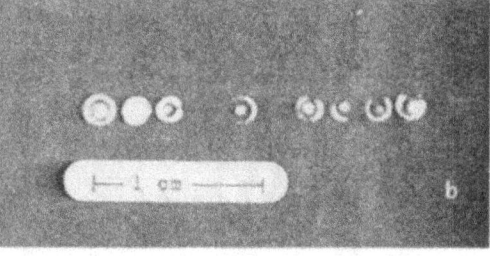

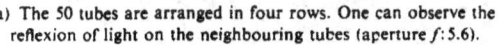

a) The 50 tubes are arranged in four rows. One can observe the reflexion of light on the neighbouring tubes (aperture f: 5.6).

b) The tubes were painted black to avoid reflexions. This time they are arranged in three rows of 17 (aperture f: 8).

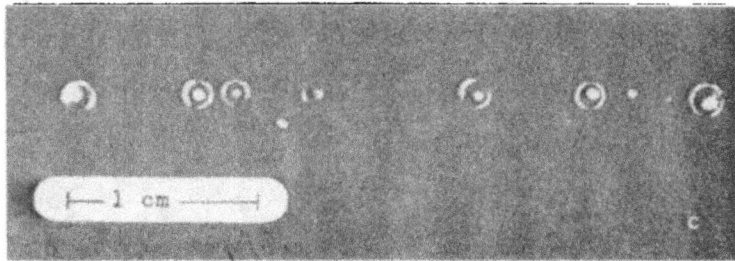

c) The tubes were painted black to avoid reflexions. This time they are arranged in three rows of 17 (aperture f: 8).

Fig. 4. Examples of tracks in the second chamber; 90 mm lenses, Kodak S0265 film.

The interesting feature is that the sparks are well localized in a single tube. With alternate layers of tubes oriented in different directions, one could thus obtain a stereo view of the track.

In order to get rid of the energy-robbing by the bright sparks, we built the second system with closed glass tubes. It is in all respects similar to the original glass-tube hodoscope[4] except that the electric field is longitudinal to the glass tubes, which presents a considerable advantage for small tubes. The wire electrodes are in the open air and the operation is quite easy.

A system of 50 tubes was tried in a 400 MeV/c PS beam.

We applied to the chamber 16 kV pulses of approximately 1 μs length.

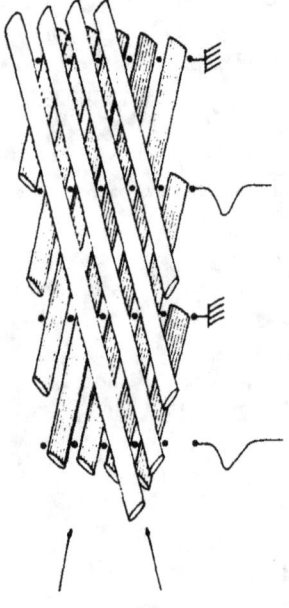

Fig. 5. Possible tube arrangement for a stereo view.

Fig. 4 shows some examples of tracks. The illumination of the neighbouring tubes in fig. 4a is due to light reflexions; by painting the tubes in black we have a perfect localization (fig. 4b and c).

By having stacks at different angles, one could obtain a stereo view (fig. 5); we have observed the possibility of flashing the tubes even at 45° with respect to the electric field.

3. Conclusions

The chamber operates with a relatively low voltage ($E = 8$ kV/cm) for any volume since one can arrange long tubes in a stack of several planes of wire electrodes.

Using tiny tubes one can expect a very high space resolution; a very good track efficiency can be also reached by increasing the gas pressure.

The discharges in the tubes are very bright (pictures were taken with an aperture of f: 8) and can easily be seen by a normal TV system.

It seems to us that, although of limited applications, such a device can find its use for instance in satellite experiments where it is much easier to operate than a normal spark chamber and also in cases where the large average density can be an advantage, such as in the detection of γ-rays.

We wish to thank Mr C. Nichols for the production of the glass tubes and Messrs G. Million and R. Bouclier for their technical help.

References

[1] G. Charpak, A. Breskin and F. Piuz, Nucl. Instr. and Meth. **100** (1972) 157.
[2] A. Breskin, G. Charpak, P. Darriulat, K. Geissler, M. Holder and K. Tittel, CERN NP Internal Report 72-21 (1972).
[3] A. Breskin, G. Charpak and K. Geissler, Nucl. Instr. and Meth. **107** (1973) 361.
[4] M. H. Conversi and A. Gozzini, Nuovo Cimento **2** (1955) 189.

142   Nuclear Instruments and Methods in Physics Research A269 (1988) 142–148
North-Holland, Amsterdam

STUDIES OF LIGHT EMISSION BY CONTINUOUSLY SENSITIVE AVALANCHE CHAMBERS

G. CHARPAK, W. DOMINIK, J.P. FABRE, J. GAUDAEN, F. SAULI and M. SUZUKI

CERN, Geneva, Switzerland

Received 10 December 1987

The optimal conditions for the optical recording of images of electron avalanches between parallel meshes have been studied. The emission spectra of gas mixtures have been investigated, where triethylamine (TEA), tetrakis(dimethylamine)ethylene (TMAE), and nitrogen, are used as the photon-emitting agents. For a given charge gain, the photon intensity decreases with electric field. This favours amplification between parallel meshes instead of wires. The use of intensified CCD cameras permits the recording of the local energy loss along the tracks.

1. Introduction

In various articles [1–3] we have described the results obtained in gaseous detectors by optical imaging of the avalanches produced by electrons multiplying in a high electric field. These electrons are produced by ionizing reactions, and a very wide range of applications can be foreseen if the light produced by the molecules excited in the gaseous medium is sufficiently intense to be easily imaged.

In the last version of our detectors [3] (fig. 1) the avalanches were produced between parallel grids, thus avoiding the delicate construction of wire planes and giving an isotropic response with respect to the orientation in the electrode plane. These gaps between the parallel grids can easily be arranged in successive layers for a multistep operation. This permits the detection of single photoelectrons and has been successfully used for Cherenkov ring imaging by charge-sensitive readout, with triethylamine (TEA) [4] as the photoionizable vapour. When the gaps are sufficiently wide, it is found that such chambers are free from photon feedback problems and permit gains above 10^6 to be easily reached. Similar results have also been found with tetrakis(dimethylamine)ethylene (TMAE) as the photoionizable gas [5].

However, we have observed that in order to detect minimum-ionizing particles at atmospheric pressure using optical readout, one single amplification step is enough, since the δ-rays along the tracks produce enough high-density ionization spots to give a good representation of the particle trajectory (as illustrated by fig. 2), which is sufficient for many applications. However, for the imaging of avalanches produced by single electrons, the multistep operation would be necessary with our

present optics and image intensifiers. As a consequence, we have oriented our research in several directions.

A charge-coupled device (CCD) camera located after the two-stage image intensifier has enabled us to exploit the dynamic range of the light-emission intensity, since this method has the precious property of displaying the local energy loss along the tracks. Using photomultipliers, we have compared the energy resolution of X-rays detected by charge collection and by light detection.

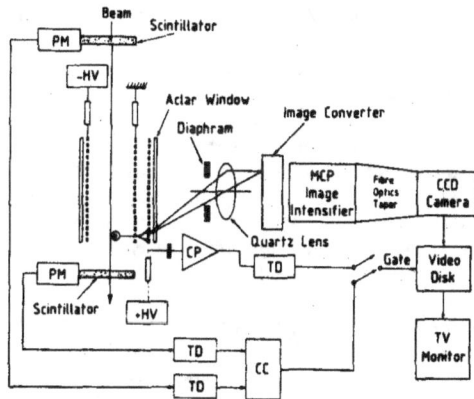

Fig. 1. Experimental setup. The ionization electrons produced in a gaseous volume drift to a multiplying gap made of two parallel meshes set 9 mm apart. This setup is the same as the one in ref. [3], but the camera placed after the image intensifier has been replaced by a CCD camera. The chamber is closed by an Aclar window transparent to light with λ > 200 nm. PM: photomultiplier; HV: high-voltage supply; CP: charge-sensitive preamplifier; TD: timing discriminator; CC: coincidence circuit.

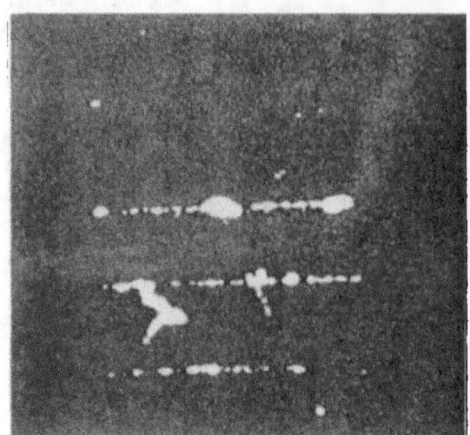

Fig. 2. Example of minimum-ionizing tracks in the drift space represented in fig. 1. The visible points are due to the avalanches from short-range δ-rays in the gas at atmospheric pressure: argon (91%) + CH$_4$ (7%) + TEA (2%).

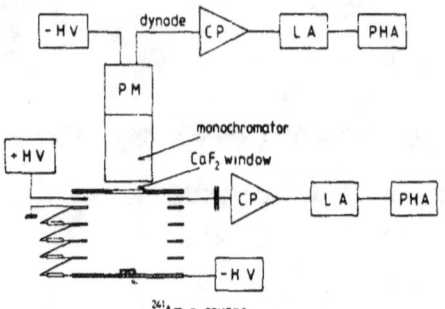

Fig. 3. Experimental set up for studying the characteristics of the light emitted in a gaseous avalanche between the grids. Alpha-particles from ^{241}Am are stopped in the gas, and the ionization electrons are drifted to a multiplying gap. The light output is measured with a photomultiplier, type RCA 31000 M, and compared with the amount of charge collected on the electrodes. The spectrum can be analysed by a spectrophotometer, type ISA H.20UV, with a grating of 1200 grooves per millimetre. PM: photomultiplier; HV: high-voltage supply; CP: charge-sensitive preamplifier; LA: linear amplifier; PHA: pulse-height analyser.

Our first satisfactory results have been obtained with TEA vapour mixed with various gases. This vapour emits photons in a spectrum centred at 280 nm, and we have shown that about one photon per electron is produced in an avalanche [3,6,7].

We have made a quantitative study of the emission spectrum of TEA mixed with some of the various gases that could be used in our detector. In order to facilitate the solution of optical problems, we orientated our efforts towards finding a vapour that emits in a spectral domain closer to visible light. We focused our interest on the properties of TMAE because of its importance as a low-threshold photoionizable vapour, and indeed found that it emits visible radiation (λ > 400 nm) when it is excited by electron avalanches.

However, TMAE is not very easy to handle, and we studied the conversion of the UV light emitted either by TEA alone, or by other gas mixtures not containing TEA or TMAE, in thin layers of a wavelength shifter (WLS) placed against the final anode grid. We found that this conversion is relatively simple and could considerably improve the light yield from a counter using TEA as the radiating vapour: in this case the shift in wavelength makes the light collection easier, thus enabling it to be improved by a factor of 20.

2. Study of the emission of light by some gas mixtures

We have used the device illustrated in fig. 3: α-particles from an ^{241}Am source are stopped in a gas mixture. The ionization electrons are drifted by an electric field to a multiplying gap between two grids, which are

either 1 mm or 1.5 mm apart according to which of the two versions of the device is being used. The photons emitted by the avalanche are analysed by a spectrophotometer or directly detected by a photomultiplier after traversal of a CaF$_2$ window. Fig. 4 shows the photon spectrum in a mixture containing TMAE vapour. The spectrum, peaked at about 480 nm, is in agreement with the fluorescence spectrum of TMAE vapour excited by UV photons (250–390 nm) [8].

The charge signals from the multiplying gap are analysed in parallel with the light signals. Using a charge preamplifier of 200 μs decay constant, we measured the real charge gain $e^{\alpha d}$.

The emission of light as a function of gain reveals

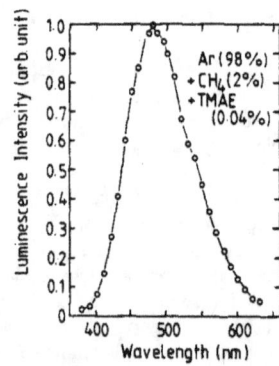

Fig. 4. Spectrum of light emission of avalanches in TMAE. Gas mixtures: argon (97%) + CH$_4$ (3%) + TMAE (0.4 Torr).

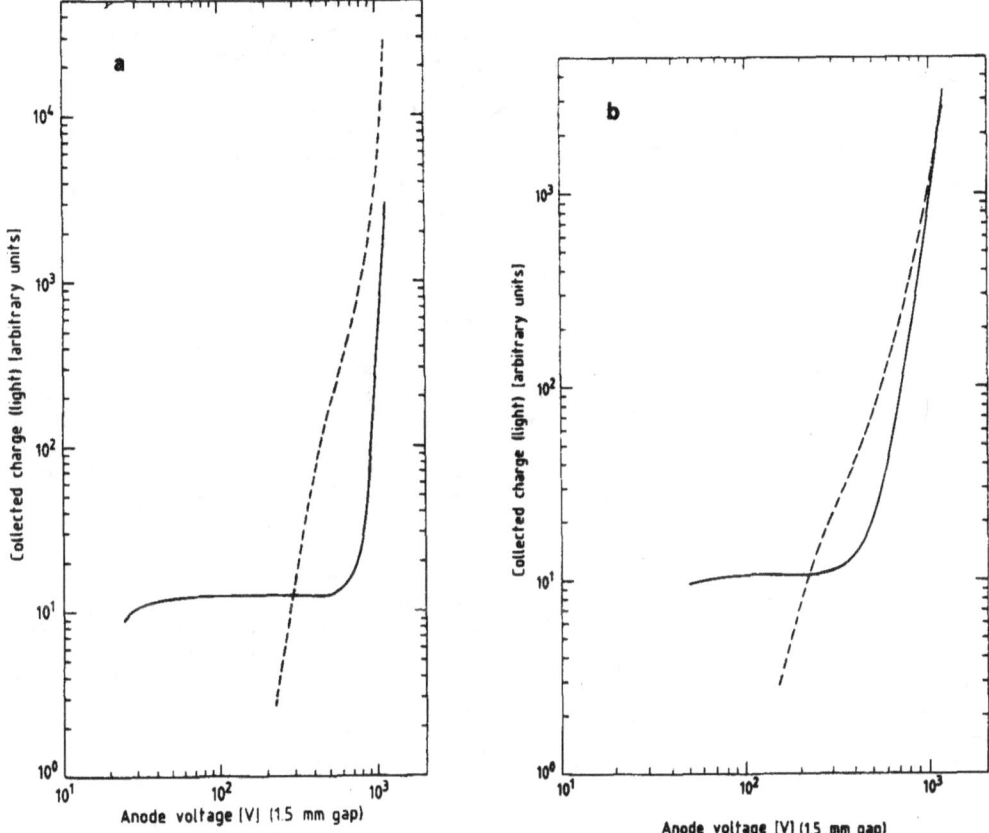

Fig. 5. Emission of light by avalanches in a uniform field. Width of the amplification gap, $d = 1.5$ mm. Solid curves: collected charge; dashed curves: collected light. Gas mixtures: (a) argon (98%) + TEA (2%); (b) argon (97%) + CH$_4$ (3%) + TMAE (0.4 Torr). We observe the emission of light before any charge amplification sets in.

interesting features. In mixtures containing TEA or TMAE, we observe light emission before any multiplication occurs (fig. 5). We can thus have gas scintillation produced by a drifting electron, as in pure noble gases. This observation has direct consequences on our detector and explains some of its features. We have seen that for the same charge gain in a counter with TEA or TMAE, we have a linear increase of photons commensurate with the gap size. We have also observed the same increase when there is no charge multiplication. If some scintillation light is emitted without ionization by atoms in the excited state, this is normal: for the same charge gain the ionization mean free path in the wider gap is longer, so that within this path there can be more inelastic collisions giving rise to scintillation light. We expect that the intensity of the scintillation photons will be proportional to the total voltage drop [7], which is larger in a wider gap. In the multiwire chambers with

very thin anode wires, the electrons reach a high energy in a very narrow region around the wire, and the relative amount of excitation may be smaller. We found the parallel-grid structure, with wide gaps, to be the most favourable for photon emission, with the clear advantage of being isotropic in the plane of the chamber.

There is no proportionality between charge gain and photon gain (fig. 6). The relative amount of light is variable between TEA, N$_2$, and TMAE. Light output is normalized to the unit of the primary charge in the drift space and corrected on the spectral sensitivity of the photomultiplier (bialkali photocathode on the quartz window). Because the spectral regions are different for these emitters – about 280 nm for TEA, 340 nm for N$_2$, and 480 nm for TMAE – the correction is necessary in order to make a proper comparison of the light yields. However, the experiments with TMAE were done at

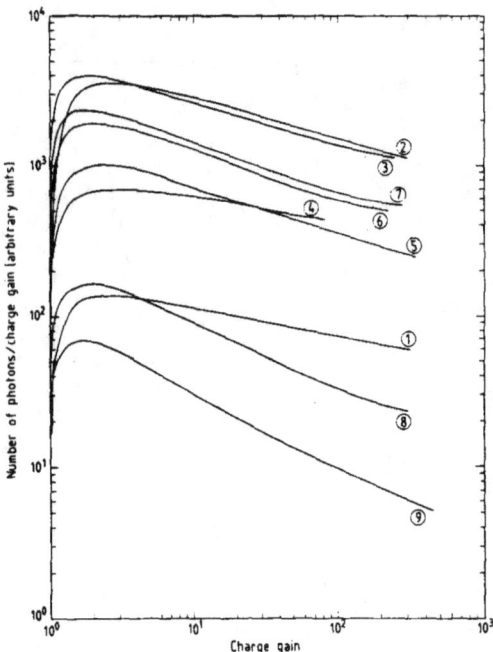

Fig. 6. Relation between charge gain and light gain. Amplification gap width $d = 1.5$ mm: (1) Ar (95.5%) + N_2 (2.5%) + CH_4 (2%); (2) Ar (98%) + N_2 (2%); (3) Ar (98%) + TEA (2%); (4) Ar + TMAE (0.4 Torr). Amplification gap width $d = 1$ mm: (5) He (87%) + CH_4 (11%) + TEA (2%); (6) Ar (87%) + CH_4 (11%) + TEA (2%); (7) Ar (98%) + TEA (2%); (8) Ar (97%) + CH_4 (3%) + TMAE (0.4 Torr); (9) CH_4 + TMAE (0.4 Torr).

$23°C$, where the vapour pressure is only 0.3 Torr. Experiments with nitrogen added to argon (fig. 6) show a very good light yield comparable to TEA. This gas, however, is not as favourable for electron multiplication.

3. Use of an intensified CCD camera

With respect to the setup of ref. [1], we simply changed the type of camera [9]. The image from the chamber is transferred to the first image intensifier via an objective $F/4.5$ for UV light or $F/1$ for visible light. The active section of our system is composed of two image intensifiers. The first one is a proximity-focused, single-stage diode of 25 mm photocathode diameter. It is essentially a wavelength converter with high quantum efficiency of the photocathode (bialkali) and relatively small gain (~ 20 photons per photon). The gain is achieved by electron acceleration in the high electric field between the photocathode and the phosphor screen-anode. The fast-X3 type phosphor is deposited

on the fibre optics backplate, allowing a very efficient coupling to the next stage of the light intensification. The second stage is a multichannel plate intensifier of 22 mm active diameter with an S20 photocathode and a P20 phosphor layer on the fibre-optics plate. This structure allows a gain of ~ 30000 photons per photon, thus making it the main light intensifier of the chain.

Since the CCD that we use has a diagonal (7 mm) smaller than the diameter of the backplate of the second image intensifier, we employ fibre optics taper for the CCD attachment to the system (in order to collect light efficiently without distortion of the image). This CCD (Thomson 7852 CDA-80) has 144 lines of 208 pixels; it produces a standard video signal which can be displayed on a monitor and fed to a digitizer card (Data Translation DT 2851) which interfaces to an IBM–PC/AT computer. This allows images produced by the chamber and viewed by the optoelectronic camera to be stored and analysed by the computer. In order to conserve disk space and to accelerate the storage process, the images are compressed using a run-length-encoding algorithm. This algorithm replaces runs of identical pixel values by a single pixel value followed by its repeat count. In our case a typical compacted image takes about 15 kbytes. The digitizer has a precision of 8 bits, which allows an accurate measurement of the light intensity over a large dynamic range, thanks to the excellent linearity of the optoelectronic chain.

The light emitted by the avalanches is proportional to the primary charge, and from the computerized images we can expect information on the local energy loss.

If we compare the electric pulses from the charges collected on the anode grid with the light pulses from a

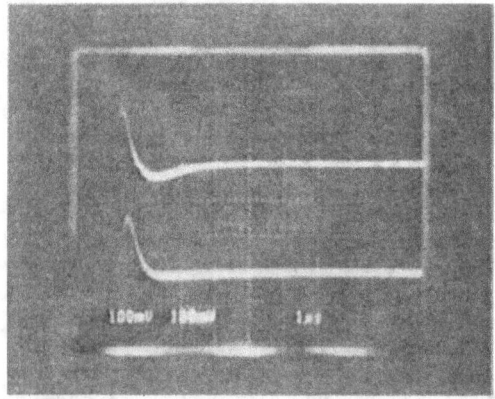

Fig. 7. Light pulses from an X-ray source. The 5.9 keV X-rays from ^{55}Fe are producing light pulses which are detected by a photomultiplier placed 100 cm from the chamber window. Gas mixture: argon (88%) + CH_4 (10%) + TEA (2%). Upper trace: collected light, lower trace: collected charge.

photomultiplier viewing the chamber at a distance of 1 m, we indeed observe that the energy resolution is just as good in both cases (fig. 7). With 5.9 keV X-rays, the spots that make up the image of each avalanche are between 1 and 2 mm wide, and the centroid – as we could verify using a well-collimated source – is accurate to within 250 μm (rms).

With α-particles the chamber displays some of its intrinsic qualities. We made the study using various gas mixtures. Fig. 8a and 8b illustrate the pictures obtained

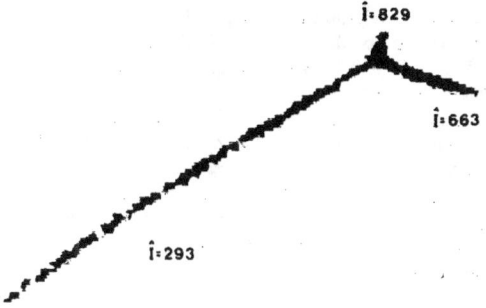

Fig. 9. A scattering event in helium. The α-particle scattered on a helium nucleus. The CCD gives the average ionization in each prong and the ionization loss along the initial tracks. Such a device would easily give a full image of the slow neutron capture on ^{3}He, permitting great accuracy on the interaction point and total rejection of background.

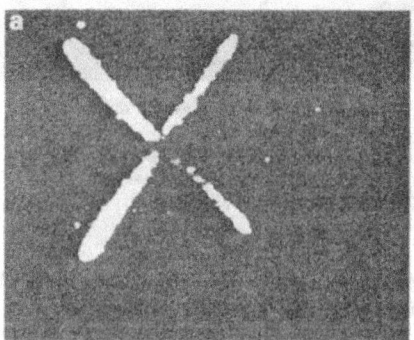

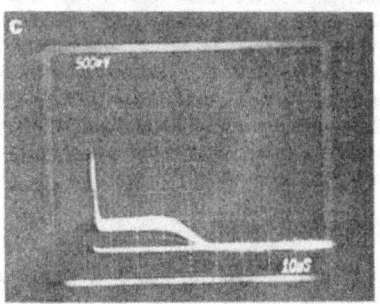

Fig. 8. Alpha-particle tracks. The α's from an ^{241}Am source are collimated along four axes at an angle of 90°. Gas mixtures: (a) argon (88%) + CH$_4$ (10%) + TEA (2%); (b) helium (92%) + CH$_4$ (6%) + TEA (2%) (the inset (c) shows the collected charge).

with argon and helium, respectively, as the main gas carriers. The interesting thing is that the energy loss along the track reflects what is expected qualitatively from the Bragg curve. In fig. 9 we have selected an event where the α-particle is scattered in the gas. The average ionization along the tracks is obtained from the CCD.

We have tried to observe tracks with TMAE with a single-step amplification in the device shown in fig. 1, and the results confirm the relative scarcity of light. The tracks are very diffuse. But if we use a multistep structure with two parallel gaps separated by a transfer space of 25 mm, we observe cosmic-ray events (fig. 10); this figure shows a heavily ionizing one. With TEA mixtures we had to use an UV objective of aperture 1/4.5, for

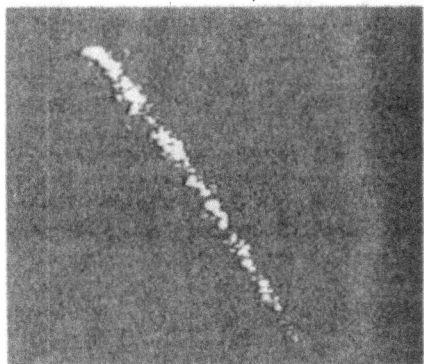

Fig. 10. A cosmic-ray event in a multistep TMAE chamber:
– Gas mixture: argon (75%) + CH$_4$ (25%) + TMAE (2.5 Torr).
– Multistep operation: two parallel gaps of 4 mm separated by a 25 mm transfer gap.
– Optical aperture: 1/4.5.

reasons of availability. With the TMAE mixture the aperture can be 1/1. The gain of nearly 20 partly compensates for the loss of light. Although one can build large-aperture optics for the TEA emission domain (but at a high cost), it is clear that the short wavelength is a limitation. For this reason we have concentrated our efforts on converting the UV light into visible light by using WLSs placed against the anode grid, thus making it possible to widen the choice of filling gases. The choice is dictated by the requirements of the physical application envisaged for the chamber.

4. Conversion of UV light

The parallel-grid geometry is very convenient for the wavelength shifting of light very near to the emission point. The majority of the excited atoms are at a distance of a few ionization mean free paths from the last grid, i.e. a fraction of a millimetre. The loss of resolution due to the distance of these excited atoms from the light converter is thus small if the WLS is placed against the grid. We tested a well-known WLS: sodium salicylate. It was deposited on a Mylar foil and covered with a 50 μm thick Aclar *) foil placed against the anode mesh of the amplification gap. The absorption length of our Aclar foil is 230 μm for photons between 250 and 350 nm. Such a WLS structure will separate the sodium salicylate powder from the detector gas. Sodium salicylate is known to convert photons in the 200–350 nm range, to visible light, with almost 100% efficiency [10]. The result is spectacular. With α-particles and a gaseous mixture containing TEA we could see the tracks with the naked eye, after only a few minutes for accommodating in the darkness.

We could record the tracks with a mixture containing only argon, methane, and nitrogen, without TEA (fig. 11 shows an example of such a track).

5. Some applications

The fact that imaging with CCD cameras records the local energy loss brings great power to this readout method. No electronic readout can deliver such complete information on a pattern of energy deposition in a gas.

5.1. High-density detectors

With heterogeneous drift spaces, one can envisage that such detectors could be valuable candidates for proton decay studies. The quality of the pattern is very

* Aclar: polychlorotrifluoroethylene (PCTFE).

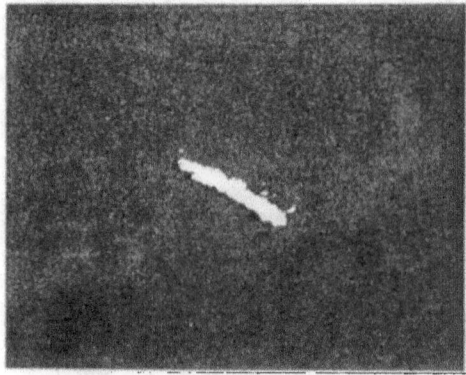

Fig. 11. An alpha track with wavelength shifter:
- Single step as in fig. 1.
- Gas mixture: argon (95%) + CH_4 (2%) + N_2 (3%).
- Sodium salicylate layer against the last grid.
- Objective Leitz Noctilux 1/1 : 50 mm.

favourable for background rejection and particle identification. The local energy-loss information is an additional asset. We are preparing tests with a large-size detector, of 15 radiation lengths depth, in high-energy electron beams. Preliminary results without a CCD camera have given encouraging pictures [3].

5.2. Slow-neutron imaging detectors

Our experiments show that helium is a very convenient filling gas: ^{3}He could be used. The interesting point is that we could get a full image of the two prongs recoiling after the neutron absorption, thus determining the origin with an accuracy of a fraction of a millimetre. One could imagine a spherical drift chamber analogous to the one used for X-rays, filled with ^{3}He + TEA and another gas such as Ar or CH_4 in order to control, at will, the range of the recoil nuclei [11]. The electrons can be transferred to a multiplying gap between two grids. The recoil pattern would be preserved by the drifting, and there would be no parallax error because of the radial field in the drift space. Such a method would be slow since it requires image analysis by a computer, but it would have the unique properties of accuracy and noise rejection.

5.3. Liquid optical detectors

It has been shown [12] that it is easy to extract electrons from liquid TMP and amplify them between two parallel grids. This observation can complement the old observation by Dolgoshein et al. [13] that electrons can be extracted from some liquefied or solid noble elements. Our study with α-particles was concentrated on vapours compatible with the existence of the liquid phase. We have observed α tracks in mixtures of He + TEA + TMP or TMS. The ionization loss of α-particles in the gas phase at atmospheric pressure is of the same order as the energy loss of minimum-ionizing particles in liquids. For a liquid-argon detector, one has to find a gas phase that emits light. The candidates can be nitrogen, or methane – which is then also a sizeable fraction of the liquid phase. Our test with the wavelength shifters shows that we can visualize α tracks in the gas phase in mixtures of argon plus methane and nitrogen (fig. 11).

References

[1] M. Suzuki, P. Strock, F. Sauli and G. Charpak, Nucl. Instr. and Meth. A254 (1987) 556.

[2] M. Suzuki, A. Breskin, G. Charpak, E. Daubie, W. Dominik, J.P. Fabre, J. Gaudaen, F. Sauli, D. Sauvage, P. Strock and T. Zeludziewicz, Nucl. Instr. and Meth. A263 (1988) 237.

[3] G. Charpak, J.P. Fabre, F. Sauli, M. Suzuki and W. Dominik, Nucl. Instr. and Meth. 258 (1987) 177.

[4] R. Bouclier, G. Charpak, A. Cattai, G. Million, A. Peisert, J.C. Santiard, F. Sauli, G. Coutrakon, J.R. Hubbard, Ph. Mangeot, J. Mullie, J. Tichit, H. Glass, J. Kirz and R. McCarthy, Nucl. Instr. and Meth. 205 (1983) 403.

[5] G. Charpak and F. Sauli, Nucl. Instr. and Meth. 225 (1984) 627.

[6] M. Suzuki, CERN EP Internal Report 87-02 (1987).

[7] A. Policarpo, Phys. Scr. 23 (1981) 539.

[8] Y. Nakato, M. Ozaki and H. Tsubomura, J. Phys. Chem. 76 (1972) 2105.

[9] J. Dupont, J.P. Fabre, P. Fonte, J. Gaudaen and M. Suzuki, internal report CERN/EF/INSTR/87-1 (1987).

[10] M. Seya and F. Masuda, Sci. Light 12 (1963) 9; J. Berlman, Handbook of fluorescence spectra of aromatic molecules (Academic Press, New York, 1977).

[11] G. Charpak, G. Demierre, K. Kahn, J.C. Santiard and F. Sauli, Nucl. Instr. and Meth. 141 (19770 449.

[12] D. Anderson, G. Charpak, R. Holroyd and D. Lamb, Nucl. Instr. and Meth. A261 (1987) 445.

[13] B.A. Dolgoshein et al., Sov. J. Part. Nucl. 4 (1973) 70.

658

Nuclear Instruments and Methods in Physics Research A283 (1989) 658-664
North-Holland, Amsterdam

BEAM TEST OF AN IMAGING HIGH-DENSITY PROJECTION CHAMBER

P. FONTE [1], A. BRESKIN [2], G. CHARPAK [1], W. DOMINIK [1]* and F. SAULI [1]

[1] *CERN, Geneva, Switzerland*
[2] *Weizmann Institute, Rehovot, Israel*

A beam test of a high-density projection chamber with optical readout is presented. The device consists of a prototype of the DELPHI HPC calorimeter on which a parallel-plate, light-emitting structure was installed, replacing the original multiwire proportional chamber readout system. It produces detailed images of the energy deposited by electromagnetic showers; hadronic interactions are easily discriminated from these. The computerized readout system gives full quantitative information on the events, showing good energy and position resolutions.

1. Introduction

The detector described in this paper is an optical readout calorimeter, which we call an imaging high-density projection chamber (IHPC). It consists of a high-density drift volume [1], terminating in an imaging chamber [2-6]. The images of the events are detected by an image intensifier coupled to a charge-coupled device (CCD) camera.

We performed a beam test with this device in order to investigate the ability of imaging chambers to detect electromagnetic showers and hadronic interactions. We also benefited from the high granularity of the device (about 30 000 pixels) to get a detailed view of the event structure in a gas sampling calorimeter. With further analysis we will attempt to improve the calorimetric measurements by using statistical information eventually contained in the shower topology.

In this paper the test is described and the first quantitative results on basic parameters are presented. Early results with a small prototype of this kind were presented in ref. [5].

2. The detector

The high-density drift volume (called the "1 mm prototype") where the showers develop was built by the CERN DELPHI HPC Group using the "lead-wire ribbon" technique [7] (see fig. 1). Its dimensions are $60 \times 44 \times 27$ cm^3, presenting 10.5 radiation lengths (X_0) and 0.34 interaction lengths to the incoming particles (along the 44 cm side), and it contains about 110 kg of lead.

* On leave of absence from the Institute of Experimental Physics, University of Warsaw, Poland

0168-9002/89/$03.50 © Elsevier Science Publishers B.V.
(North-Holland Physics Publishing Division)

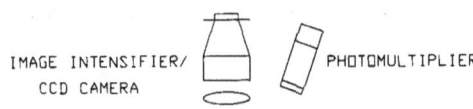

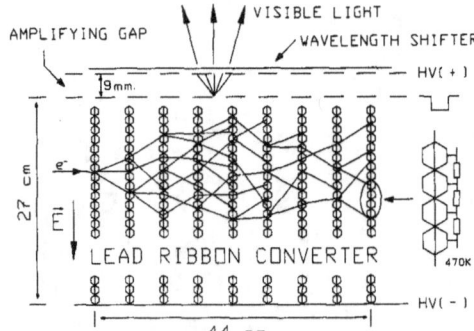

Fig. 1. Schematic view of the IHPC.

The 40 gas gaps sample the showers at intervals of $0.25 X_0$.

Ionization electrons, created in the gas gaps by charged particles in the electromagnetic shower, drift towards a parallel-plate chamber where they are amplified. Charge multiplication occurs between two parallel meshes, 9 mm apart. The drift time over the 27 cm long region is 8.2 μs, with a drift field of 600 V/cm. The gas mixture used [argon + triethylamine (TEA) vapour at 5°C (95%) and CH$_4$ (5%)] fluoresces under the impact of the electrons in the avalanches, emitting UV light [8,9]. This light is proximity-focused onto a thin sheet of plastic wavelength shifter $^+$, placed in contact with the upper mesh of the amplifying gap. This converts the UV

$^+$ Courtesy of M. Bourdinaud, Saclay, France.

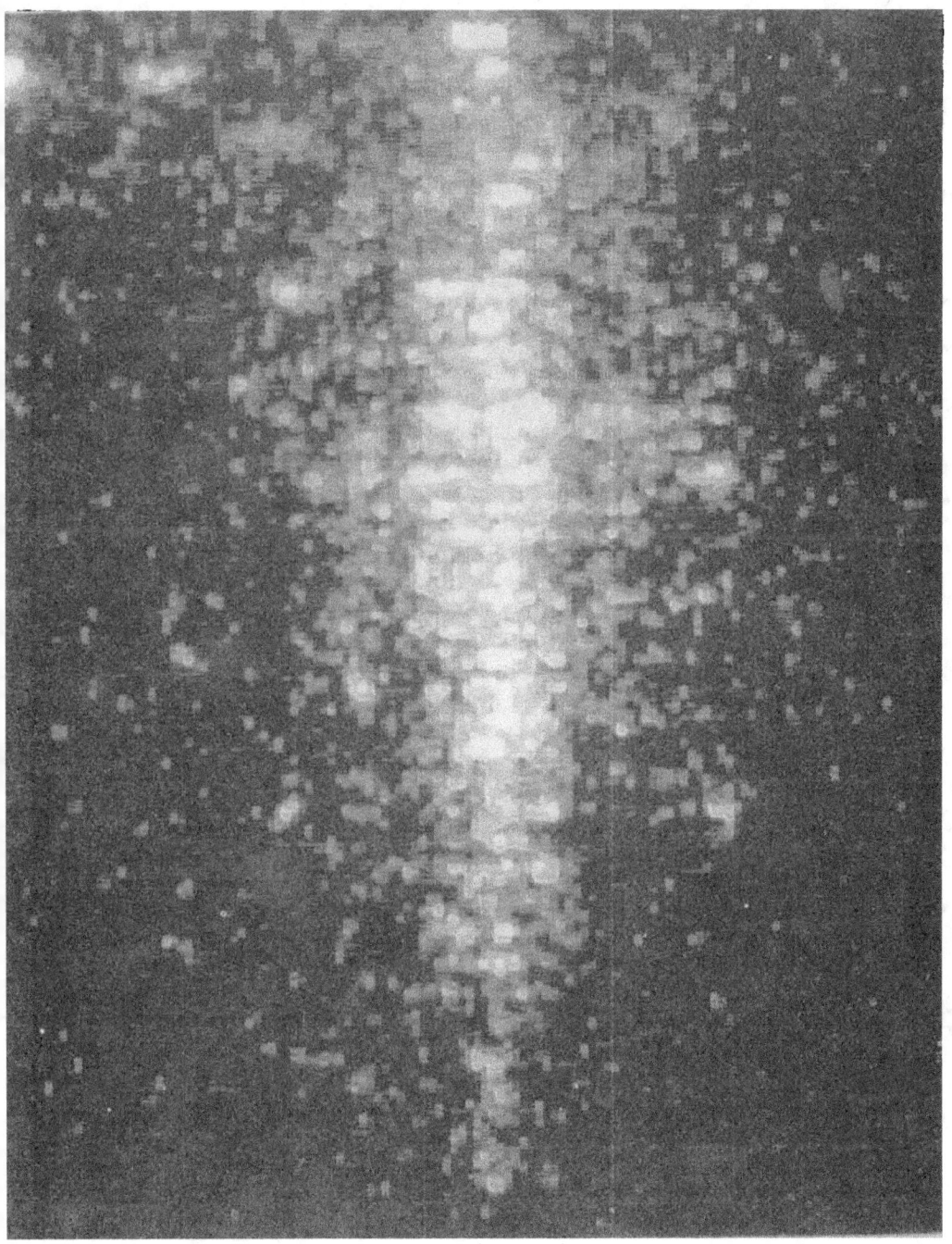

Fig. 2. Example of an 8 GeV/c electromagnetic shower.

IX. DIVERSE CHAMBERS

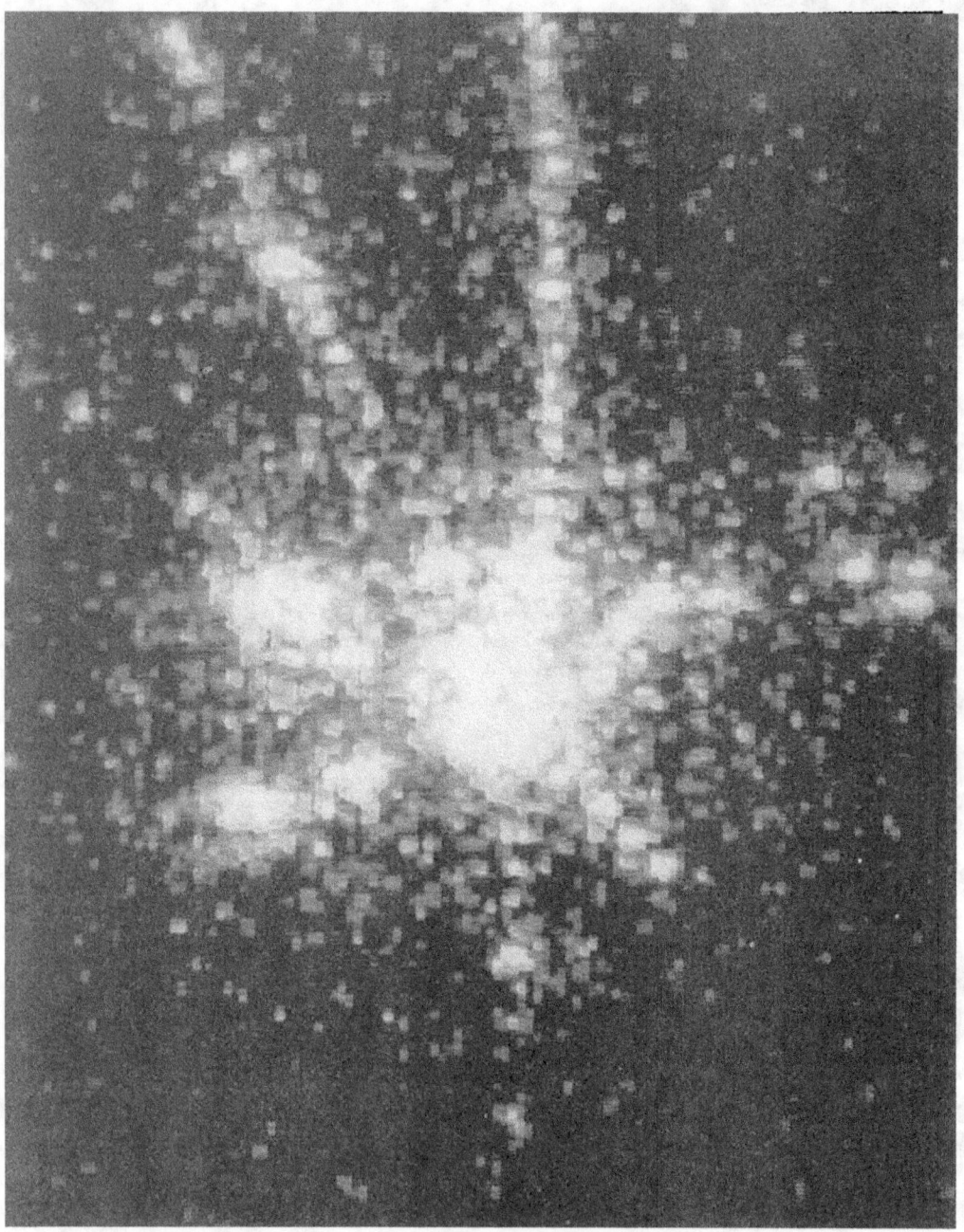

Fig. 3. Example of a 5 GeV/c hadronic interaction.

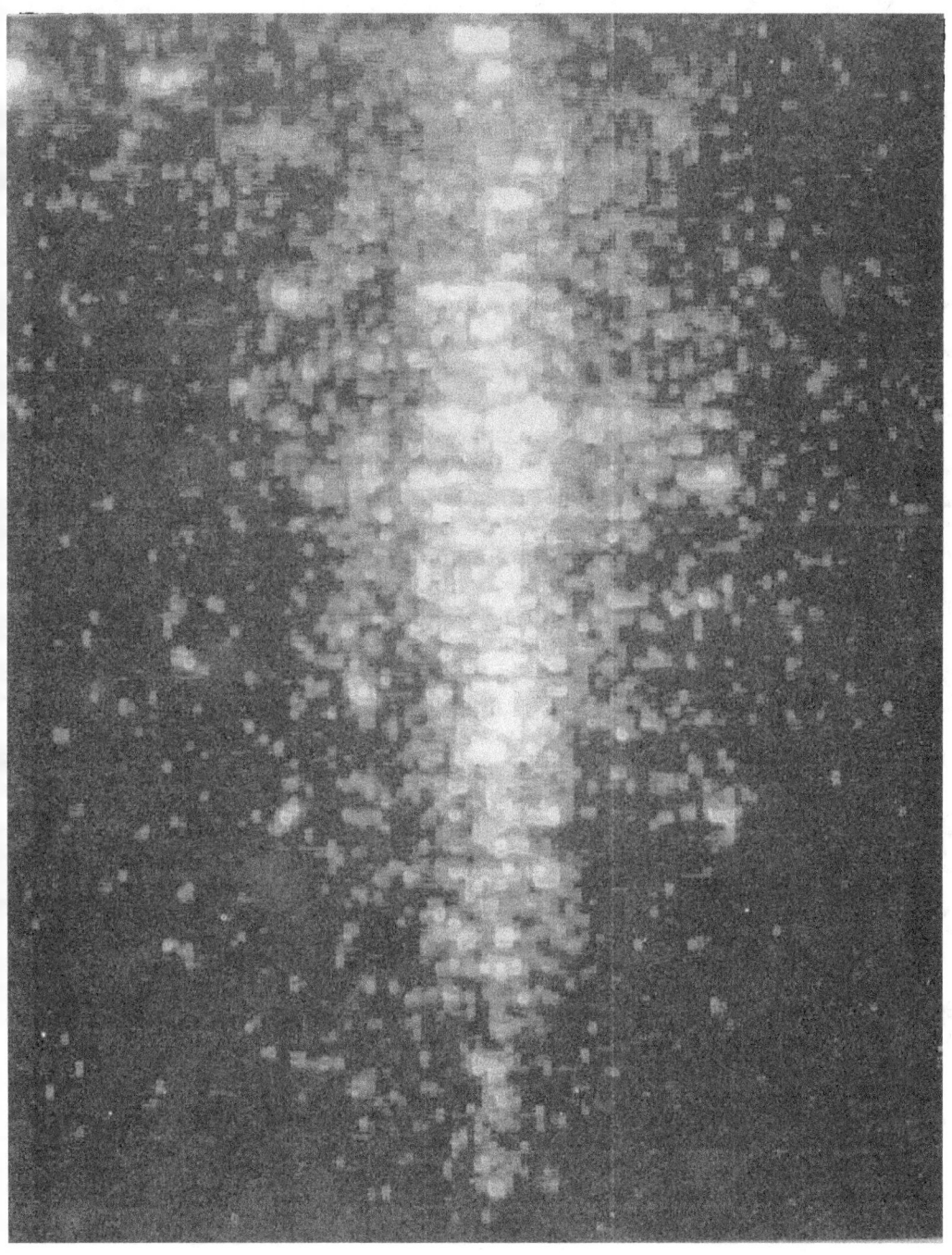

Fig. 2. Example of an 8 GeV/c electromagnetic shower.

IX. DIVERSE CHAMBERS

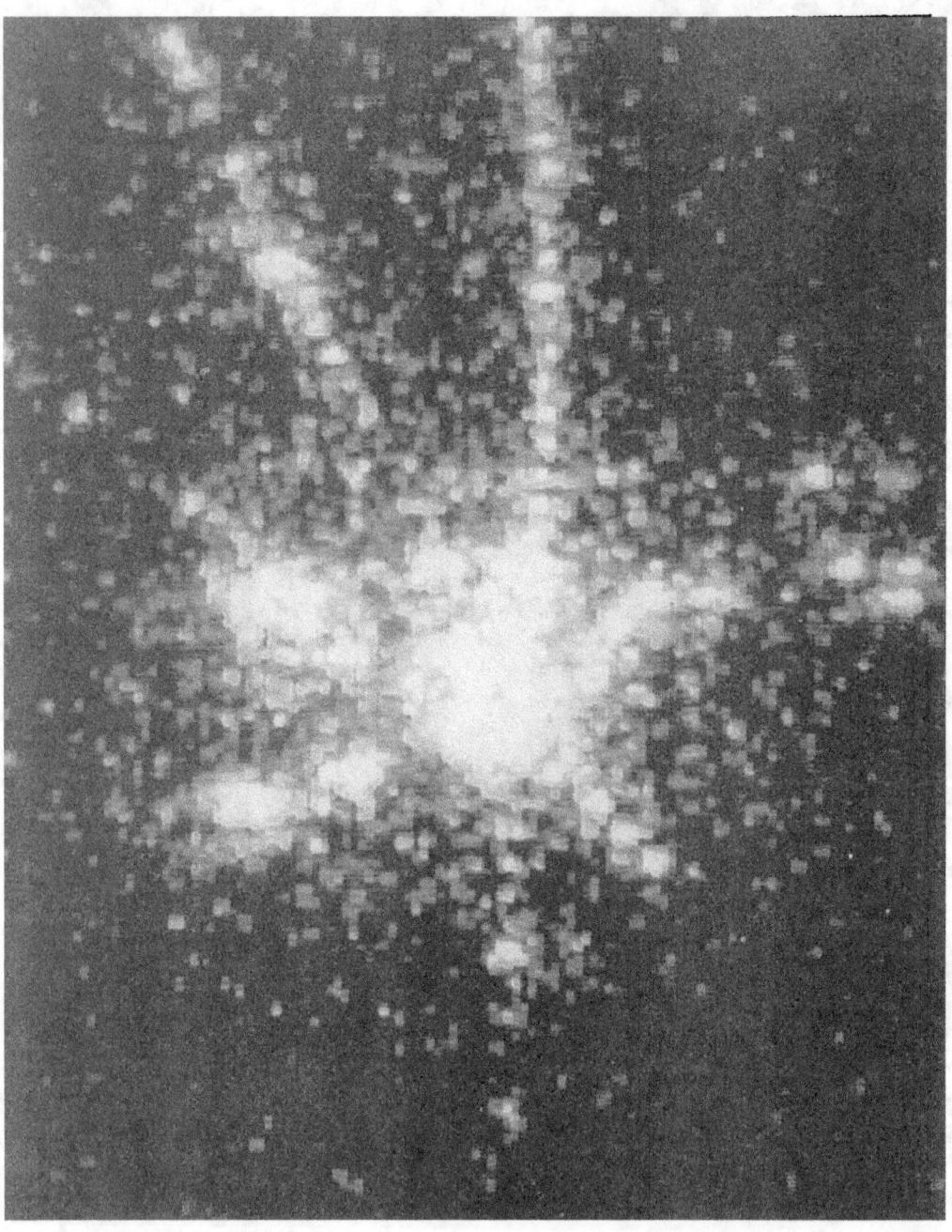

Fig. 3. Example of a 5 GeV/c hadronic interaction.

light to visible violet light, peaked at 420 nm – the region of highest sensitivity for many photocathodes, and of full transparency for standard optics (windows, lenses, etc.). The volume is closed by a double window of Mylar, with a flow of pure argon in between.

The events are imaged by an intensified CCD camera * constructed from a gateable microchannel-plate image intensifier, an active image taper and a CCD having 208×144 pixels (details of the camera are given in ref. [10]). The 208 pixels were imaged along the 44 cm side of the detector (longitudinal beam direction), yielding a granularity of 2.1 mm per pixel. Along the transverse beam direction, the granularity is 2.3 mm per pixel, covering 33 cm. The driver electronics for the CCD provides a standard video signal that can be fed into standard video equipment.

3. The test setup

The detector was installed in an unseparated beam, composed mainly of pions and electrons, at the CERN proton synchrotron (PS) East Hall facility. Two threshold Cherenkov counters, having an efficiency on the level of 99%, provided e/π discrimination. A narrow beam in the transverse direction was defined by triggering with a 2 mm wide, 2 cm high scintillator, placed vertically. The angular aperture of the beam is geometrically bound to be ≤ 10 mrad.

Owing to the natural alpha radioactivity of the lead in the converter (which induces sparking), the amplifying gap had to be kept at a reduced gain, and a square pulse of 1.2 kV and 10 µs width was applied whenever the full gain was required in order to detect an event. The pulse height chosen was such that the chamber still operates in proportional mode, the readout being completely insensitive to any electronic noise induced by this pulse. A 10 µs gate pulse was also applied to open the image intensifier in front of the CCD, thus avoiding any light-noise that would be superimposed on the event during the 20 ms CCD integration time.

A photomultiplier (PM) assembled near the CCD camera gave a time-discriminated signal proportional to the total light coming from the successive horizontal layers of the drift volume, thus providing information about the shower development in the z-direction. Using this depth-discriminated information and a beam of cosmic muons crossing the chamber along the drift direction, the drift attenuation was measured to be 50% after the maximum drift length of 27 cm.

The video signal was read by a MATROX-VIP640 VME-bus image digitizer having an accuracy of 8 bits in

---

a 512×128 matrix (thus losing 16 CCD lines). For the PM, a LeCroy-ICA2261 CAMAC module provided 10-bit resolution on 320 channels sampled at 30.4 ns intervals (equivalent to 1 mm per channel because the last 50 buckets lie outside the drift time). The data acquisition system (CERN-VALET-PLUS VME-bus microcomputer and CAMAC) allowed a maximum of four events per spill to be recorded on tape, the event size being 66 Kbytes.

Examples of an 8 GeV/c electromagnetic shower and a 5 GeV/c pion (or proton) interaction can be seen in figs. 2 and 3; their obvious difference implies that the e/π identification seems to be very good. These pictures lose much of their quality in the printing process, compared with the beautiful high-resolution colour-coded pictures that can be seen with a video monitor.

4. First results

The energy resolution of the IHPC for electromagnetic showers (fig. 4) was calculated either from the PM or from CCD signals. The CCD signal, forming an image, gives information about the image content (total light) and also about the image surface. The image surface is calculated by counting the number of pixels above the pedestal, the sum being considered as a

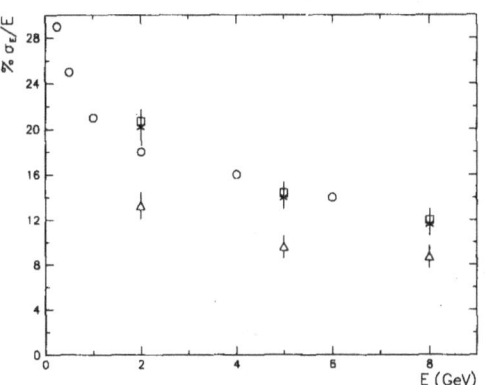

Fig. 4. Energy resolution. When the incident energy is estimated by the total light emitted by an event (photomultiplier and image content data), the IHPC energy resolution is comparable with the energy resolution measured in the same converter with MWPC readout (DELPHI HPC). If the image surface is considered as a measure of the energy, there is an improvement in the energy resolution, particularly at low energies. The image surface is calculated by counting the number of pixels above the pedestal-subtracting threshold, and the image content (total light) is calculated by summing all the pixel values after pedestal subtraction. Key: ○ DELPHI HPC; ● IHPC (image content); △ IHPC (image surface); □ IHPC (photomultiplier).

---

* Designed and assembled by J. Dupont and J.-P. Fabre, Experimental Facilities Division, CERN, Switzerland.

IX. DIVERSE CHAMBERS

measure of the energy. The image content is the sum of all pixel values after pedestal subtraction.

At 5 and 8 GeV the energy resolution determined from the image content and the PM (total light) is comparable with the values measured by the DELPHI Group in the same converter with a multiwire proportional chamber (MWPC) readout [11]. The poorer resolution of the IHPC at 2 GeV is attributable to a lack of gain in the single amplifying gap. In fact, a separate run with a 30% superior gain yielded a resolution of 18% at 2 GeV instead of the 20% presented in fig. 4. The PM and CCD readouts show strictly the same behaviour. It is clear that a short converter such as this one will not follow a $1/\sqrt{E}$ law; a deviation from the $14\%/\sqrt{E}$ law followed by the points at lowest energy is already visible at 2 GeV for the DELPHI data (see ref. [7]). For all energies, but particularly for low energy, the image surface provides a better resolution (13% at 2 GeV) than does the total light (20% at 2 GeV). In all cases the resolution is sharply degraded by any increase in the pedestal-subtracting threshold.

The signals from the CCD and the PM are strongly correlated, as can be seen in fig. 5a, the clusters from 2, 5 and 8 GeV events being clearly separated. Their centroids are marked with a cross. In fig. 5b a scatter plot of the image content versus image surface is shown. The image surface data show a tendency to saturate at higher energies – a tendency probably worsened by the non-containment of the showers.

In order to reconstruct the projection of the line of flight of the incident particle on the image plane (shower axis), the following procedure was used:

(i) the centroids of every pixel row perpendicular to the shower axis were calculated and their dispersion (σ) about the average was calculated for the 5 GeV data;

(ii) the calculated dispersions were used to weight a linear least-squares fit to the pixel-row centroids for every event, yielding the angle and the impact coordinate;

(iii) all the lines were superimposed for the 5 GeV data, and the point of smaller transverse dispersion was defined as the reference plane for the position measurement. A particle position is thus defined as the intersection of the axis with the reference plane. This plane lies at $4.1X_0$ and was used for all energies. It does not change significantly from 2 to 8 GeV.

Other procedures, such as using the weighted mean of all centres of gravity to estimate the impact point, or

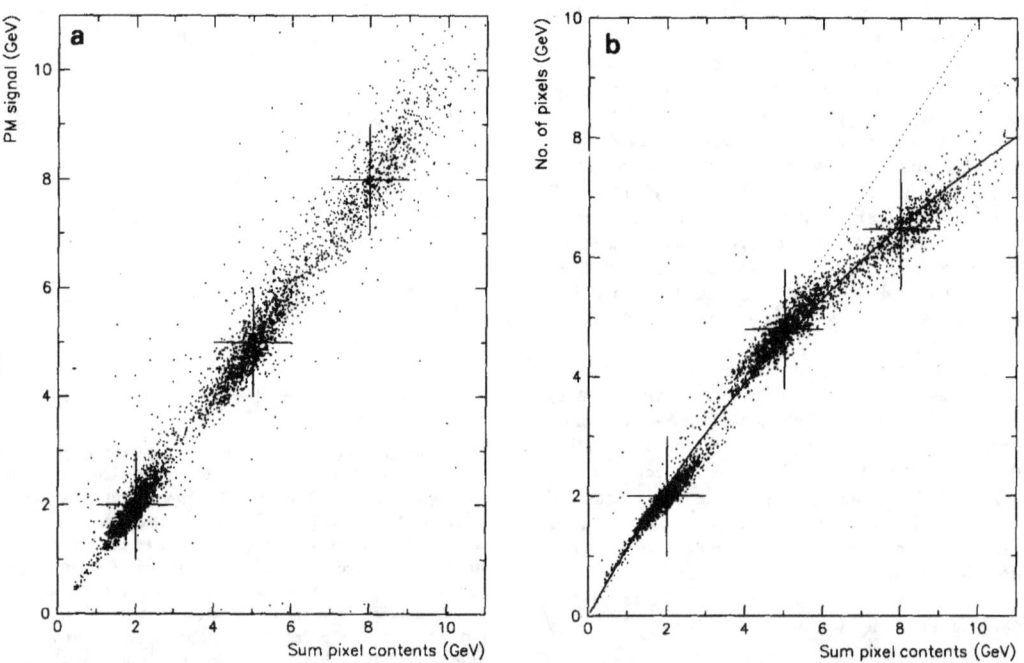

Fig. 5. (a) Photomultiplier versus image content (both quantities are converted to a GeV scale). The good correlation between these two independent forms of readout is evident. (b) Image surface versus image content (both quantities are converted to a GeV scale). The image surface grows more slowly than the image content (total light) with increasing incident energy.

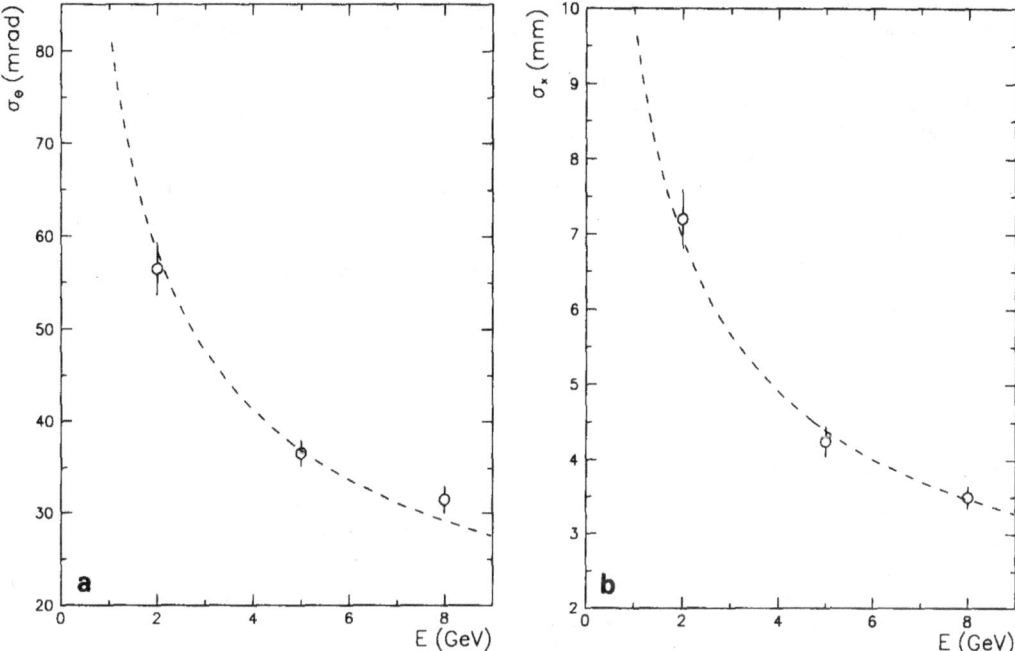

Fig. 6. (a) Angular resolution: this follows an (83 mrad/$\sqrt{E}$) law. (b) Position resolution: fitted by 9.8 mm/$\sqrt{E}$.

using variable weights for every row, proportional to the total light in the row, produced worse results.

The results from the procedure we used can be seen in figs. 6a and 6b. The angular resolution (σ_θ) is about 32 mrad at 8 GeV and seems to follow an (83 mrad/$\sqrt{E}$) law in the energy range studied. This angular resolution is worse by about a factor of 2 than the value of 38

mrad obtained by the DELPHI group [7] at 1 GeV. A possible reason for this might be the purely two-dimensional nature of the information in the IHPC image, compared with the three-dimensional of the DELPHI HPC.

A position resolution of 9.8 mm/$\sqrt{E}$ (fig. 6b) can be compared with the resolutions achieved by using MWPC

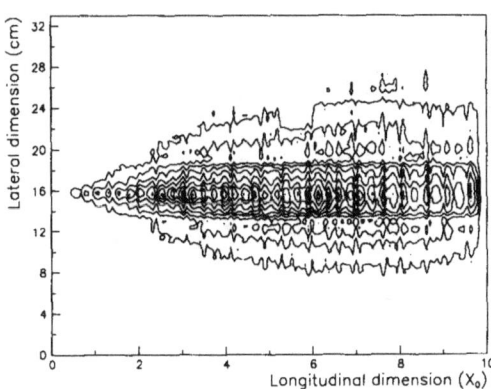

Fig. 7. Contour plot of an average of 8 GeV/c showers. The energy leakage through the end of the converter is clearly visible.

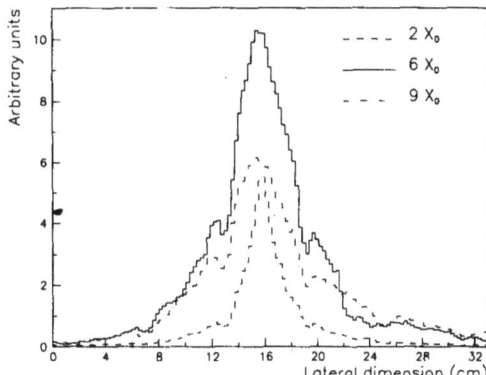

Fig. 8. Transverse cuts on the average image of 8 GeV/c showers at 2, 6 and 9 radiation lengths. The two indentations (dead zones) are explained in the text.

IX. DIVERSE CHAMBERS

techniques for shower sampling [12,13]. The effects of the beam dispersion for both the angular and the position resolution measurements lie within the error bars.

A contour plot of an average of 400 shower images at 8 GeV is shown in fig. 7. The shower broadens in the first $4X_0$, after which it keeps essentially the same width up to $9X_0$. Concentric contours over the central axis mark the centres of the gas gaps. This figure clearly shows that the showers are not contained. In fig. 8 the transverse image profile of the average of 8 GeV shower images is plotted for 2, 6 and 9 radiation lengths. The two indentations on each side of the peak correspond to mechanical supports inside the drift volume (dead zones).

5. Conclusions

The beam test of an imaging high-density projection chamber in the energy range from 2 to 8 GeV shows that the characteristics of the device seem to be close to those obtained with conventional MWPC readout for the same converter (only $10.5X_0$): 20% energy resolution (σ_E/E) at 2 GeV; 9.8 mm/$\sqrt{E}$ position resolution; 83 mrad/$\sqrt{E}$ angular resolution. The high granularity offers the possibility to carry out some nonconventional analysis; for instance, by considering the total surface of the shower's image as a measure of the incident energy, the energy resolution is improved to 13% at 2 GeV. More elaborate algorithms, which rely on the fine detail of the images, are being studied for improved particle identification and increased accuracy of electromagnetic calorimetric measurements.

The IHPC shows promise in applications where the rare or complex phenomena being studied need a high data redundancy (the cost per analog channel is re-markably low). Also, it may be suitable for covering large surfaces when the showers are still contained in a fairly big number of pixels.

Acknowledgements

We wish to thank O. Ullalland and H.G. Fischer for lending us the converter and for their assistance during the transformation work, and M. Bourdinaud for providing the wavelength shifter. The technical assistance of R. Bouclier, G. Million and J.C. Santiard was very much appreciated.

References

[1] H.G. Fischer and O. Ullaland, IEEE Trans. Nucl. Sci. NS-27 (1980) 78.
[2] R.S. Gilmore et al., Nucl. Instr. and Meth. 206 (1983) 189.
[3] D.M. Potter, Nucl. Instr. and Meth. 228 (1984) 56.
[4] T.K. Gooch et al., Nucl. Instr. and Meth. A241 (1985) 363.
[5] G. Charpak et al., Nucl. Instr. and Meth. A258 (1987) 177.
[6] G. Charpak et al., IEEE Trans. Nucl. Sci. NS-35 (1988) 483.
[7] CERN-DELPHI 85-19 GEN-22 (1985).
[8] M. Suzuki et al., Nucl. Instr. and Meth. A254 (1987) 556.
[9] V. Peskov et al., preprint CERN-EP/88-167, Nucl. Instr. and Meth. A277 (1989) 547.
[10] A. Breskin et al., Nucl. Instr. Meth. A273 (1988) 798.
[11] H.B. Crawley et al., Test results from the 1 mm prototype at DESY, Iowa State University report, February 1985, unpublished.
[12] R. Bouclier et al., Nucl. Instr. and Meth. A267 (1988) 69.
[13] E. Gabathuler et al., Nucl. Instr. and Meth. 157 (1978) 47.

3. FILMLESS SPARK CHAMBERS:
1962–1967

3. FILMLESS SPARK CHAMBERS: 1962–1967

NUCLEAR INSTRUMENTS AND METHODS 15 (1962) 318–322; NORTH-HOLLAND PUBLISHING CO.

LOCATION OF THE POSITION OF A SPARK IN A SPARK CHAMBER

G. CHARPAK[†]

CERN, Geneva, Switzerland

Received 24 January 1962

With special electrodes having the structure of a distributed delay-line, it is shown, by measuring the delay between the arrival of the signals following a spark at two opposite ends of the electrodes, that the initial position of the spark can be determined. A spark chamber with electrodes giving a delay of 200 nanoseconds has been constructed.

Up to 1960 different types of spark chambers were considered or developed:

chambers with sparks perpendicular to the plane of the electrodes[1];

chambers with the spark following the track of the particle[2,3];

chambers with the track made visible by proportional amplification by a pulsed field[2,4,7], or by a microwave[1];

chambers with tracks parallel to transparent electrodes made visible by the discharge all along the track[3].

Of these types, the first one is being widely developed because it is particularly suitable for use with high-energy accelerators [Cork and Cronin, Ref.[1]].

The ultimate aim of a spark chamber is to locate, with the maximum accuracy, the position of a particle trajectory in space. This is usually done by scanning the two stereo-pictures of the discharge in the chamber. This is not a very big problem when one is looking at rare events where the spark chamber is triggered by a well-defined logic system controlled by scintillation counters. In some cases, however, as in the measurement of an angular distribution of a single particle, one may face the problem of scanning tens of thousands of pictures.

Several ways of obtaining the data immediately have already been envisaged. For instance, at CERN, Gelerntner[5] studied the possibility of using the vidicon television camera to replace the usual photographic film and so allow the automatic treatment of the data by computers.

Some rather simple ways of locating the position of the spark in a spark counter can be envisaged. They are based on the generation, by the spark, either of sound waves which propagate in the gas or in the electrodes, or of electromagnetic signals. The possibility of using the velocity of sound in the gas has already been demonstrated[6].

In this paper we will show that one can use the finite time of propagation of the electric signal along the electrodes to measure the position of the spark in a very short time. One of the electrodes used in the spark chamber has the structure of a distributed delay line (fig. 1).

The inner part of the delay line is a plane conductor sandwiched between two insulating plates. The outer part consists of either copper wire or thin aluminium adhesive tape wrapped in a spiral around these plates. The sparks are made between this spiral and a plane electrode to which is applied a pulsed high voltage. The inductance L and

† On leave from the Centre National de la Recherche Scientifique, Paris.

[1] An extensive bibliography about spark chambers can be found in the following proceedings: 1960 Int. Conf. on Instrumentation for High Energy Physics, Berkeley, Sept. 1960 and 1961 Argonne Symposium on Nuclear Chambers, RSI May 1961.

[2] G. Charpak, J. Phys. et le Radium 18 (1957) 539; and Proceedings of the International Conference on Mesons and Recently Discovered Particles, Venice, 28 Sept. 1957.

[3] S. Fukui and S. Myamoto, Nuovo Cimento 11 (1959) 113.

[4] F. P. Arecchi, G. Cavalleri, E. Gatti and G. Radaelli, Energia Nucleare 7 (1960) 865; and 8 (1961) 539.

[5] H. Gelernter, Nuovo Cimento 22 (1961) 631.

[6] H. W. Fulbright and D. Kohler, University of Rochester Report NYo/9540.

318

capacitance C per centimetre can be chosen so as to reach the wanted delay ($\sim \sqrt{LC}$ per cm of delay line) and the wanted impedance ($\sim \sqrt{L/C}$). We

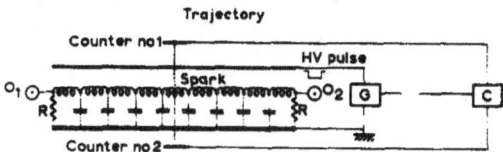

Fig. 1. Schematic diagram of the spark chamber with a delay-line electrode.

In this arrangement, the pulse generator G is triggered from the coincidence circuit C connected to the scintillation counters 1 and 2.

O_1 and O_2 are the two opposite outputs of the electrode.

have built three such lines. All were 40 cm long and 20 cm wide, and had the characteristics shown in table 1.

TABLE 1

Electrode No.	Thickness (cm)	Total delay (μs)	Impedance (Ω)	Internal resistance (Ω)
1	0.6	0.2	40	5
2	0.6	0.5	100	5
3	2.5	1.5	250	40

No. 1: In electrode No. 1 the winding consisted of 2 turns/cm of aluminium tape 3 mm wide and 0.1 mm thick, on two 3 mm thick glass plates, which sandwiched the inner electrode.

No. 2: Electrode No. 2 was built by winding 4 t/cm of copper wire on the same structure.

No. 3: Electrode No. 3 was built by winding 20 t/cm of copper wire on a 2.5 cm thick aluminium plate insulated from the winding by a $\frac{1}{10}$ mm layer of mylar.

Tests were made on these electrodes with artificial sparks produced by bringing a piece of metal connected to the pulsed high-voltage electrode close to the winding. The pictures a, b, c show that these electrodes were working as delay lines. The HV pulser was triggered by a 50 c/s pulse generator which also started the time base. These tests also showed that two sparks can be simultaneously produced and well separated. This was done by bringing two pieces of metal, which were connected to the HV electrode, close to the external windings at two different positions. Spark chambers where

then built with electrodes Nos. 2 and 3. The negative HV pulse was triggered by a coincidence produced by cosmic rays in two scintillation counters placed above and below the chamber. They worked satisfactorily as spark chambers, when filled with impure argon, with an interelectrode gap of 0.6 cm, and with pulses of about 5 kilovolts and 150 μs or 300 μs duration. At first, the time base of the oscilloscope (Tektronix 517) was triggered from the pulse coming from the coincidence circuit or from the leading edge of the HV pulse. Both of these methods gave poor definitions of the zero time because of the large time jitter between the application of the pulse and the occurrence of the spark. This can be cured by taking as the zero time a fast pulse which results from the occurrence of the spark. For instance, one could take the pulse from a photomultiplier viewing the chamber, or, as we did, by using the positive rise in the HV negative pulse, because this rise is connected with the occurrence of the breakdown. This removed the jitter and pulses were observed to be delayed by an amount proportional to their distance from the output of the line. On picture d one can see the result of the experiments done with delay line no. 1. Moving the telescope position by 15 cm changes the time of arrival of the pulses by about 75 mμs as expected.

The following factors limit the accuracy of such a method.

The number of n of turns per cm introduces an intrinsic uncertainty of n^{-1} cm. There is, however, some freedom in choosing this parameter. The time accuracy of time converters is now such that a nanosecond precision is easy to attain, giving for line no. 1 a precision of 2 mm. This is certainly not a serious limitation. The most serious one could come from the change in the rise-time of the pulse as it goes along the line. This can be seen on pictures a, with the artificial sparks along the delaying electrode no. 3. It will introduce a non-linearity in the relation between time and position. There exists an alternative method to use the information from the delay electrodes that has several considerable advantages. The time interval between the pulses arriving at both ends of the delay line is measured, a suitable delay being added at one end to a avoid

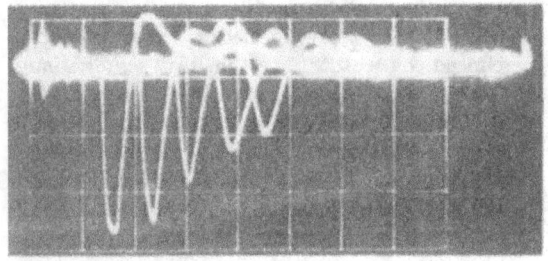

(a) 1 division = 200 mµs. Electrode No. 3. Five successive artificial sparks, 5 cm apart.

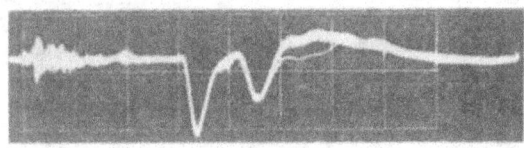

(b) 1 division = 200 mµs. Electrode No. 3. Two simultaneous sparks, distant by about 7 cm.

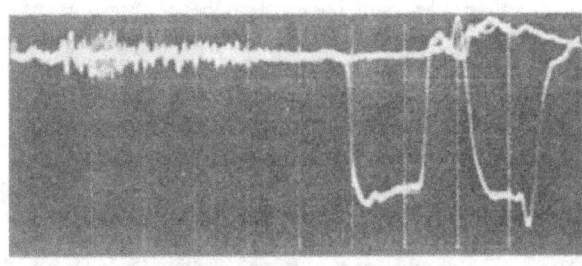

(c) 1 division = 100 mµs. Electrode No. 1. Two successive artificial sparks at the two ends of the line, distant by 40 cm.

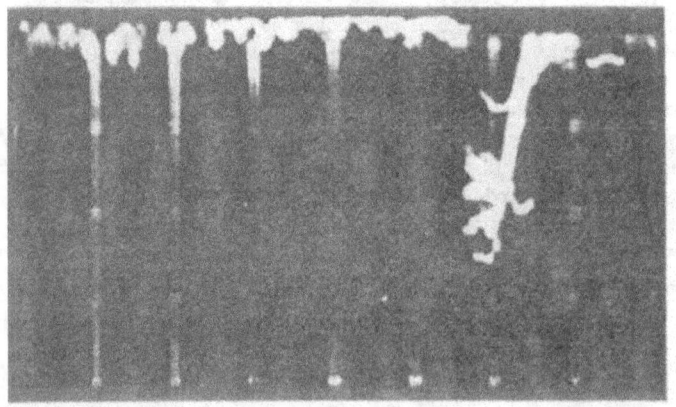

(d) 1 division = 100 mµs. Electrode No. 1. Sparks triggered by cosmic-ray telescope at a given position. Spurious sparks give displaced pulses. Telescope width 5 cm.

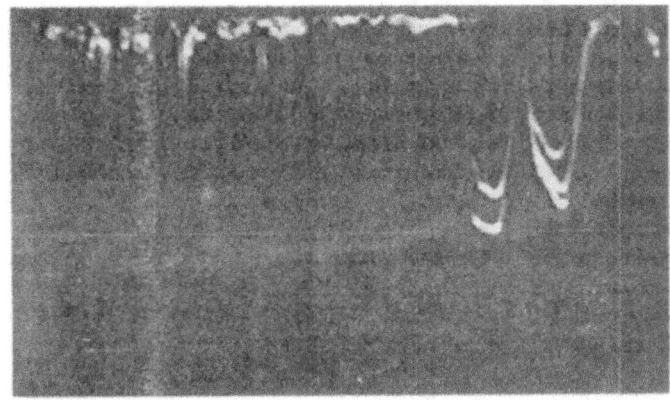

(e) 1 division = 100 mµs. Electrode No. 1. Sparks triggered by cosmic-ray telescope at two successive positions, distant by 15 cm.

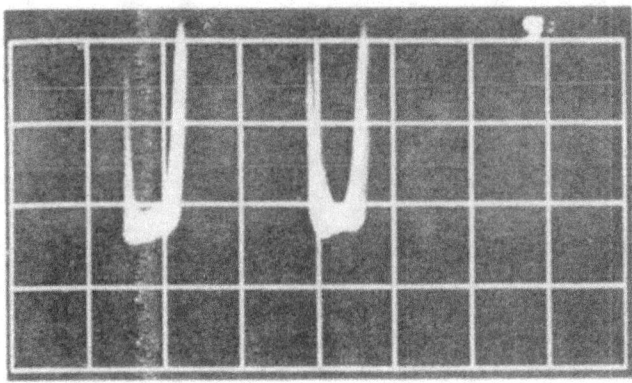

(f) 1 division = 100 mµs. Electrode No. 1. Sparks produced by an α source at two successive positions distant by about 20 cm. Notice the time dilatation by a factor two with respect to the preceding cases, due to the triggering system. The picture represents 10 sparks at each position.

The time goes from right to left. From (a) to (e) the time base is started by the positive rise of the negative HV pulse. At (f) the time base is started by the pulse arriving at one end of the electrode, while the pulse from the opposite end is delayed by 0.5 µs. The gas filling is impure argon. The applied voltage pulses are about 5 kV.

negative times. The time differences corresponding to two different positions is thus multiplied by a factor of two. The only time-jitter left is that of the time measuring instrument and can be reduced below a nanosecond without difficulty. It is probably also the good method when one makes use of the velocity of sound in the gas or the velocity of ultrasound in the electrode, to determine the location of the spark.

Pictures f illustrate this method. We used a strong α-particle source at different positions in the chamber and triggered the HV by a 50 c/s pulse generator. Each picture is a ten-pulse exposure.

In conclusion, we see that by this method one should be able to locate the position of a track with the usually wanted accuracy of about 1 mm, within a time shorter than a microsecond. This fact has an essential importance if one wants to use the information provided by this type of chamber to trigger other spark chambers, because the resolving time of spark chambers can be easily varied by means of of the clearing field or by means of oxygen-controlled admixture to bring it to any value below 10 µs with still almost 100% efficiency[1,2]). The signals provided by the chamber are several kilo-volts high with very good rise-time. They can be used for analogic or digital handling of the information such as distances or angles. They can also indicate the number of sparks occurring, if this is the information of interest in the experiment. The occurrence of several sparks should not limit the use of the signals coming at the two ends of the delay line. One can remove the confusion arising from the occurrence of several pulses at each input by requiring, by electronic logic circuits, that the sum of the times of propagation of any pair of pulses from opposite outputs be a constant equal to the total delay of the line. One can thus hope to obtain signatures of some specific rare events like the occurrence of the disintegration of a Λ particle.

This work was performed with the technical help of Mr. Bouclier. Dr. Massam was very helpful in the preparation of the manuscript.

I wish to acknowledge interesting discussions with Drs. Farley, Muller, Bernard-Levy, Professor Garwin and the encouragement of Professors Bernardini and Preiswerk.

NUCLEAR INSTRUMENTS AND METHODS 24 (1963) 501-502; NORTH-HOLLAND PUBLISHING CO.

LETTERS TO THE EDITOR

A NEW METHOD FOR DETERMINING THE POSITION OF A SPARK IN A SPARK CHAMBER BY MEASUREMENT OF CURRENTS

G. CHARPAK, J. FAVIER* and L. MASSONNET

CERN, Geneva

Received 19 September 1963

The interest attached to the automatic handling of a great number of events obtained from spark chambers has lead to several proposals and attempts to locate the spark in a spark chamber without taking a picture[1-5].

The measurement of the propagation time of the sound has proved to be able to reach an accuracy limited only by the size of the spark itself, and is already being successfully used in experiments.

We propose a new method that can replace some of the others or be complementary to them.

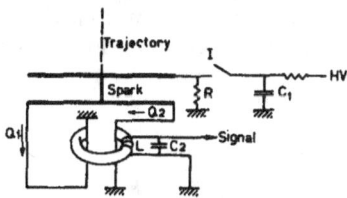

Fig. 1. Principle of the method. The charge Q of the storage capacitor is divided between two channels, in a proportion depending on the relative impedance of the two paths, which is a function of the spark position. The difference between the two charges, Q_1 and Q_2, is measured, for instance, with a ferrite transformer.

The principle is illustrated by fig. 1. Fundamentally, in a spark chamber, the charge from a condenser C_1 charged at voltage V discharges to ground through the conductive channel produced by the spark. If the grounded plate is connected to ground at two opposite points the current will be divided between the two different channels, according to their relative impedance, which is a function of the position of the spark.

Several methods can be used to measure the

* Visitor from Laboratoire Joliot-Curie, Faculté des Sciences, Orsay.

difference between the charges which flow in the two paths. A very simple method is to use a ferrite transformer with the earth currents passing through separate primary windings connected in opposition. An induced current is picked up from a secondary coil of inductance wound around the core. The signal is small and some serious pick-up from the neighbouring spark is to be expected. This is overcome by connecting a capacity C_2 in parallel with the coil. If the ringing period $2\pi\sqrt{(LC_2)}$ is very large compared to the duration of the signal, namely the duration of the spark, a damped oscillation is induced of initial amplitude proportional to $\int B \, dt$. As B is proportional to I, if the core is not saturated, the amplitude of the signal is exactly proportional to the net charge which flowed to ground through the central hole of the core.

This principle has been tried successfully both with artificial sparks produced between a spike and a plane electrode and with a spark chamber triggered by muons from the CERN Synchrocyclotron.

The coil L is made of 100 turns of thin wire wound around a core (Philips 300502). With a capacity $C_2 = 0.047 \, \mu F$, the signal is about 2V. After a simple filter the signal could be sent directly to a pulse height analyser. Fig. 2 shows the response obtained when the spark position varied. The dimensions of the electrodes are $50 \times 20 \, cm^2$, the ground electrode being made of stainless steel

[1] H. Gelernter, Nuovo Cimento 22 (1961) 631.
[2] G. Charpak, Nucl. Instr. and Meth. 15 (1962) 318.
[3] B. C. Maglić and F. A. Kirsten, Nucl. Instr. and Meth. 17 (1962) 49.
[4] H. Fulbright and D. Kohler, University of Rochester, Report NYO-9540.
[5] K. Krienen, Nucl. Instr. and Meth. 16 (1962) 262.

mesh*. The ground leads are fixed at the opposite ends of the larger dimension.

A slight non-linearity is observed when the distance of the spark from one end is of the order of the width of the electrode. Also a small dependence

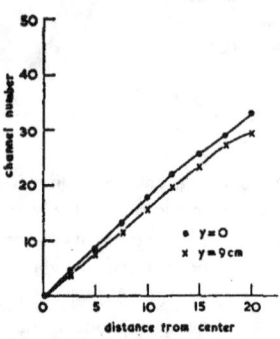

Fig. 2. Variation of the signal magnitude with position. The sparks are produced by triggering the chamber on a beam of muons of 5 mm width traversing the chamber. The chamber is moved along its longitudinal dimension at two different lateral positions. Energy of the spark 0.02 joule.

on lateral positions is observed. Three or four leads can thus be necessary to have in an analogous way the two-dimensional position of a spark. For a steady current the problem reduces to that of finding a solution of the Laplace two-dimensional equation $\nabla V = 0$, which shall be such that either V has the value zero at the position where the charges are brought to ground or else $\partial V/\partial n = 0$ at every point of the boundary. In the case of high-frequency signals as such induced by the spark, the situation is more complicated since the impedance can be essentially of inductive origin and the transmission-line properties of the chamber have to be taken into account.

A good way to avoid non-linearity effects is to extend the ground electrodes at distances which are long with respect to the width. The nature of the electrode plays an important role. It is vital to the method that the impedance of the ground leads be negligible with respect to that of the electrode itself. For this reason we found it very practical to use a mesh electrode made of thin wires.

* This mesh is obtained from: Gantois, 40 Bd. Richard Lenoir, Paris and Michel, 9 Fg. de Colmar à Mulhouse, Ht. Rhin. It is made of wires of 0.05 mm, at distances of 0.5 mm.

With thin aluminium electrodes (0.0025 cm) we found it necessary to have very short leads, made of copper braid, and going to a copper ground through the central holes of two separated ferrite cores. The secondary windings around the two cores are adjusted so as to have signals of equal values when the same current is flowing through their central hole. Then they are put in opposition. With electrodes of great thickness or of high electric conductivity the problem of making ground connections of an impedance low compared to the intrinsic impedances of the electrode becomes difficult. However, the skin effect due to the very short duration of the current flow limits the influence of thickness. Wire chambers, or chambers made with printed circuits, are very favourable for the use of the method. In the case of the use of ferrite memory core to store the information relative to the energized wire, the method offers the supplementary information as to the position of the spark along the wire. The information can be obtained within times shorter than the memory time of normal spark chambers by choosing properly the values of L and C_2. It is thus possible to use the information from this type of chamber to trigger eventually other spark cham-

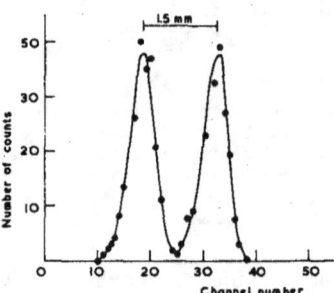

Fig. 3. Sensitivity of the method. Pulse height distribution from two artificial sparks at two points, 1.5 mm apart. Energy of the sparks 0.02 joule.

bers from which another type of information is wanted.

The accuracy of the method is very high. Fig. 3 shows the variation in pulse height for two sparks separated by 1.5 mm. It is fundamental that the total current flow be of constant magnitude. However, by using two separate cores and two additional windings with addition of the amplitudes, one obtains a signal proportional to $Q_1 + Q_2$, one eliminates this source of error. Simple analogic dividing circuits now exist to do this operation.

If several sparks are present the use of more than three channels allows the elimination of this type of event by requiring the correct relation between the currents.

This work was accomplished with the technical help of Messrs. R. Bouclier and G. Million. We are grateful to the Laboratoire Joliot-Curie, Orsay, for partial financial support.

Volume 16, number 1   PHYSICS LETTERS   1 May 1965

STUDY OF TWO-HOLE STATES IN LIGHT NUCLEI
BY MEANS OF (π^+, 2p) REACTIONS

G. CHARPAK, G. GREGOIRE *, L. MASSONNET, J. SAUDINOS **

CERN, Geneva

and

J. FAVIER, M. GUSAKOW, M. JEAN

Institut de Physique nucléaire, Orsay

Received 26 March 1965

It has been pointed out [1] that pion absorption in light nuclei offers an interesting possibility of studying the properties of nuclear two-hole states which are a logical consequence of the validity of the shell model description for light nuclei, strikingly demonstrated by the results of (p, 2p)

* On leave from University of Louvain.
** On leave from CEN, Saclay.

reactions [2]. In particular, positive pion absorption in flight provides us with a mechanism for ejecting two protons out of a nucleus leaving a neutron-proton hole behind.

Provided the initial pion energy is well defined, measurements of the summed energy spectra of the two ejected protons are expected to show in the high-energy end a structure corresponding to different binding energies of the

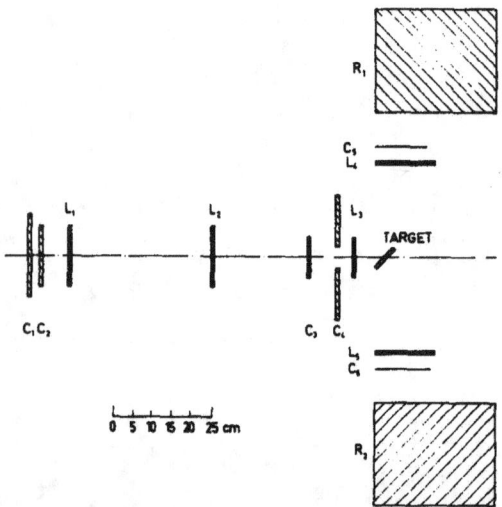

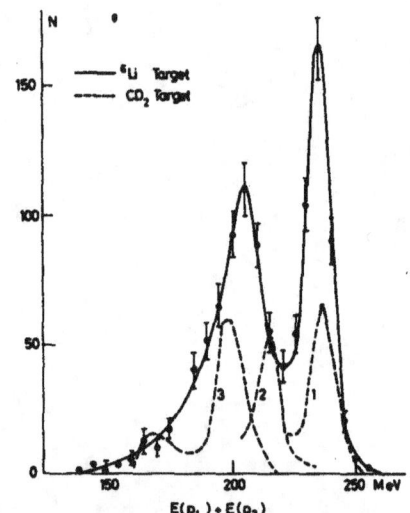

Fig. 1. Experimental set-up. $L_1 L_2 L_3 L_4 L_5$ are spark chambers with automatic read-out of the coordinates. $R_1 R_2$ are 25-gap range chambers with automatic read-out of the spark pattern. $C_1 C_2 C_3 C_4$: scintillation counters defining the pion beam. $C_5 C_6$: scintillation counters detecting the protons.

Fig. 2. Sum of the proton energies from a thick Li^6 target (cylinder of 3×3 cm^2). Dashed curves: $1 - CD_2$ target under the same conditions, at the same pion energy: 95 MeV. 2 and 3 - CD_2 target with reduction of the pion energy by 20 MeV and 40 MeV, respectively. Total energy interval 125 to 275 MeV.

initial neutron-proton pair. Thus, the two-hole excitations, if they show up, must be given by the energy separations of the peaks of this structure.

Our initial study has been directed towards very light elements in which two-hole states are likely to be sufficiently long-lived to be seen, and of which one has a good knowledge of one-hole states. We have thus investigated the $(\pi^+, 2p)$ reaction on Li^6, Li^7, C^{12} and N^{14}. Among them Li^6 is of special interest, since the reaction $\pi^+ + Li^6 \rightarrow He^4 + 2p$ can, in principle, lead to excited states of the α particle with isospin $T = 2$, as well as $T = 0$ and 1. This is of importance in view of recent controversial discussions about the existence of H^4, $n_0{}^4$ and Li^4 which are possible isobaric analogues of the eventual excited states of He^4 [3, 4].

Experiment. A beam of 106 MeV pions from the CERN Synchro-cyclotron is bombarding targets of Li^6, Li^7, CD_2, CH_2, and liquid nitrogen. The beam is defined by the counters C_1, C_2, C_3 and a counter C_4 with a hole of diameter 4 cm, in anticoincidence (fig. 1). The protons are detected by two counters, C_5 and C_6 of 15×15 cm, defining the accepted solid angle of about 0.3 sr. A $(\pi + 2p)$ event is defined by $C_1 C_2 C_3 \bar{C}_4 C_5 C_6$. The addition of an energy selection on the proton counters, eliminating the pulses from minimum ionizing particles, changes the counting rate by less than 3%. The position of the three charged particles is measured with spark chambers L_1, L_2, L_3, L_4 and L_5 (fig. 1), giving the coordinates by means of the measurement of the currents flowing to two opposite channels in the ground and high-voltage electrodes [5]. A detailed paper on the techniques used in this experiment will appear elsewhere.

Accuracies of the order of 1 mm are obtained over the whole surface of the chamber. The range of the two protons is measured in the chambers R_1 and R_2, each consisting of 25 elements separated by 4 mm of lucite. The current flowing in each chamber is used to trigger a memory element. The pattern produced in a range chamber is reproduced in a digitized way, together with the information about the coordinates of the particle, on a magnetic tape. Further processing of the data by the computer is done with different requirements which reject tracks with more than one missing gap or more than one spurious spark after the end of the track.

The most serious troubles came from the signals induced in L_4 and L_5 by the range chambers. Although the use of a new technique has

introduced some particular difficulties at this stage of the experiment, interesting results appear.

Because of the chosen kinematics, the recoil energy of the residual nucleus is negligible, and the distribution of the sum of the energies of the two protons reflects the distribution in the binding energy of the residual nucleus.

Results. Fig. 2 illustrates the results obtained with several targets. The resolution of 10 MeV for a total available energy of about 235 MeV (the pions are degraded to 94 MeV when reacting in the target) comes from the beam (6 MeV), the chambers (3 MeV per gap) and the target thickness (cylinder of 3 cm diameter and 3 cm length for the Li, equivalent longitudinal weight for the others).

The resolution is still too poor to resolve the structure in most of our targets but in the reaction $Li^6(\pi^+, 2p)He^4$ the structure expected from the two-hole interpretation of the final nuclear states is clearly displayed.

The energy calibration is checked with the CD_2 target at three π^+ energies, different by steps of about 20 MeV (fig. 2). The peak at 30 MeV has a width greater than the intrinsic width at this energy, and can be composite. We have repeated the experiment with a thin target of Li^6(3 mm), giving a slightly better energy resolution. Also by reducing the absorbers in front of the range chambers the total energy acceptance shifted from 125-275 MeV to 90-250 MeV. This reduces the contribution from the reaction to the ground state and allows the eventual appearance of peaks with the corresponding $E_x < 30$ MeV. A

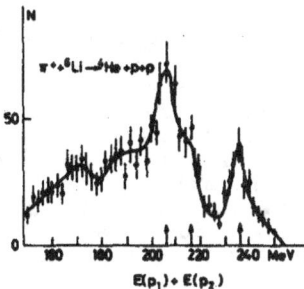

Fig. 3. Sum of the proton energies from a thin Li^6 target (3 mm thick). Pion energy at the target: 100 MeV. Total energy window: 90 to 250 MeV. The highest energy peak is cut down since only protons of rather equal energies can be accepted in this energy region.

55

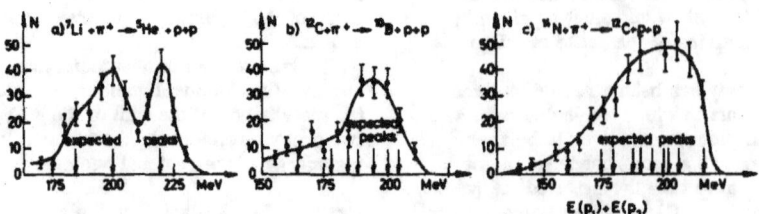

Fig. 4. Sum of the proton energies from targets of Li^7, C^{12} and N^{14}.

bump at E_x = 20 MeV also appears (fig. 3).

These two peaks compare well with the two-hole excitations expected (20 MeV for the $p^{-1}s^{-1}$ and 25-35 MeV for the s^{-2}) from the one hole excitation. Thus, it seems likely that the excited states of He^4 which we obtain can be interpreted in terms of two-hole states from Li^6, although it is probably not correct to consider them as entirely pure because of the centre-of-mass motion complications. The $p^{-1}s^{-1}$ state is quite close to the 19.24 MeV and 21.24 MeV levels measured by Donovan et al. [6], and can tentatively be identified with them, in which case they would have odd parity.

The two-hole excitations are also displayed, although not clearly, in the energy spectrum from Li^7 (fig. 4). The peak at 220 MeV is precisely centred at the energy expected, both from the Li^7 and He^5 mass difference, and from the p^{-1} hole excitation energy in Li^7 and Li^6. Thus, it can be associated to the He^5 ground state. In the same way the peak at 200 MeV and, maybe, the bump at 185 MeV, which lie in regions expected from the one-hole known excitation energies, can be interpreted as corresponding to the $p^{-1}s^{-1}$ and s^{-2} states, respectively. The excitation energy, 20 MeV, of the $p^{-1}s^{-1}$ state fits well with the 20 MeV known state of He^5 [2], which must be an odd-parity state if this assignment is correct.

The situation is not so clear for the C^{12} and N^{14} energy spectra which show only a very wide bump in each case. But the bumps lie precisely in regions where many two-hole states are expected due to the known multiplicity of p^{-1} hole states (peaks expected from 205 MeV to 165 MeV for C^{12} spectra and from 220 MeV to 160 MeV for N^{14} spectra). The absence of structure in the observed bumps is probably not entirely due to our relatively poor energy resolution, but also comes from the important natural widths of the hole states.

The cross section at 80^0 is 0.26 mb/sr for deuterium and about 0.05 mb/sr^2/MeV for Li^6. Counting rates of about 3 events/m were obtained with beam intensities of only 6000 π^+/sec and the large solid-angle obtained with the range chambers was an essential advantage.

We wish to acknowledge the technical assistance of Mr. Bouclier and Mr. Million who have had to solve, continuously, difficult problems, bound to the use of a new technique. Thanks are also due to Mr. Lindsay and Mr. Scott who designed a simple 20-channel analogue-to-digital converter, allowing the efficient use of the current division method for the localization in spark chambers. We are indebted to Dr. Ericson for many discussions, and thanks are due to Professor Preiswerk for his continuous encouragement. Mr. Cozzika and Mr. Krivine, from the Faculté des Sciences d'Orsay, collaborated in the runs at different stages of the experiment. We are indebted to the Laboratoire Joliot-Curie of the Faculté des Sciences d'Orsay for financial support.

References

1. M. Jean, Proc. Conf. on Direct interaction. Padua 1962 (Gordon and Breach, Science Publishers, Inc. 1963) p. 46.
   T. Ericson, ibid, p. 39.
2. M. Riou, Information on nuclear structure from (p, 2p) and other knock-on reactions, Conf. on Correlations of particles emitted in nuclear reactions, October 1964, Gatlinburg, U.S.A.
3. R. C. Cohen, A. D. Kanaris, S. Margulies and J. L. Rosen, Physics Letters 14 (1965) 242.
4. M. J. Beniston, B. Krishnamurtly, R. Levi Setti and M. Raymund, Phys. Rev. Letters 13 (1964) 553.
5. G. Charpak, J. Favier and L. Massonet, Nucl. Instr. and Meths. 24 (1963) 501.
6. P. F. Donovan, J. V. Kane, J. F. Mollenauer and P. D. Parker, Phys. Rev. Letters 14 (1965) 14.

* * * * *

56

4. WIRE CHAMBERS AND DRIFT CHAMBERS: 1968–1994

4. WIRE CHAMBERS AND DRIFT CHAMBERS: 1968–1994

NUCLEAR INSTRUMENTS A D METHODS 62 (1968) 262–268; © NORTH-HOLLAND

THE USE OF MULTIWIRE PROPORTIONAL COUNTERS
TO SELECT AND LOCALIZE CHARGED PARTICLES

G. CHARPAK, R. BOUCLIER, T. BRESSANI, J. FAVIER and Č. ZUPANČIČ

CERN, Geneva, Switzerland

Received 27 February 1968

Properties of chambers made of planes of independent wires placed between two plane electrodes have been investigated. A direct voltage is applied to the wires. It has been checked that each wire works as an independent proportional counter down to separations of 0.1 cm between wires.

Counting rates of 10^8/wire are easily reached; time resolutions of the order of 100 nsec have been obtained in some gases; it is possible to measure the position of the tracks between the wires using the time delay of the pulses; energy resolution comparable to the one obtained with the best cylindrical chambers is observed; the chambers operate in strong magnetic fields.

1. Introduction

Proportional counters with electrodes consisting of many parallel wires connected in parallel have been used for some years, for special applications. We have investigated the properties of chambers made up of a plane of independent wires placed between two plane electrodes. Our observations show that such chambers offer properties that can make them more advantageous than wire chambers or scintillation hodoscopes for many applications.

2. Construction

Wires of stainless steel, 4×10^{-3} cm in diameter, are stretched between two planes of stainless-steel mesh, made from wires of 5×10^{-3} cm diameter, 5×10^{-2} cm apart. The distance between the mesh and the wires is 0.75 cm. We studied the properties of chambers with wire separation $a = 0.1, 0.2, 0.3$ and 1.0 cm. A strip of metal placed at 0.1 cm from the wires, at the same potential (fig. 1), plays the same role as the guard rings in cylindrical proportional chambers. It protects the wires against breakdown along the dielectrics. It is

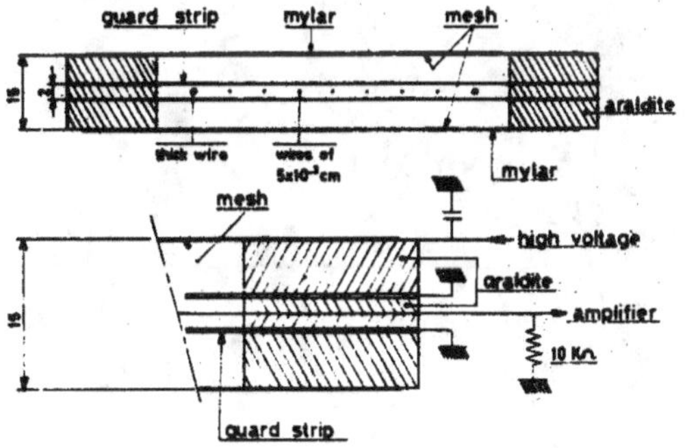

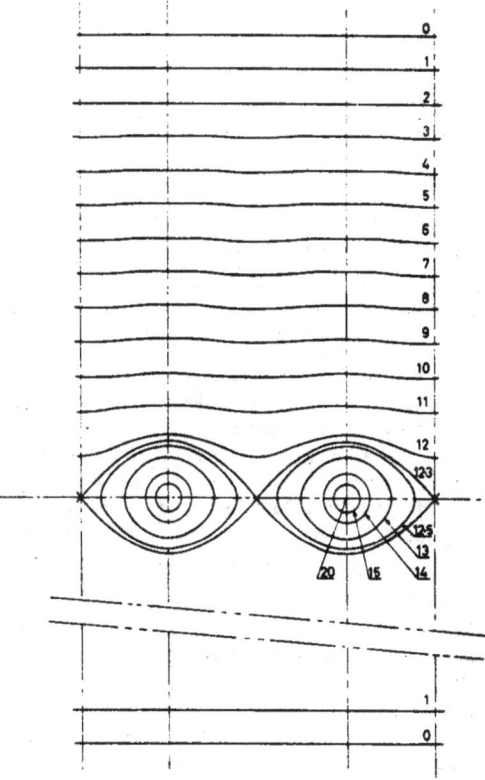

Fig. 2. Equipotentials in a chamber.
Wires of 4×10^{-3} cm diameter, 0.3 cm separation, and 1.5 cm total thickness. 20 V applied between the wires and the external mesh. Results from an analogic method.

Fig. 1. Some details of the construction of the multiwire chambers.
A copper shield protects the wires at their output from the chamber and contains the solid state amplifiers.

262

important to have the last wire on each side much thicker than the other ones in order to avoid a too high gradient on these wires. Each wire is connected to an amplifier with an input impedance of about 10 kΩ.

The chamber is flushed at atmospheric pressure by a flow of ordinary argon bubbling through an organic liquid at 0° C: ethyl alcohol, or n-pentane or heptane. A negative constant voltage is applied to the external electrodes.

3. Properties of the chamber

3.1. AMPLIFICATION

At the distances from the wires at which amplification occurs, the equipotentials are concentric to the wires. We thus expect the amplification to behave exactly as in a cylindrical proportional counter. Fig. 2 shows the distribution of these equipotentials for a chamber with a separation $a = 0.3$ cm between the

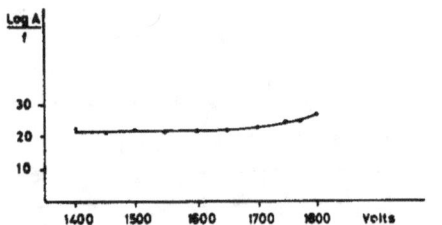

Fig. 3. Relation between the amplification factor A in the gas and the applied voltage V.
In cylindrical chambers log A is proportional to f, eq. (1). Argon-alcohol filling; wire diameter 4×10^{-3} cm; wire spacing 0.3 cm. $A \approx 50$ at 1400 V; $A \approx 5000$ at 1800 V.

wires. We have studied the amplification factor of such a chamber as a function of applied voltage. The pulses from the wires are amplified by a factor of 200 by a simple three-transistor amplifier of 10 kΩ input impedance. In cylindrical chambers the amplification factor A is given by a formula obtained under simple assumptions[1], which shows that log A is proportional to

$$f = V^{\frac{1}{2}}[(V^{\frac{1}{2}}/V_s^{\frac{1}{2}}) - 1], \qquad (1)$$

where V is the voltage applied between the wires and the external electrode, and V_s is the threshold at which amplification begins. We used as calibration source ^{55}Fe, which emits a 5.9 keV X-ray line. The variation of pulse height in argon-alcohol over a large range of voltages (fig. 3) shows that the behaviour of the wire is, indeed, that expected for a cylindrical counter. The threshold for proportional amplification is at $V_s = 1100$ V. At 1700 V the counter may enter into the semiproportional region, but as long as we are interested in counting, rather than in linearity, this is still

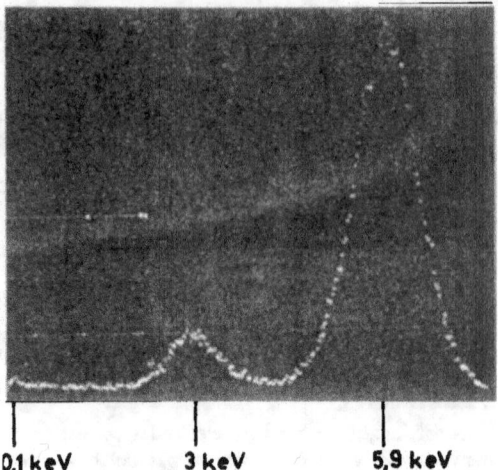

0,1 keV   3 keV   5,9 keV

Fig. 4. Pulse-height spectrum from one wire.
5.9 keV line from ^{55}Fe and escape line of 3 keV. Argon-pentane filling; wire diameter 5×10^{-3} cm; wire spacing 0.3 cm; 3750 V. Pulse-height on the wire: 5 mV at 6 keV, on a 10 pF capacitance.

acceptable. At 1800 V we obtain a 5.5 mV pulse directly from the wire, on a 30 pF load. This corresponds to an amplification factor of about 5000. At 1800 V the critical radius at which the amplification starts is, to a very good approximation[1]):

$$r_0 = \rho V/V_s = 3.3 \times 10^{-3} \text{ cm},$$

ρ being the wire radius. We thus see that it is a region where the equipotentials are exactly concentric to the wire.

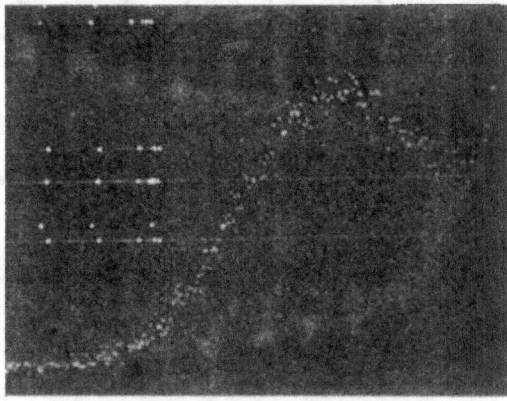

Fig. 5. Pulse-height spectrum from one wire.
β-rays from ^{90}Sr collimated on the wire. Argon-pentane filling; wire diameter 4×10^{-3} cm; wire spacing 0.3 cm; hv = 3000 V, peak at about 2 keV energy loss. Only the lowest energy part is displayed.

3.2. ENERGY RESOLUTION AND EFFICIENCY FOR DETECTING CHARGED PARTICLES

Many data have been accumulated about the properties of the different gases suitable for proportional counters. We have concentrated our efforts on selecting a gas giving the maximum amplification, so that the

minimum gain is required from the amplifiers. We found that ordinary argon, bubbling through n-pentane or heptane cooled at 0° C, gave very satisfactory results. With n-pentane we could reach pulses of 100 mV directly on the wire without entering into the Geiger region. Fig. 4 shows the energy distribution obtained

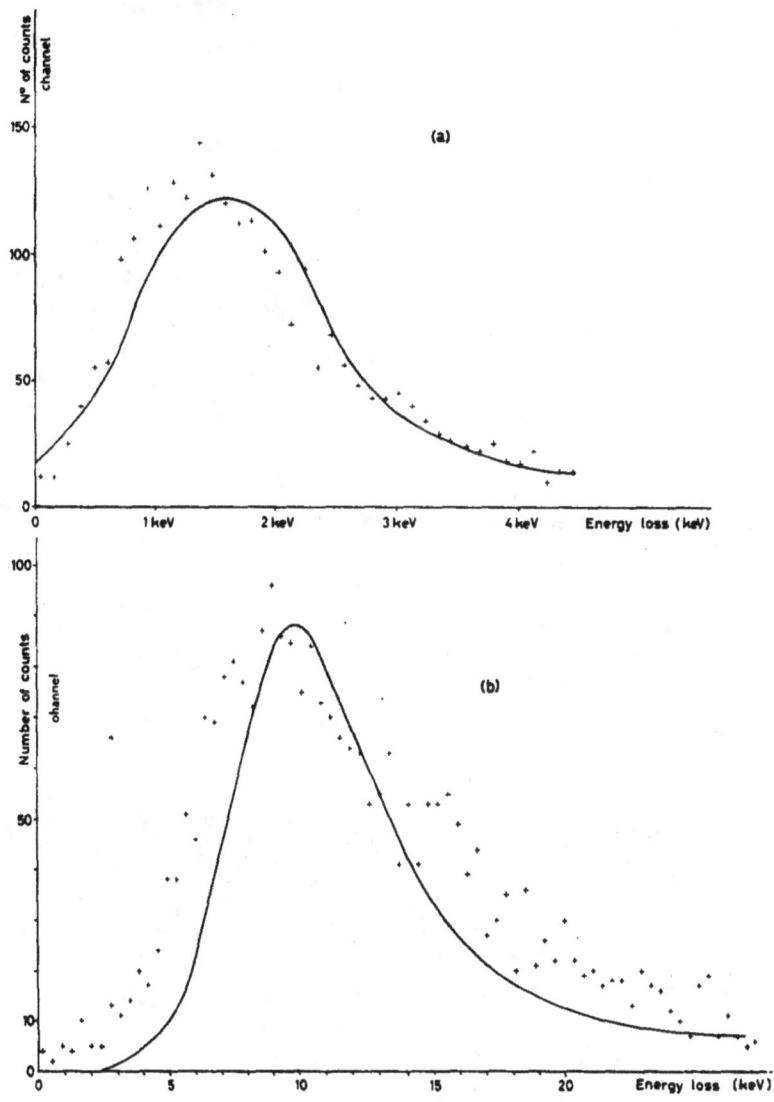

Fig. 6. Energy loss distributions by pions and protons of 370 MeV/*c*.
2.4 × 10⁻³ g/cm² of argon-alcohol. Experimental points and theoretical curve corrected for finite resolution: 30% at 6 keV varying as $E^{-\frac{1}{2}}$. a. Pions: Mean energy loss 3.6 keV; b. Protons: Mean energy loss 15.3 keV.

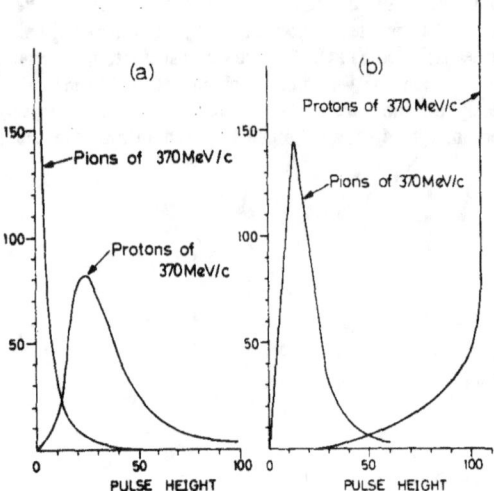

Fig. 7. Energy loss distributions by pions and protons of 370 MeV/c.
Argon-alcohol filling; hv: 1509 V (a), 1629 V (b).

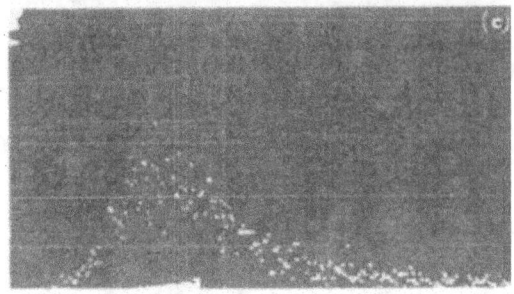

from 5.9 keV X-rays, with a simple three-transistor amplifier. With a refined amplifier the resolution is 15%.

Fig. 5 shows the energy loss distribution from β-rays traversing the counter. We see that the threshold of detection is well below the minimum energy loss. Energy resolution is important in that respect, because a poor energy resolution gives rise to an increase of low-energy pulses.

We have studied the energy lost in the counter by pions and protons of 370 MeV/c from the CERN synchro-cyclotron, in argon-alcohol mixtures, where the energy resolution was poorer (25 to 30% at 6 keV). Fig. 6 shows the observed spectrum, together with the theoretical energy loss distribution, corrected for the

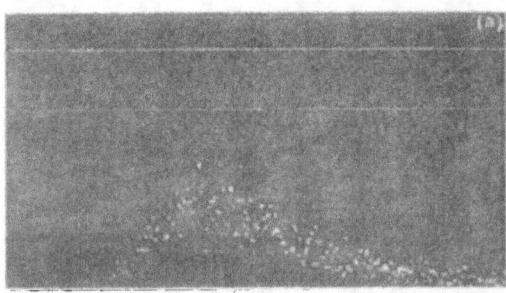

Fig. 8. Pulse height as a function of distance from the wire. Beam of protons of 370 MeV/c, 0.06 cm wide (fwhm). Argon-alcohol filling; hv = 1509 V; wire diameter 4×10^{-3}; wire spacing 0.2 cm. Beam on the wire (a). Beam respectively 0.05 cm (b), 0.075 cm (c), 0.1 cm (d) and 0.15 cm (e) from the wire.

finite resolution. We see how essential it is to have a threshold of detection in the region of 100 eV if one wishes to be close to 100% efficiency for minimum ionizing particles.

Fig. 7 shows the separation of the energy loss distributions for the pions and protons of 370 MeV/*c* at two different voltages.

3.3. LOCALIZATION

If we work in the region of proportional amplification, where the propagation of the discharge by photons plays no role, we should expect each wire to be independent of the others. We have verified this property with a beam of particles, pions and protons of 370 MeV/*c*, collimated to 0.06 cm (fwhm) by means of small counters in coincidence. Fig. 8 shows the spectrum obtained from such a beam of protons passing in the chamber near a wire, and the variation of the spectrum as we vary the distance from the wire. Fig. 9 shows the variation of the efficiency as a function of the distance from the wire, for wire separation of 0.3 cm and 0.2 cm. We see that the sharpness in the variation of the efficiency is exactly equal to the definition of the beam width: 0.06 cm. The counting rate at large distances from the wire is due to accidental coincidences between the beam telescope and the chamber (gate width 5 μsec). We may then conclude that the accuracy of localization can be brought down, at least to this limit. We have repeated these measurements with a wire separation of 0.1 cm. An efficiency of only 95% was reached at the position of the wire with the same

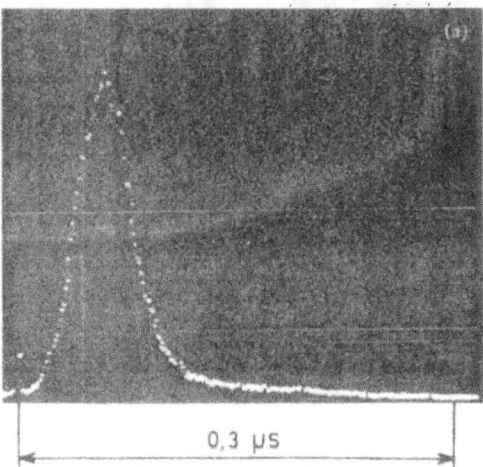

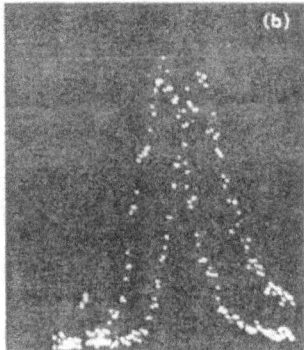

Fig. 10. Delay in the pulse as a function of the distance wire-particle.
Argon-n-pentane filling; Wire of 0.004 cm; Distance between wires: 0.3 cm; hv = 2240 V.
a. Collimated beam of β-rays ($\sim$ 1 mm width) centred on the wire; fwhm = 40 nsec.
b. Effect of a displacement by 0.1 cm from the wire.

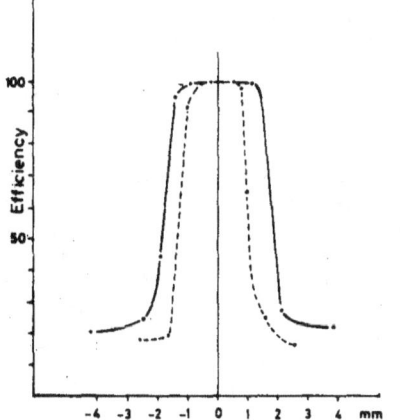

Fig. 9. Efficiency of detection as a function of position. Beam of protons of 370 MeV/*c*. Argon-alcohol filling; hv = 1509 V; wire diameter 4×10^{-3} cm; Wire spacing 0.3 cm (full line), 0.2 cm (dashed line).

fall-off sharpness and the 5% loss is probably due to the imperfect collimation of the beam. Similar results were obtained with pions.

3.4. COUNTING RATE AND TIME RESOLUTION

We have counted up to 250000 particles per second on a wire, the limit being set by our amplifiers. More than 10^7 β-rays were traversing the chamber, 20×5 cm^2, with no visible interference between the wires. The resolution time of the chamber is in principle determined by the jitter in the arrival time at the wire of electrons

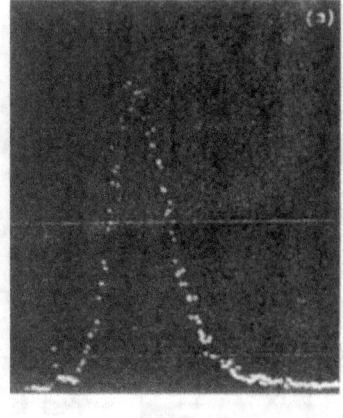
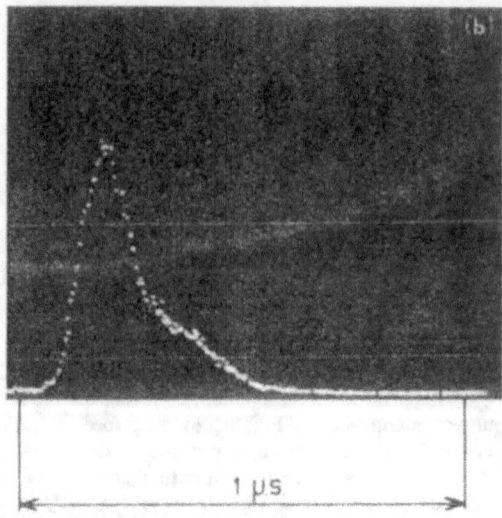

Fig. 11. Delay in the pulse as a function of the distance wire-particle.
Argon-alcohol filling; Wires of 0.004 cm; Distance between wires: 0.3 cm; hv = 1600 V.
a. Collimated beam of β-rays; b. Uncollimated beam; c. Effect of displacement by 0.1 cm from the wire of the collimated beam.

liberated in the gas. We may expect this time to vary as a function of the chamber geometry, of the electronics, of the gas mixture, and of the voltage. Our preliminary investigations show that with no precaution one obtains about 0.5 μsec, as in spark chambers. Using a source of β-rays collimated down to 0.1 cm width, we observe a very interesting property. We measure the time interval between the traversal of the chamber by the electron and the detection on the wire.

We see that at the wire the width (fwhm) of the time distribution can reach 40 nsec (fig. 10). But, as expected, this delay is shifted as we go away from the wire, since the distance to the wire is increased and since the gradients decrease further away from the wire. In the worst case of the argon-alcohol mixture (fig. 11), the maximum time delay is of the order of 0.4 μsec. Some additional jitter is introduced there by the fact that the

pulses are smaller than with n-pentane, and the time response of our electronics is not independent from the pulse height.

The variation in the delay with the distance from the wire may be exploited to give the position of the particles between the two wires. It seems to us that this may lead to spatial resolutions much better than any reached so far, and which may be of great interest when very high energy particles have to be localized. It is likely that a time resolution much smaller than the one for spark chambers will be attained after appropriate research on the gas mixtures and on the detecting electronics.

3.5. OPERATION IN A MAGNETIC FIELD

We have tested the chambers in a magnetic field at 7500 G, parallel and orthogonal to the wires, with a beam of collimated 5.9 keV X-rays. The displacement of the spatial distribution was found to be inferior to 0.2 mm. The energy resolution was not altered.

4. Conclusion

The properties of the multiwire proportional chambers can be summarized as follows:

- Each wire can amplify the initial energy loss of a particle in a thin layer of gas, of the order of 1 cm, to such an extent that minimum ionizing particles are detected with an efficiency close to 100%.
- With argon-n-pentane and argon-heptane mixtures, high amplification is possible, making easy the amplification by rudimentary solid-state amplifiers.
- With wires that are 0.1, 0.2, 0.3 and 1.0 cm apart, we have observed a good localization of the detection on each single wire.
- Any number of simultaneous particles can be detected.
- Resolution times below 0.4 μsec are readily obtained.
- Localization of the position between the wires is possible, making use of the arrival time of the pulse.
- Counting rates of the order of 2.5×10^5/sec per wire have been observed.
- Selection between particles with different ionization powers is possible.
- The chambers can be operated in strong magnetic fields.

These observations give us confidence that this type of instrument deserves a very detailed study since it can in some respects replace classical wire chambers or hodoscopes, or be a useful complementary tool, for instance as a fast decision-making chamber to trigger spark chambers. It is an ideal anticoincidence counter in front of gamma or neutron detectors, because of its very low efficiency. Since it does not require a trigger from a scintillation counter it has considerable advantages in the measurement of the spatial distribution of X-rays, γ-rays, or neutrons with the eventual association of proper radiation converters.

We wish to thank Messrs. G. Million and J. M. Fillot for their technical help. We benefitted greatly from the support given to us by Mr. G. Muratori and his group, who demonstrated that the large-scale production of this type of chamber is possible.

Messrs. G. Amato and J. P. Papis were of great help in the research into very low-cost amplifiers and were successful in this respect. They showed that less than two dollars of equipment per wire was sufficient to bring the pulses to a level close to 1 V, where their utilization by logic circuits is easy.

We are pleased to thank Prof. P. Preiswerk for his encouragement.

Reference

[1] S. C. Curran and J. D. Craggs, *Counting tubes* (Butterworths, London, 1949) p. 32.

NUCLEAR INSTRUMENTS AND METHODS 65 (1968) 217–220; © NORTH-HOLLAND

SOME READ-OUT SYSTEMS FOR PROPORTIONAL MULTIWIRE CHAMBERS

G. CHARPAK, R. BOUCLIER, T. BRESSANI, J. FAVIER and Č. ZUPANČIČ

CERN, Geneva, Switzerland

Received 29 July 1968

Several methods are discussed, which reduce substantially the number of amplifiers required for the readout of a proportional multiwire chamber.

1. The plane of wires is built as a lumped delay line. The difference in the arrival time of the pulse at the two ends of the line determines the position of the active wire.

2. The wires are connected to a resistive attenuator network. The ratio of the pulse heights arriving at the two ends localizes the active wire.

3. The wires can be connected in different subgroups by transformer coupling.

1. Introduction

In a preceding article[1]), we have described the properties of proportional wire chambers. These chambers have many advantages over conventional wire spark chambers, but their drawback for many applications is the necessity of having one amplifier for each wire. Whilst this is the best way of exploiting the properties of the chambers to the fullest extent, we have searched for an alternative that would allow a substantial reduction in the amount of electronic equipment necessary to exploit the chambers.

We describe here three methods which could in many cases lead to simple and cheap read-out systems.

2. The lumped delay-line method

The method is similar to one of the earliest principles put forward for the automatic read-out of spark chambers[2]): the electrode is built as a delay line. This is achieved, for instance, by separating the wires by self-inductances L (fig. 1). Then by measuring the delay in the arrival of the pulse at the two ends A and B,

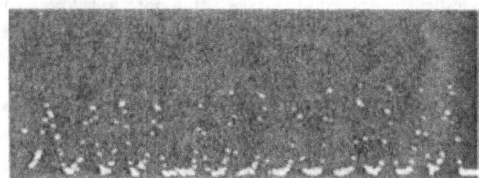

Fig. 2. Delay between the arrival of the pulse at the two ends of the wire plane. $L = 2\,\mu$H. A collimated 5.9 keV X-ray source of ^{55}Fe is displaced from wire to wire. The average distance between two adjacent peaks is about 10 nsec.

we can determine the wire carrying the pulse. With wires of 40 cm length and self-inductances of $L = 2\,\mu$H we observe a delay between each wire of 5 nsec. Fig. 2 shows the delay between the pulses at A and B, when a collimated source of 5.9 keV X-rays from ^{55}Fe is displaced from wire to wire.

The maximum number of wires which can be read in this way is strongly dependent on the quality of the line and of the electronics.

3. The pulse division method

Apart from being delayed, the pulses at the two ends of the delay-line are also attenuated in such a way that they vary with the position of the active wire. This can, in fact, be exploited to measure the origin of the pulse. It is clearly not necessary to have an inductive attenuator. By simply separating the wires by small resistances, say 30 Ω, we obtain sufficient attenuation at each end with very little change in the pulse shape.

For a 40 cm wire of 0.04 mm dia., the capacity to ground is about 10 pF and the impedance to ground, ρ, is roughly 10 kΩ. The line then behaves as a variable attenuator (fig. 3a).

This method is very similar to the current division method used in spark chambers[3]). In the latter case, however, the total amount of charge can be made close to a constant, and the difference in the pulse

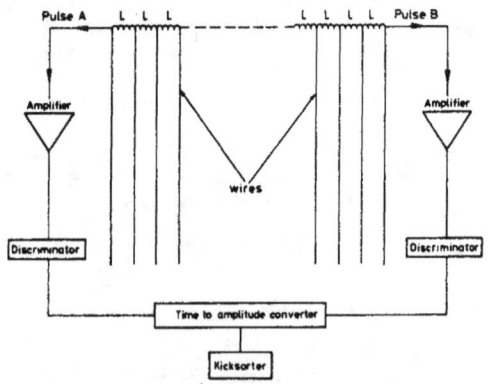

Fig. 1. The plane of wires is built as a lumped delay-line. Self-inductances L are connected to the wires.

217

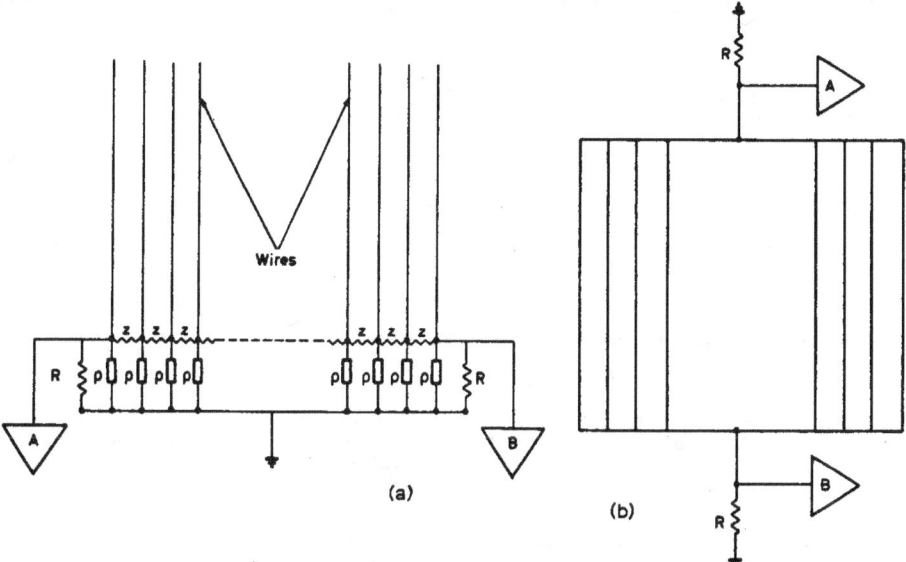

Fig. 3. The pulse division method. a. The wires are separated by resistances z. ρ is the impedance introduced by the stray capacity. The ratio of the pulses at A and B is a function of the position of the active wire; b. The wires are connected together. The ratio of the pulse-heights at A and B gives the position of the avalanche if the wire is resistive enough.

heights at the two ends of the electrode gives a sufficient accuracy. With proportional chambers, the spread in pulse heights can be very large. If the wires are separated by resistances of constant value z, the attenuation is constant for each cell. The pulse heights at each end are, to a good approximation:

$$A = V_0 \exp\left(-kz/z_0\right), \quad B = V_0 \exp\left\{-(l-k)z/z_0\right\},$$

where V_0 is the initial pulse height, k is the rank of the active wire, l is the total number of wires. The value of z_0 as a function of the characteristic parameters z, ρ, the capacitance of the wire, etc., is given in the textbooks on communication circuits.

If we take the ratio

$$A/B = \exp\left\{(l-2k)z/z_0\right\},$$

we eliminate the initial pulseheight and obtain a function depending only on the wire from which the pulses originate. Fig. 4 shows the variations of log (A/B) obtained with wires of 40 cm, $z = 30\ \Omega$ and $R = 1\ k\Omega$. The wires are clearly separated. The number of wires which can be decoded by this method depends on the linearity and on the signal-to-noise ratio of the amplifiers.

A variant of this method is to measure the position of the avalanche on wires by connecting them together

and measuring the ratio of the pulses at the two ends. This method has been successfully applied to the wire of a cylindrical proportional counter[4]), with an accuracy of 1.2 mm, for a total length of 30 cm. This indicates that with the preceding method, at least 250 wires could be decoded under the same conditions of pulse-height and signal-to-noise ratio.

4. A coding system

A proposal was made by Pizer[5]) to use for wire chambers a method of grouping the wires into different subgroups in such a way as to code the position of each wire by two numbers (a, b): a is the number of the group, and b is the position in the group.

Each wire passes successively through the central hole of two cores defining the group and the position in the group, respectively. A secondary winding on the core gives a signal indicating that a wire belonging to a given subgroup has transported the current from a spark. Such a method is possible because the coupling between the wires passing through the same core is mainly capacitive and produces relatively small pulses in the neighbouring wires.

In the case of a proportional counter, the currents are so low that it would be necessary to wind the primary wires around the core. However, the coupling

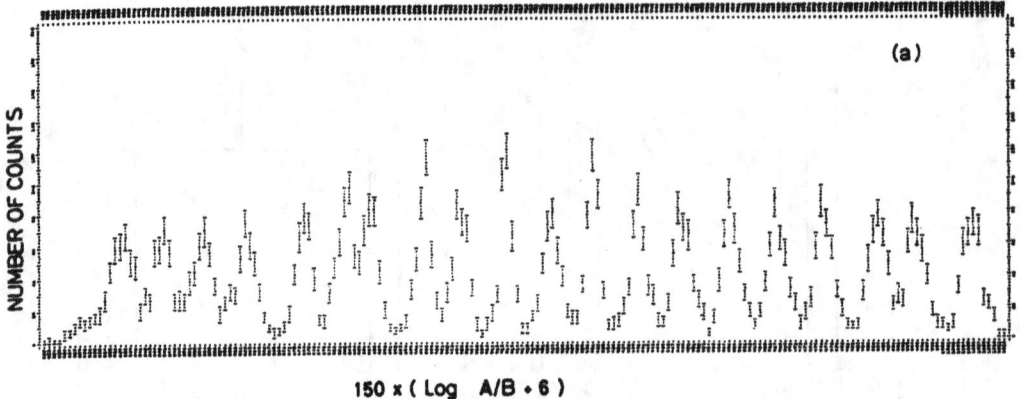

Fig. 4. Ratio A/B of the pulse-heights at the two ends of the resistive line a. a. Distribution of log (A/B) for a group of 20 wires $z = 30\,\Omega$, $\rho \approx 10$, $R = 1\,k\Omega$, 40 cm long wires separated by 3 mm. The A/B value for the first wire is 2.04 and for the 20th wire is 0.66. A collimated 5.9 keV X-ray source is displaced from wire to wire.

thus produced would be so large that it would be equivalent to connecting the wires together. The pulses from a given wire are then immediately spread among the members of a group. It is impossible to use non-linear elements such as diodes to limit the flow of currents in a given direction, because of the low level of the pulses.

We have circumvented this difficulty by the scheme represented in fig. 5. The pulses from the wires go through the primary of a ferrite core transformer before being sent to an amplifier common to all wires of a given position in the different groups. The secondaries

of the transformers belonging to one group are connected together. As a consequence, the parasitic feed-through pulses have the opposite sign from the direct signal. By using amplifiers sensitive to only one polarity of pulse, we eliminate the parasitic signals.

Fig. 6 shows the signals obtained on the primary and the secondary winding of the transformer, together with a parasitic feed-through pulse coming from a wire belonging to another group. This latter signal is 20

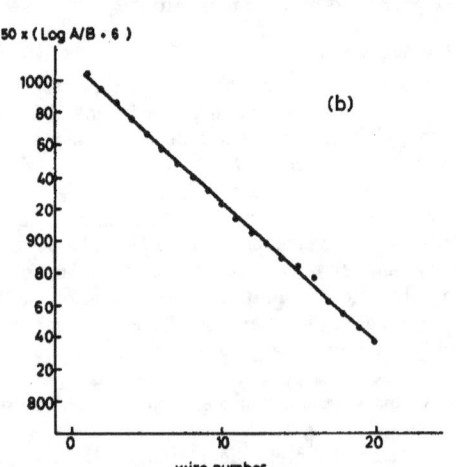

Fig. 4b. Variation of the average value of log (A/B) for the 20 wires under the same conditions.

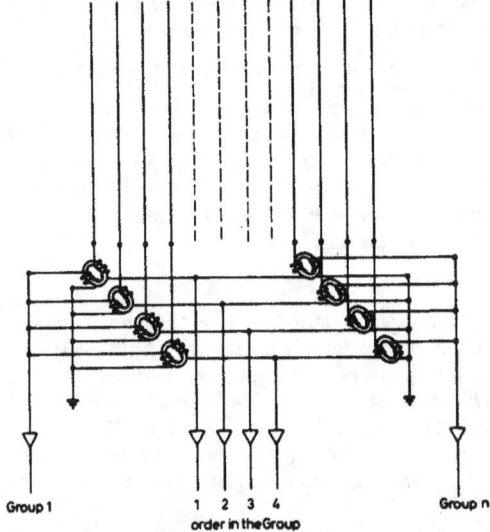

Fig. 5. Coupling of the wires to two different channels, through ferrite transformers.

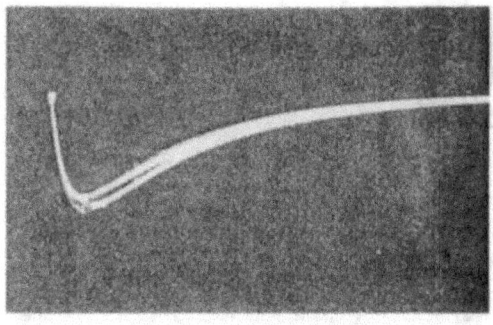

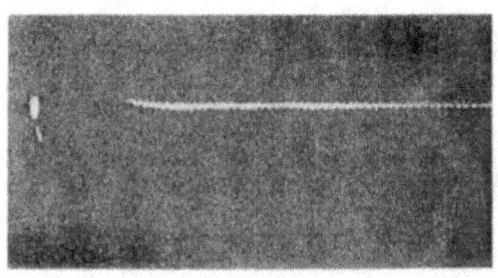

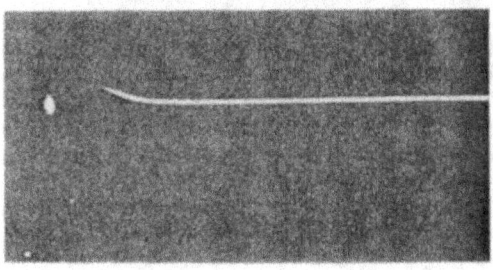

Fig. 6. Pulses obtained from the transformers. Ferrite core LTT2003 with 10 turns at the primary and 15 turns at the secondary. Source of collimated 5.9 keV X-rays. Horizontal scale: 0.5 μsec/cm. a. Pulses on the primary. Vert. 1 V/cm; b. Pulse from the corresponding secondary winding. Vert. 0.5 V/cm; c. Pulse from a secondary winding, induced by the detection in a neighbouring wire, belonging to another group. Vert. 0.1 V/cm.

times smaller than the direct pulse. In principle, such a coding system allows the reading of N wires with only $2N^{\frac{1}{2}}$ reading elements.

5. Conclusions

The three methods which are described in this article show that several relatively economic methods are available to read-out the position of an active wire in a multiwire proportional chamber. They are still at a preliminary stage, and much development work is required to make them operational.

One of the limitations of these methods is that they cannot be applied when many wires are counting simultaneously. Another limitation comes from the fact that when wires are grouped together, the capacity increases and this makes the signal-to-noise ratio worse. These methods are promising especially for simple problems, such as the detection of heavily ionizing particles in the focal plane of spectrometers. They can be combined together. For instance, a chamber can be split into n elements where each element is sufficiently small for one of the analogue decodings we have described. Then the use of the coding by transformer coupling indicates which element has been activated by a particle. This adds n amplifiers and a pattern unit with n inputs to the analogue read-out system.

We wish to thank Mr. G. Million for technical assistance. We are indebted to Mr. J. M. Fillot for his collaboration in the work on the pulse division method. Thanks are due to Professor P. Preiswerk for his continued encouragement.

References

[1] G. Charpak, R. Bouclier, T. Bressani, J. Favier and Č. Zupančič, Nucl Instr. and Meth. **62** (1968) 262.
[2] G. Charpak, Nucl. Instr. and Meth. **15** (1962) 318.
[3] G. Charpak, J. Favier and L. Massonnet, Nucl. Instr. and Meth. **24** (1963) 501.
[4] W. R. Kuhlmann, K. H. Lauterjung, B. Schimmer and K. Sistemich, Nucl. Instr. and Methods **40** (1966) 118.
[5] I. Pizer, in Proc. informal meeting on *Film-less spark chamber techniques and associated computer use* CERN report 64-30 (1964) p. 111.

NUCLEAR INSTRUMENTS AND METHODS 80 (1970) 13-34; © NORTH-HOLLAND PUBLISHING CO.

SOME DEVELOPMENTS IN THE OPERATION OF MULTIWIRE PROPORTIONAL CHAMBERS

G. CHARPAK, D. RAHM* and H. STEINER†

CERN, Geneva, Switzerland

Received 13 November 1969

Discussion of the limits of proportional amplification in multi-wire proportional chambers. Chambers with 2 mm wire spacing can give pulses of more than 200 mV on 20 pF loads with 25 nsec maximum time jitter.

Some examples of operation beyond the proportional region are given.

Multigrid chambers permit an increased gain for smaller wire separation and have multiple applications.

Better time and space resolution can be obtained from the time-position correlation. Resolutions of 7 nsec (fwhm) and 0.2 mm (fwhm) have been observed.

The use of the positive induced pulse permits the two-dimensional readout from single gaps.

The effects of magnetic fields up to 45 kG have been measured. Some new lines of research are discussed.

1. Introduction

Since the first CERN work[1] on multiwire proportional chambers (MWPC), the properties of these chambers have been described in several publications, and their use in experiments is currently under active study[2-10].

In this article we wish to report on some recent progress in the development of the chambers. We also take this opportunity to enumerate various properties and applications of the chambers as they appear from our own work or from that of other groups working in this field. This will lead to discussion of the following topics:

– Limits of amplification in the proportional region. Influence of the gas and of the construction.
– Limits in time resolution.
– Use of the position and drift-time correlation to improve the time and space resolution.
– Effects of magnetic fields up to 45 kG.
– Two-dimensional read-out from single gaps.
– Applications for X-rays or neutron mappings.
– Simple method of coding for the read-out when single particles are counted.
– Properties of multigrid chambers.
– Operation of multiwire chambers beyond the proportional region, mainly in the spark region.
– Some problems in the construction of chambers.
– New lines of research.

2. Field distribution and chamber parameters

In the first article[1]), we presented the field distribu-

tion in a multiwire chamber as it was obtained with the conductive paper method.

Although correct qualitative understanding can thus be obtained, large uncertainties arise in the field values for the central region. It is of interest for the quantitative understanding of the observations to obtain more exact values.

If we assume an infinite chamber, we can calculate the potential at any point of the chamber. This is done by solving the Laplace equation and adjusting the solutions to the potentials of the electrodes.

G. A. Erskine (CERN) has written a programme giving the field distributions in the case of finite-sized wires in symmetrical and non-symmetrical configurations. Fig. 1 illustrates some of his results. They permit the computation of the mechanical tolerances in the construction of the chambers, for given limits in the uniformity of the amplification.

We assume an infinite assembly of wires of diameter d, spacing s, and the distance from the wires to the mesh is L (fig. 2a). We centre the coordinate system on one wire, and x is in the plane of the wires and y is the perpendicular coordinate. By symmetry, all the wires have the same charge q per unit length. In the case where the wires are equally spaced and infinitely thin, the potential is given in many textbooks[+]:

$$V = q \ln [\sin^2(\pi x/s) + \sinh^2(\pi y/s)].$$

Along the symmetry lines $x = 0$, $y = 0$, $x = s/2$, simple formulae hold:

$$V(0,y) = 2q \ln \sinh \frac{\pi y}{s} \to E_y = \frac{2q\pi}{s} \coth \frac{\pi y}{s},$$

$$V(x,0) = 2q \ln \sin \frac{\pi x}{s} \to E_x = \frac{2q\pi}{s} \cot \frac{\pi x}{s},$$

$$V(s/2,y) = 2q \ln \cosh \frac{\pi y}{s} \to E_y = \frac{2q\pi}{s} \tanh \frac{\pi y}{s}.$$

* Permanent address: Brookhaven National Laboratory, Upton, LI., New York, U.S.A.
† Permanent address: Lawrence Radiation Laboratory, Univ. of California, Berkeley, Calif., U.S.A.
+ For instance, Morse and Feshbach, *Methods of theoretical physics* (McGraw-Hill, New York, 1953) p. 1236.

13

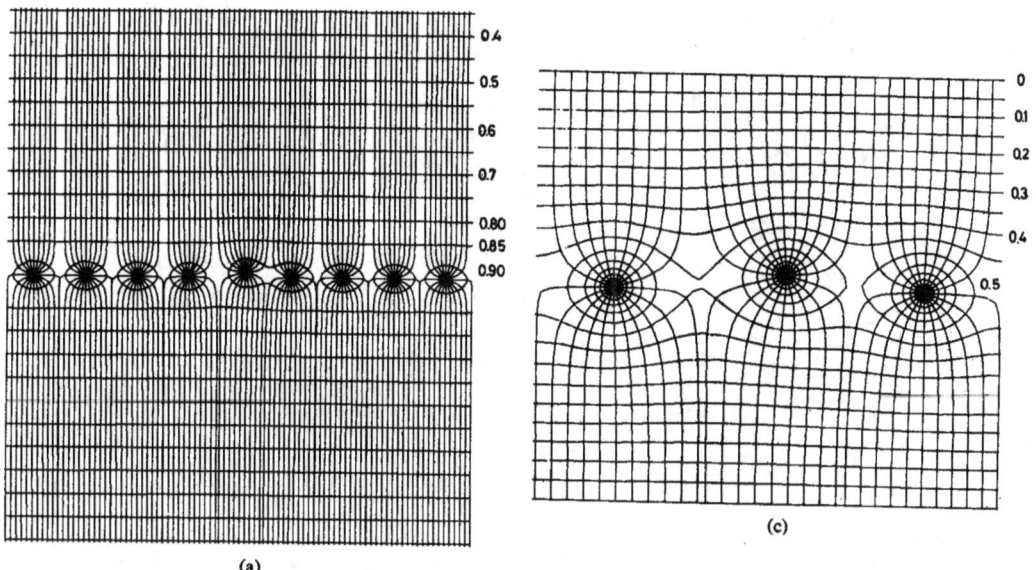

(a)

(c)

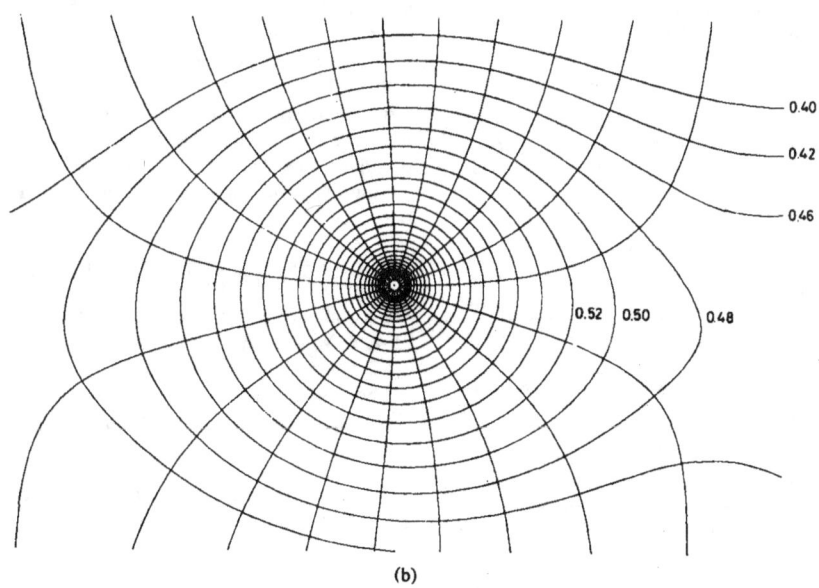

(b)

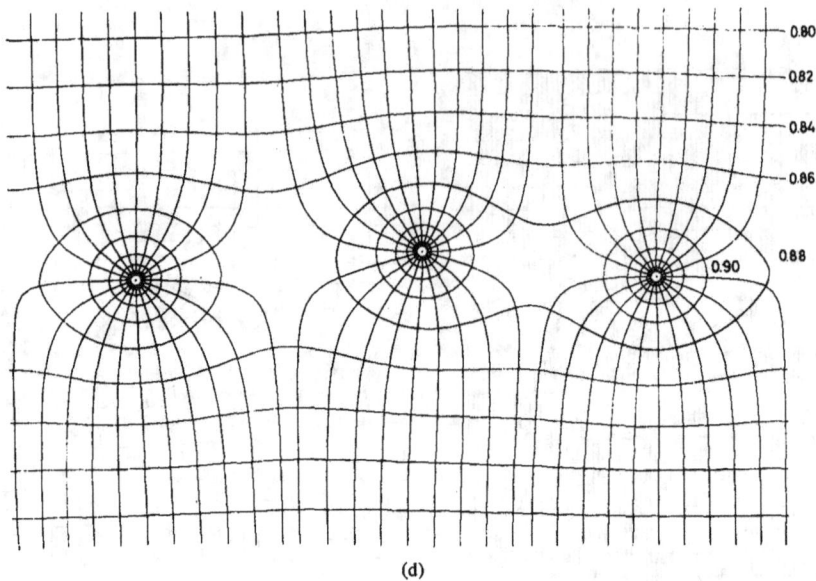

(d)

Fig. 1. Equipotentials in a wire chamber. The distributions have been obtained by solving the Laplace equations. Potential 1 on the wires, 0 on the outer electrodes. (a) Wires of 40 μm, $s = 1$ mm, $L = 8$ mm. One wire is displaced by 10% in the X and Y directions. (b) Same parameters. Enlarged view of potentials around the displaced wire. (c) Wires of 40 μm, $s = 3$ mm, $L = 4$ mm. One wire is displaced by 10% in the X and Y directions. (d) Same parameters. Enlarged view around the displaced wire.

These formulae are valid only for infinitely thin wires, but for a wire thickness of 100 μm and a spacing of 1 mm, $\sin(\pi y/s)$ and $\sinh(\pi y/s)$ differ only by 8×10^{-3},

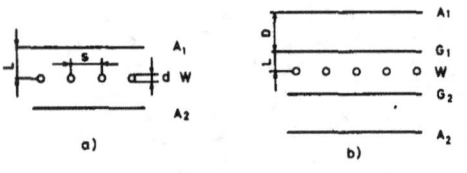

a)

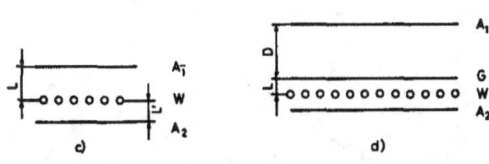

c)

b)

d)

Fig. 2. Construction parameters of the chambers used in this work: a) the symmetrical bi-electrode chamber, b) the symmetrical bi-triode chamber, c) the asymmetrical bi-electrode chamber, d) the drift chamber. The outer electrodes A_1A_2 are made of mesh of 50 μm stainless steel, with 82% transparency. The grids G are made of mesh of 82% or 90% transparency, or of wires of 0.1 mm, 1 mm apart, parallel or orthogonal to the central wires W.

and substituting an equipotential at this radius will not make an appreciable change in the field distribution.

At large distances the field becomes uniform, e.g. at $y = s/2$ the tanh and coth are equal to $\pm 9\%$, and at $y = 1.25$ they are equal to one part in 10^3, so the field has become very uniform and we can eventually put in a flat grid without perturbing the field. We can now compute q if we know the dimensions and the applied potential V_0:

$$q = \frac{V_0}{2[\ln \sinh(\pi L/s) - \ln \sinh(\pi d/2s)]}.$$

For our typical chamber with an applied potential of 4000 V, the field at the wire is 2.2×10^5 V/cm, for $s = 2$ mm and $L = 8$ mm.

The field distribution is, to a reasonable approximation, one that varies as $1/r$ near the wire, and is uniform for $y > s/2$.

3. Factors governing the proportional amplification
3.1. WIRE SPACING AND DIAMETER

The amplification of a proportional counter increases exponentially with the applied voltage above a threshold V_0. The threshold occurs when an electron can gain enough energy between collisions to ionize a molecule in the next collision. The multiplication takes place near the wire where the field is very high. The field varies as $1/r$ near the wire, so that as we reduce the wire diameter, we can reduce the applied voltage while still keeping the same field at the surface of the wire. We have found it mechanically difficult, however, to use wires smaller than 20 μm in diameter.

For a given wire diameter and applied voltage, the field at the wire (and thus the gain) will depend on the wire spacing. This is clear from the calculation of $C = q/V_0$, the capacitance per unit length with respect to the electrodes, as a function of wire parameters currently used in the chambers (see table 1).

TABLE 1

Capacitance, in picofarads/metre, of a wire in the central plane with respect to the outer electrodes for $L = 7$ mm.

Wire diameter	$s = 3$ mm	C per metre $s = 2$ mm	$s = 1$ mm
20 μm	4.97 pF	3.85 pF	2.25 pF
30 μm	5.21 pF	3.96 pF	2.28 pF
40 μm	5.30 pF	4.04 pF	2.31 pF
50 μm	5.41 pF	4.11 pF	2.34 pF

The capacitance we have computed above is that involved in charging the wire with respect to the high voltage. The capacitance for the *signal* is the capacitance of the wire to ground and is significantly higher. With careful design to minimize end effects, the capacitance can be kept down to 20 pF for a 1 m chamber with 2 mm spacing.

From table 1 we see that in going from 3 mm to 2 mm to 1 mm spacing with 20 μm wires, the voltage must be increased in the ratios 1, 1.3, 2.2 in order to keep the same charge on the wire, i.e. the same amplification. We will thus meet problems of voltage breakdown if we want to increase the spatial resolution by increasing the number of wires. If the gap L is reduced to 3 mm, the comparable capacitances are 7.94, 6.81 and 4.56 pF with the ratios being 1, 1.17, 1.74. We see that we can operate with considerably lower over-all voltages as L is decreased, at the expense of a somewhat higher capacitance.

3.2. SPACE CHARGE LIMITATION

Proportional chambers typically give signals measured in millivolts (into a load consisting of 15 pF in parallel with 10 kΩ) for minimum ionizing particles. Larger signals can be obtained by raising the applied voltage, but as the voltage is raised one finds that:

firstly, the strict proportionality between deposited ionization and signal is lost; and secondly, further increases in voltage lead to corona or sparking. For very close wire spacings, breakdown may occur even before the semi-proportional region is reached.

Hanna, Kirkwood and Pontecorvo[11] observed that the output pulse from a cylindrical proportional chamber ceases to be strictly proportional to the initial ionization when the absolute value of the output charge exceeds some critical value Q. For their counter, they found that this would correspond to losing 10^8 eV in the gas of the counter if there were no amplification, or to the creation of $\approx 5 \times 10^6$ ion pairs. Thus the upper limit on the amplification for proportional operation was $10^8/E$, where E is the energy, in electron volts, lost by the primary event. This type of a limit can be understood as being due to a space charge effect. When the charge of the positive ion cloud of the avalanche produces a significant modification in the local electric field, saturation sets in. This limit occurs when the charge of the ion cloud is comparable to the local stored charge on the wire. If the original ionization is distributed along the wire, then the limit would be higher than for the case of a low-energy γ-ray conversion electron where all the ionization is localized over a small length (1 mm for 6 keV).

For our typical chamber, the charge/cm corresponds to 1.2×10^{-10} C/cm, while 5×10^6 ions is a charge of 0.8×10^{-12} C, or 7% of the charge contained on 1 mm of the wire. These numbers are consistent with the space-charge hypothesis.

A typical practical chamber will have a capacitance to ground of ≈ 15 pF, so the maximum signal voltage for the 5×10^6 ions would be ≈ 40 mV. As discussed in section 4, we would only see about $\frac{1}{10}$ of this signal because of the usual differentiating network used in a practical amplifier (10 kΩ in parallel with the 15 pF). In practice, however, we have obtained larger signals than this, in fact up to 50 mV at the limit of proportional amplification and 200 mV just before breakdown. This corresponds to $\approx 2 \times 10^8$ ions in the avalanche.

There is no difference between the gas amplification mechanism of a multiwire chamber and that of a cylindrical chamber. However, as we reduce the distance s between the wires, we find it more and more difficult to reach the maximum gain. For chambers made of wires of diameters 20 μm, 30 μm, and 40 μm, and with a spacing $s = 3$ mm, we easily reach 200 mV or even 400 mV before breakdown. With a spacing of $s = 2$ mm and wires of 20 μm, we still reach this limit. With 30 μm wires we obtain 30 mV pulses, and with a spacing of $s = 1$ mm and wires of 20 μm we reach 20 mV as a limit.

These observations are restricted to our particular conditions of construction and gas filling, usually argon-isobutane (75%, 25%) or argon-carbon dioxyde, and a 7 mm gap. But they show a general trend; whenever we cannot increase the voltage to reach the limit that we expect, it is because breakdown occurs in some "weak points" of the chamber.

If we keep increasing the voltage beyond this limit, we can still obtain higher pulses, up to 0.4 V in some cases, but then secondary effects set in which propagate new avalanches. This will be discussed in a separate section.

3.3. GAP LENGTH AND GAS COMPOSITION

As we saw in section 3.1, decreasing the gap length decreases the total voltage required; however, the ionization deposited by a minimum ionizing particle will also decrease, while the capacitance has increased, so we would need a higher amplification. Many experiments with high voltage indicate that the maximum field that can be held by a gap increases as $1/\sqrt{L}$, where L is the gap length. Indeed, we found that we could obtain much higher amplification without breakdown in a chamber with 20 μm wires spaced 1 mm part, when the gap length was reduced from 7 mm to 2 mm.

The total gap of 14 mm employed by us at the beginning can certainly be reduced even when the chambers are used to detect minimum ionizing particles, because time resolution measurements indicate that with our electronics, electrons deposited further than 2 mm from the wire play little role in the trigger. This conclusion depends upon the sensitivity of the electronics. With highly ionizing particles, such as protons, and for applications in the focal plane of spectrometers where resolutions of 0.5 mm are to be reached, gaps of 1.5 mm would be sufficient and would allow higher fields.

Wilkinson[12] states that any gas will work in a proportional counter. He is perhaps a little optimistic, but certainly almost any gas will work. We have restricted ourselves to argon plus organic quenching agents. In our flat multiwire chambers we have a relatively dirty environment, and we need gases that give us high gains without any great sensitivity to impurities. We have found that argon plus any of the following gases will permit operation with our typical chambers: methane, propane, isobutane, household Butagaz, pentane, hexane, alcohol, ethylene, acetylene, and carbon dioxide. We did some relatively limited testing on these

mixtures, but we found the best results with argon plus 15% to 20% isobutane mixtures. We were able to achieve higher gains and to operate in modes that were not easily accessible with other gases.

In order to get high gains we need to operate with high voltages, and the limitation usually comes from sparking across the gap. With a quenching agent, the hold-off voltage increases more than the amount the voltage must be increased in order to compensate for the reduced gain. Pure argon has a low threshold voltage (≈ 1800 V in our typical chamber with $s = 2$ mm), but it also has a low flashover voltage so that the signal one can obtain is small. With pure isobutane, the working voltage is ≈ 8000 V and an appropriate mixture will work at any voltage in between.

It is important to have a construction of the chambers that does not create spots of fields higher than the average, since the breakdown limit will be reached at these places.

Gases of high density are frequently desirable, and it is interesting to compare the energy loss in different gases used in proportional counters, as shown in table 2.

TABLE 2

Gas	Energy loss in keV per cm, at TPN, of 1.3 MeV electrons
H	0.34
He	0.32
Ne	1.44
A	2.50
Kr	4.9
Xe	10.5
CO_2	3.3
CH_4	1.5
Iso-C_4H_{10}	5.3

Isobutane is advantageous, since it is cheaper than the heavy noble gases such as Kr and Xe, and has a density that should allow a reduction of thickness by about a factor of 2 compared to argon. Tests with pure isobutane show that high gains can easily be obtained.

For reasons of physics, in a particular experiment one gas or another may be preferred; for example, one may want high γ-conversion efficiency, or low multiple scattering. The gas mixture can be modified as needed. For mechanical reasons it is usually easier to operate at one atmosphere, which is something of a limitation.

We have not examined all reasonable combinations of gases, and can only report on a small sampling. Carbon dioxide is one of the best admixtures for long-range operation, since it does not polymerize under the influence of avalanches, as do the organic gases. A more systematic survey of many mixtures is needed and might turn up some agreeable surprises.

3.4. SEPARATION OF THE AMPLIFICATION AND DRIFT REGIONS

If we place a control grid close to the wires, in a region where the equipotentials are essentially parallel to the wire-plane, we can keep all the fields around the wires unchanged while we change the fields in the drift region between high-voltage electrode and the grid.

This was tried in cylindrical chambers[13]) in order to increase the drift velocity of the electrons. With this method we, too, can increase the drift velocity, which may be desirable when the distance between the wires is very great; we can also use an intermediate control grid when the wires are so close that excessively high potentials would be required in normal structures. With this method we were able to reach the expected limit set by space charge, under conditions where this was not possible with the normal construction.

The control grid can be made of wires parallel or orthogonal to the central wires, or by a mesh.

The transparency to electrons can be smaller or greater than the geometrical transparency, depending on the voltage applied to the different electrodes. This problem has been studied extensively by Buneman et al.[14]) in connection with the gridded ionization chambers. These authors calculate the efficiency of the electric shielding of the collecting electrode by means of wire screens and the transparency of the screen to the drifting electrons.

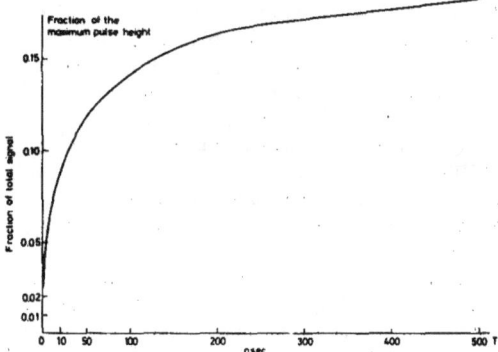

Fig. 4. Time development of a pulse. This is the computed pulse shape induced by the motion of a localized cluster of positive ions leaving the wire at $T = 0$; $d = 20 \mu m$, $L = 7$ mm.

With grids made of wires or of a mesh of 90% transparency, we observe a weak dependence of the amplification factor on the field applied in the drift space, and a collecting efficiency nearly equal to the geometrical transparency.

Fig. 3 shows the pulses obtained directly with 5.9 keV X-rays, on wires of 40 μm, with 2 mm wire spacing and a grid G at 2 mm. Potentials of only 3 kV on the screen and 4 kV on the external electrode were necessary, while 10 kV would have given the same amplification without the screen.

The reduction of the field in the drift space does not even degrade the time resolution, since the drift velocity saturates and then even decreases with increasing fields.

Such multigrid structures have many other applications, as we will discuss later.

4. Time development of pulses

There are two contributions to the effective rise-time of a pulse. One is the time it takes to collect the electrons liberated in the gas, the second is the motion of the ions away from the central wire. Let us consider the second contribution first.

We operate chambers with typical multiplications of 10^4 to 10^5, and since the mean free path of electrons in the gas is much less than a typical wire radius, almost all the ion pairs are formed very near the wire. The contribution of the electrons to the pulse is very small since most of them travel through a small fraction of the total potential. It is the motion of the positive ions away from the central wire that induces the charge on the central wire. Calculations for a chamber with 7 mm wire to mesh spacing, central wires of 20 μm spaced at 2 mm, and with 3000 V applied, show that

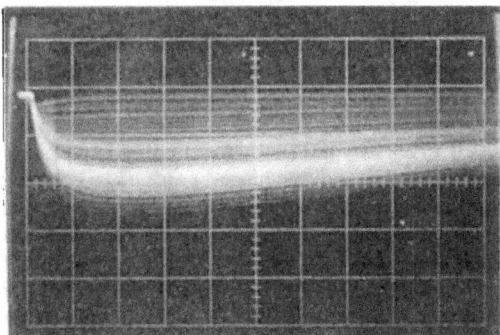

Fig. 3. Pulses obtained directly from a wire of a multigrid chamber. Type (b) of fig. 2: $L = 2$ mm, $D = 5$ mm, (hv) $G_1 = $ (hv) $G_2 = 3$ kV, (hv) $A_1 = $ (hv) $A_2 = 4$ kV, $d = 40 \mu m$, $s = 2$ mm. Pulses at the limit of proportional operations; X-rays of 5.9 keV. Vertical scale: 100 mV/cm, horizontal scale: 100 nsec/cm. Similar pulses are obtained with $d = 20 \mu m$, $s = 1$ mm.

the collected charge (for times less than 1 μsec) increases as $\ln(t + t_0)/t_0$, as in a cylindrical chamber[12]), where t_0 is about 1.8 nsec. This gives a very sharp rise (in the first 10 nsec we collect as much as is collected in the next 90 nsec) followed by a very slow increase, as seen in fig. 4. For this chamber it takes about 100–150 μsec for all the ions to reach the grid. This entire period of movement of the ions induces charge on the central wire. At 10 nsec we have collected $\approx 8\%$ of the total possible signal, at 100 nsec $\approx 14\%$ and at 500 nsec $\approx 18\%$. If one uses a differentiating circuit at the input to this amplifier, then only the initial sharp rise is seen. The above figures depend, of course, on the exact configuration of the chamber, the applied voltage, and the ionic mobility.

The other contribution, the collection time of the electrons, depends on the thickness of the chamber and the type of radiation. For low-energy X-rays where the conversion electrons may have a range of a mm or less, then the charges are spatially concentrated and will all arrive at the central wire within a few nsec of each other. For particles passing through the chamber and normal to the central plane, the electrons are distributed throughout the depth of the chamber and have to be collected from as far away as 7 mm. This collection time is about 150–200 nsec in our typical chamber, and so the pulse appears to rise more or less linearly (being a superposition of a number of fast rises) during this collection time, followed by a very much slower rise until the ions reach the mesh.

5. Induced pulses, bi-dimensional readout

The normal proportional pulses in MWPC's are

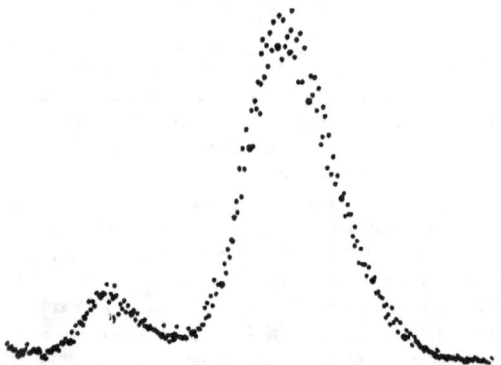

Fig. 5. Pulse-height spectrum of the positive pulses induced on a hv electrode: the chamber is of type shown in fig. 2d. Positive signal induced in the lower electrode A_2. Distance $L = 2$ mm. The energy resolution is poorer than on the wires, X-rays of 5.9 keV.

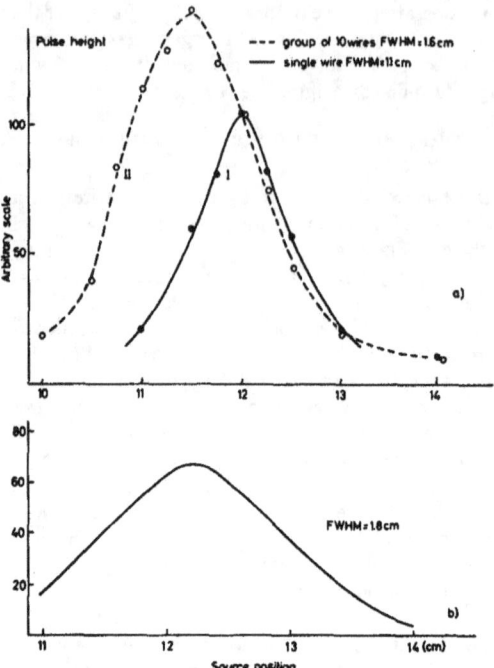

Fig. 6. Variation of the height of the induced pulses as a function of source position. a) Distance from central wire to sensing wires = 2 mm; I: pulse on one single wire, II: pulse on a group of 10 wires connected together. b) Distance from central wire to sensing wire = 7 mm.

caused by the motion of positive ions away from a wire. This same motion induces signals of opposite polarity on neighbouring wires and on the high-voltage electrodes. When the electrons from the primary ionizing event are all collected by one wire, that wire will produce a negative pulse, whereas the adjacent wires and the negative high-voltage electrode will have only positive pulses. At first we were somewhat surprised to see *only* positive induced pulses on the adjacent wires. For our conditions, proportional amplification takes place at distances from the wire that are small compared to its radius. If the ion cloud (or avalanche) were to be produced mainly on one side of a wire, one would expect the adjacent wire on that side to have a positive pulse, but the adjacent wire on the other side to have a negative signal. Instead we see positive pulses everywhere. It seems that the ion cloud is essentially symmetric about the central wire. Fig. 5 shows the pulse-height spectrum of the positively induced pulses on the lower electrode A_2 (see fig. 2).

While the sheath spreads completely around the

wire, its extent along the wire is limited. We have made a chamber with an intermediate grid of orthogonal wires spaced 1 mm part and located 1.5 mm above and below the central plane (type d, fig. 2). The voltage applied to this grid controls the amplification in the chamber. The space between the orthogonal wires and the external high-voltage electrode serves as a drift space for the ion pairs produced therein (see also the section on drift chambers). With this chamber, we can locate the position of the avalanche along the central wire. The amplitude of the pulses on the orthogonal wires varies with their distances to the avalanches.

Fig. 6a shows the pulse height of the signal induced on a single wire and a group of 10 wires in parallel when the source is moved. Fig. 6b shows the pulse-height variation on the high-voltage wires orthogonal to the central wires for a "normal chamber" with $L = 7$ mm. We see that the sensitivity to the displacements is higher with a closer wire plane. For a given sensitivity

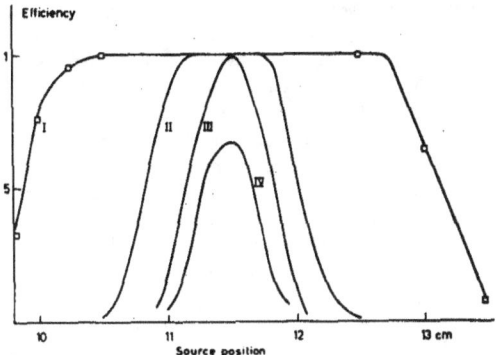

Fig. 8. Number of wires sensitive to the induced signal: Distance from central wire to sensing wire = 2 mm. For a given sensitivity on the "sensing" wire, the number of wires detecting the induced signal varies with their pulse height. For a given pulse height, the number of wires varies with the sensitivity. We triggered with only those pulses on the central wire which correspond to a 5.9 keV peak from ^{55}Fe; I: threshold at 60 (arbitrary unit), II: threshold at 150, III: threshold at !80, IV: threshold at 200.

of the circuits detecting the positive pulses on the "sensing wires", the number of wires detecting an avalanche will depend on the avalanche size.

Fig. 7 shows the pulse height of the positive signal on a wire, in coincidence with a signal of a sharply defined pulse on the central wire.

Fig. 8 shows the variation in the number of "sensing wires" detecting the avalanche as a function of their sensitivity.

The amplitude of the induced positive pulses is proportional to that of the negative signal on the central wire. Typically, the positive adjacent wire pulses are about $\frac{1}{5}$ to $\frac{1}{10}$ as large as the pulse on the central wire, depending on the geometry. The "normal" and the induced pulses are also completely correlated in time, since it is the same motion of the same ions that is responsible for both the positive and the negative signals.

When the electrons from the primary ionization process are shared among several adjacent wires, as for example when one is dealing with an inclined track, then one wire will receive electrons first, and the first avalanche will start there. This will induce positive signals on the neighbouring wires. As the electrons from further out drift in, they in turn will start avalanches on some of these neighbours, thus turning a pulse that originally started out positive into a negative one. In general when there is such a superposition of induced and normal pulses, it is the negative "normal" pulse which dominates.

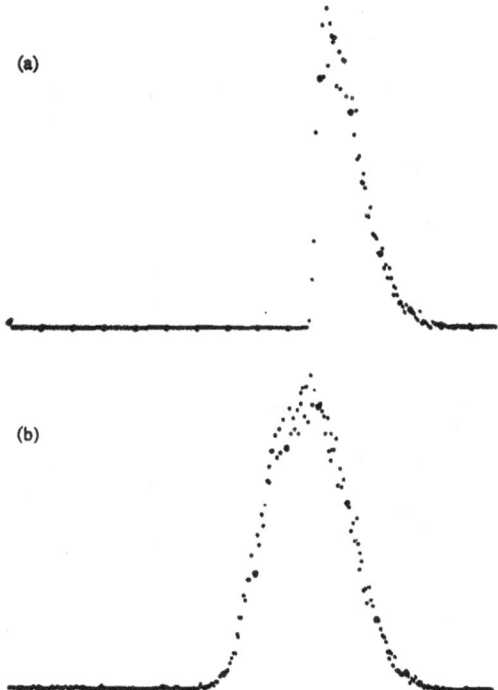

(a)

(b)

Fig. 7. Direct and indirect pulse. (a) Spectrum of pulses produced by a 5.9 keV source, after suppression of pulses below a given value by a discriminator. (b) Spectrum of positive pulses on a wire of a screen at 2 mm, orthogonal to the central wires, in coincidence with the above pulses. The positive pulses are about 5 times smaller than the corresponding negative ones.

We found it possible to use the pulses from the high-voltage electrodes of a chamber 60×60 cm^2. For very large chambers, charge amplifiers may have to be used.

There are many interesting and useful applications of these induced pulses:

1. They are suited to fast time measurement. When several wires collect charges, these will come at different times, but the rise of the pulse on the external electrode will begin with the fastest one.

2. When energy is shared among several wires, the signal on the external electrode is proportional to the total energy.

3. One can determine the orthogonal coordinate by using for the high voltage or for the control grids, wires whose directions are perpendicular to the central wires. In many cases information about the orthogonal coordinates could be obtained by simply using two ordinary chambers. However, there is an application where the method discussed here may have enormous advantages: measurement of the spatial distribution of γ-rays (or slow neutrons) where the ionization produced remains in only one chamber. There may also be an application for eliminating the ambiguities that arise when several particles traverse a chamber. Here use could be made of the very tight time and amplitude correlation between the normal and the induced pulses.

4. The digitization of the pulse heights can be accomplished by counting the number of wires with induced pulses greater than some threshold value. This method of digitization has the advantage of being simple and fast.

5. The number of amplifiers needed to transmit the pulse information from the wires in the chamber could in certain cases be reduced very significantly by suitable coding of the information from groups of wires. In principle it is possible to read out a coordinate on N wires with $2\sqrt{N}$ amplifiers. A complicated method was introduced by Charpak et al.[1]) to code a single plane of wires. The use of the induced signal offers a much simpler solution. If the hv electrode is separated into strips or wires facing groups of the central wires, they will transmit an induced signal whenever any wire of a group is collecting an avalanche. The central wires can be grouped in another way, thus giving a bi-dimensional coding. It must, however, be stressed that, for the time being, the fact that the induced pulses are smaller than the normal pulses may make some of the above-mentioned applications somewhat difficult.

6. Time resolution

In the operation of the chambers, several "times" are involved: the variable time delay between the pas-

sage of the particle and the arrival of the pulse at the logic system, the dead-time, and the read-out time. Let us consider each of these in turn.

6.1. MAXIMUM TIME JITTER

Two principal factors are responsible for the delay time in MWPC: 1) the time it takes to collect the first few electrons produced when a particle ionizes the gas in the chamber; 2) the time it takes to amplify the pulse and to transmit it to the logic system. It takes one electron about 250 nsec to go one centimeter in the type of chamber we have been using. Clearly, the time of arrival of the electrons in the amplifying region near the wire will depend on where in the chamber the primary ionization was produced. It will also depend on the voltage applied to the electrodes, the gas mixture used, the wire diameter and wire spacing, and the energy loss of the particles traversing the chamber. With ionizing particles traversing chambers having 2 mm wire spacing, the time needed to collect the first few electrons can usually be kept below about 25 nsec. One usually has to add 15–20 nsec of external delay in the amplifying system. Thus total delays smaller than 50 nsec are feasible. It is the fluctuations or jitter of this delay-time that determines the resolving time of a chamber.

In chambers with 2 mm wire spacing, the position of the primary ionization relative to the wire is uncertain to at least one millimetre. In chambers of this type one might naively expect jitter times of about 25 nsec. This is found to be case experimentally.

The actual measurements of the delay time and the jitter time were made by observing the time difference between pulses produced by a β-source in a proportional chamber and in a scintillation counter. These time differences were then converted to pulse heights and displayed on a pulse-height analyser. Some of the results obtained for different wire spacings, different gases, and different source conditions are shown in fig. 9. We see that with 3 mm wire spacing, the maximum time jitter is less than 32 nsec, with 2 mm wire spacing it is less than 24 nsec, and with 1 mm it is less than 18 nsec. These values include small contributions from our electronic circuits, which were not designed to make very fast time measurements. We conclude that without pushing voltages too high or otherwise operating chambers under marginal conditions, experiments can now be undertaken safely with a time resolution of 30 nsec for 2 mm spacing with 100% efficiency for minimum ionizing particles.

6.2. TIME DIFFERENCES FROM ADJACENT WIRES

Occasionally an ionizing particle will produce pulses

in several adjacent wires. This can happen for a variety
of reasons: the track may have been inclined with
respect to the normal to the wire plane, δ-rays were
produced, the electrons diffused laterally during the
collection process, etc. We have found that there is
normally a large ($\approx$ 100 nsec) time difference between
pulses from adjacent wires. Some results on the time
distribution of pulses from adjacent wires are shown

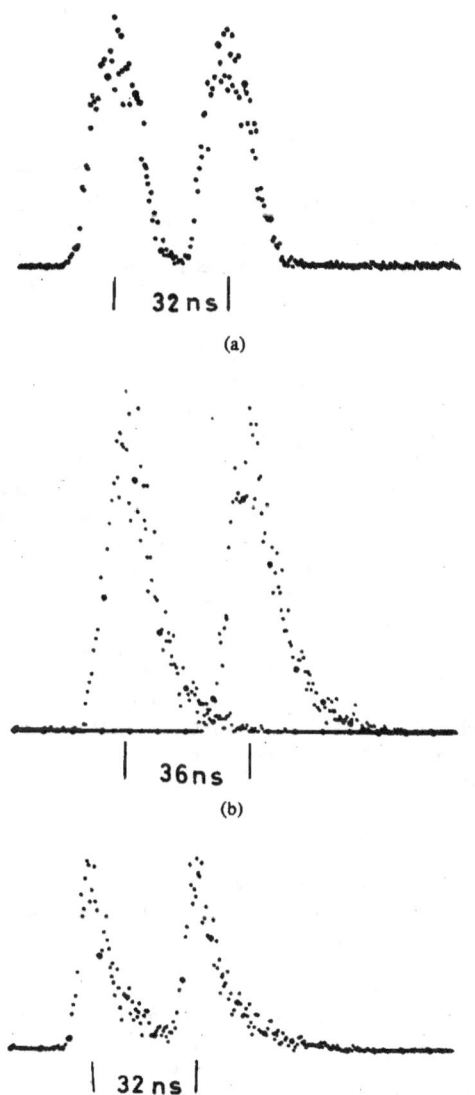

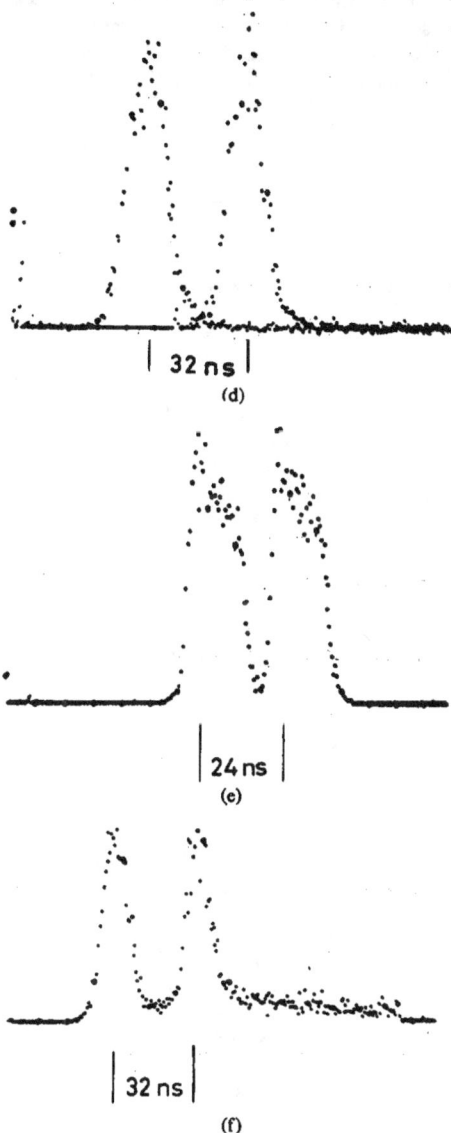

Fig. 9. Maximum time jitter. Distribution of time decays between
the passage of the particle and the detection in the chamber.
Each photo shows two spectra taken under identical conditions
but separated by an artificial delay as indicated: 3 mm spacing,
gap $L = 7$ mm; (a) Argon + isobutane (20/80) hv = 6000 V,
(b) argon + CO_2 (50/50) hv = 5000 V, (c) CO_2 + isobutane
(90/10), (d) methane (100) hv = 7200 V, 2 mm spacing, $L = 7$ mm;
(e) argon + isobutane (80/20) hv = 5500 V. 1 mm spacing, $L = 2$
mm; (f) argon + isobutane (66/33) hv = 3900 V.

in figs. 10 (normal tracks) and 11 (inclined tracks). When several adjacent wires have pulses, it is almost always the wire that produces the first pulse that is closest to the track. Often it is desirable to record only

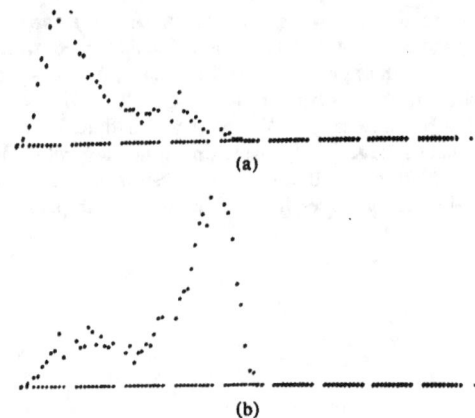

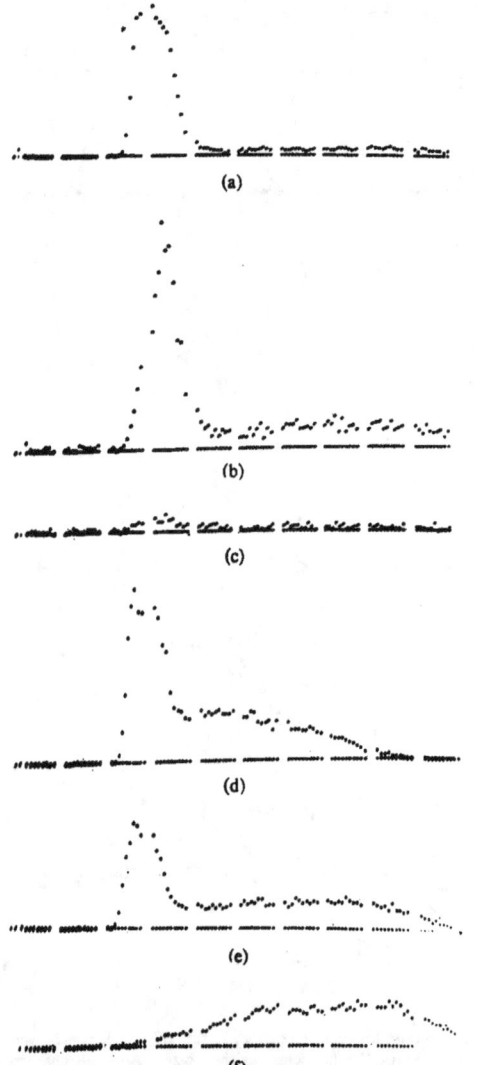

Fig. 11. Time relations between pulses from neighbouring wires. Time difference between wires. The pulses from the two neighbouring wires go to the start and stop of a time-to-amplitude converter. Time scale = 2 nsec/channel; (a) beam at 30° collimated on wire no. 2, (b) beam at 30° collimated on wire no. 1.

this first pulse and to ignore the later-arriving pulses from the other wires. To do this one can use gating circuits which allow only the first of several pulses to pass through. By keeping the width of this gate as short as possible, one can minimize the number of events in which multiple pulses from adjacent wires are recorded.

Generally speaking, the problem of multiple pulses is not serious for particles that traverse the chamber perpendicular to the central wire plane. Less than about 5% of these events produce multiple pulses under a wide range of operating conditions. Of course, care must be taken to prevent the induced pulses from causing trouble. For example, sometimes circuits are sensitive to the negative overshoot from a positive pulse. Fortunately this overshoot is usually delayed with respect to the leading edge so that narrow time-gating eliminates it.

6.3. DEAD-TIME

The dead-time of MWPC's is influenced by both internal and external factors. The internal limitations are due to the time it takes to collect *all* the electrons from the primary ionizing event, and also to a much lesser extent to the time it takes for the ion avalanche to move away from the wire. In a typical chamber with 7 mm gaps, the electron collection time is 200 nsec. This can be reduced by making the gap smaller. Owing to the statistical nature of the energy-loss process, there are large variations in the way in which the electrons are initially distributed along the track of the particle.

Fig. 10. Time relations between pulses from neighbouring wires. Beam of β-particles collimated on wire no. 1. Time scale = 2 nsec/channel. a. Beam perpendicular to the chamber. Time delay between a scintillator and wire no. 1 (a), no. 2 (b), no. 3 (c); b. beam at 30°, wire no. 1 (d), no. 2 (e), no. 3 (f). The intensity scales of no. 2 and no. 3 are four times greater than those of no. 1.

It is difficult to distinguish the collection of "new" electrons produced by a second ionizing event, from that of any "old" electrons that are still drifting through the chamber toward the *same wire*.

The major external limitation has to do with the time characteristics of the amplifiers. Differentiation times of 50 nsec are typical of those used in our tests. Thus the amplifier will require about 150 nsec before it is ready to accept another pulse without significant shift in voltage levels. It should be emphasized that we are always speaking of the effective *dead-time per wire*.

After combining all the factors that influence the dead-time, we find that the *total* dead-time per wire need not exceed a few hundred nanoseconds.

6.4. READ-OUT TIME

The read-out time probably imposes the most serious limitation on the event rates attainable in MWPC's for large numbers of wires; information transfer to computers will be inevitable. It is the time required for this information transfer (and processing) that will limit the counting rates in many experiments. Much work still needs to be done on developing fast read-out systems for experiments involving many wires. Especially important in this respect will be the development of systems that will allow coupling of MWPC's to fast decision-making logic systems for use as triggers for other detectors (spark chambers, streamer chambers, etc.) or to control information

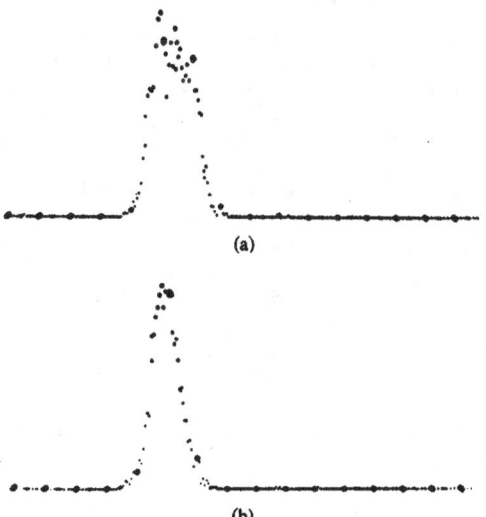

(a)

(b)

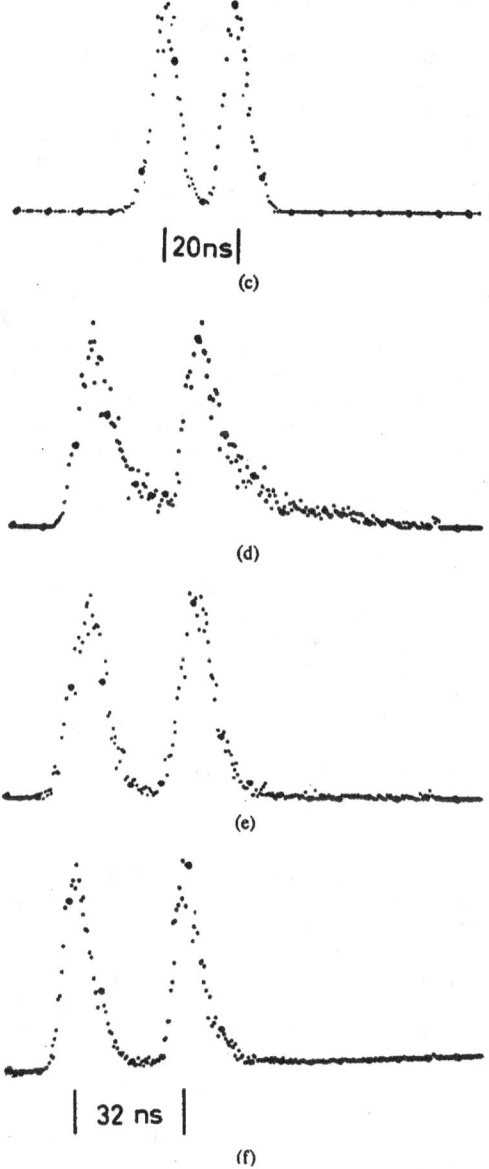

|20ns|

(c)

(d)

(e)

| 32 ns |

(f)

Fig. 12. Improvement of the time resolution with several staggered chambers. (a, b, c) $s = 2$ mm, chambers A, AB, ABC; (d, e, f) $s = 3$ mm, chambers A, B, AB; beam of β-rays.

transfer. For relatively small numbers of wires, fast logic systems already exist that can be used for this purpose.

6.5. TIME RESOLUTION IMPROVEMENT USING SEVERAL CHAMBERS

In order to bring about a further reduction of the jitter time, we have looked at the first proportional pulse when several chambers are stacked in series. With two chambers (2 mm wire spacing) suitably staggered so that the distance between wires was 1 mm when they were viewed normal to the wire planes, we obtained a significant improvement in the time resolution (see fig. 12). With a well-collimated β-source we obtained a total jitter time of 18 nsec. When the source was less well collimated, the width at the base was about 24 nsec (as with a single chamber) but the full width at half-maximum (fwhm) was only 12 nsec, which should be compared with fwhm = 20 nsec for a single chamber. With three chambers we obtained 16 nsec at the base and 8 nsec fwhm. Further improvement can be expected if still more chambers are used. It should be possible to achieve time resolutions with multiple chambers quite comparable to those usually attainable with scintillation counters. Still better time resolution might well be possible if correlations are made between the time of arrival of the pulses in several chambers whose wires are shifted relative to each other. Such ideas have already been applied to large scintillators viewed by several photomultiplier tubes[15], and the extension to MWPC's is worth investigating. Our experiments with drift chambers (see section 9) where time resolutions of 5 nsec have been attained (see fig. 16), make us optimistic that further improvements are possible.

7. Space resolution

As might be expected, the space resolution attainable with MWPC is limited mainly by the wire spacing. For particles traversing the chamber at an angle, the thickness of the gap also plays a role. Chambers which wire spacings of 2 mm or more work reliably without straining mechanical tolerances, high-voltage requirements, gas mixtures, or other factors which influence their performance. As can be seen in fig. 13, the high-voltage plateaux are large and flat. Below 2 mm, life becomes more difficult, and various tricks have been tried. We have seen that with multigrid chambers or with small gap chambers, resolutions of 1 mm can be attained. Some other rather obvious ways to improve the spatial resolution are to use staggered chambers, and/or to make use of timing information, as is done in the drift chamber.

If one tries to localize tracks to much better than 1 mm, one has to worry about the effects associated with the gap size. The lateral diffusion of the electrons

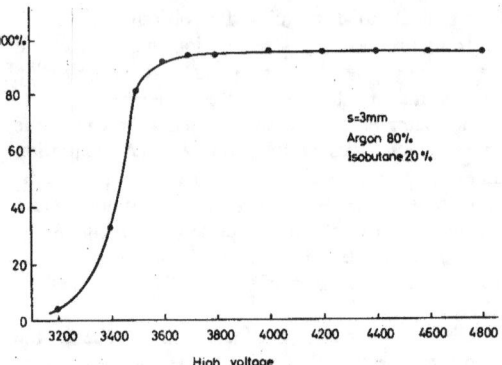

Fig. 13. High-voltage plateau. Number of pulses from a wire which exceed a predetermined threshold as a function of high voltage. Beam of β-rays collimated on the wire; argon-isobutane (80/20); $s = 3$ mm. The 3% loss on the plateau is only apparent. It corresponds exactly to the counting of γ-rays from the collimated β-source by the monitoring scintillation counter.

in the gap before they are collected at the central wire, the angle of the particle relative to the chamber, the distribution in size and position of the ion-electron pairs produced in primary ionization process, and δ-rays, can introduce incertainties in the position determination of comparable magnitude. Cylindrical proportional counters have been operated successfully with high-pressure gases[16,17] or with liquid argon[18,19]. We can expect that this should also be the case for the MWPC. In this case, much smaller gaps and wire spacings may be possible[20].

8. Multigrid structures

For many of the applications discussed in the different sections of this article, we used structures with many grids and different configurations. They were of the types represented in fig. 2. The first type (fig. 2b) was used to separate the amplification region from the drift region, as discussed in the previous section, in order to obtain higher gains at lower total voltages. The second type (fig. 2d) was used to obtain a large gap $A_1 G$, working at moderate voltages. This facilitated the sampling of the energy loss of particles parallel to the electrodes, as described by Amato et al.[6]). It can also be used to absorb X-rays with a high efficiency. The electrons are drifted to the amplification region through the grid G. Usually the energy is shared among several wires, but by using the induced pulse on the grid G or on the electrode A_2, one obtains the total energy.

Fig. 14 shows the variation in the delay of the arrival of the pulses as a function of the applied drift

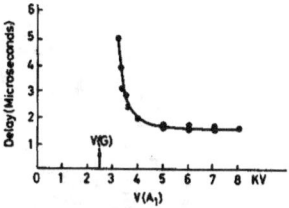

Fig. 14. Drift delay as a function of field. Beam of particles parallel to the electrodes in the drift space of a chamber of type shown in fig. 2d. $D = 10$ cm. Drift field $[V(A_1) - V(G)]/10$ kV/cm. The beam traverses the chamber at 4.3 cm above G. Drift time from G to W ≈ 0.5 μsec. Argon-progane (95/5).

voltages. We see that in argon-propane, 200 V/cm give a saturation of the drift velocity at atmospheric pressure.

In the third type (fig. 2c), the electrode A_2 close to the wire plane is at a potential corresponding to the equipotential symmetric with respect to W. It is made of a plane of wires for studying the induced pulse. Our results show that one can foresee an electrode A made of printed "sensing electrodes" of a shape adapted to particular problems. For instance, concentric printed circles would give the angular distribution around an axis.

The effect of a grid in symmetric chambers of type (b) and of asymmetric chambers of type (c) is represented in fig. 15a, b, c.

9. Drift chamber

This method has been developed by T. Bressani et al.[7]. A drift space was added to a conventional multiwire proportional chamber. The electrons liberated in the gas of the drift space move under the influence of a uniform electric field, pass through a grid into the proportional chamber, and are detected there. The time taken to cross the drift space gives a measurement of the position. This method of coordinate determination should have an accuracy limited by:

1. the inherent width of the track due to δ-rays and scattering;
2. diffusion of the electrons in their flight;
3. timing inaccuracies.

At present, effects (2) and (3) dominate.

Fig. 15. Characteristics of multigrid chambers. 5.9 keV X-ray source; symmetrical chamber of type shown in fig. 2b, $L = 2$ mm, $D = 5$ mm. (a) Pulse height as a function of the control grid potential: $V(G_1) = V(G_2)$; (b) pulse height as a function of the outer electrode potential: $V(A_1) = V(A_2)$. Asymmetrical chamber of type shown in fig. 2c, $L' = 2$ mm, $L = 7$ mm. (c) Pulse height as a function of the closer electrode potential: $V(A_2)$.

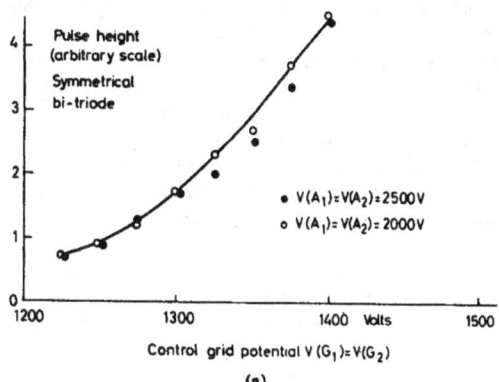

(a)

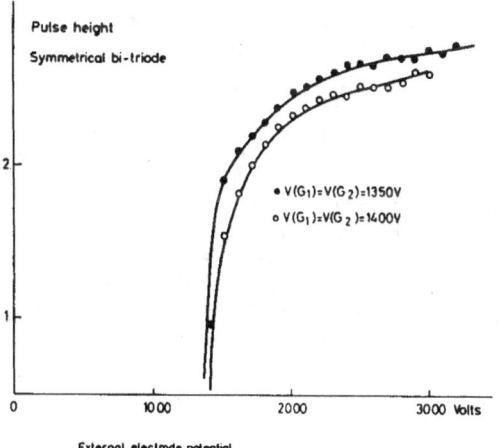

(b)

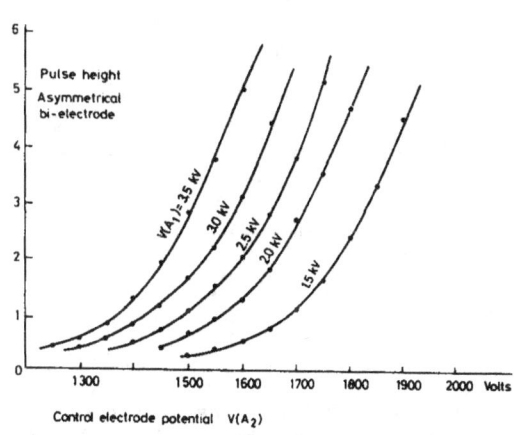

(c)

The chamber is of the type illustrated in fig. 2d. The drift space is 3 cm high × 12 cm × 12 cm. The proportional chamber is 1.4 cm high, with the wires in the centre. A grid with a transparency of 90% separates the two sections. The proportional chamber was constructed with the wires spaced at 2 mm intervals.

The beam passed through the drift space, leaving a trail of ions and electrons. The electrons moved perpendicularly to the beam under the influence of the drift-space field, passed through the grid, and were detected in the proportional chamber. Scintillation counters, 2 mm high, defined the beam and provided the zero of time T_0. The time from T_0 until signals appeared in the proportional chamber was digitized. The chamber was divided into three parts, each of which consisted of five adjacent wires in parallel.

The resolution was determined by measuring the width of the distribution of the quantity ΔT defined as:

$$\Delta T = \tfrac{1}{2}[T(1) + T(3)] - T(2),$$

where $T(1)$ is the time-of-flight to the first detector, $T(2)$ that of the middle detector. Thus, the single chamber was treated as though it were three independent chambers.

Each set of wires had an amplifier whose output went to a discriminator, the output of which provided the stop signal for a time-to-pulse-height converter. The start signal was given by the scintillation coincidence counters. The output pulses from the time-to-amplitude

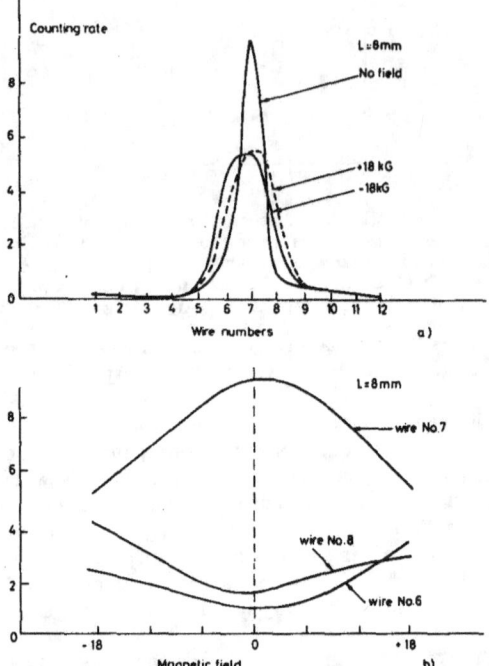

Fig. 17. Effect of a magnetic field: field parallel to the wires; $s = 2$ mm, $L = 8$ mm, hv = 4030 V; argon-isobutane (80/20); collimated beam of 5.9 keV X-rays, on wire no. 7. a) Counting rate in 12 adjacent wires at zero field and at 18 kG; b) counting rates in the three central wires as a function of the field.

converters were digitized and recorded on magnetic tape.

The drift velocities were measured by displacing the chamber a known distance, and measuring the corresponding changes in drift time. The gas used in these experiments was 3% propane—97% argon. At 100 V/cm drift field, the velocity was 2.5×10^6 cm/sec and the space resolution near the grid was ± 0.1 mm, and at 2 cm distance, ± 0.2 mm. At 600 V/cm the velocity was 4×10^6 cm/sec, the space resolution near the grid was ± 0.1 mm, and ± 0.13 mm at 1.4 cm. Increasing the drift voltage to 800 V/cm reduced the drift velocity by about 10% without changing the resolution significantly. The best time resolutions corresponded to 5 nsec (fig. 16).

The main contribution to the width of the resolution seems to have been the diffusion of electrons during their drift. It is expected that if we were to go to still higher drift voltages the diffusion of the swarm would be reduced, and since in most organic gas—argon

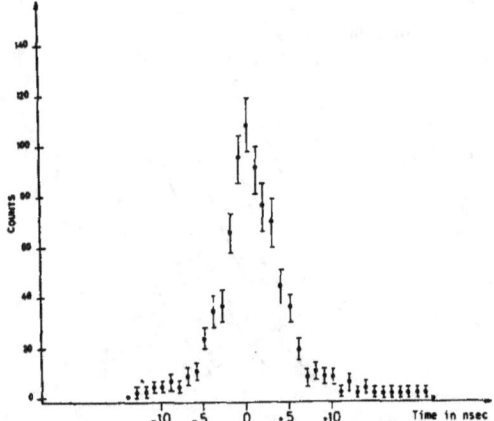

Fig. 16. Time-space correlation in a drift chamber. Beam of particles parallel to electrodes in the drift space of a chamber of type shown in fig. 2d. Distribution of $\Delta T = \tfrac{1}{2}(T_1 + T_3) - T_2$, where T_1, T_2, T_3 are the times of detection on three groups of wires, 1 cm apart; fwhm ≈ 5 nsec.

mixtures the drift velocity eventually becomes constant or even decreases, we would expect an improvement in measuring the space resolution. This should be pushed to a much higher field than we used. Other gas mixtures might have more favourable ratios of diffusion-to-drift velocity.

10. Effects of magnetic fields

We have tested the performance of MWPC's in magnetic fields up to 45 kG. When the wires are parallel to the direction of the field, the space resolution suffers in proportion to the applied field. The over-all efficiency of the chamber remains unchanged. We observed no effect when the electric field of the chamber was parallel to the magnetic field. The amplifiers worked well in the magnetic fields used in our tests.

Tests were made with chambers having wire spacing of 2 mm, gaps of 4 and 8 mm, irradiated by a well-collimated ^{55}Fe source. Such a source produces in the gas, photoelectrons of a range smaller than 1 mm in the gas. In fig. 17 we show the counting rate in each of 12 adjacent wires, with and without a magnetic field of ± 18 kG for a chamber having an 8 mm gap. Without the field, the source is quite well centred on wire 7. However, when the field is turned on, the counting rate in wire 7 decreases to about 60% of its value at zero field, whereas the counting rates in wires 6 and 8 show a marked increase. The electrons drift in the direction $\mathbf{E} \times \mathbf{B}$ before they are collected at the central wire. The drift distance is proportional to the time it takes to collect the electrons. Those originating farthest from the central wire plane drift the greatest distance. The direction of this drift is opposite for electrons produced on either sides of the central wire plane. Empirically we find that the drift velocity varies linearly with the field for $B \gtrsim 3$ kG, as is shown by the approximate expression:

$$\omega = 0.9\,B(\text{kG}) \times 10^5 \text{ cm/sec.}$$

Fig. 17b shows the counting rates in wires 6, 7 and 8 as a function of the magnetic field. Again one sees clearly how the counting rate in wire 7 decreases, while the counting rates wires 6 and 8 increase. The total counting rate remains constant. The slight asymmetry in counting rate for the two field directions shown in the curves in fig. 17a stems from the fact that slightly more X-rays convert in the first half of the chamber than in the second half. With a chamber having a 4 mm gap, the counting rate on the "central" wire as a function of the magnetic field up to 45 kG is shown in fig. 18. The results are completely consistent with the results obtained with larger gaps and smaller fields.

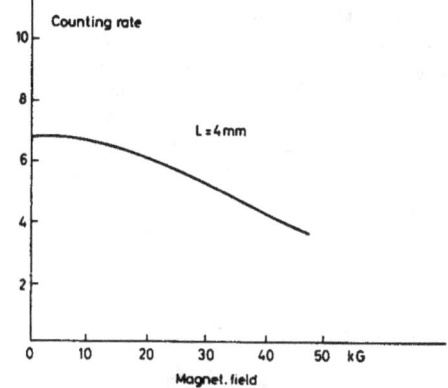

Fig. 18. Effect of a magnetic field. Field parallel to the wires, produced by superconducting coils; $s = 2$ mm, $L = 4$ mm, hv = 2580 V; argon-isobutane (80/20); collimated beam of 5.9 keV X-rays; counting rate as a function of the field.

The conclusions to be drawn from these studies is that for charged particles and normal chambers the loss of space resolution in magnetic fields should not be very serious. With a typical collection time of ≈ 25 nsec for the electrons responsible for the leading edge of the proportional pulse, the displacements in 20 kG will be about $\frac{1}{4}$ mm. In smaller fields the displacement will decrease proportionately. For drift chambers, however, the magnetic field effects will be important.

11. Operation beyond the proportional region

While working with a chamber with 40 μm wires spaced 3 mm apart, we observed peculiar output signals as we increased the applied voltage. We could obtain signals of several volts when we used a mixture of argon and isobutane (80% and 20% as read by the pressure drop flowmeters calibrated for air). These phenomena were investigated for a variety of wire diameters and spacings, but we have only scratched the surface as far as understanding or cataloging is concerned. We will summarize our principal findings.

11.1. INFLUENCE OF WIRE SIZE

We have not tried wires of diameters below 20 μm because of the difficulties of mechanical handling and soldering. With gold-plated molybdenum wires of 20 μm, we have obtained good proportional pulses for many spacings, including 1 mm. With wire of this size, as we increase the over-all voltage we get into continuous discharges before very large pulses occur. For 30 μm wires spaced at 3 mm and with 7 mm wire to mesh, we can obtain semi-proportional pulses of

≈ 200 mV for 6 keV γ-rays, but again we have a continuous discharge before obtaining large pulses. With 40 μm wires we begin to see new phenomena. With a spacing of 12 mm between wires and 7 mm from wire to mesh, we obtained a discharge that gave pulses of several volts and was sensitive to ultraviolet light. These pulses did not discharge the entire chamber and so the dead-time was small. With a spacing of 6 mm between wires and an 8 mm wire to mesh distance in a chamber 19 × 19 cm and with 4000 V applied, we could obtain signals of 0.8 mA directly from a wire into a 50 Ω load. Fig. 19a shows the pulses obtained using a 6 keV γ-source, and fig. 19b those using a β-source (^{186}Ru).

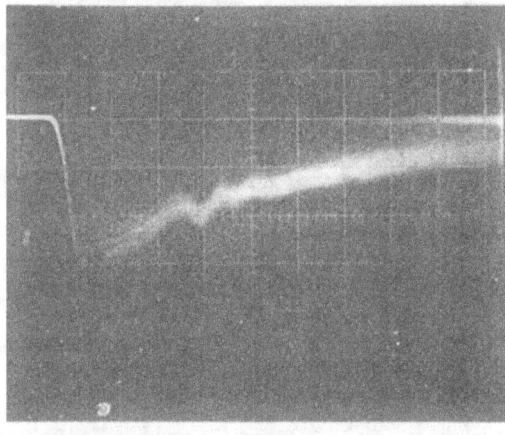

(a)

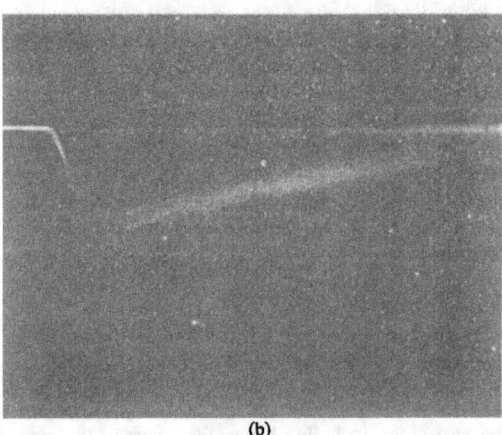

(b)

Fig. 19. Operation beyond the proportional region; $s = 6$ cm, $L = 8$ mm, $d = 100$ μm. Chamber of 19 cm × 19 cm, 4 kV, argon-isobutane (80/20); Signal into 50 Ω; vert. scale 20 mV/div., horiz. scale 10 nsec/div. (a) Pulses from 6 keV X-rays, (b) pulses from β-rays.

The signal, like that from a Geiger-Müller tube, was essentially independent of the energy deposited, and it was not a discharge that dumped all the stored energy. The energy dumped was limited and there was no obvious dead-time. The rise-time for the current is less than 5 nsec, but the time jitter was about 70 nsec, as could be expected for such large spacings. The gas gains involved are of the order of 10^8.

11.2. SPARK MODE OF OPERATION

A chamber 15 cm square with 7 mm wire to mesh spacing was equipped with 11 wires, 100 μm in diameter, spaced at 3 mm, and 2 edge wires of 300 μm, spaced 1.5 mm from the adjacent wires. Proportional amplification was seen, and as the voltage was further increased the chamber became a spark counter, with a bright flash accompanied by a clearly audible noise. (Even if it were useful for nothing else, it would make a good demonstration counter since it requires only a dc supply with no other electronics.)

The localization of the spark along the trajectory was quite good. Fig. 20a, b, c shows the pattern of flashes when a line source of 6 keV γ-rays was placed at different orientations from the wires. The edge wires were sensitive and collected electrons from a rather long distance.

Fig. 20d shows a photograph of the chamber with a patterned brass absorber located behind the chamber, and the pattern of sparks with the light off and the high voltage on. The gas mixture was 80% argon and 20% isobutane, with 6400 V applied. The spark discharged the entire chamber in about 20 nsec with a delay-time of 4 to 20 μsec after the passage of the particle. The capacitance dumped was 180 pF, and the series resistor had to be 1.5 MΩ to avoid burning out a wire. This chamber worked for many hours with ≈ 10^7 sparks without any trouble. When CO_2 was substituted for isobutane, the chamber worked at first, but burned out a wire after a few minutes.

Another chamber with 8 mm mesh to wires distance was tried with 100 μm wires at 3 mm spacing. It also worked, but not as stably as the 7 mm chamber. At wire separations of 1, 2, and 4 mm (all 100 μm wires) we could *not* obtain spark counter action. However, we were able to make a triode chamber with 2 mm between the central wires and the screen of orthogonal wires. The central wires of 40 μm were spaced at 2 mm. It worked as a spark counter. The energy discharged was less, and the sparks were smaller and less audible.

We have not pushed these spark counters to the point of reliable operation, but are merely indicating what looks like a promising field for investigation.

(a)

(d)

(b)

(e)

Fig. 20. Operation in the spark region. Same chamber as in fig. 19, 6400 V. (a, b and c) Collimated beam of 6 keV X-rays oriented at different angles, (d) patterned brass absorber in front of a 6 keV X- ray source, (e) the same picture taken with light.

12. Construction problems

The construction problems are in several respects different from those of wire spark chambers.

1. It is essential that no regions should exist with fields higher than those existing around the amplifying wires. If such regions exist they could cause breakdown, which would give rise to local pulses that are many orders of magnitude greater than the smallest proportional pulses and induce, through capacitive coupling, undesirable parasitic pulses. This is why capacitive filtering on the high-voltage electrode is often necessary in order to reduce the effect of corona breakdown that often occurs at edges of electrodes.

2. Guard strips offer useful protection against un-

(c)

wanted breakdown by collecting surface leakage currents. Solutions different from the one used by us can be adopted: for instance, a high-voltage electrode made of thick wires orthogonal to the central wires, and not extending to the edge of the frame; or thin wires or printed strips placed very close to the wires, along the dielectrics. The advantage of having these guard strips is that if accidental breakdown occurs, there is less risk of the chamber being damaged.

3. Attention must be given to the last wires. Successively thicker wires may be insufficient. The best solution is probably to have the last wire coming close to a flat metal strip, which could be a printed circuit in the median plane of the frame.

4. Special problems arise when large-size chambers are built. These have been studied in detail at CERN, by the CERN-Heidelberg group[21]). The configuration with wires in the median plane of the two high-voltage electrodes is not a stable one. The electric forces tend to displace the thin wires in alternative directions. Simple calculations give the following relation between the tensile force T on the wires, the length l of the wires and the charge per unit length q:

$$T = q^2 l^2 / s^2.$$

For a 2 m chamber, at 3 kV this corresponds to a tension of about 100 g that exceeds the strength of a 20 μm molybdenum wire.

If the tension is too small the wires move out of the median plane and give rise to sparks.

To prevent these displacements it appeared practical to place strong strands of nylon every 10 cm, perpendicular to the central wires, thus preventing any displacement outside the medium plane*. Chambers of 1 m in length were therefore built using molybdenum wires of 20 μm, with a tension of only 12 g, whereas the breakdown tension is 50 g. These chambers worked properly. The advantage of this low tension is that the mechanical forces on the frames are lower than in wire spark chambers where 100 g is usually applied to each wire. In spark chambers, wires of very small diameter are impossible because they could be destroyed by the energy of the sparks.

5. Proportional chambers with aluminized mylar electrodes can be constructed. This also is unpractical for spark chambers. This fact, together with the fact that thinner wires can be used, should in essential cases constitute an important gain in the weight factor with respect to the thinnest wire spark chambers.

* A loss of efficiency appears around the nylon wires. This can be corrected by another string of metal stretched against the nylon wire and set at a potential equal to or higher than the equipotential corresponding to the plane of that string.

6. It is of great importance to avoid spurious breakdown in the chambers. One should operate the chambers at the minimum voltage compatible with 100% efficiency. An additional safeguard can be achieved by having the high-voltage electrode made of wires connected to the high-voltage supply through an individual high resistance. In this way, only the very small local capacity gets discharged in a spark. We have operated such chambers with a strip of conductive rubber serving as a resistance which isolated one wire from another. A large reduction in the spark energy was thus obtained.

We have operated chambers for months in an intense beam, without any signs of destruction of wires or amplifiers.

13. Some applications

13.1. BEAM PROFILE ANALYZERS

In high-energy physics the first application of the chambers has been to measure the space distribution of secondary particle beams. For example, at CERN several groups have used such chambers to obtain various parameters of a beam's phase space in a single machine burst. The advantages of such devices for rapid beam-tuning are obvious.

13.2. HODOSCOPES

The chambers can be used to replace scintillation hodoscopes whenever time resolutions of the order of 25 nsec are sufficient. With staggered chambers it is possible to improve significantly both the space and time resolutions.

A brief comment is in order here about how to process the pulses from the individual wires. The detailed arrangement of logic elements will be different for different experiments. However, it seems likely that in many systems the electronic circuits associated with each wire should perform part or all of the following functions: amplification, fast signals for logical decision-making, delayed signals for gating by decision-making circuits, gates, storage. A great variety of electronic methods already exists which could be used to deal with pulses from the wires. Unfortunately, they are all still quite expensive. For small systems of less than a few hundred wires the existing circuits are practical. On the other hand, when large systems containing many thousands of wires are involved, the cost of these circuits may become prohibitive. The success of large-scale systems is thus very closely tied to the development of less expensive electronic devices capable of performing the desired functions reliably. Systems with

5000 wires and 50000 wires are under study, but will not be discussed in this article.

13.3. ANTICOINCIDENCE COUNTERS

The low mass of the chambers is at least one order of magnitude smaller than that of scintillation counters (of a thickness sufficient to detect minimum ionizing particles with 100% efficiency).

This makes them ideal anticoincidence counters in front of gamma or neutron detectors.

13.4. DETECTORS IN THE FOCAL PLANE OF SPECTROMETERS; USE OF CODING METHODS

Such chambers have a considerable advantage over spark chambers since they do not need an external trigger. They are particularly well suited for the detection of very low energy particles. With an additional grid, accuracies of less than 1 mm can be achieved, especially if heavily ionizing particles are to be detected. When only a single particle has to be measured, as is often the case, a simple coding method such as the one described in section 5 can well be applied.

13.5. DECISION-MAKING FOR SPARK CHAMBER TRIGGERING

This is probably one of the most important applications of the MWPC. The measurement of the number of charged particles traversing a counter, or rough kinematical measurements, can be made within the memory-time of spark chambers. Such systems have recently been developed at CERN by Amato et al.[22] (a series of chambers gives the position and the angle of trajectories in times shorter than 0.4 μsec at rates up to 10^5/sec) and by Bemporad et al.[8] (to give the number of particles).

13.6. X-RAYS AND NEUTRON MAPPING; MEDICAL APPLICATIONS

Spark chambers have been used in the past to obtain the spatial distribution of γ-rays emitted by human bodies containing radioisotopes. The great difficulty was the triggering. The electrons released in the chambers by the γ-rays cannot escape from the chamber to be detected by a triggering scintillation counter. This was overcome in a clever way by Lansiart et al.[23], who built a self-triggering spark chamber, making use of proportional amplification between an electrode and a grid that was transparent to electrons. The weakness of this method is that controlled amplification between plane electrodes is a very touchy technique that does not lend itself easily to industrialization. It seems that proportional chambers open up an interesting field of research for such applications. The data appear in a digitized way, which lends itself to more direct methods of analysis than the photographic ones.

Two problems arise with respect to these applications:

1. the two coordinates of the avalanches have to be measured simultaneously;

2. the efficiency of gaseous layers is small for thin layers.

The first problem can be solved by using the induced signal on a grid made of perpendicular wires, as discussed in section 5. The second problem can be solved by using a separate drift space, at high pressures, for the absorption of the γ-rays. Even large thicknesses, such as 10 cm of argon or xenon at 10 atm, are feasible, and will lead to detectors with an efficiency somewhat lower but still comparable to scintillation counters for X-rays in the few hundred keV region. This is of particular interest, since it opens up the possibility of building large-surface detectors, which are needed for some problems.

Recently, Dolgoshein[24] reported that electrons liberated in liquid argon can be drifted out from the liquid argon into the gas. Layers of liquid argon with a MWPC to localize the extracted electrons would then be an excellent γ-ray spatial detector.

A similar problem arises for neutron detection. The particles liberated in the gases can be detected, and the induced pulse method permits two-dimensional read-out.

13.7. THREE-DIMENSIONAL AVALANCHE DIGITIZED CHAMBER

The work undertaken with drift chambers has shown that it is possible to measure the coordinate perpendicular to the electrode plane. The work on the induced signals has shown that two dimensions can be obtained from a gap containing a single layer of amplifying wires. These features can be combined together to give a three-dimensional chamber. Let us consider a structure of type d, fig. 2, with grid G made of wires orthogonal to the wires W, while A_2 is made of wires at some angle to these wires. If the zero time is given by a scintillation counter, the time of arrival of the pulse at each wire gives the Z coordinate, while the coordinates X and Y are determined by the amplifying wires W and the electrodes G and A_2 with wires collecting the induced pulse.

From our experience we may say that such a method is difficult because of the small size of the induced signal. In order to achieve good accuracy, small spacing between the amplifying wires is necessary, and this leads to small energy loss for minimum ionizing particles parallel to the electrode. This situation could be improved considerably by applying a very high pulsed field in the drift space in coincidence with the passage of the particles. This is the method used in streamer chambers. However, in the application discussed here, streamers need never be formed. If one amplifies the ionization by a value somewhat below the critical value of 10^8 at which streamer propagation starts, only small additional amplification is required from the wires, and good spatial resolutions are likely. There are other advantages: the avalanches can be more confined than streamers; the control of their growth is less critical, since the propagation velocity of avalanches is 10 times smaller; the chamber will be insensitive to particles arriving after the field is pulsed. The building of such a chamber is certainly a difficult task, but so is the evaluation of streamer chamber pictures when large numbers are involved. At high pressures such a chamber may have advantages over streamer chambers, since less gaseous amplification is required in the uniform field region.

14. Conclusions

The experience gained in the last year has confirmed that the MWPC can be made a reliable tool with the following characteristics for the detection of minimum ionizing particles:

- Space resolution: full width of 2 mm (with multigrid chambers, 1 mm is easy to achieve).
- Time resolution: 25 nsec for 100% efficiency for 2 mm wire spacing. 18 nsec for 100% efficiency for 1 mm wire spacing.
- By using the correlation between position and time, space resolutions of 0.2 mm (fwhm) have been achieved, as well as a time resolution of 5 nsec (fwhm).
- The study of the positive pulses induced by the motion of positive ions shows that two coordinates can be measured from a single gap, and that the signals in the cathode can be of practical use to measure total energy losses and to give signals for fast timing or for coding purposes.
- A scheme for a three-dimensional digitized chamber is suggested incorporating many new features observed in MWPC.
- Operation of the multiwire chamber beyond the proportional region shows some promising results.

We acknowledge with thank the help of physicists at CERN, who have been working on the same problems and from whom we have gathered much information. Among them we should mention particularly members of the CERN-Heidelberg Group and the CERN-ETH Group.

We should like to thank G. A. Erskine for calculating the field in chambers with displacements of the grid and individual wires, and for providing extensive computer-generated field plots. Particular thanks are due to Mr. G. Amato and his electronics group for their constant help in solving the problems encountered in the various measurements described in this article. Mr. Y. Desclais made valuable contributions to the research on the multigrid structures and time resolution of the chambers. Mr. R. Bouclier has contributed to most of the measurements and tests involved in this article while Mr. G. Million built several prototypes.

We are grateful to Mr. G. Muratori and R. Benoit, for many inventive contributions to the construction of the prototypes.

References

[1] G. Charpak, R. Bouclier, T. Bressani, J. Favier and Č. Zupančič, Nucl. Instr. and Meth. 62 (1968) 235; 65 (1968) 217.
[2] G. Charpak, Evolution des techniques de chambres à étincelles en 1968, Proc. Int. Symp. Nuclear electronics 3 (Versailles, Sept. 1968) p. 1.
[3] J. Fisher and S. Shibata, Proc. Int. Symp. Nuclear electronics 3 (Versailles, Sept. 1968) p. 3.
[4] E. Epple and D. Decker, Nucl. Instr. and Meth. 66 (1968) 70.
[5] R. Allemand, CEN Grenoble, private communications.
[6] G. Amato, R. Bouclier, G. Charpak, D. Rahm and H. Steiner, Some research on the time, space, and energy resolution with multiwire proportional chambers, Proc. Dubna Meeting Filmless and streamer chambers (April 1969).
[7] T. Bressani, G. Charpak, D. Rahm and Č. Zupančič, Track localization by means of a drift chamber, Proc. Dubna Meeting Filmless and streamer chambers (April 1969).
[8] C. Bemporad, W. Beusch, A. C. Melissinos, E. Schuller, P. Astbury and S. G. Lee, Performance of a system of proportional wire chambers, to be published in Nucl. Instr. and Meth.
[9] L. S. Koester, R. M. Brown, T. Clark, S. Segler and R. Taylor, Wire proportional counter arrays with fast digital arithmetic for decision-making, Contribution Ispra Conf. Nuclear electronics (May 1969).
[10] G. Amato and G. Petrucci, CERN Report 68–33 (1968).
[11] G. C. Hanna, D. H. Kirkwood and B. Pontecorvo, Phys. Rev. 75 (1949) 985.
[12] D. H. Wilkinson, Ionization chambers and counters (Cambridge University Press, Cambridge, 1950).
[13] S. A. Korff, Electron and nuclear counters, 2nd ed. (D. van Nostrand Co., New Jersey, 1955) p. 186.
[14] D. Buneman, T. E. Cranshaw and J. A. Harvey, Can. J. Res. A-27 (1949) 191.
[15] G. Charpak, L. Dick and L. Feuvrais, Nucl. Instr. and Meth. 15 (1962) 323.
[16] H. W. Fulbright and J. C. D. Milton, Phys. Rev. 76 (1949) 1271; 82 (1951) 274.
[17] R. S. Wilson, L. E. Beghian, C. H. Collie and H. H. Halban, Rev. Sci. Instr. 21 (1950) 699.

[18]) G. Hutchinson, Nature 162 (1948) 610.

[19]) N. Davidson and A. Larsh, Phys. Rev. 74 (1948) 220; 77 (1950) 706.

[20]) L. Alvarez, Lawrence Radiation Laboratory Int. Note (Nov. 1968).

[21]) T. Trippe, CERN NP Int. Report 69–18 (June 1969); J. H. Dieperink, K. Kleinknecht, P. Steffen and F. Vannucci, CERN NP Int. Report 69–25 (Aug. 1969).

[22]) G. Amato, E. Chesi and H. Steiner, Fast read-out of spatial and angular distributions of charged particles with multiwire chambers, in preparation.

[23]) A. Lansiart, J. P. Morucci and G. Roux, IEE Trans. Nucl. Sci. NS-13, no. 3 (June 1966) 393.

[24]) B. A. Dolgoshein, Proc. Dubna Meeting *Filmless and streamer chambers* (April 1969).

NUCLEAR INSTRUMENTS AND METHODS 88 (1970) 149–161; © NORTH-HOLLAND PUBLISHING CO.

INVESTIGATION OF SOME PROPERTIES OF MULTIWIRE PROPORTIONAL CHAMBERS

R. BOUCLIER, G. CHARPAK, Z. DIMČOVSKI, G. FISCHER and F. SAULI

CERN, Geneva, Switzerland

G. COIGNET

Institut de Physique Nucléaire, Orsay, France

G. FLÜGGE

II. Institut für Experimentalphysik, Hamburg, Germany

Received 23 June 1970

This article describes a systematic study of the efficiency, space resolution and multiparticle separation in multiwire proportional chambers. For a variety of gas mixtures, results were obtained as functions of high voltage, time resolution and incidence angle of particles. A new gas, argon + isobutane + freon-13 B1 permits a considerable gain in amplification without entering into the Geiger or spark region.

1. Introduction

Since the observation of the independent detection of proportional pulses by each wire in a multiwire chamber, the properties of this new device have been studied by several groups[1]).

Among the physical properties of multiwire chambers which have to be known with accuracy, we may list the following:

— efficiency
— time resolution
— space resolution
— multiparticle separation (cluster size).

These quantities, however, depend on quite a number of parameters such as:

— mechanical construction and dimensions
— wire electronics
— gas mixture
— high voltage
— timing (width of memory gate)
— angle of incidence of particles.

In this paper we want to concentrate on the last four parameters, reporting results obtained on chambers of moderate size with fixed geometry and electronics.

As far as mechanical construction and electronics are concerned, today one encounters some major problems on the following points:

1. The electrostatic forces introduce stability problems[2,3]) which do not exist in pulsed chambers. This leads to difficulties in the construction of large chambers which we are not going to discuss in this article.

2. The construction of electronic circuits of low cost to treat the information flowing from the chambers is still a field of research. Several prototype circuits have been designed and the construction of the circuits equipping about 5000 wires has been undertaken by several groups. It is too early now to judge the respective merits of these circuits which are different in many respects.

Concerning the problem of gas mixture, one may state that practically any gas will give rise to proportional amplification around a wire, even if it presents strong electron attachment. It is sufficient that some electrons liberated along a track survive for a time long enough to be drifted to the region of high gradients around the wire, where the mean free path for ionization is, in general, much smaller than the mean free path for capture. However, considerable differences arise among the gases where some given property is required like a large amplification or a linear response to the energy losses, a high efficiency for small gaps or a good amplification with small wire spacings. Some relevant properties of the gases in this respect are the density, the value of the characteristic energy levels for inelastic collisions with electrons, the position of the photon absorption bands, the probability of electron attachment, or the variation of the first Townsend coefficient with the electric field.

The considerable possible choice of gases makes an exhaustive study impossible. We have therefore limited our study of gases to the wellknown mixtures of argon + CO_2 and argon + isobutane. In addition we will present the properties of some special gas mixtures, which increase the limit of proportional amplification set by space charge up to a factor of hundred. Although it may be too early to decide whether it would be wise to adopt these gases in large systems, they offer quite interesting properties for various applications.

2. Experimental procedure

2.1. CHAMBERS, ELECTRONICS, EXPERIMENTAL SET-UP

The chambers used in our measurements have a useful area of 10×10 cm^2 and are fairly similar to those constructed before[4]).

The negative high voltage is applied to two stainless steel mesh planes 16 mm apart from each other; gold-plated molybdenum wires of 20 μm diameter are located in the median plane at 2 mm spacing; guard strips held at ground potential 1 mm from the wires provide protection against breakdown at the dielectric frame made of araldite; 120 μm Mylar windows are placed 2 mm outside the mesh planes.

The gas mixtures were controlled by Rotameter* type flowmeters. All the proportions of the gas mixtures quoted in this article are corrected for the differences of density, pressure and viscosity and represent relative volumes+.

We have performed test measurements both in the laboratory, using a β-source, and in a beam of $\sim$ 350 MeV/c charged particles at the CERN synchro-cyclotron.

In the laboratory tests, pulse heights were measured using a Tektronix probe with $R = 10$ MΩ and $C = 7$ pF, the wires being terminated with 100 kΩ.

Measurements of time spectra and plateau curves were made comparing the pulses from ten wires grouped together to the pulses from a 2 mm wide scintillator also irradiated by the source. In this case the wire pulses were amplified by a factor of about

---

* Manufactured by Wisag, Zurich.
+ It should be pointed out that the reading of the rotameters has to be corrected by large factors for gases other than air.

300 in a current amplifier stage with $\sim$ 1 mV threshold and 30 Ω input impedance.

In the beam tests, a set-up of four chambers and two beam-defining scintillators was used (fig. 1).

The second chamber could be turned about a vertical axis to allow for incidence angles of the beam different from 90°. The wires were oriented vertically in all four chambers.

The signals were transported after an emitter follower over 4 m of flat mutiwire cable to the electronics.

The amplifying stage[†] had a gain of about 1000 and a threshold variable between 0.5 and 30 mV. (The initial low level of 150 μV could not be maintained due to noise picked by by the flat cables.)

The amplifier was followed by a gate normally open and a flip flop memory. An event having occurred, the memory gate is closed with an adjustable time delay between 0 and 170 nsec. This type of logics, which avoids introducing a delay line for every wire, has been described by Pagès[5]), and has been implemented using a fast gate generated by the scintillation counters.

The information contained in the memory is then read out to a PDP 8 computer and written on tape. After the read-out cycle of about 10 msec, a memory reset pulse is provided and the memory gate opened again.

2.2. EVALUATION OF EXPERIMENTAL RESULTS: EFFICIENCY AND SPACE RESOLUTION

The experimental information obtained in the beam tests (section 2.1, fig. 1) and stored on tape was processed by means of a track finding program.

Given the barycentres and cluster sizes of up to three clusters per chamber and event (as defined by the

---

† Manufactured by SAIP, Paris.

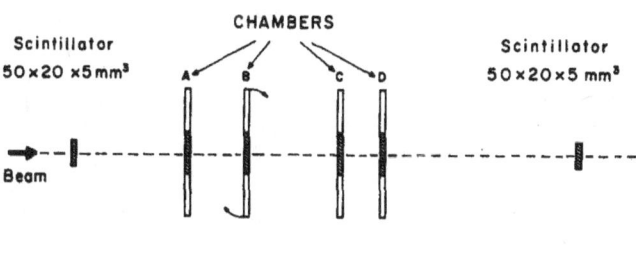

Fig. 1. Experimental set-up for beam tests.

scintillator coincidence), we fitted a straight line through three of the chambers and looked for the position of the "missing" cluster in the fourth one, thus getting efficiency and space resolution for all chambers in turn.

Clearly, efficiency and space resolution are somewhat interconnected in this method, depending on where one introduces cuts both in the χ^2 of the fitted line and in the distance between fit value and observed wire coordinates in the missing chamber. The distribution of these residuals gives, on the other hand, a measure for the space resolution of the chamber.

For tracks orthogonal to the chambers one expects, in principle, that only one wire is detecting the particles where their distance from the wire is less than $\pm\frac{1}{2}s$ (s = wire distance = 2 mm). This corresponds to a rectangular efficiency distribution around the wire with a variance

$$\sigma_{\text{wire}} = s/\sqrt{12} = 0.577 \text{ mm}.$$

The error distribution of the linear fit through the remaining three chambers is, on the other hand, Gauss-like with a variance which is also proportional to σ_{wire}:

$$\sigma_{\text{fit}} = \sigma_{\text{wire}} \cdot a,$$

where a is a correction factor < 1 which depends only on the number and geometrical arrangement of the chambers[6]).

Consequently, the measured residual distribution being a superposition of the efficiency distribution with the fitting error, we expect a variance

$$\sigma_{\text{meas}} = \sigma_{\text{wire}}(1+a^2)^{\frac{1}{2}}.$$

Instead of σ_{meas} we prefer to present the normalized value

$$\sigma_{\text{norm}} = \sigma_{\text{meas}}/(1+a^2)^{\frac{1}{2}},$$

which should be independent of the position and number of chambers and, in the ideal case, should be equal to $\sigma_{\text{wire}} = 0.577$ mm.

3. Investigation of some properties of argon + isobutane and argon + CO$_2$ fillings

The first mixture offers the following advantages. The isobutane has a high density. The energy loss is 5.6 keV/cm at TPN, against 2.5 keV/cm for argon and 3.3 keV/cm for CO_2. The amplification can be pushed to the highest values so far observed in cylindrical proportional counters, about 10^6. The variation of the gain as a function of voltage is comparably small. This gives a relative decrease in the importance of the

wire diameter irregularities or any other cause introducing a local change in the electric field.

The second mixture has several advantages which have led us to choose it as a standard filling gas: low cost, no safety problem, no danger of polymerization endangering the long-term properties of the chambers, and possibility of high-pressure bottles filled with the correct gas mixture.

3.1. PULSE HEIGHT AND PLATEAU

Because of the practical interest of the two gases, we present in fig. 2 the variation of the height of the pulses collected on the wire as a function of the applied potentials*. The decrease in the slope of the curves can reasonably be attributed to the onset of space charge effects. It is accompanied by a loss in the linearity of response with respect to the initial energy loss. The

* It should be stressed that these pulses are produced by the 5.9 keV ^{55}Fe X-ray source.

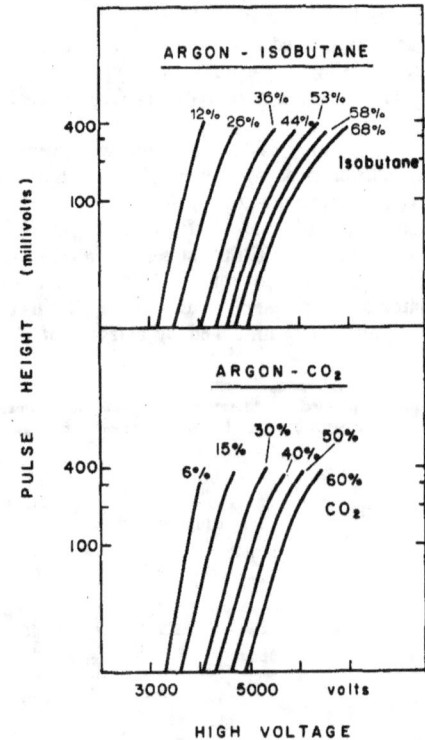

Fig. 2. Pulse height as a function of high voltage and gas concentration for argon + isobutane and argon + CO$_2$ mixtures using X-rays of 5.9 keV. Load 100 kΩ, 35 pF.

energy losses of charged particles being spread over a large spectrum, the highest pulses set the limit which can be reached. They may lead to occasional break-downs or Geiger-Müller pulses. This is another reason for having the smallest gap compatible with a good efficiency, since any unnecessary increase in the gap width leads to unnecessarily high pulses. However, although a thickness of only 2 mm is, in principle, sufficient to have an efficiency above 99%, the diffi-culties of having a constant gap of this value leads one to prefer large gaps, at least for the large chambers usually required in high-energy physics.

In fig. 3 we indicate the position of the efficiency plateau (region of constant efficiency for fast electrons up to the end of the proportional region) and the breakdown voltage as a function of gas concentrations.

In order to maintain a reasonable plateau at moderate high voltages, we chose fixed concentrations of 27% CO_2 and 34% isobutane, in order to perform more refined beam-tests of efficiency, time and spatial resolutions.

3.2. STUDY OF SPACE RESOLUTION, EFFICIENCY AND TIME RESOLUTION

Apparently the simple considerations concerning space resolution presented in section 2 can only be a first approximation to the actual conditions. First of all, it shows up that even with trajectories orthogonal to the chambers the number of clusters with more than one wire touched cannot be neglected. This is in part due to the angular spread of the beam, but especially the big clusters (up to 10 wires touched in a few cases) have to be attributed to the production of δ-rays, which at the same time tend to result in asymmetric clusters with respect to the particle trajectory.

An indication of this effect is given in fig. 4, which shows the distribution of the residuals in the second chamber for a wide time gate. (All results quoted below were obtained by introducing a cut at $\chi^2 = 3.5$ for the fitted line in order to exclude fits through spurious events.

The rather big variance is mainly due to the tails of the distribution. We find about 2% of the events outside the 3 σ limit, up to ± 4 wire distances. Cutting at 3 σ, the variance is reduced to 0.59 mm in fair agreement with the expected value.

Looking more closely, now, at the events with large deviations from the fit, we see that all clusters with more than four wires touched are to be found in the tails of the distribution with residuals larger than 3 σ. This again favours the explanation of the big clusters by means of δ-rays.

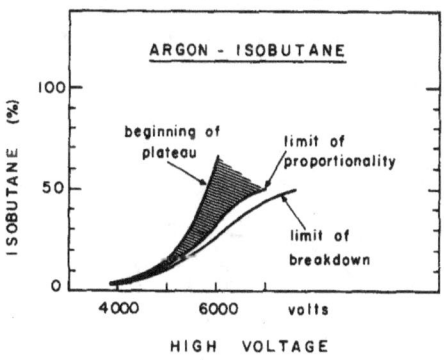

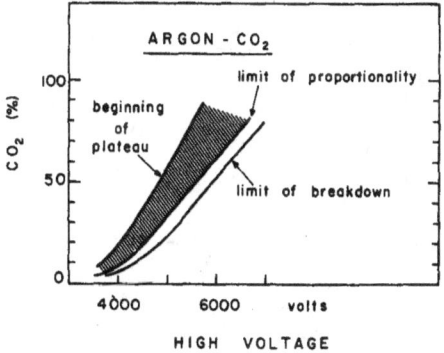

Fig. 3. Position of high-voltage plateau and breakdown voltage as a function of isobutane and CO_2 concentrations; threshold 1 mV.

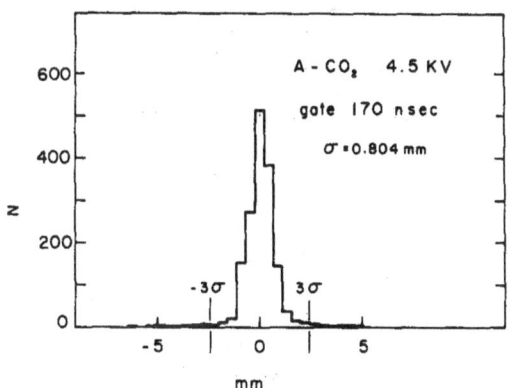

Fig. 4. Distribution of normalized residuals between cluster position and fitted line (three chamber fit).

The results quoted above have direct influence on the numerical value of the efficiency: taking into account all clusters up to ± 8 mm from the fitted trajectory, we find an inefficiency of less than 5×10^{-4} in the middle of the high-voltage plateau, whereas we loose about 2% of particles applying a cut at ± 1.8 mm.

Another interesting aspect is the dependence of space resolution and efficiency on time resolution, i.e. on the width of the memory gate. As shown in fig. 5, the efficiency defined without cut stays constant down to gate widths of about 40 nsec, whereas the number of clusters outside the 3 σ limit goes down and reaches a minimum of about 1% at the same time limit. The variance behaves accordingly, the flat minimum at $\sigma_{norm} = 0.6$ mm is close to the expected value σ_{wire}.

Looking for the percentage of clusters with two and more than two wires touched, we see in fig. 6 a corresponding behaviour with timing.

We conclude that one can reach both good efficiency and good space resolution with a time gate in the order of the time resolution of the chamber. In applications where efficiencies of more than 99% are needed, however, one has to take care of the tails of the cluster distribution around the actual trajectory. Finally, we should mention that the space resolution depends only weakly on the high voltage applied to the chambers. For the gas mixtures used in our tests, it is slightly better at higher voltages with a gradient of $\Delta\sigma/\sigma \approx 3\%$ per 200 V.

3.3. SPACE RESOLUTION AND EFFICIENCY FOR INCLINED TRACKS

For inclined trajectories, we expect that several wires will collect electrons liberated along the track. At large gate widths the mean number of wires touched should be given by the projection of the track on the chamber plane. As shown in fig. 7, there is indeed a strong correlation between cluster size and angle of incidence α. (In the following, α is always defined with respect to the normal on the chamber plane.)

At small gate widths, however, the mean number of wires touched ($\bar{n}$) decreases and approaches a limit at one wire for extremely narrow gates, fig. 8.

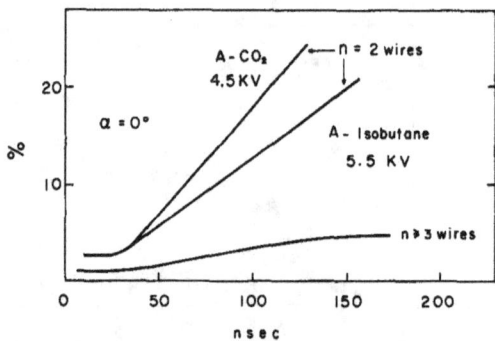

Fig. 6. Percentage of clusters with two and more than two wires touched as a function of gate width at $\alpha = 0°$.

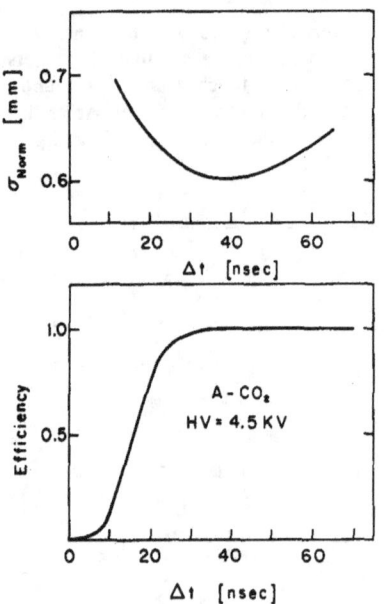

Fig. 5. Efficiency and spatial resolution as a function of time resolution.

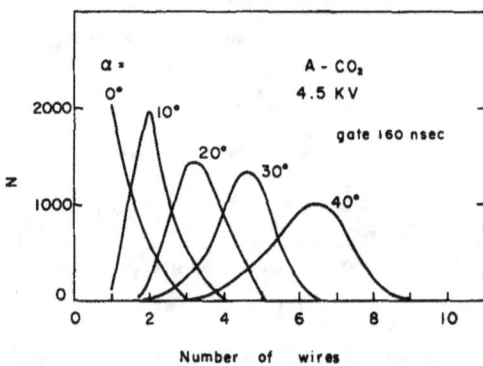

Fig. 7. Distribution of cluster size as a function of angle α.

In addition, we observe a slight dependence of $\bar{n}$ on the high voltage applied to the chamber as shown in fig. 9.

Taking into account the efficiency limit indicated in fig. 8, it appears that even for trajectories inclined at $\alpha = 40°$ only about two wires are touched with narrow time gates. This is a serious advantage over spark chambers, especially in a magnetic field where tracks of any inclination can be expected.

It is very interesting, now, to observe the behaviour of the space resolution for inclined tracks. In principle, one could expect an improved accuracy due to the large numbers of multiwire clusters, especially for wide gates. This improvement depends, however, much on the exact symmetry of the clusters relative to the

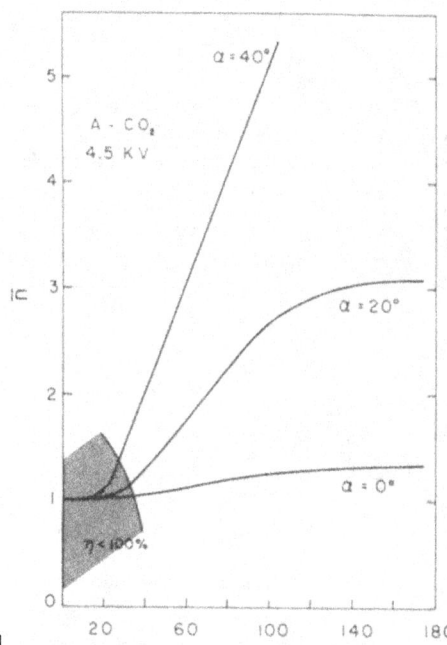

Fig. 8

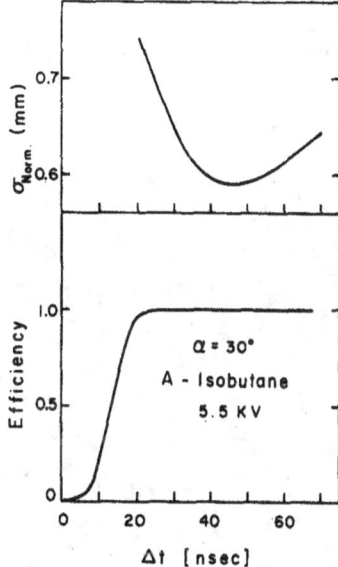

Fig. 10. Efficiency and spatial resolution as a function of time resolution at $\alpha = 30°$.

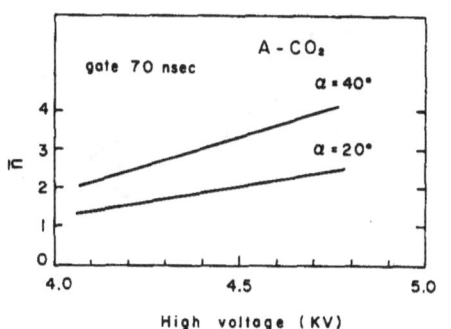

Fig. 9. Mean number of wires touched as a function of angle and high voltage.

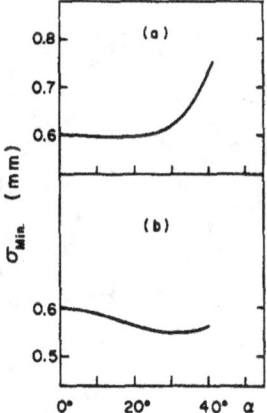

Fig. 11. a. Minimum variance in the chamber plane as a function of angle α. b. Minimum variance projected into a plane orthogonal to the particle trajectories as a function of angle α.

trajectory and is counteracted by the reduced counting probability for the outer wires of big clusters, which may result in an additional "jump" of the barycentre by $\pm \frac{1}{2}$ wire distances.

In the actual measurements, we again find a dependence of σ_{norm} (defined *in* the plane of the chamber) on the gate width with a flat minimum near the limit of the efficiency plateau, fig. 10.

Plotting σ_{min} against the angle of inclination we obtain the results of fig. 11a. The space resolution is about constant between $\alpha = 0°$ and $\alpha = 30°$ and gets worse by about 20% between 30° and 40°. Projecting the variance on a plane perpendicular to the particle

trajectory, we obtain a roughly constant space resolution up to $\alpha = 40°$ due to the additional cosine factor (fig. 11b).

Again the situation is more favourable than in a spark chamber, where the resolution rapidly degrades with angle.

As far as the efficiency is concerned, we find about the same values for inclined as for orthogonal tracks. At $\alpha = 40°$, however, the minimum inefficiency without cut at 3σ is slightly larger, about 10^{-3} instead of 5×10^{-4}.

3.4. TIME RESOLUTION

As can be seen in the curves already presented in figs. 5 and 10, one has to allow for a certain minimum gate width to maintain a high efficiency. This is due to the time resolution of the chambers plus an additional fluctuation from the wire electronics.

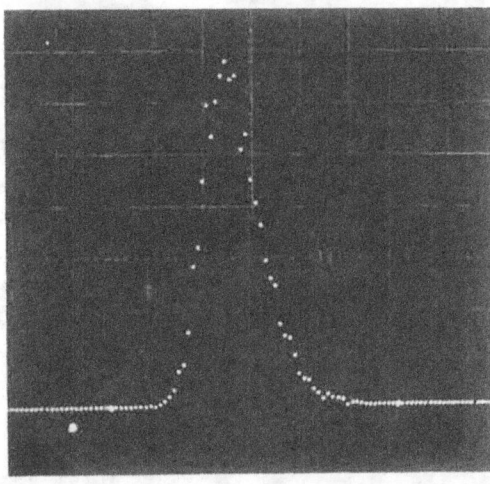

(a)

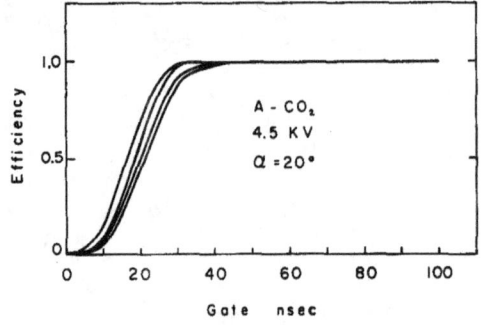

Fig. 13. Efficiency as a function of gate width for a system of four chambers.

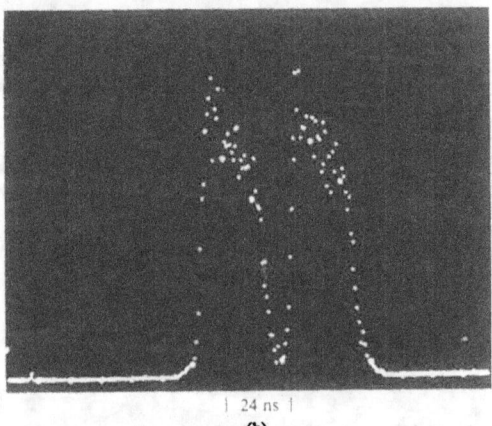

| 24 ns |

(b)

Fig. 12. a. Time spectrum for argon (73%) + CO_2 (27%) at HV = 4.5 kV, b. for argon (67%) + isobutane (33%) at HV = 5.5 kV.

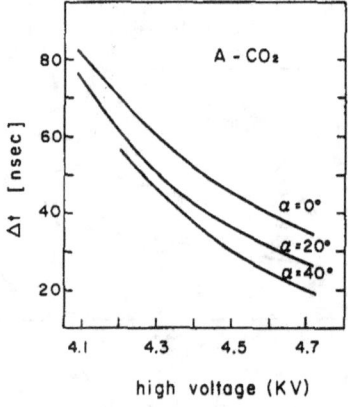

Fig. 14. Time resolution as a function of angle and high voltage.

The time spectra attained in the laboratory tests (section 2) and shown in figs. 12a, b, indicate a width of about 25 nsec for 2 mm wire spacing.

For a system with many wires, however, it is difficult to keep the electronic jitter from wire to wire to a negligible value. In fig. 13 we present a typical efficiency curve for four chambers as a function of gate width. Apparently one has to take into account a jitter of 5 to 10 nsec between different chambers in our case.

We may define the gate width at which the efficiency plateau is reached as the time resolution of a chamber. Plotting this time as a function of high voltage and angle (fig. 14), we observe a particularly steep dependence on high voltage. This makes the comparison of different gases rather difficult. Taking, for instance, the values obtained in the middle of the high voltage plateaux at 4.5 kV and 5.5 kV, respectively, we do not find a difference between argon+CO_2 and argon+ +isobutane fillings.

4. Properties of the chambers with a new, remarkable gas filling

4.1. INFLUENCE OF ELECTRONEGATIVE GASES ON THE CHAMBER PERFORMANCE

Electronegative gases have been used in the past in cylindrical counters and more recently in multiwire proportional chambers by Grunberg et al.[7]). As long as the mean free path for electron attachment is larger than the mean free path for ionization, the avalanches can grow. Two side effects appear: firstly, the pulse height is dependent on the position of the track with respect to the wire; secondly, Van Zonen et al.[8]) have shown that with ethyl bromide added to argon, the efficiency drops to zero at a distance from the wire controlled by the amount of ethyl bromide. Grunberg et al. have used this gas to limit the efficiency to narrow cylinders around the wires. This was of importance for the particular application they had in view, namely the detection of charged particles in the focal plane of a spectrometer where the trajectories are very inclined on the chambers – with ethyl bromide they could limit the counting to a single wire even in this case.

We have investigated the properties of several gases mixed with different types of freon. For a special mixture which we found of particular importance, it seems possible to neutralize the effect of space charge to such an extent that an additional factor of 100 can be gained in the pulse height, without entering into the regions where photon propagation sets in. Our main problem was to maintain this property of neutraliza-

tion of the positive ions by the heavy negative ions, and still to have an efficiency close to 100%.

This particularly favourable gas is a mixture of argon, isobutane and freon-13 B1 (CF_3 Br). It was found in a line of research covering binary or ternary mixture of argon, helium, isobutane, CO_2, freon-12 ($CCl_2 F_2$), freon-13 ($CClF_3$), freon-13 B1 (CF_3 Br), and freon-12 B1 ($CClF_2$ Br).

With this mixture we observed a striking phenomenon: for some given proportions, it was possible to push the gain to unusual values. Even the smallest pulses, corresponding to probably just one ion pair, exceed 50 mV on the wires, corresponding to gains of 10^8 without entering into Geiger operation. The iso-

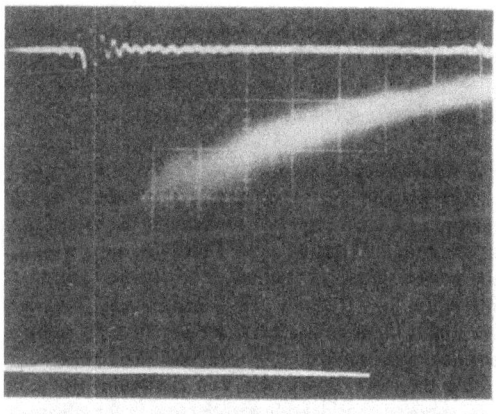

(a)

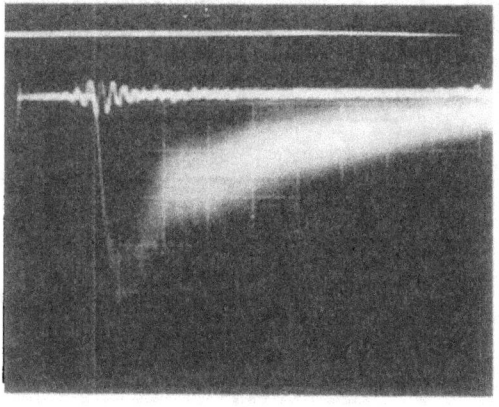

(b)

Fig. 15. Pulse on a wire after an emitter follower. Mixture argon + isobutane + freon-13 B1 (65.5/34/0.46). Load impedance 1500 Ω. a. α = 0°; b. α = 30°. Vert.: 100 mV/cm; Hor.: 50 nsec/cm.

butane plays an essential role also. Below a certain concentration one cannot reach the space charge saturation and still have a sufficient plateau. It contributes, together with the freon, to the quenching of the photons to such an extent that at the end of the plateau one usually does not observe sparks but the appearance of very small pulses, the origin of which is not quite clear.

Since this mixture appears so far to be the only one with which we could reach nearly 100 per cent efficiency together with this special type of operation, we have made a more systematic study of its properties, as described below.

4.2. SOME SPECIFIC DETECTION PROPERTIES OF ARGON + ISOBUTANE + FREON-13 B1 MIXTURES

For the results given below we used a fixed isobutane concentration of 34%. The chambers used, the experimental set-up, and the data handling were the same as described in section 2.

4.2.1. *Pulse height and efficiency*

Perhaps the most drastic effect obtained by the addition of small amounts of freon-13 B1 is the gain in the minimum pulse height. Figs. 15a, b show pulses on a wire after an emitter follower for angles of incidence $\alpha = 0°$ and $30°$, respectively, and for minimum ionizing particles. We see that no pulse is smaller than 50 mV.

We therefore expect that the electronic threshold can be raised considerably without loss in the efficiency. Fig. 16 shows the high-voltage plateau for thresholds of 1.5 and 10 mV (freon concentration 0.46%). Even at 30 mV threshold we find a plateau length of more than 700 V.

It is interesting to observe the difference in pulse

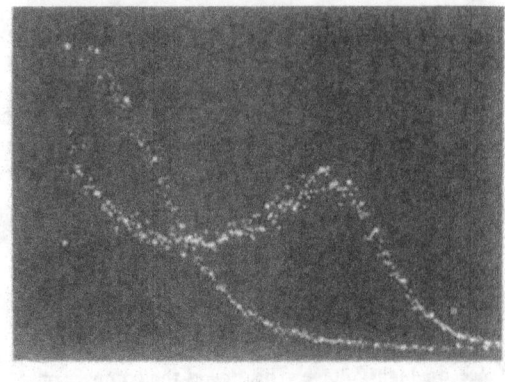

(a)

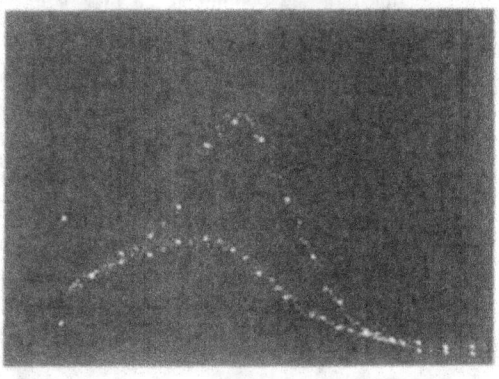

(b)

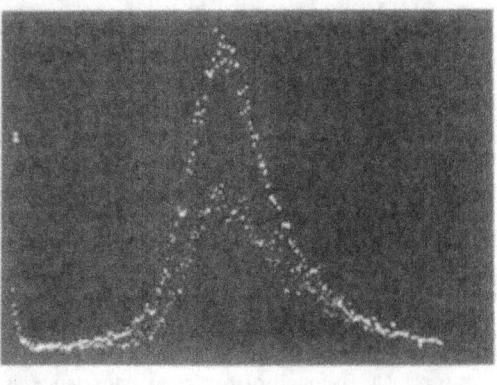

(c)

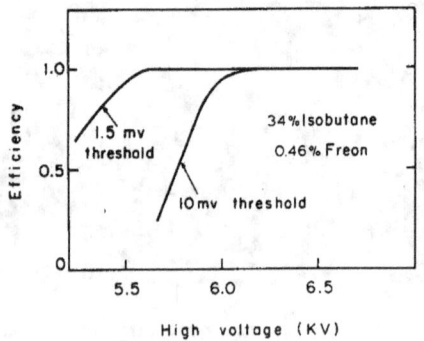

Fig. 16. High-voltage plateau for different electronic thresholds. Mixture: argon + isobutane + freon-13 B1 (65.5/34/0.46).

Fig. 17. Pulse height spectra on a wire by fast electrons and 5.9 keV X-rays. Mixture: argon + isobutane + freon-13 B1 (66/31/0.46). Gap width of chamber 5 mm, wire spacing 2 mm. HV = 4100 V (a), 4300 V (b), 4000 V (c). At 4.5 kV the pulses saturate.

height for a minimum ionizing particle and a 5.9 keV X-ray source. Whereas we see a strong difference for a high voltage of 4.1 kV, the pulse height distributions cannot be distinguished at 4.5 kV (fig. 17a, b, c). In fig. 18 we present efficiency plateaux as a function of the freon concentration. It should be noted (fig. 19), that with orthogonal tracks one cannot reach full efficiency for freon concentrations larger than 0.6%, although there is still an efficiency plateau.

This effect can be explained by the reduction of the sensitive region around the wires due to the freon admixture (section 4.1). This shows up in the fact that even at large freon concentrations one can reach full efficiency again with inclined tracks, the angle of inclination being strictly correlated to the freon content (fig. 20).

4.2.2. Space and time resolution

The results of the preceding section suggest that by reducing the efficient region around the wires, the freon

acts similarly to a narrow time gate. Correspondingly, we expect that some effect will show up in the cluster size, especially for inclined tracks.

The distribution of cluster size is indeed strongly affected by the freon concentration, as shown in fig. 21.

The mean number of wires is plotted in fig. 22 as a function of freon concentration and angle. We see that for the largest concentrations compatible with full efficiency at $\alpha = 0°$, we have less than three wires touched even for angles of incidence $\alpha = 40°$ and large gate widths.

In order to give a rough picture of the correspondence between freon concentration and timing, we may compare the experimental results obtained, for instance, on efficiency and mean number of wires touched as a function of both freon content and gate width. This comparison is somewhat difficult, because the time curves are obtained with constant high voltage, whereas we have to change the high voltage according to the freon concentration in order to stay in the plateau (fig. 18).

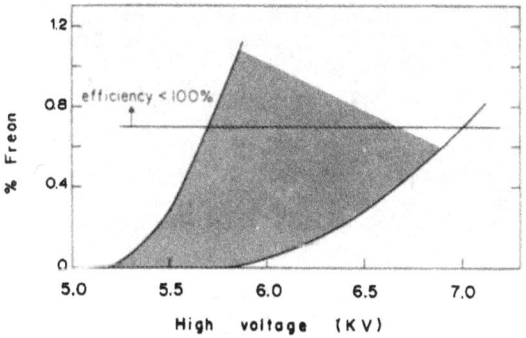

Fig. 18. Position of the high-voltage plateau as a function of freon concentration.

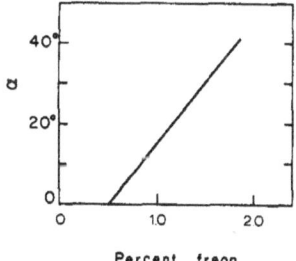

Fig. 20. Angle of inclination α for 100% efficiency as a function of freon concentration.

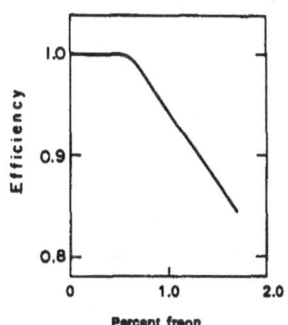

Fig. 19. Efficiency in the plateau as a function of freon concentration.

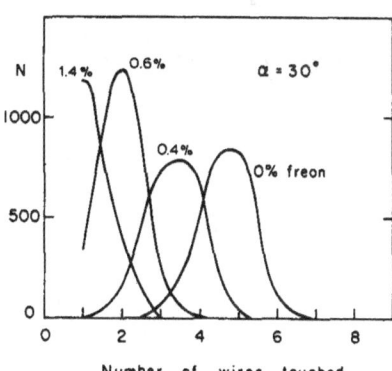

Fig. 21. Distribution of cluster size at $\alpha = 30°$ as a function of freon concentration.

However, we find a rather close correspondence between the two variables, as shown in the shaded region of fig. 23.

In view of this correspondence, we expect no big change of the spatial resolution with freon concentration. Indeed, we find no improvement of the resolution for big concentrations, exactly as in the case of narrow gate widths.

The time resolution, however, shows a difference as compared to the normal gas mixtures. Both in the time spectra obtained in the laboratory tests (compare figs. 12, 24, 25) and in the efficiency curves on the beam we find some degradation of the resolution by about 15 nsec due to a flat tail in the time distribution

(fig. 24). This is true only if we work in the centre of the high-voltage plateau and with concentrations giving full efficiency.

5. Investigations of some different freon mixtures

Strong differences are observed between the different types of freon gases. Our initial interest in these gases was their high density leading to high specific ionization that can possibly be of use for the construction of multiwire proportional chambers with high time and space resolutions. Our first observations have shown a variety of properties some of which seem to us interesting to present.

5.1. ARGON + FREON-12 MIXTURES

According to the proportion of added freon, we observe the same phenomena as mentioned by Grunberg et al. with a mixture of argon and ethyl-bromide, but at much lower concentrations. The sensitive region around each wire can be controlled at will. We have made tests with chambers with wire spacings $s = 5, 2, 1$ mm. With $s = 2$ mm, the efficiency is still near 100% with as much as 30% freon. But the pulse height distribution is strongly distorted by a shift towards low-voltage pulses. This is what we expect if the electrons liberated far from the wire have a strong chance to be captured before reaching the region of multiplication.

The time jitter shows a distribution with a fwhm of 25 nsec, without a sharp limit in the region of long delays (fig. 25a).

If the discrimination level of the pulses is increased, by a circuit in parallel to the time delay measuring circuit, an interesting phenomenon appears: the

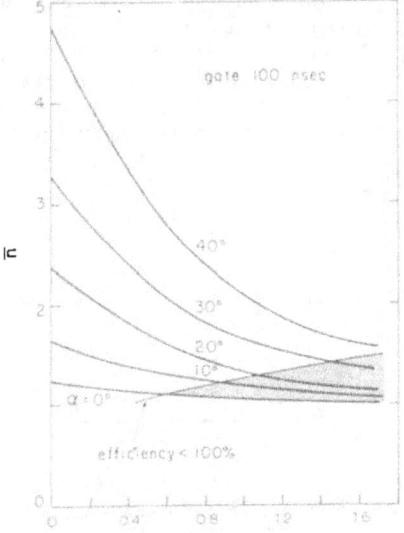

Percent freon

Fig. 22. Mean number of wires touched as a function of freon concentration and angle of incidence.

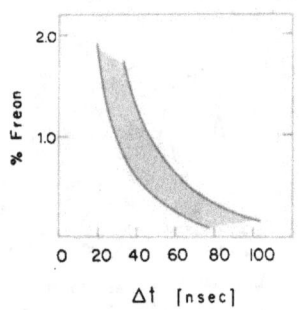

Δt [nsec]

Fig. 23. Correspondence between freon content and gate width.

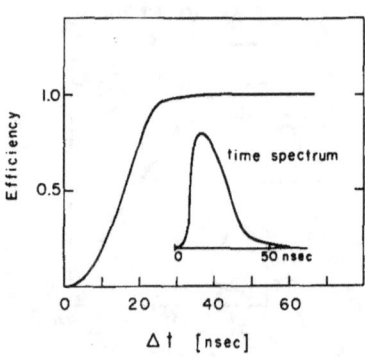

Δt [nsec]

Fig. 24. Efficiency curve as a function of gate width and time spectrum for argon + isobutane + freon mixture.

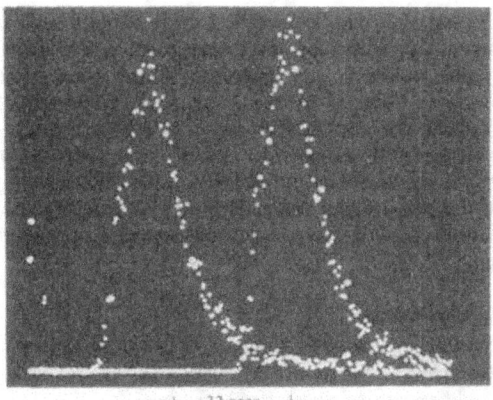

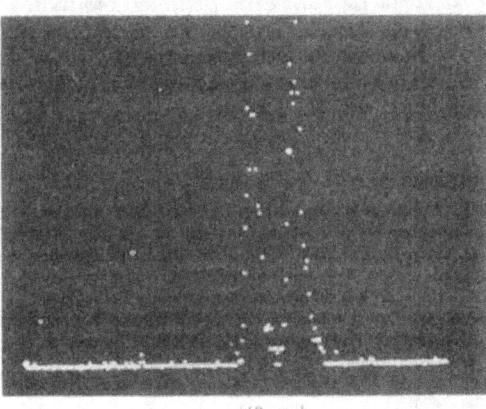

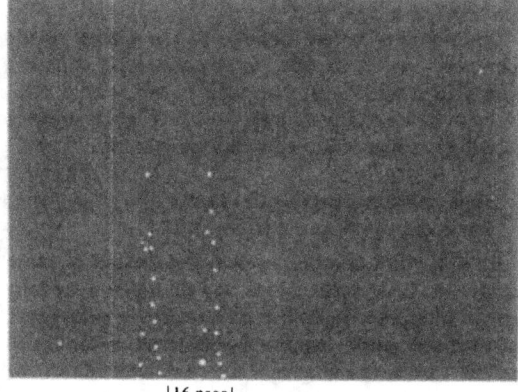

|16 nsec|

Fig. 26. Time jitter spectrum with helium+freon-13 (92/8). HV = 7100 V, efficiency 20%, calibration peak at 16 nsec, fwhm = 2.5 nsec.

Fig. 25. Time jitter spectra. (a) Argon+freon-12 (72/25),HV = 4500 V, efficiency 99%, calibration peak at 32 nsec, fwhm = 25 nsec. (b) Same conditions, threshold raised, efficiency 12%, calibration peak at 32 nsec, fwhm = 3.7 nsec. (c) Same conditions, threshold raised, efficiency 0.7%, calibration peak at 10 nsec, fwhm = 2.4 nsec.

resolution time improves and the efficiency drops, showing that a strong correlation exists between the position of the track and the pulse height, as expected from the electronegative character of the freon, and showing once again that the main contribution to the time jitter is the variation of the distance between wire and trajectory. The striking fact is that resolutions of 2.5 nsec (fwhm) can thus be reached, if the efficiency is reduced to about 1% (figs. 25b, c).

At this point one may wonder whether freon fillings should be considered in future plans for development, with high spatial resolution, along the same line of research that led to the study of liquid argon[9]). If the good time resolution is due to the fact that only a limited region of say, 50 μm around the wire is sensitive, then by placing a wire every 50 μm one gains back the efficiency. The density can be made high enough, at reasonably low temperatures, so that the specific ionization is sufficient to ensure enough primary ions. It may well be that the ultimate spatial resolution is worse than with liquid argon, but the technical problems are less formidable and it is proved that operation in the proportional mode is feasible.

This has in itself possible interesting applications. In an intense beam of particles a system of such chambers permits, by a matrix of wires in coincidence, the selection of beams with a considerable angular definition. It is an ideal massless collimator.

If we attempt to go beyond the normal proportional amplification with this gas, we do not observe stable operation.

5.2. Helium + freon-13 B1

The chamber enters rapidly into a Geiger mode operation – pulses rising to values of several volts but with a slow and linear rise time bigger than 1 μsec, but with an excellent time resolution. Fig. 26 shows a time distribution obtained with this mixture.

6. Some additional experimental tests with argon + isobutane + freon-13 B1

An important question concerning the use of freon admixtures is the behaviour of the chambers after long term irradiation of operation with very high pulse rates, because one could imagine polymerization or disintegration processes changing the characteristics of the chambers.

To test this, we have exposed a chamber (0.46% freon) for six hours to a high intensity β-source of 3.5×10^{10} particles/h. After this operation, which corresponds to a 300 days' irradiation by a typical CERN-PS secondary beam, we could not perceive any change in the chamber characteristics.

Another test was performed on chambers with different wire spacings. We find that we can operate chambers up to 3 mm wire spacing, although the region of freon concentrations providing 100% efficiency becomes rather small in this case (fig. 27). On the other hand, for 1 mm spacing one has to push the high voltage very high to obtain good efficiency.

In order to evaluate the influence of space charge on the pulse height limitation we did the following measurements: we compared the pulse height obtained directly on the wire for a 5.9 keV X-ray orthogonal to the chamber and a fast electron travelling parallel to a wire of 10 cm length. We obtained pulses of an average height of 0.4 V and 15 V, respectively, compared to energy losses of 5.9 keV for the X-ray and 20 keV for

the fast electron. Apparently, for the X-ray a serious loss is introduced by space charge effects even with freon. However, our ignorance of the amount of electrons liberated by the 3 keV photoelectrons and reaching the amplification region forbids quantitative statements. It seems to us still of interest that 15 V pulses can be obtained directly from the wires working in the proportional mode. This may be useful for the construction of cheap multiwire detectors in special cases.

7. Conclusions

The results presented above show that most of the characteristic features of multiwire proportional chambers can be understood quantitatively.

The space resolution appears to be constant up to angles of incidence of 40°, with a variance $\sigma \approx 0.3\,s$, where s is the wire spacing.

The cluster size can be controlled by the gate width and reduced to values giving good multiple track resolution even at large angles. The study of a variety of gas mixtures has shown remarkable features. In particular, the space charge limit can be shifted by a factor of a hundred by the addition of appropriate quantities of freon-13 B1 to argon + isobutane. At the same time the cluster size is reduced, thus improving the multitrack resolution especially for inclined tracks.

We are indebted to Mr. L. McCulloch for the set-up of the electronic logics and the group led by Mr. G. Amato for their help in the electronics construction.

We thank Mr. L. Dumps, Mr. G. Million and Mr. L. Naumann for their technical help.

We acknowledge many stimulating discussions with Dr. J. Favier and Dr. A. Minten.

References

1) G. Charpak et al., Nucl. Instr. and Meth. **62** (1968) 235. For a complete list of references updated to 1 March 1970, see G. Charpak, Evolution of the automatic spark chambers, to appear in Ann. Rev. Nucl. Sci. **20** (1970).
2) T. Trippe, CERN NP Internal Report 69-18 (June 1969).
3) P. Steffen and F. Vannucci, CERN NP Internal Report 69-29 (Oct. 1969); J. H. Dieperink et al., CERN NP Internal Report 69-28 (Oct. 1969).
4) G. Charpak, D. Rahm and H. Steiner, Nucl. Instr. and Meth. **80** (1970) 13.
5) R. Pagès, Nucl. Instr. and Meth. **85** (1970) 211.
6) P. C. Ziffra and M. J. Moravcsik, UCRL 8523.
7) G. Grunberg, L. Cohen and L. Mathieu, Nucl. Instr. and Meth. **78** (1970) 102.
8) D. Van Zoonen and G. Prast, Jr., Appl. Sci. Res. **3 B** (1954) 1.
9) E. Stephen et al., Preprint, UCRL 19254 (Oct. 1969).

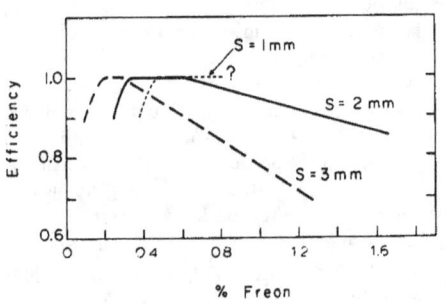

Fig. 27. Efficiency in the high-voltage plateau as a function concentration for different wire spacings (5 mV threshold of freon on amplifier inputs).

NUCLEAR INSTRUMENTS AND METHODS 94 (1971) 151–155; © NORTH-HOLLAND PUBLISHING CO.

HIGH-ENERGY CHARGED PARTICLES SEPARATION BY MEANS OF A CASCADE OF MULTIWIRE PROPORTIONAL CHAMBERS

Z. DIMČOVSKI, J. FAVIER, G. CHARPAK and G. AMATO

CERN, Geneva, Switzerland

Received 4 January 1971

The energy losses of charged particles traversing 30 proportional chambers are measured simultaneously. The relativistic rise of ionization is investigated in several gases. A maximum increase of 45% is observed in a mixture of argon–propane, where Landau theory predicts an increase of 78%. The application of these measurements for particle separation is discussed.

1. Introduction

It is well known that the energy spectrum resulting from the passage of a mono-energetic beam of charged particles through a thin absorber is not Gaussian but has a characteristic "Landau" tail on the side of the high-energy losses. This dissymmetry makes difficult the use of the relativistic rise of the energy losses as a method of distinguishing particles, because of the considerable overlap between spectra of different particles. However, a method of statistical treatment (N simultaneous measurements for a single particle, with $N \gg 1$), or sampling the ionization losses in a succession of proportional detectors, allows the difficulty due to the "Landau" shape[1]) to be overcome.

Landau has given the theoretical shape of the ionization loss distribution using some approximations (atomic electrons considered as free and no upper limit in the energy transfer). More recent theories have tried to improve the treatment; Vavilov[2]) has introduced a cut-off in the energy transfer, and Blunck et al.[3]) have taken into account the electron binding energies and the Bremsstrahlung losses, using some other approximations.

One has, however, to be careful when using these theories because of the disagreement with experiment found by several authors[4,5]) concerning the widths of the distribution losses and the slope of the relativistic increase itself. These facts render the simulation by Monte Carlo method hazardous and the results too optimistic.

Fig. 1 shows the experimental energy resolution, defined by the fwhm, obtained some time ago by West[6]) in several gases. It is clear that a too small energy loss results in a very poor resolution, but that the resolution stops improving rapidly after energy losses greater than 10 keV and then follows the expected, nearly logarithmic law. To overcome the difficulty inherent to the non-Gaussian distribution of the total energy losses, some authors have tried to measure directly the number of primary collisions in the gas. The most recent

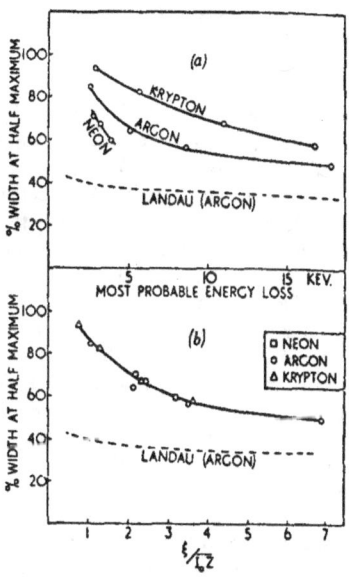

Fig. 1. Width at half-maximum of the energy loss distribution for minimum ionizing particles as a function of the most probable energy loss (West, ref. 6).

attempt by Davidenko et al.[7]) was made with streamer chambers. They chose helium at reduced pressure, where the density of ionization is small enough so that the avalanches from different δ rays do not interact. A relativistic rise of 45% was observed. However, this method of counting the number of streamers is very slow, and the time resolution of such a detector is inherently of the order of several microseconds.

The method of statistical treatment, first described by Alikhanov et al.[8]) in 1956, is to measure simultaneously the energy loss in several successive proportional counters, and to obtain an energy loss distribution for each particle. This distribution is used to obtain information about the ionization power. The

151

same authors propose to use several methods in order to obtain the average ionization power from the experimental distribution of pulse heights: a "universal method" which is similar to the maximum likelihood method; a cut-off method; and the logarithmic mean method. This latter method consists in taking the average of the logarithm of the pulses. It is attractive since it is very easy to obtain the logarithm of a pulse by analogic circuits within very short times.

Similar methods were tried by other authors using a small number of chambers, and these are discussed in detail in the articles of Ramana-Murthy and Demeester[4]), who show that the likelihood ratio method is the best way of extracting the maximum of information.

2. Construction of the detector and experimental set-up

We have built a detector consisting of 30 chambers. Each chamber is made of a wire plane with wires of 40 μm thickness spaced 5 mm apart, placed between two high-voltage planes made of wires 100 μm thick. The high-voltage planes of two successive chambers are independent and are 0.2 cm apart. This allows a fine adjustment of the amplification in each gap. The amplification of each wire is submitted to fluctuation due to a geometrical dissymmetry of the chamber, variation of the wire radii, impurities, etc. We have measured a variation of $\pm 5\%$ in the amplification from one wire to another, in agreement with preliminary calculations[9]). The sensitive surface is 10×10 cm^2. All the sensitive wires are connected together to a linear amplifier. The amplitude of the pulses from each chamber is digitized, recorded in a PDP 8 on-line, and stored on magnetic tape. Fig. 2 shows the diagram of this experimental set-up.

3. Treatment of the energy loss distribution

What is important is to know experimentally the distribution corresponding to each particle or, for a given particle, each momentum. If $\phi(E)$ is the probability of losing E for a particle of a given momentum, while $\psi(E)$ is the distribution corresponding to another particle with the same momentum, or to the same particle at another momentum, the maximum likelihood ratio method (described in detail in ref. 4) calculates the ratio for a set of N measurements

$$K = \prod_{i=1}^{N} \phi(E_i) \bigg/ \prod_{i=1}^{N} \psi(E_i).$$

$K < 1$ if the particle fits the distribution ψ, $K > 1$ if the particle fits the distribution ϕ.

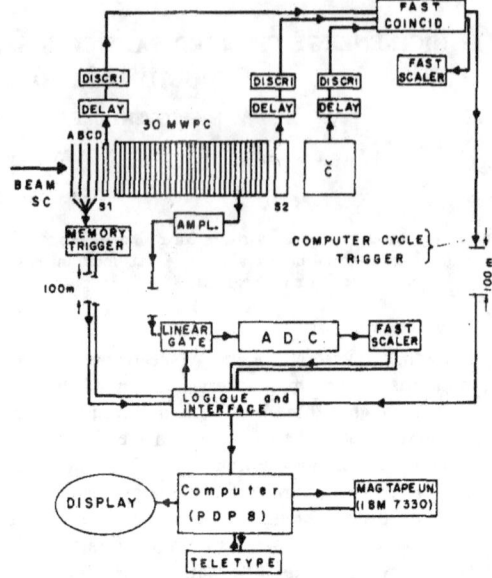

Fig. 2. Experimental set-up. Each chamber is equipped with an ADC for pulse-height measurements.

Such a method requires treatment by a computer. We present our experimental results treated by four methods:

1. the average of the 30 successive pulses;.
2. the average of the 10 smallest pulses out of 30 [this method is similar to the one of Igo and Eisberg[10]) with three chambers];
3. the average of the logarithm of the pulses;
4. the likelihood ratio method.

The respective merits of these methods are illustrated by the value of the coefficient D/L, where D is the separation between the peaks and L is the fwhm of the distributions of pulses produced by pions and electrons of 374 MeV/c.

4. Experimental results

The following mixtures have been investigated in different relative proportions: Ar–CO$_2$, Ar–iso C$_4$H$_{10}$, Ar–C$_3$H$_8$, He–iso C$_4$H$_{10}$, He–Ne–iso C$_4$H$_{10}$. The maximum relativistic increase of 45% has been observed in an argon–propane mixture in the proportion 95%–5%[11]).

Figs. 3a and 4a show the distribution of energy losses of pions and electrons in the individual chambers for a gas filling of 95% argon and 5% propane, assuming 1.2 cm to be the traversed gas length of an average chamber. Comparison is made with the theory of

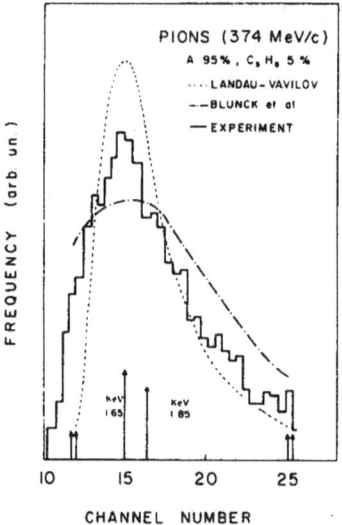

Fig. 3. Experimental and theoretical energy loss distribution for pions in 1.2 cm of argon–propane (95%–5%); $p = 374$ MeV/c. The Landau curve is corrected for the finite energy resolution of the counters (fwhm = 20% at 6 keV, varies like $E^{-\frac{1}{2}}$).

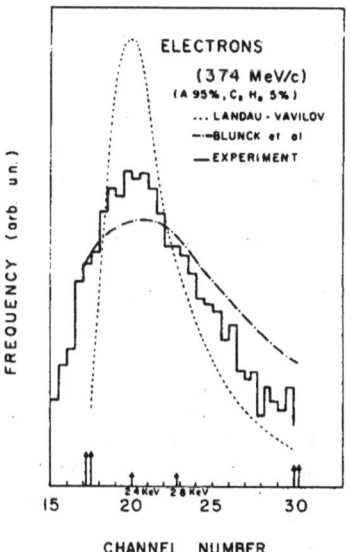

Fig. 4. Experimental and theoretical energy loss distribution for electrons in 1.2 cm of argon–propane (95%–5%); $p = 374$ MeV/c. The Landau curve is corrected for the finite energy resolution of the counters (fwhm = 20% at 6 keV, varies like $E^{-\frac{1}{2}}$).

Vavilov and that of Blunck and Leisegang (figs. 3 and 4). The finite energy resolution of the counter ($\approx 20\%$ at 6 keV, assuming $E^{-\frac{1}{2}}$ dependence) has been folded in the Landau curve, but not in the Blunck curve.

We see that neither of these theories fits the results very well. However, the applicability of the sampling method is independent of the agreement with any theory, the observed experimental distributions being taken as the functions ϕ and ψ.

Fig. 5 gives the results of the different methods of data reduction. We see that the average method yields $D'L \approx 1.1$, the average of the 10 smallest pulses $D'L \approx 1.2$, the average of the logarithms ≈ 1.3, and the maximum likelihood ratio ≈ 1.4. The 5% contamination of electrons in the pion beam is clearly visible.

The relativistic rise (i.e. the ratio of the most probable energy losses for electrons over pions) is 1.45. It has also been measured at a momentum of 219 MeV/c, where the ratio of energy loss for electrons and pions is 1.15.

Assuming that the 374 MeV/c electrons have reached the plateau value, we see that the experimental value (1.45) is considerably lower than that predicted by Landau's theory corrected for the density effect (1.78). The reason for this disagreement, found by several authors[4-6]), is not known; what is, however, obvious is that the theoretical predictions concerning the distributions of energy losses verified to 1% in solids fail in small thicknesses of gases. Therefore systematic measurements (which were not the aim of this work) over a large range of particle momenta and different gases are desirable in order to give some more hints as to the nature of this effect.

5. Results obtained with other gases

Similar measurements have been made with the mixtures quoted above where a lower value of the relativistic increase has systematically been reached.

The argon-isobutane mixture shows a striking difference: the relativistic rise is only of 10% but also the shape of the curve is changed (fig. 6). The width of the distribution of energy losses is greater with electrons. Both the high-energy side and the low-energy side show tails. This latter fact is of interest for the maximum likelihood method which is sensitive to the shape of the distribution function. The origin of this effect of broadening is not understood by us.

6. Correlation between chambers

The method is based on the independence of the energy losses in the different chambers. To check this hypothesis we have requested that the pulse in one

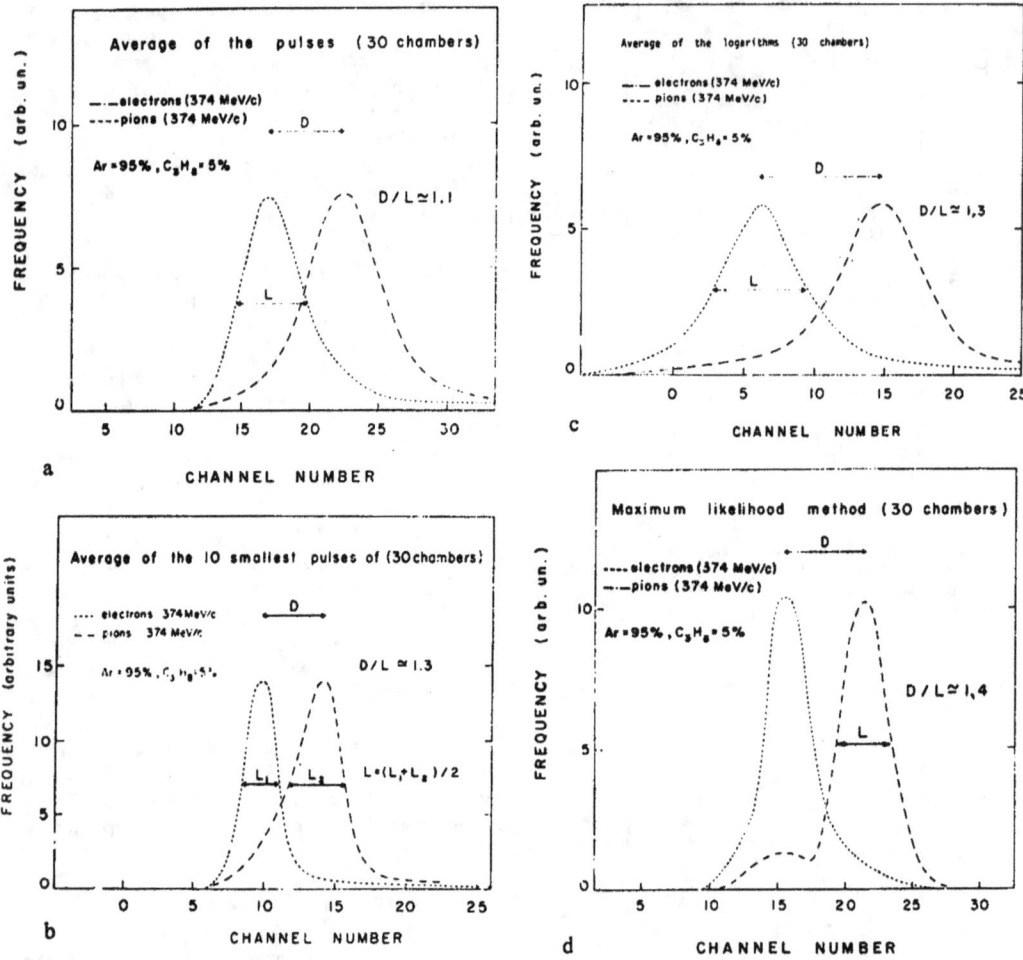

Fig. 5. Electron-pion separation by different data-reduction methods. Argon–propane; 374 MeV/c: a) average of 30 chambers; b) average of the 10 smallest of the 30 pulses; c) logarithmic average of the 30 pulses; d) maximum likelihood method; the contamination of 5% of electrons in the pion beam is clearly visible.

chamber be higher than given values and have measured the distributions in the neighbouring chambers. No sizable effect is observed.

7. Discussion of the results

One can expect that a considerable increase in accuracy can still be gained by increasing the number of chambers. For illustration, we present the pion-electron separation at 374 MeV/c, obtained with 300 chambers (fig. 7) using a simulation by the experiment (10 events and 30 chambers are equivalent to 1 event

and 300 chambers because of the statistical nature of the phenomenon and the absence of correlation between chambers). However, it is clear from fig. 1 that the optimum thickness at atmospheric pressure is not 1 cm; it would be rather 5 cm to 10 cm. With a 5 atm argon–propane filling, the resolution illustrated by fig. 7 could be obtained with 100 layers. Because of the relatively low amount of material involved, this method can be of great interest for the separation of high-momentum particles in competition with Čerenkov counters.

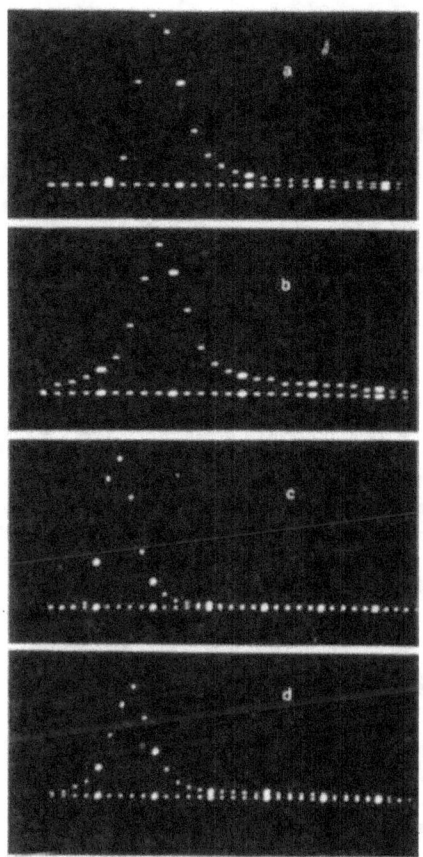

Fig. 6. Electron-pion separation with argon–isobutane. Curves a and b: π and e at 374 MeV/c (argon 13%, isobutane 87%); curves c, d; π^- and e$^-$ at 375 MeV/c (argon 50%, isobutane 50%).

Using the experimental results obtained with argon–propane, table 1 shows the number of chambers at atmospheric pressure, required in order to separate, with a 10% overlap, particles of varying energy losses.

TABLE 1

N = number of layers of 1 cm of argon–propane and TPN necessary to give a separation, at 90% confidence level, of two particles with ionization power D_1 and D_2. $\varepsilon = (D_2 - D_1)/D_1$ = relative difference of the most probable energy loss of two particles.

ε	40%	30%	20%	10%	5%
N	35	50	90	200	400

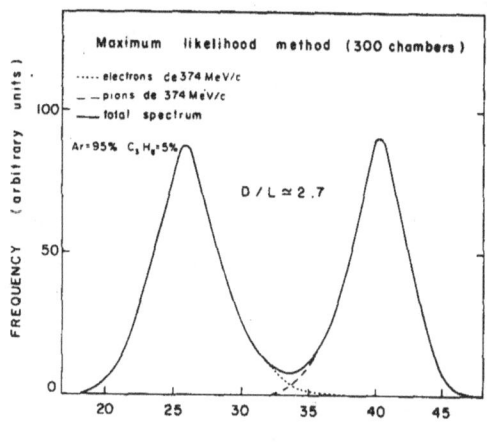

Fig. 7. Electron-pion separation with 300 chambers simulated by the experiment (10 successive particles in the 30 chambers). At 10 atmospheres, or chambers of 10 cm, this separation would be reached by less than 100 chambers (see fig. 1).

8. Conclusion

This investigation of the sampling method shows that already with 30 chambers a satisfactory separation can be obtained for the high-energy particles, i.e. for particles between the minimum ionization and the region of the Fermi plateau.

We wish to thank Prof. H. Steiner who participated in the initiation of this work. We are indebted to Drs. A. Stirling, C. Serre and V. Chabaud for many illuminating discussions. We are grateful to Messrs. R. Bouclier, S. Cairanti and J. C. Santiard for their technical assistance.

References

1) L. Landau, J. Phys. 6, no. 4 (1944) 201.
2) P. V. Vavilov, JETP 5 (1957) 749.
3) O. Blunck and K. Westphall, Z. Physik 130 (1951) 641.
4) P. V. Ramana–Murthy and G. M. Demeester, Nucl. Instr. and Meth. 56 (1967) 93.
   P. V. Ramana–Murthy, Nucl. Instr. and Meth. 63 (1968) 77.
5) T. E. Cranshaw, Progr. Nucl. Phys. (1952).
6) D. West, Proc. Roy. Soc. A66 (1953) 306.
7) V. A. Davidenko, B. A. Dolgoshein, V. K. Seminov and S. V. Samov, Nucl. Instr. and Meth. 67 (1969) 325.
8) A. I. Alikhanov, V. A. Lubimov and G. P. Elisejev, Proc. Symp. *High-energy physics* (CERN, Geneva, 1956) p. 87.
9) Z. Dimčovski, CERN NP Internal Report 70-16 (1970).
10) G. Igo and R. M. Eisberg, Rev. Sci. Instr. 25 (1954) 450.
11) Z. Dimčovski, Thesis (Université de Grenoble, 1970) and CERN NP Internal Report 70-30 (1970).

NUCLEAR INSTRUMENTS AND METHODS 115 (1974) 235–244; © NORTH-HOLLAND

PROPORTIONAL CHAMBERS FOR A 50 000-WIRE DETECTOR

R. BOUCLIER, G. CHARPAK, E. CHESI, L. DUMPS, H. G. FISCHER, H. J. HILKE, P. G. INNOCENTI,
G. MAURIN, A. MINTEN, L. NAUMANN, F. PIUZ, J. C. SANTIARD and O. ULLALAND*

CERN, Geneva, Switzerland

Received 28 June 1973

A multiparticle spectrometer at the CERN Intersecting Storage Rings is equipped with proportional chambers. In this paper we describe the construction of 20 modular 100×200 cm^2 wire chambers, the electronics for a 50k wire system, and the performance of the chambers.

1. Introduction

A multiparticle spectrometer [Split-Field Magnet, SFM[1])] has been constructed for the CERN Intersecting Storage Rings. The instrument will permit the study of multiparticle final states in proton–proton collisions.

The spectrometer consists of a magnet[2]) and a detector for charged particle trajectories which should have the following properties:

* Visitor from the University of Bergen, Norway.

1) the detector must operate in a 12 kG magnetic field;

2) the detector has to be efficient for several particles passing simultaneously, irrespective of the distance between tracks and their angle of incidence;

3) because of the geometrical dimensions involved, additional trigger counters were considered as impractical; therefore the chambers should be self-triggering.

A detector consisting of proportional wire chambers was considered to fulfil these requirements.

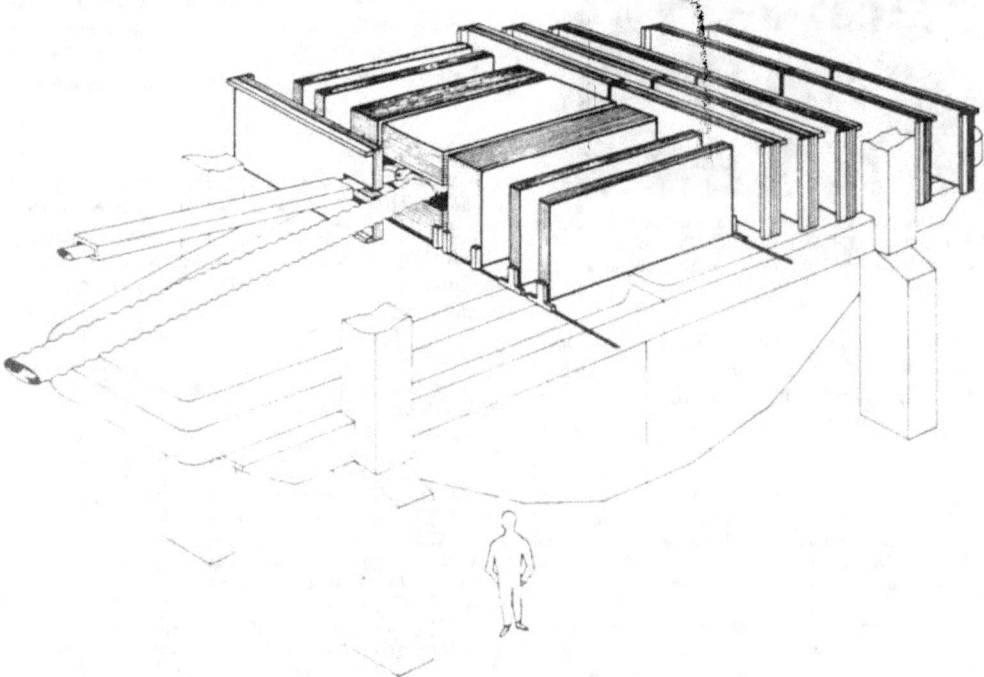

Fig. 1. Artist's view of the Split-Field Magnet detector.

In this paper we describe a system of 20 chambers and 50 000 channels of electronics constructed to date. We give a report on the mechanical design and the construction process, on the electronics circuits and the layout in a 50k system, on the special technique of high-voltage read-out, and on the performance of the chambers. We give an account of some difficulties which occurred in the project, and we will also report methods and procedures non-evident from general laws of science. We will withhold, however, witchcraft, arcana, and secret incantations[3]).

2. Design criteria

The basic constraints on the chamber construction are imposed by the magnet geometry and by the detector layout for the experiments foreseen. On the other hand, the design has to fulfil certain minimum mechanical and electric stability requirements in order to ensure safe operation in the inaccessible and radiative ISR environment. Moreover, the construction and assembly must be simple and, if possible, modular, in view of the large number of units to be built.

From the physics point of view, the SFM detector should serve as a general-purpose instrument offering a wide spectrum of experimental possibilities extending over the full phase space. Therefore, the ideal detector would cover 4π of solid angle and would give a maximum of measured points per particle trajectory. In order to approach this goal, we have chosen a box-shaped detector layout[1]) with (fig. 1):

1) a "central" part consisting of four densely packed chamber units;

2) two "forward" parts with a total of twenty-four chamber units of two wire planes each.

The detector is fully contained inside the magnetic field. The standard size of the chambers, as given by the magnet gap, is 100×200 cm^2. Cut-outs in the forward chambers minimize the acceptance loss around the incoming and outgoing beam tubes.

In view of the acceptance requirements, the "classical" chamber construction with strong frames and stretched wire high-voltage electrodes would lead to an unacceptable loss. Instead, we have developed minimum frame, self-supporting chambers. The hv electrodes are made of thin silver layers sprayed on flat, very light sandwich plates, which consist of a core of plastic foam covered with reinforced epoxy foils. In this way, the frame thickness is reduced to about 5 mm independent of chamber size. The gain in efficient surface has to be paid for, however, by a corresponding increase of material traversed by the

particles. Therefore, the weight of the sandwich has to be matched with the requirement for mechanical stability.

3. Mechanical tolerances

A figure of merit for large proportional chambers is the plateau length. This is defined as the span of voltage between reaching 100% efficiency and the set-in of excessive noise rate or breakdown. The plateau limits are determined by the electrical field on the sense wire. In a good approximation, this field value is connected to the geometrical chamber constants by

$$E(r_0) = \frac{(s/\pi r_0)\,V_0}{L-(s/\pi)\ln(2\pi r_0/s)},$$

where

r_0 : wire radius,
L : gap width,
s : wire spacing,
V_0 : chamber operating voltage.

For given field strength, we find a linear relationship between plateau voltage V_0 and gap L, with

$$\frac{\Delta V_0}{\Delta L} = \frac{E r_0 \pi}{s}.$$

In our case, a change in gap width of 0.1 mm would result in a 50 V shift of plateau. This imposes limits on mechanical gap distortions, especially since experience shows that large chambers cannot be expected to give safe plateau lengths of more than 200 to 300 V with our electronics.

There are some mechanical and electrical reasons for gap distortion:

1) *Production tolerances:* The fabrication of 2 m^2 sandwich plates with both thickness and flatness tolerances of 0.1 to 0.2 mm is on the limits of production techiques.

2) *Bending due to uniform loads such as gas overpressure:* The bending due to surface loads is growing with the fourth power of the chamber dimensions, i.e.*

$$W = \alpha \frac{qa^4}{E_f(t_f\,t_c^2/2)},$$

where

q : pressure,
a : smallest chamber dimension,
α : rectangularity parameter, $\alpha \approx 0.01$ for our case,
E_f : modulus of elasticity of sandwich facing,

* Honeycomb Sandwich Design, Hexel Products Inc., Dublin, Calif., U.S.A.

t_f : facing thickness,
t_c : core thickness.

Notice, that for foam cores, the elastic properties of the core material can be neglected for all practical purposes.

3) *Temperature gradients:* A temperature difference ΔT between the two facings of a sandwich plate yields a deflection of

$$W = \beta_f \frac{a^2}{8t_c} \Delta T,$$

where β_f is the linear temperature expansion coefficient.

Since plastic sandwiches represent ideal heat shields, it becomes important to ensure good thermal equilibrium.

4) *Bending moments due to wire tension and external loads:* External moments transmitted via the frame structure lead to deflections roughly proportional to the square of the chamber dimension:

$$W \sim \frac{ka^2(1+2L/t_c)}{4E_f t_c t_f},$$

where k is the external force per cm frame.

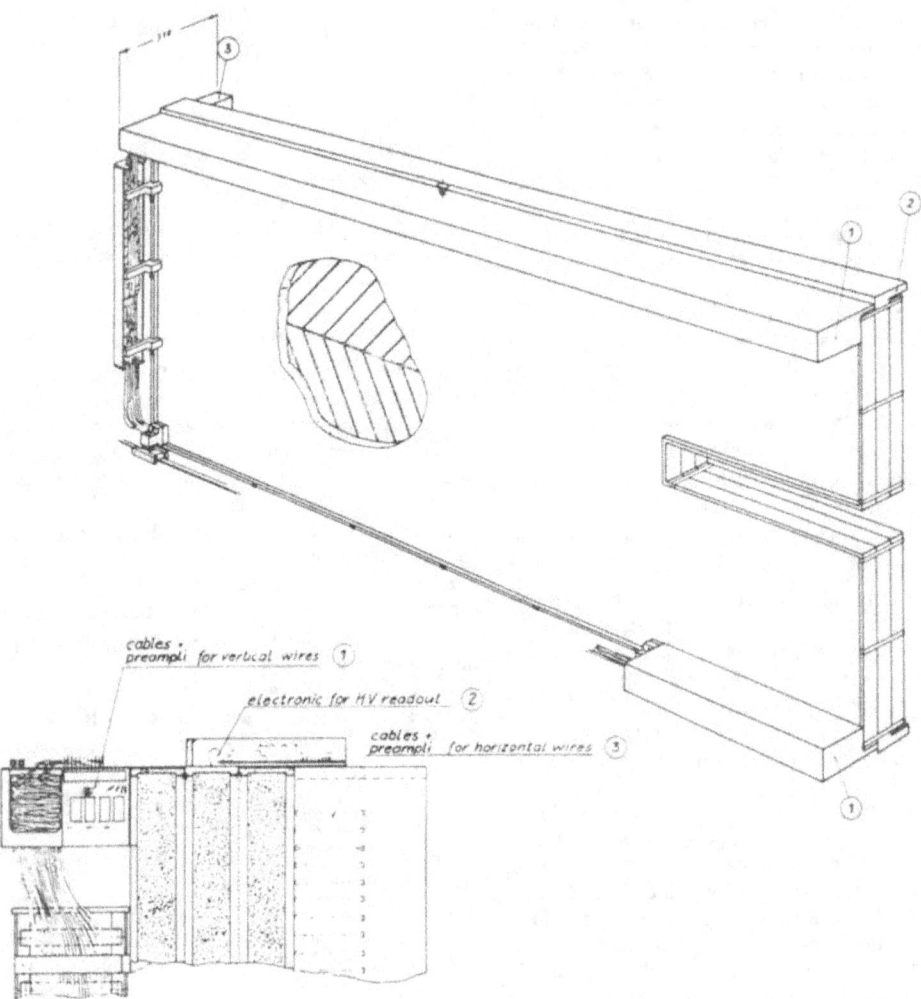

cables · preampli for vertical wires ①

electronic for HV readout ②

cables · preampli for horizontal wires ③

Fig. 2. Split-Field Magnet proportional wire chamber.

5) *Electrostatic forces on the wires:* As is well known, the electrostatic forces between the sense wires and between each wire and the hv electrodes tend to displace the wires from the equilibrium position. This can partially be compensated for by support lines stretched perpendicular to the sense wires.

4. Chamber construction

4.1. CHAMBER CHARACTERISTICS

We describe here the construction of the 20 "forward" detector chambers. Each standard chamber unit has an effective area of 102.4×204.8 cm^2 and consists of three sandwich panels, enclosing two sense wire planes (fig. 2). The wire spacing is 2 mm, wire diameter 20 μm*, wire directions are horizontal and vertical. The gap width is 8 mm. The hv electrodes are made of a 5 μm silver layer sprayed on the sandwich surfaces. On both faces of the central plate, the metallization is arranged in strips following a fishbone pattern (fig. 2). The strip width is 5.6 cm, the angle 30°. By read-out of these hv strips, we obtain two independent inclined coordinates which are used to eliminate ambiguities in track recognition.

4.2. SANDWICH PLATES

For uniform mass distribution and low Z, we have preferred plastic foam to honeycomb as core material. We use a 22 mm thick polymethacrylimid† (PMI) foam core with a weight of 30 kg/m^3. Because of the hv read-out, the facing material has to be insulating. For stability, fibre-glass reinforced epoxy (Vetronite) offers the best elastic properties. Vetronite sheets⁺ of 0.2 mm thickness are glued with epoxy to the foam surfaces ‡. One sandwich panel amounts to 0.2 g/cm^2, yielding a total of 0.6 g/cm^2 per chamber unit, which is about 1.7% of radiation length and about 1% of interaction length for the materials involved.

The mechanical tolerances were investigated on several chamber units and were found to be in reasonable agreement with the design estimates given above:

1) The production tolerances stay within the specified limits of ± 0.2 mm for flatness and thickness.

2) For uniform loads, we find that a gas overpressure of 0.1 g/cm^2 yields a sagitta of 0.1 mm. Therefore, a special gas pumping system is used to minimize overpressure effects.

3) The sagitta due to the temperature differences is

* Gold-plated tungsten, Lumalampa, Stockholm, Sweden.
† Rohacell: Röhm and Haas, Darmstadt, W. Germany.
⁺ Isola-Werke, Breitenbach, Switzerland.
‡ Contraves AG, Zürich, Switzerland.

of the order of 0.2 mm/°C. Special care has to be taken to avoid temperature gradients.

4) The biggest deformations are produced by the bending moments from the stretched sense wires. The total deflection amounts to about 1 mm from the vertical wire planes. Not much can be done to bring this back into the tolerances, since it would mean a five times increase in either core or facing thickness, which is intolerable. Instead, we add spacers in the chamber centre, which – for stability – subdivide the chamber dimensions at the expense of a small, localized efficiency loss.

4.3. SUPPORT LINES AND SPACERS

The construction of the support lines follows the method we adopted in solid frame chambers and described elsewhere[4]. We use one support line every 50 cm, which makes a total of three lines on the horizontal and one line on the vertical plane. With a compensating potential of 1.5 kV, the efficiency loss is ± 1 mm around each line. The support lines are stretched with 3.5 kg. For the large spans required in our chambers, this tension would not be sufficient to compensate the electrostatic forces on the sense wires. Therefore, the support lines are fixed to the spacer columns, which are introduced to minimize the gap distortions.

Each chamber has three spacers in both the horizontal and the vertical gap, positioned at the crossing points of the respective support lines. Fig. 3 shows a cut through one of these spacer positions. By means of a catgut passing through the chamber and the spacer columns, an external force of up to 20 kg can be exerted to bring the gaps to the nominal value.

4.4. FRAMES, PRINTED CIRCUITS AND WIRING

The frames consist of "Permaglass" fibre-glass

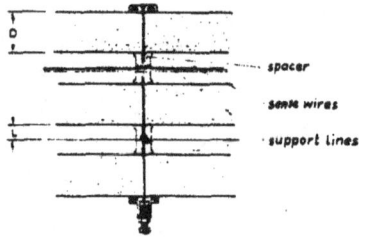

D : sandwich thickness
L : gap width

Fig. 3. Cut through a Split-Field Magnet proportional chamber at the position of a spacer and crossing support lines.

epoxy*, 5 mm thick (fig. 2). The sandwich panels are glued into a 22×2 mm^2 slot, which is machined into the frame as well as the grooves taking the O-rings and the guard strips. The sense wires are fixed on the face of the profile, at right angles to the wire direction. They are stretched with a pneumatic wiring machine. Small pistons, driven by compressed air, act on 64 wires and stress them to 50 g, independent of the elastic elongation. The wires are then bent through 90° over the frame edge and soldered onto printed circuits. The circuit board carries at the same time the connector pins[+] fitting into the preamplifier electronics cards (fig. 2).

4.5. CLAMPING, ADJUSTMENT AND CHAMBER SUPPORT

The three sandwich plates making up one chamber unit are held together by simple stainless steel clamps (fig. 2) acting on the frame edges. With one clamp every 25 cm, together with the O-ring sealing, excellent stability and gas tightness are achieved. Chambers have worked with gas flow cut for several days without noticeable deterioration.

The wire directions are adjusted orthogonal to each other during wiring and kept stable by centring pins fastened to the central sandwich. Accuracies of ± 0.2 mrad in orthogonality and ± 0.2 mm in wire positioning are reached.

Each chamber is supported on two adjustable chariots rolling on a flat rail on the magnet pole face; verticality is ensured by one top support wheel. The chambers can be easily rolled into and out of the magnet gap.

4.6. SOME SPECIAL CONSTRUCTION PROBLEMS

One of the most critical parts of the construction is the vulnerability of the sense wires in the frame region, where they are bent over the frame edge and acted upon directly by the O-ring seal of the opposite frame. This is especially true for the cut-outs, where wires pass for a long distance on the frame surface. To avoid wires breaking, all frame edges are rounded off and the sense wires are protected by a layer of 0.1 mm adhesive tape.

As is well known[4]), careful cleaning of the sense wires is needed in order to avoid discharges. All wires are wiped and brushed with amylacetate and alcohol before soldering.

The hv electrodes are extremely delicate and liable to deteriorate during the chamber construction.

Therefore electrodes are sprayed at the last possible moment, i.e. before wiring.

5. Electronics

5.1. CIRCUIT DIAGRAM

Because of space, power and radiation problems, it appeared advantageous to separate chambers from electronics; therefore the main part of electronics is located outside the inaccessible storage ring area.

The signal from a sense wire, typically a few millivolts, has to be amplified, shaped, delayed and stored. The signal produces the functions FOR (fast OR), MOR (memory OR) and DATA. The external trigger and control logics provides the functions STROBE for storing, READ for read-out and CLEAR for memory clearing. The circuit diagram is given in fig. 4, both for the technical and the functional configuration.

In the following we briefly discuss the different elements of the chain and, where necessary, their technical realization.

5.1.1. Preamplifier

The preamplifier is located on the proportional chamber. The amplifier cards (8 channels per card) are plugged directly in to the printed board on which the wires are soldered. Half an MC 1035 with 1.8 kΩ input impedance is used for each channel, providing an amplification of about 10, with a differential output.

5.1.2. Cable

The differential signal is transferred by twisted pair cables (64 channels per cable)* over a distance of 66.5 m to the main electronics. The cable has an impedance of 100 Ω, a resistance of 0.2 Ω/m and a response to a step input giving 30 ns rise-time (0 to 75%) after 66.5 m.

5.1.3. Discriminator

The cable ends on a line receiver amplifier with adjustable threshold by an external voltage V_R. It is realized by half an SN 75108, producing a TTL logic level. The output branches off to a monostable and to a fan-in to form a fast OR (FOR).

5.1.4. Monostable

The monostable, represented by an SN 74121 and tuned with an external RC circuit with 2% tolerance, produces a delay of nominally 450 ms. The delay can

* Permali, Maxeville, France.
[+] System Berg, by Aumann Co., Zürich, Switzerland.

* L. M. Ericsson, Älvsjö, Sweden.

be varied between 420 and 600 ns by an external voltage V_D.

5.1.5. *Memory gate and memory*

The STROBE function, produced by the trigger and properly timed to the monostable delay, transfers the signal into the memory flip–flop (SN 7474N). The memory contents are used to provide a level (MOR) for further slow decision, and to feed the input of the read-out gates (SN 7403 S3) to a 32-way data bus.

5.1.6. *Read-out*

The read-out[5]) is done by sequential scanning of the MOR functions. Groups with non-set MOR are skipped, groups with MOR set are called via the READ gate and read. The end of a read-out cycle (or a negative slow decision) is terminated by a memory CLEAR.

5.2. GROUPING

The *spatial* grouping of the electronics is characterized by high-density packing, resulting in 512 chan-

nels housed in a "crate", and 10 crates = 5120 channels in a standard rack.

The *logical* grouping consists of

STROBE, CLEAR in blocks of 512 channels;
FOR in blocks of 256 channels;
MOR in blocks of 32 channels;
DATA in 32 parallel lines.

5.3. ELECTRONICS PERFORMANCE

We discuss the performance with respect to three criteria.

5.3.1. *Threshold*

On a single channel, the sensitivity for operation free from oscillations is ≈ 0.5 mV at the input of the preamplifier. It is determined mainly by the switching characteristics of the SN 75108. When many channels are packed together, threshold for safe operation rises to 1 mV, due to common grounds and supplies and to the natural spread of integrated circuit properties.

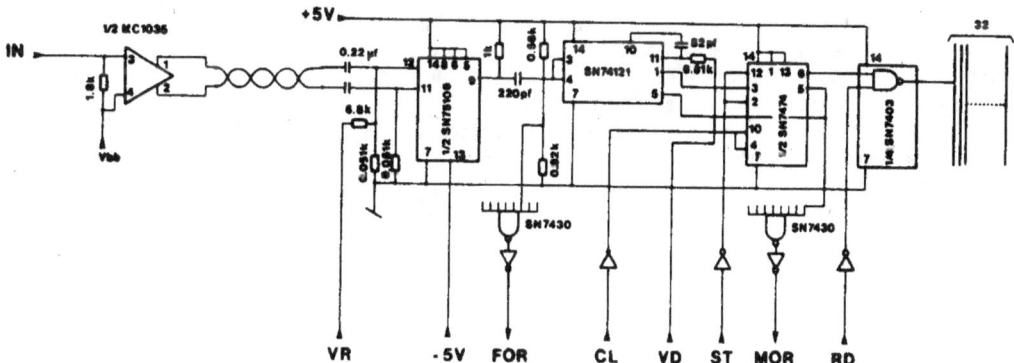

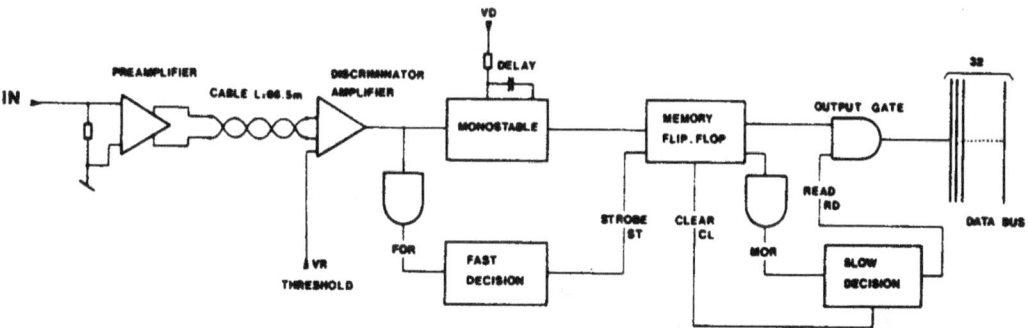

Fig. 4. Electronics diagram for one channel of proportional chamber electronics.

5.3.2. *Cross talk*

Electronic cross talk between adjacent channels, measured by applying a voltage pulse 100 ns wide to the input of the preamplifiers and by looking at the output of the SN 75108, is 14 dB, at a threshold of 2 mV. In actual chamber operation the cross-talk problem is helped by the fact that positive signals appear on the chamber wires adjacent to the wire hit by the avalanche.

5.3.3. *Time resolution*

The electronic time resolution is shown in fig. 5a for 512 channels, measured at the MOR level after pulsing all the chamber wires simultaneously. The experimental value of approximately 40 ns for full electronic efficiency can be decomposed into the following contributions:

- dispersion of transit-time and rise-time of MC 1035: 5 ns;
- dispersion of transit-time and of attenuation (depending on rise-time) in the twisted pair cables: 10 ns;
- spread of monostable time constant, due to propagation and component tolerances: 23.5 ns (fig. 5b).

In addition, for real chamber pulses a further deterioration of electronic time resolution is caused by:

- spread of propagation delay through line receiver, depending on input pulse height: 10 ns;
- monostable trigger, depending on discriminator output pulse width: 5 ns.

6. High-voltage read-out

In order to avoid additional material and cost by introducing diagonal wire planes, the two inner hv planes of each chamber are subdivided into strips forming a fishbone pattern (fig. 2). Each strip is coupled via a 1000 pF/6.3 kV capacitor to a preamplifier on the chamber edge.

The principle of the hv read-out is described elsewhere[6]. The ion drift in the electric field induces a negative pulse on the hit wire, and positive pulses on the hv electrode and the neighbouring wires. Adjacent hv strips show signals of the same sign, but smaller amplitude, i.e. position determination is therefore connected to amplitude discrimination. A higher amplification of about 70 is required from the preamplifier which is obtained by one circuit MC 1035 per channel, again with differential output and connected to the standard chain of cables and electronics.

Special care is necessary in the connection of the preamplifiers to the chamber ground in order to avoid reinjection of signals between distant preamplifiers: the propagation delay across the chamber causes a phase shift in the cross talk leading to oscillation. Shielding against external electromagnetic noise is equally important.

7. Chamber operation and performance

Up to now, ten chambers with 17 000 wires have been completely tested. We give here a short account of their basic working characteristics. All results given

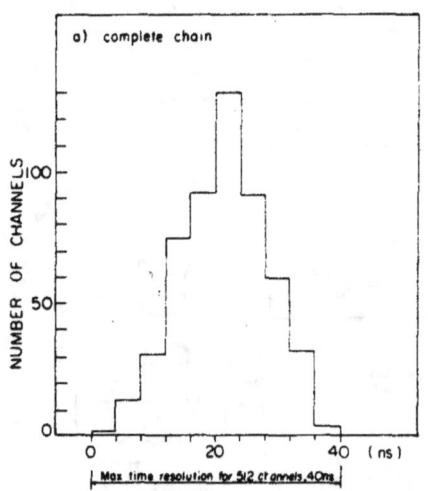
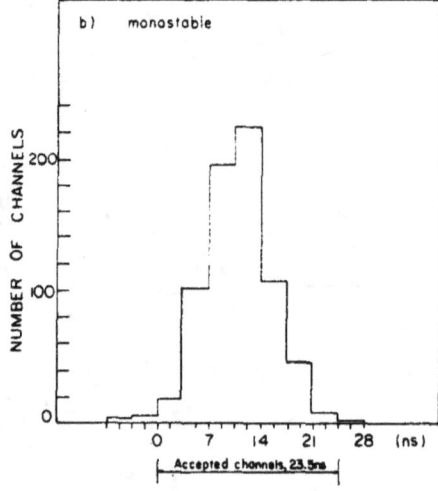

Fig. 5. a) Electronic time resolution for 512 channels. b) Time spread of monostables for 720 channels.

are obtained with our standard gas mixture ("magic" gas: 67.6% argon, 25% isobutane, 0.4% freon, 7% methylal)[7]).

7.1. WIRE READ-OUT CHARACTERISTICS

The wire read-out electronics allows safe operation down to thresholds of 1 mV for a fully equipped chamber. In fig. 6a, we present efficiency curves obtained on the horizontal (2 m long) wires for thresholds between 1 and 4 mV. At low threshold values, an apparent flattening in the approach to full efficiency is seen. This effect has been shown to be due to the freon admixture[8]). For continuous operation the electronic threshold is set to 2 mV. The time resolution of chamber and electronics, including 66.5 m of twisted-pair cable, is 90 ns on the "FAST OR" level and 110 ns on the "STROBE" level, taken over the full chamber.

7.2. PLATEAU SPREAD

In view of the mechanical stability problems mentioned above, the total spread of the beginning of the plateau, defined as $\varepsilon \geq 99.98\%$, is one of the most important chamber parameters. For the ten chambers investigated, we have found a maximum spread of ± 100 V, with respect to a mean value of 5.1 kV. Gas flow effects[8]) contribute about ± 30 V to this spread. The remaining ± 70 V are due to mechanical gap distortions. The corresponding maximum gap asymmetry is ± 0.3 mm, which is to be expected from the production tolerances.

7.3. NOISE, DISCHARGE AND PLATEAU LENGTH

After a "formation" period of about one week, the chambers reach stable operating conditions, as far as noise and breakdown characteristics are concerned. We define stable chamber operation by the noise rate of 1 count/s m per wire. Given proper cleaning, it is generally no problem to reach this noise level for all wires in a chamber. We do not have to exchange wires because of excessive noise rates. However, we have only a 50 to 100 V safety before reaching the beginning of the breakdown, which we define by a rate of 10 times the normal noise on a wire. From experience with a 5000-wire system[4]), working for two years already at the ISR, we expect the chambers to improve steadily in this respect during a period of several months. Improvement is also to be expected from a change in gas composition[8]).

7.4. IRRADIATION EFFECTS

By irradiating the chambers for a few minutes with local β-flux densities between 10^3 and $10^5/cm^2$ s, we have found that:

1) there is a general slight increase of noise rate which decays with a time constant of some minutes;

2) a few single wires show irradiation-induced

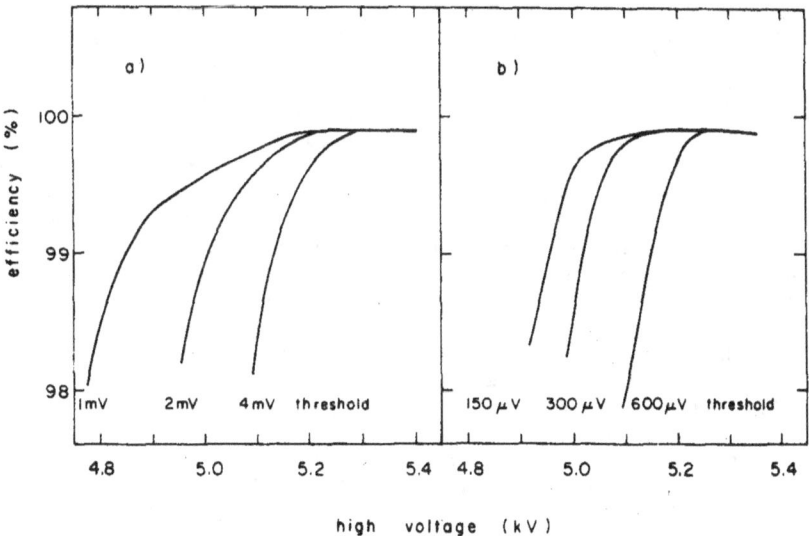

Fig. 6. Efficiency curves for different electronic thresholds: a) for wire read-out, b) for hv read-out.

discharges, which often also disappear in a few minutes' time:

3) in the case of real surface defects, stable and reproducible discharges on groups of wires are observed.

The three effects, observed with β-sources depend strongly on the surface quality of the hv electrodes. Faulty spots and surface degeneration, which occur during the production process, are readily detected by their irradiation effects. It was for this reason that we metallize the electrodes during the final production stage.

7.5. HV READ-OUT CHARACTERISTICS

There are two main problems connected to hv read-out:

1) Due to the low electronics threshold required, the system is more sensitive against oscillations and external noise; with the electronics layout and grounding precautions mentioned, we have reached safe operation down to effective thresholds of 150 μV. In fig. 6b we show efficiency curves for hv read-out which demonstrate that in the threshold region between 200 μV and 800 μV we reach full efficiency for hv plateau values, comparable to those obtained on the wire read-out (fig. 6a).

2) Due to the wide space distribution of the induced signal, and due to cross talk between strips, it becomes very important to optimize the resolution.
A good feeling on space resolution is obtained by counting the number of adjacent strips which are read-out in each event (cluster size). This depends on the following parameters:
- chamber high voltage;
- electronics threshold;
- timing (strobe width);
- geometrical position of the event relative to the hv strips.
In fig. 7, we present the percentages of single, double and triple clusters as a function of source position, scanning across several strips. At an operating potential of 5.0 kV, we are slightly below plateau for hv read-out ($\varepsilon = 97.5\%$) and the number of triple clusters is in the few per cent region. Taking into account the finite collimation of the source, we notice a sharp correlation between double clusters and source position whenever we cross the border between two strips. This is to be expected from the distribution of induced charge. If we go up to 5.3 kV chamber voltage, we obtain full efficiency, but the bulk of events gives triple clusters.

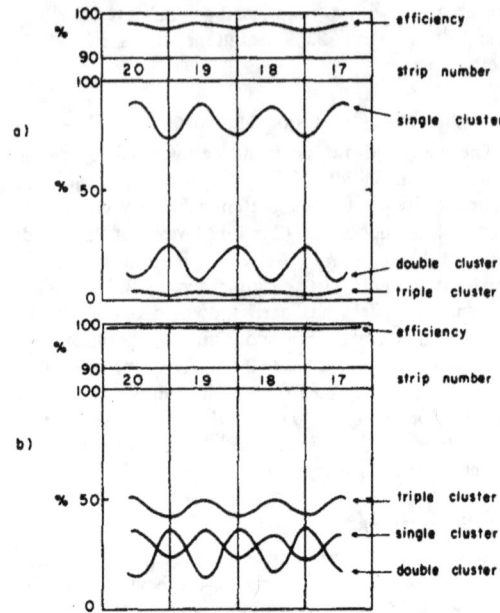

Fig. 7. Space variation of efficiency and cluster size for hv read-out: a) hv = 5.0 kV, b) hv = 5.3 kV.

Generally the operating voltage is limited by noise and the wire read-out electronics. For a chosen efficiency, threshold and strobe width can be used to minimize the cluster size. By careful adjustment, we have reached 99% read-out efficiency on each hv plane with less than 5% triple-cluster events.

We wish to thank the following persons for their continuous collaboration during the project: G. Dinkel and the Wire Chamber Workshop for the construction of the chambers; E. Albrecht, M. Blas, G. Cerutti, G. Gallay and G. Million, our group technicians; R. Alberganti and G. Di Tore for electronics; L. McCulloch for the read-out system; V. Chabaud for on-line test programs.

We are indebted to Dr F. Sauli for many contributions during the early phase of the project.

References

1) A. Minten (Ed.), *The Split-Field Magnet facility*, CERN-SFM Note-4, rev, October 1972.
2) J. Billan, R. Perin and V. Sergo, Contribution to the 4th Int. Conf. on *Magnet Technology*, Brookhaven, and CERN-ISR-MA/72-53 (1972).
3) K. B. Burns, B. R. Grummon, T. A. Nunamaker, L. W. Mo and S. C. Wright, The Enrico Fermi Institute, Chicago,

preprint EFI 72-37 (1972). The authors were the first to question metaphysical aspects of proportional chambers.

4) G. Charpak, H. G. Fischer, A. Minten, L. Naumann, F. Sauli, G. Flügge, Ch. Gottfried and R. Tirler, Nucl. Instr. and Meth. **97** (1971) 377.

5) J. Lindsay, L. McCulloch and H. I. Pizer, CERN NP Internal Report 71-16 (1971).

6) H. G. Fischer and J. Plch, Nucl. Instr. and Meth. **100** (1972) 515.

7) R. Bouclier, G. Charpak, Z. Dimcovski, H. G. Fischer, F. Sauli, G. Coignet and G. Flügge, Nucl. Instr. and Meth. **88** (1970) 149; G. Charpak, H. G. Fischer, C. R. Gruhn, A. Minten, F. Sauli, G. Plch and G. Flügge, Nucl. Instr. and Meth. **99** (1972) 279.

8) H. G. Fischer, H. J. Hilke, F. Piuz and O. Ullaland, Contribution to the Int. Conf. on *Instrumentation for High-Energy Physics*, Frascati (1973).

CERN
SERVICE D'INFORMATION
SCIENTIFIQUE

NUCLEAR INSTRUMENTS AND METHODS 148 (1978) 471-482 ; © NORTH-HOLLAND PUBLISHING CO.

PROGRESS IN HIGH-ACCURACY PROPORTIONAL CHAMBERS

G. CHARPAK, G. PETERSEN, A. POLICARPO* and F. SAULI

CERN, Geneva, Switzerland

Received 3 August 1977

For ionizing events punctually localized in the volume of a proportional chamber, the mean position of the corresponding avalanches are well defined. By reading the position of the ion cloud, using pulses induced in cathode strips arising from absorption of low-energy X-rays, the position accuracies obtained are: along the anode wire $\sigma \sim 35\,\mu m$, and on the direction orthogonal $\sigma \sim 150\,\mu m$ with 2 mm wire spacing. The intrinsic position accuracy of the method of measurement is much better.

1. Introduction

In a recent article[1], the properties of the proportional chambers' read-out, based on a direct measurement of the centroid of the charges induced on cathode strips, have been described.

Among the properties which make this approach attractive in particle physics we may mention that:

- The time resolution of the read-out method can be 50 ns. Cheap amplifiers permit in such a short time to collect enough charge from the cathode strips to permit a high resolution in the centroid's position.
- The accuracy in the measurement of avalanches along the wire was found to be far better than the width of the particle beam (200 μm fwhm) used in the experiment.
- Interpolation between the wires can be obtained for inclined charged tracks when ionization electrons are shared between two wires or more.
- The pulse height is obtained on two orthogonal electrodes for every track, thus sometimes permitting the removal of ambiguities for multiparticle hits.

This investigation was pursued with two aims: to study some limiting factors in the accuracy of position measurements with multiwire proportional chambers (MWPC); to arrive at a better understanding of the mechanism of pulse formation in the chambers.

2. Experimental set-up

The MWPC used in these experiments has been described in the previous paper[1]. The gold-plated tungsten anode wires are 2 mm apart and have a diameter of 10 μm. The cathode planes, at a dis-

tance of 8 mm, are made of 50 μm diameter wires, 500 μm apart, connected in group of six wires. Each group of six cathode wires connected together will be called a strip. In one cathode plane the wires are parallel to the anode wires; in the other they are orthogonal.

The gas fillings used were: argon or xenon (55%), isobutane (38%), and methylal (7%); and the same filling using argon and in addition freon-13 B1 (0.5%), i.e. the so-called magic mixture[2]. Except for input pulse polarity the same amplifiers* were used to handle negative pulses from the anode wire and positive pulses, either on the strips or in both anode wires adjacent to the one that collected the electrons (fig. 1a). Characteristics of these amplifiers have been reported before[1], an adjustable decay time facility being added, such that the output of the amplifiers could be made to have similar shapes. For a short track from 1.5 keV X-rays the pulses had a full duration $\leqslant 100$ ns, peaking at about 30 ns.

The pulses from the amplifiers were fed into analog-to-digital converters (ADCs), whose gate was opened during a chosen time by the main negative pulse from the anode wire a that collected the electron from the avalanche. Most of the experiments were made using 1.5 keV Al K X-rays from an X-ray tube and 5.9 keV X-rays from a ^{55}Fe source. Data collected from the ADC are fed into a computer.

3. Preliminary considerations

Let x, y, z be a system of reference, y along the anode wire a that collects the electrons from a detected event, and z normal to the plane of the

---

* On leave from Departamento de Fisica, Universidade de Coimbra, Portugal.

* Developed at CERN by J. C. Santiard, and manufactured by CIT-Alcatel.

a)

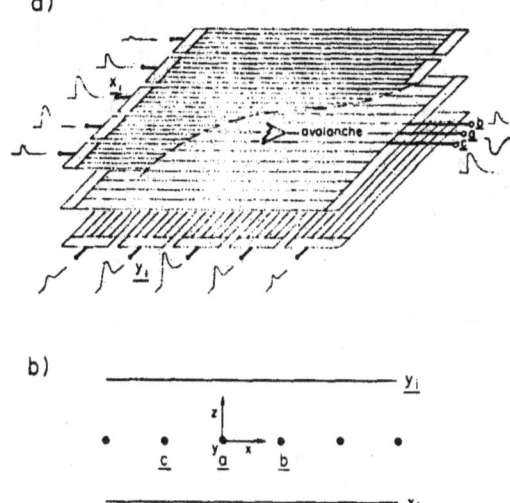

b)

Fig. 1. (a) Diagram of the MWPC showing the position of strips x_i and y_i, main anode wire a, and neighbouring wires b and c, as well as the pulses on these electrodes. (b) System of reference.

anode wires (fig. 1b). At $x = \pm 2$ mm are situated the neighbouring wires b and c. At $z = -4$ mm is situated the set of strips that will be referred to as x_i (its x coordinate is x_i) and at $z = +4$ mm the set of strips y_i (its y coordinate is y_i).

Let us use majuscules to designate the ADC readings associated with the different electrodes that surround the avalanche, i.e. A, B, C, X_i, and Y_i. The usual centre-of-gravity technique defines positions using the expressions $\sum X_i x_i / \sum X_i$ and $\sum Y_i y_i / \sum Y_i$ that will be called x' and y', respectively. These expressions arise from the assumption that x_i and y_i strips read "the avalanche position" and one expects them to be true only if the accuracies involved are not better than the dimensions associated with the ion motion responsible for pulse formation. A more careful look at the problem is then essential if one wants to reach the full accuracy inherent in the system if this accuracy is better than those dimensions.

3.1. INDUCED SIGNALS

The charge variation in any electrode from the movement of a positive point charge q between positions 1 and 2 in the chamber is the same as the charge variation that arises from the movement of a negative point charge $-q$ between po-

sitions 2 and 1. Assume that t_0, t_1, and t_2 are, respectively, the times of creation of an ion pair in the avalanche, the collection time of the electron, and the collection time of the positive ion, and that the gates are opened before t_0 and closed at some time t_3 ($t_1 < t_3 < t_2$). Then if an ion pair is created in any position in the chamber the motion of the electron will correspond simply to the inverse motion of a positive charge, as far as induced charges are concerned. Only movements of positive charges will then be considered and one sees that for times t_3 the induced charges can be obtained by considering the motion of a positive ion from an initial position at the anode surface to a final position that is the same as the position of the positive ion at t_3.

Let us now look at the motion of a point charge in a chamber. Assume that the positive point charge is very near a particular electrode. Then essentially all field lines leave q and end up on this particular electrode, the induced charges being $-q$ on this electrode and zero on all others. If q is moved slightly away from this electrode some field lines leave this electrode and end up on the other surrounding electrodes, i.e. in absolute values, there is a decrease of induced charge on the electrode nearby and an increase of induced charge on all other electrodes, the total induced charge al-

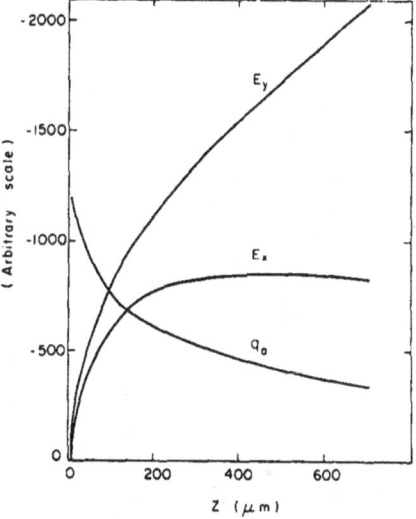

Fig. 2. Induced charges E_Y and E_X on the cathodes and induced charge q_a on the anode, as a function of the z coordinate of the "ion pin".

ways being $-q$. If q is now moved very near to another electrode, the reverse process sets in, i.e. the induced charges in absolute value will decrease on all other electrodes and increase tending to q on this particular one.

Then the initial motion of a positive point charge away from the anode surface should give rise to positive pulses both in x_i and y_i strips, and to explain this experimentally observed fact *it is not necessary that the avalanche surrounds the wire in a more or less symmetrical way.*

To verify this effect, and to obtain orders of magnitude of the most important parameters, measurements were made using a bidimensional scaled-up ($\times 100$) simulation of the chamber using conductive paper. The ion was simulated by a thin pin that could be moved over the paper and then the system behaves as if the ion was an infinite uniformly charged line. Having the "ion pin" on positive potential and keeping all electrodes at ground, for a fixed current through the ion pin (corresponding to a fixed ion charge), one was able to measure the currents through the electrodes (corresponding to the induced charges on the electrodes), for any position of the pin in the table.

Let us define $E_x = \sum X_i$ and $E_y = \sum Y_i$. Fig. 2 shows the variation of E_x and E_y with the z coordinate of the "ion pin" for $x = 0$, and up to very large distances, about $500\,\mu m$, the total induced charge, in absolute values, increases both in x_i and y_i strips, although the motion of the positive charge goes away from the x_i set; only for larger distances E_y keeps on increasing and E_x decreases. From this figure it also arises that z can be determined by using E_x and e_y information. On the same figure and in another arbitrary scale, the variation of the induced charge on the anode (curve q_a) is also shown as a function of z, decreasing always when z increases.

From what was said before, and apart from the electrode that collects the electrons, one detects essentially the induced charges corresponding to the position of the positive ions at the time t_3 when the gate closes.

Let us assume that near the anode the field is radial. This means that at the time t_3 the ions will be somewhere on a cylinder centred at the wire and with radius $r = r(t_3)$. To have information on the induced charges in the surrounding electrodes for this case the "ion pin" was located on different positions on a circle around the anode. In fig. 3 full curves show E_x vs z for several r values, as indicated in μm near the curves. In the same figure, but on another arbitrary scale, the induced charge q_a on the anode (the dashed curves) is also represented as a function of z. The two most important results are that, for a fixed distance of the ion to the wire, E_x is roughly a linear function of z, and the induced charge on the anode is constant. The ratio E_y/E_x varies more with z than E_x, for example. Also it is independent of the ion charge. It is a monotonous function of z and a good way to determine it. One then expects the same information as x to be directly available from the ratio of induced pulses on the neighbouring wires, B/C.

What does $x' = \sum X_i x_i / \sum X_i$ mean? It is the centre of gravity of the charge induced on x_i strips and it is clear that it is not only a function of x, but also of x and z, for a particular configuration.

Let us designate by q_0 the ion charge and by $q_i (i = 1, ..., n)$ the charge on the anode wires i arising from the presence of the ion in the real chamber. We shall consider a fictitious chamber, like the real one but without anode wires. The presence of a certain $q_i (i = 0, ..., n)$, located at the same position where it was in the real chamber, gives rise to a certain charge distribution on the cathode of this fictitious chamber. In the real

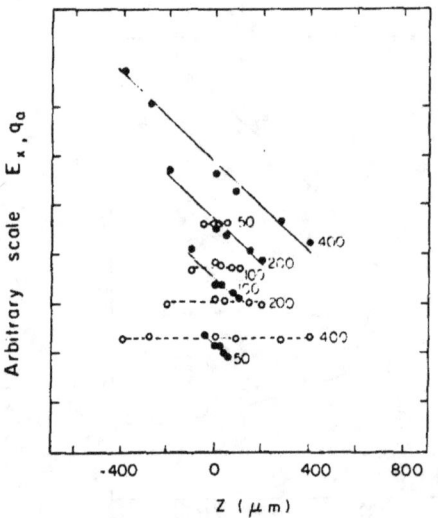

Fig. 3. Induced charges E_X (full curve) on a cathode and q_a (dashed curve) on the anode, as a function of the z coordinate of the "ion pin", for different values (indicated near the curves in μm) of the distance between the "ion pin" and the anode.

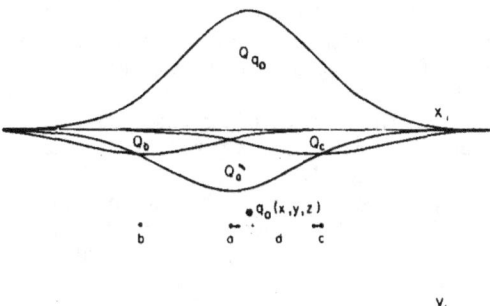

Fig. 4. Distribution of induced charges on x_i strips arising from the individual charges located on b, a, q_0, and c. The x' value corresponds to the centre of gravity of all induced charge distributions that are represented.

chamber the total charge distribution is the superposition of these individual charge distributions.

In fig. 4 are represented the induced charge distributions on the cathode x_i arising from q_0 and q_1, assuming that only neighbouring wires make a significant contribution. To calculate x' each individual charge distribution can be replaced by the total charge Q_i in it, centred in its maximum.

As we have seen, $q_b \approx \alpha + \beta x$, $q_c \approx \alpha - \beta x$, and $q_a = \gamma$; α, β, and γ being constants when one considers positions of the ion on a small circle centred in the anode. Now $Q_b = K(\alpha + \beta x)$, $Q_c = K(\alpha - \beta x)$, $Q_a = K\gamma$, and $Q_{q_0} = f(z)$. Then

$$x' = \frac{Q_{q_0}x + Q_c d - Q_b d}{Q_a + Q_b + Q_c + Q_{q_0}} = \frac{[f(z) - 2K\beta d] x}{K\gamma + 2K\alpha + f(z)},$$

where d is the distance between anode wires. For small values of x, x' is then an approximately linear function of x, because for small x, $f(z) \approx$ constant. From the data obtained in the simulation experiment $x \approx x'$. It seems that we have then at our disposal a simple way of determining the ion location, through the ratio E_y/E_x, B/C, and x'; considering the y coordinate, of course, $y' = y$.

3.2. DIFFUSION EFFECTS

Let us now consider a punctual cloud of electrons created in some point inside the chamber. In a first approximation let us neglect diffusion effects, the size of the avalanche distribution, and the finite bandwidth of the electronic system. The punctual electron cloud drifts along a field line, which at the wire surface makes an angle θ with the z axis. The punctual positive charge – the

cloud of positive ion arising from the avalanche – migrates backwards along the same field line and at $t = t_3$ has coordinates x, y, z. To each field line in the chamber, for a certain t_3, corresponds then a certain point: the characteristics of the spatial resolution of the chamber would then be defined by the accuracy on the determination of x, y, and z; by the particular line of force configuration; and by the geometric characteristics associated with the interaction of the particles in the filling gas.

If one now considers diffusion effects, in a simple approximation, the field in the chamber has two distinct regions, being uniform in one of them and radial near the wire. A punctual cloud of n electrons made very near the cathode, in our geometry, has a $\sigma_{\bar{x}}$ (r.m.s. of the mean position of the electron cloud) of $\sim 200/\sqrt{n}\ \mu$m when it reaches the radial field region. If there was no diffusion in this region, shall we say with a radius of $\sim 500\ \mu$m, then that $\sigma_{\bar{x}}$ would imply a r.m.s. of θ given by $\sigma_{\bar{\theta}_x} \approx 2°$ if $n \approx 100$.

In the region where the field is radial, if the usual formula for diffusion in uniform fields is generalized to that geometry, then the σ_θ arising from diffusion of one electron from a certain point at a distance R_N from the centre of the wire, R_0 being its radius, is

$$\sigma_\theta \approx \left(2 \frac{e D_T}{\mu a} \ln \frac{R_N}{R_0} \right)^{\frac{1}{2}},$$

on the assumption that $e D_T/\mu - D_T$ and μ being, respectively, the transverse diffusion coefficient[3]) and the electron mobility – is a constant in that region, a being defined by $E(r) = a/r$. Putting $e D_T/\mu \approx 10$ eV one gets an upper limit for σ_θ of about 30°; and with $n \approx 100$, $\sigma_{\bar{\theta}} \approx 3°$.

Another effect that contributes to the spread in the mean position of the cloud of positive ions is the statistics associated with the process of avalanche build-up. This effect can be estimated, because the statistical fluctuations in charge gain associated with one electron have been investigated[4]). Taking 0.7 for the relative variance of charge amplification for one electron, then the contribution to $\sigma_{\bar{\theta}}$ arising from this process is $\sim 2°$ for $n \approx 100$ electrons. One should then expect a good definition of $\bar{\theta}$ although the azimuthal dimension of the avalanche, arising only from diffusion, would be given by $\sigma_\theta \approx 30°$. Of course, an unknown contribution for the size of the avalanche arises from photon propagation and depends very much on the nature of the gas filling.

3.3. Data handling

Let us consider an ADC i and let P_i be its reading corresponding to a detected event. For a certain gatewidth, with no source bombarding the chamber, let p_i be the ADC reading and σ_{p_i} the r.m.s. of p_i. With $P'_i = P_i - \bar{p}_i$, $E_x = \sum P'_i$ for the ADC connected to strip x_i, and the same for E_y, A, B and C.

To decrease the contribution of the noise σ_{p_i} in the computation of x' and y', a bias b is subtracted from all readings associated with x_i and y_i strips, i.e.

$$x' = \frac{\sum (P'_i - b) x_i}{\sum (P'_i - b)},$$

where the quantity $P'_i - b$ is put to zero if it is negative. But an effect has to be taken into account: consider a charge q placed in a fixed position in the chamber; a change of the value of q implies, if b is kept fixed, a variation in x'. A first-order correction to this effect was made by putting b proportional to the anode pulse. The level of b was such that in the calculation of x' or y', four or five strips were read. Because a finite number of strips is read and they are not necessarily symmetrically distributed relatively to the centre of gravity of the induced charge distribution, one expects that the centre of gravity calculated using only these strips is slightly different from the real centre of gravity. This is, of course, a general effect that applies to both x' and y'.

Concerning the electronic noise contribution to x' and y', if $\max(P'_i)$ is the biggest strip reading, typically $\sigma_{p_i}/\max(P'_i) \approx 1/500$ and calculations showed a noise contribution to $\sigma_{x'}$ and $\sigma_{y'}$ of the order of $10\ \mu m$.

Let us now consider another effect that was analyzed. Essentially three types of events are detected in the chamber: (1) all electrons are collected at wire a; (2) all electrons are collected in a neighbouring wire (b or c); (3) the electrons split between wire a and one of the neighbouring wires. We point out that although the gates are opened by pulses from a, an event of type (2) can still open the gates through overshoot of the induced positive pulse on a. But those events were rejected by requiring that a positive pulse should be detected on b and c. This requirement also will reject some events of type (3): consider an event of type (3), for which electrons are split between a and b. In b there will be a negative pulse from the avalanche that surrounds it and a positive pulse induced from the avalanche on wire a. If the positive pulse is smaller than the negative one, the requirement stated above, that a positive pulse should be detected on a neighbouring wire, rejects that event. The rejecting of the remaining events of type (3), anyway very few, will be considered later (see section 4.1).

Finally, one should remark that the finite bandwidth of the reading system will imply that slow charge variations, such as those that would arise from the motion of ions in regions of lower electric field, will not be detected. This means that it will only be possible to detect induced charges, which correspond to the positions of the positive ion, until a certain distance from the anode wire.

4. Experimental results
4.1. X-rays (normal gas mixtures)

Low-energy X-rays provide a convenient probe to study the processes that were mentioned before, since the ranges of the photoelectrons and eventually of Auger electrons can be small. In the argon mixture, with a ^{55}Fe X-ray source, ranges are of the order of $100\ \mu m$, and Al X-rays have ranges of about $20\ \mu m$. At a distance of $\sim 500\ \mu m$ from the wire the θ uncertainty from the range effect is $\sim 2°$ for these X-rays.

Using a mixture xenon + isobutane + methylal with high xenon content if the chamber is uniformly irradiated with Al X-rays, those are mostly absorbed in one side of the chamber and then this provides a simple check of a correlation between the point of interaction of an X-ray and the corresponding final position of the positive ions from the avalanche. In fig. 5a the ratio E_Y/E_X is plotted vs B/C, each point being associated with the interaction point of an X-ray, when bombarding the chamber from the x_i side with a uniform beam of Al X-rays. If the xenon content is decreased (by a factor of 4) then the absorption of the gas mixture to the X-rays is much smaller and fig. 6b shows clearly this effect.

E_Y/E_X then effectively provides z information. In the conditions of this figure, since ion pairs are being created by the incident X-rays all over the chamber, all θ angles are fed. From the preliminary considerations one expects that all positive ions from the avalanches will lie on a circle centred on the anode wire. And this would exactly give rise to fig. 5b. Also, because both B/C and x' should be approximately linear functions of the x coordinate of the positive ions, one would expect

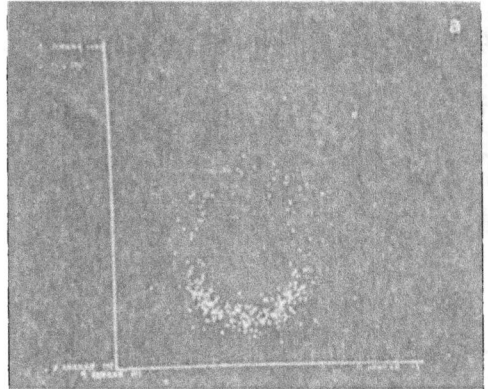

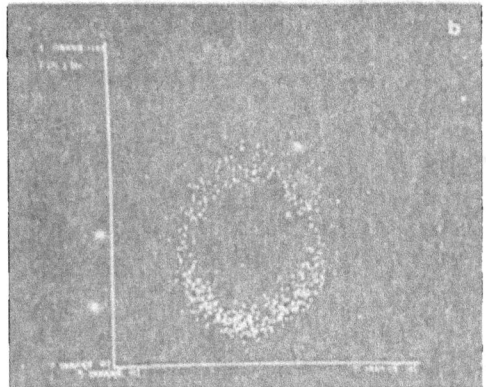

Fig. 5. E_y/E_x vs B/C for Al X-rays in a xenon + isobutane + methylal mixture, with rejection R (see section 4.1): (a) high xenon concentration; (b) low xenon concentration.

more uniform than with the xenon mixture. The chamber was bombarded with the collimated slice of X-rays, this being orthogonal to the plane of the chamber and parallel to the anode wires. In this way all X-ray interactions take place in a plane $x = x_b$ (see fig. 7, beam 0) and it is then clear that one is feeding mainly two θ angles, θ and $\pi - \theta$. Fig. 8a shows E_y/E_x vs B/C for three positions of the collimator separated by $500\,\mu m$, for the ^{55}Fe source. The same events are represented in a B/C histogram, fig. 8b, which shows that positioning between wires is possible, although the avalanche is made on only one wire. Fig. 8c displays again the same events, but as an x' position spectrum, showing a clear loss of resolution. A factor that contributes to the loss of resolution using x' is the much larger dynamic range associated with all strips electronics as compared with the electronics reading neighbouring wires, together with non-linearities.

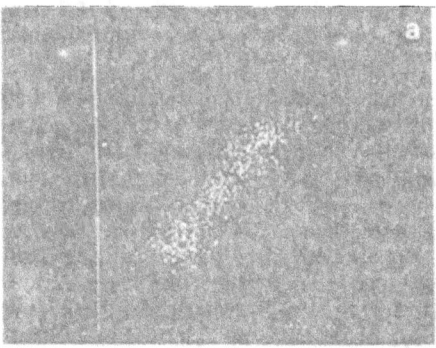

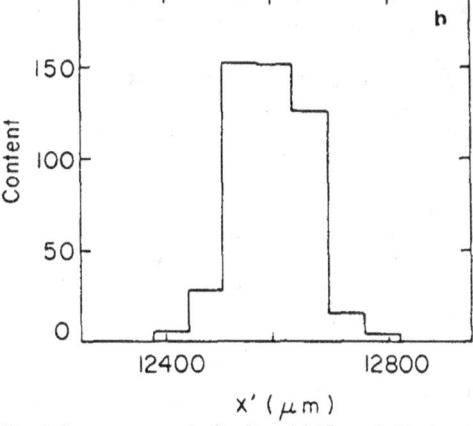

Fig. 6. Same events as in fig. 5b: (a) B/C vs x'; (b) x' position spectrum (rejection R).

a linear relationship between B/C and x', which is shown in fig. 6a for the same events as fig. 5b. The same events again, plotted as an x' position spectra, are shown in fig. 6b, the usual representation of a "wire" in a MWPC.

In fig. 6a some of the events are very far from satisfying the linear relationship B/C vs x'. These are the remaining events of type (3) as described in section 3.3, and as one sees from the figure they can easily be rejected in the data-handling procedure – this will be denoted rejection R.

The chamber was irradiated with collimated beams, trying to reach well-defined θ angles. This was done with both the Al X-ray beam and the ^{55}Fe source with a $10\,\mu m$ wide collimation, of 3 mm depth and using an argon + isobutane + methylal mixture in the chamber, such that the absorption of the X-rays in the chamber is much

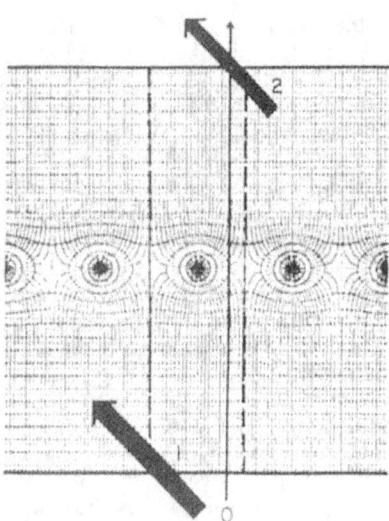

Fig. 7. Field lines and equipotentials are represented. The dashed lines limit the cell associated with the main anode wire. 0, 1, and 2 represent collimated X-ray beams as described in the text.

The results reported so far were obtained with a 100 ns gatewidth, started by the anode pulse. Bombarding the chamber uniformly with X-rays, a few runs were made with different gatewidths, displaying E_γ / E_x vs B/C, i.e. the ring of positive ions. Between 40 ns and 200 ns no clear variation of the ring was observed; for shorter gatewidths the ring would lose definition. This is the result of both the finite bandwidth of the electronics and the bad signal-to-noise ratio associated with shorter gatewidth.

It is of practical interest to know the relationship between B/C or x' and the x_b position of the X-ray bombarding beam, as well as the positioning accuracy between wires thus obtained. Of course, this data contains the many contributions associated with range of photoelectrons, diffusion, avalanche build-up, and the geometry of the electric field lines.

Using Al X-rays the position capabilities on the direction orthogonal to the anode wires are shown in fig. 9, where B/C (full curve) and x' (dashed curve) are plotted vs x_b, the position of the colli-

Fig. 8. (a) E_γ / E_x vs B/C for three positions of a collimated ^{55}Fe X-ray beam source separated by 500 μm, using an argon mixture (rejection R). (b) The same events represented in a B/C histogram. (c) The same events represented in an x' position spectrum.

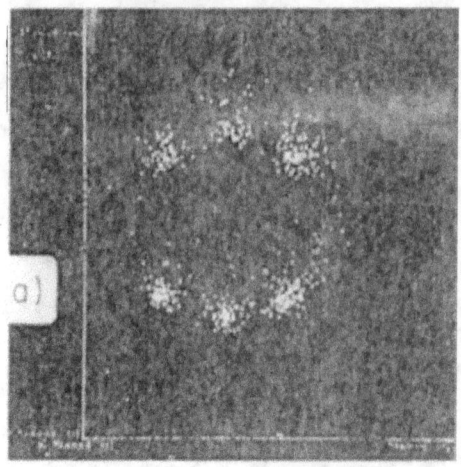

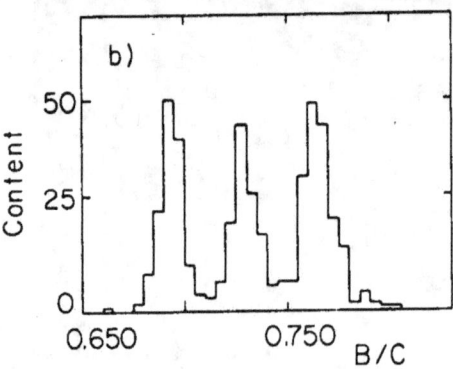

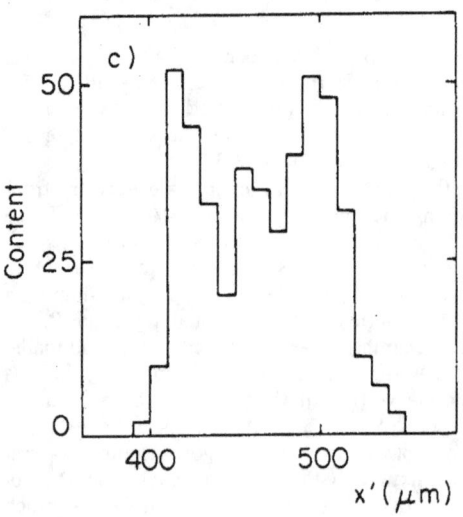

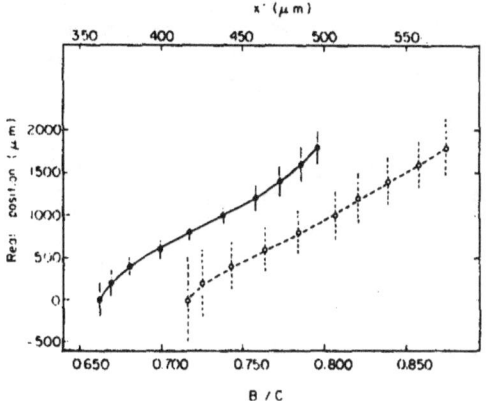

Fig. 9. Position of the X-ray beam in the direction orthogonal to the anode wires vs B/C (full curve) and vs x' (dashed curve). The bars indicate $\pm\sigma$ in μm (rejection R).

mated X-ray beam. The bars associated with every data point are $2\sigma_{xh}$ in μm. The mean position accuracy on the direction orthogonal to the anode wires was then $\sigma_{xh} \approx 150\,\mu$m using B/C information and $\sigma_{xh} \approx 300\,\mu$m using x' information. Using the ^{55}Fe X-rays source the mean position accuracy was also $\sigma_{xh} \approx 150\,\mu$m using B/C.

It is also of practical interest to know the position resolution along the anode wires' direction. Using the same collimator as for the x measurements, the thin slice of Al X-rays was positioned orthogonal to the anode, such that all X-ray interactions take place in a $y = y_h$ plane. A y' histogram, corresponding to three positions separated by 200 μm, in a xenon mixture, is shown in fig. 10. For the same mixture y_h is plotted as a function of

$\bar{y}'$ in fig. 11, the error bars being $2\sigma_{yb}$. Then, using xenon, $\sigma_{yb} \approx 35\,\mu$m. The same type of measurements with argon gave $\sigma_{yb} \approx 45\,\mu$m. Although from the considerations in section 3.3 one should not expect $y' = y_b$, data of fig. 12 show that an approximately linear relationship is obtained, even when a relatively wide range of y_b values is concerned.

4.2. X-RAYS (MAGIC MIXTURE)

To study the effect of higher gain and larger avalanches the chamber was operated with the so-called magic mixture. The results can be compared with the ones obtained with the normal mixtures, because, owing to the higher gain in the chamber, electronic-noise contributions are smaller with the magic mixture and thus the data reflect the contribution from the physical processes.

In fig. 12 the rings displayed, obtained with an Al X-ray source, irradiating the chamber uniformly, correspond, respectively, to the following conditions:

a) normal argon mixture, $h\nu = 2.8$ kV;
b) magic mixture, $h\nu = 2.8$ kV (transition between proportional and full charge saturation region);
c) magic mixture, $h\nu = 3.2$ kV (full charge saturation).

The rings displayed in this figure can be compared directly and one sees that on going from the proportional to the full charge saturation region, for most of the events there is a collapse of the ring. Nevertheless fig. 12c shows that for some events, those that are very dispersed, there are very big

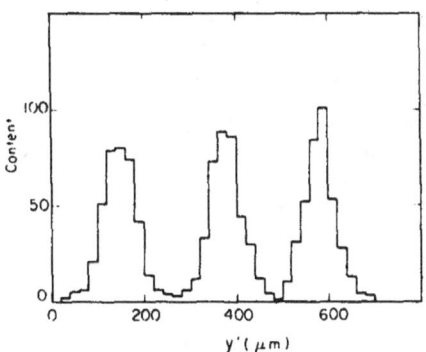

Fig. 10. y' (in μm) histogram for three positions of the collimated beam with 200 μm separation, in a xenon mixture (rejection R).

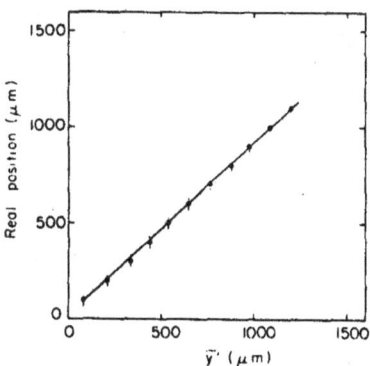

Fig. 11. Position of the X-ray beam along the direction of the anode wires vs y' (in μm) in a xenon mixture. The bars indicate $\pm\sigma$ in μm (rejection R).

asymmetries, both in E_Y/E_X and B/C. These events are not associated with any misbehaviour of the electronic system, for example, as one can see in fig. 13, where B/C vs x' is displayed for the same events.

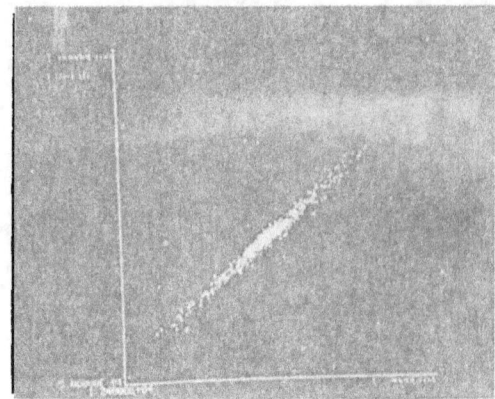

Fig. 13. Same events as in fig. 12c, but displayed as B/C vs x'.

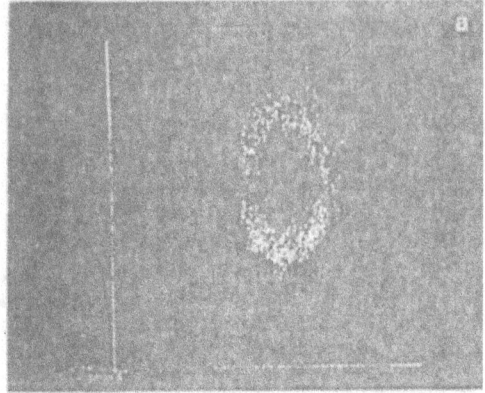

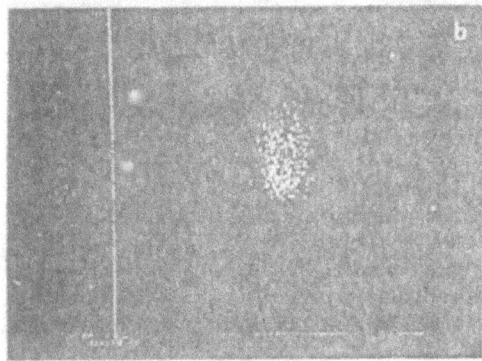

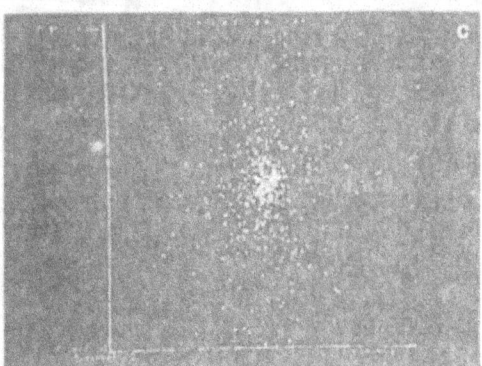

Fig. 12. Chamber irradiated uniformly with Al X-rays. Displays of E_Y/E_X vs B/C. (a) Normal argon mixture, hv = 2.8 kV (rejection R); (b) magic mixture, hv = 2.8 kV; (c) magic mixture, hv = 3.2 kV.

As said before, these data correspond to a uniform irradiation of the chamber and therefore it is not absolutely clear that there is a correlation between the point of interaction of an X-ray in the chamber and the position of the positive ion cloud that is read by the chamber. To test this effect the thin collimated Al X-ray beam was positioned in the same conditions as those of fig. 8, and fig. 14 shows E_Y/E_X vs B/C displays for: (a) beam at $-500 \mu m$; (b) beam at $0 \mu m$; and (c) beam at $+500 \mu m$, in the condition of full charge saturation. The correlation is then clear, and from B/C histograms it was found that there is an overlap in about 7% of the events when the beams at $-500 \mu m$ and $+500 \mu m$ are considered. This result should be compared with the data of fig. 8b.

To demonstrate that through better data analysis improvements are possible, an ^{55}Fe X-ray source was collimated using a $100 \mu m$ collimator and the thin beam thus obtained was positioned parallel to the anode wires, in two different positions, 1 and 2, as indicated in fig. 7. In this way, the regions of interactions of X-rays are relatively well defined in the corners of the cell that is associated with wire a.

From beam 1 let us say that the angles that are fed are between θ_1 and θ_2. Then from beam 2, one is feeding angles between $\theta_1 + \pi$ and $\theta_2 + \pi$. Fig. 15 shows the events from a usual ring plot, i.e. E_Y/E_X vs B/C, projected in a line that makes the angle $\frac{1}{2}(\theta_1 + \theta_2)$ with the z axis, the full line corresponding to beam 1 and the dashed line corresponding to beam 2. This histogram shows that events from beam 1 and 2 overlap in about

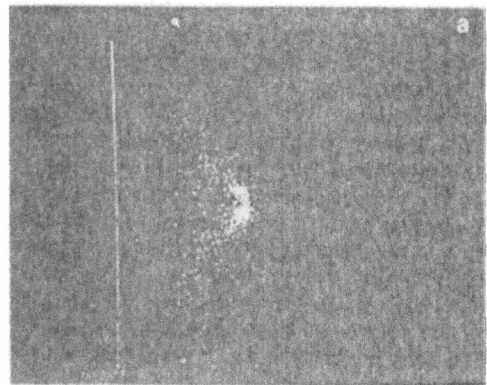

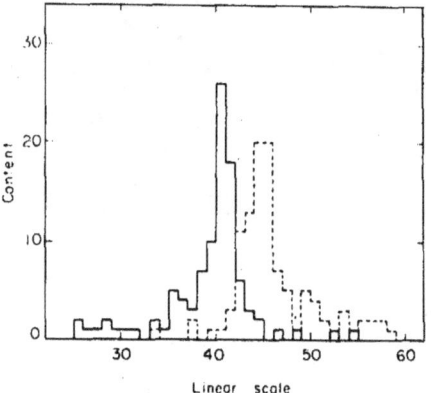

Fig. 15. Display of the asymmetry obtained with the magic mixture with full charge saturation (see text).

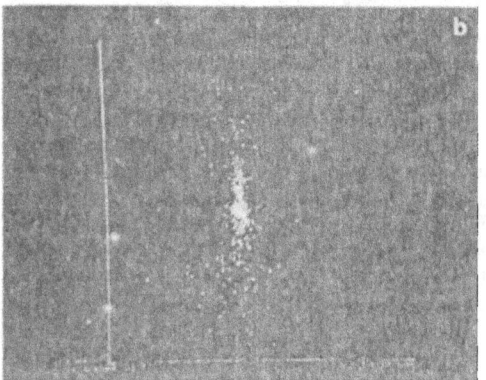

Fig. 13 indicates that $\sigma_{x'}$ is small. An upper limit to this quantity is obtained if one selects events on a very narrow band of B/C and displays the corresponding x' histogram, shown in fig. 16. This features $\sigma_{x'} = 6\ \mu m$ and, of course, the same upper limit is expected for $\sigma_{y'}$. Then $\sigma_x = \sigma_y \approx 6\ \mu m$ (see section 3.1).

The resolution along the y direction was tested with an Al X-ray beam. With the chamber working in the transition region between proportional and full charge saturation (as in fig. 12b), the measured value of σ_{y_b} was 50 μm; with full charge saturation (as in fig. 12c) $\sigma_{y_b} = 100\ \mu m$.

In general, comparing these results with those of section 4.1, the effect that a few electrons lead to charge saturation, together with electron recombination, could be responsible for worse resolutions along the y direction as well as for some loss

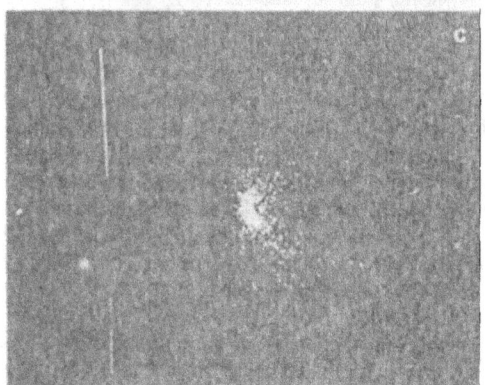

Fig. 14. E_Y/E_X vs B/C for three positions of a collimated Al X-ray beam, using magic mixture with charge saturation. (a) Beam at $-500\ \mu m$; (b) beam at $0\ \mu m$; (c) beam at $+500\ \mu m$.

10% of the cases only, even in this full charge saturation region. If only B/C information is used the overlap is much larger.

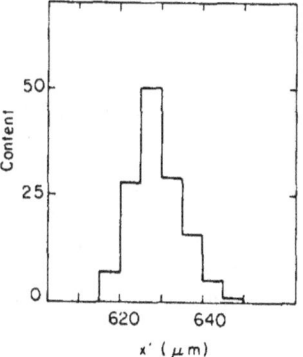

Fig. 16. The accuracy of determination of x' (magic mixture).

Fig. 17. Radiograph of the word CERN (dimensions of the word 1.5×4 mm²), using avalanches on one wire; 2 mm wire spacing.

of memory of the field lines that transport the electron cloud. The fact that for most events both E_Y/E_X and B/C show very little asymmetry suggests that there is also avalanche propagation. Nevertheless, the remaining events, associated with large asymmetries, should not be forgotten.

4.3. A DEMONSTRATION

The word "CERN" was cut out from a thin copper plate, the dimension of the word being 1.5×4 mm² and each letter being 50 μm thick. The mask obtained in this way was positioned in front of the chamber and parallel to it, the longer dimension of the word being parallel to the anode wires, the main anode wire a being centred with the mask. This mask was bombarded with an Al X-ray beam, the source being in vacuum at about 30 cm from the mask. Avalanches occurred then only on wire a, and the corresponding B/C vs y' plot is shown in fig. 17, using a normal argon mixture.

5. Discussion

The general features revealed by the experimental data obtained are in qualitative agreement with the expected behaviour of the ion pairs in the chamber and the consequent pulse formation. No quantitative approach was made, as the calculations of the induced charges in the chamber are still under way. Provided these are known as arising from a point charge, then the information available concerning diffusion of the charge carriers and the statistical fluctuations associated with the avalanche build-up, allow a proper understand-

ing of the process. Concerning the effects of photon propagation and space charge, the situation is reversed in the sense that experimental data will provide information on these effects. That the methods used are able to provide this information is clear, where, for example, one compares the "rings" of normal and magic mixtures. Another parameter of great importance, the time dependence (almost ignored in this paper), will help to probe the process of avalanche build-up. To what extent one can improve the energy resolution of proportional counters by using both the anode pulse and the "size" of the avalanche remains to be seen. On the other hand, improvement of the practical energy resolution obtained with MWPCs will be obtained simply because $\bar{\theta}$ is measured. It is clear from fig. 7 that one expects higher charge gains in regions $\bar{\theta} \approx 0°$ and 180° (regions of higher electric fields), the charge gain decreasing smoothly and reaching a minimum for $\bar{\theta} \approx 90°$ and 270°. Indeed it was found that, for a certain energy deposit in the chamber, the mean anode pulse-height decreases smoothly when $\bar{\theta}$ goes from 0° to 90°, as one should expect (the same behaviour for the other quadrants), being at $\bar{\theta} \approx 90°$ about 20% smaller than at $\bar{\theta} \approx 0°$, a non-negligible variation! A θ correction, associated with the anode pulse height, converts essentially the "bad" field line geometry inherent in a MWPC into the "good" cylindrical proportional counter geometry.

The position capabilities of the system are, in principle, limited only by the electronics noise and the fluctuations in the distribution of the charges in the avalanche. $\sigma_x \approx \sigma_y \approx 6$ μm was measured. Practical limits arise through the choice of pedestals p_i and bias b (see section 3.3), electronics non-linearities, and the integration effect over a strip width. A different effect has to do with the details of the interactions of the detected particle in the gas, the field-line distribution in the chamber and diffusion effects. For example, a few keV can be deposited in a chamber, the region of energy deposition being a few cm in some cases, a few μm in other cases. These processes have nothing to do with the capabilities of the detector itself.

Finally, the following comment can be made: better results can be obtained with a more developed analysis of the data. A very example was shown in section 4.2, a projection on the line $\frac{1}{2}(\theta_1 + \theta_2)$ being a better way of distinguishing events that, for example, use only B/C information. Another example is clear in figs. 8a and 8b.

Events are separated in fig. 8b using only the B/C information, therefore ignoring the E_Y/E_X information; in this case $\sigma \approx 150\,\mu$m in the direction orthogonal to the anode wires is really an upper limit of the system capabilities working with X-rays for which the energy deposition region is a few tens of μm. In these conditions, in the direction of the anode wires, $\sigma \approx 35\,\mu$m.

References

[1] A. Breskin, G. Charpak, C. Demierre, S. Majewski, A. Policarpo, F. Sauli and J. C. Santiard, Nucl. Instr. and Meth. 143 (1977) 29.

[2] R. Bouclier, G. Charpak, Z. Dimčowski, G. Fischer, F. Sauli, G. Coignet and G. Flügge, Nucl. Instr. and Meth. 88 (1970) 149.

[3] J. J. Lowke and J. H. Parker, Jr., Phys. Rev. 181 (1969) 302.

[4] G. D. Alkhazov, Nucl. Instr. and Meth. 89 (1970) 155.

NUCLEAR INSTRUMENTS AND METHODS 167 (1979) 455-464; © NORTH-HOLLAND PUBLISHING CO.

HIGH-ACCURACY LOCALIZATION OF MINIMUM IONIZING PARTICLES USING THE CATHODE-INDUCED CHARGE CENTRE-OF-GRAVITY READ-OUT

G. CHARPAK, G. MELCHART, G. PETERSEN* and F. SAULI

CERN, Geneva, Switzerland

Received 17 August 1979

A measurement of the analogue centre of gravity of the induced charge distribution on cathode planes of multiwire proportional chambers allows the localization, in two dimensions, of the ionizing event. This paper describes the localization properties of such a method in the detection of minimum ionizing particles, and in particular the effect of ionization statistics and δ electrons. For a beam perpendicular to the chamber plane, accuracies of around 60 μm rms are obtained in the direction along the anode wires, and of 200 μm in the direction perpendicular to them. A separate measurement of the anodic drift-time allows improvement of the last measurement reducing it to 120 μm rms. The multitrack coupling ambiguity is also discussed.

1. Introduction

Localization of ionizing radiation in multiwire proportional chambers through the measurement of the distribution of charges induced on cathode strips, and subsequent calculation of their centre of gravity[1-3], allows position accuracies approaching the limits set by the spatial extension of the ionization electrons[4] to be obtained. Because it requires the handling of a large amount of analogic information, the centre-of-gravity method has found so far a limited number of applications in cases where its bi-dimensional localization properties are essential, as in soft X-ray detection[5-7], or where the added wealth of information considerably improves the data quality, as in shower detectors[8,9]. However, with the rapid progress in microelectronics, relatively cheap and fast analogue-to-digital converters (ADC) are becoming available thus solving the major technical problem of the centre-of-gravity approach; large detection systems based on this principle can be envisaged that allow the space accuracy of present-day drift chamber systems to be attained with however the good resolution times of multiwire proportional chambers. Bi-dimensional read-out from a single detection layer, as given by the method, is very attractive also for use in detectors having cylindrical geometry, where alternative methods for obtaining the axial coordinate (delay lines, current division or small stereo angle between adjacent anode planes) offer a limited localization accuracy. A large detector of this kind is indeed already under construction[10].

The development work on the induced charge read-out has also resulted in a deepening of the understanding of several phenomena connected with the avalanche growth and distribution in proportional counters with as a result an improvement in some of the chamber's performances, as for example coordinate interpolation between wires in X-ray detection[11-17] and removal of the right-left ambiguity in multiwire and drift chambers for charged particle detection[18,19].

As most of the previous work on the subject has been devoted to the detection of soft X-ray radiation, we have continued and extended our investigation to the information obtainable with the centre-of-gravity read-out when detecting minimum ionizing particles, in various conditions of beam incidence, resolution time, and operational voltages. The results obtained so far are presented and analysed in the next sections.

2. Experimental

The set-up used to study the localization properties of the induced charge read-out for minimum ionizing particles consisted in three aligned chambers, identical both in structure and in the associated electronics to the ones described in ref. 12. Each chamber had an anode plane made of gold-plated tungsten wires, 15 μm in diameter and 2.54 mm apart ($\frac{1}{10}$ in., to match the ouput connector pitch), between two orthogonal cathode planes of Cu–Be wires 100 μm in diameter and 1.27 mm ($\frac{1}{20}$ in.) apart. The anode to cathode distance was 5 mm and the active surface of the chamber 100×100 mm^2. All wires were positioned and soldered with good mechanical accuracy, so that the

---

* Present address: Niels Bohr Institute, University of Copenhagen, Denmark.

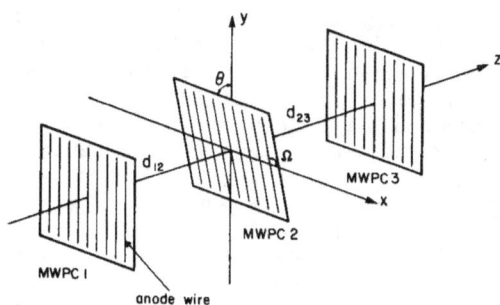

Fig. 1. The experimental arrangement of three identical chambers aligned in a medium-energy beam. The central detector can be rotated around both x and y to study the angular dependence of performances.

tolerance of each element in respect to a support frame reference was better than $\pm 15 \, \mu m$. The three chambers were mounted on individual turntables allowing both displacements and rotations with micrometer reading. Fig. 1 shows the arrangement, as well as the definition of the reference system, with the beam along the z direction; the position of the central anode plane is characterized by the polar angles θ and Ω. The system has been operated at CERN on a non-separated test beam, with a momentum around $1.5 \, \text{GeV} \, c^{-1}$; a set of scintillation counters provided both the time reference for the passage of the tracks and the geometrical definition of the beam.

As was found in previous work, the optimum cathode strip width for high-accuracy read-out corresponds to about one unit of anode-cathode distance[12]; we have therefore merged together into individual measuring channels groups of four adjacent cathode wires. For convenience, the chamber was operated with grounded cathodes; for calibration and timing purposes, the overall anodic signal was also recorded. For the set of these chambers, a total of 120 pulse-height measuring channels have been implemented on the cathode strips. three more channels being dedicated to the anode planes' signal. As far as the electronics are concerned, we have used on each channel a simple charge preamplifier mounted on the chamber and having a sensitivity of about $250 \, \text{mV} \, \text{pC}^{-1}$, a rise-time of 10 ns and a decay-time of 30 ns*; the amplified signals were then transmitted via coaxial cables to a CAMAC-based 10-bit charge-integrating ADC†.

---

* Developed at CERN by J. C. Santiard and produced with a thick-film hybrid technique by CIT-Alcatel, France.
† Le Croy 12-channel ADC Type 2249A.

Operating the converter with a 50 ns gate, we obtained a full-scale sensitivity of around 1 pC, as referred to the preamplifier input (on the proportional chamber). The overall electronics noise of the system in operational conditions was corresponding to about 2.5 ADC channels rms.

The ADC gating functions could be realized either by a common signal initiated by the scintillation counters' coincidence, or chamber by chamber by the signal detected in the corresponding anode plane, suitably discriminated and shaped. The second method produced, as expected, the best results at short gating times (50 ns or below) as it maintains a close phasing with the cathode signals in the given chamber (the typical resolution time of a chamber of this kind is around 30 ns). It could only be used however in the case where a single hit was expected in each chamber.

In addition to the cathode and anode pulse heights, the anodic signal time, with respect to the scintillation counters, was also recorded for each chamber to provide additional information. A small on-line computer was used to organize the data acquisition and transfer to magnetic tape, and to perform sample analysis to control the performance of the chambers.

As a gas filling we have used an argon–isobutane–methylal mixture in the volume concentrations 67–30–3; the operational voltage for the

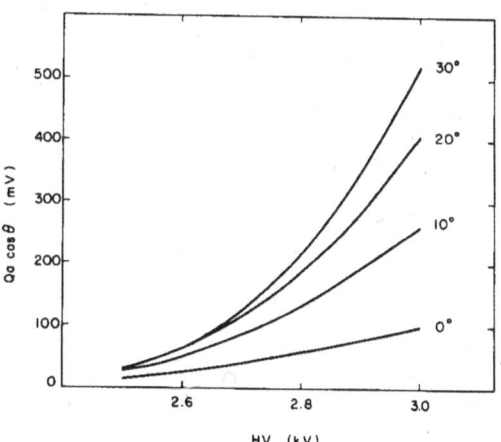

Fig. 2. Most probable detected pulse-height for minimum ionizing particles in the central chamber, as a function of anodic voltage and for several values of the rotation angle θ (see fig. 1). The measured pulse heights have been normalized to a unit track length; space charge gain saturation is very large for small angles of incidence and large voltages.

chambers was situated between 2.8 and 3.0 kV. As will be discussed later, the best results in terms of accuracy have been obtained at high operational voltages. This was not however simply a result of the larger detected pulses (improving the signal-to-noise ratio). We definitely had an effect connected with the space-charge proportional gain saturation suppressing the influence of the large charge deposition asymmetries due to δ-rays (see Section 4). The presence of large space-charge effects is apparent from fig. 2 which shows, as a function of anodic voltage, the normalized pulse height (signal per unit track length) for minimum ionizing particles and several values of the central chamber rotation angle θ (see fig. 1). The increasing deviation from proportionality at large voltages and angles was already observed in similar conditions and interpreted to be a consequence of a reduced mutual influence of avalanches when they are spread over a larger section of the anode wire[20]).

3. A system calibration and data handling

The practical possibility of recording space accuracies in the 50 μm range depends mostly on an absolute knowledge of the strips' pulse-height distributions at the level of 1% or better, during data taking. Because of the large number of channels involved, we have preferred to use cheap elements with relatively large dispersions in conjunction with a careful calibration procedure, as against the possibility of individual hardware regulations. After several trials, the following calibration procedure was retained as providing a long-term reproducibility matching our needs. To begin with, all pedestal levels in the ADC (channel content when gating the unit without analogue input) were set around the same level, channel 20 out of 1024. Next, each individual pulse-height measuring channel (including the preamplifier, cable and ADC) was pulsed with the same reference signal obtained by discharging a fixed capacitor from a constant voltage source; this reference signal, set at about the middle of the sensitivity scale, provided, after pedestal subtraction, the normalization factor between channels. Before and during the data acquisition runs, then, a common signal from a variable amplitude pulse generator was simultaneously applied to all channels by means of a bus strip capacitively coupled to the preamplifier's input on the printed circuit board. Of course, large relative dispersions were expected for the value of the

coupling capacitance that remains however fixed for a given channel. Increasing the generator pulse height in known steps (we have used a multiple calibrated attenuator, with 6 dB steps) for each channel a relative linearity curve could be obtained that, after pedestal subtraction and normalization to the previously described value, provided an absolute and zeroed charge-to-channel content calibration. As it appeared, six or seven amplitude values and a linear interpolation between adjacent steps were sufficient to guarantee the desired accuracy over several weeks of run. We should notice here that gain variations that uniformly affect the system (such as those due to moderate temperature changes) do not affect the determination of the centre-of-gravity of the pulse-height distribution.

Let us then denote by q_i the effective charge signal on strip S_i, as obtained from the raw data using the described calibration curves; we define the centre of gravity of a cluster of adjacent induced charges as the quantity:

$$\bar{s} = \sum (q_i - b) S_i / Q, \qquad Q = \sum (q_i - b)$$
$$\text{for } q_i - b > 0, \tag{1}$$

where b is a bias level and the sum extends only to positive values of $q_i - b$ measured in a group of adjacent strips. A proper choice of the bias level allows reduction of the influence of pick-up and electronics noise in the centre-of-gravity determination and will be described later. A typical induced charge distribution on a cathode plane, as recorded for a single minimum ionizing track, is shown in

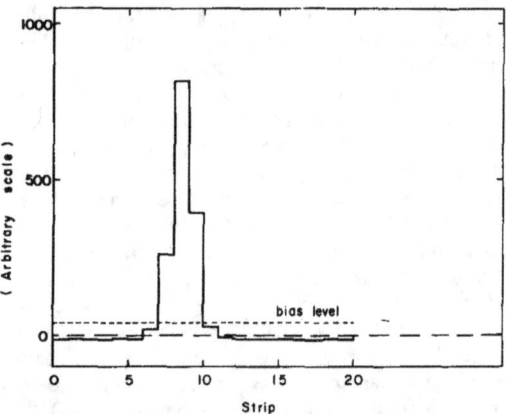

Fig. 3. An example of measured pulse-height distribution on the cathode strips. Five adjacent channels record an induced charge above zero; the choice of the bias level is discussed in the text; see also the next figure.

fig. 3 as well as the zero (pedestal) and bias levels. Negative amplitudes in the tail of the distribution, corresponding to ADC channel contents below the pedestal, are due to the unavoidable capacitive coupling in the proportional chamber between anode and cathode planes; we considered as significant for the analysis only positive amplitudes. In this definition, the events shown in fig. 3 have a cluster size corresponding to five strips, of which three are above the bias level. It appeared that for single track events and in the y direction (strips perpendicular to the anode wires) around 13% of the events have this configuration, the largest majority having four significant strips; in the x direction, owing to the regularity in the pattern of facing wires, essentially all events had four significant strips.

Raw events were accepted for reconstruction under the following restrictions:

– none of the ADC channels was overflowing;

– the extremes of the cathode pulse-height distribution did not include the first or the last two strips in each plane, to avoid edge effects;

– the number of significant strips in each plane was smaller or equal to six, to eliminate multitrack events. At the particle flux at which we have been operating, the chance of having two tracks within the resolution time (50–100 ns) close enough to escape the last selection in both planes was negligible. We also had the possibility in the analysis to reject events that provided avalanches in more than one adjacent wire, a kind of event clearly identified by the charge distribution along the x direction as will be discussed later.

For each accepted event, a straight line fit through the three pairs of computed centers of gravity provided the best estimate of the particle trajectory, and the variance of the residuals gave the positioning errors. As a rule, we measured the intrinsic dispersion of the set of three chambers operated in identical conditions and then, when changing the operational parameters only in the central one, we computed the new variance subtracting in a Gaussian sense the fixed contributions of two outer chambers.

As mentioned before, introduction of a bias level in the calculation of the centre of gravity [expression (1)] substantially improved the quality and the stability of the accuracy measurements. As is clear from fig. 3, the effect of the subtraction is to eliminate the contribution of low content channels where noise can be predominant; on the other

hand, choosing too high a bias level would result in a loss of information. We found the best results choosing a level proportional to the total charge measured in the analysed event on the significant strips for each plane:

$$b = k \sum q_i, \qquad (2)$$

the value of k being determined experimentally as follows. Data obtained in running conditions where we expected the best localization accuracies (i.e., three identical chambers operated at high values of the anodic potential and normal to the beam) were repeatedly analysed for increasing values of the constant k, computing the position accuracy in each case. The points with error bars in fig. 4 show the standard deviation of the position accuracy as a function of k; the result is almost constant for values of k between 5×10^{-3} and 2.5×10^{-2}, while it deteriorates quickly outside those limits. In the figure we also present the results of a simplified computer simulation of the signal induction process that takes into account the discrete cathode strip width as well as the actual measured distributions, without (dashed curve) and with (full curve) inclusion of the experimentally measured electronics noise contribution (2.5 ADC channels rms as mentioned before). The agreement between calculation and measurement is rather good and confirms the correctness of our procedure. For the whole of the following analysis, we assumed a value of k of 0.025; the bias level indicated in fig. 3 corresponds to this choice. A more detailed mathematical analysis of the induction process has been developed by Erskine[21]).

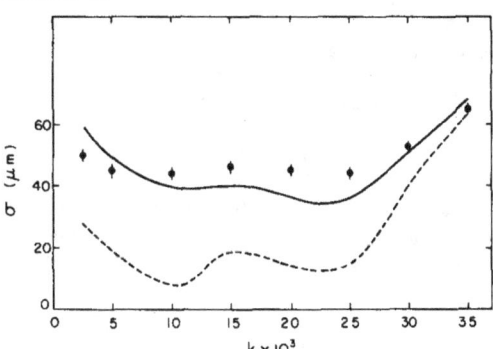

Fig. 4. Dependence of the reconstructed position accuracy on the choice of the bias level constant k [see expression (2)]. Points with error bars represent the measured values, while the curves are the result of a model calculation that does not (broken line) or does (full line) include the electronic noise contribution.

4. Experimental results

4.1. INTRODUCTION

As was expected, the localization properties of a chamber using the induced charge read-out are better for the coordinate measured along the anode wires (the y direction) than for the perpendicular coordinate (the x direction). In the last case indeed the discrete structure of the anodes is bound to introduce a quantization effect in the distribution of induced charges; it was in fact for a long time admitted that at best one could obtain a localization corresponding to the spacing of the anode wires, thus implicitly assuming a symmetric distribution of each avalanche around the wires. Small right-left asymmetries were observed[2]; however that could only be explained by a certain degree of localization of the avalanches around the wires; further work proved that this was indeed the case, at least at moderate proportional gains[12,13,15] and that the x coordinate could be determined, for very localized energy depositions as given by soft X-rays, with almost the same accuracy as for the "continuous" y coordinate[12]. This is a direct outcome of the fact that, given the multiwire proportional chamber field structure, in most of the volume there is a unique association between the x coordinate of the photo-electric conversion point and the radial direction of approach of the ionization electrons to the wire (the exception being a small cylindrical region just around the anodes).

Charged particles however do not produce a single ionization centre, but instead a succession of clusters statistically distributed along the particles' trajectory. Ionization electrons therefore arrive at the anode at different times and with different directions of approach; at least one can hope to distinguish, from the measured distribution of the induced charges, the right and left side of the anode wires.

We will describe separately in what follows the measurements realized along the two directions in the central chamber; the two outside reference chambers were always operated under the conditions that guarantee the best localization accuracy.

4.2. LOCALIZATION IN THE y DIRECTION, PARALLEL TO THE ANODE WIRES

The three chambers were mounted initially with parallel anode wires; a rotation of the central chamber by an angle θ around x allowed the study of the effect of the spread along the anode wires of the

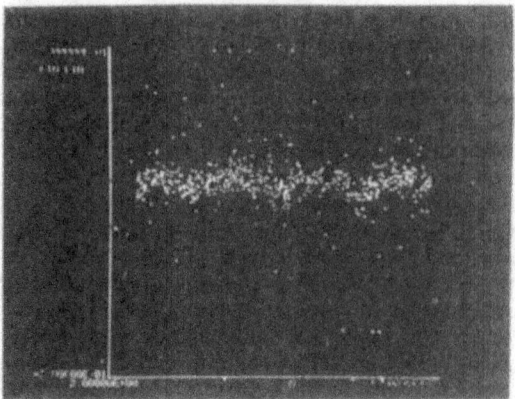

Fig. 5. Experimental distribution of the residuals in the y direction, for a uniform irradiation of the chamber. The vertical dispersion has a fwhm around 100 μm, and its uniformity along the horizontal axis (the y coordinate) is a check of the consistency of the amplitude calibration procedure.

avalanches produced by ionizing tracks. A very good check of the uniformity of response and correctness of calibration of the system is a plot of the reconstruction residuals (i.e., the difference between computed and measured coordinates) as a function of the coordinate itself. Fig. 5 shows the distribution of the y residuals as a function of y in the central chamber, for a sample of events uniformly distributed over the active surface of the chamber. The distribution is consistently centred around zero, and its projection on the vertical axis (the usual accuracy distribution) has a rms width around 45 μm, as already shown in fig. 4. A defective channel or a wrong calibration, introducing a positioning error of the same order, would clearly appear as a discontinuity in the distribution.

The production in the primary collisions of heavily ionizing δ electrons is expected to smear the physical signal, thus implying a worsening of the localization accuracy. The effect is clearly demonstrated in fig. 6 where, always for tracks perpendicular to the chamber plane ($\theta = 0$), the y residuals' distribution has been plotted for events providing an overall charge deposit [the quantity Q in expression (1)] contained in three different regions of the energy loss spectrum. Events associated with large energy losses, region III, exhibit substantial tails at large residual values as compared with average (region II) and low (region I) energy losses. The effect is even more pronounced for inclined tracks, since in this case the asymmetry in the energy loss

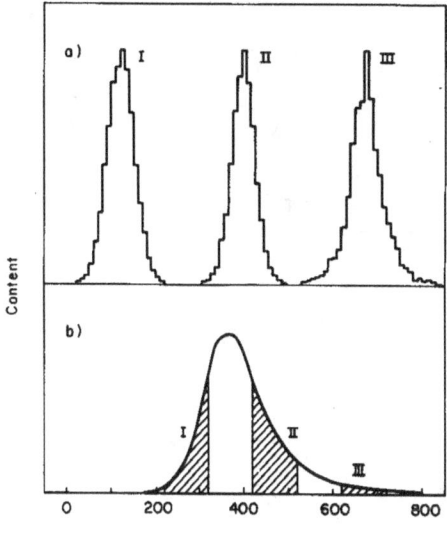

Fig. 6. Dependence of the residuals' distribution on the energy loss, measured for tracks perpendicular to the middle chamber along the y direction. Residuals have been plotted in (a) for three different regions of energy loss, which are shown in (b). The large tailing effect introduced by δ electrons is clear.

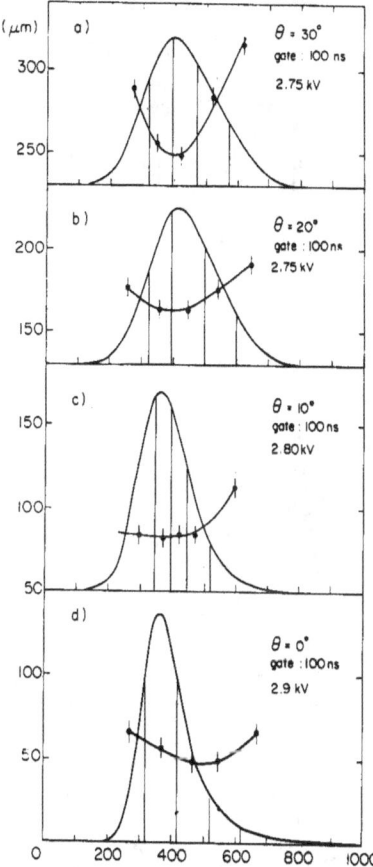

Fig. 7. Dependence of the measured localization accuracy along the y direction on the actual energy loss, for several angles of incidence of the beam. The accuracy is given by the points with error bars, computed for the slice in energy loss shown in the background. Notice that the horizontal amplitude scale is arbitrary and different for each angular setting; relative values of the actual peak amplitude were given in fig. 2 as a function of angle and anodic voltage. Analogue gating length is 100 ns.

produced on the two sides of the anode plane by a δ electron has a large effect in the determination of the centre of gravity. This is well demonstrated in fig. 7, where we have plotted the computed rms of the accuracy distribution for several values of the rotation angle θ ($\theta = 0°$ meaning a beam perpendicular to the chamber plane). The accuracy has been computed for different regions of the energy-loss distribution, shown in the background of each plot*. At large angles, not only does the average accuracy become worse but also the dependence on the energy deposit is increased.

We have observed that at high amplification space-charge effects can partially mask the quoted energy-loss dependent accuracy by suppressing the effects of high-density ionization produced by δ electrons. This effect is demonstrated by the two plots of fig. 8, obtained in identical conditions except for the anodic voltage.

The ADC gating length (the time during which the induced charge distribution is sampled) is also

---

* Notice that the horizontal amplitude scale is arbitrary and different in all plots; the relative values of the peak amplitudes were shown in fig. 2 as a function of anodic potential and incidence angle.

important in defining the achievable accuracy; fig. 9 shows the effect of reducing the gate from 100 ns (fig. 7) to 50 ns; both the average and the extreme accuracies are improved for inclined tracks, as a direct result of restricting the measurement to the section of the ionizing trail closer to the anode (in our gas mixture, the average drift time is around 20 ns mm^{-1}). Short gating times and large anodic potentials are therefore useful to reduce the effects of the ionization statistics on the localization properties of the detector.

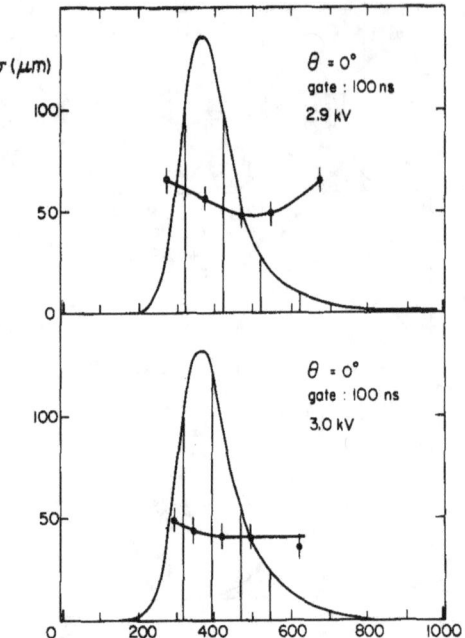

Fig. 8. The same measurement as in the previous figure, for normal tracks at two values of the anodic potential. Space-charge saturation effects clearly suppress the δ electrons' contribution at large energy losses.

4.3. LOCALIZATION IN THE *x* DIRECTION, PERPENDICULAR TO THE ANODE WIRES

The central chamber was rotated for this measurement by 90° around the *z* axis (see fig. 1), to keep the good beam localization provided by the two outer reference chambers. As expected and for the reasons discussed in section 1 the localization properties along *x* are strongly dominated by the discrete anode wire spacing. Fig. 10 shows the measured centre of gravity in the *x* direction (vertical axis), as a function of the beam real position as given by the reference chambers (horizontal axis) for a uniform irradiation over a section including two anode wires. The clustering effect around the anodes' position is clear, although one can distinguish a certain amount of interpolation events due to charge sharing between adjacent wires, and a moderate slope of the clustering due to the right-left effect. This is better seen in a projection of the scatter plot on the vertical axis, given in fig. 11: the two peaks correspond to tracks crossing the anode plane respectively at the right and left of a wire.

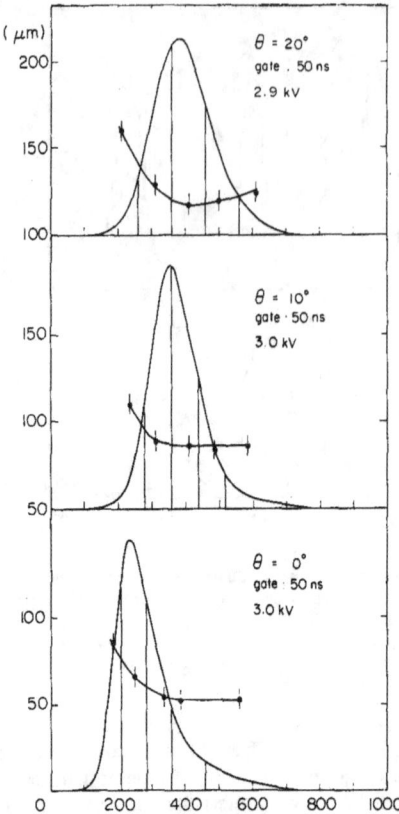

Fig. 9. Measured dependence of accuracy on pulse height and incidence angle, for a 50 ns analogue gating. Comparison with fig. 7 shows the gain in accuracy resulting from the electronic selection of a shorter segment of the energy deposit.

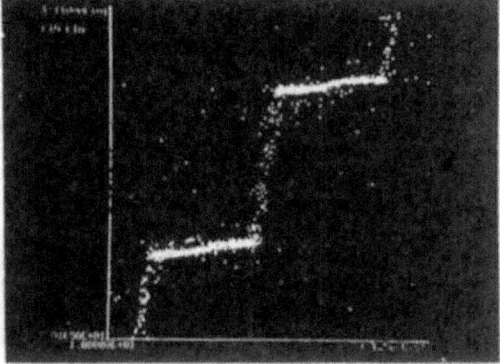

Fig. 10. Correlation between the true (horizontal) and measured coordinate along the *x* direction, for a uniform irradiation perpendicular to the chamber. The quantizing effect of the anode wires on the centre of gravity is apparent.

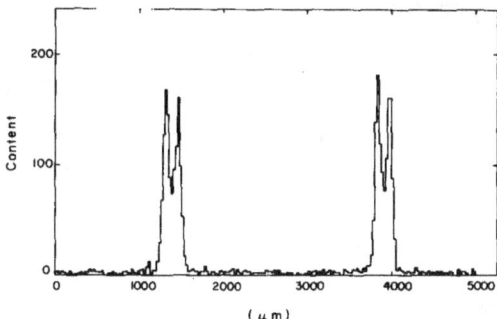

Fig. 11. Projection of the scatter plot of the previous figure on the vertical axis, showing clearly the right-left separation as two partially overlapping peaks. The uniform background is due to tracks being at small angle with the perpendicular to the wires, and producing coordinate interpolation due to charge sharing.

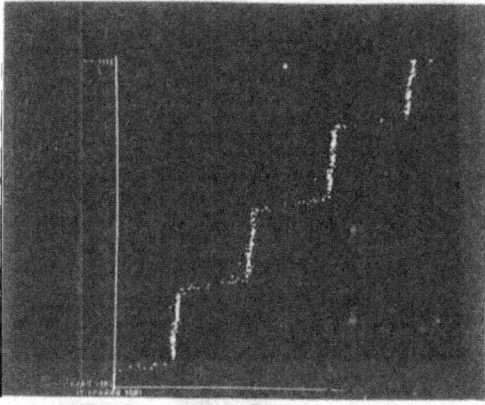

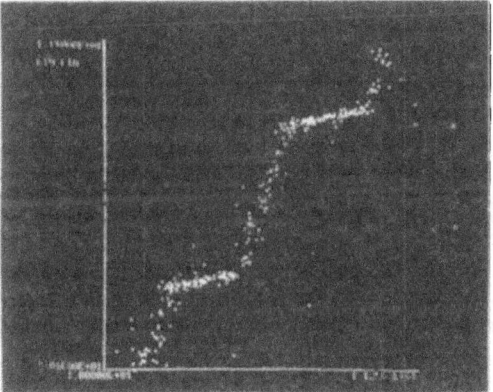

Fig. 12. Same as for fig. 10, for a small (12°) incidence angle. Coordinate interpolation is more pronounced.

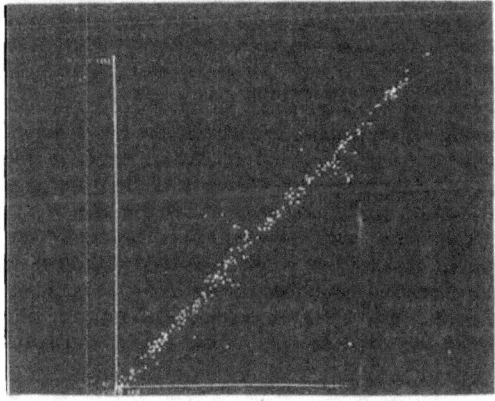

Fig. 13. Deconvolution of the centre-of-gravity distribution using an experimentally-determined correlation function. The original distribution is shown in (a), and the deconvoluted or linearized one in (b). Notice the inversion of the horizontal and vertical axes as compared with fig. 12. The resulting accuracy in the x direction has a standard deviation of about 200 μm (for 2.54 mm anode wire spacing).

The small overlap is due to tracks crossing the anode-wire region; δ-ray production and electron diffusion obviously slightly smear the right-left separation at the boundary.

For tracks perpendicular to the chamber plane, only a small number of events result in charge sharing (mainly because of the finite beam divergency) but the effect is relevant at larger incident angle. In fig. 12 is shown for example, the correlation between the measured and the real coordinate in the x direction, for a chamber rotation of 12°; at increasingly large angles all tracks produce charge sharing and the quantizing effect of the anode wires gradually disappears. This effect was already observed and analysed in detail in previous work[3]).

A closer look at fig. 10 shows that the distribution of the centre of gravity has a small slope on each side of the middle, thus suggesting a certain amount of localization between the wires. We have therefore used the function correlating the real and measured values of the coordinate x as measured with soft X-rays[12]) to deconvolute the clusters; figs. 13a and b show, respectively, the correlation between real and measured coordinates before and after the deconvolution. Owing to statistical fluctuations in the energy deposit, however, the positioning accuracy for charged particles in the x direction is not as good as the one that could be obtained with X-rays. Moreover, the compression of the centres of gravity resulting from the discrete posi-

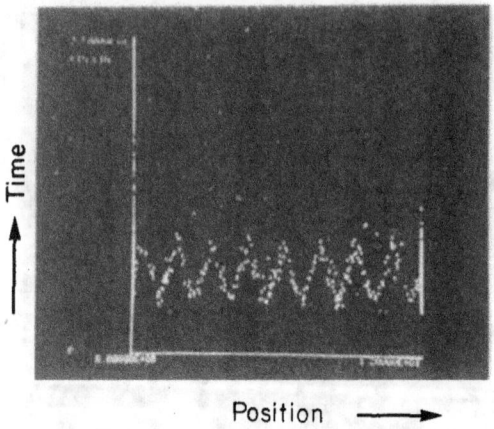

Fig. 14. Correlation between the measured drift-time on the anode plane (vertical axis) and the real track position along the x direction. The anode wire position corresponds to the lower edges.

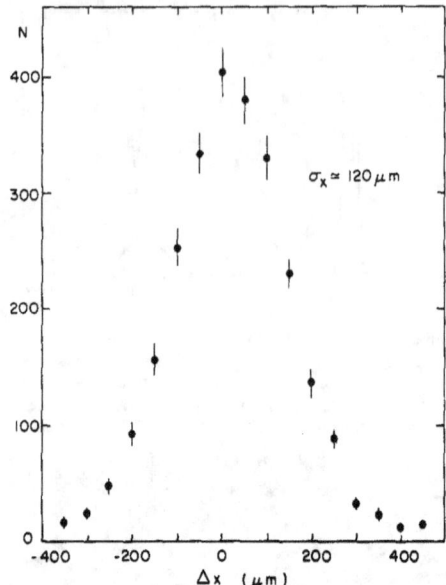

Fig. 15. Combining the right-left identification given by the centre-of-gravity read-out with the drift-time measurement, one can obtain an accuracy distribution in the x direction (perpendicular to the anode wires) having a standard deviation of 120 μm.

tioning of the anodes emphasizes the errors due to small non-linearities and calibration errors in the pulse-height measuring channels; at best, we have obtained a positioning error of 200 μm rms.

This result can, however, be substantially improved making use of the measured drift-time to the anodes as mentioned before. In fig. 14 is shown a scatter plot of the measured drift-time on the anode wires as a function of the real coordinate (as

given by the collimating chambers). The analysis-procedure is then the following: a cut in the centre-of-gravity distribution (like the one shown in fig.

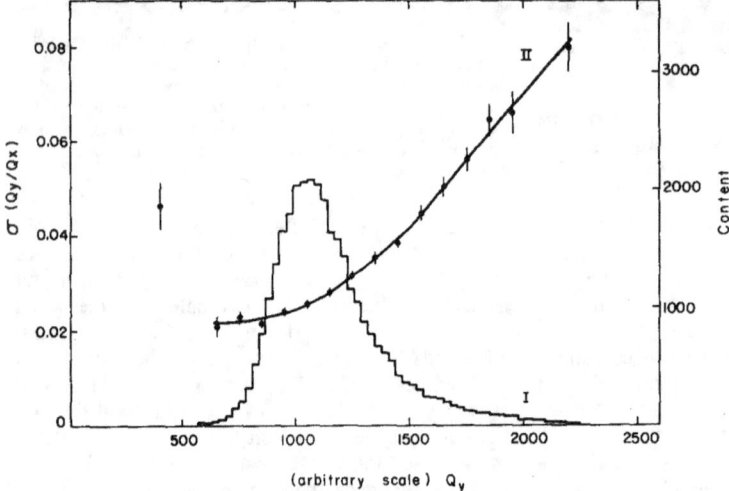

Fig. 16. Standard deviation of the ratio of measured total pulse heights on the two cathode planes, Q_y/Q_x, as a function of Q_y (points with error bars). The background histogram represents the Q_y distribution itself. The small value of the dispersion in the ratio allows in most cases an ambiguity-free correlation of the measured x and y coordinates in the case of multitrack events.

11) separates tracks crossing the chamber on the two sides of the anode wires, and the x coordinate is obtained adding to the actual wire position a quantity

$$\Delta x = \pm w(t - t_0), \tag{3}$$

where $(t - t_0)$ is the measured drift-time and w the electron drift velocity. In the gas mixture we have used, the drift velocity is almost independent of the electric field, As demonstrated in previous work on drift chambers[20]), and has a value of around $5.2 \text{ cm } \mu s^{-1}$. The sign of expression (3) is of course chosen according to the side of the crossing. The distribution of the residuals measured using this procedure in the central chamber is shown in fig. 15 having a standard deviation of $125 \mu m$; this is comparable with the value obtained in high-accuracy drift chambers for tracks close to the anode wires, where primary ionization statistics dominates the localization properties of the detector[20,4]).

4.4. DOUBLE TRACK AMBIGUITY

In the induced charge read-out method the two coordinates in a plane are measured independently and therefore the coupling is unique only in the case of single track events. However, the strict correlation that exists between the total charge measured on the two cathodes, coupled with the large dispersion in the energy loss of the individual events due to Landau fluctuations, allows in many cases a correct pairing of coordinates, as already suggested by the authors of ref. 3. Fig. 16 shows indeed the dispersion of the measured ratio of total charge detected on the two cathode planes, Q_y/Q_x, as a function of Q_y. The dispersion is contained within 2 to 4% for most of the events, while it increases strongly in the tail of the large energy losses because of the asymmetric contribution of the δ-rays. Being limited in the total beam flux, we could not obtain a large enough sample of multiple tracks to directly check the efficiency of the coupling. However, using the measured dispersions for single tracks we have computed the expected multitrack coupling efficiency using a Monte-Carlo generation of multiple events, each with the appro-

priate amplitude dispersion. The results show that in 95% of the cases the coordinates of two-track events can be correctly paired, as far as they are sufficiently separated in space to induce distinct charge distributions on both cathode planes. As from fig. 3, this minimum distance corresponds to 5 strip widths (or 25 mm) in both directions.

References

[1] G. Charpak, D. Rahm and M. Steiner, Nucl. Instr. and Meth. 80 (1970) 13.

[2] G. Charpak and F. Sauli, Nucl. Instr. and Meth. 113 (1973) 381.

[3] A. Breskin, G. Charpak, D. Demierre, S. Majewski, A. Policarpo, F. Sauli and J. C. Santiard, Nucl. Instr. and Meth. 143 (1977) 29.

[4] F. Sauli, Nucl. Instr. and Meth. 156 (1978) 147.

[5] G. Charpak, C. Demierre, R. Kahn, J. C. Santiard and F. Sauli, Nucl. Instr. and Meth. 141 (1977) 449.

[6] N. I. Kochelev and V. I. Telnov, Nucl. Instr. and Meth. 154 (1978) 407.

[7] C. Bolon, M. Deutsch, R. Lanza, G. Quigley and A. Rich, IEEE Trans. Nucl. Sci. NS-26 (1979) 146.

[8] Z. Dimcovski, N. Phinney, A. L. S. Angelis, J. Bibby, A. Coates, A. M. Segar and J. S. Wallace-Hadrill, Nucl. Instr. and Meth. 156 (1978) 123.

[9] E. Gabathuler, A. M. Osborne, R. W. Clifft and M. Sproston, Nucl. Instr. and Meth. 157 (1978) 47.

[10] CELLO, Proposal for a 4π Magnetic Detector for PETRA (Hamburg, Oct., 1976).

[11] G. Charpak, G. Petersen, A. Policarpo and F. Sauli, IEEE Trans. Nucl. Sci. NS-25 (1978) 122.

[12] G. Charpak, G. Petersen, A. Policarpo and F. Sauli, Nucl. Instr. and Meth. 148 (1978) 471.

[13] J. Fisher, H. Okuno and A. H. Walenta, IEEE Trans. Nucl. Sci. NS-25 (1978) 794.

[14] J. Fisher, H. Okuno and A. H. Walenta, Nucl. Instr. and Meth. 151 (1978) 451.

[15] J. J. Harris and E. Mathieson, Nucl. Instr. and Meth. 154 (1978) 183.

[16] E. Mathieson and J. J. Harris, Nucl. Instr. and Meth. 154 (1978) 189.

[17] E. Mathieson and J. J. Harris, Nucl. Instr. and Meth. 159 (1978) 483.

[18] A. H. Walenta, Nucl. Instr. and Meth. 151 (1978) 461.

[19] A. Breskin, G. Charpak and F. Sauli, Nucl. Instr. and Meth. 151 (1978) 473.

[20] A. Breskin, G. Charpak, F. Sauli, M. Atkinson and G. Schultz, Nucl. Instr. and Meth. 124 (1975) 189.

[21] G. A. Erskine, Charges and current induced by moving ions in multiwire chambers (in preparation).

JOURNAL DE PHYSIQUE

Colloque C3, supplément au n° 6, Tome 39, Juin 1978, page C3-7

LOCALISATION BIDIMENSIONNELLE DES POSITIONS DES TRAJECTOIRES DE PARTICULES DANS DES CHAMBRES PROPORTIONNELLES AVEC UNE GRANDE RÉSOLUTION SPATIALE ET TEMPORELLE

G. CHARPAK, A. POLICARPO, A. PETERSEN,
F. SAULI et J. C. SANTIARD

C.E.R.N., Genève, Suisse

Résumé. — En mesurant le centroïde des distributions de charges introduites sur des bandes de cathode perpendiculaires et parallèles aux fils d'anodes il est possible d'aboutir à une méthode de localisation présentant des avantages considérables.

— Le temps de résolution, peut être réduit à 30 nanosecondes.

— La mesure de la coordonnée le long des fils a une précision limitée simplement par la distribution physique des charges. La contribution de la méthode de lecture est inférieure à 7 microns.

— La coordonnée perpendiculaire aux fils peut être plus précise que la séparation entre les fils même pour les trajectoires orthogonales au plan des chambres.

— Les ambiguïtés relatives aux trajectoires multiples peuvent être levées dans la majorité des cas avec une seule chambre.

— La mesure simultanée des coordonnées de nombreuses trajectoires est possible avec un pouvoir de séparation de l'ordre de 0,5 cm.

Abstract. — By measuring the centroid of the charges induced on cathods of MWPC two dimensional localization can be obtained with advantageous properties.

— Resolution time of 30 ns
— The accuracy of the coordinate along the anode wires is limited only by the physical fluctuations in the centroid position. The read out contributes by 7 µm at most to the errors.
— Interpolation between the anode wires is obtained over a wide angular range.

1. Introduction. — Les avalanches produites sur les fils d'anodes des chambres proportionnelles induisent des charges positives sur les électrodes voisines du fil détecteur. Si ces charges sont collectées par des bandes cathodiques ou des fils cathodiques, il était connu dès l'origine des chambres proportionnelles que le centroïde de la distribution des charges induites coïncidait, en première approximation, avec la projection du centre de l'avalanche sur le plan de mesure [1]. Ceci permet la mesure des deux coordonnées d'une avalanche en utilisant des fils ou des bandes d'anode ayant des orientations différentes.

De nombreux travaux ayant en vue en général la mesure des positions de rayonnements neutres, γ ou neutrons, ont permis la mise au point de méthodes analogiques permettant d'obtenir avec précision la coordonnée parallèle aux fils d'anode. Ces méthodes reposent en général sur les retards de propagation dans des électrodes spéciales ou des différences d'intensités de charges collectées aux extrémités d'électrodes présentant une impédance finie.

Les mérites respectifs de ces méthodes sont analysées en détail par Radeka [2]. Le mérite commun à toutes ces méthodes est

d'avoir montré qu'une précision remarquable, de l'ordre de 100 µm, pouvait être atteinte dans la détermination de la position du centroïde électrique d'une avalanche le long des fils d'anode. Leur précision est en général limitée à une valeur comprise entre 10^{-2} et 10^{-3} de la longueur totale de l'électrode. Ceci explique que pour les chambres de grande taille courantes en physique des hautes énergies leur intérêt ait été limité. Le temps de résolution de toutes ces méthodes est également médiocre, de l'ordre de la microseconde, ce qui constitue une dégradation d'un ordre ou deux ordres de grandeurs de la résolution intrinsèque des chambres proportionnelles.

Pour une chambre détectant des rayons X nous avons mis au point une méthode de lecture directe du centroïde des avalanches [3]. Puis nous avons étudié l'adaptation de cette méthode à la physique des particules et avons mis en évidence des caractéristiques particulièrement intéressantes dans ce domaine et décrites en détail par ailleurs [4].

2. La méthode du centroïde dans la localisation des particules chargées. — Lorsqu'une avalanche se produit autour du fil d'anode a (Fig. 1), des avalan-

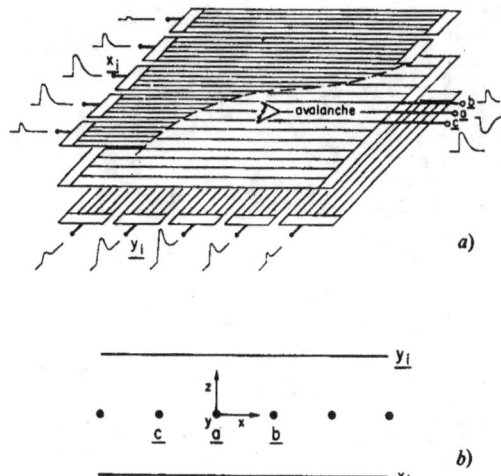

FIG. 1. — Détermination des coordonnées d'une avalanche par les signaux induits sur les électrodes voisines. La hauteur des impulsions induites est mesurée et stockée et le centroïde calculé. Les deux cathodes sont faites de bandes parallèles et orthogonales aux fils d'anode.

ches sont induites dans les fils voisins, b, et c, et sur les bandes de coordonnées x_i et y_i, où i désigne le numéro d'ordre des différentes bandes, x est parallèle et y est orthogonal aux fils d'anode.

Lorsqu'un signal est produit sur le plan d'anode les signaux X_i et Y_i recueillis et amplifiés sur les bandes x_i et y_i sont, après un délai contrôlé par des lignes à retard, acceptés ou refusés dans un convertisseur d'amplitude digital et l'information sur leur hauteur stockée dans un ordinateur (Fig. 1).

Nous nous sommes efforcés d'étudier quelle était la taille minimum de la porte d'entrée dans le convertisseur compatible avec une bonne résolution spatiale.

Nous avons observé qu'avec une porte de 30 ns et un amplificateur de faible coût [1] une précision considérable pouvait être obtenue dans la détermination de la position spatiale des trajectoires de particules au minimum d'ionisation.

Nos résultats peuvent être résumés ainsi :

a) La résolution de la position le long du fil est bien meilleure que 200 μm en largeur totale à mi-hauteur (LTMH).

b) Pour des traces inclinées par rapport au plan des chambres la résolution spatiale pour la coordonnée perpendiculaire aux fils d'anode est meilleure que celle qui est déterminée par la distance entre les fils. Il est possible d'interpoler entre les fils dans une grande gamme angulaire. On dispose ainsi, avec un seul plan anodique, de chambres de haute précision bidimensionnelles.

[1] Amplificateur type 1148, réalisé par Alcatel, Paris. Bruit à l'entrée : 2×10^{-15} coulomb.

c) La précision spatiale est peu altérée pour les traces inclinées car la largeur étroite de la porte permet de ne collecter que les électrons produits très près du plan d'anode (< 1,5 mm par exemple pour 30 ns de largeur de porte).

d) La résolution est quasi indépendante de la dimension des chambres car les mesures effectuées pour le calcul du centroïde de charge sont localisées dans un maximum de 5 bandes.

Nous avons poursuivi les travaux pour déterminer la précision ultime que l'on pourrait atteindre par cette méthode et ce sont quelques-uns des résultats obtenus que nous voulons présenter à cette conférence.

3. Mesures de haute précision sur la position des avalanches. — 3.1 AMPLIFICATION GAZEUSE MODÉRÉE. — Nous avons utilisé une amplification gazeuse modérée voisine de 10^4, pour laquelle les chambres sont loin des conditions de saturation par effet de charge d'espace. Dans ces mesures nous avons irradié les chambres avec des rayons X, provenant de l'excitation de la couche K de l'aluminium, d'énergie voisine de 1,4 keV. Cette énergie est voisine de la perte probable de particules relativistes dans 1 cm d'argon. Elle conduit donc aux mêmes conditions de gain des chambres et de l'électronique qu'avec ces particules.

Cette énergie est également voisine de celle des rayons δ qui accompagnent ces trajectoires et qui sont responsables de la limitation que l'on peut attendre dans la précision des mesures de position des trajectoires dans les gaz.

Le parcours des photoélectrons produits par ces rayons X dépend des gaz. Nous avons effectué des mesures dans des mélanges à la pression atmosphérique d'argon, isobutane et méthylal, avec ou sans fréon, et également en remplaçant l'argon par le xénon pour réduire le parcours des photoélectrons, qui dans ce cas est de l'ordre de 30 μm seulement. Nous avons mesuré, pour chaque photon X détecté sur un fil d'anode a, les impulsions X_i et Y_i et également les impulsions B et C sur les fils voisins b et c.

Certains de nos résultats ont été assez inattendus. Ils montrent qu'il est possible de déterminer avec une précision de quelques degrés l'azimut des centroïdes des avalanches produites au voisinage d'un fil. La figure 2 montre la distribution des points obtenus lors d'une irradiation uniforme, dans un diagramme où l'abscisse est le rapport B/C entre les impulsions induites sur les fils voisins et où l'ordonnée est le rapport entre les charges totales induites sur les deux cathodes, $X/Y = \Sigma X_i / \Sigma Y_i$.

Si l'on utilise un faisceau collimaté par une fente de 10 μm le diagramme de la figure 3a montre clairement les trois faisceaux séparés. La distribution des valeurs de B/C et du centroïde X (Figs. 3b et 3c) montre un certain avantage de précision

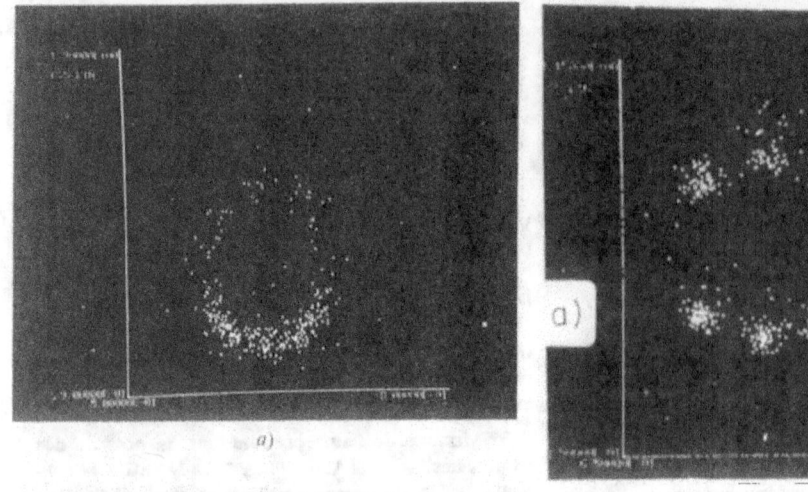

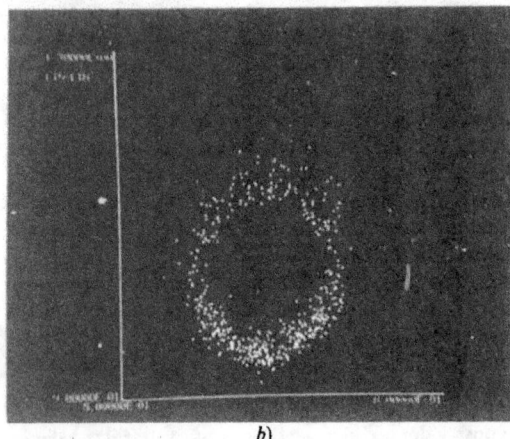

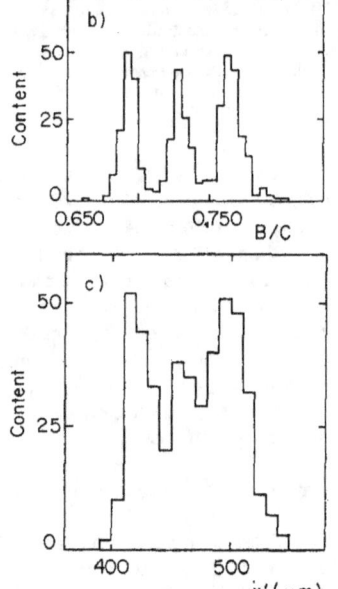

FIG. 2. — Distribution du centroïde des ions positifs pour une irradiation uniforme : gaz : xénon + isobutane + méthylal ; rayons X : 1,5 keV. *a)* Irradiation uniforme avec beaucoup de xénon. *b)* Irradiation uniforme avec peu de xénon. On voit nettement l'effet sur les différences d'absorption des deux moitiés de la chambre de part et d'autre des fils d'anode.

FIG. 3. — Distribution du centroïde des ions positifs pour des faisceaux parallèles aux fils : même gaz ; rayons X de 5,9 keV. Collimateur de 100 μm, 3 positions de faisceau distantes de 500 μm sur le fil et de part et d'autre du fil. *a)* Y/X en fonction de B/C. *b)* Les mêmes événements sur un histogramme en B/C. *c)* Les mêmes événements sur un histogramme en X' (centroïde projeté sur le plan mesurant la coordonnée X).

obtenu en mesurant B/C. Elle montre que la résolution spatiale obtenue dans l'interpolation des positions entre les fils est voisine de $\sigma = 150$ μm. La figure 4 montre les distributions de la coordonnée parallèle à un fil d'anode avec 3 faisceaux distants de 200 μm. On voit que la résolution est de $\sigma = 35$ μm.

Cette précision est limitée par le parcours des photoélectrons et non pas par la méthode de lecture comme on va le voir. Il nous a semblé intéressant de réaliser une image bidimensionnelle en rayon X, avec un masque dont la taille était plus petite que la distance entre les fils d'anode. On voit ainsi, pour la première fois à notre connaissance, une image continue qui n'est pas quantifiée par la distance entre les fils d'anode (Fig. 4).

Ceci peut présenter un intérêt considérable pour les applications où l'on veut une très grande résolution spatiale avec des rayons X d'énergie voisine de 1 keV, ce qui est le cas dans certains problèmes en astrophysique.

3.2 GRANDE AMPLIFICATION GAZEUSE. — Dans ces mesures nous avons ajouté du fréon et fonctionné dans les conditions dites du *gaz magique*,

FIG. 4. — Réponse continue bidimensionnelle. Lettres percées dans un écran de cuivre. Hauteur totale 1,5 mm. Largeur 4 mm. Fentes de 50 μm. Rayons X de 1,5 keV. Distance entre fils anodiques 2 mm. La totalité de l'image est obtenue sur un seul fil.

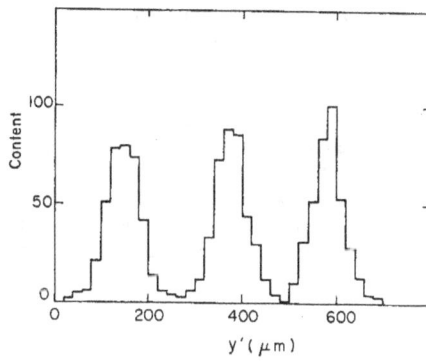

FIG. 5. — Précision dans la coordonnée parallèle au fil. Rayons X de 1,5 keV. Collimateurs de 10 μm distants de 200 μm. Mélange xénon + isobutane + méthylal σ ≈ 35 μm. En remplaçant le xénon par l'argon σ ≈ 45 μm.

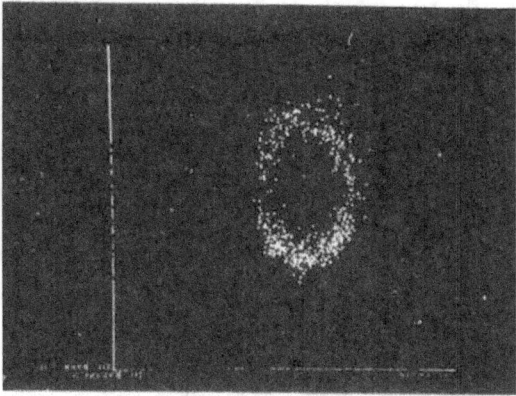

a)

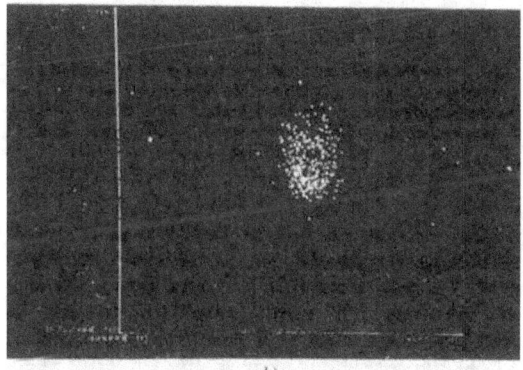

b)

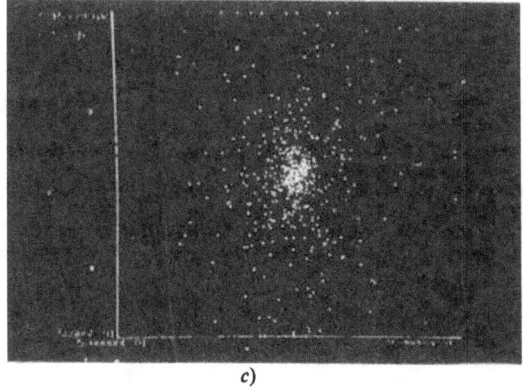

c)

FIG. 6. — Effet du facteur d'amplification. a) Mêmes conditions que figure 3 avec argon. b) Gaz magique — amplification modérée. c) Gaz magique — amplification saturée.

courantes en physique des hautes énergies. On sait que dans ces conditions le fonctionnement n'est plus en régime proportionnel. Il y a un effet de saturation de la hauteur des impulsions sous l'influence de la charge d'espace.

Les diagrammes 6b et 6c montrent que les centroïdes de distribution spatiale de l'avalanche se resserrent sur le fil. Il y a probablement une propagation de l'avalanche autour du fil et une propagation simultanée limitée le long du fil.

Si on mesure le centroïde des avalanches le long du fil on trouve d'ailleurs dans ce cas une dégradation qui montre que le fonctionnement avec le *gaz magique* n'est pas favorable aux mesures de haute précision ($\sigma < 100$ μm).

Le fonctionnement en *gaz magique* est un fonctionnement du type Geiger, mais avec une limitation de la propagation le long du fil, similaire à celui qui a été observé de façon plus nette avec des fils de diamètre plus gros (~ 100 μm), où la propagation atteint quelques millimètres [5].

Si l'on sélectionne, dans ce mode de fonctionnement, les événements donnant lieu à un rapport B/C bien déterminé, et que l'on observe la distribution du centroïde des avalanches on note une distribution étroite (Fig. 7), qui correspond à une précision $\sigma = 7 \mu$m.

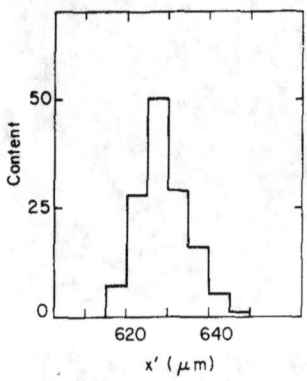

Fig. 7. — Limite supérieure de la précision de la détermination des coordonnées par la méthode du centroïde. Gaz magique. Evénements sélectionnés pour un rapport B/C donné. Distribution des centroïdes dans la direction orthogonale aux fils. $\sigma < 10 \mu$m.

Cette mesure donne la limite supérieure des imprécisions introduites par notre système expérimental. La majeure partie de cette erreur vient du bruit introduit par notre amplificateur qui est 10^4 charges électroniques à l'entrée. Une amélioration considérable serait aisée si elle s'avérait nécessaire pour la mesure de la position spatiale des phénomènes dont la localisation aurait une signification à ce niveau de précision.

Pour les phénomènes dont la localisation nous intéresse au premier chef, à savoir la position des trajectoires de particules relativistes nous pensons que la limitation de précision, au voisinage de 30 μm, viendra des phénomènes physiques intrinsèques tel que le parcours des rayons δ. Dans une chambre d'épaisseur d'un cm environ 5 % des traces donnent lieu à un rayon δ de parcours supérieur à 50 μm. Toutefois, avec des particules très ionisantes la situation est totalement différente et des précisions au niveau du micron pour le centroïde sont peut-être possibles.

4. **Conclusion.** — Nous avons montré qu'avec des méthodes simples il était possible de mesurer avec une précision voisine de quelques microns, la position spatiale des avalanches produites autour d'un fil d'anode de chambre proportionnelle.

La précision des mesures est en général limitée par les fluctuations physiques liées au phénomène étudié. Pour des particules très ionisantes, comme les ions lourds, il est probable que des progrès considérables dans la précision sont possibles par la méthode du centroïde.

Nous avons également montré qu'il était possible de mesurer dans certains cas la coordonnée de particules entre les fils d'anode d'une chambre sans qu'il y ait partage des électrons d'ionisation entre plusieurs fils. Ceci permet d'obtenir un détecteur bidimensionnel ayant une réponse quasi continue.

Bibliographie

[1] CHARPAK, G., RAHM, D. and STEINER, H., *Nucl. Instrum. Methods* **80** (1970) 13.
[2] RADEKA, V., *IEEE Trans. Nucl. Sci.* NS21 (1974), No. 1, 51.
[3] CHARPAK, G., HAJDUK, Z., JEAVONS, A., STUBBS, R. and KAHN, R., *Nucl. Instrum. Methods* **122** (1974) 307.
[4] BRESKIN, A., CHARPAK, G., DEMIERRE, C., MAJEWSKI, S., POLICARPO, A., SAULI F. and SANTIARD, J. C., *Nucl. Instrum. Methods* **143** (1977) 29-39.
[5] BREHIN, S., DIAMANT BERGER, A., MAREL, G., TARTE, G., TURLAY, R., CHARPAK, G. et SAULI, F., *Nucl. Instrum. Methods* **123** (1975) 225.

Nuclear Instruments and Methods in Physics Research A 346 (1994) 506–509
North-Holland

**NUCLEAR
INSTRUMENTS
& METHODS
IN PHYSICS
RESEARCH**
Section A

A high-rate, high-resolution asymmetric wire chamber with microstrip readout

G. Charpak, I. Crotty, Y. Giomataris *, L. Ropelewski, M.C.S. Williams

CERN, Geneva, Switzerland

(Received 18 January 1994)

We have investigated the properties of an asymmetric wire chamber with cathode strip readout. The use of a small gap between the anode plane and the cathode plane and of an electric field configuration provide fast removal of the positive ions produced in the avalanche process and restrict the area of the induced signal on strips engraved on the cathode plane. We can thus contemplate using digital readout for these strips and still have good position resolution, or use analog readout for high spatial resolution. The detector can operate at very high gas gain (10^5); thus with a small gap of 2 mm we can achieve full efficiency for minimum ionizing particles. We have measured no gain modification at a rate of 2×10^5 particles mm^{-2} s^{-1} at a gain of 10^4 with a distance of 600 μm between anodes and cathodes and 500 μm between anode and cathode field wires.

1. Introduction

The use of high-granularity position-sensitive detectors is a challenge for charged particle tracking in future high-luminosity particle accelerators. Many applications of such detectors are also under discussion in nuclear medicine, biology, and other domains. The maximum rate capability of these devices and their ageing resistance in strong irradiation environments is a fundamental issue. This demand is partially covered by the development of silicon microstrip detectors and recent progress in microstrip gaseous chambers (MSGCs) [1–4]. The MSGCs have attracted considerable interest for use at future high luminosity colliders owing to their high rate and high accuracy capabilities. Their application is, however, limited to small surface coverage.

The asymmetric chamber consists of an anode wire plane mounted close to the cathode plane with engraved pick-up strips running orthogonal to the anode wire direction. This allows a very short path for the positive ions created in the avalanche, thus reducing the collection time of the positive ions and increasing the rate capability of the device by reducing the space-charge effect. A larger gap on the other side of the anode wire plane allows a sufficient number of primary electron–ion pairs to be generated.

Asymmetric wire chambers have been considered since the beginning of the development of wire chambers [5]. Many groups have investigated this structure for various

applications [6]. For instance, in order to increase the size of signals induced on pads, a gap as small as 0.4 mm has been used for a photodetector [7].

Our main objective is to optimize the rate capabilities of these types of chambers. We are also interested in the position resolution of the microstrip readout. The position can be derived by two means, the first being with analog techniques, where the charge on each strip is measured and the position given by the centre of gravity or some similar algorithm [8–10]. This is possible at low-rate experiments (such as LEP experiments) as many channels can be multiplexed together to a few analog-to-digital converter (ADC) channels. This method is limited at high-luminosity colliders since the channels cannot be multiplexed owing to the high rate. Thus we are also interested in digital readout. If we can identify the strip with the biggest charge we can get a resolution limited by the strip width (i.e. $\sigma = 144$ μm for our 500 μm width strips). This can be further improved if we can identify the case when two strips have comparable charges; thus even with digital readout we can foresee a resolution better than 100 μm. In order to design optimal electronics it is necessary to investigate the charge sharing between both strips.

2. Description of the chamber

Fig. 1 shows the cross-section of the chamber. The anode plane consists of alternating thin 7 μm diameter anode wires and 20 μm field wires on a 500 μm pitch. The plane of pick-up strips is located 600 μm below, and has strips on a 500 μm pitch orthogonal to the anode

* Corresponding author. Tel. +41 22 767 8165, fax +41 22 783 0600.

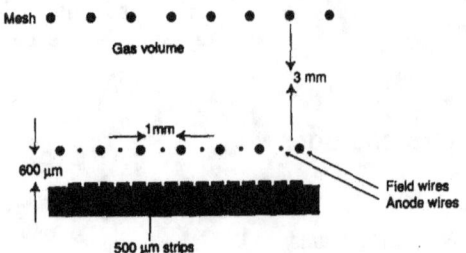

Fig. 1. Schematic of the cross-section of the asymmetric wire chamber.

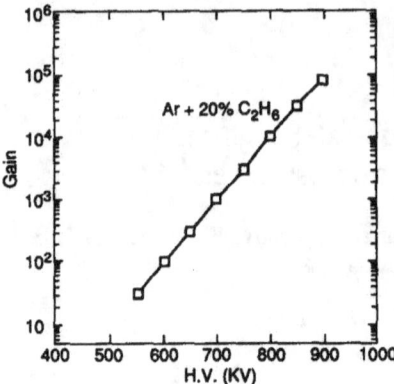

Fig. 3. Useful gain observed as a function of the high voltage applied on the anode wires. For this measurement the voltage applied on the mesh was 1600 V and on the field wires it was 100 V.

wires. Cathode strips are printed on a ceramic plate of a thickness that could, in principle, be reduced to a few hundred microns. Strips are connected to the ground through the preamplifier. There is a gas gap of 3 mm on the other side, closed by a metallic mesh foil. This plane is connected to a negative voltage, which is set at a typical value of 1500 V. Both the anode wires and the cathode strips are read out with standard preamplifiers having 300 mV/pC sensitivity. The gas filling was 80% Ar and 20% C_2H_6.

3. Results and measurements

A typical signal from the anode obtained with a minimum ionizing particle normal to the anode plane is shown in Fig. 2. Similar signal shape was obtained from the cathode strips with reverse polarity. The ratio between the collected charge on the neighbouring strip to that collected on the wire was measured to be 40%. The rise time of the signal is 15 ns including some contribution of the preamplifier time constant. Clearly the response of the detector is

satisfactory; with a more sophisticated current preamplifier and a suitable shaping it can be faster. The gain obtained as a function of the high voltage applied on the wires is shown in Fig. 3. At the value of 1000 V on the anode wires, with 50 V on the field shaping wires the useful gain was measured to be 10^5. At that gain the detector is able to detect single electrons; therefore it needs only a modest number of primary electrons produced per minimum ionizing particle and the sensitive gap can be reduced. A narrow gap is an important advantage, since the position resolution of the detector for inclined tracks is improved.

The behaviour of the chamber was tested with an ^{55}Fe radioactive source. The pulse-height distribution obtained is shown in Fig. 4. The expected peak at 5.9 keV is seen to be well separated from the Ar escape peak; it is rather uniform over the sensitive area of the detector with $\sigma = 6\%$. This is a satisfactory result for most applications and it can be improved with a careful choice of the appropriate gas and electronic chain.

To check the rate capability of the device, the chamber

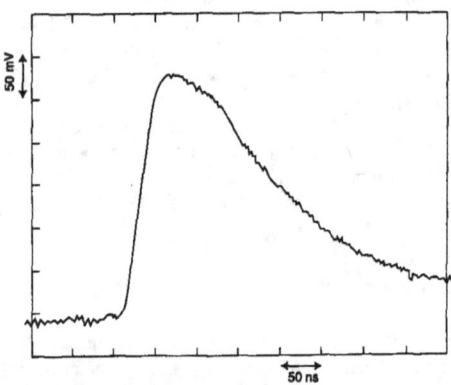

Fig. 2. A typical pulse after amplification observed with a digital oscilloscope. The pulse amplitude (50 mV per division) is plotted against time (50 ns per division).

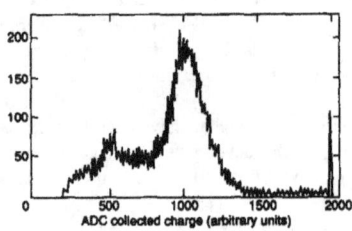

Fig. 4. Pulse-height distribution obtained for 5.9 keV ^{55}Fe X-rays (large peak) and the Ar escape peak (small peak).

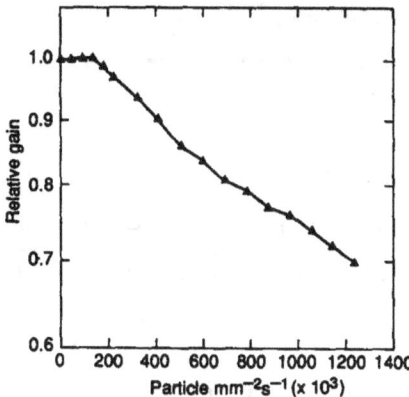

Fig. 5. The relative gain (with respect to the maximum gain) as a function of the flux of X-rays converted inside the detector. The measurement was performed at a gain of the chamber of 10^4.

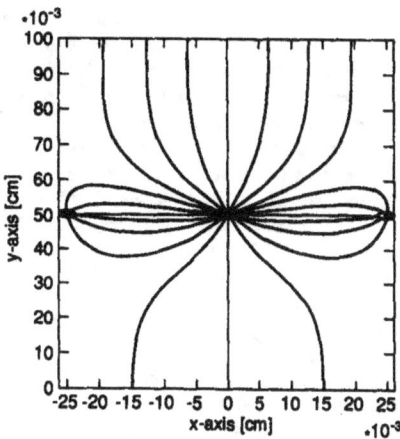

Fig. 6. Simulation of the field lines in a drift cell.

was exposed to strong radiation by an X-ray gun with a variable flux of X-rays at the energy of 8 keV. The gas filling in this case was Ar 90% and DME (dimethylether) 10%. The X-ray flux and the total current seen by the anode wires were monitored during measurements. The test was performed at the gas gain of 10^4. Fig. 5 shows the normalized gain as a function of the number of particles per mm^2 hitting the detector. The number of X-rays irradiating the detector was measured at the same time as the gain and the total current seen by the anode wires. The measurement was performed at a gain of 10^4. The gain remains stable up to a particle rate of 2×10^5 mm^{-2} s^{-1} and there is a loss of only 30% at 10^6 mm^{-2} s^{-1}. The maximum rate of operation of this device is more than one order of magnitude higher than that obtained with conventional proportional chambers. It reflects the effect of the fast evacuation of the positive ions through the small gap. This effect is illustrated in Fig. 6, which is a plot of the electric field lines around the anode wires. The major part of the positive ion charge ($\approx 80\%$), created closed to the anode wires, is collected by the field wires or cathode strips.

4. Conclusions and outlook

The microstrip asymmetric gaseous detector has been shown to be a promising device for use in large-area tracking devices in very-high-radiation environments. It operates up to 10^6 mm^{-2} s^{-1}. Our tests have shown that a significant increase in rate capability can be obtained by a wire structure that permits a more rapid evacuation of the positive ions from the avalanches. The rate capability

which we have observed, 3×10^5 mm^{-2} s^{-1}, has to be compared with tests giving a few 10^6 obtained with MS-GCs, but at a gain of only 10^3. Thus the rate capabilities are comparable. A similar work was done by a group at Siegen [11]. The structure of our chamber allows stereo tracking if anode wires are read out. Although the accuracy is limited by the wire spacing it can give valuable information, especially to remove ambiguity. In addition, the time resolution is significantly better than the MSGC. With the parameter chosen in the present design the rise time is 15 ns and the time resolution is expected to be of the order of a few nanoseconds. It is thus well adapted to the need of the future high-luminosity proton collider experiments.

The size of the detector can be large. Using narrow spacers, made for instance of insulating wires of a few hundred microns every 5 cm, the distance between anode wires and the cathode strip plane can be kept very constant, without significant efficiency loss. The chamber also allows a certain flexibility in the readout system, for instance microstrips placed on the back of a semiconducting cathode can be used.

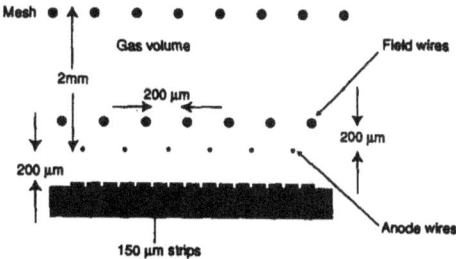

Fig. 7. Schematic of the cross-section of an optimized asymmetric wire chamber. The chamber design parameters are extended in order to achieve higher performance.

A schematic of an improved microstrip asymmetric chamber is illustrated in Fig. 7. The small cathode-anode gap is reduced to 200 μm by using insulating spacers, whilst the large ionization gap is 2 mm. The strip size is also rescaled to 200 μm. Furthermore, the anode spacing has to be reduced to increase the performances of the chamber, which is technically hard to achieve with the present design. It can, however, be accomplished by putting the field wires in a second plane, at a distance of 200 μm above the anode plane. This new configuration also allows faster positive-ion charge evacuation through a smaller gap and reduces that part of the ions drifting in the larger gap. The simulation program shows that almost 99% of the space charge is diffusing and is collected on the field wires or the cathode strips. A rough estimate of the ion drift time through the small gap (100 μm) shows that it can be of the order of 100 ns for an optimized gas mixture. It will permit an improvement of the output signal, a very-high rate capability of this detector, and a fine granularity giving a precise stereo spatial point of the incident charged particle.

Acknowledgements

We would like to thank R. Bouclier for having built the first prototype and M. Price for the preparation of the wire frame.

References

[1] A. Oed et al., Nucl. Instr. and Meth. A263 (1988) 351.
[2] F. Angelini et al., Nucl. Instr. and Meth. A283 (1989) 755.
[3] M.H.J. Gijberts et al., Nucl. Instr. and Meth. A313 (1992) 377.
[4] R. Bouclier et al., Nucl. Instr. and Meth. A 323 (1992) 236.
[5] G. Charpak, D. Rahm and H. Steiner, Nucl. Instr. and Meth. 80 (1970) 13.
[6] M. Atac, Nucl. Instr. and Meth. 176 (1980) 1.
[7] R. Arnold et al., Nucl. Instr. and Meth. A 314 (1992) 465.
[8] G. Charpak et al., Nucl. Instr. and Meth. 148 (1978) 471.
[9] A.H. Walenta, Nucl. Instr. and Meth. 151 (1978) 462.
[10] V. Radeka and R.B. Boie, Nucl. Instr. and Meth. 178 (1980) 543.
[11] E. Roderburg et al., Nucl. Instr. and Meth. A 323 (1992) 140.

Nuclear Instruments and Methods 217 (1983) 265–271
North-Holland Publishing Company

A THIN MULTIWIRE CHAMBER OPERATING IN THE HIGH MULTIPLICATION MODE

S. MAJEWSKI [1], G. CHARPAK [2], A. BRESKIN [3] and G. MIKENBERG [3]

[1] *Physikalisches Institut der Univ. Heidelberg, Fed. Rep. Germany*
[2] *CERN, Geneva, Switzerland*
[3] *The Weizmann Institute of Science, Rehovot, Israel*

The idea of a thin multiwire chamber with thick wires as a possible sampling detector for calorimeters was checked out in practice. Two small test detectors with gas sample thicknesses of 2 and 3 mm were constructed and tested in the laboratory. Avoiding argon, several pure hydrocarbons or binary mixtures with CO_2 were found to quench photons efficiently enough to make it possible to have charge multiplication factors of the order of 10^7. Efficiency of operation in this probably limited streamer mode for minimum ionizing electrons and with a single avalanche (streamer) response was found to be higher than 98%, with pulse-height resolutions down to 60% fwhm. The effect of gap-length variation is quite small: a gap change of up to 25% results in less than a 20% change in the mean charge released in a streamer.

1. Introduction

At present there exist many different structures of multiwire chambers that operate in the limited streamer or saturated modes. Some of the applications are in the domain of sampling counters in calorimeters. The advantage of such a sampling counter in comparison with the standard multiwire proportional chamber (MWPC) lies in the very high, saturated amplitude response for each particle in a shower.

But the disadvantage of most of the limited streamer counters operational at present is in the size of their individual cells, which is of the order of 1 cm × 1 cm (see refs. 2 and 3). Such large sizes are dictated by the range of photons produced in an avalanche. Firstly, photons limit the attainable amplifications by the photon-feedback mechanism taking place at cathodes. Secondly, photons from one avalanche may reach the neighbouring cells and produce secondary avalanches, thus deteriorating both the amplitude and the position resolutions of a detector. The second effect can be cured by introducing walls between cells which simply absorb photons. But there still remains a problem of photons propagating along wires. Here the idea of a resistive cathode coating was found helpful, locally desensitizing the detector around an original streamer avalanche. However, this solution probably cannot be accepted for electromagnetic calorimeters at higher energies where the density of shower particles becomes high. The best and simplest solution could be to use a gas that absorbs its own photons well enough. For example, avoiding or strongly limiting the amount of argon in a gas mixture should give a good result.

The other limit on the chamber thickness comes from practical reasons related to the necessity of keeping a sufficient level of electrode parallelism compatible with the assumed energy (amplitude) resolution of the counter over its surface. The minimum practically attainable thickness is very much dependent on the active surface of the detector. With this restriction, voltages necessary to operate the counter filled with strongly self-quenching gas in the limited streamer or saturated modes could be prohibitively high.

Thin sampling counters have the advantage of limiting the longitudinal size of a calorimeter and hence the lateral extension of showers and of permitting the construction of calorimeters with a large number of sampling layers without increasing the overall longitudinal size of a calorimeter. Moreover they should also lower the relative importance of delta rays. This was the reason why we decided to develop thin test counters [1]. We concentrated our work on medium-size detectors, aiming at a total thickness of 2–3 mm only.

2. Chamber structure

The sampling test-chamber structure is presented in fig. 1. The chamber has 50 μm diameter, sense (anode) wires and aluminium or graphite-coated mylar foils as cathodes. Originally we started with a 1 mm gap chamber having thick field wires of 100 μm diameter. Then we tried a much simpler version of a detector without field wires and a gap of 1 and 1.5 mm. We found the operation of this detector quite satisfactory as will be shown in §4. In the case of a detector with field wires the original idea was that by having a large fraction of the positive charges on the 50 μm sense wires (SW)

VIII. CALORIMETERS

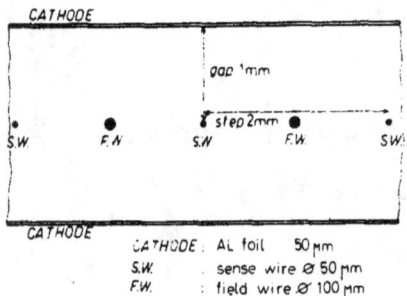

Fig. 1. Test chamber structure. Gap $L = 1$ mm; distance between 50 μm diameter stainless-steel anode wires: $s = 2$ mm. Field wires: gold-plated molybdenum, 100 μm diameter. External cathodes made of 50 μm thick aluminium foil.

produced by a large voltage difference between them and the neighbouring 100 μm field wires (FW) (fig. 2), the amplification properties would be much less sensitive to the variation of the gap thickness, which is much harder to maintain constant than the parallelism of the wires. This expectation only partly turned out to be true because, as we have found (§4.2.), the dominating factor in amplitude stability is the very high level of saturation of space charge in each avalanche, and even the simple MWPC structure was much less influenced by gap-length variations than normally operating multiwire chambers. The high-resistivity graphite-layer-coated mylar foils (> 500 kΩ per square) were tried in view of a possible readout method using pads outside the gas volume, following the method proposed by Iarocci and his col-

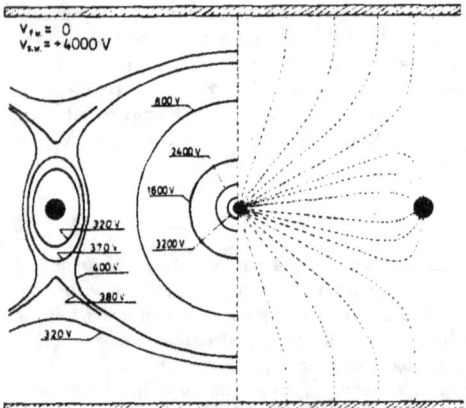

Fig. 2. Geometry of electrostatic field distribution in the structure of fig. 1. External cathodes are at the ground potential. Anode sense wires at +4.00 kV, field wires at 0 kV (resistive paper simulation).

laborators [4]. Here it was found that while sparking in the structure with field wires (due to an overvoltage or induced by radiation of high intensity) takes place between sense and field wires without any noticeable damaging effect, in the case of the structure without field wires sparks are obviously directed to the cathodes, and the graphite layer can be locally damaged by energetic sparks. The best protection against this effect is obtained by limiting the capacitance that is discharged in the spark, for example by completely avoiding the anode readout from the sense wires and thus eliminating decoupling capacitors from the wires. The other capacitance that should be limited is the capacitance of the cathode to pads (on ground potential). For example for a 10×10 cm^2 pad coupled through 100 μm thick mylar foil to the cathode, the capacitance obtained is equal to about 4 nF.

3. Gas mixtures

3.1. Gas choice

As it was pointed out in the introduction, in choosing the gas mixture we were guided by the idea of avoiding argon as a main source of ultraviolet photons in typical gas mixtures. Therefore we tried pure hydrocarbons and their mixtures with some neutral gases like CO_2, nitrogen and/or electronegative compounds such as freon-12, freon-13 and ethyl bromide. Our test of a given gas consisted in trying to obtain the highest possible amplitudes measured directly from the chamber and on a 50 Ω load resistor at the input to the scope channel. From the experience gathered by other authors using the streamer mode of operation we defined somewhat arbitrarily our goal in finding gases that could give us amplitudes of the order of 20–50 mV on 50 Ω load for a minimum ionizing particle passing perpendicularly through the detector. An additional criterion applied was to obtain the smallest possible amplitude jitter.

3.2. Gaseous hydrocarbons

We tested the following pure hydrocarbons: methane, ethylene, ethane, propane, isobutane, and neopentane (dimethylpropane). We found that only the last three can operate in the required high amplification conditions, with improving properties from propane to neopentane. In methane and ethylene the limit of amplification was due to source induced sparking at values of amplification smaller by two orders of magnitude, and in ethane maximum amplitudes were about 5 mV on 50 Ω. Because propane and isobutane are known to polymerize in gaseous detectors (though when in mixtures with argon) we tested these gases with the admixture of a well-known neutralizing agent – methylal – and found no change in operation. A small admixture (50 ppm) of water vapour was also found to prevent propane polymerization in an argon/propane mixture [5]. In a short test run we found that operating conditions for both propane and isobutane are almost unchanged, and in the case of isobutane are even slightly improved, by a 0.5 to 2% water vapour admixture. In a paper presented at this conference [6] it is shown that a small 0.1 to 0.2% admixture of hydrogen gas to argon/ethane and argon/isobutane mixtures slows down the polymerization process. We also checked the effect of additional admixtures of freon-12, freon-13 and ethyl bromide, in small quantities in order to limit the active thickness of the counter. Although the amplitude resolution was worse, it is possible that in the final application this could result in better resolution of the calorimeter, as will be discussed in §6. Finally, it was found that some small quantities of argon, up to 10% can be added to the gas without effecting in a drastic way the "magic" properties of the hydrocarbons while lowering the working voltage. Nevertheless we decided not to admix argon to our mixtures. At that stage our best practical gas was a mixture of isobutane with several (5 to 8)% of methylal and about 0.5 to 1.0% of freon-12.

3.3. Mixtures with CO_2

When considering practical problems of the gases, we came to the safety problem considerations, related to the explosive and inflammable nature of the gases used. It was decided to try binary mixtures of hydrocarbons with neutral gases like CO_2 and nitrogen. We found the maximum acceptable admixture of CO_2 in isobutane to be about 10–15% only. Then we extended our search to some hydrocarbons that exist as liquids in normal temperature and pressure conditions, but have high vapour pressure. We have checked only two substances up to now: n-hexane and n-pentane. Carbon dioxide, even with the highest possible admixture of n-hexane of about 15% (at 20°C), and nitrogen, both compounds did not operate in the high amplification mode. But CO_2 with n-pentane, which is close in structure to the gas neopentane used before, and which has very high vapour pressure at room temperature (fig. 3, [7]) of about a half of an atmosphere, operated very well. Most of the results presented here were obtained with this simple, quite cheap, but only to a slight extent, safer mixture [8].

Trying to understand why we could operate in the high amplification mode only with some hydrocarbons, we noticed (although we were certainly not the first to do so) some possible phenomenological rules on the basis of our scarce experimental data. Firstly, in all hydrocarbons fulfilling our requirement of high amplification the molecule has at least two, if not several, CH_3 atom groups. Then because ethane did not meet the

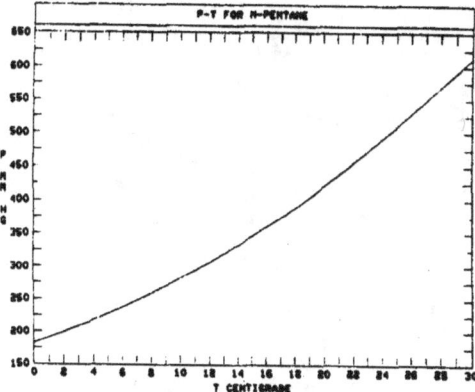

Fig. 3. Dependence of the saturated vapour pressure of n-pentane on temperature in the range of 0 to 30°C (on the basis of values from ref. 7).

amplification criteria, though it has the simplest molecular structure with two CH_3 groups, H_3C--CH_3, the second important factor is certainly the ionization potential. Its value decreases from ethane (11.1 eV) to neopentane (10.35 eV). Finally the third observation concerning liquid hydrocarbons is that the vapour pressure should be high, because below some concentrations in CO_2 the quenching is not sufficient. On the basis of these rather obvious observations, some new candidates to mix with CO_2 have been chosen and will be checked.

4. Operation of the detector

4.1. Amplitude spectra

We found that both structures described in §2 operate in the high amplification mode. Fig. 4 shows examples of pulses obtained in the structure with field wires and with pure neopentane gas filling. Peak amplitudes up to 50 mV on a 50 Ω load were obtained with a 90% fwhm pulse-height dispersion and with close to 100% efficiency for minimum ionizing electrons (fig.5). This implies a charge multiplication factor of the order of 10^7 (here we see only an electronic, fast component). In fig. 6 the pusle-height spectrum obtained in the same structure with CO_2 bubbled through n-pentane at 17°C is presented. The noticeable improvement in amplitude resolution is observed with a fwhm of 60%. In both spectra the high amplitude tail, typical of Landau distribution (fig. 7) is suppressed, indicating that there is a high saturation level of charge multiplication. With about 1% of freon-12 added to isobutane/8% methylal gas mixture an increase in attainable mean amplitude to about 100 mV/50 Ω was possible with very narrow (< 5

VIII. CALORIMETERS

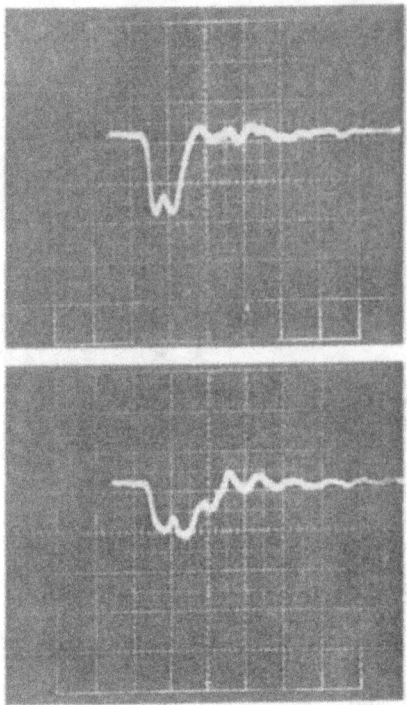

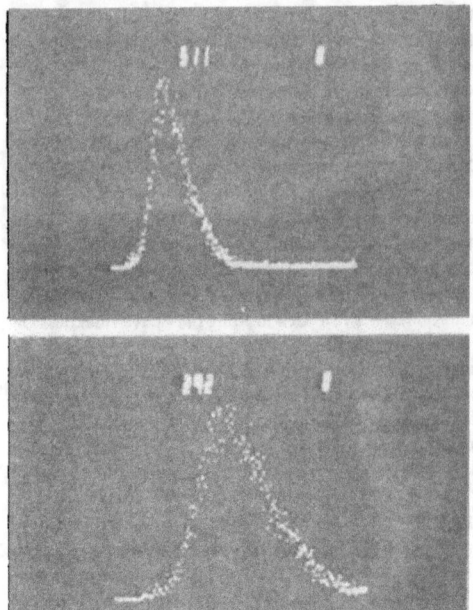

Fig. 5. An amplitude spectrum from a 2 cm wide (10 anode wires) section of the chamber. Field wires at zero potential. Collimated beam of ^{106}Ru minimum ionizing electrons. Pure neoptane gas at +4.00 kV. A fwhm of 90% was measured. (The lower picture is an enlarged view of the top one.)

Fig. 4. Two examples of pulses from minimum-ionizing electrons measured directly from a 2 cm section of the chamber on a 50 Ω load, as seen on the storage scope screen. Conditions: pure neopentane gas, working voltage: +4.00 kV. A decoupling capacitor of 470 pF was used. Horizontal scale: 5 ns/large division, vertical scale: 20 mV/large division.

ns fwhm) pulses (fig.8) but at the expense of a drop in efficiency to 93%, which nevertheless can still be accepted for hadron-calorimeter applications.

The structure without field wires was checked with anode wires of different diameters: 50, 60, and 100 μm and with a gap length of 1 and 1.5 mm. With 100 μm wires high amplitudes were also possible but at higher voltages and with not as good pulse-height resolution as for 50 μm wires. The version of the structure with the gap length increased to 1.5 mm was checked for the reasons explained in the next subsection. Fig. 9 shows a pulse-height spectrum measured in this detector, with 50 μm wires and for a 50/50% mixture of CO_2/n-pentane. A fwhm of 65% was obtained at about a maximum working voltage of 4.3 kV and of 80% at 3.7 kV, which we still find satisfactory. No difference was observed in detector operation whether it was equipped with Al foils or mylar foils with resistive graphite coating, except for the spark-induced damage in the graphite layer of the

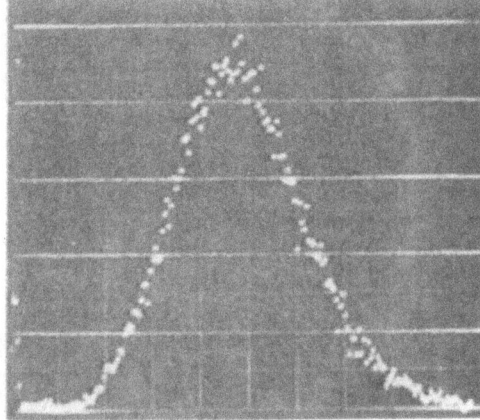

Fig. 6. An amplitude spectrum for CO_2 gas bubbled through n-pentane liquid at 17°C, working voltage +3.60 kV. The fwhm of the distribution is 60%.

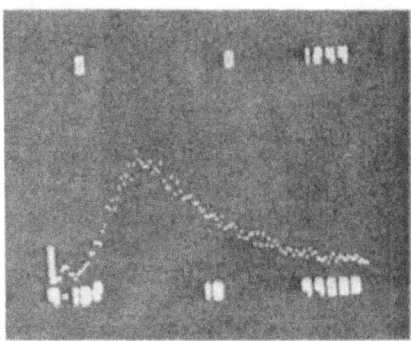

Fig. 7. An amplitude spectrum for the mixture of 62% argon + 29% isobutane + 9% methylal. Working voltage: +1.80 kV (about maximum before sparking). Most probable amplitude about 0.25 pC.

structure without field wires, due to erroneous overvoltages, mentioned in §2. Similar results were obtained at a later stage with the amount of n-pentane in the gas mixture reduced to 30%.

4.2. Gap uniformity

The obvious practical problem of our thin-gap detector is in the level of parallelism necessary to keep the local charge multiplication variations within acceptable limits. In order to test the sensitivity of the detector to gap length changes, the measurements of pulse-height spectra were performed when decreasing locally over about 1 cm^2 the gap length on one side of the anode plane and for minimum ionization electrons passing just in the middle of the region (no edge effects). For a 200 μm push, i.e. when changing a gap from 1.0 mm to 0.80 mm (which means by 20%) the change in the mean

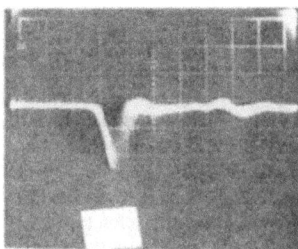

Fig. 8. Puses directly on the oscilloscope screen for the mixture of isobutane, methylal, and freon-12 (see text) at +3.40 kV. Horizontal scale: 5 ns/div., vertical scale: 50 mV/div.

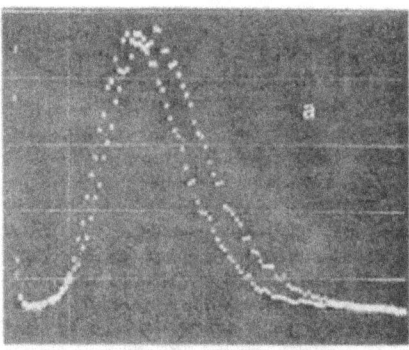

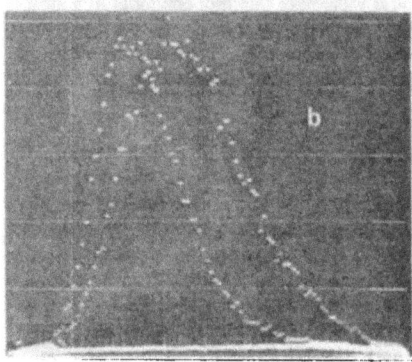

Fig. 10. Check of the effect fo gap squeezing. One of the cathode foils was pushed inwards so as to squeeze the gap distance from 1 mm to (a), 0.80 mm; (b), 0.60 mm. In both cases two overlapping amplitude spectra of minimum ionizing electrons are presented for the two positions: zero and shifted one. Structure with field wires. Conditions: gas $CO^2/20°C$ n-pentane vapour, working voltage +2.60 kV.

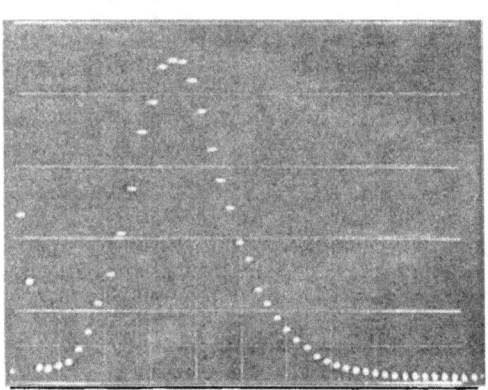

Fig. 9. Amplitude spectrum for the $CO_2/20°C$ n-pentane vapour in the structure with 50 μm diameter anode wires and no field wires. Gap length is equal to 1.5 mm. Working voltage: +4.30 kV.

VIII. CALORIMETERS

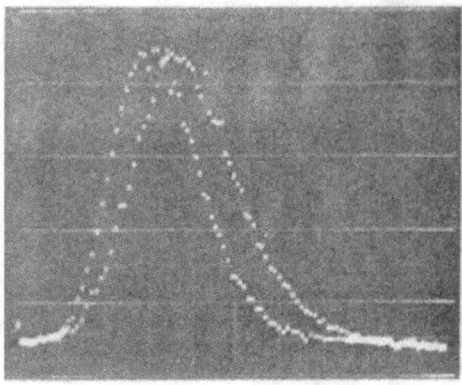

Fig. 11. Check of the same kind as fig. 10a and in the same gas, also with 50 μm anode wires but in the structure without field wires and with a gap length of 1.5 mm. Working voltage: +3.90 kV.

Fig. 12. Time resolution of the detector as measured for neopentane at +4.00 kV. Two spectra were recorded with 25 ns relative shift. The chamber output was connected through a decoupling capacitor and a short $Z = 50 \, \Omega$ cable to the input of the fast differentiating discriminator of Brahy and Rossa [10]. The threshold was adjusted at -3 mV level. The total time jitter was of about 20 ns. Note a characteristic drift-distance/time dependence contribution to the curve shape, as in the case of drift chambers.

pulse height was about 15% in the structure with field wires, with 50 μm diameter anode wires at +2.6 kV and for a 50/50% mixture of CO_2/n-pentane (fig. 10a). In the same conditions even the 400 μm push resulted only in about a 40% change and the chamber was still not discharging (fig. 10b.) As expected, the structure without field wires was found to be more, but not drastically more, sensitive to such a gap squeezing. For a 1.5 mm gap length, the same anode wires, and the same gas, a 200 μm push produced a 20% change in pulse height at 3.9 kV (fig. 11). For a comparison in a standard multi-wire proportional detector with 20 μm diameter sense wires and a gap length of 8 mm operating at a gain of around 10^6, a 200 μm cathode displacement will result in a change of about 17% [9], which falls within the same range of values as for our detector.

5. Time resolution

The time resolution of the chamber was measured only for a pure neopentane gas and a structure with field wires. All the pulses in coincidence with a scintillation counter were detected within a 20 ns time window (fig. 12). The fast differentiating discriminator [10] was used to discriminate chamber pulses. From the measured spectrum having a typical drift-time shape with a spike in the origin one can estimate the time resolution of the many-samples calorimeter detector for big showers to be below 1 ns. The 30 ns wide linear gate to ADC channels recording the shower distribution will be sufficient. Thus the expected timing properties of such a calorimeter could be comparable to those of a lead-glass/photomultiplier one.

6. Conclusions and discussion of properties

We will summarize here the interesting properties of the operation of the tested thin multiwire structures with a new family of "magic" gases:

(i) Very high gas multiplication factors of the order of 10^7, characterized by the high stage of saturation resulting in a very simple readout without noise problems and in suppression of Landau tail effects in the amplitude distribution.

(ii) Relative insensitivity to the gap-length fluctuations, making these counters no more difficult to prepare mechanically than standard multiwire chambers.

(iii) Electronic pulses short in time and with fast rise-time revealing good timing properties of the detector. Also, owing to the small gap length and much higher than typical electric fields ions are removed much faster from the gas volume, making this counter less sensitive to count-rate problems than other streamer chambers.

This counter seems then to be a good candidate to sample the number and distribution in space of shower particles in the compact and fast calorimeter modules of modest sizes. Typical thicknesses in present-day electro-magnetic calorimeters of the lead/multiwire chamber sandwich type are 2 mm for lead and 10 mm for a chamber. When using 2 mm thick chambers, the thickness of the calorimeter (taking into account the readout) will be decreased by a factor of 2. As an example, a 20 radiation lengths thick calorimeter of this type will occupy only 25 cm instead of 50 cm. On the other hand, given the maximum space to be occupied one can afford finer sampling thus improving the energy resolution.

Finally, having limited the sensitive gas volume (also by adding some electronegative admixtures to the gas) one could hope for a reduction of track-length fluctuations, which provide the main contribution to the energy resolution of gas sampling calorimeters [11].

In the works of Atac et al. [3] it was found that in their case the optimum energy resolution of a sampling calorimeter is obtained just before reaching the limited streamer mode in 1 cm thick sampling chambers, though one could expect a better resolution when operating in the streamer mode. The explanation of this effect could lie in the fact that the streamer mode was never fully attained, especially in the region of high intensity of tracks in the core of a shower and in a percentage of cases a shower particle gives no streamer response but a smaller saturated avalanche pulse. The way to improve it, as proposed by Atac et al. is to use smaller cell sizes with a better quenching gas, and our detector meets these requirements. Furthermore, in the recent measurements performed on the beam at DESY we found that up to the angle of 45° between the beam axis and detector surface, each beam particle gives a single streamer response.

To check these expectations experimentally a small-size test electromagnetic calorimeter of 20 radiation lengths is under construction.

The authors would like to thank Messrs R. Dabrowski, R. Benoit and M. Shua, from the University of Warsaw, CERN and the Weizmann Institute of Science (WIS), respectively, for their skilful work in the preparation of several test detectors. The participation of Miss Juliana Cohen and Mr Peter Fink from WIS in some measurements is also acknowledged. One of us (SM) would like to thank Professor Uzy Smilansky, Chairman of the Department of Nuclear Physics, for the invitation to visit WIS, and the staff of that Department for a fruitful stay.

References

[1] S. Majewski and G. Charpak, EP Internal Report 82-02, CERN (1982).

[2] G. Battistoni et al., paper presented at the Int. Conf. on Instrumentation for Colliding Beam Physics, Stanford (February 1982).

[3] M. Atac et al., Gas Sampling Calorimeter Studies in Proportional, Saturated Avalanche and Streamer Modes, FN-371 (1982); M. Atac et al., Nucl. Instr. and Meth. 205 (1983) 113.

[4] G. Battistoni et al., Nucl. Instr. and Meth. 152 (1978) 423, Nucl. Instr. and Meth. 176 (1980) 297; submitted to Nucl. Instr. and Meth.

[5] F. Schneider, CERN, private communication.

[6] H. Sipilä and M.L. Järvinen, these Proceedings, p. 298.

[7] R.C. Weast, ed., Handbook of Chemistry and Physics, 63rd ed., 1982–1983 (CRC Press, Boca Raton, Florida) p. D-208.

[8] J.H. Burgoyne et al., Fuel 27 (1948) 118.

[9] F. Sauli, CERN 77-09 (1977).

[10] D. Brahy and E. Rossa, Nucl. Instr. and Meth. 192 (1982) 359.

[11] H.G. Fischer, Nucl. Instr. and Meth. 156 (1978) 81.

VIII. CALORIMETERS

Nuclear Instruments and Methods in Physics Research 224 (1984) 315–317
North-Holland, Amsterdam

Letter to the Editor

A SIMPLE "VERNIER" METHOD FOR IMPROVING THE ACCURACY OF COORDINATE READOUT IN LARGE WIRE CHAMBERS

D.F. ANDERSON *, H.K. ARVELA **, A. BRESKIN *** and G. CHARPAK

CERN, Geneva, Switzerland

Received 30 January 1984

A method is proposed to improve position accuracy in large wire chambers. A localization of an avalanche is obtained simply by printing a series of double wedges on a narrow cathode strip parallel to the anode wire of a wire chamber. This method seems to be well adapted as a "vernier" technique: a gross method provides the address of a wedge subsection which then gives, with a greater accuracy, the position within the subsection.

The 1σ localization accuracy of an avalanche along a wire, in large-area wire chambers, is limited with most of present read-out techniques to about 1% of the wire length. We propose a "vernier" method to improve the coordinate read-out considerably, using a series of wedged cathode strips located in parallel to the sensing wires.

Varieties of wedge and strip methods can be used to obtain the coordinates of charge deposition on electrodes [1]. In multiwire proportional chambers, the work of Allemand et al. [2] has shown that it is possible to obtain easily the coordinates along the wires, within the limits of accuracy permitted by the physical extension of the avalanches. They have reached, with X-rays of 8 keV, an accuracy of 0.2 mm (fwhm), for a chamber of 6 cm length. The implementation of this method is in some respects easier than charge division along the wires, since it relies only on the geometry of the printed wedges which can be easily adjusted.

Several variations of this method [3] have been introduced which permit two-dimensional read-out to be obtained from single planes. This method has been extended to detectors other than wire chambers [4]. Fig. 1 illustrates the wedge-and-strip cathode discussed in ref. [4]. The cathode surface is subdivided into three domains, A, B, and C. The induced signals vary, as a function of events' two-dimensional position, according to the relations

$$x \propto Q_A/(Q_A + Q_B + Q_C),$$

* Visitor from Fermi National Accelerator Laboratory, Batavia, Ill., USA.
** Present address: Physics Dept., University of Turku, Turku, Finland.
*** Permanent address: The Weizmann Institute of Science, Rehovot, Israel.

$$y \propto Q_C/(Q_A + Q_B + Q_C),$$

where Q_A is the induced charge measured on the A electrode, etc.

In a search for a "vernier" method permitting the coordinates of avalanches along a wire in large wire chambers to be obtained with a high accuracy we have been attracted by the potential advantage of this approach. A series of double-wedge strips on a narrow circuit facing an anode wire can give with accuracy the coordinate of an avalanche along the strip that faces a portion of the wire. A gross method, such as charge division along the anode, provides the address of the strip, while the ratio between the charge induced on one

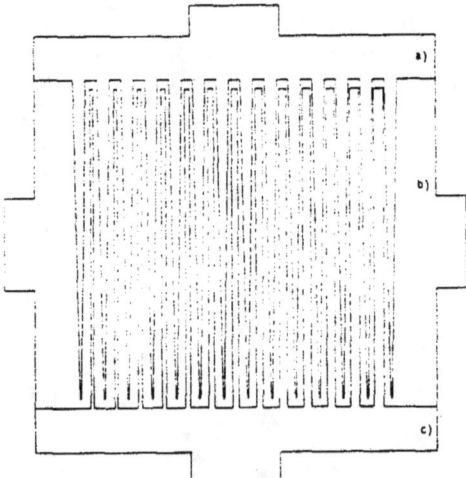

Fig. 1. Two-dimensional wedge and strip method (see text).

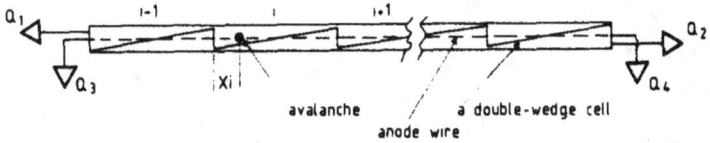

$$X_i = Q_2/Q_1 + Q_2$$
$$i = Q_4/Q_3 + Q_4 \quad \text{(or by other gross method)}$$

Fig. 2. Principle of the "vernier" method. A charge is induced on a double wedge subsection of a read-out strip. The position along the subsection is accurately obtained by the ratio of charges. The gross subsection address location can be performed by current or charge division along the wire.

wedge to the total charge induced on both wedges provides the localization along the particular strip (fig. 2). We have checked this idea by building a simple proportional counter in a tube with a square cross-section (11×11 mm^2). A printed circuit read-out strip, 9 mm wide, was glued along one side, carrying double wedges of various lengths from 10–30 cm length. The detector is shown schematically in fig. 3.

The detector was operated with an argon–isobutane gas mixture (80/20) in a proportional mode. X-ray photons of 5.9 keV from a ^{55}Fe source, collimated to 0.5 mm, could enter the counter through some window openings, as shown in fig. 3.

With simple electronics, we have obtained the following accuracies:

0.8 mm (fwhm)   for 10 cm wedge length,
1.2 mm (fwhm)   for 20 cm wedge length,
1.9 mm (fwhm)   for 30 cm wedge length.

The position spectra are shown in fig. 4. These numbers are not corrected for the source collimation width or for any systematic errors due to a displacement of the source off the centre of the counter, which may strongly influence the resolution. The resolution is quite constant along the strips, except for a minor degradation (30–40%) at distances of a few mm from the end owing to geometric problems in the design when going from one section to another. Linearity is also slightly degraded at the edges for the same reason. The results shown in fig. 4 correspond to information extracted from individual strips. When interconnecting all three strips externally to the same amplifier the resolution was degraded by about 5–10% owing to increased noise due to increased capacitance. The capacitance of a long chain of strips may deteriorate the present resolution.

We can thus envisage a printed strip made of a series of short wedges in series. The read-out of the charges at both ends gives the position along a wedge. Various methods, such as charge division, along the wire, can be used to localize the wedges which have detected the induced charge. This technique is possible for large-size detectors of several metres length and may offer resolutions of the order of 1 mm along the wire.

For smaller detectors of the type planned for some vertex detectors, one can think of a succession of strips giving alternatively a gross coordinate and a precise interpolated coordinate. A redundancy in the number of coordinates read-out would permit one to solve easily the ambiguity problems encountered when the avalanches are close to the junction points of successive wedges. This problem may also be solved by using two independent read-out strips, with displaced sections, per wire cell.

For detectors based on individual proportional tubes this method would permit the good accuracy obtained from the measurement of the drift time to be combined with the good localization of the avalanches along wires.

In the present study, the wedged read-out strips were placed inside the sensitive volume of the detector. With high-resistive cathode surfaces, which are known to be transparent to fast pulses, the strips might be placed as external pads and provide a simple and accurate localization technique.

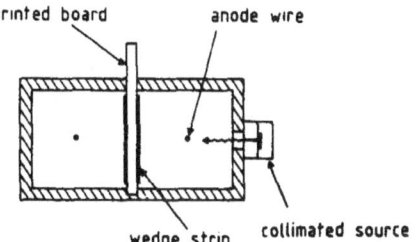

Fig. 3. A schematic view of the test detector. Only one cell was used.

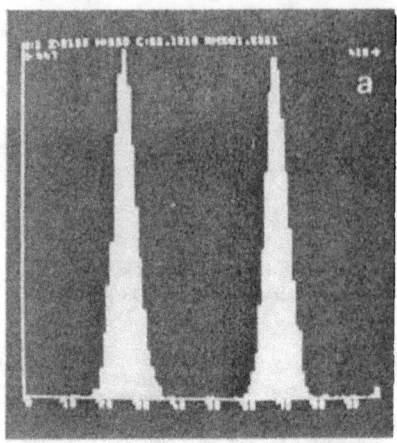

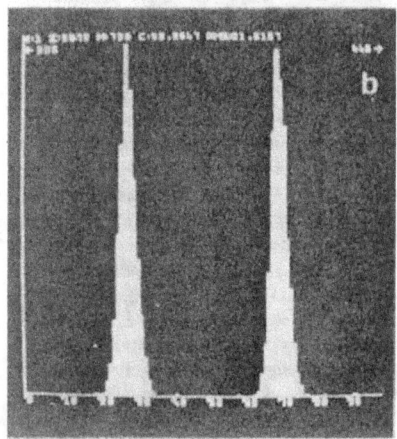

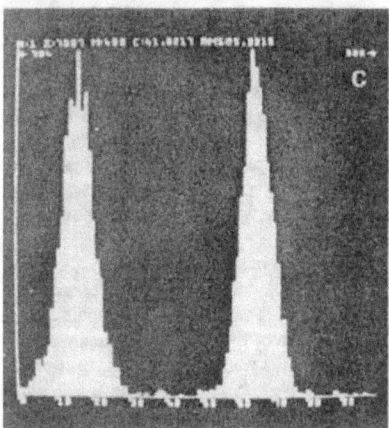

Fig. 4. Position distribution of induced charges for: (a) 100 mm long strip (5 mm between peaks), fwhm = 0.8 mm; (b) 200 mm long strip (10 mm between peaks), fwhm = 1.2 mm; (c) 300 mm long strip (10 mm between peaks), fwhm = 1.9 mm.

References

[1] H.O. Anger, Instr. Soc. Am. Trans. 5 (1966) 311.
[2] R. Allemand and G. Thomas, Nucl. Instr. and Methods 137 (1976) 141.
[3] C. Martin, P. Jelinsky, M. Lampton, R.F. Malina and H.O. Anger, Rev. Sci. Instr. 52 (1981) 1067.
[4] O.H.W. Siegmund, S. Clothier, J. Thornton, J. Lemen, R. Harper, I.M. Mason and J.C. Culhane, IEEE Trans. Nucl. Sci. NS-30 (1983) 503.

NUCLEAR INSTRUMENTS AND METHODS 108 (1973) 413–426; © NORTH-HOLLAND PUBLISHING CO.

HIGH-ACCURACY DRIFT CHAMBERS AND THEIR USE IN STRONG MAGNETIC FIELDS

G. CHARPAK and F. SAULI

CERN, Geneva, Switzerland

W. DUINKER

Utrecht State University, Holland

Received 16 January 1973

A method is proposed that allows the construction of drift multiwire proportional chambers with adjustable electric field suitable for the detection of high-energy charged particles in strong magnetic fields. Typical measured accuracies go from 50 to 200 μm, varying with the drift length, at 16 kG. The simple mechanical construction allows the wire distance to be adapted to the expected counting rate, even in the same chamber. Furthermore, the advantages of a current division method to obtain both orthogonal coordinates from the same wire are discussed.

1. Introduction

The correlation between the position of an ionized track produced by a charged particle and the time of appearance of an electric pulse at the wire of a proportional chamber can be used to measure the distance of the trajectory from the wires[1,2]).

Several types of localization detectors have been imagined based on this principle. Let us describe two of them.

1) One keeps the structure of a normal multiwire proportional chamber, and measures the time distance from the wire. The electric field between the wires is however not uniform, and is in fact zero in the centre of the separation between two wires; since the drift velocity of electrons is in general a function of the

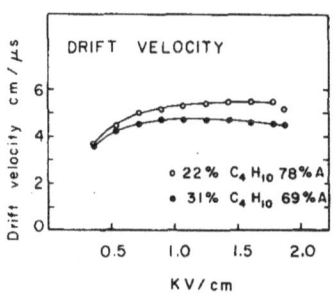

Fig. 1. Drift velocity as a function of the electric field, at atmospheric pressures, for two mixtures in different proportions of argon and isobutane. The drift velocity approaches a constant for fields larger than ≈ 500 V/cm. (From ref. 3; notice that the proportion of isobutane given in that paper was wrong, owing to an improper calibration.)

field, a complicated space-time relationship may result with long tails in the time distribution. In some gas mixtures, however, the drift velocity approaches a constant value for moderately large fields, as illustrated in fig. 1 (from ref. 3). Furthermore, addition of a corrective "field" wire[4-6]) between two sense wires increases the electric field in the critical region. As will be seen in the next chapter, however, the method is effective only if the ratio of the cathodes to wire distance is close to one; for larger wire separation, the electric field becomes too small to saturate the drift velocity.

A difficulty of such a method is the right–left ambiguity; one does not know from which side of a wire the electrons are collected.

The ambiguity can be solved in several ways; the simplest one that comes to mind is to use two chambers displaced by half a wire distance. However, problems may occur for inclined tracks since the wire distance cannot be much larger than the average chamber thickness, which is of the order of 1 cm.

A group from the Heidelberg University has used a simple method to solve the right–left ambiguity[4-6]). The single sense wire is replaced by a triplet of wires. Two thin wires close to each other and electrically separated are used to amplify the avalanches. They are separated by a thick wire which plays the role of an electric shield; the distance between the wires of a triplet is 1 mm. The triplets are 2 cm apart and are separated by a field wire which increases the drift field between the wires and reduces the spread of velocities along the drift path. Large chambers having this structure, have been successfully operated, with a

413

typical accuracy of localization of 0.35 mm (standard deviation) for a maximum drift of 10 mm. Some lack of linearity in the space–time relation was observed, that was corrected by having a non–linear clock.

We have recently proposed another method for solving the right–left ambiguity[7]. It was shown that if two wires of 20 μm are placed at a distance between 0.1 and 0.2 mm apart, the avalanche surrounds only one wire. This is demonstrated by the strong asymmetry introduced in the positive pulse induced at the neighbouring wire: the pulse is much higher in the wire which is on the same side of the collecting wire as the track. The electrostatic repulsion between the wires causes, however, electrostatic instabilities for long wires, and these have to be mechanically tightened together at regular distances.

2) One can build a special structure where a drift space is provided with a uniform field optimized for the best resolution. The electrons are collected through a grid to one or many amplifying sense wires where the time of arrival is measured.

Such a chamber of small dimension was tried at CERN in 1969[8]. It proved that an accuracy of the order of 100 μm could be obtained over a few centimetres, and that with two drift chambers the time of passage of a particle in a chamber could be measured within 5 ns or better.

A group at Saclay extended this method. They found a gas in which it is possible to drift the electrons over distances as big as 25 cm, with a reduction of accuracy which was almost unobservable – at least within their accuracy of measurements, of the order of 0.3 mm[9]).

The two experiences of Heidelberg and Saclay have clearly demonstrated the potential of drift chambers and their ease of operation. Two remarks can be made about these two methods of exploiting the space–time correlation.

a) With chambers where the electrons are drifted in non-uniform fields there exist limits to the accuracy of measurements which may be very far from the intrinsic limit of the method. This may be even more true if the chambers are placed in magnetic fields.

In a structure with wire triplets placed at distances of 1 mm there should exist variations in the position accuracy around these wires. This is of no importance as long as accuracies of the order of the wire spacing in the triplet are aimed at. But we shall see that much greater accuracies are within reach by the drift method.

b) With chambers with a single amplifying wire and a large drift space it seems difficult to go beyond the dimensions already reached at Saclay, namely 25 cm. This drift length gives rise to a time resolution for accidentals of the order of 6 μs, which may be prohibitive in many cases.

2. A method of obtaining uniform drift fields with a multiwire structure

We will show in this article that it is possible to construct chambers having a roughly uniform drift field, still keeping the structure of a normal multiwire proportional chamber.

The idea is to have the cathodes made of parallel equidistant wires placed at increasing potentials, to generate a well-controlled electric field along the drift space. This is shown schematically in fig. 2. The potential on the cathode wires grows uniformly from a minimum (V_m) in front of the amplifying wire, at left in the figure, to a maximum (V_M) at the end of the drift space. Addition of a field wire in the centre of the separation between two sense wires makes it possible

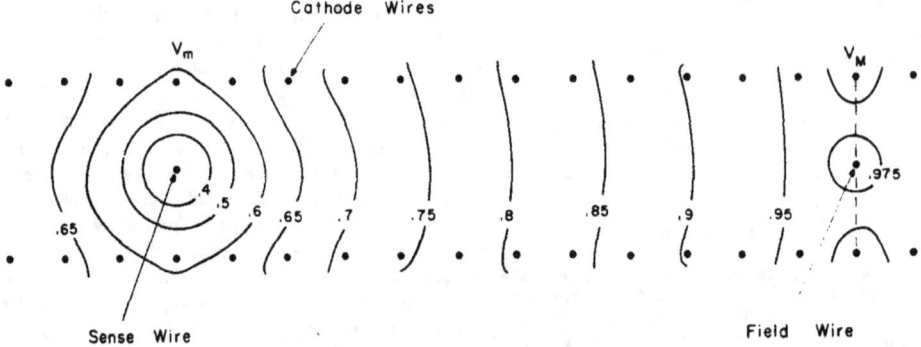

Fig. 2. Principle of construction of the adjustable field drift chamber. The cathode wire spacing is 2 mm, the gap 6 mm in total, the distance between two sense wires 48 mm. The equipotentials are shown for $V_m = 0.58$ and $V_M = 1.0$, in relative units.

to eliminate the critical low field region. In the figure, the equipotentials (measured with a conductive paper method) are shown for a uniformly increasing potential applied to the cathode wires, going from $V_m = 0.58$ to $V_M = 1.0$.

The value of the electric field, for $V_M = 1$ kV, on a central cut across the chamber (parallel to the cathode wires) is shown in fig. 3, curve A. For a comparison, curve B shows the same quantity in the absence of the field wire, and curve C the field in a chamber of similar structure but having all cathode wires (and the field wire) at the same potential, as in a standard MWPC. Since in the real case (see the next section) $V_M \geqslant 3.2$ kV, one can see that for the described structure the field is nowhere smaller than ≈ 600 V/cm, resulting in a constant drift velocity if the indicated gas mixture is used (see fig. 1). This is not the case for the structures whose field is represented by curves B and C. One can obviously vary V_m and V_M independently, in order to optimize the gain and the drift properties of the chamber.

The structure is very flexible, and the sense wire distance can be varied at will according to the experimental needs (mainly the expected counting rate per unit area); in principle one can also have a variable wire distance in the same chamber, since the operating voltage is mostly determined by V_m. The mechanical construction presents no problems; all cathode wires having the same potential are connected on a printed circuit to one of a set of bus-lines, and the voltage distribution can be organized at one end of the chamber, for example by means of a resistor network. Two power supplies can be used to vary V_m and V_M independently.

We have constructed several drift chambers with the described structure, and with the following parameters:

sense wire diameter:	20 μm,
sense wire length:	35 mm to 30 cm,
distance between sense wires:	48 mm,
distance between cathodes:	6 mm,
cathode wire diameter:	100 μm,
cathode wire spacing:	2 mm.

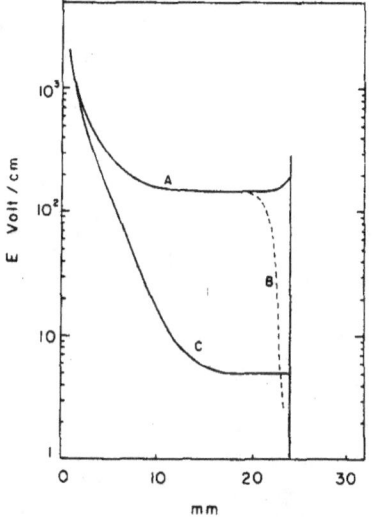

Fig. 3. Electric field along a central cut of a structure like the one shown in fig. 2: A) using an increasing potential on the cathode wires as described in the text, and with the field wire connected to V_M; B) as (A), but without the field wire; C) applying the same potential to both cathode and field wires, as in a normal multiwire proportional chamber.

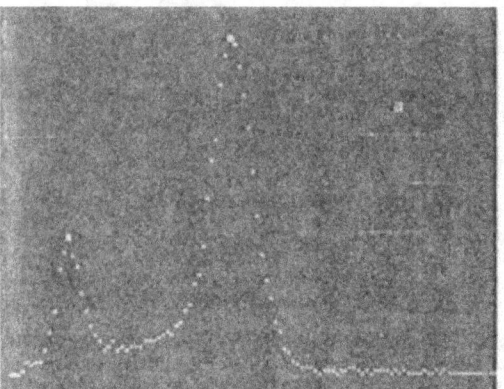
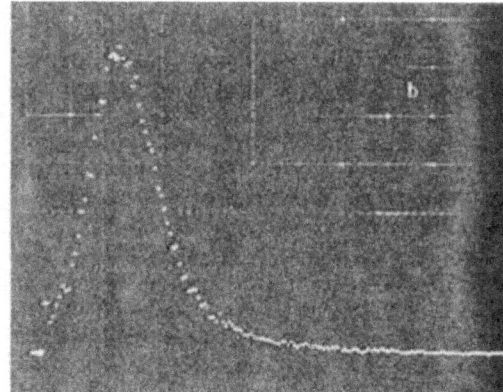

Fig. 4. Pulse-height spectrum in the drift chamber, operating with an argon (75%) and isobutane (25%) mixture and with $V_m = 1.5$ kV, $V_M = 3.2$ kV: a) for a ^{55}Fe X-ray source, 20 mV/division; b) for a 1 GeV/c momentum pion beam, 5 mV/division.

In the first instance, we have preferred as a gas filling, mixtures in various proportions of argon and isobutane since, as already mentioned, in this gas the drift velocity is only slowly varying with the field for fields larger than ≈ 600 V/cm. The linearity between space and time should then be preserved even if the electric field is not perfectly constant. This gas may not be the best, however, for reducing the electrons' diffusion which is the cause of the decrease of accuracy with the drift distance[9]).

The structure of the electric field is such as to make a drift chamber a rather good proportional counter. Figs. 4a and 4b show, as an example, the pulse-height spectra obtained with an ^{55}Fe X-ray source and in a beam of fast (1 GeV/c momentum) particles; the horizontal scales are 20 and 5 mV/div. respectively. A resolution of 17% fwhm is observed on the 5.9 keV X-ray. A threshold of detection of −3 mV was found to be sufficient to guarantee full efficiency for relativistic particles.

3. Experimental method of measuring very high accuracies

In order to study the space-time relationship and the accuracy of the described drift chambers, we had to find a method by which a beam of charged particles could be precisely defined. This was done using the space-time correlation in two small single-wire collimation drift chambers similar in design to the chamber under study. The chambers were placed in a high-energy (1 GeV/c momentum) pion beam; a couple of accurately timed scintillation counters defined the trigger and the zero time reference. A given delay was imposed on the detection of a pulse from the two collimating chambers, and events were accepted only if they did fall within a very narrow coincidence time.

Two narrow regions were thus selected on each side of the detecting wire, allowing in principle four distinct track directions. With a delay on the collimating chambers around 20 ns, two regions about 1 mm on each side of the sense wires were selected, and owing to the small divergence of the beam only two parallel beams were in fact accepted.

Taking into account the dispersion on the zero time definition (fwhm 0.8 ns) and the width of the time coincidence (1.5 ns), and for a typical drift time of 22 ns/mm (see next paragraph), we could then obtain an electronic accuracy in the definition of the two monitor beams of about 30 μm (standard deviation). The quoted definition was kept constant through all measurements described in the following.

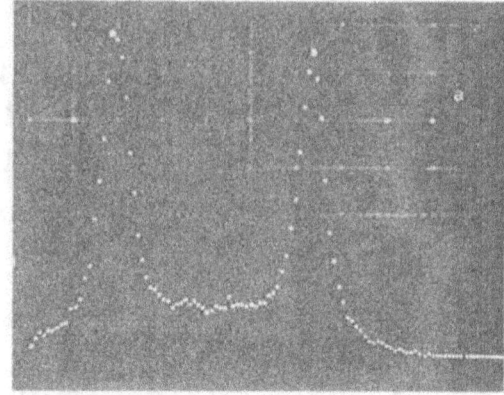
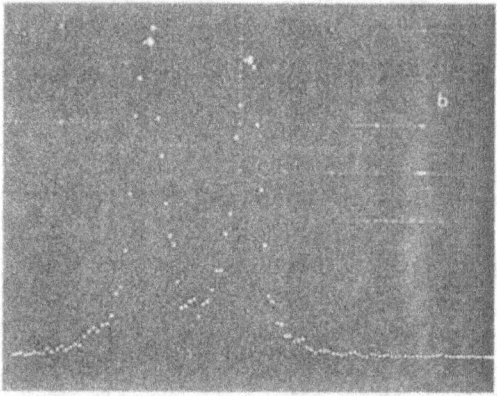

Fig. 5. Time difference between two single-wire drift chambers, as a function of their alignment on a beam. The time scale is 20 ns/division (or 2 ns/channel), and the three pictures represent the spectra obtained in three positions one mm apart from each other. In the position providing the spectrum (c), the two chambers are centred on the mean direction of the beam.

The drift chamber under study was then placed between the two collimating ones and displaced on a high-precision optical bench for accurate scanning.

We shall mention here that the time coincidence between two single-wire drift chambers is a very effective method of aligning chambers on beams, or vice versa. The time difference between the unselected pulses from the two chambers is indeed centred at zero with a dispersion corresponding to the electronic jitter and the beam divergence only if the two wires are aligned to the central direction of the beam. A small offset would produce two peaks, symmetric around the zero value, plus a constant rate in between. This is seen in figs. 5a, b, and c; here the time difference between the two collimation chambers is shown for three positions, in steps of 1 mm. The horizontal scale is 20 ns/div. Assuming an accuracy in the peak determination of 2 ns, and for a distance of 10 cm between the chambers, this corresponds to a definition of the mean beam direction of about 1 mrad.

4. Efficiency, linearity and accuracy of the drift chamber

With the method outlined in the previous section,

one then obtains from the drift chamber under study two peaks in the time distribution, corresponding to the two selected collimated beams. This is shown in fig. 6, where the time spectrum of the chamber on the

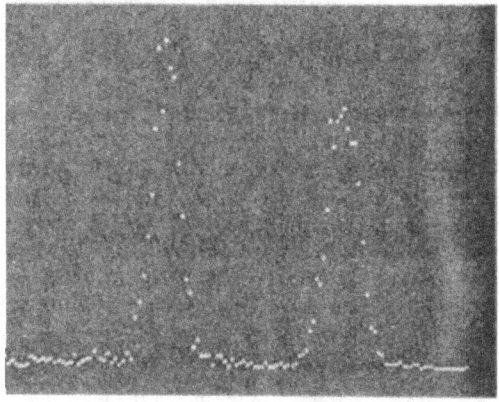

Fig. 6. Time spectrum of pulses provided by a drift chamber, detecting two narrow beams about 2 mm apart, collimated with the method described in the text. The drift lengths are 10 and 12 mm, respectively. Horizontal scale: 1 ns/channel.

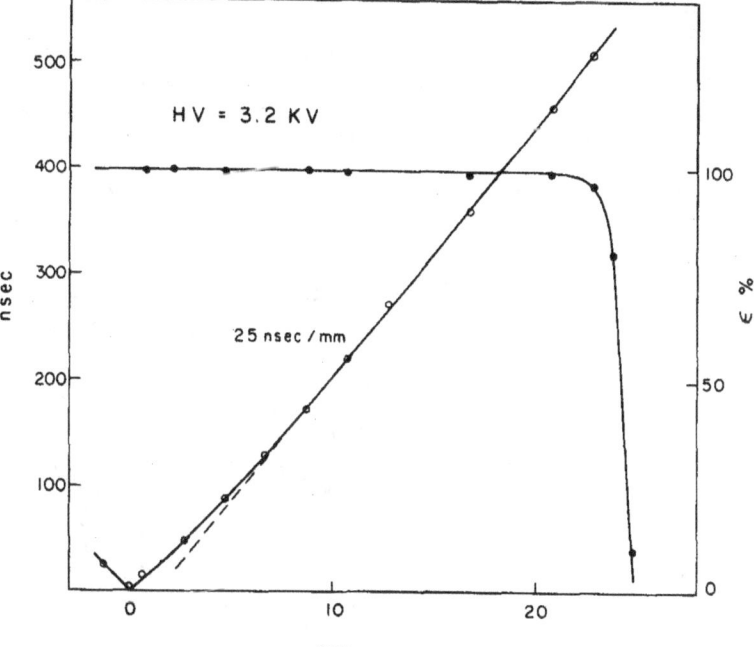

mm

Fig. 7. Measured space–time relationship and efficiency in a drift chamber, in a high-energy beam perpendicular to the chamber as a function of the drift length. The drift velocity is constant for most of the chamber length, with the exception of the region very near to the amplifying wire where the electric field rises by several orders of magnitude.

collimated beams is given; the distance between the two peaks corresponds to about 2 mm, and the drift length is 10 and 12 mm, respectively. In fig. 7 the measured space–time relationship is shown, as well as the efficiency through the 24 mm of drift space on one side of a sense wire. For this measurement the following conditions were set:

- gas mixture: argon 75%, isobutane 25%,
- high voltages: $V_m = 1.5$ kV, $V_M = 3.2$ kV,
- threshold of detection on the drift chamber pulses: -3 mV.

It can be seen that the drift velocity is constant through most of the chamber, and only slightly increasing near to the sense wire. The efficiency is constant at about 99.6% and shows a steep drop at the limiting region, proving that no smearing effects are produced, at least for a beam perpendicular to the chamber.

The accuracy in the response can be inferred from the width of the time distributions such as those in fig. 6. At least five factors contribute to the measured width:

1) the intrinsic physical jitter in the three chambers owing to the thermal diffusion of the primary electrons during the drift time;
2) the uncertainty in the zero time definition as given by the scintillation counters;
3) the width of the time coincidence used to select the collimated beams;
4) the time jitter introduced by the simple threshold detection of pulses in the three drift chambers;
5) the multiple scattering in the mylar windows (two per chamber, 12 μm thick each) and in the wires.

The raw data were corrected only for the dispersions

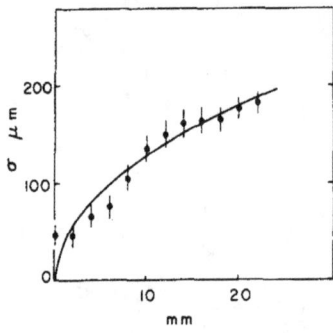

Fig. 8. Measured accuracy in the coordinate determination for a drift chamber (standard deviation), as a function of the drift length. The full line represents the diffusion eq. (1), for a coefficient $D = 0.65$ m²/s.

introduced by effects (2) and (3), accounting for about 30 μm (standard deviation, see section 3). Also, the intrinsic jitter of the three chambers was assumed to be equal for a drift of 1 mm in the central one, and then the (fixed) contribution of the two collimation chambers was subtracted in the Gaussian sense ($\sigma = 35$ μm each). For example, at 2 mm drift length the measured raw value of 75 μm is corrected to 45 μm. The result is shown in fig. 8, where the corrected accuracy of the central chamber is plotted as a function of the drift space; it increases from a minumim of 40 μm close to the wire, to a maximum of 180 μm. The full curve in the drawing represents the classical diffusion equation[10]:

$$\sigma_x = \sqrt{(2Dt)} = \sqrt{[(2D/w) x]}, \qquad (1)$$

where x is the average drift length and w the drift velocity. For the curve shown, the coefficient of diffusion D equals 0.32 m²/s.

A slight decrease in accuracy is observed close to zero; this was expected as a consequence of the statistical fluctuations in the position of the primary electrons. Indeed, the average distance between primary electrons produced by a charged particle is about 300 μm, and this introduces an additional jitter when the track is very close to the sense wire. Multiple Coulomb scattering on the wire may also contribute.

Extrapolation of the observed points with eq. (1) to longer drift spaces shows that a 10 cm drift, for example, should allow an accuracy of about 400 μm in the gas used.

5. Operation in a magnetic field

We have investigated in great detail the effect of strong magnetic fields on the drift of electrons. The worst case is of course the one in which the magnetic field has a component parallel to the wires, since the charges drifting towards the wires are lifted to the outside of the chamber and the track may be lost. We have observed that already at 5 kG the efficiency of detection decreases towards zero after about 12–13 mm of drift.

The lateral displacement of electrons drifting in an electric field E for a time t is given by[10]:

$$\Delta x = \frac{E}{B} \frac{v^2 \tau^2}{1 + v^2 \tau^2} t, \qquad (2)$$

where $v = eB/m$ is the Larmor frequency, B the magnetic induction perpendicular to E, and τ the mean free time for collision of electrons. At a given B, it can be seen that Δx increases with τ. Hence, one way of extending the efficiency region of a drift chamber in a magnetic

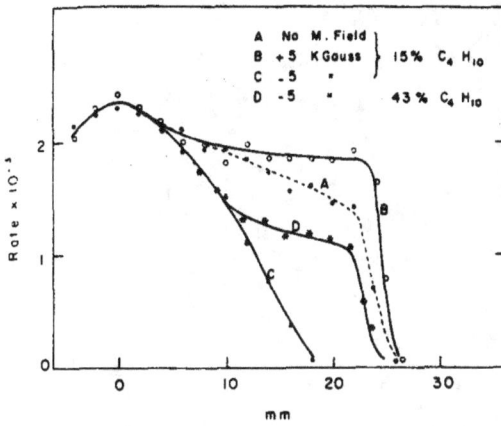

Fig. 9. Counting rate in a drift chamber, given by a collimated ^{55}Fe X-ray source, as a function of the drift length: A) in normal conditions, without magnetic field; B) in a magnetic field of +5 kG, parallel to the wires; C) in a magnetic field of −5 kG; D) as in (C), but with an increased amount of isobutane.

field is to decrease the mean free path (e.g. increase the gas density).

In a preliminary set of measurements, we have used for convenience a localized collimated ^{55}Fe X-ray source to scan the chamber, with and without a magnetic field, and defined the detection efficiency through the counting rate. The main results are presented in fig. 9. We observe a behaviour which at first is surprising: the efficiency of collection depends on the direction of the magnetic field (curves B and C, as compared with the rate measured without magnetic field: A). This comes from the fact that photons emitted by the source are mainly absorbed in the first few millimetres of the gas layer in the chamber; if then the magnetic field has the right polarity, the drifting

electrons are displaced towards the sense wire and they still reach it, at least for moderate fields. The effect of reducing the mean free path, increasing the gas density, can be seen by comparison of curve C (15% isobutane, 85% argon) and D (43% isobutane). Some improvement is observed, but only for relatively low magnetic fields.

A more general solution has been found which allows the use of drift chambers in much stronger fields*. By using separate voltage distribution lines on the two cathodes of a chamber, the electric field direction can be tilted so as to produce a component of electric force parallel and opposite to the Lorentz force. Let E be the electric field in a direction parallel to the cathode planes and B the magnetic induction perpendicular to E. One can easily obtain the tilting angle α necessary to oblige the electrons to drift parallel to the cathode planes:

$$\alpha = \arctan\frac{Bw}{E}. \qquad (3)$$

With the previously given values of the drift velocity w and of E, one obtains for a magnetic field of 15 kG, $\alpha \approx 45°$. A variable tilting can easily be produced by using separate hv distributors on the two cathode wire planes.

In fig. 10 the equipotentials in a drift chamber with the described structure, and about 50° tilting of the electric field all across the chamber are shown. The tilt corresponds to a shift by two cathode wire spacings, on each side of the symmetry axis, of the potentials V_M and V_m as defined in fig. 2.

As before, $V_M = 1$ is the maximum potential, applied

* Suggested to us by O. Guildemeister.

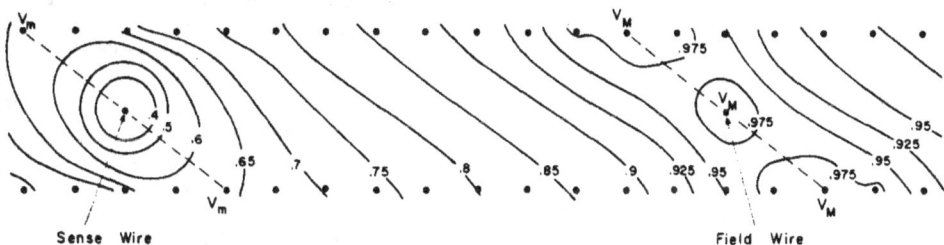

Fig. 10. Electric field in a drift chamber, where the voltages applied to the cathode wires are chosen so as to tilt the equipotential by a given angle. In the figure, the tilt is of about 50° which, according to eq. (3), makes it possible to drift the electrons properly towards the sense wire in a magnetic field of 18 kG perpendicular to the drawing.

Fig. 11. Three small single-wire drift chambers, with adjustable electric field as described, installed inside a magnet for the measurements in a strong magnetic field.

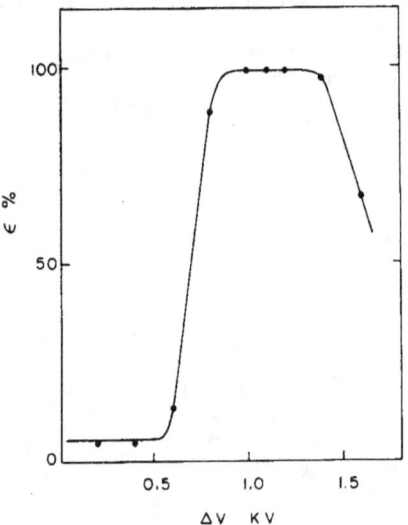

Fig. 12. Efficiency of a drift chamber operating in 15.6 kG, as a function of the voltage difference between two facing cathode wires (proportional to the tilting angle of the electric field) for a 20 mm drift length.

to the field wire as well, and $V_m = 0.64$ is the minimum; the voltage of the cathode wires is uniformly increasing between V_m and V_M. Calculation of the field across a central cut would give a curve essentially equal to curve A in fig. 3.

The set of three drift chambers used for measurements in a strong magnetic field can be seen in fig. 11; the sense wires were only 35 mm long, since the useful gap of the available magnet was 10 cm. Fairly uniform fields up to 16 kG could be obtained. To adjust the tilting angle, we used separate power supplies on the two sides of the cathode wires; as a general rule, we kept constant the mean potential between two facing cathode wires, changing the voltage difference ΔV. For a given drift distance, then, the efficiency versus ΔV was measured. As expected, the voltage difference giving the best efficiency is the same all through the chamber; fig. 12 illustrates the measurement for a drift of 20 mm. The comfortable length of the plateau makes of the tilt angle regulation a not very critical one.

The measured linearity and efficiency for detection of high-energy charged particles in a magnetic field of 15.6 kG parallel to the wires are presented in fig. 13 for $\Delta V = 1.1$ kV. For comparison, the broken line shows

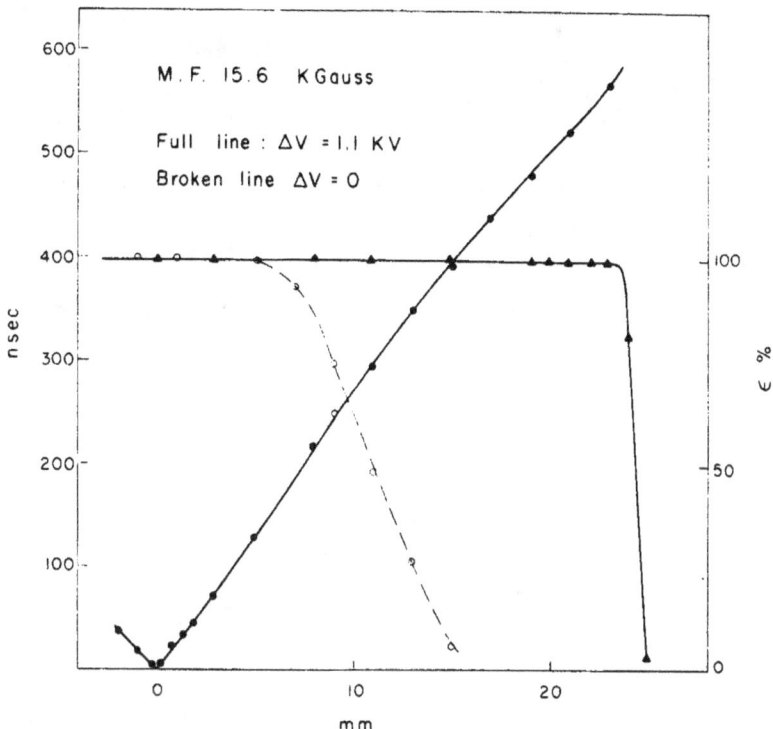

Fig. 13. Space–time relationship and efficiency for a drift chamber operating in 15.6 kG, with $\Delta V = 1.1$ kV (full lines). For comparison, the broken line shows the efficiency for $\Delta V = 0$, i.e. without any tilt of the electric field.

the efficiency measured with $\Delta V = 0$, i.e. without tilting the electric field. In fig. 14, the space-time relationship is shown in detail around zero; it can be seen that, although the slope of the curve is slowly changing close to zero, a smooth behaviour is still observed.

For the same magnetic field value, the measured accuracies as a function of the drift space are shown in fig. 15, where the full curve represents the diffusion eq. (1) with the same value of D as in fig. 8. Contrary to what one would expect[10]), the accuracy is slightly worse in the magnetic field; however, since a different set of chambers was used for the two measurements, the contribution of multiple Coulomb scattering on the beam, non-subtracted, could have been different.

6. Open problems and possible solutions for the use of adjustable field drift chambers

We have seen that the most attractive feature of the drift chamber with adjustable electric field is the high accuracy. However, several factors that contribute to the stability of the results will need further invest-

igations; let us mention some of them:

a) Temperature affect. In principle the drift velocity is very slowly affected by the gas temperature, since the electron energy in a strong electric field is about two orders of magnitude higher than the thermal energy[10]). We have not observed any fluctuation within a range of $\pm 10\,^{\circ}\mathrm{C}$ for the argon–isobutane mixture used.

b) Change in the gas composition. We have observed that a 10% variation in the isobutane content results in about 30% change in the mean pulse height, and about 1% change in the drift velocity. If one does not want to lose accuracy, a 0.5% control of the gas percentage is probably necessary.

c) Stability of the high voltage. Our measurements show that a 1% change in V_M results in about 0.3% change of the drift velocity. Looking back to fig. 1, this may mean that we were still too near to the knee before the saturation of w; to keep the maximum delay (about 500 ns) stable within ± 1 ns implies a stability on V_M of about ± 15 V. Higher values of the field in the drift region may reduce the effect.

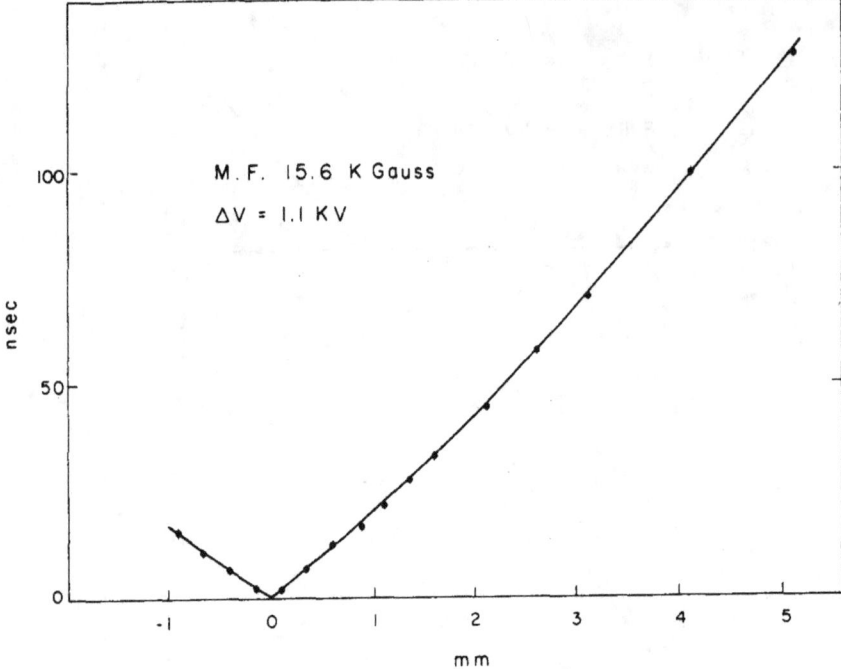

Fig. 14. Detailed measurement of the drift time near to the amplifying wire, for straight tracks.

d) Angular dependence. Because of the small size of our first chambers, we could not explore the dependence of the space-time relationship on the angle of incidence (all measurements were performed with a perpendicular beam). Probably a slightly different relation holds for large angles, and one may be obliged

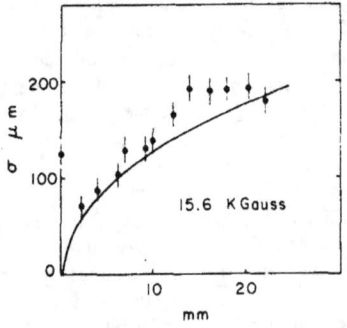

Fig. 15. Space localization accuracy of a drift chamber at 15.6 kG. The full line represents the diffusion eq. (1) with the same value of D as in fig. 8.

to correct the raw data with a given law during the software track reconstruction.

e) Electrostatic interactions. The electric field inside a chamber may be sensitive to external components or electrodes, especially if the cathode wire spacing is large. Some screening may be necessary to construct multigap thin drift chambers, or perhaps one can build the electrodes with conductive strips replacing the wires* (e.g. by vacuum evaporating aluminium on a thin insulating sheath).

Several procedures can be envisaged for controlling the drift velocity continuously or on a sampling basis. For example, the unbiased distribution in time of the pulses provided by any wire makes it possible to infer the maximum drift time. By comparison, one can quickly discover variations in the drift velocity.

More refined checks can be imagined when several drift chambers are used to detect a track; we shall mention some in the next section.

f) Large-size chambers. The mechanics of an adjustable field drift chamber is not more complicated than

* This is the solution used at Saclay, ref. 9.

for a normal multiwire proportional chamber. Because of the large wire spacing, one does not expect electrostatic instability of the wires; for a set of 20 μm tungsten wires, 50 mm apart, the length at which electrostatic instabilities should occur is about 250 cm. It is clear, however, that for large dimensions of the chamber, corrections will have to be introduced to the raw data because of the finite propagation time of the signals along the wires.

g) Multiple track resolution. We could not yet measure the minimum separation between two tracks which allows them to be resolved. We can, however, try to guess its value using available informations. Two factors play a role in defining the minimum separation: the electronic resolution and the dead-time of the amplifying wire owing to previous avalanche. The limit on the electronic resolution seems to us the width of the pulse delivered from the wire amplifier to the following discriminator and timing unit; using a low input impedence amplifier, about 50 Ω, we got an occupation time not exceeding 50 ns on straight beam tracks. This means that two tracks could be resolved if their distance, along a direction perpendicular to the wire, is larger than or equal to 2 mm.

The physical dead-time, owing to the local modification of the field around an avalanche due to a previous track, may be a much more serious problem although limited in space extension. Operating with the very high gains of the magic gas, Makowski and Sadoulet have measured dead-times as long as 10 μs on a localized spot of the wire, 0.2–0.3 mm long[11]. The situation may be rather different when working, as we are, at a much lower gain. The fact that proportional amplification is obtained in the range between minimum ionizing particles and 5.9 keV photons (see figs. 4a and 4b) proves that the saturation due to space charge, if existing, is very limited.

7. Discussion of a few schemes of utilization of drift chambers

We have already discussed the question of the right–left ambiguity in drift chambers. Although the methods of solving the problem proposed in refs. 4–7 are very attractive, they may lead to mechanical difficulties or they can spoil the high accuracy we were able to measure. One can think of alternative schemes in which the ambiguity is resolved by a combination of drift and/or proportional chambers. Elegant solutions can be imagined whenever the divergence of the tracks in a given chamber is very small, as in beams or magnetic spectrometers:

- A normal multiwire proportional chamber, with a wire spacing that is small compared to the drift space, can be placed in front of and very near to the drift chamber. Either each wire is used as a detector or, in cases where only one particle has to be detected, it is sufficient to connect together all the wires corresponding to the right and to the left drift spaces, respectively.
- A second drift chamber of the same type is used, staggered by half a wire spacing*. This configuration not only allows the resolution of the ambiguity, but since for a given angle the sum of the two drift times $T_1 + T_2 = $ const, a continuous check of the drift velocity is possible. Also, particles accidentally crossing the chambers at a time different from the one of the real event will not satisfy the time condition and can be disregarded. From the previous measurements, we can estimate an accuracy in the time correlation of about 5 ns and this will be the resolution against accidentals.

One may also think of using mean timer circuits, such as those used for scintillation counters[12]), to obtain accurate timing informations. This can be a precious property, for example at the CERN Intersecting Storage Rings, where the proportional chambers alone have too bad a time resolution to distinguish between beam–beam and beam–gas interactions using a time-of-flight measurement.

In case one gets two tracks on the same drift region, if the information of the existence of a double track is given, e.g. by pulse height in scintillation counters, the independent measurement of T_1 and T_2 still gives

---

* Two other groups are now envisaging a similar approach: at CERN, C. Rubbia (private communication); at Heidelberg, J. Heintze and A. H. Walenta, contribution to the Tirrenia Meeting, September 1972.

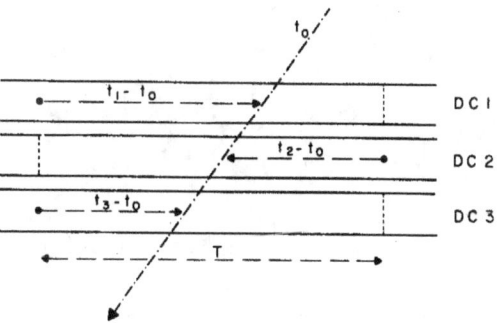

Fig. 16. Stacking of three equal drift chambers, which allows a complete resolution of the right–left ambiguity for any angle of incidence, and also provides the time t_0 of the passage of the particle as described in the text.

the distances of the two tracks from the respective wires.

In the general case of tracks having large angular spread, three chambers can be used to solve the ambiguities. By a proper choice of the respective positions of the wires, one can also eliminate the need for having a zero time definition given by other detectors. This can be seen in fig. 16: three chambers with parallel wires are stacked so that the central one has the wires displaced by half a wire distance with respect to the other two. Suppose a particle traverses the chambers, at any angle, at the unknown time t_0, and let the drift times in the three chambers be $t_1 - t_0$, $t_2 - t_0$, and $t_3 - t_0$. The following relation holds (for constant drift velocity):

$$t_0 - t_2 = \tfrac{1}{4}[(t_1 - t_2) + (t_3 - t_2)] - \tfrac{1}{2}T, \qquad (4)$$

where T is the maximum drift time (time to drift along half the wire distance). Hence, by using the central

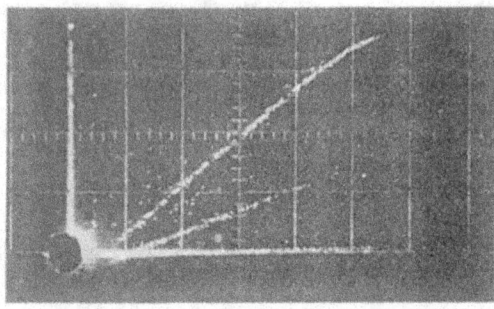

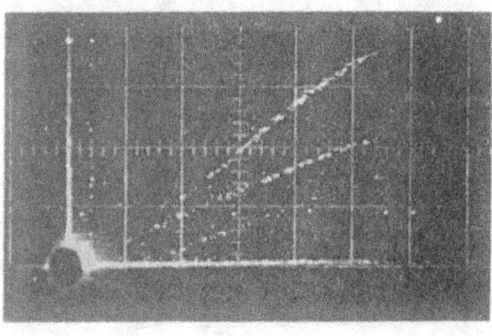

(b)

Fig. 17. Time correlation in three drift chambers. Vertical axis: $t_2 - t_0$; horizontal axis: $(t_1 - t_2) + (t_3 - t_2)$ (see text). Picture (a): beam perpendicular to the chamber. Picture (b): beam at 45° to the chamber. Top line: Signal in the three chambers. Lower lines: Signal in the first or third chamber missing. The correlation is linear and independent of the angle.

chamber as a time definition, and recording $t_1 - t_2$ and $t_3 - t_2$, one obtains the zero time $t_0 - t_2$ and then a unique track reconstruction (fig. 17). With the measured time accuracies (section 3) for each drift time, one may expect to define the time reference within about 4 ns. It is also possible with three such chambers to select a given sagitta in a magnetic field, thus permitting a rapid momentum selection. Again, with the measured average 100 μm accuracy per chamber, with three chambers 10 cm apart, a 1 GeV/c momentum should be measured at about 2%.

8. Two-dimensional read-out of drift chambers

Since a drift chamber of the type we have described leads to a relatively moderate number of sense wires, even for large surfaces, it becomes highly interesting to investigate any method, however elaborate, of obtaining the position of the avalanche along the wire. The obvious advantage is that if many particles have to be detected, the fundamental ambiguities arising from an independent measurement of the two orthogonal coordinates diappear. It is known from different measurements that the avalanche produced on the sense wires is very localized, within perhaps 0.1 mm[13].

If the two ends of the wire are then grounded, an amount of charge proportional to the position of the avalanche will flow out of each end. The ratio of the charge flowing out on one side, to the total charge, gives then the coordinate along the wire. The method is often called "current" division; it was used in spark chambers[14,15] and gave accuracies of the order of 1 mm by simply measuring the difference of the charges flowing at the end, with the help of pulse transformers. This method has also been used with other detectors; the pioneering work with single-wire proportional counters was done by Kuhlman et al.[16]). The difficulty with proportional counters in contrast to spark chambers is that the total charge is not constant and varies within a broad spectrum; it is then necessary to normalize the split charges to the total charge. This is done most easily by taking the ratio of the charges flowing at each end; which, for a given position, is independent from the absolute pulse height.

The difficulty of such a method is that the impedance of the wire has to be great with respect to the impedance of the device measuring the charges at the end of the wires, in order to achieve a good sensitivity.

By using chrome–nickel wires with $\rho = 40 \ \Omega/cm$, Kuhlman et al. achieved accuracies of the order of 1.5 mm over 30 cm. Accuracies as low as 0.5 mm have been measured by Hough[17]).

Let us mention a very related method where the

position is obtained by measuring the change of the rise-times of the pulses as a function of the origin of the avalanche – there, also, very resistive wires have to be used (80 kΩ/cm). Spectacular accuracies of the order of 75 μm have been obtained with highly ionizing particles[13]).

Spectrometers using proportional chambers with current division read-out over 0.7 m length have been built, and a typical 1.6 mm accuracy was measured[18]). However, again high resistivity wires were used, and times of the order of 1 ms were required to measure the position. For high counting rates this may be prohibitive.

It seems to us difficult to adopt such methods for a fast large-scale detector. Therefore it appeared worthwhile to investigate the possibility of using low-resistance wires, and we carried out some preliminary tests.

1) With normal tungsten wires of 20 μm (resistance = = 1.8 Ω/cm) we have measured the ratio V_1/V_2 of the amplitudes of the pulses arriving at each end using amplifiers with low input impedance (of about 2 Ω)* and a specially built fast divider†. The total conversion time to obtain the ratio V_1/V_2 was about 5 μs.

We obtained an accuracy corresponding to $\sigma = 4$ mm, which is quite encouraging. Such an accuracy could be quite sufficient for the removal, in most cases, of the orthogonal ambiguities.

* Designed at CERN by H. Verveij.
† Built at CERN by J. Olsfors.

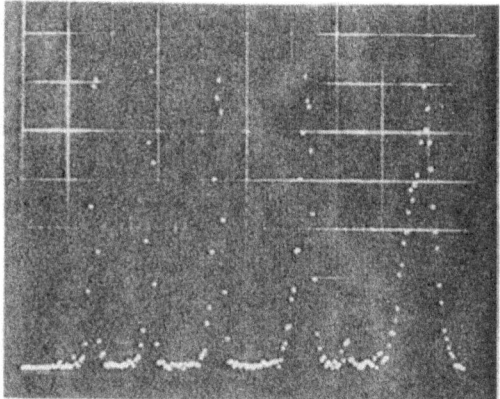

Fig. 18. Position accuracy obtained using the current division method. The pulses provided by a collimated ^{55}Fe source were read out on each end of a 50 μm constantan amplifying wire (3 Ω/cm) on 2 Ω input impedance amplifiers. The time spectrum shows the ratio V_1/V_2 of the two amplitudes, for five positions 2 cm apart from each other.

2) We used a wire of constantan, 50 μm in diameter, with a moderate resistivity, namely 3 Ω/cm; the accuracy we obtained in a first simple test over a length of 20 cm was $\sigma = 2.5$ mm. This can be seen in fig. 18 where the ratio of the signals on the two sides, V_1/V_2, is shown for five positions of a collimated ^{55}Fe source, 2 cm apart from each other. Notice that the scale is obviously not linear with the distance, and that the accuracy in space is about constant with $\sigma = 2.5$ mm. It seems to us that this is the line of research to be actively pursued, since it would lead to an ideal detector for cylindrical drift chambers in magnetic fields. The resistive wire does not have to be the wire collecting the avalanches; it can as well be a cathode wire, and the current division can be done on the induced pulses. Localization detectors using pulses induced in auxiliary electrodes have already been built[19]). In this case the mechanical requirements, such as the strength or the diameter of the read-out wire are much simplified.

The task of producing a simple and cheap electronics for fully exploiting the high potential performances of the drift chambers is not a simple one. However, it seems that if one is satisfied with the accuracies obtained by spark chambers, 0.3 mm in space and 0.5 μs in time, a detector can be built at a cost as low as or lower than that of spark chambers with still considerable advantages: continuous operation, much lower dead-time, higher efficiency for single and multi-track events, and easier operation in magnetic fields.

For the construction of detectors with a large volume, the combination of proportional wire chambers and of drift chambers may be the ideal solution. The first should rapidly provide information such as the multiplicity and participate in the fast event selection, while the second give the accurate localization.

For the future developments on the high-accuracy detectors required for the exploitation of the very high-energy machines, the drift chamber offers a most promising solution. It is already clear that with a succession of drift chambers, trajectory positioning with accuracies much better than 0.1 mm is within reach without introducing a prohibitive amount of material in the beam.

The authors wish to thank the Omega and the CERN–Columbia–Rockefeller groups at CERN for stimulating discussions and technical support.

References

1) G. Charpak, R. Bouclier, T. Bressani, J. Favier and C. Zupancic, Nucl. Instr. and Meth. 62 (1968) 235.

[2] G. Charpak, D. Rahm and H. Steiner, Nucl. Instr. and Meth. **80** (1970) 13.

[3] G. Charpak, Some research on the multiwire proportional chamber, Proc. Intern. Conf. *Instrumentation for high-energy physics* (Dubna, Sept. 1970) p. 227.

[4] A. H. Walenta, J. Heintze and B. Schürlein, Nucl. Instr. and Meth. **92** (1971) 373.

[5] A. H. Walenta, Two-dimensional read-out of drift chambers, Paper submitted to the 16th Intern. Conf. *High-energy physics* (Chicago, Aug. 1972).

[6] A. H. Walenta, A system of large drift chambers, Paper submitted to the 16th Intern. Conf. *High-energy physics* (Chicago, Aug. 1972).

[7] G. Charpak and F. Sauli, A possible solution to the right-left ambiguity in drift chambers, submitted to Nucl. Instr. and Meth.

[8] T. Bressani, G. Charpak, D. Rahm and Č. Zupančič, Track localization by means of a drift chamber, Proc. Intern. Sem. *Filmless spark and streamer chambers*, Dubna, 1969 (JINR, Dubna, 1969) p. 275.

[9] R. Chaminade, J. C. Duchazeaubeneix, J. M. Fontaine, D. Garetta, C. Laspalles, J. Saudinos and M. van den Bossche,

Note CEA-N-1522 (C.E.N., Saclay, 1971) p. 160; and R. Chaminade, J. C. Duchazeaubeneix, C. Laspalles and J. Saudinos, Localisation de particules par compteur à migration, submitted to Nucl. Instr. and Meth. (Jan. 1973).

[10] J. Townsend, *Electrons in gases* (Hutchinson's Scientific Publ., Place, Year).

[11] B. Makowski and B. Sadoulet, Space charge effects in multiwire proportional chambers, to be published at CERN.

[12] G. Charpak, L. Dick and L. Feuvrais, Nucl. Instr. and Meth. **15** (1962) 323.

[13] C. J. Borkowski and M. K. Kopp, IEEE Trans. Nucl. Sci. **17**, no. 3 (1970) 340.

[14] G. Charpak, J. Favier and L. Massonnet, Nucl. Instr. and Meth. **24** (1963) 501.

[15] J. Saudinos, G. Vallais and C. Laspalles, Nucl. Instr. and Meth. **46** (1967) 229.

[16] W. R. Kuhlmann et al., Nucl. Instr. and Meth. **40** (1966) 118.

[17] J. Hough, Nucl. Instr. and Meth. **105** (1972) 323.

[18] G. L. Miller et al., Nucl. Instr. and Meth. **91** (1971) 389.

[19] J. Hough and R. W. P. Drever, Nucl. Instr. and Meth. **103** (1972) 365.

NUCLEAR INSTRUMENTS AND METHODS 119 (1974) 1-5; © NORTH-HOLLAND PUBLISHING CO.

TWO-DIMENSIONAL DRIFT CHAMBERS

A. BRESKIN, G. CHARPAK, F. SAULI and J. C. SANTIARD

CERN, Geneva, Switzerland

Received 27 December 1973

By using simple thin delay lines, of a diameter smaller than 2 mm, and parallel to the sense wire of a drift chamber it is possible to measure the position of the avalanche along the sense wire. With a drift chamber of 150 cm length accuracies between 2 mm and 3 mm along the wire are obtained. The delay line can be placed in the cathode or can be used between two sense wires very closely spaced in order to solve the right–left ambiguity.

1. Introduction

Recent developments[1,2] have shown that multiwire drift chambers can be built with attractive properties: accuracy of the order of 100 μm, even for inclined tracks; perfect linearity within this accuracy range[2]); satisfactory operation in magnetic fields, without loss of accuracy[1,2]. An important source of progress has been the use of cathode planes made of wires or strips parallel to the sense wires, adjusted at different potentials such as to give a uniform field in most of the volume of the chambers.

In many applications envisaged in high-energy physics, at least three planes with different drift directions are necessary to determine a point in space, in order to solve two distinct types of ambiguities:

1) the right–left ambiguity – the measurement of the time of drift does not tell us on which side of the wire the particle passed;

2) the ambiguity arising when several particles cross the chamber simultaneously.

To solve the right–left ambiguity, a Heidelberg group has used a triplet of wires instead of one single sense wire: namely, two sense wires separated by a shielding wire[3]. It has also been shown that two sense wires placed at a short distance, for example 100 μm, give rise to a perfect separation between right and left tracks[4]). Our latest investigations[2]) on this subject show that, probably because of space charge, the following effect is observed when the two close sense wires are electrically independent: the first avalanche on one wire paralyses the second wire locally in such a way that even for inclined tracks there are no coincident pulses on the two wires. It is thus a very attractive way, despite the mechanical complication, for the removal of the right–left ambiguity in drift chambers.

To solve the ambiguity arising when many particles cross a plane, several methods can be envisaged. The most natural one is to use chambers oriented in three

directions. This, however, requires a computation time which is far from negligible when one deals with high multiplicities. Since this appears to be the real bottleneck in many experiments, chambers in which the two coordinates can be obtained from each wire would be very valuable. Some preliminary research[1,5]) by us and another group has shown that indeed accuracies of the order of 1 cm could be obtained in chambers of 1 m length with a current division method. The difficulties encountered in these measurements have led us to investigate another approach based on delay lines parallel to the sense wires. We have attempted several practical approaches[6,7]).

A strip line, of 4 mm width and 0.1 mm thickness, is placed on the cathode, parallel to the sense wire. The induced pulse propagates to each end of the line. By measuring the difference in the times of arrival at the two ends of the line, the position of the avalanche along the wire can be obtained. The first results, reported at the 1973 Frascati Conference on Instrumentation for High-Energy Physics[6]), showed that an accuracy of

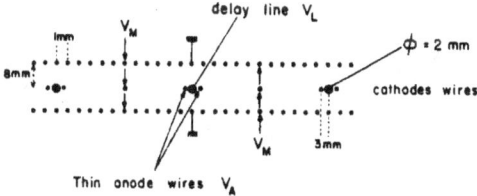

Fig. 1. Construction of a drift chamber with no right–left ambiguity and with two-dimensional read-out. The potential rises gradually from 0 to V_M to ensure a constant drift velocity. The delay line is placed between two sense wires. The gain in the chamber is controlled by V_A and the electrostatic stability is controlled by V_L. The useful length of the chamber is 1.20 m, while the length of the delay lines is 1.5 m. The distance between two adjacent groups of sense wires is 5 cm (figure not drawn to scale).

1

$\sigma = 2$ mm could easily be obtained in a 20 cm chamber. This method was much easier to implement than the current division method.

However, it appeared that it was difficult to extend this method for large dimensions. The resistance of the line was 150 Ω, twice the impedance of the line; we had thus a non-negligible attenuation.

We then found another construction method which gives much better results. The line is made simply with a thin wire wound around an insulated wire. It was very easy to build delay lines with characteristic impedances varying from 100 Ω to 1000 Ω and with delays varying from 0.7 ns/cm to 5 ns/cm and negligible attenuation. We have tried such a delay line, 20 cm long, in a chamber and the results are reported elsewhere[7]). The accuracy was varying from $\sigma = 2$ to $\sigma = 4$ mm depending on the position along the wire. The delay line was placed in the cathode plane 3 mm away from the sense wire.

We have finally built a drift chamber with a length of 1.20 m, with no right–left ambiguity, and with the measurement of the coordinate along the wire. The delay line is placed between two sense wires as shown in fig. 1. The mechanical characteristics of this chamber are the following:

1) Sense wire diameter: 20 μm.

2) Distance between the axis of the delay line and the two sense wires: 3 mm; this rather larger distance is imposed on us by the use of an existing chamber. It can indeed be smaller as shown by the experiment reported in ref. 3. Fig. 2 shows the loss of

TABLE 1

Electrical characteristics of two investigated delay lines.

Characteristics	Delay line no. 1	Delay line no. 2
Length	150 cm	150 cm
Diameter	1.9 mm	1.7 mm
Characteristic impedance Z_0	510 Ω	1.3 kΩ
Ohmic resistance	25 Ω	160 Ω
External diameter of enamelled wire	0.18 mm	0.10 mm
Attenuation/m	$\approx 50\%$	$\approx 50\%$
Rise-time (10% to 50%)	7 ns	15 ns
Delay	70 ns/m	130 ns/m

efficiency in the region of the delay line, as measured with a collimated source of 5.9 keV X-rays.

3) Distance between the groups of sense wires and delay lines: 5 cm.

4) Chamber thickness: 1.6 cm.

The drift fields could be adjusted by varying V_M (fig. 1). The gain in the chamber and the mechanical stability could be controlled by varying V_A and V_L, as in ref. 3.

We tried two delay lines with the electrical characteristics mentioned in table 1.

We wish to report the first results of our measurements, since they show that for many applications the problem of the two-dimensional read-out in drift chambers can be solved in a simple way.

2. Experimental results

A collimated source at 5.9 keV is used to trigger a pulse on the sense wires; in all measurements, the time difference between the two ends of the line has been measured in order to obtain a doubled sensitivity.

Fig. 3a shows the signal in a sense wire, and figs 3b, c and d the signals induced at each end of the delay line no. 1 for different positions of the collimated source along the line, and on the same impedance (500 Ω). A delay of about 0.7 ns/cm is measured in the line. Notice that the induced signal is narrower than the one detected on the sense wire on the same load impedance; this is probably due to the larger electric fields and hence the larger drift velocity of the positive ions, in the region around the sense wire facing the delay lines.

The size of one induced pulse is of the same order as the one of the direct signal. It depends on the voltage applied to the delay line. With the line grounded, it reaches as much as 80% of the negative pulse. With the

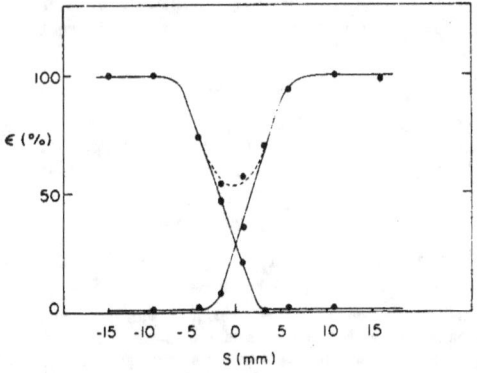

Fig. 2. Efficiency in the region of the delay line deduced from the counting rate detected using a collimated source of 5.9 keV X-rays. The hole is of the same size as the sense wire spacing. The X-rays being totally absorbed by the delay lines, this figure gives an exaggerated size to the efficiency hole.

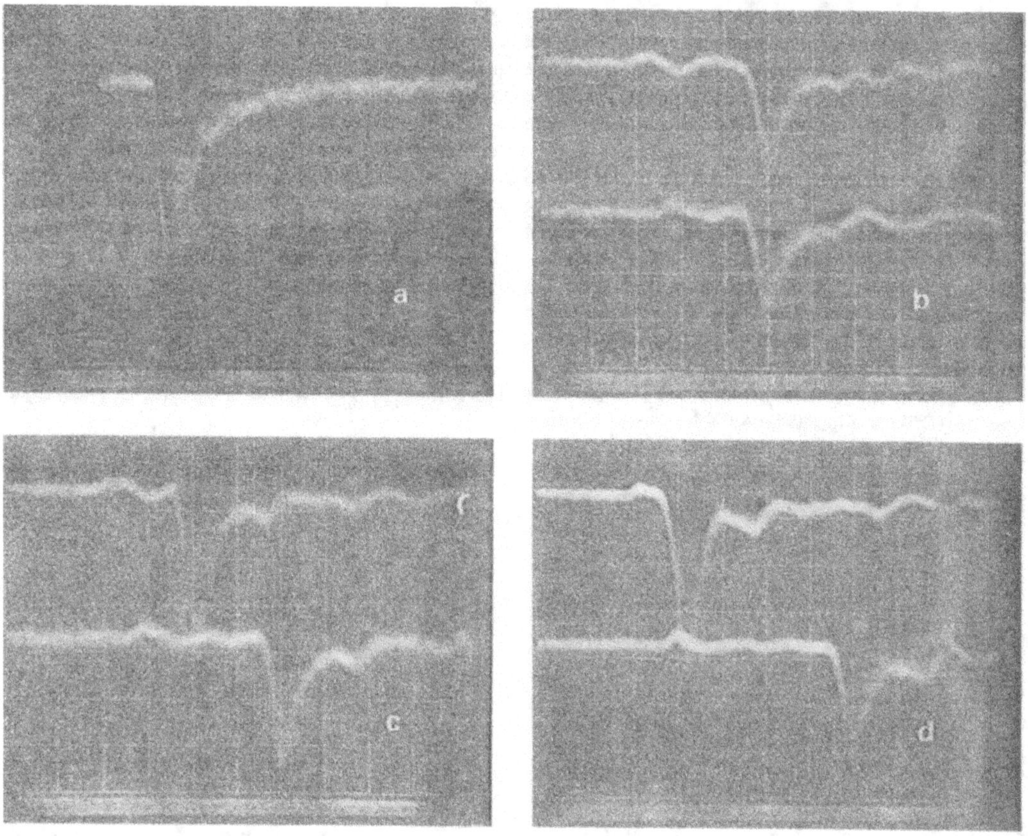

Fig. 3. Signals induced on the delay line no. 1. (a) Signal induced on the sense wire by a 5.9 keV X-ray source, on a 510 Ω load. (b, c, d) Signals induced at the end of the delay line no. 1 for three different source positions: in the centre of the chamber, i.e. 60 cm from the edge (b), at 90 cm (c), and at 120 cm (d) from one edge of the chamber. Horizontal scale: 20 ns/div; vertical scale: 5 mV/div. (The polarity of the signals in figs b, c, and d is inverted on the scope.)

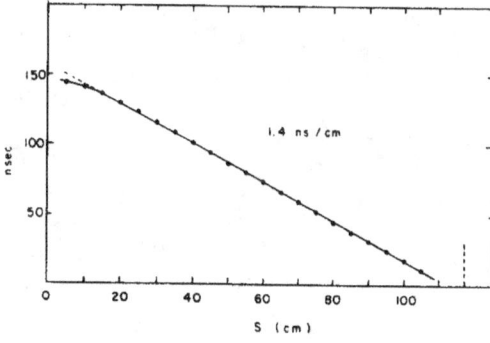

Fig. 4. Time–position relation. Delay line no. 1: 0.7 ns/cm. The time represents the difference between the arrival times at each end of the line. The end effect is due to a mechanical irregularity in the winding of the line.

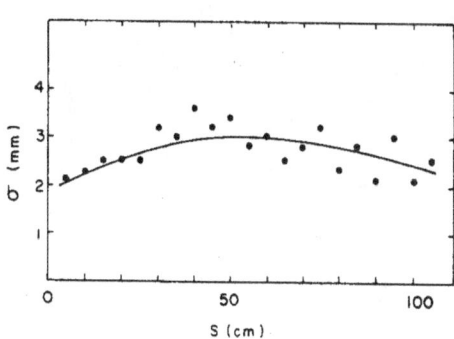

Fig. 5. Accuracy as a function of position of the collimated ^{55}Fe source.

delay line at a potential less negative with respect to the sense wire, the pulses are reduced but are still a sizeable fraction of the sense-wire pulse. The attenuation of the pulses is about 50% per metre of propagation in the delay line and the rise-time of the induced pulses is close to 7 ns.

The pulse arriving at each end of the line is shaped by a discriminator with 3 mV threshold and the time is measured with a time-to-amplitude converter. Fig. 4 shows the response of the chamber, and fig. 5 the accuracy as a function of position. It varies from $\sigma = 3$ mm in the middle to 2 mm at the edges. No effort has been made to improve the accuracy by using zero-crossing methods.

Fig. 6 shows the same results obtained with the delay line no. 2, with a propagation time twice as large: 1.3 ns/cm. The results are very similar. Fig. 7 shows the time–space relation over all of the length of the chamber, and fig. 8 shows the time distributions corresponding to two positions of the source 2 cm apart. It is our best result: $\sigma = 1.5$ mm.

We have made some preliminary study of delay lines with a ceramic tube with an external diameter of 2 mm. A 25 cm delay line of this kind exhibits better properties than the one built with the usual plastic insulation. The delay is 5 ns/cm and the rise-time is improved. Fig. 8 illustrates the quality of such lines, as measured with a pulse generator.

3. Conclusion

With delay lines of small diameter, parallel to the sense wires of drift chambers, we have obtained over a

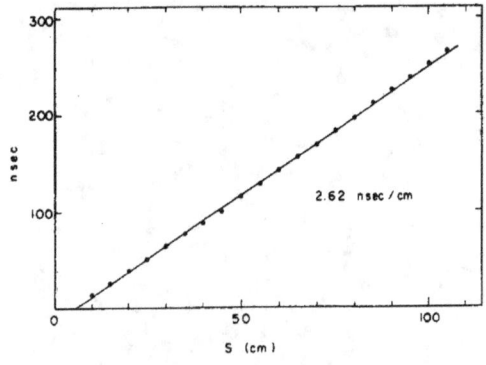

Fig. 7. Time-position relation. Delay line no. 2: 1.3 ns/cm.

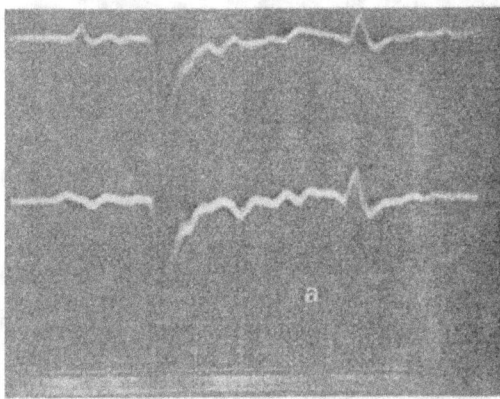

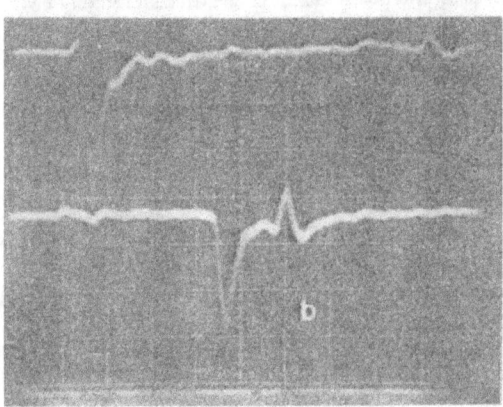

Fig. 6. Signals induced on the delay line no. 2: two positions of the collimated source, about 25 cm apart. Horizontal scale: 20 ns/div; vertical scale: 5 mV/mm (inverted polarity).

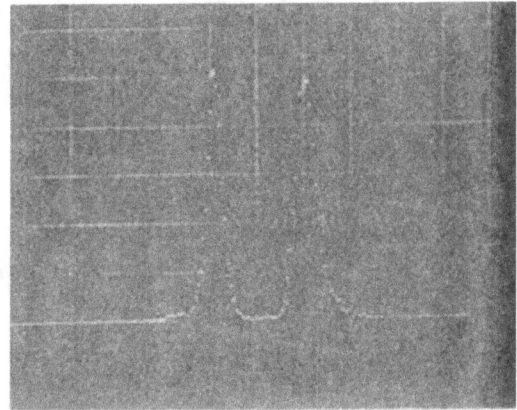

Fig. 8. Time distribution. Example of a distribution in the time difference between the two pulses arriving at each end of delay line no. 2. X-ray source: 5.9 keV. Two positions 2 cm apart: $\sigma = 1.5$ mm.

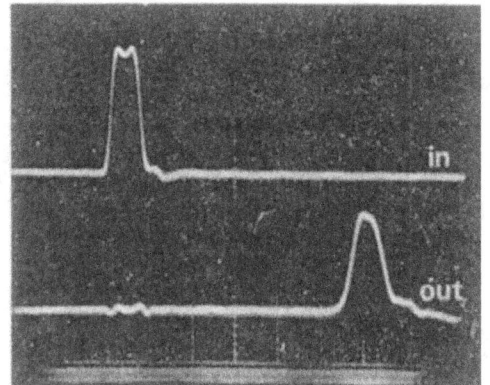

Fig. 9. Properties of a narrow 25 cm long delay line with a ceramic insulation, on a pulse generator. We see an improvement in the properties: the delay is 5 ns/cm. Scale: 20 ns/div.

length of 150 cm an accuracy of $\sigma = 2$–3 mm. These results are still not as accurate as those obtained by other authors using cathode-coupled delay lines[8]), or cathodes used as a delay line[9]). The special geometry of our drift chambers imposes severe geometrical limitations on the size of the delay line. We feel, however, that at this stage of our work we have achieved a development which can be of interest in several applications.

1) Construction of very large chambers almost free of any ambiguity. With the delay line placed between two sense wires, the measurement is free from the right–left ambiguity. The measurement of the coordinates along the wires ·eliminates the ambiguity for many simultaneous tracks detected by different wires.

Such chambers would represent considerable advantages from the point of view of computation time for complex events. They would, however, have some

inaccuracy in the region of the triplet of sense wires and delay lines. Since our construction permits spacing the sense wires by large distances, say 10 cm, with drift voltages not exceeding 6 kV, this region of inaccuracy is still a small fraction of the total surface.

However, in the case where one wants to have everywhere the same accuracy close to 100 μm, the solution is the couple of sense wires at 100 μm, with the delay line in the cathode plane parallel to the sense wire chambers[2,7]).

2) Construction of cylindrical chambers. The construction of cylindrical proportional chambers is a rather formidable problem. If an accuracy of the order of 2 mm is sufficient for the coordinate parallel to the generatrix, the proposed method is perfectly adapted to cylindrical geometry.

References

1) G. Charpak, F. Sauli and R. Duinker, Nucl. Instr. and Meth. **108** (1973) 413.
2) A. Breskin, G. Charpak, F. Sauli, W. Duinker, G. Shultz, N. Trautner and B. Gabioud, Further results on the operation of high-accuracy drift chambers (submitted to Nucl. Instr. and Meth.).
3) A. H. Walenta, J. Heintze and B. Schürlein, Nucl. Instr. and Meth. **92** (1972) 373.
4) G. Charpak and F. Sauli, Nucl. Instr. and Meth. **107** (1973) 371.
5) H. Foeth, R. Hammarström and C. Rubbia, Nucl. Instr. and Meth. **109** (1973) 512.
6) G. Charpak and F. Sauli, Proc. Intern. Conf. on *High-energy physics*, Frascati, 1973 (Lab. Naz. CNEN, Frascati, 1973) p. 246.
7) G. Charpak, F. Sauli and J. C. Santiard, Two-dimensional drift chambers, CERN NP Internal Report 73-16 (1973).
8) R. Grove, I. Ko, B. Leskovar and V. Perez–Mendez, Nucl. Instr. and Meth. **99** (1972) 381.
9) D. M. Lee, S. E. Sobotka and H. A. Thiessen, Nucl. Instr. and Meth. **109** (1973) 421.

Volume 57B, number 1 PHYSICS LETTERS 9 June 1975

CHANNELING OF 1.1 GeV/c PROTONS AND PIONS

O. FICH, J.A. GOLOVCHENKO, K.O. NIELSEN and E. UGGERHØJ

Institute of Physics, University of Aarhus, Denmark

and

G. CHARPAK and F. SAULI

CERN, Geneva, Switzerland

Received 7 April 1975

Channeling effects have been observed for 1.1 GeV/c protons and pions transmitted through a 1 mm thick germanium single crystal.

For 1.1 GeV/c protons and π^+, we have observed strong channeling effects in small-angle scattering experiments where the particles are transmitted through a germanium single crystal. Such effects may be anticipated from theory, and by extrapolating experimental results for heavy charged particles in the MeV region where correlated scattering events in crystals are known to yield pronounced directional dependences for various interactions between the penetrating particles and the crystal target, i.e. wide-angle Rutherford scattering, X-ray excitation, nuclear-reaction probabilities, and stopping power [1]. Generally one observes a reduction in yield for small-impact parameter reactions when positive particles are incident at angles smaller than a critical angle to low-index crystal axes and planes. When a classical view is applicable to negative particles, increases in such yields are found under corresponding conditions.

In the present experiment, we have observed the channeling of positive particles in a thick crystal as an increase in forward-scattering probability under channeling conditions; forward-scattering probability meaning the probability for particles being scattered by the crystal through angles of less than 1.0 mrad. The experimental set-up is illustrated schematically in fig. 1. A secondary beam produced by the CERN 30 GeV/c proton synchrotron contained about equal parts of protons and positive pions at 1.1 ± 0.03 GeV/c. The rather big angular divergence of the beam, ~ 10 and 15 mrad, in the vertical and horizontal planes, respectively, together with large surface-area detectors

EXPERIMENTAL SET-UP

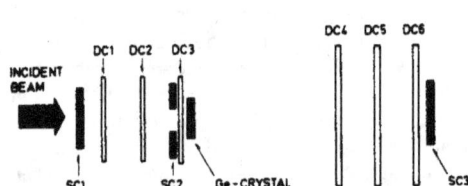

Fig. 1. Schematic illustration of experimental set-up.

facilitated simultaneous observation of a wide range of incident angles relative to the crystal. Protons and pions were distinguished by the measurement of their time-of-flight between scintillation counters SC1 and SC3; counter SC2 in anticoincidence was used to restrict the accepted beam tracks to the crystal size (1 cm^2). A set of high-accuracy, position-sensitive drift chambers [2] (DC1 to DC6) was used to record, on each event, the space coordinates of the ingoing and outgoing tracks. The intrinsic position accuracy of each detector is ~ ±150 μm; however, multiple scattering on windows and wires caused the over-all angular resolution to be ~ 0.7 mrad.

The 1 mm thick germanium crystal is located directly behind DC3. This crystal has been chosen to have a low dislocation density as measured by etch pit and X-ray anomalous transmission measurements *;

* Crystal samples were prepared at the Danish Technical University. The goniometer and beam tube were manufactured at the Niels Bohr Institute, Copenhagen.

90

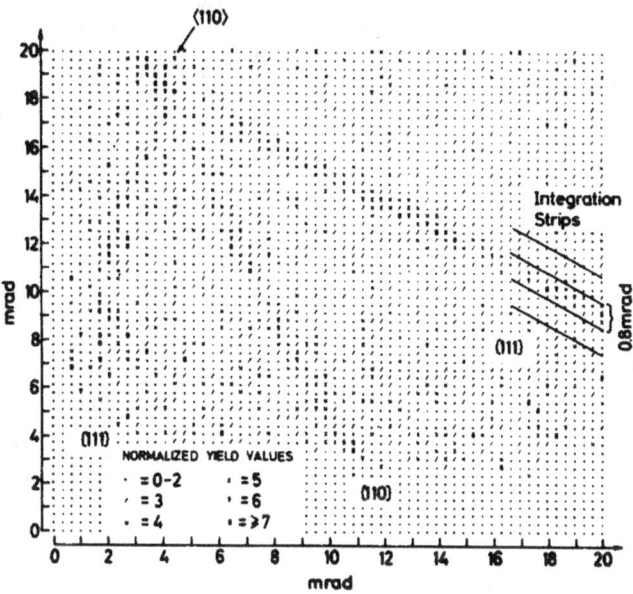

Fig. 2. a) Scatter plot of forward-scattered beam as a function of angle of incidence for 1.1 GeV/c protons.
b) Integrated planar forward-scattering intensities for 1.1 GeV/c protons and π^+ parallel to the {111} plane.

consequently its mosaic spread is negligible. The sample was positioned in the beam with the ⟨110⟩ direction aligned approximately (within ~ 1°) with the beam axis. The crystal could then be tilted through a restricted range by a one-axis goniometer. Finally, raw data were recorded on magnetic tape and later analysis determined incident and scattering angles for each event.

Preliminary results are shown in figs. 2a and 2b. Fig. 2a is a normalized plot of the incident angular distribution of those protons that have been scattered through angles smaller than 1.0 mrad obtained from several experimental runs. The presence of increased forward-scattering yield for directions of incidence parallel to {111} and {110} planes and the ⟨110⟩ axis is evident. The angles between the planes are found to be in agreement with those expected. Similar results have been obtained for π^+. In order to improve statistical accuracy, these data have been integrated parallel to the {111} plane in strips, as indicated in fig. 2a. The integrated result is shown in fig. 2b for both protons and π^+. An increase of about a factor of two in forward scattering, due to planar channeling,

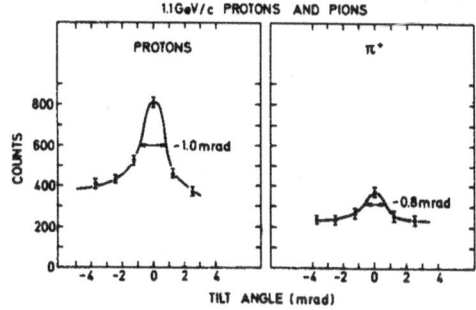

is obtained. The angular widths of the peaks shown in figs. 2a and 2b essentially represent the angular resolution of the detecting system. Based upon theoretical considerations, a reasonable estimate for the angular half-wdith of planes is given for positive particles by [3]

$$\psi_p = \left(\frac{4Z_2 e^2}{pv} \, N d_p \sqrt{3} a \right)^{1/2} ,$$

where Z_2 is the atomic number of the target atoms,

91

p the relativistic momentum, v the velocity of the incident particles, N the number of atoms per unit volume, d_p the planar spacing, and a is the Thomas-Fermi screening length. For the $\{111\}$ planes in the present case, one finds $2\psi_p \sim 0.5$ mrad. It is clear that improved resolution in future experiments is needed.

Unfortunately, owing to experimental restrictions, we were unable in this preliminary work to place the axial $\langle 110 \rangle$ direction well within the angular divergence of the incident beam; consequently, the data we have obtained for that direction are not well suited for quantitative analysis. Experiments are under way however, to obtain satisfactory results for the axial case, both for positive and negative pions and protons, with the anticipation of even stronger effects than have been observed for the planes. We are investigating now whether the channeling effect gives rise to anomalous energy losses in detectors made of crystals (germanium, silicium, CoI ...). In this relativistic region, one expects a difference of about a factor of two in the ionization loss between channeled and non-channeled particles.

If it were true, new methods of selecting the angle of high energy particles trajectories, or separating positive from negative particles would be available.

References

[1] For a general review, see D. Gemmell, Rev. Mod. Phys. 46 (1974) 129.
    The theory of channeling has been discussed most thoroughly by J. Lindhard, K. Danske Vidensk. Selsk. Mat.-Fys. Medd. 34 No. 14 (1965), and J. Lindhard, P. Lervig and V. Nielsen, Nuclear Phys. A96 (1967) 481.
[2] A. Breskin et al., Nucl. Instr. Methods 119 (1974) 9.
[3] S. Kjaer Andersen et al., Phys. Rev. B8 (1973) 4913.

92

NUCLEAR INSTRUMENTS AND METHODS 119 (1974) 7–8; © NORTH-HOLLAND PUBLISHING CO.

LOW-PRESSURE DRIFT CHAMBERS FOR HEAVILY IONIZING PARTICLES

A. BRESKIN, G. CHARPAK and F. SAULI

CERN, Geneva, Switzerland

Received 28 January 1974

Drift chambers have been operated at pressures of 20 mm, 43 mm, and 87 mm Hg of methylal vapour. At the lowest pressure an accuracy of 1.5 mm (fwhm) is obtained with α-particles.

1. Introduction

Drift chambers will probably become an important tool in high-energy physics because of their inherent advantages: accuracy of the order of 100 μm, simplicity of coordinates read-out leading to cheap, large-surface detectors, time resolutions of the order of 5 ns, when proper use is made of the correlation in several chambers.

The first measurements on drift chambers at atmospheric pressures were made with α-particles[1], and yielded indeed accuracies at least as good as those with minimum-ionizing particles. However, it is clear that in that field, low-pressure chambers present important advantages. We have made some preliminary measurements which confirm the potential interest of drift chambers for heavily ionizing particles.

2. Experimental results

We have placed a drift chamber with 13 cm of drift space (fig. 1) in a vacuum box. This box could be filled with methylal $(OCH_3)_2CH_2$ vapour at variable pressure. We have chosen methylal since it is known to be a good filling for low-pressure proportional counters. Multiwire proportional chambers have been used successfully at pressures as low as 2 mm Hg with other vapours[2] and indeed a wide choice of gases is open,

all leading to a satisfactory operation of the drift chamber.

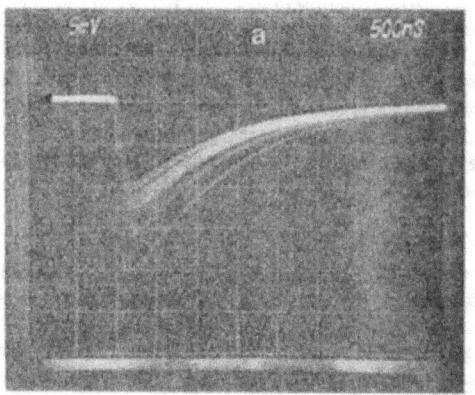

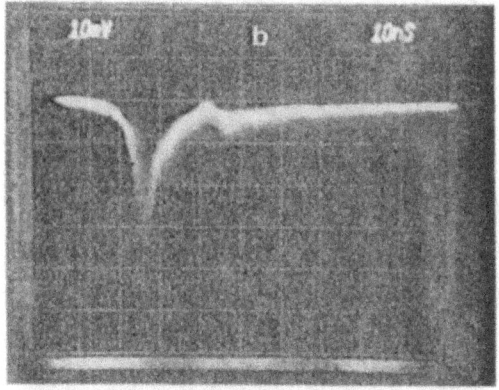

Fig. 2. Typical pulses obtained with α-particles of americium; the oscilloscope was triggered on a solid-state detector. a) $P = 20$ mm Hg, HV2 = 810 V, $E = 143$ V/cm using an emitter-follower, input impedance 2000 Ω. b) $P = 87$ mm Hg, HV2 = 1290 V, $E = 143$ V/cm, pulses are directly measured on the scope on 50 Ω impedance.

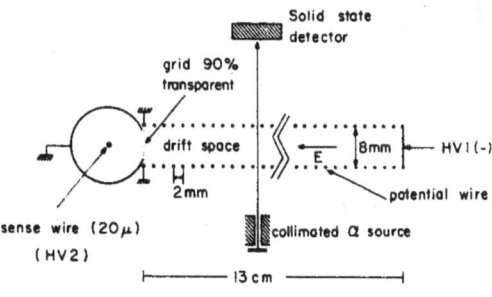

Fig. 1. Schema of the drift chamber.

A solid-state detector is used to define the time of passage of the α-particle crossing the chamber.

A source of americium provides the beam of α-particles, with a collimator of 0.1 mm width. We have worked at the three following pressures of methylal vapour: 20 mm Hg, 43 mm Hg, 87 mm Hg. We were prevented from going to lower pressures by a lack of stability in our pumping device, incompatible with the shortness of the plateau.

Figs 2a and 2b show the typical pulses obtained at a pressure of 20 mm Hg and at a pressure of 87 mm Hg. Fig. 3 shows the distribution of the time of arrival of the pulses, for two positions at 5 mm distance. The fwhm is 1.5 mm. It is still far from being as good as the best results so far obtained at atmospheric pressures[3] but many factors would have to be improved in our experiment to find out the origin of the broadening.

Fig. 4 shows the drift velocity as a function of electric field for the three pressures. We have not reached a saturation of the velocity, which is certainly a source of increased jitter. One can notice that the velocity is a function of the reduced pressure E/p, as expected.

3. Conclusions

Our work shows that drift chambers can be used at pressures of methylal as low as 20 mm Hg. We tried also to operate the chamber with the mixture[3] of argon 67%, isobutane 30%, methylal 3%. The chamber works with this gas, but we found it difficult, under our conditions, to keep a constant composition of the gas mixture. Low-pressure operation presents some impor-

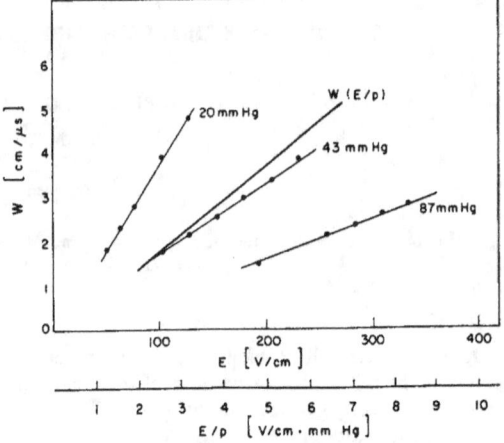

Fig. 4. Drift velocity as a function of electric field E and the reduced electric field E/p for $P = 20$, 43, and 87 mm Hg.

tant advantages whenever very thin windows are needed, as for instance in the case of slow electrons and for nuclear-physics experiments with heavy ions. In this case, the detector has to be placed in a very low-pressure vessel, and the problem of thin windows is essential. Low-pressure operation can permit a discrimination between heavily ionizing particles and a strong background of X-rays or minimum-ionizing particles, as in some experiments on the primary cosmic radiation.

An optimization of the gas and of the operating conditions could lead to an important improvement in accuracy; so far we have reached fwhm = 1.5 mm.

We think that in the domain of low-energy nuclear physics, and particularly for heavy ions, an interesting field of research is open which can lead to important progress in the accuracy and cost of detection.

We would like to thank Messrs A. Scharding and J. Sénégal for their technical help.

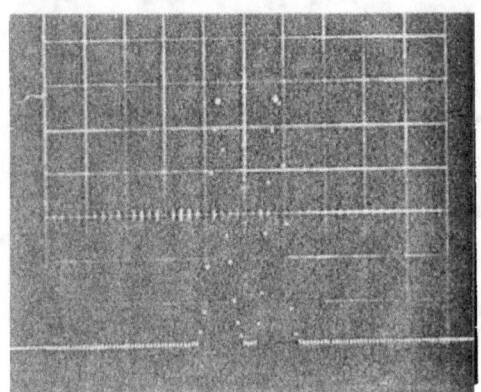

Fig. 3. Distribution of the time of arrival of pulses for two positions at 5 mm distance. fwhm ≈ 1.5 mm.

References

1) G. Charpak, J. Phys. (France) **30** (1969) 62.
2) F. Binon, V. V. Bobyr, P. Duteil, M. Gouanère, L. Hugon, M. Spighel and J.-P. Stroot, Nucl. Instr. and Meth. **94** (1971) 17.
3) A. Breskin, G. Charpak, B. Gabioud, F. Sauli, N. Trautner, W. Duinker and G. Schultz, Further results on the operation of high-accuracy drift chambers (submitted to Nucl. Instr. and Meth.).

NUCLEAR INSTRUMENTS AND METHODS 151 (1978) 473-476 ; © NORTH-HOLLAND PUBLISHING

A SOLUTION TO THE RIGHT–LEFT AMBIGUITY IN DRIFT CHAMBERS

A. BRESKIN*, G. CHARPAK and F. SAULI

CERN, 1211 Geneva 23, Switzerland

Received 12 September 1977

Using the asymmetry in the charge distributions around the anode wires, measured from the cathode-induced pulses, full separation is obtained for tracks drifting from the two sides of a wire in a drift chamber.

Drift chambers play a growing role as high-accuracy versatile particle detectors. However, measuring the instant of pulse detection at a sense wire leads to an ambiguity, since it is impossible to know from which side of the wire the electrons liberated by the ionizing particle were drifting. Among the solutions adopted to lift the ambiguity, we may quote:
- two sense wires separated by a shielding wire[1]);
- two sense wires close to each other[2]) (spaced by ≈ 200 μm);
- the use of three chambers at 120°, which solves in a mechanically economical way the problem of right–left and many multiparticle ambiguities[3]); however this leads to an increase of computing time.

We propose a simple method, requiring no mechanical change in the chamber structure, where a suitable cathode wire pick-up permits the unambiguous assignment of the detecting side of the

* On leave from the Weizmann Institute of Science, Rehovot, Israel.

anode wire; the cost increase is rather small for the electronics equipment, and can be negligible if large sense wire spacings of the order of 10 cm are used, as is practical with the graded cathode construction[4]).

It has been recently shown that the avalanche which occurs on an anode wire of a multiwire proportional chamber (MWPC)[5,6]) or a cylindrical proportional counter[7]) has an asymmetrical structure, reflected by a strong correlation between the position of the primary ionization and the distribution of the induced charges on the surrounding electrodes. We have found that this effect might be successfully applied to solve the right–left ambiguity in drift chambers, i.e. we have noticed a strong correlation between the "side" of the primary ionization, the avalanche, and the distribution of the induced pulses on the cathode wires of a drift chamber.

We have done our measurements on a $170 \times 170 \text{ mm}^2$ multiwire drift chamber, of the type described in ref. 8, constructed as follows: anode wires of 20 μm spaced by 50 mm, 50 μm ca-

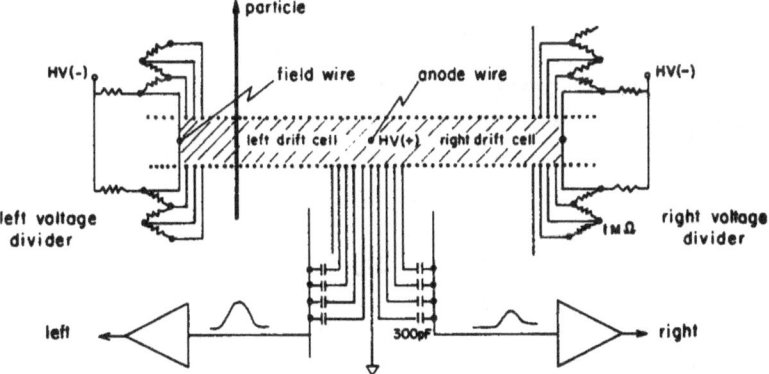

Fig. 1. Cross-section of a graded cathode drift chamber. The left and right cells have, in this case, independent voltage dividers. The induced charges can be analysed from single cathode wires or from groups of several wires.

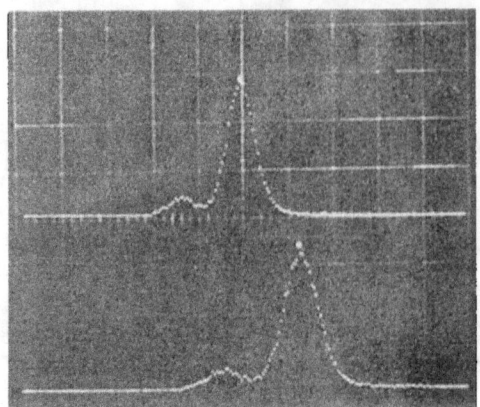

Fig. 2. Pulse-height spectrum of ^{55}Fe X-rays as analysed on the first cathode wire on each side of the anode; bottom: pulses induced on the event side, top: pulses on the opposite side; anode potential: +1600 V.

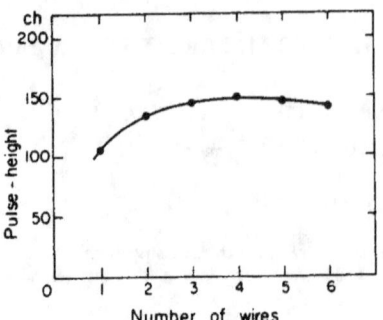

Fig. 3. Induced charge as function of the number of cathode wires connected. The maximum is reached when the first four wires are coupled together.

thode wires 1 mm apart, cathode planes spaced by 6 mm. A uniform electric field has been installed along the drift space by means of a resistive voltage divider at each side of the anode wire.

The chamber has been operated with a standard drift chamber gas (argon 67%, isobutane 30%, methylal 3%) using ^{55}Fe X-rays and β particles. The drift field has been set to 1.2 kV/cm and the anode potential varied between 1.6–1.7 kV.

The pulses induced on each of the cathode wires in the neighbourhood of the anode wire

have been measured through 300 pF capacitors; low noise current amplifiers (Alcatel type LS-CA-II) have been used, connected to individual wires or to groups of several wires as shown in fig. 1.

Fig. 2 shows the energy spectrum of ^{55}Fe X-ray induced pulses as measured on the first left and right cathode wires, when the source is at one side of the anode wire. One may remark from the merging of the 6 keV and 3 keV escape lines that the chamber operates already in the saturation region. The pulses induced on the wire situated at the side where the ionization occurred are about 28% higher than those induced on the opposite one.

The measured induced charge as a function of

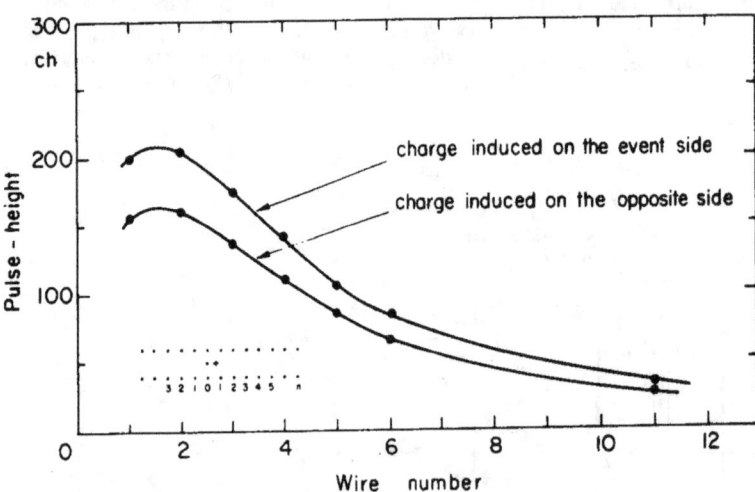

charge induced on the event side

charge induced on the opposite side

Fig. 4. Induced charge on individual cathode wires. The fwhm of the induced charge distribution is about 10 mm; it is larger than in the case of MWPC.

the number of cathode wires grouped together is plotted in fig. 3. The maximum is reached for the group of the first four wires, and then drops down owing to the increasing capacitance when a larger

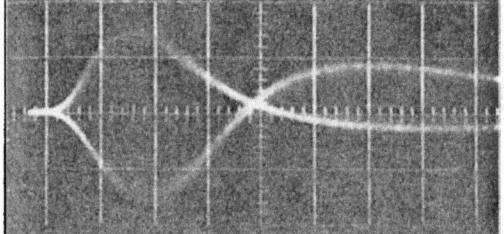

Fig. 5. Pulses from the left and right cathode wires processed by a differential amplifier and leading to output pulses of different polarities which depend on the event side; time scale: 50 nsec/div.

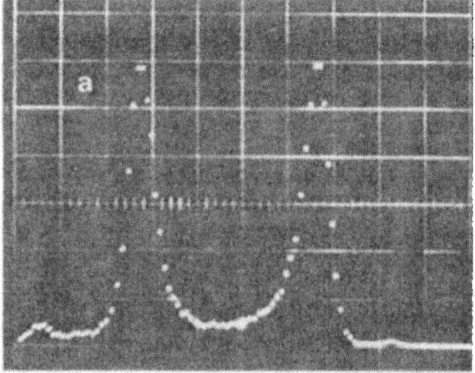

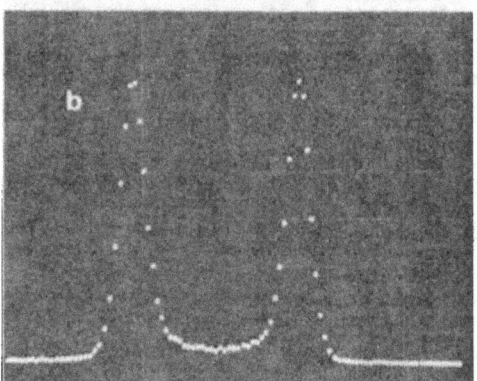

Fig. 6. Spectra corresponding to the ratio of the induced pulses when the chamber is irradiated homogeneously with: a) β particles, b) 6 keV X-rays; the two peaks are related to left or right events, and the area between them corresponds to β-rays crossing both sides or photo-electrons overlapping both sides of the wire; anode potential: +1.7 kV.

number of wires is used. The pulse height on the four-wire group is about 43% higher than that on the first wire. The induced charge on single wires is plotted in fig. 4 for both left and right wires. It can be seen that the ratio between left and right induced charges is maintained constant even at wires situated far away from the anode wire.

The simplest way of taking advantage of the left–right asymmetry is that of using a differential amplifier providing a positive or a negative signal according to the "side" on which the particle crossed the chamber (see fig. 5).

A ratio between left and right induced charges represents even better the left–right separation, as shown in fig. 6, where the ratio between left and right induced charges on a group of the first six cathode wires at each side of the anode wire has been plotted, when the total surface of the chamber is irradiated by β particles or X-rays. The small area between the two peaks that correspond to left and right events, is due to β particles traversing both sides of the wire or photoelectrons overlapping both sides of the anode wire.

An X-ray source collimated to about 0.4 mm has been displaced along the drift region. No difference between the ratio of left–right pulses has been observed for events which occur at distances between about 4 and 25 mm from the anode wire, where a complete separation is achieved (fig. 7). Fig. 8, shows the distribution when the source is "on the wire" and 1 mm to the right.

The information about the "side" of the event may be achieved even when using a single cathode wire as shown in fig. 9, where the ratio be-

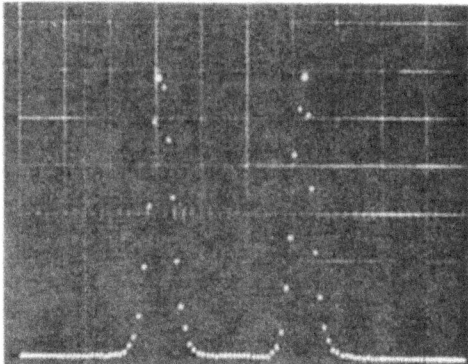

Fig. 7. Ratio of the induced pulses when the collimated X-ray source is placed 10 mm from the anode wire on its left or right side; anode potential: +1.7 kV.

A. BRESKIN et al.

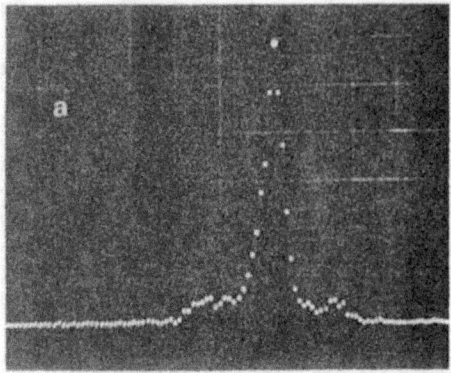

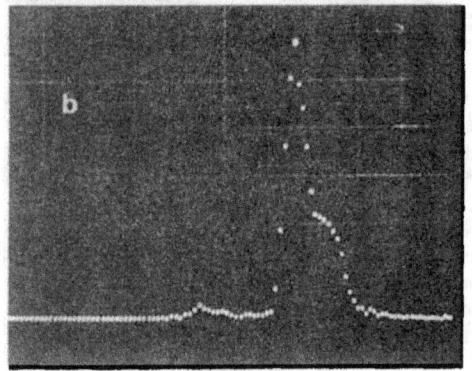

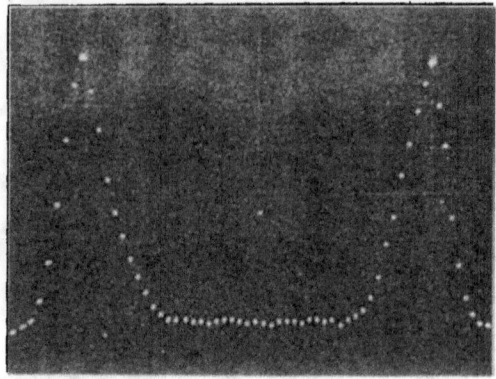

Fig. 9. Ratio of the induced pulses measured when using only one cathode wire on each side and at 6 mm from the anode; anode potential: +1.7 kV.

Fig. 8. Ratio of the induced pulses when the collimated source is: a) in front of the anode wire, b) 1 mm on the right of the wire; anode potential: +1.7 kV.

tween the "sixth" wires on each side is shown. The pulse height in that case is about 28% of the one corresponding to the six-wire group (see figs. 3 and 4).

In the chamber we used here, three drift cells were connected together, thus increasing the source capacitance; taking this into account, and the fact that a good separation has been achieved even with a single distant wire (fig. 9), we may infer that full right–left separation should be possible at least up to chamber sizes of around 2 m.

The method might also be applied to other types of drift chambers such as those having cathode planes at a constant potential. In such a case, one may use two narrow strips on one of the cathodes, one at each side of the anode wire, or the field wires.

This simple improvement makes the drift chamber even more attractive as a simple, large size, low cost, high accuracy particle detector.

References

[1] A. H. Walenta, J. Heintze and B. Schürlein, Nucl. Instr. and Meth. 92 (1971) 373.
[2] G. Charpak and F. Sauli, Nucl. Instr. and Meth. 107 (1973) 371.
[3] D. C. Cheng, W. A. Kozanecki, R. L. Piccioni, C. Rubbia, R. L. Sulak, H. J. Weedon and J. Wittaker, Nucl. Instr. and Meth. 117 (1974) 157.
[4] G. Charpak, F. Sauli and W. Duinker, Nucl. Instr. and Meth. 108 (1973) 413.
[5] A. Breskin, G. Charpak, C. Demierre, S. Majewski, A. Policarpo, F. Sauli and J. C. Santiard, Nucl. Instr. and Meth. 143 (1977) 29.
[6] G. Charpak, G. Petersen, A. Policarpo and F. Sauli, Nucl. Instr. and Meth. 148 (1978) 471.
[7] A. Breskin and G. Petersen, private communication.
[8] A. Breskin, G. Charpak, B. Gabioud, F. Sauli, N. Trautner, W. Duinker and G. Schultz, Nucl. Instr. and Meth. 119 (1974) 9.

NUCLEAR INSTRUMENTS AND METHODS 126 (1975) 381–389; © NORTH-HOLLAND PUBLISHING

THE SCINTILLATING DRIFT CHAMBER: A NEW TOOL FOR HIGH-ACCURACY, VERY-HIGH-RATE PARTICLE LOCALIZATION

G. CHARPAK, S. MAJEWSKI* and F. SAULI

CERN, Geneva, Switzerland

Received 15 April 1975

A drift chamber is described where a photomultiplier detects the photons emitted by the electrons drifting in a strong electric field in a short gap at the end of the drift space. Accuracies below 300 μm are obtained. The absence of space charge leads to considerable high-rate capabilities; at 2×10^6/s mm^2 no change of characteristics is observed.

Self-triggering modes have been investigated and various applications are discussed.

1. Introduction

A new type of particle detector has been recently developed by Policarpo et al.[1]). It is based on the fact that electrons drifting in gases under the influence of electric fields may produce detectable radiative excitation of the atoms of the gas without producing any charge multiplication.

When ionizing particles are absorbed in gases, the recoil electrons may excite some atomic levels which give rise to a photon emission detectable by a photomultiplier with a suitable wavelength shifter. This primary scintillation pulse is independent from the electric field and is of a relatively low intensity. If an appropriate electric field is applied, then additional photons are produced. These phenomena have been studied by many authors for 25 years. An extensive review of the work prior to 1964 has been done by Teyssier and Blanc[2]).

Many authors have studied this effect in noble gases or in noble gases mixed with nitrogen[3–6]). An extremely detailed study of the optical phenomena involved in gas-proportional scintillation counters has been done most recently by Thiess and Miley[7]). Measurements are reported in pure He, Ne, and Ar, in mixtures of the three gases, as well as mixtures with N_2, O_2, H_2, CO, Kr, and Xe additives for partial pressures of the second gas, from one part per million to 10%.

The real breakthrough in the field of particle detection came from the work in ref. 1. The authors had undertaken a systematic study of the light emitted by the avalanches of a proportional counter, the so-called scintillation proportional counters. Such counters had originally been built either to study the mechanism of light emission in avalanches[8]) or to build a detector with a high sensitivity[9]). It was proved indeed[9]) that the photomultiplier permits an efficiency of detection as good as the one obtained by the amplification of the induced charge on the anode. However, the progress in the electronics in the last decade clearly made the method less attractive.

The Portuguese group[1]) made a systematic study of the factors optimizing the detection of the emitted light: proper gas mixtures, wavelength shifters; and finally they reached a stage in 1972 where without any charge amplification in the gas a considerable light output was obtained, thus leading to a new class of very interesting detectors. The electrons were drifted to a spherical anode of 3 mm diameter and the electric field was below the value for charge amplification.

The improvement over the operation in a charge-amplifying mode can be summarized by the following figures:

- The amount of light per unit energy loss is one hundred times higher than in a sodium-iodide crystal.
- The suppression of the fluctuations proper to the charge-amplification process leads to a serious improvement of energy resolution. The authors obtained 8.5% (fwhm) for 5.9 keV X-rays instead of 15% with charge amplification.
- The absence of space-charge effects leads to a drastic improvement in the high-rate capability. At 10^5 counts/s no shift, within 0.4%, is observed in the pulse height[10]).

A series of further developments with low-field drift spaces and a region of light amplification between parallel grids confirmed the remarkable properties of this type of particle detection and demonstrated the geometrical flexibility of the method. Palmer and Brady[11]) applied fields of 3473 V/cm between two grids, in Xe at atmospheric pressure, and reported a

---

* On leave from the Institute of Experimental Physics, Warsaw University, Warsaw, Poland.

width of 494 eV for 5.9 keV X-rays. Conde et al.[12] reported a resolution of 1.9% for α-rays of 8.5 MeV, with fields of 4860 V/cm in xenon at 1260 torr.

2. The scintillating drift chamber

We have built a detector where the position of the particles is measured by the time of drift of electrons in a uniform electric field, as in ordinary drift chambers[13], but where the detection of the drifting electrons is obtained from the flash of light they produce when traversing a small gap of relatively intense field between two parallel transparent electrodes. The attractive prospective feature of such a device over the classical drift chamber is not a possible gain in accuracy (which is to a large extent governed by the diffusion of electrons), but a gain in the rate capability which is usually limited by the space charge in the avalanches[14]. Figs. 1a and 1b show an artist's sketch and a schematic view of the scintillating drift chamber. The field in the drift region is produced by two sets of parallel wires, 100 μm in diameter and 1 mm apart, connected to a linearly increasing potential from − HV1 to ground; the drift space is 1 cm thick and 3 cm long. The electrons liberated in the gas by a charged particle or by photon conversion drift in a uniform electric field and enter the light-producing gap of 1 mm, between a plane of wires 50 μm thick, 0.5 mm apart, and a wire mesh having 95% optical transparency. The positive potential HV2, applied to the mesh, provides the necessary acceleration of electrons to produce a light flash. The light is then detected by a 56 AVP photomultiplier, after a convenient wavelength in 1 mg/cm² of p-terphenyl deposited on a quartz window.

As gas fillings we have tried several pure noble gases, and mixtures between them or with nitrogen. As expected, we obtained the best light-emitting properties with xenon, although a mixture of 90% argon and 10% N_2 was also satisfactory. Most of the results hereafter described have been obtained with xenon, purified through calcium at 320°; the observed drift velocity however, of about 0.7 cm/μs at $E = 1$ kV/cm, appears rather low for high-flux application. Addition of 10% N_2 increases the velocity by about a factor of 2 with, however, a reduced scintillation efficiency.

3. Observation of the light flashes

3.1. Time structure of the amplified light pulses

We have studied the structure of the light pulses produced by X-rays, α-particles, and minimum-ionizing particles connecting a fast storage oscilloscope* to the photomultiplier output. Figs. 2a end 2b show the photomultiplier output for minimum-

* Tektronix Transient Digitizer R 7912.

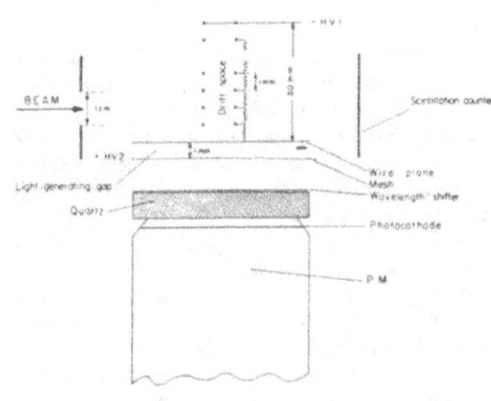

(a)   (b)

Fig. 1. The scintillating drift chamber. (a) Artist's view of the instrument. (b) Principle of construction: The electrons liberated in the drift space migrate to the light-generating gap under the influence of an electric field produced by wires or strips at a growing potential. The flash of UV light produced in the light-generating gap by a strong electric field is detected by a photomultiplier after conversion in a wavelength shifter. The origin of time is given by an external scintillation hodoscope, or by a scintillator viewed by the same photomultiplier, or by fast primary scintillation in the gas.

ionizing particles, at two potentials of the amplifying grid (HV2 = 600 and 1000 V, respectively). The single photon pile-up structure is apparent; the width of the pulse at the normal operating voltage, about 200 ns at the basis, is partly due to the decay time of the metastable states in the gas (see below). A conventional scintillation-counter hodoscope was used to define the position of the tracks in a high-energy beam, as well as to provide the trigger to the oscilloscope; the particles responsible for the light pulses shown in fig. 2 had traversed the drift region of the chamber about 10 mm above the light-producing space.

In fig. 3a we show, for a comparison, the detected pulse for a 5.9 keV photon (emitted from ^{55}Fe), and in fig. 3b the corresponding pulse-height distribution. Our energy resolution, about 20%, is much worse than the one obtained, for example, by the authors of ref. 1 (8.4% fwhm). However, in our case, the light-emitting gap is much smaller and the electric field far from homogeneous; with thin mylar windows the purity of the gas is not as high as in the devices of ref. 1, but we have been rather interested in working under the conditions of purity common with ordinary charge-amplification drift chambers. The large tail on the low-energy side is probably due to X-rays absorbed close to the wires producing the drifting field and for which

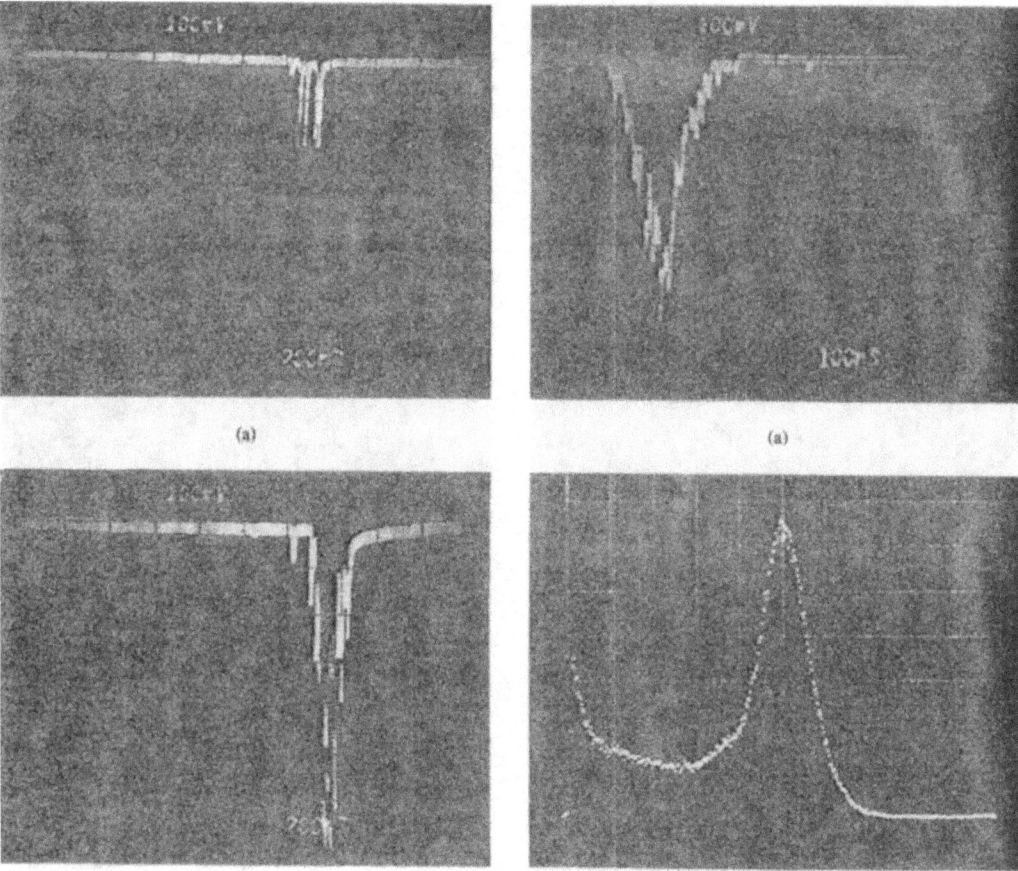

(a)

(b)

Fig. 2. Time structure of the delayed flash of light for minimum-ionizing particles. Gas filling: Xe, 760 mm Hg. (a) HV2 = 600 V/mm. (b) HV2 = 1200 V/mm. The scale is 200 ns/div. and 100 mV/div.

(a)

(b)

Fig. 3. Light flash from 5.9 keV X-rays. (a) Time-structure – same conditions as in fig. 2b. (b) Pulse-height distribution of the detected pulses. The observed energy resolution, about 20% fwhm, is due to bad geometry and impurities in the gas filling.

part of the electrons are lost in the walls during the drift to the light-amplifying gap.

3.2. POSSIBILITIES OF SELF-TRIGGERING OF DRIFT CHAMBERS

The time structure of the light pulses in xenon exhibits three types of pulses:

a) a prompt component with a short decay time;

b) a prompt component with a long decay time;

c) a delayed component, with a delay proportional to the distance between the particle and the light-generating gap.

The intensity of the first two components is a function of the nature of the particle and of its ionization power.

Fig. 4a shows a case where the fast and delayed

components are both present, for minimum-ionizing tracks. Fig. 4b shows a time distribution of the first two components. We have observed that, using pure nitrogen as the filling gas, only the first component with very short decay time persists; it is probably due to scintillation in the mylar window. If the distribution of this component in N_2 is subtracted from the distribution in Xe, for an equal amount of particles, only the second component is left: fig. 4c shows that the result has an exponential distribution $e^{-t/\tau}$, with $\tau = 30$ ns. This lifetime is shorter than the one reported for the metastable states in Xe, and would confirm that we work under conditions of purity much worse than those of refs. 1, 11, and 12. The metastable states responsible for the photon emission are probably de-excited by non-radiative collisions with impurities. In our case this may be an advantage, since the occupation time of the pulses would be too large under conditions of high purity.

The efficiency of detection of the fast pulse is clearly a function of the chamber geometry and of the ionizing power of the particles. We have measured, with our rather bad light-collection geometry, about 15% efficiency for the prompt pulse for minimum-ionizing particles, and 100% efficiency for a few MeV stopping α-particles. These results show that self-triggering of the scintillating drift chamber is possible.

It should also be noticed that for minimum-ionizing

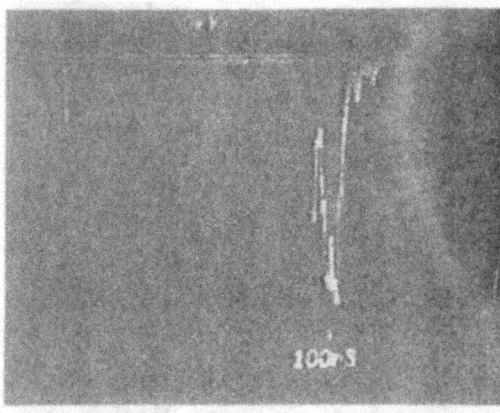

(a)

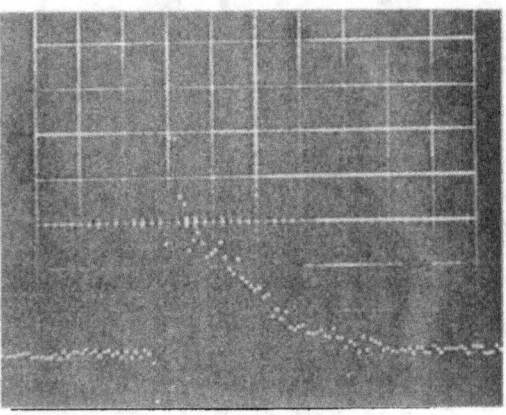

(c)

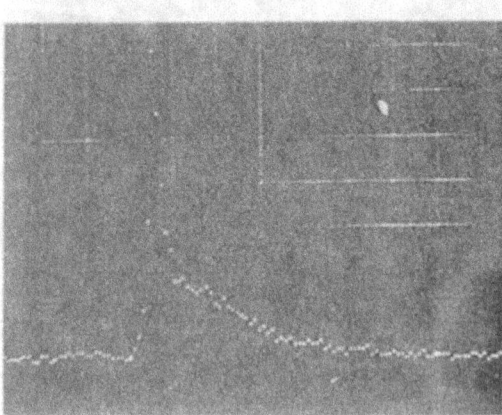

(b)

Fig. 4. (a) Fast and delayed components in the light flashes for minimum-ionizing particles. A low-intensity light pulse coincident with the passage of the particle is detected in about 15% of the events. (b) Time distribution of the fast components. (c) Time distribution after subtraction of the faster component obtained with N_2 filling; the lifetime of the decaying metastable atoms in Xe appears to be about 30 ns.

particles a fast, thin scintillator placed in front of the chamber and viewed by the same photomultiplier will give rise, on an intermediate dynode, to a large fast pulse which could give the origin of time.

For X-rays, the work of ref. 15 shows that under conditions of proper light-collection efficiency, and high gas purity, photons of 22 keV can be detected with 100% efficiency by the fast scintillation. We have thus an extremely simple one-dimensional X-ray localization detector.

4. Light intensity and efficiency for particle detection

4.1. LIGHT INTENSITY AS A FUNCTION OF THE FIELD

We have measured the light output for photons of 22 keV and α-particles of a few MeV from a ^{241}Am source traversing the chamber. Figs. 5a and 5b show the variation of the light-pulse intensity as a function of the electric field for the two radiations. Although the energy loss is different by two orders of magnitude, we observe the same law of growth. In order to ascertain that no charge amplification is causing this increase, the last electrode in the chamber was connected to a low-noise charge amplifier and the electrons collected from the α-particle trail were directly detected. The collected charge is independent of the voltage applied to the light-generating gap, as shown by the broken curve in fig. 5b.

4.2. EFFICIENCY FOR MINIMUM-IONIZING PARTICLES

We have installed the scintillating drift chamber in a minimum-ionizing charged-particle beam, as part of an existing set of scintillation counters and drift chambers. By means of the scintillation counters we could define a narrow circular beam, 1 cm in diameter, for efficiency studies, and, using the information of the standard high-accuracy drift chambers[13,14]), study the intrinsic accuracy of the new detector. Since, as mentioned above, our conditions were still not optimized in terms of light collection, to get a good efficiency plateau we had to detect the light flashes with the lowest possible electronic threshold, just above the single photoelectron noise. We found that the use of a small integration constant on the signal, around 20 ns, improves the signal-to-noise ratio without detectable losses in the accuracy*. In fig. 6 the efficiency plateau of the scintillating drift chamber is shown for minimum-ionizing particles traversing the centre of the drift region, in pure xenon and in a 90–10 mixture of xenon and nitrogen, as a function of

---

* An ORTEC 454 timing filter amplifier was used for these measurements.

the light-producing grid voltage HV2. The convenience of pure xenon is evident in terms of light output. With the drift voltage at 1 kV/cm, the drift time in the two gases was of 136 ns/mm and of 66 ns/mm, respectively.

4.3. LOCALIZATION ACCURACY OF MINIMUM-IONIZING PARTICLES

Using an electronic selection of tracks in two high-accuracy conventional drift chambers, in much the same way as described in ref. 14, we have measured the intrinsic accuracy of the new detector. An example is shown in fig. 7, where the integral time spectrum of the detector is given for a beam, as defined by scintilla-

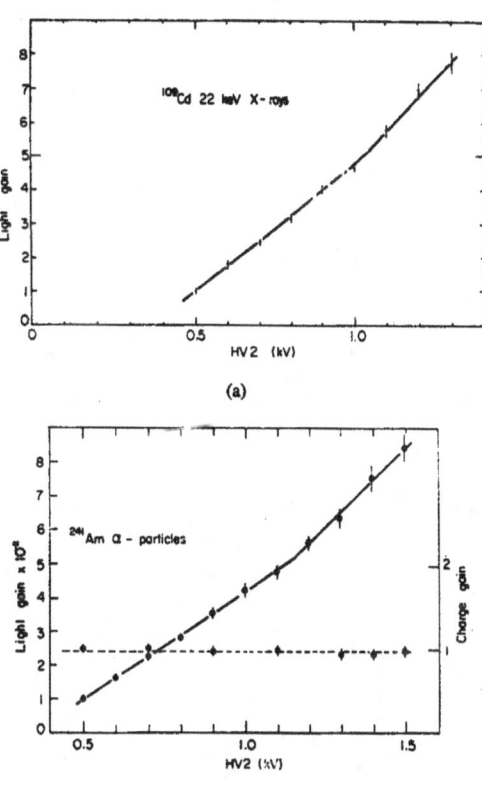

(a)

(b)

Fig. 5. Light intensity as a function of the electric field for (a) 22 keV X-rays, and (b) α-particles losing a few MeV, stopping in the gap. For both measurements the gas filling was pure Xe, and the drift voltage HV1 = 1 kV/cm. The light gain is given in arbitrary units; notice that there is a scale factor of about 100 between the two figures. In fig. 5b, the charge gain for α-particles is also shown, as measured by a charge amplifier connected to the lower grid in the chamber. There is obviously no charge multiplication in the region of interest.

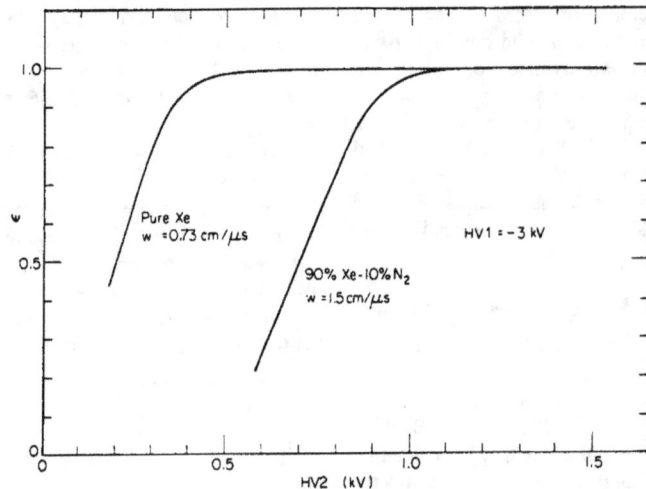

Fig. 6. Efficiency plateaus for a collimated minimum-ionizing-particle beam, and two different gas fillings. Pure Xe is obviously superior in terms of light output, but the drift velocity is lower.

tion counters, 10 mm in diameter (upper trace), and for two well-collimated beams 3 mm apart. Deducing from the measured width the known intrinsic accuracy of the standard drift chambers (100 μm standard deviation), we have obtained the intrinsic accuracy of the light drift chamber as a function of the drift distance s, as shown in fig. 8. The parabolic fit to the data shows the classical relationship for electron diffusion in gases $\sigma = (2\,Ds/w)^{\frac{1}{2}}$, where D is the diffusion coefficient and

w the drift velocity. With $w = 0.73$ cm/μs, we deduce from the measurement $D = 160$ cm^2/s.

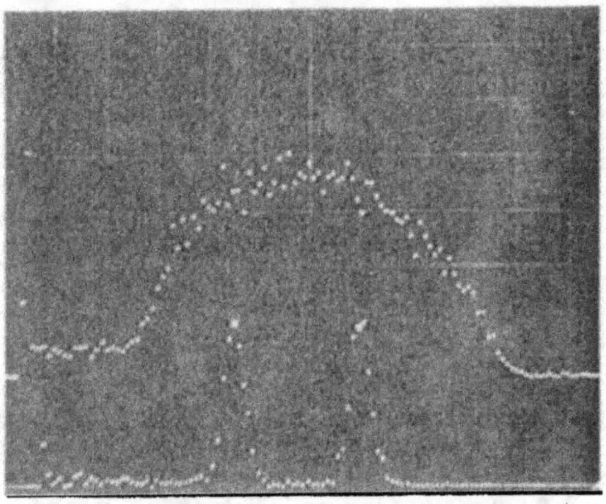

5. Rate effects

A serious limit to the rate capability of conventional proportional counters is known to be due to space-charge effects produced by the avalanches; for conventional proportional chambers, the efficiency begins to fall at about 10^4 counts/s mm^2 [14]).

The main interest of the scintillating drift chamber is the absence of space-charge effects characteristic of avalanches. We checked the stability of operation at

Fig. 7. Localization in the scintillating drift chamber. Top distribution: integral time spectrum for a circular beam of 1 cm diameter. Bottom peaks: two collimated beams 3 mm apart, as defined by ordinary drift chambers in coincidence. The time scale is 16 ns/channel.

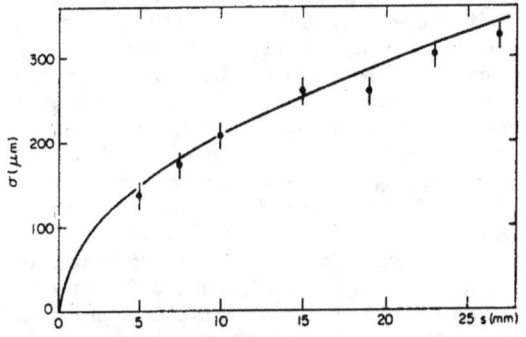

Fig. 8. Accuracy of the scintillating drift chamber for minimum-ionizing particles as a function of the drift distance s. The accuracy varies with position according to the law $\sigma = (2\,Ds/w)^{\frac{1}{2}}$, where $D = 160$ cm^2/s (Xe filling).

high fluxes in two ways. First, we measured the efficiencies of detection on a beam in the presence of a strong ^{55}Fe source irradiating the scintillating chamber; no change of efficiency or drift time could be observed at the maximum rate we could achieve, about 5×10^5 counts/s over a surface of about 1 cm^2. Fig. 9 shows a characteristic display of a single trigger under these conditions of measurements.

Secondly, we measured the single-rate plateaux with an 8 keV well-collimated 1 mm^2 beam from an X-ray generator; at fixed threshold of detection, any effect similar to the space-charge effect in proportional counters would appear as a shift in the plateau position with rate.

At the maximum rate we could achieve, of 5×10^5 counts/s with a duty cycle of about 20% (the instant rate is well above 10^6 counts/s), no decrease is detectable in the pulse height, as seen in an oscilloscope display. As shown in fig. 10, the plateau is constant in shape over three orders of magnitude of the rate. Our limitation at higher rates would come from the electronics; it is clear that the occupation time is a natural limit in the method. However, the space resolution we have obtained shows that only a fraction of the 1 mm light-generating gap is useful for giving 100% efficiency; a smaller gap is probably feasible. Also, the lifetime of the scintillating gas can be controlled by impurities. However, even at this stage, at least two orders of magnitude have been gained over the classical drift chambers or multiwire proportional chambers. Therefore this kind of chamber should find applications in high-intensity beams.

6. Applications

Several features of such chambers open wide fields of applications:

– The absence of amplifying wires, which have to be straight, gives to the detector a great flexibility as far as the shape is considered. Cylindrical or spherical drift paths for the electrons can easily be achieved.
– The self-triggered mode can be of fundamental importance for the localization of neutral particles: neutrons or X-rays. In ref. 15 it is shown that 22 keV X-rays can be detected with 100% efficiency by the fast scintillation. It is of importance also for the detection of heavily ionizing particles that cannot traverse a chamber to reach a second chamber or a scintillator giving the zero time; for the focal plane of spectrometers for heavy ions, this can be an ideal solution. We have worked in this mode with an α source of americium, collimated to a width of

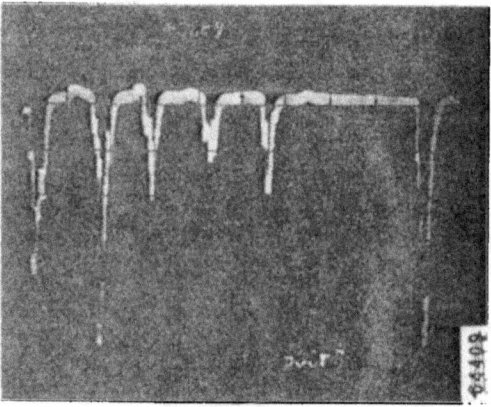

Fig. 9. Rate effects. Accidentals accompanying a single pulse from a low-intensity minimum-ionizing beam when the chamber is exposed to a 5 mCi 5.9 keV source producing about 0.5×10^6 counts/s over a surface of 1 cm^2. No change in calibration or efficiency is observed for the collimated beam; the electronics problem of separating signals from background is evident.

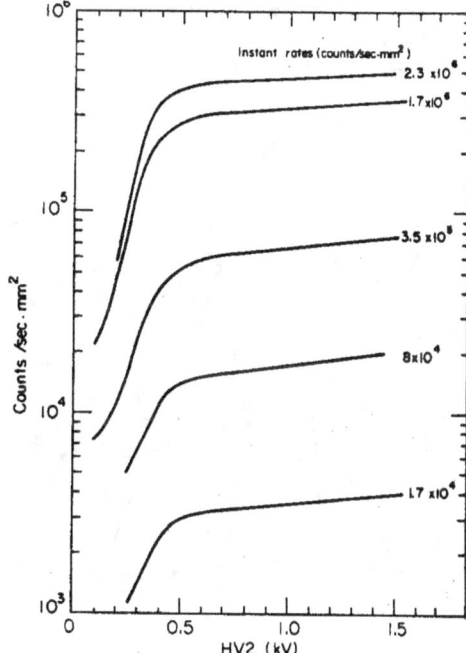

Fig. 10. Singles plateaus for an X-ray generator beam of 8 keV, collimated to 1 mm^2, with a 20% duty cycle, for increasing intensities. No effect is observed on the plateau positions at instant rates $>2 \times 10^6$/s mm^2.

about 1 mm, and obtained 100% efficiency in the self-triggered mode with a time distribution corresponding exactly to the width of the beam (fig. 11). For heavily ionizing particles the light output is sufficient, even in gases other than xenon; the measurement shown in the figure has been made in pure argon.

- For focal planes of particle spectrometers it may in general be an elegant solution if high rates have to be reached which are incompatible with proportional chambers.

For the localization of minimum-ionizing particles in high-intensity beams, the scintillating drift chamber appears to be an ideal instrument. The long drift time is not incompatible with the high rate; if the particles traverse two chambers with opposite drift directions, the sum of the drift times has to be a constant, and makes it possible to correlate the position of several particles detected during the maximum drift time (~ 1 μs/cm of drift). The time resolution of such a device is about 50 ns. With three chambers at 60°, the sum of the drift times should be a constant and the two coordinates can be obtained. We know that the widening of the track after a drift of a few centimetres is still below 1 mm and hodoscopes with a 1 mm width could easily be built, with a unique drift chamber, with the light-amplifying gap separated into sections of 1 mm each, viewed by separate photomultipliers.

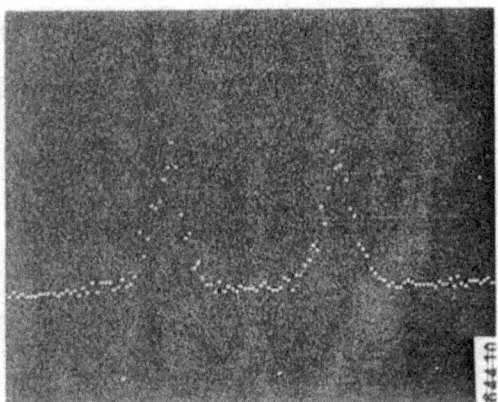

Fig. 11. Self-triggering on the fast primary scintillation for heavily ionizing particles and argon filling. Spectra of the time distance between the fast and the delayed light flashes are shown for two positions of a collimated α-ray source (^{241}Am) 5 mm apart. Even in unpurified argon the efficiency of the method is close to 100%, owing to the large light output.

7. Two-dimensional read-out

For many applications it is essential to have the information on all coordinates from a single chamber. While this is now common with ordinary proportional chambers, it was shown that this can also be achieved with drift chambers. The current division on the wires or the pulses induced on a delay line parallel to the amplifying wires permits the localization of the avalanches along the wire.

Several methods specific to the described detection system can be envisaged.

- The situation is analogous to the one in scintillation counters; the centre of gravity of the light pulse can be obtained by viewing it with several photomultipliers. The advantage over scintillators is that the amount and the duration of the light pulse can be controlled [16]).
- The thickness of the light-generating gap can be made variable. The light-pulse duration is then a function of the position.
- If the light-amplifying gap is sufficiently thick, the first detected photons may trigger a system of voltage modulation on the gap, depending on the position. For instance, a pulse propagating along the gap would modulate the light output with a delay depending on position. This is quite analogous to the method used to localize the light flash on the photocathode of a single photomultiplier [17]).
- A succession of light-generating gaps at variable distance may provide a handful of new methods.
- The use of microchannel multianode photomultipliers also opens up new ways of localization.

This rapid survey just aims at underlining the great wealth of research directions opened by this new type of particle detection.

We wish to thank Mr R. Benoit and Mr N. Greguric for the construction of the chamber. We are grateful to Mr R. Bouclier and Mr G. Million for their technical help.

References

[1] A. J. P. L. Policarpo, M. A. F. Alves, M. C. M. dos Santos and M. J. T. Carvalho, Nucl. Instr. and Meth. **102** (1972) 337.
[2] J. L. Teyssier and D. Blanc, L'Onde Electrique **44** (1964) 458.
[3] J. L. Teyssier, D. Blanc, J. Brunet and A. Godeau, Nucl. Instr. and Meth. **30** (1964) 331.
[4] J. L. Teyssier, C. Blanc and J. Brunet, Nucl. Instr. and Meth. **33** (1965) 359.
[5] F. Labernede, J. Galy, D. Blanc and J. Teyssier, Intern. Symp. on *Nuclear electronics* (Versailles, 1968; La Documentation Française, Paris) vol. I, p. 10.

6) G. L. Braglia, L. Gabba, F. Giusiano, G. De Munari and G. Mambriani, Nuovo Cimento **56B** (1968) 283.

7) P. E. Thiess and G. H. Miley, IEEE Trans. Nucl. Sci. NS-21 (1974) 125.

8) L. Colli, Phys. Rev. **95** (1953) 892.

9) G. Charpak and C. A. Renard, J. Phys. Radium 17 (1956) 585.

10) M. A. F. Alves, A. J. P. Policarpo and M. C. M. dos Santos, Nucl. Instr. and Meth. **111** (1973) 413.

11) P. E. Palmer and L. A. Brady, Nucl. Instr. and Meth. **116** (1974) 587.

12) C. A. N. Conde, M. C. M. dos Santos, M. Fatima, A. Ferreira and Celia A. Sousa, An argon gas scintillation counter with uniform electric field; Paper presented at the 14th *Scintillation and semiconductor counter* Symp. (Washington, December 1974).

13) G. Charpak, F. Sauli and W. Duinker, Nucl. Instr. and Meth. **108** (1973) 413.

14) A. Breskin, G. Charpak, F. Sauli, M. Atkinson and G. Schultz, Recent observations and measurements with high accuracy drift chambers (CERN, 15 September 1974), Nucl. Instr. and Meth. **124** (1975) 189.

15) M. Alice, F. Alves, A. J. P. L. Policarpo, M. Salete and S. C. P. Leite, The energy resolution and window area capabilities of the gas proportional scintillation counter; Paper presented to the 14th *Scintillation and semiconductor counter* Symp. (Washington, December 1974).

16) H. O. Anger, Radioisotope cameras, in *Instrumentation in nuclear medicine* (G. J. Hine and J. A. Sorenson, eds.; Academic Press Inc., New York, 1967), vol. 1, p. 514.

17) G. Charpak, Nucl. Instr. and Meth. **48** (1967) 151.

5. MULTISTEP AVALANCHE CHAMBERS: 1978–1990

5. MULTISTEP AVALANCHE CHAMBERS: 1978–1990

Volume 78B, number 4   PHYSICS LETTERS   9 October 1978

THE MULTISTEP AVALANCHE CHAMBER:
A NEW HIGH-RATE, HIGH-ACCURACY GASEOUS DETECTOR

G. CHARPAK and F. SAULI
CERN, Geneva, Switzerland

Received 14 September 1978

A new particle detector relying on an unusual avalanche mechanism mainly mediated by UV photons permits multistage amplification of ionization electrons. Single electrons are detectable. Particle fluxes, orders of magnitude more intense than in wire chambers, are acceptable. Applications can be foreseen for Čerenkov light imaging, radio-chromatography, slow neutron and X-ray detection.

Progress in elementary particle physics is very sensitive to the evolution of detectors. The introduction in this field, 10 years ago [1,2], of multiwire proportional and drift chambers (MWC's) resulted in a widespread use of these instruments and, quite naturally, a stage has been reached where some experiments are limited by the properties of this detector in terms of accuracy and rate capability. Very characteristic in this respect is the recent discovery of the upsilon particle [3]: a primary beam, one order of magnitude less intense than the available one, had to be used, mainly because of the rate limitation in the multiwire chambers.

The maximum tolerable rate in MWC's is determined by two factors: the accidental counts at high fluxes, due to the electron collection time in the gas (typical value 150 ns), and the space-charge field distortions induced by the positive ions produced in the avalanches, with a consequent loss of efficiency [4]. To overcome these difficulties a scheme has been conceived where only a fraction f of the events, suitably selected, give rise to the full avalanche multiplication [5]. The principle of operation is illustrated schematically in fig. 1a. Electrons released by an ionizing encounter in the upper part of the structure traverse a preamplifying gap A_1 where the initial charge is multiplied by a factor M_1; the resulting electron swarm is then drifted in the gas by an electric field to a gate G which transmits charges only during the short time τ of application of an appropriate pulse. The second am-

plifying element A_2, of gain M_2, is then reached by the transmitted electrons. Typically, a total gain $M_1 \times M_2 \approx 10^6$ is desirable since it brings the amplification to the level reached by good MWC's; if, for instance, $M_1 \approx M_2 \approx 10^3$ can be obtained, such a scheme would reduce by a factor M_1 the ion accumulation in A_2 due to the direct ionizations, and by a factor f the space charge due to accepted events. The resolution time is essentially defined by the jitter of the signal detected in A_2.

Two methods of solving the problem of preamplification were investigated, one based on secondary emission from low-density materials, and the other on a process of gaseous multiplication and transfer [5]. Further work on the second method has led us to the observation of an unusual multiplication process that makes it possible to go far beyond the initial aim and may give rise to a new class of detectors.

As shown by the authors of ref. [5], when in a gas a region of high electric field is separated by a wire grid or mesh from a region of lower field, under suitable conditions charge multiplication can take place with considerable electron leakage into the low-field region (fig. 1b). In the original design, a preamplification and transfer element (PAT) was associated with a gated MWC following the scheme of fig. 1a. Examination of the electric field structure in the PAT element shows that a number of field lines, equal to the ratio of the fields, escape into the lower region. Electrons multiply-

523

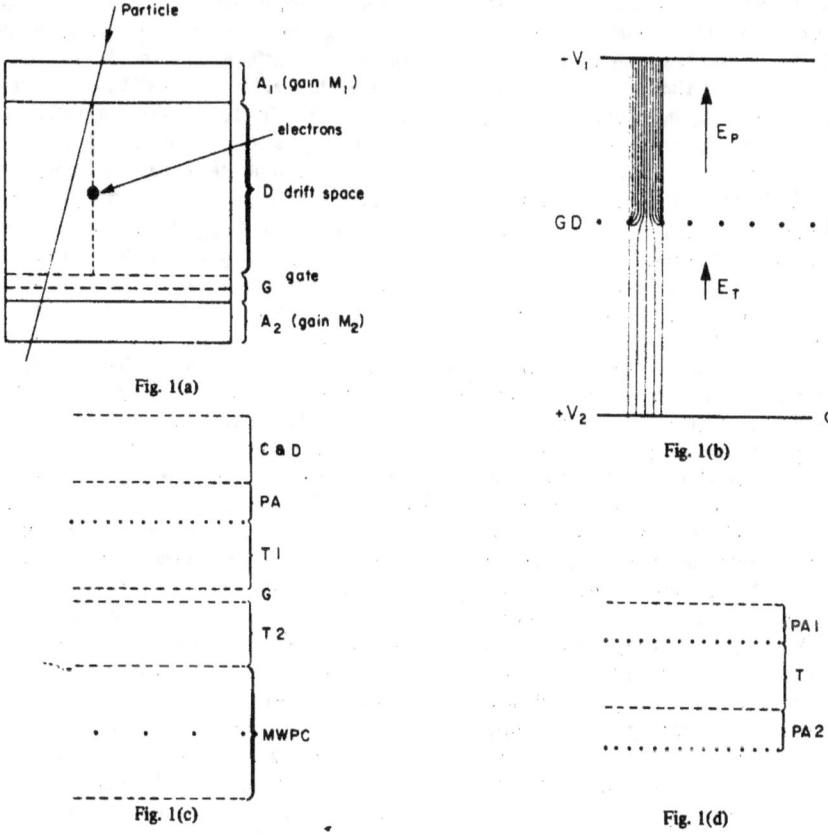

Fig. 1. (a) Principle of the gated multistep detector. Electrons liberated by an ionizing event in A_1 are amplified by a factor M_1 and transferred to A_2 (gain M_2) under the control of the gate G. (b) The gaseous preamplification and transfer element. Avalanche multiplication takes place in the high field between grids a and b, and a fraction of the electrons is transferred into the lower field region by the mechanism described in the text. (c) Construction scheme of a gated multistep chamber, associating a preamplification and transfer element with a conventional multiwire proportional chamber. For X-ray detection a conversion and drift space has been added. (d) A preamplification and transfer module followed by a parallel-plate element constitutes a true multistep avalanche chamber.

ing through a conventional avalanche process along these particular lines can be transferred into the lower field; for a typical field ratio of five to one about 20% of the charges may leak. One would, however, expect a strong position dependence of the transmission. To justify the remarkably good experimental uniformity of response and the energy resolution of the PAT element, the authors of ref. [5] had to invoke an uncomfortably large electron diffusion coefficient in the gases used (argon + alcohol and argon + carbon dioxide

+ alcohol) such as to obliterate the structure of the multiplying wires. Further observations have led us to suggest an entirely different avalanche propagation mechanism.

(i) Effective preamplification factors (inclusive of the transfer efficiency) above 500 have been obtained in argon to which had been added small quantities of ethyl alcohol, benzene, or acetone. Attempts to use argon + carbon dioxide or argon + hydrocarbons have failed owing to spark breakdown at very modest values of preamplification.

(ii) Mixtures of xenon with the above-mentioned vapours produced no detectable preamplification; the addition of even small quantities of argon, however, restored the PAT behaviour. The preamplification and transfer can be obtained again in xenon, however, using in addition a vapour with very low ionization potential, as, for example, triethylamine.

(iii) Study of the localization properties of the structures for minimum ionizing particles has shown a uniform, non-modulated efficiency of detection.

(iv) A good energy resolution was measured for well-defined charge injections into the PAT structure (~20% FWHM for 5.9 keV X-rays converted in a drift space above the preamplification region), approaching the values measured in the best single-wire proportional counters where transfer efficiency is not of concern.

The following mechanism is suggested to explain the previous observations. Electrons liberated in the gas by an ionizing event increase their energy in the high-field region and induce secondary light emission by inelastic collisions with the argon atoms. For the field values with which we are concerned, most of these photons are emitted in a wide band centred around 1250 Å (9.85 eV) [6], an energy high enough to efficiently photo-ionize the benzene and acetone molecules, and (although marginally) ethyl alcohol. Simple calculations, taking into account the known values of the photo-ionization cross section for the three vapours [7-9], provide a mean free path for photons above the ionization threshold of a few hundred micrometers in mixtures containing around two percent of quencher (atmospheric pressure). Since the secondary light emission of xenon is centred at much longer wavelengths (1500 to 1950 Å) [10], the secondary photo-ionization process cannot take place in these vapours, as observed. However, the use of triethylamine with an ionization potential of 7.5 eV permits the photo-ionization, and the PAT mechanism is restored.

A large secondary light yield followed by an efficient photo-ionization within a few hundred micrometers seems to explain all the previous observations. However, at this point we cannot completely disregard a conventional avalanche multiplication as partly contributing to the process; also, it should be noted that essentially nothing is known about secondary photon emission in noble gases mixed with organic vapours, which forbids a precise calculation on such a photon-mediated avalanche spread.

Several construction schemes have been tried in order to exploit the properties of the gaseous preamplification and transfer mechanism. Fig. 1c shows a PAT element followed by a gated MWC, as in the original design, while in fig. 1d two identical amplification elements are separated by a transfer and drift space constituting a true multistep avalanche chamber. The best stability of operation and energy resolutions have so far been obtained when operating in argon with about 2% acetone, at atmospheric pressure. Typically we have used 3 mm multiplying gaps, followed by a transfer region of 5 mm; the preamplification grid (central electrode of the structure, fig. 1b) was made either by a grid of 50 μm diameter wires, 500 μm apart, or by a stainless-steel mesh 85% transparent, as for the other electrodes. The addition of an absorption and drift space above the PAT element allowed known quantities of charge (from soft X-ray conversions) to be moved into the multiplication volume, and a multiwire proportional chamber with two-dimensional coordinate read-out using the induced cathode pulse method [11], mounted below the structure, permitted efficiency and localization studies. At fields around 6 kV/cm, multiplication takes place in the preamplification gap, and part of the electrons continue their trip in the multiple structure. At electrical transparencies around 20%, effective preamplification factors above 400 can be obtained, implying a real gain in the preamplification gap around 2×10^3. At this level the signal on the preamplification grid can be detected, as shown in the upper trace of fig. 2a. The lower trace shows the signal on the MWC anode (adding a further gain around 10^3), detected after a delay corresponding to the electron drift time between the PAT and MWC elements. The movement of the ions in the last avalanche also induces a positive feedback signal in the preamplification grid. The built-in controllable gaseous delay greatly simplifies the electronics of the read-out system, eliminating the need for a delay element on each cathode strip of the MWC. By itself, this seems a sufficient reward for the increased mechanical complexity over a conventional MWC. The operation of the pulsed electronic gate is shown in fig. 2b. The two oscilloscope tracks show, respectively, the pulse detected on the MWC anode in conditions of full transparency (upper trace) and of pulsed transmission in coincidence with the passage of the electron swarm through the gate (lower trace). A residual induction of the pulsed gate appears,

525

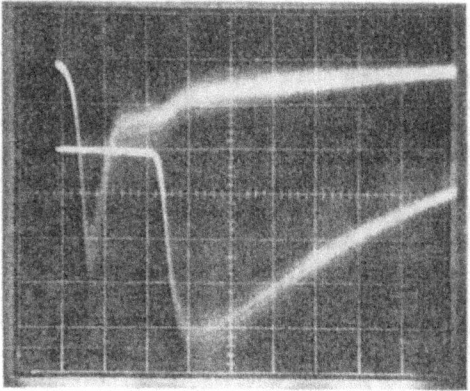

Fig. 2(a)

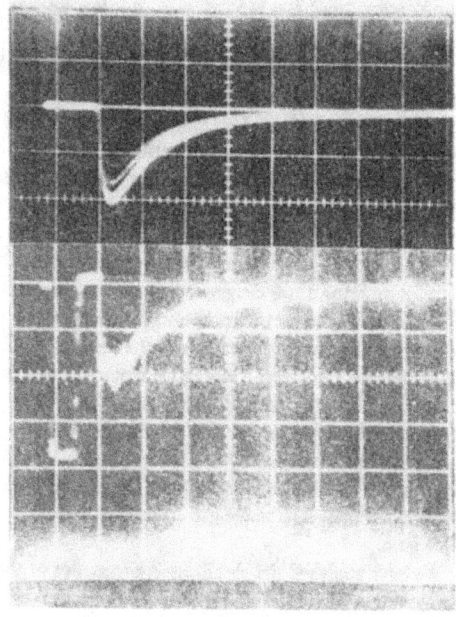

Fig. 2. (a) Pulse detected in the structure of fig. 1c on ^{55}Fe X-rays on the preamplification grid (upper track 4 mV/div) and on the multiwire chamber (lower track, 100 mV/div). Amplifier sensitivity 300 mV/pC; horizontal scale 200 ns/div. (b) Multiwire chamber electronic gating. Upper trace: dc operation (gate open); lower trace, pulsed operation: electrons are transmitted only during the gating period. Horizontal scale: 500 ns/div.

Fig. 2(b)

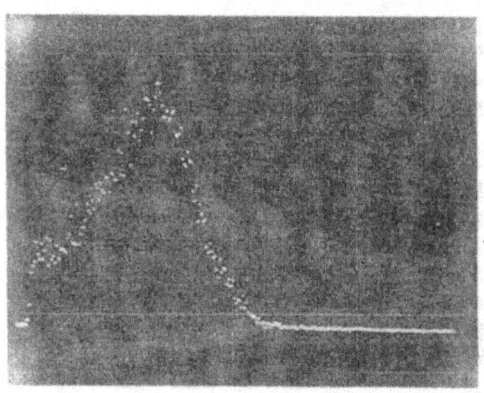

Fig. 3(a)

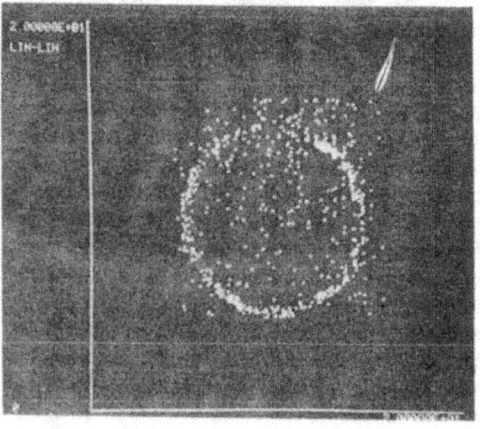

Fig. 3(b)

Fig. 3. (a) Pulse-height spectrum for single electrons, photoproduced by an UV lamp on the upper cathode of the structure shown in fig. 1c, and detected on the MWC anode. The peak value corresponds to about 1 pC on the chamber. (b) Simulated Čerenkov light ring, obtained by illuminating the upper cathode of the chamber in fig. 1c with an UV lamp through a ring mask, 50 mm in diameter. Sparse points correspond to background; exposure time ~1 min.

526

comfortably preceding in time, however, the appearance of the real charge signal.

Over-all multiplication factors above 10^7 can be obtained in the combined structure, although in a fully charge-saturated regime. This allows the detection of single electrons; fig. 3a shows the pulse-height spectrum for single electrons photoproduced on the upper cathode of the structure by an ultraviolet lamp. It should be noted in this context that in the gas used, the MWC alone cannot be operated at gains above 10^4 without entering a full Geiger regime. We explain this observation by the fact that, although there may be a large amount of hard photons from the head of the final avalanche, they have a negligible chance of propagating back into the preamplifying gap; they are instead reabsorbed within the MWC itself and give rise to a charge gain reduced by the factor M_1. This is the key reason for the stable operation of multistep structures, as compared to single-gap structures with the same gain.

We have analysed the localization properties of the PAT element on minimum ionizing particles, making use of a bidimensional induced cathode pulse read-out on the terminal MWC (see fig. 1c). Two standard chambers with the same read-out, permitting track localization within 60 μm rms [11], were used in line for monitoring purposes. Full efficiency of detection could be obtained, both with and without the drift and conversion space attachment; the fact that in the latter case the efficiency is uniform in the direction perpendicular to the preamplifying wire grid is the major argument in favour of a photon-mediated avalanche spread. The coordinate localization along this direction is instead slightly modulated by the charge scraping operated by the wires; the amplitude of the modulation implies a 50 μm displacement of the measured versus real position, and the wavelength corresponds to the distance of the multiplying wires (500 μm). Despite this, a position accuracy around 150 μm rms has been measured in this direction, which corresponds to the best determined coordinate in the MWC along its anode wires. In the perpendicular direction, where the anode wire spacing of 2.5 mm would normally introduce a large quantizing effect [11], it appears that the large avalanche spread leads to an effective charge interpolation, and an accuracy of 250 μm rms has been measured.

A complete parallel-plate avalanche chamber can be constructed with one transfer and two amplification elements, as shown in fig. 1d. Not only is this structure mechanically much simpler than the previous one (there are no thin wires, as in the MWC), but also it allows the best time resolutions on charged particles to be obtained. Operating in argon acetone, a resolution of 10 ns FWHM has been measured for fast electrons traversing the structure. Localization in this case is obtained by recording the distribution of induced charges on the last electrodes; contrary to the normal MWC case, the electron swarm may keep drifting across as many gaps as desired, allowing coordinate measurements along an arbitrary number of directions in a succession of gaps with wires oriented at different angles. Such multidimensional read-out has been used, for example, in high-accuracy X-ray detectors associated with channel plate electron multipliers [12].

In conclusion, we have shown that it is possible to multiply electrons, and to efficiently and uniformly transfer them into a succession of gaps in order to obtain a variety of functions. All electrodes in such a structure can be made with either thick wires or meshes, thus eliminating one major source of breakdown in large detectors. The large gains, good accuracies, and time resolutions that can be obtained with the multistep avalanche chambers (MSC's) allow their use in all applications where MWC's were used so far, but with a rate capability (in the gated version) at least two orders of magnitude higher for selected events. Besides, some new applications are allowed by the specific properties of the MSC:

The single-electron detection allows applications to be foreseen in the imaging of ultraviolet Čerenkov radiation [13] where the velocity of relativistic particles is measured from the opening of the emitted light cone. Using the bidimensional imaging properties of our chambers, we have simulated a Čerenkov ring (as could be obtained with suitable optics) by simple collimation of an UV light source over a cathode of the MSC (fig. 3b). Of course in this simulation, photon conversions are separated time-wise, which simplifies the pattern recognition: the diameter of the ring is 50 mm, with 1 mm thickness.

In radio-chromatography applications, the parallel-plate characteristics of the preamplifying gap should ensure a good localization accuracy, despite the long range of emitted electrons, since the largest avalanche starts from the entrance point in the gap of the electron.

527

Volume 78B, number 4 PHYSICS LETTERS 9 October 1978

For slow neutron detection, the use of a conversion foil with large cross section, e.g. made of gadolinium, as the first electrode, should, for the same reasons, allow very good localization.

In spherical drift chambers [14] a preamplifying gap at the end of the drift space should permit selection of X-ray conversions taking place within the drift space, thus in some cases eliminating the need for using xenon to reduce the background of photons converted in the transfer space or in the MWC.

References

[1] G. Charpak, R. Bouclier, T. Bressani, J. Favier and C. Zupancic, Nucl. Instr. Meth. 62 (1968) 262.
[2] G. Charpak, D. Rahm and H. Steiner, Nucl. Instr. Meth. 80 (1970) 13.
[3] S.W. Herb et al., Phys. Rev. Lett. 39 (1977) 252.
[4] F. Sauli, Principles of operation of multiwire proportional and drift chambers, CERN 77-09 (1977).
[5] G. Charpak et al., New approaches to high-rate particle detectors, CERN 78-05 (1978).
[6] J.R. Bennet and A.J.L. Collinson, J. Phys. B2 (1969) 571.
[7] J.H. Carver and P. Mitchell, J. Sci. Instr. 41 (1964) 55.
[8] J.C. Person, J. Chem. Phys. 43 (1965) 2553.
[9] J.C. Person and P.P. Nicole, J. Chem. Phys. 55 (1971) 3390.
[10] R.D. Andersen, E.A. Seimann and A. Peacock, Nucl. Instr. Meth. 140 (1977) 371.
[11] G. Charpak, G. Petersen, A. Policarpo and F. Sauli, Nucl. Instr. Meth. 148 (1978) 471.
[12] E. Kellog, P. Henry, S. Murray, L. Van Spegbroeck and P. Bjornholm, Rev. Sci. Instr. 47 (1976) 282.
[13] J. Seguinot and T. Ypsilantis, Nucl. Instr. Meth. 142 (1977) 377.
[14] G. Charpak, C. Demierre, R. Kahn, J.C. Santiard and F. Sauli, Nucl. Instr. Meth. 141 (1977) 449.

CERN 78-05
Experimental Physics
Division
14 June 1978

ORGANISATION EUROPÉENNE POUR LA RECHERCHE NUCLÉAIRE

CERN EUROPEAN ORGANIZATION FOR NUCLEAR RESEARCH

NEW APPROACHES TO HIGH-RATE PARTICLE DETECTORS

G. Charpak, G. Melchart, G. Petersen and F. Sauli,
CERN, Geneva, Switzerland

E. Bourdinaud and P. Blumenfeld,
DPhPE, CEN Saclay, Gif-sur-Yvette, France

C. Duchazeaubeneix and A. Garin,
DPhME, CEN Saclay, Gif-sur-Yvette, France

S. Majewski and R. Walczak,
Institute of Experimental Physics,
University of Warsaw, Poland

ABSTRACT

Methods are described for overcoming the limitation, due to space charge, in the detection efficiency of multiwire proportional chambers operated in high particle fluxes. By dividing the gas amplification process into two separate steps, a time interval allows for a fast selection of the detected events. The two schemes of preamplification investigated are a) secondary emission by low-density surfaces followed by electron acceleration in vacuum and detection in a multiwire chamber, and b) gaseous amplification by avalanche formation in a nearly uniform field between wire grids. Tests have shown the latter method to be capable of good gain with high resolution, which should permit the reduction of the space-charge limit of wire chambers by several orders of magnitude.

CONTENTS

1. INTRODUCTION

In recent years, several experiments have been seriously limited in their data-taking rate by the degradation of multiwire proportional chamber (MWPC) performances at high fluxes. There are two main factors that contribute to determining the maximum rates at which this class of detectors can be employed efficiently: the finite resolution and memory times, and the decrease of gain due to space-charge build-up. Although the resolution time in a good MWPC with 2 mm anode wire spacing is of about 30 nsec, the amount of accidental coincidences is mainly determined by the collection time of electrons from the drift region. For an 8 mm gap, as used in most large-size chambers, some electrons are collected, delayed by as much as 200 nsec with respect to the fast ones.

In other words, at rates around 10^7 sec^{-1} the accidentals represent the majority of the detected events even if validation gates as short as the chamber resolution are used. The addition of electronegative vapours in the gas, by reducing the effective memory times, improves the issue only slightly[1]. The space-charge build-up is an even more severe limitation to the operation at high rates[2]. The positive ions produced in the avalanche process migrate only slowly to the cathodes, and the charge accumulation that follows strongly modifies the electric field in the counter and hence its performance. At a particle rate of 10^4 mm^{-1} sec^{-1}, a gain reduction exceeding 50% has been measured at usual values of the multiplication factor[3], with a consequent decrease in the detection efficiency. Since, at least in first approximation, the gain reduction depends exponentially on the multiplication factor, the operation of counters at low gains would improve their rate capability, but in practice this implies the use of prohibitively low threshold electronic discrimination.

In the following we will describe a possible way of overcoming the quoted rate limitations in MWPCs by a drastic change in their mode of operation.

2. PRINCIPLE OF THE METHOD

With reference to Fig. 1, the principle of the method is as follows. The amplification process leading to an over-all gain M is split into two independent steps: an amplification by a factor M_1 in the region A_1, and by M_2 in A_2, such that $M = M_1 \cdot M_2$. The two regions A_1 and A_2 are separated by a drift region D, such that electrons produced in A_1 reach A_2 after a time T_D.

At the far end of region D a physical gate G (which we will describe later) may collect the electrons from D or let them be transferred to A_2. Such a physical gate plays the following role: if the event is not acceptable the electrons produced in A_1 will be collected at the end of D and not permitted to enter A_2. If the event is validated by external detectors and rapid electronic analysis, a gate of delay equal to T_D and of suitable width will transfer the electrons to A_2, which in this study, is a MWPC.

When a charged particle traverses the detector, a series of electrons are produced in A_1, A_2, and D, and the following phenomena occur:

Detection in the multiwire chamber A_2: The gain is reduced by a factor M_1 with respect to a normally operating chamber. Owing to the above-mentioned very fast exponential dependence of the positive space-charge effects from the gain, the limits in the direct chamber detection will be encountered at fluxes several orders of magnitude larger than normal, depending on the actual value of M_2. Conditions are chosen such that pulses from direct detection of charged tracks are below the discrimination threshold of the read-out electronics of A_2. We want to accept in A_2 only selected events, which may be rare, and these have been preamplified in such a way by A_1 as to be detectable in A_2.

Multiplication in the preamplification region A_1: This provides a preamplification by a factor M_1 to all tracks. Obviously this is a key part of the scheme, and in the following section we will describe two possible ways of implementing this function.

Selection of valid events: The drift space D serves a double purpose. It delays the swarm of electrons coming from A_1 by a constant time, thus allowing a logical decision to be made on the acceptance of the tracks, and it contains a controlled electronic gate at its end to forbid all but the electrons from selected events to enter the proportional chamber A_2. In conditions that are typical for the operation of gas counters, the delay T_D can be of the order of 200 nsec for a 10 mm long drift length. Classic solutions exist for implementing the electronic gate G, either by using a pulsed inversion of electric field between two meshes or by applying alternated potentials to adjacent wires so as to collect all electrons. The electronic transparency is then restored by a voltage pulse that brings the mesh to an equipotential[*] (Fig. 2).

We have experimented with both solutions. In the case illustrated by Fig. 2a, two fine wire meshes, 1.6 mm apart and having 85% optical transparency, were rendered completely opaque to electrons by using a small inverted continuous field of about 60 V cm^{-1} ($V_0 = -10$ V); by applying a 100 nsec voltage pulse V_0' of about 80 V, of the opposite polarity, a 70% electronic transparency has been measured during the pulsing time. For the scheme depicted in Fig. 2b, and with 1 mm wire spacing in the control grid, a difference of potential of about 400 V appeared to be necessary in order to stop all migrating electrons.

A fundamental limitation in the use of pulsed electric gates resides, of course, in the amount of residual noise pick-up in the active elements of the chamber at the instant of decision. With a small (10 × 10 cm^2) chamber, we have so far achieved a complete recovery of the amplifiers in about 200 nsec using both pulsing methods described, and this corresponds roughly to the drift time from the gate G to the anodes of chamber A_2 (see Fig. 1). Symmetrical pulsing of the alternate wires in the scheme of Fig. 2b reduces the pick-up problems significantly. There is certainly more to be done to bring the level of parasitic pulses to a value where cathode strips, for instance, could easily be read out for a comfortable use of the centre-of-gravity method of coordinate determination. Since our tests were performed without strict construction precautions, their results are quite encouraging.

The time resolution can be set either by the opening time of the gate or by the intrinsic resolution of the proportional chamber A_2. Its minimum value is set by the fluctuation in timing introduced by A_1 and D for particles at different spatial positions. On the assumption (which may be rather optimistic) that the opening time corresponds to the acceptance into the proportional chamber of all (and only) the charges produced by the preamplification element A_1, the total charge liberated in A_2 at a particle flux N, if n is the validated rate, is proportional to

$$N M_2 \left(1 + \frac{n}{N} f M_1 \right) ,$$

where f is the ratio of primary charges liberated in region A_1 to those liberated in A_2. If the fraction of accepted events is small, the reduction in the total liberated charge is therefore proportional to M_2.

3. DIFFERENT APPROACHES TO THE PREAMPLIFICATION PROBLEM

We will describe two distinct approaches to the most delicate problem of this undertaking. Both yield promising results and may find distinct applications in different fields.

3.1 Preamplification by secondary emission and electron acceleration in vacuum

Charged particles crossing low-density KCl or CsI surfaces[5-7] emit secondary electrons. These surfaces are prepared under special conditions; essentially, evaporation of successive layers under a low pressure of argon. The density of the deposits is of the order of 1% of the density of the respective solid, and a typical geometrical thickness of 100 μm results in an equivalent thickness of only 1 μm of dense material.

The secondary electrons ejected from such a surface must be extracted under vacuum by an electric field, typically of 20 kV/cm. Recent progress has shown that with minimum ionizing particles it is possible to eject, on the average, around 5-6 electrons and reach 90% efficiency for the detection of the relativistic particles by detecting the secondary electrons[6,7]. However, this factor of 5 is quite insufficient for our purpose and is subject to large fluctuations. We obtain the additional gain by using the extraction field to accelerate the secondary electrons in vacuum and inject them into a proportional chamber after traversal of a thin window. Let R be the residual range of the electrons after entering the chamber. For 10 keV we may expect a ratio of at least two orders of magnitude between the amount of energy deposited by the primary particle in a gas layer of thickness R and the amount of energy deposited by the secondary electrons entering the chamber with this energy.

To check the feasibility of such a scheme, we have constructed the chamber illustrated in Fig. 3. Ten layers of CsI have been evaporated on a thin aluminium support, providing a total thickness of 150 μm of low-density secondary emitter. At a distance of 33 mm, a 2.5 μm thick mylar window separates the acceleration region from a drift space D followed by a conventional MWPC. The entrance window is coated on both sides by a vacuum-evaporated, 0.5 μm aluminium layer, and is supported by a thin wire mesh.

The pressure in the acceleration space is reduced to about 10^{-3} Torr by continuous pumping, and under these conditions up to 30 kV could be applied across the gap without breakdown. This voltage is used to extract the electrons from CsI and accelerate them towards the entrance window of the gas detector. Tests were made with minimum ionizing electrons from a ^{106}Ru source placed above the secondary-emitter layer; fast electrons from the source traversed the chamber and were detected in a pair of scintillation counters, providing the trigger signal.

At low values of the accelerating potential V_A, only a direct signal is observed in the proportional chamber, corresponding to the electron collection from the chamber itself and from the drift space. At a threshold value of about 13 kV, the detection of a pulse begins with the expected delay of about 700 nsec, corresponding to 34 mm of drift in the gas. A further increase in V_A results both in an increase in the delayed signal pulse height and in a decrease in the drift time corresponding to the longer range of the secondary electrons in the chambers.

Analysis of the height of the delayed pulse offers interesting information. Figure 4 shows a pulse-height distribution of the delayed pulses for various accelerating voltages; the average energy loss in the peak, at 25 kV, corresponds to about 10 keV. Since for every electron we expect an energy of about this value, the result implies that we are obtaining essentially single electrons from the secondary emitter. Previous work[7] has shown that for electrons crossing a layer of 125 μm at 45°, the average number of electrons is only 1 at 10 kV/cm and reaches 2.5 at 25 kV/cm. We are applying about 7 kV/cm when the acceleration voltage is 23 kV. It is thus not surprising that we extract only one electron and the detection efficiency is 10% at 23 keV. However, our results are encouraging in the sense that we have demonstrated that a single electron produced by a secondary-emission surface can be detected with almost 100% efficiency. This may have applications that are different from the one originally foreseen. By increasing the extraction field and increasing the layer thickness, we have several possibilities of increasing the efficiency.

Measuring the delayed pulse time-distribution at a fixed threshold of detection in the proportional chamber, we observed the spectra shown in Fig. 5. Just above the threshold value in the accelerating potential, the delayed peak has a narrow time-spread of 20 nsec FWHM, corresponding to electronics jitter and geometrical changes in the drift length (due to the window bulging). At higher values of V_A a tail develops towards shorter times, indicating that the residual range of accelerated electrons in the gas approaches a few millimetres or so. In other words, to obtain a good resolution time we should not take the front end of the pulse but the falling edge, corresponding to the entrance position, which is fixed. In all pictures, the background of events at low times corresponds to accidental triggers on the normal ionization trail.

3.2 Gaseous preamplification

Charge amplification factors can easily be obtained in gaseous detectors; the problem is, however, to transfer all or a fraction of the multiplied charge to another part of the structure.

During the early days of multiwire chambers, hybrid systems were designed where the primary ionization in a gas was preamplified in a single-gap wire chamber and the electrons then transferred to a spark chamber[8-10]. The aim was to use the cheaper and at that time routine technology of spark chamber read-out. A schematic view of the preamplification element in a hybrid chamber is shown in Fig. 6 [9]; it differs from a conventional MWPC in that the electric field is not symmetric around the multiplying electrode. Electrons were supposedly multiplying when reaching the neighbourhood of the wires from the upper high field region, and some of them were escaping in the lower field region where they could trigger a new detection element. The transfer mechanism was attributed, at the time, either to some diffusion process during the avalanche multiplication around the wires or to a photon-induced spread. A preliminary set of measurements done in this direction, using a multiwire chamber in place of the spark chamber of the original design, led to very disappointing results: although some preamplified pulses were detected, both the efficiency and the energy resolution were very bad. Addition of a drift space in front of the preamplification region completely changed the issue, and allowed us to understand the real mechanism of preamplification and transfer of charges. Figure 7 shows the equipotentials and field lines around the central electrode structure, such as the one of Fig. 6, for 35 μm thick wires, 0.5 mm apart, and for a field E_1 (upper region) that is three times larger than E_2 (lower region). There are no lines of force leaving the wires towards the lower field region, and if all the avalanche process takes place in the vicinity of the wires, it would require, for electrons to leak in the lower region, a diffusion against the field by several hundred microns, which is an unlikely process. There remains the possibility of charge spread by photons emitted in the avalanches, in which case very bad localization properties would be expected from this kind of detector. In the framework of the studies on bidimensional read-out of proportional chambers using the cathode-induced pulses[11,12], we have verified that actually this is not the case even for very poorly quenched gases, and avalanches remain very well localized as long as the gains are not too large. Similar results have been obtained by other authors[13]. Figure 8 shows the angular distribution, around a 20 μm anode wire of a proportional counter, of the ions in an avalanche started by a well-localized cluster of ionization electrons. The technique used is described in Ref. 12. We see that at gains of the order of 1000, well above the value we want to reach in A_1, the avalanches do not propagate around the wire, which excludes a propagation mediated by long-range photons.

These facts all led to the following understanding: the hybrid chamber is nothing but a parallel-plate counter, with some escaping field lines. Multiplication takes place more or less uniformly in the whole of the high field region, and the transfer efficiency depends mainly on the ratio of fields E_1/E_2. It is clear, then, why the addition of a drift region before entering the high field space improves the resolution dramatically: the quantizing structure of the "transparent" multiplication region is obliterated by the increase in electron diffusion when approaching the grid. The correctness of this interpretation is confirmed by a recently published work[14] where a parallel-plate proportional chamber is coupled to a spark chamber, and relies precisely on the use of electron avalanches in a uniform field transferred to a spark gap where the conditions of stability and the plateau of counting rates of β particles are much improved by the injection of a large cluster of electrons at a precise distance from the anode.

We have designed and tested a chamber, showing very clearly the phenomena of preamplification and transfer, with a high efficiency for 6 keV X-rays and fast electrons (Fig. 9). The main construction parameters are as follows. A diffusion region is separated from the preamplification region by a wire mesh, 85% transparent; the 40 μm diameter preamplifying wires are 500 μm apart. A transfer and delay space of 3.5 mm follows, and charges are then drifted into a conventional MWPC. In this preliminary design the gating facility was omitted. Low-price charge amplifiers, having a sensitivity of 250 mV/pC and a time constant of 3 μsec [*)], were connected both to the preamplification grid and to a group of anodes in the proportional chamber. A suitable choice of the operating potentials then allows the structure to be operated as a conventional MWPC with an additional drift space on one side. When, however, the potential difference HV3-HV2 is increased, at a quite well defined threshold value, preamplification if observed for charges produced above the upper wire grid. Figure 10 shows two phases of this transition in the detection of a ^{55}Fe 5.9 keV X-ray source uniformly irradiating the sensitive volume of the chamber. In Fig. 10a the preamplified pulses begin to appear, separated from the normal ones (which correspond to photons converted below the wire grid), and in Fig. 10b a preamplification factor of about 25 is obtained; this value increases

---

*) The amplifiers were developed at CERN by J.-C. Santiard, and are produced in a thick film hybrid technique by CIT-Alcatel, France.

very quickly with the potential difference HV3-HV2. So far, the highest preamplification factors have been observed with a gas filling of pure argon bubbled through ethyl alcohol at 0°C; the addition of a small amount of CO_2 (2 to 5%) greatly improves the time resolution of the detector, as we will see later, but reduces the maximum safe preamplification before breakdown.

Figure 11 is a summary of the measured preamplification factors for two values of the MWPC anodic potential and a fixed voltage difference in the transfer and diffusion region, and as a function of the voltage applied to the preamplifying grid. All measurements have been done in the quoted argon-alcohol gas mixture; the error bars in the drawing represent the FWHM of the detected 5.9 keV line. The energy resolution is surprisingly good in these conditions; it varies from 25% at the lowest gains to about 15% at preamplification factors of 100. Examples of pulse-height distributions are given in Figs. 12a and 12b for preamplification factors of 7 and 40, respectively; at the lowest gain, the direct 5.9 keV and 3 keV escape peaks are still visible (Fig. 12a), while they have been suppressed in the spectra of Fig. 12b. The uneven distance between the peaks at the higher gain shows that the space-charge limited the proportionality region approaches, which may partly explain the remarkable resolutions obtained.

At preamplification factors of about 100, a charge signal begins to be detectable directly on the multiplying grid (see Figs. 13). The upper trace shows the signal detected on the pre-amplification grid, and the lower one the charge on the proportional chamber, operated in a pure ionization chamber mode (Fig. 13a) and at increasing anodic potentials (Fig. 13b and 13c). The scope sensitivities are 4 mV and 200 nsec per division in both Fig. 13a and Fig. 13b; in Fig. 13c the conditions are set for full charge multiplication in the MWPC (in this case, the lower trace sensitivity is decreased to 100 mV/division). The influence of the positive ion feedback on the preamplified pulse is clearly visible.

A very important question that is still pending is the actual value of the transfer efficiency of the preamplification element. All we said so far in terms of preamplification refers, of course, to the fraction of charges that manage to reach the MWPC, and not to the total number produced by the preamplification process. The following interpretation is suggested for the peculiar shape of the direct preamplified pulse shown by the upper traces in Figs. 13. A fast swarm of multiplying electrons traverse the preamplification region, and part of them leave it, thus explaining the fast increase and subsequent decrease of the detected signal. Slow positive ions are left behind which induce the slowly decreasing tail (because of the time constant of the amplifier). If this is a correct interpretation, the transfer efficiency would appear to be the ratio of the two fast-induced signals of opposite polarity, about 50% as from the pictures. Such a high transfer efficiency is surprisingly good. On the other hand, Fig. 13a shows that the charges reaching the anodes of the MWPC, operated in a pure collection mode, induce a pulse height which is about one quarter of the preamplification one; the transfer efficiency would then be about 25%. Further analysis of the properties of charge induction in this complex structure is clearly necessary, taking into account the ion velocities, the electronics time response, and the cathode transparency.

A simple threshold discrimination on the direct preamplified pulse and on the delayed pulse in the MWPC allows both efficiency curves and time delay measurements to be obtained. Figure 14 shows a spectrum of the time difference between delayed and fast pulses, measured with a uniform irradiation of the chamber with a 5.9 keV source, at a preamplification factor of about 100. The FWHM of the distribution is 8 nsec.

Some preliminary measurements have been done using also a ^{106}Ru β emitter in an arrangement similar to the one illustrated in Fig. 3. In Fig. 15 the pulse-height spectra of fast electrons and 5.9 keV X-rays are compared, after a preamplification factor of about 100, as detected in the MWPC. The MWPC efficiency as a function of the anodic potential has been measured and is shown in Fig. 16 for several values of the preamplification grid potential, at a fixed discrimination threshold, corresponding to about 0.1 pC on the chamber. Owing to large multiple scattering and γ accidentals, efficiency measurements realized with β emitters cannot reach 100%, but the presence of a comfortable plateau shows that full efficiency is obtained.

Interestingly enough, if the potential in the diffusion region is reduced below HV2 so as to multiply only charges produced in the preamplification region, full efficiency of detection can still be obtained although the plateaux are reduced in length, as shown in Fig. 17; this suggests that for charged tracks that produce distributed charges, the diffusion space may be suppressed with obvious advantages. In the figure, the efficiency of the system for direct detection of tracks in the MWPC (i.e. without preamplification, HV2 = 0) is also shown, and a comfortable region still exists where full rejection of direct signals is possible.

The time-spread in the detection of the preamplified electron pulse was also measured using the proportional chamber. Because of the large variance in the pulse height, due to Landau fluctuations in the energy loss, good timing is obviously more difficult than it is for X-rays since the typical rise-time of the signals is about 100 nsec (see Fig. 13c). The best result obtained so far is shown in Fig. 18, and has been obtained adding $\sim$ 3% CO_2 in the argon-alcohol gas mixture to increase the drift velocity of the electrons and decrease their diffusion; the FWHM of the distribution (time difference between the scintillation counter signal and the delayed, preamplified pulse in the chamber) is about 20 nsec. A 50 nsec wide electronic gate, of the kind discussed in the previous section, would guarantee full efficiency of detection, and represents the time resolution of the system. Further work is in progress to improve this parameter.

4. CONCLUSION. FUTURE PROSPECTS AND APPLICATIONS

Our investigations have shown that the amplification process leading to the detection of particles in a MWPC can be separated into two successive steps, delayed by a time interval permitting the acceptance of events only after some fast logic decision.

Two approaches have been investigated: the first based on secondary emission from a low-density surface and subsequent electron acceleration; the second one based on gaseous amplification.

This last approach has led to a satisfactory solution of the problem of preamplification. Gains up to 500 are easily reached, leading to a perfect separation of the signals from the direct beam and the preamplified electron cluster. The reduction in the space-charge limit, with a proper use of gating grids, should be about proportional to the preamplification factor. Even for moderate rates the approach of preliminary amplification offers attractive features.

The delay of the signal with respect to the passage of the particle is obtained from the chamber without requiring the painful and sometimes expensive methods of electronics or cables.

The wire spacing of a large MWPC has usually to be kept above 2 mm for mechanical reasons and because of the difficulty of reaching proper amplification fields without entering into unstable operation conditions. The pregain of 500 permits a considerable release in the construction constraints, and it becomes as easy to detect minimum ionizing particles as to detect α particles!

An important application, which can readily be foreseen for this technique of preamplification gap associated with a wire chamber, is the reduction of the thickness of the sensitive gas layer in a chamber. It is well known that when one aims at imaging the distribution of electrons emitted by planar sources placed against a wire chamber, the finite range of the electrons in the gas limits the spatial resolution. This is the case for paper chromatography of radioactive sources, or for thermal neutron conversion in gadolinium foils. While it is very difficult to imagine wire chambers of 1 mm thickness, it seems much easier to have a preamplifying space where the sensitive layer is very thin, thus reducing to a negligible value the parallax errors for inclined tracks.

The method used for secondary-electron emission and subsequent acceleration is still far from giving the proper preamplification factor. The different factors which can be optimized do not seem to promise more than one order of magnitude gain in this factor, and full efficiency for minimum ionizing particles is so far not guaranteed. However, here also several promising applications emerge from the preliminary results.

For thermal neutrons, reactions in ^{10}B and ^{6}Li lead, with a very high cross-section, to nuclear fragments. These fragments cannot emerge from layers that are thick enough to obtain full efficiency. However, if low-density surfaces can be made with these substances, hundreds of secondary electrons are produced by the heavily ionizing fragments[5] and the efficiency can then reach 100%. The absence of parallax, which is due to the small thickness of the layers, the low sensitivity to γ-rays, and the high accuracy of the MWPC, makes this approach to slow neutron detection extremely attractive when compared to all other existing techniques.

Another possible application worth mentioning is connected with the detection of the electrons in electron microscopes. Several groups have considered using drift chambers or multiwire chambers for this purpose. A secondary-emission layer is attractive for two reasons: it is very thin and the accuracy does not depend on the angle; and it has to be in vacuum on the emission side, which makes it mechanically much more easy for electron microscopes since the impinging electrons have also to travel in vacuum.

We are thus faced with a task that is of considerable interest, and which will require much development if we are to exploit all the potential promises of these first results.

REFERENCES

1) M. Breidenbach, F. Sauli and R. Tirler, Nuclear Instrum. Methods $\underline{108}$, 23 (1973).

2) B. Sadoulet and B. Makowski, Space charge effects in multiwire proportional counters, CERN DPH II/PHYS 73-3 (1973).

3) A. Breskin, G. Charpak, F. Sauli, M. Atkinson and G. Schultz, Nuclear Instrum. Methods $\underline{124}$, 189 (1975).

4) See, for example, L.B. Loeb, Basic processes of gaseous electronics (University of California Press, Berkeley, 1961), p. 22 and p. 400.
Also, J.L. Pack and A.V. Phelps, Phys. Rev. $\underline{121}$, 798 (1961).

5) M.P. Lorikian, Nuclear Instrum. Methods $\underline{122}$, 377 (1974).

6) J.C. Faivre, H. Fanet, A. Garin, J.P. Robert, M. Rouger and J. Saudinos, IEEE Trans. Nuclear Sci. $\underline{NS-24}$, 299 (1977).

7) J.C. Faivre, H. Fanet, A. Garin, J.P. Robert, M. Rouger and J. Saudinos, Compte-rendu d'activité, Note CEA-N-2026, p. 262 (1978).

8) J. Fischer and S. Shibata, Proc. Internat. Symp. on Nuclear Electronics, Versailles, 1968 (Documentation française, Paris, 1969), Vol. 3, p. 2-1.

9) J. Fischer and S. Shibata, Nuclear Instrum. Methods $\underline{101}$, 401 (1972).

10) V. Böhmer, Nuclear Instrum. Methods $\underline{107}$, 157 (1973).

11) G. Charpak, G. Petersen, A. Policarpo and F. Sauli, Nuclear Instrum. Methods $\underline{148}$, 471 (1978).

12) G. Charpak, G. Petersen, A. Policarpo and F. Sauli, IEEE Trans. Nuclear Sci. $\underline{NS-25}$, 122 (1978).

13) J. Fischer, H. Okuno and A.H. Walenta, Spatial distribution of the avalanches in proportional counters, submitted to Nuclear Instrum. Methods (August 1977). Also BNL 23163.

14) T. Aoyama and T. Watanabe, Nuclear Instrum. Methods $\underline{150}$, 203 (1978).

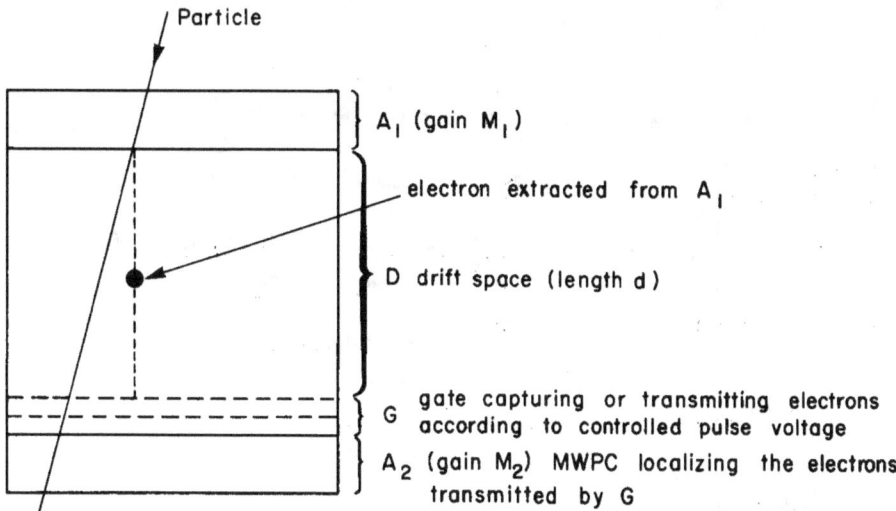

Fig. 1 Principle of gated multistep detector. The amplification M_2 of the MWPC A_2 is decreased by a factor M_1 with respect to normal gain $M = M_1 \cdot M_2$. The electrons liberated in the gap A_1 are preamplified by a factor M_1 and then transferred to A_2, after drifting in the space D for a time T_D, under the control of the gate G. The density per unit length of the electrons produced in A_1 is much higher than the density of ionization produced in the gas of D or A_2; conditions are set such that A_2 detects only the preamplified electron bunch. The electronic gate G opens only for preselected events; if their number is much smaller than the over-all particle flux, the space-charge reduction is roughly equal to M_1.

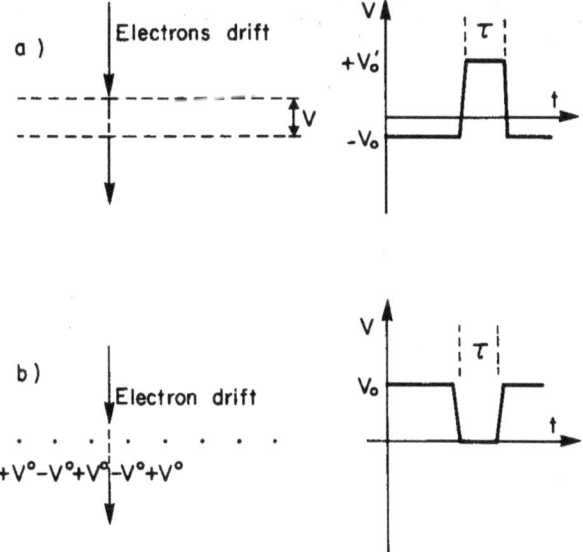

Fig. 2 Two possible solutions for the electric gating function G. In Fig. 2a two fine wire meshes are normally polarized, so as to stop all drifting electrons. Inversion of the applied potential for a time τ allows the electrons to cross the gap during the gating time. In Fig. 2b, instead, the gating function is obtained by a single grid of parallel wires at alternate potentials, transparency is restored by a voltage pulse that brings all wires to the same potential. Both methods have been tested in a prototype chamber with encouraging results.

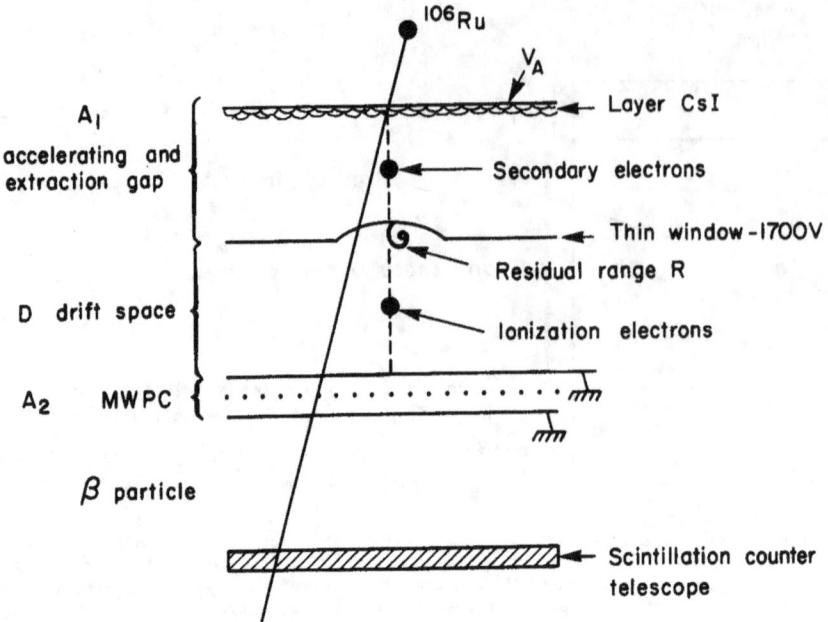

Fig. 3 Preamplification by secondary emission and acceleration of electrons. A secondary emission layer of CsI produces electrons in vacuum (10^{-3} Torr) when traversed by charged particles. A voltage up to 30 kV can be applied to the accelerating gap; the emitted electrons therefore receive enough energy to penetrate in the drift space, through a window of 2.5 μm aluminized mylar. The density of ionization of the electrons with the residual range R is much higher than the density along the primary trajectory. CsI layer of 150 μm, 1% density, on 40 μm of Al. Vacuum gap: 33 mm. Total drift length: 34 mm.

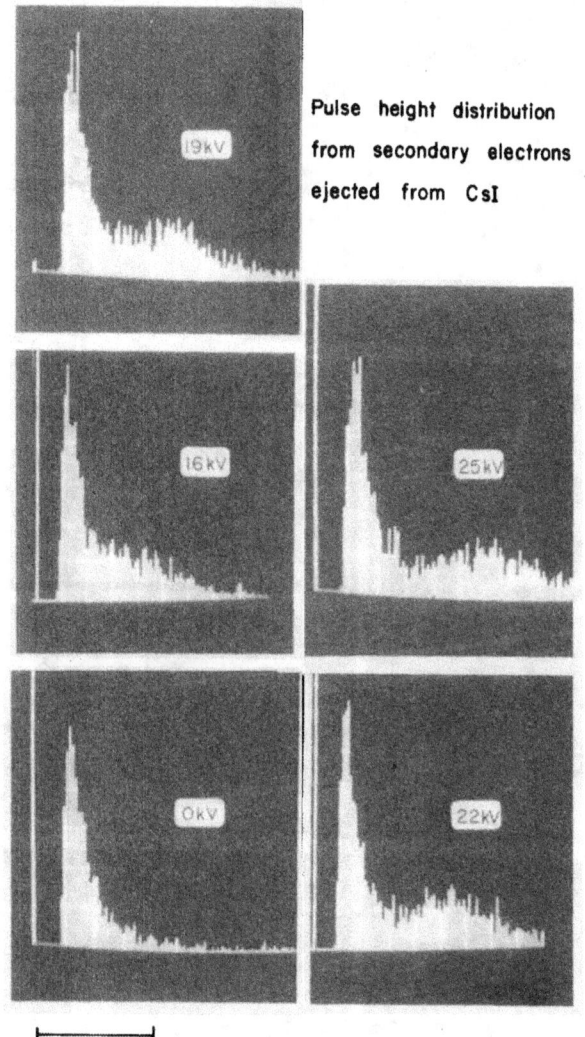

Pulse height distribution from secondary electrons ejected from CsI

6 keV

Fig. 4 Pulse-height spectra of the delayed pulse observed in the chamber of Fig. 3, at variable accelerating potentials, and for secondary electrons ejected by β particles. The peak at 25 keV corresponds to an energy loss of 10 keV compatible with the residual energy of a single accelerated electron.

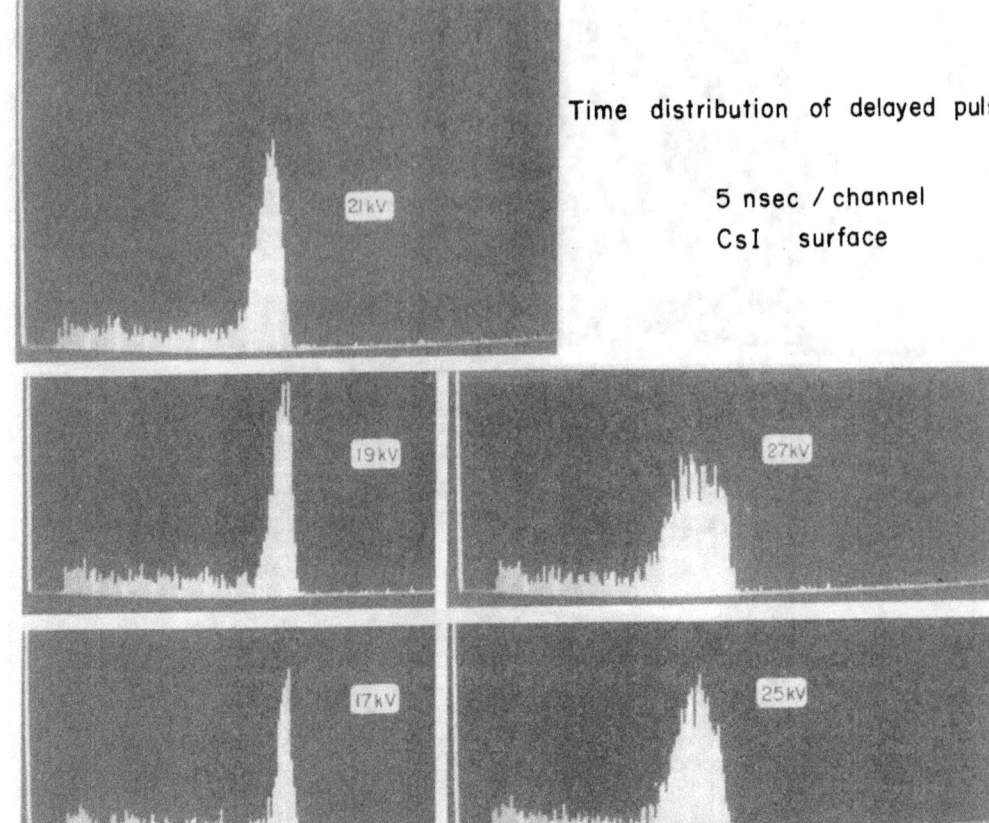

Time distribution of delayed pulses

5 nsec / channel

CsI surface

100 nsec

Fig. 5 Time delay of the secondary electrons measured in the chamber shown in Fig. 3. With the increase of the acceleration voltage V_A the secondary electrons penetrate more deeply into the drift space and are detected at shorter times. The horizontal scale corresponds to 5 nsec per division, with arbitrary origin; the background of events outside the peaks is generated by accidental triggers on the primary ionization electrons.

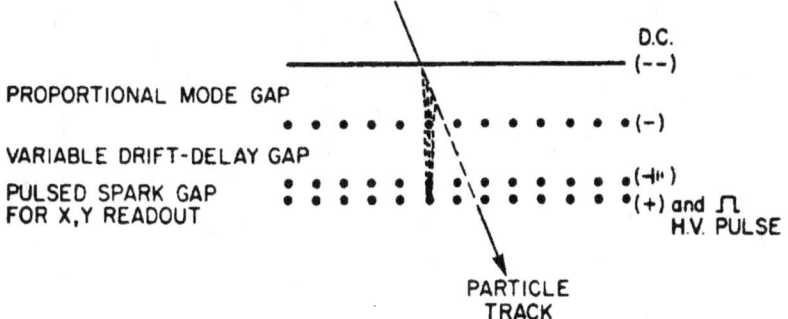

Fig. 6 The hybrid proportional and spark chamber. This figure, taken from Ref. 9, illustrates the supposed mechanism. Primary electrons give rise to an avalanche around the wires of the first plane; the electrons from the avalanche are transferred to the spark chamber via a drift space.

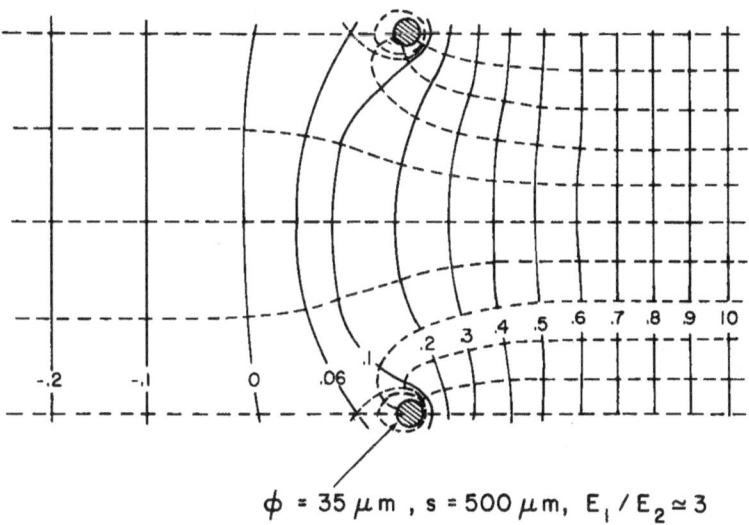

$\phi = 35 \, \mu m$, $s = 500 \, \mu m$, $E_1 / E_2 \simeq 3$

Fig. 7 Field lines and equipotentials around the preamplifying wires of a structure such as the one shown in Fig. 6. A set of parallel 35 μm diameter wires, 500 μm apart, separate a high-field region (E_1) from a transfer region (E_2). The ratio of fields in the two sections is a factor of three. As described in the text, the structure acts simply as a parallel-grid proportional counter with leaking-out field lines.

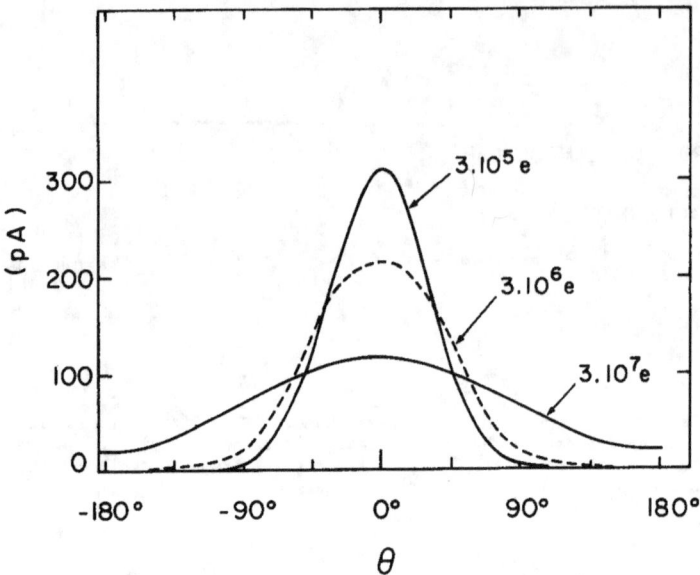

Fig. 8 Angular distribution of the ions around an anode wire. The angular distribution of the ion densities following an avalanche started by localized electrons in a cylindrical counter with wire thickness = 20 μm, cathode diameter = 1 cm, and gas filling: argon (94%), CO_2 (6%), bubbling through ethyl alcohol at 0°C. The curves represent the distribution for three values of the collected charge. Initial electron charge about 300 e.

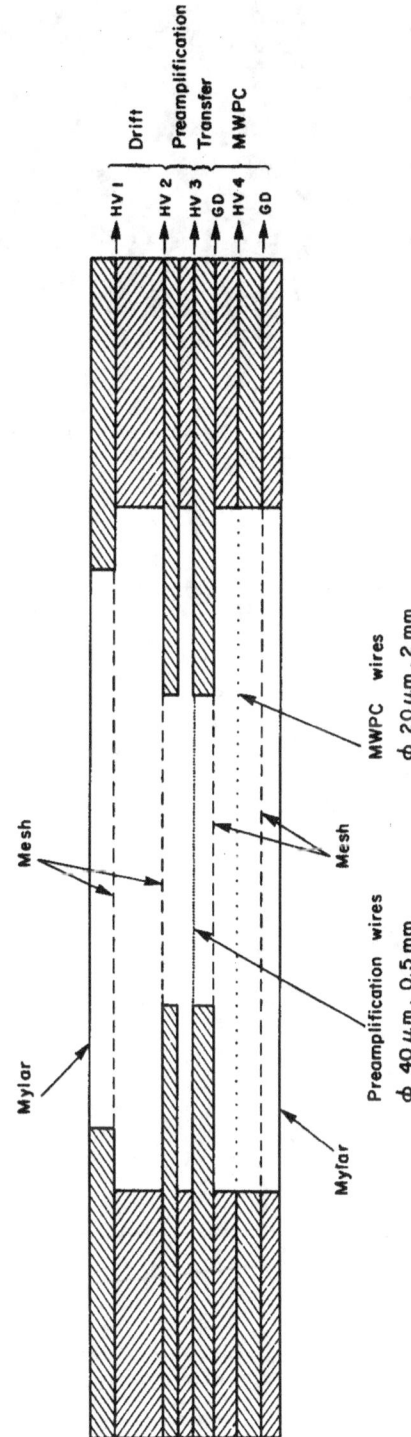

Fig. 9 Schematic cross-section of a test multi-step proportional chamber, using the hybrid chamber concept to implement the preamplification function. Electrons are multiplied in the preamplification gap, and some of them are transferred to the MWPC. The addition of a drift space in front of the preamplification gap greatly improves the chamber energy resolution for neutral radiation, although it appears that it is not essential for the efficient detection of fast charged particles.

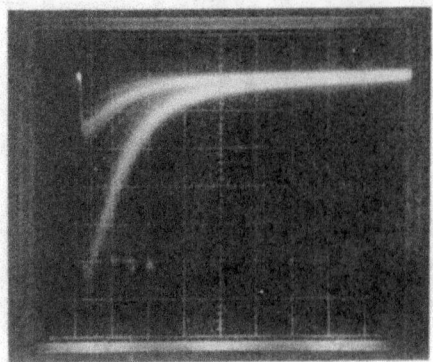 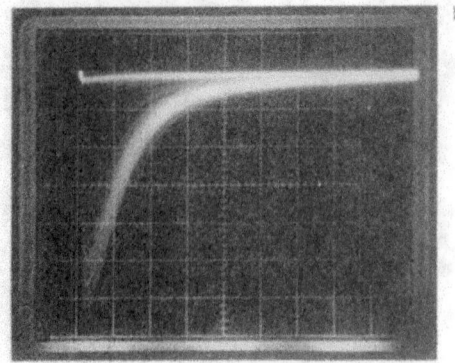

Fig. 10 Separation of direct and preamplified pulses induced by 5.9 keV X-rays, detected directly in the MWPC and after preamplification by a factor close to 25. Vertical scale: 20 mV/div. (a) and 200 mV/div. (b); horizontal scale 1 μsec/div. A charge amplifier with a sensitivity of 250 mV/pC and a 2 μsec time constant has been connected to the chamber.

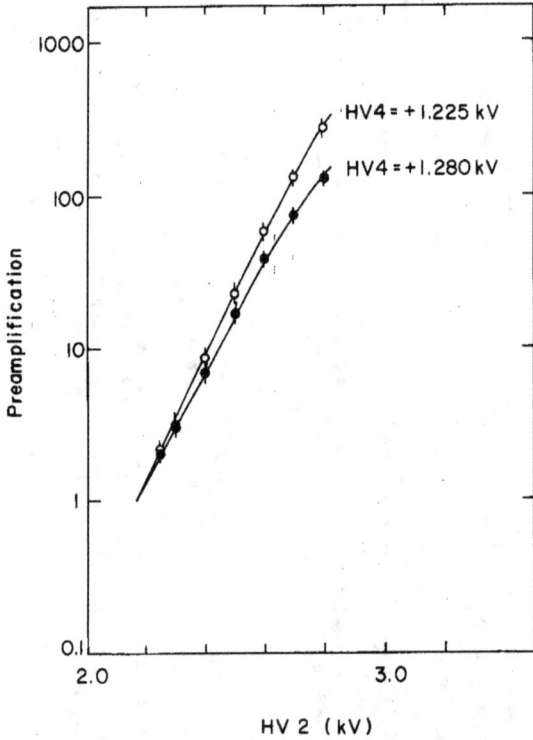

Fig. 11 Measured preamplification factor for ^{55}Fe X-rays as a function of the preamplification grid voltage for two values of the MWPC potential. The error bars around the measured points represent the FWHM of the pulse-height distribution of the detected 5.9 keV line. The chamber shown in Fig. 9 was operated in a gas mixture obtained by bubbling pure argon through ethyl alcohol at $0°C$.

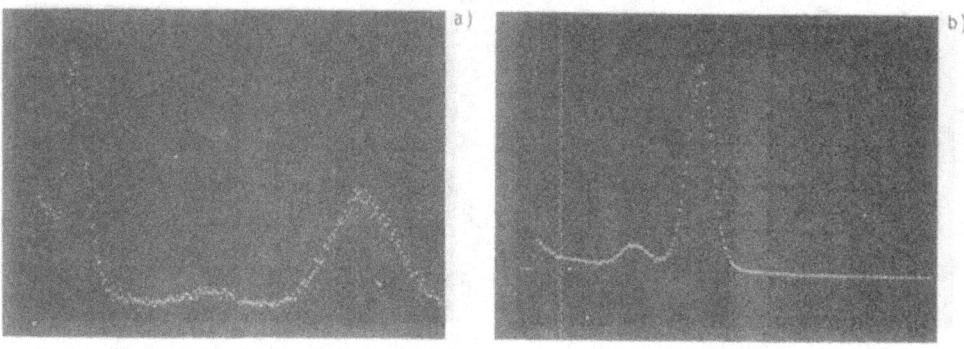

Fig. 12 Pulse-height spectra from ^{55}Fe X-rays, measured just above threshold for preamplification (a), and after preamplification by a factor of about 40 (b). The horizontal scale is arbitrary and is different in the two pictures.

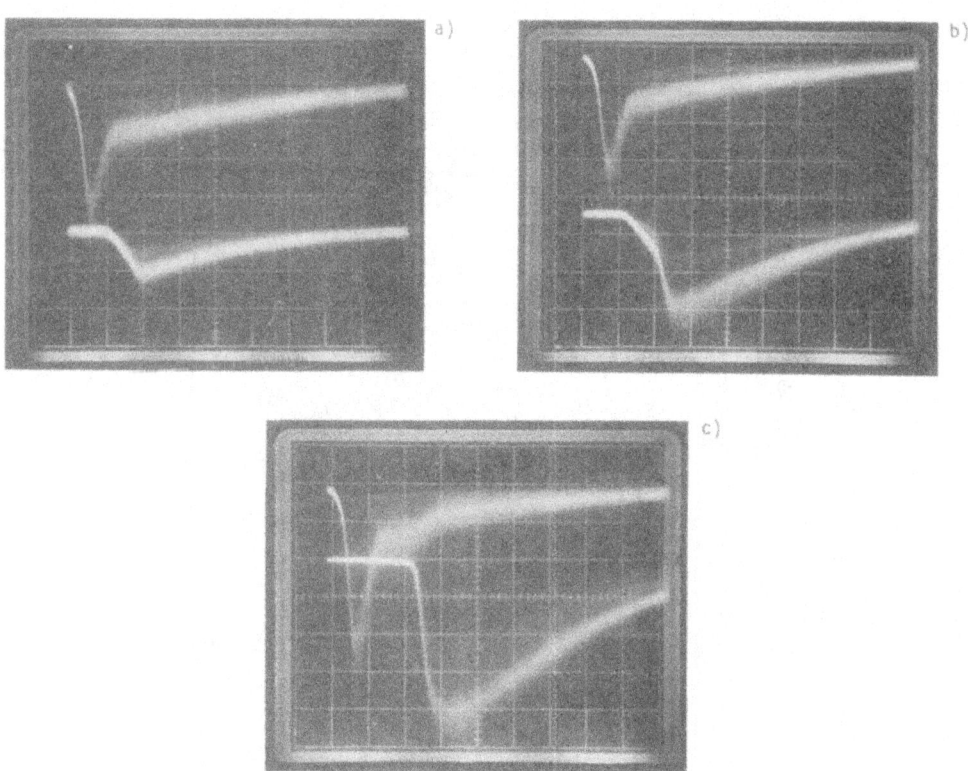

Fig. 13 At high preamplification factors (about 100) a signal can be detected directly on the preamplification grid, as shown in the upper traces in all pictures. The lower traces show the charge detected in coincidence on the MWPC operated in a pure collection mode (a), at the beginning of proportional multiplication (b), and in full operation (c). The vertical sensitivities are 4 mV/div. in all pictures, except for the lower trace in (c) where it has been decreased to 100 mV/div.; the horizontal scale is 200 nsec/div. The charge amplifier described in the text has been used to obtain the pictures.

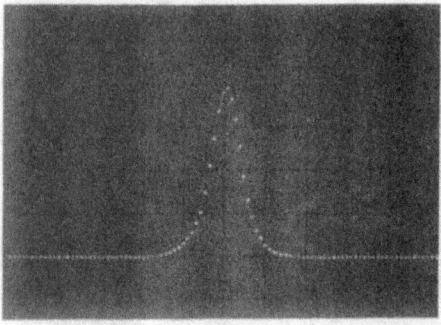 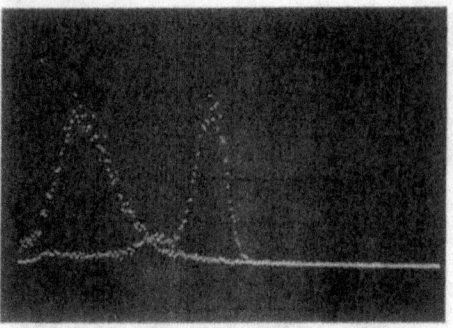

Fig. 14 Time jitter measured at fixed discrimination threshold ($\sim$ 0.1 pC on the
chamber) between the fast preamplified pulse
(upper trace in Figs. 13) and the delayed
pulse detected in the proportional chamber,
for ^{55}Fe X-rays. The horizontal scale is
10 nsec/div.

Fig. 15 Comparison of pulse-height spectra
on the preamplified charge at a preamplification factor of about 100, for ^{55}Fe X-rays
and minimum ionizing electrons.

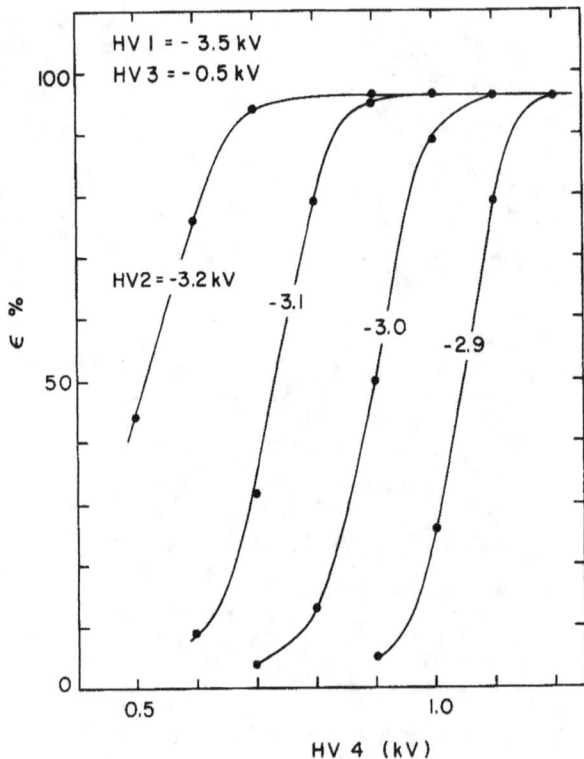

Fig. 16 Detection efficiency for fast electrons, at a fixed discrimination threshold on
the proportional chamber ($\sim$ 0.1 pC at the wires), as a function of anodic potential, and
for several values of the preamplification grid voltage. When using a radioactive β emitter for this kind of measurement, it is usually impossible to reach 100% efficiency.

262

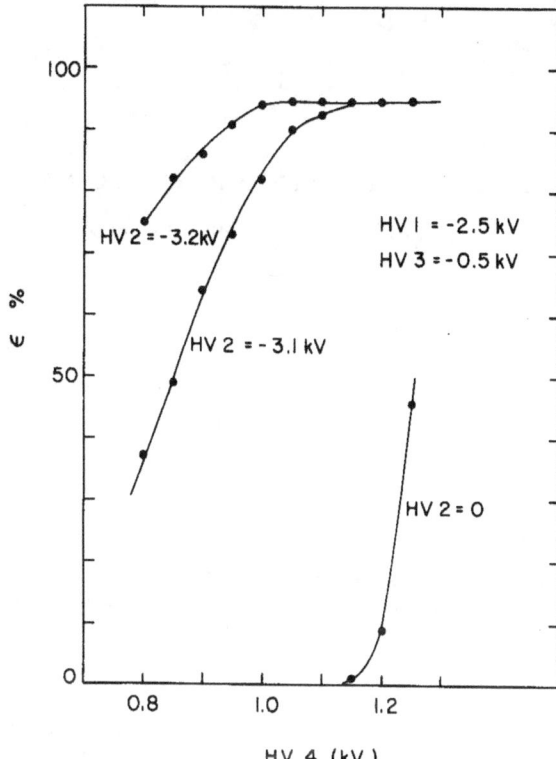

Fig. 17 Efficiency for the detection of fast electrons, in the conditions of Fig. 16 except for an inversion of electric field in the upper drift region (see Fig. 9). It appears that for charged particles full efficiency can be obtained on the preamplified tracks even without the addition of the first drift section. The efficiency of detection of the direct (non-preamplified) section of the track is also shown (curve $HV_2 = 0$).

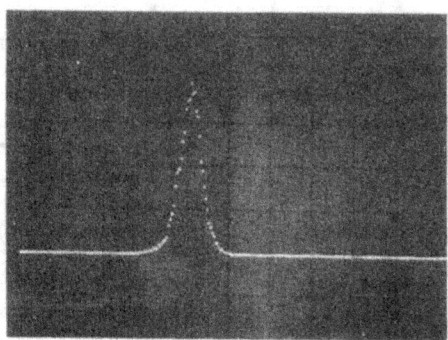

Fig. 18 Time resolution for charged-particle detection. Time jitter distribution between detection of ^{106}Ru β-particles by a scintillation telescope and by the MWPC after preamplification. Mixture of argon-CO_2 (97%, 3%) bubbling through alcohol at 0°C. FWHM = 20 nsec. Horizontal scale 50 nsec/div.

Nuclear Instruments and Methods 178 (1980) 11–25
© North-Holland Publishing Company

HIGH FLUX OPERATION OF THE GATED MULTISTEP AVALANCHE CHAMBER

A. BRESKIN *, G. CHARPAK, S. MAJEWSKI **, G. MELCHART, A. PEISERT **, F. SAULI

CERN, Geneva, Switzerland

F. MATHY

CEN, Saclay, France

and

G. PETERSEN

Niels Bohr Institute, Copenhagen, Denmark

Received 28 May 1980

A multiple gaseous detector based on the preamplification and transfer mechanism has been tested in a high radiation flux environment. The device, called the gated multistep avalanche chamber, can detect and localize with good space and time accuracies selected tracks in minimum ionizing particle fluxes exceeding 10^5 s^{-1} mm^{-2}. The general operation properties, as well as some of the problems encountered are discussed.

1. Introduction

The original motivation for the development of the multistep avalanche chamber was an attempt to overcome the limited flux capability of the multiwire proportional chamber (MWPC), caused by the space-charge effects of positive ions [1]. The idea was to realize a composite structure in which the overall gas proportional gain M, needed for detection of minimum ionizing particles, could be obtained as a product of two independent amplification elements of effective gain M_1 and M_2, respectively. Under the control of an electronic shutter or gate, however, only selected events could receive the full gain $M_1 M_2 = M$. The number of positive ions produced per unit time in the second amplifying element at a particle flux R is then:

$$nRM_2(1 + fM_1),\tag{1}$$

where n is the number of primary charges liberated in the drift space of the two amplifying elements by the ionizing particle (assumed for simplicity to be the

same in the two detection elements) and f the ratio between selected and total particle flux. Obviously, for $f \ll 1$, the reduction of positive charges in the second element equals M_1, as compared with a normal counter having gain M. For an overall gain of around 10^5, suitable for minimum ionizing particle detection, and assuming $M_1 \simeq M_2$, a positive space-charge reduction by a factor of 100 is realized at a triggering rate f of 1%. Following Hendricks [2], the gain reduction in a proportional counter at a flux R depends exponentially on the total number of ions produced; in the quoted example, we should therefore, in principle, increase the flux capability of a gated chamber by two orders of magnitude. At high rates the device can be used, of course, only for track sampling, either random or under a triggering condition imposed by accessory equipment.

The practical realization of the proposed structure depends entirely on the possibility of implementing a region of gaseous amplification with the transfer of a sizeable fraction of the electron avalanches to the following sections. Much work has been devoted to the understanding of the preamplification and transfer mechanism [3–5], and has led to a new family of gaseous detectors with peculiar properties, finding applications already in the ring imaging of Cherenkov photons [6–8] and radio-chromatography imaging

* On leave from: The Weizmann Institute of Sciences, Rehovoth, Israel.
** On leave from Institute of Experimental Physics, Warsaw University, Poland.

11

[9]. The present paper presents the preliminary results obtained in the operation of a gated multistep avalanche chamber (GMSC), in the detection of minimum ionizing particles both at low and high fluxes.

2. Principles of construction and operation

We have built a GMSC around an existing 100 × 100 mm² active area mechanical framework, although as we will see later, only a small fraction of the chamber (about 20 × 10 mm²) was activated for the high rate beam measurements. The structure of the GMSC used is shown in fig. 1. The electrode identified by HV1 consists of a 10 μm thick aluminium foil, while electrodes HV2, HV3 and GND are implemented with stainless steel crossed-wire grids, of 50 μm wire diameter and 500 μm pitch. The gate electrode and the readout electrode (HV6) are two meshes of copper–beryllium 50 μm diameter wires, 500 μm apart and perpendicular to each other. The wires in the gate electrode are connected alternately to two independent high-voltage supplies, HV4 and HV5, and a pulsed operation of the gate can be obtained through a pair of capacitors as shown. The wire grid between the gate and the readout electrode is grounded all

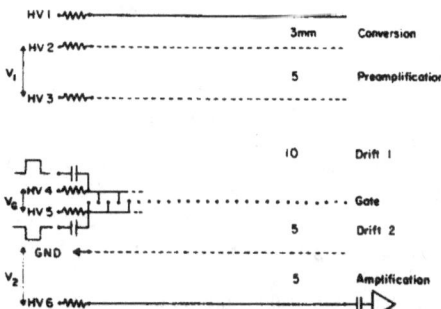

Fig. 1. Schematics of the gated multistep avalanche chamber (GMSC). Electrodes identified by the potentials HV2, HV3, and GND are crossed wire meshes, having 50 μm wire diameter and 500 μm pitch. Electrodes HV4–HV5 and HV6 are planes of parallel wires, with diameter and pitch as before, perpendicular to each other; the particular layout of the gate electrode allows pairs of adjacent wires to be dc or pulse offset. Electrode HV1 is a 10 μm thick aluminium foil. By a proper choice of potentials, charges liberated in the conversion region are preamplified and transferred to drift region 1; if the gate electrode is equipotential at the time the electron cloud approaches, the charge is transferred to the second element of amplification, equipped with readout electronics.

around the chamber edges on a large copperized plate, and the quality of this grounding appears to be the major element in limiting the capacitive pick-up of the gating pulse (several hundred volts) by the amplifier on the readout electrode. For the same purpose, a symmetric bipolar gate pulsing was preferred to a simpler unipolar pulsing; indeed, a capacitive pick-up almost two orders of magnitude smaller than in a previously described chamber [4] was obtained.

The structure shown contains as gaseous amplifying elements two parallel-plate avalanche chambers, where high electric fields are required for electron multiplication (around 7 kV cm⁻¹). Spontaneous breakdown at the gap edges has to be avoided in order to attain good working conditions; we have found that in most cases a single thin insulating strip (mylar or Kapton, 50–100 μm thick and a few mm wide) extending into the gap at the edges between the anode and cathode is sufficient to prevent edge breakdown. The good quality of the insulating frame surface, machined fibre glass in the present case, appears to be essential for correct behaviour. The best way to avoid this kind of breakdown is however a physical increase in the gap thickness at the edges; methods to implement such a structure have been developed and will be described in a more specialized paper [10].

With a proper choice of the grids' potential, to be detailed in the following sections, electrons produced in the conversion space of the GMSC by an ionizing radiation can be preamplified and transferred into the first drift space. An identical value for the potentials of the alternate sets of wires in the gate, either dc (HV4 = HV5) or pulsed as shown in fig. 1 then allows the transfer of electrons to the drift region 2 and to the second step of amplification for selected events. Simple field inversion in the conversion region (|HV1| < |HV2|) also allows the first section to operate as a parallel-plate counter; despite the obvious disadvantage of such a configuration, i.e. a gain dependent on the position of the primary pair production within the gap, we have found that better time resolutions can thus be obtained, as will be described later in the "conversion off" condition. Localization was performed by analysing the pulse-height distribution on the last set of wires, after suitable amplification and analogue-to-digital conversion. Details of the amplifiers and analogue-to-digital converters (ADCs) will not be given here, as they were identical to the MWPC cathode readout system already described [11,12] except for the polarity of the detected signals; the typical charge amplifier's sensitivity was about 1 V pC^{-1}, with an output noise around 5 mV rms.

A small on-line computer was used for the data storage and retrieval during the beam tests. The gas mixture for all the described measurements was argon—acetone in a 98—2 volume concentration.

For the beam tests, the GMSC was installed between two scintillation counters providing, in coincidence, the time reference for the generation of the gating pulse, at a rate conveniently reduced by a preset electronics paralysis. Also, for a survey of the localization properties of the device, we have installed in the line a conventional MWPC having 2.54 mm wire spacing and used in the shift mode, i.e. recording the time of detection as referred to the trigger in a time-to-digital converter. Owing to the small beam spot size, about 100 mm^2, only two of the MWPC wires were illuminated, and this resulted in a very severe rate limitation due to space-charge build-up: indeed, the localization accuracy of the GMSC could only be measured at low particle fluxes (below 10^3 s^{-1} mm^{-2}, see section 6).

3. The gating function

Figure 2 shows the absolute gain as a function of electric field measured in a 5 mm thick parallel-plate structure, identical to the two amplification elements of the GMSC. As discussed in detail in ref. [4], pre-amplification factors in excess of 10^4 can be obtained at fields around 7 kV/cm, with a subsequent transfer of part of the electron avalanche in the lower field region; the transfer efficiency is, in a first approximation, equal to the fields' ratio. On the other hand, the gating function requires the generation of an electric field between the alternate wires which is sufficiently large to defeat the existing one, i.e. a difference of potential increasing with the value of the transfer field, the obvious limit being spontaneous breakdown betweeen adjacent wires. We have found good operational conditions setting the transfer field in both drift regions at around 1.3 kV/cm; as shown in fig. 3, a full blocking of electrons transfer through the gate is then obtained at a potential difference between alternate wires of $V_G = |HV4 - HV5| \simeq$ 180 V. The average gate potential $\frac{1}{2}$(HV4 + HV5) can instead be varied over a large range without modifying the overall gate transparency (fig. 4); indeed, losses of transmission from one region appear to be compensated by gains in the transfer to the next.

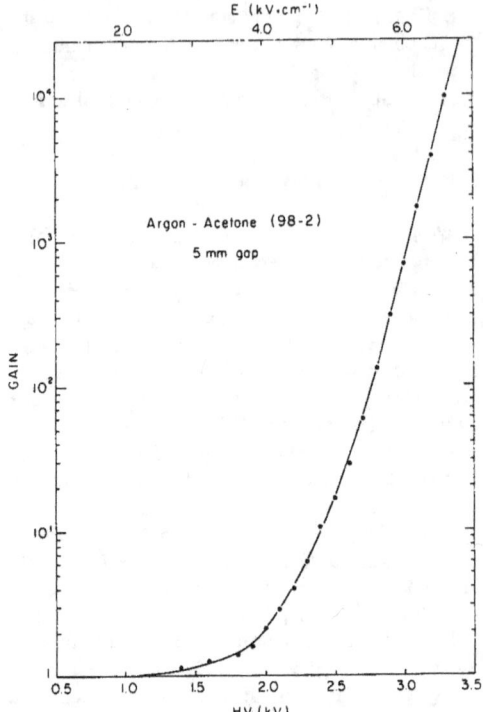

Fig. 2. The absolute gain in a 5 mm thick parallel-plate chamber, operated at atmospheric pressure in a 98—2 mixture of argon—acetone, as a function of the applied field.

Detailed measurements of the grid transparency of the electrons were given in ref. [4]. In all the measurements to be described we have adopted the following setting of the potentials: HV3 = −2 kV, $\frac{1}{2}$(HV4 + HV5) = −0.7 kV. The condition HV4 = HV5 = −0.7 kV corresponds then to a dc open gate, while full dc gate blocking is obtained at |HV4 − HV5| = 200 V.

In the dc blocked condition, the gate transparency can be restored by the application of a symmetric bipolar pulse with a total amplitude approaching the existing dc offset. Fig. 5 shows the simple avalanche transistor circuit used to generate the gating pulses *; cable shaping is used to adjust the length of the gate and two symmetric pulses are obtained through a pair of circuits with identical and inverted transformers. The identity of the shape of the symmetric pulses and that of the loads is, as mentioned, one of the pick-up

* Developed at CERN by J.C. Santiard and M. Kudla.

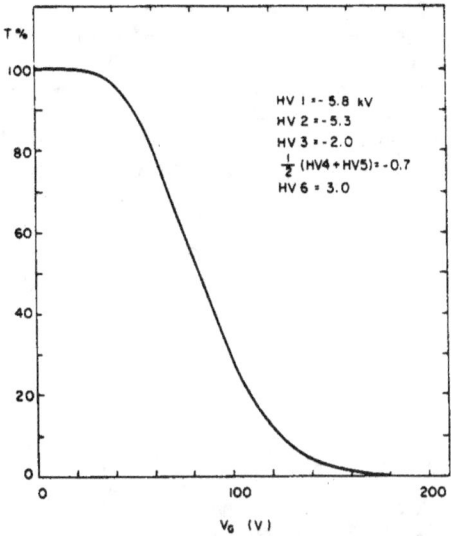

Fig. 3. Transparency for electrons of the gate electrode, as a function of the difference of potential $V_G = (HV4-HV5)$ between adjacent pairs of wires (500 μm pitch). The voltage difference for blocking, about 180 V, depends of course on the existing drift field, 1.4 kV/cm as can be deduced from the values indicated in the inset.

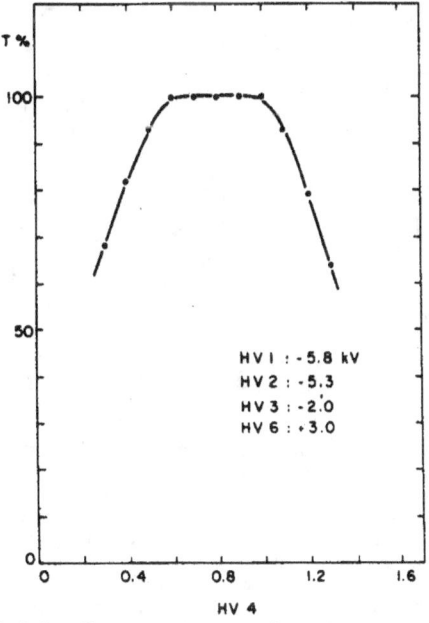

Fig. 4. Overall system transparency, detected at the last electrode, for an equipotential gate as a function of its voltage; a large region of uniformity exists.

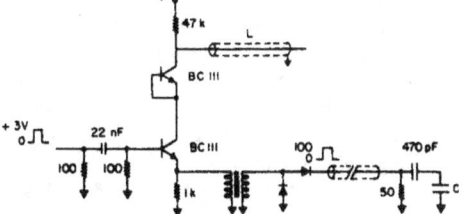

Fig. 5. Two identical avalanche transistor circuits are used to generate the pair of symmetric gating pulses (one only is shown in the figure; in the second the transformer winding and the output diodes are inverted). The pulse duration can be adjusted varying the length of the shaping cable L. The gate electrode is represented by the capacitive load C; the diodes in the output protect the transistors from accidental overvoltage in case of spark breakdown in the chamber.

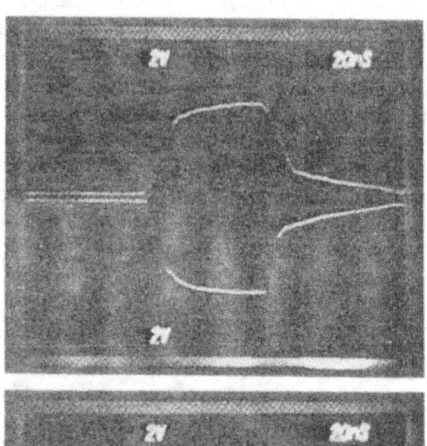

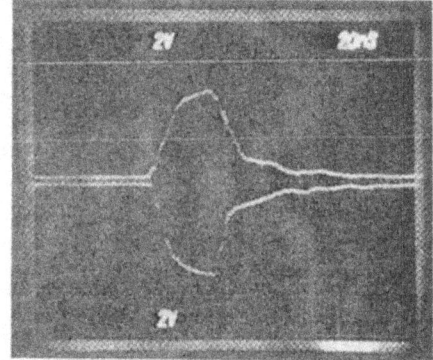

Fig. 6. Shape observed on the chamber of the gating pulses, for 60 and 30 ns hsaping constants, respectively. The vertical axis is attenuated on the scope by a factor of 20.

reducing factors; fig. 6 shows examples of the gating pulses for 60 and 30 ns shaping lengths, respectively. A 20X attenuation probe was used to realize the pictures on the oscilloscope, the actual peak voltages at the chamber being ±100 V as required to counterbalance the dc offset. Rise- and fall-times of the pulses are purposely rather slow to reduce pick-up.

Under pulsed operation, the gate transmission depends on the gating pulse length when this is too short to let the full electron avalanche through. As will be documented in a subsequent paper [13], the longitudinal avalanche extension in the described conditions has a standard deviation of about 300 μm; with a typical value of electron drift-time of about 20 ns mm^{-1}, we expect to start losing transmission for gate lengths below 20–30 ns. This is indeed the case, as shown in fig. 7. The pulsed-gate transparency was measured by irradiating the GMSC with a ^{55}Fe 5.9 keV X-ray source, at sufficiently high preamplification (around 10^4) so as to be able to detect the preamplified pulse directly on electrode HV3; the discriminated signal could then be used to trigger the gate generator at the proper time. The ratio of the observed pulse height on the last electrode HV6 under gating conditions, to the one measured at dc open gate determines the transmission. As apparent from the figure, a 30 ns fwhm gate is sufficient to guarantee full transmission of the drifting electron cloud.

The picture in fig. 8(a) shows the signals detected on the first preamplification electrode (upper trace,

5 mV/division) and on the second amplification gap under pulsed-gate transmission (lower trace, 50 mV/division). As there is not enough screening between the gate and the preamplification electrode, a large pick-up swing is observed there at the time of pulsing; the coupling to the main amplification grid instead is visible only at a much increased scope sensitivity (fig. 8(b), 5 mV/division). Notice also that, owing to the drift-time of the electrons from the gate to the avalanche amplification detection, of around 200 ns, the residual pick-up is well separated in time from the

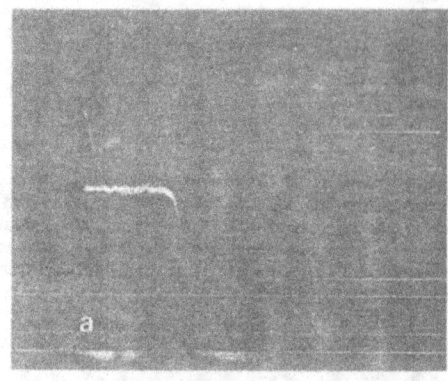
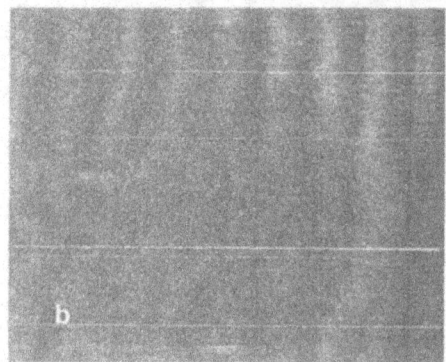

Fig. 8. Pick-up of the gating pulse on the readout electrode. In (a) the upper trace (with 5 mV/division sensitivity) shows the signal detected on the preamplification element for a 5.9 keV X-ray, used to trigger the gate pulse, and the lower trace (500 mV/division) the charge detected on the last electrode after gating and amplification. Since the upper meshes in fig. 1 are not screened from the gate electrode, a large pick-up swing appears at the time of pulsing on the preamplification element. The pick-up on the main amplification grid is however much smaller, as shown in (b) on a more sensitive scale (5 mV/division), and is easily discriminated.

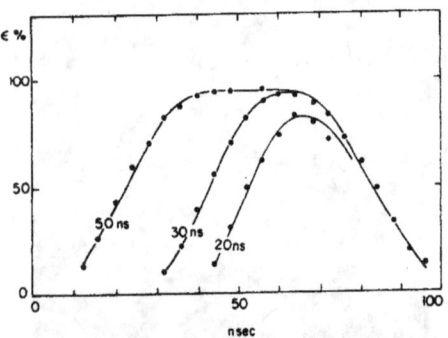

Fig. 7. Gate transmission in the pulsed mode of operation as a function of the gate delay, relative to the migrating electron cloud and for several shaping constants. The measurement was performed by irradiating the GMSC with a ^{55}Fe X-ray source and detecting the preamplified pulse to generate the gate pulse. The transmission efficiency is defined as the ratio of the detected pulse height in the last electrode for a gated and for a dc open operation.

physical signal and constitutes no problem in the detection.

4. Rate-dependent phenomena

Using the beam of a collimated 8 keV X-ray generator we could increase the ionizing radiation flux to values largely in excess of the ones considered at the limit for correct MWPC operation (around 10^4 s^{-1} mm^{-2} [14]). No decrease in the average pulse height was observed in the gated operation, thus confirming the validity of the starting assumptions. However, an unexpected problem appeared, which may set an ultimate limit to the rate capability of the multistep chamber: sudden gas breakdown in the region of the irradiation, intervening at gas gains which are lower the higher the flux. It is suspected that the cause of the rate-dependent breakdown around the source position is the local field increase induced by the space charge of the positive ions at the cathode surface. The practical implication of the observation is a limit on the maximum safe operational voltage, and therefore on the gas gain, for a given flux. Obtaining full detection efficiency depends then on the capability of the wire electronics to discriminate properly between the charge signals at a chamber gain below the critical one.

It is observed that in a multistep device, breakdown always occurs in the section with the higher charge densities, i.e. the last one for elements with similar gain. In fig. 9 the breakdown voltage in the second element, V_2, is shown as a function of the radiation flux (8 keV X-rays detected in the conversion region) both in the dc open gate condition and in pulsed-gate operation. The measurement bars in the second case correspond to two gate opening frequencies, given as a percentage of the total rate [the factor f in expression (1)]. Clearly, in the gated operation at high fluxes the operational voltage in the second amplification element can be higher by more than 500 V than in the dc open condition, corresponding to a gain factor of almost two orders of magnitude (see fig. 2). Notice also that the breakdown voltage depends very little on the gate opening ratio; at low fluxes indeed, where it was possible to approach a 100% ratio (the limit being set by the recovery time, about 100 μs, of the avalanche transistor circuit), the working voltage in the pulsed operation is still about 300 V higher than in the dc open mode. Since in this case most of the converted X-rays

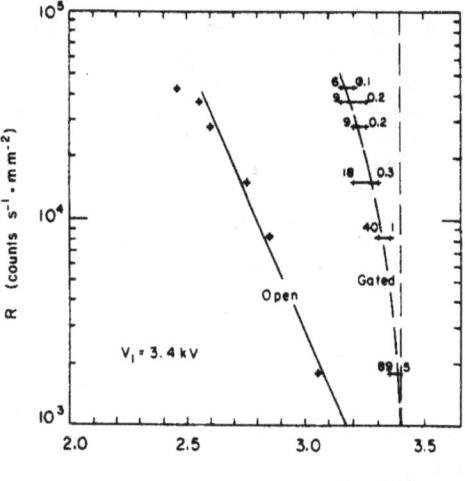

Fig. 9. The spark breakdown voltage in the second amplification element, for a fixed preamplification, as a function of the detected 8 keV X-ray flux. Measurements are shown both for the dc open gate condition, and for the pulsed gate operation; in the last case, the small figures on the two sides of the measurement bars indicate the gate opening frequency (as a percentage of the total flux detected on the preamplification element). The gated operation allows the attainment of larger chamber gains (by almost two orders of magnitude, see fig. 2) and also enable opening frequencies approaching 100%, thus indicating that breakdown is caused by feedback of positive ions.

receive the full amplification, the observation strongly suggests that the mechanism responsible for the breakdown is a secondary extraction mechanism at the first cathode induced by positive ions. In the gated operation the ions generated in the second element of amplification cannot reach the first, since generally they find the gate closed in their backward drift (even if the gate would randomly open at the proper time, its length of around 30 ns only lets through a minor fraction of the slowly migrating ions).

A systematic investigation of this phenomenon is under way, and its results will be published later [13]; in what follows, we will just consider the rate-dependent breakdown as a phenomenological limitation to the working voltage of the device.

5. High flux beam measurements

The GMSC has been installed and operated at CERN in a high-intensity minimum ionizing charged particle beam, having an average surface at the detector position of about 10 mm². Because of intrinsic limitations in the rate capability of the pair of scintillation counters used for beam monitoring, a maximum intensity of about 10⁷ counts/2 s spill was used, providing an average flux in the detector of 2×10^5 s⁻¹ mm⁻² (assuming a uniform illumination over a surface of 25 mm²; as will be shown in the following, see for example fig. 24 in section 6, this is a rather optimistic assumption). The beam trigger itself was provided by the coincidence of the two scintillation counters, one of which had a surface (16 mm²) barely exceeding the beam spot size. The gating frequency was reduced to the maximum repetition rate allowed by the avalanche circuit, about 10⁴ s⁻¹, by a 100 μs paralysis in the main coincidence. Owing to the small beam size, and in order to reduce the pick-up and noise problems, only a 20 mm wide strip

in the gate electrode was pulsed open, while the surface of the gate (100 × 100 mm²) was dc polarized. All efficiency measurements of the detectors were then realized electronically using the logic sum of the discriminated output of twenty adjacent wires in the last amplification element; the average discrimination threshold on each wire had a value of 0.05 pC, corresponding to a 50 mV discrimination level at the output of the preamplifier having 1 V pC⁻¹ sensitivity.

As already mentioned, the statistical fluctuations in the production of primary ion pairs tend to increase the avalanche size in the case of detection of charged particles, as compared with X-ray detection. This is apparent from fig. 10, showing the measured efficiency plateau in the described conditions and at moderate beam flux, both for a dc open gate and for the operation at several pulsed gate widths. The efficiency is given for constant preamplification and

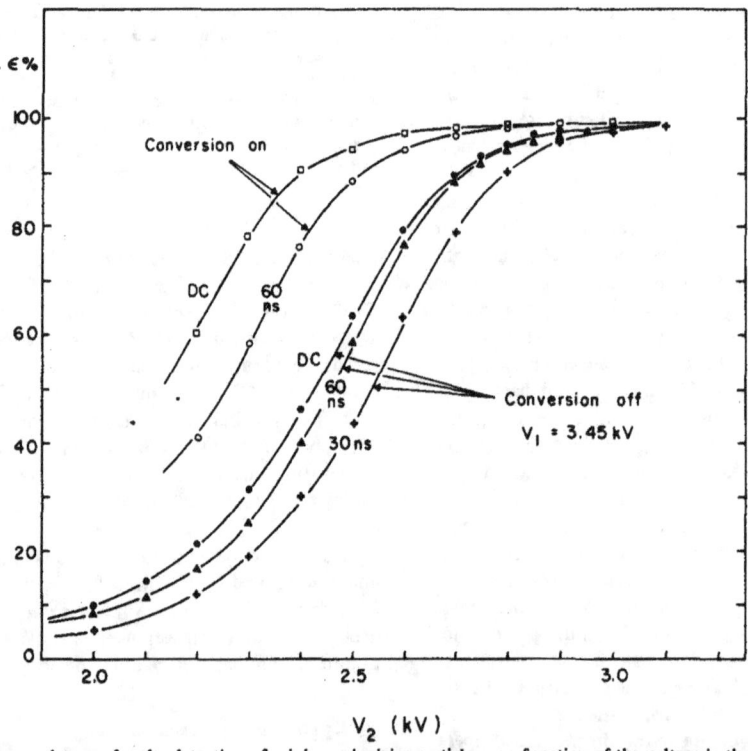

Fig. 10. Efficiency plateaux for the detection of minimum ionizing particles, as a function of the voltage in the second amplification element, for several operating conditions. The conversion on mode corresponds to detection of the ionization produced in the conversion space, while in the conversion off only the charge produced within the preamplification element is multiplied. The two modes can be simply selected by a proper adjustment of HV1 (see fig. 1).

transfer voltages as a function of the second amplification voltage; setting $|HV1| > |HV2|$ (see fig. 1) includes in the detection the ionization produced in the upper region (conversion on), while $|HV1| < |HV2|$ defines the conversion off condition. The last mode of operation, as we will see in more detail, results in better time resolution and reduced memory for the detector, but has the obvious disadvantage of requiring higher chamber gains to detect the few electrons produced close to the upper surface of the pre-amplification region and receiving therefore full amplification. In our operating conditions, indeed, the average distance between primary ion pairs, 300 μm, is comparable with the mean distance of multiplication in the gap, about 500 μm for a gain of 10^4. Charges created within the preamplification volume receive therefore an exponentially decreasing multiplication for an increasing distance from the upper cathode. As a result the efficiency plateau is reached with difficulty, especially for short gate lengths; the maximum measured efficiency in the conversion off condition is indeed 98%, while it approaches 99.5% in the conversion on operation. The difference in the detected charge profile for the two conditions is apparent from the pulse-height spectra shown in figs. 11(a) and (b) (conversion on and off, respectively). Notice that the second spectrum was measured at a much higher chamber gain than the first; the bad detection statistics, corresponding essentially to the amplification of single electrons produced at variable depths in the gap, is apparent.

Since the avalanche propagation time slightly decreases with an increasing chamber gain, the pulsed-gate timing has to be adjusted at each value of gain to reach full efficiency. Typical gate delay curves in the conversion off case are shown in fig. 12, for several values of V_2; the width of the coincidence reflects the width of the avalanche at a given discrimination level, and determines the memory of the detector, i.e. the chance to detect uncorrelated tracks at high fluxes. At the highest value of V_2, where full efficiency is reached, the memory appears to be about 100 ns fwhm, which is certainly not very appealing for high flux operation (although still smaller than for conventional MWPCs). Possible ways of reducing the memory will be discussed in the last section. Notice that the resolution time of the chamber, i.e. the time jitter of the detected tracks, is much smaller than the memory, and a good recording of the time of detection should allow the improvement of background rejection. The best resolution times in the GMSC have

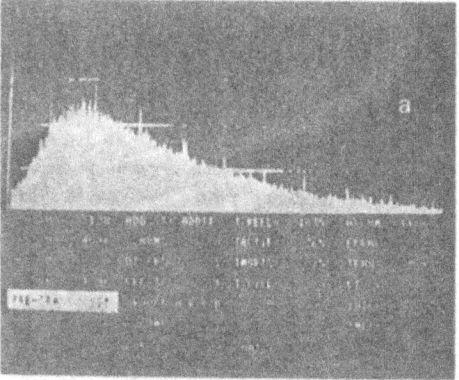

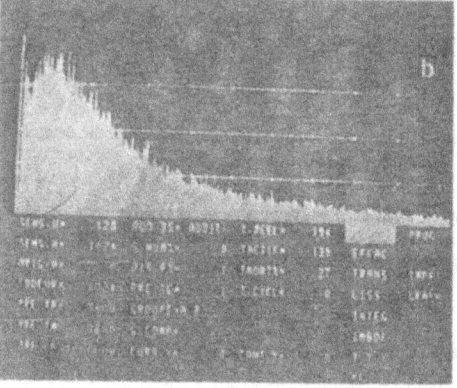

Fig. 11. Pulse-height spectra for minimum ionizing particles, in the fully gated operation of the GMSC, in the conversion on (a) and off (b) conditions. The chamber gain in the second case has been increased by about a factor of 10.

been measured operating in the conversion off condition with a 30 ns gate, and discriminating the detected signal after a double differentiation (a simple method of peak-time measurement). A typical time spectrum obtained in these conditions is shown in fig. 13 and has a 10 ns fwhm; the presence of tails is again a manifestation of the large non-Gaussian fluctuations due both to the primary pair statistics and to the avalanche size.

The efficiency plateaux shown in fig. 10 were obtained at a moderate beam flux (below 10^3 s^{-1} mm^{-2}). As already discussed in section 4, spontaneous breakdown on the beam spot is observed at high gains and particle rates, thus limiting the maximum operational voltage of the detector. The effect is particularly large in the dc open gate condition, as

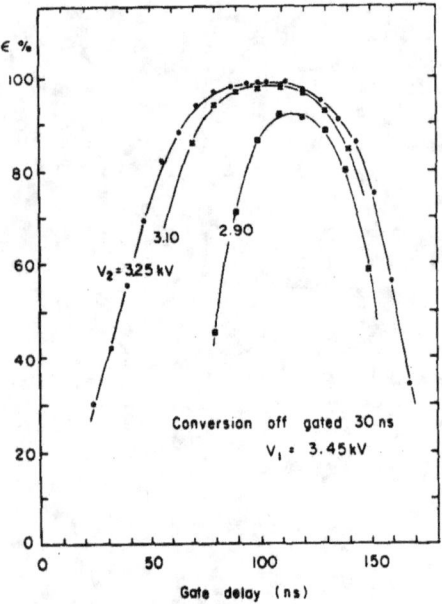

Fig. 12. Efficiency versus gate delay in the detection of minimum ionizing particles, in the conversion off condition and for several amplification voltages.

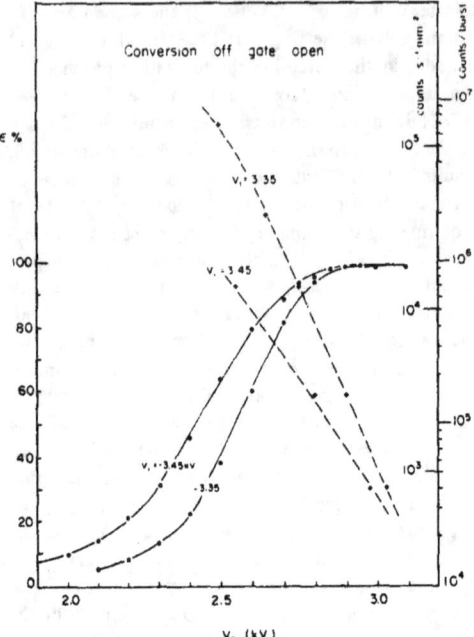

Fig. 14. Efficiency for detection of minimum ionizing particles, measured at small rates and for two values of preamplification (full lines and left vertical scale), and breakdown voltage as a function of the particle rate (broken lines and right vertical scales). Operating conditions for the GMSC: conversion off and gate dc open.

already discussed. In fig. 14 typical efficiency curves (measured at very low rate) and flux-dependent breakdown voltages are shown in the conversion off, open gate operation; for rates above 10^3 s^{-1} mm^{-2} full efficiency cannot be reached, owing to break-

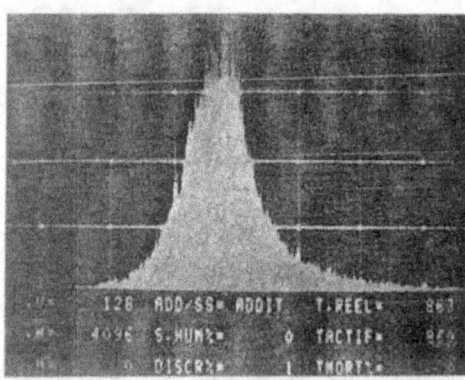

Fig. 13. Time resolution of the GMSC for fast particles, measured with a double differentiation circuit on the signal detected on the last electrode. The horizontal scale is 10 ns/division.

down. Two sets of points are measured for different preamplification factors.

Much higher gains can be reached before breakdown at a given flux in the fully gated operation of the GMSC. This is shown in figs. 15 and 16 for the conversion on and off operation, respectively. As for the previous figure, efficiency plateaux are given as a function of V_2 for several values of V_1, and the breakdown voltage in the second amplifying element is shown as a function of particles' flux. The experimental bars for the breakdown voltage cover the possible range of the gating frequency, from 10^4 s^{-1} to zero; the last point corresponds obviously to the natural breakdown of the single gap without transfers. At large fluxes, this intrinsic limit is also reached in the first amplifying element, and the figures close to the experimental bars indicate the value of V_1 at which the measurement was performed; at each rate, the proper efficiency plateau has to be considered. For example, in fig. 16, at a flux of 2×10^5 s^{-1}

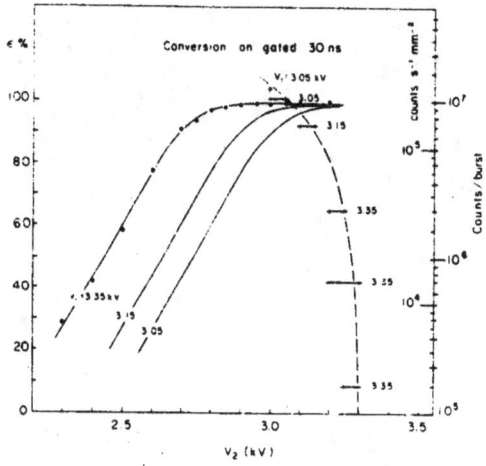

Fig. 15. Same as for fig. 14, for the conversion on and 30 ns gate operation. The sparking point bars represent the measurement of breakdown, for a given flux, at two extremes of gate-opening frequencies (10 kHz and zero). The value of the preamplification voltage for each measurement is also shown, and the reading has to be combined with the corresponding efficiency plateau for a given flux.

mm^{-2}, V_1 can be increased up to 3.25 kV, and breakdown occurs at $V_2 = 3.05$ kV; the corresponding maximum efficiency is then around 90% as read on the $V_1 = 3.25$ kV curve. No noticeable difference in

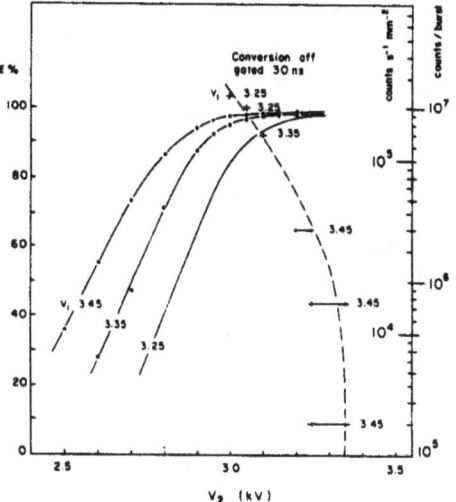

Fig. 16. Same as fig. 15, for a gated, conversion off operation.

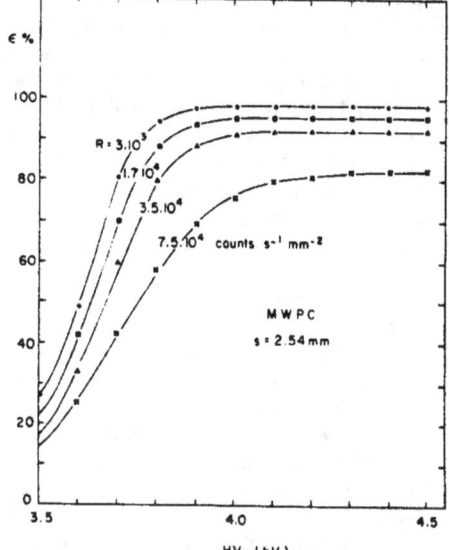

Fig. 17. Efficiency plateau for different beam fluxes measured in the MWPC used as a monitor for the measurement of accuracy. The efficiency reduction due to space charge is visible and has limited the use of the chamber to rates below about 10^4 s^{-1} mm^{-2}.

the efficiency plateau shape was observed at increasing particle fluxes, proving that space-charge gain reduction, such as that observed in normal MWPCs, is absent in the GMSC, as already mentioned. We had indeed mounted in the same beam a conventional MWPC, having 2.54 mm anode wire spacing, to be used as a positional accuracy monitor (see the next section): typical efficiency plateaux obtained in the MWPC at increasing beam fluxes as a function of anodic voltage are shown in fig. 17. The maximum efficiency is already reduced to 90% at a flux of 3.5×10^4 s^{-1} mm^{-2}, and this has limited our study of the GMSC localization properties to moderate beam rates.

6. Localization properties of the GMSC

The large avalanche spread in our particular low-quenched gas mixture is responsible for the observed preamplification and transfer properties of the multistep avalanche chamber; in the gas mixture used in the present work, 2% acetone in argon, the standard deviation of the electron avalanche for one step of multiplication is about 300 μm [6]. For a double-step multiplication process as in the GMSC, also on account of diffusion in the drift regions, a transverse charge extension at the last electrode having about 700 μm standard deviation is observed [13]. As a consequence, several wires in the last plane share the detected charge and, as noticed in a similar case for MWPC, the position accuracy obtained with a centre-of-gravity measurement is better than the wire spacing itself [11,12]. When recording the fast component of the negative signal on the electrode, however, a compensation mechanism similar to the one observed on the anodes of MWPCs takes place. Indeed, electrons collected by one wire, and producing there a negative signal, induce a positive charge on the adjacent wires that overlaps with and partially compensates the direct electron signal; as a result, the spatial extension of the fast induced signal is considerably smaller (about a factor of two) than the true electron avalanche width. A typical induced signal pulse-height distribution for a single event, measured on the last electrode of a GMSC with the method outlined in section 2, is shown in fig. 18; each horizontal bin corresponds to one wire spacing, 500 μm, and the distribution has a standard deviation of about 0.4 mm. The variance of the avalanche width, measured on a large sample of events, is very small, $\frac{1}{25}$ of a bin size, or 20 μm; the analysis of the avalanche width should be a powerful tool in the case of multiparticle or jet events. The reduction of the measured over the physical width is particularly interesting if the two-track

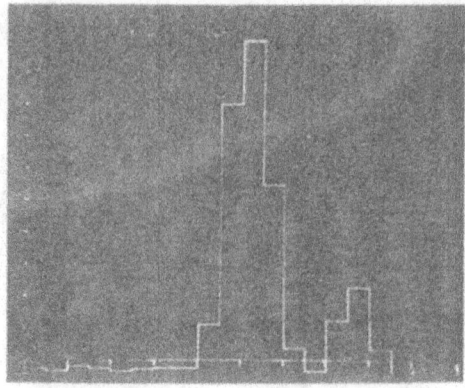

Fig. 19. Same as for fig. 18, for a double-track event; the distance between the two centres of gravity corresponds to about 2.5 mm.

separation allowed by the detector is considered; from the figure, we would expect to identify two tracks about 1.5 to 2 mm apart. As an example, fig. 19 shows the pulse-height distribution measured for an event having two simultaneous tracks 2.5 mm apart and recorded during the high-flux measurements previously described. Fig. 20 shows instead an event that may correspond to two tracks 1 mm apart, and with the pulse-height spectra partially overlapping. The analysis of this kind of event with a single detector is however difficult and has not been

Fig. 18. Pulse-height distribution for a single event measured in the GMSC on the last electrode after subtraction of the ADC pedestal. The channel width corresponds to the wire spacing, i.e. 500 μm.

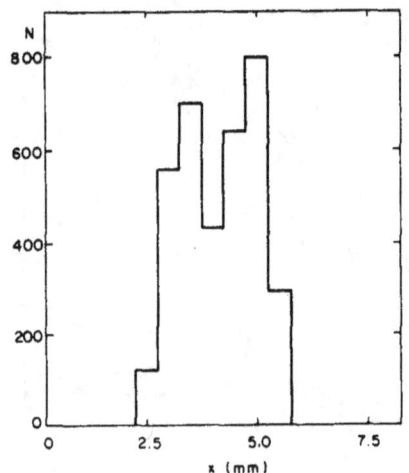

Fig. 20. Pulse-height distribution for an event probably corresponding to two tracks 1.5 mm apart.

attempted yet, especially in the case when the position distortion due to the partial overlap of the distribution of the induced signals. Notice that in the pulse-height spectra of figs. 18 to 20 the ADC pedestal, a presettable dc level at the input of the converters, has been subtracted following the procedure outlined in ref. [12]; this results in measured amplitudes below the base line for all channels adjacent to the ones of the peak, because of the opposite polarity signals induced on the GMSC wires by the charge-compensating mechanism described above. In computing the centre of gravity of the distribution, only the cluster of adjacent significative channels above the base line have been taken into account.

The localization properties of the GMSC have been checked using a collimated ^{55}Fe X-ray beam and operating the device in the self-gated mode. The centre-of-gravity distribution as measured for two beam positions 3 mm apart is shown in fig. 21; the physical width of the collimated beam, about 400 μm in diameter, gives of course the major contribution to the measured dispersion. The observed width of the distribution, having a standard deviation of about 200 μm, can only be justified by an intrinsic localization accuracy in the 100 μm region, thus proving that, at least for a localized charge cluster such as the one provided by 5.9 keV X-rays, the large avalanche spread takes place in a rather symmetric way.

The position accuracy in the detection of charged particles is normally best analysed using a triplet of identical detectors. This could not be done however in our test run, and we have installed on-line a normal MWPC with 2.54 mm anode wire spacing, used in the drift mode, as positional reference. The time of detection of the beam particles triggering the GMSC system was recorded in a set of time-to-digital converters, and assumed to be proportional to the distance of the track from the MWPC wires; owing to the small beam size, only two wires in the MWPC were actually read-out. A linear dependence of the drift-time on the distance was assumed, with a slope of 20 ns mm^{-1}, as often verified in similar conditions for tracks perpendicular to the anode wire plane, with intrinsic localization accuracies equal to or better than 100 μm [15]. Obviously, the same time of drift is measured for tracks on both sides and at equal distances from an anode wire (the right–left ambiguity). Fig. 22 shows the correlation plot between the coordinates of the beam tracks measured through the centre-of-gravity calculation in the GMSC (horizontal axis) and the drift-time measured on the two wires of the MWPC (vertical axis). Data for the GMSC correspond to the conversion off, 30 ns gate operating condition; as already emphasized, this correlated measurement could only be performed at low beam flux

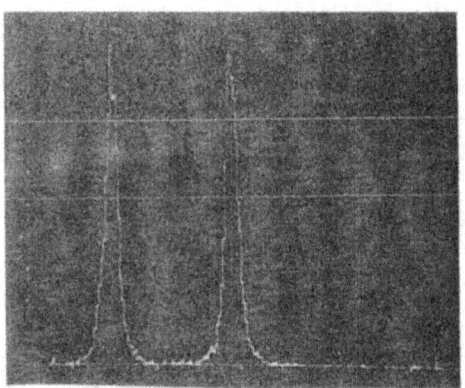

Fig. 21. Centre-of-gravity distribution measured in the GMSC for two positions of a collimated 5.9 keV X-ray source, 3 mm apart. The two peaks have a standard deviation of about 200 μm, which also includes the source collimation.

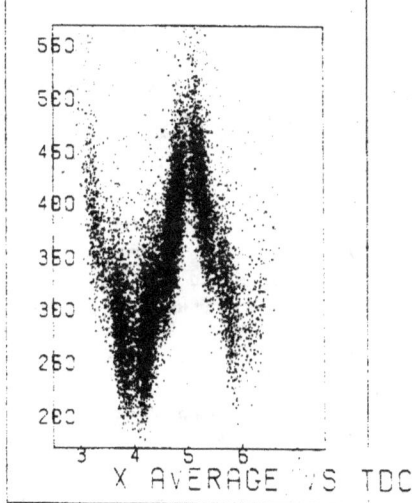

Fig. 22. Correlation plot between the centre of gravity of the signals induced on the last electrode of the GMSC, horizontal axis, and the time of drift measured on two adjacent wires of an MWPC for minimum ionizing particles.

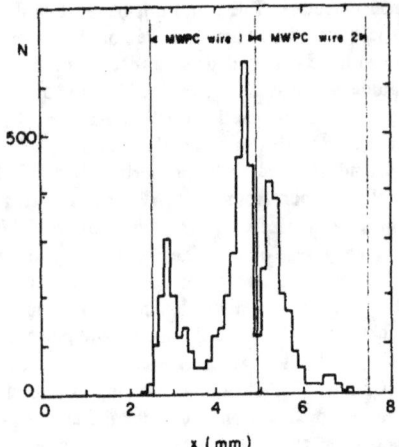

Fig. 23. Projection histogram of a horizontal slice 2 ns wide in the plot of fig. 22, for a time of drift in about the middle of the range. The standard deviation of the peaks, corresponding to the different right-left possibilities of the MWPC wires, is about 200 µm, including the localization accuracies of both the GMSC and of the MWPC used in the drift anode.

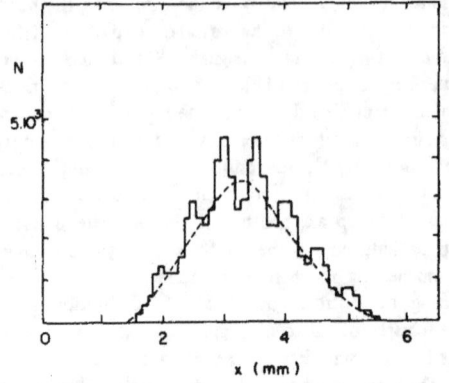

Fig. 24. Histogram of the distribution of the centres of gravity measured for a large unselected sample of tracks in the GMSC; a modulation effect with a 500 µm period and about ±50 µm amplitude appears over the Gaussian-like beam distribution.

($\sim 10^3$ s^{-1} mm^{-2}) because of the rate limitations in the MWPC. The gating frequency for the GMSC was in this case limited to about 200 s^{-1} by the data acquisition link.

The localization accuracy in the GMSC is clearly good enough to separate the right and left tracks in the MWPC, at least for events not too close to the anode wires. A projection of the two-dimensional scatter plot for events in a slice 2 ns wide around the centre of the drift-time distribution is shown in fig. 23; each peak has a standard deviation of around 200 µm, which includes of course both the contributions of the MWPC and of the GMSC. From the previous considerations, it is inferred that the localization accuracy of the GMSC for minimum ionizing particles is about 150 µm.

When plotting the centre-of-gravity distribution for unselected events in the GMSC, fig. 24, a modulation effect appears with an amplitude corresponding to about ±50 µm (too small to be disclosed by the previous correlation). The modulation is probably due to the discrete structure of the multiplying grids or of the readout electrodes; a similar behaviour has already been noticed and is reported in ref. [4]. Further study of the modulation effect is however required before attempting a nonlinear correction of

the data with the aim of improving the localization accuracy of the device.

7. Summary of the results and future developments

We have shown that, at least for chambers of moderate size, the problems connected with capacitive pick-up of the gating pulse from the sensitive electrodes can be solved by a proper grounding and by the use of a symmetric bipolar pair of gates. The residual pick-up is not much higher than the amplifier noise, and furthermore is several hundred nanoseconds in advance of the physical signal. Full efficiency of detection can be obtained for minimum ionizing particles in various operating conditions, the more stringent one corresponding to the conversion off, 30 ns gate width; the resolution time in this case is of about 10 ns fwhm, with a memory around 100 ns. No gain reduction due to space charge is observed at increasing radiation fluxes; however, spontaneous breakdown in the beam region appears at a well-defined gain for a given flux, thus limiting the maximum operational voltage. Strong evidence suggests that breakdown is due to a mechanism of secondary extraction at the cathodes due to the charge density of the positive ions. With the electronics sensitivity used, about 0.05 pC at the wires, full efficiency of detection for minimum ionizing particles can be reached up to maximum fluxes slightly in excess of

$10^5 s^{-1} mm^{-2}$, which is an order of magnitude more than in a conventional MWPC.

May we emphasize here that breakdown in a multi-step evalanche chamber has no dramatic consequences, except for a short dead-time due to the recovery time of the power supply (that can be reduced to a few milliseconds); all electrodes in the structure are indeed realized with thick wire grids or meshes, contrary to the MWPC case, and the electronics can be protected by a pair of fast clamping diodes at the input, at least for chambers of moderate capacity.

The localization properties of the GMSC are rather unique, because of the large avalanche spread in the gas (about 1 mm fwhm). Charge interpolation is automatically performed, and a position accuracy of 150 μm standard deviation has been measured by recording the induced charge distribution on the last anode wires, 500 μm apart. Obviously, simpler digital on-wire electronics can be used, similar to the systems developed for MWPCs, and we would then expect an accuracy comparable with the wire spacing. This last parameter can be varied over a large range, since in the parallel-plate avalanche mode of amplification the diameter of the wires and their distance play no role (as long as their surface field is not too large).

Although this has not been implemented in the chamber described, two-dimensional or multi-dimensional localization can be obtained in multistep devices by transferring some of the electron charge in the last avalanche to another electrode, at an angle to the previous one; the transfer properties of such structures are well understood. In particular, the last electrode in the chamber can be made with a printed circuit, where any convenient readout pattern can be implemented.

In the formation of the induced signal on a set of adjacent wires, a compensation phenomenon appears, reducing the measured fast-charge distribution to about half of the physical electron avalanche size: this is particularly convenient to improve the two-track separation that has been shown to be around 2 mm. Also, the presence of signals of opposite polarity on the tails of the induced signal distribution sharpens the transition from a significative signal to the background level, thus eliminating the common problem of low-level tails in the centre-of-gravity calculation.

The major drawback of the GMSC lies in its long memory time, around 50 ns, due to the extended longitudinal avalanche size. It is obviously easy to reduce this size, by increasing the amount of quencher, as shown for example by fig. 2 of ref. [6]; however, the uniform transfer properties of the pre-amplification element are quickly degraded when the avalanche size is reduced below the transfer grid wire spacing. With the meshes we have used, having 0.5 mm pitch, the minimum avalanche size for correct behaviour has around 200 μm standard deviation. We are currently investigating the possibility of using finer grids or micromeshes, which would allow a good transfer for smaller avalanche size.

It seems also desirable to improve the GMSC time resolution over the measured one, 10 ns fwhm. The reduction in the avalanche size is expected to reduce its fluctuations, but the intrinsic limit of the present design is given by the statistics of the primary ion-pair production in the gap. Use of heavier gases (xenon replacing argon) or of pressures higher than atmospheric should, by the increase of the ionization density, remove part of the statistical fluctuation, and we are investigating this approach.

Several parameters need however to be improved to make the GMSC a useful device, for example in selective high flux beam localization. We are, in particular, investigating in detail the breakdown mechanism, and its dependence on the gas and the nature of the electrodes lowering the discrimination threshold (which was not particularly low for the described measurement) could be another way of improving the rate capability of the GMSC, by allowing operation at a lower voltage.

References

[1] G. Charpak, G. Melchart, G. Petersen, F. Sauli, E. Bourdinaud, P. Blumenfeld, C. Duchazeaubeneix, A. Garin, S. Majewski and R. Walczak, CERN 78-05 (1978).
[2] R.W. Hendricks, Rev. Sci. Instr. 40 (1969(1216.
[3] G. Charpak and F. Sauli, Phys. Lett. 78B (1978) 523.
[4] A. Breskin, G. Charpak, S. Majewski, G. Melchart, G. Petersen and F. Sauli, Nucl. Instr. and Meth. 161 (1979) 19.
[5] G. Charpak, G. Melchart, G. Petersen and F. Sauli, IEEE Trans. Nucl. Sci. NS-26 (1979) 186.
[6] G. Melchart, G. Charpak and F. Sauli, IEEE Trans. Nucl. Sci. NS-27 (1980) 124.
[7] G. Charpak, S. Majewski, G. Melchart, F. Sauli and T. Ypsilantis, Nucl. Instr. and Meth. 164 (1979) 419.
[8] J. Séguinot, J. Tocqueville and T. Ypsilantis, Nucl. Instr. and Meth. 173 (1980) 283.
[9] G. Petersen, G. Charpak, G. Melchart and F. Sauli,

CERN-EP/80-39 (1980), presented at the Wire Chamber Conf., Vienna (February, 1980).

[10] R. Bouclier, G. Charpak, G. Million, J.C. Santiard and F. Sauli, to be published.

[11] G. Charpak, G. Petersen, A. Policarpo and F. Sauli, Nucl. Instr. and Meth. 148 (1978) 471.

[12] G. Charpak, G. Melchart, G. Petersen and F. Sauli, Nucl. Instr. and Meth. 167 (1979) 455.

[13] G. Charpak, F. Mathy, A. Peisert and F. Sauli, to be published.

[14] F. Sauli, CERN 77-09 (1977) and refs. therein.

[15] G. Charpak, F. Sauli and W. Duinker, Nucl. Instr. and Meth. 108 (1973) 413.

Nuclear Instruments and Methods in Physics Research 220 (1984) 349–355
North-Holland, Amsterdam

ON THE LOW-PRESSURE OPERATION OF MULTISTEP AVALANCHE CHAMBERS

A. BRESKIN

The Weizmann Institute of Science, Rehovot, Israel

G. CHARPAK

CERN, Geneva, Switzerland

S. MAJEWSKI

Physikalisches Institut der Universität Heidelberg, Fed. Rep. Germany

Received 31 August 1983

It is demonstrated that in the double-step parallel-plate structure filled with pure low-pressure isobutane or methane, the gas preamplification-and-transfer mechanism occurs with an efficiency close to 100%, probably due to the electron diffusion process. Amplification in successive steps in isobutane and methane at pressures of 2–10 Torr is shown to be possible. An increase in total gain of 10–80 as compared to maximum amplifications attainable in a single-step structure is observed with 5.5 MeV alpha particles and single electrons. The possible applications of multistep low-pressure counters for single-electron (photoelectron) imaging are discussed.

1. Introduction

The idea behind our attempt to test multistep avalanche chamber structures at low gas pressures was to try to increase the maximum attainable gain of low-pressure detectors in view of their applications to single-electron detection and imaging. Low-pressure gaseous detectors, widely used for the detection of heavily ionizing particles, have been proved to detect efficiently single electrons [1]. They have several advantages in comparison with detectors operating at normal gas pressures, in cases where the detection process does not rely on direct ionization of a gas by charged particles but on secondary electrons produced by a photoelectric effect on photocathodes or in photosensitive gases:

1) High extraction efficiency of photoelectrons from a photocathode surface to a gas, due to an increased reduced electric field E/p at the cathode surface.
2) High gains of up to 10^7 [1] obtained in pure hydrocarbons, resulting in a saturated avalanche response for a single electron.
3) Subnanosecond time resolution.
4) Very low sensitivity to relativistic charged particles and to low-energy gamma rays, due to low gas density.
5) High counting rate capability [2], due to much faster drift velocities of positive ions in the high reduced fields resulting in the reduction of space charge and

in short electronic occupation time. (Examples of the fast ion-collection times are shown in fig. 1.)

Detectors such as parallel-plate avalanche chambers (PPACs) or multiwire proportional chambers (MWPCs), operating at pressures of a few Torr, are at present used as single-electron detection and imaging devices coupled to solid [3,4] or liquid [5] photocathodes for photon-detection applications.

Multistep avalanche chambers (MSACs) [6] have been proved to be successful detection devices at normal gas pressures for minimum ionizing particles at high rates [7] and as photosensitive detectors for UV photons in Cherenkov ring imaging [8].

The preamplification of an initial charge and the transfer of the primary avalanche to a second amplification step, as shown in fig. 2, have been the subject of an intense study during the last few years [6,7,9,10]. Using pure electrostatic considerations it was proved that an efficient charge transfer from the preamplification region to the transfer region ($E_p \gg E_t$) can only occur if the avalanche has a lateral spread of the same order as or exceeding the wire spacing of electrode b [6]. At normal pressures and with standard gas mixtures the electron diffusion, in a simple avalanche model, is often too small to allow an efficient transfer. Extended photon-mediated avalanches can be obtained in some binary gas mixtures, similar to the one used in Geiger counters, where the ionization potential of one of the molecules is lower than the energy of the excited state

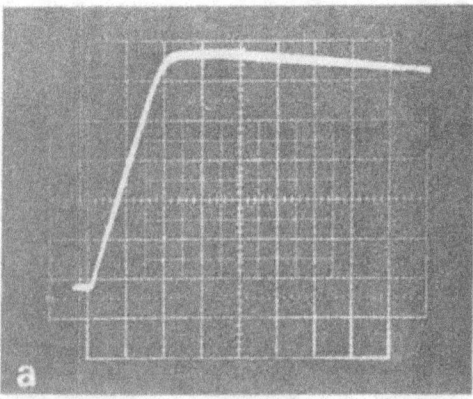

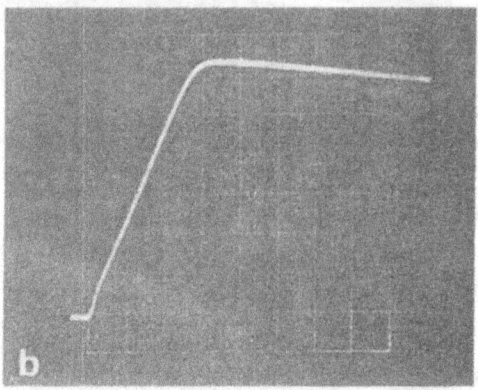

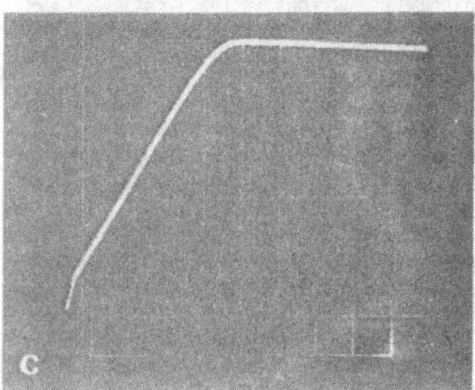

Fig. 1. Examples of α-particle pulses, after a charge-sensitive preamplifier, in a single-step structure filled with low-pressure isobutane: upper trace (a) at 2 Torr, middle (b) at 5 Torr, and lower (c) at 10 Torr. Horizontal scale: 500 ns/div. Note the difference in the rise-time, corresponding to total ion-collection times. At 2 Torr the electronic and ionic components of the pulse are confused.

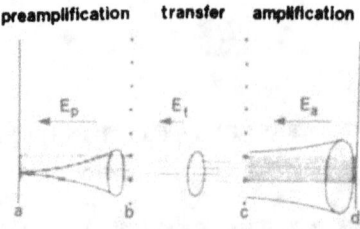

Fig. 2. A double PPAC structure used for the measurements. Preamplification and amplification regions are 3.3 mm thick and the transfer region is 8.4 mm thick. Electrode a is a copper plate, while all other electrodes are made out of 93% transparent stainless-steel mesh. α-particles' tracks from the ^{241}Am source are approximately perpendicular to the electrodes. UV photons enter through a CaF$_2$ window and produce photoelectrons at the electrode a, initiating two-step avalanches. The initial avalanche can spread laterally in the transfer region owing to diffusion when drifting in a low electric field E_t.

of the other atom or molecule [6]. At low gas pressures electron diffusion leads to wide avalanches [11] and we show that the multistep operation can take place efficiently even in organic gases such as isobutane or methane, with an increase in total gain of 10–80 compared to single-step operation.

Several of the interesting features of MSACs may be of importance for the application to single-electron detection:

1) Higher amplification in comparison with a single-step PPAC or a MWPC.
2) Gating possibility on selected events that offer a better counting-rate capability [7].
3) Reduction of the photon feedback mechanism at the cathodes and in the gas, since photoelectrons produced by photons coming from the avalanche are absorbed in the transfer region and are not subject to the full two-step amplification.
4) A possibility of reducing the positive-ion feedback effect on the cathodes by stopping backward going positive ions with a gating electrode.
5) Unambiguous read-out of multiple events with multielectrode structures by passing the avalanche electrons through successive electrodes that have wires oriented in various directions [6].

In the present work we show the first results of the low-pressure operation of MSACs and discuss possible applications.

2. Experimental conditions

We have studied the operation of two different double-step structures. In both cases the first step is a 3.3 mm thick parallel-plate detector with its anode made

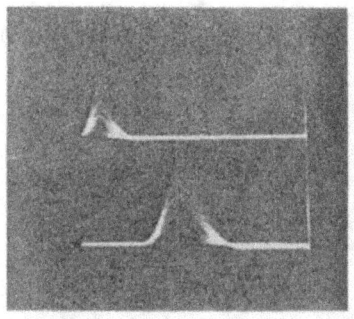

Fig. 3. Demonstration of the high efficiency of a double-step amplification process in 5 Torr of isobutane. Electronically shaped pulses from electrodes a (upper trace) and d (lower trace) from fig. 2, are displayed. The oscilloscope trace was triggered by pulses from the first step, electrode a. $E_p/p = 430$, $E_t/p = 4$, $E_a/p = 240$ V·cm^{-1}·Torr^{-1}.

out of stainless-steel 50 μm diameter wire mesh, with a cell size of 500 μm, and an 8.4 mm thick transfer region followed by an electrode made of the same kind of mesh. The second amplification step is either a 3.3 mm thick parallel-plate counter or a multiwire chamber having a 2 × 3 mm gap with 10 μm diameter anode wires 1 mm apart. Since both structures show very similar behaviour, only that concerning the double-step avalanche

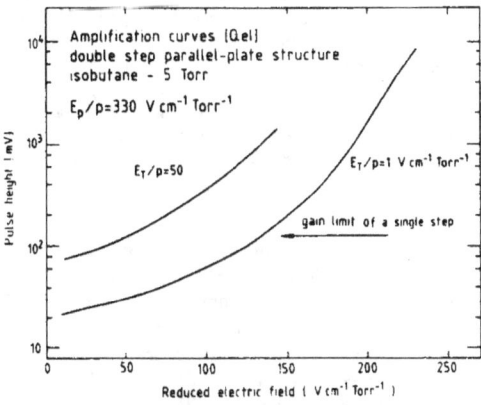

Fig. 4. Amplification curves (electronic component, Q_{el}, only) measured with α-particles in a double PPAC detector filled with 5 Torr of isobutane. The curves were measured at a constant preamplification field of 330 V·cm^{-1}·Torr^{-1} and with a variable E_a/p field of up to 230 V·cm^{-1}·Torr^{-1}. Two extreme cases were analysed with a very low ($E_t/p = 1$ V·cm^{-1}·Torr^{-1}) and a high ($E_t/p = 50$ V·cm^{-1}·Torr^{-1}) electric field in the transfer region. Notice the difference in the maximum attainable amplification factor (just before breakdown) in the two cases.

chamber is presented in this work.

The measurements were performed at pressure values of 2–10 Torr of isobutane and methane. The detectors were operated either with a ^{241}Am source, placed inside the chamber, providing 5.5 MeV α-particle tracks perpendicular to the detector surface, or with single photoelectrons produced at the first cathode (the cathode of the first PPAC) by UV photons. The UV photons from an external light source entered the chamber through a CaF$_2$ window.

3. Experimental results

Our first aim was to check if an electron swarm can be efficiently transferred from a first preamplifying step

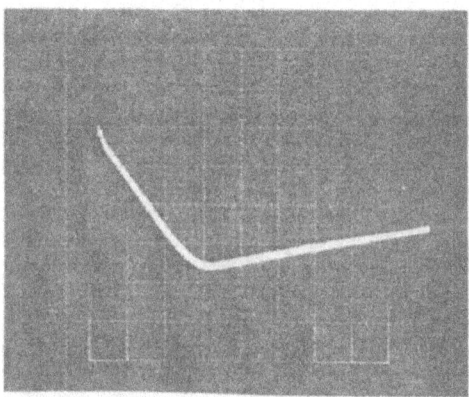

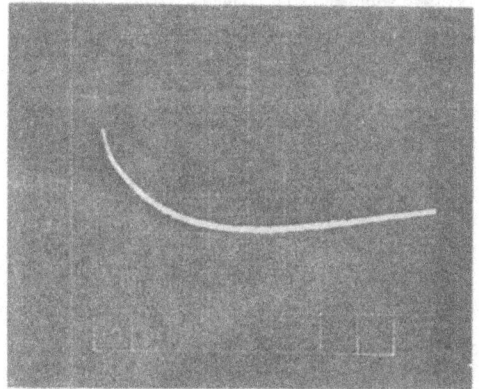

Fig. 5. Examples of α-particle pulses in a single-step (upper trace, 5 mV/div and 500 ns/div) and a double-step (lower trace, 200 mV/div and 500 ns/div) structure in 5 Torr isobutane as measured directly on a 1 MΩ load. In the first case the detector was operated with an E_p/p value of 330 V·cm^{-1}·Torr^{-1}. In the second case the field values were as follows: $E_p/p = 330$, $E_t/p = 1.2$, $E_a/p = 305$ V·cm^{-1}·Torr^{-1}. Notice the shape of the ion component in the double-step mode, corresponding to a saturated-gain mode.

to the second step at a low gas pressure. We found that in both gases tested an efficient transfer mechanism takes place. The transfer efficiency is shown in fig. 3 for single electrons in a double PPAC structure filled with isobutane at 5 Torr. The upper oscilloscope trace shows electronically shaped pulses from the first stage, the lower one from the second stage (electrodes a and d respectively, from fig. 2). The time base was triggered by the pulses from the first step. As is seen, the efficiency of the transfer is close to 100% (a small effect of missing pulses can be explained by edge effects due to lack of precise collimation of the UV-photon beam). It means that for each avalanche produced in the first PPAC a substantial part of the electron cloud is transferred to the following region of a lower electrostatic field. The electrons are then transferred to the second amplifying step, where the electrical field is higher. The main condition required for a proper transfer mechanism is that the transversal size of an avalanche produced in the preamplifier step should be larger than the wire mesh spacing (500 μm in our case) of electrode b. It is likely that the reason for the broad size of avalanches in low-pressure isobutane (1–2 mm fwhm [11]) is related to electron diffusion at low pressure.

We have proceeded to search for the optimum operating conditions, trying to reach higher amplifications than in a single-step structure. Our first observation is that in a pressure range of 2–10 Torr, isobutane works much better than methane. The highest gains were achieved with isobutane at a pressure of about 5 Torr. In methane reasonable amplification factors, though lower than with isobutane, could be reached only above 10 Torr. In all cases we have observed an increase in the total gain in the double structure as compared with the

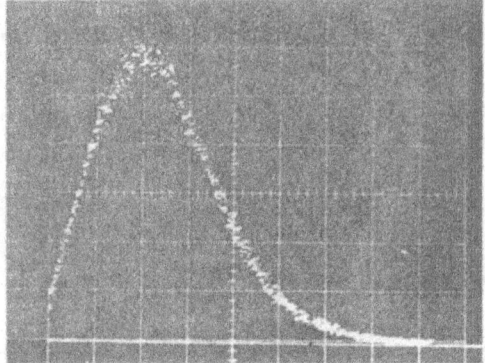

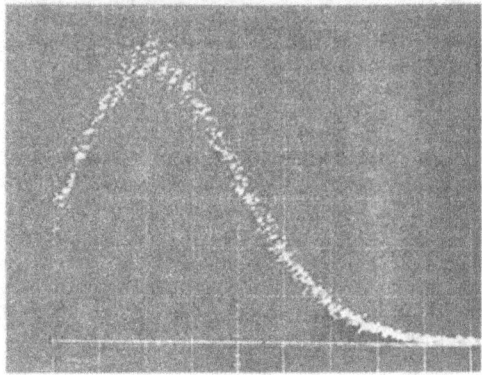

Fig. 7. Pulse-height spectra from single photoelectrons as measured on electrodes a and d in the conditions specified in the caption to fig. 3. The spectrum measured at the electrode d was not gated by the pulses from the electrode a and many pulses of low amplitude are due to avalanches started by photoelectrons produced at the meshes b and c. The total increase in gain of the double-step structure is of the order of 30. The charge at the peak of the spectrum measured at the electrode d is about 3 pC.

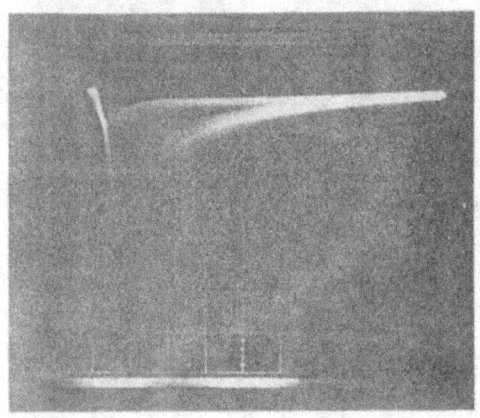

Fig. 6. Current pulses from α-particles measured directly on a 200 Ω load in a double-step detector filled with 5 Torr of isobutane. $E_p/p = 330$, $E_t/p = 1.2$, and $E_a/p = 215$ V·cm^{-1}· Torr^{-1}. Scales: 40 mV/div and 200 ns/div.

single structure. This gain improvement factor was variable and could reach up to the 80, depending on the operating conditions. Apart from the pressure dependence it was found that the maximum possible gain in a double-step structure increases when the electric field strength in the transfer region is decreased. Assuming that the spark limit at given conditions is related to a space-charge density in an individual avalanche in the second detector step, one can explain the observed effect by an enhancement of the lateral avalanche size due to diffusion of electrons drifting in the transfer region. Another interpretation can be related to reduction of positive-ion feedback; there is a limit of amplification due to positive-ion feedback at the cathode of the first step, and the percentage of ions moving back to the first stage is about linearly dependent on the field

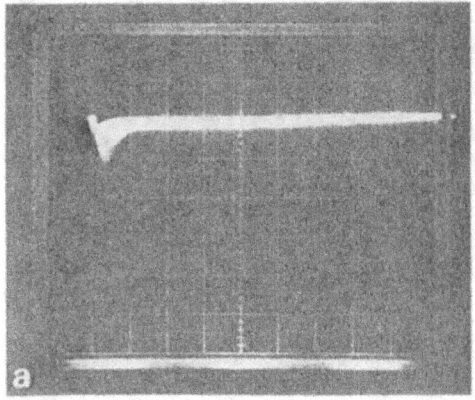

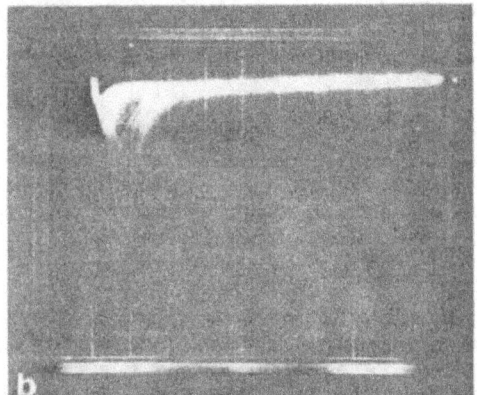

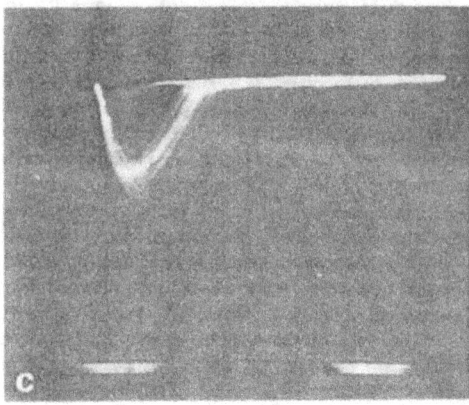

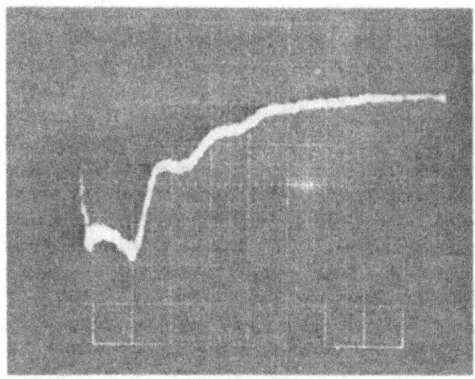

Fig. 9. Example of an α-particle current pulse from a low-input impedance amplifier as observed at the limit of amplification in a single-step detector filled with isobutane at 2 Torr. $E_a/p = 750$ V·cm^{-1}·Torr^{-1}. Horizontal scale: 500 ns/div.

intensity in the transfer region if the other parameters are kept constant.

3.1. Results of measurements in isobutane

Fig. 4 shows the amplification curves for two extreme values of the transfer field E_t. The total gain is increased by almost two orders of magnitude at the low transfer field; notice the limit of amplification of a single-step PPAC shown in the figure. Examples of charge pulses taken at 5 Torr with α-particles in a single step and a double step operating with a transfer field of 2 V·cm^{-1}·Torr^{-1} are shown in fig. 5. Current pulses

Fig. 8. Transition to the "high amplitude mode" in a double-step PPAC detector with 10 Torr of isobutane, measured with α-particles directly on a 200 Ω load. $E_p/p = 212$, $E_t/p = 2.5$ V·cm^{-1}·Torr^{-1}. Horizontal scale: 500 ns/div. Photographs taken at 3 working points: (a) normal proportional mode pulses, $E_a/p = 9$ V·cm^{-1}·Torr^{-1}; vertical scale 10 mV/div., $E_a/p = 18$ V·cm^{-1}·Torr; vertical scale: 20 mV/div., (c) $E_a/p = 27$ V·cm^{-1}·Torr^{-1}; vertical scale: 100 mV/div. Notice the low amplification field in the amplifier region.

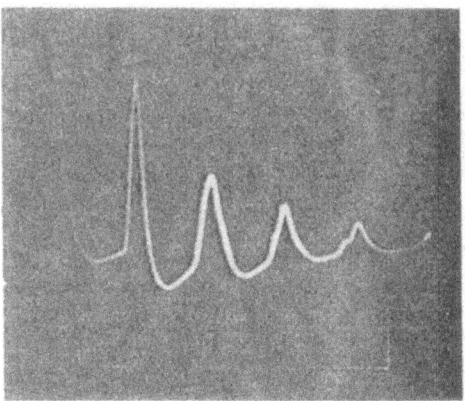

Fig. 10. Demonstration of the positive-ion feedback process taking place in a double-step PPAC counter in methane at 2 Torr. Differentiated single photoelectron pulses from a charge amplifier. $E_p/p = 680$, $E_t/p = 17$, $E_a/p = 36$ V·cm^{-1}·Torr^{-1}. Horizontal scale: 20 μs/div.

of α-particles in the double-step mode are shown in fig. 6. In fig. 7 are presented two spectra obtained at 5 Torr in a two-step structure, measured on electrodes a and d (fig. 2) with single photoelectrons. The saturated response is observed in both stages. At the reduced transfer field of about 2 $V \cdot cm^{-1} \cdot Torr^{-1}$ the two-step amplification is about 30 times higher than the maximum amplification attainable in a single step. At a pressure of 10 Torr, the double-step gain increase is only of about 10. At this pressure and with a reduced field of 2 $V \cdot cm^{-1} \cdot Torr^{-1}$ the chamber gets into a "very high amplitude" mode with pulses reaching 1 mA peak current (fig. 8). We have not explored further the properties of this "magic" mode of operation.

At 2 Torr a gain increase factor of about 30 was obtained in a wide range of reduced transfer fields from 3 to 30 $V \cdot cm^{-1} \cdot Torr^{-1}$. The drift velocity of positive

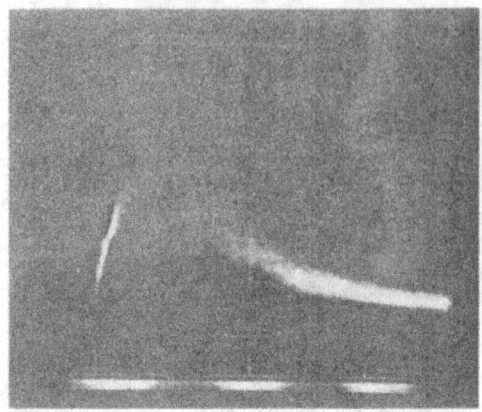

Fig. 12. Photon-feedback mechanism as seen on a differentiated output pulse from a charge amplifier connected to the single-step counter filled with 4 Torr of methane. Horizontal scale: 200 ns/div.

ions is already so high that one cannot distinguish between electronic and ionic components of a pulse (fig. 1a). The absolute amplification in a single step is limited by positive-ion feedback, clearly visible in fig. 9.

3.2. Measurements in methane

In methane, as was already mentioned, the low amplification limit in the range of pressures from 2 to 10 Torr is due to secondary processes at the cathodes. The strong positive-ion feedback mechanism in the two-step PPAC structure at 2 Torr is shown in fig. 10 for single photoelectrons. Fig. 11 shows the same phenomenon in the same structure, but at 4 Torr for α-particles in the cases of a weak and strong transfer field. The observed difference in the feedback time constants is due to the different drift velocity of positive ions in the two cases. In a single-step structure we could also observe the photon feedback mechanism, as can be seen in the oscilloscope picture of fig. 12 (appearing there as an after-pulse). In contradiction with these results, in a short measurement done at 20 Torr we have observed much better operation and safely attained high amplifications. The effect of the higher pressure is probably in slowing down methane positive ions of high mobility and then in limiting their efficiency of secondary emission when striking a cathode. A gain increase factor 2–3 at most was found in these conditions, which could mean that methane is not a good choice for a double-step structure.

4. Summary and possible applications

With these first results we have demonstrated that a double-step process is possible in low-pressure gases,

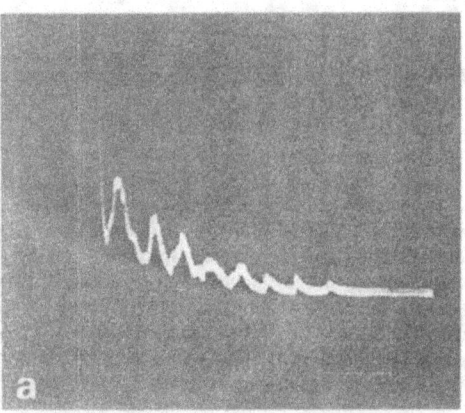

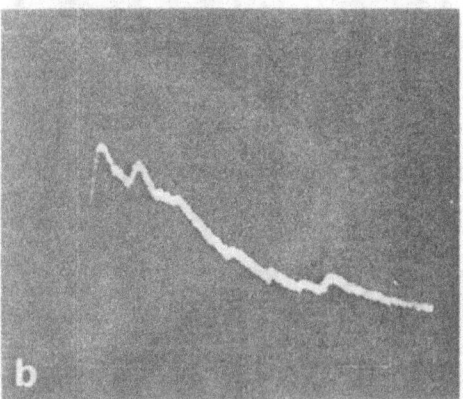

Fig. 11. The feedback mechanism as observed with single photoelectrons at the limit of amplification in a double-step detector with a 4 Torr methane filling. $E_p/p = 375$ $V \cdot cm^{-1} \cdot Torr^{-1}$. Case a: $E_t/p = 4$ $V \cdot cm^{-1} \cdot Torr^{-1}$. Case b: $E_t/p = 50$ $V \cdot cm^{-1} \cdot Torr^{-1}$. Horizontal scale: 50 μs/div.

most probably due to electron diffusion in the first amplification process. In the case of low-pressure isobutane one can achieve excellent performances with very high amplification factors of up to 10^7. The double-step structure enables one to gain 1 to 2 orders of magnitude in the total amplification factor. The "very high amplitude" mode obtained in isobutane should be further investigated because of possible advantages of operation with saturated amplitudes.

One could envisage several future applications of low-pressure multistep detectors based mainly on their high amplification, their insensitivity to relativistic particles, and their imaging and gating capabilities. The gating could be fast or slow: fast, when avalanche produced in the first stage are allowed to enter the second amplification region only if an electronic gate pulse is applied; slow, when one wants to prevent positive ions from reaching the cathode. This could be used in the case of gas photodiodes filled with methane or isobutane, when positive ions destroy the bialkali photocathode [3]. The idea in this case will be to close a gate each time after an electronic pulse is detected and recorded.

One could think of another specific application of such a structure in dc and gated operation for Cherenkov ring imaging either with photosensitive gases or liquid photocathodes. The vapour pressure of the presently used photosensitive compound tetrakis(dimethylamine)ethylene (TMAE) added to the main gas is very low even at room temperature (0.35 Torr at 20°C) [12]. One could work with very low pressures of gas, preserving the same quantum efficiency while the chamber is insensitive to passing charged particles and having all the other above-listed advantages of low-pressure double-step detectors.

Some other studies of low-pressure double-step parallel-plate structures, or parallel-plate counters followed by MWPCs, were also independently performed for heavy ion applications [13,14].

We would like to thank R. Bouclier and G. Million for technical help.

References

[1] A. Breskin, Nucl. Instr. and Meth. 196 (1982) 11;
A. Breskin, in: Detectors in heavy ion physics, ed., W. von Oertzen, Lecture Notes in Physics, vol. 178 (Springer, Berlin, 1983) p. 44.

[2] A. Breskin, R. Chechik and N. Zwang, IEEE Trans. Nucl. Sci. NS-27 (1980) 133.

[3] G. Charpak, W. Dominik, F. Sauli and S. Majewski, IEEE Trans. Nucl. Sci. NS-30 (1983) 134.

[4] S. Majewski, W. Kusmierz and F. Sauli, Wire Chamber Conf., Vienna (1983).

[5] D.F. Anderson, Phys. Lett. B 118 (1982) 230.
D.F. Anderson, R. Bouclier, G. Charpak and S. Majewski, Nucl. Instr. and Meth. 217 (1983) 217.

[6] For example: A. Breskin, G. Charpak, S. Majewski, G. Melchart, G. Petersen and F. Sauli, Nucl. Instr. and Meth. 161 (1979) 19.

[7] A. Breskin, G. Charpak, S. Majewski, G. Melchart, A. Peisert, F. Sauli, F. Mathy and G. Petersen, Nucl. Instr. and Meth. 178 (1980) 11.

[8] For example: R. Bouclier et al., Nucl. Instr. and Meth. 205 (1983) 403.

[9] G. Charpak, G. Melchart, G. Petersen, F. Sauli, E. Bourdinaud, P. Blumenfeld, C. Duchazeaubeneix, A. Garin, S. Majewski and R. Walczak, CERN 78-05 (1978).

[10] G. Charpak and F. Sauli, Phys. Lett. 78B (1978) 523.

[11] A. Breskin, R. Chechik, I. Levin and N. Zwang, Nucl. Instr. and Meth. 217 (1983) 107.

[12] D.F. Anderson, IEEE Trans. Nucl. Sci. NS-28 (1981) 842.

[13] H. Stelzer, in: Detectors in heavy ion physics, ed., W. von Oertzen, Lecture Notes in Physics, vol. 178 (Springer, Berlin, 1983) p. 25.

[14] A. Breskin, R. Chechik, Z. Fraenkel, I. Tserruya and N. Zwang, in press.

Nuclear Instruments and Methods in Physics Research A283 (1989) 673–678
North-Holland, Amsterdam

A LINEAR DRIFT CHAMBER READOUT OF A GATED MULTISTEP AVALANCHE CHAMBER

N. SOLOMEY [1], R. BOUCLIER [2], G. CHARPAK [2], I. GOUZ [2]*, D.C. LAMB [2]** and F. SAULI [2]

[1] University of Geneva, Switzerland
[2] CERN, Geneva, Switzerland

A classical drift chamber with linear field cage wires and highly segmented cathode strips is elegantly combined with a gated multistep avalanche chamber. The drift chamber has a linear drift-time to drift-distance relation over 95% of its length; it provides a flexible readout method that can be tailored to the expected event multiplicity of a given experiment; the readout time is considerably faster in comparison to that of similar devices that can handle large multiplicities; and the readout techniques are well known in high-energy physics. The gated multistep avalanche chamber with a thin conversion gap amplifies the desired signals, in a given time slice, above the ungated minimum-ionizing background, making such a device suitable for use in high-intensity experiments, whilst the gas drift delay and gating allows time for the experiment to generate a trigger and to read out selectively only the events of interest. The separation of the gas amplification into three sections permits large gains, which are necessary for VUV photon detection in ring-imaging Cherenkov techniques, thus reducing the gain needed in each parallel-plate avalanche gap and completely eliminating electrical discharges.

1. Introduction

The future of high-energy particle physics requires detectors that are capable of handling high-intensity particle fluxes and events with high multiplicities. One type of gaseous detector that is capable of handling these requirements is the multistep avalanche chamber. Such devices have been shown to perform well under high-intensity particle fluxes [1]; however, the readout methods had limitations that could prevent their use in certain experiments. The original multistep detector had a multiwire proportional chamber (MWPC) readout [2] that was limited to low-multiplicity events. Camera readout of the secondary light emission from excitation of the molecules by electron impact has also been used [3], but this is limited by the camera acquisition rate of 25 Hz. Electronic readout by charge-collecting pads also works [4]; it is limited only by the cost, which is determined by the size and number of pads. Multistep detectors have found limited use in high-energy physics experiments because of their small size and frequent electrical discharges.

Preliminary results from a multistep avalanche chamber with a linear drift-chamber readout are presented. An attempt was made to read out a parallel-plate avalanche chamber (PPAC) with a drift chamber [5], but the drift-time to drift-distance relation was nonlinear. This problem has been resolved by finding the

proper electric field that permits a smooth and efficient transfer of the electrons from the multistep chamber into the drift chamber. The detector has possible applications for high-multiplicity charged-particle tracking and for single-photoelectron detection, as in Cherenkov ring-imaging. The detector has a 100 ns charge-collection time coupled with a 14 µs drift chamber, making it considerably faster and less expensive than other readout methods. Measurements indicate that the position resolution for single-photon detection can be better than 2.4 mm FWHM; the position resolution for charged particles should be much better. Because in the detector the amplification is done in three steps, this device is very suitable for efficient single-photon detection – as is demonstrated by a peaked pulse-height spectrum – and makes it possible to eliminate completely the electrical discharges in the parallel-plate amplification regions. The separation of the amplification by sufficiently large distances greatly reduces feedback.

2. Detector design

The detector structure is shown in fig. 1. The detector was tested either with ultraviolet (VUV) photons or with beta particles, entering the gas volume from the left. In the first case a quartz window was used, and the VUV photons were absorbed in the conversion gap by the photosensitive gas. In the latter case, we used a Mylar window when testing minimum-ionizing beta particles, which ionized the gas as they traversed the gap. It is planned to use tetrakis(dimethylamine)ethyl-

* On leave of absence from IHEP, Serpukhov, USSR.
** On leave of absence from the University of Illinois, Urbana, IL, USA.

0168-9002/89/$03.50 © Elsevier Science Publishers B.V.
(North-Holland Physics Publishing Division)

IX. DIVERSE CHAMBERS

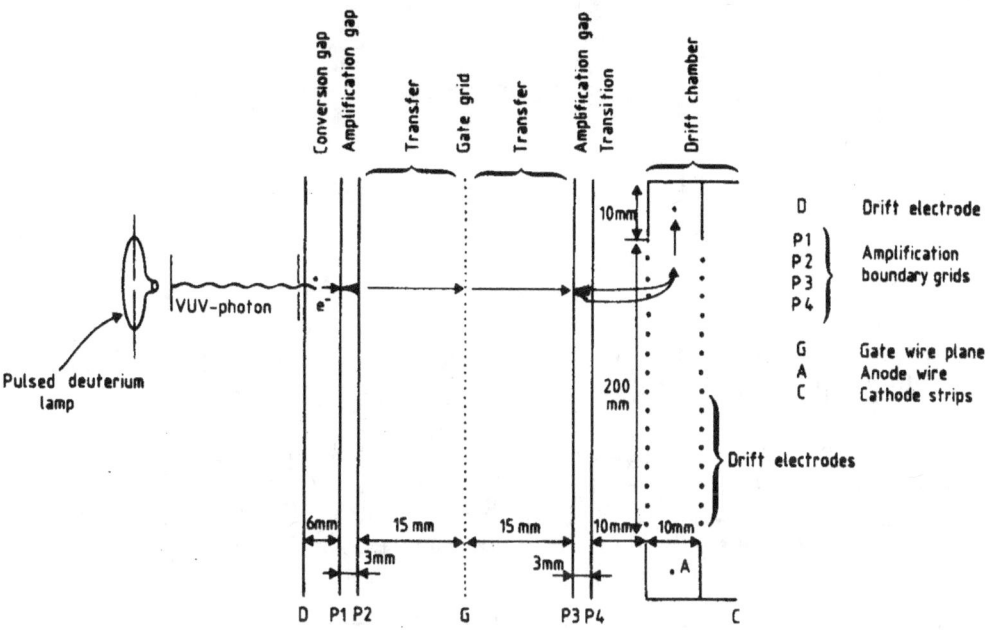

Fig. 1. Detector design showing the mesh planes as solid lines, the gating grid as a dotted line, and the drift-chamber readout.

ene (TMAE) as the photosensitive gas. The efficiency of the photon detection is determined by both the thickness of the conversion gap and the concentration of the TMAE. The original aim of this development was to test the idea of such a device as a prototype for a ring-imaging Cherenkov (RICH) light-detector for high-intensity fluxes of particles. The requirement for the fast RICH was that it had to have a 100 ns collection time, which meant that the conversion gap had to be 6 mm so that the electrons could all be collected within the 100 ns window. Any RICH detector requires a high single-photon detection efficiency, and the aim was to have a conversion gap with one conversion length (66% detection efficiency). This demands a high TMAE concentration, which can only be achieved by heating the TMAE to 42°C. To prevent condensation of the TMAE onto the walls of the detector, it was planned to heat the chamber to 50°C. In determining the thickness of the conversion gap, it was necessary to take into account the planned gas mixture's drift velocities as a function of the electric field, and the interaction length of the photosensitive gas.

The electrodes are stainless-steel meshes made of 50 μm diameter wires with 0.5 mm pitch. The voltages on the electrodes are such that the electrons produced in the conversion gap are moved to the right (see fig. 1), where they enter a 3 mm amplification gap. In this latter gap, the higher electric field accelerates the electrons to an energy that is sufficient to ionize the gas

even more, as in a PPAC. The electrodes are made of meshes that are transparent, so that some of the electrons produced can be transferred to another amplification gap. The percentage of electrons transferred depends, with good approximation, on the percentage of field lines leaving the amplification region and entering the transfer gap; this is just the ratio of the electric field in the transfer gap to that in the amplification gap [1]. Transfer efficiencies of 10% to 20% are reasonable. In the centre of the transfer gap there is a wire grid that allows gating of the events of interest. The transfer distance to this gate grid is large (15 mm), which means that with a gas-drift delay this would allow time for an experiment to generate a trigger and thus open the gate for only those events of interest. The transfer distance after the gate is 15 mm, but it could be much smaller. In this second transfer gap, the background signal originates from the ionization deposited by the charged particles crossing the gap, but which are not associated with the event of interest. Making this second transfer gap smaller would reduce the background signal. The unwanted ionization from the particles that are out of time with the event, and the gated signal from the conversion gap then enter a second amplification gap of 3 mm, identical to the first one. Each of these amplification gaps can easily reach a gain of 1000 to 10 000, with 10% transfer efficiency, making an effective gain of 100 to 1000 in each gap. The purpose of the multistep chamber is not to amplify single photoelectrons suffi-

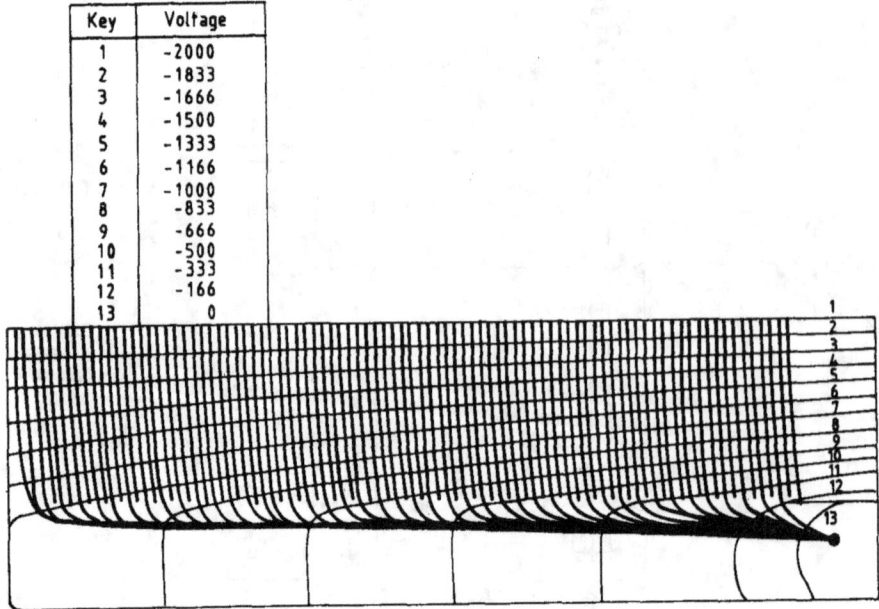

Key	Voltage
1	-2000
2	-1833
3	-1666
4	-1500
5	-1333
6	-1166
7	-1000
8	-833
9	-666
10	-500
11	-333
12	-166
13	0

Fig. 2. Electric-field calculations for the transition region from the multistep chamber into the drift chamber. The dark lines are the force lines, and the lighter ones are the lines of equipotentials. The values of the equipotentials are given in the key. There is an offset voltage of +30 V across the drift chamber, with more on the top electrode.

ciently to detect them, but is only to amplify them above the background ionization. This ionization can also be reduced by proper selection of the gas mixture. For charged-particle tracking, a gas mixture of argon (90%) + methane (10%) was used, and any further reduction of the background was not possible. However, since TMAE provides the single photoelectron, the carrier gas could be a low-ionization gas such as helium, with a small amount of quencher added. Drift chambers [6], as well as PPACs [7], have been made to work with mixtures containing mostly helium, and so do not present a problem.

After the signal has been amplified in the second amplification gap it enters the transition region, where the electrons are transferred from the multistep chamber into the drift chamber (see fig. 2) in such a way that the timing information can be used to determine the coordinate perpendicular to the anode wire. It is also important for the timing information to be linear so as to make the data-taking and analysis easier. The drift chamber was constructed using a standard method: field-wire electrodes every 2 mm, and a 10 mm thick drift chamber with 100 mm drift distance in both directions to the 50 µm diameter, 200 mm long, anode wires. The anode wires were surrounded by ground electrodes on three sides and open to the drift chamber on the fourth. At the back of each anode wire there was placed

an array of cathode strips, 2.5 mm in pitch; these were used to determine the coordinate of the signal along the anode wire. For simplicity, it was necessary to have the cathode pads at ground potential, which required all the multistep electrodes to be at negative high voltage. The correct selection of the electric field, required to make a uniform transfer from the multistep chamber to the drift chamber, was made by electric-field calculations. These calculations also showed that the drift-chamber electrodes had to have an electric field across the drift-chamber gap to help turn the lines of force; otherwise, many force lines from long drift distances would not terminate on the anode wire. Additional losses of electrons occur in the transition region because not all the electric field lines enter the drift chamber: a simple comparison of the number of entering field lines with the total number gave 8% transparency. Other, more detailed, calculations than ours were made [8] by determining the field density on both sides of the drift-chamber electrodes, and arrived at 2% transparency.

3. Measurements

The detector was first operated with a mixture of argon (90%) + methane (10%). The operating voltages used are listed in table 1, where the symbols refer to

IX. DIVERSE CHAMBERS

those in fig. 1. The drift-time to drift-distance relation shown in fig. 3 was measured using a collimated [106]Ru beta source. A scintillator and a photomultiplier tube (PMT) provided the Start signal, and the cathode strip provided the Stop. A measurement of the pulse height as a function of a drift distance in the drift chamber (fig. 3) shows a reduction of the pulse height. This is because the transfer efficiency out of the second parallel-plate amplification region varied as a function of the drift distance, caused by the electric field variation in the transition region. The electrodes have their highest voltage in the centre of the drift chamber, whereas at the edges they are close to ground potential. It is near the edges that the electric field in the transition region is highest and the transfer efficiency out of the multistep chamber is greatest. A pulse-height spectrum is shown in fig. 4; it was taken at a 7 cm drift distance. The black curve in the figure is the spectrum with the gate open, and the underlying white curve that with the gate closed; in both cases the spectrum was taken in coincidence with a PMT trigger. From a comparison of the two spectra in fig. 4 it can be seen that a threshold can be set to eliminate the sensitivity to background particles. The pulse height for the gate-closed signal can be made smaller by a factor of 8, by reducing the transfer gap after the gate plane to 2 mm.

As a prelude to testing the detector as a RICH counter, it was operated in the laboratory, using single photoelectrons with a gas mixture of helium (95%) + ethane (5%) bubbling through liquid TMAE at 42°C. The detector was operated in an insulated hot box kept at 50°C, and the necessary operating voltages were those listed in table 1. Ultraviolet photons were produced by a deuterium spark lamp, which provides a

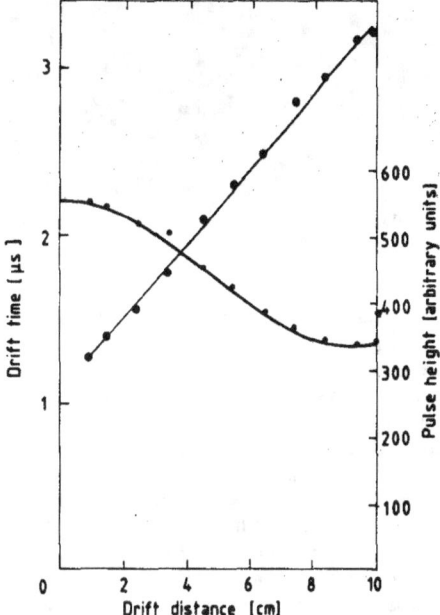

Fig. 3. The straight line indicates the drift-time to drift-distance relation using argon (90%)+methane (10%). The pulse height is shown as a function of drift distance using a beta source.

trigger when the lamp pulses. The VUV light was seen to be emitted over a 25 ns time period and was collimated down to single photoelectrons by using a 30 cm tube with 200 μm diameter holes at both ends. Single-photoelectron signals on the cathode strips were observed to be as large as 40 mV, using a simple current amplifier with sensitivity of 10–20 mV/pC, and produce a pulse-height spectrum that starts to peak above the sensitivity threshold of the multichannel analyser, as can be seen in fig. 5. A VUV photon absorber, transparent to VUV light above 190 nm, was placed between the chamber and the VUV lamp; the rate of single-photon detection was reduced, but the pulse-height spec-

Table 1

Voltages used for operation. Symbols refer to those of fig. 1. The gate-plane voltage was for the condition of 'gate open'; the 'gate closed' condition had ±100 V on alternating wires.

Electrodes	Voltages	
	Ar (90%)+ methane (10%)	He (95%)+ ethane (5%) &42°C TMAE
D	−12 600	−5600
P1	−12 500	−5000
P2	−8700	−3700
G	−8050	−3200
P3	−7400	−2700
P4	−4000	−1500
Drift chain	−3000 to 0	−1000 to 0
Offset across drift chamber added on entrance electrodes to dc	+30	+10
Anode	+1500	+900

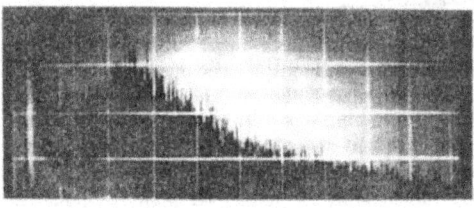

Fig. 4. The black (white) spectrum is the pulse-height distribution for betas from a source at [106]Ru with the gate open (closed).

Fig. 5. Pulse-height spectrum for single photoelectrons in a gas mixture of helium (95%) + ethane (5%) through TMAE at 42°C.

trum remained unchanged, indicating that the spectrum measured was indeed that of single photons. For these measurements, the two PPACs were operated at more than 400 V below their sparking threshold, completely eliminating any electrical discharges. The hot TMAE gas, or the chamber, appears to generate a 1 kHz noise, similar to a thermal photocathode noise in a PMT. This noise was not a problem because the events were triggered by the pulse signals from the VUV lamp, and this noise would only be noticeable if the gate was left open. The pulse-height spectrum of this hot noise is similar to the single-photon spectrum but with an extended exponential tail. At present, the pulse-height difference in the helium-gas mixture, between single photoelectrons, minimum-ionizing particles gated with the event, and background particles, is not known. There are many places where the minimum-ionizing background can be reduced by making the size of the transfer gap after the gate smaller, and by adding an additional gate plane before entering the drift chamber.

The drift-time to drift-distance relation, shown in fig. 6, was measured with single photoelectrons in the helium-gas mixture by using the VUV lamp trigger pulse as the Start and the cathode strip signal as the Stop. The position resolution for single photoelectrons was measured using the deuterium VUV lamp and the collimating telescope in order to produce a small spot of VUV light on the quartz entrance window of the chamber. Fig. 7 shows the time spectrum of the single photoelectrons at 5 cm drift distance. The horizontal scale is 710 ns per division, and using the drift-time to drift-distance relation of fig. 6 gives 2.4 mm FWHM (1 mm sigma) position resolution. The actual position resolution is probably better than this, because the VUV spot was 200 μm and the VUV light was emitted by the lamp during 25 ns (FWHM). The main limiting factor on the position resolution with this type of detector is the fluctuation in the arrival time of the signal on the anode, which is caused by the different conversion depths of the VUV photons in the gas. The naïve approach would be to say that the position resolution should be the thickness of the conversion gap. However, two other factors are important: first, the probability for a VUV photon interaction is exponential; and sec-

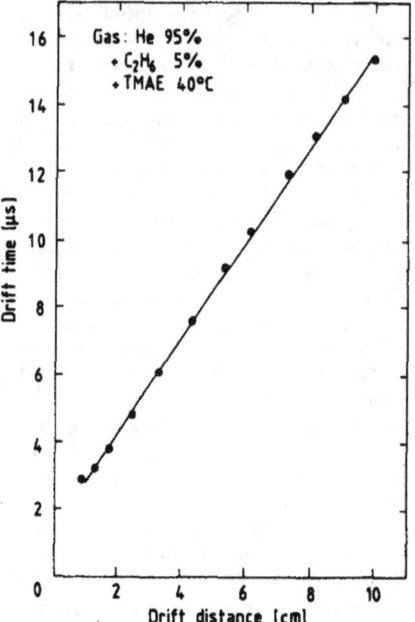

Fig. 6. Drift-time to drift-distance relation for single photoelectrons using the same gas mixture as for fig. 5.

ondly, the drift velocity in the conversion gap is much faster than in the drift chamber. The voltages in the drift chamber were set to be three times slower than in the conversion gap, so that the 6 mm conversion gap would map into a 2 mm error in the drift chamber's timing measurement. This effect was well demonstrated when the position resolution was measured for different electric fields in the conversion gap, and is shown in fig.

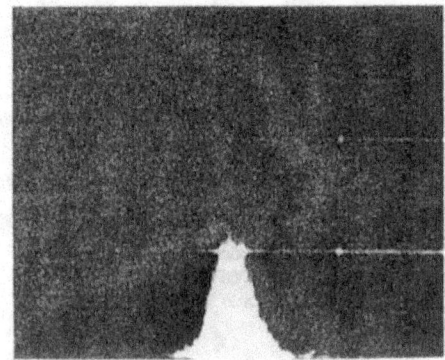

Fig. 7. Timing distribution for a 200 μm spot of VUV light at 5 cm drift distance. Each box is 710 ns and, using the drift velocity from fig. 6, the position resolution is 2.4 mm FWHM.

IX. DIVERSE CHAMBERS

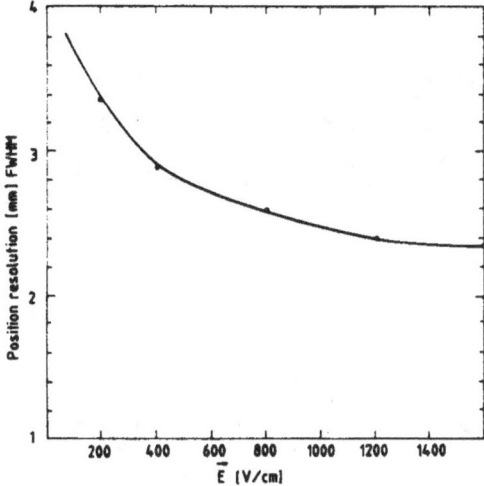

Fig. 8. Position resolution as a function of the electric field in the conversion gap.

8. It can be seen that the position resolution is best when the drift velocity saturates to its maximum in the conversion gap.

4. Conclusion

The principles of operation of this preamplified drift chamber have been shown to work, and warrant further investigation. Many more measurements are necessary before such a detector can be seriously considered for use in a high-energy physics experiment. A detailed study of the position resolution over the entire drift distance is necessary, for both charged-particle tracking and single photoelectrons. Uniform detection efficiency, again over the whole drift distance, is necessary, and an investigation into the efficiency variations of the transfer into the drift chamber must be tackled. Small dead regions in the transfer are not a problem, since the parallel-plate amplification spreads the signal out over 2 mm, depending upon the gas mixture, but large dead regions of poorer transfer efficiency would mean a

larger dynamic range of the signal to be detected. At present the detector has been operated in dc mode, with either gate open or gate closed, but in the future the pulsing will be required to make the detector suitable for use in high-intensity experiments. Pulsing limitations and induced signals must be looked into. Studies at high intensities are necessary in order to determine the actual upper limit with such a detector. Its use in or near a magnetic field cannot be guaranteed without further calculation of electric- and magnetic-field effects on the lines of force.

This preamplified drift chamber has many possible uses in experiments that have high-intensity particle fluxes and require high readout rates, e.g. in charged-particle tracking and for VUV photon detection, such as in ring-imaging of Cherenkov light or BaF_2 scintillation light. The fact that an electrical signal can be taken from the second parallel-plate amplification gap makes this drift chamber the first to be able to provide its own Start trigger, even for X-ray sources. Since the readout methods of the drift chamber are diverse, it is not clear which electronics is best suited for a given application, and this could be an interesting topic for further study. At present, ≈ 100 ns charge-collection times have been chosen, because this is the memory time of the MWPC in the experiment, but in principle shorter collection times are possible, down to the 10 ns rise time of the high-voltage gate pulser. For faster VUV photon collection, higher TMAE temperatures will be necessary, or possibly other liquid or solid photocathodes will have to be found.

References

[1] A. Breskin et al., Nucl. Instr. and Meth. 178 (1980) 11.
[2] G. Charpak et al., CERN 78-05 (1978).
[3] M. Suzuki et al., Nucl. Instr. and Meth. A263 (1988) 237.
[4] P. Fischer et al., IEEE Trans. Nucl. Sci. NS-35 (1988) 432.
[5] N. Solomey et al., Nucl. Instr. and Meth. A271 (1988) 423.
[6] N. Solomey and E.N. May, Argonne National Laboratory Internal Report ANL-HEP-PDK-92 (1984).
[7] G. Charpak et al., Nucl. Instr. and Meth. 205 (1983) 403.
[8] W. Beusch, CERN, private communication.

Nuclear Instruments and Methods in Physics Research A274 (1989) 275–290
North-Holland, Amsterdam

GASEOUS DETECTORS WITH PARALLEL ELECTRODES AND ANODE MESH PLANES

G. CHARPAK [1]), W. DOMINIK [1])*, J.C. SANTIARD [1]), F. SAULI [1]) and N. SOLOMEY [2])

[1]) CERN, Geneva, Switzerland
[2]) University of Geneva, Geneva, Switzerland

Received 5 July 1988

The amplification of electron avalanches in a uniform electric field between parallel electrodes can be used in single or multistep structures for a great variety of applications in fields such as charged-particle tracking, X-ray imaging, ultraviolet photon detection for ring imaging of Cherenkov light, beta autoradiography, and gamma-ray astrophysics. In this article some observed properties of parallel-electrode structures are analysed and compared with those of wire chambers. Parallel electrodes have advantages over wire chambers in their better energy resolution, better timing resolution, and the fact that they are easier to construct and are more durable. It is now possible to construct proportional gaseous detectors without the need to string wires, with readout options using either electronic charge signals or optical light imaging. For either of these methods the position and energy of the avalanches can easily be determined.

1. Introduction

The multiplication of charges in a gaseous medium through a succession of inelastic ionizing collisions building up a Townsend avalanche is one of the oldest detection techniques in nuclear physics [1,2]. Detectors based on amplification around wires have long been preferred since they permit larger gains before breakdown, which was a decisive advantage in the early days. However, it was a routinely used technique in the field of heavily ionizing particles, because there the required gains were much smaller. The limitation to the amplification came mainly from the onset of disruptive sparks when the avalanche reached a given threshold, this threshold being much lower than with wires. In a wire chamber the electric field is intense only around the wires and after the avalanche grows beyond a certain level; the decreasing electric field away from the wire is not favourable for the propagation of avalanches by ultraviolet (UV) photons which leads to disruptive streamers.

However, electronics has evolved, providing low-noise, low-cost, spark-resistant amplifiers permitting work with much smaller gains and greater safety. More importantly the understanding of the parallel-electrode structures has increased. It is now possible to use multistep structures with gains so large that avalanches from single electrons liberated by UV photons can be detected with nearly 100% efficiency. Even the single-gap structures have evolved, with a better understanding of

* On leave of absence from the University of Warsaw, Poland.

factors limiting the gain. The readout of the position of avalanches in parallel gaps has also progressed. It can be done in two dimensions by a variety of electrical methods and very conveniently by observation of the light emitted by the excited atoms in the avalanche. This last technique can provide an advantage in some applications, in particular when events of high complexity have to be analysed.

After the initial work on multistep chambers in 1979 [3,4] several groups started to adopt amplification between parallel gaps for soft X-ray imaging, demonstrating sizeable advantages over proportional wire chambers, with essentially the same energy resolution. There is at present some confusion concerning the limitation in energy resolution or local counting rates. In this work we present some observations on the properties of the amplifying gaps between parallel electrodes, with particular emphasis on the construction where the anode plane is made of a mesh or wires since this is the condition for electron transfer to successive gaps. The parallel-plate avalanche chambers (PPACs) constructed in this way work in conditions equivalent to uniform field amplification.

Some observations are presented here on the relationship between charge amplification and light production using multiwire proportional chambers (MWPCs) and uniform electric fields such as in PPACs. A study of energy resolution, timing resolution, and feedback was performed in an effort to determine which of these amplification techniques was most suitable for certain needs. A study of light production in different electric fields, use of anode meshes under conditions of symmetric electric fields similar to those applied to wire planes in MWPCs, and a description of the setup and operation of a single-gap chamber that has two steps of amplification – in the uniform gap and on the wires of the limiting anode grid – is presented. Although the work described here deals mostly with the properties of single-gap amplification structures, it has implications for multi-gap structures; these have the advantage of permitting the gating of the events and can thus be used in high-rate environments.

2. Properties of a single amplification gap limited by mesh planes

When a voltage is applied to a mesh plane in order to produce a uniform electric field, the field is uniform in most of the gap but is increased near the wires. The strength and distribution of the field lines depend on the wire diameter, the wire spacing, and the electric field on either side of the mesh. The case of a grid made of parallel wires has been extensively described in many articles on wire chambers [5]. Therefore, we limit ourselves to the case of wire meshes with wires stretched and interwoven in two orthogonal directions: these mesh structures have a mechanical advantage over wires in that they are quicker to make and stronger. We have worked with anode planes made of meshes and wires, in single-step and multistep modes, in order to make comparisons.

Gaseous chambers containing vapours such as triethylamine (TEA) or tetrakis (dimethylamine)ethylene (TMAE) can be made to produce observable light in the UV and visible regions, respectively [6–9]. This light production is the result of deexcitation of excited states formed by electron impact collision with the molecules during the amplification process [7]. Previously it was shown that the ratio of the amount of light produced to the amount of charge present was a function of the charge gain [8]. These relationships have a peak at low charge gain (5–10) indicating that to produce light most efficiently it is best to have a low charge gain close to the peak of this curve (see fig. 1). In this work these observations were extended and also now include measurements with intense electric fields and the effects of their geometry.

2.1. Multistep chamber with an optical-fibre readout

A multistep chamber [3,4] is a device combining a PPAC with either another PPAC or a MWPC. If the PPACs are constructed using meshes, the electrons produced by the Townsend avalanche mechanism in one PPAC can be transferred out through the mesh plane and drifted into a second amplification step, which is either another PPAC or a MWPC; higher gains can thus

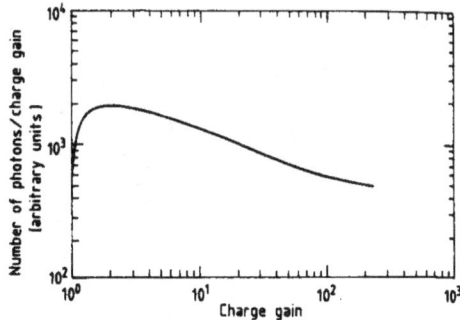

Fig. 1. Relation between charge gain and light gain. Amplification gap width $d = 1.5$ mm, for a gas mixture of 88% argon, 10% methane, and 2% TEA [8].

be achieved. These types of devices can be of interest in high-energy physics because they permit high-rate operation ($> 10^7$ Hz) when gated; this also allows event selection from a high background [5].

Since the electrons produced are drifted perpendicularly to the mesh planes, such a device can be used with complex events, having a large multiplicity, provided the means for reading out such a large multiplicity can be found. One method of handling these events is by optical imaging of the secondary light produced from the charge-multiplication process. The first multistep light chamber [6] had a geometry identical to that of the first multistep chambers used with charge-readout methods; they both used a MWPC as the second step of amplification, and initially it was assumed that most of the light was produced in the MWPC final amplification step. However, after some measurements with a multistep light chamber composed of a PPAC as the first step of amplification and a MWPC as the second, there was an indication that the light was mainly produced in the PPAC and not in the MWPC. The chamber construction is shown in fig. 2. It was a multistep chamber with the first amplification gap being a PPAC made with meshes, and the second amplification gap was either another PPAC made only of meshes or a classical MWPC. The chamber had thin Aclar * windows to allow X-rays to penetrate, and to permit UV light produced in the chamber to be imaged. Located on the outside of the exit window was an array of ten wavelength shifting optical fibres **, 1 mm in diameter; these shifted TEA emission to 450 nm and were connected to a photomultiplier tube (PMT); the amount of light produced by the chamber, measured by this system, is shown in fig. 3. Two types of measurements were made: first the gain in the first step of amplifica-

---

* Polychlorotrifluoroethylene (PCTFE).

** Fibres provided by M. Bourdinaud, CEA, Saclay (France).

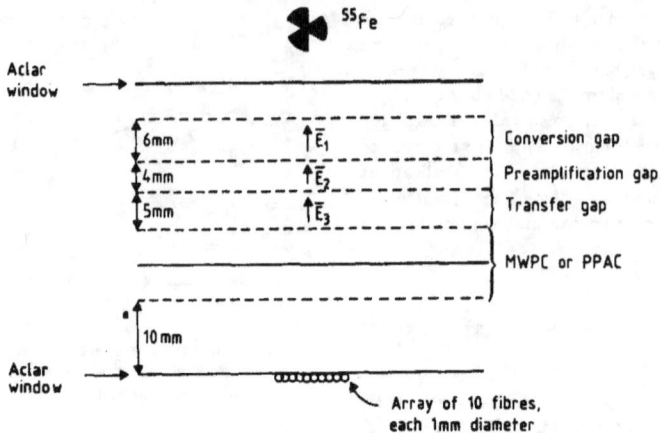

Fig. 2. Multistep chamber construction with either a PPAC or a MWPC as the second step of amplification, and an optical-fibre readout.

tion was kept constant, while that in the second MWPC step was varied; this was done for several different gains in the first step of amplification and the results are shown as the solid curves in fig. 3. The second type of measurement was done keeping the gain in the second MWPC step constant while that in the first step, a PPAC, was varied; the result is shown as the dashed curve in the figure. From a comparison of these curves, the fact that the slope of the dashed curve is much steeper indicates that light was produced most efficiently when the gain in the first amplification step was increased.

The light source could be determined by using only one optical fibre connected to the PMT, and moving the radioactive source perpendicularly to the optical fibre in order to measure the light distribution. This was done with only the first amplification step turned on (dashed curve in fig. 4), and with only the MWPC turned on (solid curve in fig. 4). The widths of these two curves are due to the different distances of the optical fibres

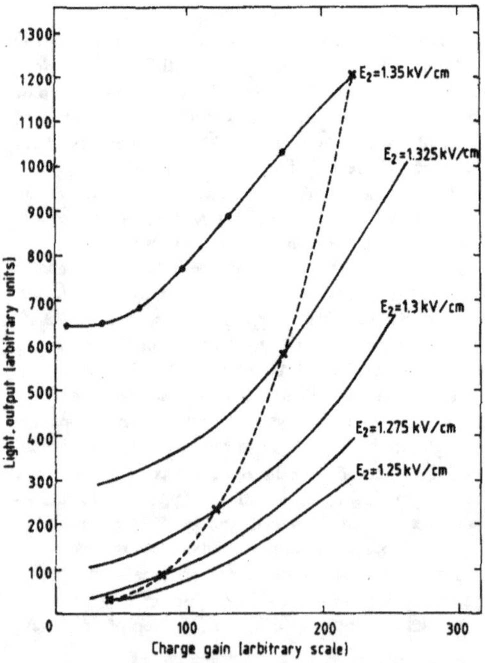

Fig. 3. The multistep chamber with a MWPC as the second step of amplification, showing the light output as a function of the charge gain. See text for an explanation of the curves.

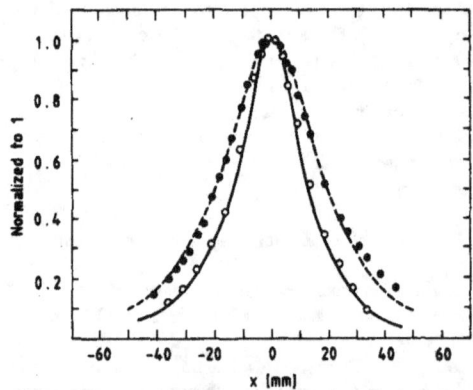

Fig. 4. Light distributions as viewed with the optical fibres. See text for an explanation of the curves, which have been normalized to each other.

from the light source. When the chamber was operated as a multistep light chamber, with a high gain in the first stage, the distribution of light after normalization was identical to that shown by the dashed curve in fig. 4; this indicates that most of the light was produced in the first step of amplification, the PPAC, and not very much in the second step, the MWPC. With the experience obtained in using optical fibres for these tests, it is now possible to suggest using an array of optical fibres close to the light source in order to determine the position of the avalanche. Both dimensions could be obtained by having two layers of optical fibres perpendicular to each other, their readout being achieved by using a linear charge-coupled device or a pin diode string. If the optical fibres are placed close to the light source, then the distribution of the light will be much smaller than shown in fig. 4. Although this method would only be able to handle the same multiplicity as the MWPC charge readout, it does have other advantages of the light-readout methods, which can be useful in some applications. When using optical-fibre readout in a gas scintillation detector for astronomy applications [10], difficulties arise from insufficient light; here, with stimulated light emission by electron multiplication, this is not a problem.

This work was our first indication that the light yield was dramatically greater in a uniform electric field amplification process, such as that taking place in a PPAC. After determining that the MWPC was hindering the maximum possible amount of light, the MWPC

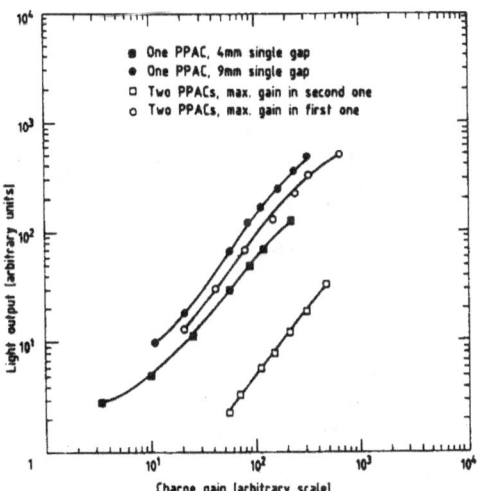

Fig. 5. Light output as a function of charge gain, using either a single PPAC with gaps of different thicknesses, or a multistep chamber with two PPACs. Comparisons should be made only between the two curves with full data points and between the two with empty data points.

was replaced by a second PPAC using meshes. To understand the limiting factors in the maximum light achievable, a comparison of light output for different charge gains was performed, using a single-gap PPAC, and multistep chambers with two PPACs. The results are shown in fig. 5. Here it is seen that the maximum charge gain achieved has some upper limit dependent upon the gas mixture, but independent of the method used to obtain it. The most striking feature of this figure is that the light produced with a multistep chamber using two PPACs is at a maximum when the gain in the first PPAC is at a maximum, with the second step at a lower gain; the light was then observed to originate almost completely from the second PPAC. This should be compared with the lower light output, when the chamber was operated with maximum gain in the second PPAC and moderate gain in the first. Under these conditions, even though the final charge gains were identical, 200 times less light was observed than with the more favourable operating conditions. This effect is related to the decaying tail in the light-to-charge ratio at high charge gains.

2.2. Light and charge measurements with a single amplification gap

Fig. 6 shows the field lines in a gap limiting by meshes, made of 50 µm diameter wires spaces every 500 µm, which are common and commercially available *. The electric fields are calculated in three dimensions using the actual geometry of a perfect mesh. The black dots are the wires of the mesh, perpendicular to the plane of the paper; this plane is half way between two of the mesh wires that are parallel to the plane of the paper (and thus perpendicular to the other wires). Fig. 6a shows a mesh plane with nonsymmetric fields, as is commonly found in multistep structures; fig. 6b shows an anode mesh plane with symmetric electric fields (similar to those fields used in MWPCs, but here with a mesh replacing the wire plane). It is clear that if the electric field is large enough for inelastic collisions to occur in the uniform field region, such collisions will also occur close to the wires, with an increased probability per unit length since the fields are slightly higher there. The gain may depend upon the different paths taken by the lines of force guiding the electrons, which may differ as a function of the initial position of an initially converted electron. The gain may be variable, thus spoiling energy resolution if such a gap is used for X-ray imaging when compared to an amplification gap with a perfectly uniform electric field. The electrons produced in the uniform field may either go to the wire at various angles, thus traversing different electric fields,

---

* Manufactured by Gantois, Saint-Dié, France.

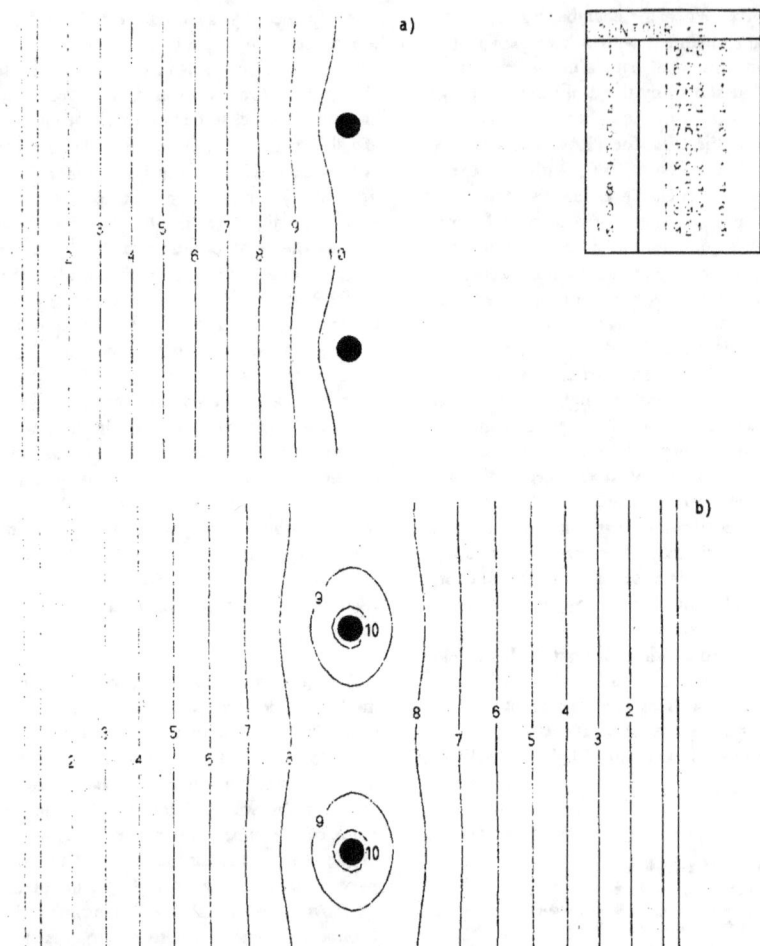

Fig. 6. (a) Equipotentials for a mesh plane with the amplification gap on the right, and zero electric field on the left; these are the conditions referred to as PPAC mode. (b) Equipotentials for a mesh plane with a symmetric electric field similar to those in MWPCs; in the case of a wire plane, the ovals would be circles. The slices shown in both these figures are half way between the mesh wires parallel to the plane, and the solid circles are the mesh wires perpendicular to the plane. Notice should be taken of how the electric field in the gap is reduced when a symmetric electric field is applied around the anode mesh plane.

or be transferred into the region beyond, which has a different electric field. This may affect the electronic gain and also the photon yield. This is best illustrated by the following experiments, performed on a double gap where the electronic signal is collected on the anode mesh and there is a simultaneous measurement of the light intensities with a photomultiplier.

The experimental system consisted of a ^{55}Fe X-ray source placed in front of the chamber. The chamber had thin Aclar windows, transparent to light of wavelength larger than 280 nm, on both sides so that the UV photons emitted from the excited TEA molecules could

be observed with a PMT that had a quartz window. The chamber was constructed with mesh planes mounted onto fibre-glass-epoxy frames. The schematic of the chamber construction and experimental setup is shown in fig. 7. The electrode mesh planes were constructed out of 50 μm diameter stainless-steel wires spaced 500 μm apart. The anode plane was changed according to the tests being performed, but it was either the same as the electrode mesh planes, or another mesh made of 30 μm stainless-steel wires with the same spacing, or wires of different diameter and spacing. The thickness of the conversion gap was 17 mm; the thickness of the amplifi-

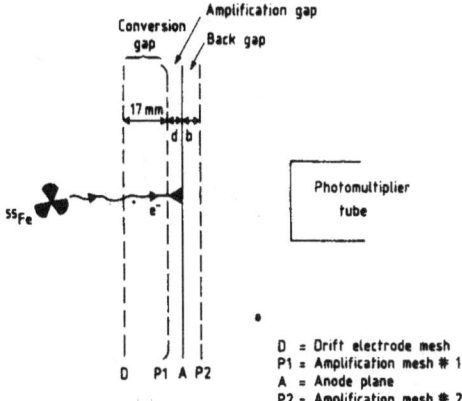

Fig. 7. Experimental setup for the measurement of light and charge from the test chamber: the charge signal was taken from the plane labelled A and the light signal was recorded by the PMT.

cation gap and the back gap could be changed to investigate the dependence of charge and light output. The gas mixtures used were mainly 90% argon and 10% methane, or 90% argon, 8% methane, and 2% TEA; also other gases were occasionally used.

The initial tests were started after calculating the electric field for an anode mesh plane at high voltage between two cathode ground planes; these electric-field conditions are similar to those in the MWPC, but with the wire plane replaced by a mesh plane. We were interested in measuring the relation between the charge and light produced as a function of the electric fields in the two gaps. In the past, the PPACs with mesh electrodes were operated with a high electric field on one side of the charge-collection mesh and zero electric field on the other; these conditions are referred to as antisymmetric fields. Fig. 8a shows the electronic charge signal and fig. 8b the light signal as the electric field is changed from an antisymmetric to a symmetric field around the mesh anode. The results are at first sight surprising: whilst the charge signal increases when the field is made symmetric, the light signal decreases. The explanation lies in the fact that practically no light is produced in regions of intense electric fields such as those found around wires, and a careful study of fig. 6 provides the answer. The upper half of fig. 6 shows the antisymmetric case and the lower half the symmetric case. It is clearly seen from a comparison of these two figures that the electric field in the gap decreases when the electric field around the mesh plane is made symmetric. With the electric field lower in the gap and higher on the mesh wires, there is less amplification in the gap. Even though the charge signal increases, the light production is less, indicating that the latter comes

mostly from the gap and not from amplification processes in intense electric fields. Another example of this effect will be described later in the paper. This work also shows that an anode wire plane may be replaced by an anode mesh, which is much stronger and easier to manufacture, and the detector still acts like a proportional wire chamber, as shown by measurements of pulses produced by X-rays (fig. 9). Other experiments have proposed the use of metal strips or pads, for the last electrode, to carry out the charge collection [11,12]. There have been some difficulties with this technique owing to charge buildup on the insulators between strips or pads, and spark discharges to the charge-sensitive pads destroying the electronics. After seeing that it was possible to use anode mesh planes as the charge-collection electrode in a symmetric electric field we are led to suggest the use of cathode pads as the back electrode (see fig. 10). This would have the advantage of a highly segmented pad readout for high-multiplicity events, as in the other cases, but it would not create the difficulties encountered when the pads are used as the last electrode for charge collection as mentioned above.

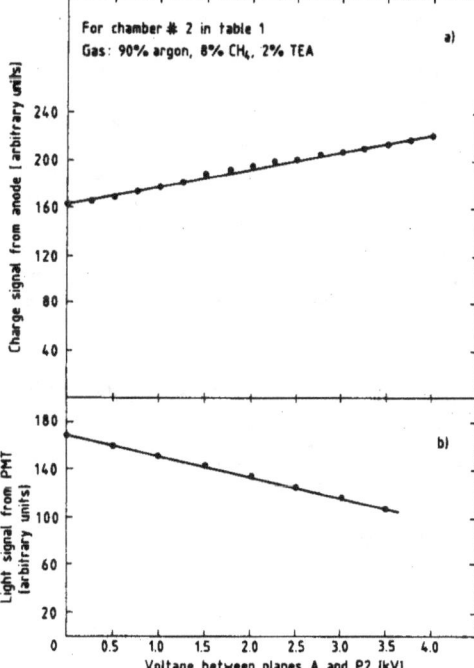

Fig. 8. Measurement of the charge signal (a) and the light signal (b), as the electric field around the anode mesh plane is made symmetric.

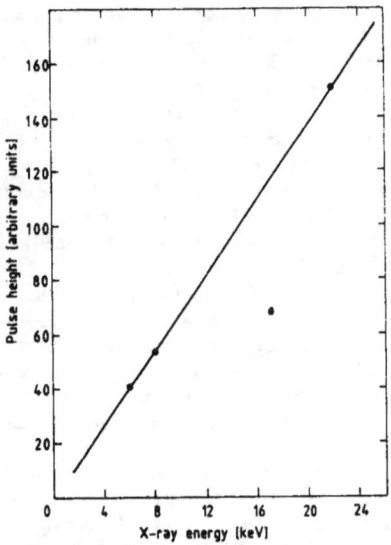

Fig. 9. The charge-signal pulse height for different energy X-rays, showing that the anode mesh plane with a symmetric electric field is proportional to energy.

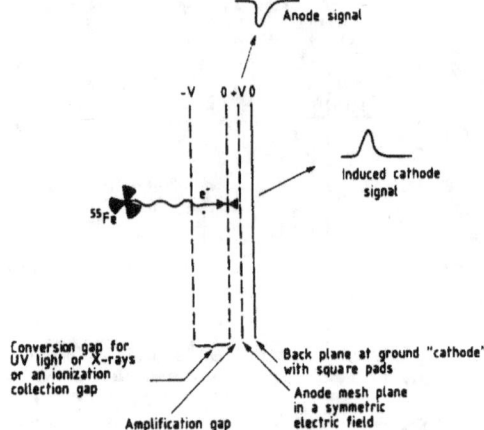

Fig. 10. A suggested chamber design for the anode mesh plane with symmetric electric field and a back plane made of cathode strips or pads to localize the shower.

Table 1 lists the many different chambers which were constructed to determine the optimal conditions for light and charge measurements. The data in table 1 were obtained at the maximum safe operating voltage for the given conditions, and the symbols refer to those in fig. 7. With one chamber we also investigated different gas mixtures under conditions of symmetric and antisymmetric electric fields around the anode mesh, and the results are listed in table 2. This has clear consequences for developments which aim at obtaining the maximum amount of light. It is more advantageous to split the amplification process into two steps: a first step aiming at the maximum electric charge multiplica-

tion which can be attained under stable conditions with respect to sparking, followed by a second step with a modest gain to optimize the light yield [13] (see fig. 11).

We have found that there is more light with larger gaps between meshes, since there are then no regions of intense electric fields, and this is why the imaging optical chambers are now much easier to operate. In the early times when the final step of amplification was a wire chamber it was operated at extreme gains, such as those found in limited streamer mode, in order to observe satisfactory tracks. Conclusions drawn from these data indicate that the optimum chambers for light-imaging readout are with parallel electrodes made of meshes, and the best conditions for electronic charge readout are with wires having the spacing of the classical MWPC. However, for those experiments where it may be desirable to construct large chambers out of

Table 1

Results of light and charge measurements with different chambers and operation modes [a]

Chamber				E	Pulse height (arbitrary units)			
Number	Wire diameter (μm)	Mesh of wires	d	(V/mm)	Charge signal		Ligth signal	
					ASM [b]	SM [c]	ASM	SM
1	30	mesh	2	850	150	200	250	112
2	30	mesh	5	1000	160	220	168	108
3	50	mesh	3	1100	120	120	375	225
4	50	wires every 0.5 mm	3	1100	175	350	80	42
5	20	wires every 2 mm	4	660	1000	1000	2	4

[a] Compare down columns, and in rows only ASM and SM below Charge signal and Light signal, respectively. The gas is 90% argon, 8% CH_4, and 2% TEA.
[b] ASM = antisymmetric electric field around anode plane.
[c] SM = symmetric electric field around anode plane.

Table 2
Results of charge and light measurements with different gas mixtures [a]

Gas	E (V/mm)	R [b]	
		Charge	Light
90% Ar + 10% CH$_4$	1240	2.0	–
98% Ar + 2% TEA	690	1.1	0.55
95% Ar + 5% TEA	825	1.5	0.60
90% Ar + 8% CH$_4$ + 2% TEA	1100	1.0	0.60
97.5% Ar + 2.5% isobutane	720	1.3	–
90% Ar + 10% isobutane	1250	1.7	–
80% Ar + 20% isobutane	1600	3.0	–
80% Ar + 18% isobutane + 2% TEA	1330	3.2	–

[a] The chamber used for these measurements was number 3 in table 1.
[b] R = pulse height SM/pulse height ASM, where SM is the symmetric electric field around anode plane and ASM is the antisymmetric field around anode plane; – means no light produced.

meshes and still use electronic readout methods, because of the easier construction, it is still possible to do this if lower pulse-height signals are acceptable. Also the best gases for charge signals seem to be those of the MWPC.

It was suggested that the amount of light being observed was influenced by shadowing of the wires; in MWPCs most of the light could have been shadowed by the wires if the intense electric field concentrated the avalanche on one side of the wire, such that the light produced was hidden from the view of the PMT or camera. We have measured the amount of light being emitted on both sides of the chamber for MWPCs as well as PPACs and found no difference. We also measured the amount of light at varying angles with respect to the normal to the exit window and found the light emission to be isotropic.

2.3. Energy resolution

The initial work with multistep avalanche chambers [3,4] had a surprisingly good energy resolution obtained from the first amplifying gap. Fig. 12 illustrates this initial structure, which consisted of a conversion space for the X-ray photons, an amplifying gap between parallel grids, a drift space where the electrons from the

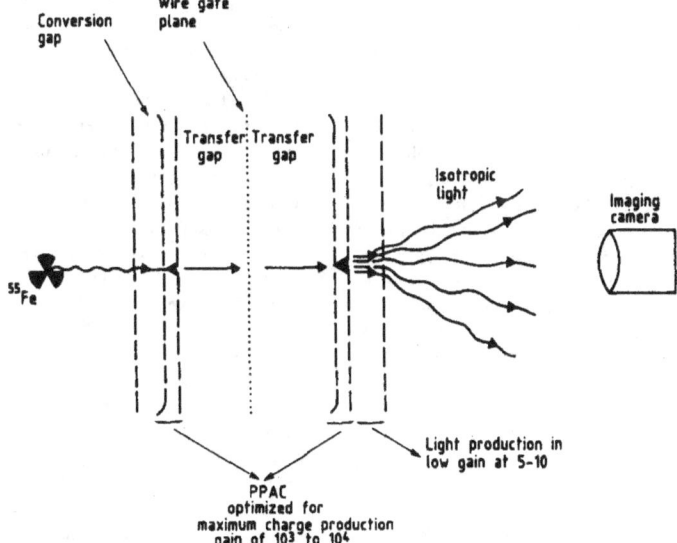

Fig. 11. A multistep chamber followed by a light production gap which is optimum for maximum charge production and maximum light emission. This structure is the one which was used in the low-pressure Cherenkov counter of ref. [13].

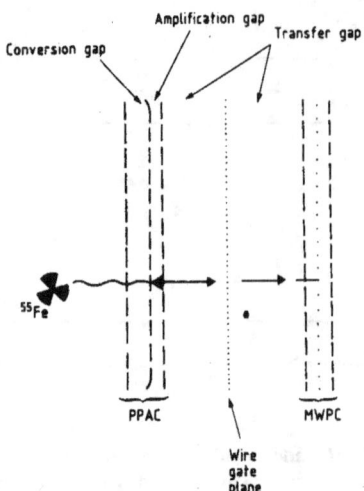

Fig. 12. A schematic of the original multistep chamber with MWPC readout.

amplifying gap were transferred and, finally, a multi-wire chamber. The resolution was roughly the same as that of wire chambers, i.e. 18% FWHM for 5.9 keV X-rays. It was not the best which could be reached by a careful tuning of all the parameters contributing to the resolution, but it was sufficient for a number of applications. The fluctuations in uniform electric fields have been analysed by Alkhazov [14], and the attainable energy resolution should, in principle, be better than with single-wire proportional chambers.

The charge-signal energy resolution is shown in fig. 13a for our single-gap chamber filled with 90% argon, 8% methane (CH_4), and 2% TEA. It has 14% FWHM for the PPAC mode and 18% when the anode mesh plane is operated with a symmetric electric field. The light-signal energy resolution, shown in figs. 13b and 13c, is 18% for the PPAC and 20% for the symmetric electric field mode. In the PPAC mode the charge-signal energy resolution is practically 1% short of the value reached by those groups who have strived to attain the best energy resolution, even though in our tests we have used low-cost electronics. These results are in agreement with Alkhazov's calculations that uniform electric fields produce better energy resolution. The energy resolution of our light measurements is worse than the charge-signal energy resolution, as would be expected from an additional fluctuation in the light-production mechanism on top of the charge fluctuations.

In this respect we can mention other work advocating the use of the "Penning imaging counter", which was a normal multistep chamber filled with a Penning gas [15]. Such a mixture was used in the original multistep chamber, when it was thought that a Penning

mixture was important for the uniform transfer of charge from the amplifying gap to the transfer gap. Indeed, studies with a LiF window between these two gaps showed that with these mixtures the transfer mediated by photons does exist. However, further studies showed that the transfer was also uniform with mixtures not exhibiting the properties of a Penning mixture, where the electrons produced in the avalanche simply moved along the lines of force going from one gap to the other.

It is also worth mentioning that the transfer of charge does not always lead to a loss of energy resolution. This may be due to the fact that the lateral spread of the avalanche is very favourable for proportional amplification at the anode mesh of the final stage. The anode mesh limits the effect of space charge from the positive ions, which is strongest when they are in concentrated avalanches such as those found around wires. Such an effect was observed with the spherical drift chambers [16], where the energy resolution from the charge gain improved after the ionization electrons had drifted by as much as 15 cm, with respect to that obtained by direct detection in the wire chamber.

2.4. Feedback measurements from mesh planes

A useful way of reducing the secondary effects produced by the UV photons emitted from TEA vapour was found when the stainless-steel mesh wires were gold-plated. The chamber used in these tests was similar to that shown in fig. 7, with a 3 mm amplification gap but with a gold-plated drift electrode. Since the electronics used to observe the charge signal was not fast enough, it was not possible to measure the photon feedback from a mesh plane other than that of the drift electrode. The time delay between the drift electrode and the anode plane was 180 ns. The electronics circuit used to measure the feedback is shown in fig. 14; the main idea in the measurement of the feedback was to determine for each event the time until the next event. The threshold of the electronics for detection was on the level of 5 primary electrons; hence our measurement was not sensitive to secondary single photoelectrons. If all of the observed signals were originating from the source, then the time spectrum should be flat, owing to the random time process of the decaying source. If there was feedback from the TEA molecules in the gas there should be an observed increase in the time spectrum for times corresponding to gas drift distances when compared to the longer times which are purely due to statistics. For feedback from the mesh plane the time spectrum should have a bump related to the drift-time needed for the electrons extracted from the mesh plane to drift back to the anode mesh. The threshold of electron extraction for stainless steel is 4.5 eV, lower than that for gold, which is 5.1 eV., The light-emission spectrum of TEA was previously measured [17], and is

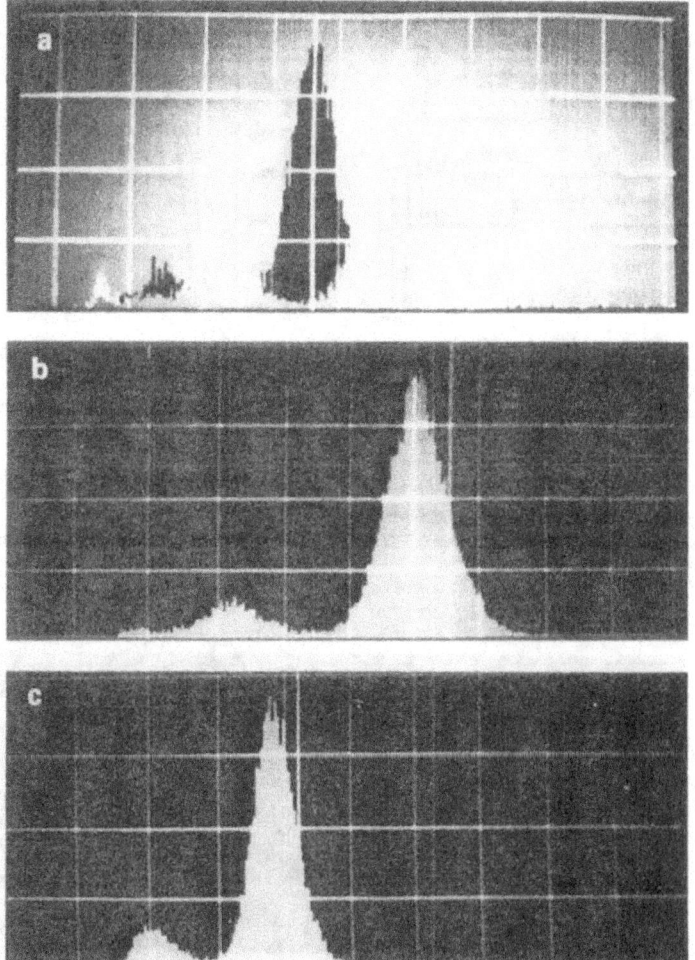

Fig. 13. (a) The energy resolution of the charge signal for ^{55}Fe X-rays: the black spectrum is 14% FWHM for PPAC operation, and the white spectrum is 18% FWHM with a symmetric electric field around the anode mesh plane. (b) The energy resolution of the light signal is 18% FWHM with ^{55}Fe for PPAC operation. (c) The energy resolution of the light signal is 20% FWHM with ^{55}Fe for symmetric electric field around the anode mesh plane.

shown in fig. 15. Because less photons emitted from TEA reach the threshold of gold than that of stainless steel, we observed a sharp decrease in feedback pulses, from 0.7% with stainless-steel meshes to 0.05% for gold-plated meshes, for the same observed charge pulse height, as shown in figs. 16a and 16b respectively.

Such a dramatic drop in the number of photoelectron feedback pulses from the gold-plated mesh and the lack of observable feedback from the TEA molecules in the conversion gap show that the main source of photon feedback in avalanches is photon emission by the TEA

and not the vacuum ultraviolet (VUV) photons from the Ar or other gases used in the detector. Feedback effects, caused by the reabsorption of photons produced in the avalanche, may also affect the energy resolution. If the feedback effects are reduced, higher gains and better energy resolution could be achieved. Since the secondary effects leading to feedback or discharges are dependent on the light intensity, it is most favourable that the first step of electron multiplication be reached under conditions of large electric fields. Also for any chamber where it is wished to observe light; regions of

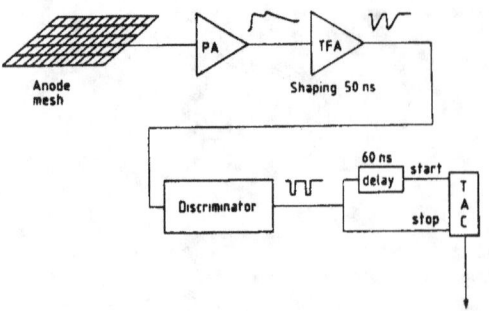

Fig. 14. Schematic of electronics used for measuring the amount of feedback pulses due to the reabsorption of UV photons emitted in the avalanche. PA is a preamplifier, TFA a timing filter amplifier, and TAC a time-to-amplitude converter.

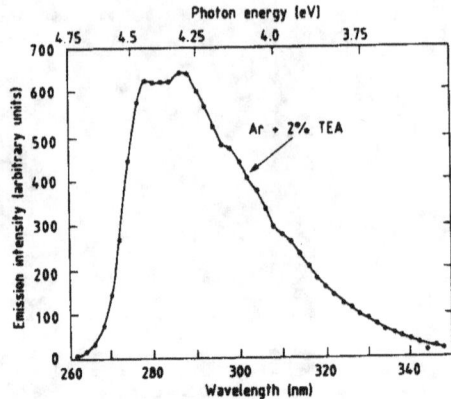

Fig. 15. Light-emission spectrum from mixtures containing TEA [17].

2.5. Combining uniform-field amplification with wire amplification

We wished to construct a chamber that had two steps of amplification in a single gap. Our first attempt

intense electric fields such as those found around wires should be eliminated to avoid feedback from VUV photons.

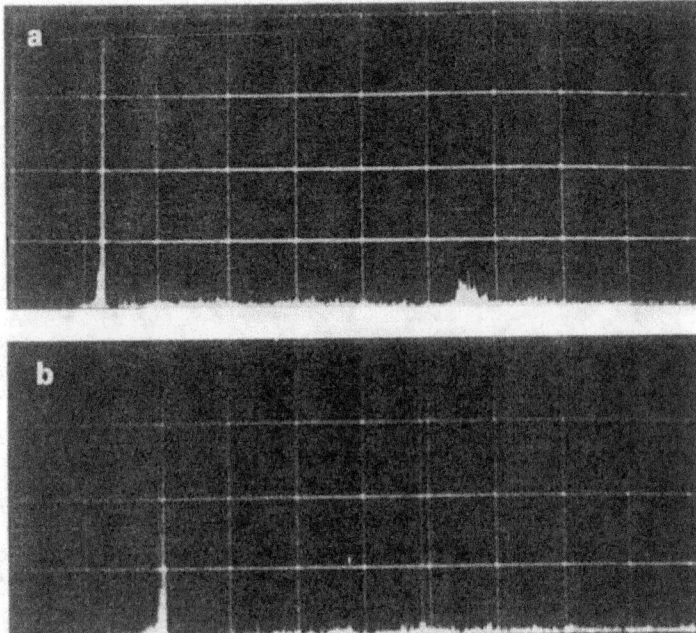

Fig. 16. Time spectra showing the time between pulses in the chamber. The large peak on the left of both figures indicates those events that have no stop for a given start within the range setting of the time-to-amplitude converter. This peak is highly suppressed so as to enhanace the background level. (a) The feedback pulses from the stainless-steel mesh plane; (b) the feedback pulses with the gold-plated mesh plane. The measured feedback from the mesh plane in (a) is 0.7%, clearly seen above the background, and the feedback in (b) is 0.05%.

was a chamber that used an anode plane of wires with 10 μm diameter and 1 mm spacing; the chamber construction is shown in fig. 7; the amplification gap thickness is $d = 3$ mm and the back-gap thickness is $b = 5$ mm. The aim was to have uniform electric field amplification in the gap and additional amplification on the wires. For the detection of X-rays or single UV photons it is possible to remove the parallax error by obtaining a knowledge of the depth of conversion.

The charge gain in an amplifying gap between parallel electrodes, $G = e^{\alpha x}$, is a function of x, the distance between the initial electron and the anode, and of α, the first Townsend coefficient (a function of the electric field and the gas). If the initial electrons from a drift space are injected into the gap through a cathode grid, this gain is constant. If the initial conversion electrons are created inside the amplifying gap the gain will be variable. This variable gain may be exploited to advantage in some applications: it is useful to enhance those events that occur very near to the cathode, such as in autoradiography [18] or slow neutron imaging [19] with gadolinium converting foils. It is also an advantage for applications where the pulse heights of the avalanches from low-energy X-rays are used to obtain information about the depth of absorption in order to overcome parallax error [20].

The most direct proof of combined amplification came when we used an 8 keV X-ray generator, collimated to a 300 μm diameter beam, which was directed at an angle of 45° to the mesh planes (see fig. 17). For this measurement only one anode wire signal was read out. The X-ray beam was moved in 0.5 mm steps so as to change the average depth of conversion with respect to the readout wire, and the average pulse height was recorded. The geometry was taken into consideration and the depth of conversion calculated for each point. The pulse height as a function of the depth of conversion is shown in fig. 18. Although this was not the largest amplification possible in the uniform electric field, it was the best linearity that could be achieved with the electronics at hand. Depth of conversion was measured away from the anode plane so that the largest

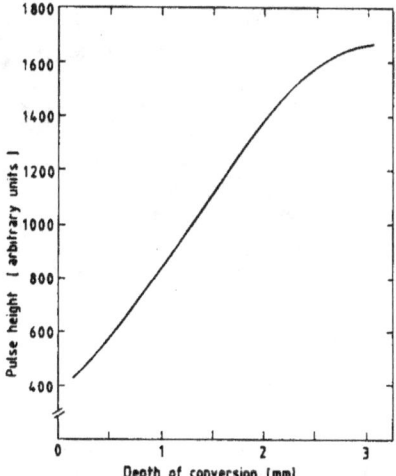

Fig. 18. A measurement of the pulse height as a function of the depth of conversion, using the X-ray beam collimated to 300 μm as described in fig. 17.

pulse height occurred when the X-ray converted 3 mm away from the anode plane. The saturation of the amplifier is starting to dominate at large pulse heights.

Table 3 lists the voltages used on this special chamber when it is operated in classical multiwire chamber mode, in PPAC mode, and in a combined mode. The standard gas mixture of 90% argon and 10% methane was used in the experiments described below. By applying a low symmetric electric field around the anode wire plane, the chamber operated as a classical MWPC, exhibiting 20% FWHM energy resolution for ^{55}Fe. At slightly higher symmetric electric fields, amplification in the gap started to become more important, and with small increases in the electric field the energy resolution quickly degraded. After seeing this degradation it was obvious that the same effect must have been seen 20 years ago when other experiments, including some of the present authors, tried to make MWPCs with very small wire spacing. Our conclusions are that care must be taken, when operating an MWPC with 1 mm wire

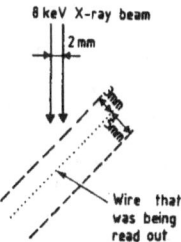

Fig. 17. Experimental setup with the X-ray beam to measure pulse height as a function of the depth of conversion.

Table 3
Electric fields [V] used in timing measurements

	Mode of operation		
	MWPC	PPAC	Combined
P1 [a]	− 2600	− 4000	− 3340
A [a]	0	0	0
P2 [a]	− 4100	0	− 2650

[a] P1 (P2) = amplification mesh No. 1 (No. 2) and A = anode plane.

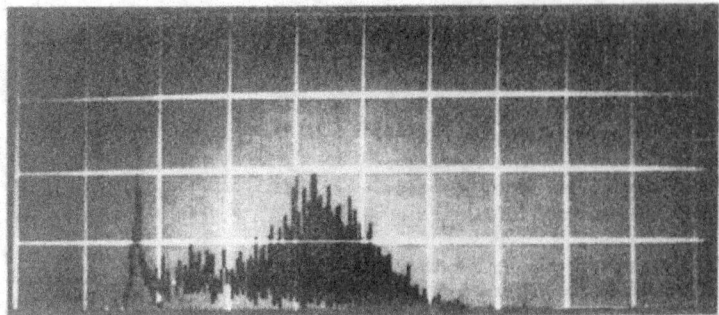

Fig. 19. The black (light, superimposed) spectrum is the pulse-height distribution with the conversion gap on (off); this is additional proof of the amplification in the gap and on the wires.

spacing, in the choice of the gap distance between the anode and the cathode planes, as well as in the use of the correct gas mixtures, so that uniform field amplification combined with wire amplification does not destroy the energy resolution at high gains. Parallel-plate amplification was when there was an electric field across the gap in front of the anode wire plane and zero electric field across the back gap, but it was far from being pure PPAC operation because the thin wires always added some contribution to the amplification. These two modes of amplification were combined by first finding the proper operating voltage for classical MWPC mode and then adding amplification in the front gap by increasing the electric field across this gap. There was an additional conversion gap preceding the uniform field amplification gap. When the conversion gap had a reversed voltage across it, so that the electrons produced there were not collected by the anode wire plane, then the dominant signal came only from those X-rays that converted in the uniform field amplification gap. The conversion gap was also operated so as to drift the electrons towards the anode wire plane; in this case almost all of the electrons had the same amount of amplification in the uniform electric field. The pulse-height spectra for combined amplification mode with the conversion drift on and off are shown in fig. 19. The difference in pulse-height spectra for these modes is a second proof of combined amplification. Differences in the amplification factors of wires and of uniform fields were between 3 and 100, but the amplification in the gap was probably higher. When the conversion gap was turned off, the good energy resolution was destroyed by the varying amplification coming from the different depths of conversion for the X-rays. The charge amplifier used had a sensitivity of 230 mV/pC, as calibrated with the capacitance of the chamber included. Average maximum pulse heights in proportional mode were 820 mV for ^{55}Fe X-rays, which indicated a gain above 10^5.

Using a gas mixture of 90% argon, 8% methane, and 2% TEA vapour, the light output was measured with a photomultiplier. Large amounts of light were seen when the chamber was operated in PPAC mode and in combined mode, but very little light was observed when the chamber was operated in classical MWPC mode. This agrees with the measurements with the other chambers described in section 2.2, which are summarized in table 1. The amount of light was measured when the chamber was operated at maximum gain in the gap for combined

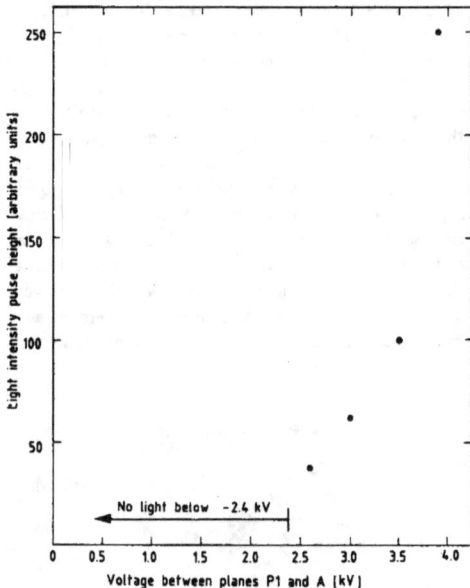

Fig. 20. Using the combined-amplification chamber, the amplification in the gap was lowered and that from the wires increased, keeping the same charge pulse height; the amount of light decreased, proving that light comes only from the uniform field amplification region.

amplification mode and the charge-signal pulse height noted. The electric field across the front amplification gap was lowered, but that across the back gap was raised, in order to keep the same charge-signal pulse height as before, and the amount of light was measured. This was continued as much as possible and the amount of light output as a function of electric field across the front gap was measured, shown in fig. 20. It is seen from this figure that as the electric field across the amplification gap was decreased, the amount of light also decreased, even though the charge signal remained the same. This indicates that the light is produced mainly from the uniform electric field region and that there is very little light production in intense electric fields.

2.6. Timing measurements

The test chamber with the combined amplification described in section 2.5 was operated without the conversion region, using a ^{106}Ru beta source. Others who have worked with low-pressure MWPCs [21] have noticed that the timing resolution was extremely good. They attributed this to the property of two steps of amplification, one from the uniform-field amplification and another from the wires or mesh. Their good timing resolution and electric-field calculations were their main arguments for two-step combined amplification. We were interested in knowing if our combined amplification chamber described above would have better timing resolution than a normal PPAC or a MWPC with 1 mm wire spacing. The experimental setup for the timing measurements consisted of a time-to-amplitude converter; the start signal was provided by a scintillator-photomultiplier at the back of the chamber and the stop signal by the anode-wire plane. Simple threshold timing electronics was used for the anode wire signal [22], and a standard discriminator for the scintillator signal. The timing measurements were made using a gas mixture of 90% argon and 10% methane for classical MWPC, PPAC, and combined-amplification modes with the voltages given in table 3. The measured timing resolutions in FWHM are: 34 ns for the classical MWPC mode, 26 ns for the PPAC mode, and 42 ns for the combined-amplification mode. These measurements have the expected timing resolution for the MWPC mode, but it is our conclusion that the PPAC mode is better for timing measurements than the combined-amplification one.

It was known that low-pressure operation has a large transverse diffusion as the avalanche develops [23], and so it was postulated that this might be the origin of the good timing resolution in the low-pressure MWPC mode. To test this hypothesis, helium or neon gas was used in our chamber, because the diffusion for these gases is larger than for argon [24]. Also, from this same

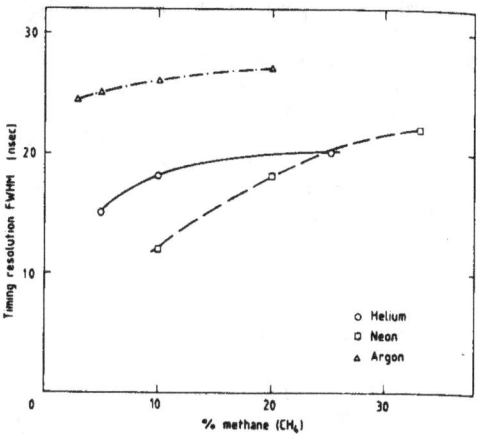

Fig. 21. Timing measurements for different gases and different percentages of methane for the chamber operated in the PPAC mode. The curves should not be extended downwards since the PPAC mode was not possible with lower percentages of quencher.

report it is known that as the percentage of quencher is decreased the diffusion increases. Measurements for timing resolution with either argon, neon, or helium, combined with different percentages of methane, are shown in fig. 21. The best timing resolution, i.e. 12 ns FWHM, was obtained in the PPAC mode with 90% neon and 10% methane, as shown in fig. 22. All the measured points in fig. 20 are for the same observed pulse height, and the curves cannot be extrapolated because the PPAC mode could not be reached at lower concentrations of methane for this chamber. These measurements encouraged our belief that the good timing resolution was originating from the larger dispersion of the avalanche. To confirm this belief we used a gas mixture of argon and TEA, where the transverse spread

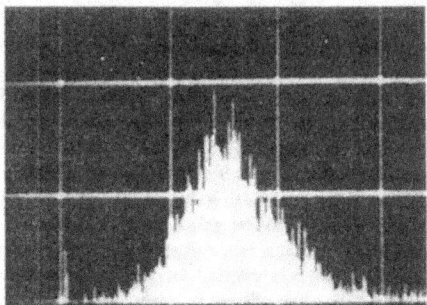

Fig. 22. The best timing measurement of 12 ns FWHM for a gas mixture of 90% neon and 10% methane. The horizontal scale is 17 ns per box.

of the avalanche is known, smaller for a higher TEA concentration [25]. We measured an improvement in the timing resolution as the percentage of TEA was decreased; however it was not better than any of our measurements with helium or neon. This improvement in timing resolution can mainly be attributed to a large avalanche size, which helps to remove the geometry effects of the anode signal plane. A side effect of this might be an increase in the spot size of the light produced for easier optical imaging.

3. Summary and conclusion

The process of amplification in gaseous wire chambers is sufficiently well documented that it was not discussed here, and we refer the reader to the many places where its description may be found (see for instance refs. [26,27], and references therein). The process of gaseous amplification in a uniform electric field, and characteristics of this amplification, can be found in refs. [1,2], and a detailed description of the fluctuations in gaseous amplification was given by Alkhazov [14].

Above some minimum electric field, depending on the gas, the drifting electrons are accelerated and then undergo inelastic collisions. These collisions, if sufficiently energetic, may lead to the liberation of free electrons by ionizing collisions, or to the emission of photons by the excited atoms or molecules if the electric field is low [8], and more often to a mixture of these two processes. Gaseous amplification processes produce photons which can be generated by atomic excitation or complex molecular effects; we do not want to analyse these here, referring the reader to the work where they are used in gaseous detectors with good energy resolution [28]. The emission of photons plays a great role in the maximum amplification attainable in the avalanches since they are the main contribution to the propagation of discharges; they generate fresh electrons outside the initial avalanche, either in the gas or on the electrodes. This generation of further avalanches leads to streamers or sparks if the gain is extremely high. In photosensitive gases, such as TMAE or TEA, feedback effects are enhanced because of their low ionization potential. Avalanches reaching a size of the order of 10^8 electrons are usually followed by a spark. However, it should be emphasized that such high gains are usually impossible to reach. Even at modest gains, such as 10^4 or 10^5, it always happens that a rare event, produced for instance by an alpha particle emitted from a contaminant gas such as radon, often present in gases, or a very ionizing reaction produced by cosmic rays, leads to a destructive spark. A non-negligible point, in the progress of parallel-plate amplification structures, has been the availability of cheap amplifiers, well protected against such occasionally large energy sparks.

We have made a comparison between uniform field amplification and MWPCs to study energy resolution, timing resolution, feedback, and maximum charge and light production. The best energy resolution and timing resolution were found in uniform electric fields in a PPAC structure. Feedback from the mesh planes was reduced by an order of magnitude in mixtures containing TEA by gold-plating the mesh planes. Although the wire chamber is a popular device it seems that the future of high-intensity beams, or those events covering a large surface with high multiplicity, can better be dealt with by using multigap structures with either charge-signal or light-imaging readout. The past few years have seen the light-imaging chambers develop from a novel idea into many possible applications. Although the work presented here was an academic study aimed at optimizing the light production for imaging, without any particular application in mind, it is now clear that light is best produced in uniform electric fields that lack regions of high electric fields such as those around wires. Meshes made of 50 µm diameter wires 500 µm apart are favourable. Use of wire-grid planes or meshes with smaller wires produces less light. The PPACs are more favourable for light emission because they have no concentrated intense electric-field regions. A two-step process of first producing maximum charge, which is then transferred into a large gap suitable to produce the maximum light yield, can aid in light-imaging applications. In this work we have also constructed a chamber that has combined the uniform field amplification process with that of wire amplification. This type of chamber can be used to remove parallax errors in some applications such as the imaging of X-rays or VUV photons.

References

[1] S.C. Curran and J.D. Craggs, Counting Tubes (Butterworth, London, 1949).
[2] P. Rice-Evans, Spark, Streamer, Proportional and Drift Chambers (Richelieu, London, 1974).
[3] G. Charpak et al., New approaches in high-rate particle detectors, CERN 78-05 (1978).
[4] G. Charpak and F. Sauli, Phys. Lett. 78B (1978) 523.
[5] A. Breskin et al., Nucl. Instr. and Meth. 178 (1980) 11.
[6] M. Suzuki et al., Nucl. Instr. and Meth. A263 (1988) 237.
[7] G. Charpak et al., Nucl. Instr. and Meth. A258 (1987) 177.
[8] G. Charpak et al., Nucl. Instr. and Meth. A269 (1988) 142.
[9] D. Sauvage, A. Breskin and R. Chechik, Weizmann Inst. preprint WIS-88/8/Feb-PH., Nucl. Instr. and Meth., in press.

[10] B. Sadoulet et al., Gas scintillation drift chambers with wave shifter fiber readout, to appear in Proc. Workshop on Nuclear Spectroscopy of Astrophysical Sources.

[11] A. Peisert and F. Sauli, Nucl. Instr. and Meth. A247 (1986) 453.

[12] P. Fischer et al., Pad readout for gas detectors using 128-channel integrated preamplifiers, Univ. Heidelberg preprint HD-PY 87/13, presented at the IEEE Nuclear Science Symp., San Fancisco, 1987.

[13] A. Breskin et al., Proc. London Conf. on Position-Sensitive Detectors, London, 1987, Nucl. Instr. and Meth. A273 (1988) 798, Weizmann Inst. preprint WIS-87/08/Sept-PH.

[14] G.D. Alkhazov, Nucl. Instr. and Meth. 89 (1970) 155.

[15] H.E. Schwarz and I.M. Mason, IEEE Trans. Nucl. Sci. NS-32 (1985) 516.

[16] R. Kahn et al., Nucl. Instr. and Meth. 172 (1980) 337.

[17] M. Suzuki et al., Nucl. Instr. and Meth. A254 (1987) 556.

[18] G. Petersen et al., Nucl. Instr. and Meth. 176 (1980) 239.

[19] G. Melchart et al., Nucl. Instr. and Meth. 180 (1981) 613.

[20] A. Breskin, G. Charpak and J.C. Santiard, Nucl. Instr. and Meth. 195 (1982) 469.

[21] A. Breskin et al., IEEE Trans. Nucl. Sci. NS-27 (1980) 133.

[22] J.C. Santiard, Multipurpose low-noise charge amplifier, CERN EP Internal Report 85-05 (1985).

[23] R. Chechik and A. Breskin, Nucl. Instr. and Meth. A264 (1988) 251.

[24] A. Peisert and F. Sauli, Drift and diffusion of electrons in gases, CERN 84-08 (1984).

[25] R. Bouclier et al., Nucl. Instr. and Meth. 205 (1983) 403.

[26] F. Sauli, Principles of operation of multiwire proportional and drift chambers, CERN 77-09 (1977).

[27] W. Franzen and C.W. Cochran, Pulse ionization chambers and proportional counters (Wiley, New York, 1956).

[28] A.J.P.L. Policarpo, Phys. Scripta 23 (1981) 539.

THE MULTISTEP CHAMBER AS A HARDWARE EVENT BUFFER

P. ASTIER *, G. CHARPAK, W. DOMINIK ** and F. SAULI

CERN, Geneva, Switzerland

Presented by P. Astier

We study a multistep chamber optically readout used as a buffered tracking detector for a tagged neutrino beam.

1. Introduction

We have investigated the possibility of using multistep chambers as part of a tagged neutrino beam. Such a setup can be shortly described as follows: in a standard narrow band beam, one equips the end of the decay tunnel to measure the position and the energy of the charged secondaries from meson decays, and to identify muons and electrons. This tunnel is followed by about 200 m of shielding and a neutrino calorimeter. This kind of facility aims at detecting, for each neutrino interaction, the accompanying particle(s) produced in the meson decay: the flavor of the neutrino is thus tagged on an event-by-event basis. The physics opened by such a setup is described elsewhere [1], we just quote here the main constraints it should fulfill to equip the neutrino Narrow Band Beam at the CERN SPS (assuming a 2 s long spill):

- The flux reaches 10^5 to 10^6 m.i.p./s cm² at the end of the decay tunnel; it falls very rapidly when going off axis. This flux is on the very high side of MWPC's capability considering pileup and space-charge effects. To get an acceptable tagging efficiency, the apparatus should cover a 1.2 m radius disk with possibly a 15 cm hole or dead region to let the meson beam through.
- The total rate on the tagging station is a few hundred MHz. As it is hopeless to record the whole information, the tagger should be read out only when a neutrino interaction occurs. The particle(s) produced with the neutrino hit the tagger long before the trigger signal reaches it; the tagging system should thus have a memory of the order of 2 μs allowing the neutrinos to reach the calorimeter, the trigger to be

formed and sent back. Delaying thousands of channels of the tracking system with cables or electronic devices is not very appealing.

With its high rate capability and its built-in gaseous delay, the multistep chamber [2] is an appropriate choice for the tracking part of the system. It provides also a low sensitive mass (a few g/m² of sensitive area), an interesting feature in a highly radioactive environment; multistep chambers finally accomodate a two-dimensional optical readout which practically suppresses pileup problems.

In this paper, we do not aim to propose a final design of a would-be tagging system. We only study a possible tracking subdetector to contribute both to the physics goal and to detector development.

2. Apparatus

2.1. Chamber

Our test chamber is a multistep parallel plate avalanche chamber made of successive crossed wire grids (50 μm wires spaced by 500 μm, 10 × 10 cm² in size) stretched on glass fiber insulating frames. The grids are held at suitable potentials which define regions with different functions in the gas volume (fig. 1):

- A conversion region in which electrons released by charged particles (or X-rays in some lab tests) are collected;
- the first stage: a high-field region where the collected electrons are preamplified in the parallel plate mode,
- a low-field drift (or transfer) region where part of the preamplified electrons are tranferred and drift. The transferred fraction can be estimated by the electric fields ratio: $E_{transfer}/E_{preamp}$ [3] and is around 10% for our voltage settings;
- the second stage is again a high-field region where the preamplified and transferred electrons are finally

---

* Now at Laboratoire de Physique Nucléaire et de Hautes Energies, IN2P3-CNRS/Universités Paris VI-VII, 4 place Jussieu, F-75252 Paris Cedex 05, France.
** On leave of absence from Warsaw University, Poland.

0168-9002/90/$03.50 © 1990 – Elsevier Science Publishers B.V. (North-Holland)

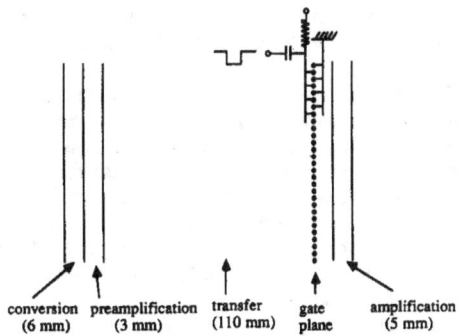

Fig. 1. Schematic transverse view of the chamber.

amplified. The gating electrode (or gate), located just before the second stage cathode can block or let the drifting electrons through.

The delay provided by the chamber is fixed by the transfer section length (11 cm) and the drift velocity (around 5 cm/μs for most of the gas mixtures): 2.2 μs are available for the trigger signal to come back and command the opening of the gate. This gate is made of 50 μm parallel wires spaced by 1 mm; odd and even wires are connected together to form two groups of wires whose voltage difference drives the gate transparency (fig. 2): the gate is normally closed (150 V is used as a blocking voltage) but when a trigger comes, a 150 V pulse brings the difference near zero and the electron cloud passes through to the second stage. In this mode of operation, the positive ions from the avalanches in the second stage are collected on the gate and do not disturb the electric field upwards. The gate potential is to be chosen to equate the electric fields upwards and downwards in order to maintain straight

Transparency

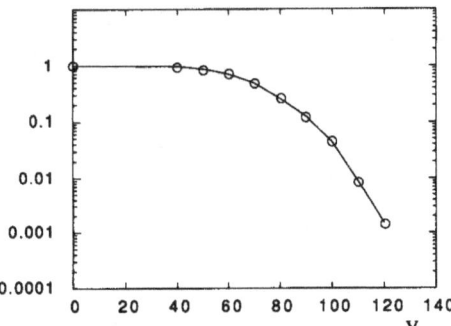

Fig. 2. Gate transparency as a function of voltage difference between the two groups of wires. The 30 V long plateau around zero difference allows a rough setting of the opening pulse amplitude.

field lines in the gate vicinity. We used a home-made pulser with 5 ns rise time and 10 ns fall time. Minimum workable width was 30 ns.

2.2. Gas filling and readout

It has been shown that the addition of some vapours (among which some photosensitive compounds and water) in the gas mixture results, under avalanche conditions, in copious light emission in or close to the visible region [4]. The amount of light produced increases with vapour concentration up to a plateau in the range of 1 photon per electron. Among such vapours tetrakis(dimethylamine)ethylene (TMAE) and triethylamine (TEA) have been extensively studied [5]. TMAE emits in the green, around 480 nm, but has the practical inconvenience of a low vapour pressure (0.4 Torr at room temperature) which obliges to heat the detector to reach an acceptable scintillation yield. The relatively high vapour pressure of TEA (52 Torr at 20°C) is counterbalanced by its emission in the close ultraviolet range (peaked at 280 nm). We used TEA as light emitter (in an argon–methane mixture) for practical convenience (no heating necessary and TEA does not react with air), and we converted its emission to visible, placing a thin wavelength- shifter (WLS) foil against the last grid [6]. As UV light is emitted close to this grid, the re-emitted light is not too much spread out by this trick. To allow also direct imaging of UV light, the chamber window was made of Aclar (polychlorotrifluoroethylene or PCTFE), a plastic reasonably transparent in the close UV region.

The emitted light can be detected by a commercially available solid-state Charge Coupled Device (CCD) video camera, preceded by an objective and an image intensifier. The intensifier spectral sensitivity must match the emitted wavelengths as well as the lens and chamber window transparencies. This requirement restricts the choice for UV readout but is relaxed if one uses the wavelength shifter foil. The light loss in the conversion (about one half due to isotropic emission of shifted light) is then more than compensated by the greater aperture of available visible light objectives ($f/1$ for visible light and $f/4.5$ quartz optics for UV light). The image intensifiers now include a gating capability down to a few nanoseconds, and reduce the recorded photocathode noise to practically zero.

If CCDs reach the mega-pixel range, their maximum readout speed capability is nowadays 50 Hz when driven by standard video electronics, and a factor of 2 or 3 faster with a dedicated driving scheme. This would probably be sufficient for the neutrino tagging application. But as some particle physics applications need both fine-grained and fast readout [7], some improvement in this field is anyway to be expected in the coming years.

VII. NEW DETECTION TECHNIQUES

3. Operation characteristics

3.1. Charge and light yields

The charge gain was measured using a two-channel charge amplifier connected to the cathodes of the amplifying stages, with an integration time around 1 μs. Thus the charge signal recorded does not include the positive ions contribution. It is related to the input charge by $Q_{measured} \simeq (M/\ln M)Q_{input}$ [8], where M is the actual multiplication factor and Q_{input} the charge of electrons released in the conversion space. The ratio $Q_{measured}/Q_{input}$ is often called "practical gain" and we will call it gain in what follows.

The light yield was measured using an RCA 8850 photomultiplier tube calibrated on the single and double photoelectron peaks of the output charge spectrum. As the inferred number of photons emitted relies on a computation using solid angle acceptance, photocathode area and quantum efficiency, and assumes phototube linearity over two decades, the results are to be taken within a factor of 2. The phototube pulse shape does not show any evidence for long lived excited states of TEA resulting in delayed light emission.

The first-stage gain can reach 10^4 in an Ar–CH$_4$–TEA mixture for any methane concentration between 0 and 17% and 2% TEA. The gain slope (with respect to high voltage) decreases with increasing methane concentration. The light output was found to stay proportional to charge (around 1.5 photon per "practical" electron) when varying both high voltage and methane concentration. The energy resolution (measured at 5.9 keV) was found around 25% (FWHM) using charge collection and 30% using light; these resolutions include some 10% gain variations on the chamber area. The second stage has similar characteristics when operated alone; its gain is limited to a few hundred when the whole chamber is on in open-gate (or dc) mode, and reaches 3×10^3 in pulsed (or ac) mode. Taking into account the 10% transferred charge fraction the overall gain can reach more than 10^6. The practical high-voltage limitation arises from sparking, a universal tendency of parallel plate avalanche chambers, which can be reduced by a careful design insisting upon the grids beeing parallel to a high accuracy.

3.2. Delay characteristics

The neutrino tagging application needs an accurate definition of the chamber delay; in case of poor timing at the end of the transfer, a wide gating pulse would have to be applied, and many events would then be transferred to the second stage and read out, increasing the difficulty to associate the right candidate with the downstream neutrino interaction. Long term fluctuations of the delay can certainly be monitored, using e.g.

laser-induced ionization; statistical fluctuations from event to event must be taken into account in the final design.

One measurement was carried out using 5.9 keV X-rays, each cathode being connected to a preamplifier–discriminator [9], the outputs of which fed a time-to-amplitude converter, the first stage as the t_0 and the second as the stop. The chamber was operated dc in order to measure its intrinsic delay and not depend on the gate circuit timing properties. The time jitter is 14 ns FWHM for a nominal delay of 2.45 μs. This measurement using a localized charge deposition accounts for the electron cloud spread in the transfer. Variations of the mean delay over the chamber area were found under 3 ns.

In the case of minimum ionizing particles, the electron cloud used for detection is as long as the conversion section (6 mm in our case); the light emission will hence last at least as long as the time necessary to collect this cloud (~ 120 ns). We measured the delay spectrum of the chamber for m.i.p. using scintillators slabs for the t_0 and a phototube to measure the time of the light pulse, its discriminated output being used as the stop. The peak is then 35 ns (FWHM) wide, still measured in dc mode. The same measurement as before gives about the same precision. The charge delocalization clearly worsens the delay accuracy with respect to 5.9 keV X-rays. This measurements refer to the leading edge jitter of the light or charge pulse and not to their duration.

The figure of merit of this chamber used as a hardware event buffer is the curve of chamber efficiency (using light output) versus gating pulse delay: it describes the probability to detect a particle off-time with respect to the neutrino interaction. Considering the cloud length (120 ns) and the gating pulse width (50 ns), one should expect a ~ 150 ns efficiency curve width. The measurement shown in fig. 3 exhibits a 250 ns width which includes an acquisition acceptance time of

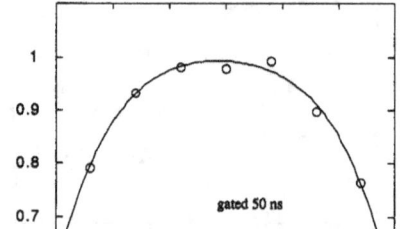

Fig. 3. Chamber efficiency versus variation of the delay of the gating pulse around its nominal value.

100 nS. The actual memory time is thus of the order of 150 ns as could be checked practically in the test beam described below when tuning the gate delay. Narrowing the gate under 50 ns decreases the efficiency without reducing much the memory time. Reducing the conversion gap thickness to 3 mm would narrow the memory (down to about 100 ns) and would adapt the gate and cloud length. In the sketched tagged neutrino beam, the gate would then select about 25 candidates per neutrino interaction. A precise time measurement (even with a poor granularity) would then be necessary to sort out the right one, which must anyway fulfill the kinematical constraints of the meson decay (mass and momentum).

3.3. Position accuracy

The position accuracy was measured in the CERN PS T7 test beam, using the setup sketched in fig 4. The vertical position of incident 5 GeV pions was measured using two microstrips planes having 60 μm pitch. With horizontal gate wires, we expect the worst position resolution to be obtained in the vertical direction because of unavoidable electrostatic distortions around the wires of the gate. The second stage of the chamber was imaged in a mirror by a gateable image intensifier coupled to a CCD camera [10]. The mirror reflectivity ranges from 0.80 at 250 nm to 0.87 in the visible range. We used a 105 mm f/4.5 quartz lens for UV readout (i.e. without a WLS foil), and a 50 mm f/1 for visible light readout; both were used at full aperture. A photo-tube (RTC 8850) was set up beside the camera to monitor in a simple way the light output. The distance from chamber to camera and phototube was 70 cm. We tested three operation modes described below.

The trigger signal was the coincidence of three scintillators aligned with the microstrips. It was used to initiate the acquisition process, and delayed by about 2 μs to trigger the gates of the chamber and of the image

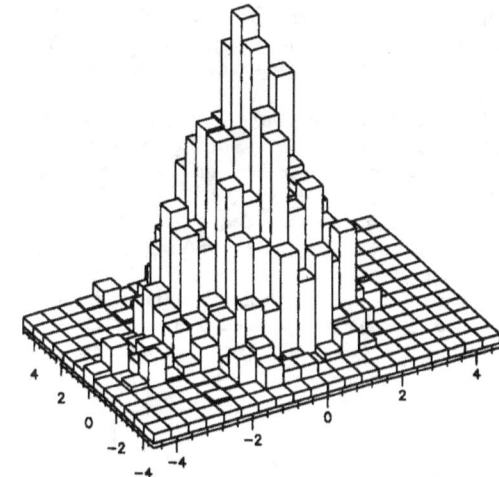

Fig. 5. A typical light spot drawn as a lego plot. The height represents the pixel brightness; the horizontal scale is in mm measured on the chamber. The small satellites around the spot are probably due to backscattered electrons on the phosphor screen of the image intensifier [12].

intensifier. The major noise source in the images appeared to be due to the light emitted by the numerous particles going through the first stage of the chamber during the 20 ms video cycle, rather than photocathode noise which can be easily cut out while preserving the signal. Gated operation of the image intensifier (which consists in driving the photocathode bias) cuts both of them. A 30 ms inhibition was introduced in the trigger to avoid recording more than one track within the same video cycle (20 ms).

Data acquisition was performed by means of a Valet-plus system [11] reading out and writing on tape the ADCs of the microstrips, the ADC of the monitoring phototube, and the video signal from the camera (consisting essentially of a chain of electric pulses proportional to the pixel charge content) digitized on a commercially available 8 bit digitizer. The digitized video output was zero suppressed by software because the light spots hit typically a few hundred pixels among the 30 000 pixels of the CCD (see fig. 5). The acquisition rate was limited by the available memory in the digitizer (4 images) and the speed of the zero suppression loop. With a hardware zero suppression at digitization time, the acquisition rate would be limited by CCD scanning speed, which allows at least 50 images/s. The video signal was split between the digitizer and a TV screen which allowed to monitor by eye the chamber operation.

The vertical track positions in the microstrips and in the chamber were computed by using a center-of-gravity method. This computation involved approximately 3

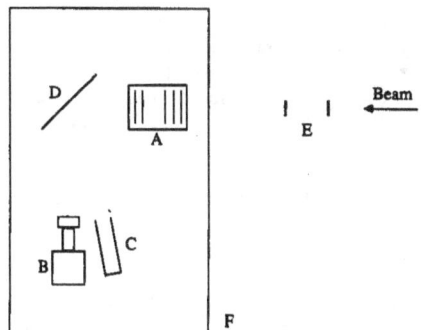

Fig. 4. Sketch of the test beam setup for position accuracy measurement. A: chamber, B: camera, C: phototube, D: mirror (aluminized mylar foil), E: microstrips planes, F: dark box.

VII. NEW DETECTION TECHNIQUES

strips per microstrips plane and 20 CCD lines. The residual distribution (fig. 6a) exhibits a 0.8 mm resolution (FWHM) for the three gas and readout conditions tested (see table 1).

We used the root-mean-square of the light spatial distribution as a spot size estimator (fig. 6b). The mean values quoted in the table show that using the WLS enlarges the spots by 20 to 30%. The double track separation capability of the device (depending on a specific pattern recognition algorithm) can certainly lie in the range of 2 to 3 mm.

The light yield was measured with the phototube and expressed in number of photoelectrons (fig. 6c). The numbers in table 1 show an increase when using methane

Table 1
Data for the three gas and readout conditions tested

Gas filling	Position accuracy (FWHM) [mm]	Light spot size (σ) [mm]	Light output (number of photoelectrons in PMT)
Ar methane (7%) TEA (2%) No WLS	0.8	1.25	–
Ar methane (7%) TEA (2%) With WLS	0.8	1.6	1200
Ar TEA (5%) With WLS	0.8	1.4	660

as the main quencher; actually, adding methane enables higher gains before breakdown under beam conditions.

The inefficiency was estimated as the fraction of events having an integrated intensity of the digitized video signal under a given threshold (10% of the mean signal from m.i.p.). The raw result is 5%, from which 3% are due to trigger falling in a "dead" part of video cycle. The intrinsic inefficiency of the chamber can then be evaluated to 2%. The dead time in the video cycle can be removed by using a more sophisticated scheme for driving the CCD than the one used in the video mode.

4. Conclusions

If one discards the direct UV readout for practical reasons (need for a specially ordered image intensifier with enhanced UV sensitivity and UV lens), the chamber is to be operated with shifted light readout and Ar–methane–TEA gas filling. If the camera is positioned three times farther from the chamber (i.e. 2 m), the total collected light will decrease by a factor of 10 and lie in the manageable range of 100 photoelectrons, regarding efficiency and resolution; the light collected per pixel will not vary (as long as the spot hits a few pixels). The spatial resolution of our setup is certainly not limitated by the readout granularity but by the scale of electrostatic distortions around the meshes and the gate. Reducing the wire pitch of the gate to 0.5 mm seems hard to achieve for an area in the square meter range. The figure obtained here (0.8 mm) is anyway more than sufficient for the neutrino tagging application.

The memory time can reasonably hoped to be in the 100 ns range or even below with a 3 mm conversion. This is too long for the neutrino tagging, but this experiment would anyway need a precise timing measurement to associate particles from the same decay

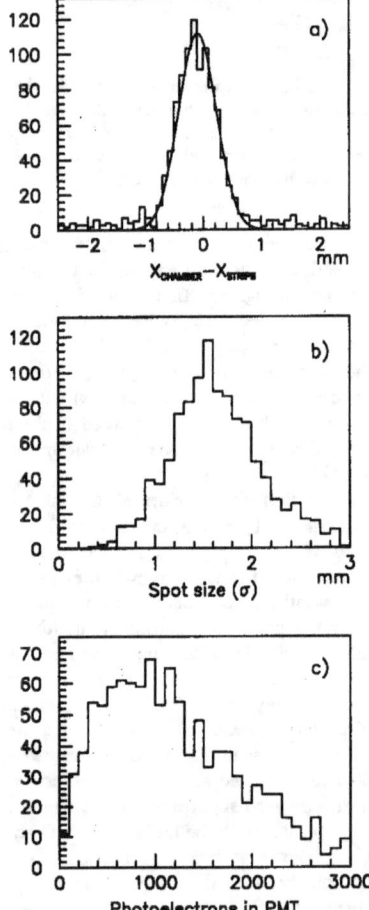

Fig. 6. Distributions from test beam analysis with Ar–methane–TEA gas filling and WLS foil: position residual (a), rms of spots (b), light output in number of photoelectrons in PMT (c).

(especially for the K_{e3} decay) in order to reduce the mistag rate (i.e. wrong particle(s) associated with a neutrino having interacted). This timing could then be used to refine the selection performed by the gating sheme.

Acknowledgements

This work was made possible by the technical help from R. Bouclier, J. Dupont and J.C. Santiard.

References

[1] Proposal for neutrino experiments at the IHEP accelerator using a tagged neutrino beam and a liquid argon detector, INFM(Pisa), IHEP(Zeuthen), JINR(Dubna) and IHEP (Serpukhov).

[2] A. Breskin et al., Nucl. Instr. and Meth. 178 (1980) 11, and references therein.

[3] A. Breskin et al., Nucl. Instr. and Meth. 161 (1979) 19.

[4] P. Astier et al., IEEE Trans. Nucl. Sci. NS-36 (1989) 300.

[5] M. Suzuki et al., Nucl. Instr. and Meth. A254 (1987) 556; G. Charpak et al, Nucl. Instr. and Meth. A269 (1988) 142.

[6] 100 μm of doped polystyrene, made by M. Bourdinaud, DPhPE, CEN Saclay, France.

[7] See e.g. D'Ambrosio et al., CERN-EP 89–44.

[8] J.R. Hubbard et al., Nucl. Instr. and Meth. 176 (1980) 233.

[9] J.C. Santiard, EP Internal Report 82–04.

[10] The CCD used was a 144×208 pixels matrix of size 4.32×5.82 mm^2 (Thomson 7852) driven by standard video electronics. We used a multichannel plate gatable image intensifier, 18 mm in diameter with a S20 photocathode and P36 fast phosphore screen (RTC XX 1410/SP 41721.160); A DEP electrostatic demagnifier reduces the image from 18 to 7 mm to match the diagonal size of the CCD.

[11] The Valet system is an acquisition package, developed at CERN, operating on a 68000 processor (or its younger brothers) sitting in a VME crate interfaced with CAMAC.

[12] For the study of noise in image intensifiers see e.g. the contribution by C. Angelini et al., these Proceedings (4th Pisa Meeting on Advanced Detectors, Isola d'Elba, Italy, 1989), Nucl. Instr. and Meth. A289 (1990) 356.

VII. NEW DETECTION TECHNIQUES

6. IMAGING OF VACUUM ULTRA VIOLET PHOTONS. APPLICATON TO CERENKOV AND GAMMA IMAGING

6. IMAGING OF VACUUM ULTRA VIOLET PHOTONS. APPLICATION TO CERENKOV AND GAMMA IMAGING

NUCLEAR INSTRUMENTS AND METHODS 164 (1979) 419-433; © NORTH-HOLLAND PUBLISHING CO.

DETECTION OF FAR-ULTRAVIOLET PHOTONS WITH THE MULTISTEP AVALANCHE CHAMBER. APPLICATION TO CHERENKOV LIGHT IMAGING AND TO SOME PROBLEMS IN HIGH-ENERGY PHYSICS

G. CHARPAK, S. MAJEWSKI*, G. MELCHART, F. SAULI

CERN, Geneva, Switzerland

and

T. YPSILANTIS

DPhPE, CEN Saclay, Gif-sur-Yvette, France

Received 17 April 1979

We have designed and operated a multistep avalanche chamber capable of detecting single photons in the vacuum ultraviolet wavelenght (130–160 nm). The device has been used to image photons generated from charged particles by Cherenkov effect in solid an gaseous radiators, and providing a ring pattern with a radius related to the particle's velocity; experimental values are given for the angle and velocity resolution obtained. Possible applications of the method for particle identification are discussed.

1. Introduction: preamplification and transfer of ionization electrons in gases

It has been shown recently[1,2] that with a proper choice of binary gas mixtures, $A + B$, a new variety of particle detectors can be built, based on the development of electron avalanches in uniform electric fields through a mechanism quite different from the one occurring in most proportional counters. The name multistep avalanche chamber (MSC) has been proposed for the new device.

In the carrier gas A, which according to our tests can be any of the noble gases, an electric field is applied with a strength such that free electrons experience non-ionizing inelastic collisions. These collisions lead, most probably via the production of dimers, i.e. molecular bound states of an excited atom and an atom in the ground state, to the emission of light quanta in a wide band centred around an energy characteristic of each gas. The process has been studied extensively[3-6] and is referred to as secondary light emission in the works on scintillating proportional counters[7].

When a small quantity of a gas B having an appropriate ionization potential is added to A, an efficient process of photoelectric reconversion takes place leading to an unusual photon-mediated avalanche growth[1]. In fig. 1 we have represented the

measured secondary light emission continua of several pure noble gases at atmospheric pressure[6]), as well as the ionization potentials of vapours that can be used in conjunction. Since the absorption cross-section of all these vapours above the ionization threshold is around 50 Mb, the mean free path of photons is around 300 μm for a few per cent vapour addition at normal conditions: quantum efficiencies are typically around 50%. An electron avalanche developing under these conditions has several unusual properties, such as a large lateral

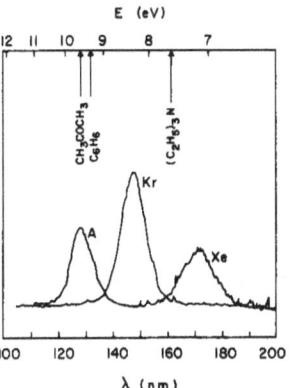

Fig. 1. Secondary photoemission spectra in several noble gases, measured at electric field values similar to the ones encountered in the multistep chamber (from ref. 6). The ionization potentials of acetone, benzene and triethylamine (TEA) are also shown.

* On leave from the Institute of Experimental Physics, University of Warsaw, Poland.

spread that tends to obliterate any quantizing effect of discrete electrode structures (grids or wires), and the ability to "jump" through potential wells as observed in the development of the MSC. Consider indeed the three-electrode structure of fig. 2, where a region P of high field E_P is followed by a region T of lower field E_T; as shown in fig. 2a, a fraction of the lines of force roughly equal to E_T/E_P escapes to the lower section. As represented schematically in fig. 2b, an electron accelerated in P induces the emission of a photon; a subsequent photoionization may then generate a new electron and the process continues, somehow resembling the usual avalanche multiplication. Photons released close to the central electrode may of course be reconverted in the region T, thus enhancing the purely electrostatic electron leak from P to T determined by the ratio of the fields and favoured by the lateral charge spread. The experimentally observed preamplification and transfer yields have been well reproduced using a Monte Carlo simulation; fig. 2c shows, for example, the computed lateral avalanche spread obtained assuming a mean free path for both photon emission and reabsorption around 200 μm.

The possibility to transfer and drift a sizeable fraction of the preamplified electrons has led to the following development in the art of particle detec-

tion: the swarm of electrons can be transmitted through a grid to a new element where a second step of amplification occurs: a multiwire chamber, or another parallel-electrode avalanche chamber. Insertion in the drift space of a gating structure permits then to control the passage of the swarm and to admit to the second step only those electrons associated with preselected events. The delay introduced by the drift-time allows this type of decision with fast auxiliary counters for instance; this was the first motivation of the research. It permits the restriction of the total amplification to a selected class of events, thus considerably reducing the detrimental space-charge effects due to the accumulation of positive ions at high particle fluxes[8]).

However the interest of the multistep approach appears to be broader:
1) The fact that the second step may also be a parallel-grid counter opens the way to detectors without any thin amplifying wires. The difference with a single-step parallel-plate counter with the same total gain stems from the fact that the feedback mechanism mainly responsible for the discharge propagation in parallel-plate structures, namely the emission of ultraviolet (UV) light by the head of an avalanche and the subsequent release of photoelectrons at the cathode, is strongly suppressed by the transfer gap.
2) The fact that the additive B is highly efficient for photoelectric conversion makes it an ideal detector for UV light of energy higher than its ionization threshold and lower than the cutoff of the window placed at the entrance of the structure. The intention of the present article is to illustrate some applications of the MSC in the last field.

2. The multistep avalanche chamber as a single photon detector

It has already been demonstrated by the authors of refs. 1 and 2 that in a multiple structure, where a preamplification and transfer element is followed by a conventional multiwire proportional chamber, large enough gains can be obtained to allow single photoelectron detection. Fig. 3 (from the quoted works) shows the pulse-height spectrum recorded when illuminating the structure with an UV light source. The measured charge (about 1 pC peak) implied an effective chamber gain in excess of 10^6 and allowed detection and localization of single

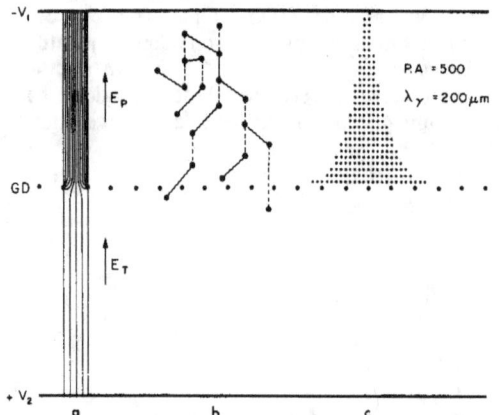

Fig. 2. Photon-mediated preamplification and transfer mechanism in gases, in a three-electrode structure. Electrons produced in the upper region experience inelastic collisions with the molecules of the main constituent, A, of the gas mixture, which lead to the emission of photons; some of them are reconverted to electrons by photoionization of component B. There results a large, concentration-controlled lateral spread of the avalanche, which permits obliteration of the quantizing effects of the central electrode wires. Transfer of charges from the upper to the lower region is controlled by the ratio of the fields.

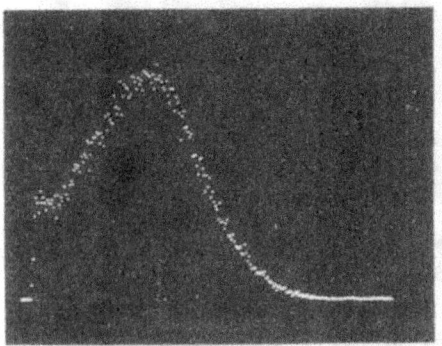

Fig. 3. Pulse-height spectrum for single photoelectrons generated by a UV lamp in a multistep avalanche chamber (MSC) (from ref. 1).

photons using the well-established bidimensional MWPC read-out techniques. Although not directly measured by the authors, the quantum efficiency of the device could be estimated from that known for the additive. In fig. 4 we have represented the transmission properties of some commercially available window materials for far UV wavelengths*, as

* ORIEL Optik GmbH, Darmstadt, West Germany.

well as the quantum efficiencies of two additives used, benzene (C_6H_6) [9] and triethylamine [TEA, $(C_2H_5)_3N$] [10]. For example, using benzene in argon with a thin LiF window allows average quantum efficiencies of around 25% in the photon energy range 9.2–10.6 eV (taking into account the window transmittance), as far as each converted photoelectron can be detected with full efficiency.

To study the detection and localization properties of the MSC for multiple photon events in this energy region, whe have constructed the device represented in fig. 5. A spark-plug is fired with an external high-voltage pulse, generating a burst of photons in an energy spectrum characteristic of the filling gas and whose quantity can be controlled by the discharge energy. The fraction of the photon spectrum that traverses the window can be absorbed in the chamber gas. Although we could have used an electronic technique for charge localization[2,11], it appeared that an optical recording on photographic plates was the most suited for a quick survey of simultaneous multiphoton events. The reamplification and transfer element of the MSC was therefore followed by a triggered spark chamber, as shown in the figure; the triggering time, in respect to the main plug sparking, was obviously

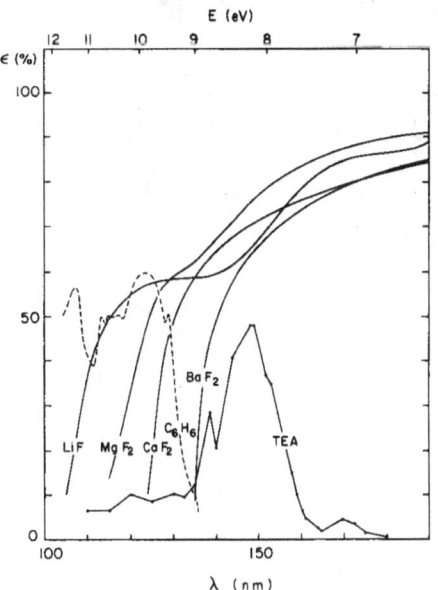

Fig. 4. Compilation of the quantum efficiency of benzene[9]) and TEA [10]), and of the window transparency for 5 mm thick crystals as a function of photon wavelength.

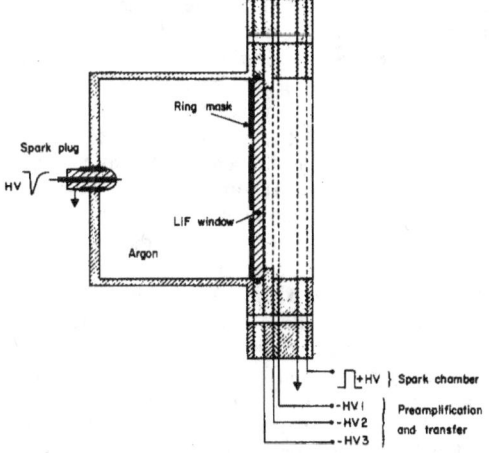

Fig. 5. Testing the imaging capability in the vacuum UV for multiphoton events with the MSC. Bursts of photons are generated in argon by the triggered spark plug, and a mask in front of the detector allows the simulation of a ring image. The MSC in this case consists of a preamplification and transfer element followed by a triggered spark chamber. The gas filling that provides the best multiphoton efficiency appeared to be helium or helium–neon (Henogal) in conjunction with 2–5% of acetone, benzene or TEA.

determined by the electron's drift-time in the MSC. We may mention here that a device of this kind, named hybrid chamber, was operated long ago by Fisher and Shibata[12]), although not for the same purposes. To obtain the best multitrack efficiencies the spark chamber had individually terminated 50 μm diameter wire electrodes, and a high-voltage square pulse about 100 ns long was applied from a low impedance source at a time corresponding to the arrival of the preamplified electron swarm in the neighbourhood of the grounded electrode in the spark chamber. It was found that a transfer space at least 10 mm thick is necessary to avoid parasitic firing in the preamplifying gap because of capacitive pulse couplings. Using argon as gas filling in the spark-plug volume we could generate a burst of photons with a wavelength spectrum mostly below 150 nm, suited for this study[13]). As filling for the MSC, we have tried several combinations of noble gases and photoionizing vapours, under the constraints imposed by the described preamplification and transfer mechanism; it appeared that the best multitrack efficiency in the spark chamber could be obtained using helium of helium–neon together with acetone or TEA (a well-known outcome of previous track chamber work[14]). Although the secondary emission spectra of helium and neon are not known in our field conditions, their general behaviour under condensed discharge conditions suggests that photon emission takes place mostly around 80 nm [13]), thus satisfying the described MSC operation constraint. Use of a low-density gas is also convenient in order to decrease the direct contribution of the ionization in the detector due to charged particles traversing the structure when one wants to detect, for example, the photons emitted by the particle by Cherenkov effect in a radiator (as we will see in the following sections).

The pictures in fig. 6 show the images obtained in these conditions, with a mask placed in front of the chamber window to simulate a ring image, at an increasing number of photons in the burst. In fig. 6c, the photoionization charge is so large as to result in a transition from the individual spark behaviour to a continuous glow regime. The ring mask diameter is 35 mm with an opening of 0.1 mm; analysis of a succession of images like the ones shown confirmed the ability of the device to record multiphoton events with intrinsic accuracies better than a millimeter.

The charge signal induced on the preamplifying grid by single photoelectrons is far too small, at the

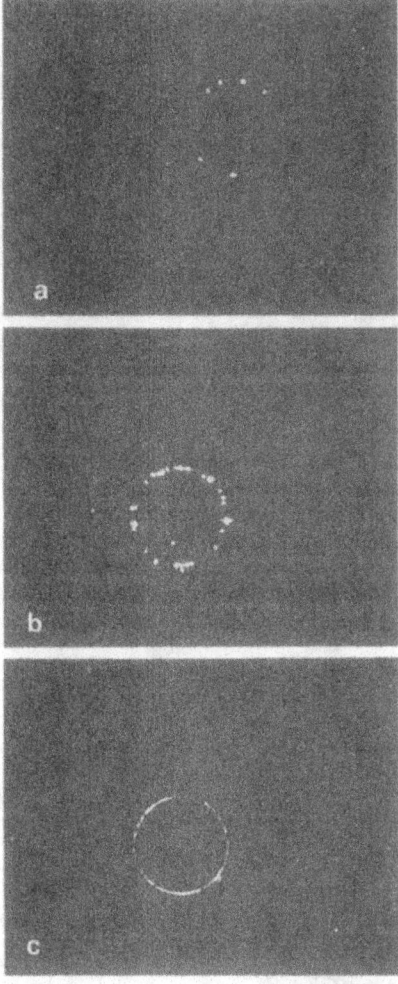

Fig. 6. Ring images obtained by a photographic recording of single events, at an increasingly large number of photons generated by the spark plug of fig. 5. The ring diameter and width are 35 and 0.5 mm, respectively (polaroid film, 3000 ASA, opening 5.6).

present stage, to allow a self-triggered operation of the chamber. However, this can be realized using a strongly ionizing radiation source, as for example X-rays of several keV. In this case, the discriminated signal from the preamplification grid can be used to generate at the proper time the high-voltage pulse applied to the spark-chamber element and allows therefore the imaging of neutral radiation. Pioneering work on this matter was realized several years ago by Böhmer, using the hybrid chambers[15]). The picture in fig. 7a shows the image recorded with the chamber illustrated in fig. 5 operated in the self-triggered mode and detecting 5.9 keV X-rays. In this case, the UV photon pulser was replaced by a non-collimated ^{55}Fe source, and the ring mask replaced by a 5 mm thick aluminium absorber with cross-shaped cutouts, with 2 mm wide openings, having arms of 40 and 20 mm length, respectively.

Increasing the voltage in the last section of the chamber, one can eventually get a dc operation of the system; fig. 7b shows an example of the image obtained, with the same source and geometry, operating the device as a multistep spark counter. This mode of operation, different from that for the classic spark counter in that the total amplification process is separated in two steps, seems to have a considerably better stability of operation, an observation already made (although in a different con-

Fig. 7. Imaging soft X-rays (5.9 keV) with the multistep spark chamber. In (a), the image is obtained operating the device in the self-triggered mode, the hv pulse to the spark chamber being initiated by the signal detected on the preamplification element. In (b), instead, the same image is obtained operating the chamber as a dc spark counter.

text) by Aoyama and Watanabe[16]). Obviously, the general background level in the dc mode is larger than for the pulsed operation, as one can see comparing figs. 7a and b; this is due both to the larger values of the applied voltages and to the cosmic radiation (the total exposure time to obtain the image in fig. 7b was about 5 min).

3. Imaging of the Cherenkov light produced in gases

3.1. INTRODUCTION

A recent analysis[17]) has shown that great advantages in Cherenkov detection techniques could be obtained by exploiting the quanta emitted in the far UV region where photoionization gaseous counters could be used, as they offer the possibility of detection and imaging of individual photons. Several groups have been working on this subject[18-20]), the key problem being to obtain large enough gains in a proportional chamber that maintains a good quantum efficiency; this is not a trivial issue, since large amplification factors can normally be obtained in proportional counters using organic quenchers that suppress the photon-propagated discharge or breakdown mechanism. As mentioned in the Introduction, the MSC approach defeats this otherwise fundamental limitation, allowing gains in excess of 10^6 to be obtained in gas mixtures with large quantum efficiencies in the far UV.

A simple optical arrangement, making use of a spherical mirror of radius r, permits the formation, on a spherical surface of radius $\frac{1}{2}r$, of a ring image corresponding to all photons emitted in a transparent medium at an angle θ_c with the radiating particle trajectory. The radius, on the image surface, of the ring is given by

$$R = f\theta_c, \tag{1}$$

where f is the mirror focal length ($2f = r$). For $r \gg R$, a flat detection plane approximates well the spherical image surface and in this case the previous expression can be written as

$$R = f \tan\theta_c. \tag{1'}$$

Since we will need them in the following analysis, let us here briefly summarize some relevant expressions concerning Cherenkov light emission[21,17]). Charged particles with velocity β (referred to the speed of light) emit Cherenkov radiation in a medium having the index of refraction n at an angle θ_c given by

$$\cos \theta_c = \frac{1}{n\beta}, \tag{2}$$

above a threshold velocity given by $\beta = n^{-1}$. The refraction index is in general a function of photon frequency or energy, $n = n(E)$. The number of photons emitted in the energy range $E_2 - E_1$ by a radiator of length L is given by:

$$N = \frac{2\pi\alpha L}{hc} \int_{E_1}^{E_2} \left[1 - \frac{1}{[\beta n(E)]^2} \right] dE. \tag{3}$$

In computing the average number of detected photons, of course, one has to take into account all sources of energy-dependent efficiencies, like the absorption and reflection losses in the optical system, as well as the detection efficiencies. Assuming, in the energy interval $E_2 - E_1$, an average value n and ε for the refraction index and the over-all efficiencies, one can write

$$N = N_0 L \sin^2 \theta_c, \qquad N_0 = 370(E_2 - E_1)\varepsilon, \tag{3'}$$

(L in cm and E in eV). The velocity resolution obtainable from a ring radius measurement can be estimated by differentiation of expression (1') and substitution into eq. (2):

$$\frac{\Delta\beta}{\beta} = \sin^2 \theta_c \frac{\Delta R}{R}. \tag{4}$$

Furthermore, if all sources of dispersion are Gaussian-like, the radius resolution when measuring N photons, assuming that the centre of the ring image is known, will be given by:

$$\left. \frac{\Delta R}{R} \right|_N = \frac{1}{\sqrt{N}} \left. \frac{\Delta R}{R} \right|_1, \tag{5}$$

the right-hand resolution corresponding to that for a single detected photon. A naive statistical argu-

ment based on the available degrees of freedom suggests that in case the centre point is unknown, and if the radius is computed from a fit to events with three or more photon points, the previous expression can be replaced by

$$\left. \frac{\Delta R}{R} \right|_N = \frac{1}{\sqrt{(N-2)}} \left. \frac{\Delta R}{R} \right|_1, \quad N \geq 3. \tag{5'}$$

For small values of the Cherenkov angle, taking $\sin \Delta\theta_c \simeq \Delta\theta_c$, and combining eqs. (1), (3), and (5):

$$\Delta\theta_c \simeq \frac{1}{\sqrt{(N_0 L)}} \left. \frac{\Delta R}{R} \right|_1, \tag{5''}$$

($\Delta\theta_c$ in rad). When approximation (5'') holds, one can see that the angular resolution is constant for a given geometry, i.e. independent of the Cherenkov angle. In practice, the major sources of dispersion in the radius measurement are the chromatic aberrations in the radiator and windows and the positioning errors in the photon imaging device. The first effect is particularly important in the far UV domain, since the index of refraction varies rapidly in this region.

3.2. EXPERIMENTAL OBSERVATION OF CHERENKOV RINGS

We have constructed and tested a gas Cherenkov ring imaging detector, mounting the described MSC on a radiator tube, about 1 m long, operated in argon at a pressure slightly above the atmospheric one (fig. 8). The focal length of the reflecting mirror, $f = 93$ cm, matched the distance between the mirror and the MSC. High-purity argon (3 ppm impurities, mostly water and oxygen) was used as radiator to minimize photon absorption; the mirror was coated with magnesium fluoride to provide a

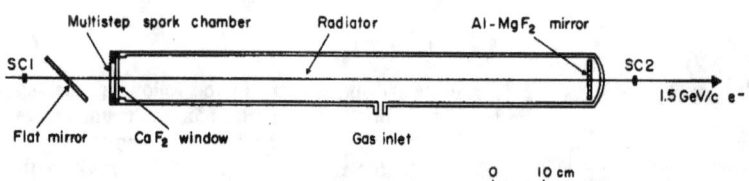

Fig. 8. Cherenkov ring imaging with a gas radiator (argon at normal conditions). The optical system (mirror and window) is optimized for good response in the far UV region of sensitivity of the detector, around 150 nm. The MSC detector itself is the same as the one illustrated in fig. 5, and photographic recording of individual events is obtained by a fast camera as represented.

good reflectivity in the far UV (75% at 150 nm as given by the manufacturer). The MSC itself was operated with about 5% TEA in commercial grade helium; the system was installed in a 1.5 GeV/c non-separated charged particle beam at the CERN Proton Synchrotron, equipped with a gas threshold Cherenkov counter to identify the electrons (the only particles above threshold for this momentum). Several hundred pictures have been taken with a fast camera under these conditions. Because of the properties of the image-forming optics, of course, the direct ionization spark is located in the geometrical centre of the ring only for those particles whose trajectory coincides with the central axis of the system. Fig. 9 shows an integrated image over ~30 events ang gives an idea of the dispersion and the acceptance of the system; the Cherenkov photon ring appears clearly around the charged particle points, with a diameter of about 50 mm.

About 400 individual pictures have been hand-analysed by simple projection on a screen and with the following acceptance criteria:

- only centred events (i.e. with a spark in the fiducial volume corresponding to the beam collimation) were considered;
- the coordinates of all sparks, whatever their position, were measured in a reference system relative to this centre;
- a least-squares fit to a circle (i.e. not using the centre constraint) provided the best estimate of

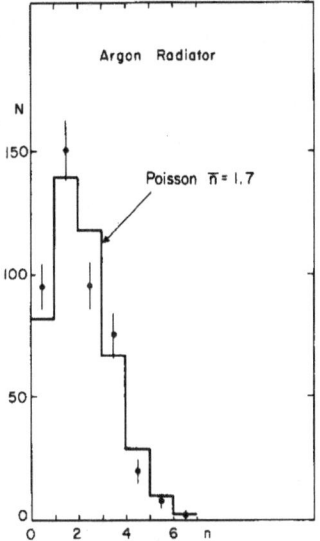

Fig. 10. Measured distribution of the number of photons (points with error bars) as recorded with the detector of fig. 8 on 1.5 GeV/c electrons. The histogram represents a Poisson distribution for the same number of events, computed for an average of 1.7 photons.

the ring radius R for events having three or more photon points.

The measured photon number distribution is shown in fig. 10 (points with error bars), together with the Poisson distribution for $n = 1.7$ normalized to the same total number of events that represents the best fit to the data. Fig. 11 instead shows the measured ring radius distribution, obtained as described for events containing three or more photons (about one hundred out of the total). The continuous curve is a Gaussian fit to the distribution with an average radius of 24.42 mm and a standard deviation of 2.05 mm (or about 8%).

Analysis of individual pictures reveals very little – if any – background that could be due to direct scintillation of argon in the sensitivity range of the MSC (an observation also made by the authors of ref. 18).

3.3. DISCUSSION OF THE DATA

From the measured value of the average ring radius, $\bar{R} = (24.42 \pm 0.22)$ mm, and expression (1′), one can compute the average Cherenkov angle to be $\bar{\theta}_c = 1.50° \pm 0.02°$ and, from expression (2) for $\beta = 1$, the corresponding average index of refraction for argon in the detected wavelength domain,

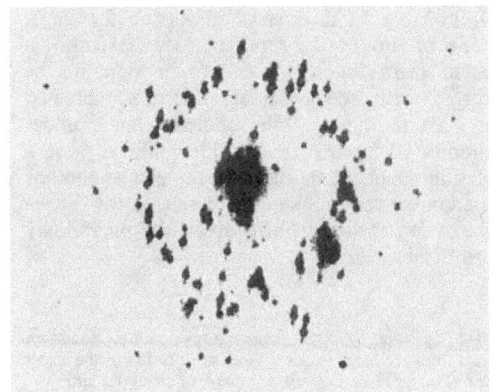

Fig. 9. Integrated image over ~30 events obtained with the apparatus shown in the previous figure. The Cherenkov ring pattern appears clearly, with a radius of about 25 mm; the central spot corresponds to the overlap of the sparks developing on the direct ionization of the beam particles (1.5 GeV/c electrons). The small amount of off-ring counts proves that direct scintillation in the radiator is almost negligible.

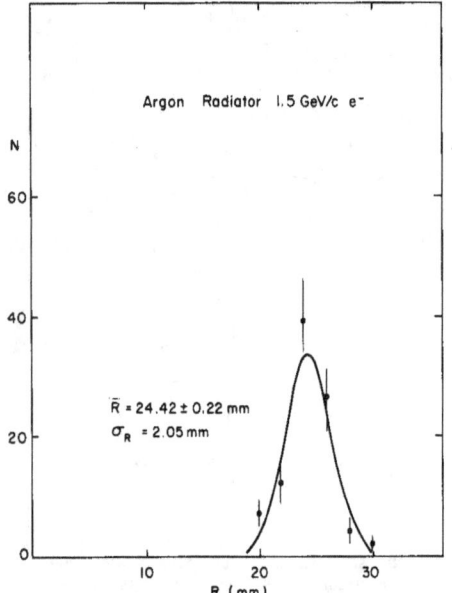

Fig. 11. Measured distribution of the average ring radius (points with error bars), obtained from events having three or more photon points. The full curve represents a Gaussian fit with average value and standard deviation as indicated.

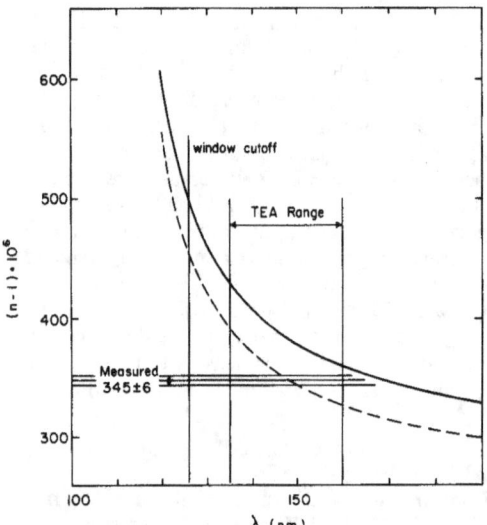

Fig. 12. Comparison between the computed index of refraction for argon[22]) in the far UV, at 0 °C, 760 mm Hg (full curve) and in our conditions at 20 °C, 770 mm Hg (broken curve), and the average value provided by our measured Cherenkov ring radius; the detector sensitivity range is shown, peaked around 150 nm.

$(\bar{n}-1) = (345 \pm 6) \times 10^{-6}$. This result is compared in fig. 12 with the computed dependence of $(n-1)$ for argon in the vacuum UV region[22]), as given in the quoted reference at 0 °C, 760 mm Hg and as computed in our conditions (20 °C, 770 mm Hg) with an approximated density dependent relationship. The approximate limits of sensitivity of our detector, using TEA and a calcium fluoride window, are also shown in the figure, the maximum quantum efficiency being around 150 nm (see fig. 4).

In practice, several independent sources of dispersion limit the achievable radius resolution. In table 1, we summarize the estimated value of the standard deviation of the single photon dispersion, assuming the central point to be known, for the following contributions:

- the chromatic aberrations in the gas, or the change in the refraction index with the detected photon energy in the sensitivity region; the known dependence of $n-1$ from wavelength (fig. 12) has to be folded in with the quantum efficiency of the detector, fig. 4; this effect is obviously particularly large in the vacuum UV domain;
- the multiple scattering of the charged particle in the radiator producing a dispersion in the angle of the emitted photons, estimated as an average over the radiator length;
- the accuracy of localization of the detector itself and mainly of the spark chamber and associated optics.

As from table 1, an over-all dispersion for single photons of around 1.2 mm standard deviation is expected from the three sources, a value to be compared with the observed 2.05 mm, obtained from a fit to mostly three photon events, under conditions where expression (5′) would suggest a dispersion identical to the one for single photons with a known centre. We cannot explain the discrepancy so far, insisting however on the preliminary nature of our results.

TABLE 1

Standard deviation of the single photon radius dispersion expected from the three major sources of diffusion in the argon radiator. The resulting over-all dispersion is about 1.2 nm.

Source	σ_R (mm)
Chromatic aberration	0.67
Multiple scattering	0.94
Detector	0.50

TABLE 2

Average estimated efficiencies for photon transmission due to the indicated effects in the argon radiator case. The over-all optical efficiency (not taking into account the detector quantum efficiency) is around 32%.

Source	ε
Window transmittance	0.65
Gas transmittance	0.86
Mirror reflectivity	0.75
Mesh transparency	0.77

The expected number of detected photons can be computed from expression (3), using approximate values for the efficiencies involved. Table 2 summarizes our estimates of the major losses, as due to the following effects:

- window transmittance at wavelengths around the detector sensitivivity region;
- gas absorption, assuming 3 ppm of impurities with the oxygen cross-section around 150 nm (around 14 Mb);
- mirror reflectivity, given by the manufacturer to be around 75%;
- absorption in the first wire mesh of the detector, due both to its optical transparency (85%) and to be probability for the photoelectron to be captured by the mesh itself, if produced too close to the window; we have assumed a 30 μm dead layer for a 330 μm absorption length in the gas.

To these factors one has to add the integrated detector quantum efficiency; from fig. 4 we estimate the average efficiency-energy product for TEA to be around 0.37 eV. From expression (3') and the data of table 2, one gets therefore $N_0 \simeq 44 \, \text{cm}^{-1}$ and, for $L = 93$ cm and $\theta_c = 1.50°$, $N \simeq 2.8$ photons, larger than the measured value ($\overline{N} = 1.7$, see fig. 10). The more likely explanation for the discrepancy seems to us a larger residual oxygen contamination in the Cherenkov radiator, that could not for practical reasons be evacuated before the argon filling (since the CaF_2 window would not withstand the overpressure). A photon yield lower by a factor of two than the expected one was observed also by Chapman et al. under similar operating conditions[18]). We cannot of course exclude an unknown source of inefficiency in the chamber itself.

4. Use of fluoride crystals as radiators

The UV transparent layer used as entrance window for the photon detector is itself a Cherenkov radiator for particles in the appropriate velocity range. In fig. 13 we summarize the known values of the index of refraction for several fluoride cyrstals[23,24]) (the corresponding UV cut-off was shown in fig. 4). It is interesting to note that in the TEA sensitivity region, below 160 nm, all crystals (with the possible exception of magnesium fluoride) have an index that allows total reflection for the Cherenkov light emitted by particles perpendicular to their surface above a critical velocity. From simple optics and expression (2), total reflection is encountered at a critical velocity given by

$$\beta = \left[n \cos \arcsin \frac{1}{n} \right]^{-1}, \qquad (6)$$

a condition that can only be satisfied for $n \geq \sqrt{2}$. For example, LiF, whose index of refraction at the TEA ionization threshold is 1.47, outputs detectable UV Cherenkov radiation for normal tracks only in the velocity range $0.68 \leq \beta \leq 0.93$, a peculiarity that may offer interesting possibilities of application (see section 5).

As in the case of a gaseous radiator, one has to make sure that direct scintillation in the radiator is not a substantial source of background. We have checked this point by mounting a commercially available UV sensitive photomultiplier with a magnesium fluoride window 1 mm thick* in a non-separated charged particle beam. The sensitivity

* EMI Gencom Inc Type G26H315.

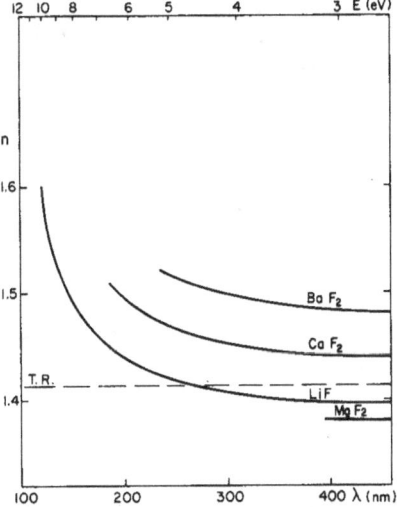

Fig. 13. Known dependence of the refraction index in the far UV for several fluoride crystals[23,24]).

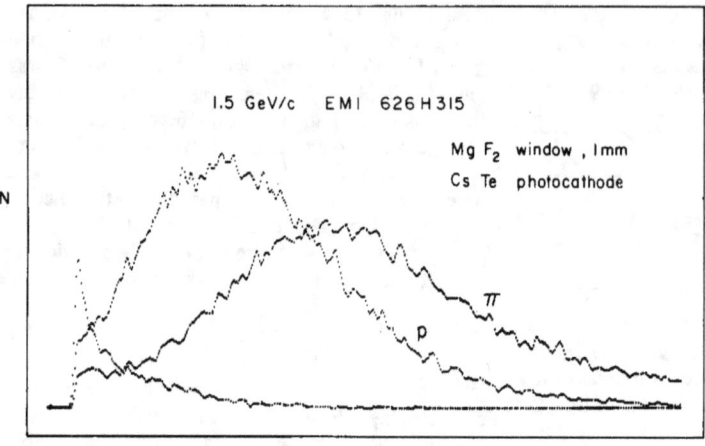

Pulse Height

Fig. 14. Pulse-height spectrum measured on a 1 mm thick MgF$_2$ window with a UV-sensitive phototube, for 1.5 GeV/c protons and pions traversing the system so that the Cherenkov light emitted by the crystal could be directly detected on the photocathode. The smaller peak corresponds to the same total exposure to the beam, having turned the system by 180°.

range of the photocathode (CsTe) extends from about 120 to 300 nm, a region where the index of refraction of MgF$_2$ is mostly below the total reflection limit. Fig. 14 shows the measured pulse-height spectrum for 1.5 GeV protons and pions, traversing the window in the direction that allows the Cherenkov light to be detected on the photocathode, as well as the spectrum measured inverting the tube direction (for the same particle flux, lowest spectrum in the figure). Obviously, the extraction of direct secondary electrons from the photocathode or from the first dynodes may contribute to the last

spectrum, giving evidence that direct scintillation, if any, is very small even in the extended sensitivity region of the tube as compared with TEA. Quite a different behaviour is expected in the visible range, where all fluorides are known to scintillate[25]).

Since a large number of photons is emitted by such materials above threshold (they have a large index and therefore a large Cherenkov angle), thin radiators may be used thus allowing the simplification of the ring imaging optics. Fig. 15 shows the arrangement we have tested; it consists of a 5 mm thick LiF window, used as radiator, 7.5 mm apart

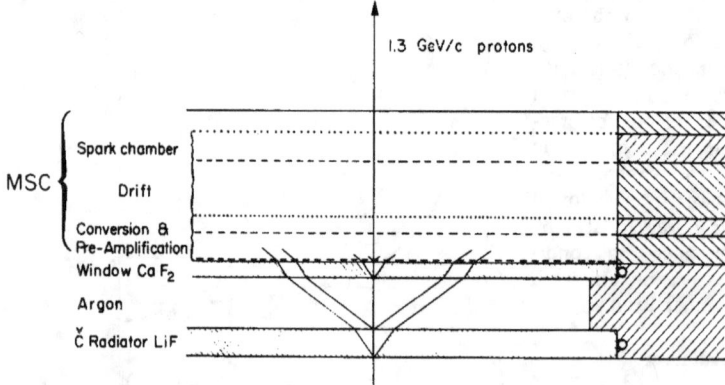

Fig. 15. The MSC used for imaging the photons emitted by Cherenkov effect in a 5 mm thick LiF radiator. The chamber window is a 3 mm thick CaF$_2$ crystal. The ring image forms directly in the UV-sensitive detector, with a dispersion that depends on the radiator thickness.

from an MSC with optical spark recording as described in section 2. The chamber window is a 3 mm thick CaF_2 crystal. The ring image is here obtained by refraction of the light cone in the argon gap between the two windows, a considerable simplification in the optics which, however, has the disadvantage of providing an extra intrinsic dispersion due to the radiator width.

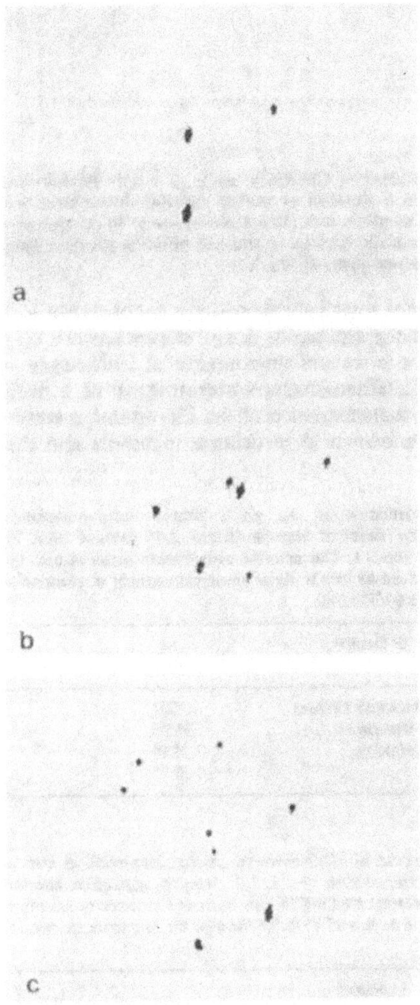

As for the gas radiator, several hundred pictures were taken in this geometry and a least squares fit to the measured photon points (if three or more) provided the best estimate of the image radius; some examples of individual events are given in fig. 16. Again, the centre point could not be used (other than for consistency checks) since photons radiated in the MSC window are also detected and blur the charged track coordinate measurement. Figs. 17 and 18 show, respectively, the measured distribution of the number of photons and of the reconstructed ring radius (for $N \geq 3$) for 1.3 and 1.5 GeV/c protons traversing the detector. The corresponding Cherenkov angles are 34.05° and 37.75°, well matched by the assumption of an average index of refraction for LiF of 1.50 (fig. 19) consistent with the known values (see fig. 13). In fig. 19 we have also represented the standard deviations in the angle estimate corresponding to the measured dispersion in R; for the restricted class of events (three or more photons) the equivalent velocity resolution is about 2% [see eq. (4)]. Table 3 summarizes an estimate of the main effects contributing to the single photon dispersion, computed

Fig. 16. Examples of single events obtained with the detector illustrated in the previous figure, for 1.3 GeV/c protons. The central point, corresponding to the charged particle trajectory is often accompanied by photon points emitted in the MSC window.

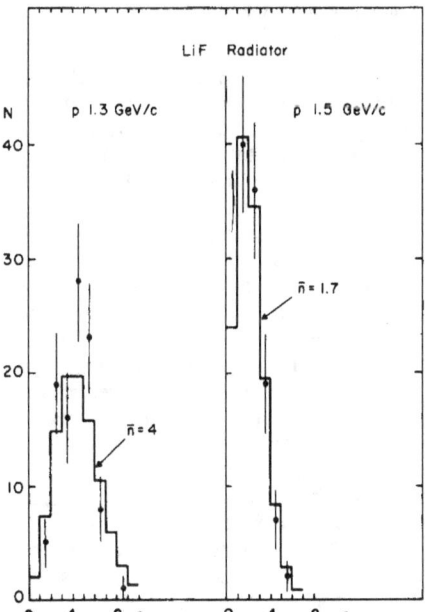

Fig. 17. Measured distribution in the number of photons obtained with the detector shown in fig. 15, for 1.3 and 1.5 GeV/c protons (point with error bars). The histogram corresponds to a Poisson distribution with the indicated average.

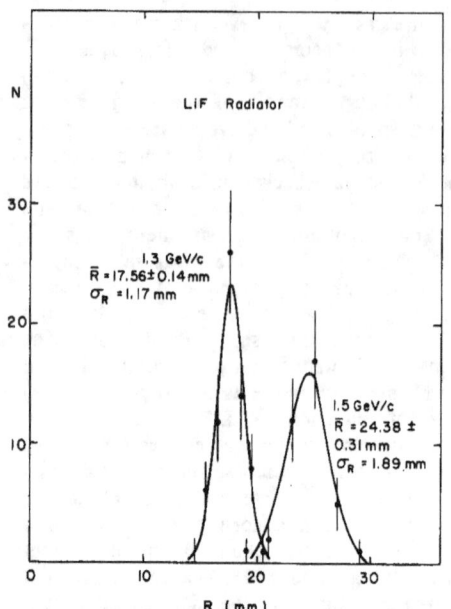

Fig. 18. Measured distribution of the average ring radius (points with error bars). The full curves represent a Gaussian fit with the indicated average values and standard deviations.

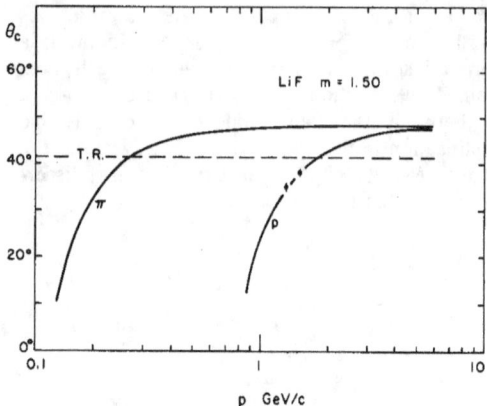

Fig. 19. Computed Cherenkov angle in a LiF radiator with $n = 1.50$, as a function of particle momentum for pions and protons. The points with error bars correspond to the measured values at 1.3 and 1.5 GeV/c, obtained from the previous figure and the known geometry (fig. 15).

for 1.3 GeV/c protons; the over-all single photon dispersion corresponds to about 2.4 mm standard deviation. As the majority of the reconstructed events contains 4 or 5 photons at this momentum (see fig. 17) the experimentally measured dispersion of 1.17 mm is consistent with expression (5′).

The expected number of photons can be evaluated as for the previous case, taking into account reflection and absorption losses in the two windows; the contributions of the main sources of efficiency are given in table 4, for 1.3 and 1.5 GeV/c momentum, respectively. For the estimate of the reflection losses on the crystal-gas interfaces we have used Fresnel's law for P polarized Cherenkov radiation; the average window absorption and quantum efficiency have been obtained from fig. 4 as in the previous case. From expression (3′), therefore, we compute an average number of photons $N = 8.5$ and $N = 6.6$ for the two momenta. As for the case of the gas radiator, we are missing more than a factor of two in the photon yield, the discrepancy being particularly large for the higher momentum measurement, where however the large beam divergence and some misalignment may explain the losses, the Cherenkov angle being rather close to

the one for total reflection in the radiator. Since in this case the gas purity is not of concern, we tend to believe in an unknown source of inefficiency in the MSC; alternatively, we may think of a badly degenerate transmission of the LiF crystal, a material that is known to be delicate to handle and that

TABLE 3

Standard deviation of the single photon radius dispersion expected for different reasons in the LiF radiator case for 1.3 GeV/c protons. The detector contribution includes also the spread introduced by the finite absorption length of photons in the gas (about 330 μm).

Source	σ_R (mm)
Radiator thickness (5 mm)	1.02
Chromatic aberration	1.99
Multiple scattering	0.26
Detector	0.77

TABLE 4

Average estimated efficiencies for photon transmission due to the indicated sources in the LiF radiator case. The over-all optical efficiency (not taking into account the detector quantum efficiency) is 40% and 26% for the two beam momenta, respectively.

Source	ε	
	1.3 GeV/c	1.5 GeV/c
Window transmittance	0.60	0.50
Interface reflections	0.98	0.88
Mesh transparency	0.70	0.60

can substantially change its transmission properties at wavelengths below 200 nm [26]). This point obviously requires further careful investigation.

5. Future prospects and applications

Detection of the Cherenkov rings by a photographic device, convenient for a rapid survey of the technique, may not be adapted to the present experimental requirements. On the other hand, the large number of simultaneous coordinates to be measured prevents the use of conventional bidimensional multiwire chambers with either cathode-induced pulses or delay line read-outs because of poor multitrack and ambiguity resolution. Multicell structures with individual needle counters have been proposed[17]) but they do not appear very simple to build. We are presently implementing a digital system having as a last step in the detector a multiwire chamber with current division read-out on the anodes where one expects, for moderate chamber sizes with 1 mm wire spacing, localization accuracies around 1 mm in both directions[27]). The ambiguity-free multihit capability remains, however, to be verified, since it could be worse than the simple wire spacing because of the large avalanche spread intrinsic in the MSC operation. It seems to us that coupling of an optical device of the kind described to a television-like scanning digitizer (like the plumbicon systems extensively used for spark chamber assemblies) may still offer distinctive advantages whenever large particle multiplicities and moderate data acquisition rates are expected.

Cherenkov light imaging using the MSC is of course restricted to the use of radiators that are transparent in the wavelength range between 120 and 160 nm. We have summarized in table 5 the properties of some of them, giving the estimated value of the index of refraction[22,23]) at 150 nm and the corresponding Cherenkov threshold [in terms of the velocity β or of $\gamma = (1 - \beta^2)^{-1}$]. For gases, the values are given at 0 °C, 1 atm. Based on the previous results, one can estimate the expected resolution of an imaging counter; as an example in fig. 20 we have represented the momentum dependence of the Cherenkov angle and of the ring radius for pions, kaons, and protons, obtained with a 50 cm long, 10 atm argon radiator. The width of the strips, one standard deviation each side of the average, has been estimated assuming a 10% relative radius accuracy determined by a single photon (an optimistic extrapolation of our previous results), combined with expression (5), and an average photon number given by eq. (3), also indicated in the figure. The angular resolution is almost constant in all the range, corresponding to about 0.1°, as can also be inferred from the approximated expression (5').

For a Poisson-like distribution in the number of photons around the average, and requiring at least three photons detected to define a circle, we have estimated the constant inefficiency lines shown in the figure. Under these conditions, a kaon–proton separation by more than three standard deviations can be foreseen from detection threshold (7.5 and

TABLE 5

Refraction index around 150 nm and corresponding Cherenkov threshold for several materials transparent in the vacuum UV range. For solid fluorides and liquid neon the value given is an extrapolation of known values at longer wavelength (ref. 23); for gases, the computed values of ref. 22 are given, at 0°, 1 atm. For liquefied gases, the density-dependent expressions given in the same references have been used.

Radiator	n or $(n-1) \times 10^{-4}$	β_T	γ_T
LiF	1.50	0.68	1.36
CaF$_2$	1.57	0.63	1.28
Ne (liquid)	1.10	0.92	2.49
He (liquid)	1.03		
Ne	0.749		81.7
A	3.785		36.4
Kr	6.806		27.1
e	23.898		14.5

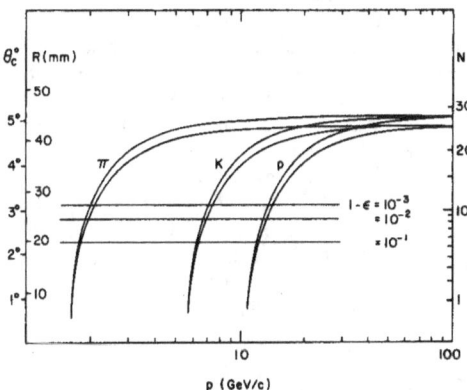

Fig. 20. Extrapolation of our results to a 10 atm, 50 cm long argon radiator using the MSC Cherenkov imaging device. The strips show the expected radius and angle resolution (one standard deviation around the average) for an $f = 50$ cm optics, as well as the expected number of photons N and the corresponding inefficiency levels if three of more photons are required for an event to be reconstructed.

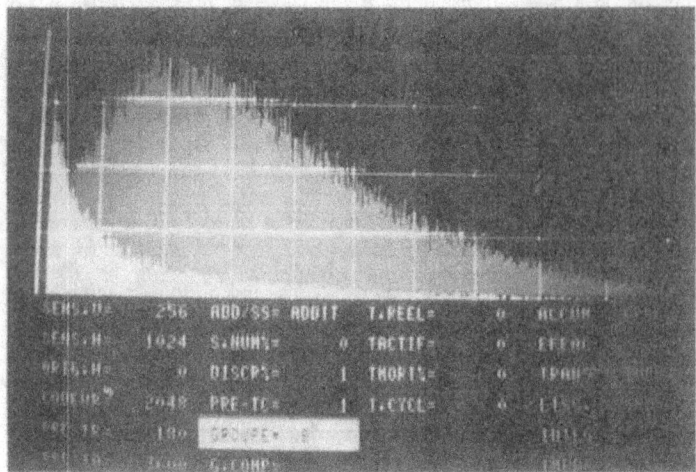

Fig. 21. Pulse-height spectra obtained in an MSC having a proportional chamber in the last element with a 5 mm thick LiF radiator under Cherenkov emission conditions (larger peak, protons at 1.5 GeV/c) and total reflection conditions (smaller peak, pions at 1.5 GeV/c). The residual signal, in the second case, that coincides with the spectrum obtained for both particles turning the detector by 180° in respect to the beam, is given both by direct ionization in the gas of the MSC and by scintillation in the radiator. Removing the sensitive part of the detector from the beam, one may obtain virtual insensitivity for particles having a velocity above the total reflection limit.

14 GeV/c for 10^{-3} inefficiency, for kaon and proton, respectively) up to about 28 GeV/c, an energy domain where kaon–proton identification is rather difficult using other techniques.

We may point out here that, contrary to photomultipliers, the MSC has a very little sensitivity to magnetic fields (only a small image shift may result at the strongest fields), and can in principle be operated at high pressures as fas as the quenching vapour does not condense (its absolute amount should be maintained constant in the mixture, irrespective of the pressure, to keep a value of the absorption coefficient compatible with the MSC operation). One can therefore foresee the use of the gas detector even simply as replacement of a photomultiplier in threshold operated devices, in the presence of strong magnetic fields.

The use of fluoride radiators, as we have seen, allows the detection of Cherenkov photons emitted by particles above threshold velocities $\beta \simeq 0.7$; this may be a handy device to identify short-lived, low-momentum particles in a detector only a few mm thick. Another attractive possibility is offered by the total reflection property, for particles normal to the radiator surface, above a critical velocity ($\beta \simeq 0.93$ for LiF, see section 4), which makes the detector insensitive to high-energy particles. One

may think of using this peculiarity to search for or to identify rare, low-energy particles produced in a high flux of very relativistic tracks. We show in fig. 21 a simple pulse-height measurement, obtained with an MSC similar to the one illustrated in fig. 15 except that the terminal spark chamber was replaced by a conventional multiwire proportional chamber. The two pulse-height spectra show the measured yield under Cherenkov emission conditions in a 5 mm thick LiF radiator (larger peak, protons at 1.5 GeV/c), and under total reflection conditions (smaller peak, pions of the same momentum). The residual signal in the second case, which coincides with the spectrum obtained for both masses inverting the detector as compared to the beam direction, is given both by the direct ionization in the gas of the detector and by scintillation in the radiator. Obviously, the direct charge signal can be eliminated if one removes the MSC from the beam and uses a reflecting mirror.

Note added in proof: Further measurements of the authors of ref. 10 seem to indicate that the triethylamine quantum efficiency (shown in fig. 4) could be overestimated by at least a factor of 2. This would explain the discrepancy between the computed and observed number of photons discussed in sect. 3.3 and 4.

References

[1] G. Charpak and F. Sauli, Phys. Lett. 78B (1978) 523.

[2] A. Breskin, G. Charpak, S. Majewski, G. Melchart, G. Petersen and F. Sauli, Nucl. Instr. and Meth. 161 (1979) 79.

[3] J. R. Bennet and A. J. L. Collinson, J. Phys. B2 (1969) 571.

[4] A. Gedanken, J. Jortner, B. Raz and A. Szöke, J. Chem. Phys. 57 (1972) 3456.

[5] R. D. Andresen, E. A. Leiman and A. Peacock, Nucl. Instr. and Meth. 140 (1977) 371.

[6] M. Suzuki and S. Kubota, submitted to Nucl. Instr. and Meth.

[7] A. Policarpo, Space Sci. Instr. 3 (1977) 77, and references therein.

[8] G. Charpak, G. Melchart, G. Petersen, F. Sauli, E. Bourdinaud, P. Blumenfeld, C. Duchazeaubeneix, A. Garin, S. Majewski and R. Walczak, CERN 78-05 (1978).

[9] J. C. Person and P. P. Nicole, ANL-7760 (1970).

[10] G. Comby and T. Ypsilantis, private communication.

[11] G. Charpak, G. Petersen, A. J. P. Policarpo and F. Sauli, Nucl. Instr. and Meth. 148 (1978) 471.

[12] J. Fisher and S. Shibata, Nucl. Instr. and Meth. 401 (1972) 157.

[13] Y. Tanaka, A. S. Jursa and F. J. LeBlanc, J. Opt. Soc. Am. 48 (1958) 304.

[14] G. Charpak and L. Massonnet, Rev. Sci. Instr. 34 (1963) 664.

[15] V. Böhmer, Dissertation (Un. Karlsruhe, July 1972).

[16] T. Aoyama and T. Watanabe, Nucl. Instr. and Meth. 150 (1978) 203.

[17] J. Seguinot and T. Ypsilantis, Nucl. Instr. and Meth. 142 (1977) 377.

[18] J. Chapman, D. Meyer and R. Thun, Nucl. Instr. and Meth. 158 (1979) 187.

[19] R. S. Gilmore, J. Malos, D. J. Bardsley, F. A. Lovett, J. P. Melot, R. J. Tapper, D. I. Giddings, L. Lintern, J. A. G. Morris, P. H. Sharp and P. D. Wroath, Nucl. Instr. and Meth. 157 (1978) 507.

[20] S. Durkin, A. Honma and D. W. G. S. Leith, SLAC-PUB-2186 (1978).

[21] J. Litt and R. Meunier, Ann. Rev. Nucl. Sci. 23 (1973) 1.

[22] P. W. Langhoff and M. Karplus, J. Opt. Soc. Am. 59 (1969) 863.

[23] D. E. Gray (ed.), American Institute of Physics Handbook (McGraw-Hill, New York, 1963).

[24] D. M. Roessler and W. C. Walker, J. Opt. Soc. Am. 57 (1967) 835.

[25] See, for example, K. Przibram, Irradiation colours and luminescence (Pergamon Press, London, 1956) p. 177 ff.

[26] D. A. Patterson and W. Vaughan, J. Opt. Soc. Am. 53 (1963) 851.

[27] C. W. Fabjan, J. Lindsay, F. Piuz, F. Ranjard, E. Rosso, A. Rudge, S. Serednyakov, W. J. Willis, H. B. Jensen and J. O. Petersen, Nucl. Instr. and Meth. 156 (1978) 267.

Nuclear Instruments and Methods 205 (1983) 403–423
North-Holland Publishing Company

PROGRESS IN CHERENKOV RING IMAGING:
Part 1. Detection and localization of photons with the multistep proportional chamber

R. BOUCLIER, G. CHARPAK, A. CATTAI, G. MILLION, A. PEISERT,
J.C. SANTIARD and F. SAULI

CERN, Geneva, Switzerland

G. COUTRAKON, J.R. HUBBARD, Ph. MANGEOT, J. MULLIE and J. TICHIT

CEN, Saclay, France

H. GLASS, J. KIRZ and R. McCARTHY

State University of New York, Stony Brook, NY, U.S.A.

Received 2 July 1982

The multistep proportional chamber, operated with a photosensitive gas filling, makes it possible to obtain stable multiplication factors in excess of 10^6 and can be used for the detection of single photoelectrons released in the gas. The efficiency and localization properties of the device in the detection of vacuum ultraviolet photons are discussed here, in view of its use for particle identification exploiting the Cherenkov ring-imaging method.

1. Introduction

Cherenkov ring imaging using multiwire proportional gas counters as photon detectors was introduced several years ago as a very promising technique for particle identification at high energies [1]. Various methods for photon detection and localization have since been devised and are described in the literature; for a complete bibliography see, for example, Coutrakon et al. [2]. Two major problems confront the experimenter in this field. On the one hand, the high-frequency cut-off of the best vacuum ultraviolet windows demands the use in the detector of photoionizing vapours with the lowest possible ionization potential, generally of rather delicate manipulation. Moreover, electronic detection of single photoelectrons in a gaseous counter requires stable and large multiplication factors, of 10^6 or so, to be attained and this in a gas mixture that, because of its intrinsic large photosensitivity, is particularly prone to secondary processes.

In the course of the development of a large Cherenkov ring-imaging device for experiment E605 at Fermilab, we have built and tested various configuratins of photon detectors and studied extensively their localization properties. Most of these measurements have been realized using a multistep proportional chamber as detector, and triethylamine (TEA) as the photo-ionizing vapour in addition to a noble gas carrier; the main

results of our investigation are described here. In a separate paper, we will describe the beam test of a ring-imaging detector prototype and its particle-localization properties [2]. The reasons for our choice of TEA as against the possibility of using even lower ionization potential vapours [such as tetrakis(dimethylamine)ethylene (TMAE)] are also discussed there.

2. Avalanche-size distribution generated by single electrons in proportional counters

The avalanche multiplication characteristics of gaseous proportional counters in the detection of single photoelectrons have been the subject of extensive research, both from the theoretical and the experimental standpoint. Knowledge of the fluctuations in the size of avalanches started by one electron is indeed essential in order to estimate the limiting energy resolution of proportional counters (see for example Curran and Craggs [3]). In a uniform field geometry, and assuming that each electron has a constant probability of undergoing ionizing collisions, simple probability considerations allow the computation of the following experimental avalanche-size distribution [4]:

$$P(n) \simeq \frac{1}{\bar{n}} e^{-n/\bar{n}}, \tag{1}$$

where $\bar{n}$ is the average value of the avalanche size. The expression implies that the maximum probability corresponds to no multiplication at all. Such a pulse-height distribution is indeed observed at moderate multiplication factors. At high gains, however, the experimental distribution ceases to be exponentially decreasing and develops a distinct peak around the average with a consequent reduction of its variance [3]. It appears that the shape of the distribution depends more on the field strength (and therefore on the value of the first Townsend coefficient) than on the total gain, proving that the peak is not a space-charge-related effect; this is apparent in fig. 1, where the pulse-height distributions for avalanches generated by single electrons in methylal are measured at increasing values of field, but at correspondingly smaller values of the gap so as to keep the effective total charge roughly constant [5].

Several theoretical models have been proposed to explain the observed avalanche-size distribution at high fields [6–9]. They generally agree in suggesting a functional form of the Polya type with one parameter b:

$$P(n) = \frac{b^b}{\bar{n}\Gamma(b)} \left(\frac{n}{\bar{n}}\right)^{b-1} e^{-b(n/\bar{n})} \qquad (2)$$

that reduces to the exponential (1) for $b = 1$. Typical observed values of b at high gains vary between 1.5 and 2.

The practical implications of a peaked pulse-height distribution as compared to an exponential one are obvious. In the case where the counter's signals are simply discriminated and scaled, the appearance of a peak in the input pulses' distribution implies that for a low enough threshold setting a constant efficiency plateau can be measured when increasing the operational voltage above the minimum for detection: this would not be the case for an exponentially decreasing pulse-height distribution. Examples of efficiency plateau obtained in the peaked condition will be shown in section 5.1. Moreover, when using a centre-of-gravity readout to improve the localization accuracy, a peaked distribution is useful since it limits the required dynamic range for the amplitude-measuring electronics and reduces the dispersions connected with gain linearity, uniformity, and signal pileups.

While a peaked avalanche distribution seems to be a general characteristic of high gains, this does not imply that one can reach these operational conditions in all gases; moreover, especially in photosensitive gas mixtures such as the ones we are interested in, various secondary processes connected with photon and ion feedback limit the gain to about 10^5 or so before discharge sets in. For a single electron avalanche, this is somewhat at the lower edge of the amount of charge that can be handled by the simple and cheap electronic circuits currently used in the work on proportional chambers.

Large stable gains can indeed be obtained in multi-wire proportional chambers, sometimes in excess of 10^7, by a careful choice of the gas mixture that normally includes photon-absorbing hydrocarbons and electronegative vapours for an effective quenching of secondary processes leading to discharge [10]. Such an approach is obviously not possible when the goal is to preserve good quantum efficiency in a wavelength interval as large as possible.

Operating a conventional multiwire proportional chamber with various concentrations of photosensitive vapours added to noble gases, we have identified two main mechanisms that limit the stable gain to values around 10^5. The first effect, which appears when using a few percent of TEA in argon or helium, is the onset of a typical Geiger discharge propagating along the anode wires irradiated and sometimes extending across several adjacent wires. Indeed, one expects this to happen in a gas where the mean free path for photoelectric absorption of photons, emitted in the avalanches in the energy domain where the additive has large quantum efficiency, is around 1 mm *. As an example, fig. 2 shows the transition from proportional to Geiger regimes observed in a multiwire proportional chamber with 2 mm wire spacing, operated with 2% TEA in helium. The maximum proportional gain is limited to about 10^5, and the transition is observed when increasing the anodic voltage by a few tens of volts.

Use in a proportional chamber of photosensitive vapours having much lower vapour pressure, such as TMAE, leads to a different secondary process. Because of their long range in the gas, several centimetres or so, ultraviolet photons emitted in the avalanche can reconvert rather far from the origin or can reach the surrounding electrode and extract there secondary electrons; a second avalanche develops in this case at a well-defined time interval (the time it takes for the electrons to drift back to the anode). An example is shown in fig. 3, as observed in a chamber operated with 1‰ TMAE in argon at gains slightly in excess of 10^4. Increasing the gain further, new generations of secondary avalanches appear and very quickly a diverging condition is reached leading to discharge. Such a behaviour is perhaps more insidious than the previous one, since it leads to spurious avalanches well separated spatially from the primary thus spoiling the localization properties of the detector or increasing its rate of accidentals. Only recently, by a careful design of the detector geometry and using pure methane, instead of a noble gas, gains in excess of 10^5 have been recorded using TMAE as the photo-ionizing agent [11].

The problems quoted have frustrated most early attempts to use conventional multiwire proportional

---

* The total absorption cross-section of TEA is about 20 Mb at 1450 Å, which implies a 1 mm absorption length for 2% TEA at that wavelength.

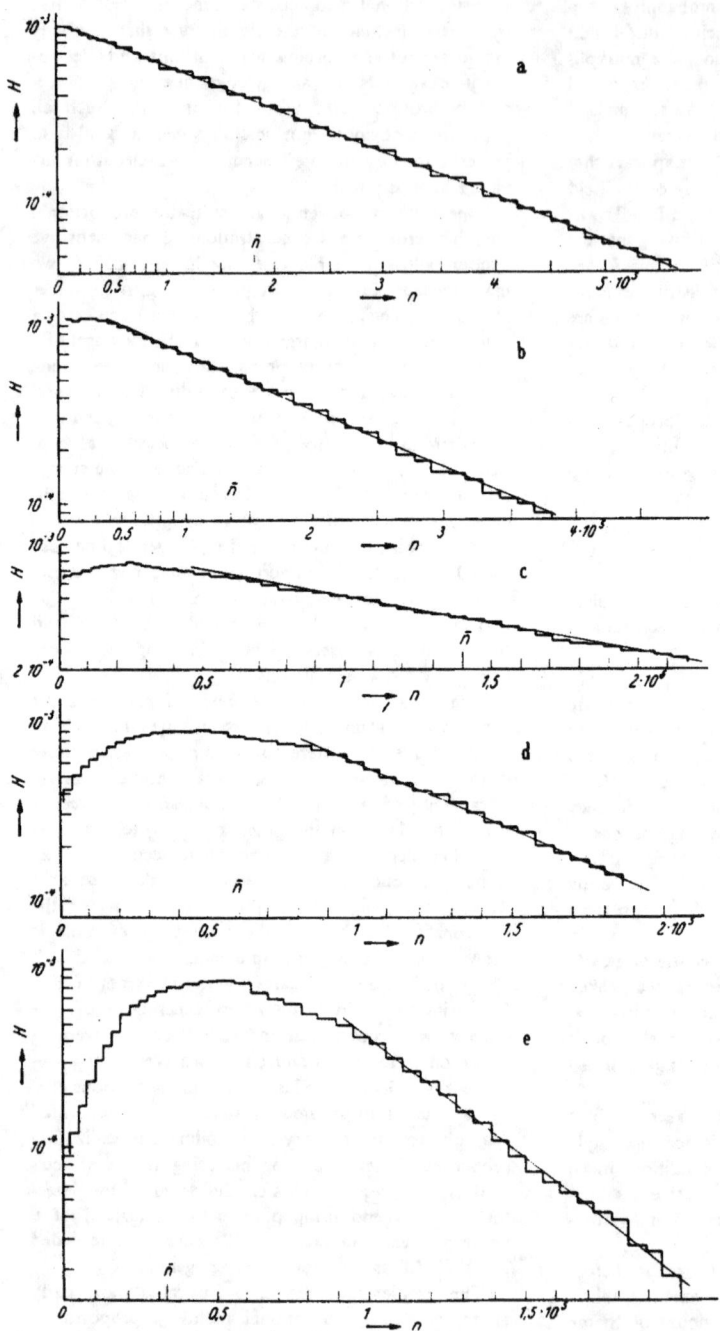

Fig. 1. Avalanche-size distributions for single electrons multiplying in methylal at increasing values of the reduced field [(a)–(e)]; to keep the average size roughly constant, the gap width has been reduced correspondingly. The transition from exponential to peaked distribution is evident. (From ref. 5.)

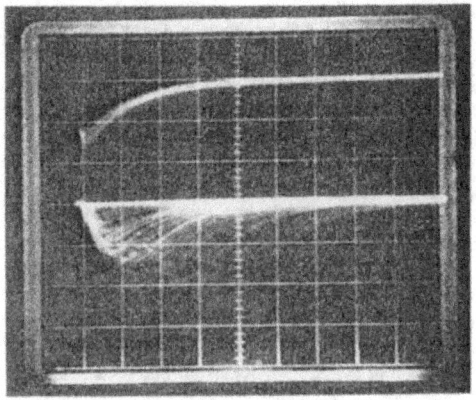

Fig. 2. Transition from proportional (top trace) to Geiger regime (lower trace) observed in a multiwire proportional chamber operoated with 2% TEA in helium, at gains around 10^5.

chambers for efficient detection of ultraviolet photons. During the development of the multistep proportional chamber [12] it was observed that separation of the overall amplification process into two distinct elements, each well below the critical region, allowed large stable gains, of 10^6 or more, to be attained in photosensitive gas mixtures, thus permitting efficient detection of single photoelectrons. The first Cherenkov ring images in a gaseous detectors were indeed observed using a multistep device [13].

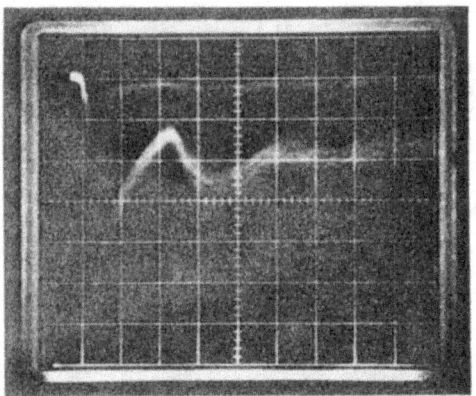

Fig. 3. Secondary avalanches generated by electron photoproduction on the cathodes. The chamber was operated with 1‰ TMAE in argon at gains slightly in excess of 10^4.

3. The multistep proportional chamber as a photon detector

The operational principle and the properties of multistep proportional chambers have been described extensively elsewhere [12–16] and only a brief summary will be given here.

Under the action of a strong uniform field between two electrodes, electrons experience various inelastic processes leading to the growth of an avalanche. For some gas mixtures, mainly the ones containing a pure noble gas with the addition of a little alcohol, acetone, or other organic vapours, the avalanche spreads spatially during the growth more than would be expected from the low-field electron diffusion coefficients in the given gas mixture; typically, in a 4 mm gap operated with 1% acetone in argon at a gain around 10^4, the front of the electron avalanche has a width of around 2 mm fwhm. As the ionization potential of the quoted additives is generally smaller than the excitation potential of the noble gas and of the energy of photons emitted by secondary scintillation, a photon-propagated or a Penning-type diffusion mechanism have been suggested to contribute to the avalanche spread as well as the direct ionization process [12–15]. One practical consequence of the large avalanche extension is, however, that the electric field in a multielectrode structure can be arranged in such a way as to transfer a uniform fraction of the electron avalanche into a subsequent element. A scheme of the electrode configuration in a multistep chamber is shown in fig. 4. A region of moderate electric field, named conversion or drift space, is separated by a wire mesh from a region where higher fields can be applied in a parallel-plate geometry, the so-called preamplification space. Electrons produced within the conversion space by an ionizing event drift towards and into the high field region where they multiply in an avalanche process. Electrons in the avalanche located within the field-line tubes connecting the pre-

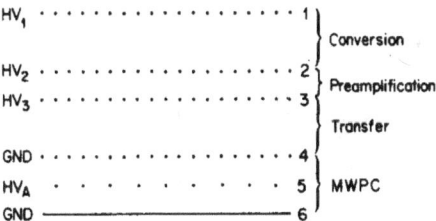

Fig. 4. Schematics of the multistep proportional chamber. Charges produced by an ionizing event in the conversion region drift into the high-field preamplification space where an avalanche grows. A fraction of the electrons in the avalanche head proceed into the transfer region and receive further amplification in the multiwire proportional chamber.

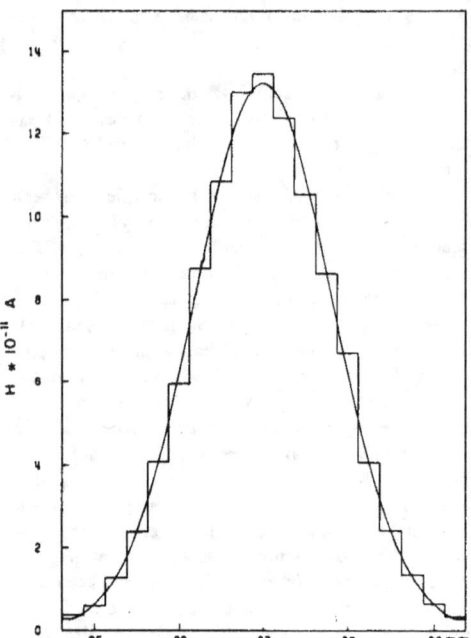

Fig. 5. Measured transverse distribution of electron charge in an avalanche developing in a 4 mm thick gap at gains around 10^4, in a gas mixture containing 0.5% TEA in argon [17].

amplification region to the lower field transfer region are injected there and proceed towards the following amplification element, a conventional multiwire proportional chamber as in the drawing.

A typical distribution of electron charge as measured for an avalanche developing in a 4 mm thick gap in a mixture of 0.5% TEA in argon is shown in fig. 5 [17], and has a Gaussian shape with a standard deviation of 800 μm. In fig. 6, the measured r.m.s. of the avalanche width in the same geometry is plotted as a function of the TEA concentration in argon. Somewhat smaller spreads have been obserrved in mixtures of TEA with helium (see section 5.3), but a systematic measurement of avalanche size in this case has not been made.

As far as the transverse avalanche spread in the preamplification region is comparable with the pitch of the wires constituting the transfer grid, the charge-transfer efficiency has only a little dependence on the average position of the original avalanche and equals roughly the ratio between the transfer and preamplification field strengths. Fig. 7 shows, for example, the computed transfer efficiency for several avalanche sizes, as a function of average position and for a wire mesh having 500 μm pitch. Clearly, for avalanches having 200 μm r.m.s. or more, the transfer efficiency is almost independent of the position.

The fraction of electrons drifting in the transfer region enters into the second amplification element, a conventional multiwire proportional chamber, see fig. 4; as the field there is generally higher, almost all charge is transferred. It was observed in the early works on the multistep chamber that in the double structure one could reach safely overall amplifications largely exceeding the ones that could be obtained in a single device when operated with a photosensitive gas mixture. As an example, fig. 8 shows the pulse-height distributions for single photoelectrons, as measured in a multistep proportional chamber operated with 1.5% TEA in argon at

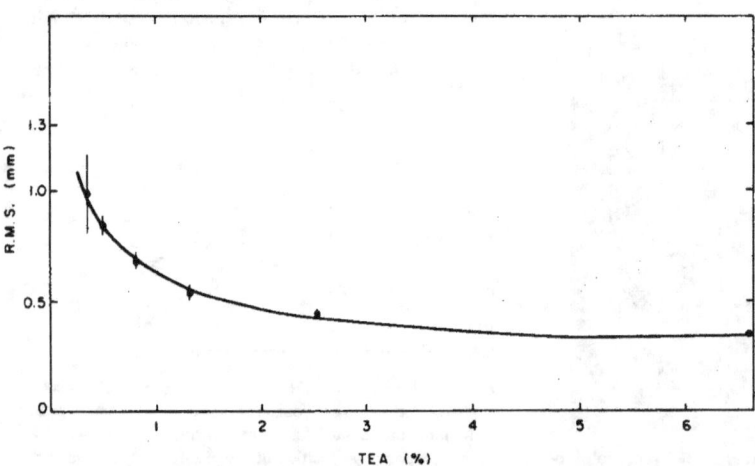

Fig. 6. Transverse avalanche size in a 4 mm gap, at gains around 10^4, measured as a function of TEA concentration in argon. A constant source-width contribution, estimated to be about 200 μm r.m.s. has not been subtracted from the data [17].

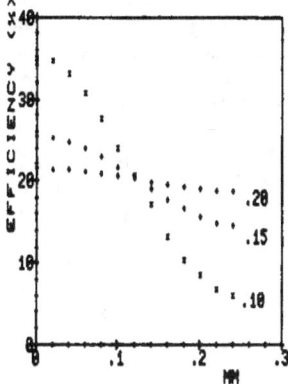

Fig. 7. Efficiency of charge transfer from the preamplification into the transfer region, computed for several avalanche sizes (their standard deviation is indicated on the plots, in millimetres). The ordinate represents the distance of the avalanche centre from the boundary between two wires in the transfer mesh, with 0.5 mm pitch. For an avalanche size exceeding 200 μm r.m.s., the transfer efficiency tends to be uniform and equal to the ratio of fields in the transfer and preamplification regions (20% in the example).

increasing values of the overall gain [17]; the amplitude scale is given in terms of detected charge on the anode wires (1 pC corresponds to a charge gain of about 6×10^6 in the detector). The peaked structure of the distribution typical of large gains is apparent, as discussed in the previous section.

In such a structure full detection efficiency for single electrons can be achieved with relatively low sensitivity

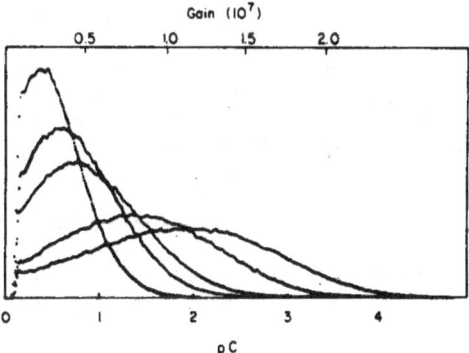

Fig. 8. Pulse-height distribution in the detection of single electrons recorded with a multistep proportional chamber operated with 1.5% TEA in argon at increasing values of the overall gain. The ordinate is given in terms of the detected charge; 1 pC corresponds to a gain of about 6×10^6. The peaked structure at large gains is apparent.

electronics, and localization by the centre-of-gravity method is greatly facilitated by the peaked signals with reduced dynamic range (see section 5).

4. Construction and testing of the photon detector

4.1. The multistep proportional chamber

We have built and tested several prototype photon detectors based on the multistep proportional chamber principle. Most of the results described here have been obtained with a detector which has an active area of 200×200 mm^2 and whose major parameters are listed in table 1 (which refers to fig. 4). For the actual Cherenkov ring imager, described in ref. 2, some parameters have been slightly modified, as for example the multiwire proportional chamber gap thickness, which was reduced to 3.2 mm to improve the separation of adjacent clusters. A set of machined fibre-glass frames, holding the various electrodes, are stacked and held together with a conventional O-rings and bolting system, see fig. 9. A ribbed aluminium frame on one side, and the calcium fluoride window mount on the other side ensure the necessary rigidity of the structure. To eliminate the possibility of water vapour permeation into the gas of the detector, which could affect its quantum efficiency, a double mylar window configuration has been adopted with the outgoing gas from the active volume flowing in the interface between the two sheets before the exhaust.

All passive electrodes, i.e. without readout electronics, are implemented with a stainless steel mesh having crossed wires 50 μm thick at 500 μm pitch. The meshes are stretched over the frames and soldered on suitable copperized-board insets; electrical contact is provided by extensions of the board itself.

The multiwire proportional chamber consists of three meshes of parallel wires; copper–beryllium wires 50 μm in diameter and 1.27 mm apart for the cathodes, and gold-plated tungsten wires, 20 μm in diameter with 2 mm spacing for the anode plane. The two cathode

Table 1
Construction parameters of the detector.

Electrode	Type	Pitch (mm)	Diameter (μm)	Gap (mm)
1	mesh	0.5	50	
2	mesh	0.5	50	6
3	mesh	0.5	50	4
4	wires	1.27	50	10
5	wires	2.0	20	5
6	wires	1.27	50	5

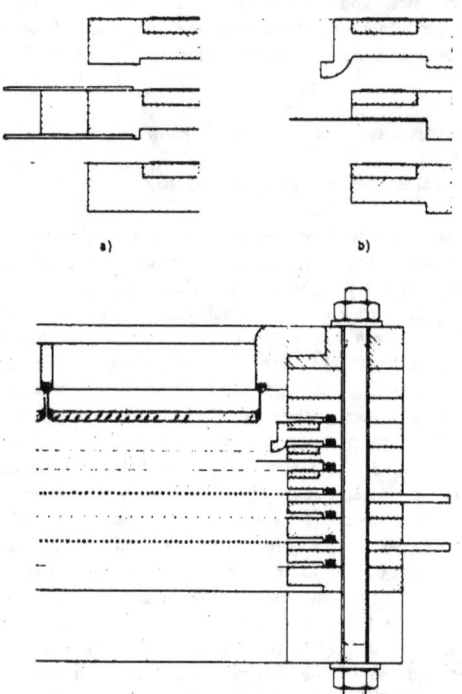

Fig. 9. Schematic cross-section of the multistep proportional chamber used for the test. A stack of fibre-glass frames holds the various electrodes that constitute the structure; on one side of the stack is mounted the CaF$_2$ window, while on the other a double mylar foil guarantees the gas-tightness. The insets show two possible solutions to the edge-breakdown problems encountered in the preamplification gap: an insulating foil inset (a) and a gap reducer (b).

planes are perpendicular to each other, and the anode wires mounted at 45° to the cathode. Although each wire output is accessible on a multipin connector for testing purposes, for most of the measurements four adjacent cathode wires are grouped together on each measuring channel, constituting strips about 5 mm wide. The anodes, on the other hand, are read out individually.

The wire orientation chosen ensures a redundant measurement of the coordinates for each point, thus partly overcoming the ambiguous coordinate coupling in the case of several close and simultaneous points (see below).

The construction of the preamplification gap deserves particular mention. Since one wants to attain here very high values of electric field in a parallel-plate geometry, precautions against edge breakdown have to be taken. We have experimented with various configura-

tions that help to prevent the spontaneous breakdown at the frame edges before reaching a useful preamplification gain. The more effective geometries are shown in the insets of fig. 29. In inset (a) two thin insulating strips have been inserted on each side of the gap, in close contact with the electrodes; 100 μm thick mylar or kapton strips, extending 4–5 mm into the gap, appear to be sufficient to prevent edge breakdown up to very high fields, probably just because they more than double the possible conduction channels on the frame surface. A more effective protection is given, however, by the geometry shown, as well as in inset (b), in the complete cross-section, that provides a physical increase of the gap thickness at the edge when the chamber is assembled. A specially machined fibre-glass frame, the "pusher", has insulating extrusions that reduce the gap thickness upon tightening the chamber by squeezing down the upper grid all around its edges. A reduction of 1/3 to 1/4 of the original gap appears sufficient to allow the highest fields to be reached without edge breakdown. Although mechanically more difficult to implement, this solution gave the more reproducible results and should be preferred whenever possible.

Because of its parallel-plate geometry, the preamplification element has a rather critical gain dependence on the gap thickness and uniformity of the electrodes. It has been computed, and experimentally verified, that the gain variation due to a decrease δ in the gap thickness l can be expressed by [15]:

$$M(l - \delta) = M(l) e^{k\delta}, \quad M(l) = e^{\alpha l}, \tag{3}$$

where k is a gas constant describing the first Townsend coefficient α at increasing fields, in a local linear approximation:

$$\alpha(E) \simeq AE - k. \tag{4}$$

We have measured the amplification factor, for a 4 mm thick gap, in a helium–triethylamine mixture (1.5% TEA in helium), see fig. 10. The parameter k, taken from expression (4) is about 8.1 mm^{-1}; from expression (3), therefore, one can see that a gain increase by a factor of two is obtained for a reduction of 90 μm in the 4 mm gap. While a 40–50 μm gap tolerance is not difficult to achieve in a small detector, like the one described here, both mechanical non-uniformities and electrostatic gap distortions are increasingly hard to control in large detectors and some gap-restoring or support-line device has to be envisaged with a geometry similar to the one adopted at the frame's edge.

4.2. Construction of the calcium fluoride windows

Detection of photons in the vacuum ultraviolet domain (VUV) requires the use of a fluoride crystal window. Despite its lower cut-off (about 10 eV) we have preferred calcium to lithium fluoride because of the

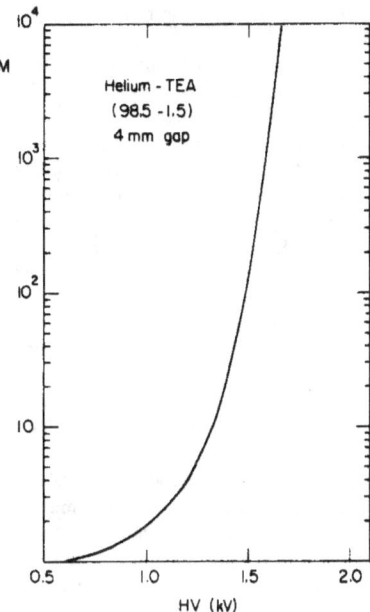

Fig. 10. Multiplication factor measured in a 4 mm thick parallel field gap for 1.5% TEA in helium.

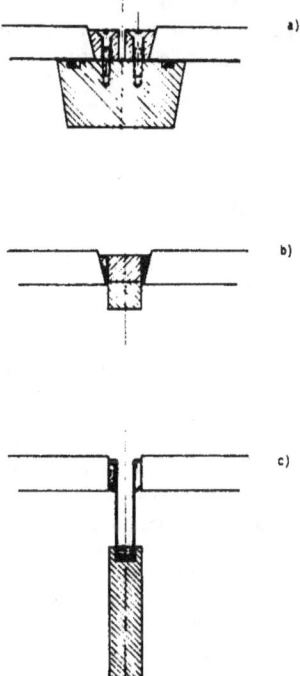

Fig. 11. Three possible ways of mounting CaF$_2$ crystals on thin frames, in order to implement large-surface composite windows. The scheme shown in (a) which uses thin compression bars and O-rings, was found to be not sufficiently gas-tight. Glueing with soft silicon rubber (b) was mechanically acceptable but showed large permeability to TEA; the third scheme (c), which couples the rigid support tot he crystals through a flexible frame, gave satisfactory results.

stability in transmission characteristics that are not degraded by prolonged exposure to the atmosphere. Calcium-fluoride crystals with VUV transmission properties can be purchased up to large diameters, about 270 mm [*]; in the perspective, however, of having to build a larger surface detector, we have preferred to experiment with smaller crystals mounted on a support frame, a solution that turns out to be cheaper for a given surface. To cover the 200 × 200 mm^2 active area of the detector, we have used four 4 mm thick crystals and tried several methods of mouting on a support that has the necessary stiffness to withstand the mechanical stresses and introducing a minimum dead space for photon transmission. A first attempt, based on the use of conventional rubber O-rings compressed on the crystal edges by tiny bars (fig. 11a), gave rather poor results in terms of gas-tightness and also presented severe problems of electrical insulation between the metal support bars and the first mesh in the detector (that has the highest potential in the chamber). Directly glueing the crystals on the support frame, fig. 11b, was more satisfactory and indeed most of the results described in the next section and in ref. 2 were obtained using a composite window of this kind. The choice of the glue has, however, to satisfy somewhat conflicting requirements. Because of the different thermal-expansion coefficients of the crystal and

of the support, the coupling should have sufficient elasticity to avoid the buildup of excessive stresses; on the other hand, the purity requirements of the radiator side in the ring-imaging set-up demand a very small outgassing of all materials employed. It appears that most vacuum-grade epoxies (such as for example Torr Seal [*] and Epotek [**]) have a very high grade of hardness and their use has resulted in localized or extended cleavage of the crystals. A limited success was obtained using a softer two-component silicon compound [***].

---

[*] Torr Seal, produced by Varian Vacuum Division, Palo Alto, Ca., U.S.A.
[**] Epotek H77, produced by Epoxy Technology Inc., Billerica, Mass., U.S.A.
[***] CaF$_4$, produced by Rhone Poulenc, Div. Silicones, Paris, France.

---

[*] Produced by Harshaw Chemical Co., Solon, Ohiol. U.S.A.

which has the desired mechanical characteristics to avoid local stresses but appeared in the long range to be permeable to TEA thus partly spoiling the purity of the gas in the Cherenkov radiator.

The last mounting scheme, fig. 11c, seems to have solved the above-mentioned problems. Each individual window is first glued along its edges to a thin (0.1 mm) stainless steel square-shaped tube, which has a size slightly exceeding the perimeter of the crystal. The framed elements are then glued into a supporting frame that has suitable grooves as shown in the figure. Both bondings are realized with vacuum-grade epoxy, thus ensuring a very low outgassing and no permeation to foreign molecules. The tensional stress due to temperature variations is thus avoided by the flexible coupling; indeed, part of the curing process requires a temperature above 100°C to be reached without damage to the crystals. The picture of an assembled four-crystal window, before being mounted on the detector, is shown in fig. 12. For structural reasons, the main cross-like support is machined out of a brass sheet; because of the way the crystals are mounted, however, it is not possible to electrically insulate the support from the first mesh in the detector. The metal frame is therefore glued to an insulating fibre-glass frame, visible in the pictue, that can then bolted to the radiator body with a teflon separation frame that ensures the insulation.

4.3. A vacuum ultraviolet timing photon source

For laboratory testing of the detector, and in order to study its localization properties, we have developed a VUV photon source exploiting the light-emission properties of noble gases excited by ionizing radiation. As schematically indicated in fig. 13, a gas-tight insulating envelope with a CaF_2 window contains an α-emitting source. A semitransparent mesh, mounted on the inside surface of the crystal, allows a difference of potential to be applied across the gas volume, the source holder being grounded. The presence of an electric field allows not only the ionization produced in each decay to be collected and detected, but also to increase largely the photon yield through a process of secondary scintillation. Even using very thin collimators, 100 μm or so wide, between the source and the chamber, average yields of several photons per disintegration can be raelized. We have mostly used krypton at atmospheric pressure as a gas filling, since its secondary emission occurs in a broad preak centred at around 1450 Å [18] almost coinciding with the peak quantum efficiency of TEA [13,19]. The source assembly can be applied over the detector window, after evacuating a small cylindrical flange, which is sealed both on the window and on the source side by O-rings. Collimators can be introduced within this volume to study localization properties.

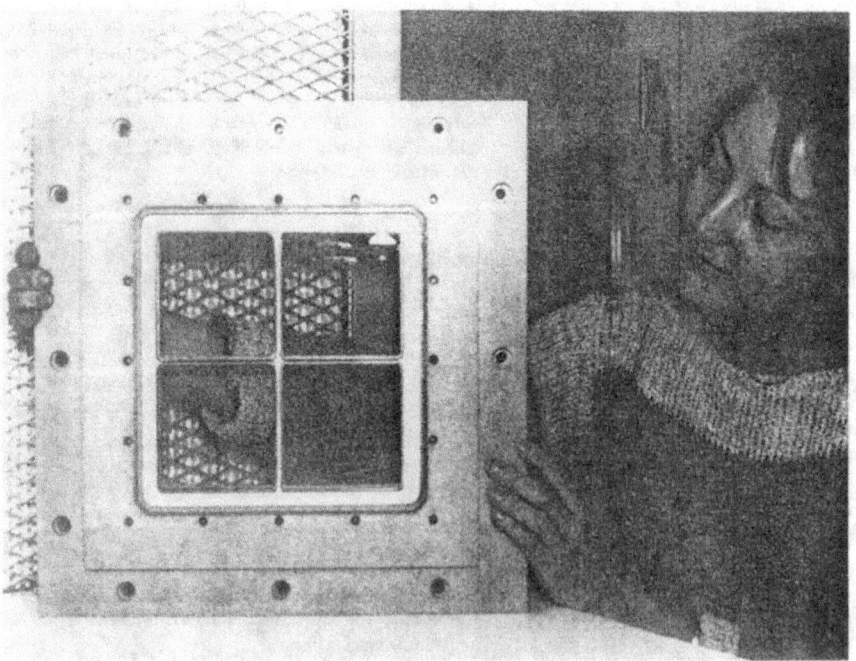

Fig. 12. A four-crystal mount used for testing the photon detector; each crystal is 100×100 mm^2 and 4 mm thick.

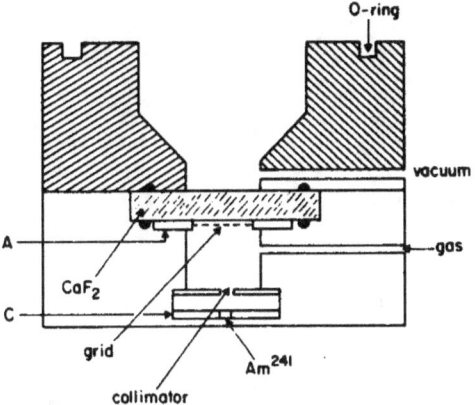

Fig. 13. The VUV photon source used for measuring efficiency and localization properties of the detector. 5 MeV α particles emitted by the ^{241}Am source stop in the gas between two electrodes producing both ion pairs and scintillation photons; using krypton as the filling, the photon emission is centred around 1450 Å at peak TEA quantum efficiency. The charge signal, detected on the top mesh, can be used for coincidence measurements; moreover, the photon emissin can be largely increased by applying a suitable difference of potential.

Because of the long lifetime of the excited states in krypton and of the collection time of the electrons in the source, the photon emission occurs on each disintegration over about 1 μs, preventing any study of the timing properties of the detector with this device. Efficiency measurements can instead be implemented by using the charge signal from the source in coincidence with the detector output; only relative efficiencies can be measured, however, unless the number of emitted photons is separately measured with a calibrated detector. With careful outgassing of the source and keeping a small flow of gas in it (1 cm^3 or so per minute) we could achieve a sufficient long-term stability in the photon yield to allow a comparison of efficiency plateaux measured with various detector geometries and gas fillings.

4.4. Pulse-height recording and calibrations

Charges collected on each anode wire and on groups of cathodes are preamplified on the detector and transmitted, through coaxial or twisted pair cables, to linear receivers followed by gated charge-to-digital converters. From the resulting pulse-height distribution, after some corrections to be described, the coordinates are computed using a centre-of-gravity algorithm [20]. Various configurations of amplifiers and receivers have been tried; a basic requirement was that they be cheap and compact enough so as to allow their use in large quantities. The scheme that has been used for part of the

measurements described here and for the prototype ring-imaging detector is shown in fig. 14. It consists of a charge preamplifier with a twisted-pair differential output (fig. 14a), followed by a linear differential receiver buffered for a low impedance output (fig. 14b). Only the circuit used for the cathode channels is shown, which has a full dynamic range (about −4 V at the receivers' output) for positive input charges; the anode channels are identical in design, but use complementary transistors in the preamplifier and an attenuation network (shown in the inset) to compensate for a larger input charge. We have adopted for the charge preamplifiers an RC shaping constant of around 400 ns. The charge-to-digital converter's gating pulse is shorter than the input signal and overlaps with it so that the recorded charge corresponds to the integral of the input during the gate length. Such a configuration has been preferred to a peak-sensing converter since it allows the overall sensitivity of the system to be modified by changing the gate length. The relatively long decay constant of the amplifiers allows the charge measurement to be less dependent on the relative timing between signal and gating pulse; indeed in our application the photon absorption length in the conversion space, typically 2 mm, introduces a jitter of about 100 ns in the detection time.

The sensitivity of the system, from the input to the receiver's output (on 50 Ω), is 400 and 150 mV pC^{-1} for the cathode and anode channels, respectively. With the typical sensitivities of commercial charge-to-digital converters, which is about 250 pC full scale, an average anodic input charge of 1 pC (see fig. 8) is recorded roughly in the middle of the dynamic range for a 50 ns gate.

The average electronics noise of the chain referred to its input corresponds to about 10^{-2} pC r.m.s., and is small compared to the average detected charge. The preamplifiers are organized on printed-circuit boards by groups of eight, the individual cards being plugged into a multiple connector board distributing the wires' signals. The connection is direct for the grounded cathode-wire plane, while low-leak high-voltage capacitors mounted on the mother board couple the anode wires to the corresponding amplifiers. The operating voltage is provided to the anode wires through individual 1 MΩ protection resistors, all connected to a common hv bus on the card. Protection against overload or accidental breakdown at the input is provided by three fast-switching diodes; no damage to the circuit results from a 500 pF, 5 kV discharge at the input, which largely exceeds the expected overload in the case of spark breakdown in the detector.

At the receiver side, an analogue sum of signals over 24 adjacent channels is used for monitoring; it is used also to allow self-triggering of the detector when using a calibration source.

For digital recording of the charge, we have used

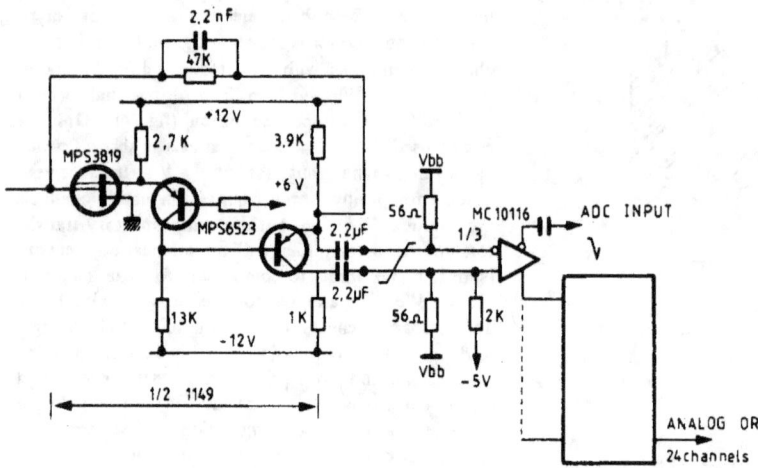

Fig. 14. Electronic circuit of the charge preamplifier and receiver for the cathode (positive) signals; the anodic circuit uses complementary transistors.

Fig. 15. The multistep proportional photon detector used for the measurements; it has a 200×200 mm^2 active area, fully equipped with charge preamplifiers on each anode wire and on cathode wire strips. To help reconstruction, the two cathode planes are perpendicular to each other and the anode wires are mounted at 45° to the cathodes.

commercial high-density CAMÁC-based charge-to-digital converters * connected to a small on-line computer ** for data handling and monitoring.

An assembled multistep proportional chamber complete with the preamplifiers' electronics, used for laboratory testing, is shown in fig. 15. It has a 200 × 200 mm² active surface, with a composite CaF₂ window as described. With two orthogonal cathode planes having 5 mm wide strips readout (indeed, 4 wires at 1/20 of an inch grouped together), and anode wires with 2 mm spacing inclined at 45° and individual readout, the detector requires around 230 pulse-height measuring channels to be fully equipped.

To reduce the unit cost of the amplifiers, the channel gain is not individually adjusted to the required tolerance. Instead, we have used a software calibration procedure for the complete system, which compensates for the gain differences and long-term drifts in individual channels. At a conveniently low operational voltage, to avoid saturation, the detector is uniformly irradiated with a ⁵⁵Fe 5.9 keV X-ray source, and the pulse-height distributions are recorded using the receivers' analogue sum output as the triggering source. After pedestal subtraction, the pulse-height spectra on all measurement channels are plotted, under the restrictive condition that a single anode has a significative pulse height in order to avoid the charge-sharing events (see the next section) that would produce a poor pulse-height resolution. For the cathode strips, on which the pulse height is obviously position-dependent, only the channel containing the highest pulse height in the distribution in each event is plotted; this ensures a good enough energy resolution for both the anode and cathode channels (fig. 16). Once the pulse-height spectra have been collected with sufficient statistics for all channels, the average pulse height on the 5.9 keV peak is taken as representative of the channel gain; this includes all sources of gain variation, from the chamber to the charge-to-digital converter. The table of gains thus compiled is then used as a calibration constant to reduce all channels' gains to the same normalized value. We have found that this procedure is reproducible and reduces the corrected channel-to-channel dispersion to 5% or less.

A convenient way of monitoring the residual positioning errors connected with all dispersions and non-linearities of the system is to plot, again for a uniform illumination of the detector with X-rays, the difference Δ between a measured coordinate, and the value of the same coordinate as computed from the two other measured projections; a plot of Δy as a function of y is presented in fig. 17. The projection of the distribution

<space holder> * LeCroy A.D.C. type 2249 A.
** Hewlett Packard HP 2100.

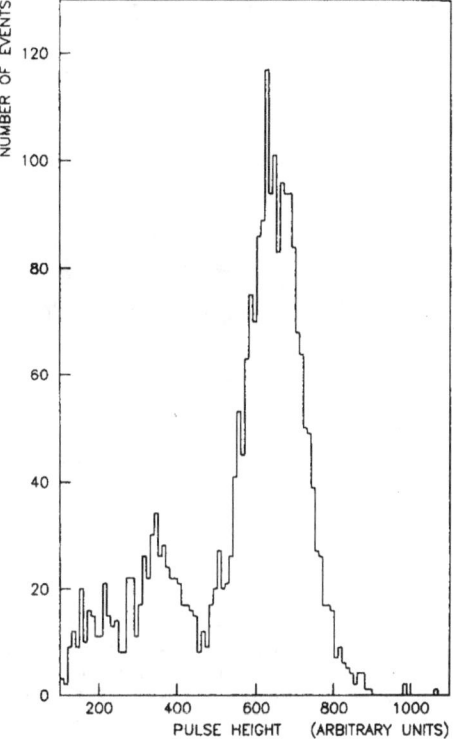

Fig. 16. As part of the calibration procedure, we record the pulse-height distribution for ⁵⁵Fe 5.9 keV X-rays for all channels. The figure shows a typical spectrum obtained on an anode wire, operating the detector at low gain with argon–TEA. A similar distribution is obtained for the cathode strips, under the restrictive condition that the strip contained the computed centre of gravity of the induced cluster.

on the ordinate has a Gaussian shape with 250 μm r.m.s.; this represents the overall localization error due mostly to residual non-linearities, gain variations, and electronics noise.

One should remember that the centre-of-gravity localization procedure is not very sensitive to local gain variations; this is because the distribution is normalized to itself. In fig. 18, we have simulated the effect of a 50% gain variation on one cathode channel, corresponding to the 5 mm wide strip centred around y = 56 mm. The oscillation that results has a maximum width of 1 mm; the positioning error due to a single-channel gain variation is therefore about 20 μm for each variation.

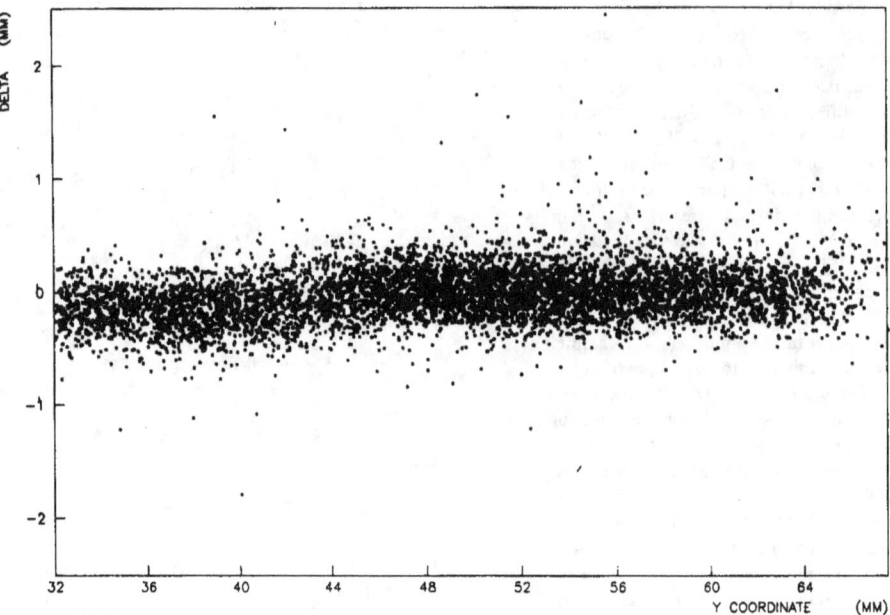

Fig. 17. Scatter plot of the difference between the directly measured coordinate y and its value computed from the other pair of coordinates, as a function of y itself. The deviations from linearity of the plot and its width represent the precision of the centre of gravity calibration. The projected distribution has 250 μm r.m.s.

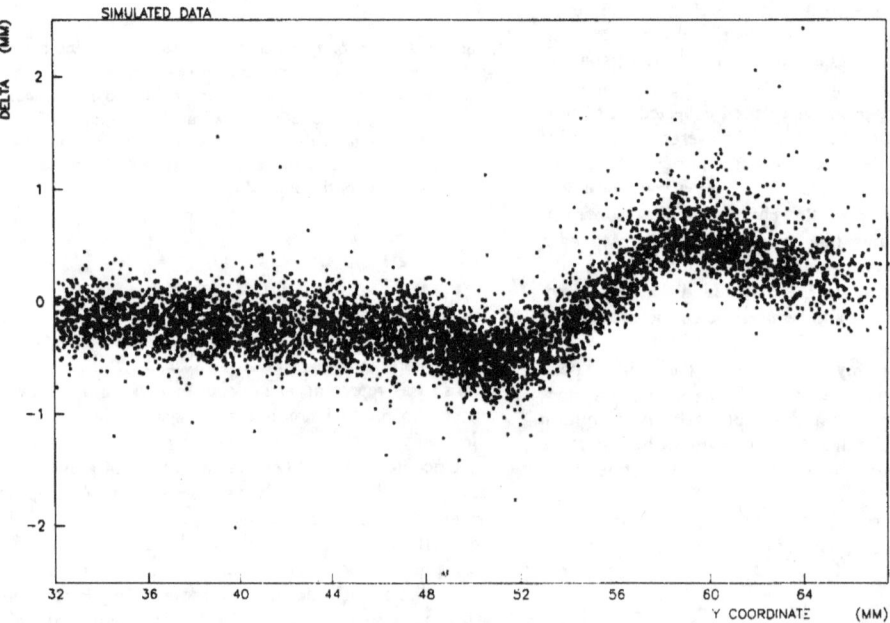

Fig. 18. Simulated effect of a variation by 50% of the gain on one of the cathode amplifiers mounted on the strip centred around $y = 56$ mm.

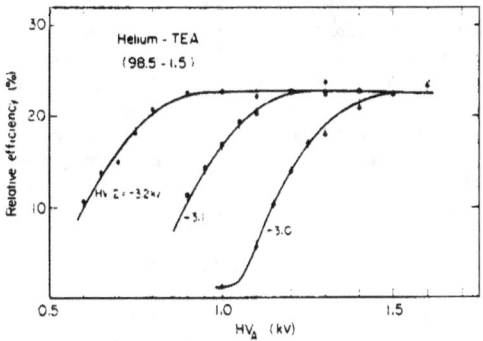

Fig. 19. Relative efficiency plateaux in the detection of VUV photons with the multistep proportional chamber, as a function of the anodic potential and for three values of the preamplification voltage. The gas mixture is 1.5% TEA in helium.

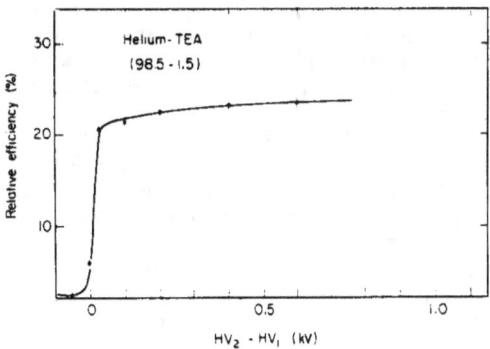

Fig. 21. Relative detection efficiency as a function of the drift voltage. A slight increase towards the higher fields may be due to increasing collection efficiency for photoelectrons produced very close to the window.

5. Detection and localization of vacuum ultraviolet photons

5.1. Detection efficiency

We have used the timing VUV source described in section 4.3 both to study the detection efficiency of the chamber and to analyse its localization properties. To perform the first kind of measurement, the analogue summing output from an anode receiver is applied to a faster discriminator with adjustable threshold. The detector efficiency is then defined as the ratio between the number of discriminated pulses in coincidence with the timing trigger provided by the source and the number of triggers. This is of course a relative measurement, containing no information about the absolute quantum efficiency of the device. However, adjusting the source voltage and collimation so as to measure a plateau with typically 20% detection efficiency, one can make sure

that the results correspond to the detection of a single photoelectron. Indeed, for a Poisson-like distribution in the number of detected photons, at the quoted total-ef-

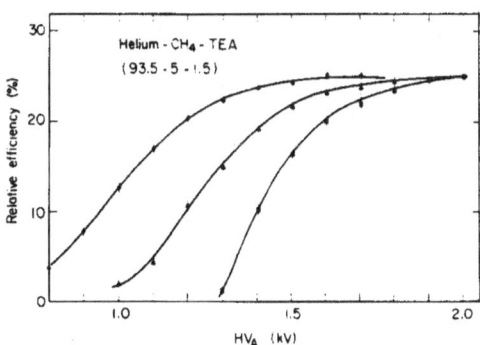

Fig. 20. Relative efficiency plateaux in the detection of VUV photons, for a gas mixture containing 1.5% TEA and 5% methane in helium.

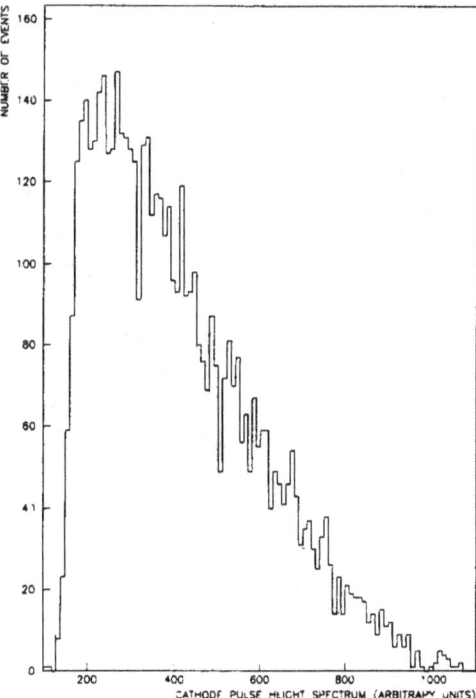

Fig. 22. Pulse-height spectrum for single photoelectrons recorded on the cathodes when operating the detectors with 1.5% TEA in helium. For this measurement, a set of 10-bit charge-to-digital converters was used.

ficiency level the number of events with two or more simultaneous photons is about 2%. This argument holds only if a constant efficiency plateau is found at increasing gain of the chamber; this is indeed the case for a gas filling containing 1.5% TEA in helium, as shown in fig. 19. The relative efficiency is given as a function of the anodic voltage, and for three values of the pre-amplification gap voltage; the transfer potential, HV3 in fig. 4, was kept constant and equal to -1 kV. The discrimination threshold was set at 0.1 pC, which is comfortably higher than the electronics noise and corresponds to about 5×10^5 electrons (see also fig. 8). Essentially the same behaviour is observed when adding a small quantity of methane to the mixture, although the working potentials are slightly higher (fig. 20).

Both quoted measurements were obtained by keeping a constant drift field in the conversion gap, about 1 kV cm^{-1}; it is interesting, however, to look at the behaviour of the relative efficiency at fixed gain, as a function of the drift potential (fig. 21). After a sudden increase above zero field, where electron collection begins, there is an indication of a slight increase of efficiency over several hundred volts until a plateau is reached at around 600 V (1 kV cm^{-1} in the 6 mm thick conversion gap). This may be an indication of either a slight electronegativity of the gas mixture, decreasing at high fields, or of a poor collection from the region between the first mesh and the window at low fields.

5.2. Pulse-height correlation and cluster size

Using a timing VUV photon source and the complete charge-recording system, we have studied the pulse-height and cluster-size distribution on anode wires and cathode strips. For each event, and after the pedestal subtraction and gain correction as described, the sum of recorded charges for the cluster of cathode-induced signals is computed, as well as the total anodic charge

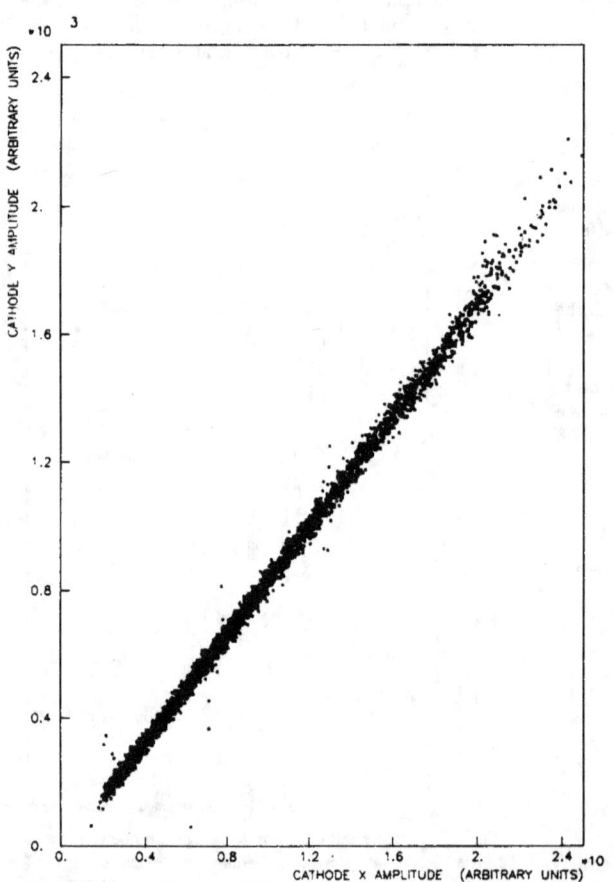

Fig. 23. Correlation between the overall charge detected in each cluster by the two cathode planes.

(because of charge interpolation, se below, a large fraction of the event interests two adjacent anode wires). A typical integrated cathode charge spectra, for a single photoelectron, is shown in fig. 22.

Owing to the extended dynamic range of the detected charge, a fraction of the events (about 2%) contains one or more channels exceeding the linear range of the converters; for these saturated events an improved centre-of-gravity algorithm has been developed.

Because they originate in the same physical process, there is a correlation between the integral anode and cathode charges, as shown in figs. 23 and 24 for the cathode vs cathode and cathode vs anode total detected charges, respectively. The correlation can be exploited to recognize correctly the triplet of clusters corresponding to each detected photon in the case of multiple events with ambiguous geometrical correlations. The relative amplitude difference between the two cathodes has a width of 5% fwhm (see fig. 25).

Another powerful constraint used to disentangle spatially close events is given by the knowledge of the width of the induced cluster. Fig. 26 shows the normalized charge distribution around the computed centre of gravity, for a large number of events; the distribution is very well represented by a Gaussian with a fwhm of 8 mm (or two cathode strips width). In the case of two close photons, partly overlapping in one projection, a fit with two Gaussian functions of known width (four free parameters) provides a good reconstruction of the original event, see fig. 27. For the event in the figure, the distance between the two photons in the projection is about 8 mm.

The double-cluster resolution is obviously a function of the gap thickness in the MWPC, which is 5 mm for the chamber described here; to improve this parameter, we have reduced the gap to 3.2 mm in the finally adopted geometry of the Cherenkov ring-imaging detector.

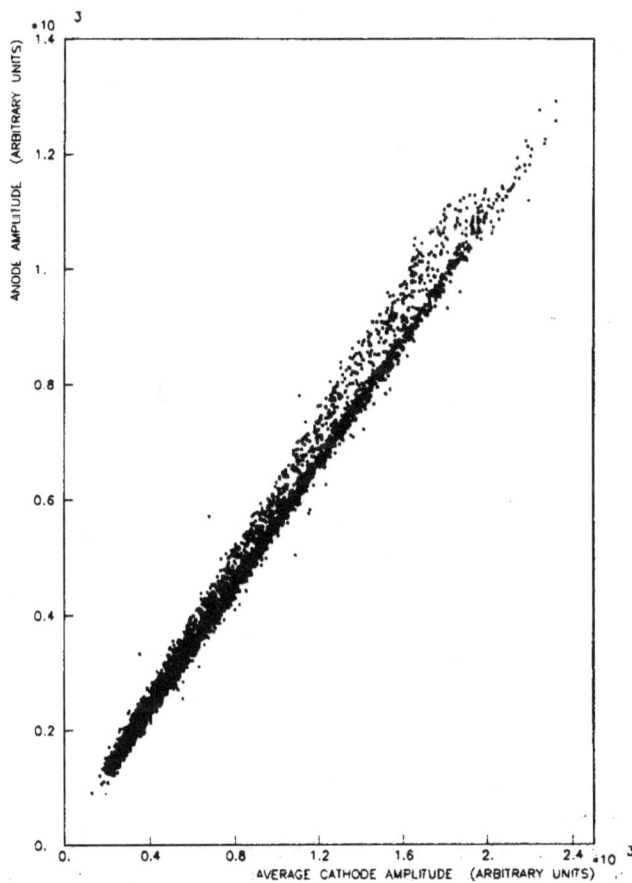

Fig. 24. Correlation between the charge detected on the anode and on one cathode plane.

5.3. Localization accuracy

Using a narrow slit collimator on the photon source, we have recorded and analysed the centre-of-gravity distributions for various relative orientations between the source and the detector. Indeed, while one expects a uniform dependence of the centre of gravity on the real position for the coordinate measured in the direction parallel to the anode wires, for the perpendicular direction the quantizing effect of the anodes normally leads to discrete values corresponding to the wire spacing. In the multistep detector, however, as described in section 3, the lateral spread of the avalanche in the preamplification element results for most events in a charge sharing between two or more adjacent anodes with, as a consequence, an effective coordinate interpolation.

Measurements of avalanche size as a function of the TEA concentration in argon were given in fig. 6; in order, however, to estimate the probability of charge sharing, one has to take into account the detailed mechanism of the mutual signal induction between adjacent anode wires. In the MWPC geometry of our detector, the positive signal induced on the neighbouring wires by an avalanche developing around a given anode represents about 10% of the main signal, however with opposite sign thus partly cancelling the signal due to the original shared charge. In fig. 28 we show the computed probability of charge sharing, i.e. the relative number of events resulting in two adjacent anodes with significative (negative) pulse height, as a function of the physical avalanche size and for a uniform exposure of the detector to the source. With reference to fig. 6, in a 1.5%

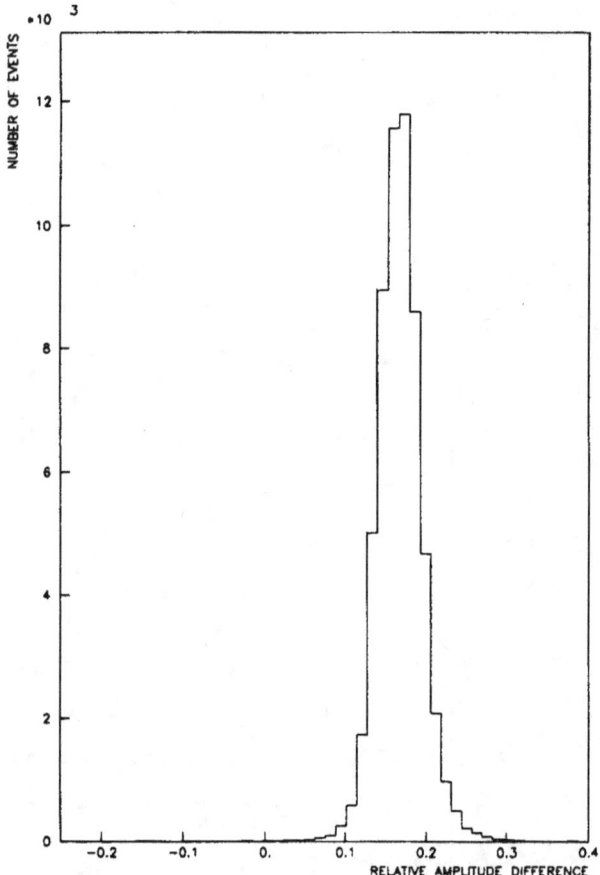

Fig. 25. Projection of the relative amplitude difference between the charge detected in the two cathodes; it has a fwhm of about 5%. The correlation has been exploited to couple multihit events properly in the case of clusters overlapping or ambiguous geometrical reconstruction.

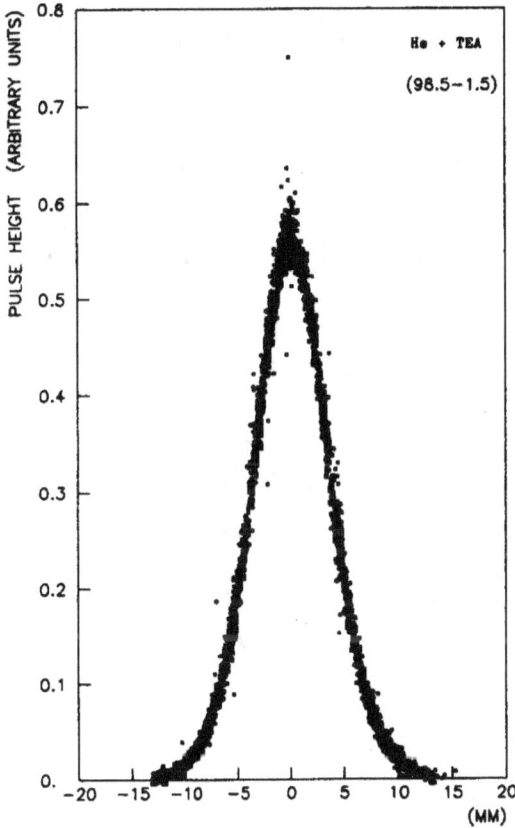

Fig. 26. Normalized charge distribution around the computed centre of gravity of cathode clusters. The distribution is roughly Gaussian with 8 mm fwhm.

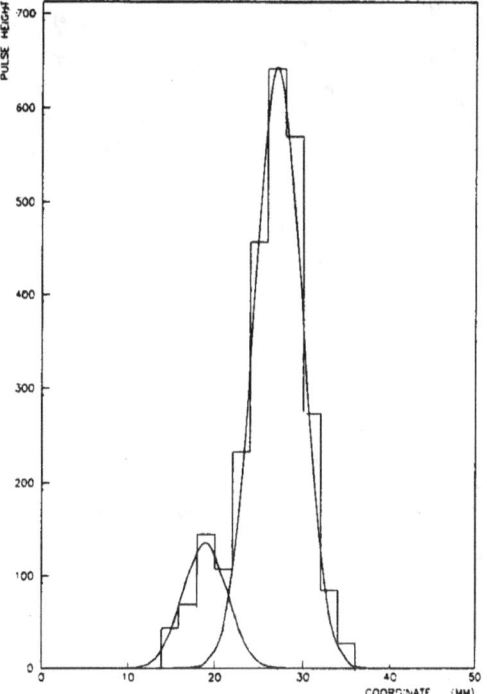

Fig. 27. An example of a double-cluster event, as seen on one cathode, and the Gaussian fit to two distributions with a known standard deviation (as from fig. 28).

Subtracting the intrinsic dispersions of the localization method, which account for about 150 μm (as indicated in section 4.4), and the known source width

concentration of TEA in argon, the avalanche has about 600 μm r.m.s., and this results in roughly 80% of the events being of the charge-interpolating kind, as indeed observed for a uniform illumination of the detector with the photon source. The effect is very obvious when plotting the centre-of-gravity distribution with the source collimated in the direction perpendicular to the anodes (fig. 29a). The dsitribution is indeed realtively uniform, the small oscillations being due to varius geometrical shadows within the source collimation (owing, for example, to photon absorption in the grids); in the absence of charge interpolation, the distribution would appear as a succession of peaks 2 mm apart (for comparison, see fig. 32). A projection of the distribution in the direction parallel to the anode wires is shown in fig. 29b; it has a standard deviation of about 450 μm, which represents the experimental localization accuracy for this direction.

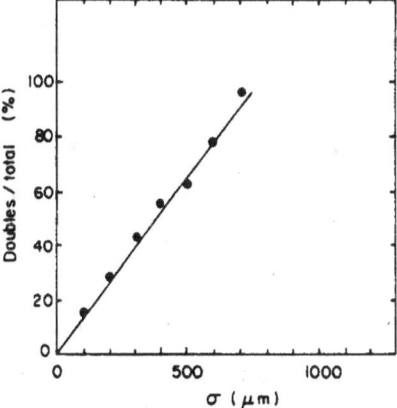

Fig. 28. Computed ratio of charge sharing vs total number of events for a multistep detector with 2 mm anode wire spacing, as a function of avalanche size in the preamplification gap.

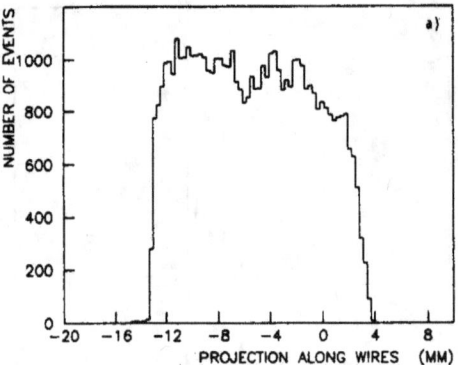

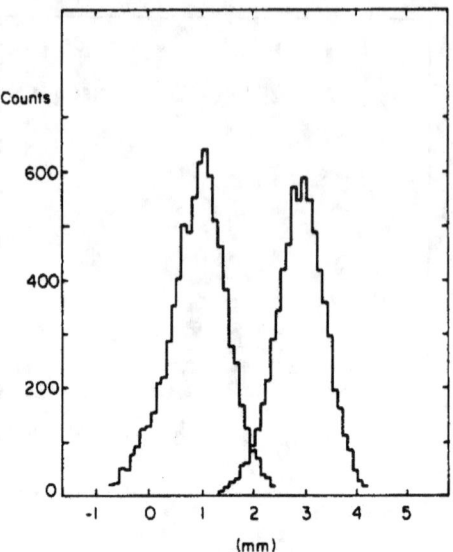

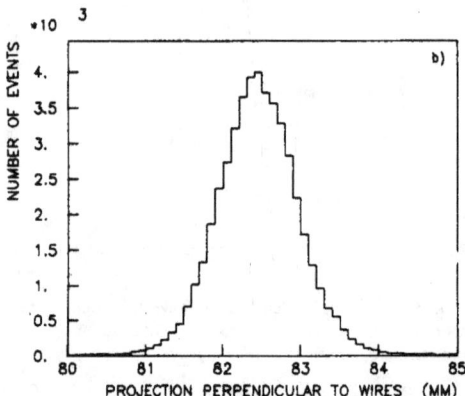

Fig. 29. Distribution of the computed centre of gravity in the detection of photons emitted from a source collimated in the direction perpendicular to the anode wires. In (a) the projection along the wires shows almost no structure owing to the large amount (80%) of charge-sharing events, in (b) the projection along the collimation direction provides the localization accuracy perpendicular to the anodes, which is 450 μm r.m.s. Gas filling: 1.5% TEA in argon.

Fig. 30. Centre-of-gravity distribution for a source collimated in the direction parallel to the anode wires, for two positions 1 mm away from each side of the wire. The distributions have a width of about 500 μm r.m.s.

(150 μm), it is apparent that the residual value of about 400 μm r.m.s. represents the physical dispersion connected with the detection of a single photoelectron in the structure and is most likely due to the statistical fluctuation of the average position of the charge in the first amplification element. Indeed, it has been observed that a decrease in the avalanche size (obtained by increasing the TEA concentration) results in a better localization accuracy in the direction parallel to the anodes, but obviously the charge-interpolation probability decreases at the expense of the accuracy in the perpendicular direction (see below).

To understand better the interpolation process and the resulting localization properties, we have mounted

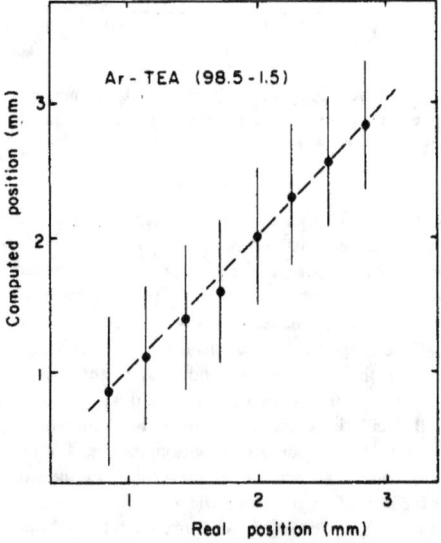

Fig. 31. Average computed centre of gravity and dispersion as a function of the actual collimated source position, for the coordinate perpendicular to the anode wires. Owing to the fluctuations in the charge-sharing process the localization error, 500 μm r.m.s., is larger than for the other direction.

the collimated photon source on a mechanical scanner with micrometer reading, and recorded the centre of gravity when displacing the source by known amounts along the direction perpendicular to the anodes. A typical distribution measured for two positions of the collimator 1 mm away from each side of a wire are shown in fig. 30; they have a width of 500 μm r.m.s. The results are summarized in fig. 31, again for the argon–TEA 98.5–1.5 mixture; the correlation is obviously linear, showing the effectiveness of the interpolation mechanism, with an average dispersion of 500 μm. Subtracting again the source-width and electronics-dis-

persion contribution, the residual value of about 450 μm represents the intrinsic localization accuracy in the direction perpendicular to the anode wires, which is slightly worse than that for the parallel coordinate because of the fluctuation in the charge-sharing process.

We have repeated the same measurements in mixtures of helium and TEA, where the avalanche size appears to be smaller than for identical concentrations of TEA in argon. The effect of the reduction of the charge-interpolating events is visible in fig. 32a, which is obtained using the same geometry as for fig. 30; about 50% of all avalanches are detected by a single anode wire, and the discrete anode-wire structure appears clearly in the distribution. From the quoted charge-sharing ratio, and from fig. 28, one infers an avalanche width of about 400 μm r.m.s. for these conditions; a direct measurement of the avalanche size in helium–TEA has, however, not been performed. Projection of the distribution in the direction parallel to the anodes, fig. 32b, provides a value of about 350 μm r.m.s. for the accuracy along this coordinate. As expected, the physical dispersions on the avalanche centre of gravity are smaller for the smaller avalanche; subtracting as before the intrinsic resolution and the source–collimator width, we infer a residual fluctuation of about 280 μm (as compared to the 400 μm previously quoted).

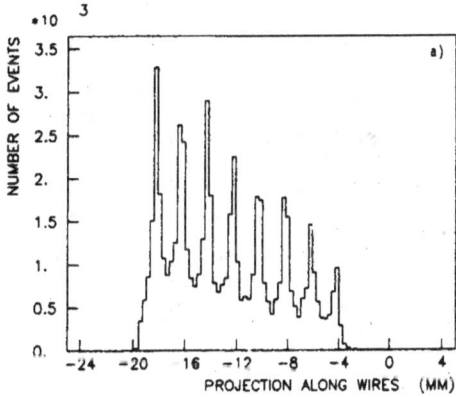

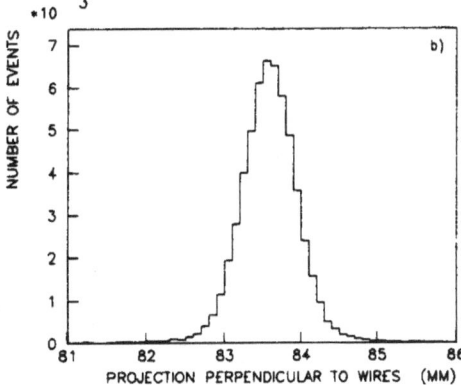

Fig. 32. Distribution of the centre of gravity in the detection of photons for a source collimated in the directin perpendicular to the anodes, using 1.5% TEA in helium. The smaller avalanche size in this case implies a reduced number of charge-sharing events (around 50%). In (a) the peaked structure corresponding to the anode wire positions is obvious. Because of the smaller fluctuations in the avalanche shape, however, the localization accuracy in the direction parallel to the wires which is about 350 μm r.m.s. (b) is better than for the previous case.

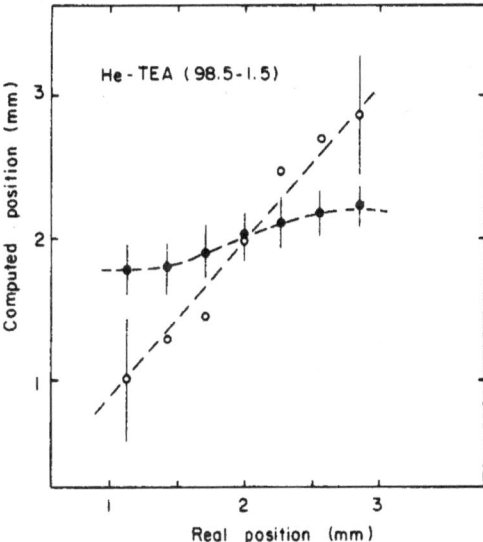

Fig. 33. Computed centre of gravity along the coordinate perpendicular to the anode wires, as a function of the real source position. Open circles represent charge-sharing events, and follow a more or less linear relationship to the real position, while full points represent events where a single anode provided a significant charge and cluster around the actual wire position.

A mechanical scanning along the direction perpendicular to the anode wires produces the results illustrated in fig. 33. In this case, we have separately plotted the centre of gravity for events interesting a single anode wire (full points) and charge-sharing events (open circles). While for the second class a more less linear correlation is observed, the single event obviously cluster around a position corresponding to the anode wire. A small right–left dependence is observed, although of course the dispersions in the position determination proevent the identification of individual events.

The fact that, depending on the avalanche fluctuations, an event in a given position may or may not result in charge-sharing obviously increases the overall localization error for the coordinate perpendicular to the anode wires. It can be seen from fig. 33 that this error varies between about 300 μm r.m.s. for a source position facing the wire up to about 600 μm at the edges.

Based on the quoted results it would seem convenient, in order to improve the localization accuracy when using a given number of readout channels, to increase the anode wire density, for example by a factor of two, grouping two adjacent wires on the same pulse-height measuring channel, and operating at a smaller avalanche size so as to reduce the dispersions in its centre of gravity still taking advantage of the charge-interpolation process.

References

[1] J. Seguinot and T. Ypsilantis. Nucl. Instr. and Meth. 142 (1977) 377.

[2] G. Coutrakon, J.R. Hubbard, Ph. Mangeot, J. Mullie, J. Tichit, R. Bouclier, G. Charpak, G. Million, A. Peisert, J.C. Santiard, F. Sauli, H. Glass, J. Kirz and R. McCarty, Nucl. Instr. and Meth., to be submitted.

[3] S.C. Curran and J.D. Craggs, Counting tubes (Butterworth, London, 1941).

[4] H.S. Snyder, Phys. Rev. 72 (1947) 181.

[5] H. Schlumbohm, Z. Physik. 151 (1958) 563.

[6] T. Byrne, Proc. Soc. Edinburgh Sect. A 66 (1962) 33.

[7] A. Lansiart and P. Morucci, J. Phys. Radium 23, Suppl. 6 (1966) 102A.

[8] W. Legler, Brit. J. Appl. Phys. 18 (1967) 1275.

[9] G.D. Alkhazov, Nucl. Instr. and Meth. 89 (1970) 155.

[10] F. Sauli, CERN 77-09 (1977).

[11] E. Barrelet, T. Ekelöf, B. Lund-Jensen, J. Séguinot, J. Tocqueville, M. Urban and T. Ypsilantis, submitted to Nucl. Instr. and Meth . (1982).

[12] G. Charpak and F. Sauli, Phys. Lett. 78B (1978) 523.

[13] G. Charpak, S. Majewski, G. Melchart, F. Sauli and T. Ypsilantis, Nucl. Instr. and Meth. 164 (1979) 419.

[14] A. Breskin, G. Charpak, S. Majewski, G. Melchart, A. Peisert, F. Sauli, F. Mathy and G. Petersen, Nucl. Instr. and Meth. 178 (1980) 11.

[15] J.R. Hubbard, G. Coutrakon, M. Cribier, Ph. Mangeot, H. Martin, J.Mullie, S. Palanque and J. Pelle, Nucl. Instr. and Meth. 176 (1980) 293.

[16] G. Charpak, A. Peisert, F. Sauli, A. Cavestro, M. Vascon and G. Zanella, Nucl. Instr. and Meth. 180 (1981) 387.

[17] A. Cattai, Thesis, University of Trieste, Italy (July 1981).

[18] M. Suzuki and S. Kubota, Nucl. Instr. and Meth. 164 (1979) 197.

[19] J. Séguinot, J. Tocqueville and T. Ypsilantis, Nucl. Instr. and Meth. 173 (1980) 283.

[20] G. Charpak and F. Sauli, Nucl. Instr. and Meth. 113 (1973) 381.

Nuclear Instruments and Methods 216 (1983) 79–91
North-Holland Publishing Company

PROGRESS IN CHERENKOV RING IMAGING
Part 2: Identification of charged hadrons at 200 GeV/c

Ph. MANGEOT, G. COUTRAKON *, J.R. HUBBARD, J. MULLIÉ, J. TICHIT and A. ZADRA
DPhPE, CEN Saclay, Gif-sur-Yvette, France

R. BOUCLIER, G. CHARPAK, J. MILLION, A. PEISERT, J.C. SANTIARD and F. SAULI
CERN, Geneva, Switzerland

C.N. BROWN
FNAL, Batavia, Illinois, USA

D. FINLEY, H. GLASS, J. KIRZ and R.L. McCARTHY
SUNY, Stony Brook, New York, USA

Received 11 April 1983

We have used a ring-imaging Cherenkov detector to separate π's, K's, and antiprotons in a 200 GeV/c beam at Fermilab. This device was built as a prototype for a large-aperture counter now in operation in Fermilab experiment E605. The radiator consisted of 8 m of atmospheric-pressure helium gas. The photon detector was a multistep proportional chamber. Cherenkov photons near 8 eV were detected by photoionization of triethylamine (TEA) vapor in the chamber. An average of 2.5 to 2.7 Cherenkov photons were observed per event, corresponding to a figure of merit $N_0 \approx 45$ per cm. A single-photon radius uncertainty of 0.47 mm was obtained with a helium/TEA/CH$_4$ gas mixture in the photon detector. The rms uncertainty in the determination of the Cherenkov angle was $\Delta\Theta_c/\Theta_{max} = 0.006$, corresponding to one-standard-deviation π/K separation at 500 GeV/c. At 200 GeV/c, the particle identification efficiency in a beam containing 95.2% π', 4.3% K', and 0.5% antiprotons was 92% for the π's, 83% for the K's, and 90% for the antiprotons.

1. Introduction

A ring-imaging Cherenkov detector has been used to separate π^-, K$^-$, and antiprotons in a 200 GeV/c beam at Fermilab. The ring-image was formed in the entrance-plane of the radiator by an 8 m focal-length spherical mirror. The photon detector was a 20×20 cm^2 multistep proportional chamber containing triethylamine (TEA) vapor as the photo-sensitive component. Some details of the construction, operation, and performance of the multistep chamber as a detector of photons in the vacuum ultraviolet (VUV) region were presented in Part 1 of this paper [1].

2. Cherenkov ring-imaging technique

Cherenkov ring-imaging was first proposed by Roberts [2] as a technique for accurate measurement of particle velocities and directions. When a charged par-

ticle passes through a medium of index of refraction n at a velocity ($v = \beta c$) greater than the velocity (c/n) of light in the same medium, photons are emitted at an angle Θ_C whose cosine is the ratio of the velocity of the light to that of the charged particle:

$$\cos\Theta_C = 1/\beta n. \tag{1}$$

Photons emitted at various points along a particle's straight-line trajectory, when reflected from a spherical mirror of radius R, are focused onto a circle of radius

$$r = f\tan\Theta_C \tag{2}$$

in the focal plane, a distance $f = R/2$ from the mirror. If the particle is inclined at an angle Θ_D to the optical axis, the center of the circle will be displaced a distance

$$\Delta q = f\tan\Theta_D \tag{3}$$

from the axis. Thus, measurement of the center of the Cherenkov circle determines the particle direction. Inversely, external measurement of the particle direction determines the center of the Cherenkov circle, so that *each* detected photon gives a measure of the Cherenkov

* Also at SUNY, Stony Brook, USA.

0167-5087/83/0000–0000/$03.00 © 1983 North-Holland

ring radius, and therefore of the particle velocity. In our application the particle momentum and direction are both determined externally, so a single detected photon yields a measurement of the particle mass.

3. Design considerations

The work reported here was undertaken to develop a prototype for a detector to identify π's, K's, and protons of momentum 100 to 400 GeV/c for Fermilab experiment E605. This is a high-luminosity experiment designed to study high-P_t single particles and particle pairs. The Cherenkov counter is downstream of two momentum-analyzing magnets. It has a 2×2 m^2 entrance plane and accepts particles with wide angular dispersion (120 mrad $\times$ 60 mrad).

The detector must have large acceptance, good velocity resolution ($\gamma_{max} \approx 800$), and reasonably good time resolution (≤ 1 μs), but it is only expected to identify at most two particles per event.

3.1. Choice of the photo-detection technique

Cherenkov ring-imaging was first demonstrated experimentally using image-intensifiers operating at visible wavelengths [3–8]. This approach is unsuitable for our purposes because of the limited acceptance and high cost of these devices. The detector surface can be reduced by optical subtraction, as in the spot-focusing Cherenkov counter [9,10], but this technique is limited to beams with small divergence.

In order to obtain a large detection surface compatible with our high rates, we use a proportional wire chamber containing photo-sensitive vapor, as suggested by Séguinot and Ypsilantis [11]. Difficulties were encountered in the development of this technique because the known compounds with reasonable vapor pressures are sensitive only in the vacuum ultraviolet (VUV) region. This fact sharply limits the choice of radiators and windows, since most materials absorb in the VUV region. Furthermore, chromatic dispersion increases for these higher photon energies. These problems have been brought under control by the development of detectors using vapors with very low emission potentials. The further problem of efficient detection and localization of the single photoelectrons produced has been resolved by use of a multistep structure for the proportional chamber, as described in ref. 1.

3.2. Choice of the photo-sensitive vapor

The early work on Cherenkov ring-imaging with gaseous detectors [12–14] used acetone (ionization potential 9.69 eV) or benzene (9.24 eV) as the photosensitive vapor. Lithium fluoride crystals, which are highly

hygroscopic and difficult to handle, were required for the windows. The introduction of triethylamine (TEA) vapor, with a photo-ionization threshold at 7.5 eV, provided considerable simplification. Crystals of CaF$_2$ or MgF$_2$, transparent to about 10 eV and more stable than LiF, can be used with TEA. The vapor pressure of TEA is sufficiently high (40 Torr at 15°C) that photons are completely absorbed in a few mm of gas. Considerable effort has been invested in the development of photon detectors using TEA vapor [15–18]. Noble gases are normally used as radiators in order to minimize the chromatic dispersion.

We chose TEA vapor for our detector, even though several compounds with much lower ionization potentials are now available [19–22]. These new compounds can be used with quartz windows (transparency cut-off around 7.5 eV for UV grade), instead of the fragile fluoride crystals required for TEA, but they have the disadvantage of very low vapor pressures and correspondingly long absorption lengths. Tetrakis dimethyl-amino-ethylene (TMAE), for example, has a photoionization threshold of 5.4 eV, but a vapor pressure of only 0.35 Torr and an absorption length [23] of about 20 mm at 20°C. Such a long absorption length implies a timing jitter of a μs or more and significant loss in spatial resolution due to parallax. With TEA vapor at 50% saturation at 15°C, the absorption length is only 1 mm [16], and both timing jitter and parallax are acceptably small.

3.3. Choice of the detector gas mixture

Once the photosensitive vapor has been selected, we must choose an appropriate carrier gas. Usually a noble gas is used to avoid photon capture by the carrier gas. We have used helium as the principal component of our gas mixture, rather than the more common argon, for two reasons: first, because charged particles ionize much less in helium, so we avoid large pulses which can mask the smaller photon pulses and which can lead to chamber instabilities. Second, because the excitation and ionization levels are much higher in helium than in argon, and we can obtain higher gains before breakdown occurs.

We use TEA at 50% saturation to avoid condensation in the photon chamber. The TEA temperature was 15°C. Thus, the first gas mixture used was

He(97.5%) + TEA(2.5%).

A second gas mixture was formed by adding methane (CH$_4$) to the above mixture, yielding

He(90%) + TEA(2.5%) + CH$_4$(7.5%).

The methane was added to improve resolution by the absorption of high-energy photons, as discussed in section 6.2. (The transmission of 7.5% methane in our gas

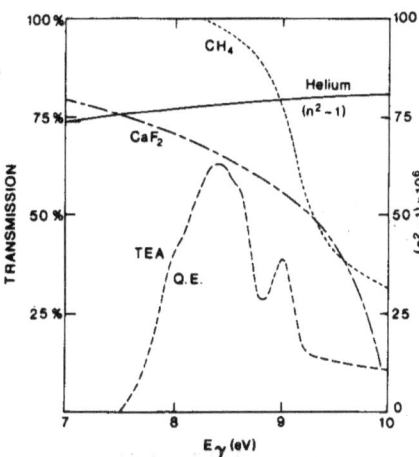

Fig. 1. The helium index of refraction [26] at STP, the quantum efficiency of TEA [16,22], the transmission of CaF_2, and the transmission of the methane in our $He/TEA/CH_4$ gas mixture [24,25], as a function of photon energy.

mixture, calculated from measured absorption coefficients [24,25], is shown in fig. 1.)

3.4. Choice of radiator gas

Helium was chosen for the radiator gas in order to limit chromatic dispersion and allow particle identification at high energy. Atmospheric-pressure helium at 0°C (STP) has an index of refraction of 1.0000386 for photons of 8.6 eV (near the peak of the TEA quantum efficiency). The energy dependence of the index of refraction [26], as well as the quantum efficiency of

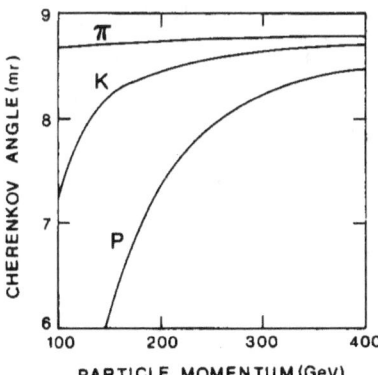

Fig. 2. Cherenkov angle for different momenta and particle types. The refractive index of helium used here was 1.0000386 at STP for 8.6 eV photons [26].

TEA [16,22], are shown in fig. 1. Fig. 2 shows the Cherenkov angle for various particle types as a function of momentum. The Cherenkov threshold for protons is around 100 GeV/c, and we can hope to separate π's and K's out to 400 GeV/c.

4. Experimental set-up

Our prototype particle identifier is shown schematically in fig. 3. The radiator vessel was a stainless-steel tube with a spherical mirror mounted at one end and the photon detector at the other end. The spherical mirror, at the downstream end, was accessible through a bolted cover in order to allow the proper alignment. A CaF_2 window and the multistep photon detector were mounted with conventional O-ring seals at the upstream end. A control tube mounted above the radiator tube could be used to measure the transmission of the radiator gas. Drift chambers measured the particle trajectory, and scintillation counters provided the trigger.

4.1. Radiator

The radiator tube was 8 m long and 30 cm in diameter. Gas purity was an important consideration in the design, because the photon absorption cross sections of oxygen and water are large in our photon energy range. The stainless-steel vessel was cleaned, then vacuum-baked at 100°C before the initial use at CERN. The radiator gas was obtained as boil-off from a liquid helium dewar. Gas flow was maintained at one volume change per hour during most data runs. The gas flowed out through the control tube to the recovery system.

4.2. Control tube

The control tube was 6 m long and 10 cm in diameter. It was used to monitor continuously the outflowing gas purity by measuring its transmission for VUV photons. A VUV photon source was mounted at one end of this tube, and a solar-blind photomultiplier with a MgF_2 window at the other end. The photon source, described in detail in ref. 1, used secondary emission in krypton [27], which matches closely the TEA quantum efficiency response. The photomultiplier response was measured while the helium flowed through the tube. Then the control tube was isolated from the radiator and evacuated, and the photomultiplier response was measured again. The ratio of the response with and without gas in the tube gave a reliable measurement of the transmision of the radiator gas: 87% during the run without methane, 81% during the run with methane in the photon chamber.

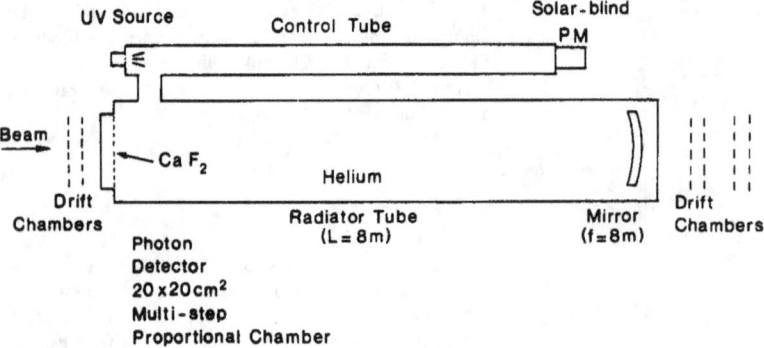

Fig. 3. Experimental set-up.

4.3. Spherical mirror

The mirror was manufactured by vacuum depositon over a pyrex substrate of a reflecting layer of aluminum 822 Å thick, coated with 165 Å of MgF_2 to enhance reflectivity and to avoid oxidation. Its reflectivity (fig. 4) was measured to be $(70 \pm 4)\%$ over the region of high TEA quantum efficiency. The focal length was (794 ± 1.2) cm, so the mirror was positioned 794 cm from the conversion gap of the photon detector.

4.4. Calcium fluoride window

A square array of four 4 mm thick CaF_2 crystals separated the radiator from the detector. The transmission of the CaF_2, measured with a monochromator, is shown in fig. 1. The average transmission for the He/TEA run was 64%, obtained by weighting the measured transmission by the quantum response of TEA in the energy range 7.5 to 10 eV. For the $He/TEA/CH_4$ run, the average transmission was also weighted by the effective transmission of the methane, yielding 65% for this run. The crystals were mounted on a brass frame to match the CaF_2 thermal expansion (for details, see ref. 1). The 1 cm wide arms formed a dead-space for photons (an additional loss of 9%), but not for beam tracks. The effective window transparency was thus $0.64 \times 0.91 = 58\%$ for the He/TEA run, and $0.65 \times 0.91 = 59\%$ for the $He/TEA/CH_4$ run.

4.5. Multistep proportional chamber

The 20×20 cm² multistep proportional chamber is shown schematically in fig. 5. Cherenkov photons pass through the CaF_2 window and ionize TEA molecules in the conversion gap (C), producing single photoelectrons. The photoelectrons are drifted into the preamplification gap (PA), where each photoelectron initiates an

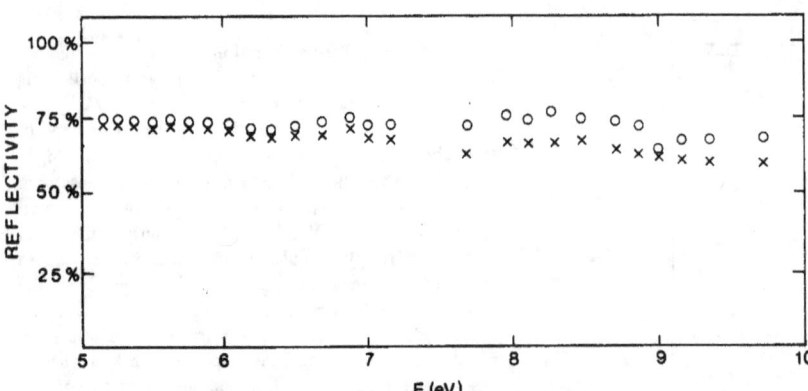

Fig. 4. Reflectivity versus photon energy. Measurements were made with an ultraviolet monochromator. The circles correspond to source intensity measurements made before inserting the mirror and crosses correspond to source intensity measurements made after the mirror was withdrawn, indicating some intensity fluctuations.

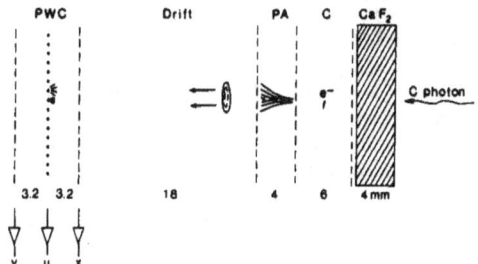

Fig. 5. Photon detector structure.

avalanche in the strong electric field. A fraction of the electrons in each avalanche, typically about 10^3, emerge into the drift space and are transferred to the proportional wire chamber (PWC). A second amplification takes place around the anode wires, bringing the total charge to about 10^7, which permits easy electronic detection.

The conversion and amplification gaps were formed of three stainless-steel meshes of 50 μm diameter wires, with a 500 μm pitch. The first mesh was in contact with the CaF_2 window. The optical transparency of this mesh was 81%. Some photoelectrons produced in the conversion gap are lost because of electric field distortions near the 100 μm thick meshes. Conversions which take place in the first 50 μm or so of the detector can be lost due to field distortions in that region. The mean free path of photons in 2.5% TEA is 1 mm, implying a 5% loss from these first 50 μm. Furthermore, some photoelectrons fail to enter the preamplification gap, and are lost, collected on the second mesh. We estimate a loss of 8% of the photoelectrons at the second mesh for our operating conditions. The conversion gap was 6 mm thick (6 mean free paths), so there was no loss due to photon escape. The effective transmission of the conversion gap was therefore $0.18 \times 0.95 \times 0.92 = 71\%$.

The PWC half-gap width was 3.2 mm. This narrow gap was chosen to reduce the width of the cathode pulses, and thereby improve the multiple-hit resolution. The cathode planes, made with 50 μm diameter, $\frac{1}{2}$mm pitch wires, were perpendicular to each other and parallel to the detector frames, defining two orthogonal coordinates X and Y. The anode plane was implemented with 20 μm wires, 2 mm apart, mounted at a 45° angle with respect to the other electrodes, and provided an inclined coordinate U. All three planes were read out every 2 mm into analog-to-digital converters (ADCs). There were 96 channels per plane which covered most of the active surface of the detector.

The multistep proportional chamber was operated with two gas mixtures, He/TEA and He/TEA/CH$_4$, as discussed in section 3.3. The drift fields in the conversion and transfer gaps were 100 V/mm for both

gas mixtures. With He/TEA the PA gap field was below 600 V/mm, and the PWC anode–cathode voltage difference was 1100 V. With He/TEA/CH$_4$ the PA gap required 1000 V/mm, and the PWC operated at 2000 V. We experienced some difficulties with chamber stability, and the higher voltages necessary with He/TEA/CH$_4$ increased these difficulties.

Further details of the construction and performance of the multistep proportional chamber used as a photon detector are given in ref. 1.

4.6. Electronics

Signals from the 96 anode wires and 2×96 cathode strips were amplified on the chamber by discrete-component amplifiers with a gain of 125 mV/pC, a time constant of 300 ns, and a noise level equivalent to less than 5×10^4 electrons. The pulses were transmitted in differential mode to the control room (a distance of 30 m), where they were referenced to ground and further amplified ($\times 6$) by MECL 10115 receivers. These receivers had trimmers which allowed a channel-by-channel adjustment of the gain. The signals were fed to charge-sensitive ADCs (LeCroy model 2249) with input impedance of 50 Ω. Standard NIM logic units were used to trigger the system and provide the gates for the ADCs. A 75 ns gate was used, delayed by 800 ns to compensate the drift time of the electrons. The digitizing time of the ADCs was 60 μs.

4.7. Particle beams

The complete prototype particle identifier was first tested in a secondary electron beam (about 8 GeV/c momentum) at CERN. This run provided information on the detection efficiency and multihit resolution, but the radius resolution was poor because of multiple Coulomb scattering in the windows and in the gas.

A second test run, with a hadron beam, was necessary to demonstrate the particle-identification possibilities of the device. This run was performed in the Meson Laboratory at Fermilab with an unseparated 200 GeV/c negative beam. Results from this second run are presented below.

4.8. Drift chambers

Two sets of high-accuracy drift chambers, before and after the radiator, were used to measure the trajectory of the incident particles. A third set of drift chambers was installed to determine drift-chamber resolution. Each of the six planes of drift chambers contained four wires, one wire every 5 cm, with one TDC per wire. The drift chambers were positioned so that the beam was measured on a single wire of each chamber.

4.9. Data acquisition

A coincidence of four scintillation counters, two before and two after the Cherenkov counter, defined a particle event and provided the trigger for readout of the drift chambers and the multistep proportional chamber. All data were read from standard CAMAC units by a Hewlett-Packard mini-computer. The raw data recorded on magnetic tape included 24 TDC values from the drift chambers and 288 ADC values from the photon chamber. The data acquisition rate was computer-limited to 40 events per second. The quality of the data was controlled by an on-line program that included a simple reconstruction of individual events.

5. Data analysis

5.1. Pulse-height distributions

The ADC pulse-height information from a typical event is shown in fig. 6. The cathodes defined the orthogonal coordinates X and Y, and the anodes provided the inclined coordinate U, with $U = (X + Y)/\sqrt{2}$. The preamplified charge from a single photoelectron usually fell on one or two adjacent wires (see ref. 1), while the induced charge on each cathode plane was spread over six or seven 2 mm channels.

The pulse-height spectra on the cathode X and Y planes for the two runs are shown in fig. 7. The value of the pulse height for each entry in the histogram is the sum of the ADC contents of each channel in a cluster of cathode strips above threshold (20 counts) multiplied by a conversion factor 2.2×10^{-4} pC per ADC count. The pulse-height threshold for the cluster was 200 counts, corresponding to a minimum detectable signal of 3×10^5 electrons.

Fig. 8 shows the correlation in cathode X and Y pulse heights for single photo-electrons. The cathode X pulse heights were 20% larger than the cathode Y pulse heights.

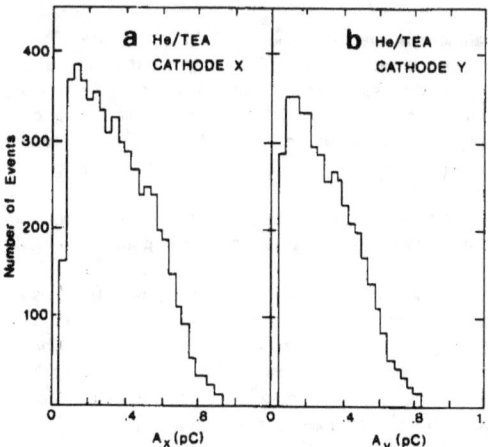

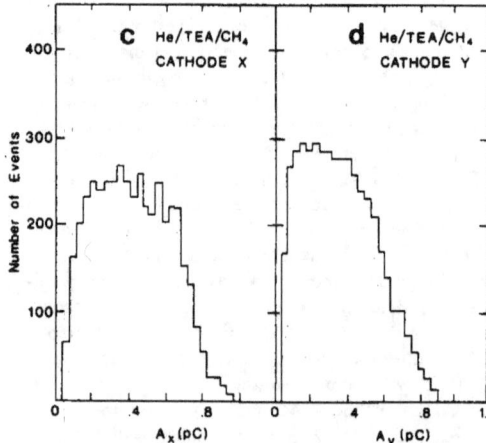

Fig. 7. Cathode X and Y pulse height distributions for He/TEA (a) and (b) and He/TEA/CH$_4$ (c) and (d). The amplitude is the sum of the ADC values for a cluster of channels above threshold. The conversion from ADC value to picocoulombs is discussed in the text. Only clean photon candidates were included in these plots.

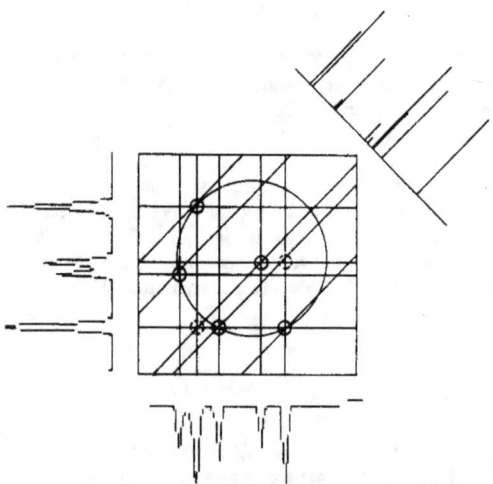

Fig. 6. An example of photon reconstruction for an event with four photons and a beam track. The beam track is not at the center of the Cherenkov ring, which is determined by the particle direction. Two ghost points appear, but they are easily eliminated because all three coordinates are required for other, real points.

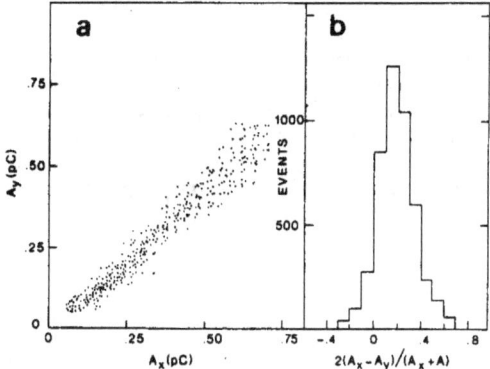

Fig. 8. Correlation between the cathode X and Y amplitudes. (a) Low statistics correlation plot A_x versus A_y for unambiguous photon hits. (b) High statistics distribution showing the cathode X amplitudes to be 20% larger than the cathode Y amplitudes.

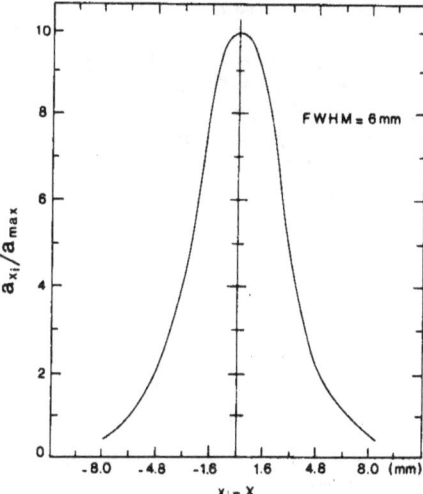

Fig. 9. Pulse height distribution of individual cathode strips. Here a_{xi} is the amplitude of the pulse from the strip with coordinate x_i, a_{max} is the maximum pulse height in the group of strips, and X is the coordinate calculated for the group of strips by the center of gravity method.

because the preamplified charge arrived at the anode wire closer to cathode X than to cathode Y.

5.2. Coordinate calculation

The pulse-height information was used to calculate coordinates by the center-of-gravity method for each group of wires hit. For the ith group of wires in the X plane

$$X_i = \sum_k x_k a_{xk} / \sum_k a_{xk},$$

$$A_{xi} = \sum_k a_{xk},$$

where a_{xk} is the ADC pulse height on the wire or strip at position x_k, and X_i and X_{xi} are the center-of-gravity coordinate and its amplitude. The distribution of cathode pulse heights about the calculated coordinate position is shown in fig. 9. The minimum separation for which two coordinates of equal pulse height can be resolved is 6 mm. For the anodes, the minimum separation is 4 mm.

When multiple peaks were apparent in a given group of wires, multiple coordinates were calculated by a rough partition of the measured pulse heights. Some coordinates were missed or poorly determined at this stage, because of overlapping pulse-height distributions.

5.3. Point reconstruction

An initial list of point candidates was obtained by selecting all coordinate triplets (X_i, Y_j, U_k) with good spatial correlation, unless their amplitudes differed by a very large factor. Explicitly, we selected triplets satisfy-

ing the following three requirements: (a) correlated in space to better than 2 mm, (b) cathode amplitudes equal within a factor 4, and (c) cathode and anode amplitudes equal within a factor 9.

Some points did not satisfy the above criteria because of unresolved coordinates. In order to recover these points, a second class of candidates was defined as follows: For each coordinate pair (X_i, Y_j), (X_i, U_j), or (Y_i, U_j), the expected value of the third coordinate U_k, Y_k, or X_k was calculated; if the contents of the ADCs near the calculated coordinate were large enough to match the amplitudes of the original pair, then a new point candidate was created.

A final list of points was selected by removing poorer candidates from this combined list until a "best" fit to the overall event was found. Candidate points were most likely to be eliminated if all three coordinates were used elsewhere in the candidate list, and if the amplitudes of the three coordinates were very different.

When the final list was established, final coordinates were determined as follows: (1) A coordinate was considered unambiguous, and its measured value was retained, if it was used in a single point in the final list. (2) If a point had only one ambiguous coordinate, that coordinate was recalculated from the two unambiguous coordinates. (3) If more than one coordinate was ambiguous, we attempted to recalculate these ambiguous coordinates by removing from the center-of-gravity calculation the contribution from previously resolved points.

5.4. Recalculation of ambiguous coordinates

This last technique can be illustrated by considering the specific example in fig. 10. The X and Y coordinates of photon 2 of this event are both ambiguous, so the position of that photon is poorly determined. But, the ambiguous coordinate X_1 and its amplitude A_{x1} can be expressed as the sum of the contributions from photons 1 and 2:

$$X_1 = \frac{X(1) A_x(1) + X(2) A_x(2)}{A_x(1) + A_x(2)},$$

$$A_{x1} = A_x(1) + A_x(2).$$

The X coordinate of photon 1, $X(1)$, can be calculated from the unambiguous coordinate pair (Y_1, U_1)

$$X(1) = \sqrt{2}\, U(1) - Y(1) = \sqrt{2}\, U_1 - Y_1.$$

Furthermore, the X and Y amplitudes due to this photon are closely correlated, so

$$A_x(1) \approx A_y(1) = A_{yi},$$

Therefore, the X amplitude due to photon 2, $A_x(2)$, can be calculated from the coordinates amplitudes:

$$A_x(2) = A_{x1} - A_x(1) \approx A_{x1} - A_{y1}.$$

Now these values can be used to calculate the X coordinate of photon 2:

$$X(2) = \frac{X_1 A_{x1} - X(1) A_x(1)}{A_x(2)}.$$

We calculate $Y(2)$ in a similar way. Then we verify that the triplet $[X(2), Y(2), U_2]$ is well-correlated in space before accepting this point for further calculations.

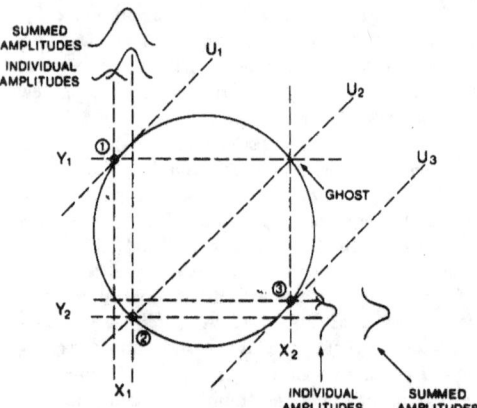

Fig. 10. Example of two pulse overlap problem. The circles mark the position of the photon hits. The coordinates were calculated by the center of gravity method and are labeled on the sides. Photon 2 in this figure has an ambiguous coordinate in X and Y.

5.5. System resolution

The drift chamber measurements of the angles of the particle trajectory were used to determine the center of the photon ring. For each point candidate, a ring radius was calculated (the distance from the ring center). Single-photon radii in the region of the π peak are plotted in fig. 11 for the two data runs. The distribution

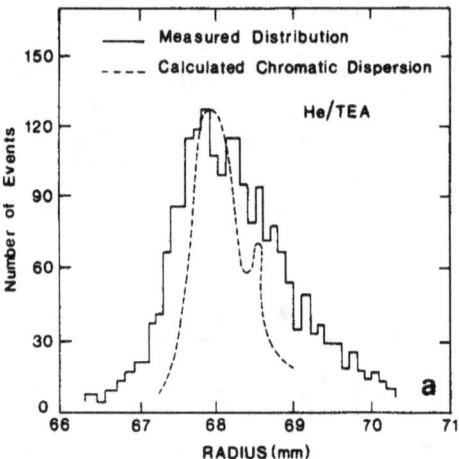

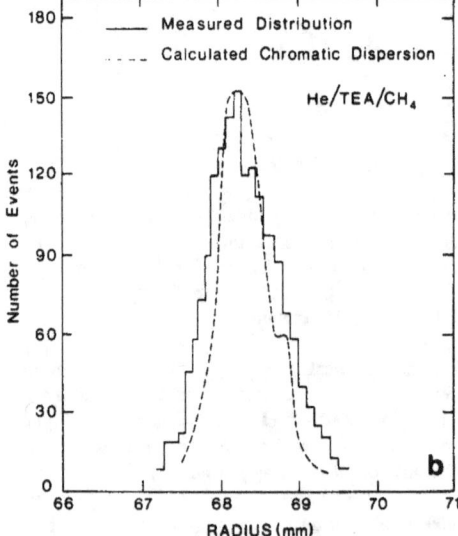

Fig. 11. The measured one photon radius distribution and the expected chromatic dispersion (a) with He/TEA in the chamber, and (b) with He/TEA/CH$_4$. The chromatic dispersion was calculated from eq. (10), using the helium index of refraction in fig. 1 and the TEA quantum response in fig. 15.

of single-photon radii for all 200 GeV/c π's should give the system resolution function. The fwhm of this distribution is 1.5 mm for the He/TEA data and 1.1 mm for He/TEA/CH$_4$. These distributions are not Gaussian. They are convolutions of the detector resolution with the chromatic dispersion of the helium radiator (see section 6.2). The equivalent rms widths are 0.64 mm and 0.47 mm, respectively.

5.6. Photon selection

The final point list contained beam tracks as well as photon candidates. Indeed, the incident particles traversed the multistep chamber near its center, leaving a trail of ionization. This beam track was often detected along with the Cherenkov photons, even though the photons reach the chamber with a 53 ns delay. The beam track was located easily by interpolation from its drift-chamber coordinates. With the beam track removed, the remaining points were the photon candidates.

Background points were them removed by a clustering technique. For a multi-photon event, all photon radii were required to lie within 3 mm of each other. When this was not true, the cluster with the most photons satisfying this criterion was used. When two ore more clusters had the same number of photons, the cluster with the smallest range of radii was used. Background points were removed efficiently by the cluster cuts except in the case of zero-photon events, where any background point was interpreted as a single photon (1% of all events).

The distributions of the number of photons selected by the cluster technique are shown in fig. 12. For the He/TEA run, an average of 2.64 photons were found per event; for the HE/TEA/CH$_4$ run, the average was 2.48 photons per event.

These distributions are not perfectly Poissonian, however, even after shifting background events down to the zero-photon bin. Limited two-photon resolution and other program inefficiencies shift many higher-multiplicity events down to lower bins, with a net loss of about 8% of all photons. The lowest multiplicity bins, which determine the number of unidentified events, correspond to the higher averages, 2.9 photons per event for He/TEA and 2.7 photons per event for HE/TEA/CH$_4$.

5.7. Particle identification

The distribution of average photon radii is shown in fig. 13. The antiproton peak is well separated from the combined π–K peak. This latter region is shown in more detail in fig. 14. The superior resolving power of the He/TEA/CH$_4$ gas mixture is evident here. The rms width of the π peak was 0.53 mm for the He/TEA run, and 0.42 mm for the He/TEA/CH$_4$ run, for a π–K radius difference of 2.64 mm. These widths are smaller than the widths of the single-photon distributions, but larger than expected for a $N^{1/2}$ improvement.

For the He/TEA data run, 1% of the π's and 25% of the K's are ambiguous between the π and K hypotheses.

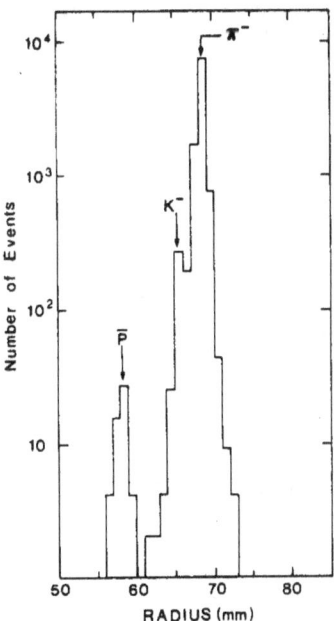

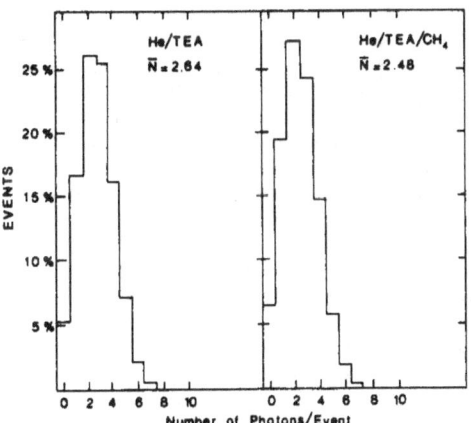

Fig. 12. Histogram of number of events versus number of photons per event, with and without CH$_4$.

Fig. 13. Radius distribution with 1 mm bin size, showing the π^-, K$^-$ and $\bar{p}$ peaks. The radius shown here is the average of all photons in the event.

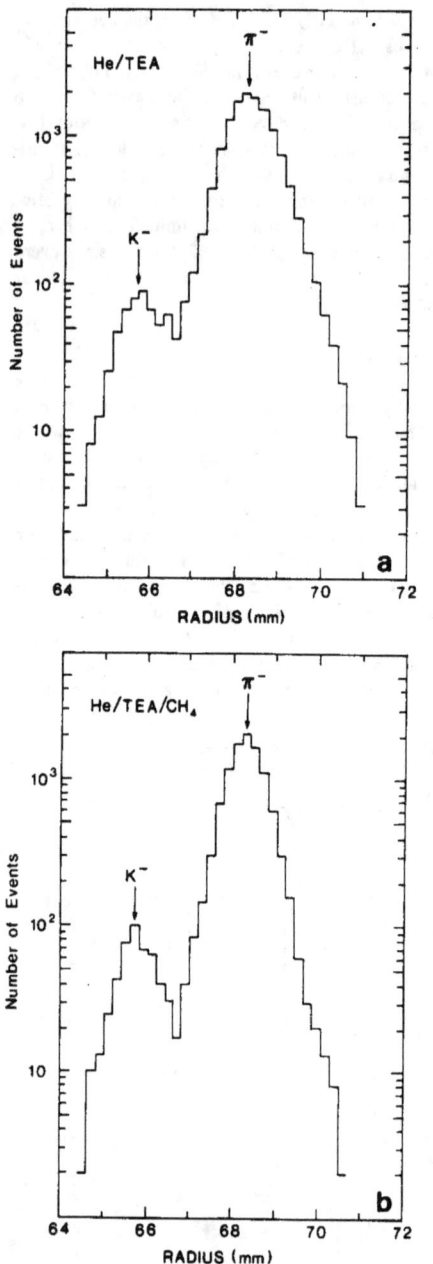

Fig. 14. Radius distributions with 200 μm bin size, (a) for the He/TEA gas mixture, and (b) for He/TEA/CH₄.

If we assign these events to the most probable hypothesis, then 11% of the K's and only 0.4% of the π's are misidentified (0.8% of all events). Zero-photon events

(including background events which appear in the one-photon bin) account for 6% of the π's, 7% of the K's, and 9% of the antiprotons. The particle identification efficiency is thus 93% for the π's, 68% for the K's, and 91% for the antiprotons.

For the He/TEA/CH₄ run, π–K ambiguities are only half as frequent. Less than 1% of the π's and 9% of the K's are ambiguous between the two hypothesis. Assigning these events to the most probable hypothesis, 4% of the K's and 0.2% of the π's are misidentified (0.4% of all events). Zero-photon events account for 7% of the π's, 8% of the K's, and 10% of the antiprotons. The identification efficiency is thus 92% for the π's, 83% for the K's, and 90% for the antiprotons.

5.8. Beam composition

The measured Cherenkov ring radii were 68.3, 65.7, and 57.5 mm for π^-, K^-, and antiprotons, respectively. The average index of refraction of the radiator gas was thus 1.0000372, reflecting the atmospheric conditions during these data runs. The 200 GeV/c beam composition was determined to be 95.2% π^-, 4.3% K^-, and 0.5% antiprotons.

6. Cherenkov detector system evaluation

6.1. Figure of merit

Cherenkov systems are often characterized by their figure-of-merit, N_0, defined such that the number of photons of Cherenkov angle Θ_C detected from a radiator of length L is

$$N = N_0 L \sin^2 \Theta_C .$$

The experimental figure-of-merit can be calculated directly from the ring-image radius and the average number of detected photons. For the He/TEA run, $N = 2.64$ detected photons per event and the ring radius is 68.3 mm, so $N_0 = 45$ per cm. For the He/TEA/CH₄ run, $N = 2.48$ and $r = 68.3$ mm, so $N_0 = 42$ per cm.

The figure-of-merit is supposed to represent the system characteristics independent of the radiator used in a particular experiment. Correcting the above values for the transmission of the radiator gas in each run, we find $N_0 = 45/0.87 = 52$ per cm for the He/TEA run, and $N_0 = 42/0.81 = 52$ per cm for the He/TEA/CH₄ run. We expect a larger N_0 value for the He/TEA data, since the methane in the He/TEA/CH₄ gas mixture removes some of the photons. The equality of the corrected values is undoubtedly due to the lower pulse heights in the He/TEA run (see fig. 7). In fact, the Y-cathode pulse-height distribution from the He/TEA/CH₄ run seems to indicate that the chamber was not completely efficient even for these data.

The experimental figure-of-merit can be compared to the value expected on the basis of the physical properties of the detector system, obtained by integrating the detector response over the energy spectrum of the Cherenkov emission. Since Cherenkov photons are emitted in a flat energy spectrum, the number of photons *produced* in an energy interval dE is given by

$$dN = KL \sin^2\Theta_c dE, \tag{4}$$

where $K = \alpha/hc = 370$ photons per cm per eV. The number of photons *detected* is

$$dN = KL \sin^2\Theta_c \epsilon(E) dE, \tag{5}$$

where $\epsilon(E)$ is the detection efficiency for photons of energy E. If we neglect a small energy dependence of the Cherenkov angle (see section 6.2), we can express the total number of detected photons as

$$N = KL \sin^2\Theta_c \int \epsilon(E) dE = N_0 L \sin^2\Theta_c. \tag{6}$$

The efficiency $\epsilon(E)$ is the product of the efficiencies (at photon energy E) of all the components of the system – reflectivity of the mirror, transmission of the radiator gas, the CaF_2 window, and the conversion gap of the multistep chamber, quantum efficiency of the TEA, single photoelectron efficiency of the chamber, and pattern-recognition efficiency of the analysis program:

$$\epsilon(E) = R_m T_r T_w T_c Q_{tea} \epsilon_{pe} \epsilon_{pr}. \tag{7}$$

The main energy dependence comes from Q_{tea}, the TEA quantum efficiency. The form of the TEA quantum response [16] is shown in fig. 1. The value at the maximum has been measured to be $Q_{max} = 0.65$ [22].

We have used the functional dependences in fig. 1 to perform the numerical integration indicated by eq. (6). The results are shown in table 1. The average mirror reflectivity (R_m) was 70%. The effective window transmission (T_w) was 58% for the He/TEA run and 59% for the He/TEA/CH₄ run, as discussed in section 4. The effective transmission of the conversion gap T_c) was 71% for He/TEA. The average methane transmission in the He/TEA/CH₄ gas mixture was 83% (see fig. 15). This reduced T_c to 59% for the He/TEA/CH₄ run. The remaining terms – transmission of the radiator gas, photoelectron detection efficiency, and pattern-recognition efficiency – are rather dependent on the particular experimental conditions. Ignoring them for the moment, we find *maximum* expected values of N_0 of 73 per cm for He/TEA and 61 per cm for He/TEA/CH₄.

In order to calculate N_0 for this particular experiment, we must include the measured radiator transmission ($T_r = 87\%$ for the He/TEA run and 81% for the He/TEA/CH₄ run), and the analysis efficiency ($\epsilon_{pr} = 92\%$). Then, if we assume the photoelectron detection efficiency (ϵ_{pe}) to be 100%, we find expected values of $N_0 = 58$ per cm for the He/TEA run, and $N_0 = 45$ per

Table 1
System efficiencies

	He/TEA	He/TEA/CH₄
Radiator gas transmission	0.87	0.81
Mirror reflectivity	0.70	0.70
CaF₂ crystal transmission	0.64	0.65
CaF₂ window frame	0.91	0.91
First wire mesh transparency	0.81	0.81
Conversions in first 50 μm of detector	0.95	0.95
Detector carrier gas transmission	1.00	0.83
Second wire mesh transfer efficiency	0.92	0.92
Multistep chamber photoelectron efficiency	0.78	0.93
Analysis program efficiency	0.92	0.92
TEA quantum efficiency $\int Q_{tea}(E)dE$	0.68 eV	0.68 eV
System efficiency $\int \epsilon(E)dE$	0.121 eV	0.114 eV

cm for the He/TEA/CH₄ run. These calculated figures-of-merit are still larger than the experimental values, especially for the He/TEA run, indicating that the multistep chamber was not completely efficient, as already suggested above.

We can use the expected and experimental values of N_0 to estimate the photoelectron efficiency in the two

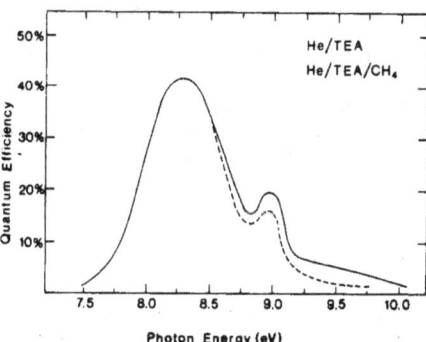

Fig. 15. Quantum efficiency curves versus photon energy with and without CH₄ in the multistep chamber. These curves include the TEA quantum efficiency [16,22], the transmission of the CaF₂ crystals, and the absorption in the methane [24,25]. Further corrections can be found in table 1.

runs. For the He/TEA run, we find $\epsilon_{pe} = 45/58 = 78\%$. For the He/TEA/CH$_4$ run, we find $\epsilon_{pe} = 42/45 = 93\%$. These values seem compatible with the pulse-height distributions in fig. 7.

6.2. Photon detector spatial resolution

The system resolution, determined from the single-photon radius distributions (see section 5.5), depends in part on the choice of radiator for this particular experiment. From eqs. (1) and (2), the ring radius is given by

$$r = f\left(\beta^2 n^2 - 1\right)^{1/2}. \tag{8}$$

But the index of refraction, and thus the ring radius, depends on the photon energy, as shown in fig. 1 for our helium radiator. The range of photon energies over which our detector is sensitive is given by the efficiency $\epsilon(E)$, which depends principally on the TEA quantum efficiency, as discussed above.

The expected dispersion is given by

$$\frac{dN}{dr} = \frac{dN}{dE} \Big/ \frac{dr}{dE} = \frac{dN}{dE} \Big/ \left(\frac{dr}{dn}\frac{dn}{dE}\right). \tag{9}$$

Differentiating eq. (8) and using eq. (5), we obtain

$$\frac{dN}{dr} = \frac{K\epsilon L \sin^3\Theta_c}{\beta(dn/dE)}, \tag{10}$$

where dn/dE, the chromatic dispersion of the helium index of refraction at photon energy E, can be read from the curve in fig. 1. In eq. (10), ϵ, Θ_C, and dn/dE are all functions of the ring radius, $r = r[n(E)]$, through their dependence on the photon energy E.

The dispersion thus calculated is shown on the single-photon radius distributions in fig. 11. The methane in the He/TEA/CH$_4$ gas mixture cuts off high-energy photons, as shown in fig. 15, and eliminates any potentially long tails in the radius distribution. The rms chromatic dispersion (calculated in the region where the distribution exceeds 5% of the peak values) is found to be 0.34 mm for He/TEA and 0.29 mm for He/TEA/CH$_4$.

The detector resolution is found from the deconvolution of the system resolution and the chromatic dispersion. We find $\sigma = 0.54$ mm for the detector operating with He/TEA, and $\sigma = 0.37$ mm with He/TEA/CH$_4$.

7. Conclusions

We have identified particles in a 200 GeV/c beam containing 95.2% π^-, 4.3% K$^-$, and 0.5% antiprotons with an efficiency of 92% for the π's, 83% for the K's (68% with He/TEA in the photon chamber), and 90% for the antiprotons. The multistep chamber detected photons with a spatial resolution of 0.54 mm with He/TEA and 0.37 mm with He/TEA/CH$_4$. The system resolution (including the effect of chromatic dispersion in the radiator gas) was 0.64 mm for the He/TEA run and 0.47 mm for the He/TEA/CH$_4$ run. The experimental figure-of-merit of the ring-imaging system was $N_0 \approx 45$ per cm. A ring-imaging systems based on this prototype is now in operation in Fermilab experiment E605 [28].

We would like to thank T. Ypsilantis, H. Jöstlein, A. Cattai, A. Breskin, and R. Praca for their help in various stages of this experiment. We would also like to acknowledge the support of the National Science Foundation, the Commissariat à l'Energie Atomique, and CERN.

References

[1] R. Bouclier, G. Charpak, A. Cattai, G. Million, A. Peisert, J.C. Santiard, F. Sauli, G. Coutrakon, J.R. Hubbard, Ph. Mangeot, J. Mullié, J. Tichit, H. Glass, J. Kirz and R.L. McCarthy, Nul. Instr. and Meth. 205 (1983) 403.

[2] A. Roberts, Nucl. Instr. and Meth. 9 (1960) 55.

[3] M.M. Butslov, M.N. Medvedev, I.V. Chuvilo and M.V. Sheshunov, Nucl. Instr. and Meth. 20 (1963) 263.

[4] G.T. Reynolds, J.R. Waters and S.K. Poultney, Nucl. Instr. and Meth. 20 (1963) 267.

[5] D.M. Binnie, M.R. Jane, J.A. Newth, D.C. Potter and J. Walters, Nucl. Instr. and Meth. 21 (1963) 81.

[6] R. Iredale, G.W. Hinder, A.G. Parham and D.J. Ryden, IEEE Trans. Nucl. Sci. NS-13 (1966) 399.

[7] R. Geise, O. Gildemeister, W. Paul and B. Schuster, Nucl. Instr. and Meth. 88 (1973) 83.

[8] B. Robinson, Phys. Scripta 23 (1981) 716.

[9] M. Benot, J.M. Howie, J. Litt and R. Meunier, Nucl. Instr. and Meth. 111 (1973) 397.

[10] M. Benot, J.C. Bertrand, A. Maurer and R. Meunier, Nucl. Instr. and Meth. 165 (1979) 439.

[11] J. Séguino and T. Ypsilantis, Nucl. Instr. and Meth. 142 (1977) 377.

[12] R.S. Gilmore, J. Malos, D.T. Bardsley, F.A. Lovett, J.P. Melot, R.J. Tapper, D.I. Giddings, L. Lintern, J.A.G. Morris, P.H. Sharp and P.D. Wroath, Nucl. Instr. and Meth. 157 (1978) 507.

[13] S. Durkin, A. Honma and D.W.G.S. Leith, Proc. 1978 Isabelle Summer Workshop BNL 50885 (1979) 120.

[14] J. Chapman, D. Meyer and R. Thun, Nucl. Instr. and Meth. 158 (1979) 387.

[15] G. Charpak, S. Majewski, G. Melchart, F. Sauli and T. Ypsilantis, Nucl. Instr. and Meth. 164 (1979) 405.

[16] J. Séguinot, J. Tocqueville and T. Ypsilantis, Nucl. Instr. and Meth. 173 (1980) 283; D. Solomon and A.A. Scala, J. Chem. Phys. 62 (1975) 1469.

[17] G. Comby, Ph. Mangeot, J. Tichit, H. de Lignières, J.F. Chalot and P. Monfray, Nucl. Instr. and Meth. 174 (1980) 77.

[18] G. Comby, Ph. Mangeot, J.L. Auguères, S. Claudet, J.F. Chalot, J. Tichit, H. de Lignières and A. Zadra, Nucl. Instr. and Meth. 174 (1980) 93.

[19] F. Sauli, Phys. Scripta 23 (1981) 526.

[20] T. Ekelof, J. Séguinot, J. Toqueville and T. Ypsilantis, Phys. Scripta 23 (1981) 718.

[21] G. Charpak, A. Peisert, F. Sauli, A. Cavestro, M. Vascon and G. Zanella, Nucl. Instr. and Meth. 180 (1981) 387.

[22] E. Barrelet, T. Ekelof, B. Lund-Jensen, J. Séguinot, J. Tocqueville, M. Urban and T. Ypsilantis, Nucl. Instr. and Meth. 200 (1982) 219.

[23] K. Fransson, private communication.

[24] L.G. Christophorou, Atomic and Molecular Radiation Physics (Wiley, New York, 1971) p. 165.

[25] P.G. Wilkinson and H.L. Johnston, J. Chem. Phys. 18 (1950) 190.

[26] M.C.E. Huber and G. Tondello, J. Opt. Soc. Am. 64 (1974) 390.

[27] A. Gedanken et al., J. Chem. Phys. 57 (1972) 3456.

[28] H. Glass et al., submitted to IEEE Trans. Nucl. Sci.

Nuclear Instruments and Methods in Physics Research 225 (1984) 627–635
North-Holland, Amsterdam

Section VI. Particle identification

USE OF TMAE IN A MULTISTEP PROPORTIONAL CHAMBER FOR CHERENKOV RING IMAGING AND OTHER APPLICATIONS

G. CHARPAK and F. SAULI

CERN, Geneva, Switzerland

We tested a multistep proportional chamber of 20×20 cm^2 active area, conceived to localize single vacuum ultraviolet photons. The detector has a capability of imaging large multiplicities of Cherenkov rings with a high density of detected photoelectrons and uses tetrakis(dimethylamine)ethylene, with a threshold of 5.4 eV, as the photoionizing vapour. Time slicing in a conversion gap reduces the multiplicity at a given instant of time by a large factor and permits the use of simple methods of localization. Separation of the amplifying gaps solves the problems of photon feedback encountered, at high gains, with single-step detectors. We also discuss an application of this chamber for the detection of superheavy monopoles if they have a speed in the range of the electron drift velocity.

1. Introduction

Efficiency and localization properties of gas proportional chambers in the detection of ultraviolet (UV) photons have recently been extensively studied in view of their use for Cherenkov ring imaging; a technique that permits identification of fast charged particles in a wide range of velocities [1,2]. A new device, the multi-step proportional chamber, operated with a photosensitive gas filling, allows the attainment of the large amplification factors (10^6 or more) necessary in order to conveniently detect and localize single electrons released in the gas by the photoelectric effect [3]. Efficiency and localization properties of the device, when operated with triethylamine (TEA, $E_i = 7.5$ eV) as the photosensitive vapour, have been extensively studied [4,5].

Because of its good time and space resolutions (50 ns and 350 μm, respectively), the detector is particularly suitable for use in high-flux, very high momentum set-ups; two large-area multistep proportional chambers are indeed mounted on a 16 m long helium radiator in the first operational ring-imaging particle identifier, installed in experiment E605 at Fermilab. Results of the test runs have been reported and are up to expectations [6,7].

Use of TEA is very convenient, because of its high vapour pressure (55 Torr at 20°C), which allows both high proportional gains and good time resolutions to be obtained. It implies, however, the use of calcium or magnesium fluoride windows and particular care concerning the purity of the radiating medium, in order to avoid absorption losses. The choice of radiators is indeed restricted to pure noble gases and fluoride crystals. The use in the detector of an induced-charge cathode read-out method for localization limits, moreover, the maximum number of simultaneous photons in a ring that can be unambiguously detected to around 5.

A vapour having much lower ionization potential, tetrakis(dimethylamine)-ethylene (TMAE, $E_i = 5.4$ eV) was introduced several years ago [8] and has been adopted since in the time-projection chamber (TPC) design for Cherenkov ring imaging [9,10]. Opening up the possibility of using a large variety of radiators and a fused silica window, photon detectors using TMAE are very attractive for large-area, intermediate momentum particle identifiers. The TPC approach is favoured by the very low vapour pressure of TMAE (0.5×10^{-5} Torr at 20°C), which implies a rather thick detection layer of about 4 to 5 cm. This leads, however, to serious background problems when charged particles are also allowed to traverse the detector. The long drift-times due to the lateral motion towards the detecting wires can also be a nuisance for high-rate experiments.

Moreover, severe problems of photon feedback have been met, because of the long path of photons emitted in the avalanches; limiting the maximum practical gain to 10^5 or less (with obvious electronics implications) and requiring a particular design of the anode structure to limit spurious points [10].

We present, in what follows, preliminary results obtained using a multistep proportional chamber, redesigned in order to allow operation with TMAE; they suggest that photon feedback can be completely suppressed by the double-gain structure and that by recording both timing and amplitude information, the intrinsic limitation of the bi-dimensional cathode read-out method in handling many simultaneous points as in a ring image can by and large be overcome.

VI. PARTICLE IDENTIFICATION

2. The multistep proportional chamber using TMAE

The original multistep proportional chamber design, conceived for the use of a high concentration of photosensitive vapour, had to be modified for operation with TMAE because of the much larger absorption length for photons, around 30 mm at room temperature. We have correspondingly increased the thickness of the conversion region to 50 mm and, to reduce photon feedback

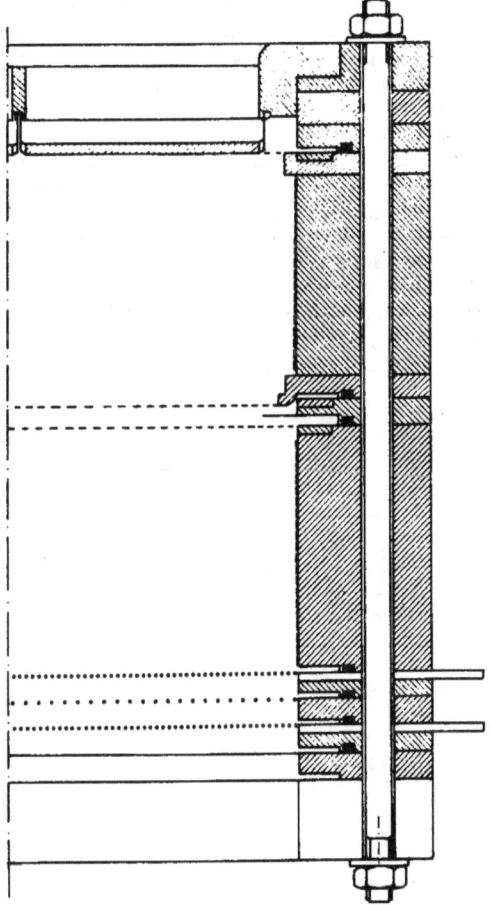

Fig. 1. The multistep proportional chamber adapted for the use of TMAE. Both the conversion and the transfer spaces have been increased in thickness to 50 mm, to cope with the photon absorption length. Charges produced in the upper gas layer (e.g. by photoelectric effect from a UV photon source mounted above the CaF$_2$ window) drift towards the central thin-gap parallel-plate preamplifying structure. A fraction of the electrons in the avalanche enter the transfer region and reach the lower MWPC, where further amplification occurs. Localization is realized, recording both pulse-height distribution and time on anode and cathode wires in the MWPC.

(see below), also the thickness of the transfer region to the same value (see fig. 1). The thin-gap preamplifying structure and the multiwire proportional chamber (MWPC) were left unmodified; a field cage with graded potentials all around the edges was added to guarantee a uniform field in the thick regions. A drift field is created in the conversion region by applying a suitable potential between the upper mesh (in contact with the window) and the first mesh of the preamplifying gap; similarly a transfer field is created between the preamplifying gap and the MWPC end structure.

The operation of the multistep proportional chamber is briefly described as follows. A charge released in the upper gas layer is caused to drift into a region of very high field, the preamplification space, where avalanche multiplication occurs. This region is realized using two well-stretched meshes, 4 mm apart, with a particular profile on the frame side to avoid edge breakdown (see ref. [4]). A fraction of the total electron charge produced in the avalanche, typically 10–20%, is injected into the transfer region and proceeds towards the second element of amplification; a conventional thin-gap MWPC with bi-dimensional cathode read-out.

Separating the overall multiplication process into two steps, as described in the early works on the subject, allows combined gains to be reached far in excess of the ones that could be reached in each element separately. This is partly due to the fact that photons emitted in the last avalanche have a very small probability of getting all the way through into the conversion region, with the consequence of suppressing secondary effects that lead to a discharge. While using TEA (or indeed any photosensitive vapour in high concentration), the most relevant outcome of the mechanism is that it allows large gains; in TMAE one expects the suppression of the so-called "satellites" or spurious electron signals observed in Cherenkov ring imagers around the main event [9,10].

As gas filling for the detector, we have tried various combinations of argon and helium with methane, ethane and isobutane, always saturated with TMAE at 20°C. The highest gains and best localization properties have been obtained with He–C$_4$H$_{10}$–TMAE; a convenient gas mixture because of its low drift velocity (saturated around 2 cm/μs, see below) and low specific ionization for charged particles.

3. Experimental set-up

The multistep proportional chamber used for the present work had the structure shown in fig. 1 and an active area of 200 × 200 mm^2. In the MWPC section, the two cathode planes perpendicular to each other, are made of copper–beryllium wires, of 100 μm diameter, 1 mm apart; the anode wire plane is realized with 20 μm gold-plated tungsten wires, 2 mm apart, mounted at 45° with respect to the cathodes. All anode wires and pairs of cathode wires were equipped with individual charge amplifiers for timing and pulse-height recording.

All other electrodes in the structure are realized with a good-quality crossed-wire stainless-steel mesh having 0.5 mm pitch. The preamplifying grids, stretched originally over a 5 mm thick frame, are pushed inwards by a suitably shaped frame to reduce the gap to 4 mm, except along the edges, so as to avoid edge sparking [4].

The gas volume is delimited on one side by the UV transparent window (a mosaic of four CaF_2 crystals mounted as described in ref. [4]), and on the other side by two foils of 50 μm thick Mylar–Aclar, 5 mm apart; the filling gas is made to flow between the two foils before evacuation, thus preventing the pollution of the inside volume from the outside atmosphere by diffusion through the thin window.

Gain and localization properties of the detector were studied by irradiating the chamber from the thin window side with X-ray sources, or through the CaF_2 window with a photon source. This source, described in more detail in ref. [4], consists of a layer of gas between two grids traversed by α particles emitted by an internal radioactive source. Photons are emitted, with a spec-

trum characteristic of the gas filling, either by direct primary scintillation or by secondary scintillation of the drifting ionization electrons if the electric field is increased. A charge amplifier on one mesh is used to detect the ionization, thus allowing precise timing of the decay for triggering purposes. Using Xe as the gas filling, scintillation occurs around 7.5 eV, appropriate for detection with TMAE.

For part of the measurements we have also used a pulsed N_2 gas laser beam, entering the detector through the window and inclined at about 45°. Ionization is produced in this case not directly (the laser emits at 337 nm, or 3.7 eV), but by multiple excitations; by varying the beam intensity, we could cover a very wide range of specific ionization, from 10^6 down to a few electrons per centimetre.

Each anode wire and pairs of adjacent cathode wires were equipped with charge amplifiers having sensitivities of 200 and 600 mV/pC, respectively. Optimized to perform localization by the centre-of-gravity method, these amplifiers were relatively slow and not optimal for getting timing information. Pulse height and time could be recorded on each channel using a linear splitter on the amplifier output, with a set of CAMAC-based multiple analog-to-digital (ADC) and (after discrimination)

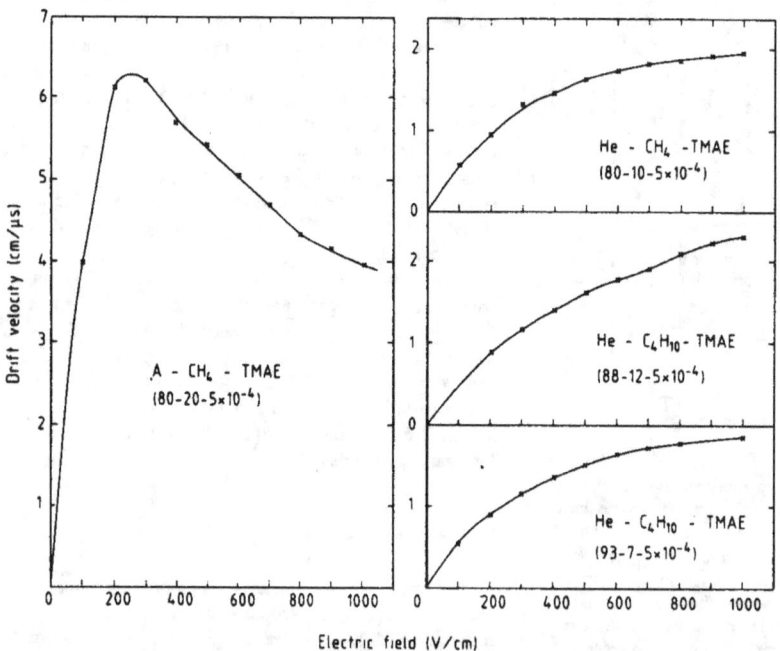

Fig. 2. Electron drift velocity plotted against field for several gas mixtures containing TMAE at saturated vapour pressure (20 °C). Mixtures with helium are attractive, because of the low saturated velocity, improving the vertical separation for a given accuracy in the drift time measurement and because of the low ionization density produced by ionizing tracks.

VI. PARTICLE IDENTIFICATION

time-to-digital (TDC) modules. The gating and timing signals were provided by the discriminated charge signal on the UV xenon source, or, when using the N_2 laser, by a photodiode pick-up on the main beam.

As gas filling, we tested several combinations of argon or helium with methane or isobutane; the mixture was bubbled through a small volume of TMAE at room temperature. To avoid the problems connected with the chemical activity of TMAE (it combines with oxygen producing oily pollutants), the complete gas system, down to the chamber input, could be evacuated and purged with a pure noble gas. To estimate the TMAE concentration in the mixture, we assumed a saturated vapour pressure at room temperature. During about a month of continuous operation, we had no problems with the TMAE manipulation system, except for a slight opacity on the outer Mylar layer probably due to the described chemical reaction with oxygen molecules diffusing through the thin outer window.

Fig. 2 shows measured values of drift velocity as a function of field in several gas mixture containing TMAE; they were realized with the method outlined in section 5. It appears that the addition of TMAE does not sensitively modify the drift velocity curves, at least for the mixtures we have investigated.

4. Experimental results obtained with X-ray and UV sources

Gain curves were measured, irradiating the drift volume in the chamber with a 5.9 keV X-ray source and recording the pulse-height spectra on the anode plane. For most of the measurements, we kept the drift and transfer fields at constant values; about 800 V/cm. In the transfer region, this is somewhat at the lower and of the useful range and implies a small transfer efficiency (around 15%); use of larger fields, however, would imply reaching excessive potentials on the upper grid (because of the peculiar composite window mounting, the maximum safe potential that an be applied to this electrode is around 10 kV).

A typical gain curve obtained for a He–CH_4–TMAE mixture is shown in fig. 3. The MWPC potential is set so as to obtain a gain of about 3.5×10^3 in this element and increasing the difference of potential between the 4 mm apart preamplifying grids, the overall gain is increased to its maximum safe values of about 7×10^5. In fig. 4, obtained in a He–C_4H_{10}–TMAE mixture, two curves are shown for different initial settings of the MWPC gain; the maximum overall gain that can be reached is about the same. Notice the deviation from a pure exponential behaviour at the highest gains, probably due to space–charge effects.

Although definitely smaller than the gains observed when using TEA in a similar structure, the quoted

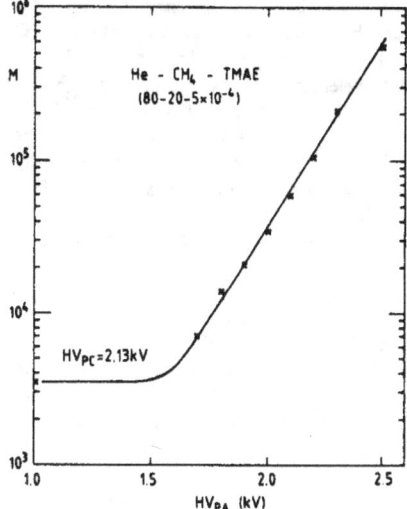

Fig. 3. Amplification curve in the multistep proportional chamber as a function of the potential difference between the preamplification grids (4 mm apart) for a fixed gain in the MWPC (3×10^3). Gas filling He–CH_4–TMAE, in the percentages (80–10–5×10^{-4}).

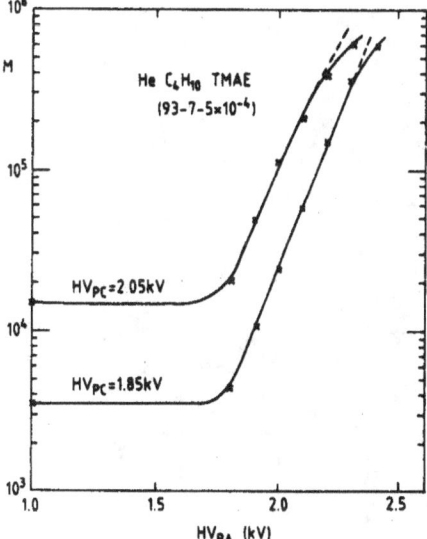

Fig. 4. Amplification curves in the multistep proportional chamber, as a function of the preamplification voltage, for two values of gain in the MWPC (3.5×10^3 and 1.5×10^4). About the same maximum gain is reached before discharge (~ 7×10^5). Gas filling He–C_4H_{10}–TMAE, in the percentages (88–12–5×10^{-4}).

multiplication factor above 5×10^5 exceeds by an order of magnitude what could be obtained using the MWPC alone in the same gas mixture The repartition of gains between the two elements implies that a photon emitted in the last avalanche and reconverted within the MWPC or transfer regions should not be detected. Photons emitted in the first avalanche are, of course, greatly reduced in number due to the moderate gain, about 10^3, of the preamplification region (as is clear from figs. 2 and 3 and taking into account the quoted transfer efficiency).

We have used the timing UV source, described in section 3, to study the detection efficiency for single photons of the chamber. Mounted on the CaF_2 window side and suitably collimated, the source was operated in the primary scintillation mode (i.e. with a very small collection potential). Evidence of single-photoelectron operation is obtained both from the value of the re-

corded efficiency (around 20%) and by simple oscilloscope observation of detected pulses; because of the long absorption length (30 mm), pulses due to several individual photoelectrons are generally well separated in time, see figs. 5a and b. The procedure was then as follows: starting with a source intensity such as to produce on one wire the signal structure shown in fig. 5, the source collimation was further increased so as to have single pulses within the drift time in about 20% of the events. This guarantees that multiple photoelectron pulses within the time window are very unlikely.

In such conditions, the drift time spectrum of detected pulses was recorded, the time zero being provided by the charge signal on the UV source. Although detection of this slow signal is not free from time jitter,

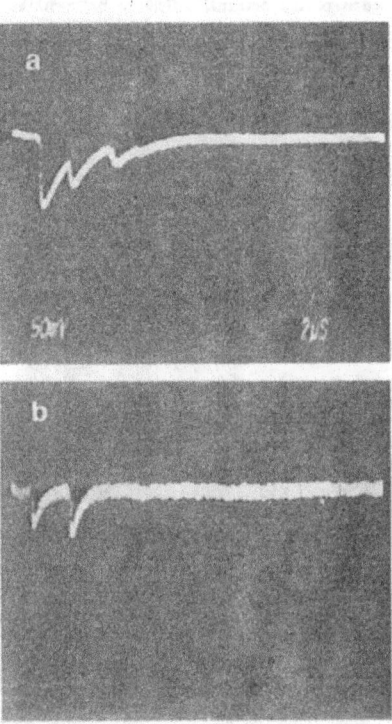

Fig. 5. Multiphoton events detected on a wire of the MWPC; photoelectrons are produced in the conversion volume by a collimated UV source, described in the text. Because of the long (30 mm) absorption length, photoelectrons are well separated in time at low source intensities; a single photoelectron condition can easily be reached by collimation.

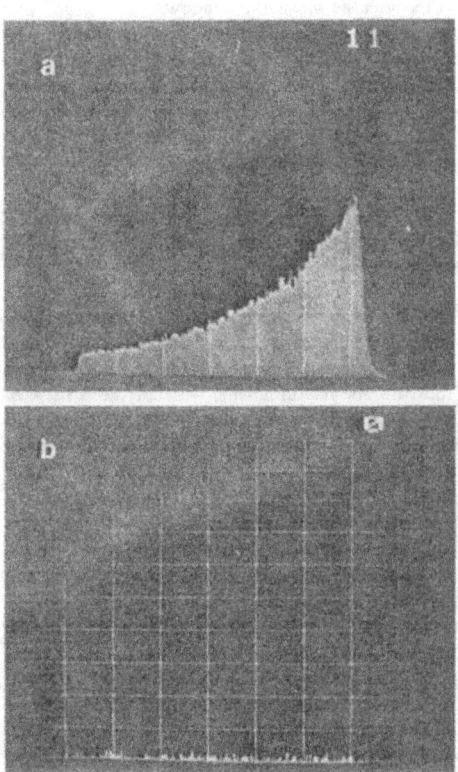

Fig. 6. (a) Time spectrum of single detected photoelectrons recorded on an anode wire (time zero provided by the source). Longer times correspond to the photons converted close to the entrance window; the cut-off at the shortest time corresponds to the preamplification region; (b) time spectrum recorded during the same time on a wire 16 mm from the one facing the source. The absence of time-correlated feedback photoelectrons is clear.

VI. PARTICLE IDENTIFICATION

operating the source without the secondary scintillation enhancement guarantees at least that all photons are emitted by primary scintillation within a very short time. Fig. 6a shows an example of a recorded time spectrum; the long lines correspond, of course, to conversions close to the CaF$_2$ window and the shape of the spectrum provides the absorption mean free path of ~ 30 mm compatible with the known TMAE cross section for room temperature operation. Notice the sharp transitions on both sides of the spectrum, the first evidence that timing is rather well preserved despite the multiple avalanching process.

The time spectrum in fig. 6a has been recorded on a single anode wire, facing the collimated source; to obtain information on satellites or background counts, we have recorded for the same amount of time the spectrum on a wire 16 mm apart (fig. 6b). The residual counts observed do not have the time structure one would expect in the case of photon-induced background and are indeed compatible with a purely accidental distribution (events were accumulated over several minutes, with a single-electron sensitivity). In other words, we have seen no evidence of secondary, photon-induced background signals, as anticipated because of the double gain structure of the detector.

Pulse-height spectra for single photoelectrons are not as good in TMAE as they were for the TEA operation, where much higher gains could be obtained. An exam-

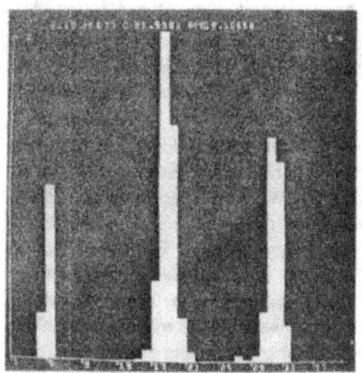

Fig. 8. Charge profile recorded for one photoelectron event on the MWPC; each bin represents an anode wire or a cathode strip, all 2 mm apart. The clusters correspond to the anode and the two cathode planes, from left to right. In most of the events two adjacent anode wires detect the avalanche, because of the dispersion produced in the preamplification, thus allowing interpolation. On cathodes, the fwhm of the induced signal is about 4 mm.

ple is shown in fig. 7; a peaked structure begins to appear (counts in the lower channels are due to amplifier noise), but the distribution remains essentially exponential.

The picture in fig. 8 shows a single-electron event as recorded on the pulse-height measuring system; each bin in the picture represents a 2 mm wide electrode and the three distributions correspond respectively to the anode and the two cathode planes. Because of the long absorption length for photons in the drift volume and because of the angular divergency of the source, we could not measure the localization accuracy in the MWPC plane; everything indicates, however, that it should not be very far from what was measured in the TEA operation (350–400 μm standard deviation, see ref. [4]). Charge is shared between two adjacent anode wires (2 mm apart) in most of the events, giving an idea of the spatial extension of the preamplified charge and allowing interpolation.

5. Experimental results obtained with the N$_2$ laser beam

In order to study the timing properties of the detector, we have used as the source of ionization a pulsed N$_2$ laser having very high energy density and small-beam cross section and divergence *. This kind of laser is currently used in the high energy physics community for

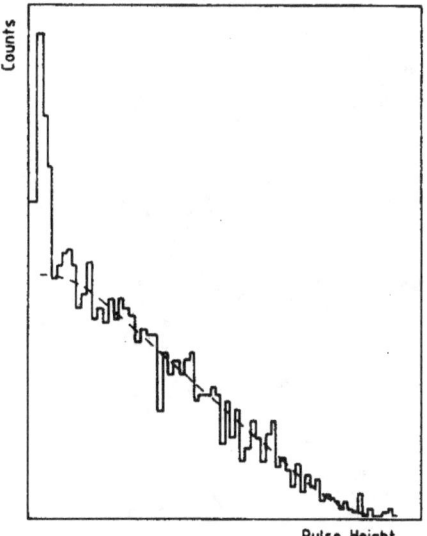

Fig. 7. Single photoelectron pulse-height spectrum at high gain (5 × 10^5). The shape is still close to an exponential, as against the peaked structure observed using TEA at higher gains (the peak in the lower channels is due to the noise of the amplifiers.

---

* MOPA-400, produced by Multilasers SA (Geneva).

calibration purposes, since it allows the creation, by multiple excitation processes, of ion pair trails with densities comparable with those produced by fast charged particles in gas along a line with sub-millimetre cross section.

In our detector, laser-induced ionization densities were dramatically enhanced by a factor of 10^4 or so, when adding TMAE. We could indeed detect charge signals and measure drift velocities operating the chamber at unity gain. The estimated charge density as produced by the 200 μJ, 0.5 mm^2 section, pulsed beam was about 10^6 pairs per centimetre. Moreover, strong light emission at green wavelengths was observed along the beam in the chamber (N_2 lasers emit at 337 nm). This can be due either to some process of multiple excitation and subsequent decay, or to direct fluorescence (TMAE is known to have a strong absorption in the blue and re-emission around 500 nm [11]).

The beam could be attenuated using neutral density filters down to the point of obtaining single electrons at an average distance exceeding the wire spacing (a condition verified with the procedure described in the previous section).

We have recorded the drift time of the detected charge on a number of adjacent anode wires and fitted a straight line through the points to obtain information on timing accuracy. An example is shown in fig. 9, as recorded for a large ionization density on a group of 8 adjacent anode wires; the abscissa is the wire position (in 2 mm steps) and the ordinate the computed vertical coordinate. The localization accuracy at these large charge densities (about 10 × minimum ionizing) is about 50 μm for an average drift over 7 cm. The slope of the correlation (given the angle of crossing of the beam), of course, provides the drift velocity; the curves in fig. 2 were indeed measured in this way.

Reducing the beam intensity so as to be close to the single electron per wire condition, the method provides the time dispersion corresponding to the detection of one electron (figs. 10a and b). Several physical sources contribute to the dispersion, such as the diffusion of electrons before and after preamplification, the statistical fluctuations in the size and shape of the preamplified avalanche, the electronics slewing due to the large dynamic range of the pulses. Howver, in our set-up, the largest contribution is purely geometrical, owing to the angle of crossing (45°) and the 2 mm anode wire spacing; charges are created along 2 mm vertically. Not surprisingly, the resulting time localization accuracy for one electron is rather modest in our set-up, having a fwhm of 100 ns or 4 mm (fig. 11). This is to be considered an upper limit; the real intrinsic dispersion due to all physical effects being probably about 2 mm fwhm. The geometrical dispersion can be avoided, of

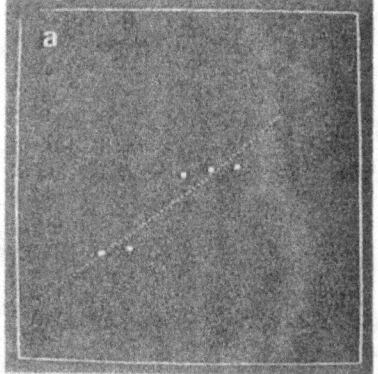

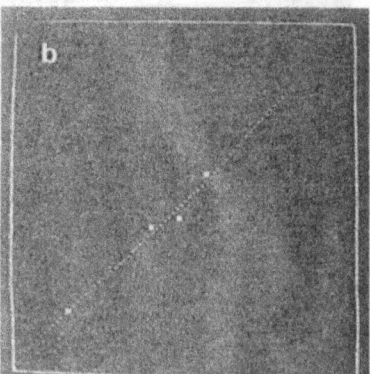

Fig. 10. (a) and (b). Drift time recorded for two events on a set of eight adjacent anode wires, operating the laser beam at very low intensity so as to reach a single photoelectron per wire condition. This dispersion of the recorded time around the straight-line fit is due to both physical and geometrical factors.

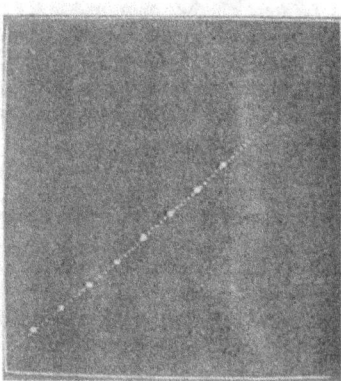

Fig. 9. Recorded drift time on eight adjacent anode wires, 2 mm apart, for an ionization track produced in the conversion region by a N_2 laser at 45°.

VI. PARTICLE IDENTIFICATION

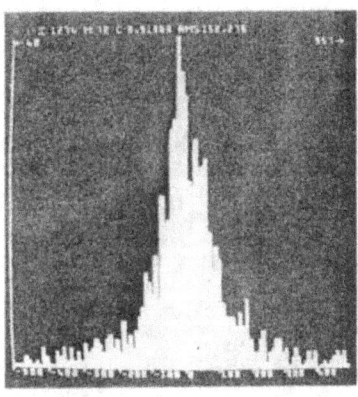

Fig. 11. Distribution of the measured drift time for single electrons, around the average track position (see fig. 10). The distribution has 100 ns fwhm, or 4 mm; most of the width is due to geometry (laser beam at 45° with respect to the anode wire plane). Gas filling: $He-C_4H_{10}-TMAE$.

course, by setting the laser beam along a line parallel to the anode plane; a condition that could not be realized in our set-up.

6. Applications to Cherenkov ring imaging

We have shown that the multistep proportional chamber, in the modified version described, operates satisfactorily in various gas mixtures containing TMAE and allows the detection and localization of single photoelectrons produced by quanta of energy above the photoionization threshold of 5.4 eV. Using isobutane in the gas filling, photons are absorbed above ~ 7.3 eV, which roughly corresponds to a UV grade quartz window cut-off. The spectral range of sensitivity can be extended, of course, by using methane or ethane with a fluoride window.

The possibility of recording both the pulse height and the drift time of detected pulses offers two advantages. Firstly, roughly knowing the conversion depth, one can correct the parallax error intrinsic in the detection of photons non-perpendicular to the window surface; our results indicate that a depth accuracy of the order of 2 mm can be achieved with the corresponding angular accuracy. Moreover, one can largely overcome the major drawback of the centre-of-gravity localization method, i.e. its inability to handle events with too many simultaneous photons. For a typical cathode-induced profile of 4 mm fwhm and with Cherenkov rings of 60 mm radius, the reconstruction algorithm quickly becomes inefficient for events consisting of more than 5–6 points, because of the overlaps and ambiguities. The time spread of pulses in the TMAE detector essentially

allows the recording of pulse height in many independent time slices. With charge amplifiers having 100 ns shaping constants and using the low-drift-velocity $He-C_4H_{10}-TMAE$ mixture (2 cm/μs) one can subdivide the conversion volume into about 20 time slices; thus correspondingly increasing the multiphoton handling capability of the device. This point requires, of course, further studies with a complete detector.

As far as the electronics is concerned, a system of charged coupled devices or of flash ADCs, such as the ones developed for time projection or imaging chambers, with 40–50 ns sampling rates would be adequate.

7. The multistep proportional chamber as a detector of monopoles

The detector we have described has excellent single-electron imaging capability and low intrinsic noise, as shown, for example, in figs. 8 and 9. In a search for exotic particles having an ionization power far below that of minimum ionizing particles, the addition of a thicker drift volume would increase the likelihood of recognizing an anomalous energy loss. For slow, superheavy monopoles, an additional factor makes a drift detector quite attractive; namely the fact that their expected velocity at the earth's surface is about the same as the drift velocity of electrons in gases. A monopole traversing the detector at that speed would generate a very peculiar charge signal, very short or corresponding to twice the maximum drift time, depending on the direction of motion. A pair of multistep proportional chambers mounted back to back, or a single central MWPC with preamplification and drift on both sides, would provide a unique electronics signature for a slow monopole candidate, easily recognizable at the trigger level.

There remains the question of estimating the ionization power of slow monopoles. Recent calculations by Drell et al. [12] have shed light on the matter. According to those authors, monopoles having masses of the order of 10^{16} GeV, as predicted in some supersymmetry models, are expected to increase substantially their ionization above a threshold velocity of about 10^{-4}; this threshold can be even lower, 3×10^{-5} or so, in heavy noble gases. Since this value corresponds to the velocity of a free falling monopole on the earth's surface, the described detector seems ideal.

Using a drift volume 50 cm thick, even one electron released (and detected) every 25 mm would provide a recognizable track, at 1% minimum ionizing. In addition, one can adjust the electrons' velocity to be about 10^6 or 3×10^5 cm/s. The observation of an aligned trail of electrons at very low density, having an arrival time distribution corresponding to a very anomalous drift velocity, would provide a unique signature for a slow moving particle. There exists nowadays sufficient experience in the field of gaseous detectors to make a project with a sensitive area of ~ 1000 m^2, which is not too unrealistic – at least from the technical point of view.

References

[1] J. Séguinot an T. Ypsilantis, Nucl. Instr. and Meth. 142 (1977) 377.

[2] T. Ypsilantis, Phys. Scripta 23 (1981) 370.

[3] G. Charpak and F. Sauli, Phys. Lett. 78B (1978) 523.

[4] R. Bouclier, G. Charpak, A. Cattai, G. Million, A. Peisert, J.C. Santiard, F. Sauli, G. Coutrakon, J.R. Hubbard, Ph. Mangeot, J. Mullie, J. Tichit, H. Glass, J. Kirz and R. McCarthy, Nucl. Instr. and Meth. 205 (1983) 403.

[5] G. Coutrakon, M. Cribier, J.R. Hubbard, Ph. Mangeot, J. Mullie, J. Tichit, R. Bouclier, A. Breskin, G. Charpak, J. Million, A. Peisert, J.C. Santiard, F. Sauli, C.N. Brown, D. Finley, H. Glass, J. Kirz and R.L. McCarthy, IEEE Trans. Nucl. Sci. NS-29 (1982) 323.

[6] H. Glass, M. Adams, A. Bastin, G. Coutrakon, D. Jaffe, J. Kirz, R. McCarthy, J.R. Hubbard, Ph. Mangeot, J. Mullie, A. Peisert, J. Tichit, R. Bouclier, G. Charpak, J.C. Santiard, F. Sauli, J. Crittenden, Y. Hsiung, D. Kaplan, C. Brown, S. Childress, D. Finley, A. Ito, A. Jonckheere, H. Jostlein, L. Lederman, R. Orava, S. Smith, K. Sugano, K. Ueno, A. Maki, Y. Hemmi, K. Miyake, T. Nakamura, N. Sasao, Y. Sakai, R. Gray, R. Plaag, J. Rothberg, J. Rutherfoord and K. Young, IEEE Trans. Nucl. Sci. NI-30 (1983) 30.

[7] M. Adams, A. Bastin, G. Coutrakon, H. Glass, D. Jaffe, J. Kirz, R. McCarthy, J.R. Hubbard, Ph. Mangeot, J. Mullie, A. Peisert, J. Tichit, R. Bouclier, G. Charpak, J.C. Santiard, F. Sauli, J. Crittenden, Y. Hsiung, D. Kaplan, C. Brown, S. Childress, D. Finley, A. Ito, A. Jonckheere, H. Jostlein, L. Lederman, R. Orava, S. Smith, K. Sugano, K. Ueno, A. Maki, Y. Hemmi, K. Miyake, T. Nakamura, N. Sasao, Y. Sakai, R. Gray, R. Plaag, J. Rothberg, J. Rutherfoord and K. Young, Proc. Wire Chamber Conference, Vienna (1983) Nucl. Instr. and Meth. 217 (1983) 237.

[8] D. Anderson, IEEE Trans. Nucl. Sci. NS-28 (1981) 842.

[9] E. Barrelet, T. Ekelöf, B. Lund-Jensen, J. Séguinot, J. Tocqueville, M. Urban and T. Ypsilantis, Nucl. Instr. and Meth. 200 (1982) 219.

[10] T. Ypsilantis, Advances in Cherenkov Counters, this Conference.

[11] Y. nakato, M. Ozaki and H. Tsubomura, J. Phys. Chem. 76 (1972) 2105.

[12] S. Drell, N. Kroll, M. Mueller, S. Parker and M. Ruderman, Phys. Rev. Lett. 50 (1983) 644.

VI. PARTICLE IDENTIFICATION

Nuclear Instruments and Methods in Physics Research A277 (1989) 537–546
North-Holland, Amsterdam

ETHYL FERROCENE IN GAS, CONDENSED, OR ADSORBED PHASES: THREE TYPES OF PHOTOSENSITIVE ELEMENTS FOR USE IN GASEOUS DETECTORS

G. CHARPAK, V. PESKOV, F. SAULI and D. SCIGOCKI

CERN, Geneva, Switzerland

Received 5 December 1988

We have investigated the properties of an organometallic compound, ethyl ferrocene (EF), which we propose to use as the photosensitive element in gaseous detectors, both in the gas (vapour) phase and, in condensed or adsorbed layers, as photocathodes. The big advantage of EF is that it is easy to handle, as it is not reactive to oxygen. The sensitivity for the detection of BaF_2 fast emission was measured with EF vapour and was found to be lower by a factor of close to 1.5 compared with TMAE vapour measured under the condition of full light absorption. Adsorbed or condensed layers of EF used as photocathodes in a gaseous detector achieved an efficiency that was lower by a factor of 4 to 10, depending on the experimental conditions.

1. Introduction

The concept of using a photosensitive element in gaseous detectors has a long history [1]. The first idea for this type of device came from Séguinot and Ypsilantis [2], who proposed a ring-imaging Cherenkov detector. A wire chamber filled with photosensitive vapour was elaborated independently by Bogomolov et al. [3].

Further, and successful, developments made it possible to use triethylamine (TEA) vapour (ionization potential $E_i = 7.5$ eV) as the photosensitive element [4,5]. The second idea, put forward by Peskov [6] and Policarpo [7], was for an imaging gas scintillation proportional chamber (GSPC); a realization of this idea was carried out by Charpak et al. [8]. The technique of using an organic vapour called tetrakis(dimethylamine) ethylene (TMAE), with $E_i = 5.36$ eV, as a photosensitive element in a GSPC was first employed by Anderson [9].

The proposal to couple a dense inorganic scintillator, BaF_2, with a low-pressure multiwire proportional chamber (MWPC) came from Anderson et al. [10–12], who named the device the solid-scintillator proportional counter (SSPC). The properties of BaF_2 are: high density (4.9 g/cm^3); short radiation length (2.05 cm); a fast ultraviolet (UV) scintillation component with a decay-time of 600 ps [13]; a photon yield that is independent of temperature [14,15]; and resistance to radiation damage up to more than 10^7 rad [16,17]. Adding these advantages to those of a low-pressure chamber (which resulted in a very good time-resolution [18] and low sensitivity to direct ionization) made the SSPC a most attractive device for many applications [19–21]. Until recently, the only photosensitive compound that had

been used with MWPCs to detect, with good efficiency, the fast component emitted by BaF_2, was TMAE. The major disadvantage of TMAE is that it is very reactive to oxygen. Furthermore, this compound has proved to be corrosive for many materials.

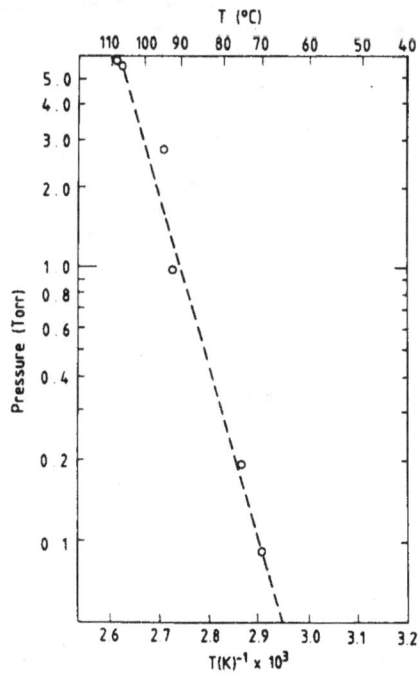

Fig. 1. EF vapour pressure as a function of temperature [22].

The aim of this work is to investigate the properties of an organometallic compound, ethyl ferrocene (EF), which we propose to use as the photosensitive element in gaseous detectors, either in the vapour phase or as a photocathode made with an adsorbed or a condensed layer of EF. We are carefully studying the possibility of using EF as the photosensitive element for the detection of BaF_2 fast emission in an SSPC.

2. General properties of ethyl ferrocene

Ethyl ferrocene (chemical formula $Fe(C_5H_4)_2C_2H_5$; $E_i = 6.2$ eV) is a brown liquid with a faint odour. It is relatively easy to produce, is not toxic, and is very stable. It does not interact chemically with oxygen, or with gases or materials commonly used in gaseous detectors [22], and can therefore be manipulated in air. On the other hand, the vapour pressure of EF, which is given in fig. 1 as a function of temperature [22], is very low compared with that of TMAE (0.1 Torr compared with 8.5 Torr, at 70 °C [22]).

An important characteristic of photosensitive gases is the light emission generated during avalanches. On the one hand, this can produce photon feedback in the chamber [23]; on the other hand it can be used for the optical readout of wire chambers [24]. For these reasons, both the spectrum and the intensity of EF light emission were measured [25]. In the wavelength region between 200 and 800 nm, the main contribution to the emission shows a peak at 310 nm, the region belonging to OH^-. This emission, which is probably due to small traces of water in the EF or in the experimental setup, is rather strong (see fig. 2), i.e. at the level of that of an argon + TMAE mixture. When we increased the proportion of water vapour in the gas mixture, the light emission increased and reached the level of that of an argon + TEA mixture, which is the maximum known at

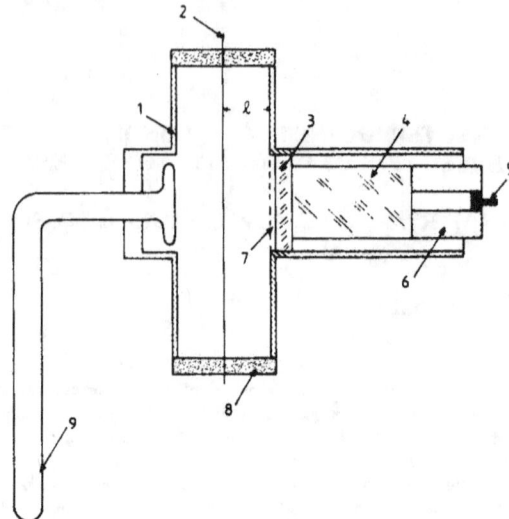

Fig. 3. Setup of the single-wire counter [26]. (1) Body of the counter; (2) anode wire; (3) CaF_2 window; (4) BaF_2 crystal; (5) lead collimator; (6) ^{241}Am gamma source; (7) cathode mesh; (8) ceramics; (9) cathode that can be cooled; l is the absorption gap.

present [24]. But when the water traces were removed from the EF and from the whole system by heating at 90 °C during a few days, this emission dropped by an order of magnitude (curve 4): with the elimination of water traces, the EF vapour becomes a very good quencher.

3. Experimental results

3.1. Experimental setup

3.1.1. Single-wire counters

We have used single-wire counters. The first counter consists of a cylindrical cathode, 40 mm in diameter, made from stainless steel and having a central anode wire of 50 μm diameter (fig. 3). Two transverse flanges were added in the middle of the cylinder. One of the flanges was equipped with a CaF_2 window; the other had joined to it a Cu cathode which could be independently cooled. In one setup (fig. 3) a $15 \times 15 \times 40$ mm^3 BaF_2 crystal was placed in contact with the CaF_2 window. This crystal was irradiated, through a lead collimator, by 59 keV gammas from a ^{241}Am source. In a second setup (see ref. [26]), a Jobin–Yvon H20 UVL monochromator was joined to the flange that had the CaF_2 window. The counter was flushed with a gas mixture consisting of argon and EF or TMAE vapour at a total pressure of 1 atm. The counter and the gas system could be heated up to 90 °C.

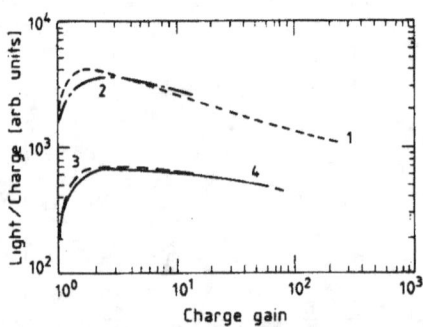

Fig. 2. Light-to-charge ratio as a function of charge gain. (1) Argon + 50 Torr of TEA [24]; (2) argon + EF + 2 Torr of water vapour; (3) argon + 0.5 Torr of TMAE [24]; (4) argon + EF.

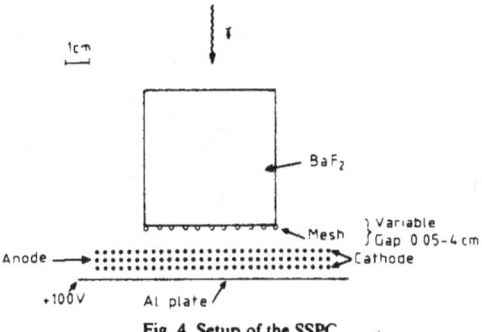

Fig. 4. Setup of the SSPC.

The second counter was mostly used for taking measurements of ageing. We assumed that the ageing characteristics can be largely dependent on the contamination of the gas and on the purity of the whole gas system. For this reason we tried to make these measurements under the cleanest conditions. A sealed counter was used, with a stainless-steel cathode, 16 mm in diameter, and a molybdenum anode having 0.1 mm diameter wires. The counter was equipped with an LiF_2 window of 2 mm diameter. Before sealing, the counter was pumped up to 5×10^{-6} Torr. It was then heated up to $200°C$ over a period of a few days, after which the EF vapour was cryopumped into the counter and spectral clean argon, at 300 Torr, was added. The counter worked in the Geiger mode. During the measurements it was kept at $70°C$ and was irradiated by vacuum ultraviolet (VUV) light in the spectral region between 105 and 200 nm.

3.1.2. The solid-scintillator proportional counter

The setup of the SSPC used for these measurements is shown in fig. 4. It consists of an aluminium vessel containing a $5 \times 5 \times 5$ cm³ BaF_2 crystal that is in contact with a MWPC. The latter has an absorption gap l that can be varied between 0.05 and 4 cm. The face of the crystal that is in contact with the chamber is covered by a stainless-steel mesh made of wires of 50 μm diameter, 500 μm apart. A negative voltage is applied to this mesh to repel the photoelectrons liberated on the crystal or in the conversion gap. The MWPC has two cathode planes made of 100 μm diameter wires, 1 mm apart, and orthogonal to each other, and an anode plane of 15 μm diameter wires. There is a gap of 3 mm between each of these three planes. Two types of anode planes were used: one with wires 1 mm apart at low pressure, and a second with wires 2 mm apart at atmospheric pressure. The anode is connected to a positive potential and the cathodes to ground. An aluminium plate is placed 3 mm after the last cathode plane and is grounded or connected to a positive voltage (+100 V).

The detector was assembled inside a thermostatic box, where the temperature could be varied between 5 and 80°C. We made two kinds of experimental tests: at low pressure (≤ 30 mbar), and at higher pressures (between 0.4 and 1 atm). For the low-pressure measurements a reservoir containing liquid EF was coupled to the detector and could be heated independently. Before these measurements, the detector was pumped up to 5×10^{-5} Torr and then EF vapour was introduced into it. The liquid EF was pumped for only a few minutes before each vapour filling and was not especially purified. In some cases, 30 mbar of CH_4 or a few millibars of isobutane, ethane, tetramethyl pentane (TMP), tetramethyl silane (TMS), or neopentane (NP), were added in order to allow higher gains in the detector. For comparison, some measurements were also performed with the chamber filled with TMAE vapour and with 30 mbar of CH_4. At higher pressure, preliminary measurements were made under the same conditions of chamber operation but with the additive gas at different pressures. We used CH_4 at 0.5 atm or a gas mixture composed of $He + CH_4$ at 1 atm. In a second phase, the setup of the experiment was modified: instead of using a closed system, the $He + CH_4$ gas mixture, at atmospheric pressure, was continuously flushed through the detector after being bubbled in a bottle of liquid EF. For the measurements, the BaF_2 crystal was irradiated by collimated gamma sources [~ 1 MeV gammas from ^{60}Co (60 μCi) or 59 keV gammas from ^{241}Am (100 μCi)] placed 50 mm from the crystal. We measured the number of counts produced with the anode plane of the MWPC through an Ortec 142 preamplifier followed by an Ortec 450 research amplifier.

3.2. Results

3.2.1. EF vapour

(a) *Measurements with the single-wire counters.* Fig. 5a shows our rough estimate of the behaviour of the efficiency of EF vapour obtained from our measurements at $T_{EF} \simeq 70°C$, for an absorption gap $l_{EF} = 16$ mm in the spectral region 105 to 200 nm, measured with the first single-wire counter described in subsection 3.1.1. For the case of full absorption, we expected an efficiency of more than 25%.

For comparison, fig. 5b gives the quantum efficiency of TMAE vapour ($E_i = 5.36$ eV) measured between 150 and 250 nm [27] (curve 3), the BaF_2 fast emission spectrum [10] (curve 1), and the estimated quantum efficiency of EF (curve 2). In the region of the first peak at shorter wavelengths (180–200 nm), which according to calculations gives the largest contribution to the SSPC's response, the integral quantum yield of EF is estimated to be about one half the one of TMAE.

To verify this estimate experimentally, we measured the count rate N_{EF} using the single-wire counter in the

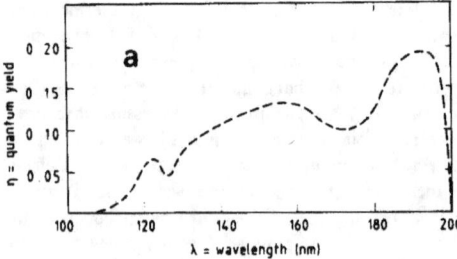

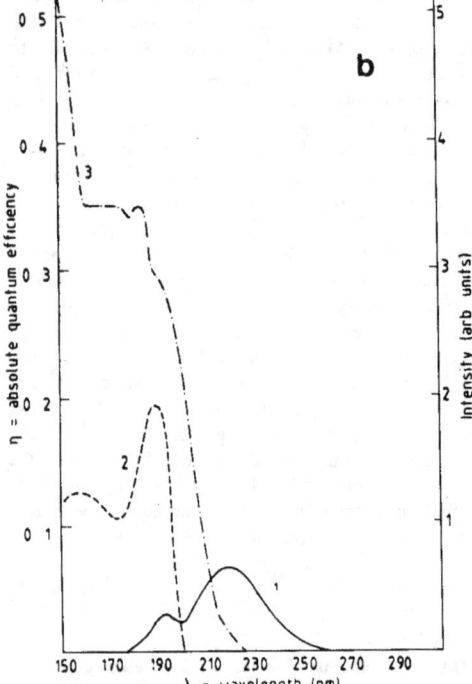

Fig. 5. (a) Estimated quantum efficiency of EF vapour as a function of wavelength. This result was obtained with a vapour pressure corresponding to the liquid at $T_{EF} \simeq 70\,^{\circ}$C and $l_{EF} = 16$ mm. (b) Comparison of the efficiency of EF and TMAE vapours for BaF_2 photon emission: (1) BaF_2 fast emission spectrum; (2) quantum efficiency of EF in the same conditions as in (a); (3) quantum efficiency of TMAE vapour for full light absorption [27].

configuration shown in fig. 3, in the Geiger regime. The advantage of the Geiger mode is that the pulse-height spectrum has a sharp peak and so its efficiency for single-electron detection is close to 100%. This allows a very precise comparison to be made between TMAE (in conditions of full light absorption) and EF vapour efficiencies for BaF_2 emission. Table 1 shows the results of this comparison. The BaF_2 crystal was irradiated with

Table 1
Efficiency of EF vapour (see text) with 59 keV gammas from a ^{241}Am source

Detector	Temperature T_{EF} [$^{\circ}$C]	Gap l_{EF} [mm]	N_{TMAE}/N_{EF}
Single-wire	80	36	1.5 [a]
SSPC	52	40	7 [b]
	80	35	1.5 [b]

[a] With $l_{TMAE} = 36$ mm and T_{TMAE} 27°C.
[b] With $l_{TMAE} = 40$ mm and $T_{TMAE} = 30\,^{\circ}$C.

59 keV gammas from a ^{241}Am source. Because of the small amount of energy deposited in the BaF_2, the average number of photoelectrons n_{pe} produced in the chamber is < 1. Therefore, the ratio N_{TMAE}/N_{EF} gives the relative quantum efficiency of TMAE and EF. The result is slightly better than our preliminary estimate based on figs. 5.

In other measurements we used the sealed counter described in subsection 3.1.1 to estimate the effect of ageing on the properties of EF vapour under clean conditions in the gas system. The single-wire counter worked in the Geiger mode, and information on the ageing was obtained by measuring the gas gain versus the applied voltage. The total charge collected in our experiment is estimated to be $\sim 10^{-3}$ C, without any change observed in the gas gain. Therefore the ageing characteristics of EF vapour are at least not worse than those of TMAE (see ref. [28]).

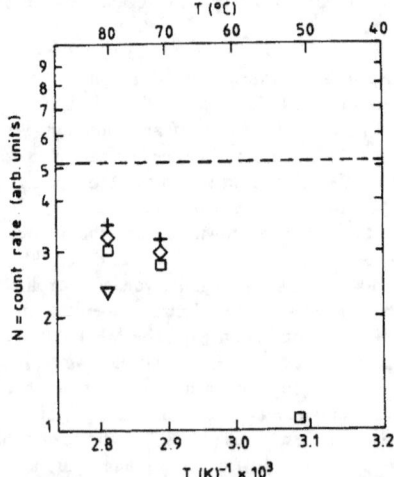

Fig. 6 Count rate measured in a BaF_2 counter (SSPC) filled with EF vapour, irradiated by 59 keV gammas from ^{241}Am, as a function of temperature and for different gaps: $l_{EF} = 35$ mm (+); 25 mm ($\Diamond$); 15 mm ($\Box$); 10 mm ($\triangledown$). Dashed line is TMAE vapour, $l_{TMAE} = 35$ mm.

(b) *Measurements with the SSPC.* All the measurements using the SSPC with EF vapour were made at low pressure, using the detector shown in fig. 4. To avoid condensation, the detector was kept at a temperature higher ($\sim 10\,^\circ$C more) than that of the bottle of EF liquid.

With this wire chamber we can operate in the proportional mode only. Since, as above, $n_{pe} < 1$, the pulse-height spectrum was exponential-like. In this case, owing to the discrimination level in the recording electronics, the measured number of counts can be less than in the Geiger mode. To verify this, we made two experimental checks: first, we checked whether we had a counting plateau; secondly, we measured N_{TMAE}/N_{EF} in conditions of gain large enough to detect single photoelectrons. Table 1 gives the results.

The highest efficiency with the SSPC confirmed our result obtained with the single-wire counter under similar conditions, and indicates that, even in the proportional mode, the efficiency of the SSPC filled with EF vapour is close to that of TMAE. Fig. 6 shows N_{EF} as a function of the temperature and for different l_{EF}. The points show clearly the increase of efficiency with the vapour pressure. The best result is $\sim (67 \pm 5)\%$ of the efficiency obtained with TMAE vapour in conditions of full light absorption; it was measured with $T_{EF} = 80\,^\circ$C and $l_{EF} \simeq 35$ mm.

To check once more that, for $T_{EF} \simeq 70$–$80\,^\circ$C, we worked closed to full absorption, we inverted the polarity in the drift region and made measurements for different l_{EF}. Photoelectrons produced in the drift region cannot reach the wire chamber, and only those created inside the chamber gave a signal. From this measurement we have inferred that at an EF vapour pressure corresponding to a temperature of $70\,^\circ$C, the mean free path of VUV BaF_2 photons is about 1 cm. We do not know if this is due to real absorption in the EF vapour or to electronegative impurities in it.

3.2.2. Condensed layer of EF

(a) *Measurements with the single-wire counters.* Recently, the possibility of using liquid or solid organic photocathodes for gaseous detectors with a good sensitivity in the spectral region between 105 and 300 nm was successfully investigated [24]. We tried to perform the same measurement for condensed EF, using the setup described in ref. [26]. In the present experiment, the single-wire counter was heated to $40\,^\circ$C and the external part of the Cu cathode was cooled to $0\,^\circ$C (for details, see ref. [26]). The measurements were made in the Geiger mode, at atmospheric pressure. In this case we can be sure that the detection efficiency for single photoelectrons is about $\sim 100\%$ (see above).

Fig. 7 (curve 2) shows the quantum efficiency measured between 180 and 250 nm for a thin condensed layer of EF deposited on the copper cathode. For

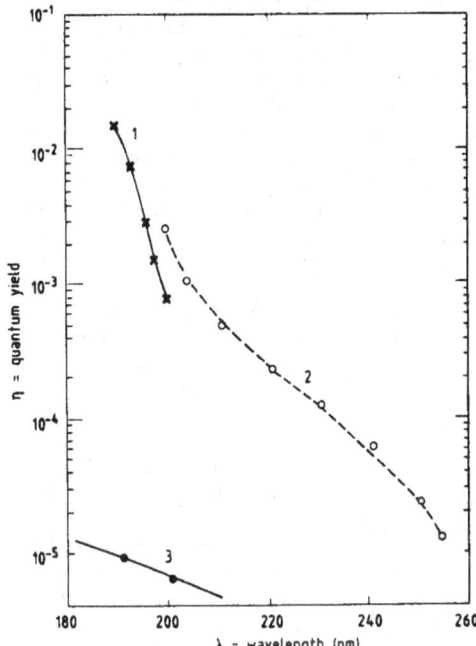

Fig. 7. Quantum efficiency measured with a single-wire counter [26]; (1) EF vapour, corresponding to the liquid EF, at $T_{EF} = 40\,^\circ$C and $l_{EF} = 15$ mm; (2) thin layer of EF condensed on a copper cathode; (3) clean copper cathode.

comparison, the curve of EF vapour at $T_{EF} = 40\,^\circ$C for $l_{EF} = 15$ mm (curve 1) and that of a clean copper cathode (curve 3) were added. As can be seen, the condensed layer of EF is sensitive to radiation at wavelengths longer than the photoionization threshold in the gas phase. The efficiency of this layer, measured with a monochromator at $\lambda = 240$ nm is shown, as a function of cooling duration, in fig. 8. In this case, to condense on the cathode, we kept its temperature at a few degrees less than that of the whole counter. These results indicate that above a certain thickness of the condensed layer, the efficiency stays unchanged. The simplest explanation is that after a delay of five hours we reached a thickness that was sufficient to achieve full absorption for the UV light. In this case, the efficiency of photoelectrons creation in a liquid layer would be constant. This opens up the possibility of interesting applications, e.g. EF could be used as a liquid photocathode.

(b) *Measurements with the SSPC.* All the measurements done to study a condensed layer of EF in the SSPC described in subsection 3.1.2. (fig. 4), were made at low pressure.

To test liquid EF as the photosensitive element in the SSPC, we began by filling the detector, kept at

65°C, with EF at a vapour pressure corresponding to a liquid temperature of 55°C and adding 30 mbar of CH_4 or a few millibars of other additive gases. We then cooled the detector slowly until it reached room temperature. We expected the EF to be condensed uniformly in the detector, thus producing a photocathode deposited on the BaF_2 crystals. Under these conditions the detector worked successfully and we measured the count rate produced by the gamma irradiation from a ^{60}Co or ^{241}Am source. Using methane as the additive gas, and with a gap $l = 5$ mm, we then found that at room temperature the count rate was only 1.5 lower than the rate measured when the EF was in the vapour phase and at a pressure corresponding to a liquid temperature of 55°C. To verify that this efficiency came from the condensed liquid layer and not from the vapour, we brought the temperature down to ~ 10°C: the signal was the same as before (see fig. 9) whereas, according to our estimate based on calculations taking into account the EF vapour pressure, it should have dropped by a factor of more than 5.

To check this by other means, we measured the count rate produced by scintillation of BaF_2 irradiated by the ^{60}Co source as a function of T_{EF}. For these measurements, the chamber and the whole gas system were kept at the same temperature, and the temperature of the EF container was a few degrees lower. Then we introduced the EF vapour and 30 mbar of CH_4 into the chamber, as described earlier. The measured count rates, as a function of T_{EF}, are given in fig. 9. As can be seen, when T_{EF} goes from 55 to 10°C, the count rate drops by about a factor of 6, which is in good agreement with our estimate. But when T_{EF} goes from 20 to 5°C, the count rate drops by about a factor of 2.5. This is less than we expected from calculations (about a factor of 10). We suggest that this discrepancy is due to a contribution to photoelectron creation in a thin layer of EF adsorbed everywhere, including on the BaF_2 crystal (see subsection 3.2.3).

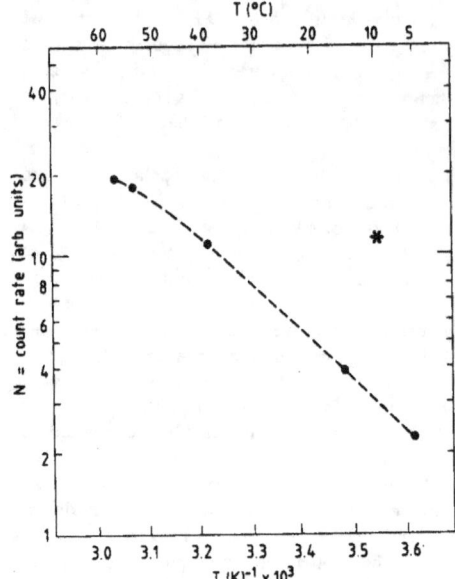

Fig. 9. Count rate measured in the SSPC irradiated by gammas from ^{60}Co for $l_{EF} = 5$ mm; (dots) EF vapour (and probably adsorbed layer); (asterisk) layer of condensed EF.

We also compared the efficiency when different gases were added in the chamber: 30 mbar of CH_4, or a few millibars of TMP, TMS, or NP. After cooling form 65°C to room temperature, we measured an equivalent efficiency for the EF layer with all these gases, and found that it was lower by a factor of 2 compared with that obtained with the EF vapour corresponding to $T_{EF} \approx 65°C$ with $l_{EF} = 5$ mm. These results differ from those obtained with a condensed liquid layer of TMAE [26], where the quantum efficiency of the photocathode increased when the additives TMP, TMS, or NP were used.

Other results were obtained with isobutane or ethane (at a pressure of a few Torr) as the additive gases in the SSPC. With ethane, the efficiency of the condensed layer of EF was 2.5 times lower and with isobutane 3 times lower than what was achieved with the other gases. Therefore, isobutane and ethane would seem to deteriorate the efficiency of the condensed liquid layer of EF. Similar results were obtained with a liquid layer of TMAE [26].

We next increase T_{EF} from 65 to 80°C before filling the detector and then cooled it to room temperature. With 30 mbar of methane as the additive gas we did not observe any improvement in the efficiency of the condensed layer. This is in good agreement with the results obtained with the single-wire counter (see section 3.2.2(a0)).

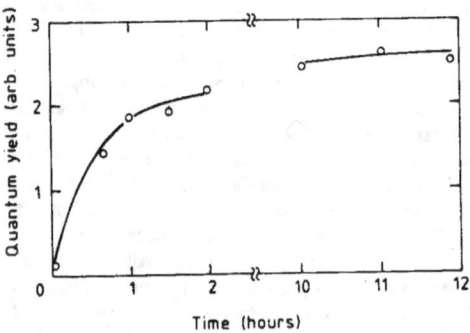

Fig. 8. Efficiency of the layer of condensed EF as a function of the time from the start of condensation, at constant temperature and gain.

The maximum efficiency reached by the condensed layer of EF for the BaF_2 fast component was estimated to be $\sim (10 \pm 3)\%$ of that obtained with TMAE vapour under the condition of full light absorption.

3.2.3. Adsorbed layer of EF

We have discovered that when EF vapour, at a pressure corresponding to room temperature, is introduced into an SSPC kept at the same temperature, an efficient and stable layer of EF is adsorbed on the BaF_2 crystal. This adsorbed layer of EF can be used as a photocathode to detect the fast component of the BaF_2 scintillation.

The measurements with an adsorbed layer of EF deposited on a BaF_2 crystal were made, with the same SSPC, at low and up to atmospheric pressure.

(a) *Experiments at low pressure.* At a temperature of $26\,°C$, with a gap of 5 mm, and with 30 mbar of methane as the additive gas, we measured an efficiency equivalent to that obtained with the condensed layer of EF, i.e. $(10 \pm 3)\%$, under similar conditions (see above). In principle, the total efficiency can be divided into two parts: one part coming from the adsorbed layer itself, and the other being due to the influence of EF vapour at room temperature with a gap $l_{EF} = 5$ mm. We estimated the latter by heating the detector to $78\,°C$, which strongly reduces the effect of the adsorption and adding EF vapour corresponding to a liquid temperature of $27\,°C$. The efficiency measured was lower, by a factor of ~ 3, than that with the detector at room temperature. This means that not more than 30% of the signal comes from the vapour. But even this residual efficiency is probably still due to the electron extraction from the surface, because the EF vapour pressure at room temperature is so low ($\sim 10^{-4}$ Torr, see fig. 1) that the EF gas efficiency in a gap of 5 mm would be much lower than the efficiency measured under these conditions.

(b) *Experiments up to atmospheric pressure.* The gain of gaseous detectors depends essentially on the reduced electric field per unit of pressure E/P. At low pressures it is possible to achieve a very high value of E/P and, as a consequence, a good gain with low electric fields, typically about a few hundred volts per centimetre. Higher fields are needed (approximately a few kilovolts per centimetre) at atmospheric pressure in order to reach similar multiplication. The efficiency of extraction of photoelectrons produced in a liquid photocathode placed in a gaseous detector increases with the electric field [29], and it has been observed in ref. [28], for quencher gases, that it is not strongly dependent on the pressure. This means that low pressure is not favourable for liquid photocathode efficiency because it is not possible to apply a very high electric field without discharges. To test whether the comportment of an adsorbed layer of EF is equivalent to that of a condensed layer, we tried to improve the efficiency of the EF layer adsorbed at room temperature on BaF_2 by increasing the pressure in the SSPC.

As a first step we ensured that the experimental conditions were similar to those at low pressure with a closed system. After vacuum pumping and the introduction of EF vapour at room temperature, ~ 0.5 atm of pure CH_4 or a gas mixture composed of $He + CH_4$ at atmospheric pressure were added in the SSPC. We found that the sensitivity of the layer was not disturbed and continued to be very stable. In comparison with TMAE vapour in the condition of full light absorption, the efficiency of the EF layer measured with $l_{EF} = 5$ mm was

$(25 \pm 2)\%$ with CH_4 at 0.5 atm,
$(20 \pm 2)\%$ with CH_4 at 0.4 atm,
$(10 \pm 3)\%$ with $He + 10\%\ CH_4$ at 1 atm.

The difference in sensitivity observed when the gas and the pressure were changed is correlated with the electric field applied. As we expected, the efficiency seems to increase with the electric field, which indicates that it comes essentially from the layer adsorbed on the BaF_2 crystal and not from the vapour. Some tests were made to verify this. First, we varied l between 5 and 40 mm; the number of counts remained unchanged. If the EF vapour contributed strongly to the signal, the number of counts should increase linearly with l. Another indication was given by reversing the field applied in the gap $l = 5$ mm. Under these conditions the number of counts decrease by a factor of close to 5. If the vapour gives a large contribution, the signal should drop by a factor of ~ 2, because then the only sensitive volume of the detector is within the thickness of the chamber, which is equal to 6 mm. Major proof of the negligible contribution of EF vapour came from the cooling of the detector, under working conditions, from 27 to $2\,°C$. According to the EF vapour pressure (see fig. 1), the sensitivity, which must decrease linearly with it, should have dropped by more than one order of magnitude. We measured a decrease of only ~ 2 for the number of counts under these conditions. Taking into account the tests made at low pressure, all these experiments demonstrated clearly that the major contribution to the sensitivity indeed comes from the EF layer adsorbed on the BaF_2 crystal and not from the vapour.

The second important series of measurements made at atmospheric pressure concerned the stability of the layer when the gas mixture was flushed through the detector. The gas mixture, composed of helium and different concentrations of CH_4 was bubbled through a bottle of liquid EF before reaching the SSPC. The detector and the bottle of EF were kept at room temperature. The efficiencies measured compared with those achieved with TMAE vapour at full light absorption were $\sim (25 \pm 2)\%$ with $He + 50\%\ CH_4$, $\sim (10 \pm 3)\%$ with $He + 10\%\ CH_4$. The stability in time was tested for several days, and no modification in the efficiency of the adsorbed EF layer within the above-mentioned interval of error was observed. In these measurements, the stability of the gas gain was estimated to be ~ 30%. We expect that the stability in the accuracy of this type of operation would essentially be disturbed by small variations in the outside pressure and the detector temperature. In future, this would be compensated by electronic regulation of the gain according to the high voltage applied in the chamber.

4. Discussion and conclusion

Our investigation proves that EF can be used as a replacement for TMAE vapour as the photosensitive element in gaseous detectors, with three possibilities:

(1) Under the condition of full light absorption, *EF vapour* has an efficiency comparable to that of TMAE vapour: our measurement showed it to be only $(67 \pm 5)\%$ of it. The EF vapour pressure is, however, much lower than that of TMAE. For practical applications, particularly for good time resolution compatible with high detection efficiency for VUV photons from BaF_2 emission, it is necessary to reach a higher temperature ($\sim 100\,^\circ C$ with EF compared with $40-50\,^\circ C$ with TMAE) in order to decrease the mean free path of VUV photons and obtain a thin absorption gap in SSPCs. A prototype chamber – a multistep parallel-plate avalanche chamber (PPAC) designed to work at high temperature ($\sim 100\,^\circ C$) – is under development. With large detectors, for example an electromagnetic (e.m.) calorimeter, it may be a technical challenge to use this type of SSPC. However, the disadvantage of higher temperature could be compensated by the good properties of noninteraction with air and with the standard materials used in the construction of the chambers, and by the simplicity of manipulating EF compared with TMAE. Ethyl ferrocene vapour may be an attractive alternative to TMAE where small-size detectors are commonly used, such as in positron-emission tomography (PET) [19,20].

(2) We have demonstrated the capability of *condensed layers of EF* deposited, as photocathodes, on a metallic cathode in a single-wire counter and on the surface of a BaF_2 crystal in an SSPC. The preliminary measurements of the quantum efficiency of the condensed layer of EF gave results comparable to those obtained with different substances, such as TMAE, in ref. [26]. When the layer was deposited on a BaF_2 crystal in an SSPC, we measured a sensitivity, at low pressure, of ~ $(10 \pm 3)\%$ of that obtained with TMAE vapour under the condition of full light absorption. The efficiency of liquid photocathodes in gaseous detectors usually increases according to the applied electric field [26]. This means that, at atmospheric pressure, a condensed layer of EF in an SSPC would reach even higher sensitivity. To verify this and to study the properties of liquid photocathodes, more extensive investigations will be necessary. With its property of noninteraction with air, EF used as a liquid photocathode in a gaseous detector should find a wide range of applications.

(3) The third possibility studied was to use an *adsorbed layer of EF* as a photocathode for SSPCs. We discovered that an efficient layer of EF is adsorbed on the BaF_2 crystal in the SSPC when the detector is filled, at room temperature, with EF vapour. We found that the efficiency of this layer, for the BaF_2 component, is ~ $(10 \pm 3)\%$ of that of TMAE vapour (under the condition of full light absorption) at low pressure, and reaches ~ $(25 \pm 2)\%$ at a higher pressure. In large detectors (e.g. high-energy e.m. calorimeters) based on SSPCs – which are being considered for high-energy machines [20,21] – the lower efficiency of the EF adsorbed layer compared to the vapour is not really a limitation. Again, the sensitivity of the EF photocathode depends on the electric field applied near the cathode (i.e. it increases with the electric field) and therefore on the type of gas mixture used and on the pressure. The stability of the layer seemed to be rather good, even when we flushed the gas mixture through the SSPC at atmospheric pressure.

For applications where the speed of the detector is an essential factor, it is interesting to detect only the signal that comes from the fast component (decay time 600 ps) of BaF_2 emission, and not the one from the slow component (decay time 620 ns). In fact, if the spectral efficiency cutoff of EF vapour is around 200 nm, the tail of efficiency between 200 and 280 nm observed for condensed liquid EF (see fig. 7) can in principle show some sensitivity for the slow component of BaF_2 emission peak at 310 nm. Unfortunately, we cannot estimate it correctly because we did not measure the quantum efficiency of a liquid EF layer for $\lambda < 200$ nm. If we assume that here the efficiency is constant, then, as a pessimistic estimate, the contribution of the slow component of BaF_2 emission is not more than 10% of the photoelectrons detected with our electronics (1 μs integration time).

In SSPCs with TMAE vapour as the photosensitive element, it is necessary to work at a high temperature ($40-50\,^\circ C$) in order to decrease the thickness of the absorption gap. With an EF adsorbed layer at room temperature, a considerable fraction of all the photoelectrons is produced on the surface of the BaF_2. It is then possible to eliminate the conversion gap in the SSPC, which may allow more compact devices and better time resolution to be obtained because the jitter coming from the conversion position of VUV photons in this gap disappears. With drift space in which to collect photoelectrons, a direct amplification could probably start at the surface of the crystal, and the detector would then work in parallel-plate avalanche mode.

The concept of the SSPC is based historically on the association of BaF_2 scintillators and low-pressure MWPCs. This type of chamber was used essentially for two reasons. First, low pressure allows minimization of direct ionization, which adds to the signal coming from the BaF_2 VUV light. The second reason is its good timing properties compared with those of MWPCs at atmospheric pressure; the SSPC achieves time resolutions that are two orders of magnitude better [18], and the positive-ion drift velocity is only a few microseconds as compared to ~ 100 µs at atmospheric pressure. In low-pressure MWPCs, this latter property leads to much less sensitivity to the space-charge effect produced by the drift of positive ions, which at high rates decreases the gain of the chambers. The PPACs at atmospheric pressure also have good timing properties – in particular, the drift-time of the positive ions can be only a few microseconds. Furthermore, with an adsorbed layer of EF as the photosensitive element in a PPAC, only photoelectrons produced near the crystal are detected because only the photoelectrons produced near the BaF_2 surface receive the total amplification. Then again, in the PPAC, only a small area close to the face of the crystal may be sensitive to direct ionization that would add to the signal coming from the photoelectrons produced in the layer. This means that even at atmospheric pressure, a gas mixture with a low sensitivity to direct ionization could give a negligible direct contribution from ionizing particles. For example, with a gas mixture based on helium, where approximately only eight ion pairs per centimetre are produced by minimum-ionizing particles (MIPs) – and assuming that a gas gap of only 1 mm after the crystal is the sensitive detection volume – the level of this contribution will be less than one electron per MIP crossing the SSPC. In BaF_2, the energy deposited by one MIP is ~ 6.5 MeV/cm. In one radiation length (X_0 = 2.05 cm), 13 MeV will be deposited, which may produced ~ 13 photoelectrons. (Under these conditions the efficiency of the EF adsorption layer is estimated to be one photoelectron per MeV deposited in BaF_2, while that of TMAE is 10 photoelectrons per MeV.) The contribution from direction ionization will be small (less than 10%) compared with the signal from the adsorbed layer.

Our observation shows that in single-wire counters working in the Geiger mode, the effect of ageing on the properties of EF is, at least, not worse than on TMAE (in the case of wire chambers [28]). In the case of PPACs, which in the future will be used with the adsorbed layer of EF as the photocathode, we expect the ageing effect to be even less noticeable because laterally the avalanches would be larger; this would reduce the local charge density, and the gas gain would be less disturbed by the nonuniformity on the mesh surface than would be the case with wire chambers.

If the ageing characteristics are really favourable, then the use of an adsorbed layer of EF at room temperature as the photosensitive element in PPACs at atmospheric pressure, may be simple and attractive solution for calorimetry in high-energy physics, particularly compared with TMAE vapour and low-pressure MWPCs. However, further extensive research will be required.

Acknowledgements

This work is part of the research on BaF_2 calorimetry within the framework of the LAA project.

We than Prof. W. Schmidt for his contribution to discussions, and R. Bouclier, G. Million and I. Crotty for technical assistance.

We are indebted to G. Aleksandrov, of the Institute for Chemical Physics of the USSR Academy of Sciences, Moscow, for supplying us with our first samples of ethyl ferrocene.

We also wish to thank P. Müller and J. Pfyffer of the Organic Chemistry Laboratory, University of Geneva, who kindly synthesized for us more than 0.1 litre of ethyl ferrocene within a very short time.

References

[1] D.F. Anderson, IEEE Trans. Nucl. Sci. NS-32 (1) (1985) 495.
[2] J. Séguinot and T. Ypsilantis, Nucl. Instr. and Meth. 142 (1977) 377.
[3] G.D. Bogomolov, Yu.V. Dubrovskii and V.D. Peskov, Instr. Exp. Tech. 21 (1978) 779.
[4] G. Charpak et al., Nucl. Instr. and Meth. 164 (1979) 419.
[5] V.D. Peskov, Instr. Exp. Tech. 23 (1980) 507.
[6] V.D. Peskov, Proc. All-Union Conf. on Vacuum Ultraviolet, "VUV-78", Leningrad (1978) (Acad. Sci. USSR, Leningrad, 1978) p. 99.
[7] A. Policarpo, Nucl. Instr. and Meth. 153 (1978) 389.
[8] G. Charpak, A. Policarpo and F. Sauli, IEEE Trans. Nucl. Sci. NS-27 (1980) 212.
[9] D.F. Anderson, Nucl. Instr. and Meth. 178 (1980) 125.
[10] D.F. Anderson, Phys. Lett. 118 (1982) 230.
[11] D.F. Anderson, R. Bouclier, G. Charpak and S. Majewski, Nucl. Instr. and Meth. 217 (1983) 217.
[12] D.F. Anderson, G. Charpak, Ch. von Gagern and S. Majewski, Nucl. Instr. and Meth. 225 (1984) 8.
[13] M. Laval, M. Moszynski, R. Allemand, E. Cormorèche, P. Guinet, R. Odru and J. Vacher, Nucl. Instr. and Meth. 206 (1983) 169.
[14] P. Schotanus, C.W. van Eijk, R.W. Hollander and J. Pijpelink, Nucl. Instr. and Meth. A238 (1985) 564.
[15] M. Suffert and G. Charpak, CERN EP Internal Report 86-03 (1986).
[16] S. Majewski and D. Anderson, Nucl. Instr. and Meth. A241 (195) 76.
[17] A.J. Caffrey et al., IEEE Trans. Nucl. Sci. NS-33 (1986) 230.

[18] A. Breskin, Nucl. Instr. and Meth. 196 (1982) 11.

[19] P. Miné et al., IEEE Trans. Nucl. Sci. NS-34 (1987) 458.

[20] P. Miné et al., Nucl. Instr. and Meth. A269 (1988) 385.

[21] R. Bouclier et al., Nucl. Instr. and Meth. A267 (1988) 69.

[22] G. Melin, Handbuch der Anorganischen Chemie. Band 14, Teil A, Ferrocen 1 (Springer-Verlag, Heidelberg, 1974) p. 259.

[23] R. Arnold et al., to be published in Nucl. Instr. and Meth.

[24] G. Charpak et al., to be submitted to Nucl. Instr. and Meth.

[25] V. Peskov, G. Charpak, W. Dominik and F. Sauli, this issue, Nucl. Instr. and Meth. A277 (1989) 547.

[26] V. Peskov et al., Nucl. Instr. and Meth. A269 (1988) 149.

[27] R.A. Holroyd et al., Nucl. Instr. and Meth. 261 (1987) 44.

[28] C.L. Woody, IEEE Trans. Nucl. Sci. NS-35 (1988) 493.

[29] R.A. Holroyd, S. Ehrenson and J.M. Preses, J. Phys. Chem. 89 (1985) 4244.

[30] A. Zichichi, in: Report on the LAA project (9 December 1986).

[31] G. Anzivino et al., in: Report on the LAA project (25 June 1987).

798

Nuclear Instruments and Methods in Physics Research A273 (1988) 798–804
North-Holland, Amsterdam

A HIGHLY EFFICIENT LOW-PRESSURE UV-RICH DETECTOR
WITH OPTICAL AVALANCHE RECORDING

A. BRESKIN, R. CHECHIK *, Z. FRAENKEL, D. SAUVAGE, V. STEINER and I. TSERRUYA

Weizmann Institute, Rehovot, Israel

G. CHARPAK, W. DOMINIK, J.P. FABRE, J. GAUDAEN, F. SAULI and M. SUZUKI

CERN, Geneva, Switzerland

P. FISCHER, P. GLÄSSEL, H. RIES, A. SCHÖN and H.J. SPECHT

University of Heidelberg, Heidelberg, FRG

UV photons from a Cherenkov radiator are multiplied in a multistep avalanche chamber operating in a gated mode at low gas pressure (40 Torr). The gas mixture is C_2H_6–argon (80/20) and TMAE at 34°C. Visible light emitted from single photoelectron avalanches is detected by a CCD camera coupled to an image intensifier system. The detector was tested with 5 GeV/c electrons, using a CH_4 radiator gas at 1 atm. Cherenkov rings essentially free of particle background and of secondary photon feedback were obtained in this mode of operation with a mean number $n \simeq 11.5$ ($N_0 \simeq 76$ cm^{-1}). We present this new method and discuss its performance.

1. Introduction

A multistep avalanche chamber, operating at low gas pressure (LPMSC) with a photosensitive mixture is an efficient tool for the imaging of single UV-photons [1,2]. A LPMSC was proposed as a UV-photon detector for Cherenkov ring imaging (RICH), in relativistic heavy ion experiments [3], and it was recently shown that a detector of this type can be applied for recording multiple photon events [4]. In these kinds of experiments the detector has to record several simultaneous Cherenkov rings per event (~ 10 rings/event in S + Au at 200 GeV/u), in the presence of a high background consisting of high energy photons and charged particles. The main advantages in using LPMSCs are [1,2]: high gain (10^7–10^8), reduced photon feedback, low sensitivity to background radiation (low dE/dx), low self-absorption of UV-photons in the carrier gas, high rate capability and the possibility of gating the detector on events of interest.

The position sensing with wire chambers is usually done using wire readout electronics, recording the avalanche induced charges on several wire planes. However, when many simultaneous particles, complex particle events or multiple Cherenkov rings are to be recorded, the readout electronics becomes very sophisti-

cated and expensive. In order to achieve maximum redundancy, parameters like pulse height and time have to be recorded in addition to the position coordinates, using multihit electronics like flash ADCs, as in the case of drift chambers [5] or multistep chambers [3]. A genuine two-dimensional pad-structured readout is thus the natural solution and some efforts have been made in this direction, for example in endcup detectors for TPCs [6]. Some development is presently under way in applying pad readout in MWPCs or LPMSCs for RICH [7].

Another 2D localization method is the optical recording of electron avalanches. The idea of recording the light produced by inelastic collisions during the avalanche process was put forward a few years ago [8–12]. It was clear that using a "photographic" technique would have many advantages for recording complex events, but the small amount of light usually produced in the avalanche, and its unfavorable spectral properties (in most cases the emission is in the far UV-range), made this detection a non-trivial task. The possibility of shifting the UV-radiation into the visible was proposed, using an appropriate organic fluor coating of the detector anode [10]. Another way was to search for an efficient gas mixture which could yield a large amount of light at convenient wavelengths. In the last few years an active research was carried out in this domain and it was found that high light yields can be reached by mixing TEA (triethylamine) vapours with Ar, Kr and hydrocarbons. This is due to an efficient

* The Hettie H. Heineman Research Fellow.

0168-9002/88/$03.50 © Elsevier Science Publishers B.V.
(North-Holland Physics Publishing Division)

process of resonant energy transfer from the noble gas atoms to the TEA molecules. A light yield of 1–4 secondary photons was measured, at normal gas pressure, per single secondary electron produced in the detector volume [14,15]. Successful attempts were made to image particle tracks using MWPCs [16] and parallel grid chambers [14] operating at normal gas pressures with Ar/CH$_4$/TEA-80/18/2. The TEA emission being in the UV (peaked at 280 nm) [13], UV optics and a UV sensitive image intensifier had to be used, coupled to a Vidicon camera and finally to a CCD device.

A considerable step forward was made by realizing that replacing TEA by TMAE [tetrakis (dimethylamine) ethylene], which emits in the visible (peaked at 480 nm) [17], provides a light/charge yield comparable to that of TEA, for equal vapor concentrations and operation pressures [15].

The emission in the visible range provides a great advantage since standard, high aperture lenses and standard image intensifiers can be used. Moreover, TMAE has the largest quantum efficiency [18], among known photosensitive vapors, in the UV range (120–230 nm).

A systematic study of light emission from argon–TMAE and C$_2$H$_6$ or CH$_4$ mixtures was carried out and it was found that a light/charge ratio of the order of 0.1 can be reached at low gas pressures in mixtures having gain and diffusion properties suitable for the application to RICH [15]. Using LPMSCs, 10^6–10^7 secondary visible photons can be emitted per single photoelectron avalanche which is sufficient to make the detection of single photoelectrons possible and efficient.

We have built a prototype of a 3-stage gated UV-photon detector [19] and have studied the application to RICH with a 5 GeV/c electron beam at the CERN-PS. In the present setup, we recorded on the average $\bar{n} \simeq 10.2$ photoelectrons per Cherenkov ring, which corresponds to an N_0 value of 76 cm^{-1}. The localization resolution per single photon is $\sigma \simeq 2.2$ mm, mostly due to chromatic aberration and diffusion. We describe here the detector, its optical system and the experimental setup, and present preliminary results.

2. The UV detector

The UV photon detector is a multistep avalanche chamber operated in a gated mode [20], at a low gas pressure, of 40 Torr of C$_2$H$_6$/Ar (80/20) + TMAE at 34°C. The detector is schematically shown in fig. 1. It has an active area of 20×20 cm^2 and is built in a modular structure of epoxy resin and Delrin frames interslated with Viton O-rings. All the electrodes are made of stainless steel mesh with 81% transparency. Incident UV photons are converted to electrons in the conversion region; its width of 30 mm corresponds to

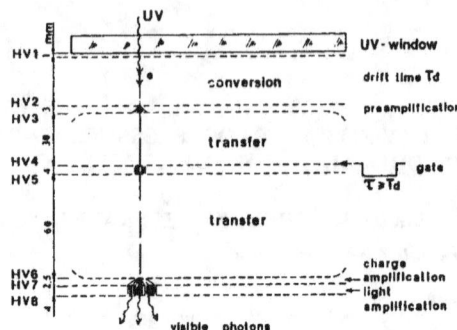

Fig. 1. Schematic diagram of the UV photon detector.

$L = 3\lambda$, λ being the photon absorption length of TMAE at 34°C [2]. The electrons drift to a first amplification stage, operated in a parallel plate avalanche mode. A fraction of the preamplified electron swarm drifts towards the gate device, formed by two meshes having between them a "normally reversed" field. The charges can be transmitted further towards the second amplification stage if the electric field is set to a correct value and direction, by applying a negative pulse on electrode 4 (see fig. 1). The gate pulse has to be applied at a proper time after the formation of the preamplified avalanche. It has to have a width larger than or equal to the drift time of the electrons in the conversion region, so as to ensure full efficiency of the photon detection. The second amplification stage is also made of two grids, operating in a parallel-plate avalanche mode. Electrons are further transferred to a third parallel mesh element. The electric field in this stage is high but below the charge amplification threshold. Electrons drifting across the gap of this element produce light by excitation of the gas molecules. This last stage is indeed responsible for an increase of the light emission by at least an order of magnitude, without further charge amplification.

3. The optical readout

3.1. Light emission from TMAE mixtures

Fig. 2 shows the light emission spectrum as recorded using a parallel grid chamber coupled to a monochromator. The gas, argon (98), CH$_4$(2), TMAE (23°C), at atmospheric pressure, was excited by α-particles from a ^{241}Am source. The spectrum, peaked at about 480 nm, is in agreement with data obtained from TMAE excitation by UV photons (250–390 nm) [17]. An example of the light/charge yield in a gas mixture used in the present work, namely C$_2$H$_6$ (80) argon (20) TMAE, is shown in fig. 3. The light yield was measured as a function of pressure and TMAE concentration by recording charge

VII. PARTICLE PHYSICS

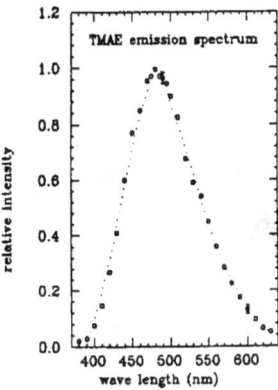

Fig. 2. Light emission spectrum of TMAE, excited by avalanche electrons in a parallel plate avalanche detector, using Ar (98%) + CH_4 (2%) + TMAE (23° C) mixture.

and light emitted from electron avalanches, initiated by 5.9 keV X-rays from a ^{55}Fe source, in a parallel grid chamber. A detailed description is given elsewhere [15]. It was found that the light yield is indeed a function of the TMAE partial pressure (temperature).

3.2. The optical readout system

The optical recording system is shown in fig. 4. It is composed of the following elements:
(1) Large aperture lens, Leitz Noctilux 1 : 1/50 mm.
(2) A gateable MCP image intensifier, Philips XX1410
  – active diameter: 18 mm
  – photocathode: S20
  – phosphor: P36, decay-time 250 ns.

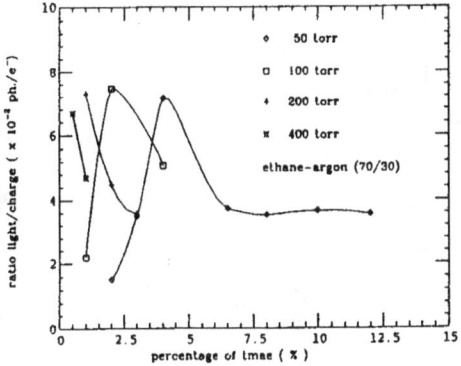

Fig. 3. Number of photons emitted, over 4π, by one electron in the avalanche, as a function of gas pressure and TMAE concentration, in C_2H_6 (70%) + Ar (30%) + TMAE mixture.

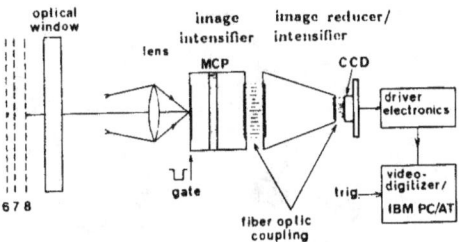

Fig. 4. Schematic view of the optical readout system.

(3) Image intensifier and reducer, DEP XX1490
  – active diameter: ϕ_{in} = 18 mm to ϕ_{out} = 7 mm
  – photocathode: S20
  – phosphor: P46, decay-time 100 ns.
(4) CCD camera, Thomson TH 7852 FO
  – active area: 4.32×5.82 mm^2
  – pixels: 144×208.
(5) Video digitizer: Data Translation DT 2851.
(6) Computer: IBM PC/AT.

The two image intensifiers and the CCD camera are coupled to each other via optical fibers. The first image intensifier can operate either in a continuous mode or in a gated mode. In the latter case, the potential between the photocathode and the MCP cathode is reversed and is set to a proper value only upon the application of a HV pulse of 280 V, triggered by a main trigger system as for the detector gate. The gating of the MCP intensifier and the use of fast phosphors considerably reduce the "noise" from light penetration and "old" events. The video digitizer runs synchronously at 25 Hz. When a trigger occurs, the corresponding frame is stored.

Taking into account the quantum efficiencies of the photocathodes and the phosphor efficiencies, the light gain of the first intensifier is 2×10^3 secondary photons per single photon (at 480 nm) at the photocathode. The light gain of the reducing intensifier is 7, this giving a total light gain of 1.4×10^4. The demagnification factor in the present experiment is of the order of 12.

4. Experimental set-up for RICH

The UV detector described in section 2 is coupled via a UV-transparent window to a gas radiator. The window is made of four slabs of 10×10 cm^2, 7 mm thick each, two of them are made of CaF_2 and two are Suprasil quartz. The radiator set-up is identical to that described in ref. [3], except for a longer focal length of 1.7 m of the spherical UV mirror mounted inside the radiator tank. The radiator operates in a flow mode with CH_4 gas at 1 atm, using a 99.95% pure gas and an "oxisorb" filter. The tank is pumped to below 10^{-5} Torr before the filling. The UV transmission of the

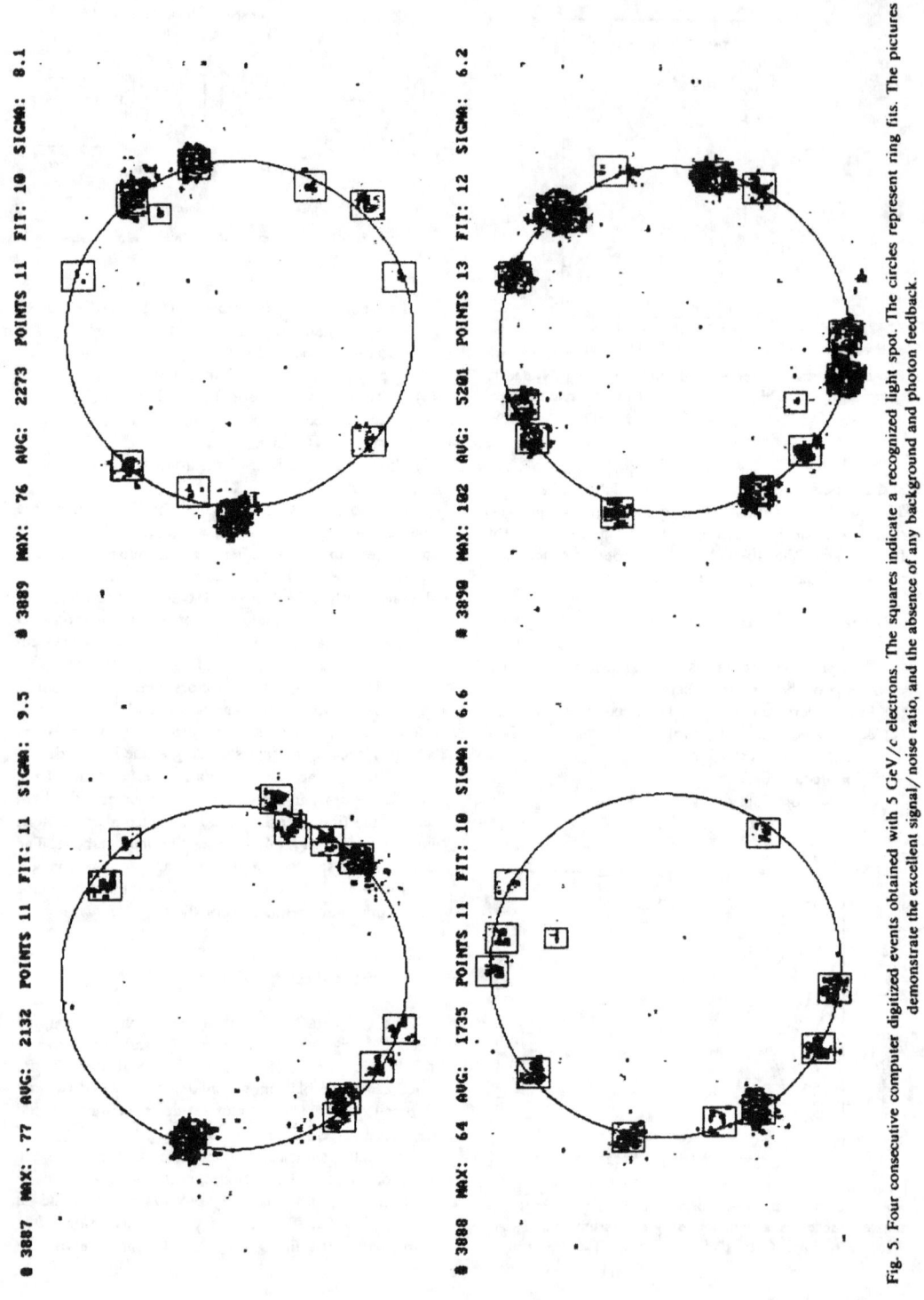

Fig. 5. Four consecutive computer digitized events obtained with 5 GeV/c electrons. The squares indicate a recognized light spot. The circles represent ring fits. The pictures demonstrate the excellent signal/noise ratio, and the absence of any background and photon feedback.

VII. PARTICLE PHYSICS

radiator gas is continuously monitored in the wave-length range of 150–220 nm. The monitor consists of a UV-monochromator coupled to a 1 m long stainless steel tube, with the radiator gas continuously flowing through it. With the present system we measured a transmission of about 0.9 in the range of 160–220 nm.

The UV detector is operated in a flow mode with 40 Torr of C_2H_6/Ar–80/20 (both 99.995% pure + oxysorb), bubbling through a liquid TMAE bubbler, which is immersed in a bath at $34°C$. At this temperature the percentage of TMAE vapors is of 2.4%. The gas flow system of the detector is described in more detail in ref. [2]. The detector and all the gas tubings are heated to about $40°C$, to prevent TMAE condensation on walls and windows of the detector. The gas is pumped out through a cold trap at $-30°C$, to prevent TMAE vapours from contaminating the pump.

The RICH prototype was tested with a beam of 5 GeV/c electrons from the PS accelerator at CERN. A pair of threshold Cherenkov counters and appropriate scintillators provided the beam trigger for the detector and image intensifier gates and for the computer digitizing system. A netgative gating pulse of 40 V was applied to electrode 4 (see fig. 1) at the proper time. The gate pulse duration was about 2 μs, to allow the full transmission of the preamplified electron avalanche.

The gating of the UV detector proved to be essential in this experiment. It made the detector blind to most of the background radiation, enabling an operation at high gains. Furthermore, it prevented the second amplification stage from detecting delayed parasitic secondary avalanches, due to photon feedback.

Within the gate period the UV detector had a gain of about 10^7; the operating voltages were: HV1 = 0, HV2 = 100 V, HV3 = 980 V, HV4 = 1320 V, HV5 = 1300 V, HV6 = 1600 V, HV7 = 2320 V, HV8 = 2890 V.

The optical readout system was coupled to the rear end of the UV detector through a 180 mm diameter quartz window, mounted at about 30 mm from the last electrode (No. 8). The image intensifier was operated in a pulsed mode, with a pulse generated from the beam trigger, with a duration varying from 10–500 μs.

With the present video digitizing system used for this test run, several rings could be detected per beam burst (480 ms). Due to the high photon yield, neither the UV-detector nor the MCP image intensifier were operated at their maximum gain. Though normally the lens was used with an aperture of $F = 1$, we could observe the photons without substantial losses up to $F = 4$.

5. Results

Examples of typical Cherenkov rings (4 consecutive events) are shown in fig. 5. One can see that due to the low pressure operation and the gating of both the

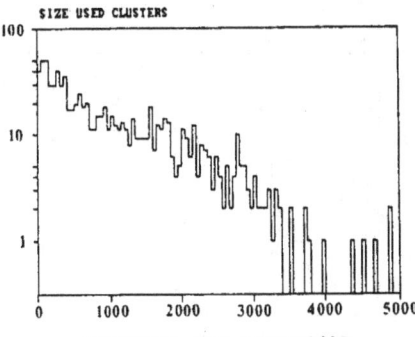

Fig. 6. Number of photons in the recognized light spots. The exponential distribution is typical of a single electron spectrum at low amplification.

detector and the image intensifier the rings are free from background and of photon feedback avalanches. Fig. 6 shows the light distribution from single photon avalanches obtained from a sample of 95 events. It has an exponential behaviour, typical of single electron avalanches at a relatively low gain, non-saturated operation mode. We have measured the absolute number of photons per avalanche, by using a calibrated green LED and replacing the optical chain, that follows the lens, by a photomultiplier. We measured a mean value of about 1000 visible photons per avalanche, at the present solid angle (demagnification factor 12), as shown in fig. 6. Taking into account the light gain of the image intensifiers we get on the average more than 10^7 photons on the CCD, per initial UV photon! The light spots are rather broad, due to the considerable size of the avalanche in the LPMSC, and spread over about 100 pixels, which makes the calculation of the centroid a rather simple task. The average signal-to-noise per pixel is of the order of 100. The size of the light spots may be

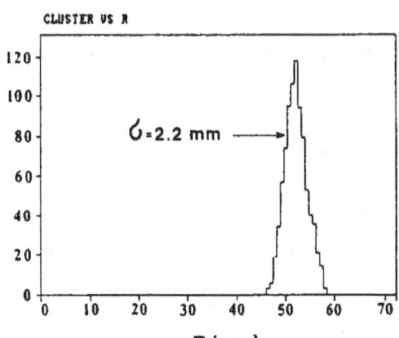

Fig. 7. A distribution of distances of light spots from the ring's center.

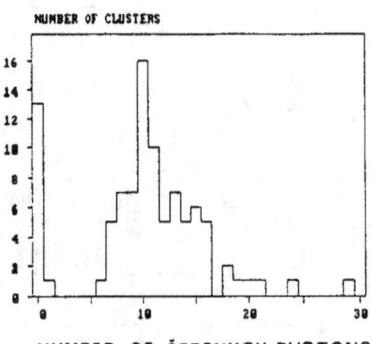

NUMBER OF CLUSTERS

NUMBER OF ČERENKOV PHOTONS

Fig. 8. Distribution of the number of recognized light spots per ring.

reduced by reducing the length of the transfer gap in the UV detector and increasing the gas pressure.

Rings with a fixed radius ($R = 52$ mm) were fitted to the data. The radial distribution of the centers of gravity of the light spots is shown on fig. 7. The distribution has an rms width of 2.2 mm, very similar to that obtained using a FADC electronic readout [3]. The major contributions to this width come from the diffusion of single electrons in the conversion gap ($\sigma \cong 1.1$ mm) and from chromatic dispersion in the radiator gas ($\sigma \cong 1.3$ mm). The contribution of the readout system is negligible due to the excellent signal-to-noise ratio.

The distribution of the number of photoelectrons is shown in fig. 8. The mean value is 10.2 photons. The minimum number of photons is 6. The "zero" events correspond to sparks in the detector – due to a bad isolation between the metal frame of the window and the last electrode, and to empty frames – due to some inefficiency of the computer trigger. When corrections for spatial pile-up are introduced, based on the distribution of the number of photons for a recognized light spot, the experimental mean value becomes 11.5 photons per ring.

6. Discussion and conclusions

We have shown that a combination of a low-pressure multistep UV-phton detector and an optical readout may be an efficient solution for recording Cherenkov rings. The optical readout provides a real unambiguous, 2D localization of single photons, which is crucial for recording multiple Cherenkov rings.

The considerable light yield of TMAE and the high detector gain enable us to reach on the average about 1000 photons per single UV photon, on the photo-

cathode of the first image intensifier – which makes the system practically free of noise.

The N_0 value of the UV-detector was calculated using the data of fig. 9. The window was made of CaF_2/Suprasil (50/50). Taking into consideration the cutoff of C_2H_6, quantum efficiency of TMAE and the transmission of Suprasil, we get a theoretical value of $\langle N_0 \rangle_{th} = 158$ cm^{-1}. However, the efficiency of the radiator and the detector system is affected by several losses:

– mirror reflectivity: 0.85,
– transmission of the radiator gas: 0.90,
– transmission of the window frame: 0.94,
– losses due to a 1 mm distance between the window and the first mesh: 0.90,
– transmission of the first mesh: 0.81,
– efficiency of the 3λ conversion gap: 0.95.

The total efficiency is therefore only of 0.5, yielding an expected value of $\langle N_0 \rangle = 79$ cm^{-1}.

Taking into account the radiator length of 1.41 m and a $\gamma_{TH} \cong 30.6$, the expected number of photoelectrons should be $\langle \bar{n} \rangle = 11.9$. We have measured experimentally an average number of 11.5 photons (after pile-up correction) corresponding to $N_0 \cong 76$ cm^{-1}. This is in very good agreement with the calculated value. With an improved design of the detector, an efficiency of at least 0.7 should be obtainable.

The localization resolution of a single light spot was measured to be of $\sigma = 2.2$ mm. With an average of 11.5 points per ring, the center – needed for particle tracking, or the radius – used for identification, will be determined with a resolution of ≈ 1.0 mm and ≈ 0.75 mm, respectively. The detector resolution which is dominated by the electron diffusion in the conversion gap can be improved by raising the TMAE temperature (smaller conversion length) and raising the gas pressure.

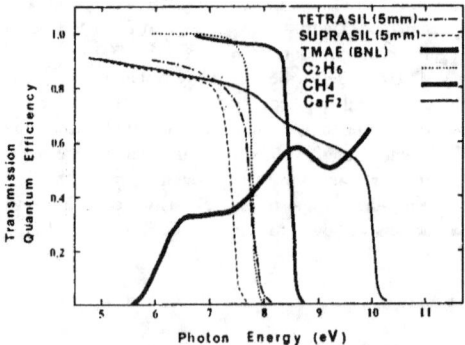

Fig. 9. Typical transmission and absorption curves for C_2H_6 and CH_4 gas, CaF_2 and two quartz windows, and the quantum efficiency curve of TMAE.

VII. PARTICLE PHYSICS

The chromatic dispersion of the radiator can be reduced by using other gases in the radiator such as freons [21].

The present optical readout system has certainly not been optimized for the final application to RICH. In the case of large area detectors, image intensifiers of larger diameters can be used in order to have a reasonable demagnification. If a larger gain is needed, another light amplification stage can be added. For most experiments the present CCD readout system has to be replaced by a faster one, in order to reduce deadtime. Several fast systems are presently being developed, as for example the readout of scintillating fibers [22].

Acknowledgements

We would like to thank Mrs J. Gil, Mr. J. Asher, Mr G. Lamade and Mr M. Klin for technical assistance. We are indebted to Mr R. Benetta, Mr J. Dupond, Mr P. Nappey and Mr S. Reynaud from the CERN/EF division for the construction of the optical readout system. We are grateful to Mr A. Braem from CERN for mirror evaporation and to Dr J. Giomataris for the preparation of high purity TMAE.

This work was supported by the Kadoorie Family Endowment Fund – a project of the Fund for Higher Education, by the BSF US-Israel Research Grant No. 85-00280/1, and by the Minerva Foundation.

References

[1] A. Breskin and R. Chechic, Nucl. Instr. and Meth. A252 (1986) 488.

[2] R. Chechik and A. Breskin, Nucl. Instr. and Meth., A264 (1988) 251.

[3] P. Glässel, H. Ries, H. Specht, A. Breskin, R. Chechik, Z. Fraenkel and I. Tserruya, CERN/HELIOS Note No. 135/March 1986.

[4] A. Drees, P. Fischer, P. Glässel, G. Lamade, H. Ries, E. Schmoetten, H.J. Specht, A. Breskin, R. Chechik, Z. Fraenkel and I. Tserruya, Nucl. Instr. and Meth., in press.

[5] H.J. Burckhart, J. Vavra, K. Zankel, U. Dusziak, D. Schaile, Q. Schaile, P. Igo-Kemenes and P. Lennert, Nucl. Instr. and Meth. A244 (1986) 416.

[6] D.R. Nygren and J.N. Marx, Phys. Today 31 (1978) 46.

[7] G.B. Coutrakon, B. Biggs and S. Dhawan, IEEE Trans. Nucl. Sci. NS-33 (1986) 205; P. Fischer, A. Drees, P.

[8] O. Siegmund, P. Sanford, I. Mason, L. Culhane, S. Kellock and R. Cockshott, IEEE Trans. Nucl. Sci. NS-28 (1981) 478.

[9] R.S. Gilmore, T.K. Gooch, W.L. Kwan, I.C. McArthur, J. Malos, J.P. Melot, R.J. Tapper and R.J. Wyley, Nucl. Instr. and Meth. 206 (1983) 189.

[10] D.M. Potter, Nucl. Instr. and Meth. 228 (1984) 56.

[11] T.K. Gooch, R.S. Gilmore, D.R.N. Jeffery, W.L. Kwan, T.J. Llewellyn, I.C. McArthur, J. Malos and R.J. Tapper, Nucl. Instr. and Meth. A241 (1985) 363.

[12] G. Charpak, Proc. Int. Symp. on Lepton and Photon Interactions at High Energies, Kyoto, 1985 (Kyoto Univ., Kyoto, 1985) p. 514.

[13] M. Suzuki, P. Strock, F. Sauli and G. Charpak, Nucl. Instr. and Meth. A254 (1987) 556.

[14] G. Charpak, J.P. Fabre, F. Sauli, M. Suzuki and W. Dominik, Nucl. Instr. and Meth. A258 (1987) 177.

[15] D. Sauvage, A. Breskin and R. Chechik, preprint WIS-88/8/Feb-PH, submitted to Nucl. Instr. and Meth.

[16] M. Suzuki, A. Breskin, G. Charpak, E. Daubie, W. Dominik, J.P. Fabre, J. Gaudean, F. Sauli, D. Sauvage, P. Strock and T. Zeludziewicz, Nucl. Instr. and Meth. A263 (1988) 237.

[17] Y. Nakato, M. Ozaki and H. Tsubomura, J. Phys. Chem. 76 (1972) 2105.

[18] R.A. Holroyd, J.M. Preses, C.L. Woody and R.A. Johnson, Paper 6408 submitted to the XXIII Int. Conf. on High Energy Physics, Berkeley (July 1986).

[19] A. Breskin, R. Chechik, D. Sauvage and W. Dominik, in preparation.

[20] A. Breskin, G. Charpak, S. Majewski, G. Melchart, G. Peterson and F. Sauli, Nucl. Instr. and Meth. 161 (1979) 19.

[21] T. Ypsilantis, private communication.

[22] R.E. Ansorge et al., Nucl. Instr. and Meth. A265 (1988) 33.

Discussion

B. Ramsay: Do you have any problems with TMAE dissociating in your detector?

A. Breskin: We were working with TMAE up to temperatures of 80°C. We got the same gain as for TMAE at 30°C, the only problem we had was damage to some supports in the detector due to the high temperature. You have to choose appropriate materials if you go to these temperatures.

596   Nuclear Instruments and Methods in Physics Research A283 (1989) 596–601
North-Holland, Amsterdam

A GAMMA RING-IMAGING TELESCOPE FOR HIGH-ENERGY PHOTON DETECTION

G. CHARPAK, W. DOMINIK *, Y. GIOMATARIS, A. GOUGAS and F. SAULI

CERN, Geneva, Switzerland

We describe experimental results obtained using a new type of detector for high-energy gamma astrophysics. The detector will measure gamma rays in the GeV range, with an angular resolution of a few mrad and a field of view of 50°. The telescope is based on the imaging of the Cherenkov light produced in a dense medium (liquid or solid) limited by a parabolic or spherical reflecting surface. The high-energy electrons produced at the start of the electromagnetic shower retain enough information about their initial direction to generate a ring-shaped photon image. This may produce hundreds or thousands of photoelectrons on an appropriate photon-imaging detector placed at the focal plane of the system, which is at the entrance surface of the converter. A prototype consisting of a liquid radiator (C_6F_{14}), a multistep parallel-plate avalanche chamber and an optical readout system was tested in a 9 GeV/c charged-particle beam. Preliminary results show that an angular resolution of a few mrad can be obtained. Furthermore, the capability of such a detector to identify e/π in the GeV range seems promising.

1. Introduction

The determination of the direction of high-energy gammas is a challenge both to particle physics and to astrophysics. In most accelerator physics experiments, the direction of a gamma is determined by the conversion point inside a calorimeter and by the interaction point lying a few meters away, and the main goal is to improve the spatial and the energy resolution. However, this is possible for short-lived particles only, as for example in the case of $\pi^0 \rightarrow 2\gamma$ disintegration or in the case of direct gamma production. It is not possible for long-lived particles such as K_L^0, K_S^0, etc. [1], nor in experiments where the interaction point is not well known, as for example in neutrino experiments. In high-energy gamma-ray astronomy, where the photon intensity is very low [2] (for instance, on the ground, the Crab pulsar intensity is of the order of 5×10^{-6} gammas/(cm² s) and at 1 GeV/c, i.e. many orders of magnitude lower than the charged-particle flux), the development of detectors with better energy resolution – and, in particular, higher angular resolution – is needed in order to improve the signal-to-noise ratio.

The ring-imaging technique [3] for high-energy γ-direction measurements relies on the fact that the direction of electrons produced during electromagnetic shower development is strongly correlated to the momentum vector of the gamma ray initiating the shower [4]. Although the direction of each individual

electron in the shower is affected by multiple scattering, the Cherenkov photons are emitted around the well-known cone. However, because of their relatively high number, and if they are registered with good efficiency and spatial accuracy, they retain much of the information on the initial direction.

In our prototype, the Cherenkov light produced in a transparent medium (C_6F_{14}) by shower electrons, and lying in the ultraviolet (UV) region, is reflected by a spherical mirror and detected by a multistep avalanche chamber [5] filled with a mixture of He (97%) and C_2H_6 (3%), doped with a photosensitive compound (TMAE at 35–45°C, which corresponds to its partial vapour pressure of 1.29–2.46 mbar) [6]. The active area of the prototype is 20×20 cm², and the overall thickness, including the radiator, the mirror and the multistep chamber, is less than 15 cm (fig. 1). The Cherenkov ring image is read out by a charge-coupled device (CCD) camera equipped with an image intensifier. Here we exploit the phenomenon of light emission during the charge-multiplication process in a parallel-plate gaseous chamber filled with a TMAE mixture. A more detailed description of the apparatus is given in ref. [7].

The proposed telescope, now under development at CERN, has an angular resolution of the order of a few mrad in the GeV range. In addition, the separation capability of photons from protons (the latter being the main source of background in an experiment in the space environment), and in particular the large aperture of this apparatus, permits a simultaneous scanning of a large number of celestial high-energy gamma-ray sources.

---

* On leave of absence from the Institute of Experimental Physics, University of Warsaw, Poland.

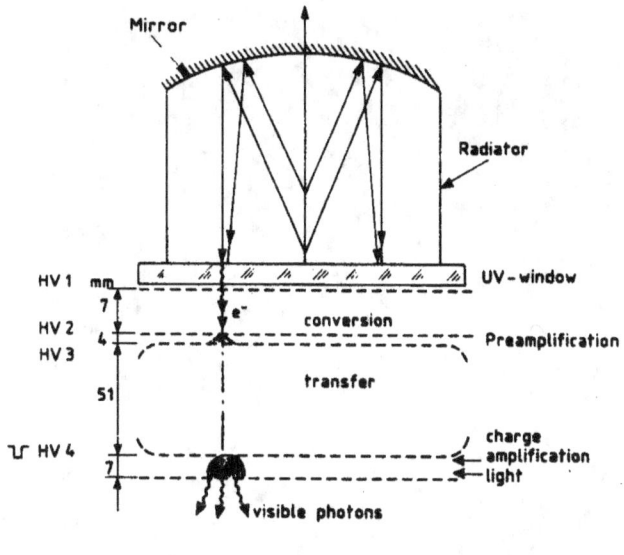

Fig. 1. Schematics of the apparatus.

2. Test results

The reconstruction of the Cherenkov rings proceeds in different and successive steps: information from the optical readout system (i.e. the image-intensified CCD camera) is fed to a frame-grabber * connected to an IBM PC port and stored on a magnetic device [7]. With the present video digitizing system, we could record 2–3 events per second. This limitation is due to the speed of our data transfer and storage system. Showers were initiated on a lead-glass plate of 1 radiation length thickness, placed in front of the gaseous detector.

We studied the performance of the detector at 9 and 10 GeV/c, using a charged-particle beam containing electrons tagged by a gas Cherenkov counter. The average size of an electromagnetic shower event was about 15 Kbytes – significantly larger than that of the single π events collected under the same conditions, which did not exceed 5 Kbytes per event. Factors contributing to the event size are: variation of the CCD thermal noise, accidental hits and background hits due to direct ionization in the gaseous detector. In the case of an electromagnetic shower event, we could observe – apart from the ring pattern – a bright spot at the point of beam

---

* DT-2851, Data Translation, Inc., Marlborough, MA, USA.

incidence on the detector, along with 5–6 hits around the core of the shower; these hits came from ionization in the multistep avalanche chamber (fig. 2). In the case of pions the picture is about the same, but there is only one single bright spot at the point of beam incidence without any other dispersed hits on the image. This background hit is easily distinguishable from the ring pattern as it is 3–4 times brighter than single-photoelectron hits (as expected from the gas mixture we used), and in addition as it lies at a fixed geometrical position for a given angle of beam incidence.

If four or more points are found in an event, a ring fit is done. The ring radius r is initially fixed to the calculated value of $r_0 = 4$ cm. In the iterative ring fit only points with a distance $\langle r - r_0 \rangle \le 1.5\sigma_r$ are considered to belong to the ring, where σ_r is the rms distance of the points from the ring. We collected a total of 680 events for which ring fits with ≥ 4 points are possible according to these criteria, and with $\sigma_r \le 2.4$ mm. This allows background hits from the fit to be discarded. Setting these conditions, we were able to analyse 75.5% of the total data sample collected with an electron trigger. The average number of hits on the ring pattern is $\langle n \rangle = 25$. The rest of the events were mainly empty owing to the acquisition problems, or they contained very few hits because of varying gain in the chamber.

VII. CHERENKOV RING-IMAGING DETECTORS

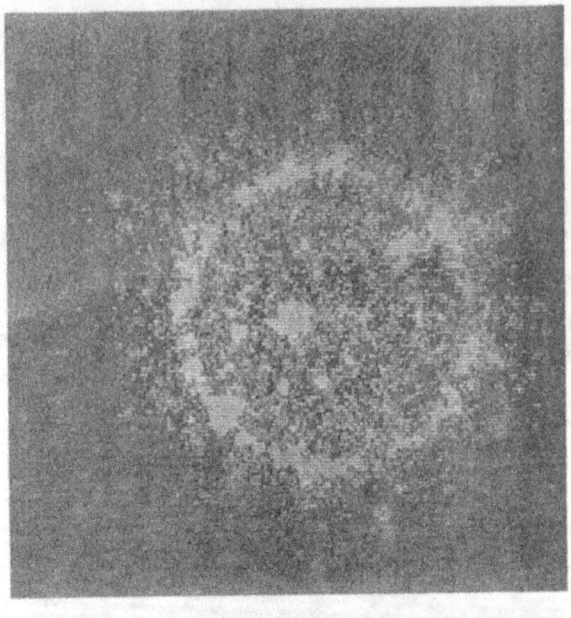

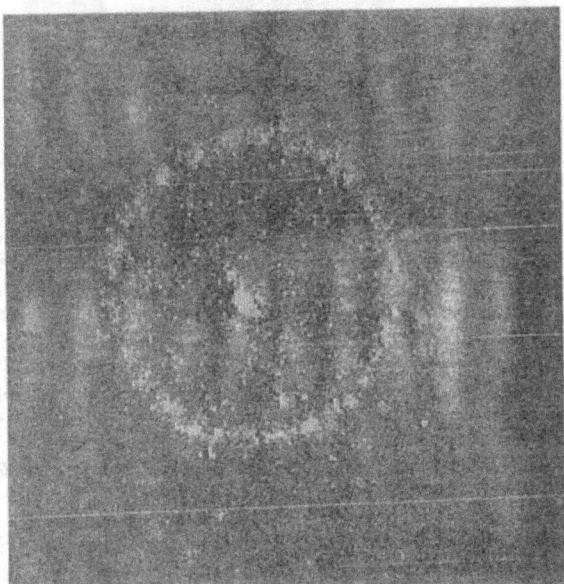

Fig. 2. Cherenkov rings from a shower initiated by an electron at 9 GeV/c (single event).

The centroids of the fitted events are overlaid in fig. 3. Apart from the ring pattern, an intense spot at the center of the image can be observed, which is due to primary ionization in the chamber, and in addition there is a second, fainter ring which is probably due to multiple reflections on the mirror. Further away a part of the ring of Cherenkov light on the CaF_2 window is visible. The angular resolution for the determination of the direction of the incident electron can be given by measuring the spread of the fitted ring centers; the

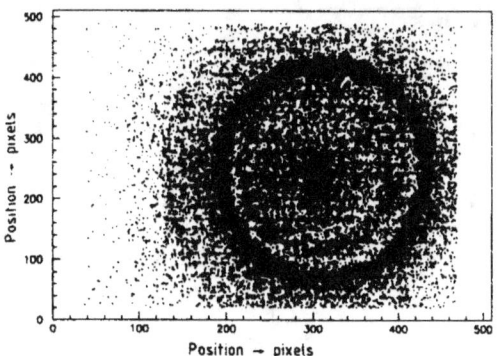

Fig. 3. Overlap of the centroids of 680 events having four or more points on the ring.

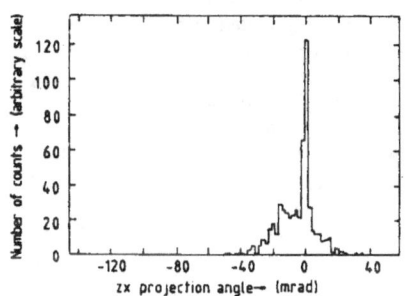

Fig. 4. Ring-center distribution x-axis. An angular resolution $\sigma_\theta \approx 4$ mrad is observed.

analysis gives $\sigma_\theta = 7.3$ mrad, which is determined mainly by the beam dispersion and multiple scattering in the shower converter. Figs. 4 and 5 show the position distribution of the fitted ring centers. The asymmetry observed on the x-plot of the position of the ring center is caused by the asymmetric distribution of background hits lying at radii that are smaller or larger than the ring radius (fig. 6). These hits are mainly due to reflections on the mirror, to accidental hits (sparking etc.) and to

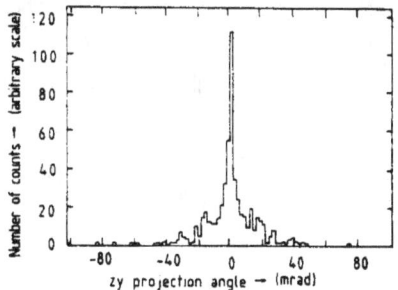

Fig. 5. Ring-center distribution y-axis. The angular resolution is $\sigma_\theta \approx 6$ mrad.

Cherenkov light produced on the CaF_2 entrance window. The distribution of hit radii with respect to the fit center is shown in fig. 7. The Gaussian-like peak has an average of $\langle r \rangle = 4$ cm and a FWHM = 2.2 mm. According to a Monte Carlo simulation, for a geometry such as that in fig. 1 and the same type of radiator the contribution of multiple scattering is expected to be 3.5 mrad at 9 GeV/c, for normal incidence of the initial electron at the center of the lead-glass converter. The beam divergence was 3.1 and 5.6 mrad along the horizontal and vertical planes, respectively. We expect that the use of another type of radiator, made of dense material such as CaF_2 or NaF, would improve the angular resolution of the apparatus, as the incident gamma or electron will initiate the shower inside the radiator. The Monte Carlo simulation for a 5.2 cm thick BaF_2 crystal gives and angular resolution of 1.5 mrad for normal incidence of a 10 GeV/c electron at the center of the crystal. However, the use of BaF_2 is not recommended for this application because of a heavy direct-scintillation component to which TMAE vapours are sensitive [8]. Another improving factor in this case will be the absence of background hits observed in the ring pattern caused by ionization in the chamber, and the resulting ambiguity will be minimized. Other factors that contribute to this uncertainty are electron diffusion

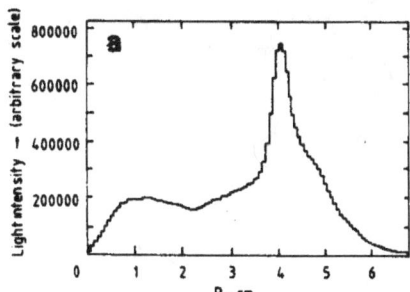

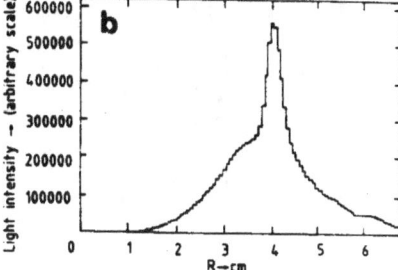

Fig. 6. Radial light distribution for (a) the right half and (b) the left half of the ring. Comparing the background around the peak, asymmetry can be seen.

VII. CHERENKOV RING-IMAGING DETECTORS

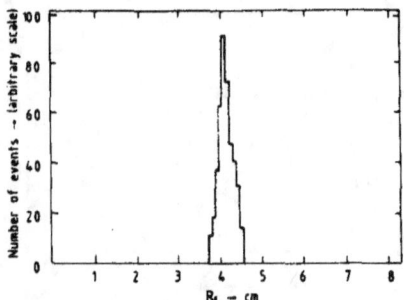

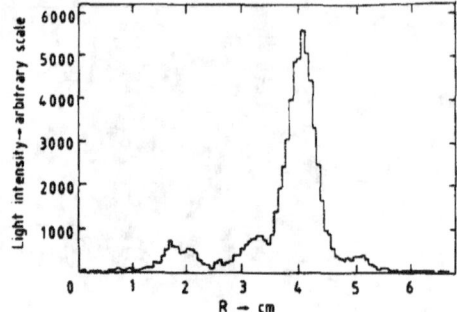

Fig. 7. Fitted radius distribution for electromagnetic shower events. As expected from Monte Carlo calculations, $\sigma_r \approx 2.2$ mm, which is higher than in the case of a single-pion event.

Fig. 9. Radial light distribution of a pion at 9 GeV/c. In this case $\sigma_r \approx 1.2$ mm.

in the conversion gap of the avalanche chamber and the error coming from the reconstruction method used by us.

According to Monte Carlo simulations [9], a spread σ_r of the individual hits results in a resolution $\sigma_0 = 1.6\sigma_r/\sqrt{n}$ for the ring center in one dimension and for n hits on the ring. In this case, we therefore expect $\sigma_0 \approx 700$ μm for the $\langle n \rangle \approx 25$ hits found per ring and $\sigma_r/r = 5.4\%$, which results in an angular resolution of $\sigma_\theta = 7.6$ mrad, comparable with the result obtained by taking the ring-center distributions. Single events with more than 50 points on the ring pattern were also detected. The inconsistency between the average number of photoelectrons of the sample and single events containing more than twice as many hits was due to the non-fully-optimized test conditions and to the statistics of the shower development. Another limitation came from the geometrical properties of the gaseous detector we used and from the TMAE temperature T, namely, the conversion stage of the multistep avalanche chamber was 7

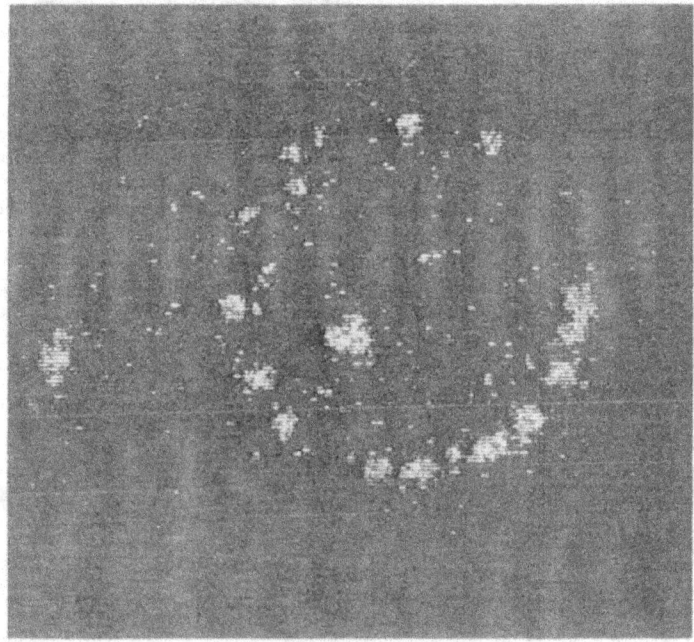

Fig. 8. Cherenkov ring from a single pion at 9 GeV/c. The pattern contains less points than in the case of showers, and the ionization hit is better localized.

mm and the TMAE temperature was always below 45°C. The absorption mean free path l of TMAE for $\lambda = 200$ nm is given by [10]

$$l[\text{mm}] = 19.4 \exp\left[-5614\left(\frac{1}{300} - \frac{1}{T}\right)\right].$$

In our case, the photon conversion depth was always ≥ 6.2 mm, resulting in a loss of $\approx 33\%$ of the UV photons. For these reasons we cannot quote a final number for N_0. An optimization of the gaseous detector will lead to an improvement in the light yield. To achieve this, we are thinking of introducing a gating electrode before the last amplifying stage and operating the chamber at pressures higher than 1 atm. Thus we expect that the light yield will increase and the diffusion be smaller, resulting in a better angular resolution.

A sample of hadron events at 9 and 10 GeV/c was collected. The trigger was provided by a set of scintillators in anticoincidence with a threshold Cherenkov counter. We detected similar ring patterns, but in this case the pattern contained less hits on the ring and the ionization spot was better localized (fig. 8). The event size for hadrons did not exceed 6 Kbytes. Another interesting characteristic of these events is that by applying the same analysis method as that used for the shower events we obtained $\sigma_r/r = 3.1\%$ (fig. 9). We are currently investigating the possibility of distinguishing between single-shower and hadron events by taking into account the integrated intensity of the pattern and the spread of individual hits. For this, a more sophisticated pattern-recognition code is being developed.

3. Conclusions

We have proved that electromagnetic shower ring-imaging is feasible. An angular resolution of ≈ 7 mrad at 9 GeV/c was obtained, which is close to the limits imposed by beam dispersion and multiple scattering of the lead-glass converter. The overall performance of the proposed scheme could be bettered by developing the chamber. This could be done by controlling the pressure and gating the last amplifying stage, and by improving the transparency of the radiator (or replacing it with a dense, transparent medium of low refraction index, which would serve both for shower initiation and as the Cherenkov radiator). Such improvements should result in minimizing the background hits observed and maximizing the number of hits on the ring pattern. Extrapolating from the results obtained for an optimized detector, we expect an angular resolution of ≈ 1 mrad for an incident photon of 10 GeV energy.

Electron–pion separation in the 1–10 GeV/c range is another interesting feature of the proposed telescope. In a subsequent paper we will present the e/π discrimination obtained with this device, along with expectations of Monte Carlo calculations.

The possibility of using the apparatus for π/K separation up to 4 GeV/c is under investigation. For this reason another prototype has been constructed and tested, and data are being analysed [11]. The first results seem to be encouraging.

Acknowledgement

We are indebted to K. Zioutas for the interest he has shown and for his useful suggestions.

References

[1] T. Ferbel (ed.), Experimental Techniques in High-Energy Physics (Addison–Wesley, Menlo Park, CA, 1987).

[2] E. Aprile (Columbia University), private communication; see also B.N. Swanenburg et al., Astrophys. J. 243 (1981) L69.

[3] J. Séguinot and T. Ypsilantis, Nucl. Instr. and Meth. 142 (19770 377.

[4] Y. Giomataris and G. Charpak, preprint CERN-EP/88-94 (1988), presented at the 20th Int. Cosmic-Ray Conf., Moscow, 1987.

[5] G. Charpak and F. Sauli, Phys. Lett. B78 (1978) 523.

[6] Y. Giomataris et al., CERN internal note DELPHI 86-17, RICH-15 (1986);
D. Anderson, Fermilab preprint FN-473 (1988), submitted to Nucl. Instr. and Meth. A.

[7] Y. Giomataris et al., preprint CERN-EP/88-96 (1988), submitted to Nucl. Instr. and Meth.

[8] P. Schotanus, C. Van Eijk and R. Hollander, Nucl. Instr. and Meth. A269 (1988) 377;
R.A. Holroyd, J.A. Preses and C.L. Woody, Nucl. Instr. and Meth. A261 (1987) 440;
S. Ekelin, TMAE Quantum Efficiency, Internal Report, Royal Institute of Technology, Stockholm (1981).

[9] A. Breskin et al., IEEE Trans. Nucl. Sci. NS-35 (1988) 404.

[10] R. Arnold et al., Nucl. Instr. and Meth. A270 (1988) 255; see also J. Séguinot, T. Ypsilantis and P. Petroff, Proposal for the Development of a Fast Ring-Imaging Cherenkov Detector with Local Readout for Use on a Hadron Collider, memorandum to A. Zichichi, LAA project (24 Sept. 1987).

[11] R. Stock et al. (NA35 Collaboration), CERN Proposal SPSC-88/30, SPSC/M438 (1988).

VII. CHERENKOV RING-IMAGING DETECTORS

Nuclear Instruments and Methods 217 (1983) 217–223
North-Holland Publishing Company

COUPLING OF A BaF$_2$ SCINTILLATOR TO A TMAE PHOTOCATHODE AND A LOW-PRESSURE WIRE CHAMBER

D.F. ANDERSON [2], R. BOUCLIER [1], G. CHARPAK [1] and S. MAJEWSKI [3]
with an Appendix by G. KNELLER [4]

[1] CERN, Geneva, Switzerland.
[2] Visitor at CERN, Geneva, Switzerland.
[3] Physikalisches Institut der Universität, Heidelberg, Fed. Rep. Germany.
[4] Lehrstuhl f. Theor. Physik - A, RWTH, Aachen, Fed. Rep. Germany.

The short wavelength component of a BaF$_2$ scintillator has been successfully coupled to a tetrakis(dimethylamino)ethylene (TMAE) photocathode and a low-pressure wire chamber. An energy resolution of 28.5% fwhm has been measured for protons losing about 18 MeV in the crystal. A timing resolution of 540 ps fwhm has been measured for 350 MeV α-particles. We foresee a new generation of calorimeters with good spatial, temporal, and energy resolution, able to work at high rates. Other possible applications are also discussed.

1. Introduction

The modern work in the field of UV light detection with a wire chamber began with the suggestion by Séguinot and Ypsilantis [1] that an admixture gas with a low photo-ionization potential I_g might be added to a wire chamber with an optical window. This was then demonstrated by Charpak et al. [2] with the use of triethylamine (TEA) in a multistep chamber, imaging Cherenkov photons. The threshold of TEA is 7.5 eV (165 nm), which requires the use of CaF$_2$ windows. The next advancement was made by Anderson [3] with the introduction of tetrakis(dimethylamino)ethylene (TMAE), with a threshold of 5.4. eV (231 nm), which allowed the use of quartz windows. The primary interest in photosensitive wire chambers has been in the detection of the light from xenon-filled gas scintillation proportional counters and Cherenkov ring imaging. Although TMAE is somewhat hampered by its low vapour pressure (0.35 Torr at 20°C) [4], no gas with a lower value of I_g has yet been found.

In the search for gases with low I_g, it was noted that the photocurrent threshold E_{th} of the material in solution is much lower than its value of I_g. A value of $E_{th} = 3.54$ eV (350 nm) has been measured for TMAE dissolved in tetramethylsilane [5]. This led to the demonstration of a liquid TMAE photocathode coupled to a low-pressure wire chamber, by Anderson [6] who also detected the UV photons from BaF$_2$. It should also be noted that the same author earlier observed the detection of the light from BaF$_2$ with a single-wire proportional counter filled with TMAE, and a mixture of argon (90%) and methane (10%) [4]. It was believed that

the light was detected by the TMAE in the gas phase, but, as will be shown by this work, this was in fact the first demonstration of a TMAE photocathode. The poor efficiency obtained in that early work was probably due to the use of a counter working at atmospheric pressure.

The reasons for the interest in coupling a high-Z scintillator to a wire chamber are many. It would, in principle, allow the detection of high-energy particles or photons (e.g. calorimetry, or positron emission tomography) without the restraints imposed by photomultiplier tubes. One could also envisage systems that are very compact, offering good timing and spatial resolution.

2. Test detector

Figs. 1a and 1b show a schematic of the test detector and a detail of the electrode structure. The detector consists of an aluminium chamber sealed with Viton O-rings. The scintillator is a BaF$_2$ crystal, 130 mm in diameter and 26 mm thick, in one of two configurations: (i) the first surface is aluminized and the second surface is in contact with a 90% transparent mesh; or (ii) the first surface is clean and the second surface is coated with a thin surface of NiCr as a transparent conductive surface. After an 8 mm drift space, there is an amplification region consisting of two 90% transparent meshes and an anode made of 20 μm wires with a 1 mm pitch. The cathode-to-anode spacing is 3 mm. After a second drift space of 4 mm, there is a temperature-controlled aluminium surface. The surface of the BaF$_2$ and that of aluminium are kept at ground potential, and the cathode meshes are connected to a com-

VI. SPECIAL DEVICES

TEST DETECTOR

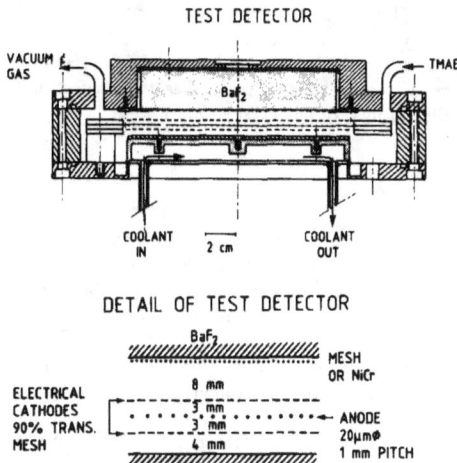

Fig. 1. (a) Schematic of test detector; (b) Detail of electrode structure of test detector.

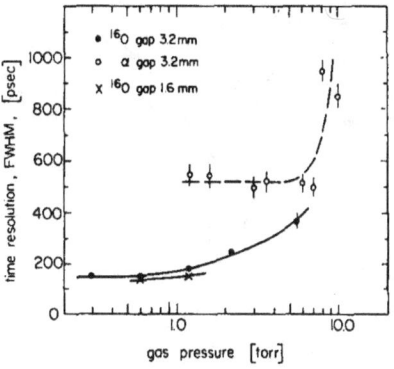

Fig. 2. Timing resolution as a function of gas pressure measured for an isobutane-filled low-pressure proportional counter. (Courtesy of Breskin et al. [7].)

mon power supply. The vacuum and gas systems are connected to one port, and a temperature-controlled flask of TMAE liquid is accessed through a second one.

In the original concept of the test detector, the photosensitive surface would be a condensed layer or TMAE on the aluminium surface. The collection and amplification regions were separated to give greater flexibility in the operation parameters. In most cases the gas filling was 3 to 9 Torr of pure isobutane. Methylal was also used as a counter gas.

The test detector has a small, thin, aluminium window to allow the entry of low-energy γ-rays. Their interaction with the BaF₂ was used to provide single photoelectrons from the cathode.

3. Advantages of low-pressure counters

There are several reasons that have led us to use low-pressure counters (see refs. 7 and 8 on the subject). The first reason is that high-pressure gas has been found to decrease the collection efficiency of photoelectrons from a photocathode. As discussed in the Appendix to this paper, the use of low-pressure organic gases gives the highest quantum efficiency from a photocathode.

A second advantage of low-pressure counters is the good timing characteristic. The dependence of time resolution with counter gas pressure is shown by the work of Breskin et al. [7] (fig. 2). As will be discussed in the next section, the component of BaF₂ to which TMAE is sensitive has only a 600 ps decay time. Thus a low-pressure counter seems to be a natural choice if the

good timing characteristics of BaF₂ are to be maintained.

Another advantage of low-pressure counters is that they work with pure "quench" gases such as isobutane or methylal. Other gases such as argon/methane mixtures, or even pure methane, produce UV photons in the gas amplification process. These photons are to be avoided since high gain is desired. With isobutane we were able to achieve gains of the order of 10^6 without photon feedback.

Low-pressure counters have very low sensitivity to minimum ionizing particles, which will lose, on the average, one electron per centimetre in 3 Torr of isobutane. Thus the photon detector is almost exclusively sensitive to the UV photons and not to the particles in the shower.

Finally, low-pressure counters have very high ion drift velocities. The ion occupation time is low and, since the photons are detected over some area, the ion density is also low. Thus BaF₂ coupled to a low-pressure counter should be able to work at very high rates.

4. BaF₂ as a scintillating crystal

The scintillation properties of BaF₂ have recently gained in interest with the discovery of the very short decay constant of one its two emission components [9]. The first scintillation component to be discovered had an emission peak at 310 nm and a decay constant of 620 ns. Later, a fast component containing about one quarter of the light was discovered, with an emission peak at 225 nm and a decay constant of only 0.6 ns. Using this fast component, Laval et al. [9] have obtained a timing resolution of 112 ps fwhm for ^{60}Co photons. They have also measured the emission spectra of BaF₂ that are

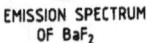
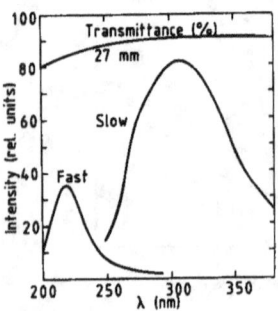

Fig. 3. Emission spectra of BaF₂ measured by Laval et al [9], and the transmission of 27 mm of BaF₂.

shown in fig. 3. The transmission of 27 mm of BaF₂ is also included. From an early measurement done with filters, the detection threshold for a TMAE photocathode was estimated to be about 4.3. eV (290 nm). Thus it is primarily sensitive to the fast component.

Before the demonstration of the detection of the light from BaF₂ with a wire chamber, BGO and NaI(Tl) were the only high-Z crystal scintillators considered for calorimetry. Some properties of BaF₂, BGO and NaI(Tl) are listed in table 1 (refs. 9–12), and it can be seen that BaF₂ has properties that fall somewhere between the other two, with the notable exceptions of decay constant and index of refraction. Its short decay component is 400 to 500 times faster, whilst its smaller index of refraction allows a larger fraction of the light to leave the crystal. The light yield of the fast component is about the same as that of BGO but is considerably less

than NaI(Tl). In general, BaF₂ is a strong, non-hygroscopic crystal, with the greatest resistance to high-energy radiation of all the fluoride listed in the Harshaw Optical Crystal Catalog. It also has the hardest UV scintillation light of any crystal known to us.

5. Performance

In this section we will discuss data taken with a parasitic test beam at Saclay. The BaF₂ crystal with the aluminized first surface was used along with the ground plane of 90% transparent mesh. The aluminium surface was maintained at 2°C, with a thin layer of TMAE liquid condensed on it. As will be shown in the next section, this is not the best configuration. The gas filling was 3 Torr of isobutane.

The first tests were made using α-particles with a maximum energy of 350 MeV. Figs. 4 and 5 show the nature of the pulses from the test detector. The first figure, with a time scale of 4 ns/division, shows the electron component of the signal. The rise-time is about 10 ns. Fig. 5, with a time scale of 0.5 μs per division, shows an ion drift time of less than 1.5 μs.

Again using α-particles, the timing resolution between the test detector and a 1 cm plastic scintillator was measured. The timing spectrum is shown in fig. 6. The time difference between the two peaks is 2 ns, and the resolution is only 540 ps fwhm.

For an energy resolution determination, 940 MeV protons were used, depositing about 18 MeV in the BaF₂ crystal. Fig. 7 shows a pulse-height spectrum taken at an instantaneous rate of about 2×10^5 counts/s. The energy resolution was measured to be 28.5% fwhm. Assuming an $E^{1/2}$ dependence on energy resolution, this implies a resolution of 3.8% fwhm (or $\sigma = 1.6\%$) at 1 GeV loss in the detector. Thus at high energies the

Table 1
Properties of three scintillators [a]

	BaF₂	BGO	NaI(Tl)
Density (g/cm³)	4.9	7.1	3.7
Radiation length (cm)	2.1	1.1	2.6
dE/dx (min.) (MeV/cm)	~ 6	8	4.8
Linear attenuation coeff. at 511 keV (cm⁻¹)	0.47	0.92	0.34
Peak emission (nm)	225 310	480	410
Decay constant (ns)	0.6 620	300	250
Index of refraction	1.56	2.15	1.85
Light yield (photons/MeV)	2×10^3 6.5×10^3	2.8×10^3	4×10^4
Hygroscopic	no	no	yes

[a] See refs. 9–12

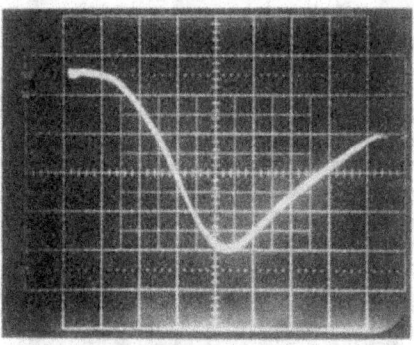

Fig. 4. Electron component of signal. Time scale is 4 ns/division.

VI. SPECIAL DEVICES

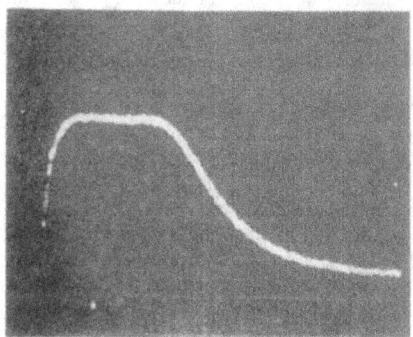

Fig. 5. Ion drift of signal. Time scale is 0.5 μs/division.

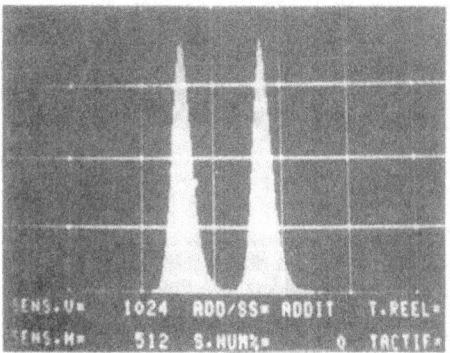

Fig. 6. Timing spectrum measured with 350 MeV α-particles. The time between peaks is 2 ns.

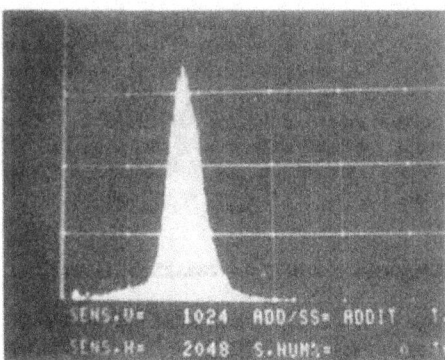

Fig. 7. Pulse-height spectrum for 940 MeV protons depositing about 18 MeV in the BaF$_2$. The energy resolution is 28.5% fwhm.

energy resolution of a BaF$_2$ calorimeter will probably not be limited by photon statistics but by physical factors.

An estimate was also made of the number of photoelectrons detected per MeV. This was done by comparing the pulse height of cosmic rays depositing about 15 MeV in the BaF$_2$, with a single photoelectron spectrum generated with 59.5 keV γ-rays from ^{241}Am. The best estimate for the configuration discussed here was 11.3 (±20%) photoelectrons per MeV or 11.3×10^3 photoelectrons per GeV. Assuming 2×10^3 photons per MeV produced in the crystal, a critical angle of 40°, and a 30% enhancement in light collected by the aliminized surface, we estimate the quantum efficiency of the condensed TMAE photocathode to be about 4%.

6. TMAE as a photocathode

After the measurements described in the previous section, it was discovered that it was not necessary to condense TMAE on a cold surface. If TMAE gas is introduced into an evacuated chamber, all metal surfaces become photosensitive within seconds. Depending on the surface and the counter temperature, it may take from ten minutes to tens of hours to remove this photosensitivity by evacuation.

Upon realizing this fact, we replaced our BaF$_2$ crystal with the second one coated with a thin layer (15–25 Å) of NiCr. This was to provide a UV transparent, conductive surface for the TMAE. For the three figures that follow (figs. 8,9, and 10), with time scales of 50 ns/division, the evacuated detector was kept at room temperature and exposed to the temperature-controlled source of TMAE liquid. The counter was then filled with 9 Torr of isobutane. Fig. 8 shows cosmic-ray events with the TMAE at 2°C. The three bumps show the arrival of the photoelectrons from (i) the electrical cathode mesh;

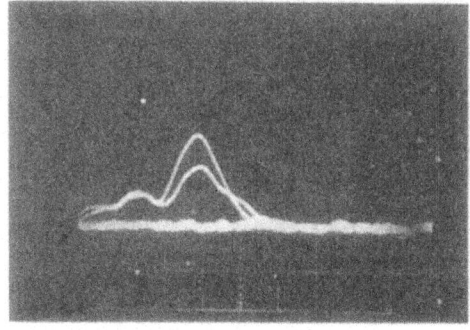

Fig. 8. The three components of electrons from cosmic rays with the TMAE at 2°C. The time scale is 50 ns/division.

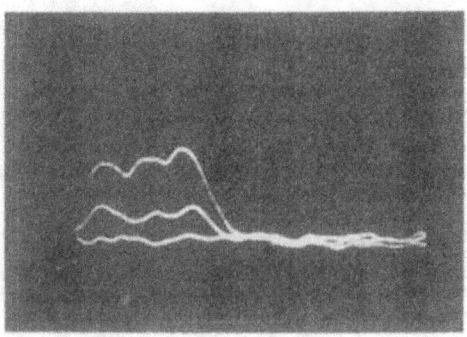

Fig. 9. The three components of electrons from cosmic rays with the TMAE at 20°C. The time scale is 50 ns/division.

(ii) the aluminium surface, and (iii) the BaF$_2$ surface (see fig. 1b). Fig. 9 shows the cosmic-ray events with the TMAE at 20°C. Finally, fig. 10 shows the response of the detector after 12 h of evacuation. The BaF$_2$ crystal is still sensitive. After another 24 h of evacuation, this effect disappeared. It should be noted that the vertical scale on these three figures has no absolute significance, since the gain of the detector is dependent on the amount of TMAE present; thus the scale was adjusted to give events of similar heights. Although these figures are representative, there was considerable variation in the ratio of the bumps seen from event to event.

On first looking at fig. 8, one notices that the contribution of the 90% transparent mesh is disproportionately high. Some of this may be because of enhanced emission due to the stronger electric field at the surface of the mesh [13]. But we also have evidence that suggests that the electrons coming from this mesh have slightly higher gain than those from the other surfaces. If this is the case, such cathodes should be

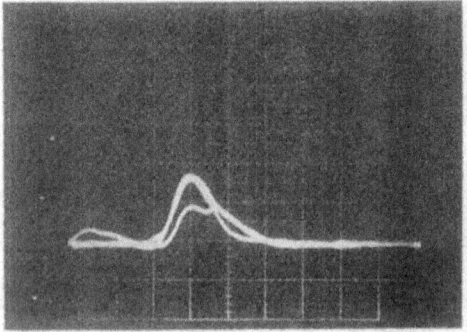

Fig. 10. Single component of electrons from the BaF$_2$ surface after 12 h of evacuation. The time scale is 50 ns/division.

avoided since they would degrade the energy resolution. Also, if it is not enhanced emission, such cathodes would seem to be detrimental where optimum timing is desired.

We now see that the use of TMAE as a photocathode is not as difficult as was first anticipated. There are no detector cooling requirements. Since the amount of TMAE absorbed on the surfaces seems to be dependent only on the pressure of the TMAE, flow counters seem to be feasible with the counter gas simply bubbled through TMAE at a controlled temperature. Also, since the TMAE can be deposited directly on the BaF$_2$ surface, good position resolution should be possible for imaging detectors using these crystals.

Unfortunately we have not determined the number of photoelectrons detected per MeV of energy loss in the BaF$_2$ in this new configuration. One would hope that with the NiCr and TMAE directly on the BaF$_2$, less light will be trapped in the crystal and thus more will be detected. We have also not been able to evaluate the dependence of quantum efficiency of the photocathode on the electric field strength. There is still much work to be done in understanding TMAE as a photocathode.

7. Cost of BaF$_2$

It is difficult to discuss costs in a publication, but when one talks of calorimetry the quantity is in tons, and costs are important (1 ton of BaF$_2$ is a cube with a 60 cm side). At present we pay $6 per cubic centimetre for cut and polished BaF$_2$ crystals, 120 mm in diameter by 25 mm thick. This is about 25% of the price for large crystals of BGO. But the potential for reduction in price is dramatic. The price for pure BaF$_2$ powder (of the order of 1 ton from Fluka, Switzerland) is $0.018 per cubic centimetre of crystal. The Ge in BGO costs $1.42 per cubic centimetre (in the form of GeO$_2$ from China).

It is also much easier to produce BaF$_2$ crystals than BGO. Six years ago, about 40 kg of BaF$_2$ were produced at CERN. The production and material costs were low, and we have found that this BaF$_2$ is within a factor of 2 of the light output of the best BaF$_2$ available. There is hope that for large instruments the cost of the BaF$_2$ will be greatly reduced over the present price for small orders.

8. Advantages of a BaF$_2$ plus TMAE calorimeter

One of the greatest advantages of BaF$_2$ plus TMAE as a calorimeter is that BaF$_2$ can be subdivided into slices of 1 or 2 radiation lengths, separated by 4 to 6 mm of low-pressure counter. This will allow the longitudinal shower development to be observed in a compact instrument with scintillator energy resolution. This

VI. SPECIAL DEVICES

would also allow lower quality material to be used, since self-absorption will be much less of a problem in 2 radiation lengths pieces than in one 22 radiation lengths piece. The insertion of thinner pieces of BaF$_2$ coupled to imaging detectors would also allow the lateral development of the shower to be monitored and would give some two-particle identification capability.

The speed of a BaF$_2$ plus TMAE calorimeter will also be a great advantage. We have demonstrated good energy resolution at a rate of 2×10^5 counts per second. With the proper electronics, this number could be substantially increased. Also, with the large energies deposited in a calorimeter, and with several BaF$_2$ crystals available for timing, the 0.5 ns timing resolution reported here should easily be attainable in a practical instrument.

The performance of this instrument in a magnetic field has not, as yet, been demonstrated. It will certainly be less sensitive than photomultipliers, and may possibly work in very high magnetic fields with only little effect.

There is also reason to hope that the price of large quantities of BaF$_2$ will be substantially reduced. This would allow the construction of instruments, at reasonable cost, that can handle very high rates whilst giving more information than other calorimeter designs.

9. Other applications

If the light detection efficiency can be improved for the BaF$_2$ plus TMAE system, many applications suggest themselves. Using thin BaF$_2$ crystals and an imaging proportional counter, imaging γ-ray detectors should be possible for γ-ray astronomy and nuclear medicine. One can also foresee applications in positron emission tomography. There are also possible applications in radiation detection and dosimetry, where the gaseous detector can be used either in the ionization mode or with gain.

We also see an application for TMAE photocathodes in Cherenkov ring imaging. Although the quantum efficiency is lower than for TMAE gas, the lower threshold allows a greater variety of radiators to be used. One possible configuration for the use of a TMAE photocathode for Cherenkov ring imaging is to have planes of wires, with collection of the photoelectrons on a single plane of anodes. This would allow the imaging of a greater number of points in an event, since electrons from different planes will arrive at different times. If we are seeing enhanced emission from wires, the quantum efficiency may be much higher than the 4% quoted above.

10. Conclusions

We have presented here a technique that offers new prospects in electromagnetic calorimetry. BaF$_2$ coupled to a TMAE photocathode has the potential of yielding calorimeters capable of high rates, energy resolution limited only by leakage, subnanosecond timing, and ability to display both longitudinal and lateral shower development. The work presented here also has applications in many other fields, such as nuclear medicine, γ-ray astronomy, and Cherenkov ring imaging. But there is a great deal more work to be done. We need to find a source of large pieces of BaF$_2$ at a moderate price. The effects of magnetic fields, collection voltage, and surface materials need to be studied. A search for photosensitive materials other than TMAE must also be pursued. But it seems clear that the use of condensed gases as a photosensitive surface has a promising future.

We are pleased to thank Messrs. J. Saudinos and J.C. Duchazeaubenex for their kind help in setting up our instrument, under very improvised conditions, in a particle beam at Saturne in Saclay.

Appendix: Extraction of photoelectrons into a gas

For the work presented above, it was considered necessary to have at least a qualitative understanding of the effect of gas pressure and composition on the collection of photoelectrons from a photocathode. To study this, we constructed a photodiode consisting of a CaF$_2$ window followed by a grounded mesh and an aluminium photocathode. The cathode-to-mesh distance was 5 mm. the photocathode was held at a negative potential and the mesh at ground. A UV light source with an intensity that varies at 100 Hz was used. The output voltage was measured at the photocathode with an ac coupled

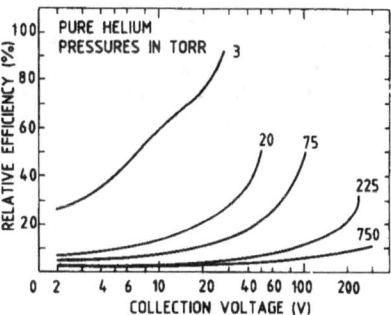

Fig. 11. Relative collection efficiency of photoelectrons as a function of collection voltage for various pressures of helium. All measurements are normalized to the vacuum measurement.

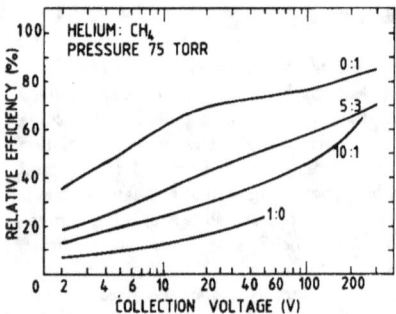

Fig. 12. Relative collection efficiency of photoelectrons as a function of collection voltage for helium/methane mixtures at 75 Torr. All measurements are normalized to the vacuum measurement.

amplifier. The photodiode was connected to a vacuum and gas-filling system.

The gases evaluated were helium, argon, methane, and isobutane. For all tests the instrument was first evacuated for several hours, and the output voltage (peak-to-peak) was measured as a function of collection voltage. All tests made with gases were then normalized to the vacuum measurement at the same collection voltage. An efficiency of the same value as the vacuum measurement would have a value of 100%.

With the exception of helium, we found that after an exposure to a gas, the vacuum measurement was not reproducible. This effect varied with gas type, pressure, and length of filling time. The effect was that of greater electron yield after exposure to a gas.

We present here two sets of measurements that are the most reproducible and show the effect of gas pressure and composition.

Fig. 11 shows the effect of helium for various pressures on the collection of photoelectrons. Again, all measurements have been normalized to the vacuum measurements. As can be seen, even 3 Torr of helium has a dramatic, negative effect. Fig. 12 shows the effect of 75 Torr of a mixture of helium and methane, going from pure helium to pure methane. We can see that the organic gas is much better for photoelectron collection.

In general, we found that organic gases have much better efficiency than noble gases, the difference being quite large. These results are consistent with measurements made with gas-filled photodiodes [14].

We would like to give a simplified explanation of the effect that we see. The atoms of noble gases, having many fewer excitation states available at low energies, are like hard balls scattering the electons elastically, with a high probability that they will be scattered back into the photocathodes and lost. The electrons have inelastic collisions with the organic molecules, and thus have a smaller probability of being scattered back into the pull of their image charges on the cathodes and lost. But whatever the explanation, the message is clear: for efficient collection of photoelectrons, use an organic gas at as low a pressure as possible.

References

[1] J. Séguinot and T. Ypsilantis, Nucl. Instr. and Meth. 142 (1977) 377.
[2] G. Charpak, S. Majewski, A. Melchart, F. Sauli and T. Ypsilantis, Nucl. Instr. and Meth. 164 (1979) 419.
[3] D.F. Anderson, Nucl. Instr. and Meth. 178 (1980) 125.
[4] D.F. Anderson, IEEE Trans. Nucl. Sci. NS-28 (1981) 842.
[5] Y. Nakato, T. Chiyoda and H. Tsubomura, Bull. Chem. Soc. Japan 47 (1974) 3001.
[6] D.F. Anderson, Phys. Lett. 118B (1983) 230.
[7] A. Breskin, R. Chechik and N. Zwang, Nucl. Instr. and Meth. 165 (1979) 125.
[8] A. Breskin, Nucl. Instr. and Meth. 196 (1982) 11.
[9] M. Laval, M. Moszyński, R. Allemand, E. Cormoreche, P. Guinet, R. Odru and J. Vacher, Nucl. Instr. and Meth. 208 (1983) 169.
[10] R. Allemand, M. Laval, J. Vacher, M. Moszyński, E. Cormoreche and R. Odru, Communication LETI/MCTE/82-245, Grenoble, France.
[11] "BGO–NaI(Tl) Comparison", paper distributed at the International Workshop on Bismuth Germanate, Princeton University (10–13 November 1982).
[12] M.R. Farukhi and C.F. Swinehart, IEEE Trans. Nucl. Sci. NS-18 (1971) 200.
[13] C.I. Coleman, Appl. Opt. 17 (1978) 1789.
[14] G. Charpak, W. Dominik, F. Sauli and S. Majewski, IEEE Trans. Nucl. Sci. NS-30 (1983) in press.

VI. SPECIAL DEVICES

Nuclear Instruments and Methods in Physics Research 228 (1984) 33–36
North-Holland, Amsterdam

TEST RESULTS OF A BaF$_2$ CALORIMETER TOWER WITH A WIRE CHAMBER READOUT

D.F. ANDERSON *, G. CHARPAK, W. KUSMIERZ **, P. PAVLOPOULOS and M. SUFFERT ***

CERN, Geneva, Switzerland

Received 23 July 1984

The results from a calorimeter tower consisting of BaF$_2$ crystals coupled to photosensitive, low-pressure wire chambers are presented. The longitudinal and lateral shower development is shown. The results for 108 MeV and 200 MeV electrons yield an energy resolution of $\sigma/E = 2.5\% \, E^{-1/2}$ (GeV).

1. Introduction

There has been a continuing effort in the development of the scintillator BaF$_2$ coupled to a photosensitive wire chamber [1–4]. This new instrument, consisting of a solid scintillator plus a proportional counter, has been given the abbreviation SSPC. The photosensitive material used to detect the BaF$_2$ scintillation has been TMAE (Tetrakis (dimethylamine)ethylene), used in the form of gas [1], liquid photocathode [2], and as an adhered surface [3,4].

As the only scintillator demonstrated to work in an SSPC, BaF$_2$ has many fortunate characteristics. It is non-hygroscopic, has a density of 4.9 g/cm^3 and a radiation length of only 2.1 cm. The short-wavelength component to which TMAE is sensitive has a decay constant of only 0.6 ns. This is about 500 times faster than NaI(Tl) or BGO.

In order to enhance the photoelectron collection efficiency [3] and reduce the sensitivity of the wire chamber to passing particles, low-pressure wire chambers have been used. (See Breskin [5,6] on the subject.) Low-pressure chambers have the added advantage of high positive ion mobility. Thus an SSPC using such a wire chamber yields an instrument with good timing and high rate capability.

In an earlier work a resolution of 28.5% fwhm was measured for protons losing about 18 MeV in an SSPC. But there has never been a measurement made with a stopping high-energy particle. In this work we report results of a test done with an SSPC calorimeter tower stopping high-energy electrons, μ^- and π^-.

---

* Visitor from Fermi National Accelerator Laboratory, Batavia, Illinois, USA.
** Visitor from the Institute of Experimental Physics, University of Warsaw, Poland.
*** Visitor from Centre de Recherches nucléaires et Université Louis-Pasteur, Strasbourg, France.

2. Tower design

Fig. 1 shows a schematic of the SSPC tower. It consists of 14 BaF$_2$ crystals, 12.6 cm in diameter, varying in thickness from 1 to 5 cm. The total thickness of BaF$_2$ is 40.5 cm, making 19.3 radiation lengths, X_0. Each crystal is preceded by its own wire chamber to maximize the position resolution at each step [4]. Table 1 gives the crystal thickness. In general, the crystals become thicker with depth, with the exception of no. 6 which was made thinner to examine the shower after 4.5 X_0. Unfortunately, owing to an oversight, the cathode grouping of this module was not made proportionately narrow. The entire tower was placed in a thin aluminium chamber so that all wire chambers share a common gas volume.

Fig. 2 shows a schematic of an SSPC module. Each crystal is coated on one surface with aluminium to provide an electrode for the next chamber and to maintain optical isolation. For the first chamber a thin aluminium plate was used as the first electrode. The other surface of the BaF$_2$ is coated with about 15 Å of NiCr to provide a conductive, u.v. transparent surface. Between the crystal surfaces and the anode are two orthogonal, cathode planes to give the x–y event location. The anode is made of 15 μm wires with a 1 mm pitch, while the cathodes are made of 100 μm wires with the same 1 mm pitch. The cathode wires are connected in groups of various widths (see table 1) and operated at ground potential. The spacing between each electrode plane is 3 mm. See ref. [4] for more details on the operation principles of an SSPC.

Before filling, the aluminium chamber was first evacuated for several days and checked for leaks. It was then exposed to liquid TMAE and allowed to come to equilibrium. The chamber was filled to 3 Torr of isobutane and the valve closed. No degradation in performance was seen during the two weeks of tests.

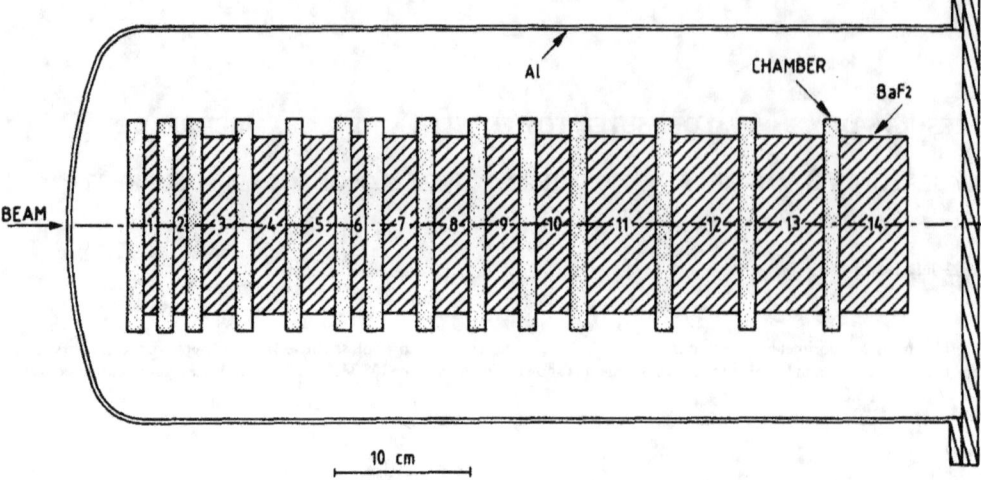

Fig. 1. Schematic diagram of the SSPC tower design.

Table 1
BaF$_2$ crystal thickness and cathode group width.

Crystal no.	Crystal thickness (cm)	Cathode grouping (cm)
1–2	1.0	0.5
3–4	2.5	1.0
5	2.5	1.5
6	1.0	1.5
7	2.5	1.5
8–10	2.5	2.0
11–12	5.0	2.0
13–14	5.0	10.0

Because of the cool environment of the test area, the TMAE temperature was only 17°C. This corresponds to a TMAE partial pressure of 0.28 Torr. An increase in TMAE pressure results in an increase of photon efficiency. Thus it would have been desirable to have been able to heat the chamber and TMAE. This will be done in future work.

3. Test results – shower development

Before putting the tower in the test beam, it was first calibrated with cosmic rays. The gains were adjusted so that the average dE/dx in each crystal was the same.

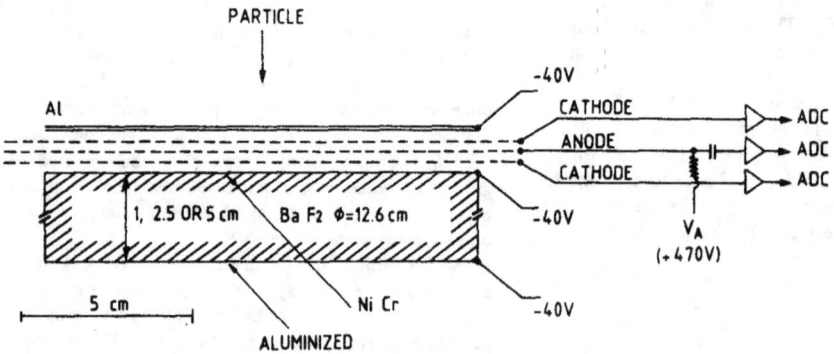

Fig. 2. Schematic diagram of the SSPC module design.

The tower was put in a test beam at the CERN synchro-cyclotron, where π^-, μ^-, and electrons were available with momenta of 108 MeV/c and 200 MeV/c.

Figs. 3 a) and b) show the energy deposit per module for 108 MeV and 200 MeV electrons. There are 2000 events in each sample. The energy deposit in module 6 seems low because it is only 1 cm thick. The longitudinal shower development can easily be seen, with the shower maximum moving deeper for increased energy. Fig. 3 c) shows the energy deposit per module for 108 MeV/c π^- and μ^-. The ratio of π^-/μ^- is about 2:1. Here the short range of these particles is obvious.

To see the lateral shower development, we used the induced pulses on the cathodes. Fig. 4 shows the x and

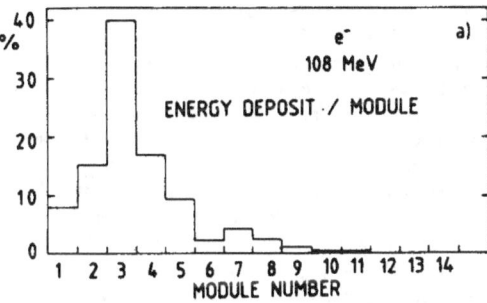

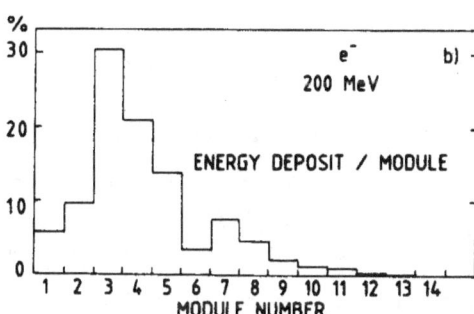

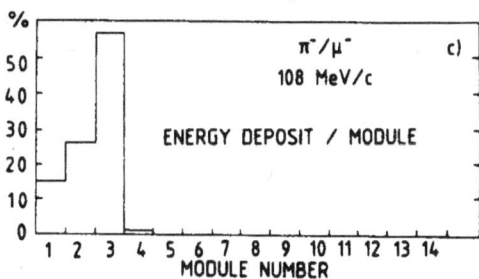

Fig. 3. Energy deposit per module for a) 108 MeV e⁻, b) 200 MeV e⁻, and c) 108 MeV/c π^-/μ^-. The energy deposit in module 6 seems low because it is 1 cm thick.

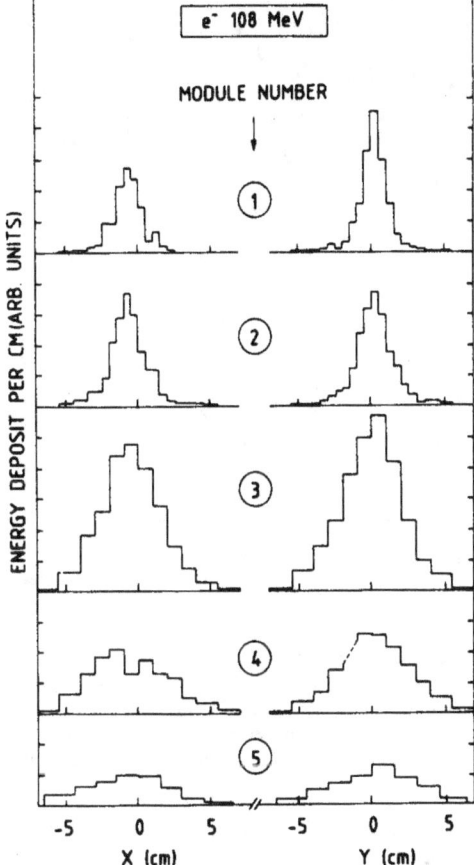

Fig. 4. x and y cathode signals for modules 1–5.

y pulse-height distribution in the first 5 modules for 108 MeV electrons. Unfortunately, owing to electronics problems during the test, we were unable to calibrate the gain of the cathode amplifiers, but the lateral development of this shower is clearly visible.

4. Test results – energy resolution

The pulse-height spectra for 108 MeV and 200 MeV electrons are shown in fig. 5. By simply dividing the widths at half maximum by the peak values, the resolutions are 30% and 20%, respectively. But this is not the true energy resolution. Because of leakage from the sides of the tower, the peak channel is not at the full energy.

In order to estimate the energy resolution of an SSPC that is large enough to totally confine the particle

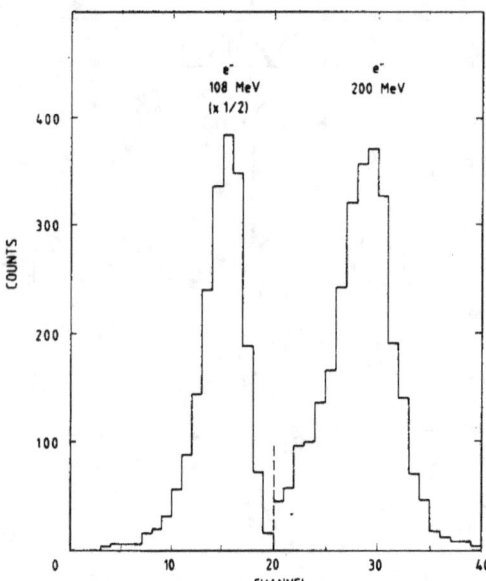

Fig. 5. Pulse-height distribution for 108 MeV e$^-$ and 200 MeV e$^-$.

energy, we used the Monte Carlo simulation, EGS [7]. Table 2 gives the experimental results and the simulation results. The simulation EGS(A) is for a tower with perfect resolution. Thus the spread is due only to energy leakage. EGS(B) is the same simulation but for an

inherent detector resolution of $\sigma/E = 2\%\ E^{-1/2}$ (GeV). Channel resolution is as defined above. Energy resolution is the energy width at half maximum divided by the full particle energy.

The simulation EGS(B) is in good agreement with the experimental results if the resolution is degraded by about 25%. A substantial fraction of the line broadening is due to lateral energy leakage. From these results it seems a resolution of $\sigma/E = 2.5\%\ E^{-1/2}$ (GeV) is a good estimate of the performance of an SSPC calorimeter.

5. Conclusion

It has been shown that a BaF$_2$ SSPC has good energy resolution for an electromagnetic calorimeter. A resolution of $\sigma/E = 2.5\%\ E^{-1/2}$ (GeV) has been determined with a TMAE pressure of only 0.28 Torr. If the entire chamber was heated to 40°C, the TMAE pressure would be increased by a factor of 5. This would increase the u.v. photon detection efficiency and improve the energy resolution.

An SSPC calorimeter also allows the longitudinal and lateral shower development to be monitored. This, combined with its good position resolution [4] and timing resolution [3], should make the SSPC a very powerful instrument for medium- and high-energy physics.

Table 2
Tabulated resolution of the experiment, and that for the EGS simulations with energy leakage, but with perfect detector resolution, (A); and a detector resolution of $\sigma/E = 2\%\ E^{-1/2}$ (GeV), (B). See text.

	180 MeV		200 MeV	
	channel (% fwhm)	energy (% fwhm)	channel (% fwhm)	energy (% fwhm)
Experiment	30	–	20	–
EGS(A)	14	13	11	10
EGS(B)	24.5	22	16	15

References

[1] D.F. Anderson, IEEE Trans. Nucl. Sci. NS-28 (1981) 842.

[2] D.F. Anderson, Phys. Lett. B 118 (1982) 230.

[3] D.F. Anderson, R. Bouclier, G. Charpak and S. Majewski, Nucl. Instr. and Meth. 217 (1983) 217.

[4] D.F. Anderson, G. Charpak, Ch. von Gagen and S. Majewski, preprint CERN-EP/83-200 submitted to Nucl. Instr. and Meth.

[5] A. Breskin, Nucl. Instr. and Meth. 196 (1982) 11.

[6] A. Breskin, in: Lecture notes in physics vol. 178, ed., W. von Oertzen (Springer, Berlin, 1983) p. 44.

[7] R.L. Ford and W.R. Nelson, The EGS code system (version 3), SLAC 210 (1978).

Nuclear Instruments and Methods in Physics Research A267 (1988) 69–86
North-Holland, Amsterdam

TEST OF AN ELECTROMAGNETIC CALORIMETER USING BaF$_2$ SCINTILLATORS AND PHOTOSENSITIVE WIRE CHAMBERS BETWEEN 1 AND 9 GeV

R. BOUCLIER [1], G. CHARPAK [1], W. GAO [1], G. MILLION [1], P. MINÉ [1], S. PAUL [3], J.C. SANTIARD [1], D. SCIGOCKI [1], N. SOLOMEY [1] and M. SUFFERT [2]

[1] CERN, Geneva, Switzerland
[2] CRN, Strasbourg, France
[3] MPI, Heidelberg, FRG

Received 28 September 1987

We describe an electromagnetic calorimeter constructed from layers of BaF$_2$ crystals, coupled to low pressure MWPCs with hot TMAE gas as the photosensitive constituent. By making use of the fast component from the BaF$_2$ scintillation, this detector is well suited for a high rate, intense radiation environment. We present the results of a test performed with our prototype in a 1–9 GeV/c beam, which gives an energy resolution better than $4\%/\sqrt{E}$, a position resolution of 1 mm, and a time resolution better than 1 ns. The detector is highly segmented, with tracking capabilities and good e/π rejection. We discuss the possible application to experiments with intense colliders.

1. Introduction

The proposition to couple BaF$_2$ crystals with a MWPC came from Anderson et al. [1–3] who proposed the name solid state proportional counter (SSPC).

The properties of barium fluoride are the following. It is dense (4.9 g/cm^3), and has a short radiation length (2.05 cm). It is also nonhygroscopic, and cheaper than BGO. Its fast ultraviolet component, discovered by Laval et al. [4] has a decay time of 600 ps, and is insensitive to temperature [5,6]. It has an excellent radiation resistance, more than 10^7 rad, which has been studied by several authors [7,8].

The photons are detected and localized in a low pressure wire chamber. Breskin [9] has extensively studied the properties of such chambers. This device, used to detect the scintillation light of the BaF$_2$, has the advantages of being cheap, naturally segmented, and not very sensitive to magnetic fields. In order to take advantage of the properties of the SSPC, it is useful to operate the MWPC at low pressure, a few Torr, for two reasons. First, the chambers are then insensitive to the ionization due to the charged particles of the shower, which would introduce large fluctuations in the signal, and spoil the energy resolution. Second, an atmospheric chamber is much slower than a low pressure one: time resolutions less than 1 ns at high energy [10,11] and 2.4 ns at low energy [12–16] have already been reported.

The electromagnetic calorimeter prototype, used in the present work, was built and tested at low energies (108 and 200 MeV) in 1984 [17]. After a brief description of the improvements we made, we will report on the experimental procedure used in a PS beam at CERN from 1 to 9 GeV during 1986. We will present results on energy measurement, spatial resolution and its application to electron/hadron separation. The time resolution and high rate behaviour, which are of extreme importance for future high luminosity hadron colliders, will be studied.

There exist two types of electromagnetic calorimeters: homogeneous and sampling. The best energy resolutions are obtained with homogeneous calorimeters, whereas sampling calorimeters optimize the spatial and angular resolutions, with the lepton–hadron separation. These two families exhibit traditionally incompatible performances. Our calorimeter can combine the advantages of the two systems.

Concluding with a discussion that summarizes our findings, and a comparison with other experiments using the same technique, we describe alternative design, concerning crystal size for the imaging properties, or MWPC construction to improve the high rate capability. Some still open problems, like the ageing of the chambers and the quantitative effect of the magnetic field are mentioned.

2. The experiment

2.1. The detector

The constituents of the detector have been described elsewhere [17] and we briefly mention the main characteristics. It is made of fourteen cylindrical barium

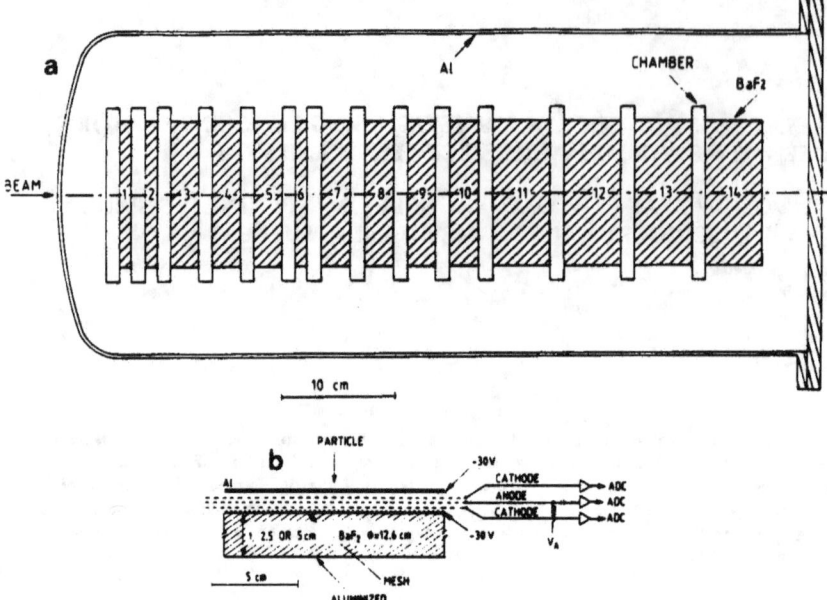

Fig. 1. Schematics of the calorimeter and details of one MWPC.

fluoride crystals, 12.6 cm diameter and 1, 2.5 or 5 cm thick. The total BaF_2 depth is 40.5 cm, which represents 19.3 radiation lengths. Each crystal is preceded by a low pressure MWPC, which detects the BaF_2 scintillation, produced by the development of the electromagnetic shower. Thanks to their low density, the chambers do not interfere with the shower development in the detector, which behaves as an homogeneous calorimeter, despite its segmentation. The longitudinal segmentation was chosen to visualize the shower profile. The whole calorimeter is placed in an 35 l aluminium tank, containing the gaseous mixture necessary for the MWPC operation.

Each crystal is aluminized on the plane face which is not in contact with the chamber and on the side, for light reflexion (fig. 1). The last plane face is covered with an stainless steel mesh, made of 50 μm diameter wires, spaced by 500 μm. This electrode system was found as efficient as the previous NiCr layer, and easier to make. A voltage of −30 V is applied to this mesh, to repel the photoelectrons liberated in the 3 mm conversion space between the crystal and the first cathode plane. The same voltage is applied to the aluminium plate after the chamber, with separates the crystal from the following MWPC. Each plane is thus optically isolated from the next one.

The MWPC has two cathode planes made of 100 μm diameter wires, orthogonal to each other, to localize in the x and y position, and an anode plane made of 15 μm wires to measure energy and time. The three planes are separated by 3 mm. The wires are 15 cm long, spaced by 1 mm. The anode is connected to a positive potential. The cathodes are connected to a virtual ground through the amplifiers and their wires are grouped into strips of different widths (0.5 cm for the minimum). The lateral profile of the shower is obtained in x and y by recording the induced signals on the strips.

2.2. Experimental working conditions

In the first low energy test performed with this detector the experimental configuration was different from the present one. The detector was at room temperature (17°C). The aluminium tank was first evacuated with a primary and secondary pumping unit. Then a valve connecting the tank to a bottle containing liquid TMAE was opened. The TMAE vapour filled all the volume after a few minutes. Due to the low TMAE pressure (0.28 Torr at 17°C), it was necessary to add 3 Torr of isobutane to have correct working conditions in the MWPC. Then the volume was closed during all the data taking. With this experimental procedure a deterioration of the gain was observed after a few hours. It was necessary to pump and refill with a new gas mixture.

For the work described in this paper, the working conditions were different on two points and as a consequence some modifications had to be made:
(1) The detector was heated to optimize the UV photon detection efficiency [6].

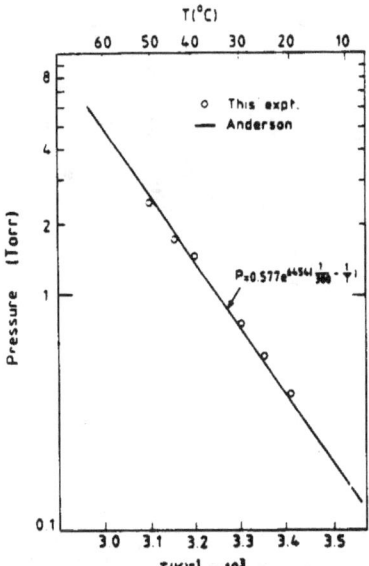

Fig. 2. TMAE vapour pressure as a function of the inverse of the absolute temperature in °C, including points from ref. [18].

(2) The gas volume was constantly renewed by a low pressure bubbling system, enabling a continuous operation of the experiment.

Increasing the TMAE vapour pressure, by heating the liquid, decreases the conversion length of the photons. For a given gas depth, the detection efficiency rises. We have measured the pressure curve with a Baratron MKS type 223BHA10. Our data points shown in fig. 2 follow the Clausius–Clapeyron relation as in ref. [18] which is compatible with use at low temperature. In our calorimeter the best temperature was estimated to be 40°C, for a gas depth of 12 mm. The UV photons mean free path is equal to 5 mm, with a TMAE pressure of 1.4 Torr, and the detection efficiency is about 10 photoelectrons for 1 MeV lost in the crystal [12,13]. The bottle of liquid was immersed in a water thermostat controlled bath regulated to better than one tenth of a degree. To avoid TMAE condensation, the tank was heated to a higher temperature (45°C) with heating belts wrapped around. The temperature of the gas volume was kept constant within 1.5°C by a thermostatic regulation driven by a thermistance placed inside the tank. The detector was also covered with insulating material.

A schematic diagram of the continuous bubbling system is shown in fig. 3. In this configuration, the tank is evacuated once by the primary and secondary pumping units, down to 10^{-5} Torr. An isobutane bottle is connected to the TMAE bubbler, kept at 40°. The isobutane bubbles through the liquid, carrying away the TMAE vapour to the detector. The other side of the tank has a microleak, or an electric valve driven by a pressure transducer, followed by a refrigerated bath to freeze the TMAE, a charcoal trap and a vacuum pump. By this method the gas volume is continuously renewed in the detector. Its pressure is set at 2 Torr (1.4 TMAE and 0.6 isobutane). We have found that adding isobutane is mandatory to ensure an external gas flow to renew TMAE at a sufficient rate (a choice of gas other than isobutane would probably also be adequate). The

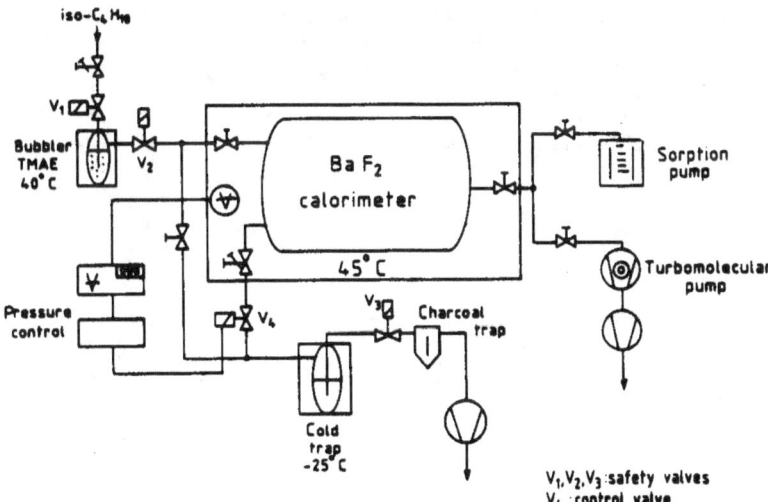

Fig. 3. Schematics of the gas continuous flowing system.

pressure transducer was the MKS type 223BHA10 used for the determination of the vapour pressure curve, precise to ±0.1 Torr. With this system we have been able to keep the gain stable within a few percent, as will be explained in detail below, and to operate the apparatus continuously without pumping and filling for three months.

Another improvement was added to the original system: the fourteen anodes had separate electrical high voltage connections, since at high energy the signals were strongly different with the longitudinal position in the shower.

2.3. The beam

Our data were collected in the T7 test beam at the CERN PS. The momentum could be varied from 1 to 9 GeV/c, the $\Delta p/p$ dispersion was 2% (FWHM) and the angular dispersion was 10 mrad. The proportion of secondary particles in the beam was momentum dependent: for the electrons 42% at 3 GeV/c, 20% at 5 GeV/c and 3% at 9 GeV/c. The end of the data taking had to be done with deuterons circulating in the PS. In that case the proportion of electrons was lower: 3% at 5 GeV/c. The burst length was 0.5 s, during which we usually sent 1500–3500 particles in the detector, except for the high rate tests described below.

Fig. 4 shows the general environment of our apparatus in the test beam. Upstream were two gas Cherenkov counters C_1 and C_2 for particle identification. The signals from their photomultipliers were recorded by ADCs. Fig. 5 shows the correlation between C_2 and the energy measured in the calorimeter.

The Cherenkov counters were followed by two drift chambers Ch_1 and Ch_2 giving the horizontal and vertical position of the incoming particles with a precision of 100 μm. They worked at atmospheric pressure with a mixture of 4% methylal, 30% isobutane and 66% argon.

Just in front of the calorimeter were placed three scintillators: P_3 was 20 × 20 cm², P_2 had the same size

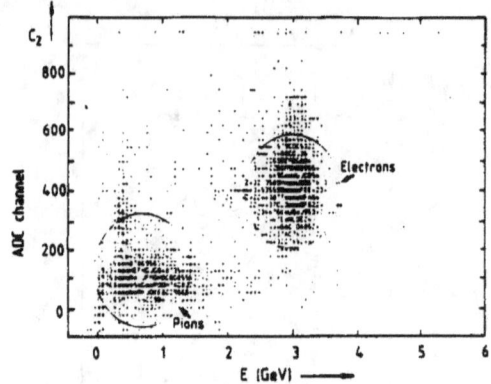

Fig. 5. Correlation between pulse height in Cherenkov counter C_2 and the energy measured in the calorimeter, for a 3 GeV/c beam.

with a 5 mm diameter hole in the center. P_1 was 2 × 2 cm². P_3 was always part of the trigger, while the signals from P_1 and P_2 were recorded on TDCs in order to select at the analysis stage the events well centered in the detector.

Downstream of the calorimeter, a 1 m long iron block stopped all the particles except muons. It was followed by two 20 × 20 cm² scintillators μ_1 and μ_2, used for tagging muon events.

Two types of data were taken: muons for calibration and electrons (with a controlled percentage of pions). The muon trigger was a coincidence of $P_3 \cdot \mu_1 \cdot \mu_2$ and the electron trigger was $P_3 \cdot P_1 \cdot C_1$. It was also possible to rotate the calorimeter by 90° and take data on vertical cosmic rays with a special two-counter trigger. This setup was used to monitor the gain stability over long periods, when the beam was off.

2.4. The acquisition electronics

Our apparatus included 256 electronic channels for the cathodes and 14 for the anodes. For the cathodes

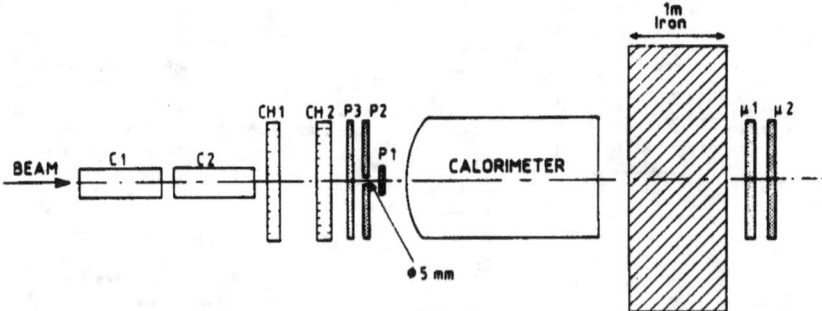

Fig. 4. Setup of our experiment in the test beam.

were used the same amplifiers and line receivers as in ref. [17] the only modification being an attenuator, by a factor of 60, to handle higher energies. Each line was recorded on one ADC. The dispersion on the gains was ± 10%, and the sensitivity was 400 mV/pC.

The 14 anode channels were improved to handle high rates up to 250 kHz and to enable timing measurements. Each channel consisted of a charge amplifier, located on the detector, followed by a 6 dB attenuator on some lines. Then the signal was split into two directions. The first one was shaped and fed into a constant fraction discriminator (Ortec 934) before going to a TDC. The second one was shaped for charge measurement, then fed into an attenuator box and finally into the ADC. Two attenuations were possible, corresponding to two types of data collection, muons or electrons. The attenuation values had to be different because the electrons deposited more energy that the muons. Also each chamber plane, relative to the others, had different amounts of energy deposited. We could compensate for the gain differences, and the longitudinal shape of the electron shower. Table 1 gives the different values of the resistors, attenuations and crystal thicknesses for all the planes. The method to determine these values will be described in the section dealing with calibration.

The total length of the signal in the MWPC was about 3 μs, including the drift of positive ions. We used a 600 ns gate on the ADCs, which was enough to include the maximum of the signal. The data acquisition was performed with a CAMAC system connected to a PDP 11/24 from Digital Equipment Corp. The behaviour of the anode channels at rates up to 250 kHz was tested with a random pulser: no significant shift was measured in the ADC data. We found an equiv-

alent noise between 100 keV and 2 MeV on the fourteen channels, the average being equal to 700 keV.

3. Energy measurements

3.1. Calibration

As mentioned in section 2.2, each plane of the calorimeter had its own high voltage power supply. We adjusted the voltages so that there were no ADC overflows on the cathode channels during a 5 GeV/c data collection run with electrons. Their range was distributed between 305 and 330 V, corresponding to a gain of the low pressure MWPC close to 4×10^3. Then we determined the value of the attenuators for the anode channels, which were different for electron and muon runs (table 1), to get a reasonable signal in the ADCs. When the beam momentum was higher than 5 GeV/c, all the fourteen high voltages were lowered by 10 V, in order to decrease the gain by a factor of 2.

The muon signal was very clean, even for the thinnest 1 cm crystal, as can be seen in fig. 6 showing the histogram of the pulse height of one muon, leaving a 6.5 MeV ionization energy. The asymmetry coming from the Landau distribution is visible. For this reason all the energy resolutions were measured in full width at half maximum (FWHM), and then converted into variance. Here we find a value $\sigma(E)/E = 28\%$. This energy resolution is good enough to use the position of the muon peak p_i in each crystal to calibrate the calorimeter. The energy given in the plane i by an electron E_i can be

Table 1
Values of resistors and attenuations for the 14 planes

Plane no.	Resistance		Attenuation	
	R_μ [kΩ]	R_e [kΩ]	Att$_\mu$	Att$_e$
1	0	1.96	1	20.6
2	0.3	1.96	4	20.6
3	0.1	3.83	2	39.3
4	0.1	3.83	2	39.3
5	0.1	3.83	2	39.3
6	0.1	3.83	2	39.3
7	0.1	3.83	2	39.3
8	0.1	3.83	2	39.3
9	0.1	3.83	2	39.3
10	0.3	3.83	4	39.3
11	0.3	3.83	4	39.3
12	0.3	3.83	4	39.3
13	0.681	1.96	7.81	20.6
14	0.3	1.96	4	20.6

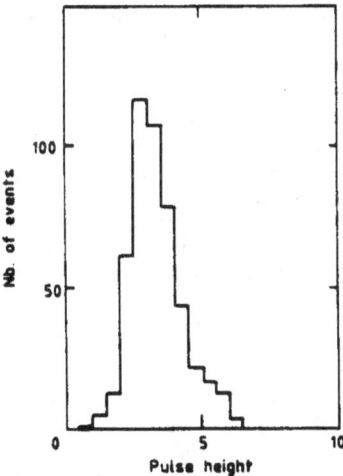

Fig. 6. Pulse height delivered in a 1 cm thick crystal by minimum ionizing muons, corresponding to an energy equal to 6.5 MeV.

calculated, knowing the crystal width L_i, the attenuation factors for the electrons A'_e and for the muons A'_μ, by the equation:

$$E_i = 6.5 L_i A'_e \mathrm{ADC}_i / A'_\mu p_i.$$

where ADC_i is the electron signal measured by the ADC connected to plane i. The total energy deposited in the calorimeter by an electron shower is simply the sum of the fourteen E_i's calculated by this method. The procedure is correct only if the response for muons and electrons is the same. It is known that this assumption is not true in sampling calorimeters [19] due to transition effects. However our apparatus is, in fact, a homogeneous calorimeter, since the MWPC is almost transparent to the shower and is expected not to suffer from such a problem. This point will be discussed in section 3.3.

The operating mode for taking data was the following: each electron run at any energy was preceded and followed by a muon run at 5 GeV/c, with the same high voltages, for the purpose of calibration and to check the gain stability. Several thousand events were recorded, giving a statistical error on the gain $\approx 0.5\%$. The main error came from the uncertainty on the determination of the peak position p_i, equal to the width of one ADC channel. It is estimated to be $1\% \pm 0.5\%$.

3.2. Gain stability

Monitoring the gain, and trying to keep it as stable as possible, is extremely important for a detector aiming at an energy resolution of the order of a few percent. A qualitative check was made during long periods with cosmic rays, the calorimeter being in the vertical position. We tested the continuous flowing system developed for this experiment, and could maintain the apparatus in working order for several months without any pumping or cleaning.

During the beam periods we collected muon calibration runs with 5000 events each. The fluctuations of the peaks for the fourteen crystals are shown in fig. 7 as a function of the time. They stay within a few percent and no general shift is detectable in 34 h.

A more precise determination was performed with a 120 000 events data collection lasting 12 h with 5 GeV/c electrons. The results are shown in fig. 8, each point is the energy peak for the whole calorimeter, calculated with 1000 events. The precision is much better than in the other measurements since the energy resolution is $\sigma(E)/E = 1.7\%$ for 5 GeV/c electrons; we can then observe short time fluctuations. One can see a modulation with a period of 20 min and an amplitude of the order of 2%. There exists another phenomenon having the same period in our setup: the heating cycle of the gas tank. This temperature was not as well regulated as the one of the liquid TMAE bottle. Also, for technical

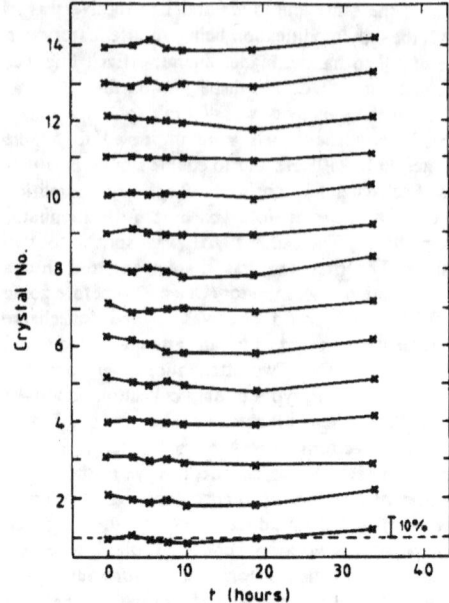

Fig. 7. Relative variation with time of the mean pulse height measured for muons in each of the fourteen crystals.

reasons the heat source was located around the aluminium cylinder and had to diffuse into the gas, where the thermistance was located which drove the thermostatic loop. Probably the temperature fluctuations of a few degrees produce modifications of the pressure and outgassing through the walls of the calorimeter, which are impossible to control. To minimize this effect in future devices, it would be useful to heat from inside and to control the temperature in and out of the tank, reducing the duration of each heating cycle.

3.3. Energy resolution and linearity

In the previous measurement [17] with 108 and 200 MeV electrons the energy resolution was $\sigma(E)/E =$

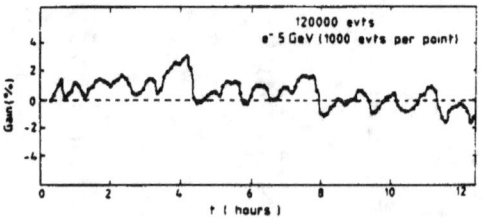

Fig. 8. Relative variation with time of the mean pulse height measured for 5 GeV/c electrons in the calorimeter.

12.7% and 8.5% respectively. In the experiment described in this paper a measurement in the same region of energy was performed with minimum ionizing muons, which deposit 263 MeV in the 40.5 cm total depth of BaF_2. We obtain 10.6% (fig. 9). Fig. 10 shows the energy spectrum given by 9 GeV/c particles. One can clearly distinguish the peak of minimum ionizing particles (π and μ), interacting pions giving a wide spectrum and the electron peak centered near 8 GeV, with a $\sigma(E)/E$ = 1.7%. A portion of the shower energy (≈ 1 GeV) leaked because of the finite size of the detector. A simulation using the EGS [21] program predicted a loss of 982 MeV through the sides of the calorimeter, and 82 MeV at the back. Thus the longitudinal leaks are negligible (less than 1%) for every energy of our test; on the other hand, the lateral leaks are very important. They were calculated by EGS, and the result displayed in fig. 11 shows that they are around 11–12% for incoming energies greater than 500 MeV. All the results presented in the following will be corrected for this effect.

The linearity of the calorimeter between 1 and 9 GeV is indicated in fig. 12. This picture demonstrates the perfect linearity of the apparatus in our energy range and as a consequence the linearity of the fast component of BaF_2. Measurements of the good linearity of a barium fluoride calorimeter have been reported between 2 and 40 GeV [20] for the slow component detected by silicon photodiodes. Our muon data at 263

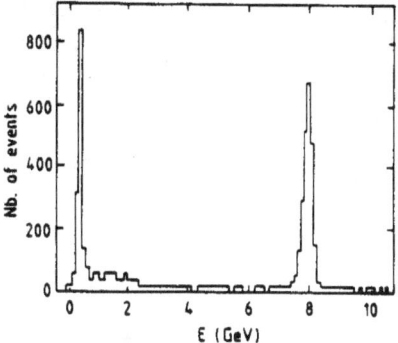

Fig. 10. Energy spectrum measured for 9 GeV/c incident particles in the calorimeter.

MeV are also plotted and fit correctly with the straight line of electrons. This confirms the assumption that muons and electrons give the same response.

Table 2 gives the measured energy resolution for different energies. It indicates the contribution coming from leaks, calculated by EGS, and the contribution from the beam momentum spread. The two last lines give the corrected resolution with its uncertainty. These results are summarized in fig. 13, the curve represents an energy dependence $\sigma(E)/E = 3.9\%/\sqrt{E}$. This determination was made after a cut in the total energy spectrum, to remove the electrons from the hadrons and muons. Additional cuts using the Cherenkov counters or the centered events traversing the 5 mm hole in P_2 did not significantly improve the resolution.

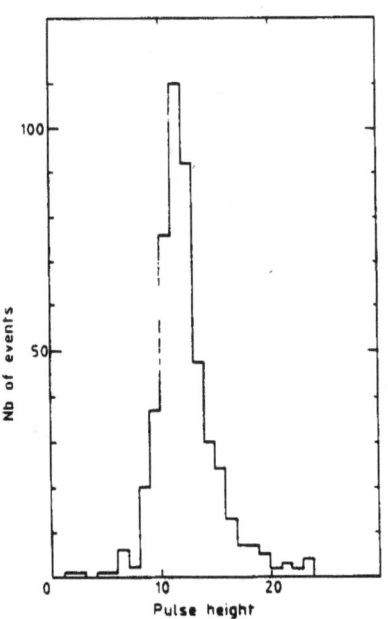

Fig. 9. Energy spectrum for muons depositing 263 MeV in the calorimeter.

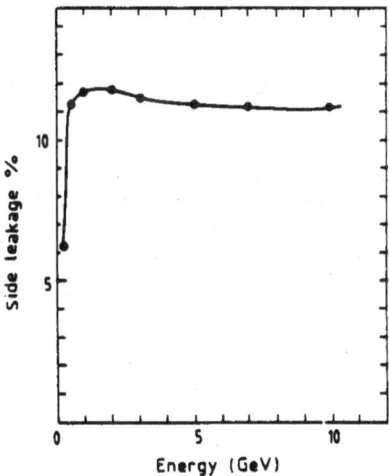

Fig. 11. Proportion of the energy of an electron shower lost by side leakage in our detector, simulated by the EGS program.

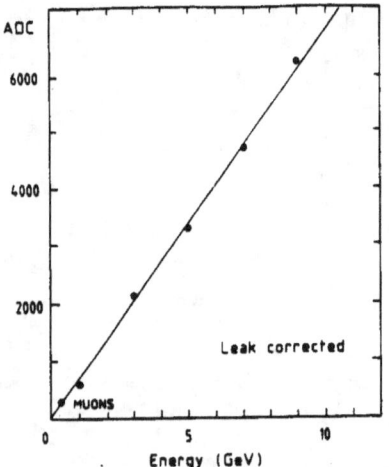

Fig. 12. Linearity response of the calorimeter for electron energies ranging from 1 to 9 GeV, corrected for side leakage. Minimum ionizing muons are also plotted.

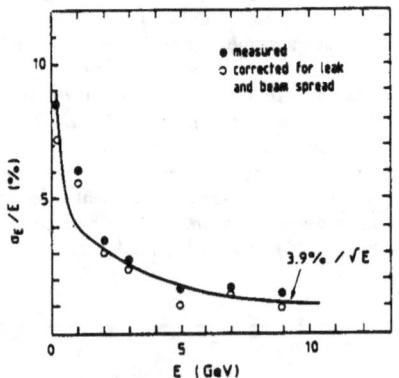

Fig. 13. Energy resolution as a function of the incoming energy.

Let us estimate some effects which contribute to the energy resolution. Some are energy independent, as for example the calibration, giving 1%, but most of them decrease with energy. The photoelectron statistics is an important factor since the quantum efficiency of the

BaF_2–TMAE coupling is small (≈ 10 photoelectrons/ MeV). Its contribution is estimated as:

$$\sigma(E)/E = \alpha/\sqrt{E},$$

where:

$$\alpha = \sqrt{\left(\sigma_p/p\right)^2 + \left(\sigma_g/g\right)^2},$$

following ref. [11]; p is the mean number of photoelectrons per GeV (≈ 10000) and g is the average gain in the chambers. The first factor follows a Gaussian law: $\sigma_p^2 = p$. The second one is due to the gain fluctuations in the chambers and is estimated to be of the same order as the first one. Then

$$\alpha = \sqrt{2}/p$$

and

$$\sigma(E)/E = 1.5\%/\sqrt{E} \ (\text{GeV}).$$

Let us now estimate the contribution coming from the direct ionization by the shower particles traversing the low pressure MWPC. As far as we know, no measurement exists of the total ionization in an isobutane–TMAE gas mixture. But these data are well known for pure isobutane: 195 ion pairs/cm atm [22]. The number of ion pairs n_s produced by the crystal scintillation is derived from the measured quantum efficiency (10 photoelectrons/MeV), and dE/dx (6.5 MeV per cm on BaF_2). For a 1 cm crystal we expect $n_s = 65$ ion pairs in the 12 mm sensitive depth of the MWPC. If we assume a total ionization equal to $n_i = 1$ ion pair in the same volume, for 1.4 Torr of TMAE added to 0.6 Torr of isobutane, we finally get the ratio:

$$R = n_i/n_s = 1/65 = 1.5 \times 10^{-2}.$$

This is an upper limit of the "pollution" from direct ionization because the crystals are 1, 2.5 or 5 cm. The typical resolution of a sampling calorimeter with MWPCs measuring the shower by ionization is $20\%/\sqrt{E}$ (GeV). Then the contribution of ionization to the total resolution is:

$$\sigma_i(E)/E = 0.2 \ R/\sqrt{E}$$

or

$$\sigma_i(E)/E = 0.003/\sqrt{E}.$$

which is much lower than the measured $3.9\%/\sqrt{E}$.

The electronic noise was found on average to equal

Table 2
Measured energy resolutions for different energies

E [GeV]	0.108	0.2	1	2	3	5	7	9
$(\sigma_E/E)_{meas}$ [%]	12.7	8.5	6.1	3.5	2.7	1.7	1.7	1.5
$(\sigma_E'/E)_{leak}$ [%]	5.9	4.7	2.03	1.44	1.62	0.97	0.82	0.72
$(\sigma_E/E)_{beam}$ [%]	0	0	0.85	0.85	0.85	0.85	0.85	0.85
$(\sigma_E/E)_{corr}$ [%]	11.2	7.1	5.7	3.1	2.4	1.1	1.5	1.0

700 keV, and its contribution is negligible for energies higher than 1 GeV (less than $0.1\%/\sqrt{E}$). We estimated the gain fluctuations at a level of 2% and this contributes to a small amount: $2\%/\sqrt{12} = 0.6\%$.

If we add quadratically the five contributions discussed above, we find for 1 GeV a value $\sigma(E) = 2\%$. The difference with the experimental data may be caused by unestimated contributions, such as, for example, some nonuniformity between the center and the edge of crystals.

4. Position measurements

4.1. Position resolution

The two drift chambers Ch_2 and Ch_2 measured the horizontal and vertical position of the incoming particle, X_1, X_2, Y_1, Y_2, respectively, with a precision of 100 μm. D represents the distance between the two drift chambers, and L the distance between Ch_2 and the average maximum of the electronic shower. One can deduce the lateral coordinates of the center of the shower, taking into account the angle of each beam particle, by the formula:

$$X = X_2 + (X_2 + X_1)L/D,$$
$$Y = Y_2 + (Y_2 + Y_1)L/D.$$

The angular spread of the beam was momentum dependent. It was non-negligible at 1 GeV/c, but was almost zero for higher momenta. In that case we, simply took $X = X_2$ and $Y = Y_2$.

In each MWPC plane we calculated the centre of

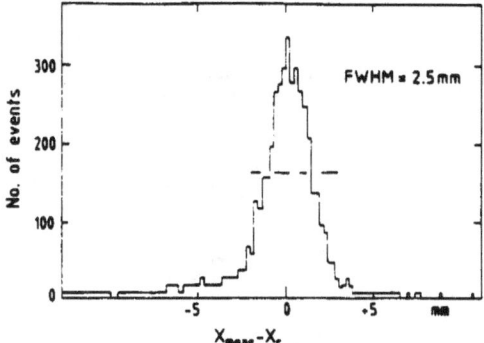

Fig. 15. Difference between the positions of a 9 GeV electron measured by a drift chamber and by the calorimeter.

gravity of the cathode signals to get the position of the shower centre X_i, Y_i. There were insufficient data to calculate an individual calibration for each cathode strip, but we checked a posteriori the absence of large fluctuations by the linearity of the position measurement and by the regularity of the lateral profiles (next subsection). Then we determined the X_c and Y_c coordinates measured by the calorimeter by the formulas:

$$X_c = \sum (X_i/\sigma x_i^2)\Big/\sum (1/\sigma x_i^2),$$
$$Y_c = \sum (Y_i/\sigma y_i^2)\Big/\sum (1/\sigma y_i^2),$$

where σx_i (σy_i) were the resolution measured in the plane i for the x (y) coordinate. These weights were chosen so as to minimize the final error in X_c and Y_c. Fig. 14 shows the correlation between X_2 and X_c for electrons at 9 GeV/c. We can observe a good linearity in a range of a few cm, indicating that the relative gains exhibit no troublesome fluctuation. Fig. 15 is a plot of the resolution distribution $X - X_c$ at the same energy; its FWHM is equal to 2.5 mm. Fig. 16 summarizes the

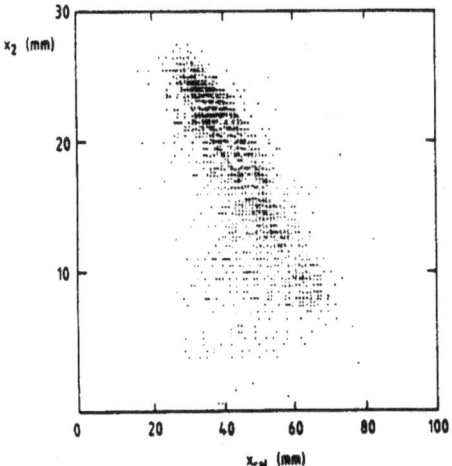

Fig. 14. Correlation between the position of a 9 GeV electron measured by a drift chamber in the beam and by the center of gravity of the shower in the calorimeter.

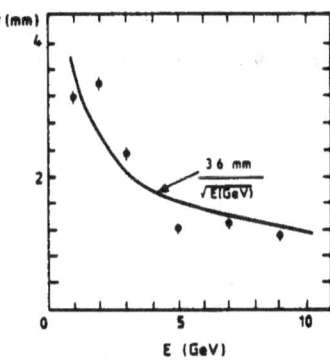

Fig. 16. Position resolution as a function of the incoming energy.

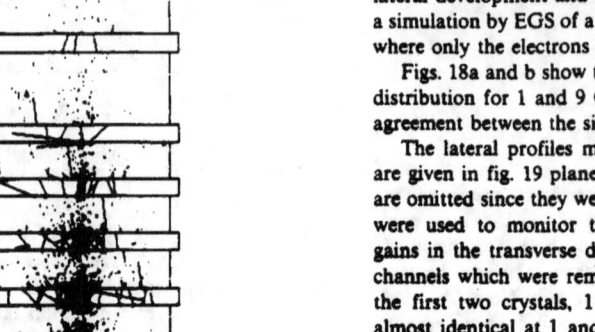

Fig. 17. Simulation of a 9 GeV electromagnetic shower in the calorimeter, displaying electron and positron tracks.

decreases from 3.5 mm at low energy, but saturates at 1 mm after 5 GeV.

4.2. Imaging the shower development

The fourteen plane segmentation of the apparatus enables an easy visualization of the longitudinal shower development, as in a sampling calorimeter. Moreover the detection by the MWPCs gives information on the lateral development and on the position. Fig. 17 shows a simulation by EGS of a 9 GeV electromagnetic shower where only the electrons and positrons are represented.

Figs. 18a and b show the relative longitudinal energy distribution for 1 and 9 GeV showers. We find a good agreement between the simulation and the data.

The lateral profiles measured by the cathode strips are given in fig. 19 plane by plane (the last two planes are omitted since they were not segmented). These plots were used to monitor the relative calibration of the gains in the transverse direction: two planes had dead channels which were removed from all calculations. In the first two crystals, 1 cm thick, the energy loss is almost identical at 1 and 9 GeV because the electrons interact on the average after one radiation length. The effect on the incoming energy becomes visible after these layers. The comparison with the profiles of 7 GeV/c muon events (fig. 20) gives an idea of the imaging possibilities of this technique.

The separation capability for two showers simultaneously present in the detector (for example two photons from a π^0) can be evaluated by the width in the first planes. It is equal to ≈ 1 cm (FWHM) in the first two planes and to ≈ 2 cm in planes 3 and 4. In fact the separation is probably better since these numbers include the beam width (1 cm at 9 GeV/c) and divergency.

4.3. Hadron versus electron rejection

results from 1 to 9 GeV, plotted in standard deviation, together with a curve $\sigma = 3.6/\sqrt{E}$ (mm/GeV). The resolution does not follow precisely this parametrization; it

In an electromagnetic calorimeter the main source of hadronic background is the interaction $\pi^- p \rightarrow \pi^0 n$ (or $n^+ n \rightarrow \pi^0 p$), which can simulate an electronic shower in the detector. These events can be rejected by two methods:

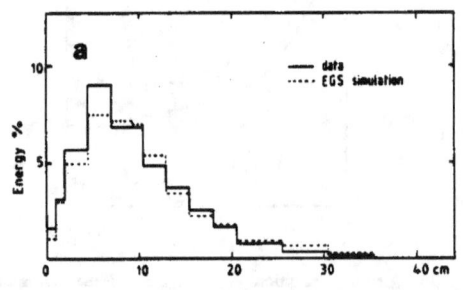
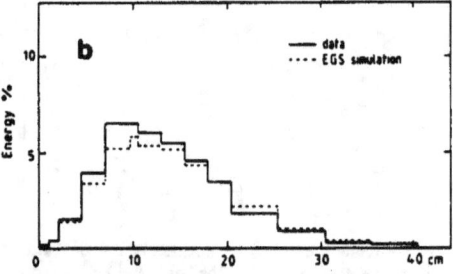

Fig. 18. Relative longitudinal energy distributions for 1 and 9 GeV showers.

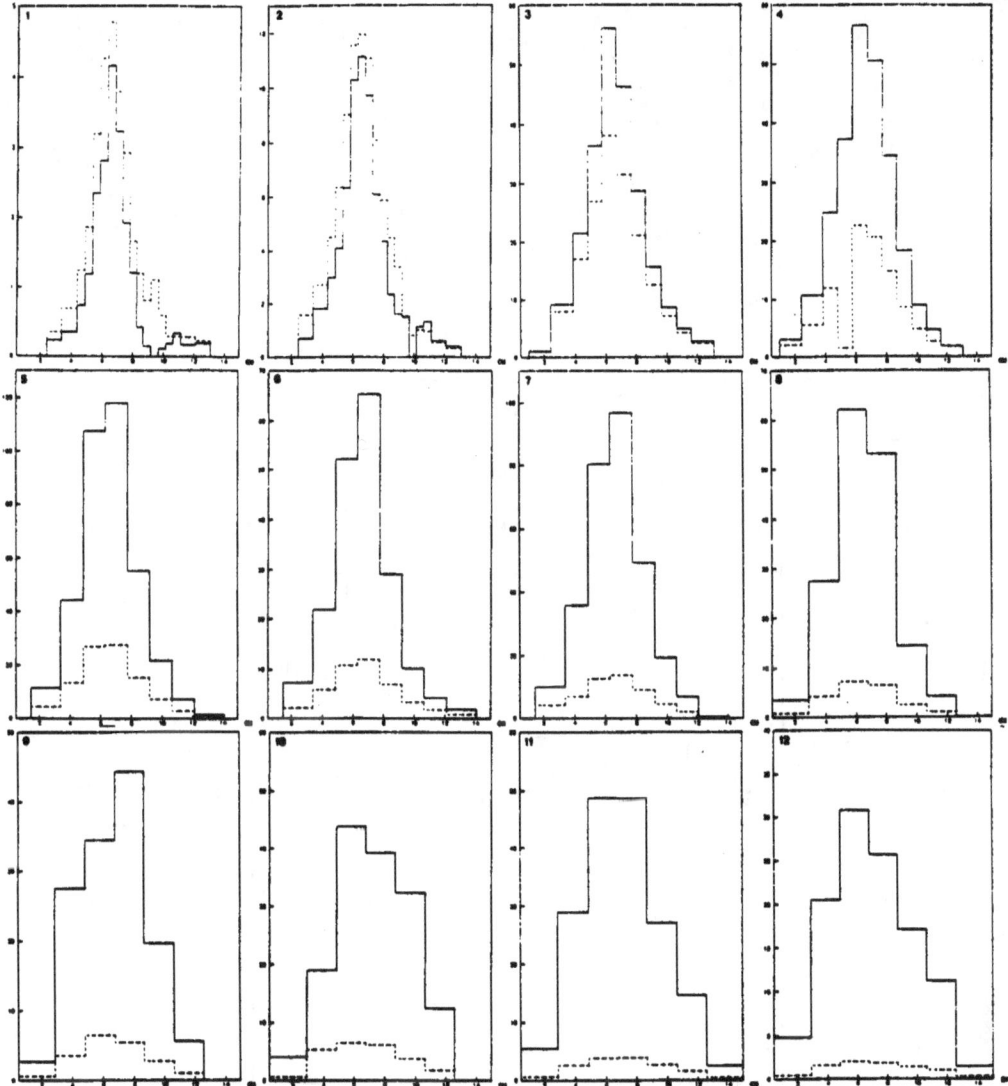

Fig. 19. Lateral energy distribution for 1 and 9 GeV/c electrons.

(a) If the detector has a good energy resolution and if the incident particle energy is known, a cut on this variable can reject the hadrons with a good precision.

(b) If the detector is well segmented longitudinally and laterally, the position of the interaction and the shape of the shower can be used to discriminate electrons and hadrons. In BaF$_2$ the radiation length is $X_0 = 2.05$ cm and the nuclear interaction length is $\lambda = 50$ cm. Of course a better rejection factor is obtained by combination of these two methods.

To study this problem we started with a data sample of 140 000 events at a momentum of 5 GeV/c. A clean sample of 2549 pions was extracted by a cut on the first Cherenkov counter signal, on which we applied criteria (a) and (b).

(a) A first energy cut consisted in the elimination of the events for which the measured energy E_m represented less than 70% of the peak energy of the electrons E_e: $E_m < 0.7E_e$. This cut eliminated 0.7% of the elec-

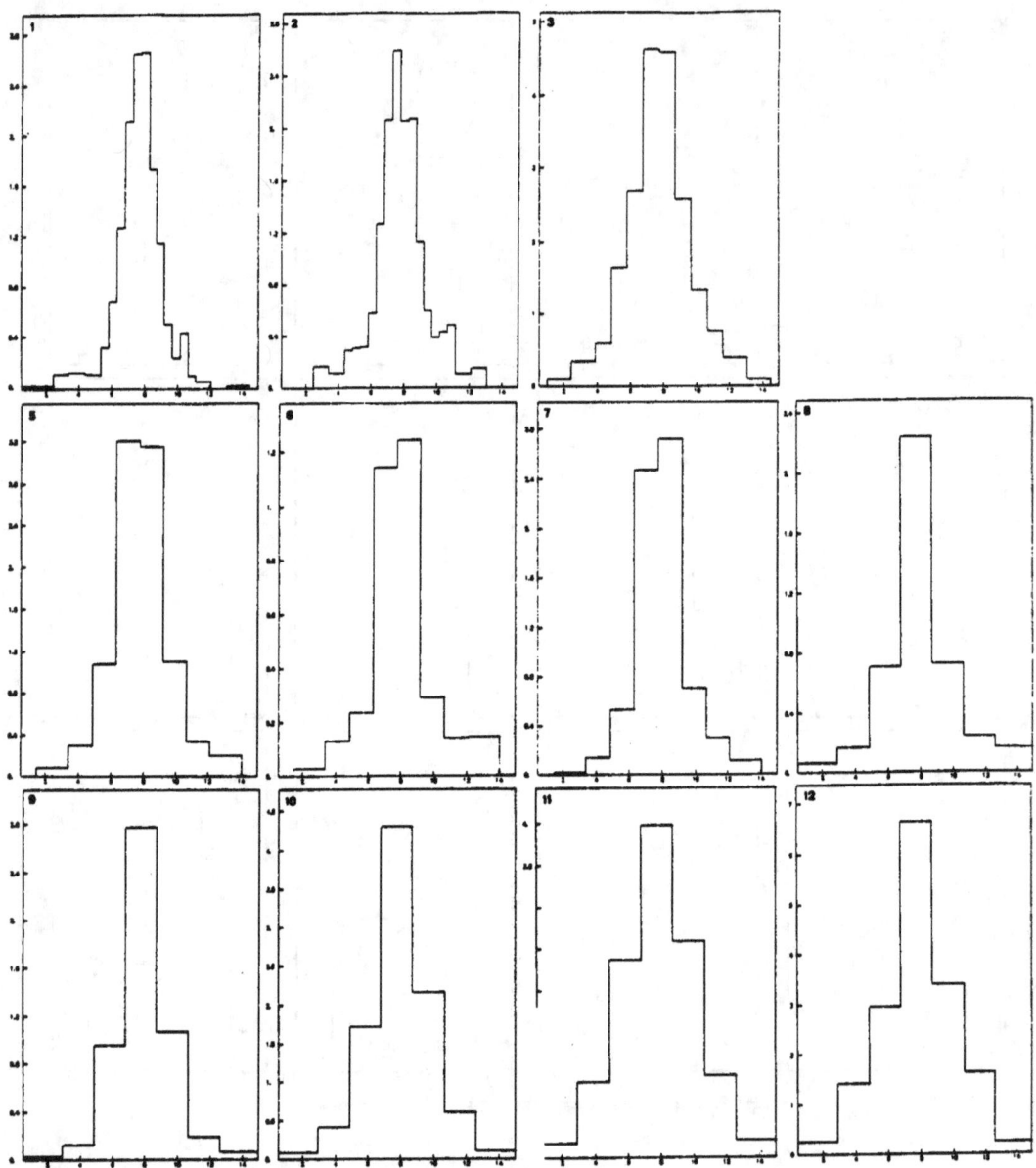

Fig. 20. Lateral energy distribution for 7 GeV/c muons.

trons and left us with 10 pion candidates which could simulate electrons out of the 2549. After a cut on the second Cherenkov counter C_2, to correct for C_1 inefficiency, the remaining numbers were respectively 7 and 2480. Thus the value of the rejection factor is:

$$r_{e/\pi} = 7/2480 = 2.8 \times 10^{-3}.$$

This energy cut was rather loose. With a sharper selection, for example $E_m < 0.85E_e$, 2% of the electrons were eliminated, but only 6 candidates remained. After a cut on C_2 we were left with 3 candidates out of 2480, and the new rejection factor was:

$$r_{e/\pi} = 3/2480 = 1.2 \times 10^{-3}.$$

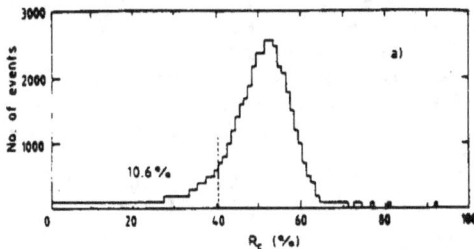
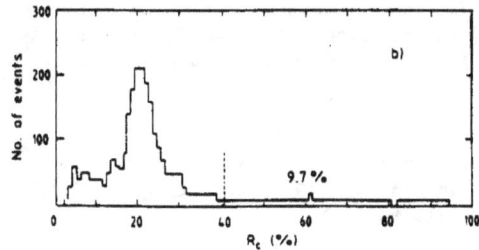

Fig. 21. Distribution of R_c, the ratio of the energy deposited in the center of the detector, over the total energy seen, for 5 GeV/c electrons (a) and pions (b).

(b) For the second cut, we used only the longitudinal information to discriminate pions and electrons. The lateral development used in some analyses [19] would certainly improve the result, but needs a detailed simulation and was jeopardized by the limited diameter of our detector.

(b1) In a simplified analysis we considered three parts in the calorimeter. The first included the four crystals at the beginning, where the electrons deposit a fraction of their energy not very far from minimum ionizing particles. In the second part, made by crystals 5–8, the shower exhibits the highest energy deposition, whereas the pions still have a low interaction probability. The third part consisted in the last five thick crystals where only the tail of the electromagnetic shower is seen. In this last section it is generally difficult to distinguish between a pion and an electron. Let us define the ratio R_c of the energy deposited in the second part, over the total energy seen in the calorimeter. The value of R_c was calculated for each event of the pion sample defined above and for a clean electron sample (37761 events) selected by a cut on the C_1 signal. In fig. 21a the percentage of eliminated electrons is given by the proportion of events at the left of the cut, 10.6%. The pions simulating electrons are at the right of the cut in fig. 21b. Their number is 9.7%, which gives a rejection factor $r_{e/\pi} = 9.7 \times 10^{-2}$. Taking into account the second Cherenkov counter did not change these data. We tested other partitions of the detector into three sections: they all gave a worse result.

(b2) In another method, still based on the shape of the shower ignoring spectrometric information, we used the complete fourteen plane segmentation. For each event we compared the signal E_i in each chamber ($i = 1–14$), to the average signal for minimum ionizing particles $\langle E_\mu \rangle_i$. If $E_i < k \langle E_\mu \rangle_i$, where k is a parameter to adjust, the particle was considered to be minimum ionizing and was rejected as a candidate to simulate an electron. In the reverse case, we assumed that the particle interacted. This method not only eliminates minimum ionizing particles, but also determines the interaction plane. The electrons interact in the beginning of the calorimeter (90% in the first three crystals), whereas the pions interact along the whole length. Since the interaction point was known for each event, the rejection factor was simply obtained by dividing the number of pion events of this type in the first crystals, which could simulate electrons, by the total number in the pion sample.

Table 3 summarizes the results obtained for $2 \leqslant k \leqslant 8$, with a cut on the first two planes ($1 X_0$) and on the first three planes ($2.5 X_0$). It gives the percentage of pions eliminated by the minimum ionizing criterium, the rejection factor $r_{e/\pi}$ and the percentage of rejected electrons. One can see that the cut on the first two planes gives a good rejection factor, but at the expense of a high percentage of rejected electrons. The best compromise is reached with a cut on the three first planes and k between 4 and 6. For example, with $k = 5$, one gets $r_{e/\pi} = 5.4 \times 10^{-2}$ and 2.8% loss of electrons. An additional cut on the second Cherenkov counter did not change this result.

It is clear that the best rejection factor can be obtained by using simultaneously the two methods (a)

Table 3
Summary of results for $2 \leqslant k \leqslant 8$

k	2	3	4	5	6	7	8
Min. ionizing pions [%]	37.2	57.6	66.9	69.7	71.4	72.6	74.1
$r_{e/\pi}$ (cut on first two planes) [%]	13	6.2	4.2	3.3	2.6	2.4	2
Rejected electrons [%]	6	15.5	25.3	36.5	46	56.4	64
$r_{e/\pi}$ (cut on first three planes) [%]	19.5	9.3	6.6	5.4	4.8	4.4	4.1
Rejected electrons [%]	0.2	0.6	1.3	2.8	4.6	7	10

and (b), when the incoming energy is measured. For example the association of (a) and (b1), with an energy cut $E_m < 70\%$, eliminated 3 pions out of 7. The result is:

$$r_{e/\pi} = 4/2480 = 1.6 \times 10^{-3}.$$

After an energy cut $E_m < 85\% E_e$, no event remained. We can give the following upper limit, with a 90% confidence level:

$$r_{e/\pi} < 2.3/2480 = 9 \times 10^{-4}.$$

5. Time measurements

5.1. Time resolution

We measured the time resolution with electron data at 5 GeV/c on the anode channels. The best result was obtained by averaging planes 4 and 5, which corresponds to the position of the maximum deposition of energy by an electromagnetic shower. The mean was calculated with weights inversely proportional to the variance, to minimize the final error.

The main component of this resolution is the jitter coming from the fluctuation of the materialization position of the UV photons in the detector. If we neutralize

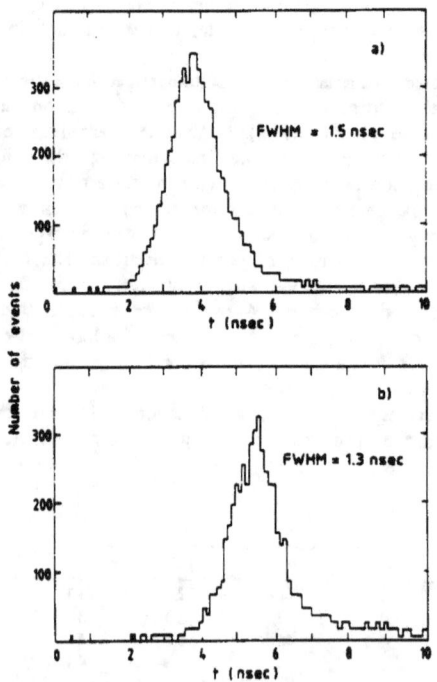

Fig. 22. Time spectra with drift space (a) and without (b).

the drift space by reversing the voltage (+ 30 V), we keep only the electrons produced in the MWPC. Thus the time resolution is improved, but the energy resolution is degraded, because a factor of 7 is lost in the signal height. We measured the time resolution in two configurations: (1) with the drift space. (2) without the drift space.

Figs. 22a and b show the results obtained in the two cases: 1.5 ± 0.1 ns and 1.3 ± 0.1 ns (FWHM). After the substraction of the jitter due to the trigger counter P_3, equal to 1.2 ns FWHM, we are left with the following results, expressed in standard deviation:
(1) $\sigma_1 = 400 \pm 200$ ps,
(2) $\sigma_2 = 170 \pm 130$ ps.

5.2. High rate capability

An important feature for an apparatus foreseen to operate in high intensity beams is its ability to work at high rates. Three parameters have to be considered in this problem:
(1) The occupation time, defined as the duration of the physical signal caused by one particle traversing the detector.
(2) The integration time, determined by the electronics as the time the gate is opened to measure the signal.
(3) The timing resolution, which is the precision of the determination of the impact of the particle.

Our prototype was tested at 5 GeV/c with a counting rate, measured by P_3, varying from 6 to 800 kHz. With the experimental working conditions described in sections 2.2 and 2.4, we had the following values for the three parameters:
(1) The occupation time was equal to the drift time of the positive ions in the MWPC, i.e. 3 μs.
(2) The integration time was chosen to be 600 ns, to completely include all of the signal from the electrons.
(3) The time resolution was better than 600 ps, as shown in the previous section.

The first parameter comes from a physical process, intrinsic to gaseous detectors. The maximum theoretical rate that the chambers can handle at a given spot without a modification of their properties is consequently 3×10^5 events per second in a continuous flux. Beyond this value the accumulation of positive ions between the electrodes decreases the gain. The quantitative effect is affected by the actual signal processing, like differentiation. It has to be measured and compared to the expected resolution and linearity in energy. In fact the practical capability depends of the segmentation of the detector. The quoted number corresponds to a maximum rate per module. In our prototype, where only the fourteen anode channels were equipped with adequate high rate electronics, one module is a 15×15 cm^2 MWPC. For a detector segmented in strips having the size of an electromagnetic shower, say 3 cm, this number would be multiplied by a factor of 5. The best configuration would be to make all measurements using rectangular pads 3×3 cm^2, for which the limit is 8.5 MHz.

The second parameter gives a theoretical limit equal to 1.5×10^6 events per second. If two events are separated by less than 600 ns, there appears a pileup effect which caused a dispersion of the measured signals in the ADCs. As before, the effect depends of the segmentation of the calorimeter. The quoted rates concern an ideal case where the events are regularly spaced in time. One should also take into account the Poisson statistics and the arrival of the beam particles in bunches.

For the third parameter, it is clearly not a limitation at high rate since it is several orders of magnitude smaller than the two others.

Our experimental study consisted in the relative measurement of the gain for electrons and muons on the fourteen anode planes, as a function of the counting rate. The beam had a mixture of electrons, hadrons and muons in variable proportion, depending upon the momentum, which was measured with the Cherenkov counters. The important cause of variation being the space charge effect in the MWPCs, the relevant parameter to characterize the rate is then the average energy deposited in the calorimeter per event. This number varied with the nature of the particles circulating in the PS (proton or deuterons). For deuterons it was equal to 0.4 GeV, and for protons 1.1 GeV. Fig. 23 gives the result as a function of the number of "minions" per seconds (equivalent minimum ionizing particles). Two measurements were made using electrons, with and without drift space. Detection of minimum ionizing muons was not possible without the drift space, because the signal was small. The space charge effect is expected to be more pronounced in the drift space, where it competes with a field lower than in the MWPC.

Fig. 23 demonstrates that without the drift space one can tolerate higher counting rates without a substantial loss in gain. In fact, in that configuration, the amount of signal is decreased by a factor of 7 in the MWPC, the photons which materialized in the 3 mm of drift space went undetected. It is interesting to plot the same gain measurement as a function of the intensity per wire unit of length. This is done in fig. 24. This intensity I is calculated for each chamber plane by the formula:

$$I = Q E_{dep}/A N E_e,$$

where E_{dep} is the energy deposited in the calorimeter per second, E_e is the beam energy (5 GeV), N is the number of wires per unit of length in a MWPC and A is the area of a shower estimated from EGS and the measured profiles (2800 mm^2 on average). The charge Q is defined by:

$$Q = \text{ADC Att } N_{pc}/G,$$

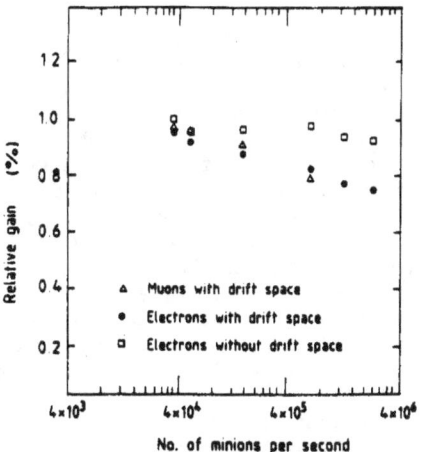

Fig. 23. Relative gain as a function of the number of minimum ionizing particles per second.

where ADC stands for the ADC output of that plane, Att for the attenuator value, N_{pc} for the number of picocoulombs per channels ($= 0.25$) and G for the electronic gain ($= 256$). The error on I comes essentially from the estimation of A. For the value I we use the average on the fourteen planes of the calorimeter, the signals being roughly equalized after the tuning of the attenuators and the MWPC high voltages. Fig. 24 shows that the gain decline begins around 10^{-10} A/mm and reaches 25% for $I \approx 10^{-9}$ A/mm in configuration (a). The common range for (a) and (b) is too narrow to conclude for a real discrepancy between the behaviours in the two conditions.

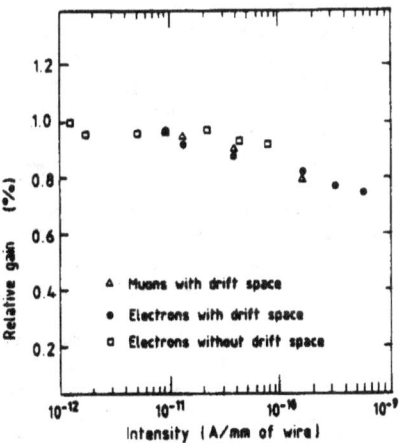

Fig. 24. Relative gain as a function of the intensity per unit of length in the wires of the MWPCs.

6. Discussion

6.1. Specific properties of this calorimeter

The tests performed between 1 and 9 GeV/c with this prototype calorimeter demonstrate that the low pressure continuous bubbling (≈ 2 Torr) and the precise temperature controls (40 and 45°C) improve the ease of operation. Continuous working over long periods is now possible with high stability for the MWPC gain, comparable to the precision in energy.

The energy resolution of $3.9\%/\sqrt{E}$ is of the same order as the performance of a good homogeneous calorimeter, lead glass for example. The result found by the authors of refs. [10,11] who used a single BaF$_2$ crystal to measure minimum ionizing particles, is in agreement with ours: they find $3.5\%\sqrt{E}$ for a deposition of 67.8 MeV. Our longitudinal segmentation of fourteen crystals has probably deteriorated the energy resolution, but gives the possibility to achieve a position resolution and an electron/hadron separation comparable to specialized sampling calorimeters.

The position resolution obtained is equal to 3.1 mm at 1 GeV/c and 1.2 mm at 9 GeV/c. The e/π rejection is $\approx 5 \times 10^{-2}$ at 5 GeV/c, without using the energy information. The lateral shape of the shower was unused in our rather simple analysis. It is known [19] that the addition of this information improves the rejection by a factor of 3–5. Higher energies are also expected to be more favourable, both for the position resolution and for the electron/hadron separation. For this last point one order of magnitude has been quoted at 100 GeV. This type of calorimeter should reach an e/π rejection factor of the order of 10^{-3} for energies higher than 100 GeV. The good energy resolution, combined with a good rejection factor, which can be obtained without magnetic field, makes this device particularly attractive for the TeV region. We have demonstrated in another study [13] that crystals as small as 5×5 mm^2 area can be used efficiently to detect low energy nuclear sources. We think that the imaging capability of the calorimeter can even be improved by a choice of small parallelepipeds instead of large cylindrical monocrystals, as discussed in section 4.4.

The timing resolution is better than 600 ps, and can reach 300 ps when the drift space is neutralized. Woody et al. [11] find the same effect: their best result (800 ps FWHM) is obtained with negative bias, i.e. without drift space in our notation. The improvement of the timing by the suppression of the drift space is paid for by a loss in detected light and consequently by a deterioration of the energy resolution. One can imagine another configuration, in which the MWPC is directly in contact with the crystal. The vapour pressure of TMAE can be increased by heating at 55°C, to get a mean free path of the photons ≈ 2.5 mm. The probability of photoelectric conversion in the chamber alone is then 90%, equal to the value at 40°C in the present configuration. However, the energy resolution is probably worse since the electron collection mechanism is different. In the actual geometry most of the electrons are produced in the drift space and they are amplified identically by the MWPC. In the other design all of the electrons would be produced in the chamber, where the gain varies with the position. Those produced near the anode wires are less amplified. This effect, which exists certainly at a low level with the present design, would introduce fluctuations in the energy measurement and would be the cause of a deterioration in energy resolution.

6.2. The high rate problem

As pointed out in the section 5.2, the high rate capability of a detector increases with its segmentation. For an electromagnetic calorimeter it is useless to go further than the size of the electromagnetic shower.

The gain loss ($\approx 25\%$ at 10^{-9} A/mm) could probably be minimized by suppressing the drift space, although our measurement (fig. 20) is not conclusive. We would then choose the configuration without the drift space, which gives the best resolution in time, at the expense of the energy resolution.

The experimental working conditions were chosen to have the highest gain in the MWPCs. The high voltage was set in each plane so that low energy depositions could be seen in the cathode strips. Attenuators were used for the anode signal for both muon and electron data collection. Another method, more delicate, could be to operate with different voltages for calibration and for normal data taking, without attenuators. It is clear that limiting the gain of the chambers could decrease the space charge effect by one order of magnitude and optimize the detector for high rates.

An improvement in electronics would probably improve the high rate behaviour. Higher sensitivity amplifiers would allow a lower gain in the chambers. They would also, with a more adapted shaping of the signal, allow a shorter gate for the ADCs compared to the 600 ns we used. A proportional reduction of the pileup would be expected.

6.3. Some other problems

The behaviour of an electromagnetic calorimeter in a strong magnetic field is an important property, since this type of detector is often placed close to the magnet in a colliding beam experiment. Recent measurements [23] exhibit a decrease in the gain of a SSPC in fields of a few kG, especially when the magnetic field is perpendicular to the drift electric field of the MWPC. The origin of this effect and its influence on the properties of the calorimeter certainly deserve a special study. The first planned application of a SSPC is a highly efficient veto counter working inside a magnet, for the AGS experiment 787 [11], designed to detect very rare kaon decays.

The exceptional radiation resistance of barium fluoride would be of little advantage if the wire chambers used to detect the light were destroyed by large fluxes of particles. The ageing due to polymerization is well known at atmospheric pressure, the maximum tolerated charge per unit length of a wire is on the order of 0.1–1 C/cm. Recent tests [24] made with TMAE and methane or isobutane give the very pessimistic limit of 10^{-4} C/cm. At low pressure the situation may be different, and traces of additives in the gas or in the material of the MWPC can dramatically change the result. We have not seen any measurable effect in our studies but we were far below the quoted numbers. For the moment this question is unsolved.

In addition to the scintillation light, there is some Cherenkov light which is produced by relativistic particles crossing the BaF_2 crystals. A simple evaluation shows that, taking account of the shorter wavelength of these UV photons and of the maximum sensitivity of TMAE at 150 nm, the numbers of photoelectrons produced by the two effects should be of the same order. An application of using this Cherenkov effect in gamma ray astronomy was proposed [25]. Yet the refractive index of BaF_2 is 1.56 and the Cherenkov angle equal to $38°$ is bigger than the limiting angle of $37°$. In principle Cherenkov photons emitted by particles running parallel to the axis of the crystals are trapped, in contrast with the scintillation photons emitted isotropically. Multiple scattering in the electromagnetic shower may change this result and it is possible than some proportion of the light we measured was in fact due to the Cherenkov effect. A high sensitivity to this additional source of signal would probably be disturbing since it can introduce nonlinearities and a variation of the efficiency with the angle, etc. An increase in light is not needed since we already have a lot of photons from the high energy shower. In any case a study of this effect by shooting beam particles at various angles through an SSPC would be useful.

6.4 Calorimetry for future accelerators

The calorimeters which will be used in future experiments on very high luminosity accelerators SSC or LHC have to satisfy a list of requirements, which we now discuss, following ref. [26].

(1) The speed of the detector is a crucial parameter if the expected luminosity is 10^{33} cm^{-2} σ^{-1} with 10^8 interactions per second. The delay between bunches is ≈ 15 ns. Our time resolution, $\sigma < 300$ ps, gives the possibility to distinguish events from different bunch crossings. It can even disentangle events coming from the same bunch crossing, within a 1 ns window, which otherwise would be superimposed.

(2) The energy resolution is important, not only for single particles, but also for jets and missing transverse momentum. The search for the Higgs boson, and new particles in the range of a few hundred GeV/c^2, requires a jet–jet mass resolution of a few percent. Our actual detector performs as good as most homogeneous electromagnetic calorimeters in this respect.

(3) The granularity of a calorimeter was initially requested for the identification of the electrons close to the jets. The quoted number for the angular separation, from the interaction point, is $\Delta\phi \approx 0.03$. We can have a very fine granularity with BaF_2 since long rods, with a 5×5 mm^2 area, are feasible and have been tested [13] at low energy.

(4) The absolute calibration must be kept and monitored within 1%. We have shown how to calibrate our detector by muons with a precision of 1%. The stability was of the order of 2% but it could be improved by a thermostatic system more precise that what we had during the test ($\pm 1.5°$C).

(5) The uniformity of the detector must be better than 1% so that it does not dominate the energy resolution at high energy. In a BaF_2 calorimeter this parameter is dependent on the variation of the amount of light emitted by each crystal and on mechanical tolerances on the MWPCs.

(6) A high density is required for compactness and the length of the calorimeter must be long enough to contain most of the electromagnetic shower. Although BaF_2 is the most dense inorganic scintillator after BGO, its density is not at the requested level (< 8 g/cm^3). A total crystal length of 40 cm represents 22 radiation lengths, and it does not include the thickness of the readout MWPCs. One possible solution is to separate the calorimeter into two parts. The first one, $(10–15)X_0$ long, would be homogeneous, as the actual prototype. It will contain 95% of the energy of the showers below 10 GeV and have the highest energy resolution. The second part would be a sampling calorimeter made from BaF_2, MWPCs and lead sheets. It will measure the tail of the high energetic showers. In this region the energy resolution is anyhow dominated by the crystal homogeneity and calibration, and can suffer a reduction of the light output. With this design it is possible to obtain more compactness, without disturbing too much the energy resolution.

(7) In consideration of the high luminosity of the accelerator it is mandatory for any detector to accept radiation doses up to 10^6 rad. BaF_2 is for the moment the only dense scintillator which stands such values. However, the behaviour of the low pressure MWPCs in these extreme conditions has to be investigated.

(8) A good linearity for events up to 5 TeV is a

useful property for any calorimeter. We have tested from 0.2 to 9 GeV, and another group [20] has tested, with a different readout scheme, to 40 GeV. No deviation from linearity was found for BaF_2.

(9) A good hermiticity, i.e. a minimum of dead regions in the calorimeter, is an important parameter for the detection of missing neutrinos. In a BaF_2 calorimeter the crystals, being nonhygroscopic, can be closely packed. No cryostat is needed in this technique and the dead space is due mainly to the passway of the cables for the MWPCs.

(10) A good e/π separation based on the pattern in a calorimeter can only be obtained by using both the longitudinal and lateral measures of the shower. Our simple analysis has shown the very good imaging capability of the detector and a reasonable expectation to reach $\approx 10^{-3}$ at 100 GeV.

7. Conclusion

We have investigated the performance of an electromagnetic calorimeter consisting of BaF_2 scintillators and wire chambers at a temperature of $40°C$. It is possible to operate this device continuously for several months at low pressure, as soon as the TMAE is renewed by a constant flowing gas system. Minimum ionizing particles are very clearly seen, with an energy resolution of $\sigma(E)/E = 10.6\%$. They are used for calibration and gain monitoring. A precise measurement of the gain fluctuations with the electrons shows an amplitude of 2%, the limiting parameter being probably our present thermostat regulation.

We find an energy resolution for electromagnetic radiation equal to $3.9\%/\sqrt{(E)}$ and calculate that it is mandatory to operate at low pressure, to avoid a degradation of this number by direct ionization of the gas in the MWPCs.

The position resolution for an electron shower, derived from the center of gravity method, is 3.5 mm at 1 GeV and 1 mm at energies higher than 5 GeV. The e/π rejection at 5 GeV/c is 5.4×10^{-2}, without using the incident energy information, and better than 9×10^{-4} when this information is used. Better performances are certainly feasible after a more refined analysis and taking into account the transverse shape. The technique enables a very good segmentation both longitudinally and laterally. Thus it combines the advantage of sampling calorimeters (imaging capability) with the characteristics of homogeneous ones (good energy resolution, homogeneity).

The time resolution we measured is better than 600 ps. With different working conditions we can obtain better than 300 ps, but the energy resolution is then degraded. A test of the calorimeter in a high intensity beam showed that the gain begins to decline by 25% for a rate of 4×10^6 minions per second, or 10^{-9} A/mm of wire. This rather good performance can probably be ameliorated by suppression fo the drift space and an upgrading of the electronics.

Clearly some questions are unsolved: behaviour in a magnetic field, ageing of the chambers due to time or irradiation, influence of the Cherenkov effect. However, a critical comparison of the characteristics of this technique, with the requirements of detectors for future colliders, already shows that the SSPC is a very good candidate for electromagnetic calorimetry.

References

[1] D.F. Anderson, Phys. Lett. B 118 (1982) 230.
[2] D.F. Anderson, R. Bouclier, G. Charpak and S. Majewski, Nucl. Instr. and Meth. 217 (1983) 217.
[3] D.F. Anderson, G. Charpak, Ch. von Gagern and S. Majewski, Nucl. Instr. and Meth. 225 (1984) 8.
[4] M. Laval, M. Moszynski, R. Allemand, E. Cormoreche, P. Guinet, R. Odru and J. Vacher, Nucl. Instr. and Meth. 206 (1983) 169.
[5] P. Schotanus, C.W.E. van Eijk, R.W. Hollander and J. Pijpelink, Nucl. Instr. and Meth. A238 (1985) 564.
[6] M. Suffert and G. Charpak, CERN-EP internal report 86-03.
[7] S. Majewski and D. Anderson, Nucl. Instr. and Meth. A241 (1985) 76.
[8] A.J. Caffrey et al., IEEE Trans. Nucl. Sci. NS-33 (1986) 230.
[9] A. Breskin, Nucl. Instr. and Meth. 196 (1982) 11.
[10] C.L. Woody, C.I. Petridou and G.C. Smith, IEEE Trans. Nucl. Sci. NS-33 (1986) 136.
[11] C.L. Woody, BNL experiment 787 technical note no. 120 (1986).
[12] P. Miné et al., IEEE Trans. Nucl. Sci. NS-34 (1987) 458.
[13] P. Miné et al., Symp. on Wire Chambers in Medical Imaging, Corsendonk, Belgium (1987).
[14] D. Scigocki, Thesis, Université de Savoie (1987).
[15] P. Schotanus, C.W.E. van Eijk, R.W. Hollander and J. Pijpelink, Nucl. Instr. and Meth. A252 (1986) 255.
[16] P. Schotanus, C.W.E. van Eijk, R.W. Hollander and J. Pijpelink, IEEE Trans. Nucl. Sci. NS-34 (1987) 272.
[17] D.F. Anderson, G. Charpak, W. Kusmierz, P. Pavlopoulos and M. Suffert, Nucl. Instr. and Meth. 228 (1984) 33.
[18] D.F. Anderson, IEEE Trans. Nucl. Sci. NS-28 (1981) 842.
[19] C.W. Fabjan, Calorimetry in high energy Physics, Techniques and concepts in high-energy physics, ed. T. Ferbel (Plenum, New York, 1985).
[20] E. Lorenz, G. Mageras and H. Vogel, Nucl. Instr. and Meth. A249 (1986) 235.
[21] W.R. Nelson et al., SLAC-265 (1985).
[22] F. Sauli, CERN-EP internal report 77-09.
[23] C.L. Woody and D.F. Anderson, Fermilab-pub 87/42.
[24] J. Va'vra, IEEE Trans. Nucl. Sci. NS-34 (1987) 486.
[25] I. Giomataris and G. Charpak, Int. Union of Pure and Applied Physics 20th Int. Cosmic Ray Conf., Moscow, USSR (1987).
[26] Report on the task force on detector R&D for the SSC, LBL SSC-SR 1021.

CERN
SERVICE D'INFORMATION
SCIENTIFIQUE

Nuclear Instruments and Methods in Physics Research 225 (1984)
North-Holland, Amsterdam

8

RECENT DEVELOPMENTS IN BaF$_2$ SCINTILLATOR COUPLED TO A LOW-PRESSURE WIRE CHAMBER

D.F. ANDERSON *, G.CHARPAK and Ch. VON GAGERN

CERN, Geneva, Switzerland

S. MAJEWSKI

Physikalisches Institut der Universität, Heidelberg, Germany

Received 30 January 1984

A mode of operation is discussed for a detector using BaF$_2$ as the scintillator, and tetrakis(dimethylamine)ethylene (TMAE) as the photosensitive agent adsorbed on a NiCr layer to form a photocathode. The position resolution of the detector with a crystal 25 mm thick and 120 mm in diameter is measured to be about 3 mm fwhm. The physics of TMAE as a liquid is also discussed.

1. Introduction

In earlier works [1–3] it has been shown that the short-wavelength component of the scintillator BaF$_2$ can be detected by a wire chamber using tetrakis(dimethylamine)ethylene (TMAE) as the photosensitive material. This has given rise to a new class of instrument, consisting of a solid scintillator plus a proportional counter (SSPC). The SSPC combines the high stopping power of a high-density solid with the versatility of a proportional counter.

As a scintillator, BaF$_2$ has many desirable features, as can be seen in table 1 which summarizes some properties of BaF$_2$, BGO, and NaI(Tl) [4–7]. The density of BaF$_2$ is 4.8 g/cm^2 and it has a radiation length of only 2.1 cm. The short-wavelength component, peaking at 225 nm, has a decay constant of only 0.6 ns. This material is also rugged and non-hygroscopic, and is readily available in sizes up to 150 mm in diameter. We believe, from contacts with the manufacturers, that the price of BaF$_2$ could soon be comparable with that of NaI(Tl).

The best results for the SSPC have been achieved with low-pressure wire chambers (see Breskin [8,9] on the subject). These detectors allow for high photoelectron collection efficiency and good timing properties that permit one to take advantage of the fast response of the BaF$_2$. Low-pressure wire chambers also have the advantage of being themselves insensitive to radiation, ensuring that only the light from the scintillator is detected in a high-energy physics environment. Finally,

* Visitor from Fermi National Accelerator Laboratory, Batavia, Ill., USA.

because of the fast ion drift in low-pressure wire chambers, an SSPC can be operated at very high rates with far less space-charge problems that one atmosphere.

In an earlier work [3] we reported on an SSPC with a BaF$_2$ crystal 25 mm thick and 120 mm in diameter. The TMAE was condensed on a cooled surface. With this detector we achieved a timing resolution of 540 ps fwhm with 350 MeV α-particles. An energy resolution of 28.5% fwhm was measured for protons losing about 18 MeV in the crystal.

In this work we carry the study of the SSPC further. We will discuss a new, simpler mode of operation that does not require cooling, as well as the imaging capability of the SSPC. The physics of TMAE as a liquid photocathode will also be discussed.

Table 1
Properties of three scintillators [a]

	BaF$_2$	BGO	NaI(Tl)
Density (g/cm^3)	4.9	7.1	3.7
Radiation length (cm)	2.1	1.1	2.6
dE/dx (min.) (MeV/cm)	~ 6	8	4.8
Linear attenuation coefficient at 511 keV (cm^{-1})	0.47	0.92	0.34
Peak emission (nm)	225 310	480	410
Decay constant (ns)	0.6 620	300	250
Index of refraction	1.56	2.15	1.85
Light yield (photons/MeV)	2×10^3 6.5×10^3	2.8×10^3	4×10^4
Hygroscopic	no	no	yes

[a] See refs. [4–7].

2. Gas versus surface

In the first work with the SSPC, the TMAE was condensed as a very thin liquid on a surface, which is not difficult owing to its low vapour pressure. The cathode surface was typically kept at 2°C. This gave good results, with an estimated quantum efficiency of about 4%. But, for the field of high-energy calorimetry, one would gladly sacrifice some of this efficiency for easier construction and operation. As was reported earlier, when TMAE is introduced in to an evacuated chamber all surfaces become photosensitive. We thus made a study of the operation of the SSPC without cooling.

The counter was rebuilt in the configuration shown in fig. 1. The surface of the BaF₂ crystal (25 mm × 120 mm diameter) was coated with a 15 Å layer of NiCr, which serves as a UV transparent conductive coating. A wire grid was placed 5 mm from the BaF₂ crystal and a copper-clad printed-circuit board was placed 5 mm from the grid. The NiCr surface and the printed circuit were kept at ground potential, while the grid was connected through a picoamperemeter to a power supply.

In operation the evacuated chamber was first filled with TMAE at various pressures. After a short time, left to allow the TMAE to reach an equilibrium, methane was introduced to bring the pressure to 9 Torr. This was to ensure that there was no charge gain under the applied voltage. The BaF₂ was then excited by a strong ^{90}Sr source. The current was measured with a collection voltage of $+30$ V and then of -30 V. At $+30$ V the electrons from the gas and from the surface were measured. At -30 V only the positive ions from the gas were collected. Thus the contribution from the surface and the gas could be determined. One motivation for these measurements is that since there is some overlap between the BaF₂ emission spectrum and the TMAE gas photoionization spectrum (fig. 2) [10,11], we wanted to verify that there was a contribution from the surface without cooling.

The results of these measurements are shown in fig. 3, where the contribution of the gas and TMAE on the surface are plotted as a function of TMAE pressure. The temperature corresponding to this pressure is also shown. As can be seen from the data, the contribution

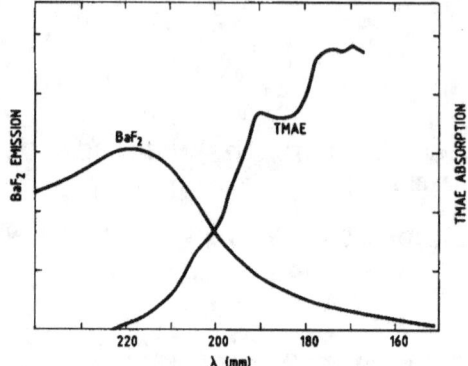

Fig. 2. Part of the emission spectrum of BaF₂ and the absorption of TMAE as a function of wavelength.

from the gas starts to saturate at about 15°C, while the surface contribution continues linearly over the region examined. For the above measurements the gas contributed about 75% of the signal at a pressure of 0.44 Torr (23°C) [1].

These measurements were made with the SSPC at 23°C. A question yet to be studied is the effect of surface temperature on the gas-to-surface ratio. From earlier tests [3] we have determined that most of the efficiency that we measure from the surface is from the NiCr surface of the BaF₂, with a small contribution from the surface of the printed-circuit board. This may be due to less light being available to this second surface.

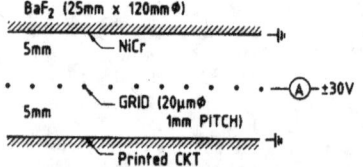

Fig. 1. Counter configuration used to measure the effect of TMAE in the gas and on the surface.

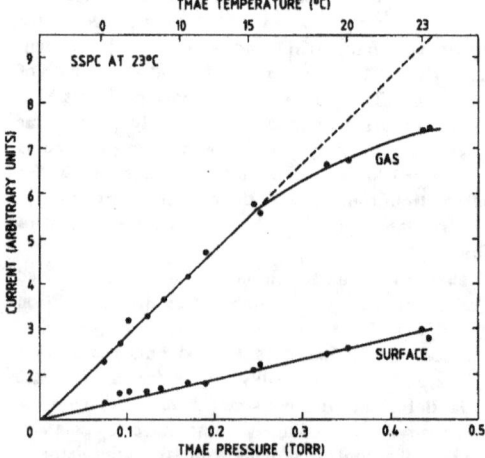

Fig. 3. Contribution of the TMAE on the counter surface and in the gas as a function of TMAE pressure. The corresponding TMAE temperature is also given.

It should be noted that in actual operation of an SSPC the charge released in the gas makes a much smaller contribution to the signal than implied above. This is because low-pressure wire chambers operate very close to a parallel-plate mode. Thus the charge from the surface undergoes a full gain of

$$G(\text{surface}) = e^{\alpha d},$$

where α^{-1} is the mean free path for charge amplification and d is the gap width. The average gain on the charge uniformly distributed in a parallel-plate counter is

$$G(\text{parallel plate}) = (1/\alpha d)\, e^{\alpha d},$$

This is because charges liberated closer to the anode will receive substantially less gain. So for a gain of 10^4 for the counter, the average gain of the charges liberated in the gas is reduced by almost an order of magnitude.

3. Imaging capability

One of the characteristics of the SSPC that makes it unique, as compared to other scintillation detectors, is its potential for high spatial resolution. To demonstrate this aspect, we rebuilt our test detector as shown in fig. 4. The BaF₂ crystal was separated from the amplification region by a 4 mm drift region to reduce ion feedback and thus allow higher gain. The cathode-to-cathode spacing was 6 mm. The second cathode was a printed circuit with a row of 5 × 5 mm² pads to allow a one-dimensional determination of the event location. Two additional 5 × 10 mm² pads were also included to allow a check that the event was at least roughly centred on the row of smaller pads. In operation, we filled the SSPC with TMAE at 25°C and 18 Torr of isobutane to increase the gain for the cathode amplifiers. The voltages were $V_c = +40$ V and $V_a = +1020$ V. It should be

noted that we have measured gains as high as 10^4 in pure TMAE.

A calculation of the light distribution on the surface of the BaF₂ crystal was made for one and two particles traversing the crystal. Fig. 5 shows the intensity as a function of radius for BaF₂ thicknesses of 2.54 cm and 10 cm. Figs. 5a and 5b are for single particles passing normally ($\theta = 0°$) through the crystal, and at $\theta = 30°$, respectively. At first the narrow distribution seems surprising. It is due to the fact that the solid angle subtended by a spot on the surface goes as ρ^{-2}, where ρ is the distance between a point of scintillating crystal and the spot on the surface. Thus the surface nearest to the point of exit (or entry) receives a large amount of light. This gives the narrow distribution, with the additional light from the thicker crystal going mostly into the wings of the distribution. Figs. 5c and 5d show the simulation with two particles, 1.5 cm apart, passing at $\theta = 0°$ and $\theta = 30°$, respectively.

In our first test with cosmic rays, we found the signals on our pads had a distribution consistent with

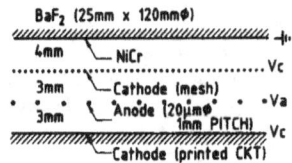

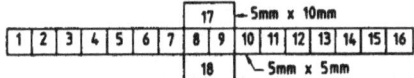

Fig. 4. Counter configuration, with details of the cathode pads, used to measure the one-dimensional position resolution of the SSPC.

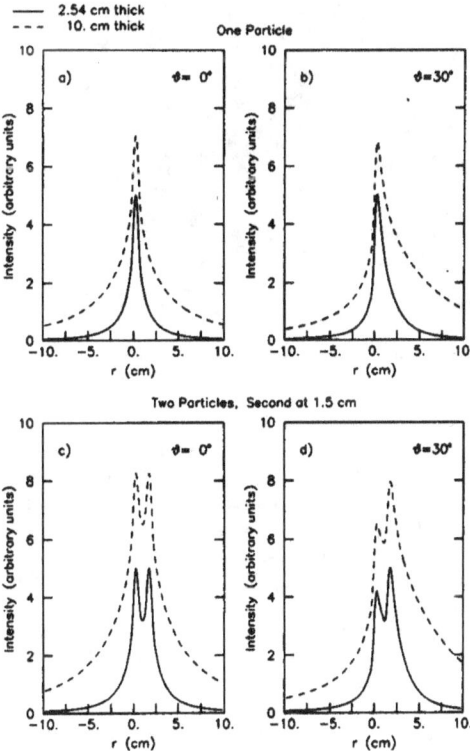

Fig. 5. Simulation of the light intensity on the surface of BaF₂ crystals 2.54 cm thick and 10 cm thick, for one particle a) normal to the surface ($\theta = 0°$), b) $\theta = 30°$; and for two particles 1.5 cm apart c) $\theta = 0°$ and d) $\theta = 30°$.

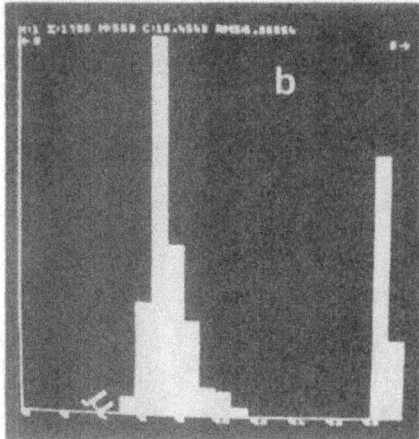

Fig. 6. Cathode signals for two cosmic-ray events. The two channels on the right are pads 17 and 18 in fig. 4.

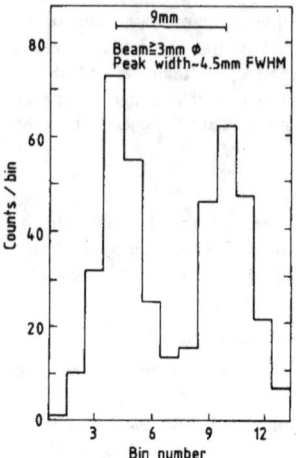

Fig. 7. Two superimposed position spectra for the SSPC, displaced by 9 mm. The beam diameter was about 3 mm and the peak widths are about 4.5 mm fwhm, implying a position resolution of about 3 mm fwhm.

our simulations. Fig. 6 shows typical pulse-height distributions on the 5×5 mm² cathode pads for two cosmic-ray events. The two channels on the right are pads 17 and 18 in fig. 4. This we feel is further evidence that most of the electrons come from the BaF₂ surface, with little contribution from the TMAE on the surface of the printed circuits.

A test of the position resolution of the detector was made with a beam of π^-, μ^-, and e⁻ at the Synchro-cyclotron at CERN. The momentum of the particles was 200 MeV/c. The particles entered the crystal with a beam size of about 3 mm. A centroid was calculated for 220 events; then the detector was displaced by 9 mm and an additional 220 events were recorded. The resulting position spectra are shown in fig. 7. The width of the peaks is about 4.5 mm fwhm. Thus the position

resolution of this detector for high-energy particles is about 3 mm fwhm.

It should be noted that, although the simulations and the measurements indicate that a good position resolution is possible for one particle, and a reasonably good two-particle separation is possible from a large crystal, the energy loss of each of two close particles cannot be accurately determined. For this, an array of BaF₂ rods with smaller cross-section would be better.

4. The liquid photocathode

Although the experimental material reported on in this paper involves TMAE that is either in the gas phase or adhering to a surface, we would like to discuss here the physics of TMAE as a liquid photocathode (LPC).

The photoionization threshold, E_{TH}, of a photosensitive material dissolved in a non-polar liquid is expected to be lower than the ionization potential, IP, in the gas phase. This threshold is for electrons collected in the solvent, which in our case is TMAE itself. The relation between E_{TH} and IP is given by [12,13]

$$E_{TH} = IP + V_0 + P_+, \qquad (1)$$

where V_0 is the ground-state energy, with respect to the vacuum, of a free electron in the liquid, and P_+ is the polarization energy of the positive ion.

In the case where V_0 is positive or zero the photoelectric threshold for the LPC, E_{LPC}, will be given by eq. (1), i.e.

$$E_{LPC} = IP + V_0 + P_+, \quad \text{for } V_0 \geq 0. \qquad (2)$$

In the common case where V_0 is negative, the photoelectron must overcome V_0 at the surface, and thus it expresses itself as a work function, cancelling its effect on the threshold in the liquid. Thus we expect

$$E_{LPC} = IP + P_+, \quad \text{for } V_0 \leqslant 0. \tag{3}$$

The value of V_0 for TMAE is in fact negative [14] and thus eq. (3) is applicable.

To evaluate P_+ for liquid TMAE, the Born equation can be used [15,16] to give

$$P_+ = (-e^2/2R)\left[1 - (1/\epsilon_{op})\right], \tag{4}$$

where ϵ_{op} is the optical dielectric constant and R is the radius of the sphere for the positive ion.

We have measured the index of refraction of TMAE, for the shortest wavelength that will traverse it, to be $n = 1.51$. Since $\epsilon_{op} = n^2$, and using the value of $R = 3.21$ Å [15], we calculate $P_+ = -1.26$ eV. Taking the value of IP $= 5.36$ [15] we arrive at $E_{LPC} = 4.10$ eV from eq. (3). This is in good agreement with the earlier experimental estimate of Anderson [2] of $E_{TH} = 4.3$ eV.

Although we believe that when V_0 is negative it does not express itself in the lowering of the threshold of the LPC, its value is not without physical effects. Firstly, within the liquid the threshold is lowered by V_0 and thus there are electrons liberated within the liquid that do not escape. Secondly, the value of V_0 will probably express itself in the quantum efficiency of the LPC. There is a thermalization range, r, for a photoelectron, which in general increases as V_0 becomes more negative [16]. Thus the electron must be liberated in a distance r', less than r, from the surface in order to still have enough energy to escape. If one assumes that the photoelectrons leave the atoms in all directions, a simple calculation shows that only about 25% of the electrons liberated within a distance r' of the surface will escape. Thus as V_0 becomes more negative, r increases and therefore one would expect the quantum efficiency to increase.

From the earlier measurements of Anderson et al. [3], the quantum efficiency of the TMAE LPC was estimated at 4%. This means that the electrons are coming from a depth great enough to stop at least 16% of the light. This implies a large r' are therefore a very negative V_0.

There are some interesting implications if r', and therefore the electron mobility μ, is large for liquid TMAE [17]. The mobility of electrons in a non-polar liquid increases rapidly with r, as does the probability that the electron escapes from the Coulomb field of the positive ion. Thus there may be an application for TMAE in liquid ionization chambers. This is certainly a subject for further study.

The above arguments also imply that with an LPC we are only seeing a small fraction of the electrons available. It is possible that, with strong electric fields,

these electrons can be removed from the TMAE, giving a much higher quantum efficiency and a threshold lowered by V_0.

5. Conclusions

It has now been shown that the SSPC has good spatial resolution as well as good energy and timing resolution. However, it would be desirable to improve the efficiency of the photocathode for application in medium–high-energy physics and in nuclear medicine. For this, work should continue on improving the light collection from the BaF₂ and on improving the electron extraction from the LPC.

The prospects of using TMAE for liquid ionization detectors also look promising. The most obvious application would be for uranium calorimeters, where its low vapour pressure, and thus low fire hazard, would be advantageous.

References

[1] D.F. Anderson, IEEE Trans. Nucl. Sci. NS-28 (1981) 842.
[2] D.F. Anderson, Phys. Lett. 118B (1982) 230.
[3] D.F. Anderson, R. Bouclier, G. Charpak and S. Majeswki in Proc. Wire chamber Conf. Vienna (1983) Nucl. Instr. and Meth. 217 (1983) 217.
[4] M. Laval, M. Moszyński, R. Allemand, E. Cormoreche, P. Guinet, R. Odru and J. Vacher, Nucl. Instr. and Meth. 208 (1983) 169.
[5] R. Allemand, M. Laval, J. Vacher, M. Moszyński, E. Cormoreche and R. Odru, Communication LETI/ MCTE/82-245, Grenoble, France (1982).
[6] BGO–NaI(Tl) comparison, paper distributed at the Int. Workshop on Bismuth germanate, Princeton University (1982).
[7] M.R. Farukki and C.F. Swinehart, IEEE Trans. Nucl. Sci. NS-18 (1971) 200.
[8] A. Breskin, Nucl. Instr. and Meth. 196 (1982) 11.
[9] A. Breskin, in Detectors in heavy ion physics, ed., W. von Oertzen, Lecture Notes in Physics (Springer, Berlin, 1983) vol. 178, p. 44.
[10] Y. Nakato, M. Ozaki and H. Tsubomura, Bull. Chem. Soc. Japan 45 (1972) 1299.
[11] J. Vacher, R. Allen, M. Laval, M. Moszyński and R. Odru, Communication LETI/85X, Grenoble, France (1983).
[12] R.A. Holroyd and R.L. Russell, J. Phys. Chem. 79 (1983) 483.
[13] Y. Nakato, M. Ozaki and H. Tsubomura, J. Phys. Chem. 76 (1972) 2105.
[14] R.A. Holroyd, private communication.
[15] Y. Nakato, T. Chiyoda and H. Tsubomura, Bull. Chem. Soc. Japan 47 (1974) 3001.
[16] R.A. Holroyd and R.L. Russell, J. Phys. Chem. 78 (1974) 2128.
[17] G.R. Freeman, in Proc. 5th Int. Congress of Radiation research, Seattle (Academic Press, New York, 1975) p. 367.

Nuclear Instruments and Methods in Physics Research A269 (1988) 149–160
North-Holland, Amsterdam

LIQUID AND SOLID ORGANIC PHOTOCATHODES

V. PESKOV [1,2], G. CHARPAK [1], P. MINÉ [1]*, F. SAULI [1], D. SCIGOCKI [1], J. SÉGUINOT [3], W.F. SCHMIDT [1,4] and T. YPSILANTIS [3]

[1] CERN, Geneva, Switzerland
[2] Institute for Physical Problems, Moscow, USSR
[3] Collège de France, Paris, France
[4] Hahn-Meitner Institute, Berlin, FRG

Received 10 December 1987

We have investigated the possibility of creating photocathodes for gaseous detectors with a high sensitivity in the photon spectral region between 105 and 300 nm. Metal cathodes covered with liquid or solid organic layers, such as tetrakis(dimethylamine)ethylene (TMAE) and tetramethyl-p-phenylenediamine (TMPD) and solutions of these substances, were studied using two different experimental setups: a proportional wire chamber and a single-wire counter. Three effects were observed. First, a thin film of adsorbed vapours led to the efficiency of the metallic cathode being increased by more than one order of magnitude at photon wavelengths up to $\lambda \approx 400$ nm. Secondly, a thick layer of liquid strongly increased the cathode efficiency Q for radiations of $\lambda < 270$ nm: e.g. with TMAE we obtained a yield of about 0.5%, for $\lambda = 235$ nm. Thirdly, we found that some solid layers, such as neopentane + TMAE, give a maximum efficiency $Q \approx 3\%$ at $\lambda = 235$ nm. The quantum yields of liquid and solid photocathodes increase with increasing applied electric field. Some applications of the liquid and solid photocathodes are discussed.

1. Introduction

The photoionization process in a gas is of intrinsic interest in understanding the propagation of electron avalanches in an electric field around a wire, i.e. the gain mechanism in proportional and Geiger counters. Practically, this process served as the basis for the development of gas-phase single UV photon detectors for Cherenkov ring-imaging [1,2] and for plasma diagnostics [3]. Data made available in a recent review [4] give relative quantum efficiencies versus photon wavelength for noble, simple organic and simple inorganic gases. Absolute quantum efficiencies are now measured for some complex organic molecules [tetrakis(dimethylamine)ethylene (TMAE), tetramethyl bi-imidazolidine (TMBI), triethylamine (TEA)] which have the lowest ionization thresholds and high efficiencies [5–8].

The photoionization process is also observed in liquids (both neat and doped), and a recent review giving relative quantum efficiencies is available [9]. In liquids, as compared with gases, there are several new related factors which tend to reduce the quantum yield: e.g. geminate recombination (due to the short electron thermalization length r_0 in liquids) and low electron mobility μ. If, in addition, transfer of the photoelectrons from the liquid to the gas phase is required (where

amplification is possible), then additional loss of efficiency may occur [10]. The threshold energy for photoionization in liquids is, however, generally lower than in gases owing to the polarization energy P_+ of the positive ion and to the lowering of the electron conduction band V_0. The relation between the gas-phase ionization energy threshold E_{gas} and the liquid threshold E_{liq} is

$$E_{liq} = E_{gas} + P_+ + V_0. \tag{1}$$

The polarization energy is always negative, whereas V_0 may be of either sign. Relation (1) will, for example, allow some photosensitivity in liquid TMAE ($V_0 = -0.26$) up to 330 nm [11], whereas for TMAE gas, the sensitivity ends at 230 nm.

A characteristic of these liquids when they are specially purified is that the electron lifetime τ is long (many µs); so if the electron mobility is high ($\mu > 10$ cm^2/V s), the electron mean free path $l_c = \mu E \tau$ will be long. As an example: for $\mu = 10$ cm^2/V s, $\tau = 10$ µs, and an electric field $E = 10$ kV/cm, the electron mean free path is $l_c = 1$ cm. Such a long mean free path would allow all photoelectrons that escape geminate recombination to drift and be collected at the liquid surface, since the liquid (or solid) photocathode is only µm (10^{-4} cm) thick. Its thickness is determined by the photon absorption length $l_{ph} = 1/n\sigma$, which is about 0.1 µm (n is the number of molecules per cubic centimetre of liquid $\approx 3 \times 10^{21}$/cm^3 and σ is the molecular

---

* Now at the Ecole Polytechnique, Palaiseau, France.

0168-9002/88/$03.50 © Elsevier Science Publishers B.V.
(North-Holland Physics Publishing Division)

absorption cross section ≈ 40 Mb). A 1 μm thick cathode therefore absorbs 99.995% of the incident photons. In this simple model the final quantum efficiency is determined by geminate recombination and the extraction probability of the electrons at the liquid–gas interface. An electric field that is large enough for efficient extraction will also reduce the geminate recombination. The condition that $l_c \gg l_{ph}$ is completely satisfied for high-mobility liquids the primary reason for their high quantum efficiency.

Anderson [12,13] proposed to use a liquid-TMAE layer as the photocathode in a wire chamber. With liquid photocathodes, one can expect to build UV detectors with good sensitivity in the wavelength range up to 300 nm, which opens up new prospects for applications in high-energy physics and in other areas of science. Subsequent investigations, however, have shown conflicting indications. The existence of liquid-TMAE photosensitivity was also demonstrated by Schotanus et al. [14] using spectral measurements. Other investigators [15,16] have failed to confirm Anderson's results, although Holroyd [8] observed an effect in TEA which could be explained as a manifestation of a liquid photocathode.

The aim of this work is to investigate liquid-TMAE photocathodes as well as other organic liquid and solid photocathodes, and to understand the conflicting results previously indicated. We have repeated the conditions of Anderson [13] based on the detection of UV scintillation light emitted from BaF$_2$, and have extended this work with a photocathode coupled to a monochromatic light source of variable wavelength and with a counter capable of efficiently detecting single photoelectrons.

2. Experiment with BaF$_2$ scintillator and a low-pressure MWPC

2.1. Experimental setup

This experiment, which followed closely the setup of Anderson [12,13] in observing a liquid-photocathode effect, was performed with the assembly shown schematically in fig. 1. This consisted of: a low-pressure ($\leq$ 3 mbar) multiwire proportional chamber (MWPC) assembled inside a thermostatic box; a gas-flow system; a liquid-cathode receptacle; a pump; and detection electronics. A BaF$_2$ crystal was coupled to the MWPC as previously described [17] and as shown in fig. 1. The aluminium cathode had a groove ("cathode bath") which held the liquid on the cathode. It was filled either by vapour condensation as the chamber was cooled, or from a liquid reservoir. An overflow bottle coupled to the grooved channel prevented accidental spillage of the liquid across the rim of the cathode bath. If there was a need to fill the chambers with vapour alone, special

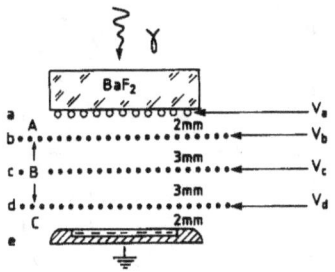

Fig. 1. Schematic drawing of the apparatus used to observe the liquid-cathode effect. It includes: a BaF$_2$ crystal; a wire plane (a) touching the crystal; wire planes (b, d), which are cathodes; wire plane (c), which is the anode of a MWPC; and finally a liquid-cathode receptacle (e). The voltages V_a, V_b, V_c, and V_d could be independently varied, whilst $V_e = 0$, so as to provide signals from region A, B, or C.

valves were used to open the vessels containing the appropriate liquids. The cathode bath ((e) of fig. 1) was earthed ($V_e = 0$), whereas the BaF$_2$ plane (a), the MWPC cathodes (b, d), and the MWPC anode (c) potentials could be varied independently. Depending on the voltages chosen, it was possible to work with or without gas amplification in spaces A and C (fig. 1). In this way, by adjusting polarity and voltages, the signals from volumes A, B, and C could be measured separately. Signals from the anode wires (c) were amplified and sent in parallel to a pulse analyser and to an oscilloscope. The thermostatic box (and the gas and liquid fill lines) could be operated between $-20\,^\circ$C and $+70\,^\circ$C.

2.2. Results with BaF$_2$ scintillator and low-pressure MWPCs

For these measurements the UV scintillation light from the BaF$_2$ crystal was excited with γ-rays from ^{60}Co (1.17 and 1.33 MeV) or by X-rays from ^{241}Am (59 keV). The number of UV photons is estimated, from the measurements of Laval et al. [18] and of Schotanus et al. [14], to be 2000/MeV of energy lost in the crystal in the wavelength interval 180 to 230 nm. As a result of the overlapping of this radiation spectrum with the TMAE-vapour quantum efficiency, the number of photoelectrons created inside a wire chamber at room temperature ($\sim 20\,^3$C) was less than one per millimetre gas gap. For this reason, when the signal from the MWPC region A or C of fig. 1 was recorded on the pulse analyser, the shape of the spectrum displayed a low pulse-height noise peak and an exponential single-electron distribution. The electronics noise peak was eliminated by a discriminator so as not to overload the analyser.

When the chamber was filled with purified TMAE vapour, with 30 mbar CH$_4$, the count rate from region

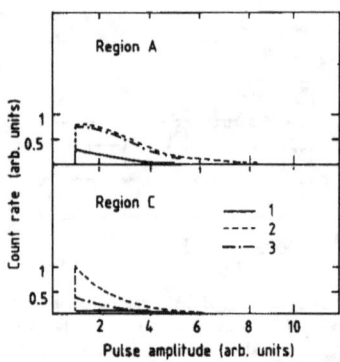

Fig. 2. Observed pulse amplitudes from regions A and C of fig. 1 when working with a cooled low-pressure wire chamber: (1) In this experiment the chamber was cooled to 4°C and filled with TMAE vapour at 4°C and 30 mbar CH_4. (2) The cathode bath was then filled with liquid TMAE at 4°C: p_{CH_4} = 30 mbar. The working voltage on the chamber was kept at the same value as for the first experiment. (3) Signals from the BaF_2 detector cooled to −12°C: p_{CH_4} = 30 mbar.

B (fig. 1) was slightly higher than the rates from regions A or C. Injection of purified TMAE liquid into the cathode bath did not substantially change the count rate from region C, indicating that the gas-phase photo-ionization was dominant. Therefore, in order to obtain clearer observations of the liquid-cathode effect, we usually proceeded by cooling the chamber. When TMAE was cooled from 20 to 4°C, the saturated vapour pressure fell from 0.46 to 0.13 mbar [13,19].

After filling the chamber with saturated TMAE vapour at 4°C, methane was added up to a pressure of 30 mbar. The working voltage of the chamber was determined mainly by the methane pressure. At 4°C, however, we were unable to obtain a gas amplification large enough to record all pulses reliably in the wire chamber (region B), but it was possible to obtain pre-amplification in the drift space and to record signals from regions A and C (see fig. 2 (solid lines)). A comparison with room temperature indicated that the count rates from regions A and C fell by a factor of approximately 5, which shows fairly good agreement with the expectation from the reduction of the gas-phase TMAE pressure.

When liquid TMAE at 4°C was injected into the cathode bath the gas conditions in the chamber did not change, because the chamber was again filled with the saturated pressure of TMAE at 4°C plus 30 mbar CH_4. So the working voltages, the gas amplifications, and the pulse-height spectrum were exactly the same as before injection. This allowed us to make an accurate comparison of the count rates from regions A and C. The results of the measurements are presented in fig. 2 (dashed line). As can be seen, when liquid TMAE (4°C) was

injected into the cathode bath, the count rate from region C rose by a factor of 7 to 10 and that from region A by a factor of 3 (because of the low condensation). This demonstrated the liquid-TMAE photocathode effect, with transfer of electrons from the liquid phase to the gas phase. It also showed that a liquid-cathode effect occurred at the BaF_2 surface. In all cases when there were vapours inside the chamber, a thin photosensitive liquid layer was formed on all the chamber surfaces, including that of the BaF_2 crystal. This can open up a range of applications for photocathodes created directly on the entrance windows of gas detectors.

In another experiment we tried to obtain a liquid photocathode by means of stronger condensation. To do this, the chamber was filled with TMAE vapour at a temperature of 20°C and then cooled down to 0°C. Here the highest count rate was from region A, indicating that photoelectron emission from the thick TMAE layer condensed on the BaF_2 crystal is dominant. The performance of the chamber at −12°C, which is below the freezing point of TMAE, was also tested. In this case, the count rate from region A changed only very slightly, whereas from region C it actually decreased (see dash–dotted line in fig. 2). This shows that emission takes place mainly from TMAE in its solid state formed on the window, and that as a result of the absorption and scattering of the radiation in this layer, the signal from region C decreases.

Similar experiments were performed with purified liquid tetramethyl pentane (TMP) in which 3% TMAE was dissolved; TMP was chosen because of its low vapour pressure (≈ 6 mbar at 0°C) and its long thermalization length r_0 (≈ 10 nm) and electron mobility μ (≈ 30 cm²/V s). All these factors decrease the recombination rate of TMAE ions and therefore increase the quantum yield.

The experimental procedures were the same: first we introduced TMP + TMAE vapour into the chamber, added CH_4 to a pressure of 30 mbar, and carried out the measurements; then we filled the cathode bath with the liquid and compared the results. The high-voltage gas amplification and pulse-height spectrum remained unchanged. The experiments were carried out at room temperature and at 0°C. At room temperature a small (~ 20%) increase in the count rate was noticeable when the bath was filled with liquid. To make the effect stand out more clearly, we cooled the chamber down to 0°C. When the liquid was placed in the cathode bath, the following changes were observed, which were not seen when the detector was filled with vapour alone: the count rate from region B remained almost the same, but the rates from regions A and C increased by a factor of 2 to 3 (see fig. 3).

A large number of other control experiments were carried out. For example, the detector was filled with a

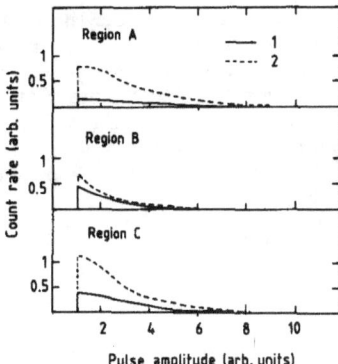

Fig. 3. Observed pulse amplitudes from regions A, B, and C of fig. 1 when (1) the MWPC was filled with TMP + TMAE vapour + 30 mbar CH$_4$ at 0°C; (2) the same gas filling was used, but with the bath (e) containing liquid.

vapour mixture of TMAE and TMP and heated to about 38°C (the gas amplification was kept constant by voltage adjustment). In this case, because of the rise in vapour pressure, the signal from region B increased, but the signals from regions A and C decreased owing to the evaporation of the thin, liquid, surface film and a consequent loss of sensitivity in these parts of the chamber.

All the experiments described were stable and easily reproduced.

2.3. Discussion and interpretation

The measurements given in the preceding subsection show conclusively the existence of a liquid-TMAE photocathode effect.

Two sources could give rise to the increase sensitivity of the chamber: the formation of a thin adsorbed film, and the extraction of photoelectrons from the liquid (thick condensed layer and cathode bath). Since the introduction of liquid into the cathode bath between 0 and 4°C increased the quantum efficiency for both TMAE and TMP + TMAE by a factor of more than 3, it could be expected that the liquid would provide a contribution that is at least three times larger than that of the adsorbed film. At room temperature, the difference between the adsorbed film and the liquid is lower. This is due to the fact that with a greater partial vapour pressure, the adsorbed film [8] is thicker, thus reducing the difference between the film and the liquid itself.

If we ignore the effect of the adsorbed film at low TMAE vapour pressure, the fact that the sensitivity increases as the cathode is filled with liquid leads us to conclude that for temperatures between 0 and 4°C the emission of electrons from the liquid is at least three

times higher than that from the ionization of the gas. Therefore at room temperature, when the TMAE vapour pressure rises by a factor of 4 to 5, the contribution to the full current by the charges that are extracted from the liquid will be at least 60% for a gap thickness of 2 mm. For a 1 cm thickness the contribution from the liquid will amount to about 15%, whilst that coming from the surface films will be lower than 15%, which is too small from the practical point of view.

In the case of TMP + TMAE, cooling to 0°C caused the counting rate to drop by less than a factor of 4, indicating the existence of a surface effect. In addition, even at room temperature the difference between the liquid and the adsorbed layer could be seen. In qualitative terms this means that a greater sensitivity was obtained with a TMP + TMAE liquid cathode than with TMAE alone.

Thus the idea of Anderson and some of his observations were confirmed. There clearly exists a photoelectric enhancement effect due to the condensation of photoionizable vapours in wire chambers used for detecting the UV component of BaF$_2$. But the absolute quantum yield of a liquid photocathode could not be estimated from this experiment, so measurements with monochromatic light and a single-wire counter were then performed.

3. Experiments with a UV monochromator and a single-wire counter

3.1. Experimental setup

The apparatus shown in fig. 4 was built for the precision measurement of the quantum yield from different photocathodes. It consists of a D$_2$-filled UV arc lamp with a MgF$_2$ window (Hamamatsu L879-01), a UV photon monochromator (Jobin–Yvon H20 UVL) with 2 nm resolution at 150 nm, and a single-wire counter which operates in such a way that single electrons can be detected with near 100% efficiency. The photon beam from the monochromator was split into a reflected beam (intensity I_R) and a transmitted beam (intensity I_T) by means of a 4 mm thick CaF$_2$ crystal ((3) in fig. 4) placed at 45° to the beam direction. The reflected beam was monitored by an RCA8850 phototube (14) whose glass was coated, in vacuum, with a 1000 nm thick film of p-terphenyl wavelength shifter ((13) in the figure). The transmitted beam I_T was parallel-focused by a CaF$_2$ lens of 100 mm focal length and 25 mm diameter. The 3 mm diameter beam entered the single-wire gas counter (6) after traversing a neutral density filter station (4) which could reduce I_T by calibrated amounts, a CaF$_2$ exit window (15), and two parallel, 81% optically transparent meshes (5) which were biased to trap photoelectrons produced by possible TMAE deposits on the CaF$_2$ window (15).

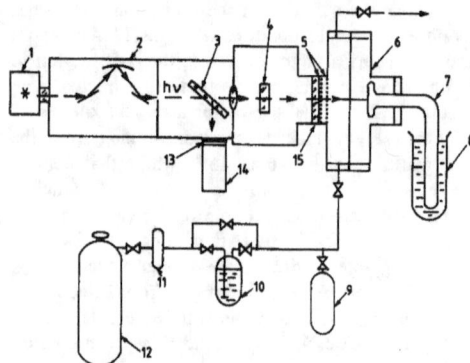

Fig. 4. Apparatus for cathode quantum efficiency measurements. The components are as follows: (1) D_2 arc lamp; (2) UV monochromator; (3) CaF_2 beam splitter; (4) filter station; (5) transparent meshes; (6) single-wire counter; (7) copper cathode; (8) cryogenic bath; (9) TMS, TMP, or neopentane liquid container; (10) TMAE liquid container; (11) gas flowmeter; (12) gas bottle supply; (13) p-terphenyl wavelength shifter deposit on (14) the photomultiplier; (15) CaF_2 exit window.

Fig. 4 shows the design of the single-wire counter. It consists of a 40 mm diameter stainless-steel cylinder with two transverse flanges placed along an axis in the middle of the cylinder. The counter was linked to the monochromator by one of the flanges. Cathodes of differing design, manufactured from various materials, could be fixed on the other flange. This figure shows the most frequently used design, which permits the copper cathode to be cooled or heated. The vapour can be condensed by cooling, and a liquid or solid photocathode obtained; by heating, it can be evaporated. A 0.2 mm diameter wire was used as an anode.

The gas supply (12) passed through a flowmeter (11) and could be bypassed through liquid TMAE (10). A supply of liquid TMP, tetramethyl silane (TMS), or neopentane (NP) was available (9) in the circuit. All these liquids and the TMAE were specially purified.

Different mixtures of gases could be flushed through the counter: methane, argon, and isobutane with (or without) TMAE, TMP, TMS, NP, or diethyl ferrocene vapours at total pressures from 0.03 to 1.6 bar.

3.2. Calibration

The system was calibrated before starting measurements with the single-wire counter. Calibration of I_T relative to I_R was achieved by replacing the single-wire counter ((6) of fig. 4) with a second photomultiplier, which was also covered with a 1000 nm thick vacuum-deposited film of p-terphenyl (identical to (13) and (14) of fig. 4). The wavelength shifter efficiency has been measured to be constant ($\pm 10\%$) over the range 100 to

300 nm [20], relative to sodium salicylate which itself is constant over this interval [21]. Measurement of the photomultiplier current ratio $R = I_T/I_R$ serves to calibrate the transmitted flux: in terms of the reflected flux, $I_T = RI_R$ for any wavelength. The count rate C of the single-wire counter is proportional to I_T and to the quantum efficiency Q (for 100% single-electron counting efficiency); thus

$$C = kI_TQ = kI_RRQ, \qquad (2)$$

where the second equality expresses the count rate of the monitored flux $I_R(\lambda)$ and the calibration ratio $R(\lambda)$. The proportionality constant k is evaluated by measuring, at wavelength λ_0, a count rate C_0 and reflected flux I_{R_0} when the single-wire counter is filled with TMAE vapour with known quantum efficiency Q_0. Thus $k = C_0/I_{R_0}R_0Q_0$, where $R_0 = R(\lambda_0)$ is from the calibration. The TMAE-vapour relative quantum efficiency versus wavelength has been measured by Ekelin and Fransson; their results were published a few years ago [6,7]. Normalization of this curve to a maximum quantum efficiency $Q_{max} = 0.67$ at $\lambda = 145$ mm leads to agreement with the Cherenkov ring-imaging data [7] which gave the Cherenkov quality factor [1] $N_0 = 95$ cm^{-1} for an argon-gas Cherenkov radiator and a TMAE-vapour-filled, single-photon detector with quartz windows. With this normalization the quantum efficiency at $\lambda_0 = 200$ nm is $Q_0 = 0.21$. The more recent absolute measurement of Q for TMAE vapour by Holroyd et al. [8] gives a value $Q_0 = 0.23$. We adopt a value $Q_0 = 0.22$ at $\lambda_0 = 200$ nm for full photon absorption.

Special care was taken to suppress short wavelength light coming from second-order diffraction and the tail of the zeroth-order (undispersed) beam. For example, when taking measurements at $\lambda = 250$ nm, light of $\lambda = 125$ nm may be present from second-order diffraction. Even though the zeroth and second orders have only $\sim 1\%$ of the intensity of the first order, the counter was much more sensitive to this shorter wavelength light; a signal from this parasitic light was approximately one order of magnitude larger than the useful signal. A filter system ((4) of fig. 4) was therefore always used to suppress this component, and the level of this light as well as of the tails of the zeroth order was regularly checked.

It should be noted that when the counter was flushed with Ar and isobutane or with isobutane alone, it was possible to achieve the Geiger regime. In this case the amplitudes of all pulses were approximately equal. This makes it possible to reach almost 100% probability in photoelectron recording, which is of extreme importance for precision measurements of the photocathode efficiency. When the counter was filled with methane, only a restricted proportional regime was possible, with a counting plateau at about 200 to 300 V. To estimate

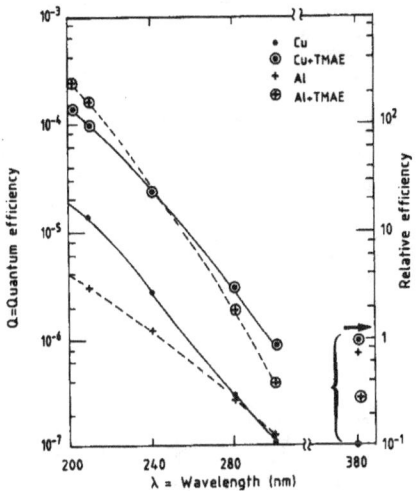

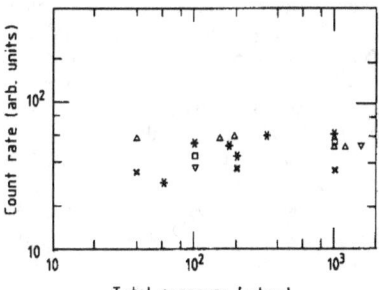

Fig. 6. The relative quantum efficiency at $\lambda = 380$ nm (unless otherwise noted) of the copper cathode (20°C) as a function of the single-wire counter gas pressure for the following gas fillings: (∗) argon + 50% methane; (□) methane;; (×) argon + 50% methane, $\lambda = 240$ nm; (△) isobutane; (▽) methane + TMAE.

Fig. 5. The quantum efficiency Q of a copper and an aluminium cathode without and with exposure to TMAE vapour ($p < 0.1$ mbar) for a period of 20 h vs wavelength. The single-wire counter gas was argon + 10% methane at a pressure of 1 bar. The cathode was at room temperature (20°C).

the efficiency of this counter, a calibration procedure was then carried out. We measured the count rate from a Geiger counter filled with isobutane + TMAE vapour, at $\lambda = 200$ nm. Then the same measurements were repeated with the counter filled with methane + TMAE vapour. In the region of the counting plateau, the two

rates agreed, showing that efficient single-electron counting in methane is possible, as previously shown by Ekelöf et al. [22].

3.3. Liquid and solid photocathodes containing TMAE

3.3.1. TMAE photocathodes

The quantum efficiency Q of an uncoated copper cathode at room temperature (20°C) was measured at atmospheric pressure in argon + 10% methane versus wavelength λ, as shown in fig. 5. The admittance of an extremely low partial pressure of TMAE ($p < 0.1$ mbar)

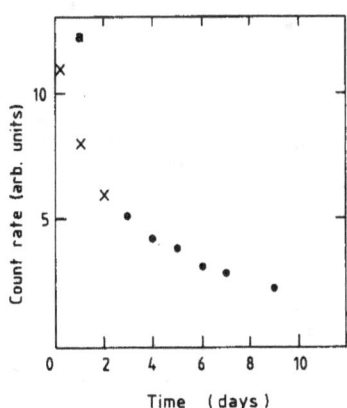

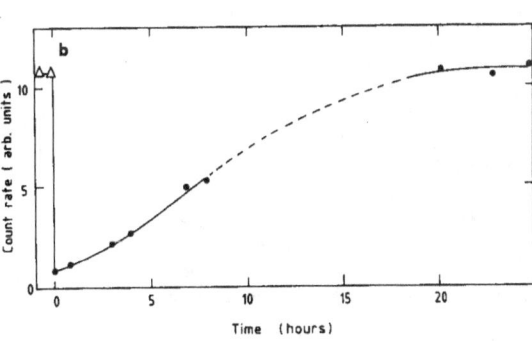

Fig. 7. (a) Measurements illustrating the destruction of a thin TMAE layer on the upper cathode as a function of time. After formation of the TMAE layer at time zero, for the whole nine days of measurements, argon + 50% methane (no TMAE) was flushed through the system at the rate of 3 l/h (counter volume = 0.25 l). For the first three days, the entire single-wire counter system was heated to 50–70°C. Measurements were at $\lambda = 380$ nm. (b) Formation of a photosensitive layer on a copper cathode (20°C) vs time. The deposit is due to TMAE impurities in the gas system. After heating the entire system for 10 h, a new, clean cathode was installed at time zero. The gas flow was 3 l/h of argon + 10% methane (without TMAE). The TMAE deposit resulted from TMAE impurities in the system. Measurements were at $\lambda = 380$ nm.

over a period of 20 h was sufficient to form a photosensitive layer with the quantum yield increasing by more than a factor of 10, even up to λ = 380 nm. This test was repeated with an initially clean aluminium cathode (see fig. 5). During the first 10 to 12 h, the results obtained with the aluminium cathode were similar to those with the copper cathode, but after 20 h the sensitivity at λ = 380 nm dropped. The quantum yield of this type of photosensitive layer did not depend on the counter pressure or on the type of gas filling (see fig. 6). The photosensitive layer is extremely tenacious, as is shown in fig. 7a. The whole single-wire counter system had to be heated to about 60°C for two days and gas-flushed (argon + 50% methane) for nine days in order to recover the initial low quantum efficiency of the copper cathode. The formation time of the TMAE cathode at 20°C is shown in fig. 7b. The previous TMAE thin-film cathode was replaced at time zero with a clean copper cathode, and then flushed with gas (without TMAE). The TMAE film cathode was formed by TMAE trapped in the gas system, and reached its previous value in about 20 h.

When the external part of the copper cathode is cooled to $t_{cath} = 0°C$, intensive TMAE condensation begins, because the gas input is at room temperature and TMAE gas is cryopumped onto the copper cathode. The quantum efficiency grows by factors of between 10 and 1000 at λ < 250 nm. A new, steeper dependence of Q on λ appears in the region of the increased quantum efficiency. At the same time, because of absorption in a comparatively thick liquid layer, the sensitivity to radiation between 290 and 400 nm is reduced by two orders of magnitude.

The same results were obtained at a TMAE pressure of $p > 0.1$ mbar. Typical data are shown in fig. 8a (curve 2). In this experiment, the gas mixture flowing in the counter contains TMAE vapour at 0.45 mbar, which is almost its room-temperature equilibrium vapour pressure. At this partial pressure the photon absorption mean free path is about 20 mm at λ = 200 nm [19], hence 86% of these photons are absorbed in the counter gas path (36 mm) before reaching the cathode. Since the TMAE vapour is ionized by radiation from λ < 230 nm downwards, the observed quantum yield in this region is due to ionization of the vapour. There is good agreement between the present measured shape of Q(λ) and the measurements of Holroyd et al. [8] and Arnold et al. [19]. This demonstrates the reliability of our calibration system.

At λ > 230 nm the gas-phase quantum efficiency is negligible and the full signal is due to the liquid cathode. When the TMAE cathode was heated to $t_{cath} = 40$ to 50°C for several minutes, the sensitivity for λ > 230 nm decreased significantly (compare curves 3 and 2 of fig. 8a), showing that the cathode effect is due to condensation. All these results are easily reproducible,

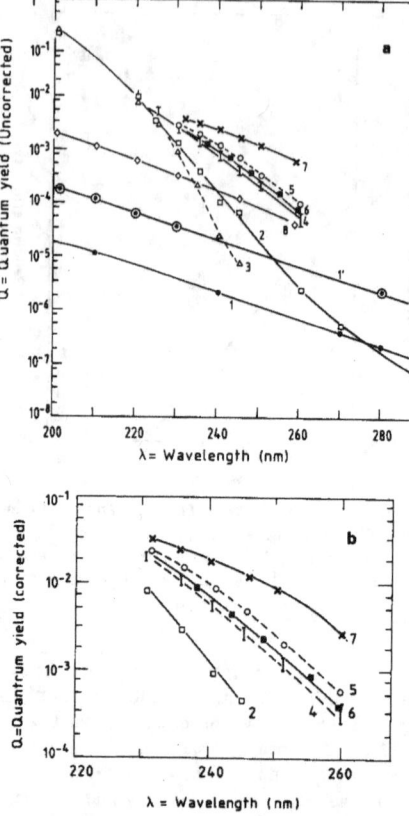

Fig. 8. (a) Experimental results of quantum efficiency Q vs photon wavelength λ for various liquid and solid cathodes deposited on a copper cathode at a pressure of 1 bar unless otherwise noted: (1) and (1′) are the full curves of fig. 5; (2) methane + TMAE, $t_{cath} = 0°C$; (3) the same as 2, but with the cathode heated for several minutes to 40–50°C; (4) methane + TMP + TMAE, $t_{cath} = -15°C$; (5) methane + TMS + TMAE, 1.5 bar, $t_{cath} = -5°C$; (6) methane + NP + TMAE, $t_{cath} = 0°C$; (7) methane + NP + TMAE, $t_{cath} < -20°C$; (8) methane + diethyl ferrocene, $t_{cath} = -40°C$. (b) Experimental results presented in fig. 8a, but corrected for light absorption in the counter gas.

although they depend very much on the deposited cathode material. Corrections are necessary to account for absorption of photons in the counter gas before they reach the cathode. This correction was measured experimentally and, depending on temperature and wavelength, amounted to a factor of about 5 to 7, which is in agreement with the absorptance measurements of Holroyd [8] and Arnold et al. [19]. The results, corrected for absorption, are presented in fig. 8b (curve 2). The corrected quantum efficiency for TMAE is 0.4% at λ = 235 nm, to be compared with Holroyd's value of

0.56% [11] as measured in a liquid ionization chamber where the photoelectrons are not extracted into the gas phase.

Several control experiments have verified that the $\lambda > 230$ nm signal originates at the cathode. They include: (1) changing the bias on the entrance meshes so as to allow photoelectrons produced at the entrance CaF_2 window to be collected (no effect was observed); (2) cooling the counting gas to $0°C$, which caused the signal to increase, as expected, owing to less photon absorption in the counter gas; (3) moving the copper cathode closer to the entrance window (and wire), which increased the signals, also as expected, owing to a reduced absorption path length; and (4) using a copper cathode with a hole, 7 mm diameter, in the centre, which caused an almost complete loss of signal. These independent measurements with a single-wire counter showed also that the sensitivity of the cathode to UV light is increased by the TMAE film. With the cathode at room temperature we observed effects that are intermediate between those of the thin and thick layers.

At $t_{cath} < -5°C$ the TMAE layer becomes solid. The quantum efficiency of this layer was approximately equal to the sensitivity of the liquid layer.

3.3.2. Photocathode with TMAE in solution

We also tested liquid photocathodes obtained from TMS + TMAE, TMP + TMAE, and NP + TMAE condensation. The electrons in these liquids have the highest drift mobility and the longest thermalization lengths r_0 of any known organic compound [9,23]; hence the maximum quantum yield can be expected from them. Before the measurements, however, it was not evident which of the three liquids was the best, since one has maximum μ, the second has maximum r_0, and the third has maximum V_0 (see table 1).

Attempts were made to form photocathodes from these mixtures when the copper cathode was at room temperature. Liquid condensation did not occur, but a thin adsorbed film was formed which had some photosensitivity; however, the liquid was three times better.

In our experiments we obtained a mixture of liquids by cooling the cathode down to $t_{cath} < 0°C$ when vapours of TMAE (with TMS, or TMP, or NP) were flushed through the counter. It is difficult to calculate

Table 1
Electron mobilities and band energies at 295 K and thermalization lengths at $h\nu - E_{liq} \geq 1$ eV [9,22]

	μ [cm^2/V s]	V_0 [eV]	r_0 [nm]
TMS	100	−0.55	~10
NP	70	−0.43	~20
TMP	24	−0.33	~10

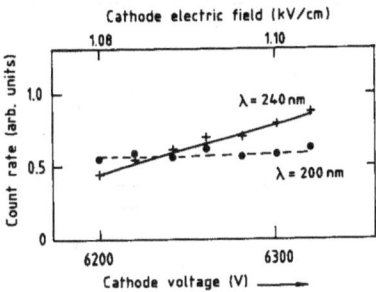

Fig. 9. Count rate vs cathode voltage (or cathode electric field) for two different photon wavelengths, $\lambda = 200$ and 240 nm. The gas flow was methane + TMS + TMAE at 20°C and 1 bar, with the copper cathode at $t_{cath} = 0°C$. At $\lambda = 200$ nm the count rate is mostly due to gas-phase photoionization of TMAE, but at $\lambda = 240$ nm the signal is mainly due to a liquid photocathode.

the TMAE concentration accurately in this kind of liquid photocathode, but we consider it to be at least equal to the ratio of partial pressures. The TMAE concentration could be slightly altered by changing the velocities at which a gas was flushed through the system. The maximum efficiencies attained are given in figs. 8a and 8b. Curves 5 and 6 of fig. 8b show the best liquid-cathode efficiencies obtained. As can be seen, the sensitivity of TMAE in NP and of TMAE in TMS gave ~ 2% at $\lambda = 235$ nm. The TMAE + TMP cathode was always a factor of 1.5 to 2 times less efficient than the TMAE + TMS cathode.

The best result is, however, given by a solid photocathode formed at $t_{cath} = < -20°C$. Curve 7 in fig. 8b shows the results for a solid mixture of TMAE + NP. The efficiency achieved in this case is ~ 3% at $\lambda = 235$ nm.

The quantum yield of the liquid or solid photocathodes increased with increasing electric field at the cathode surface, as shown in fig. 9. A large variation of the field strength at the cathode surface was not possible because of a concomitant increase of the electric field at the wire, causing breakdown or instability. Even though the surface electric field increased from 1.08 to only 1.10 kV/cm, the count rate doubled, showing a very strong field dependence at $\lambda = 240$ nm. No electric-field effect is seen at $\lambda = 200$ nm, because the signal is due to gas photoionization. A striking confirmation of this dependence was provided by replacing the solid copper cathode by a cathode of parallel 100 μm wires with 1 mm spacing. At the wire surface the electric field is calculated to be a factor of 5 higher than for a corresponding uniform cathode. The measured quantum efficiency at $\lambda = 235$ nm for a TMAE + TMS cathode was (6 ± 1)% at room temperature, which should be compared with 1% for the flat cathode also at room

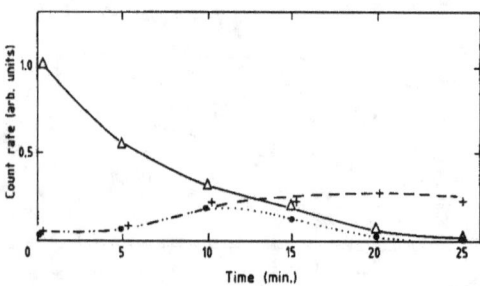

Fig. 10. Count rate vs time when TMAE is bypassed and pure methane is flushed through the single-wire counter at an accelerated rate of 2 l/min. The curve labelled (△) is for λ = 200 nm and an initial TMS (or TMP) + TMAE photocathode; (●) for λ = 245 nm and for an initial liquid TMS + TMAE photocathode; (+) for λ = 245 nm and an initial solid TMP + TMAE photocathode.

temperature. This indicates that it is possible to have high-efficiency cathodes at room temperature, given a large extraction field.

In all the measurements of figs. 8a and 8b methane was always flushed through TMAE and the other liquids (TMS or TMP or NP), and it was the main constituent of the counting gas in the single-wire counter. If, however, isobutane was used for the same purpose, the quantum yields from the liquid cathode were reduced by a factor of 2 to 3 as has been previously noted [24]. This is due to little-known effects at the cathode surface.

The stability of the liquid and solid photocathode mixtures was tested as follows. After forming the mixed cathodes as previously described pure methane was flushed through the single-wire counter by bypassing the TMAE and TMS (or TMP) reservoirs (see fig. 4), and by increasing the flow velocity, by a factor of about 40, to several litres per minute. The results are shown in fig. 10. At λ = 200 nm the signal fell quickly because it was predominantly due to the gas-phase photoionization of TMAE, whereas at λ = 245 nm the signal initially increased in both the TMS and TMP cathodes because of increased gas transparency. The intensity of the pulses from the solid-TMP cathode continued to increase, and stabilized after 20 min to a value about five times the initial value (this is the correction for photon absorption), whereas the TMS cathode reached a peak after 10 min and then decreased to a negligible value after 25 min. This shows that the liquid cathode (TMS + TMAE) evaporated, whilst the solid cathode (TMP + TMAE) remained stable for many hours.

3.4. Other photocathodes

Since solid (or liquid) photocathodes at atmospheric pressure could have many important practical appli-

cations (see section 4), we tested other substances in addition to TMAE, such as tetramethyl-p-phenylenediamine (TMPD), caesium iodide (CsI), anthracene ($C_{14}H_{10}$), diazobicyclo-octane (DABCO), and diethyl ferrocene $(C_5H_4-C_2H_5)_2$Fe, which have a low ionization potential. The best results were obtained with a TMPD cathode. The molecular structure of TMPD is related to that of TMAE, but it is solid even at room temperature and is not reactive with the oxygen in air (for short exposures). A TMPD cathode of about 1 μm thickness was obtained by vacuum evaporation on the standard copper cathode. A second TMPD cathode was fabricated by rapid evaporation of a solution of TMPD in ethyl alcohol (100%) on a warm copper-cathode substrate. Typical results are shown in fig. 11. The curves labelled 1, 2, and 3 are liquid ionization chamber measurements [25,26] of TMPD in TMS, TMP, and NP, respectively (without electron extraction into the gas). The points on the figure show our measurements of quantum efficiency versus wavelength, with electron extraction: (△) for solid TMPD; (×) for TMPD in TMP; (●) for TMPD in TMS; and (□) for TMPD in NP. We note that the solid-TMPD result is lower than that of liquid TMAE by a factor of more than 5 at λ = 235 nm. The mixed cathodes were made by flushing the cooled copper cathode covered by TMPD (from t_{cath} −5 to −40°C) with methane plus the vapours of TMS (or TMP or NP): a solution was probably formed by the absorption of these vapours in the TMPD. The measured points are, as before, somewhat lower than

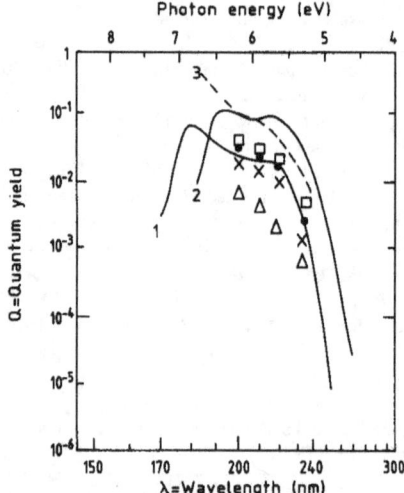

Fig. 11. Quantum efficiency of TMPD cathodes vs photon wavelength λ: (△) is for solid TMPD; (×) for TMPD in TMP; (●) for TMPD in TMS; (□) for TMPD in NP. The curves labelled (1), (2), and (3) are liquid ionization chamber measurements [25,26] of TMPD in TMS, TMP, and NP, respectively.

the liquid ionization measurements. However, some of the difference could be recuperated by increasing the electric field. The measurements of TMPD in NP were not very reproducible, because the NP has such a high vapour pressure that it tends to evaporate quickly. The cathode, however, could not be cooled too much (for better stability) because neopentane solidifies at $-20°C$. The data ($\square$) shown in fig. 11 are those considered to be the most reliable for a solution of TMPD in NP. The solid TMPD + NP cathode (at $t_{cath} < -20°C$) shows an efficiency of about 4% for $\lambda = 235$ nm, but this cathode has a spontaneous emission of about 10 to 30 photoelectrons per second.

Another solid cathode was tested: diethyl ferrocene, whose gas-phase photoionization threshold is $\lambda = 184$ nm. The result, given as curve 8 in fig. 8a, showed lower efficiency than with the TMAE liquid, but still with a factor of 80 increase over the bare copper cathode up to $\lambda = 260$ nm.

3.5. Discussion

Measurements with a single-wire counter confirm and extend, in a totally independent way, the qualitative results obtained from a wire chamber. Three effects which increase the sensitivity to photons were observed; these were due to (1) a thin adsorbed film; (2) real liquid condensation; (3) the formation of a solid layer. Let us discuss these effects separately.

3.5.1. Thin adsorbed film

Generally speaking, the effect of a thin molecular film on the efficiency of the cathode may have a simple explanation: in order to escape from the surface of the metal, the electron must overcome a barrier defined by the work function ϕ. When there is contact between a metallic surface and a liquid film possessing a conduction band energy V_0, the work function of the metal decreases if $V_0 < 0$. In this way, electrons find it easier to overcome the metal–liquid barrier than the metal–gas barrier. However, to escape from the liquid film, the electrons must overcome the conduction band energy. Depending on the relationship of these two factors, the liquid film can either increase or reduce the quantum efficiency of the cathode. In the particular case of thin TMAE and TMPD films, the metal work function is evidently lowered – owing to the predominant orientation of the adsorbed molecules – in such a way that the π-electrons enter the metal and form an electric dipole layer (for details see, for example, refs. [27] and [28]). This explains why, in the case of a copper or an aluminium cathode, we observed an increase in sensitivity, of one order of magnitude, up to 400 nm. This also explains why it is so difficult to remove the TMAE layer from a metallic surface: it is only by heating for a

period of a few days that the surface can be completely cleansed.

3.5.2. Liquid layer

If there is real liquid condensation, the sensitivity to short wavelength radiation ($\lambda < 270$ nm) rises sharply. In this case the radiation does not reach the metal because of absorption in liquid, and electrons are produced mainly through photoionization of the liquid molecule. In this way, the metallic cathode itself plays no direct part in the generation of photoelectrons [*]. The following mechanism seems plausible. It was found that for liquid cathodes with $V_0 \approx -0.2$ eV, a quantum yield Q is close to the value obtained with liquid ionization chambers. For example, the authors of refs. [11] and [12] have measured, in a liquid ionization chamber, a quantum efficiency Q_{liq} of 0.56% at $\lambda = 235$ nm in pure TMAE, 4.6% for 1% TMAE in 2,2′,4-trimethyl pentane (closely related to TMP = 2,2′,4,4′-tetramethylpentane), and of 0.2% for TMPD in trimethyl pentane. Our measurements give 0.4% for pure TMAE, 1.2% for TMAE in TMP, and 0.1% for TMPD in TMP. For liquids with $V_0 \approx -0.5$ eV, $Q = (0.1-0.3)Q_{liq}$. Indeed, at $\lambda = 235$ nm, Q_{liq} is 16% for 1% TMAE in TMS, ~3% for TMPD in TMS, and 0.15% for TMPD in NP [11,25,26]. Our measurements, with electron extraction, give 2% for TMAE in TMS, 0.3% for TMPD in TMS, and 0.05% for TMPD in NP at the same λ. This may be explained in the following way. In experiments with a liquid ionization chamber the true quantum yield of the liquids was measured. However, in the case of the liquid photocathode, the electrons have to be extracted from the liquid to the gaseous phase. To do so, the kinetic energy of their random motion (not of the drift motion, which is smaller) has to be higher than $-V_0$; in other words the electric field at the cathode must be greater than a given value E_{crit}. For example, for liquid Ar, $V_0 \approx -0.2$ eV and $E_{crit} \approx 5$ kV/cm [29]. In a recent work of Anderson et al. [30] the full extraction of electrons from liquid TMP to the gas phase has been observed with $E \approx 30$ kV/cm. In our case, however, the electric field at the cathode was $E \approx 1$ kV/cm, which is not great enough for efficient extraction. This explains why, in the case of $V_0 < -0.2$ eV, $Q < Q_{liq}$.

One way of increasing the quantum yield is therefore to apply the maximum possible electric field at the cathode. This will increase the probability of the electrons overcoming the potential barrier and reduce the likelihood of geminate recombination and trapping by

---

[*] This fact may explain results obtained with an Al cathode (see subsection 3.3 and fig. 5) if one supposes that the attachment of Al to TMAE is stronger than that of Cu, and therefore that a thicker layer of TMAE can be formed after 20 h.

impurities. In fact, in the experiments with the multi-wire cathode plane with $E \approx 5$ kV/cm, we were able to obtain $Q = 5$ to 7% for $\lambda = 235$ nm (TMAE + TMS at 20°C).

If an electron is born inside the liquid, it has the kinetic energy $K = h\nu - E_{liq}$, where E_{liq} is the ionization threshold of the liquid (see eq. (1)). If the electron is close to the surface, and if $K > -V_0$, it will have a high probability of escaping from the liquid, even in a relatively weak electric field. The thermalization length $r_0 = 10$ nm in TMS and NP is 5 to 10 times less than the photon absorption length in solutions with TMAE, so that only a small fraction of the photoelectrons can be extracted in this way.

The results obtained make it possible to link up all the disparate findings of earlier published literature which were mentioned in the Introduction. This literature may be arbitrarily divided into research of two kinds: (1) measurements without spectral resolution, and (2) measurements with spectral resolution. In the experiments without spectral resolution, efforts were made to observe the part played by a thin adsorbed TMAE film in wire chambers; this was seen from the time structure of the sequences of pulses coming from the different electrodes and the different gas gaps [15,16]. Based on our measurements with a single-wire counter, the proportion of the signal due to surface photoeffect can be estimated. To do this, it is enough to calculate the convolution of the spectrum of the BaF_2 emission [31] in the 180 to 300 nm range with the quantum efficiency of TMAE vapour and of the surface photoeffect. Our calculations show that for a wire chamber with a 1 cm gap and with methane flushing, the surface contribution would be 6% for TMAE and 12% for TMS + TMAE. This is in good agreement with our measurements mentioned in subsection 2.3. For isobutane the effect would be half as much, i.e. it would correspond to 3% and 6%. For gaps of > 1 cm the contribution would be correspondingly lower. Such a small contribution from the surface relative to the gas can only be observed under conditions of very good statistics, which was not the case in most attempts referred to above.

Let us now turn to experiments carried out with spectral resolution. In our section 1 we pointed out that in ref. [14] – where, as in our own case, a monochromator was used – the photoeffect from TMAE condensed on the surface was clearly observed. In ref. [8] the surface photoeffect was investigated on a dielectric window, so that to compare these results with our own would be incorrect. In ref. [15] the spectral measurements were carried out by means of filters, and the authors obtained the following estimates: $signal_{230\,nm}$ to $signal_{210\,nm} = 7 \times 10^{-2}$ and $signal_{250\,nm}$ to $signal_{210\,nm} = 3 \times 10^{-3}$, which coincide closely with our own measurements (see fig. 8a).

As can be seen from this cursory analysis, the results of our experiments do not conflict with any work known to us.

3.5.3. Solid photocathode

In this work we have discovered that some solid photocathodes have a sensitivity that is equal to or even better than that of the corresponding liquids. This fact has a natural explanation because, in the solid state, V_0 as well as μ is usually higher than in liquids [23]. These two factors contribute to achieving higher sensitivity for VUV. In the special case of neopentane, V_0 becomes positive in the solid phase [23,32]. This explains why, in the case of solid NP, we observe the highest efficiency: about 3% at $\lambda = 235$ nm.

4. Applications

Liquid and solid organic photocathodes can find a wide range of applications in diversified fields. The main interest of such photocathodes lies in the fact that large surfaces, in the square meter range, are relatively easy to produce without having to go to the high vacuum technique. Also, they are compatible with most of the gases used in wire chambers or in other gaseous amplifying structures. This permits position accuracies in the millimeter range coupled to time resolutions of better than a nanosecond, since the absence of parallax in the film absorbing the UV photons suppresses a major cause of position and time jitter. It is foreseen to use such cathodes to read out scintillation light from liquid argon or xenon. This would probably improve the energy resolution of liquid Ar or Xe chambers [33]. On the other hand, a solid photocathode in contact with liquid Ar would allow us to build fast calorimeters [34]. The present efficiency reached in our work, i.e. 3% at 235 nm, is far from being the maximum achievable with this method. In high-energy physics, one often encounters situations where the number of photons is large enough to make this efficiency attractive.

These cathodes can be used in plasma diagnostics, astrophysics, X-ray spectroscopy, and medical imaging.

Acknowledgements

The authors would like to tender their appreciation to R. Bouclier, G. Million and R. Saigne for their technical assistance.

References

[1] J. Séguinot and T. Ypsilantis, Nucl. Instr. and Meth. 142 (1977) 377.

[2] R. Arnold et al., Nucl. Instr. and Meth. A252 (1986) 188.

[3] G.F. Karabadjhak, V.D. Peskov and E.R. Podolyak, Nucl. Instr. and Meth. 217 (1983) 56.

[4] C.Y. Ng, Adv. Chem. Phys. 52 (1983) 263.

[5] Y. Nakato, M. Ozaki and H. Tsubomura, Bull. Chem. Soc. Japan 45 (1972) 1299.

[6] S. Ekelin, Photoionization quantum efficiency measurements of photosensitive compounds for use in Cherenkov ring-imaging detectors, Internal report, Royal Institute of Technology, Stockholm (1981).

[7] E. Barrelet et al., Nucl. Instr. and meth. 200 (1982) 219. This paper gives the TMAE and TEA relative quantum efficiencies of the preceding reference versus photon energy. The peak quantum efficiencies have been established from Cherenkov ring-imaging data as 0.67 for TMAE and 0.55 for TEA.

[8] R.A. Holroyd et al. Paper 6408, submitted to the 23rd Int. Conf. on High-Energy Physics, Berkeley (1986). This paper gives absolute quantum efficiencies and photon mean free paths for TMAE and TEA versus photon wavelength. The peak TMAE and TEA quantum efficiencies disagree with ref. [7] (e.g. 0.56 and 0.32 compared with 0.67 and 0.55, respectively). See ref. [19] for details.

[9] B.S. Yakovlev and L.V. Lukin, Adv. Chem. Phys. 60 (1985) 99.

[10] R.M. Minday, L.D. Schmidt and H.T. Davis, J. Chem. Phys. 54 (1971) 3112.

[11] R.A. Holroyd, S. Ehrenson and J.M. Preses, J. Phys. Chem. 89 (1985) 4244.

[12] D.F. Anderson, Phys. Lett. B118 (1982) 230.

[13] D.F. Anderson et al., Nucl. Instr. and Meth. 217 (1983) 217.

[14] P. Schotanus et al., Nucl. Instr. and Meth. A252 (1987) 255.

[15] J.C. Michau et al., Nucl. Instr. and Meth. A244 (1986) 565.

[16] M. Suffert and G. Charpak, CERN EP Internal Report 86-03 (1986).

[17] P. Miné et al., IEEE Trans. Nucl. Sci. NS-34(1) (1984) 458.

[18] M. Laval et al., Nucl. Instr. and Meth. 206 (1983) 169.

[19] R. Arnold et al., submitted to Nucl. Instr. and Meth. B (1987).

[20] K. Watanabe, J. Opt. Soc. Am. 43 (1953) 32.

[21] J.A.R. Samson, Techniques of VUV spectroscopy (Wiley, New York, 1967) p. 216.

[22] T. Ekelöf, J. Séguinot, J. Tocqueville and T. Ypsilantis, Phys. Scr. 23 (1981) 718.

[23] H.A. Holroyd, private communication.

[24] J.S. Edments et al., Studies of a gas-filled UV detector with semitransparent photocathode, Instrument Technology, Ltd., unpublished.

[25] H.T. Choi et al., J. Chem. Phys. 77 (1982) 6027.

[26] E.H. Böttcher, Ph.D. Thesis, Free University Berlin (1984).

[27] D.P. Woodruff and T.A. Delchar, Modern techniques of surface science (University Press, Cambridge, 1986).

[28] D.A. King and D.P. Woodruff, in: The chemical physics of solid surface and heterogeneous catalysis (Elsevier, Amsterdam, 1984) vol. 3, part B.

[29] W.F. Schmidt, IEEE Trans. Electr. Insul. EI-19 (5) (1984) 389.

[30] D.F. Anderson, G. Charpak, R.A. Holroyd and D.C. Lamb, Fermilab Pub-87/81, 2110.000 (1987) Nucl. Instr. and Meth., in press.

[31] P. Schotanus et al., IEEE Trans. Nucl. Sci. NS-34 (1987) 272.

[32] D. Grand and A. Bernas, J. Phys. Chem. 81 (1977) 1209.

[33] Proc. UCLA Conf. on Low-Level Counting in Liquid Drift Detectors, Los Angeles (1987). See discussion on X-ray energy resolution.

[34] P. Pétroff, J. Séguinot and T. Ypsilantis, Proposal CERN/LAA P23 (1987).

NEW SCINTILLATORS FOR PHOTOSENSITIVE GASEOUS DETECTORS

G. Charpak, V. Peskov and D. Scigocki

CERN, Geneva, Switzerland

and

J. Valbis

Inst. Solid-State Physics, Latvian State University, Riga, USSR

Abstract

A new family of scintillators are presented. Their properties are similar to those of barium fluoride, and the spectrum of the scintillation emission is between 140 and 300 nm. Our latest efficiency measurements of ethyl ferrocene and triethylamine liquid or caesium iodide solid photocathodes, in parallel-plate avalanche chambers (PPACs) at high electric field, are also presented. We discuss the revolutionary consequences of the combination of the new scintillators with PPACs with semitransparent photocathodes deposited on the crystals, such as high speed, high resistance to radiation damage, compacity, high gamma efficiency, and applications to tracking devices with scintillation optical fibres.

1. Introduction

The proposal to couple a dense inorganic scintillator, barium fluoride (BaF_2), with a low-pressure multiwire proportional chamber (MWPC) filled with a photosensitive vapour such as tetrakis-(dimethylamino)ethylene (TMAE) or liquid TMAE photocathode, in order to detect its light emission, came from Anderson et al. [1-3]; they named the device the solid scintillation proportional counter (SSPC). The properties of BaF_2 are: a high density (4.9 g/cm^3); a short radiation length (2.05 cm); a very fast ultraviolet scintillation component with a decay time of 600 ps [4]; a photon yield that is independent of temperature [5, 6]; and a resistance to radiation damage up to 10^5 Gy [7, 8]. Adding these advantages to those of a low-pressure chamber — resulting in a very fast collection time and good time resolution [9], a high rate capability, and a very low sensitivity to direct ionization — made the SSPC working with TMAE vapour a very attractive device for many applications, such as nuclear medicine [10, 11] and electromagnetic calorimetry [12]. But nobody succeeded in using liquid TMAE photocathodes in an SSPC, because of its low efficiency [13].

Until recently BaF_2 was the only known scintillator to possess adequate properties; on the other hand, the only photosensitive compound that had been used with MWPCs to detect the fast component emitted by BaF_2 was TMAE.

The discovery of a new vapour, ethyl ferrocene (EF), photosensitive to the BaF_2 fast emission, broke the monopole of TMAE vapour in SSPCs [14]. Also a new type of organic photocathodes, much more efficient in the VUV region than liquid TMAE, was investigated by Peskov et al. [13]; some of these could be used to detect the BaF_2 fast scintillation in an SSPC [13, 15]. A new concept of SSPC was then developed successfully: it consists of BaF_2 crystals coupled with a parallel-plate avalanche chamber (PPAC) working at room temperature and with a gas mixture at atmospheric pressure [16]. This type of detectors permits us to keep most properties of the traditional SSPC, with high-temperature, low-pressure wire chambers filled with TMAE vapour, eliminating its main disadvantages (see Section 3).

The other way to improve the SSPC would be to find new scintillators that would have properties similar to those of BaF_2, but a shorter emission wavelength, in the range where the known photo-sensitive vapours and photocathodes achieve higher efficiencies. The discovery of a scintillation peaked at 173 nm, with a decay time of ~ 6.3 ns, in lanthanum fluoride (LaF_3) crystals doped with Nd^{3+} allows its use in an SSPC filled with TMAE vapour [17]. The light yield is very small and the detection efficiency was not higher than that obtained with the BaF_2 device. The main advantages of LaF_3 over BaF_2 are its higher density (5.94 g/cm^3) and its shorter radiation length ($X_0 \simeq 1.7$ cm).

On the other hand, the mechanism by which BaF_2 emits light with a fast component (which gave it its unique properties) was not understood. Studies made by Jansons et al. [18] have now allowed them to explain this fast scintillation process. Based on this research, a new family of inorganic scintillators was discovered [18, 19]. Their properties are similar to those of BaF_2, but they emit light in a wavelength region between 140 nm and 300 nm.

In this paper we detail the characteristics of these scintillators, and discuss the implications of their use in SSPCs for a wide range of applications. In particular, we show the revolutionary consequences of the association of this new type of scintillators with a PPAC at room temperature and at atmospheric pressure, using semitransparent photocathodes as readout.

2. The new scintillators

Recently Valbis and his group investigated the luminescent properties of heavy alkali and alkaline-earth metals such as BaF_2, CsCl (caesium chloride), RbF (rubidium fluoride), etc. [18], of KF (potassium fluoride), $KMgF_3$ (potassium magnesium fluoride), $KCaF_3$ (potassium calcium fluoride), and solid solutions of KF + RbF [19]. In these crystals the energy separation between the valence band, involving mainly the halogen electronic states, and the upper cation core band is smaller than the band gap. After excitation generating holes in the core band, radiative electronic transitions from the valence band to the core band take place, giving rise to a specific emission which these authors proposed to call cross-luminescence (CRL).

The spectra and decay kinetics of CRL were studied under excitation by a 7 keV electron beam, with current density ≈ 100 $\mu A/cm^2$ in the pulsed regime. The experimental set-up, based on a Seya–Namioka-type vacuum monochromator with a concave diffraction grating (1200 grooves/mm) was used as described in Ref. [20]. The spectra were measured in a photon-counting regime, using a FEU-106-type photomultiplier tube with a magnesium fluoride (MgF_2) entrance window. The decay kinetics were studied using the correlated photon-counting method under excitation by 100 ns long electron pulses having a decay time of 2 ns, which was also the limit of the minimum measurable CRL decay times.

The fluoride crystals were grown by the Bridgman method, in graphite crucibles, in fluorizing or inert atmosphere. The chlorides and bromides were grown in fused silica crucibles, in air. For CRL measurements, small samples (about $1 \times 1 \times 0.5$ mm^3) were fixed to the cryostat's cold-finger using indium metal.

In these studies, the fast emission spectrum of BaF_2 (see Fig. 1 [18]) was explained as coming from the radiative electronic transitions between the F-2p valence band and the Ba^{2+}-5p core band. Since the ionization threshold of the Cs^+-5p core bands of caesium halides is lower than that of other alkali halides' core bands, the CRL was anticipated in the region of relatively low photon energy. Figure 2 (from Ref. [18]) presents the spectra of the fast component of the emission of two Cs halides, CsCl and CsBr. (caesium bromide). It was expected [18] that, with RbF, the whole CRL spectrum would also correspond to the transparency region of the crystal; consequently, the CRL intensity would be comparable to that in CsCl and CsF (caesium fluoride). The results confirmed these expectations (see Fig. 3 from Ref. [18]).

In order to illustrate the fact that the emission wavelength of these scintillators can be controlled, we show in Fig. 4 (from Ref. [19]) the different emission spectra of KF when the Rb^+ concentration is changed. Such wavelength shifts could be used, for instance, to obtain a better overlap of the emission spectrum with the transparency region. The CRL spectra of $KMgF_3$ and $KCaF_3$ (see Fig. 5, from Ref. [19]) exhibit two main maxima, of nearly equal intensity. The total CRL intensity is comparable to that of BaF_2; it remains constant (within 10%) in the temperature range between 80 K and 400 K. The CRL spectrum lies entirely within the transparency region. As all the scintillators described in this paper, $KMgF_3$ and $KCaF_3$ have an emission decay time (< 2 ns) comparable to that of the BaF_2 fast emission. In addition, their resistance to radiation damage [21] was also estimated to be comparable to that of BaF_2.

All these properties, combined with an emission wavelength between 140 and 300 nm, made these two scintillators an interesting alternative to BaF_2, their main disadvantages being their lower density (~ 3 g/cm^3) and long radiation length ($X_0 \approx 7$-8 cm). However, these scintillators are not unique, and others are being developed by the same group; belonging to the same family, they have the same general characteristics, but their atomic number and density are higher.

Recently, Valbis and his group studied BaF_2 crystals doped with a large proportion of hydrogen (up to one atom of H for one atom of Ba). Their preliminary measurements showed that these crystals kept the other properties the same as those of the pure BaF_2.

3. Application to SSPCs

3.1 Wire chambers filled with photosensitive vapours

All the scintillators quoted in the first section emit light in the wavelength region below 300 nm. Some of them, such as $KMgF_3$ and $KCaF_3$, have properties similar to those of BaF_2 and, in addition, have a maximum light emission between 150 and 180 nm. In this region the common photosensitive vapours used in SSPCs, such as TMAE, have achieved very high efficiencies (between 35 and 60% for TMAE, see Fig. 6); other photosensitive vapours, such as triethylamine (TEA), nitrogen monoxide (NO), toluene, ..., are sensitive to this scintillation (see Fig. 6). Some of them, e.g. TEA, have a rather high vapour pressure at room temperature and do not interact with air, simplifying their manipulation. It was thus attractive to use TEA to detect the emission of these new scintillators in SSPCs. This was done by Buzulutzkov et al. [21], who coupled a $KMgF_3$ crystal with a wire chamber filled with a gas mixture of methane (CH_4) + 7.5% TEA at one atmosphere. They measured an efficiency of ~ 4-5 photoelectrons per MeV of energy deposited in the crystal. The sensitivity is half that obtained with TMAE and BaF_2, but a major improvement is the possibility to work at room temperature, whereas the temperature needed to achieve a high enough pressure, and thus efficiency, is 40 °C with TMAE and 80 °C with EF. However, even with TEA vapour, it is still necessary to work at low pressure in order to exploit the good timing properties and insensitivity to direct ionization of traditional SSPCs. Furthermore, the BaF_2 fast emission was recently detected also with a layer of TEA adsorbed on a BaF_2 crystal in an SSPC [22], making it less interesting to use the new scintillators with TEA. It seems more attractive to use them with TMAE vapour: the efficiency would then be higher, by factors of at least 5 to 10, than with TEA (see Fig. 6). This high sensitivity would not only greatly improve the results obtained with BaF_2 in electromagnetic calorimetry [12] and in positron emission tomography (PET) [10, 11], but it would also open a new range of applications in astrophysics, industrial imaging, nuclear medicine imaging, and so on.

3.2 Parallel-plate amplification counters (PPACs)
with liquid or solid semitransparent photocathodes

The main physical limitations of the traditional SSPC, made with BaF_2 and a low-pressure wire chamber filled with TMAE, are set, on the one hand by the fast ageing effect due to polymerization observed with TMAE in a wire chamber [23], and on the other hand by the small spectral overlap between the BaF_2 fast scintillation and the TMAE quantum efficiency (see Fig. 7), which had allowed only a small part of the BaF_2 fast emission to be detected. The practical disadvantages of this detector are that it has to be operated at low pressure in order to keep the properties of this traditional type of SSPCs, and also that TMAE vapour is very reactive to air and corrosive for many materials. Besides, in order to increase the vapour pressure of TMAE in the chamber, and hence the efficiency, it is necessary to work at high temperature. The discovery of the new photosensitive vapour, ethyl ferrocene (EF) [14], interesting since it does not interact with air, did not allow us to increase the efficiency with respect to TMAE vapour, and did not solve the problem of high temperature. The solution may come from the work of Peskov et al. [13], who observed that there is an increase of up to 3 orders of magnitude of the quantum efficiency of metallic photocathodes when they are covered by liquid or solid layers of organic compounds (see Fig. 8). The shift in ionization potential, which is smaller when these organic compounds are in the liquid or solid phase than in the gaseous one [13], allows some of these compounds to be used, in an SSPC, to detect the BaF_2 fast scintillation. Semitransparent photocathodes of liquid EF, adsorbed or condensed on a BaF_2 crystal, were tested in an SSPC containing a mixture of He + CH_4 at atmospheric pressure and room temperature with good results [15]. Semitransparent photocathodes deposited directly on the surface of the scintillator crystal serve as the photosensitive element in the SSPC; this makes it possible to eliminate the gap that is needed to convert the VUV light emission in the photosentitive vapour. The low-pressure chamber can then be replaced by a PPAC working at atmospheric pressure and room temperature, directly coupled with the crystal, and keeping similar properties.

A prototype, made of a BaF_2 crystal covered by an EF (condensed or adsorbed) layer as the photocathode, coupled to a PPAC of 4 mm thickness and filled, at atmospheric pressure, with a gas mixture of helium and 100 Torr of tetramethyl silane (TMS), worked successfully at room temperature [16]. The properties of this new type of readout for SSPCs are: compacity (3–4 mm thickness), very fast collection time ($\leqslant$ 10 ns) and very good time resolution ($\leqslant$ 1 ns), negligible sensitivity to direct ionization at atmospheric pressure (with helium) and to a magnetic field [24], high gain (single-photoelectron detection), high rate capability [24] (no space-charge effect), and we did not observe any ageing effect due to polymerization. Nevertheless, the efficiency obtained with this new device is no higher than the traditional one, with a low-pressure wire chamber and TMAE vapour.

In order to improve this new SSPC, it would be attractive to find photocathodes that detect the BaF_2 fast scintillation more efficiently. A promising result was obtained with a solid photocathode, made with a new compound [16] called CPIHMB (for cyclopentadienyl-iron-hexamethyl-benzene) (see Fig. 9). It belongs to a large new family of organometallic compounds, discovered by Astruc [25]. They have good chemical stability, associated with very low ionization potential (some of them between 4.2 and 4.8 eV in the gaseous phase). A programme of measurements of the quantum efficiencies of these new promising compounds, in the gaseous and in the liquid or solid phases, is being developed.

The most revolutionary improvement for the SSPC concept is the coupling of the new scintillators, covered with semitransparent photocathodes, in the liquid or the solid phases, with PPACs working at room temperature and atmospheric pressure. The advantage of this type of counter, developed originally with BaF_2, was explained above. Moreover, our calculations show that it is very inefficient, for photon collection, to detect the scintillation light with photosensitive vapours in SSPCs, because all these crystals have a high index of refraction (n = 1.56 for BaF_2 and n $\simeq$ 1.5 for $KMgF_3$): by internal reflection, only a small part of all the photons produced by the scintillation escape out of one face of the crystal. This phenomenon is also well known when the BaF_2 is coupled to a photomultiplier (PM) — the light yield then increases by a factor of at least 2, if a good optical contact is obtained by applying an optical grease between the surface of the crystal and the PM window. In the case of liquid or solid semitransparent photocathodes, which are of course in direct contact with the surface of the crystal, the light yield increases significantly, since the difference of refraction index is smaller: typically n = 1.3 for liquid photocathodes. We estimate that the light yield would increase by a factor of 2.5 to 5 (depending on the transparency of the crystal).

These organic photocathodes achieve rather good efficiencies already in regions of relatively high wavelengths, e.g. ~ 2-7% at 235 nm (see Fig. 8). This efficiency can increase strongly in regions of shorter wavelengths; see, for example, Fig. 10, where the increase of the efficiency of an EF liquid photocathode is of at least one order of magnitude when the wavelength drops from 235 to 200 nm. In the wavelength region of maximum emission of the new scintillators (for $KMgF_3$ and $KCaF_3$ between 150 and 180 nm), the quantum efficiencies of our photocathodes would be much higher, and could probably achieve values equivalent to what can be obtained with a photomultiplier. Moreover, it was observed by Holroyd et al. [26] that the number of photoelectrons extracted from the liquid photocathode into the gaseous phase increases with the intensity of the applied electric fields. This effect was verified by our measurements in SSPCs. For example the increase of efficiency measured between 1 kV/cm and 5 kV/cm is shown in Fig. 8 for a photocathode made of methane + neopentane (NP) + TMAE, cooled at T < −20 °C; Fig. 10 shows this increase between 1 kV/cm and 7.5 kV/cm for an EF-liquid photocathode. As a consequence, in a PPAC, in which the applied electric fields are higher than ~ 10 kV/cm, the use of the liquid photocathodes optimizes the quantum efficiency of the SSPC. Even if the sensitivity is generally rather lower, it is more attractive, from the point of view of practicability (e.g. manipulation) and of stability, in SSPCs, to use solid photocathodes at room temperature than liquid ones. A good candidate would be pure tetramethyl-p-phenylenediamine (TMPD), in which a relatively high efficiency was achieved at 200 nm [13]; another could be caesium iodide (CsI), with a quantum efficiency of ~ 10% in the wavelength emission region of the new scintillators (see Fig. 6). We estimate that a CsI photocathode should achieve an efficiency equivalent to or higher than that obtained with TEA in an SSPC with $KMgF_3$.

In addition, it is estimated that PPACs are much less sensitive to the ageing effect due to polymerization observed in the majority of proportional wire counters; this is because the local charge density is much lower in a PPAC than around wires. Moreover, in order to avoid photon feedback on the photocathode and to reach a high gain in the PPAC, it is necessary to use a good quencher in the gas mixture, tetramethyl silane (TMS) for example. Next, the quenchers we introduce should have very good ageing properties; the best known so far candidate is dimethyl ether (DME). No ageing effect was observed in proportional wire chambers filled with pure DME up to 1.2 C per centimetre of wire [27].

We expect that the association of the new scintillators, covered with CsI semitransparent photo-cathodes, with a PPAC filled with helium and DME, could achieve properties and results equivalent to those obtained with BaF_2 associated with a low-pressure wire chamber filled with TMAE. In addition, there would be no ageing effect in the counter, the radiation resistance would be equivalent to that of BaF_2, and the compacity would be better: ~ 3 mm as against 1 cm.

4. Discussion

An electromagnetic calorimeter made with PPACs as described above, estimated to be very resistant to radiation damage, to have a time resolution of < 1 ns and a signal collection time of < 10 ns, with an energy resolution of ~ $4\%/\sqrt{E(GeV)}$ or better, would be very well adapted to the needs of experiments in high-luminosity accelerators, such as the SSC and the LHC. In many other applications, positron emission tomography (PET) for instance, this type of SSPC would be a great improvement, since it could cover large surfaces at a low price, with a good position resolution. If electromagnetic calorimetry is important in high-energy-physics experiments, another aspect, as essential, of the experiments planned for the high-luminosity accelerators is hadronic calorimetry. In fact, homogeneity would make it necessary to use the same technique in electromagnetic and hadronic calorimetry.

Ideally, these new detectors should possess all the following qualities:
- high speed,
- high resistance to radiation damage,
- ability to realize compensation phenomena, in order to achieve a good energy resolution,
- technical ease of building,
- compacity.

Classical sampling calorimeters made of uranium or lead and organic scintillators [28, 29], or those based on the new approach developed by Wigmans [30], with lead and scintillating fibres, are fast enough and can achieve compensation, but their radiation resistance is not high enough in the hot region of the SSC or LHC, whereas those made of uranium with a warm liquid such as TMP [30] are good on radiation resistance but not fast enough. Also, they are technically difficult to realize. To make a hadronic sampling calorimeter, it would be of advantage to use SSPCs made of a PPAC and photocathodes with inorganic scintillators, such as BaF_2 or the other ones presented in this paper, sampled with uranium or lead. Such a calorimeter would have all the qualities required above, except for the compensation, since the inorganic scintillators do not contain the hydrogen which is necessary to achieve it.

The BaF_2 crystals doped with a large proportion of hydrogen may be of interest for the purpose. Then a fine-sampling hadronic calorimeter, using uranium or lead and this 'new' crystal coupled to PPACs with semitransparent photocathodes as readout, should combine the speed and technical practicability of the uranium (or lead)/organic scintillator detectors with the good resistance to radiation damage of the uranium/warm liquid detectors. In addition, this compact calorimeter would also probably achieve compensation.

A recent step forward in the manufacturing of BaF_2 optical fibres[*] opens considerable prospects for the solution of an important problem in detector physics. While scintillating fibres present attractive

---

*) D. Winn (Fairfield University), private communication.

features for many problems in particle physics, they suffer from a major defect when they are compared to gaseous detectors, namely the difficulty to delay the information for a time long enough for the selection and gating of the events of interest to be made.

The solution that has been found for the manufacturing of BaF_2 fibres can be extended to a vast variety of scintillators. It consists in depositing the substance on a quarz fibre, by chemical vapour deposition (CVD). The author[*] advocates that BaF_2 films of 1 to 100 μm can be deposited on quartz fibres of diameter 10 μm to 500 μm. This opens the way for devices where the light coming from a bundle of such fibres is converted into electrons in a gaseous photosensitive device; there the photo-electrons are amplified, delayed, gated and multiplied again in a multistep device, with the emission of light induced by the adequate addition of a vapour such as TEA or TMAE and the detection of the image pattern in a CCD. With some of the new scintillators, the photosensitive device could consist of the thin transparent photocathodes discussed previously, with efficiencies of the order of 10%. Since a single photoelectron can easily be imaged with the multistep chamber, this may lead to the efficient detection of the energy deposited in the fibres, although the yield of light is smaller for these scintillators than in the plastic scintillation fibres investigated up to now for particle tracking or calorimetry. It is an alternative approach to a solution imagined by Sauli[**], where visible light is transformed into VUV by accelerating electrons, in vacuum, between a photocathode sensitive to visible light and a thin layer of BaF_2.

5. Conclusion

The perfect overlap between the emission spectra of the new scintillators and the quantum efficiencies of the photosensitives elements used in gaseous detectors would strongly improve the SSPC concept. In particular, an SSPC made of these scintillators covered with solid or liquid photocathodes, coupled with a PPAC working at room temperature and at one atmosphere, would reach a sensitivity that would be practically equivalent to that of a PM ($\geqslant$ 10%), keeping the properties of proportional gaseous detectors, i.e. compacity, possibility to cover large surfaces, negligible sensitivity to magnetic fields, low noise, ... In addition, this detector would be very fast (time resolution $\leqslant$ 1 ns and collection time of the signal $\leqslant$ 10 ns), with a high rate capability and a very high resistance to radiation damage.

All these characteristics should allow substantial improvements in many fields of application, at low energies (PET, for example) as well as high energies (electromagnetic and hadronic calorimetry, for instance).

We expect that a new field of applications would also be opened with the scintillating fibres made of BaF_2 and probably of the new scintillators, coupled with photosensitive gaseous detectors as readout.

This work was carried out in the framework of the LAA Project.

---

[*] D. Winn (Fairfield University), private communication.
[**] F. Sauli, Proposal for the development of an optic-electronic image delay tube, Proposal for LAA project, CERN, 6 June 1988.

References

[1] D.F. Anderson, Phys. Lett. **18** (1982) 230.

[2] D.F. Anderson, R. Bouclier, G. Charpak and S. Majewski, Nucl. Instrum. Methods **217** (1983) 217.

[3] D.F. Anderson, G. Charpak, Ch. von Gagern and S. Majewski, Nucl. Instrum. Methods **225** (1984) 8.

[4] M. Laval, M. Moszynski, R. Allemand, E. Cormorèche, P. Guinet, R. Odru and J. Vacher, Nucl. Instrum. Methods **206** (1983) 169.

[5] P. Schotanus, C.W. van Eijk, R.W. Hollander and J. Pijpelink, Nucl. Instrum. Methods **A238** (1985) 564.

[6] M. Suffert and G. Charpak, CERN EP Internal Report 86–03 (1986).

[7] S. Majewski and D. Anderson, Nucl. Instrum. Methods **A241** (1985) 76.

[8] A.J. Caffrey et al., IEEE Trans. Nucl. Sci. NS-33 (1986) 230.

[9] A. Breskin, Nucl. Instrum. Methods **196** (1982) 11.

[10] P. Miné et al., IEEE Trans. Nucl. Sci. NS-34 (1987) 458.

[11] P. Miné et al., Nucl. Instrum. Methods **A269** (1988) 385.

[12] R. Bouclier et al., Nucl. Instrum. Methods **A267** (1988) 69.

[13] V. Peskov et al., Nucl. Instrum. Methods **A269** (1988) 149.

[14] G. Charpak, V. Peskov, F. Sauli and D. Scigocki, CERN-EP Internal report 88-02 (1988).

[15] G. Charpak, V. Peskov, F. Sauli and D. Scigocki, preprint CERN-EP/88-166 (1988), submitted to Nucl. Instrum. Methods.

[16] V. Peskov, G. Charpak, F. Sauli and D. Scigocki, Organometallic photocathodes for parallel-plate and wire chambers, presented at the Wire Chamber Conference, Vienna, 1989.

[17] P. Schotanus, C.W. van Eijk and R.W. Hollander, Nucl. Instrum. Methods **A272** (1988) 913.

[18] J.L. Jansons, V.J. Krumins, Z.A. Rachko and J.A. Valbis, Phys. Stat. Sol. (b) **144** (1987) 835.

[19] J.L. Jansons, V.J. Krumins, Z.A. Rachko and J.A. Valbis, Solid State Commun. **67** (No. 2) (1988) 183.

[20] J.L. Jansons and Z.A. Rachko, Phys. Stat. Sol. (a) **53** (1979) 121.

[21] A.F. Buzulutzkov et al., Serpukhov preprint IHEP-88-167 (1988).

[22] G. Charpak, V. Peskov and D. Scigocki, Group internal report.

[23] C.L. Woody, IEEE Trans. Nucl. Sci NS-35 (1988) 493.

[24] F. Sauli, Proc. Meeting on the Future of Intermediate and High Energy Physics in Switzerland, Les Rasses, 1985 (SIN, Villingen, 1985), p. 167.

[25] D. Astruc, Accounts Chem. Res. **19** (12) (1986) 377.

[26] R.A. Holroyd, S. Erhenson and J.M. Preses, J. Phys. Chem. **89** (1985) 4244.

[27] M. Jibaly et al., The ageing of wire chambers filled with dimethylether, paper presented at the Wire Chamber Conference, Vienna, 1989.

[28] T. Åkesson et al., Nucl. Instrum. Methods **A262** (1987) 243.

[29] E. Bernardi et al., Nucl. Instrum. Methods **A262** (1987) 229.

[30] R.M. Wigmans, The Spaghetti calorimeter project at CERN, Paper presented at the Workshop on Future Directions in Detector R & D for Experiments at pp Colliders, Snowmass, 1988.

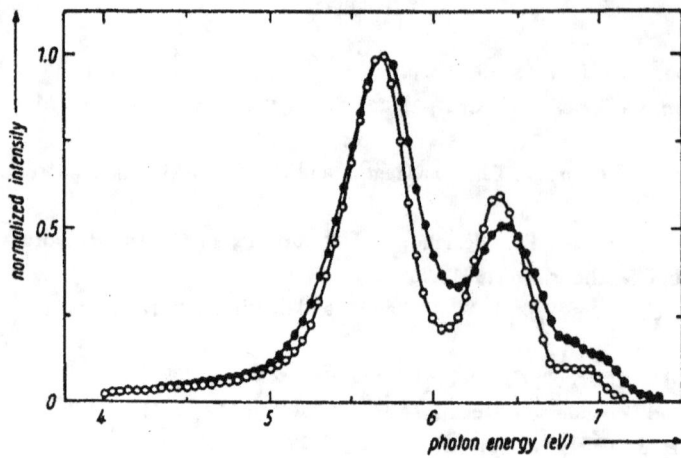

Fig. 1 Luminescence emission spectra normalized to equal maximum intensity of BaF$_2$ at T = 80 K (○) and 300 K (•).

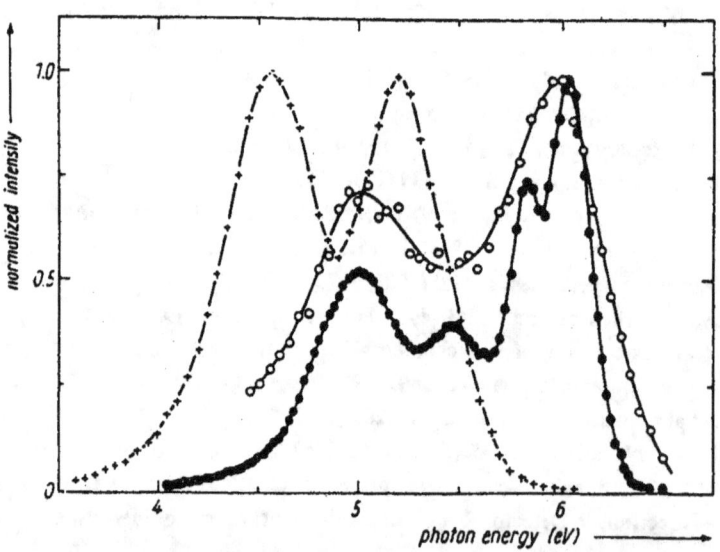

Fig. 2 Luminescence emission spectra normalized to equal maximum intensity of CsCl at 300 K (+) and CsBr at 300 K (○) and 80 K (•).

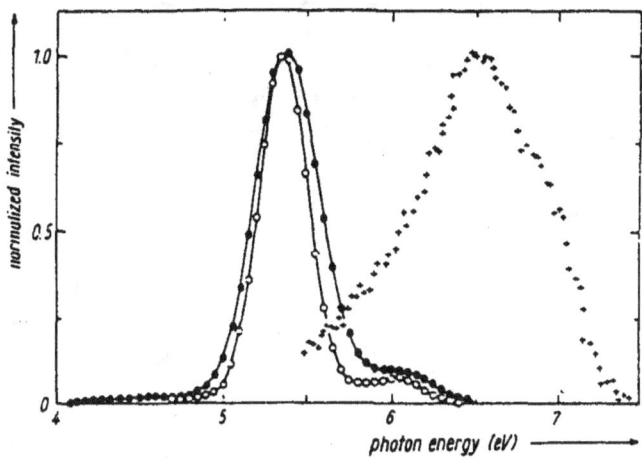

Fig. 3 Luminescence emission spectra normalized to equal maximum intensity of RbF at 80 K (○) and 300 K (•) and RbCl at 300 K (+).

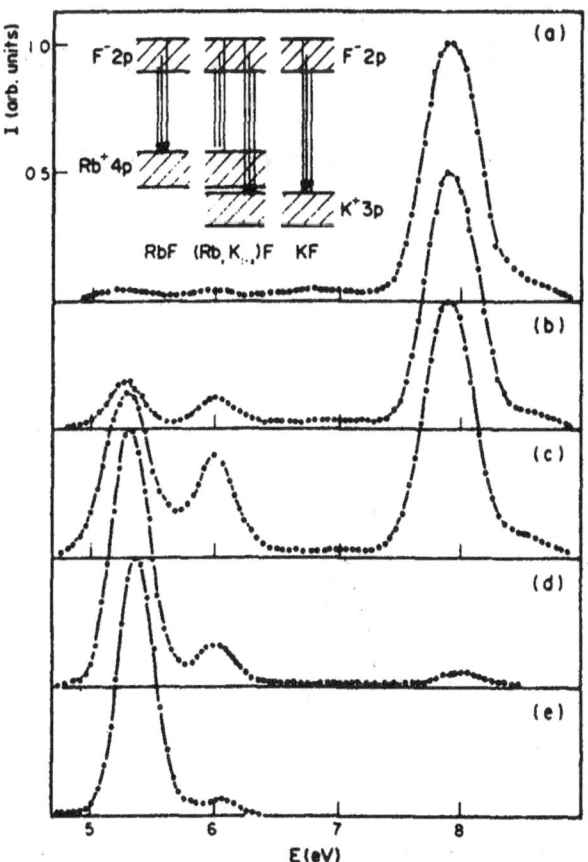

Fig. 4 Luminescence emission spectra of $K_{1-x}Rb_xF$ solid solutions, with varying x = 0 (a); 0.001 (b); 0.01 (c); 0.1 (d); 1 (e).

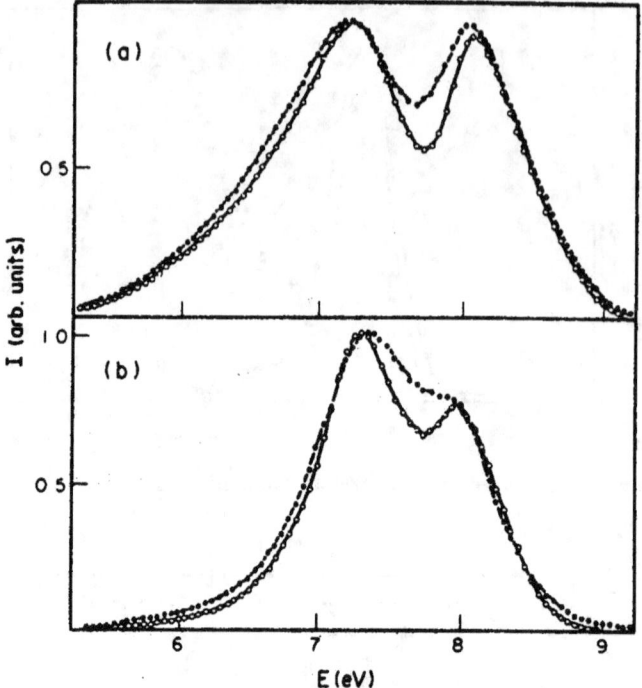

Fig. 5 Luminescence emission spectra of KMgF₃ (a) and KCaF₃ (b) at 80 K (○) and 300 K (•).

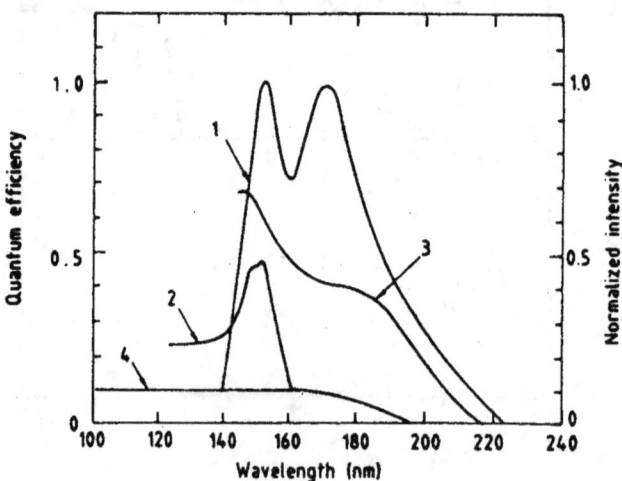

Fig. 6 Emission spectrum of KMgF₃ and quantum efficiencies of TEA and TMAE vapours and of CsI solid photocathodes as a function of photon wavelength λ. (1) KMgF₃ emission spectrum; (2) TEA vapour quantum efficiency; (3) TMAE vapour quantum efficiency; (4) CsI solid photocathode quantum efficiency.

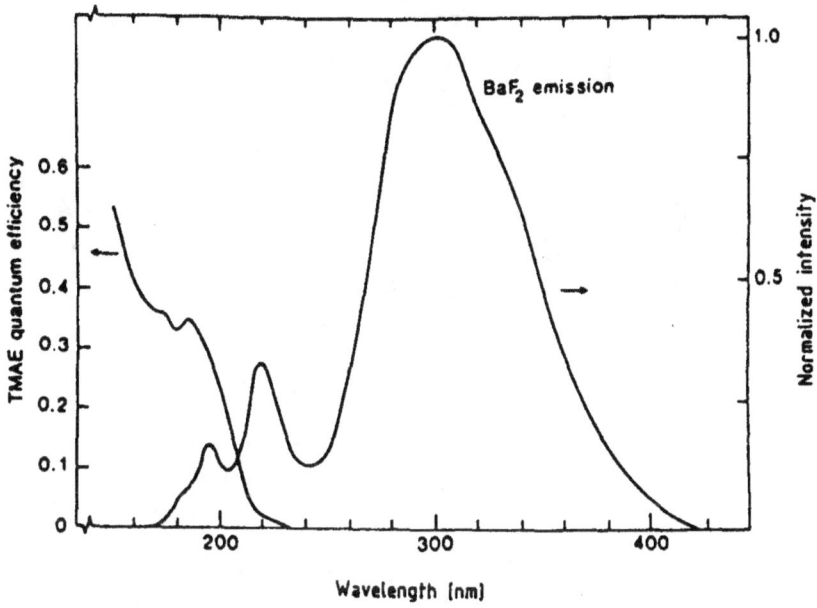

Fig. 7 BaF₂ emission spectrum and TMAE vapour quantum efficiency as a function of wavelength

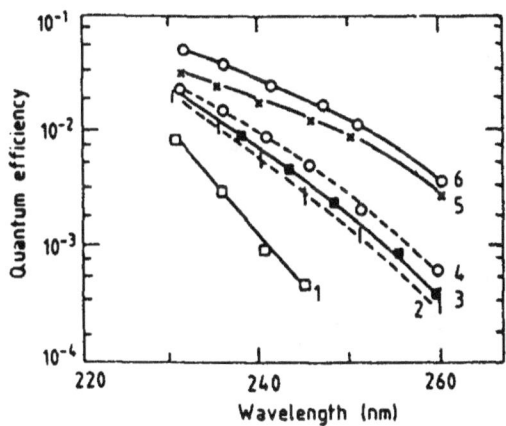

Fig. 8 The experimental quantum efficiency as a function of the photon wavelength λ for various liquid and solid photocathodes at a pressure of 1 bar, unless otherwise noted, and electric field $E = 1$ kV/cm; (1) methane + TMAE, $T_{cath} = 0$ °C; (2) methane + TMP + TMAE, $T_{cath} = -15$ °C; (3) methane + neopentane (NP) + TMAE, $T_{cath} = 0$ °C; (4) methane + TMS + TMAE, $p = 1.5$ bar, $T_{cath} = -5$ °C; (5) methane + NP + TMAE, $T_{cath} < -20$ °C; (6) same as (4) with electric field $E = 5$ kV/cm.

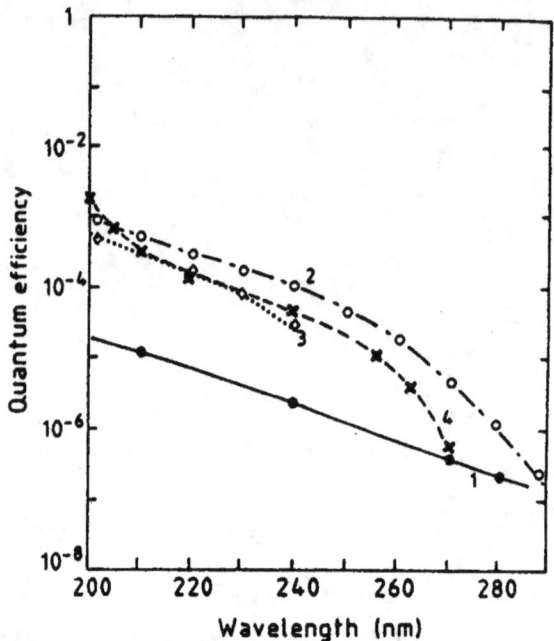

Fig. 9 Quantum efficiency as a function of wavelength for different photocathodes. (1) Clean Cu; (2) Cu + condensed CPIHMB; (3) Cu + condensed CPIHMB + TMS; (4) Cu + condensed EF.

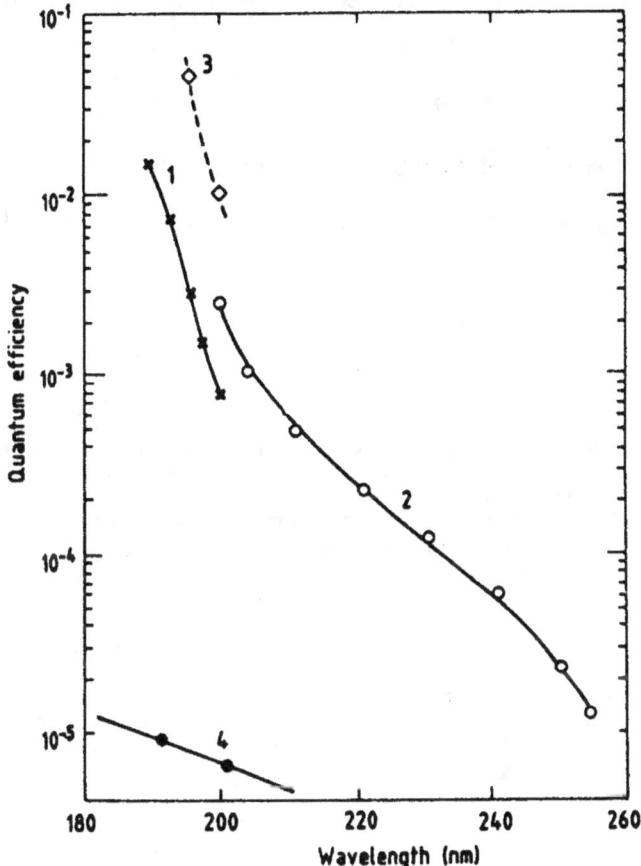

Fig. 10 Quantum efficiencies measured with a single-wire counter: (1) EF at T = 40 °C and a gap L = 15 mm; (2) thin layer of EF condensed on a copper cathode in an electric field E = 1 kV/cm; (3) same as (2), with E = 7.5 kV/cm; (4) clean copper cathode.

ORGANOMETALLIC PHOTOCATHODES FOR PARALLEL-PLATE AND WIRE CHAMBERS

V. PESKOV, G. CHARPAK, F. SAULI and D. SCIGOCKI

CERN, Geneva, Switzerland

V. DIEP and D. JANJIC

Chemistry Department, University of Geneva, Switzerland

We have investigated the possibility of using organometallic photocathodes (ferrocene, diethyl ferrocene, ethyl ferrocene, and cyclopentadienyl–iron–hexamethyl–benzene) in gaseous detectors. We found that condensed layers of these substances could increase the sensitivity of metallic cathodes to VUV radiation by several orders of magnitude. The best choice seems to be ethyl ferrocene, which can be used as the photosensitive element in the vapour phase as well as in condensed and absorbed layers. A parallel-plate chamber with this photocathode, operated at atmospheric pressure, combines a rather good time resolution (~ 10 ns or better) with a sensitivity better than 1% for wavelengths shorter than 230 nm.

1. Introduction

The molecular structure of organometallic compounds enables them to combine the desirable properties of low ionization potential and little chemical activity [1]. For this reason these compounds seem to be very attractive for use as photocathodes in gaseous detectors. We have tested four substances as candidates: ferrocene [2], diethyl ferrocene [2], ethyl ferrocene (EF) [3], and cyclopentadienyl–iron–hexamethyl–benzene (CPIHMB) [4]. The highest sensitivities in the spectral range 200–300 nm were achieved with the last two substances, on which we have therefore concentrated our investigations. In this article we give a review of our last results obtained using CPIHMB and EF as photocathode for gaseous detectors.

This work has been done in the framework of the LAA Project on Detector Research and Development [5].

2. The CPIHMB photocathode

2.1. Experiment

CPIHMB [chemical formula: $Fe(\eta\text{-}C_5H_5)\eta - (C_6Me_6)$] * is a brown powder with a vapour pressure of $\sim 10^{-6}$ Torr at room temperature, and a photoionization potential in the gas phase, $E_i = 4.5$ eV (275 nm) [1].

For this investigation, the setup was the same as that described in ref. [2]. A single-wire counter associated

---

* Me = methyl.

0168-9002/89/$03.50 © Elsevier Science Publishers B.V.
(North-Holland Physics Publishing Division)

with a cooled metallic cathode was used. This detector and the method of depositing these substances on the cooled cathode are also described in detail in ref. [2]. Different gas mixtures – CH_4, argon (90%) + CH_4 (10%), and CH_4 + tetramethyl pentane (TMP) or tetramethyl silane (TMS) vapour – were flushed at a total pressure of 1 atm through a bubbler containing CPIHMB powder, and through the counter. The bubbler, the single-wire counter, and the whole gas system were maintained at 70°C, and the temperature of the metallic cathode was kept at only a few degrees. The gas flow carried the CPIHMB vapour from the bubbler to the counter volume, where it was cryopumped onto the cooled metallic cathode. After deposition of more than a few monolayers of CPIHMB, all the measurements were made at room temperature.

2.2. Results

The quantum efficiency of the CPIHMB photocathode was measured in the range 200 to 300 nm. The results are presented in fig. 1 [4]: curve 1 is the sensitivity of a clean copper cathode; curve 2 is the efficiency of the cathode with a layer of CPIHMB; curve 3 gives the sensitivity obtained with the layer of CPIHMB when TMS vapour was present in the gas mixture. It can be seen that after deposition of the CPIHMB layer, the efficiency of the metallic cathode increases by 1 to 2 orders of magnitude in the wavelength region $\lambda < 260$ nm. As we expected (see ref. [2]), the measured photoionization threshold of CPIHMB in the condensed form $E_{i,cond} = 4.2$ eV (295 nm) was smaller than $E_{i,gas}$ in the gas phase.

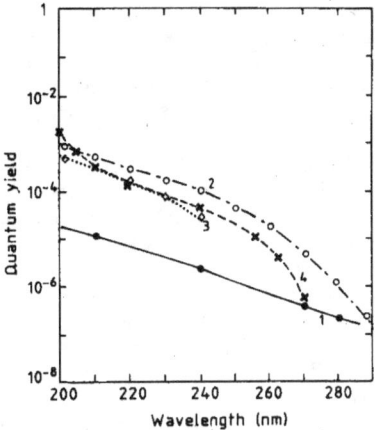

Fig. 1. Quantum yield as a function of wavelength for different photocathodes. (1) Clean Cu; (2) Cu + condensed CPIHMB; (3) Cu + condensed CPIHMB + TMS; (4) Cu + condensed EF.

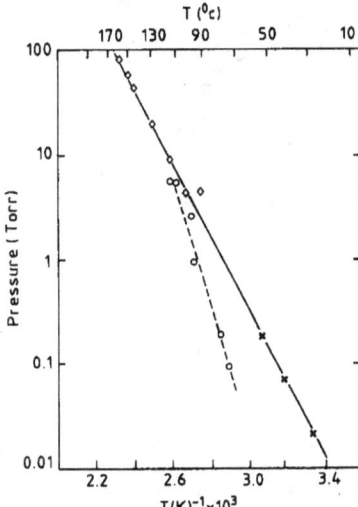

Fig. 3. EF vapour pressure as a function of temperature: ○ old measurements [6]; ◆ and × our measurements (see text).

The formation time and the stability of the CPIHMB photocathode are shown in fig. 2. In these measurements the single-wire counter was flushed with CH_4 + CPIHMB vapour, as described in ref. [2]. Cooling of the metallic cathode was started at time $t = 0$ and stopped at t_{stop}. It was then kept at room temperature, as were the counter and the whole gas system. The photocathode remained stable for a few days, even after the cooling was stopped – a property that is favourable for practical applications.

3. EF photocathode

Our result from the previous investigation of EF as a photosensitive element were presented in ref. [3]. The efficiency of EF photocathodes is shown as curve 4 in fig. 1. Although the efficiency of EF as a photocathode is slightly lower than that of CPIHMB, the fact that EF

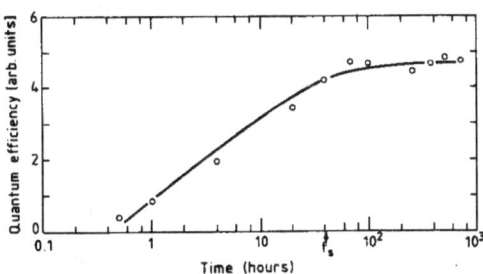

Fig. 2. Quantum efficiency of a CPIHMB photocathode as a function of the deposition time (up to t_{stop}) and of the stability of the sensitivity (as from t_{stop}).

is easy to handle because it does not react with air makes it rather attractive. We therefore tried to use it as the photosensitive element in a BaF_2 scintillator proportional counter [3]. The efficiency achieved at 1 atm (in He + 10% CH_4) for BaF_2 fast scintillation light (λ = 193 nm) was about 10% compared with TMAE gas under the condition of full absorption. The efficiency curve showed a peak at 193 nm. Previously published information [5] on vapour pressure led us to estimate that this efficiency is due to the photosensitive layer, and that the contribution from the gas can be neglected since the vapour pressure at room temperature is ~ 10^{-4} Torr. However, during the course of an experiment with an ionization chamber, we found that this pressure was higher (see fig. 1 in ref. [3]). This meant that although the absolute efficiency measured in ref. [3] was correct, it was likely that the gas contribution could not be neglected. For this reason, we made some additional investigations; the results are presented below.

3.1. Measurements of EF vapour pressure

In order to check on qualitative observations made with the above-mentioned ionization chamber, we performed our own measurements of EF vapour pressure (see fig. 3). Two methods were used: (i) In the temperature interval 80–200° C, we used the standard method of determining the boiling point of EF at a given pressure (points ◆ in fig. 3). (ii) In the interval 25–55° C, where the standard method is not applicable, we used a pressure transducer (Baratron MKS type 221A) (points × in fig. 3). The results of these measure-

XII. TECHNIQUES AND METHODS

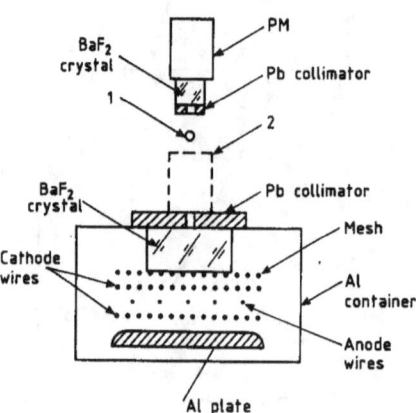

Fig. 4. Experimental set-up for the investigation of the EF photocathode: (1) shows the position of the H_2 flash lamp (or of the ^{22}Na source, for the measurements with the PM). (2) is the position of the PM for the measurement using the internal radioactivity of BaF_2.

drift-time enabled us to recognize in which part of the wire chamber the electrons were created. By applying a positive voltage on the Al plate or on the mesh on the BaF_2 crystal, it was possible to exclude the contributions from the cathode and from the adjoining drift regions.

Typical signals from the fast charge-sensitive amplifier (rise time ≈ 10 ns) are sketched in figs. 5. In each figure the solid line corresponds to average results obtained with hot EF vapour ($T_{EF} \approx 60°C$), the dashed line to EF vapour at room temperature (an absorbed layer of EF was also found [3]), whereas the dash-dotted line shows the results obtained with a condensed layer of EF; in the last case the chamber, at a temperature of $60°C$, was filled with EF vapour at the same temperature, then cooled down to $20°C$ or $0°C$. In all cases, 100 Torr of CH_4 was added to the vapour filling, and the pressure was then brought to 1 atm.

Within a division of 200 ns per division on the time scale, most of the signal is from photoelectrons col-

ments are in good agreement (fig. 3). However, our results give a much higher vapour pressure than was previously quoted (dashed line) [3,6]. We attribute this discrepancy to the fact that our earlier measurements were made in a temperature range where the accuracy of the method is poor.

Since our new results indicated that at room temperature the EF vapour contribution to the full ionization signal could not, as we had thought [3], be neglected, we started to do another set of measurements to separate clearly, in the signal, the contributions coming from the photocathode and from the vapour.

3.2. Detailed investigation of the wire chamber with an EF photocathode

3.2.1. Experiments with a pulsed VUV lamp

The wire chamber construction was described in refs. [2] and [3]. But in this latest experiment the aluminium container of the chamber had a CaF_2 window placed just near the BaF_2 crystal (see fig. 4). The wire chamber was irradiated through the BaF_2 crystal by a hydrogen flash lamp [7]. The output window of the lamp was covered by an interference filter 193-N-ID (Acton research Co.), with a maximum transmission at 193 nm and a FWHM of 18 nm. In this way, we can imitate the BaF_2 fast-component scintillation. A typical light-pulse measured by a photomultiplier (PM XP2020Q) was seen to be rather short, with a FWHM of 25 ns. In principle, the light-flash can liberate the photoelectrons from the cathode and from the EF gas. The number of photoelectrons created inside the wire chamber was about 100. Because of the short pulse duration, measuring the

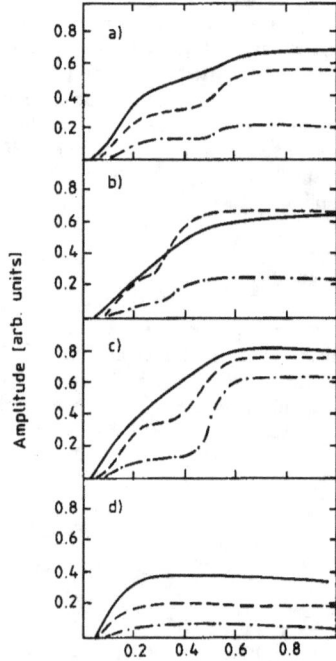

Fig. 5. Shape of pulses from the wire chamber. (a) Potential of the mesh on the BaF_2 crystal (V_{BaF_2}) and potential of the Al plate (V_{Al}) are kept at 0 V; (b) $V_{BaF} = 0$ V, $V_{Al} = -200$ V; (c) $V_{BaF_2} = -200$ V, $V_{Al} = 0$ V; (d) $V_{BaF_2} = +100$ V, $V_{Al} = +100$ V. Solid lines: EF vapour at $60°C$; dashed lines: EF at $20°C$, dash-dotted lines: condensed EF (i.e. the vapour was cooled from $60°C$ to $20°C$).

lected on the anode wire, and the influence of ions can be neglected.

Fig. 5a corresponds to the case where a positive voltage was applied to the anode wire while the potential of the Al plate on the mesh placed against the BaF_2 crystal was kept at 0 V. Since the electric field penetrated the drift region, we collected not only the electrons created in the wire chamber, but also some that came from the drift gap. Fig. 5b shows the case of a negative voltage applied to the Al plate. Here, in addition to the electrons created inside the wire chamber, we collected those from the drift space between the Al plate and the wire chamber, and also those liberated by the plate itself. The results shown in fig. 5c were obtained when a negative voltage was applied to the mesh on the BaF_2 crystal. In this case, the electrons were collected in the wire chamber itself, in the gap between the chamber and the BaF_2, and in the photosensitive layer on the surface of the crystal. When a positive voltage was applied to the mesh and to the Al plate (fig. 5d), the only electrons collected were those from the wire chamber.

Let us now analyse the shape of the pulses. In the case of a hot EF vapour, the signal grows monotonically, i.e. for several hundred ns the collection of photoelectrons, which are due to photoionization of the gas, is continuous.

When the signals were obtained at room temperature, their shape was completely different. Then the partial pressure of EF was much less, but a photosensitive layer of it was formed (adsorbed or condensed) on the BaF_2 crystal and on the Al plate. These signals clearly start off by growing sharply: this corresponds to the collection of photoelectrons from the wire chamber (including those created on the surface of the wire). They then show a slightly rising plateau, indicating that we collect a relatively small number of photoelectrons from the drift space, followed by another sharp rise corresponding to the collection of electrons liberated from the surfaces. The duration of the plateau corresponds to the drift time of the electrons in the gap; the rise time of the sharp growth that follows this plateau corresponds to the duration of the VUV pulse produced by the lamp. We can thus conclude that although we obtain a big signal from the surfaces, there exists a contribution from ionization of the EF vapour, even at room temperature.

Measurements performed using a mixture of He + CH_4, without EF vapour, and with cleaned surfaces of the wire chamber, including the Al and BaF_2, did not exhibit any signal from either the gas or the surfaces, even though the high voltage was pushed to the breakdown limit. This proves that the sharp rise in the signals shown in figs. 5b and 5c comes from the electrons liberated by the photocathode.

If we know the vapour pressure of EF as a function of temperature, it is possible, from the shape of the pulses shown in fig. 5, to estimate the absolute quantum efficiency of the EF photosensitive layer. Using the data for the absolute quantum efficiency of the detector at $T_{EF} = 70\,°C$ (see fig. 5 of ref. [3]), we can recalculate the efficiency of our wire chamber at room temperature. The efficiency due only to gas ionization, under the condition of full photoelectron collection (wire chamber and two drift spaces), is $\approx 0.7\%$ for $\lambda = 193$ nm. The amplitude of the signal from the cathode was approximately twice that of the signal from the vapour (see fig. 5). Therefore the efficiency of the adsorbed layer is in the range between 0.7% and 1.5%, increasing with the electric field. In the case of the condensed layer, the efficiency is between 1.5% and 2%. The full efficiency of the chamber (i.e. gas and surfaces) would be about 3%, which is about 10% of that of TMAE vapour under the condition of full absorption. This is in very good agreement with the data presented in ref. [3] for the same gas mixture. It should be noted that the efficiency of the EF layer increases with the electric field E and the concentration of the quencher gas. For example, in pure CH_4 at $p = 0.5$ atm and $E \approx 6$ kV/cm, the quantum efficiency of the layer for $\lambda = 193$ nm was about 6.5%, which is $\approx 22\%$ of that of TMAE.

3.2.2. Measurements with radioactive sources

All these measurements were done under the same conditions: the chamber was filled with EF vapour at room temperature, and then 100 Torr of CH_4 + He, up to 1 atm, were added.

3.2.2.1. Investigation based on the shape of the pulses.
It was expected that the wire chamber with the EF photocathode would have a high photon feedback, and that it would be impossible to achieve a high gas gain. To understand this, we studied the shape of the pulses produced by VUV emission of BaF_2 excited by radioactive sources of ^{241}Am and ^{60}Co.

In the case of the ^{241}Am source of γ-rays (59 keV), which produced single photoelectrons in the wire chamber with very small efficiency, we could not observe any photon feedback (double pulses) up to a rather higher gain ($\sim 10^5$). Inverting the polarity in the drift hardly changed the shape of the pulses at all: only their risetime became a little shorter.

The ^{60}Co source of γ-rays (of 1.17 and 1.33 MeV) produced mostly single photoelectrons, with an efficiency of 20 times larger (per incident photon). In this case, when the polarity on the mesh or the crystal was negative, we could sometimes observe double pulses, even at gains of 10^5. The second pulse was delayed by a time corresponding to the electron drift from the surface. Since the gain was 10^5, where we know that there is no feedback, the second pulse was an indication that, in the case of the ^{60}Co, two photoelectrons per photon

XII. TECHNIQUES AND METHODS

were sometimes detected from the scintillation light of the BaF$_2$, the second one coming from the photocathode itself. It was thus shown that the largest contribution to the chamber sensitivity came from the photocathode.

3.2.2.2. Measurements of the absolute efficiency of the wire chamber.

Using the radioactive sources and the experimental set-up shown in fig. 4, we made two independent measurements of the absolute efficiency of the wire chamber. In the first one, a PM (C31000M) was placed directly against the CaF$_2$ window of the wire chamber. We found that the BaF$_2$ was slightly radioactive. This contamination created bursts of VUV light. It has been shown earlier [8] that approximately half of these pulses come from the fast component ($\lambda = 193$ nm) of the light. By comparing the count rate of pulses from the PM with that from the wire chamber (the latter being delayed by the time the electrons drifted from the Al plate to the wire chamber (~ 200 ns)), it was possible to estimate the absolute efficiency of the EF adsorber layer. In our case this ratio of count rates was about 100, increasing with the electric field. Taking into account the solid angle of the PM, it can be concluded that the absolute efficiency of the adsorber layer was between 0.7% and 1%, which is consistent with previous measurements.

The number of wire chamber pulses delayed by at most 1 µs increase with decreasing temperature and also with the drift voltage. Observations made at different temperatures showed that:
– at 9.1°C, 50% of the coincidence pulses were delayed by 200 ns,
– at 2°C, 70% of the coincidence pulses were delayed by 200 ns.
So at 2°C, at least 70% of all photoelectrons come from the photocathode.

In the second measurement, the PM, with the other BaF$_2$ scintillator, was placed 10 cm away from the CaF$_2$ window of the wire chamber. The BaF$_2$ crystal in the wire chamber and the one on the PM were irradiated by a ^{22}Na source, placed between them on the common axis. They were thus irradiated simultaneously, giving two photons of ~ 0.5 MeV energy, propagating in opposite directions and producing scintillation pulses in both crystals. The pulse–amplitude spectrum of BaF$_2$ VUV emission consists in the peak produced by 0.5 MeV photons, the background being produced by ^{22}Na itself and by the internal radioactivity of the BaF$_2$. After subtraction of this background, the numbers N of coincidences (with 200 ns delay) measured at different electric fields E were the following: at $E = 200$ V/cm, $N \approx 1/200$; at $E = 400$ V/cm, $N \approx 1/130$; at $E = 2$ kV/cm, $N \approx 1/100$. Since the PM records the bursts of scintillation light from the BaF$_2$ with 100% efficiency, the value of the absolute efficiency of the EF layer, for

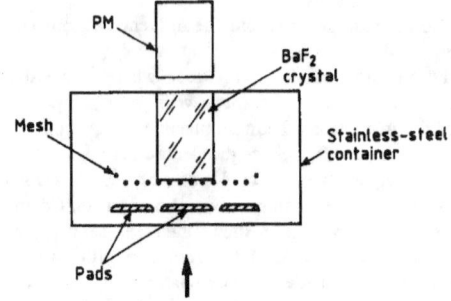

Fig. 6. Construction of the PPAC. The thick arrow shows the direction of the beam.

$\lambda = 193$ nm, can be estimated to be between 0.5% and 1%, depending on the electric field. This is again in good agreement with the above-mentioned data.

3.3. Parallel-plate avalanche chamber with the BaF$_2$ scintillation

As was shown above, the BaF$_2$ scintillation emission in the wire chamber, containing EF vapour at room temperature, produced photoelectrons due both to photoionization of EF vapour and to extraction of the electrons from the photocathode. In order to investigate the EF photocathode under "clean" experimental conditions, i.e. without any contribution from EF vapour, we constructed a parallel-plate avalanche chamber (PPAC) containing a BaF$_2$ crystal, $2.5 \times 2.5 \times 7$ cm^3. The transparency of the crystal along the axis for $\lambda = 193$ nm was about 50%. The chamber construction is presented in fig. 6. The distance between the two electrodes was 4 mm. It was filled with EF vapour at room temperature, and then 100 Torr of TMS mixed with He up to 1 atm were added. As in the case of the wire chamber, the EF vapour formed a photosensitive layer, which was deposited everywhere, including on the surface of the BaF$_2$ crystal and of the Cu cathode. The photoelectrons extracted from the surfaces produced avalanches, so that the number of electrons reaching the anode was larger by a factor N, corresponding to the gas gain. In principle, the photons emitted by the BaF$_2$ crystal can also ionize the EF vapour; but the photoelectrons that are created at a distance from the cathode larger than λ_i, the mean free path before ionization, will produce avalanches with a much smaller gas gain. For $N = 10^5$, $\lambda_i \approx 0.3$ mm; therefore only 1/10 of the gap can in principle contribute to the creation of photoelectrons due to photoionization. Under these conditions and at room temperature, we can really neglect the contribution of the gas photoionization.

3.3.1. Investigation of the BaF₂ PPAC with a VUV lamp and with radioactive sources

Measurements with the VUV flash lamp give efficiencies that are of the same order of magnitude for a layer of EF adsorbed on the BaF_2 crystal and on the Cu cathode. The chamber was stable during operation and the gas gain ($\gtrsim 10^5$) was sufficient to detect single photoelectrons.

Measurements made with an ^{241}Am source, as described in ref. [3], show that the BaF_2 fast emission detection efficiency of a PPAC, with an adsorbed layer of EF, is four times less than that of a wire chamber [3]. For a radiation of $\lambda \approx 193$ nm, the absolute efficiency of an EF layer for the given gas mixture is thus $\approx 0.7\%$. The rise time of the signals from the PPAC was ≈ 10 ns.

3.3.2. Measurements in a test beam

To check once more the efficiency of an EF layer, we also made preliminary tests in an unseparated charged-particle beam, of about 7 GeV energy, that also contained electrons. By inserting a thick absorber, we could also work only with muons. We estimated that the energy deposited by the muons in our BaF_2 crystal was approximately 49 MeV, and the maximum energy deposited by the electrons was about 0.7 GeV. The stainless-steel container of the BaF_2 PPAC was closed with a CaF_2 window, on which the PM (C31000M) was directly mounted (see fig. 6) as was done for the measurement of the absolute efficiency of the EF layer in a wire chamber (see section 3.2.2.2). This BaF_2 crystal was without Al reflector. Calculations and measurements show that this reduces by $\approx 30\%$ the intensity of the BaF_2 light entering the PPAC. When we worked with the muon beam, the efficiency of the PPAC in coincidence with the PM was about 100%. The amplitude of the signal from the PM for the fast component was 50 photoelectrons (the absolute efficiency of the PM to this radiation was about 25%). This means that in the case of the PPAC, with an efficiency of $\approx 0.7\%$ for the EF layer adsorbed on Cu, we should expect about 2 photoelectrons for 49 MeV deposited energy and about 50 photoelectrons for 0.7 GeV. Therefore, in the case of the muon beam, we expect the pulse-height distribution to have a broad maximum. For the 0.7 GeV electrons, we expect an energy resolution [9] of $\approx \sqrt{2/50} \approx 20\%$, since the trigger is a signal from the PM, corresponding to the maximum energy loss in the BaF_2. Qualitative results from the beam tests confirm these data, and we can say that the efficiency of an EF adsorbed layer was really about 1%. This means that in the case of a crystal with 90% transparency and with an Al reflector we can expect the energy resolution at 1 GeV to be about 10%. The gap between the plated is thin, only 4 mm, which makes it possible to collect light of the front and the back surface of the crystal (or on all six surfaces), thus increasing the signal and partly compensating for the loss of efficiency with respect to TMAE. We are now also preparing for another test in a beam, with a PPAC working with a condensed layer of EF. There we may expect the efficiency to be better.

We remind the reader that, in the case of TMAE under the condition of full absorption of the scintillation light from the BaF_2 crystal (with a transparency of the crystal of 75% and with an Al reflector), the energy resolution for 1 GeV was 6%, improving to $3.9\%/\sqrt{E}$ at high energies [10]. With EF, we may also expect such an improvement, which would made the calorimetry with an adsorbed layer attractive, even at the present level of sensitivity.

4. Conclusion

All the experiments described above show that the absolute efficiency of an adsorbed layer of EF is about 1% for $\lambda = 193$ nm. When the EF layer is condensed, this efficiency increases by a factor of 3 to 4, which makes such a calorimeter very useful at very high energies.

Acknowledgements

The authors would like to tender their appreciation to R. Bouclier, G. Million, I. Crotty and J.-C. Santiard for their technical assistance and their advice.

References

[1] D. Astruc, Accounts Chem. Res. 19 (12) (1986) 377..
[2] V. Peskov, G. Charpak, P. Miné, F. Sauli, D. Scigocki, J. Séguinot, W.F. Schmidt and T. Ypsilantis, Nucl. Instr. and Meth. A269 (1988) 149.
[3] G. Charpak, V. Peskov, F. Sauli and D. Scigocki, preprint CERN-EP/88-166, submitted to Nucl. Instr. and Meth. (1988).
[4] D. Astruc, G. Charpak, P. Miné, V. Peskov and D. Scigocki, Proc. Int. Workshop on Liquid-State Electronics, Berlin, 1988 (Hahn–Meitner-Institut, Berlin, 1988) p. 109.
[5] A. Zichichi et al., The LAA Project, Proc. Meeting of 15 December 1986, 22 June 1987, and 25 July 1988, CERN-LAA/88-1 (1988).
[6] G. Melin, Handbuch der Anorganischen Chemie, Band 14, Teil, Ferrocene 1 (Springer-Verlag, Heidelberg, 1974), p. 259.
[7] I.B. Berlman, O.J. Steingraber and M.J. Benson, Rev. Sci. Instr. 39 (1968) 54.
[8] K. Wisshak and F. Käppeler, Nucl. Instr. and Meth. 227 (1984) 91.
[9] P.A. Lansiart and J.P. Morucci, J. Phys. & le Radium 23 (1962) 102A; G. Woody, Brookhaven Nat. Lab. Technical Note N120 (1986).
[10] R. Bouclier, G. Charpak, W. Gao, G. Million, P. Miné, S. Paul, J.C. Santiard, D. Scigocki, N. Solomey and M. Suffert, Nucl. Instr. and Meth. A267 (1988) 69.

XII. TECHNIQUES AND METHODS

Nuclear Instruments and Methods in Physics Research A297 (1990) 133–147
North-Holland

Reflective UV photocathodes with gas-phase electron extraction: solid, liquid, and adsorbed thin films

J. Séguinot [1], G. Charpak [2], Y. Giomataris [2], V. Peskov [3], J. Tischhauser [2] and T. Ypsilantis [1]

[1] Collège de France, Paris, France
[2] CERN, Geneva, Switzerland
[3] World Laboratory, Geneva, Switzerland

Received 3 July 1990

The photoemission quantum efficiency of reflective photocathodes in methane gas has been investigated in the spectral range between 140 and 250 nm. The spectral response of solid metals and CsI, as well as of liquid and solid TMAE film, have been measured. The high quantum efficiency of CsI (35% at 170 nm) makes it attractive for BaF_2 or xenon scintillation detection. A BaF_2 crystal coupled to an ionization chamber with a reflective CsI photocathode has been successfully tested. Adsorbed TMAE films can significantly increase the quantum yields of metal and CsI (to 46% at 170 nm), making them suitable for fast RICH and other applications.

1. Introduction

Investigations of organic reflective UV photocathodes with electron extraction into a gaseous medium were reported in an earlier article [1]. The molecules studied were those of liquid tetrakis(dimethylamine)ethylene (TMAE) and of tetramethyl-p-phenylenediamine (TMPD). Most of these measurements were carried out in the spectral region of wavelength $\lambda > 230$ nm so as to exclude the masking effects of gas-phase photoionization. The quantum efficiency was always found to increase with photon energy, and extrapolation to $\lambda \approx 180$ nm indicated that large ($\sim 10\%$) efficiencies were possible (see figs. 8b and 11 in ref. [1]).

In order to verify this extrapolation, we have investigated some reflective solid/liquid/adsorbed-film photocathodes in the region between 140 and 250 nm. An apparatus was designed to deposit photocathodes by cryopumping a gas onto a cold metal substrate. After deposition, the flow of the photosensitive gas could be stopped, thus permitting photoefficiency measurements to be made with minimum interference from gas-phase photoionization. However, there always remained the effect of the equilibrium vapour pressure of the photosensitive molecule, which could be removed – in order to isolate the intrinsic photocathode effect – by making an appropriate set of measurements. The data show that the quantum yields are indeed high. One particular photocathode (solid CsI + adsorbed TMAE film) has 46% efficiency at 170 nm. Such a cathode can have important applications in the field of high-energy

physics, i.e. for fast ring-imaging Cherenkov (RICH) detectors [2], for BaF_2 calorimetry [3], and for liquid-xenon electromagnetic calorimetry [4]. In addition, it may find uses in positron emission tomography (PET), either with BaF_2 or liquid xenon.

A preliminary report of these results was presented in September 1989 at the ECFA Conference held in Barcelona and was published in the Proceedings [5].

2. Experimental set-up

The apparatus is shown in fig. 1a. It consists of a deuterium-filled low-pressure dc arc lamp with a MgF_2 window (labelled La in the figure; Hamamatsu L879-01); a UV photon monochromator (PMC; Jobin-Yvon H20VUV), and a photoefficiency test chamber (TC). The wavelength-selected photon beam is parallel-focused by an $f = 100$ mm, $\varnothing = 25$ mm CaF_2 lens (Le) placed 100 mm downstream from the monochromator exit slit. This parallel beam is then split into a reflected beam and a transmitted beam by means of a flat, 5 mm thick, CaF_2 crystal beam-splitter (BS) tilted 45° to the beam direction. The reflected beam is monitored by a photomultiplier (PMR; RCA 8850), whose glass entrance window is coated (by vacuum deposition) with a p-terphenyl layer (10^4 Å thick [6]) of wavelength shifter (WLS) protected by a 10^3 Å thick layer of MgF_2. An identical photomultiplier (PMT; RCA 8850) is positioned in such a way as to measure the transmitted photon beam (see fig. 1a). The 5 mm diameter beam

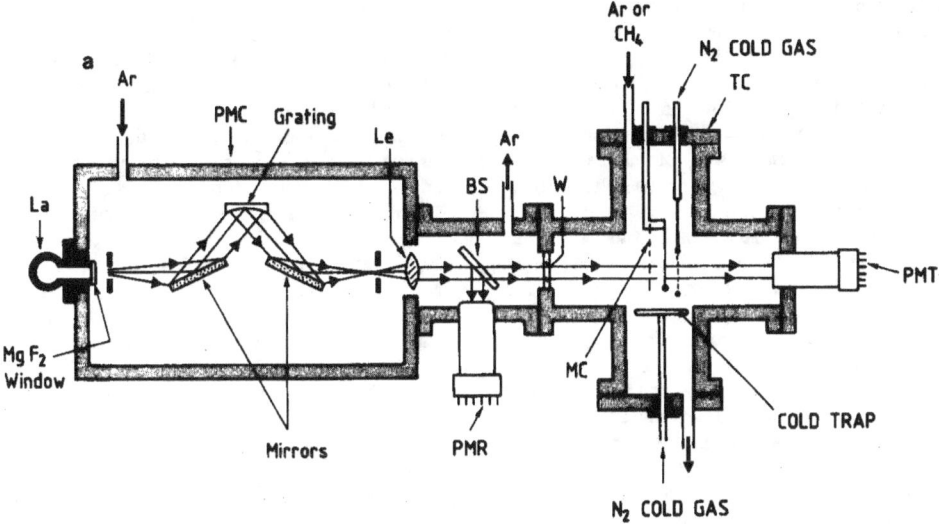

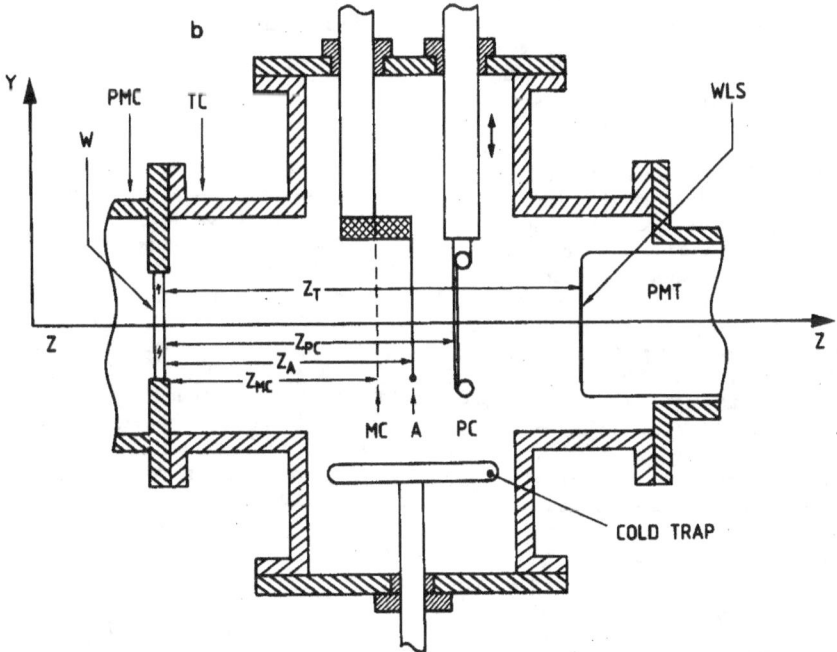

Fig. 1. (a) The experimental layout of the photon monochromator (PMC) with the D_2 UV lamp (La), the CaF_2 lens (Le), the CaF_2 beam splitter (BS), the reflected beam photomultiplier (PMR), and the CaF_2 window (W). The test chamber (TC) is shown in more detail in (b). The test chamber contains the mesh cathode (MC), the wire anode (A), the photocathode (PC) with its cooling loop, the cold trap, and the transmitted beam photomultiplier (PMT). The glass window of PMT (and PMR) is coated with a 1 μm thick vacuum-deposited film of the wavelength shifter (WLS) p-terphenyl. Not shown in the figure are the gas-flow inlet and outlet.

enters the test chamber through a flat, 5 mm thick, CaF_2 window (W), which isolates the monochromator from the TC and guarantees that the light source is unaffected by any contaminants originating in the TC.

The test chamber is shown in more detail in fig. 1b. It consists of a horizontal (z-direction) stainless-steel cylinder of 100 mm diameter with a transverse vertical (y-direction) cylinder positioned centrally to form a cross. The horizontal cylinder is linked by one end-flange to the photon monochromator (PMC), whilst the other end-flange is linked to the photomultiplier (PMT). The test chamber contains an ionization cell composed of a transparent mesh cathode (MC) plane, an anode (A) wire plane, followed by the solid photocathode (PC) plane and a cold trap. The cold trap can be cooled to a temperature much lower than that of the photocathode, so as to trap most of the impurities coming from the TC ambient-temperature walls. This facility was used to test operation of CsI photocathodes down to 160 K.

The PC plane is made of a stainless-steel (ss) plate ($\Delta y = 45$ mm high, $\Delta x = 22$ mm across, and $\Delta z = 0.5$ mm thick) with a 4 mm diameter ss tube soldered around its perimeter. The ss plate can be cooled to ≥ 120 K by flowing cold N_2 gas through the tube by means of an evacuated transfer line. Separate plates of brass, copper, or stainless steel can be clipped onto the PC plane (with good thermal contact), permitting study of these bare metals. These plates can also be transported to a separate vacuum deposition facility for coating with various low-vapour-pressure materials (CsI, Zn, etc.) for subsequent measurement in the TC. The temperature of the PC plane is continuously monitored by means of a calibrated platinum resistor. The PC plane itself can be raised vertically by $\Delta y = 50$ mm (without affecting the gas flow or the temperature), allowing the photon beam to impinge directly onto the monitor PMT. The PC potential V_{PC} is negative and variable: $0 \leq -V_{PC} \leq 3$ kV.

The anode or collector plane (A) consists of three 2 mm diameter vertical wires spaced 10 mm apart in the transverse horizontal (x) direction. The middle wire ($x = 0$) is kept at ground potential via a digital electrometer (Keithley 616A; lower current limit = 0.01 pA). It collects all electrons coming from the target region of the PC ($-5 \leq x \leq 5$ mm), whilst the two edge wires ($x = +10$ mm and $x = -10$ mm) are independently grounded so as to completely define the electrostatic configuration of the ionization cell.

The mesh cathode (MC) potential V_{MC} is also negative and variable between 0 and 2 kV. It serves to define the fiducial collection region for gas-phase photoionization, as well as to ensure that all lines of force from the cathode terminate on the collection anode.

Before assembly, all the metallic elements of the TC were cleansed by sequential washing in acetone, detergent, and distilled water, then pumped and outgassed at 200°C.

The PMC is continuously flushed with argon (quality N48; 20 ppm impurities) at atmospheric pressure. The window (W) isolates the PMC from the variable conditions in the TC. The test chamber itself is flushed independently, usually with methane (quality N35; 500 ppm impurities), or sometimes with argon (N48). The flushing gases are cleansed by passing them through a cartridge of Oxisorb (chrome trioxide, silica gel, and molecular sieves; Messer Griesham GmbH) before entering either the PMC or the TC. The TC flushing gas can be saturated at a partial vapour pressure of an organic photosensitive molecule by flow through a bubbler containing the aforesaid molecules in the liquid state. In order to prevent any risk of condensation of the photosensitive molecules onto the PMT wavelength shifter, the PMT was heated to about 50°C. The TC exhaust is discharged into air through a stainless-steel capillary (3 m long, 2 mm in diameter) so as to limit the retrodiffusion of oxygen.

The distances between the entrance window (W) and the mesh cathode (MC), the collection anode (A), the photocathode (PC), and the phototube (PMT) are $z_{MC} = 116$, $z_A = 129$, $z_{PC} = 141$, $z_T = 236$ mm, respectively.

3. Measurements

3.1. Flux calibration

A current measurement is used to determine the photon flux, because of its greater linear dynamic range compared with photon counting.

Initially, the TC is flushed with methane at atmospheric pressure, and measurements begin only when optimal photon transmission is obtained (at $\lambda = 160$ nm). The measurement cycle starts with a calibration of the transmitted photon beam relative to the reflected photon beam. The reflected and transmitted photocurrents I_R and $I_T(0)$ are measured (between 140 and 250 nm) with 1 mm monochromator slits ($\sigma_\lambda = 4.3$ nm) in order to maximize the sensitivity. This increased sensitivity is especially necessary at the longest wavelengths, where the quantum yield is small. At each wavelength λ, the photocurrents are determined by the following relations:

$$I_R = eI_0 R_{BS} Q_R G_R,$$
$$I_T(0) = eI_0(1 - R_{BS})Q_T G_T = eI_0' Q_T G_T, \tag{1}$$
$$C = I_T(0)/I_R,$$

where e is the electron charge; I_0, I_0' are the incident and transmitted photon flux respectively, Q_T, Q_R, and G_T, G_R are the transmittive and reflective photomultiplier quantum efficiencies and gains, respectively; and

R_{BS} is the reflectivity of the beam splitter. The argument (0) in the above equations means that there is no photoionizing gas in the test chamber. The wavelength dependence is implicit. The quantity C calibrates the transmitted beam current $I_T(0)$ in terms of the reflected beam current I_R. This current is measured before the beam enters the TC; hence it is independent of the gas conditions in the TC (i.e. without the argument).

When a photoionizing (PI) gas is introduced into the TC, a photoionization current $I_A(PI)$ is observed on the anode wire. It is given by the relation

$$I_A(PI) = e I_0' Q_{PI} A_{PI}(z_{MC}, z_{PC}), \qquad (2)$$

where

$$A_{PT}(z_{MC}, z_{PC}) = \exp(-z_{MC}/l_{ph}) - \exp(-z_{PC}/l_{ph}), \qquad (3)$$

and Q_{PI} is the gas-phase photoionization quantum efficiency. The quantity $A_{PI}(z_{MC}, z_{PC})$ is the absorptivity of the PI gas in the ionization cell between z_{MC} and z_{PC}. The photon absorption length (l_{ph}) is determined by measuring the current $I_T(PI)$, which reaches the PMT when the photoionizing gas is in the TC and (necessarily) with the PC in the "up" position. The equation for this current is then determined by the TC absorptivity as

$$I_T(PI) = e I_0' Q_T G_T \exp(-z_T/l_{ph})$$
$$= I_T(0) \exp(-z_T/l_{ph}), \qquad (4)$$

whilst from eq. (1), $I_T(0) = C I_R$; hence

$$l_{ph} = z_T / \ln[C I_R / I_T(PI)]. \qquad (5)$$

The transmitted photon flux I_0' may now be calculated from eq. (2) as

$$I_0' = I_A(PI) / [e Q_{PI} A_{PI}(z_{MC}, z_{PC})], \qquad (6)$$

where $I_A(PI)$ is measured directly, Q_{PI} is from the known PI efficiency (i.e. for TMAE [7,8]), A_{PI} is from eq. (3), and l_{ph} is from eq. (5).

Having evaluated l_{ph} and I_0', the product $Q_T G_T$ may then be evaluated from eq. (4). The gain G_T is independent of the wavelength and is given by the manufacturer as $g = 1.4 \times 10^5$ at the chosen operating voltage of 1.4 kV. The quantity $Q_T G_T / g$, plotted in fig. 2a, is the quantum efficiency of the PMT if the gain g is, indeed, correct. The average PM quantum efficiency is ~ 10%, not inconsistent with previous measurements [6], but with a surprising (i.e. not flat) wavelength dependence, probably due to a variation in the p-terphenyl wavelength-shifter efficiency. It must be emphasized, however, that the derived PC quantum yield depends only on the TMAE photoionization yield and not on the PMT quantum efficiency, it being used only

as an intermediate standard. A measurement of the product $Q_T G_T$ was also made in the regions above 200 nm, by using a calibrated photodiode EGG UV 360 BQ. Good agreement was obtained in the TMAE overlap region (200 to 220 nm) confirming the previous measurement.

3.2. Photocathode quantum efficiency

When measuring a photocathode that has a negligible equilibrium vapour pressure, the current $I_A(PC)$ on the wire anode (A) is given by the obvious relation

$$I_A(PC) = e I_0' Q_{PC}. \qquad (7)$$

Solving for the PC quantum efficiency, with I_0' from eq. (6), gives

$$Q_{PC} = Q_{PI} A_{PI}(z_{MC} z_{PC}) [I_A(PC)/I_A(PI)], \qquad (8)$$

where A_{PI} is from eq. (3). The PC quantum efficiency thus depends on the PI gas quantum efficiency, on the measured currents, and on the absorptivity (i.e. l_{ph}), and is independent of the PM quantum efficiency and gain.

When measuring a photocathode that has non-negligible vapour pressure, the anode signal I_A will have a component arising from gas-phase photoionization as well as from the PC. In this case, the procedure is to measure the anode current I_{Ad} with the PC in the down position (so as to intercept the beam), and the current I_{Au} when the PC is in the up position. In this latter position a current I_T may also be observed in the PMT, which is due to unabsorbed photons reaching the PMT. Note that when the PC is in the up position the ionization cell becomes much longer because the PMT cathode is also at a negative potential (-1.4 kV), and charge collection occurs between z_{MC} and z_T. It has been verified that the photoionization charge is efficiently collected in this region provided $|V_{PC}|$ is higher than 1.0 kV, as shown in fig. 2b. This current determines the photon absorption length l_{ph}, which then allows the photoeffects due to the PC and those due to the PI gas to be separated. The observed currents are

$$I_{Ad} = I_A(PC) + I_A(PI) = e I_0' \big[Q_{PC} \exp(-z_{PC}/l_{ph}) + Q_{PI} A_{PI}(z_{MC}, z_{PC}) \big],$$

$$I_{Au} = e I_0' Q_{PT} A_{PI}(z_{MC}, z_T),$$

$$I_T(PI) = e I_0' \exp(-z_T/l_{ph}) Q_T G_T. \qquad (9)$$

Combining the first two relations of eq. (9) gives the key equation for the PC quantum efficiency:

$$Q_{PC} = Q_{PI} (I_{Ad}/I_{Au}) \big[\exp(z_1/l_{ph}) - \exp(-z_2/l_{ph}) \big] + 1 - \exp(z_1/l_{ph}), \qquad (10)$$

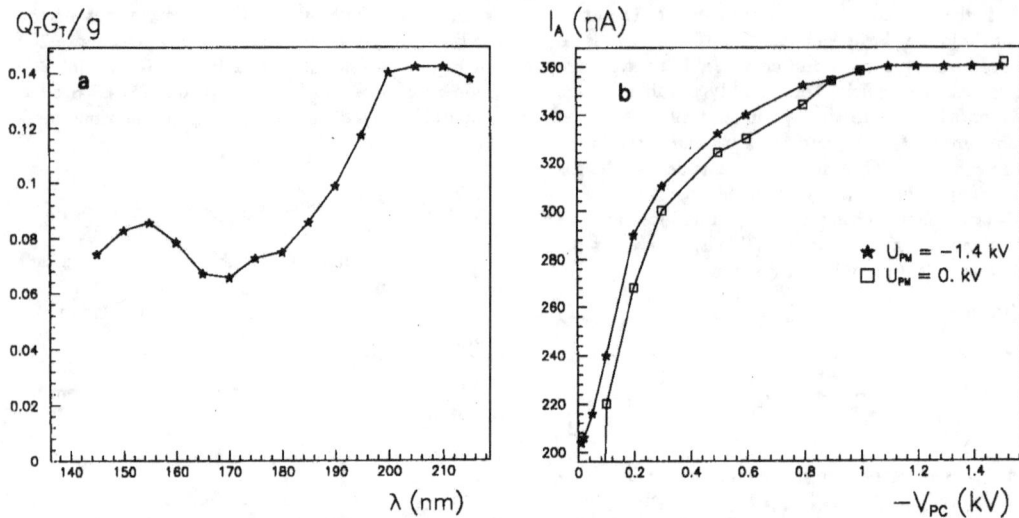

Fig. 2. (a) The measured product of quantum efficiency and gain $Q_T G_T$ of the transmitted beam photomultiplier (PMT) divided by the rated gain $g = 1.4 \times 10^5$ at 1.4 kV versus the wavelength λ. (b) The anode-collected photocathode current I_{Au} in a methane–TMAE (0.04 Torr) mixture as a function of the photocathode voltage V_{PC}, for two different PMT voltages, at $\lambda = 160$ nm.

where $z_1 = z_{PC} - z_{MC}$ and $z_2 = z_T - z_{PC}$. The third relation of eq. (9), with $I_T(0)$ from eq. (1), determines l_{ph} exactly as in eq. (5). Here, as before, the PC quantum efficiency depends on the PI gas quantum efficiency, on the measured currents, and on l_{ph}, and is independent of the PMT quantum efficiency or gain.

The final results are independent of any intensity variations in the lamp or in the monochromator throughput because all measurements are normalized to the reflected photon current I_R.

4. Experimental results

4.1. Metal photocathodes

The quantum efficiencies of bare steel, copper, and brass PC plates have been measured, and the results are shown in fig. 3. The quantum yields relative to TMAE are determined from eq. (8). As expected, the yields are low ($\leq 10^{-4}$) for $\lambda \geq 185$ nm and in the ss case reach only 0.4% at $\lambda = 145$ nm.

It was expected that a zinc (or cadmium) photocathode would have the highest yield, reaching 0.7% at 200 nm [9]. For this reason, a 2 μm layer of zinc (Ventron; impurities ≤ 10 ppm) was vacuum-deposited on a PC plate. The measured yield, shown in fig. 3, is only a factor of 2 larger than that of the copper cathode and much less than the value given in ref. [9]. However,

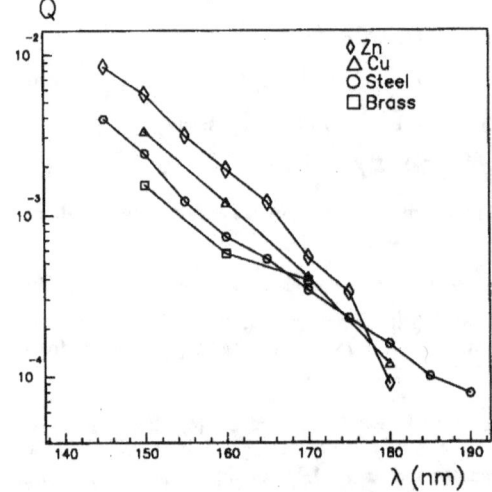

Fig. 3. The measured quantum yield Q of four bare metals versus the photon wavelength λ.

the zinc surface was exposed to air for about 10 min, hence some oxidation might have occurred.

4.2. Solid-TMAE photocathodes

A solid-TMAE photocathode was prepared by flowing clean methane through a TMAE bubbler at a tem-

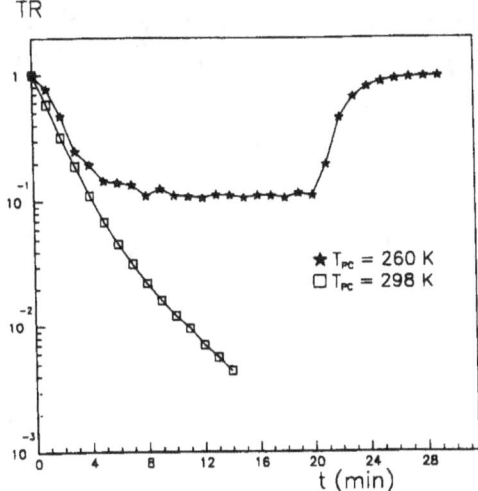

Fig. 4. The transmission TR of TMAE-laden methane gas in the TC for two temperatures of the PC plate versus gas flow-time t. In the case of $T_{PC} = 260$ K the TMAE gas injection started at $t = 0$ and stopped at $t = 20$ min. For $T_{PC} = 298$ K the TMAE gas injection was not stopped and the transmission plateaued at TR $= 7.8 \times 10^{-4}$.

perature of 293 K, and then into the test chamber with the PC plate cooled to 260 K. The TC walls and the PMT were maintained at 298 K. The melting point of TMAE is 269 K, its boiling point is 453 K, and its equilibrium vapour pressure is

$$p = p_0 \exp\left[(\Delta H/R)(T_0^{-1} - T^{-1})\right], \tag{11}$$

where the factor $\Delta H/R = 6370$ K, and $p_0 = 0.55$ Torr is the reference pressure at the reference temperature $T_0 = 300$ K [10,11]. A reduced TMAE concentration is obtained, when required, by mixing the TMAE line with a pure methane line in adjustable proportions via calibrated mass flowmeters.

The 160 nm light transmission was monitored by the PMT as a function of time after TMAE injection, as shown in fig. 4. The transmission (TR) decreased for 8 min, as the PI gas filled the TC, then became constant for 12 min, after which time the TMAE injection was stopped. The constant value TR_{PC} determines the equilibrium photon mean free path l_{PC} of the cryopumped TMAE gas, i.e. $l_{PC} = -z_t/\ln(TR_{PC})$. The corresponding equilibrium vapour pressure of TMAE is

$$p_{PC} = kT_{PC}/\sigma l_{PC}, \tag{12}$$

where k is Boltzmann's constant ($= 1.036$ mm Torr Mb/K), T_{PC} is the absolute temperature of the photocathode (and gas), and σ is the photoabsorption cross-section ($= 30$ Mb) of TMAE at $\lambda = 160$ nm.

The transmission of the same TMAE-laden gas in the TC and PC at ambient temperature (298 K) is also

shown in fig. 4. In this case the transmission plateau is very small ($\sim 7.8 \times 10^{-4}$), because the TMAE equilibrium vapour pressure is much higher without cryopumping. The mean free path l_{ph} at this temperature T, may be inferred from the slope of the curve, thus permitting the equilibrium vapour pressure of the non-cryopumped TMAE to be calculated analogously to eq. (12). The difference between these partial pressures determines the rate (dn/dt) at which TMAE molecules are cryopumped onto the cold PC, i.e.

$$\Delta p = p - p_{PC} = (k/\sigma)(T/l_{ph} - T_{PC}/l_{PC})$$

and

$$dn/dt = (dV/dt)\Delta p/kT_{PC},$$

hence

$$dn/dt = \frac{1}{\sigma} \frac{dV}{dt}\left[\frac{T/T_{PC}}{l_{ph}} - \frac{1}{l_{PC}}\right], \tag{13}$$

where dV/dt is the volume flow-rate of the gas into the test chamber. Characteristic numbers are $dV/dt = 10$ l/h, $l_{PC} = 10.7$ cm $T_{PC} = 260$ K, $l_{ph} = 3.3$ cm at $T = 298$ K, and $\sigma = 30$ Mb, hence $dn/dt = 1.4 \times 10^{18}$ molecules/min. The photocathode surface area is $2\Delta x \times \Delta y = 19.8$ cm^2, hence it has a molecular surface density of 7.1×10^{16} molecules/cm^2 min, which for $\sigma = 30$ Mb corresponds to 2.1 l_{ph} deposited per minute. The TMAE molecular weight is 200 and its estimated solid density is 0.7 g/cm^3, giving a film deposit rate of 0.34 µm/min. The deposit time was usually about 10 min, giving a thickness (≈ 3.4 µm $= 21$ l_{ph}), which is more than suffi-

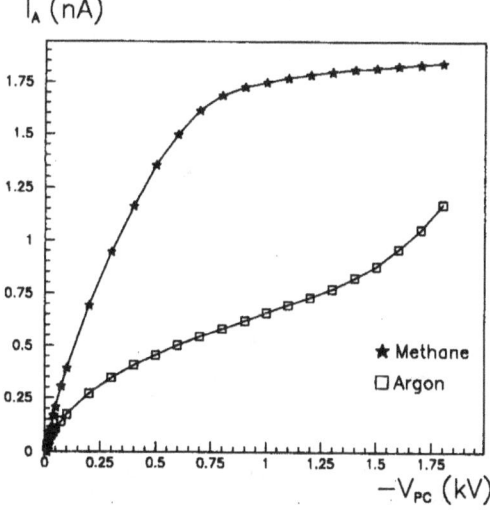

Fig. 5. The anode-collected photocathode current I_A versus the photocathode voltage V_{PC} for electron emission into methane or argon.

cient to ensure that the photons are completely absorbed in the TMAE layer before they reach the metal substrate.

The collected anode current I_A versus the PC potential $-V_{PC}$ is shown in fig. 5 for a solid photocathode with electron collection in pure methane. The current rises rapidly and becomes approximately constant between 0.8 and 1.8 kV, indicating efficient charge collection without amplification. Also shown in fig. 5 is the charge collection in pure argon gas. Note that the yield is reduced by about a factor of 3 owing to backscattering of electrons from argon, and also that amplification sets in for $|V_{PC}| > 1.5$ kV.

The quantum efficiency of four solid TMAE photocathodes (calculated from eq. (10)) is shown in fig. 6, with the TMAE gas-phase yield also shown for comparison. The curves for solid TMAE are labelled by T_{PC} (p). The data were taken under the following conditions: 260 (0), 260 (0.045), 243 (0.045) and 220 (0.045). Note that the yield increases with decreasing temperature. However, it was observed that when the TMAE gas flow was stopped, the higher yield curve, 260 (0.045), decayed rapidly to the lower curve, 260 (0), indicating that the increase in efficiency is probably due to a strong gradient of TMAE gas at the cathode surface. This was confirmed when melting a solid TMAE photocathode was allowed to melt. We then observed an

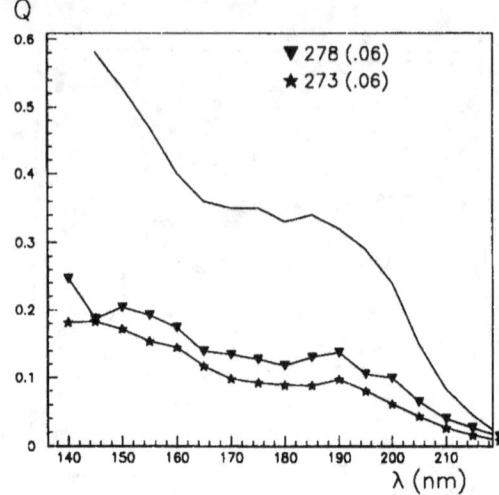

Fig. 7. The measured quantum yield Q versus the photon wavelength λ for two liquid-TMAE photocathodes. As in fig. 6, the labelling gives the photocathode temperature (K), followed by the TMAE partial pressure (Torr) in brackets. The solid line without data points is the TMAE gas-phase quantum efficiency.

increase of the current I_{Ad} by a factor of 2–3, before it fell to the quoted limit. The yields reported in this and subsequent sections are the intrinsic photocathode efficiencies, with the small gas-phase photoionization contribution removed by eq. (10).

These cathodes may be useful for fast RICH detectors (even though the best efficiency is well below the gas-phase efficiency), because the photocathode produces an isochronous (hence fast) signal and because the photon feedback is strongly suppressed owing to the low TMAE partial pressure. Better cathodes for fast RICH detectors have, however, been produced, as will be shown in subsequent sections.

4.3. Liquid-TMAE photocathodes

A liquid-TMAE photocathode was obtained by warming a deposited solid TMAE layer to $273 \leq T_{PC} \leq 278$ K; however, at this temperature the photocathode evaporated rapidly. To obtain a stable response it was necessary to flow, through the test chamber, methane that contained 0.06 Torr of TMAE, i.e. nearly its equilibrium pressure. The data are shown in fig. 7 by the curves labelled 273 (0.06) and 278 (0.06). The efficiencies are similar to those for solid TMAE, and are significantly smaller than the TMAE gas-phase efficiency, which is also shown in fig. 7 for comparison.

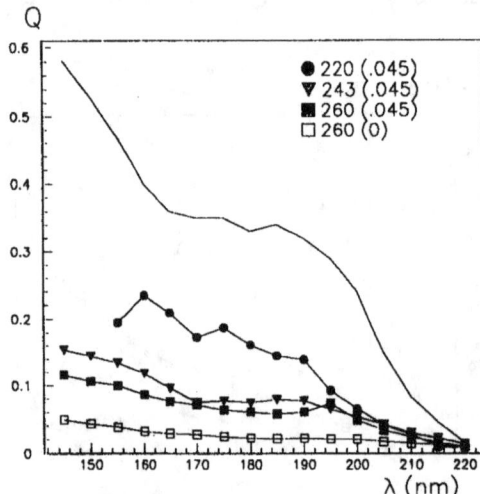

Fig. 6. The measured quantum yield Q versus the photon wavelength λ for four solid-TMAE photocathodes. The labelling in the top right-hand corner gives the photocathode temperature (K), followed by the TMAE partial pressure (Torr) in brackets. The solid line without data points is the TMAE gas-phase quantum efficiency.

4.4. Adsorbed TMAE films

In this case a bare metallic PC was kept at room temperature (298 K), and a small partial pressure (0.045 Torr) of TMAE-laden methane gas was flowed through the TC. The thickness of the adsorbed TMAE film could not be measured (it was adsorbed and not cryo-

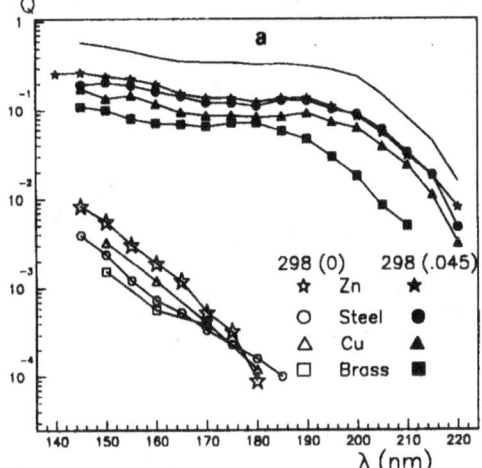

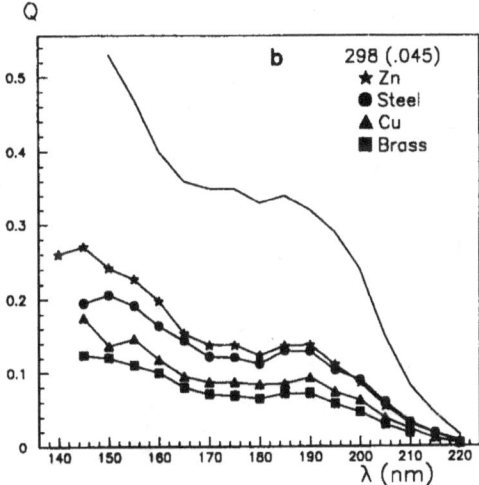

Fig. 8. (a) The measured quantum yield Q versus the photon wavelength λ for four metal cathodes with an adsorbed TMAE film. The labelling gives the photocathode temperature (K), followed by the TMAE partial pressure (Torr) in brackets. The solid line without data points is the TMAE gas-phase quantum efficiency. The open points are from fig. 3. (b) The same TMAE adsorbed film data as in (a), but on a linear scale.

pumped); however, it was so thin that the yield depended on the substrate metal. The measured photocathode yields are shown in fig. 8a on a logarithmic scale and in fig. 8b on a linear scale. The four different metal substrates – ss, copper, brass, and zinc, labelled 298 (0.045) – are shown in both fig. 8a and fig. 8b, whereas the bar-metal yield – 298 (0) – are shown only in fig. 8a. The TMAE gas-phase efficiency is also included for comparison. An efficiency of $\sim 15\%$ (at 185 nm) obtained with the ss and zinc PCs should be compared with a yield of $\sim 10^{-4}$ for the bare metals. This enormous enhancement of more than a factor of 10^3 may occur, as has been first observed by Anderson et al. [12] and confirmed by Peskov et al. [1], because the TMAE adsorbed on the metal surface forms an electric dipole layer, which facilitates the electron extraction [13,14]. The existence of a high-yield adsorbed film has also been hypothesized in order to obtain quantitative agreement with the observed photon feedback in the DELPHI barrel RICH prototype detector [15].

All yields increased by about 10% when the TMAE partial pressure was increased by 67% to 0.075 Torr.

The adsorbed TMAE film efficiencies are high; the films are relatively easy to obtain, because the PC is at room temperature with only a small TMAE partial pressure in the gas flow. The obvious advantages are fast response, suppressed photon feedback, and room-temperature operation.

4.5. Cesium iodide photocathode

Reflective CsI photocathodes with high quantum efficiency have been known since 1957, through the work of Taft and Philipp [16], and have been further developed by Carruthers [17], for space astronomy. His reflective photocathodes operate by extracting electrons into vacuum. In the present work, atmospheric pressure gas extraction is required.

To fabricate the CsI photocathode, a ss plate was mounted in a bell-jar pumped to a pressure $p \leq 10^{-6}$ Torr, and a ~ 100 nm layer of aluminium was vacuum-evaporated onto it, followed by a layer of CsI (Merck #2861; impurities of Rb, Na, K, Ba, Li, Mg, Sr, and Ca (in ppm) are 1000, 500, 500, 50, 40, 10, 5, and 1, respectively, with 0.13 ppm of other metals). The aluminium layer was deposited in order to insulate the CsI layer from impurities adsorbed on the ss plate. After CsI deposition, the bell-jar was filled with dry nitrogen at atmospheric pressure, and the ss plate was removed. It was rapidly wrapped in aluminium foil (to avoid exposure to sunlight) and stored in a silica-gel-filled glass dessicator, which was then pumped to a pressure $p \leq 10^{-3}$ Torr, sealed, and transferred to the photon monochromator laboratory. When the time came for measurement, the dessicator was opened to air, and

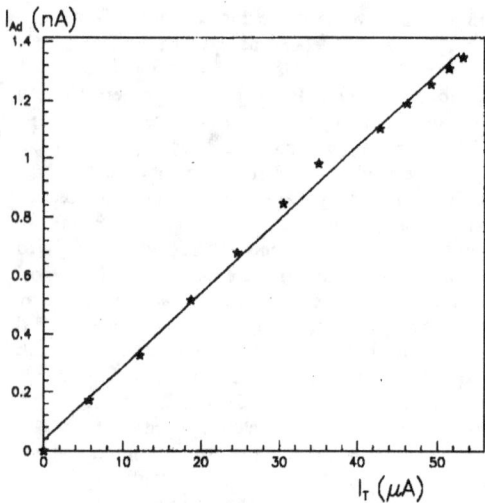

Fig. 9. The linearity of the anode-collected photocathode current I_A versus the transmitted beam monitor current I_T (with the PC plate in the up position). The PMC light intensity was varied by adjusting the height of the monochromator exit slit.

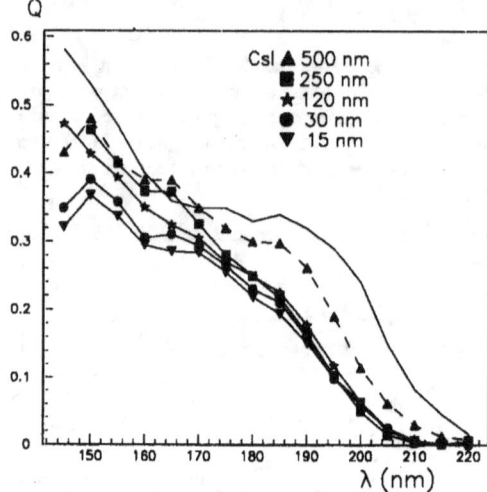

Fig. 10. The measured quantum yield Q versus the photon wavelength λ for five CsI photocathodes of different thicknesses. The solid line without data points is the TMAE gas-phase quantum efficiency.

the CsI photocathode plate was clipped manually onto the PC plane in the test chamber. As soon as the TC was closed, it was then flushed with a fast flow of clean methane gas. The total integrated air-exposure time of the CsI PC was between 10 and 20 min.

The 160 nm light transmission of the TC gas is monitored by the current I_T with the PC in the up position. Full photon transmission (at 160 nm) is obtained in about 1 hour of normal gas flow (20 1/h), but it requires about 48 h in order to obtain the full anode signal I_{Ad}. The TC and entry lines have a volume of about 3 l; hence, with the normal gas flow, about seven volume changes occur per hour. This is consistent with the observed time needed to obtain full photon transmission (i.e. to reduce the oxygen level); however, full photocathode yield requires more extensive rinsing of the CsI surface.

A check of the linearity of I_A with I_T made when the 160 nm light intensity was reduced by mechanically decreasing the height of the monochromator exit slit, is shown in fig. 9. A comparison of the currents I_{Ad} with I_T together with the measured value of $Q_T G_T = 1.12 \times 10^4$ (see fig. 2) shows that this CsI photocathode has a 3.57 times higher quantum efficiency than the p-terphenyl coated PMT. For $Q_T \cong 0.1$ (see fig. 2) then $Q_{CsI} \cong 0.36$ (at $\lambda = 160$ nm) in agreement with the direct comparison with gas phase TMAE (see below).

Five CsI photocathodes of different thicknesses (15, 30, 120, 250, and 500 nm) were deposited and measured. The yields are shown in fig. 10, with the TMAE

gas-phase efficiency included for comparison. The quantum yields (calculated from eq. (8) because CsI has negligible equilibrium vapour pressure) are large, and

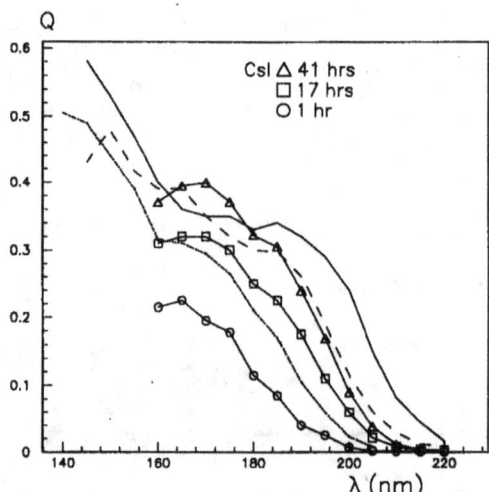

Fig. 11. The measured quantum yields Q versus the photon wavelength λ of the second 500 nm thick CsI photocathode after 1, 17, and 41 h of methane gas-flow through the TC. The solid line without data points is the TMAE gas-phase yield, the dashed line is the first 500 nm thick CsI photocathode yield, and the dash-dotted line is the Carruthers vacuum photocathode yield [17].

they are close to the TMAE gas-phase efficiency, especially at the shorter wavelengths. Note that the efficiency generally increases with increasing thickness. An (assumed) absorption cross-section $\sigma = 10$ Mb corresponds to an absorption mean free path $l_{ph} = 100$ nm, which is probably too small since the yield is still rising even for the 500 nm thick PC (i.e. $5l_{ph}$). Even so, for the quantum yield to continue to increase implies that there must be efficient electron transport from deep layers to the surface. Thicker layers have not yet been made, but will be tried out in the near future.

A second 500 nm thick CsI photocathode has been made and compared with the first one, as shown in fig. 11. The data show the evolution of the quantum yield, measured 1, 17, and 41 h after the (30 l/h) gas flow started. This second cathode is perhaps the better one, but overall they are in reasonable agreement ($\sim 10\%$). They both give a somewhat larger yield than Carruthers' 1000 nm thick vacuum photocathode [17], which is also shown in fig. 11. His photocathode was deposited, transferred, installed, and pumped without rinsing with methane gas. It therefore appears either that electron extraction in methane is more efficient than vacuum extraction (in contradiction with the Appendix of ref. [12]), or that the CsI must be rinsed to obtain its maximum yield. The yield of this cathode remained unchanged after 3 weeks' storage under reduced pure methane gas flow (1 l/h).

The best CsI photocathode yield is almost as good as that of gas-phase TMAE for $\lambda < 185$ nm. It therefore represents a serious alternative to TMAE for RICH detectors, as well as for readout of the fast component of BaF_2 scintillation.

Recently, Dangendorf et al. [18], in an independent measurement, have used a 1500 nm thick CsI reflective photocathode to detect primary and secondary scintillation photons from xenon in an X-ray imaging gas-scintillation counter. Their measured quantum efficiency is 9% at 170 nm, compared with $\sim 36\%$ in this work. Their calibration was by photon counting, with the quantum efficiency of the photomultiplier (Hamamatsu, R1460) as the reference, whereas we measure the photocurrent and use the gas-phase photoionization of TMAE [7,8] as the reference. Their photocathode was deposited directly onto a copper substrate, and only pumped to 10^{-1} Torr for outgassing. The quality of their CsI was identical to ours since they both came from the same source. We therefore believe that the above difference is caused by their photocathode not being sufficiently rinsed. As can be seen from fig. 11, the 170 nm yield increased by a factor of 2 during the 1 to 41 h period after the (30 l/h) gas flow started. We could not measure the increase in yield between 0 and 1 h because the TC gas was not yet sufficiently transparent due to residual oxygen.

When the 500 nm thick CsI photocathode was cooled

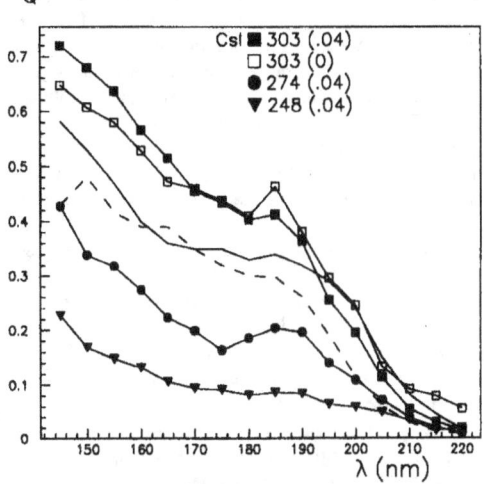

Fig. 12. The measured quantum yield Q versus the photon wavelength λ of the first 500 nm thick CsI photocathode, onto which a solid, a liquid, or an adsorbed film of TMAE has been deposited. The labelling gives the photocathode temperature (K), followed by the TMAE partial pressure (Torr) in brackets. The solid line without data points is the TMAE gas-phase quantum efficiency and the dashed line is the CsI yield.

slowly down to 160 K (the temperature of liquid xenon), the quantum yield decreased by a factor of about 100 from its room-temperature starting value. It was hypothesized that this was due to the ambient-temperature TC walls being gettered by the cold CsI cathode, which poisoned its operation. However, the photocathode always recovered its full efficiency when it returned to room temperature. To verify this supposition, the cold trap (figs. 1a and 1b) was cooled to about 110 K before the CsI cathode was cooled to 160 K. The quantum yield then decreased by only a factor of 1.2, in accordance with the hypothesis. The small residual gettering effect that remained will probably disappear with further cooling of the trap. On this basis, it appears that a CsI photocathode will operate with full efficiency at liquid-xenon temperatures.

4.6. Cesium iodide + adsorbed TMAE-film photocathodes

Because of the enormous effect that an adsorbed TMAE film has on metals (see subsection 4.4 and fig. 8a), it was decided to investigate the effect of a solid, a liquid, or an adsorbed TMAE film on a CsI photocathode. The results, with the (first) 500 nm thick CsI photocathode as the substrate, are shown in fig. 12. The solid, liquid, and adsorbed films were formed by flowing methane, mixed with 0.04 Torr TMAE, through the

TC, with the PC plate at temperatures of 248, 274, and 303 K, respectively. In all cases the yield was several times higher than the corresponding case where a metal substrate was used. In the case of the solid and liquid layers, the yield was lower than for the pure CsI photocathode. However, for the adsorbed layer it was higher – even more than the TMAE gas-phase yield (both of these curves are included in fig. 12 for comparison). Even more remarkable is the fact that the adsorbed-layer yield stayed unchanged when the TMAE partial pressure was reduced to zero, as shown by the curve labelled 303 (0), and it remained stable even after the CsI photocathode had been flushed for three weeks with pure methane. This photocathode was later exposed to air by opening the TC for one hour. Its quantum efficiency returned to the previous level after several hours of vigorous methane gas-flow (80 l/h).

The CsI photocathode with an adsorbed TMAE film has several advantages as a RICH detector, e.g. it has a high yield, it is fast, and it is solar blind. The biggest problem in a detector with gas amplification will probably be due to photon feedback. Gas-phase feedback cannot occur since the TMAE partial pressure in the gas is negligible. Photon feedback from the photocathode itself will tend to merge with the primary avalanche if the amplifying structure (wire) is near the cathode (i.e. 0.5 mm = 10 ns). The amplifying gas must, of course, be opaque to the C^* (atomically excited carbon) lines at 156 and 166 nm (i.e. 7.95 and 7.47 eV). The use of pentane or hexane, with methane as the amplifying gas, assures the requisite opacity and simultaneously defines the upper energy limit of the detector response (i.e. ~ 7 eV). The C^* emission line at 193 nm (6.42 eV) is then the only remaining source of feedback instability. It will have the same level of photon feedback as that from a triethylamine (TEA)-based photoionizing gas detector, which is sensitive only to the 156 nm line. The quantum efficiencies of TEA at 156 nm and CsI + TMAE at 193 nm are similar, and since these emission lines are equally probable it is expected that these two types of detectors will show a similar performance. A TEA pad-detector (with amplifying wires 0.5 mm from the pad plane) has recently been tested successfully [19]. It showed considerably enhanced stability to photon feedback, compared with the results obtained with a TMAE gas detector.

A pad-detector, in which the CsI + TMAE photocathode is deposited on the pad plane, is now being fabricated in order to test this cathode for applications as a RICH detector.

Tests for reading out a BaF_2 scintillating-crystal calorimeter with a CsI photocathode have been made; they will be briefly described in section 5 and more completely in ref. [20].

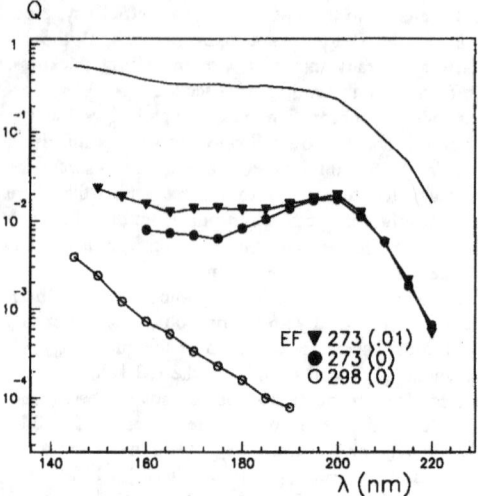

Fig. 13. The measured quantum yields Q versus the photon wavelength λ of the stainless-steel cathode with an adsorbed ethyl ferrocene (EF) film. The labelling gives the photocathode temperature (K), followed by the TMAE partial pressure (Torr) in brackets. The solid line without data points is the TMAE gas-phase quantum efficiency.

4.7. Ethyl ferrocene photocathode

Quantum-yield measurements of an ethyl ferrocene (EF) photocathode in the region $\lambda > 190$ nm have been published in a previous work by some of the present authors [21,22]. This material is photosensitive and, unlike TMAE, does not react with air. The latter property may be of interest for some applications. Such a cathode has already been used as the photosensitive element in a small BaF_2 calorimeter [23]. In the present work a new measurement of this cathode has been made over a wider wavelength range, as shown in fig. 13. The three curves are: i) for the bare ss PC plate at room temperature, labelled 298 (0); ii) for a condensed photocathode at 273 K with the EF equilibrium partial pressure of 0.01 Torr, labelled 273 (0.01), and iii) for the same cathode after the flow of EF has been stopped, i.e. 273 (0). In cases ii) and iii) the yield at the BaF_2 scintillation wavelength ($\lambda \sim 193$ nm) is about 2%, which is in agreement with previous work [22]. However, the EF yield is considerably smaller than any of the others tested here.

5. BaF_2 scintillation detected with a CsI photocathode

In order to independently confirm the large quantum yield of the CsI photocathode (see subsection 4.5), a home-made pulsed 100 kV electron accelerator was used to excite fast scintillation in a BaF$_2$ crystal, as shown in figs. 14a and 14 b. The electron source for the accelerator starts with a pulsed (15 ns FWHM) photon beam from a low-pressure H$_2$ UV lamp with CaF$_2$ windows. The pulsed light source is focused by a CaF$_2$ lens through a CaF$_2$ window and onto a CsI reflective vacuum photocathode placed at the beginning of the 40 cm long, 100 kV, accelerating structure, shown in fig. 14a. The number of photons per pulse could be regulated by an adjustable iris located in front of the uv lamp (not shown in fig. 14a). The maximum electron intensity is $\sim 10^7$ electrons per pulse having a maximum kinetic

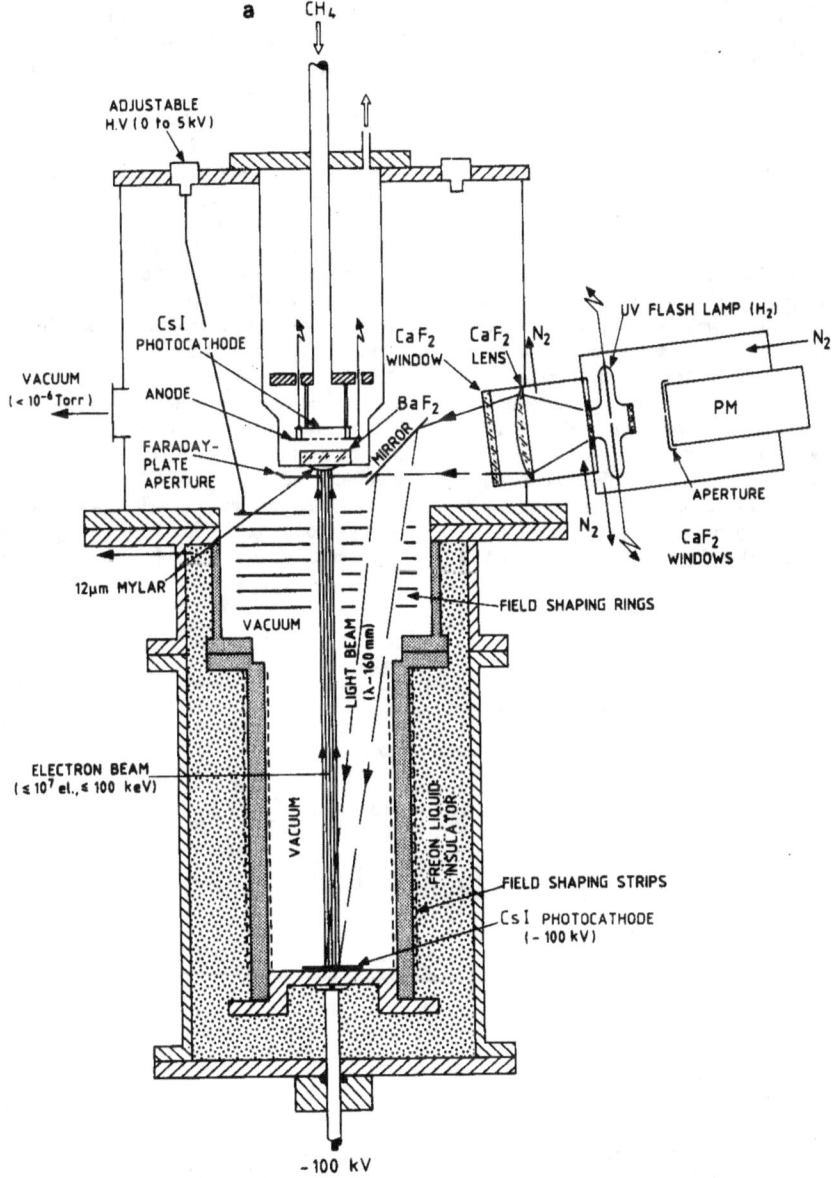

Fig. 14. (a) A schematic view of the 100 kV electron accelerator with the pulsed H$_2$ light source, the CsI photocathode, the accelerating structure, and the copper Faraday-plate collimator. (b) A schematic view of the BaF$_2$ scintillator, the CsI reflective photocathode, the collection mesh, the charge-sensitive preamplifier, and the 0.8 μs shaper.

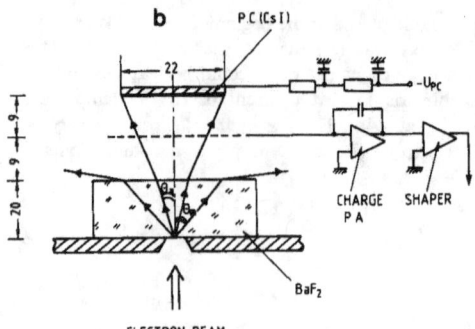

Fig. 14 (continued).

energy of 100 keV corresponding to a maximum total excitation energy of 1 TeV per pulse. The minimum measurable excitation energy is ~ 1 MeV. The number of electrons in the beam is monitored by a copper Faraday-plate collimator, which intercepts about 75% of the beam and transmits the remainder to the BaF_2 crystal. In a separate study, with a second Faraday plate (placed to intercept the collimated electron beam), it was shown that the charge measured on the Faraday-plate collimator determines the transmitted charge with a fractional rms error of about 0.3%. The electron kinetic energy is determined by the power supply voltage corrected for energy loss in a 12 μm Mylar entrance window. A more complete description of this device will be given in ref. [20].

The BaF_2 crystal was mounted to intercept the collimated electron beam, and the (second) 500 nm thick CsI photocathode was positioned to view the scintillation source, as shown in fig. 14b. The refraction-corrected solid angle subtended by the photocathode was $\Delta\Omega/4\pi = 2.44\%$. A fast signal was observed on the collection grid of a low-noise charge-sensitive preamplifier ($\langle NEC \rangle \approx 300$ electrons; 400 mV/pC) followed by a shaper (ELSCINT model CAU-N3; 0.8 μs shaping time) (see fig. 15a). The signal (normalized to a constant charge input by the Faraday-plate monitor) varied linearly with the electron kinetic energy as shown in fig. 15b. The slope of this curve, the solid angle, the crystal transmission (0.8), and the charge input (2.65 × 10^5 electrons per pulse) determine a total (4π) signal of 40 photoelectrons per MeV of deposited energy.

In a separate study of BaF_2 scintillation, a total emission efficiency of 560 photons per MeV for the fast component and 9600 photons per MeV for the slow component was measured using two different photomultiplier detectors (Hamamatsu R1460 and 1332Q). The observed pulse waveform is shown in fig. 16, where the fast and slow signals may be clearly discerned. The lifetime of the slow component was measured to be 740 ns, whereas the fast pulse reflects only the 15 ns excitation pulse width.

The measured quantum-yield spectrum of this CsI photocathode (fig. 11) has been averaged over the fast BaF_2 scintillation spectrum [24] to give an average yield $\langle Q \rangle = 5.75\%$. This yield, together with the measured

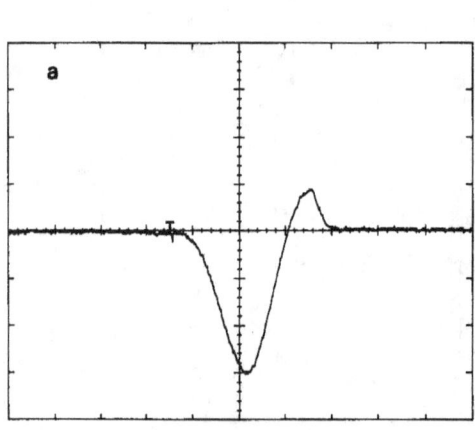

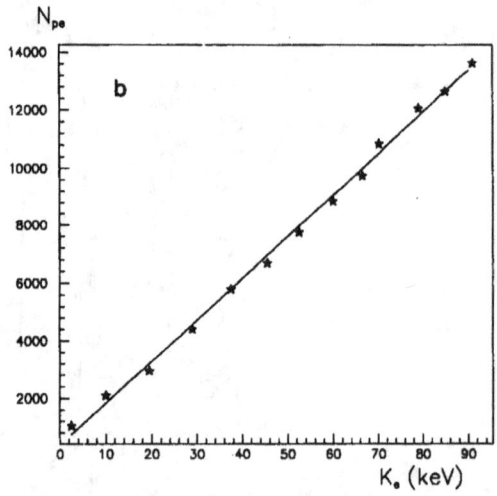

Fig. 15. (a) Typical CsI photocathode signal observed, at the output of the shaper, with a digital oscilloscope Tektronix 2440, for 5 GeV total energy deposited in the BaF_2 crystal. The horizontal and the vertical scales are 1 μs and 2 mV per division, respectively. (b) The CsI photocathode-detected signal versus the incident electron kinetic energy K_e. The signal is normalized to a constant charge input of 2.65×10^5 electrons per pulse, hence the 100 keV electron kinetic energy point corresponds to 26.5 GeV of total energy deposited in the BaF_2 crystal.

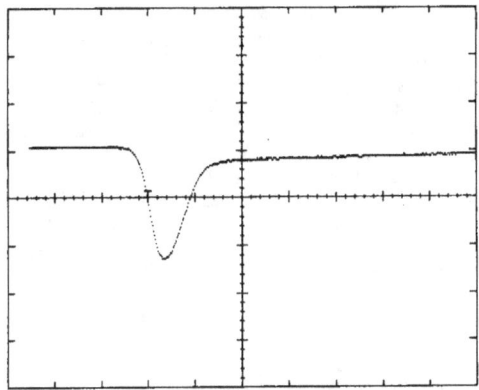

Fig. 16. The BaF_2 scintillation pulse waveform observed with a Hamamatsu R1460 photomultiplier with a Cs_2Te solar-blind photocathode. The photomultiplier pulse amplitude (1 mV per division) is shown versus time (50 ns per division). The fast pulse width reflects only the excitation pulse width (15 ns FWHM), whereas the slow pulse has a decay time of 740 ns. The Cs_2Te photocathode is more efficient for the fast pulse (193 nm), hence the relative amplitudes of the fast and slow components are biased.

fast emission yield of 560 photons per MeV, then predicts a total (4π) signal of 32 photoelectrons per MeV, assuming that no reflectivity occurs onto the other faces of the crystal, hence in rough agreement with the measured value of 40 and also confirming the large CsI yield.

Note that with a reflective cathode the maximum observable signal is limited, by total internal reflection, to about 11.6% per face, i.e. to 4.6 photoelectrons per MeV per face. A six-faced detector (or a one-faced detector with each of the other five faces mirrored) could then obtain a total of 28 photoelectrons per MeV with perfect crystal transmission. A way to overcome the loss of photons, which is due to internal reflection, could be the insertion, between the crystal and the CsI photocathode, of a liquid interface such as TMS or TMP, which would transmit the fast BaF_2 scintillation component. Such a test is under preparation.

A transmissive photocathode, deposited directly on the crystal's surface, would avoid the problem of losses due to total reflection, but it would not increase the detected signal because its quantum yield is only 10% of the reflective cathode yield [17]. It would also require a transmissive conducting film, deposited directly on the BaF_2 crystal, as a substrate for the transmissive CsI photocathode. Such a conducting film would engender additional photon absorption.

The use of the reflective CsI + TMAE adsorbed film cathode will increase the quantum yield by a factor of 2 (at 193 nm), to give a quite respectable detected signal

of 9.2 photoelectrons per MeV per face, and hence a total of 56 photoelectrons per MeV.

In comparison, the TMAE gas-phase quantum yield [7,8] has been averaged over the fast BaF_2 scintillation spectrum [24] to give $\langle Q \rangle = 9.84\%$, hence a total ($4\pi$) signal of 55 and a detectable signal of 6.4 photoelectrons per MeV per face. Therefore, the response from the CsI + TMAE adsorbed film cathode should be 44% larger than that from gas-phase TMAE.

Other ways of increasing the signal require that the scintillation spectrum be shifted to about 170 nm [25] where the CsI + TMAE quantum yield is high (~ 46%). The peak of the liquid-xenon emission spectrum is, in fact, at 170 nm, and its scintillation efficiency is enormous, i.e. about $(2-3) \times 10^4$ photons per MeV, corresponding to a potential total signal of (10–15) photoelectrons per keV. This potential signal could be fully realized because the photocathode would be in direct contact with the liquid, which means that total reflection losses could be avoided. However, the photoelectrons must be efficiently injected into the liquid, where they would drift (in an applied electric field) to a collection grid for detection.

6. Summary and conclusions

An investigation of reflective photocathodes has shown that electron emission into methane gas is as efficient as emission into vacuum, and that the use of TMAE adsorbed films can increase the electron yields significantly. These cathodes must be rinsed by gas flow in order to obtain optimal yield. A specific CsI + TMAE adsorbed-film photocathode has been developed, which has 46% quantum efficiency at 170 nm and is well adapted for fast RICH detectors because it has isochronous signals, reduced photon feedback, and room-temperature operation. This cathode also promises to allow the simple, efficient, and cheap readout of fast inorganic UV scintillators such as BaF_2, as well as of the noble liquid scintillators.

Acknowledgements

The work reported here is part of the LAA project; it was also supported by the Institut National de Physique Nucléaire et de Physique des Particules (IN$_2$P3) of France. Y.G. wishes to acknowledge the support of the National Research Centre Demokritos (NRCD), Athens. We wish to thank G. Passardi for assisting in the construction of the apparatus, and M. Bosteels for his help with the gas system. Thanks are due also to A. Brehm and C. Nichols for their excellent vacuum evaporations, to J.-P. Jobez for the design of the apparatus, and to R. Saigne and J.L. Escourrou for its assembly and modifications. J.C. Santiard provided us with the low-noise preamplifier, and gave us the benefit of his expertise.

References

[1] V. Peskov, G. Charpak, P. Miné, F. Sauli, D. Scigocki, J. Séguinot, W.F. Schmidt and T. Ypsilantis, Nucl. Instr. and Meth. A269 (1988) 149.

[2] J. Séguinot, Proc. Symp. on Particle Identification at High-Luminosity Hadron Colliders, Batavia, IL, 1989, eds. T.J. Gourlay and J.G. Morfin (Fermilab, Batavia, 1989) pp. 215 and 671;
T. Ypsilantis, ibid., p. 133 and/or preprint CERN-EP/89-150 (1989);
J. Séguinot and T. Ypsilantis, report CERN-LAA/89-1, p. 327.

[3] G. Charpak, V. Peskov and D. Scigocki, ibid., p. 139.

[4] J. Séguinot and T. Ypsilantis, ibid., p. 187.

[5] T. Ypsilantis, Proc. ECFA Study Week on Instrumentation Technology, Barcelona, 1989, eds. E. Fernandez and G. Jarlskog (CERN 89-10, Geneva, 1989) vol. 2, p. 661.

[6] P. Baillon, Y. Declais, M. Ferro-Luzzi, B. French, P. Jenni, J.-M. Perreau, J. Séguinot and T. Ypsilantis, Nucl. Instr. and Meth. 126 (1975) 13.

[7] R.A. Holroyd, J.M. Preses, C.L. Woody and R.A. Johnson, Nucl. Instr. and Meth. A261 (1987) 440.

[8] R. Arnold, P. Baillon, H.J. Besch, M. Bosteels, E. Christophel, M. Dracos, Y. Giomataris, J.L. Guyonnet, G. Passardi, P. Petroff, J. Séguinot and T. Ypsilantis, Nucl. Instr. and Meth. A270 (1988) 289.

[9] W. Kluge, in: Landolt–Börnstein (Springer, Berlin, 1959), vol. 2, part 6;
R. Schulze, Z. Phys. 92 (1934) 223.

[10] D.F. Anderson, Nucl. Instr. and Meth. 126 (1975) 13 and 178 (1980) 125;
IEEE. Trans. Nucl. Sci. NS-28 (1981) 842.

[11] Y. Giomataris, CERN-DELPHI 86-17, RICH15 (1986).

[12] D.F. Anderson, R. Bouclier, G. Charpak and S. Majewski, with Appendix by G. Kellner, Nucl. Instr. and Meth. 217 (1983) 217.

[13] D.P. Woodruff and T.A. Delchar, Modern Techniques of Surface Science (University Press, Cambridge, 1986).

[14] D.A. King and D.P. Woodruff, in: The Chemical Physics of Solid Surface and Heterogeneous Catalysis (Elsevier, Amsterdam, 1984), vol 3, part B.

[15] R. Arnold, P. Baillon, H.J. Besch, M. Bosteels, E. Christophel, M. Dracos, Y. Giomataris, J.L. Guyonnet, G. Passardi, P. Petroff, J. Séguinot, D. Toët, J. Tocqueville and T. Ypsilantis, Nucl. Instr. and Meth. A270 (1988) 255.

[16] E.A. Taft and H.R. Philipp, Phys. Chem. Solids 3 (1957) 1.

[17] G.R. Carruthers, Appl. Opt. 8 (1969) 633, 12 (1973) 2501, and 14 (1975) 1667.

[18] V. Dangendorf, A. Breskin, R. Chechik and H. Schmidt-Bocking, Weizmann Inst. preprint WIS-89/81/December/PH, presented at SPIE 1990, Symp. on Instrumentation in Astronomy, Tucson (Ariz.), 1990; also Nucl. Instr. and Meth. A289 (1990) 322.

[19] R. Arnold, Y. Giomataris, J.L. Guyonnet, J. Séguinot and T. Ypsilantis, report in preparation.

[20] J. Séguinot, M. Bosteels, Y. Giomataris, G. Passardi, V. Peskov, J. Tischhauser and T. Ypsilantis, report in preparation.

[21] G. Charpak, V. Peskov, F. Sauli and D. Scigocki, Nucl. Instr. and Meth. A270 (1988) 255.

[22] G. Charpak, D. Lamb, V. Peskov, D. Scigocki and J. Valbis, see Proc. of ref. [5], vol. 2, p. 593.

[23] V. Peskov, G. Charpak, F. Sauli, D. Scigocki, V. Diep and D. Janjic, Nucl. Instr. and Meth. A283 (1989) 786.

[24] P. Schotanus et al., IEEE Trans. Nucl. Sci. NS-34 (1987) 272.

[25] P. Schotanus et al., IEEE Trans. Nucl. Sci. NS-36 (1989) 132;
also A.F. Buzulutskov et al., Nucl. Instr. and Meth. A288 (1990) 659.

7. MISCELLANEOUS DETECTORS:
1962–1994

7. MISCELLANEOUS DETECTORS: 1962–1994

NUCLEAR INSTRUMENTS AND METHODS 15 (1962) 323–326; NORTH-HOLLAND

LOCATION OF THE POSITION OF A PARTICLE TRAJECTORY IN A SCINTILLATOR

G. CHARPAK, L. DICK and L. FEUVRAIS

CERN, Geneva, Switzerland

Received 24 January 1962

The light pulses travelling to opposite directions in a flat scintillator are collected by two photomultipliers.

1. The delay between the arrival of the two pulses, measured with a time-converter, gives the position of the traversal of the scintillator by a ionizing particle. It is found that the average propagation speed of the light in the scintillator is 10^{10} cm/s. The relation between position and delay is linear along the axis of a 40 cm long scintillator. The accuracy (full width at half height) in the determination of the position is 5 cm with a collimated strontium source.

2. Possibility of eliminating the fluctuations introduced in time-of-flight measurements with extended scintillators is discussed.

3. Addition of the pulses from the two photomultipliers eliminates to a great extend the variation of pulse height as a function of the position of the ionizing particle. The variation is less than 10%.

In a recent study[1] about the localization of sparks in a spark counter, spiral electrodes built as distributed delay lines were used in the following way: after each spark, the time interval between the signals travelling along the line to the two opposite outputs is measured and determines the position of the spark. One of the properties of this arrangement is that the time intervals are amplified by a factor of 2.

This can be of interest in problems where the propagation speeds are such that they reach the accuracy limit of the time measuring instrument. A similar method is used to determine the point where a particle crosses a scintillator (fig. 1). The

plates often used in nuclear physics, and with two pairs of photomultipliers and two time-converters, the position in space of the particle could be established.

If a trajectory crosses a scintillator at a point distant by Y cm from one end of the scintillator, the time interval between the two output pulses will change by ΔT when Y is varied by ΔY:

$$\Delta T = \frac{\Delta Y}{3 \times 10^{10}} \cdot 2 \cdot n \cdot f \tag{1}$$

where n is the refractive index (equal to 1.58 for a plastic scintillator), f is a geometrical factor expressing the fact that the average path length of the

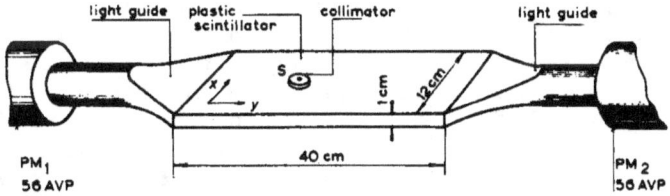

Fig. 1. View of the light guide and scintillator assembly.

light signal is emitted isotropically, and, by measuring the time interval between the arrival of the light signals at two opposite ends of a scintillator, one of the coordinates is obtained. With the thin

light is not a straight line parallel to the Y-axis. f_{min} is equal to 1; f_{max} corresponds to the reflexion

[1] G. Charpak, Location of the position of a spark in a spark counter, Nucl. Instr. and Meth. 15 (1962) 318.

at a critical angle on two pairs of orthogonal planes, and is equal to n^2.

Thus $\varDelta Y \times 10^{-10}\,s < \varDelta T < \varDelta Y \times 2.6 \times 10^{-10}\,s\,.\,(2)$

Therefore, to get a resolution of 1 cm, an accuracy of about 2×10^{-10} s. at least is needed. However, a limited resolution can lead to the very accurate determination of a distribution, when this varies smoothly across the width of the window, and if the stability of the centre of the window is good. Stability of the order of 10^{-10} s can be reached, and with two scintillators one meter apart, an angular distribution measurement to within one degree is obtained with a resolution of 2 cm leading to the measurement of time intervals of the order of 4×10^{-10} s.

The experiment was performed under the following conditions (fig. 1): a 40 cm long $\times$ 0.8 cm thick $\times$ 12 cm wide scintillator was viewed at both ends, through two short light guides, by two Radiotechnique 56 AVP photomultipliers. The pulses were fed to a time converter of the type developed by L. Dick at CERN[2]) (fig. 2). Calibration with delay lines of fixed length shows linearity on a range of 20×10^{-9} s, and gives 0.15×10^{-9} s per channel of a TMC 256-channels analyser.

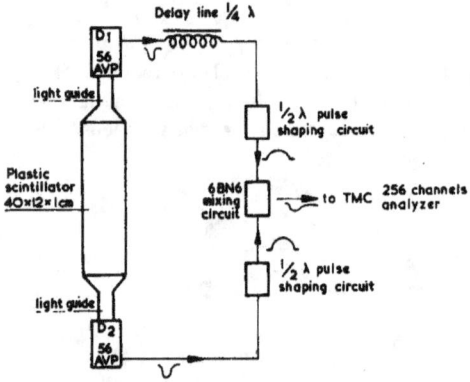

Fig. 2. Simplified diagram of the time converter.

The pulses in the scintillator were produced by a collimated β strontium source 2 mm wide. A slow coincidence circuit is used to select the pulses of the highest amplitude. The results obtained are given in fig. 3. They show that the full width at half

maximum corresponds to about 5 cm and is practically independent of position of the particle's impact point. An other result is that the average velocity of the light along this scintillator is

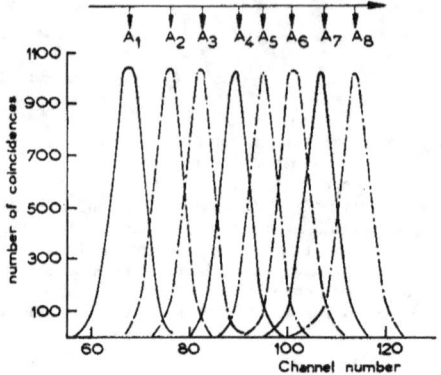

Fig. 3. Pulse height distribution. Position of strontium source. Each A_n curve corresponds to a step of 5 cm. A_1 and A_8 are at 2.5 cm from the light guides.

10^{10} cm/s, i.e. only $\frac{1}{3}$ of c, and is in agreement with inequality (2). This gives the time fluctuations to be expected when extended plastic scintillators are used to measure time-of-flight, for instance.

We also found that the displacement of the source along a transverse line, over the whole width of the scintillator, gives no appreciable shift.

Fig. 4 shows the linear relation between the time converter output pulse height and the position of the source on the Y- axis. A slight non-linearity is also observed at one end (which was to be expected since, at the ends of the scintillator, an appreciable amount of light can go directly to the photomultiplier with less reflexions and therefore a shorter optical path).

This time-converter has proved to have a stability, over a range of more than a week, of better than one channel, corresponding here to 0.7 cm;

This technique thus offers the possibility to improve the resolution and efficiency in some experiments, such as angular distributions. It allows the simultaneous scanning of big solid angles with the same spatial resolution as obtained with coun-

[2]) L. Dick, Mesure des intervalles de temps dans le domaine de 0.1 ns avec des détecteurs à scintillation, CERN Report (to be published).

ters of very small dimensions. Multidimensional-multichannel analysers provide facilities for correlated storage along two perpendicular coordinates.

The resolution being a function of the intensity

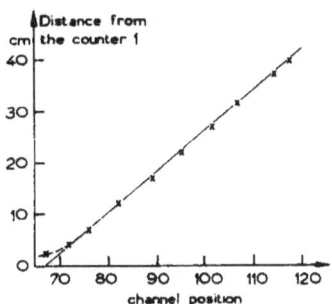

Fig. 4. Average pulse height as a function of strontium source position.

of the light pulses, pulses of standard height should allow a more precise time determination. For that reason, minimum ionising particles traversing a flat scintillator should be particularly suitable, while a continuous β-spectrum like the one used for the test is rather unfavourable. For instance, a time-of-flight measurement with pions of 200 MeV/c and a classical telescope with counters of $6 \times 6 \times 0.2$ cm gives a resolution of 5×10^{-10} s better by a factor of 2 than the resolution obtained with a continuous β-spectrum.

The simultaneous use of the light outputs from the two ends of a scintillator can help to improve the accuracy of the time-of-flight devices using extended scintillators. It is well known that the principal limitation comes, not from the electronics, but from the fluctuations of time connected with the difference in length of the light paths from the different parts of the scintillators. Advantage can be taken of this in the following way (fig. 5).

Let t_1, t_2 and t_3, t_4 be the times of propagation of the light pulses to the outputs of the two scintillators, and t_0 the time-of-flight of the particle from counter 1 to counter 2. Time-converter A measures $t_0 + t_4 - t_2$. Time-converter B measures $t_0 + t_3 - t_1$. In C, the pulses from A and B are added together; then the result is

$$2t_0 + (t_4 + t_3) - (t_2 + t_1) = 2t_0 + \text{const.}$$

if the time-converters caracteristics are the same; for scintillators with the same length, const. = 0.

The time-converters used are based on the properties of the 6BN6 gated beam tubes; the

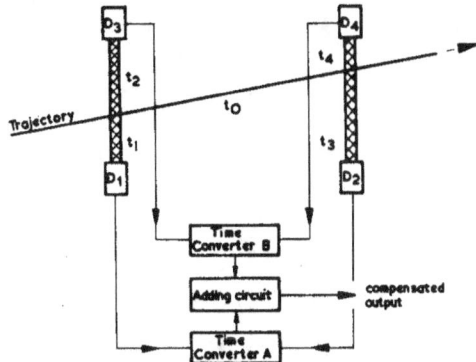

Fig. 5. Block diagram for the compensated time of flight measurement.

addition can be made simply by giving a common plate resistor to the two 6BN6 of the two time-converters.

The method described permits the scanning of a beam in a very convenient way, by coupling the two plates of an oscilloscope to the two time-converters, giving two orthogonal co-ordinates.

Other informations are provided by such a system:

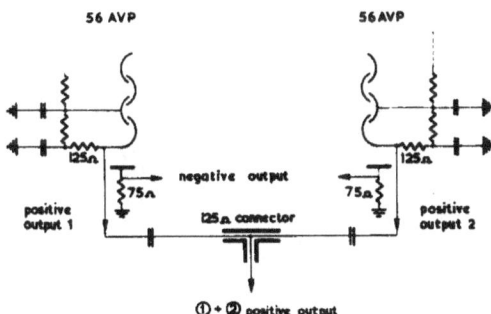

Fig. 6. Addition of the two positive outputs.

1. The ratio between the pulse heights at two opposite ends is dependant only on the position

where the light is produced, and not on the initial energy loss.

2. The sum of the two outputs is nearly constant. This fact was verified under the following condi-

within 10%, while the single output varies by a factor of two from one end of the scintillator to the other (fig. 7).

This property can be used to distinguish be-

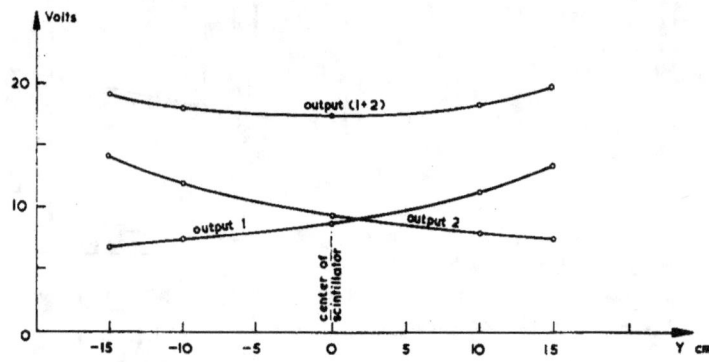

Fig. 7. Pulse height vs the position of the source along X axis.

tions: the gains of the two photomultipliers are adjusted to be the same; the pulses from the two positive outputs are added together, amplified and fed to the kick-sorter (fig. 6). The sum is constant

tween the passage through the scintillator by one or several particles, when these particles all loose the same amount of energy, as is often the case in high energy physics.

NUCLEAR INSTRUMENTS AND METHODS 48 (1967) 151–153; © NORTH-HOLLAND PUBLISHING

LETTERS TO THE EDITOR

THE LOCALIZATION OF THE POSITION OF LIGHT IMPACT ON THE PHOTOCATHODE OF A PHOTOMULTIPLIER

G. CHARPAK

CERN, Geneva, Switzerland

Received 27 October 1966

When light impinges on the transparent photocathode of a photomultiplier, electrons are extracted by photoelectric effect. They are focused and accelerated to the first dynode by several auxiliary electrodes. The photoelectrons are emitted in a direction almost orthogonal to the incident light.

Very small magnetic fields can perturb the trajectory of these very low energy electrons. We investigated the possibility of creating very localized magnetic fields in such a way that only the photoelectrons coming from a limited area of the photocathode be perturbed. For this purpose we used the set-up shown in fig. 1.

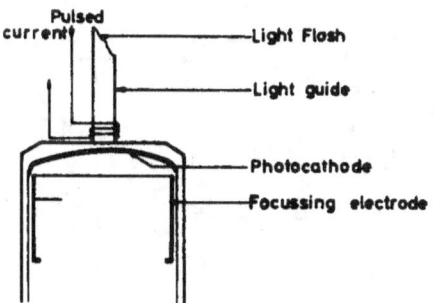

Fig. 1. Experimental set-up to study the effect of localized magnetic fields. Currents of several amperes can be sent through the seven-turn coil C at the bottom of the 6 mm wide light guide.

Light flashes produced by a light pulser are sent on a 56 AVP Radiotechnique photomultiplier, through a 6 mm dia. light guide at the bottom of which a small coil of seven turns can receive current pulses of several amperes. These current pulses can be applied simultaneously with the light flashes, or through a variable delay line. We observed (fig. 2) a strong effect when the two pulses coincide, and no effect when the current is delayed before or after the light. However, we observed that we had also a strong effect by disconnecting the return to ground of the coil.

This shows that the dominant effect comes from the application of the local electric field.

We repeated the experiment with the set-up of fig. 3. A thin transparent mesh of 6 mm dia. is placed under

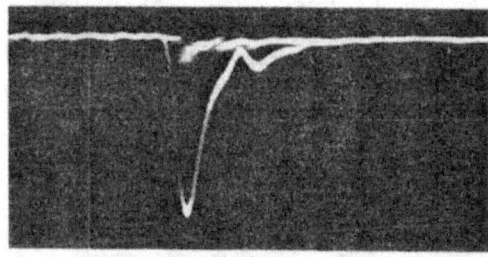

Fig. 2. Pulse of a 56 AVP Radiotechnique gated by a current pulse in the coil C. Flashes of light 25 nsec wide from a Pek flashtube. Big pulse: no current in the coil. Small pulse (at the bottom): ∼ 20 A in the coil.

the light guide G_1. Electric pulses of 70 V can be applied to this mesh. A second light guide, G_2, is placed at 1 cm from the first one and can also receive the light flashes. The following effects are observed:

A positive pulse of 70 V applied on the mesh extinguishes the pulse from the photomultiplier by a factor of 10 when the light is flashed through light guide G_1. It enhances the pulses by a factor of 4 when the light goes through the neighbouring light guide G_2.

Pulses of the opposite polarity quench the light from either guide.

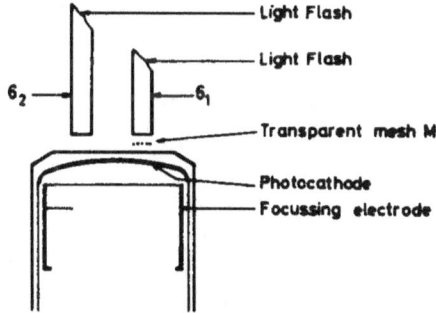

Fig. 3. Experimental set-up to study the effect of local electric fields. Two light guides 1 cm apart can receive the light pulses from a Pek flash-tube. The transparent mesh M is placed between one light guide and the photomultiplier. It can be pulsed at adjustable voltages with a variable delay.

151

Fig. 4 shows the variation of the effect as a function of the applied field.

The pictures of fig. 5 show the effect of the gating on a double light pulse (two light pulses, distant by about 25 nsec). It is possible to extinguish the pulse from one of the light flashes without disturbing the second. This should have a practical application when high initial short light-bursts overload the circuits.

The picture of fig. 5c also shows how by lengthening the gating pulse on the mesh, the second light pulse gets extinguished. The sharpness of the effect is exactly the sharpness of the falling edge of the gating pulse ($\sim$ 5 nsec). Thus by this method it is certainly possible to get a very sharp time control of the electron emission to the

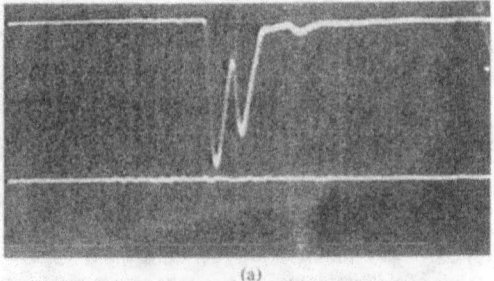

(a)

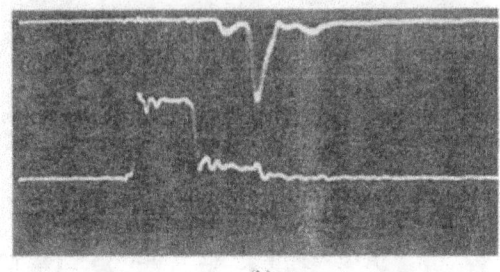

(b)

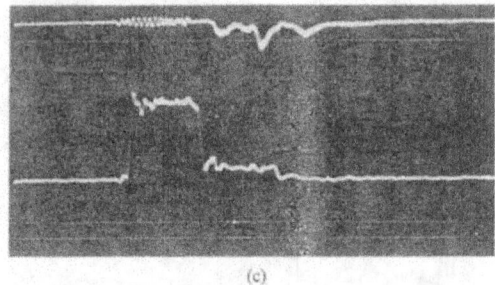

(c)

Fig. 5. Effect of a positive 70 V pulse on M. Double light-pulse. Hor. 50 nsec/div. Top signal from photomultiplier. Bottom signal from the mesh M. a. No pulse on M; b. 70 V pulse on M; c. The same pulse widened by 5 nsec.

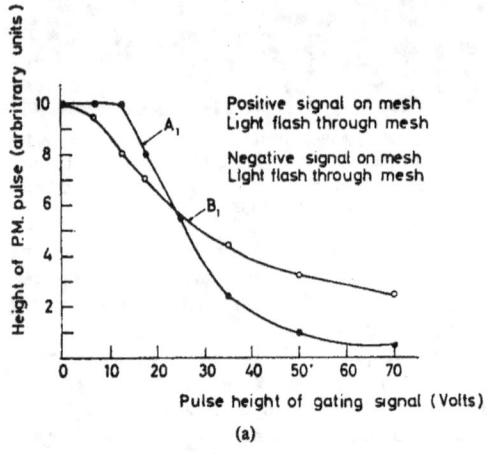

(a)

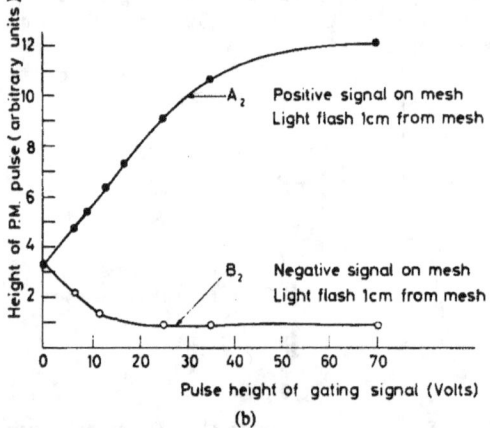

(b)

Fig. 4. Effect of the pulsed field applied to the grid M: a. flash through light guide G_1, positive or negative pulses applied to M (curves A_1 and B_1); b. flash through light guide G_2, positive or negative pulses applied to M (curves A_2 and B_2).

first dynode. Since pulses of either sign have an effect, and in some cases the pulses from neighbouring spots can be enhanced, this points to the fact that the defocusing of the local electron beam is a dominant effect.

The resistivity of the caesium-antimony cathode is very high, of the order of $10^7\,\Omega$ for a square. Even with a 1 pF capacity per cm^2, the time constant for the propagation of potentials along the cathode is very high. Onec can thus understand that local potentials can be applied for short times on different spots on the photocathode. The gating of photomultiplier tubes has already been realized by applying a reverse bias to the focusing electrode between cathode and fast dynode[1,2]. It is worth investigating how these two methods compare as far as the recovery time of the cathode after intense illumination is concerned. This method also opens up the possibility of using one single photomultiplier with several light guides coming from different light sources, if these latter are sufficiently delayed with respect to the information used for the gating decision.

I am grateful to Mr. R. Bouclier for technical help.

References

1) F. J. M. Farley and B. S. Carter, Nucl. Instr. and Meth. 28 (1964) 279.
2) K. B. Keller and B. M. K. Nefkens, Rev. Sci. Instr. 35 (1964) 1359.

NUCLEAR INSTRUMENTS AND METHODS 51 (1967) 125–128; © NORTH-HOLLAND PUBLISHING CO.

RETARDATION EFFECTS DUE TO THE LOCALIZED APPLICATION OF ELECTRIC FIELDS ON THE PHOTOCATHODE OF A PHOTOMULTIPLIER

G. CHARPAK

CERN, Geneva, Switzerland

Received 2 December 1966

It is shown that by applying electrical voltages of some hundreds of volts to transparent conductors placed against the photocathode of a photomultiplier, on the external side, two effects are observed: 1. the intensity of the electron emission can be modulated locally; 2. the photoelectrons can be slowed down and very long delays can be obtained for electrons emitted by a localized part of the photocathode.

In a preceding letter[1]) the following effect has been described: if a transparent conductor (a mesh for instance) is placed against the external wall of the photocathode of a phototube, and if electrical pulses are applied to this mesh in coincidence with the light traversing it, the intensity of the pulse obtained from the photomultiplier can be modulated by these electrical fields. In particular it is possible to gate out sharply intense short light flashes. The effect is, to a great extent, localized to the part of the photocathode facing the mesh. The effect is sharpest with positive fields, and this can be understood since these fields compensate the accelerating fields and can eventually retain the electrons. The very high resistivity of the thin photocathode ($\sim 10^7\ \Omega$ for a square) is responsible for such an effect. Even with the small distributed capacity of the photocathode the time of propagation of potentials along the photocathode is very long compared to the collection times of the electrons. However, the effect of fields of both signs shows that focalizing or defocalizing effects of the pulsed fields play a strong role.

In investigating the effect of these localized fields on the photocathode of a 58 AVP Radiotechnique, of 11 cm dia., we have observed another effect which can be of great interest. If the light flash is coming off at the end of the electric pulse applied to the mesh, the height of this pulse can be adjusted such as to have the photoelectrons slowed down and the pulse from the photomultiplier strongly delayed. I have observed delays as long as 80 nsec. Fig. 1 shows the effect of the applied voltages. The light is produced as close as possible from the falling edge. At this position, when the gating pulse is at its maximum of 180 V, the pulse height from the PM (photomultiplier) drops by a factor two with no appreciable shift in time (figs. 1b and 1c). Then by widening the gating pulse by 20 nsec the pulse shifts by the same amount, fig. 1d. This effect depends critically on the relative position of the falling edge of the gating pulse and the light pulse, on the gating field value and on the field of the focusing electrode which we adjust empirically to maximize these effects. By reducing the gating voltage the shifted pulse moves back to the original position. It is thus possible to delay the pulse from a PM by altering the width and the height of the gating field applied to the mesh. Fig. 2 shows the successive shifts of the pulse as a function of the width of the field (figs. 2b to 2g) and for two different values of the voltage at a given width (figs. 2h and 2i). We have also used two light pulses in coincidence at two different positions on the photocathode, at a distance of 5 cm from each other and equidistant from the centre of the photocathode. One of the light pulses was transmitted through a light guide with a mesh at its bottom on which could be applied the pulsed electrical field. It is possible to delay this pulse without perturbing the other one, as can be seen in fig. 3. In other words, it is possible to distinguish the position of the light impact on a photocathode by the different delays introduced in the collection of the photoelectrons from different spots of the cathode, by the application of localized fields with the help of these auxiliary transparent external electrodes.

To summarize these effects, if the time of arrival of a light signal is known, as is sometimes the case for instance with accelerators, one can determine the position of the arrival of a light signal on a photocathode from one of many effects produced by the pulsed electric fields applied to meshes placed against the photocathode: modulation of the pulse height depending strongly on the gating voltage on the mesh; delay of the pulse depending on the width, intensity and relative position of the gating voltage with respect to the light. When sources of light of sufficient time duration are available, as with caesium iodide crystals one can use the fast signal from the first photoelectrons to trigger the gating signal. We have verified this effect. Fig. 4 shows the effect of the gating pulse on the pulse

125

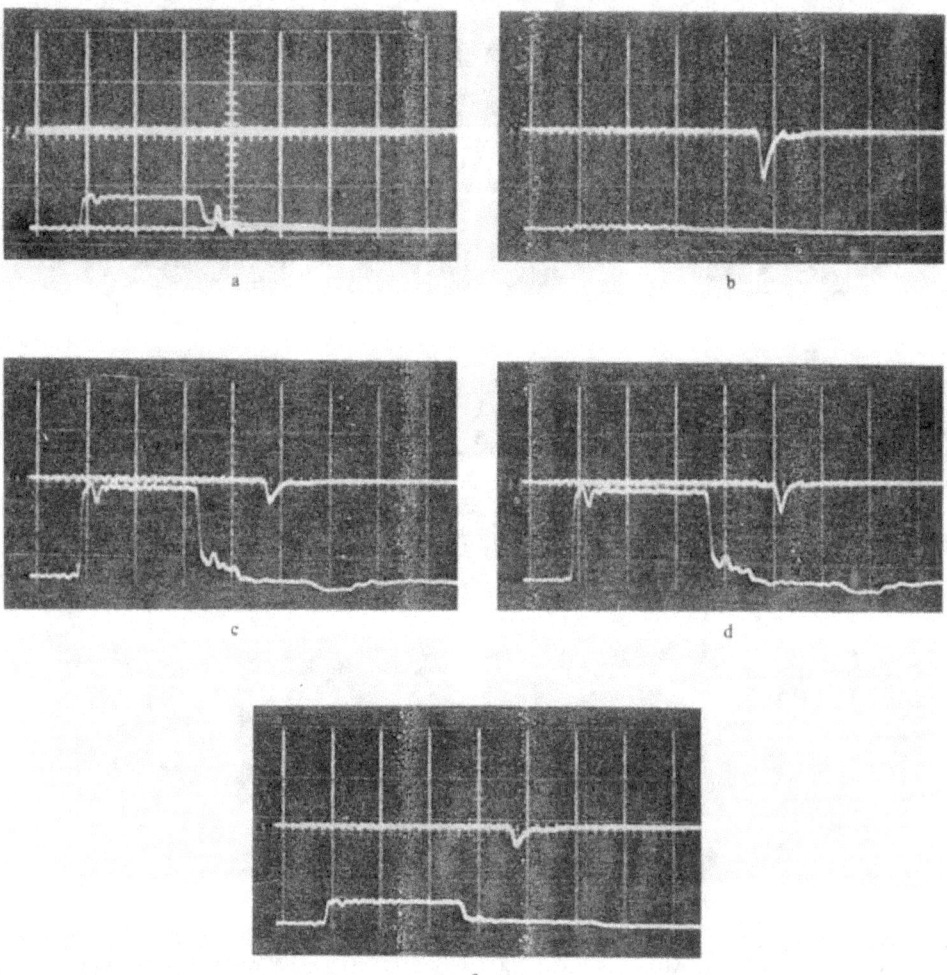

Fig. 1. *Top line*: pulses from a 58 AVP Radiotechnique, hv 2200 V. Light pulses from a PEK flash tube. Vert. 10 V/div.; Hor. 50 nsec/div. *Bottom line*: gating square pulse applied to the mesh placed between the photocathode and the light guide (dia. 6 mm). Vert. 100 V/div.; Hor. 50 nsec/div. Tektronix 585.

a. On the bottom line, superposition of the gating pulse and of the pulse applied to the flash tube, this latter delayed by additional 10 nsec to go to the oscilloscope.

b. PM pulse with no gating pulse.

c. PM pulse with +180 V gating pulse. If the light pulse was applied earlier by, say, 50 nsec, its pulse height would be reduced by a factor 40. With a −180 V gating pulse the pulse height is doubled.

d. Widening the gating pulse by 20 nsec. The light pulse shifts by 20 nsec.

e. If the preceding gating pulse is too low, the pulse shifts back to its initial position.

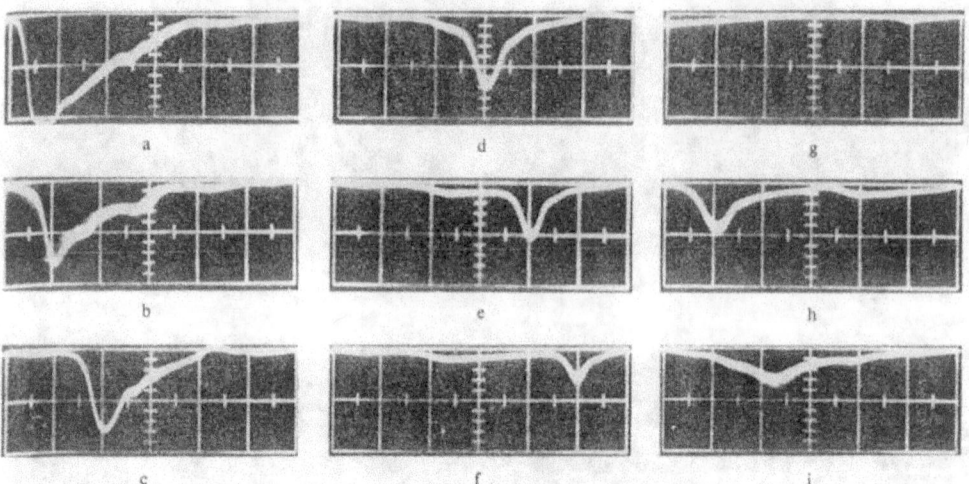

Fig. 2. Pulses from a PM 58 AVP Radiotechnique, produced by a PEK flash tube. Delays of the pulse as a function of width of the gating signal on the mesh. 10 nsec/div. Tektronix 519. No gating pulse (a); Pulse of +140 V, width 80 nsec (b); 90 nsec (c); 100 nsec (d); 110 nsec (e); 120 nsec (f); 160 nsec (g), additional 40 nsec added in the trigger signal of the oscilloscope (g); Pulse of +70 V, width 110 nsec; compared to 2e the shift is smaller by 30 nsec (h); Pulse of 110 V, width 110 nsec (i).

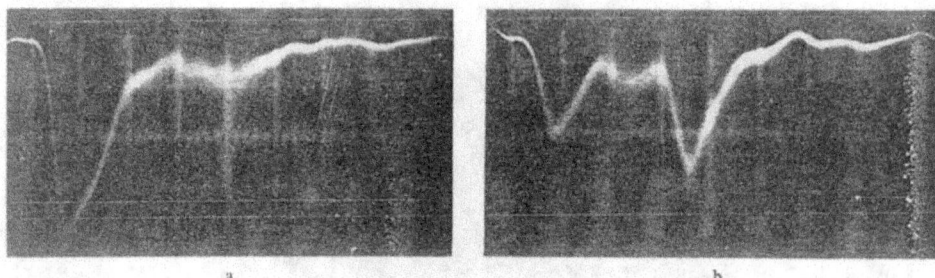

Fig. 3. Pulses from two different light guides at a distance of 5 cm from each other, placed at 5 cm from the centre of the photocathode. 10 nsec/div. Tektronix 585. a. No pulse applied to the mesh of one guide and pulse of 70 V applied to the mesh of the second. The two pulses appear simultaneously. b. The gating pulse on the mesh is brought to 140 V. The second pulse is displaced by 25 nsec.

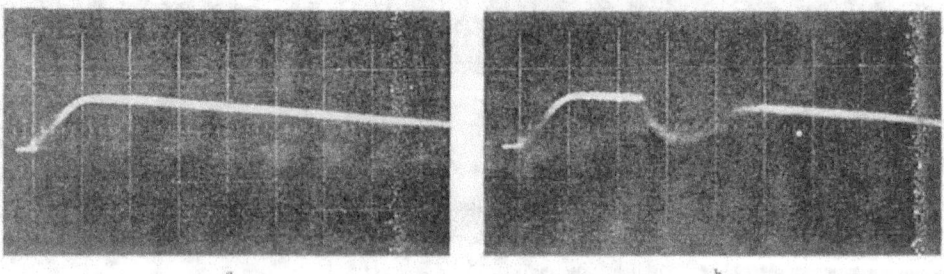

Fig. 4. Pulses obtained with a CsI crystal and a ^{60}Co source. A transparent mesh is placed under the crystal. a. No pulse on the mesh, 200 nsec/div.; b. Electric pulse on the mesh, triggered by the rising edge of the pulse and delayed.

obtained from a caesium iodide crystal with a ^{60}Co source. It is thus possible to examine a large number of caesium iodide crystals or any scintillators with sufficiently long response time, and produce a scintillator hodoscope using one single photomultiplier with many crystals.

It is a pleasure to acknowledge the technical assistance of Mr. R. Bouclier, who made several important observations during the development of this work.

Reference

[1] G. Charpak, Nucl. Instr. and Meth. 48 (1967) 151.

NUCLEAR INSTRUMENTS AND METHODS 96 (1971) 363–367; © NORTH-HOLLAND PUBLISHING CO.

MULTIWIRE CHAMBERS OPERATING IN THE GEIGER–MÜLLER MODE;
NEW SIMPLE METHOD OF PARTICLE LOCALIZATION

G. CHARPAK and F. SAULI

CERN, Geneva, Switzerland

Received 8 June 1969

Standard multiwire proportional chambers have been successfully operated in the Geiger–Müller mode using an argon-ethylbromide gas filling. Very large pulses of a standard shape can be obtained, which allow several simple electronic arrangements for the detection. It is also proposed to use the streamer propagation time along the anode wires to obtain, with suitable pick-up electrodes, the coordinates of the initial avalanche. Accuracies ranging from 4 to 10 mm (fwhm) have been measured with a 38 × 38 cm² chamber.

1. Introduction

Several groups experimenting with multiwire proportional chambers have observed that some gases can give rise to the Geiger–Müller mode of operation. The situation is different from a single wire counter, since in order to achieve localization in multiwire chambers it is necessary that the mean free path of the ultraviolet light responsible for the streamer propagation along the wire be very short, so that the discharge does not propagate from one wire to another. Very few observations have been written, and to our knowledge the only available publication reporting on the properties of different gases in multiwire chambers is the one by Stuckenberg[1]. What is still missing is a detailed description of the time and space resolution of chambers working in the Geiger–Müller mode.

In an attempt to understand the properties of the argon-ethylbromide gas used by Grunberg et al.[2], we have come to several conclusions:

1. This gas could not be used for operation in the proportional mode with large chambers. We were unable to reach more than 90% efficiency with a time resolution as large as 100 nsec, without entering the Geiger region. Some improvement has been achieved by adding a small amount of methylal [$(OCH_3)_2CH_2$]; the results will be described elsewhere.

2. Argon-ethylbromide is a perfect gas for the Geiger–Müller operation of multiwire chambers. The discharge is localized essentially on a single wire, and standard pulses of about 1 V on a 1000 Ω impedance can be obtained over a wide range of voltages.

3. It is possible to exploit the slow speed of streamer propagation along a wire to measure the position of the initial avalanche.

* This was obtained by bubbling 40% of the argon flux through C_2H_5Br at 0°C.

With a single MWPC element, one can obtain the two coordinates plus redundant information for multitrack detection, with a simplicity of read-out that seems to be matched by no other method of automatic read-out in spark chambers or proportional chambers. The accuracy, however, as far as our conditions are concerned, was limited to 10 mm (fwhm) along the wire for a 30 cm length of streamer propagation.

2. Properties of multiwire chambers filled with argon-ethylbromide

For all the following measurements, a standard MWPC was used having an active area of 38 × 38 cm², 20 μm gold-plated tungsten wires 2 mm apart on the anode, and 100 μm copper-beryllium wires for the cathode [see also Bouclier et al.[3]]. The best results, in terms of stability of the Geiger operation, have been measured with 4% C_2H_5Br in pure argon*.

Fig. 1 shows the shape of the pulses observed on different load impedances and time scales. It is of great interest to observe the pulse on a small load (1000 Ω). The time constant is very small with respect to the propagation time in a Geiger–Müller counter, and the rate of growth of the charges on the wire can be observed. We see that the pulse reaches its peak value in about 100 nsec, and then remains constant at two levels; the relative width of the two levels is a function of the source position. The streamers originating from the initial avalanche propagate to the right and the left, and when one streamer reaches the end of the wire, the rate of growth of the current suddenly decreases, thus generating the lower step.

It has been verified that only rarely two wires fire simultaneously (less than 1% of events). When this happens, the effect of the positive pulse induced by the positive ion sheath moving away from the neigh-

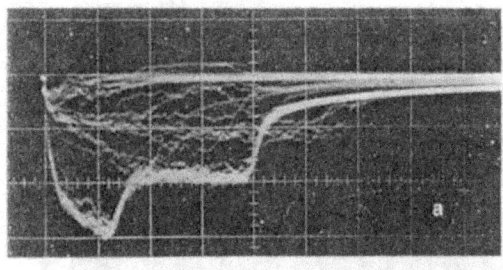

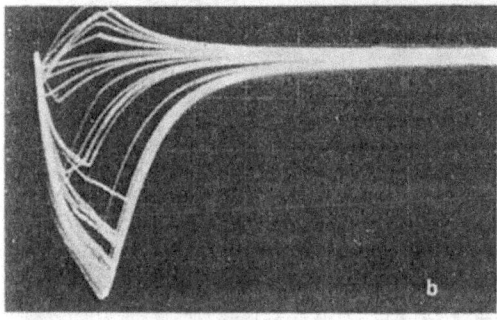

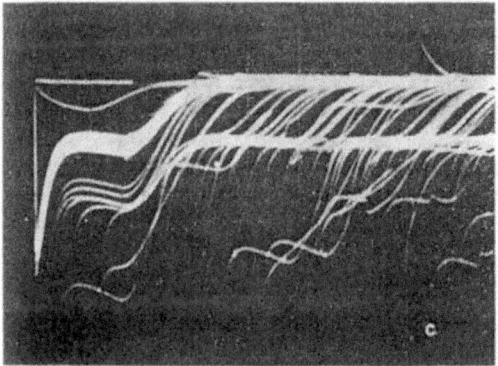

Fig. 1. Pulses obtained from the MWPC described in the text, operating in the Geiger-Müller mode. The working voltage is 3.8 kV in all pictures. (a) Single wire terminated on 1 kΩ. Horizontal: 1 μsec/cm; vertical: 500 mV/cm. (b) Single wire terminated on 100 kΩ. Horizontal: 5 μsec/cm; vertical: 10 V/cm. c) 64 wires strip. Terminated on 100 kΩ. Horizontal: 100 μsec/cm; vertical: 2 V/cm. A piling-up effect due to the large counting rate can be clearly seen.

bouring wire is to reduce the negative pulse height on the detecting wire; fig. 1b shows that only a small proportion of the pulses are smaller than the main pulse. Moreover, the pulse shape is completely independent from the initial ionization; the same standard

pulse is obtained with 5.9 keV X-rays of ^{55}Fe and with a single ion pair produced by an ultraviolet light.

Fig. 2 shows the efficiency plateau measured for minimum ionizing electrons whilst operating the MWPC in the Geiger-Müller mode, for 60 nsec resolution time and −100 mV threshold on the sense electronic. The measurement presents some difficulties, owing to the long dead-time proper of the GM operation; a low counting rate had to be used.

The dead-time of a wire is the time necessary to sweep away the positive ions; it is close to 200 μsec as seen from fig. 3, which shows the time distribution of the pulses on a wire irradiated by a strong source. A lack of detection is visible after each pulse, for about 200 μsec. The same number can be obtained looking at fig. 1c, where clearly the positive ion sheath, which generates the long constant plateau, needs about 200 μsec to reach the cathode. This is a strong limitation compared to proportional chambers; however, for large surfaces working in a low flux this can be quite acceptable.

The large amplitude and the standard shape of the pulses detected on the wires suggest the use of several simple read-out systems. In the scheme proposed, for example, by Pagès[4]), a low-threshold bistable (such as may be found in the emitter-coupled logic circuits) can be directly flipped by the pulse without amplification. If a good time-resolution is not required, even simpler systems in which the "memory" of the wire is used can be imagined; a prompt read-out from wires or groups of wires can be used for the trigger selection, and opening a gate on each wire before the end of the Geiger propagation (15 μsec for 1 m long wires) would present the selected information to the read-out.

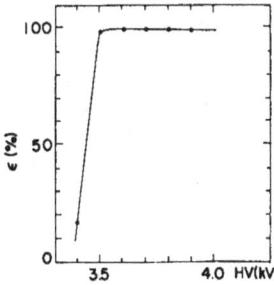

Fig. 2. Efficiency plateau of a MWPC operating in the Geiger-Müller mode, for fast electrons. The threshold of detection on each wire is −100 mV, and the time resolution (as required by a coincidence with two scintillation counters) 60 nsec. Due to the large dead-time proper to the GM operation, a low source intensity had to be used.

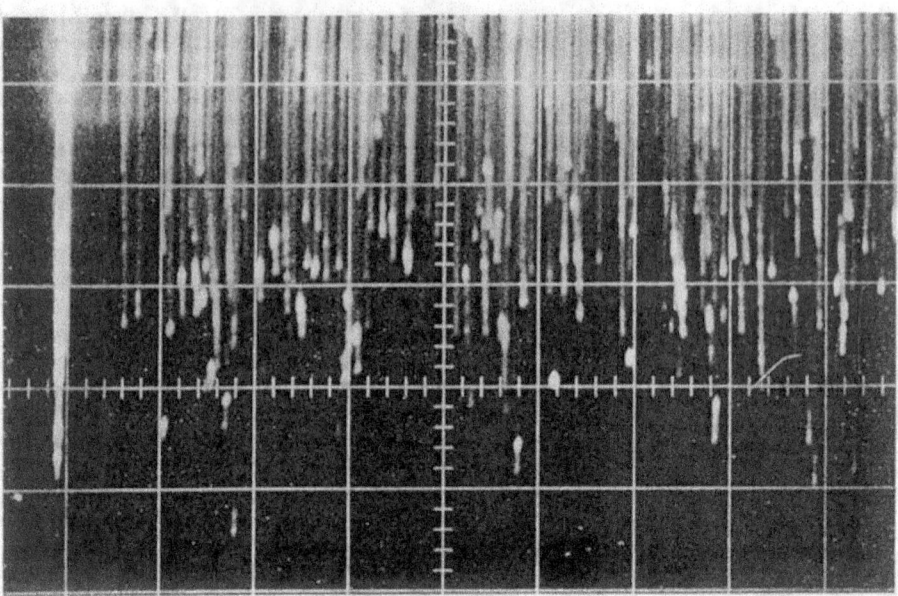

Fig. 3. Counting rate on a wire, for very high irradiation level. A lack of detection can be seen, about 200 μsec long (horizontal scale 200 μsec/cm).

For short distances we successfully tried a very simple discrete *LC* delay line (see fig. 4). The delay of each cell (20 nsec) was sufficiently constant to obtain the wire localization, at least over a range of 32 wires. The zero time is provided by a common collector, and measurement of the delayed pulse gives the wire coordinate. A simple application of this read-out scheme is given in the next section.

3. Use of the Geiger–Müller propagation time to determine the coordinates

As was already visible in fig. 1a, the information on the initial positions of the avalanche is contained in the pulse shape, directly on the sense wires. However, it can be obtained in a simpler way. If a pick-up wire or strip is set across the anode sense wires, say in the

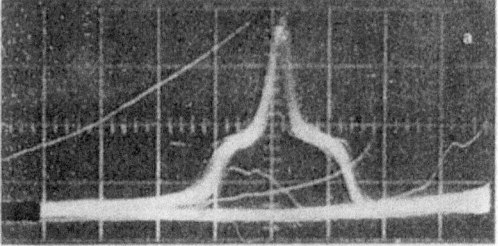

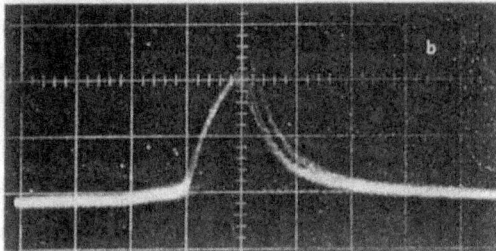

Fig. 5. The streamer front as detected by two pick-up electrodes. (a) Single wire in the hv plane, read out by a 500 pF capacitor on 2 kΩ. The scope was triggered by the prompt sense wire output, the source being at about 15 cm from the pick-up electrode. Horizontal: 1 μsec/cm; vertical: 1 mV/cm. (b) Strip of hv wires, 5 cm large, read out by 500 pF on 2 kΩ. Same conditions as in (a). Horizontal scale: 1 μsec/cm; vertical: 50 mV/cm.

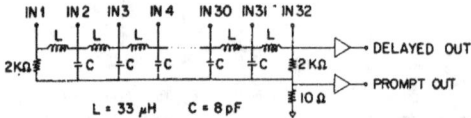

Fig. 4. Scheme of the simple discrete cell delay line used for the measurements described in the text. The delay of each cell is about 20 nsec, and a prompt as well as a delayed output is provided.

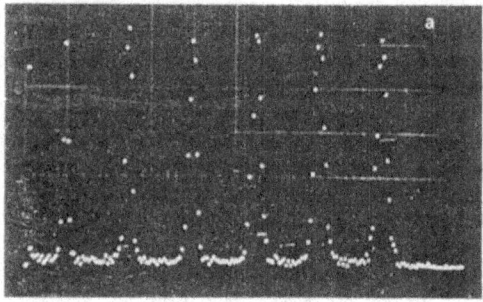

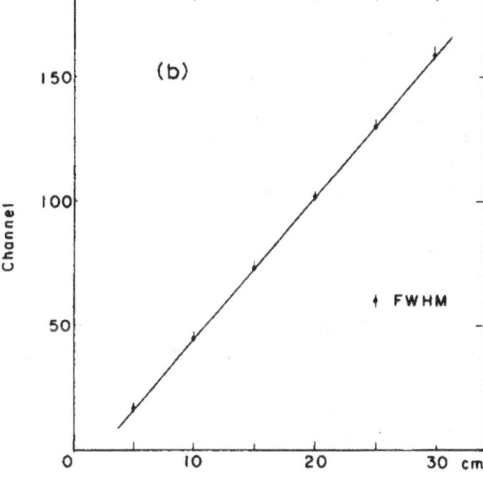

Fig. 6. (a) Time distribution of pulses, in respect of the prompt output of a sense wire, for the pick-up electrode described in fig. 5a. A well-collimated ^{55}Fe source was used, and the peaks correspond to source pick-up distances of 5, 10, 15, 20, 25 and 30 cm. (b) The peak value as well as the total width at half-height is given as a function of the source pick-up distance. The hv on the chamber is 3.7 kV.

sense wire itself, as given by an electronics sensitive to pulses higher than −100 mV.

A well-collimated ^{55}Fe source was used to measure the time distribution $t_p - t_0$. for several distances between the source and the pick-up electrode. The best results were obtained using as pick-up one of the hV wires read-out through a 500 pF capacitor. The time distributions obtained are summarized in fig. 6a for several distances, between the source and the pick-up, going from 5 to 30 cm in steps of 5 cm. In fig. 6b the peak position as well as the total width at half-height of the distributions is given, as a function of the source pick-up distance. A very good linearity is observed, corresponding to a streamer propagation time of 200 nsec/cm. The speed of propagation is of course function of the high voltage.

The width of the distribution varies as a function of the distance; it is 4 mm (fwhm) at 5 cm from the wire and reaches 10 mm at a distance of 30 cm. This lack of resolution is very probably introduced by the fluctuations in the propagation velocity of the Geiger–Müller streamer, and varies roughly with the square root of the length. This resolution should be compared to a resolution of 22 mm (fwhm) reported in a Geiger–Müller cylindrical counter using a current division method[5]).

In the measurements described, only one coordinate was obtained using the streamer propagation time; a simple geometry can, however, be imagined which gives both coordinates and also the absolute streamer speed on each event. With reference to fig. 7, a method is proposed to get the x and y coordinates, plus the

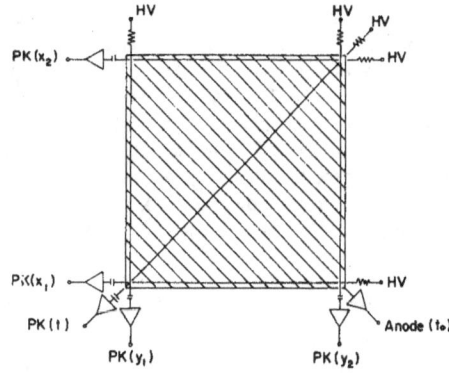

Fig. 7. Possible configuration for a chamber giving all coordinates, plus several redundances, with only five pick-up electrodes. The sense wires can be read out all together to provide the zero time of an event, and the streamer pick-up electrodes may be wires or strips on the cathode.

high-voltage plane, it will detect a sharp positive pulse when the front of the streamer intersects the pick-up electrode. As pick-up electrodes we have used: a guard strip at one end of the chamber, carefully brought to a proper negative intermediate potential (so that the streamer comes as close as possible to it); a single wire of the high-voltage cathode plane; a set of wires connected together in the high-voltage plane. Fig. 5 shows the pulses observed in the last two cases. By measuring the time of arrival of the positive pulses, t_p, we could localize the position of the initial avalanche along the wire. We used as zero-time t_0 the negative pulse on the

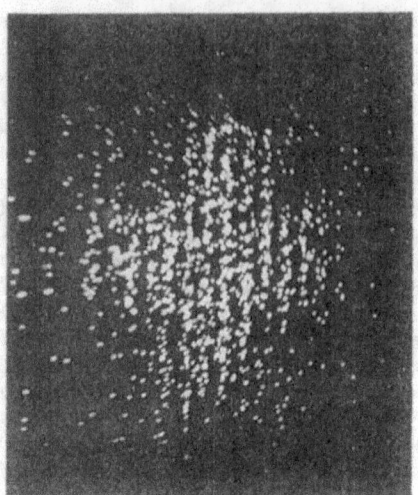

Fig. 8. Scope display obtained by the time-to-analogue conversion described in the text. A cross-shaped contour was cut out of a thick paper sheet, and the chamber irradiated with a point-like ultraviolet source. The vertical direction is given by the streamer propagation, whereas the horizontal coordinate was obtained using the discrete delay line described in fig. 4.

speed of the streamer, from t_0, and the four pick-up electrodes $PK(x_1)$, $PK(x_2)$, $PK(y_1,)$ $PK(y_2)$; the inclined pick-up wire $PK(t)$ can be used to solve the ambiguities in case of multiple tracks.

We have simulated an x–y read-out in a chamber, by using on the sense wires the discrete delay line described in section 1. Both pieces of time information, one (vertical) deduced from the streamer pick-up wire and the other (horizontal) from the delay line, were transformed into amplitude information and used to drive the x–y axes of a scope. A small cross, with 2 cm arms, was cut out of a thick sheet of paper which was used to mask the chamber. A point-like ultraviolet source was placed about 20 cm from the wire planes, and the resulting scope display is shown in fig. 8. It can be seen that the horizontal definition corresponds to the wire spacing.

4. Conclusions and summary

Chambers filled with argon-ethylbromide work well in the Geiger–Müller region. The pulses are of a height sufficient to be used directly in logic circuits. A method is described to obtain the coordinate parallel to the wires by measuring the time of propagation of the Geiger–Müller streamer along the wire; an accuracy ranging from 4 to 10 mm (fwhm) in a 38×38 cm^2 chamber has been obtained.

It would be very interesting to investigate the effect of all the parameters of a chamber in the resolution that can be obtained with the described system. In particular, the diameter and the smoothness of the surface of the sense wires may play a big role in the streamer propagation.

For some applications where dead-time and modest space resolution are acceptable, no other methods of automatic read-out of large detectors seem to have comparable advantages:

– It is as simple as the magnetostriction read-out but works in magnetic fields.

– It is self-triggered.

– Single planes can give three coordinates.

For shower detectors, the constancy of the pulse, combined with the fact that with this electronegative gas only one wire is generally firing on each track, opens up the possibility of simple methods of measuring the size of an electron avalanche as it propagates in intermediate absorbers.

It should also be observed that it would be easy to localize the position of the initial avalanche in an ordinary cylindrical Geiger counter by using either the information on the wire itself or the pulses detected on the guard rings which are common in many counters. The problem is even simpler with a glass cathode counter of the Maze type.

We wish to thank Mr. C. Grunberg for many discussions and information on the properties of chambers filled with ethylbromide.

We are indebted to Messrs. R. Bouclier and J. C. Santiard for technical help and for the performance of some of the initial measurements.

References

[1] H. J. Stuckenberg, DESY Report 69-49 (1969).
[2] C. Grunberg, L. Cohen and L. Mathieu, Nucl. Instr. and Meth. **78** (1970) 102.
[3] R. Bouclier, G. Charpak, Z. Dimčovski, G. Fischer, F. Sauli, G. Coignet and G. Flügge, Nucl. Instr. and Meth. **88** (1970) 149.
[4] R. Pagès, Nucl. Instr. and Meth. **85** (1970) 211.
[5] W. N. McDicken, Nucl. Instr. and Meth. **54** (1967) 157.

NUCLEAR INSTRUMENTS AND METHODS 124 (1975) 491–503; © NORTH-HOLLAND PUBLISHING CO.

THE HIGH-DENSITY MULTIWIRE DRIFT CHAMBER

A. P. JEAVONS, G. CHARPAK and R. J. STUBBS

Cern, Geneva, Switzerland

Received 12 December 1974

A multiwire proportional chamber, with a high-density drift space attached, has been developed as a position-sensitive detector for non-ionizing radiation.

A chamber with a 5 g/cm³ drift space, comprising a lead–bismuth matrix of 1 mm square holes on a 1.5 mm pitch has been investigated. For 0.66 MeV photons a detection efficiency of 5%, combined with a spatial resolution of 1.3 mm fwhm, is obtained. Much higher efficiencies are possible. A theoretical analysis of the detection efficiency as a function of photon energy is presented. It agrees well with experimental results.

The chamber offers new possibilities for photon and neutron detection, and as a shower detector for high-energy particles.

1. Introduction

Experimental work involving the spatial localization of ionizing particles has advanced in recent years following the development of the multiwire proportional chamber (MPC)[1]. Large systems involving many thousands of wires are now in operation. Progress in this field continues with the drift chamber[1-3], which offers the possibility of submillimetre resolution and very large (4 m × 4 m), comparatively cheap, detectors.

Increasing attention is now being paid to the use of MPCs as position-sensitive detectors for non-ionizing radiation. This is hardly surprising in view of the many possible applications. The uses of MPCs for medical X-ray imaging[4], X-ray crystallography[5] and neutron detection[6] have been reported; other fields of application include X- and gamma-ray astronomy, solid-state physics, and non-destructive material testing.

The detection of non-ionizing radiation presents two problems for the MPC:

i) The conversion of the non-ionizing radiation into ionizing particles.

ii) Obtaining X and Y information from one chamber, since, in general, the ionizing products will either not propagate to a second chamber or else will be strongly scattered before reaching it.

The second problem is relatively easily resolved. Various two-dimensional read-out techniques: pulse-rise-time[7], delay-line[8], charge-division[9] and centre-of-gravity[10] have been described, offering different combinations of cost, accuracy, sensitivity, and data rate.

To date, the first problem has received little detailed attention. Effort has centred on using the gas in the MPC as the converting material. For 8 keV X rays, gas conversion is very successful. 100% efficiency combined with submillimetre spatial resolution is possible using argon at atmospheric pressure. Recently, the introduction of the spherical multiwire drift chamber[11] has extended this efficiency and spatial resolution to the cases of X-ray imaging from a point source or from a divergent beam.

As photon energy is increased, gas conversion becomes less and less useful. Firstly, the photoelectric capture cross section falls very rapidly and, secondly, the range of the photoelectron increases. Thus efficiency and spatial resolution both degrade. By using a dense gas with a high atomic number, under pressure, the photon energy at which the chamber performs satisfactorily can be maximized. Xenon at 4 atm pressure has been used for 60 keV gamma rays to obtain an efficiency of 50% and a spatial resolution of 1 mm[4]. This would seem to be a useful limit for gaseous converters.

Carrying the idea of pressure to its logical conclusion leads to a chamber filled with a liquid. And indeed, liquid-argon- and liquid-xenon-filled chambers have been investigated. For 0.28 MeV gamma rays an efficiency of 65% and a resolution of 4 mm has been recorded[12]. However, serious technical problems exist due to the low-temperature operation, the very high liquid purity needed, and the fact that avalanche amplification is difficult to obtain. The chambers must operate in the ionization region with consequent electronic problems and cost.

As an alternative to a liquid-filled chamber, an ordinary gas-filled chamber may be equipped with a solid converter. Then the advantages of the solid – a high capture efficiency for gamma rays and short range of the resulting photo-electron – may be obtained, together with the normal operation of the gas-filled MPC. Such a hybrid chamber and its application to

491

photon energies greater than 0.1 MeV, i.e. above the gas-conversion limit, is the subject of this paper.

2. A gas–solid hybrid chamber

2.1. THE BASIC DESIGN

A dilemma arises for the detection of the ionizing products resulting from the conversion of non-ionizing radiation. If sufficient bulk material is used to give a high capture probability for the non-ionizing radiation, then generally the ionizing products will have a low probability of escaping from the material and being detected. This dilemma is neatly overcome with the noble gases. The free electrons resulting from ionization may be transported, or drifted, over a distance of many centimetres under the influence of an electric field, and then detected. Electron drifting is also possible in liquid argon and xenon. Thus chambers employing a noble gas or liquid as the converting medium can give high detection efficiencies.

The simplest way of using a solid for conversion is to utilize the walls of a chamber. This is how X and gamma rays are detected with a conventional Geiger counter. With walls made of lead, an efficiency of around 1% is obtained, over the energy range 0.1 to 2.0 MeV[13]).

To obtain a high detection efficiency with a solid, a structure is required where the solid is in thin sections, interleaved with the gas. The photoelectron will then have a high probability of escaping to the gas and creating free electrons which may be extracted with a drift field. To obtain a good spatial resolution, the range of the photoelectron must be restricted to small pockets of gas. A structure that fits these two require-

ments is a solid block, perforated with a large number of small holes close together. By making the block thick enough, the photon will have a high conversion

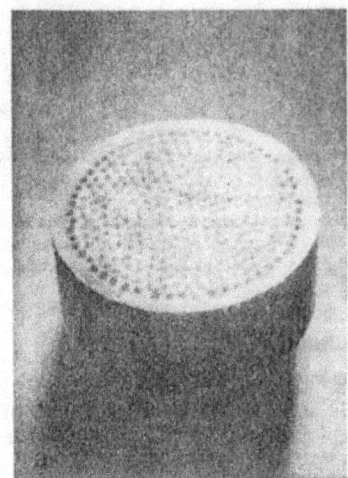

b

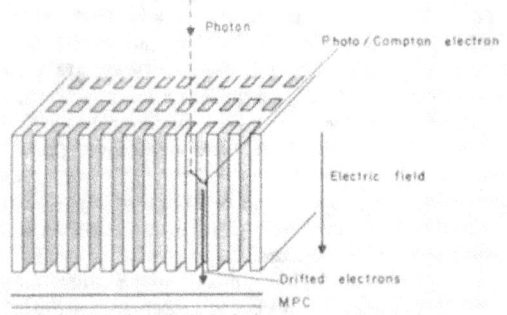

Fig. 1. Principle of the gas–solid hybrid chamber. Photons are captured in the solid bars and produce fast electrons which can escape to an adjacent hole. The free electrons resulting from gas ionization in the hole may be extracted by an electric drift field, and detected by a proportional chamber.

c

Fig. 2. Different ways of making the solid matrix: (a) a bundle of glass tubes (full size); (b), (c) a copper-plate, mylar sandwich (× 2/3).

probability. With a correct choice of hole size and spacing, the resulting photoelectron will have a high probability of escaping to a hole but a low probability of propagating to a second hole. The application of an electric field will drift the free electrons out of these holes for subsequent detection by a MPC. If the hole axis is parallel to the direction of the impinging photons, the two dimensions perpendicular to this direction may be resolved to an accuracy determined by the hole size (see fig. 1).

A somewhat similar, independent approach has been reported. An open honeycomb structure was used to provide a large surface area of lead for photons angled at 20° to the direction perpendicular to the chamber. A detection efficiency of 2.5%, combined with a spatial resolution of 5 mm, was obtained[14]).

2.2. EXPERIMENTAL TESTS

Two small prototypes were tested to assess different methods of fabricating a solid matrix and to prove that electrons could be drifted efficiently through holes in a solid.

2.2.1. Glass tubes

A matrix 4 cm in diameter and 2 cm long was made by packing together soda-glass tubes of 1.5 mm bore and 0.5 mm wall thickness (see fig.2a). It was placed in a drift space attached to a small (10 cm × 10 cm) MPC. Steel mesh grids immediately above and below the bundle of tubes provided the drift field. A source of 5.9 keV ^{55}Fe X rays was placed above the matrix and the count rate and pulse spectrum of the chamber monitored to observe changes with and without the drift voltage applied. Initially the tubes exhibited an electrostatic charging effect. An increase in count rate was observed when the drift voltage was applied, but it decayed rapidly (1–2 min). Following a dipping of the tubes in an antistatic glue, a reasonably stable, eightfold increase in count rate was obtained. Fig. 3 shows the variation in pulse height with drift voltage obtained from one anode wire. With a back-drift voltage applied (fig. 3a) good energy resolution is apparent. This is expected since the tubes will act purely as an X-ray collimator. The application of a forward-drift voltage gave a large increase in the num-

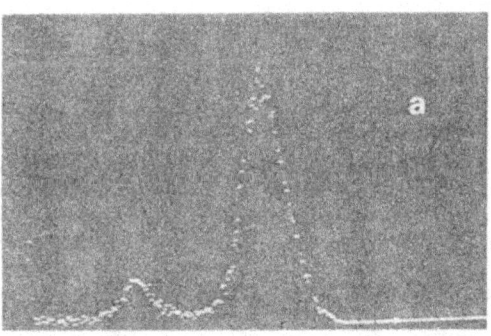
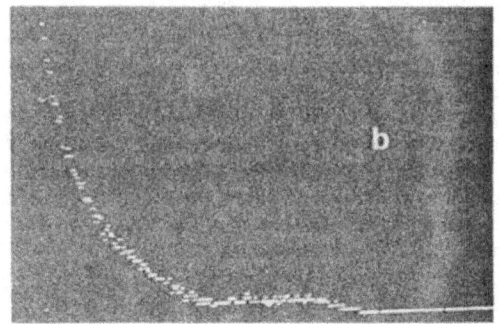

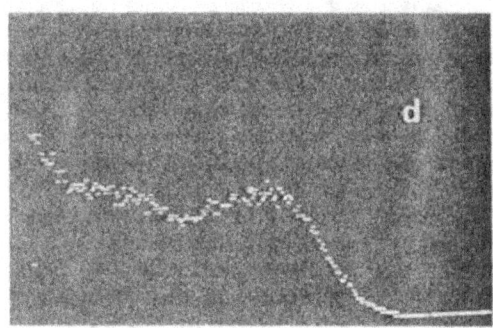

Fig. 3. The variation of pulse-height spectrum with drift voltage for the glass tubes. 5.9 keV photon source.
Back-drift voltage. The tubes act solely as an X-ray collimator. (a) 1000 V.
Forward-drift voltage. The count rate increases eightfold. (b) 1000 V; (c) 2000 V; (d) 3000 V.

ber of low-energy pulses (fig. 3b). Increasing the drift voltage did not further increase the count rate, but flattened the pulse-height spectrum, indicating an increased efficiency of electron collection (see figs. 3c and 3d).

An important result to observe is the flat spectrum of fig. 3d, which implies an almost complete loss of energy resolution. This is caused by the random amount of energy lost by any photoelectron in interacting with the wall of a glass tube.

Electrostatic charging effects were still seen, particularly if the drift voltage was decreased, when the count rate and pulse-height spectrum would take many minutes to stabilize.

2.2.2. Metal plates

To avoid electrostatic charging and to have a denser matrix, a stack of perforated metal plates, insulated from one another, was investigated. Eleven plates of copper 6 cm × 6 cm × 1 mm were interleaved with mylar sheets 0.1 mm thick. The complete stack was perforated with 2 mm diameter holes every 3.5 mm on a hexagonal pattern (see figs. 2b and 2c). The drift field within the holes was produced by connecting the plates to a linear resistor chain. Testing with the ^{55}Fe source as before gave similar results, but without any electrostatic charging problems.

In view of the completely stable performance of the metal plates compared to the glass tubes, it was decided to construct a larger chamber of metal plates to make detailed measurements of detection efficiency and spatial resolution. Before this is described, some theoretical questions will be considered.

3. Theoretical analysis

The calculation of the detection efficiency of the solid matrix is essentially the problem of calculating the probability that the fast electron produced by a photon interaction will escape from the solid. From first principles this is very difficult, due to the complex nature of the multiple scattering of the fast electron. However, some excellent theoretical [Spencer[15,16]] and experimental [Seliger[17]] work may be invoked to make the calculation relatively straightforward.

Fig. 4 illustrates a matrix of square holes. The hole size is h and the bar width between holes is b. The matrix is made of n plates each of thickness t, material density ρ and photon interaction cross section σ. Consider a line segment along a bar at a distance x from the edge of a hole. The area of this segment is $(h+2x)\,dx$. The unit cell of the matrix has an area $(h+b)^2$ and includes 4 such segments, so the area ratio

as a function of x is:

$$[4/(h+b)^2](h+2x)\,dx.$$

Let a photon impinging on the line segment have a probability $P_e(E)$ of producing an electron, energy E, which in turn has a probability $P_d(x,E)$ of escaping to a hole and being detected. $P_d(x,E)$ will steadily decrease as x increases. The value of x at which $P_d(x,E)$ goes to zero is the electron residual range, $R(E)$[18]).

The assumption is made that an electron can only escape to the nearest hole, i.e. $R(E)<b/2$. Then, if F is the fraction of electrons detected from the photons interacting anywhere in the matrix,

$$F = \frac{4}{(h+b)^2}\int_0^{E_1} P_e(E)\int_0^{R(E)} (h+2x)\,P_d(x,E)\,dx\,dE.$$

(1)

The integration limit E_1 will be discussed later.

The over-all efficiency, ε, is obtained by including the probability of photon interaction and summing for the relevant processes:

$$\varepsilon = [1 - \exp(-\sigma_T \rho n t)]\sum_i \frac{\sigma_i}{\sigma_T} F_i,$$

(2)

where $\sigma_T = \sum_i \sigma_i$.

3.1. THE ELECTRON ESCAPE PROBABILITY

To evaluate F_i, functions must be found for $P_e(E)$ and $P_d(x,E)$. Spencer[15]) used a numerical method of spatial moments to calculate $P_d(x,E)$, but he also gives an analytic expression:

$$P_d(x,E) = K\left[1 - \frac{x}{R(E)}\right]^p \exp\left\{-\frac{A}{1-[x/R(E)]}\right\}. \quad (3)$$

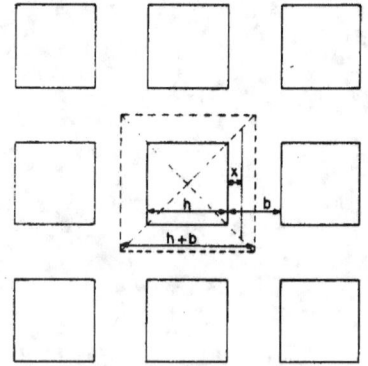

Fig. 4. Parameter details for the theoretical analysis.

TABLE 1

The values of the coefficient A in eq. (3) obtained from Seliger's data. A is reasonably constant for each material. Note the deviations from Spencer's asymptotic values.

Material	Energy (MeV)	$\dfrac{x}{R(E)}$ 0.02	0.05	0.1	0.2	0.3	0.4	0.5	0.6	0.7	Spencer
Al	0.16	–	–	3.2	3.2	3.2	3.1	2.9	2.7	2.4	1.58
	0.25	–	–	3.1	3.1	3.1	3.1	2.9	2.6	2.4	1.57
	0.34	–	–	3.1	3.1	3.1	3.1	2.9	2.7	2.5	1.56
	0.96	–	–	3.1	3.1	3.1	3.0	2.9	2.7	2.5	1.45
Ag	0.16	5.8	5.9	6.1	6.0	6.0	5.8	–	–	–	6.3
	0.25	5.5	5.7	5.8	6.0	6.2	6.3	–	–	–	6.5
	0.34	5.3	5.5	5.5	5.8	6.1	6.2	–	–	–	6.6
Pb	0.16	8.4	9.7	9.5	9.5	9.2	–	–	–	–	10.0
	0.25	8.2	8.4	8.7	8.7	8.9	–	–	–	–	10.5
	0.34	7.3	7.8	8.5	8.6	8.3	–	–	–	–	11.2
	0.96	6.7	6.8	7.6	7.7	8.0	–	–	–	–	12.3

K, A and p are constants; $R(E)$ is the residual range as defined and tabulated by Nelms[19]) of a fast electron, initial energy E, in the material.

The value of K follows directly from the boundary condition $x = 0$, $P_d(x, E) = 1$:

$$K = \exp A. \qquad (4)$$

Eq. (3) is relatively insensitive to the value of p. A will be sufficiently large that the exponential term dominates. Spencer used integral or half-integral values of p between $+1$ and -2. Here the choice

$$p = -3 \qquad (5)$$

is made to simplify the integration.

Eq. (3) has two limitations. Firstly, Spencer used it to extrapolate from a finite to an infinite number of spatial moments; i.e. it applies to the asymptotic tail of $P_d(x, E)$ at deep penetration in the solid and does not necessarily describe the full distribution. Secondly, the Spencer theory does not include range straggling to which the asymptotic tail is particularly sensitive. To investigate these inaccuracies, previous experimental results may be invoked. Seliger[17]) has accurately measured $P_d(x, E)$ for a number of materials. He used a 2π counter with an efficiency of 99% down to an electron energy of a few hundred electron volts: an experimental situation entirely analogous to the present work. It is true that the source of electrons was different: a beam impinging perpendicularly on the material, rather than a point isotropic source within the material. However, the electron velocity in the material is quickly randomized,

as is evidenced by the back-scattering probability being greater than 50% [20]). Further, Spencer points out that the electron distribution function for a parallel beam source is the same as for a point isotropic source. The two situations are very similar.

The fit of eq. (3) to Seliger's data is demonstrated by the A values in table 1. A is reasonably constant for a given material over the full range of the distribution. The deviation of A from Spencer's theoretical values for the asymptotic tail of the distribution is evident.

The integration of eq. (1) over x may now be made by substituting for $P_d(x, E)$ from eq. (3). With the change of variable

$$y = \{1 - [x/R(E)]\}^{-1}$$

and the values given by eqs. (4) and (5):

$$F = \frac{4}{(h+b)^2} \int_0^{E_1} P_e(E) R(E) \frac{1}{A^2} \times$$

$$\times \left[\{2AR(E) - [h+2R(E)](Ay+1)\} \times \right.$$

$$\left. \times \exp\{A(1-y)\}\right]_1^\infty dE. \qquad (6)$$

When $y \to \infty$, $\exp\{A(1-y)\} \to 0$. Eq. (6) requires evaluation only for $y = 1$, which gives:

$$F = \frac{4}{(h+b)^2 A^2} \times$$

$$\times \int_0^{E_1} P_e(E) R(E) [2R(E) + h(A+1)] dE;$$

$$R(E) < b/2. \qquad (7)$$

3.2. The electron energy spectrum

To complete the evaluation of F, a knowledge of the function $P_e(E)$, the probability of a photon interaction producing an electron energy E, is required. Two types of photon interaction, the photoelectric effect and Compton scattering, are relevant to the energy range considered here. Pair production starts above 1 MeV, but it is still a minor contribution to the total interaction cross section up to 2 MeV. The question of photon energies greater than 2 MeV will be the subject of a future paper.

3.2.1. The photoelectric contribution

For the case of photon capture by the photoelectric effect, $P_e(E)$ is simply a delta function at the photon energy minus the binding energy of the electron. Two possibilities, K and L + M shells, are treated separately:

$$P_e(E) = 1, \qquad E = E_{ph} - E_{K, L+M},$$
$$P_e(E) = 0, \qquad E \neq E_{ph} - E_{K, L+M}. \tag{8}$$

Putting eqs. (8) into eq. (7):

$$F_{K, L+M} = \frac{4 R_{K, L+M}}{(h+b)^2 A^2} [2 R_{K, L+M} + h(A+1)];$$
$$R_{K, L+M} < b/2, \tag{9}$$

where $R_{K, L+M} = R(E_{ph} - E_{K, L+M})$.

3.2.2. The scattering contribution

Compton scattering gives rise to a continuous spectrum of electron energy[21]). As a reasonable approximation the spectrum may be assumed flat and $P_e(E)$ constant up to the Compton edge energy, E_C. Then:

$$\int_0^{E_C} P_e(E)\,dE = 1; \quad \therefore P_e(E) = \frac{1}{E_C}. \tag{10}$$

For $R(E)$, a range–energy relationship[22]) holds:

$$R(E) = aE^q, \tag{11}$$

where the constants a and q follow from Nelms' data[19]).

Substituting eqs. (10) and (11) into eq. (7) and integrating with $E_1 = E_C$,

$$F_C = \frac{4 R_C}{(h+b)^2 A^2} \left[\frac{2 R_C}{2q+1} + \frac{h(A+1)}{q+1} \right];$$
$$R_C < b/2, \tag{12}$$

where $R_C = R(E_C)$.

Note that eqs. (9) and (12) are the same for $q = 0$ and $R_C = R_{K, L+M}$.

3.2.3. The escape-peak contribution

One further small contribution to F should be included. This arises from the de-excitation of an atom following photoelectric capture: the escape peak. Only K-shell de-excitation needs to be considered. An electron resulting from L- or M-shell de-excitation will be of too low an energy to be significant. The de-excitation will either produce an electron immediately (the Auger effect), or after a further photoelectric capture of the photon with the shell energy. The contribution F_{KE} to detection efficiency by this electron may be treated in the way described in

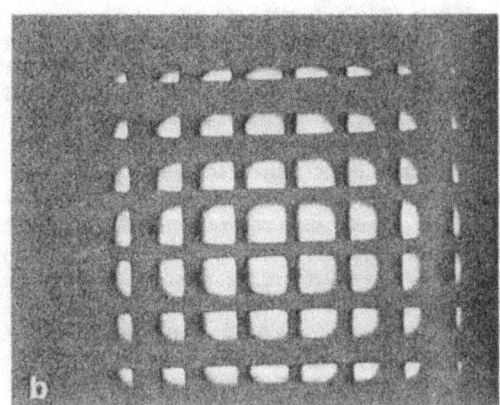

Fig. 5. The lead–bismuth plate chamber: (a) top view ($\times 1/3$); (b) close-up of the plates ($\times 6$).

section 3.2.1:

$$F_{KE} = \frac{4 R_{KE} [2 R_{KE} + h(A+1)]}{(h+b)^2 A^2},$$

where $R_{KE} = R(E_K)$, E_K being the K-shell energy.

The final stage is to sum the four contributions F_K, F_{L+M}, F_C, and F_{KE}, according to eq. (2). Photon-capture cross sections. and K to L+M shell relative probability have been tabulated by Grodstein[23]).

In the next section, the construction and operation of an experimental chamber is described. The results from the theoretical analysis will then be compared with experimental measurements.

4. An experimental chamber

4.1. CHAMBER CONSTRUCTION

Fig. 5 shows a small chamber constructed according to the ideas detailed in the previous sections.

To obtain the maximum advantage from the photoelectric effect, lead is a natural choice for the plate material. Since lead is so soft, a eutectic alloy* of lead and bismuth was chosen instead. The matrix was made from 17 plates each cast and machined to a thickness of 0.5 mm. A pattern of square holes, side 1 mm, pitch 1.5 mm, was cut in each plate over an area of 9 cm × 4 cm by spark-erosion machining. The insulation between the plates was oxidized aluminium foil. Each foil, 0.1 mm thick, was cut to the same pattern as the lead–bismuth plates by photochemical methods and then anodized to give an oxide layer 0.02 mm thick on each side. A MPC was positioned 3 mm beneath the metal stack. Orthogonal cathode planes were of 0.1 mm diameter, and the anode plane 0.01 mm diameter, gold-plated tungsten wires. Cathode and anode wire spacings were 1.5 mm, the same as the hole pitch in the plates. The anode wires were centred under the holes, and the cathode wires under the bars, of the matrix. The spacing between wire planes was 3 mm. The chamber was closed with mylar windows.

4.2. CHAMBER ELECTRONICS AND OPERATION

The chamber was operated with the cathodes at earth potential and a positive voltage of 2.6 kV on the anode wires. The usual argon–isobutane–methylal gas mixture[24]) was used. The resistor chain connected to the plates of the matrix was of 100 kΩ steps. To obtain a focusing field, the last four resistors were increased in value progressively by 20%. The electric field strength

over the gas space between the matrix and the MPC was double that at the last plate. The drift voltage was 750 V. Exceeding this value caused a big increase in the chamber noise, probably due to the porous nature of the aluminium oxide insulating layers.

Charge-sensitive preamplifiers show considerable advantages for recording pulses from MPCs. A simple three-transistor circuit can give an order of magnitude better signal-to-noise ratio than the usual voltage amplifier with a 10 kΩ input impedance. Such preamplifiers were made, similar to those described by Radeka[25]). They worked well and will be detailed elsewhere.

For pulse counting, the negative anode pulses were used, since they are typically five times larger than the positive cathode pulses. A band of twelve anode wires was connected via an isolating capacitor, to a preamplifier and then to the main amplifier, giving a gain of 60 dB. Some differentiation was applied to remove low-frequency noise. The excellent signal-to-noise ratio of the preamplifier revealed the presence of a secondary pulse about 100 μs after every main pulse. This was the arrival of the positive argon ions at the high field around the cathode wires. To prevent this pulse, and differentiation overshoots, from counting, a 200 μs one-shot was connected between the discriminator and timer-counter (see fig. 6). The consequent counting-rate limitation did not matter, since rates were never greater than a few hundred counts per second.

5. Chamber performance

The chamber was investigated to assess its electron drifting action, and its detection efficiency and spatial resolution for photons in the range 0.1–2.0 MeV. The experimentally measured efficiencies are compared to the theoretical values derived in section 3.

5.1. ELECTRON DRIFTING ACTION

Testing with an ^{55}Fe source gave results similar to those obtained with the copper matrix, described in section 2: a large increase in count rate, and a flat pulse-

---

* Trade name CERROBASE. Composition 44.5% Pb, 55.5% Bi, melting point 124 °C.

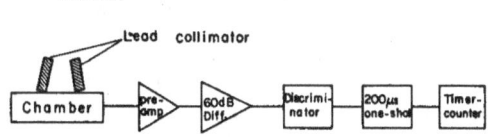

Fig. 6. The electronics arrangement for measuring detection efficiency.

height spectrum, when a forward-drift voltage was applied. At the maximum chamber voltage, 2.6 kV, the 5.9 keV line, which corresponds to two hundred ion pairs in the chamber, gave a pulse height three hundred times the discriminator setting. On the assumption that the chamber was operating in the proportional mode, one ion pair was detectable.

The chamber was next tested with 0.66 MeV photons from a ^{137}Cs source. At this energy, photon conversion by the gas is completely negligible. The source was fixed 10 cm above the matrix. The photon flux was collimated onto twelve rows of holes by two tilted lead blocks, 5 cm deep (fig. 6). This prevented spurious counts due to Compton-scattered photons entering the active area from elsewhere in the matrix. Applying the drift voltage produced a sevenfold increase in count rate and a flat pulse-height spectrum. Fig. 7 shows the count rate for a fixed drift voltage as a function of chamber voltage. The saturation at 2.6 kV suggests a detection threshold down to one ion pair, in agreement with the ^{55}Fe observations.

To assess the influence of hole size and plate thickness on the drifting action, the count rate was measured, for different drift voltages, with the plates connected to the resistor chain in singles, doubles, and triples. The count rate saturates in all three cases; very rapidly in the single-plate case (see fig. 8). The saturation values of count rate are plotted against the ratio, hole diameter to plate thickness (h/t), in fig. 9. Now, for

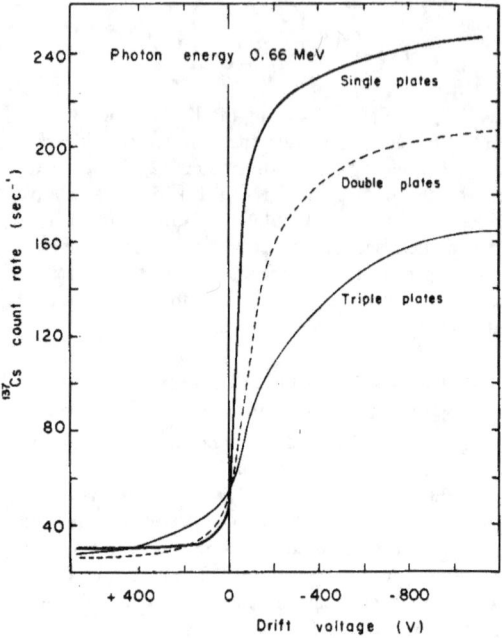

Fig. 8. Count rate as a function of drift voltage for 0.66 MeV photons. The influence of plate thickness is shown.

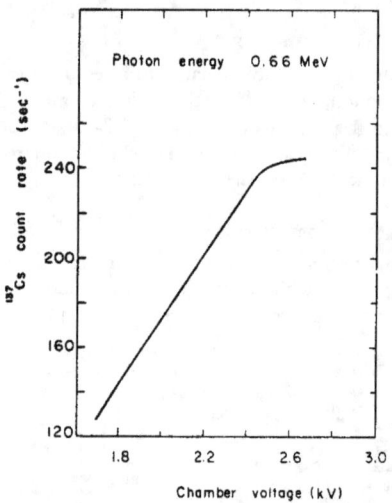

Fig. 7. Count rate as a function of chamber voltage for 0.66 MeV photons. The saturation at 2.6 kV indicates a detection level down to 1 ion pair.

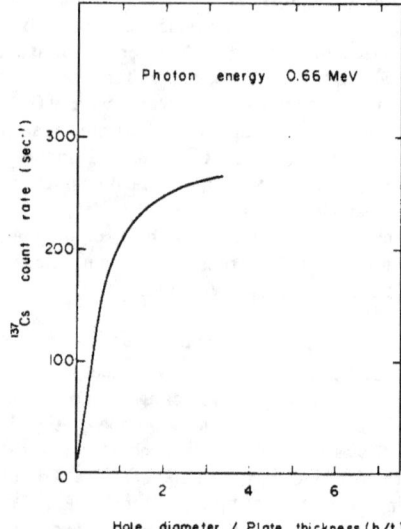

Fig. 9. The influence of plate thickness on the count-rate saturation values of fig. 8. The saturation here indicates a high probability of electron collection from the matrix.

TABLE 2

The results of the experimental measurements of detection efficiency.

Source	Photon energies (MeV)	Contribution (%)	MPC efficiency (%)	Matrix efficiency (%)	Efficiency ratio
^{57}Co	0.014	7.8			
	0.122	84.8 ⎫	0	2.2	∞
	0.136	11.4 ⎬			
	0.707	0.2 ⎭	–	–	–
^{203}Hg	0.071/0.085	12.7			
	0.279	81.5	0.24	3.33	13.9
^{22}Na	0.511	181	0.58	3.46	6.0
	1.275	100			
^{137}Cs	0.032/0.037	8.1			
	0.662	85.1	0.61	4.37	7.2
^{54}Mn	0.835	100	0.73	5.16	7.1
^{60}Co	1.173	100 ⎫	1.03	7.0	6.8
	1.332	100 ⎬			
^{88}Y	0.898	91.5			
	1.836	99.5	1.7	11.3	6.7
	2.76	0.5	–	–	–

large values of h/t, the chamber will look like a gas drift space ($h/t \to \infty$), which is well known to be capable of drifting electrons with no losses. From the saturation of fig. 9 one sees, therefore, a clear indication of efficient drifting action by the matrix, provided $h/t > 2$. Some changes were made to the resistor chain for the single-plate connection. No significant change from fig. 8 was observed. However, it is possible that the situation $h/t < 2$ could be improved by a careful optimization of resistor values.

A far more detailed experimental investigation would clearly be of interest; for example to determine the pulse-height spectrum as a function of drift voltage, resistor values, and photon energy, and to look at much thicker stacks. In general, it may be said that for a 1 cm stack the drifting action for count rate is very efficient, provided the hole diameter is at least twice the plate thickness.

5.2. THE DETECTION EFFICIENCY

The detection efficiency was measured as a function of photon energy by replacing the ^{137}Cs source used previously with each of a number of calibrated sources*, details of which appear in table 2.

* Gamma Reference Source Set, The Radiochemical Centre, Amersham, England.

5.2.1. Measurement procedure

Three measurements of count rate were made at each energy for both forward- and backward-drift fields. Subtraction separates the direct chamber and matrix contributions. "No source" background, being the sum of cosmic and natural radiation and chamber discharges, was also measured and subtracted. For the ^{22}Na source, thin aluminium discs were placed in contact with the source to ensure complete positron annihilation close to it. A 2 mm thick sheet of lucite was interposed between the chamber and the ^{137}Cs source to absorb the 0.63 MeV electron radiation.

Only the ^{54}Mn source provides monoenergetic photons. But the ^{60}Co lines are sufficiently close in energy that the source may be considered as giving monoenergetic photons at the average energy. Conveniently, the results from these two sources may be used to resolve the double lines of ^{88}Y and ^{22}Na. For ^{57}Co, the 0.707 MeV line may be ignored; a calculated gas capture correction applied for the 0.014 MeV photons and the other two energies averaged. The weak, low-energy lines of ^{203}Hg and ^{137}Cs were allowed for by extrapolating the ^{57}Co result, together with gas capture corrections.

5.2.2. Experimental results

Results are given in table 2. The detection efficiency

due to the MPC varies from 0.25% at 0.28 MeV to 1.7% at 1.8 MeV. A comparison may be made with the known efficiency of a cylindrical Geiger counter over this energy range[13]) (fig. 10). A consistent picture is seen. At high energies the MPC looks like a lead-walled Geiger counter and at low energies like an aluminium-walled counter. This is explained by the screening effect of the matrix at low energies. The attenuation length for 0.1 MeV photons in lead is 0.2 mm. Only those photons that pass through the holes in the matrix and convert in the MPC, presumably on the bottom mylar window, may be recorded. At higher energies the photons penetrate the matrix and the MPC receives Compton electrons and photoelectrons from the bottom lead–bismuth plate.

The matrix efficiency is 1.8% at 0.1 MeV, rising to 11% at 1.8 MeV. It should be noted that the total photon interaction probability for the matrix is 18% at 1.8 MeV. Thus 60% of the photons that interact are detected. The ratio matrix efficiency to MPC efficiency is roughly constant, above 0.5 MeV, at 7, a value measured previously (section 5.1). At the lower energies this ratio increases, reflecting the rapidly decreasing MPC efficiency, explained above.

In fig. 11 the theoretical calculation of the matrix detection efficiency, described in section 3, is compared with the experimental results. The four contributions: K-shell (ε_K) and L,M-shells (ε_L) photoelectric capture, Compton scattering (ε_C), and the K-shell escape peak (ε_{KE}) are shown, together with the total (ε_T). The general theoretical–experimental agreement is very pleasing. Two small discrepancies are evident: the experimental

values exceed the theoretical above 1.0 MeV and below 0.4 MeV. There are two explanations for the high-energy discrepancy. Firstly, the basic condition $R(E) < b/2$ (see section 3) of the theoretical model is violated above 0.4 MeV. In fact, one sees from Seliger's data that this condition is too severe. For lead, the electron escape probability is effectively zero at $\frac{1}{2}R(E)$. Thus the condition relaxes to $R(E) < b$, which is not violated until 0.8 MeV, just where the discrepancy begins to appear. The second factor is the onset of pair production above 1.0 MeV. At 1.8 MeV the pair-production cross section is 6% of the total. However, the electron–positron pair will average only 0.4 MeV each in kinetic energy: 1.0 MeV of energy is lost in creating the pair. And the subsequent annihilation of the positron will not contribute significantly to the efficiency: detecting the resulting 0.5 MeV photons is a second-order process. One would expect only a fraction of a per cent extra in efficiency.

The reason for the low-energy discrepancy is less obvious. It is probably due to an extra contribution from the upper surface of the top plate of the matrix. As pointed out above, the attenuation length of 0.1 MeV photons in lead is 0.2 mm. Nearly all the photon capture will be in the first plate, at this energy; and fig. 10 shows a sharp increase in the efficiency of a lead Geiger counter below 0.4 MeV.

5.3. SPATIAL RESOLUTION

The centre-of-gravity method[10]) was applied to single cathode wires to measure the spatial resolution. In fig. 12 are displayed histograms of the number of

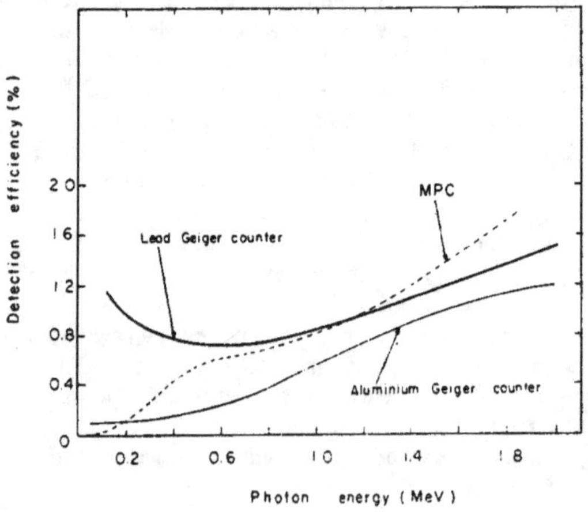

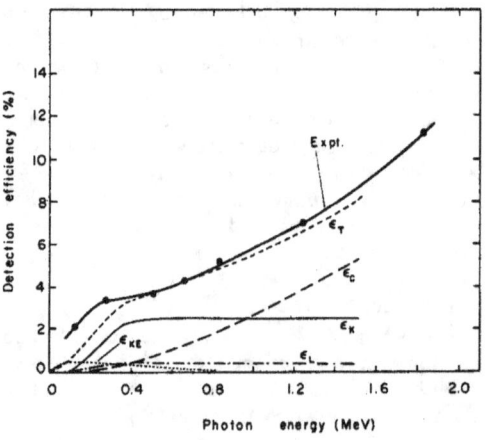

Fig. 10. A comparison of the MPC efficiency with the known efficiencies of lead and aluminium Geiger counters.

Fig. 11. A comparison of the experimental and theoretical efficiencies of the matrix. $\varepsilon_T = \varepsilon_{KE} + \varepsilon_L + \varepsilon_K + \varepsilon_C$.

events as a function of position *along* the anode wires, i.e. the direction of continuous resolution. Each tick mark represents 0.5 mm. Fig. 12a shows the result for an uncollimated 0.66 MeV ^{137}Cs source, with a back-drift voltage, i.e. the MPC background. When the forward-drift voltage is applied, the hole structure of the matrix is revealed (fig. 12b). The peaks are narrower than the hole size because of the focusing action of the drift field. The source was collimated with a 0.5 mm wide slot in a lead block 5 cm deep. The slot was positioned along a row of holes perpendicular to the anode wires. Fig. 12c shows, as expected, that the spatial resolution is basically the hole size. The envelope of the three peaks has a fwhm of 1.3 mm. Finally, in fig. 12d, one sees the result from a collimated ^{60}Co source (1.25 MeV). The resolution is lost because of the dominance of Compton scattering over the photo-electric effect at this energy. At 0.66 MeV the photo-electric-to-Compton cross-section ratio is about 1:1.5, and at 1.25 MeV, 1:3. This is an important result: it clearly demonstrates the need for photoelectric capture to obtain a good spatial resolution.

Figs. 13a–e show the sequence obtained by moving the collimated 0.66 MeV source across the matrix in half-pitch (0.75 mm) steps. The passage of the beam from hole to hole may be clearly followed. Indeed a source movement of 0.05 mm caused an observable change in the relative heights of the peaks.

6. Conclusion

The gas–solid hybrid MPC described here offers new possibilities for the spatial localization of non-ionizing radiation, over a very wide energy range.

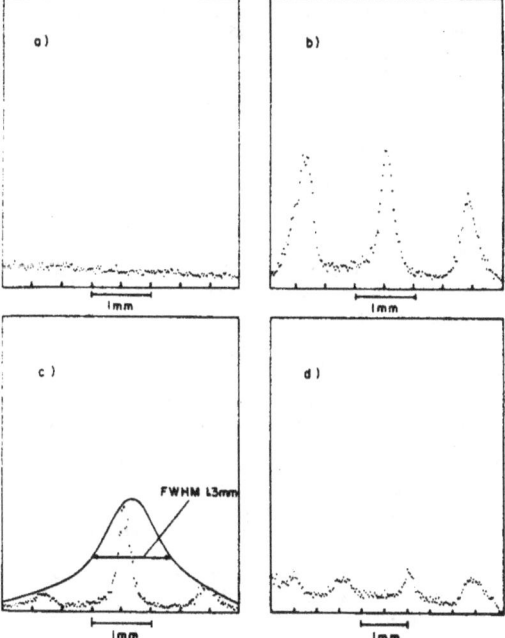

Fig. 12. Spatial resolution given by the matrix. 0.66 MeV photon source: (a) Uncollimated source, back-drift voltage; (b) Uncollimated source, forward-drift voltage; (c) Collimated source, forward-drift voltage. 1.25 MeV photon source: (d) Collimated source, forward-drift voltage. Spatial resolution is lost at high photon energies due to the predominance of Compton scattering.

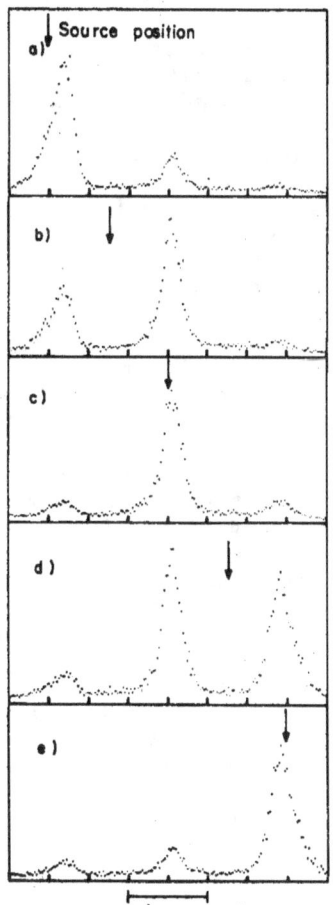

Fig. 13. (a)–(e) The passage of a collimated beam of 0.66 MeV photons from hole to hole of the matrix, in steps of 0.75 mm.

These possibilities stem from two basic properties of the solid matrix:

i) Good detection efficiency due to the high probability of interaction of the non-ionizing radiation with the solid.

ii) Good spatial resolution due to the hole size limitation to the range of the ionizing products.

Advantage may be taken of either or both of these properties, depending on the nature and energy of the radiation.

It has been shown that both properties are fundamental for detecting and localizing photons in the energy range 0.1–2.0 MeV. At 1.8 MeV, 60% of the photons interacting with the solid are detected. Spatial resolution of 1.3 mm fwhm is obtained below 1.0 MeV, where the photoelectric effect is dominant. Energy resolution is lost, however. By correct choice of matrix material and geometry and by using multiple stacks, such results should be possible for any particular energy within the 0.1–2.0 MeV range. Of particular interest is 0.5 MeV for positron work and around 0.1 MeV for general radio-isotope imaging. If the matrix is made of neutron-sensitive material, for example boron, lithium, or gadolinium, similar results should be forthcoming for thermal and epithermal neutrons. An extension to fast-neutron detection should be possible by replacing the matrix insulation with sheets of hydrogeneous material.

Using the property of range limitation alone is applicable to the case where the gas provides good detection efficiency, but insufficient spatial resolution. An obvious example is the use of ^{3}He for thermal-neutron imaging[26]. At the expense of energy resolution, the use of a matrix would remove the need for pressurization or addition of a stopping gas, such as krypton. The same argument applies to using argon for detecting 20–50 keV photons, or xenon for 50–100 keV. Whether the matrix could be made in a spherical form, for use in a spherical drift chamber, is perhaps an intriguing technological problem.

At very much higher energies (GeV) the range limitation property disappears, but the high detection efficiency remains. The gas–solid matrix becomes a high-density drift space; using dense metals such as tungsten, more than 10 g/cm^3 should be possible. One then has a shower detector that should compete very favourably with the liquid-argon[27] or lead-glass[28] detectors. By making use of drift-time as well as spatial information, three-dimensional siting of the shower vertex should be obtained. Further, if the sampling of the shower from the gas ionization is adequate, energy resolution will be regained.

A general program of work is under way to explore the varied and very promising topics outlined above.

We are greatly indebted to Messrs. K. Kull, G. Dinkel and J. Toustou for constructing the chamber, and Messrs. C. Parkman, J. C. Santiard and Z. Hajduk for assisting with the electronics.

We thank Prof. M. Peter, and Drs M. G. N. Hine, G. R. Macleod, P. Zanella and H. Davies for their enthusiastic interest and support.

References

1) G. Charpak, Ann. Rev. Nucl. Sci. 20 (1970) 195.
2) A. H. Walenta, J. Heintze and B. Schürlein, Nucl. Instr. and Meth. 92 (1971) 373.
3) J. Saudinos, J.-C. Duchazeaubeneix, C. Laspalles and R. Chaminade, Nucl. Instr. and Meth. 111 (1973) 77.
4) S. N. Kaplan, L. Kaufman, V. Perez-Mendez and K. Valentine, Nucl. Instr. and Meth. 106 (1973) 397.
5) C. Cork, D. Fehr, R. Hamlin, W. Vernon, N. H. Xuong and V. Perez-Mendez, private communication.
6) K. Valentine, S. N. Kaplan, V. Perez-Mendez and L. Kaufman, IEEE Trans. Nucl. Sci. NS-21 (1974) 178.
7) C. J. Borkowski and M. K. Kopp, IEEE Trans. Nucl. Sci. NS-17 (1970) 340.
8) V. Perez-Mendez and S. I. Parker, IEEE Trans. Nucl. Sci. NS-21 (1974) 45.
9) W. R. Kuhlmann, K. H. Lauterjung, B. Schimmer and K. Sistemich, Nucl. Instr. and Meth. 40 (1966) 118.
10) G. Charpak, A. P. Jeavons, F. Sauli and R. J. Stubbs, CERN 73-11 (1973).
11) G. Charpak, Z. Hajduk, A. P. Jeavons, R. Kahn and R. J. Stubbs, Nucl. Instr. and Meth. 122 (1974) 307.
12) S. E. Derenzo, T. F. Budinger, R. G. Smits, H. Zaklad and L. W. Alvarez, Proc. Symp. on Advanced technology arising from particle physics research, Argonne, 1973 (ANL 8080, Argonne, Ill., 1973), p. 11, 1.
13) S. C. Curran and J. D. Craggs, Counting tubes (Butterworth, London, 1949), p. 89.
14) C. B. Lim, D. Chu, L. Kaufman, V. Perez-Mendez and J. Sperinde, IEEE Trans. Nucl. Sci. NS-21 (1974) 85.
15) L. V. Spencer, Phys. Rev. 98 (1955) 1597.
16) L. V. Spencer, Energy dissipation by fast electrons, N.B.S. Monograph (Nat. Bureau of Standards, Washington, 1959).
17) H. H. Seliger, Phys. Rev. 100 (1955) 1029.
18) E. Segrè, Experimental nuclear physics (J. Wiley, New York, 1953), Vol 1, p. 292.
19) A. T. Nelms, Energy loss and range of electrons and positrons (Nat. Bureau of Standards, Washington; circ. 577, 1956).
20) H. H. Seliger, Phys. Rev. 88 (1952) 408.
21) R. D. Evans, in Handbuch der Physik (Springer Verlag, Berlin, 1958), Vol. 34, p. 268.
22) R. D. Birkhoff, in Handbuch der Physik (Springer Verlag, Berlin, 1958), Vol. 34, p. 132.
23) G. W. Grodstein, X-ray attenuation coefficients from 10 keV to 100 MeV (Nat. Bureau of Standards, Washington; circ. 583, 1957).
24) G. Charpak, H. G. Fisher, C. R. Gruhn, A. Minten, F. Sauli

and G. Plch, Nucl. Instr. and Meth. **99** (1972) 279.

[25] V. Radeka, IEEE Trans. Nucl. Sci. NS-21 (1974) 51.

[26] B. H. Meardon and D. C. Salter, Rutherford lab. rep. RHEL/R 262 (1972), p. 12.

[27] S. E. Derenzo, A. Schwemin, R. G. Smits, H. Zaklad and L. W. Alvarez, Berkeley report LBL-1791 (1973).

[28] J. Beale, F. W. Büsser, L. Camilleri, L. Di Lella, G. Gladding, A. Placci, B. G. Pope. A. M. Smith, B. Smith, J. K. Yoh and E. Zavattini; B. J. Blumenfeld and L. M. Lederman; R. L. Cool, L. Litt and S. L. Segler, Proc. Intern. Conf. on *Instrumentation for high-energy physics*, Frascati, 1973 (Lab. Naz. CNEN, Frascati, 1973), p. 415.

Nuclear Instruments and Methods 186 (1981) 613–620
North-Holland Publishing Company

THE MULTISTEP AVALANCHE CHAMBER AS A DETECTOR FOR THERMAL NEUTRONS

G. MELCHART, G. CHARPAK, F. SAULI
CERN, Geneva, Switzerland

G. PETERSEN
The Niels Bohr Institute, University of Copenhagen, Copenhagen, Denmark

and

J. JACOBÉ
Institut Laue-Langevin, Grenoble, France

Received 4 August 1980

We have investigated the performance of the multistep avalanche chamber in combination with a solid converter foil of gadolinium for detection of 2.4 Å thermal neutrons. High spatial resolution, 700 μm fwhm, reasonably good time stability, and low γ-ray sensitivity are among the promising features of this new kind of detector.

1. Introduction

Multiwire proportional chambers (MWPCs) are in current use as one- or two-dimensional position-sensitive detectors for thermal neutrons [1]. The basis for detection is the capture of a neutron, performed by one of the gas components, followed by the emission of ionizing compounds, such as protons, deuterons, α-particles, etc.

To obtain good detection efficiency combined with a reasonable position resolution and small parallax errors for neutron incident angles different from 90°, it is necessary to pressurize the detector. For a two-dimensional detector over a certain size this imposes serious mechanical problems in the construction.

Instead of using the detector gas as a conversion medium, one can use a converter foil of solid material and detect the emitted conversion electrons by means of a standard gas flow MWPC. To limit the range of the conversion electrons in the detector gas, in many cases several centimetres, one can introduce a solid collimator in front of the foil [2]. However, the collimator introduces a modulation in the detector's position response, as well as an increased γ-ray background and a decreased neutron detection efficiency is to be expected.

The detector described in this work also uses the standard MWPC and a solid converter foil. The collimation is obtained by introducing an extra gas amplification stage in the chamber, the so-called pre-amplification stage. When the conversion foil is mounted close to the outer electrode of the pre-amplification stage, it is preferentially that part of the ionization trail, a few hundred μm, which is closest to the emission point of a conversion electron that is amplified and seen by the MWPC.

2. The converter foil

A 10 μm thick foil of natural gadolinium was used as a converter; it contained about 15% of gadolinium-157, which is the isotope with the desirable converter properties. A foil enriched with this isotope would therefore have been better suited to our purpose. The conversion in gadolinium consists of an (n, γ) process, where the neutron is captured, followed by the emission of one or more conversion electrons. The internal conversion probability in gadolinium is so high that each captured neutron creates one or more electrons. The energy spectrum of the conversion electrons has a mean value of 70 keV [3].

The neutron capture cross-section is high, and around 25% of an incoming beam of 2.4 Å neutrons give rise to conversion electrons liberated from the

0029-554X/81/0000–0000/$02.50 © North-Holland

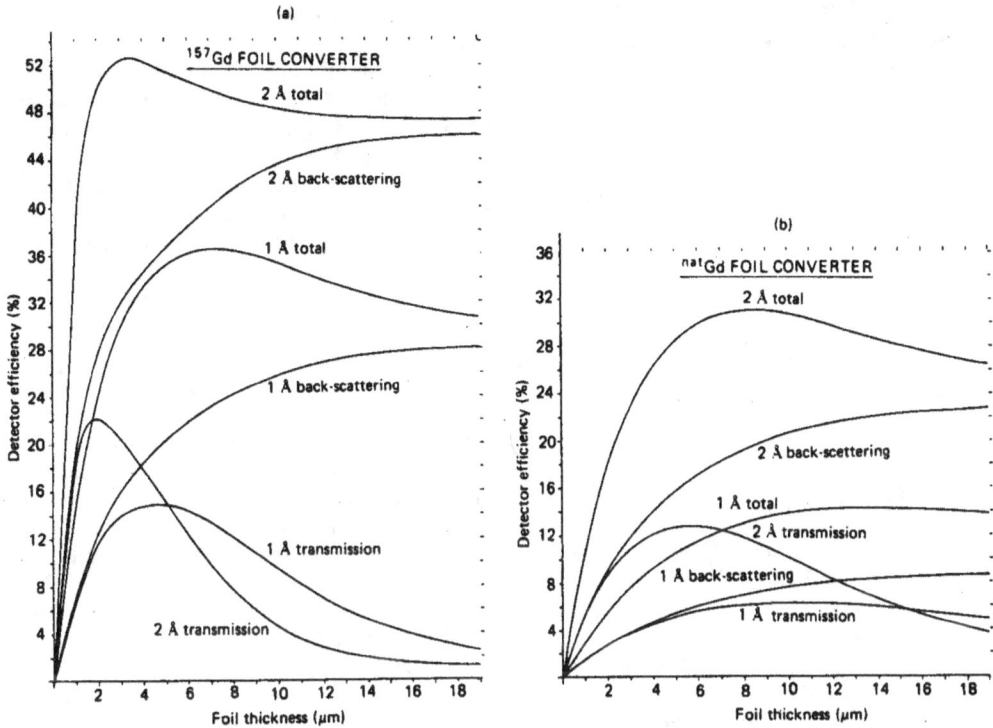

Theoretical detection efficiencies for gadolinium-157 and for natural gadolinium [2].

converter foil in back-scattering. Fig. 1 shows the theoretical detection efficiencies for gadolinium-157 and for natural gadolinium [2]. A strong dependency of the efficiency on wavelength and foil thickness for both transmission and back-scattering can be seen.

3. The multistep avalanche chamber

3.1. The photon-mediated avalanche

It has been shown [4,5] that, with a proper choice of gas mixture, it is possible to develop an avalanche between parallel grids with the following characteristics: the lateral spread of the avalanche is such that a sizeable fraction of the resulting electron cloud can be extracted into a following transfer space. The transferred fraction is nearly independent of the position of the electron initiating the avalanche. This permits proportionality between the initial ionization and final transferred charge and prevents modulation of the efficiency as a function of position.

It has been inferred that the emission and absorption of VUV photons play a role in the spread of the avalanche; however other processes, such as electron diffusion and the Penning effect, may also participate.

3.2. The collimation effect

The charge multiplication of an electron in the preamplification stage varies exponentially with its distance from the second preamplification grid. An electron cloud, obtained by preamplification and transfer of the electrons in an ionization trail produced in the preamplification region, will therefore privilege the initial electrons closest to the entrance grid limiting the preamplification gap. When the converter foil is mounted in close contact with this grid, the preamplification mechanism will serve as a kind of collimator for the conversion electrons.

The localization effect can easily be treated quantitatively, using the following simplifying assumptions: the ionization tracks follow straight lines, they are infinitely long, and they are emitted uniformly in space from a point on the entrance grid. Under these conditions one gets a simple expression for the centre of gravity distribution of preamplified and transferred electron clouds

$$I(Z) = \frac{1}{1 + \left[\frac{\ln M}{l} Z\right]^2} ,$$

where l is the distance between preamplification grids, M is the multiplication factor over the distance l, and Z is the coordinate in a direction parallel to the grids. The formula is accurate within a few percent for values of $M > 50$. In the case where there is no multiplication the expression is modified to

$$I(Z) = \frac{1}{1 + \left[\frac{2}{l} Z\right]^2} .$$

Fig. 2 shows $I(Z)$ for $l = 3$ mm and different values of M. In the case of $M = 2000$, which corresponds to the working conditions of our detector, we expect a position accuracy of about 1 mm fwhm.

In practice, the ionization tracks created by electrons from the foil converter are neither straight lines nor infinitely long. The energy spectrum of emitted electrons extends down to zero, which implies that a large amount of the tracks have a very limited range, less than a few hundred μm, in the detector gas. This improves the final position accuracy of the detector.

4. Experimental set-up

The arrangement of wire planes in the detector is shown in fig. 3. Two electrodes have been mounted on top of a standard MWPC to provide the preamplification and transfer regions. The preamplification electrode F is a mesh of stainless steel wires, 30 μm in diameter and 500 μm apart. The electrode E is a micromesh of Ni wires, 10 μm in diameter and 100 μm apart. A foil of natural gadolinium G, mounted on an aluminium ring H, 50 mm in diameter, is pressed against the outer electrode in such a way that in the preamplification region the distance between the electrodes E and F is 3 mm, whereas the distance between them at the exges of the chamber is 5 mm. We have designed the preamplification region in this way to prevent sparks from appearing at the edges during operation. The distance between the electrode E and the MWPC is 3 mm. A is a mylar window which closes the MWPC on the other side, making it gas tight.

The MWPC and associated electronics have been described elsewhere [6]. Briefly, the anode plane (electrode C) is made of gold-plated tungsten wires, 10 μm in diameter and 1.27 mm apart. The two orthogonal cathode planes B and D are made of Cu–Be wires, 100 μm in diameter and 1.27 mm apart. The anode–cathode distance is 5 mm, and the active area of the MWPC is 10×10 cm^2. The bi-dimensional read-out of the MWPC is done with the centre-of-gravity method where, for each avalanche occurring on the anode, the positive induced charge distributions on the two cathode planes are recorded. The cathode wires are connected together 4 by 4 in strips, and each strip is connected to a charge preamplifier

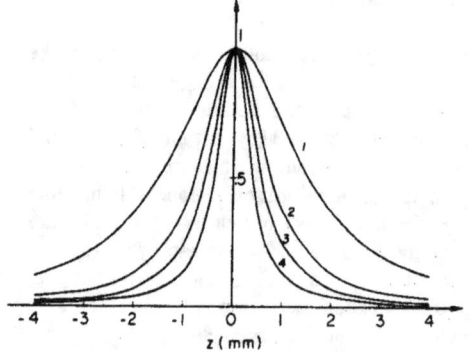

Fig. 2. Theoretical distributions of the centre of gravity of preamplified and transferred electron clouds for different values of the amplification factor M. The curves 1–4 correspond to the M values 1, 50, 200, and 2000.

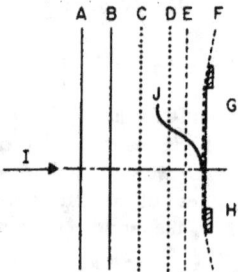

Fig. 3. Schematic view of the detector. A is a mylar window; B, C, and D constitute a standard MWPC; E and F are the preamplification electrodes; G is a gadolinium foil mounted on an aluminium ring H; I is an incoming neutron, converting to an electron that ionizes along the track J.

having a sensitivity of about 250 mV/pC, a rise-time of 10 ns, and a decay-time of 50 ns [*]; the amplified signals are then transmitted via coaxial cables to a CAMAC-based, 10-bit current integrating ADC [†].

All anode wires are connected together; the signal from this plane is recorded in an ADC channel and provides also the gate for the ADC modules after suitable discrimination and shaping. The gate width is 100 ns. A small on-line computer is used to organize the data acquisition and transfer to magnetic tape.

As a gas filling we have used a mixture of argon and acetone in the volume concentration 98–2.

5. Calibration and data handling

A detailed description of the procedure used to calibrate the chains of the preamplifier, coaxial cable, and ADC channel, is given in ref. 6. The calibration allows us to correct in software for non-linearities in the chains as well as for dispersions between them.

Let us denote by q_i the effective charge signal on strip S_i, as obtained from the raw data using the above-mentioned calibration procedure. We define the centre-of-gravity of a cluster of adjacent induced charges as the quantity:

$$\bar{S} = \sum (q_i - b) S_i / \sum (q_i - b),$$

where b is a bias level, and the sum extends only to positive values of $q_i - b$. A proper choice of bias level allows reduction of the influence of pick-up and electronics noise in the centre-of-gravity determination. In our case a value of $b = 2\%$ of Σq_i gave satisfying results.

6. Experimental results

6.1. Introduction

The detector was tested in a 2.41 Å monoenergetic neutron beam at the Institut Laue–Langevin (ILL) in Grenoble, France. The beam could be shaped to the desired profile using collimators made of 1 mm sheets of cadmium. The γ-ray background in the surroundings of the detector, measured with a dosimeter, was typically at the level of 6 mRad/h. The chamber was operated at a voltage of +2.20 kV on the anode,

and −300 V between the electrodes D and E. The cathodes are grounded; the voltage between E and F is −3.6 kV or less.

6.2. Detection efficiency

The counting rate of the detector was measured for different values of the voltage HV_F on electrode F, in a beam with an intensity of 2000 neutrons/s. The threshold for counting was set just over the electronics noise level. The result is shown in fig. 4. The background, obtained from counting with the beam off, has been subtracted before plotting the curve. The counting time was 1 min for each point; therefore the uncertainty in the measurement from statistical fluctuations is negligible. We see that no plateau is obtained. The reason is that it was not possible to operate the detector in a stable way with a preamplification voltage high enough to detect the low-range part of the ionization tracks.

It has been shown [5] that it is possible to detect single electrons after preamplification, so with a good detector one should be able to get close to 25% detection efficiency, which is the theoretical value for 2.4 Å neutrons, see fig. 1. In our case the measured

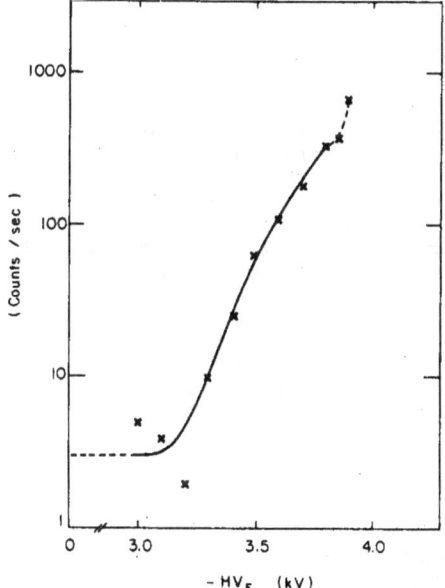

Fig. 4. Counting rate in the detector versus preamplification voltage. No plateau is obtained. The beam intensity was 2000 neutrons/s.

---

[*] Developed at CERN by J.C. Santiard and produced with a hybrid thick film technique by CIT-Alcatel, France.
[†] Lecroy, 12-channel ADC, type 2249A.

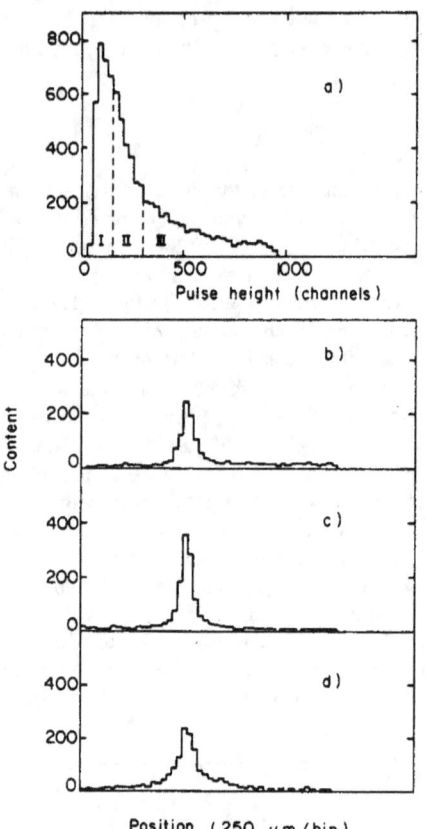

Fig. 5. (a) Typical anode pulse spectrum. The spectrum is cut at the lower side by an electronic threshold. At the higher side we cut the spectrum in order to eliminate those pulses which are too high and give rise to overflow in the ADCs. The spectrum is divided into three intervals corresponding to the events in figs. 4b, c and d. (b, c, d) Detector response in the direction along the anode wires for different intervals of the anode pulse spectrum. Bin width 250 μm.

efficiency is around 20% for HV_F equal to −3.85 kV, which was the operational voltage on the preamplification electrode during most of the tests.

6.2. The γ-ray background

On a normal pressurized MWPC for thermal neutrons, where the conversion takes place in the detector gas, one can easily discriminate against unwanted γ-ray events, as these in most cases give rise to much lower energy losses in the detector gas than the neutrons do.

Two factors contribute to a decrease of the γ sensitivity as compared with a normal proportional chamber:
– Only the electrons extracted from the entrance window (F, fig. 3) are counted.
– Only the Compton electrons back-scattered in the gadolinium foil are counted.

We wondered also if the detected pulses showed a difference between γ's and neutrons. In a detector like ours, one should not expect to see much difference, in both cases it is only a part of an ionization track, created by a conversion electron, that is detected. It turns out, however, that there is a possibility to reduce the γ-ray background, although at the price of reduced detection efficiency. Fig. 5 shows a typical anode pulse spectrum as obtained with our detector in the normal working conditions. The cut at the lower end is introduced by the electronic threshold. The cut at the higher end is set by the software program, which does not accept events that have given overflow in any of the ADC channels [6]. The beam is collimated to 350 μm in the direction along the anode wires. Figs. 5b, c and d show the detector response in that direction for each of the three intervals shown in fig. 5a. It is seen that an increase of the electronic threshold to some extent decreases the number of counts outside the peak. The γ-ray events seem to appear mainly in the low end of the anode pulse spectrum.

The sensitivity to γ-rays was also tested by a comparison between the detector responses obtained with and without a 5 cm thick lead block positioned in the neutron beam. No significant difference could be observed between the two cases, showing that the γ-rays are coming from the general background of the area. Further tests are needed to get more quantitative information about the sensitivity for different γ-ray energies.

6.3. Position accuracy

From fig. 5 it can also be seen how the position accuracy of the detector varies for different selections in the anode pulse spectrum. The bin width in the spectra is 250 μm. The best result, 650 μm fwhm, is obtained for the interval II. The pulses are here sufficiently high to exclude the contribution of the electronics noise in the centre-of-gravity determination. For the interval III the accuracy decreases; note the high tails in the spectrum. The reason is that conversion electrons, which are scattered back towards the

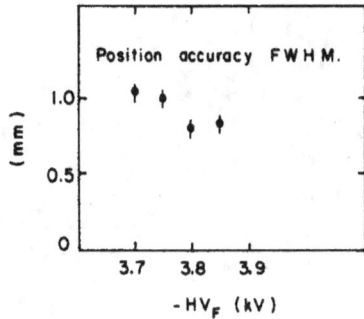

Fig. 6. Variation of the position accuracy with the pre-amplification voltage.

gadolinium foil, or emitted at angles far from the direction orthogonal to the foil, do not localize the emission point very well, and also tend to give larger pulses because of a larger ionization close to the electrode F.

The following results are obtained without any selection in the anode pulse spectrum. In all cases

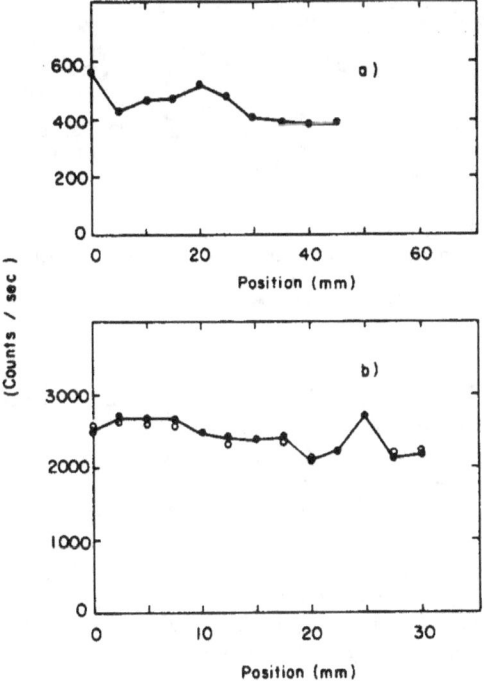

Fig. 7. Variation of the counting rate in the detector (a) along the anode wires, and (b) in the orthogonal direction, indicated by closed dots. A scan made one hour later is indicated by open dots. The time stability seems to be good.

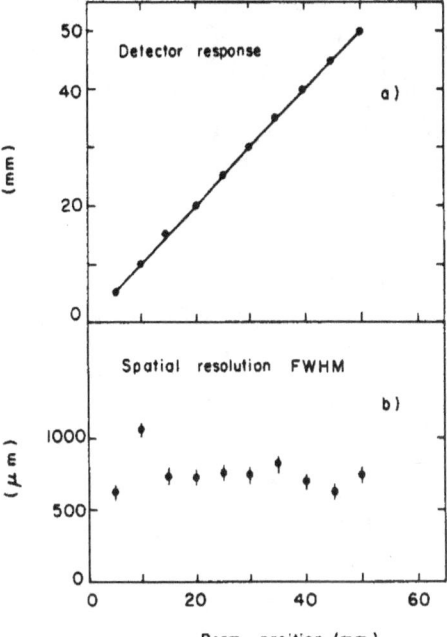

Fig. 8. The detector response for different positions of the beam in the direction along the anode wires, showing (a) the linearity and (b) the spatial resolution. A mean value of 700 μm fwhm is obtained.

where values for the position accuracy are given, these values would improve by around 100 μm if only events from the interval II were taken into account.

The position accuracy was measured for different values of HV_F. The results are shown in fig. 6, which is a plot of the fwhm of the position spectra against HV_F. We see the tendency mentioned above that the position accuracy increases with increasing pre-amplification voltage. The results shown from now on have all been obtained with $HV_F = -3.85$ kV.

6.4. Variations of detector response with position and time

For many applications it is of great importance that the detection efficiency of the chamber remains stable during several days. Variations in the efficiency for different positions in the detector can be corrected for in the software data handling, if the time stability is good. That it is possible to obtain good time stability with our detector is not obvious, since the electronic threshold for detection is not positioned well

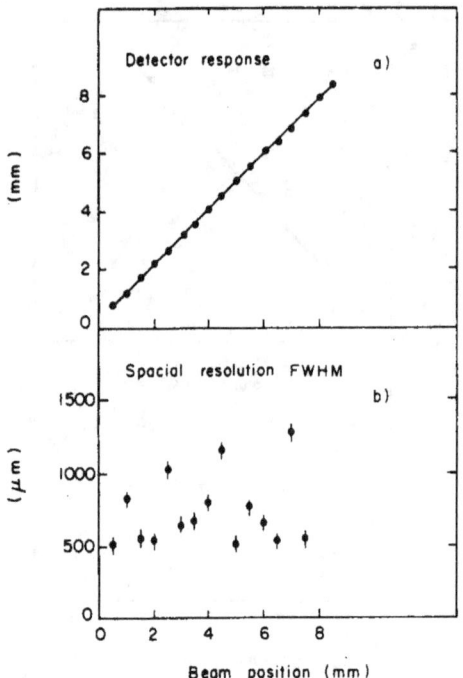

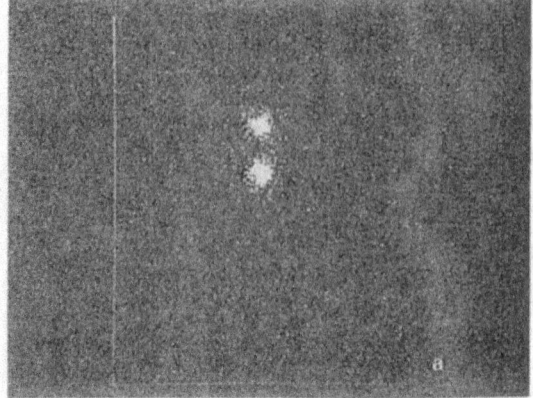

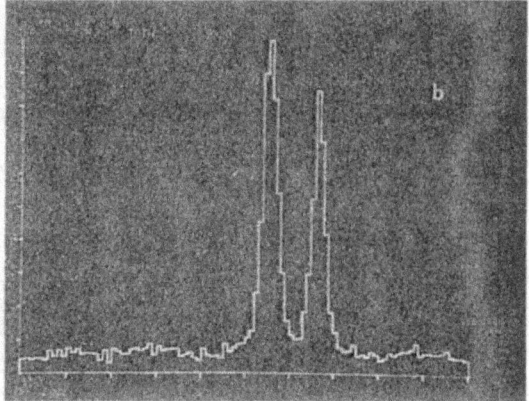

Fig. 9. The detector response for different positions of the beam in the direction orthogonal to the anode wires, showing (a) the linearity and (b) the spatial resolution. A mean value of 700 μm fwhm is obtained.

below the pulse height spectrum. We have measured the counting rate in the chambers, scanning the sensitive area in the direction along the anode wires, fig. 7a, and in the orthogonal direction, fig. 7b. The counting time was one minute for each point, giving negligible statistical fluctuations in the measurement. After one hour, one of the scans was repeated, indicated by open dots in fig. 7b. The time stability seems reasonably good, but further tests, over many days, are needed before any definite conclusion can be made.

6.5. Linearity and position response

In the direction along the anode wires the linearity of the detector response was expected to be good. That this is the case can be seen from fig. 8a, which shows a scan over the active area in steps of 5 mm. In fig. 8b are shown the corresponding position accuracies. A mean value of around 700 μm fwhm is obtained. A scan with steps of 500 μm in the ortho-

Fig. 10. The detector response for a beam collimated through two holes of 600 μm diameter and 2.5 mm apart, showing (a) the response in two dimensions and (b) in the anode wire direction. The scales are 20 mm and the bin width 250 μm.

gonal direction is shown in fig. 9. Also here we have a mean value for the accuracy of 700 μm. The anode wire structure, wire distance 1.27 mm, introduced no modulation in the linearity, fig. 9a, and only a small modulation in the value of the local accuracy, fig. 9b. This can be explained from the above-mentioned lateral spread in the electron avalanches after preamplification. For each event, avalanches will appear on more than one anode wire at the same time, and this effect improves the linearity of the detector. The scans of figs. 8 and 9 have been made with beams collimated to 350 μm.

A few tests were made with other beam profiles. Fig. 10a shows the two-dimensional response, when the beam enters the detector through two holes in a

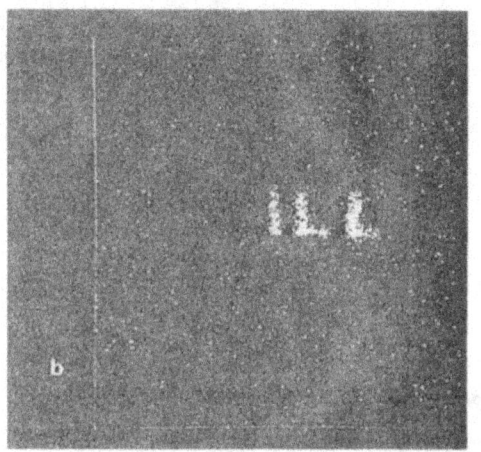

Fig. 11. (a) the pattern ILL drilled in the cadmium sheet as a series of holes, which are 600 μm in diameter. (b) The detector response for neutrons going through the holes.

cadmium sheet, 600 μm in diameter and 2.5 mm apart. In fig. 10b is shown for the same events, the detector response in the direction along the anode wires. The bin widths is 250 μm. The two peaks are completely resolved.

A cadmium sheet with a hole pattern as shown in fig. 11a was mounted in the beam. The diameter of the holes was 600 μm. The detector response is shown in fig. 11b.

7. Conclusions

The combination of the multistep avalanche chamber with a converter of gadolinium-157 turns out to be promising for thermal neutron detection. The detection of the back-scattered conversion electrons reduces the Compton electron background expected from a solid converter.

The chamber is operated with gas at atmospheric pressure, which prevents problems with outgasing, purity and high pressure. The two-dimensional spatial resolution is around 700 μm fwhm. The detection efficiency is mainly determined by the converter foil, and is measured to be around 20% for 2.4 Å neutrons. Higher values, close to 50%, are feasible, by the application of converters enriched in gadolinium-157. The preliminary tests indicate that the background of γ-rays can be kept at a reasonably low level. Also the time stability seems to be good, although further tests are needed to clarify this point completely.

G. Petersen wishes to thank the Danish Natural Science Council and the Danish Technical Science Council for support that made it possible for him to join this project.

References

[1] R. Allemand, J. Bourdel, E. Randaut, P. Corvert, K. Ibel, J. Jacobe, J.P. Collon and B. Farnaux, Nucl. Instr. and Meth. 126 (1975) 29.
[2] A.P. Jeavons, N.L. Ford, B. Lindberg and R. Sachot, IEEE Trans. Nucl. Sci. NS-25 (1978) 553.
[3] H. Rausch, F. Grass and B. Feigl, Nucl. Instr. and Meth. 46 (1967) 153.
[4] G. Charpak and F. Sauli, Phys. Lett. 78B (1978) 523.
[5] A. Breskin, G. Charpak, S. Majewski, G. Melchart, G. Petersen and F. Sauli, Nucl. Instr. and Meth. 161 (1979) 19.
[6] G. Charpak, G. Melchart, G. Petersen and F. Sauli, Nucl. Instr. and Meth. 167 (1979) 455.

Nuclear Instruments and Methods 192 (1982) 235–239
North-Holland Publishing Company

ELECTROSTATIC IMAGING OF PARTICLE TRAJECTORIES

G. CHARPAK, R. BOUCLIER, A. BRESKIN *, R. CHECHIK *

CERN, Geneva, Switzerland

and

J. LEWINER

Ecole Supérieure de Physique et de Chimie Industrielles, Paris, France

Received 4 August 1981

The ions liberated in a high-pressure gas or in some liquids can be collected, by electric fields, on the surface of insulators and can be accurately localized.

In a simulation of this method at atmospheric pressure, we applied it to α particles, with the additional amplification from a parallel grid gap. By directly measuring the static electric charges collected on mylar foils, we observe tracks of 1 mm fwhm and charge densities as low as 10^4 electrons/mm^2. The combination of multistep gated avalanche chambers with this read-out method should permit high-accuracy measurements of minimum ionizing particles. The limits of the method and some conditions for detection by liquid toners are discussed.

1. Introduction

Wire chambers and drift chambers have serious limitations in the imaging of complex configurations of charged particles close to the interaction vertex of high-energy nuclear particles inside a target.

We describe a new approach to this problem, which applies to gaseous targets and also to some specific liquid or solid targets where the ions liberated by the ionizing particles can drift along electric field lines.

From the first results reported here it is clear that our method opens up the way to very high accuracies in the measurement of particle trajectories in gases, as well as to additional potential diverse applications.

2. Principle of electrostatic imaging of ionizing particles

The initial idea is the following (fig. 1a): if an ionizing particle liberates free ions in a gas, it should be possible to attract the ions of a given sign to an insulating surface, where they are trapped and can be

* On leave from the Weizmann Institute, Israël.

revealed by the appropriate methods. Indeed, in such a device only a projection of the trajectories is obtained, but at the present stage of high-energy physics considerable interest is attached to the search for very short-lived particles, decaying within a range of tens or hundreds of microns near the interaction vertex. In this case a high resolution in one projection can be

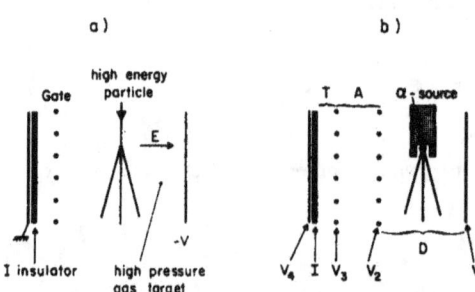

Fig. 1. Electrostatic imaging of ionizing particles. (a) High-energy particles produce a reaction in a very high pressure gaseous target, or in a liquid target. The ions are collected on the insulating foil I. (b) Schematic views of the experimental set-up. Alpha particles ionize mixtures of argon and vapours at 1 atm. The primary electrons drift to the amplification gap A, where they produce an avalanche. They are then transferred to T and collected on the insulator I.

sufficient to clarify complex configurations, while additional gaseous detectors can give the missing coordinates.

The imaging of ionizing radiations in gases or insulating liquids has been the subject of many investigations over the last 25 years, aimed at measuring X-ray spatial distributions for medical applications, or detecting β-ray distributions for use in chromatography. These methods have been extensively reviewed by one of the pioneers in this field, Boag [1]. In fact they are closely related to techniques used in xerography [2], where the latent image is an electrostatic distribution on a dielectric or a photoconductor. It should be emphasized that they all result in the detection of variable ionization densities, corresponding to an integration over a large number of particles. Even in the case where additional gaseous amplification was used in order to decrease the doses of X-ray irradiation, only integrated distributions have been reported [3].

We have investigated the use of commercial powder or liquid toners to reveal the distribution of charges collected on dielectrics from ionization of the gas by charged particles. These materials have a threshold of about 2×10^7 electrons per mm^2, thus even at pressures of 500 bar of argon, we are below the threshold.

In view of this limitation we described here a method of direct measurement of the electrostatic charges, which seems to be of adequate sensitivity. We used α particles in argon at atmospheric pressure as a simulation for minimum ionizing particles. We also used an amplifying gap, composed of two parallel grids, which permitted us to vary the final electron density [4, 5] at will, before collecting the electrons on an insulating mylar foil. The foil surface was then scanned and the charge distribution measured. With this flexible set-up we observed narrow linear charge distributions characteristic of α trajectories, and we are now in a position to make a realistic evaluation of the prospects and limits of our approach.

3. Experimental

The experimental set-up is illustrated in fig. 1b. The ^{241}Am α source is placed inside a drift space from which the ionization electrons can be extracted and transferred, by the appropriate field, to the amplification gap A. The source is grossly collimated in a direction parallel to the electrodes. The avalanche

electrons are transferred to the space T, 1 mm deep, and are collected on the 125 μm mylar foil, which is backed with a layer of conductive material. The gas filling consists of argon with a small admixture of propane or acetone, and the range of the α particles is close to 4 cm.

The transfer voltage $(V_4 - V_3)$(fig. 1b) is applied for a few seconds. Meanwhile, the pulses on the anode are monitored and show that some avalanches have been transferred. The foil I is then removed and scanned with a dedicated electrometer *, which consists of a probe shielded by a metallic housing with a 0.6 mm diameter hole, through which penetrate the lines of force of the charges deposited on the foil. The hole is at about 0.5 mm from the foil. The operation principle of the electrometer consists of bringing the probe to the same potential as the charges facing the hole, thus setting to zero the electric field between the probe and the foil. The ground of the instrument is connected to the conductive backing of the foil.

If the thickness of I is small with respect to the width of the charge distribution, we can simply calculate the voltage as if we had a capacitor with two conducting electrodes. This applies to our case very well, where we expect a charge distribution width of about 1 mm.

4. Experimental results

We have taken data, varying the range of the following parameters: the gain in A, the depth of T, and the gas composition. We report a few results which illustrate the potential power of the method and clarify its limits. Fig. 2 shows the voltage distribution observed along a mylar foil scanned in a direction roughly orthogonal to the direction of the tracks. We observe signals of about 1 mm fwhm and a few volts maximum amplitude. At some places we observe signals from six tracks (fig. 2a). By scanning successive lines, we can reconstruct the path of the tracks (fig. 2b). Fig. 3a shows the variation of charge profile along a given track; fig. 3b shows the variation of ionization along the track.

The observed width of the charge distribution is a function of several factors: (1) the intrinsic width of the avalanche, which increases with a decrease in ace-

* Electrostatic millivoltmeter model 1015. Monroe Electronics, INC, Lyndonville, NY, U.S.A.

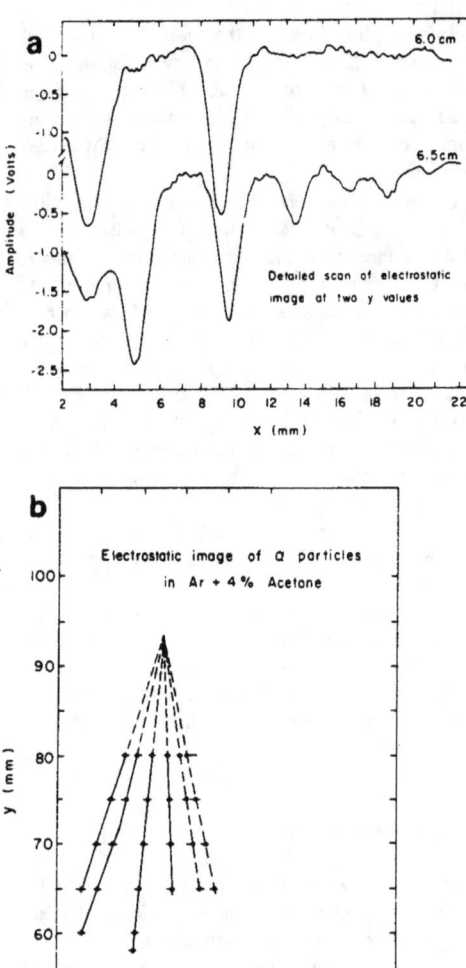

Fig. 2. Distribution of charges across tracks. (a) Charge distribution across tracks at two positions along the beam. M1015 Monroe probe at 0.5 mm from I. (b) Reconstruction of the trajectories.

tone content, as observed elsewhere [4], and increases with the depth of T; (2) the considerable space charge of the amplified electron cloud, which produces a field that exceeds the drift field, and might cause an "explosion" of the electron cloud before it reaches the insulator foil I; (3) integration due to scanning procedure.

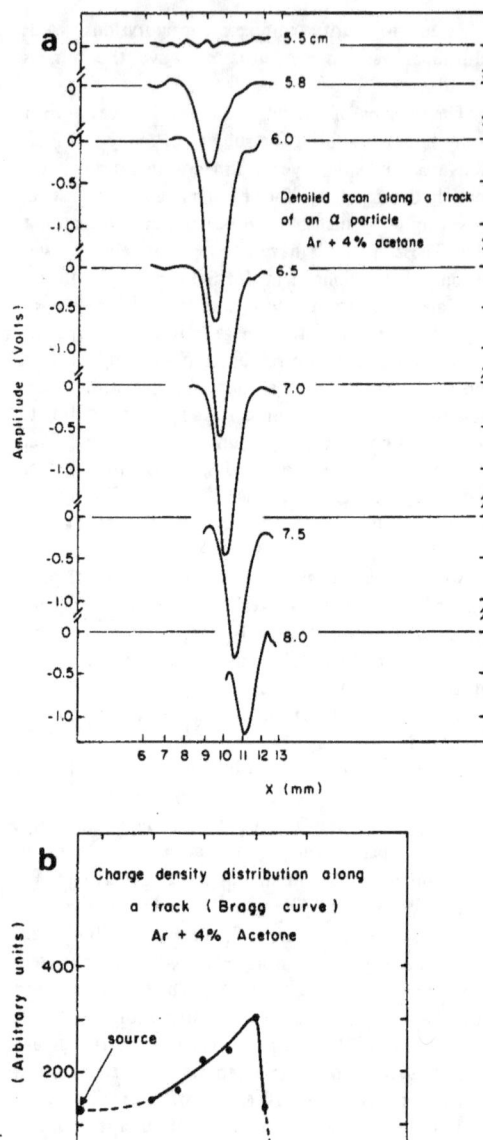

Fig. 3. Charge density along an α-track. (a) Measured charge along a track. (b) Extrapolated charge density distribution.

The distribution of charge intensity along a track shows a characteristic Bragg behaviour, as demonstrated in fig. 3b.

The data of figs. 2 and 3 were obtained with a gain in A of about 10^4, giving rise to about 30% of transferred electrons. This gives a linear charge density of about $10^7 e/cm$, where e is the electron charge. These data permit us to appreciate the limits of the method.

5. Sensitivity of electrostatic imaging

There are two sources of noise in the present method: the initial charges and dipoles existing in all good insulators, and the noise of the read-out instrument.

The initial charges can lead to very high local fields, well above 10 kV/cm. It is almost impossible to work with a mylar foil taken from the shelf. The charges can be eliminated by heating – or better, by γ-ray irradiations to levels of the order of a 10^6 rad with good conductive contact on the surface to permit the evacuation of the charges. We chose the first method, and after simple chemical degreasing, the foils were heated at 180° for 30 min. This brought the noise down to levels giving rise to signals below 5 mV in our case.

The electronic noise of the electrometer is claimed by the manufacturer to be 3 mV. This would correspond to a charge of about 2000 electrons on our foil, on the surface of 0.6×0.6 mm^2 viewed by the probe. A minimum ionizing particle in 500 atm of argon for instance, would liberate nearly this charge over a length of 0.6 mm; we are thus at the threshold for detection without any need for amplification.

We have, however, several other ways of improving the signal-to-noise ratio.

(i) A straightforward way is to reduce the capacitance of the system. By stretching a thin 10 μm foil at a distance of 250 μm from the collecting electrode, we gain a factor of close to 6 with respect to our present design.

(ii) The lateral diffusion is strongly reduced by the high pressure and by the possible adjunction of a magnetic field parallel to the electric field. The width of the electron cloud can then be brought down to levels well below 100 μm. We therefore have to reduce the width of the read-out probe accordingly, which seems possible.

(iii) Since the read-out is non-destructive we can repeat a measurement any number of times and find the centroid to an accuracy of the order of one per cen' of the width and the charges remain for a long time. We observed a loss of about 25% in 48 h proba-

bly due to air conduction produced by the radiation background and to charge compensation and transport on the foil itself. If the charge image is to be kept for a long period, better foils than mylar can be used (polypropylene, Teflon) which can also be stored in vacuum.

One of the main advantages of the method is, that in contrast to ordinary gaseous detectors where the electrons are absorbed by the electrodes after a measurement, here the electrons are kept in a memory and permit a quite different approach to the read-out. In this respect we have the same situation as in photographic emulsion, but with the essential difference that the trajectory is produced in gas and the ionization electrons can be monitored and gated with electric pulsed fields.

(iv) Finally, even with the present tools, we can hope to detect minimum ionizing particles in high-pressure gases or in liquids such as liquid argon or tetramethyl–tin.

6. Applications

The above discussion shows that our initial goal – to detect charged particles in high pressure gases or liquids, without amplification – is compatible with the physical limits of the method. The reduction in diffusion should lead to good particle resolution, and the possibility of a static centroid measurement should permit good position accuracy.

For the detection of minimum ionizing particles in gases at atmospheric pressures, it is possible to use a double-step parallel-plate amplifying gap, with an intermediate gating. Our work at CERN [5] has shown that gains of 10^6 can be achieved even at high rates. A single avalanche of 1 mm width with 10^6 electrons, collected on a 10 μm mylar foil subtended at a distance of 250 μm from our electrode, would rise, with respect to this electrode, at a potential of about 1.5 V. In other words, it is measurable even with the present conditions. We are investigating this possibility since it is of a great interest, both for the high-accuracy measurement of ionizing particles and for the imaging of VUV photons in photoionization detectors. Despite the width of the tracks introduced by the double-step amplification, we think that the peculiarity of the stored distribution should permit a high accuracy for the centroids determination.

Needless to say, for heavy ions the situation is even more favourable. The fact that the energy loss is

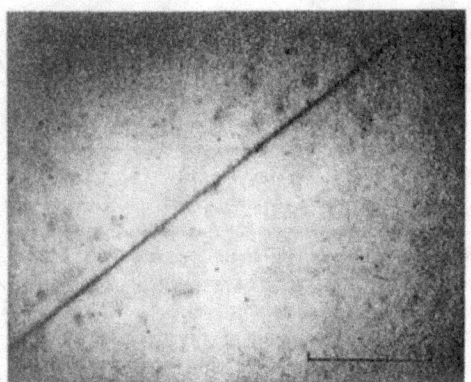

Fig. 4. Track image produced by accumulated charges from the collimated α source. An amplitude of 15 V (2×10^7 electrons/mm^2) is visible with liquid developer, Kodak MX 1112. (The scale-line is 1 mm long.)

measurable along the track is an important advantage for particle identification.

There are a few more situations where this method may be attractive. For example, interest has recently developed in avalanche streamer chambers at atmospheric or at high pressures, with low gains of about 10^5, and with the help of light amplifiers. The collection and detection of the electrons or the ions in such chambers would be an easy task. The level of ionization should then be sufficient for the use of powder or liquid developers. Also, the choice of gases may be

broadened since the emission of visible light is not required. It is indeed intriguing to compare the quality of the optical and the electrical read-out. We are undertaking to investigate this subject, and as a first step we have tried to estimate the threshold for obtaining a visible line with Kodak liquid toner MX 1112 for a linear distribution of charges on mylar. Fig. 4 shows the picture obtained with accumulated charges from an α source collimated to 0.3 mm, giving rise to a peak voltage of 15 V. We see a narrow fine line of only 0.04 mm width, with the liquid toner. At the voltages below 10 V we do not observe a line. There is probably room for much progress in this direction also, and a considerable variety of other methods of visualizing small charge distribution have already been tested or envisaged [1, 2].

We are indebted to J.-C. Santiard and R. Benoit for their help in the technical set-up of these experiments.

References

[1] J.W. Boag, Phil. Trans. R. Soc. London A292 (1979) 273.
[2] J.H. Dessauer and H.E. Clark, Xerography and related progress (Focal Press, London, 1965).
[3] K.H. Reiss, Z. Angew. Math. and Phys. 19 (1965) 1–4.
[4] G. Charpak and F. Sauli, Phys. Lett. 78B (1978) 523.
[5] A. Breskin, C. Charpak, S. Majewski, G. Melchart, A. Peisert and F. Sauli, Nucl. Instr. and Meth. 178 (1980) 11.
[6] A. Fenster and H.E. Johns, Med. Phys. 1 (1979) 262.

Nuclear Instruments and Methods 196 (1982) 555–557
North-Holland Publishing Company

Letter to the Editor

ELECTROSTATIC IMAGING OF MINIMUM IONIZING PARTICLES

G. CHARPAK, H. CZYRKOWSKI, A. FARILLA and S. MAJEWSKI *

CERN, Geneva, Switzerland

Received 23 October 1981

Charge amplification in a multistep chamber permits the electrostatic imaging of minimum ionizing particles trajectories in various gases at atmospheric pressure. In helium the single avalanches from primary electrons are easily detected.

A recent work at CERN [1] has demonstrated the feasibility of charged-particle trajectory imaging in gases by measuring the electric charges stored on insulating surfaces. This was illustrated with α-particles and with an amplification of the primary ionization by a factor of close to 10^3 with the help of a single-step parallel grid gap.

In view of the great interest attached to the accurate localization of minimum ionizing particles or single electrons liberated, for instance, by VUV radiation, we have verified the same concept of detection with a two-step gated avalanche chamber [2].

---

* Present address: Physikalisches Institut der Universität, Heidelberg, Fed. Rep. Germany.

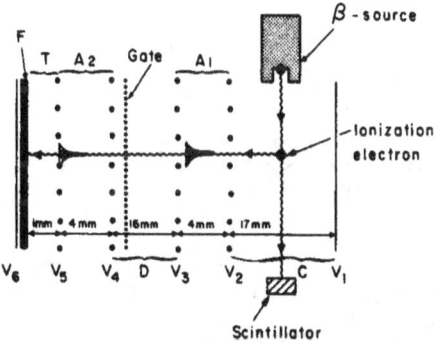

Fig. 1. Experimental set-up. The fast β-particles liberate electrons in the gas of space C. The electrons are multiplied in A_1, drift in D, and are transmitted through the control grid G to A_2, where they are again multiplied, transferred to T, and collected by F. The voltages V_1 to V_6 for the helium-methane-triethylamine gas mixture are -5400, -4400, -1800, 0, $+2600$, $+3100$ V, respectively; G is gated by pulses of ± 150 V on alternating wires. The mean voltage of G is -700 V.

Minimum ionizing electrons from a ^{106}Ru source are detected by a scintillation counter after having traversed a conversion gap C, and triggered the gating grid G

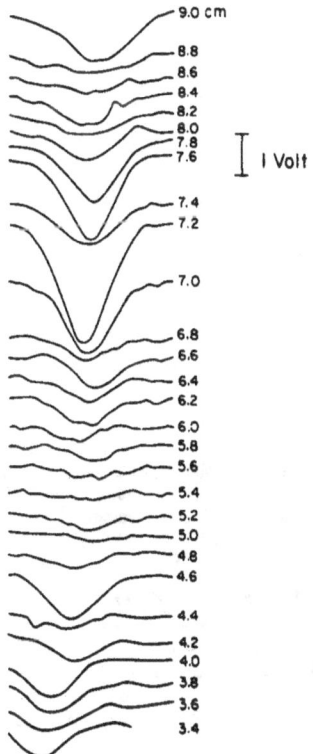

Fig. 2. Electron track in argon. Argon (97.5%)–propane (2.5%) mixture. Transverse scanning of an electron track. The width (fwhm) is 2.5 mm larger than the resolution of the scanning device ($\sim$0.6 mm).

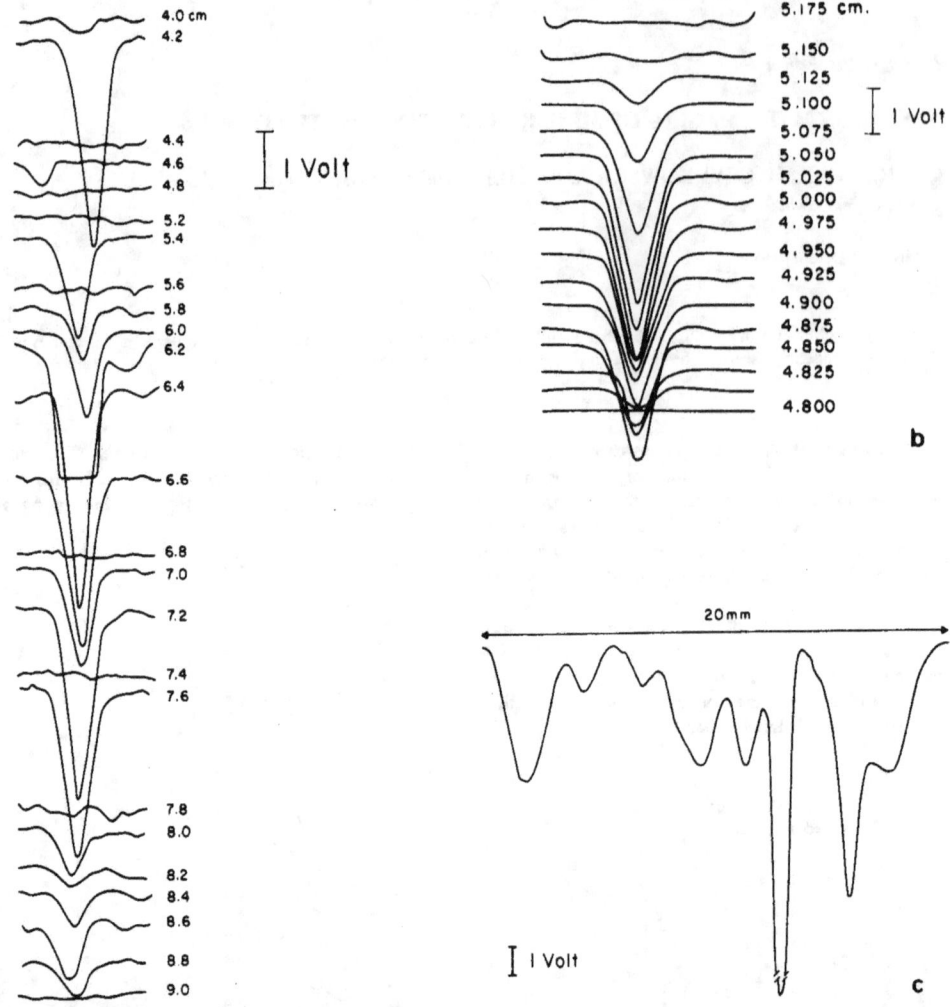

Fig. 3. Electron track in helium. Helium (93%)–methane (4%)–triethylamine (3%) mixture. (a) Transverse scanning in steps of 2 mm. Many holes are observed. (b) Transverse scanning, in steps of 0.25 mm, of an isolated avalanche. (c) Longitudinal scanning of an electron track. The longitudinal track over a length of 2 cm shows a number of separated avalanches compatible with the scanning resolution and the primary ionization.

separating two amplifying gaps A_1 and A_2 (fig. 1), with delays adjusted to the drift times in C, in D, and in A_1.

The electrons are collected on a 125 μm mylar foil F backed by a conductive layer of Sn_2O_3. The experiments were performed in two gas mixtures: argon (97.5%) + propane (2.5%), and helium (93%) + methane (4%) + triethylamine (3%).

The gain permitted by the two-step amplification is sufficient for the convenient detection of minimum ionizing particles, and even single electrons as can be seen from our results.

1) *Trajectories in argon–propane.* When we trigger the gating grid after detection of one electron in the scintillation counter, we observe (fig. 2) a rather wide track of 2.5 mm fwhm along the 5 cm of the sensitive length of the foil. The width is much larger than the resolution of our scanning device, which is the same as in ref. 1: an electrostatic probe with a 0.6 mm hole, set at about 0.5 mm from the foil.

2) *Trajectories in helium–methane–triethylamine.* This gas presents the greatest interest; the primary ionization being only 6 ions/cm in helium, the distance between

avalanches should make it possible to test the capability of detecting single avalanches. It is also used, in practice, for the imaging of VUV Cherenkov light.

This gas is favourable for high multiplications before reaching breakdown (which is a very great nuisance in our set-up, since every spark deposits a very broad splash of electrons on the mylar). Fig. 3a shows the transverse scanning of a typical electron track over the total length of 5 cm of the sensitive region of the chamber. We observe gaps due to the low primary ionization power.

Fig. 3b shows a transverse scanning of one avalanche in steps of 0.25 mm. We observe a fwhm of about 1 mm in both the lateral and the longitudinal dimension, which can be considered as the size of the image of a single avalanche in our set-up. The signal-to-noise ratio is still favourable enough to make the detection of single electrons easy, even if several electrons of a δ-ray contributed to the avalanche. Fig. 3c shows the charge distribution along an electron track, clearly displaying the primary ionization distribution. The number of avalanches along a track is compatible with the expected primary ionization of 6 ions/cm, taking into account the limited resolution of our scanning set-up.

In conclusion, this further work confirms that at the present stage of electrostatic imaging, the detection of relativistic ionizing particles, in helium mixtures at atmospheric pressures, can be made feasible by combining it with multistep amplification. The considerable accuracy of position measurement in a stored image could make this detector of interest for some specific applications.

However, the greatest prospects for this detector are still in the detection of particle trajectories in dense media, without any amplification. This requires some further new steps in the charge detection methods and is being actively investigated.

References

[1] G. Charpak, R. Bouclier, A. Breskin, R. Chechik and J. Lewiner, preprint CERN-EP/81-89 (1981) Nucl. Instr. and Meth. 192 (1982) 235.
[2] A. Breskin, G. Charpak, S. Majewski, G. Melchart, A. Peisert and F. Sauli, Nucl. Instr. and Meth. 178 (1980) 11.

Reprinted from *J. Phys. III France* 3 (1993) 2299–2304
© Les Editions de Physique

DECEMBER **1993**, PAGE 2299

Classification
Physics Abstracts
07.50 — 29.40

Détection de particules α à l'aide de charges électriques de surface

G. Charpak (^{1}), G. Cordurié (^{1}), J. Lewiner (^{1}), D. Morisseau (^{1}), M. Chabot (^{2}) et
J.-C. Santiard (^{2})

(^{1}) Laboratoire d'Electricité Générale, E.S.P.C.I., 10 rue Vauquelin, 75231 Paris Cedex 5,
France
(^{2}) CERN, Genève, Suisse.

(*Reçu le 7 juin 1993, accepté le 27 septembre 1993*)

Résumé. — La trajectoire d'une ou quelques particules α peut être visualisée en collectant sur un diélectrique les charges libérées dans un gaz le long de son trajet. La quantité de charges obtenue, sans multiplication, est faible et difficile à extraire du bruit de fond. Ce bruit de fond est dû, en particulier, aux charges ou à la polarisation résiduelles à la surface du diélectrique. Nous avons abaissé, de façon sensible, ce niveau de charges résiduelles en les compensant avec les charges créées par une source de rayons α. La lecture des charges déposées par une seule particule est alors possible avec un dispositif de mesure électrométrique à tête tournante.

Abstract. — The trajectory of one or a few α particles can be visualized by collecting on a dielectric the charges released in the gas by ionisation. The charge quantity obtained, without multiplication, is small and difficult to extract from the background noise. This background noise is, in particular, due to the residual charges or polarization near the surface of the dielectric. This residual surface charges have been significantly reduced by balancing them with charges created by a beam of α particles. Then, the reading of charges due to one particle can be performed with an electrometric measuring setup using a rotating head.

1. Introduction.

Des travaux précédents [1, 2] avaient montré qu'on pouvait détecter des particules α individuelles ou des particules au minimum d'ionisation, en collectant sur un film diélectrique les ions libérés dans un gaz le long de la trajectoire, ceci après une multiplication considérable des charges initiales au moyen d'avalanches de Townsend entre grilles parallèles. En l'absence de multiplication, les charges collectées sont faibles et les méthodes de lecture des charges réparties sur la surface isolante ne permettent pas d'extraire le signal utile du bruit de fond. Ce bruit de fond est essentiellement dû à l'inhomogénéité des charges résiduelles à la surface du diélectrique avant le dépôt des charges créées par la particule à détecter.

Il est possible de mesurer sur une feuille de diélectrique vierge, les charges de surface à l'aide d'un électromètre. On peut trouver en certains points de la surface des charges produisant des potentiels de quelques milliers de volts. Un nettoyage de la surface à l'acétone (qui enlève les traces de graisse), puis avec de l'alcool éthylique, réduit considérablement les charges résiduelles à un niveau tel que le potentiel équivalent est de l'ordre de quelques volts. Mais ce niveau de charges est encore trop élevé pour permettre de mesurer les charges déposées le long de la trajectoire d'une particule.

Nous avons mis au point une méthode qui permet d'abaisser le niveau des charges résiduelles sur la surface du diélectrique, ce qui rend possible la détection de la trace d'une seule particule α.

Une méthode de révélation par poudre n'étant pas assez sensible pour mettre en évidence la trace d'une particule, les charges sont lues à l'aide d'une sonde capacitive déplacée près de la surface du diélectrique [3].

2. Méthode.

2.1 DÉTECTION DE PARTICULES α. — Une particule α a une trajectoire de quelques centimètres dans l'air, au cours de laquelle elle ionise l'air sur son passage. En appliquant un champ électrique suffisant, les ions positifs et négatifs sont séparés. Il est alors possible de recueillir sur un film diélectrique des charges, positives ou négatives selon la polarité du champ appliqué, tout le long de la trajectoire de la particule. La trajectoire de la particule doit être parallèle à la surface du diélectrique, à quelques millimètres de distance.

Une particule α perd approximativement une énergie de 1 MeV par centimètre parcouru, elle produit ainsi une charge de l'ordre de 5×10^{-15} C/cm parcouru. En tenant compte de la diffusion, la largeur de la trace, sur la surface du diélectrique, est de l'ordre de 100 μm, ce qui donne une densité superficielle de charge de 5×10^{-13} C/cm^2. Le dispositif de lecture de charges doit donc être capable de mesurer des densités de charge de cet ordre de grandeur sur une largeur de 100 μm et une longueur de quelques centimètres.

2.2 MÉTHODE DE MESURE DES CHARGES SUPERFICIELLES. — Nous avons utilisé une méthode de mesure des charges électriques superficielles dérivée des mesures électrométriques. Au lieu de mesurer les variations de potentiel de l'électrode placée près de la surface chargée, nous mesurons le courant de charge et de décharge du condensateur formé par l'électrode et le diélectrique et son support [3, 4]. Lorsque la sonde est déplacée à vitesse constante près de la surface chargée, le courant mesuré représente la dérivée spatiale, dans la direction du balayage, de la distribution de charges.

2.3 DIMINUTION DE LA CHARGE RÉSIDUELLE DE SURFACE. — Après nettoyage de la surface du diélectrique à l'acétone et à l'alcool, des charges électriques sont encore présentes. Pour éliminer ces charges et les ramener à un niveau faible devant les charges à collecter, nous avons irradié l'air au voisinage de la surface, en champ électrique nul, avec les particules α émises par une source intense de ^{241}Am ($3,7 \times 10^7$ Bq). Les ions libérés dans l'air sont attirés par les ions, de charge opposée, piégés à la surface du diélectrique et les neutralisent. Même les charges piégées plus profondément dans le diélectrique attirent les ions de signe opposé jusqu'à ce que le champ électrique, au voisinage de la surface, s'annule. En quelques minutes, les charges disparaissent et le signal issu du système de lecture des charges ne montre plus que le bruit de fond propre du dispositif. La figure 1 montre l'enregistrement de ce bruit de fond. La variation lente du signal est uniquement due à la transmission optique du signal de la sonde vers l'amplificateur, au cours de la rotation. Ce signal, reproductible, peut être éliminé. Le bruit de fond mesuré, exprimé en tension équivalente à la surface du diélectrique, est de 3,5 mV crête. Ce niveau de bruit doit permettre de mesurer les traces de particules.

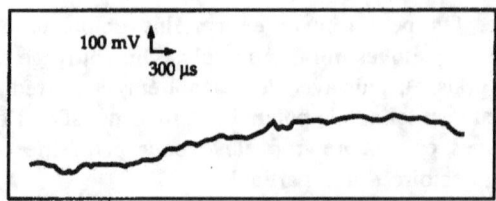

Fig. 1. — Bruit de fond du système de mesure après irradiation du diélectrique avec une source de rayons α. On observe une variation lente due au dispositif tournant et des petites fluctuations qui correspondent à une tension de surface maximum de 2 mV.

[Noise level of the measuring system after the dielectric has been irradiated with a α source. The slow variation of the signal is related to the rotating head ; fast fluctuations correspond to a 2 mV surface equivalent voltage.]

3. Dispositif expérimental.

3.1 CAPTEUR. — Le diélectrique utilisé est du téréphtalate de polyéthylène de 120 μm d'épaisseur dont une des faces est couverte d'une couche d'oxyde d'étain conductrice. La feuille est collée sur une plaque métallique mise à la masse qui sert de référence de tension. La surface du diélectrique collecte les charges produites par des particules ionisantes dans le volume de gaz compris entre la feuille et une grille métallique placée à une distance de quelques millimètres (Fig. 2). La grille est portée à un potentiel V par rapport à la masse pour recueillir sur le diélectrique les ions d'une seule polarité.

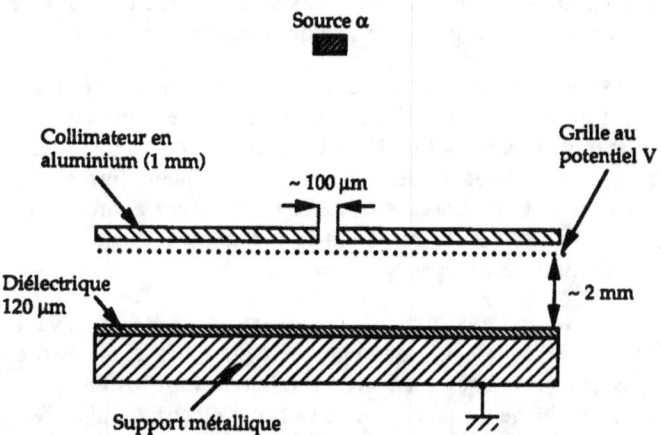

Fig. 2. — Dispositif expérimental utilisé pour déposer des charges de polarité et d'amplitude connues pour effectuer la calibration du système de mesure à tête tournante.

[Experimental setup to put on the dielectric surface a known quantity of charges.]

3.2 DISPOSITIF DE LECTURE DES CHARGES. — Les charges électriques sont lues à l'aide d'une sonde à déplacement circulaire. Dans ce dispositif, la sonde est formée d'une électrode de 100 μm de largeur et 10 mm de longueur, maintenue et entraînée par un disque conducteur, dont elle est isolée électriquement, qui sert d'anneau de garde. L'électrode est positionnée de telle façon que sa plus grande dimension est suivant un rayon du disque ; le milieu de l'électrode est à 50 mm de l'axe de rotation de l'ensemble. Au cours de son déplacement la sonde balaie une bande de 10 mm de large sur une largeur de 100 μm. La sonde se déplace à une distance voisine de 300 μm de la feuille chargée. Les changements dans le flux des lignes de champ électrique coupées par l'électrode induisent des variations de tension à la sortie de la sonde dont l'amplitude est proportionnelle au flux.

Afin d'éliminer le bruit électrique engendré par un moteur électrique, le mouvement de rotation de la sonde est produit par une turbine à air comprimé. La tête de mesure peut tourner ainsi à une vitesse angulaire voisine de 100 tours par seconde, ce qui correspond à une vitesse linéaire de l'ordre de 30 m/s.

Le signal issu du préamplificateur, qui est localisé le plus près possible de l'électrode et est en rotation avec elle, est transmis à la chaîne de mesure par couplage optique afin d'éviter le bruit que pourrait générer des contacts tournants.

Nous exploitons la nature périodique des mesures, liée à la rotation de la sonde, pour réduire le bruit du signal. Pour cela, nous utilisons une méthode d'accumulation des signaux sur plusieurs périodes de rotation de la sonde. Il en résulte une amélioration du rapport signal sur bruit d'un facteur 5 à 10.

3.3 CALIBRATION. — Préalablement aux recherches de traces de particules, la chaîne de mesure a été calibrée. Après traitement de la surface du diélectrique comme indiqué précédemment, des charges électriques sont déposées en collectant les ions produits dans l'air ionisé par les rayons α émis par une source de ^{241}Am (intensité $\approx 3,7 \times 10^7$ Bq). Si la grille (Fig. 2) est polarisée à une tension V, et si on irradie le gaz avec les rayons α collimatés par une fente de 100 μm de large, les ions sont collectés sur le diélectrique jusqu'à ce que le potentiel de la surface devienne égal au potentiel de la grille. Nous avons vérifié que ceci était vrai, à l'aide d'un électromètre, à une précision de 20 %, pour une tension de grille de 1 V, à condition d'effectuer la collecte des ions pendant une durée suffisante.

La figure 3 montre le signal obtenu pour une polarisation de la grille de − 1 V. Sur la figure 4 est présentée la courbe de calibration obtenue, c'est-à-dire l'amplitude de la tension mesurée en

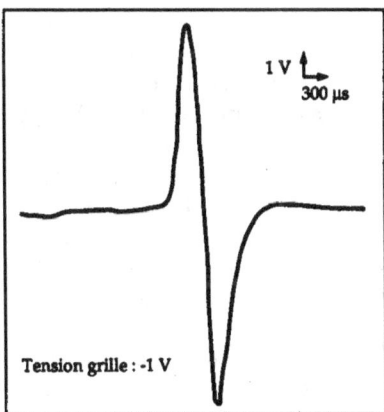

Fig. 3. — Signal obtenu pour des charges déposées sur le diélectrique à l'aide du dispositif de la figure 2, pour une tension grille de − 1 V.

[Signal obtained with the setup of figure 2, with a grid bias of − 1 V.]

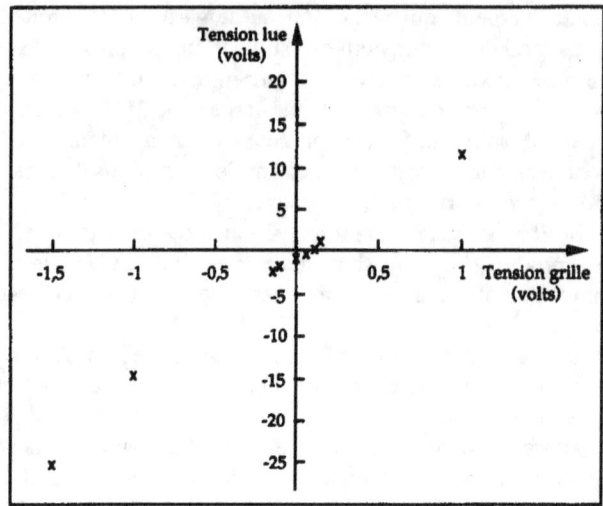

Fig. 4. — Courbe de calibration de l'appareil de lecture à tête tournante : amplitude de la tension mesurée en fonction de la tension de polarisation de la grille.

[Calibration of the measuring rotating system : measured voltage *versus* grid bias.]

fonction de la tension appliquée sur la grille. On note, pour des tensions voisines de zéro, un décalage de la courbe qui ne passe pas par l'origine. Il faut remarquer qu'en champ électrique très faible, il est possible qu'on ne puisse pas séparer efficacement les ions de signes opposés et qu'une diffusion importante étale l'image. Le signal est alors perturbé parce qu'il dépend aussi de la forme du bord de la distribution de charges. On ne considère comme valable pour la calibration que les mesures faites avec des tensions équivalentes de l'ordre de 1 V. La calibration montre qu'une charge linéaire d'une largeur de quelque centaines de micromètres se traduit par un signal de 15 V à la sortie du système de lecture.

4. Résultats expérimentaux.

Les particules α sont fournies par une source de ^{241}Am dont l'activité est de $3,7 \times 10^7$ Bq. Le montage présenté sur la figure 5 a été utilisé pour obtenir un flux de particules de 1 à 2 particules par seconde. Le collimateur est réalisé dans un bloc de laiton de 10 mm qui absorbe le rayonnement γ (60 keV) émis par la source. Les rayons α sont collimatés par une fente de 200 μm par 100 μm et sont absorbés dans un espace de dérive de 5 mm de hauteur, situé entre une grille portée à – 1 000 V et la surface du diélectrique dont le support est à la masse. Nous avons vérifié que l'intensité de la source collimatée est comprise entre 1 et 2 particules par seconde. Cette intensité permet aisément des irradiations pour lesquelles on obtiendra zéro, une ou deux particules.

Sur la figure 6 est représenté le signal obtenu avec la sonde tournante lorsqu'on détecte une trace de particule. L'amplitude du signal est de 500 mV crête à crête, ce qui correspond à une tension équivalente sur le diélectrique de 33 mV. C'est l'ordre de grandeur attendu, sachant que la capacité du diélectrique est voisine de 20 pF/cm², en supposant que la largeur de la trace est voisine de 100 μm.

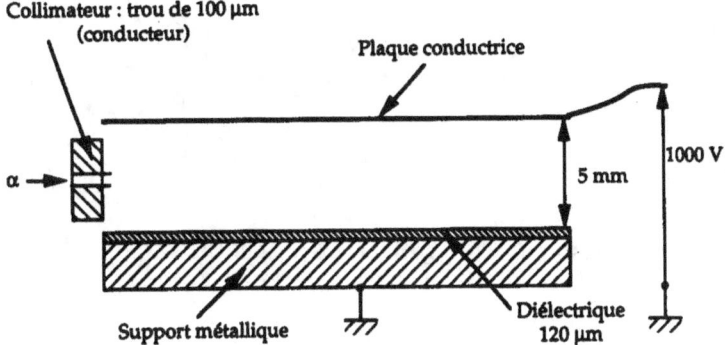

Fig. 5. — Dispositif expérimental permettant de déposer sur le diélectrique la charge produite par des rayons α individuels. Le collimateur laisse passer une ou deux particules par seconde.

[Experimental apparatus to collect on dielectric surface the charges produced along the trajectory of few α particles.]

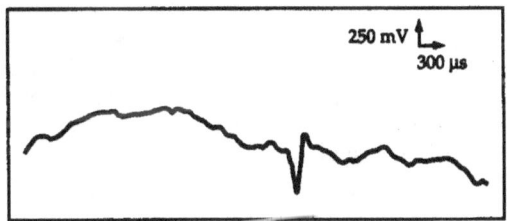

Fig. 6. — Signal enregistré pour une particule α isolée. Les charges sont collectées pendant un temps très court pour avoir zéro ou un rayon α.

[Signal obtained for one α particle. Charges are collected during a very short time to get only zero or one particle.]

5. Conclusion.

En irradiant, avec une source de particules α, en champ nul, l'air au voisinage de la surface d'un diélectrique, il est possible de neutraliser les charges résiduelles piégées dans le diélectrique ou à sa surface. Le niveau de charges résiduelles ainsi obtenu est suffisamment bas pour permettre la mesure des charges déposées lors du passage d'une seule particule.

La méthode de mesure des charges que nous avons mis en œuvre permet d'autres applications intéressantes des distributions superficielles de charges, aussi bien en physique des particules qu'en imagerie des rayons X et des rayons β.

Bibliographie

[1] Charpak G., Bouclier R., Breskin A. et Lewiner J., *Nucl. Instrum. Methods* **192** (1982) 235.
[2] Charpak G., Czyrkowski H., Farilla A. et Majewski S., *Nucl. Instrum. Methods* **196** (1982) 555.
[3] Charpak G., Cordurié G., Lewiner J. et Morisseau D., *J. Phys. III France* **3** (1993) 2149.
[4] Lewiner J. et Charpak G., US Patent n° 4567530 (1986).

Reprinted from *J. Phys. III France* 3 (1993) 2149–2161   NOVEMBER **1993**, PAGE 2149

Classification
Physics Abstracts
07.50

Méthode de mesure de distributions de charges électriques de surface

G. Charpak, G. Cordurié, J. Lewiner et D. Morisseau

Laboratoire d'Electricité Générale, E.S.P.C.I., 75231 Paris, France

(*Reçu le 25 septembre 1992, révisé le 24 mai 1993, accepté le 4 août 1993*)

Résumé. — Une méthode de mesure des distributions de charges électriques superficielles sur un diélectrique ou un photoconducteur est décrite. Les dispositifs électroniques de mesure et de balayage permettent une lecture suffisamment rapide (2 000 points par seconde) pour réaliser une image de distribution de charges. Des expériences ont été menées dans différents domaines : imagerie X ou γ, détection de particules α, suivi de profils de charges sur des supports associés à des semiconducteurs.

Abstract. — A method of measuring the charge distribution on a dielectric or semiconductor surface is described. The electronic measuring system and the sweeping apparatus are fast enough (5 lines of 400 points by second) to enable charge imaging. Experiments have been performed in various fields : X ray and γ ray imaging, detection of α particles, evolution of charge distribution on semiconductor related substrates.

1. Introduction.

La mesure de la distribution de charges électriques à la surface d'un diélectrique peut présenter un intérêt dans des domaines divers : imagerie médicale, contrôle industriel, détection de particules, contrôle de décharges électriques.

Dans le domaine de l'imagerie médicale, le film photographique est encore le support d'image le plus répandu en radiographie, mais, pour certains examens particuliers comme la mammographie, la technique photographique peut être remplacée par un procédé xérographique. Dans cette méthode, une image radiographique latente est obtenue sous la forme d'une distribution de charges électriques à la surface d'un matériau photoconducteur, distribution représentative de l'intensité du rayonnement X qui a été plus ou moins absorbé en traversant le milieu étudié. Les photons X cèdent tout ou partie de leur énergie dans le matériau en créant des paires électron-trou. Ces charges contribuent à l'augmentation de la conductivité électrique du photoconducteur ce qui permet de compenser les charges électriques préalablement déposées à la surface.

Dans le domaine industriel, on rencontre une situation analogue en contrôle non destructif. En effet, la radiographie est utilisée pour détecter la présence de défauts dans des pièces mécaniques, des soudures, etc. Le spectre des énergies des rayonnements utilisés (des rayons X aux rayons γ de quelques MeV) est beaucoup plus étendu que dans le domaine médical, et, par conséquent, les interactions rayonnement-matière peuvent être différentes.

Comme nous le verrons, la visualisation de distributions de charges à la surface d'un diélectrique permet aussi de mettre en évidence la trajectoire de particules élémentaires. Une particule ionise sur sa trajectoire le gaz ou le liquide qu'elle traverse. Ces charges sont séparées par un champ électrique et, ainsi, des charges positives ou négatives sont déposées sur un diélectrique le long de la trajectoire de la particule. La distribution de charges sur le diélectrique donne des informations sur le passage des particules et leurs trajectoires.

Dans le domaine des décharges électrostatiques, la mesure des charges déposées sur un isolant est une méthode directe de visualisation. La récente apparition des cartes « à puces » a soulevé un problème déjà rencontré au cours de la fabrication des circuits intégrés : l'accumulation de charges électriques sur le support isolant du composant peut être suffisante pour provoquer une décharge électrique qui peut détruire le circuit. Il est possible d'avoir des renseignements sur la façon dont les charges se répartissent sur la carte, ainsi que sur leur amplitude, en visualisant la distribution de charges en fonction de certains paramètres caractéristiques du matériau.

Les deux méthodes les plus utilisées pour visualiser une distribution superficielle de charges sont la révélation par une poudre chargée [1] et la mesure électrométrique [2, 3].

La révélation par poudre consiste à pulvériser sur la surface chargée une poudre dont les particules très fines, sensibles au champ électrique ou à son gradient, se déposent sur les charges de signe opposé. La concentration de particules de poudre est d'autant plus importante que la densité de charge est élevée. Généralement, la poudre ainsi distribuée est transférée sur un papier où elle est fixée par chauffage. On obtient de cette façon une image représentative de la distribution de charges initiale dont la résolution spatiale est bonne ; elle n'est pas limitée par la taille des grains de la poudre mais par les effets de déformation du champ électrique au voisinage des variations rapides de charge. Cette méthode, très utilisée en xérographie et xéroradiographie, donne une image de bonne qualité mais ne donnant que des informations qualitatives. On ne peut, en effet, par cette méthode, obtenir une mesure de la quantité de charge. Il faut remarquer, en outre, que cette méthode détruit la distribution de charges qu'on visualise, on ne peut donc pas l'utiliser pour étudier l'évolution au cours du temps des charges déposées sur un diélectrique.

L'autre méthode de mesure des charges distribuées sur une surface consiste en une mesure électrométrique des charges induites sur une électrode placée près de la surface. Cette mesure est très sensible à la distance entre l'électrode et la surface. Pour s'affranchir de cette difficulté, des systèmes de mesure à électrode vibrante [4] ou périodiquement occultée [5] permettent d'ajuster le potentiel de l'électrode pour que le champ électrique entre l'électrode et la surface soit toujours nul. Cette méthode, si elle rend l'amplitude mesurée indépendante de la distance, ne permet pas d'approcher l'électrode de la sonde tout près de la surface. De plus, la constante de temps imposée par l'asservissement est assez longue (0,1-1 s), ce qui rend la méthode peu adaptée à l'imagerie qui demande une bonne résolution et des mesures rapides.

2. Description de la méthode de mesure.

2.1 PRINCIPE. — La méthode, présentée ici, de mesure des charges électriques réparties sur la surface d'un diélectrique est dérivée des mesures électrométriques. Au lieu de mesurer les variations du potentiel de l'électrode placée près de la surface, sont mesurées les variations de la charge induite sur l'électrode lorsque celle-ci se déplace, c'est-à-dire le courant de charge ou de décharge du condensateur formé par l'électrode, d'une part, et le diélectrique et son support, d'autre part [6]. On peut montrer que, lorsque l'électrode de mesure est déplacée à vitesse constante le long de la surface chargée, le courant mesuré est presque proportionnel à la dérivée spatiale de la charge déposée. Cette méthode permet donc un balayage rapide (durée de la mesure d'un point < 200 µs) de la surface à mesurer, alors que les méthodes électrométriques classiques ont des constantes de temps assez longues (0,1-1 s).

2.1.1 *Description de la sonde de mesure.* — La sonde est composée d'une électrode formée par un fil (ou d'une lame) de cuivre isolé qui traverse une plaque conductrice qui sert d'anneau de garde (Fig. 1). La surface sensible de la sonde est la section droite de l'électrode. L'électrode est maintenue à une distance a de la surface chargée à mesurer. Le diélectrique ou le photoconducteur est déposé sur un support métallique, en laiton ou en aluminium, qui sert, à la fois, à maintenir mécaniquement le diélectrique, et d'électrode de référence pour fixer le potentiel du dispositif. L'électrode de garde, qui contribue aussi à la rigidité de la sonde, est suffisamment large pour qu'on puisse supposer, en première approximation, que les champs entre l'électrode et la surface sont perpendiculaires à celle-ci.

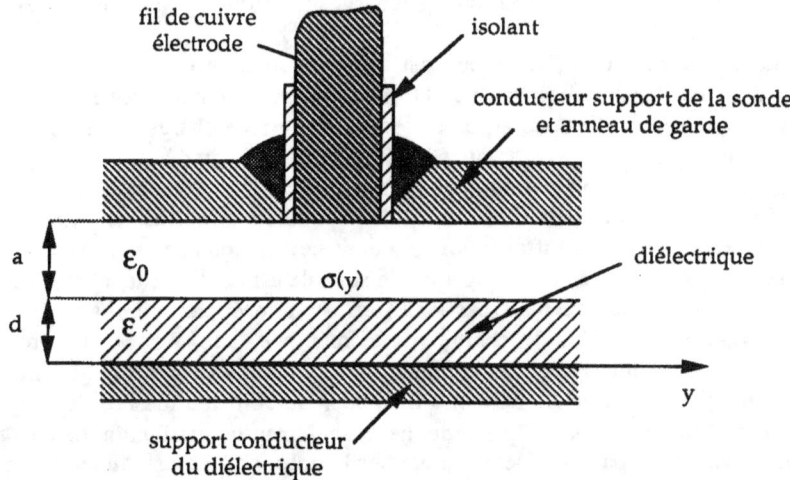

Fig. 1. — Description de l'électrode de mesure et du diélectrique chargé.

[Description of the measuring electrode and of the charged dielectrics.]

L'électrode de la sonde est reliée à un préamplificateur de type transimpédance (voir Fig. 2) dont la tension de sortie est proportionnelle au courant d'entrée, c'est-à-dire au courant qui est induit par la variation de la charge de l'électrode [7].

Si le courant venant de l'électrode, à l'entrée du circuit, est i_{in}, la tension à la sortie de l'amplificateur est :

$$V_s = - Ri_{in} \left(\frac{R_2}{R_1} + 1 \right) .$$

Les résistances R, R_1 et R_2, définies sur la figure 2, sont telles que R_1 et R_2 sont très petites devant R.

On a donc à la sortie de ce qu'on appellera par la suite la sonde, c'est-à-dire l'ensemble mobile formé de l'électrode, de l'anneau de garde et du préamplificateur, un signal V_s représentant la variation de la charge induite sur l'électrode.

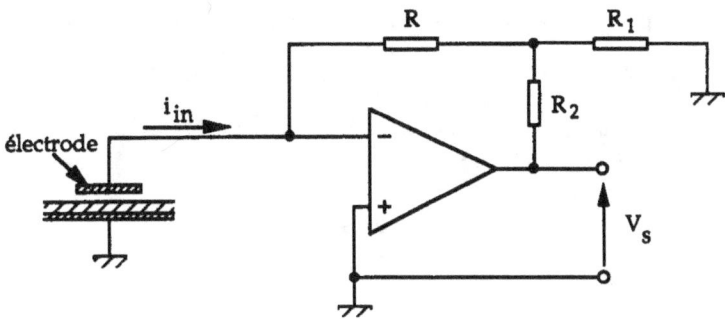

Fig. 2. — Sonde de mesure : ensemble de l'électrode et de l'amplificateur transimpédance.

[Measuring probe : electrode and transimpedance type amplifier.]

Pour lire une distribution de charges sur une feuille diélectrique, la sonde est déplacée à une vitesse élevée (1,4 m/s), constante dans la zone de mesure, à une distance constante de la surface. Le signal obtenu à la sortie de la sonde est amplifié et intégré (voir Fig. 3), puis converti numériquement pour être enregistré dans la mémoire d'un ordinateur. Il peut être stocké pour être visualisé ultérieurement ou immédiatement après un enregistrement.

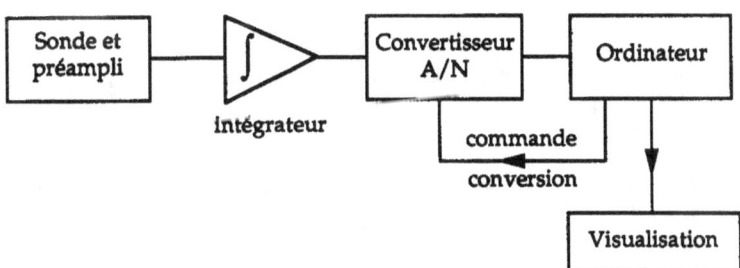

Fig. 3. — Schéma de la chaîne de mesure, d'acquisition et de visualisation des charges.

[Diagram of the measuring, data acquisition and charge visualisation setup.]

2.1.2 *Caractéristiques du système de mesure.* — La lecture d'une distribution de charges est faite ligne par ligne dans la direction Oy, avec un déplacement pas à pas dans la direction Oz. On suppose que, au cours du balayage suivant y, la variation de la densité de charge suivant z est petite et qu'on peut négliger son influence sur la mesure. A chaque balayage d'une ligne, pour une valeur de z, on mesure la variation de la charge induite sur l'électrode par la variation de densité superficielle de charge $\sigma(y)$.

Afin de faire un calcul simplifié de la tension V_s, à la sortie de la sonde, il est fait l'hypothèse, en première approximation, que les lignes de champ électrique entre la surface et l'électrode restent perpendiculaires à la surface malgré une variation de la densité de charges. Le courant de l'électrode de mesure arrive sur l'entrée inverseuse d'un amplificateur opérationnel. On peut donc considérer, en supposant l'amplificateur opérationnel parfait, que ce point est maintenu au potentiel zéro de référence. Dans ce cas on peut décrire l'ensemble de l'électrode et de l'anneau de garde par une seule électrode reliée à la masse (Fig. 2).

Les différences de potentiel V_{sf} (sonde-surface) et V_f (surface-masse) sont donc telles que :

$$V_{sf} + V_f = 0 \; .$$

Les champs électriques dans l'air $\mathbf{E}_a$ et dans le diélectrique $\mathbf{E}_d$ (Fig. 4) sont liés aux différences de potentiel V_{sf} et V_f par les relations :

$$\mathbf{E}_a = -\frac{V_{sf}}{a}\mathbf{i} \quad \text{et} \quad \mathbf{E}_d = -\frac{V_f}{d}\mathbf{i} \; .$$

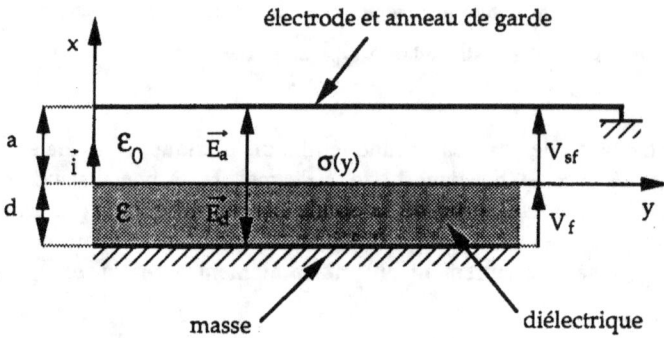

Fig. 4. — Schéma électrique équivalent de la sonde de mesure.

[Equivalent electrical circuit of the measuring probe.]

La relation de continuité à l'interface entre l'air et le diélectrique impose que :

$$\varepsilon_0\, \mathbf{E}_a \cdot \mathbf{i} - \varepsilon \mathbf{E}_d \cdot \mathbf{i} = \sigma \; .$$

On peut donc en déduire le champ électrique dans l'air, au voisinage de l'électrode de mesure :

$$\mathbf{E}_a \cdot \mathbf{i} = \frac{\sigma}{\varepsilon_0}\frac{d}{d + a\varepsilon_r} \; .$$

La densité de charge induite sur l'électrode, σ_s, se déduit de l'expression du champ. On obtient :

$$\sigma_s = -\sigma\frac{d}{d + a\varepsilon_r} \; .$$

On peut remarquer que la sonde forme avec la masse deux condensateurs en série, de surface S égale à celle de la sonde, le premier, de capacité C_{sf}, entre la sonde et la surface chargée et le second, de capacité C_f, formé par le diélectrique. La charge électrique Q_f commune aux deux condensateurs est la charge qu'on cherche à mesurer, c'est-à-dire celle qui est déposée sur la surface S du diélectrique :

$$Q_f = \iint_S \sigma(y)\,\mathrm{d}s \; .$$

Dans l'expression de σ_s, le rapport $d/(d + a\varepsilon_r)$ n'est autre que le rapport entre les capacités :

$$\eta = \frac{C_{sf}}{C_{sf} + C_f}.$$

La densité de charge induite sur l'électrode en chaque point y est reliée très simplement à la densité de charge superficielle :

$$\sigma_s(y) = - \eta \sigma(y).$$

La charge induite sur l'électrode de mesure, q_s, est donc l'intégrale de la densité superficielle de charges induite sur la surface de l'électrode :

$$q_s(y) = \iint_S \sigma_s(y) \, ds.$$

Le courant i_{in} mesuré (Fig. 2) est la variation de la charge induite sur l'électrode, c'est-à-dire :

$$i_{in} = - \frac{dq_s}{dt}.$$

La densité superficielle de charges $\sigma(y)$ ne dépend que de la variable spatiale y ; elle est indépendante du temps. En revanche, la densité superficielle de charges induite sur l'électrode, qui est le reflet de la densité superficielle de charges qui est en regard de celle-ci, dépend du temps indirectement parce que la sonde décrit « y » à une vitesse v. La dérivée par rapport au temps de $\sigma(y)$, qui n'a pas de sens physique puisque la distribution de charges est statique, peut s'exprimer en fonction de la dérivée par rapport à y, variable qui dépend du temps : $y = vt$. On peut, alors, écrire que :

$$i_{in} = \eta v \iint_S \frac{d\sigma(y)}{dy} \, ds.$$

Le signal V_s, à la sortie de la sonde, est donc proportionnel à la dérivée spatiale de la distribution de charges le long de la ligne de balayage de la sonde.

Le coefficient η dépend du rapport entre la distance sonde-surface, a, et l'épaisseur « effective » du diélectrique d/ε_r. η est inférieur à 1. On peut augmenter ce coefficient et l'approcher de 1 en diminuant la distance entre la sonde et la surface ; on augmente ainsi la sensibilité de la sonde. Les fluctuations mécaniques du système de balayage peuvent, alors, perturber les mesures de façon significative.

Sur ce principe, nous avons mis au point deux types de lecteurs de charges. Dans les deux cas, la sonde se déplace à vitesse constante, pendant la période de lecture, près de la surface à lire. L'un des systèmes utilise un déplacement linéaire de la sonde de lecture, l'autre un déplacement circulaire.

2.2 SONDE DE MESURE À DÉPLACEMENT LINÉAIRE. — Dans ce dispositif (Fig. 5), la sonde de lecture se déplace linéairement, à vitesse constante égale à 1,4 m/s, pendant la durée de la lecture. La sonde est, en effet, entraînée par un moteur et un système de poulies et de courroies qui transforment le mouvement de rotation du moteur en mouvement linéaire. La zone de balayage utilisée pour la lecture, dans ce dispositif, est celle pendant laquelle la vitesse de balayage est constante.

La lecture est faite ligne par ligne. Un autre moteur entraîne le déplacement du diélectrique et de son support sous la sonde après la lecture d'une ligne pour préparer la lecture de la ligne suivante.

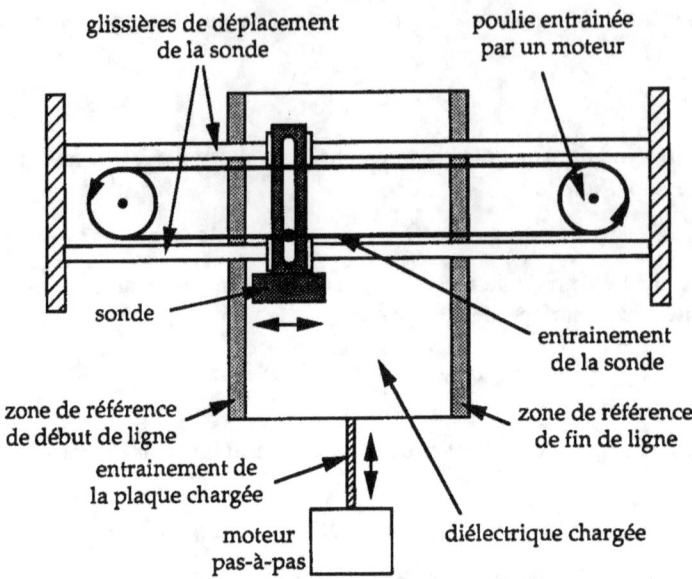

Fig. 5. — Dispositif mécanique d'entraînement rapide de la sonde et de déplacement de la plaque isolante au cours de la lecture.

[Mechanical setup for the bidirectional sweeping of the plate.]

L'intégrateur possède un système de contrôle de dérive en fonction du temps. Il est, malgré tout, préférable de faire une remise à zéro de l'intégrateur à chaque début de ligne avant de commencer la lecture. Au moment de cette remise à zéro, il est nécessaire que la sonde n'envoie aucun signal à l'intégrateur. Pour cela, une bande de quelques millimètres de large au bord du diélectrique est métallisée et portée au potentiel zéro. La remise à zéro de l'intégrateur est faite lorsque la sonde passe au-dessus de cette bande de référence. Cette référence est indispensable pour que le signal intégré le long d'une ligne représente bien la charge le long de la ligne et que les lignes soient comparables entre elles, ce qui est primordial pour former une image [7].

Le signal issu de la sonde est injecté à l'entrée de l'intégrateur après amplification de gain G. Compte tenu des relations précédentes, le signal à l'entrée de l'intégrateur est :

$$V_p = -RgG \iint_S \frac{d\sigma_s}{dt} ds = -RgG\eta v \iint_S \frac{d\sigma}{dy} ds$$

où $g = 1 + R_2/R_1$.

$R_i C_i$ étant la constante de temps de l'intégrateur réalisé avec un amplificateur opérationnel, le signal intégré est de la forme :

$$V = -\frac{1}{R_i C_i} \int V_p \, dt \; .$$

On obtient, après intégration :

$$V = \frac{RgG\eta}{R_i C_i} \iint_S \sigma(y) \, ds = \frac{RgG\eta}{R_i C_i} Q_f \; .$$

On remarque que la vitesse de déplacement de la sonde n'apparaît pas dans cette expression.

Ce dispositif intègre, non seulement le signal V_p, mais aussi les tension et courant d'offset et de polarisation propres à l'amplificateur opérationnel utilisé. Ceci se traduit par une dérive de la tension de sortie qui se superpose au signal. On peut minimiser ces effets en choisissant convenablement l'amplificateur mais ils ne sont jamais complètement annulés ; ils fluctuent, en particulier, avec la température. Ces signaux parasites sont d'autant plus gênants qu'on peut être amené à mesurer des plages de charges d'amplitude constante ; on ne peut donc s'en affranchir en filtrant les composantes basses fréquences du signal.

Pour remédier à cet inconvénient, le montage de la figure 6 a été mis au point. A chaque début de ligne l'intégrateur est remis à zéro par les interrupteurs analogiques I_1 et I_2 quand la sonde précède et lit la première moitié de la zone de référence. A ce dispositif indispensable au bon fonctionnement d'un intégrateur, a été ajouté un système d'évaluation de la dérive et d'asservissement tendant à ramener le signal de sortie de l'intégrateur vers zéro en dehors de la mesure des charges. L'interrupteur I_3 est fermé à un instant où le signal doit être nul (en dehors de la remise à zéro de l'intégrateur). Le signal, amplifié avec le gain K (négatif), commande un générateur de courant qui charge ou décharge la capacité C', suivant le sens de la dérive. Une fraction de la différence de potentiel présente aux bornes de la capacité est alors injectée sur l'entrée non inverseuse de l'intégrateur. Ce dispositif d'asservissement permet de corriger la dérive de l'intégrateur au cours du temps sans modifier le signal utile puisque la correction est faite en dehors de la mesure.

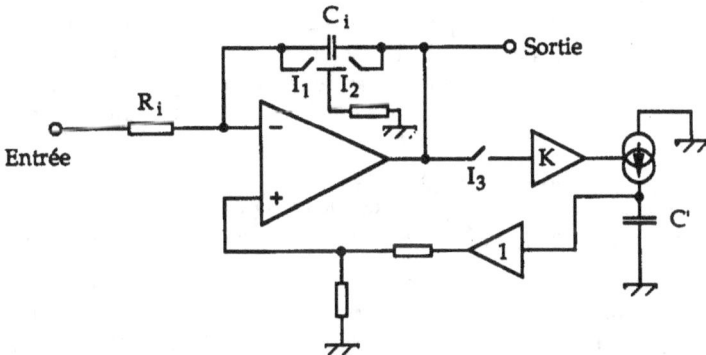

Fig. 6. — Schéma de l'intégrateur et de son système de compensation de dérive.

[Integrator circuit with its drift compensation system.]

2.3 SONDE DE MESURE À DÉPLACEMENT CIRCULAIRE. — Dans ce dispositif, la sonde est formée d'une électrode de 100 μm de largeur et de 10 mm de longueur entourée d'une plaque conductrice dont le potentiel est maintenu à zéro qui sert d'anneau de garde ; la surface de mesure est donc 1 mm². L'ensemble est entraîné dans un mouvement de rotation par une turbine à air comprimé afin d'éviter le bruit et les vibrations engendrés par un moteur électrique. La tête de mesure peut ainsi atteindre une vitesse angulaire de 100 tours par seconde. La sonde étant à une distance de l'ordre de 50 mm de l'axe, sa vitesse linéaire peut être voisine de 30 m/s. L'électrode est placée de telle sorte que sa grande dimension soit suivant le rayon. La sonde tourne à une distance de l'ordre de 300 μm de la surface à lire. L'électrode est connectée à un transistor à effet de champ à très faible niveau de bruit qui est à l'entrée d'un amplificateur. Ce dernier tourne avec la sonde ; il est donc alimenté par une pile, également entraînée dans le mouvement de rotation, pour éviter d'amener l'alimentation du circuit par des contacts tournants. Le signal issu de l'amplificateur est transmis par un coupleur optique analogique afin d'éviter les bruits de contacts.

La sortie de la sonde peut être connectée à un intégrateur, comme dans le dispositif à balayage linéaire. Mais afin de réduire le bruit et d'abaisser le seuil de détection, on exploite la nature périodique des mesures due à la rotation de la sonde. Pour cela, une méthode d'accumulation des signaux sur plusieurs rotations de la sonde est utilisée. Cette technique permet une amélioration du rapport signal sur bruit d'un facteur 5 à 10.

2.4 RÉSOLUTION ET PERFORMANCES DES SONDES. — La résolution spatiale de la sonde dépend essentiellement de la dimension de l'électrode et de la distance entre celle-ci et la surface chargée. L'amplificateur transimpédance est suffisamment rapide pour ne pas limiter la résolution par son temps de montée. Si on définit la résolution comme étant la distance entre les deux positions de la sonde qui donnent respectivement 10 % et 90 % de la variation totale du signal lorsqu'on lit un échelon de charge, la résolution spatiale de la sonde est égale à approximativement 0,7 fois le diamètre de celle-ci, si on suppose que les lignes de champ restent perpendiculaires à la surface. Dans cette hypothèse, la résolution est indépendante de la distance entre la sonde et la surface.

En réalité, au voisinage d'une discontinuité de charges, le champ électrique est déformé. La charge induite sur l'électrode dépend alors de la distance entre la sonde et la surface puisque la charge induite est fonction de la concentration des lignes de champ. Afin de calculer la forme des champs au voisinage d'une discontinuité de charges, nous avons décrit cette discontinuité de charges par une distribution périodique, en créneaux, de période spatiale assez grande pour pouvoir négliger l'influence des charges se trouvant au-delà d'une fraction de la période. Les lignes de champ ont été obtenues en représentant la distribution périodique par sa décomposition en série de Fourier. Par cette méthode nous avons obtenu des résultats comparables à ceux déjà publiés [8].

Il a été, ainsi, possible de calculer la résolution du système pour différentes configurations. La résolution peut être sensiblement plus grande que le diamètre de la sonde et elle est optimale lorsque la distance entre la sonde et la surface est à peu près égale au diamètre de la sonde.

2.4.1 Sonde à déplacement linéaire. — Dans ce dispositif, le niveau de bruit à la sortie de l'intégrateur est de l'ordre de 10 mVcc pour un gain $G = 2$. Cette valeur correspond à une variation de charge de surface de 0,04 pC, c'est-à-dire, à une densité superficielle de charge de 60 pC/cm². Définissons la tension équivalente de surface comme la différence de potentiel V_{f_0} entre un point (y, z) de la surface du diélectrique et la masse, en l'absence d'électrode au-dessus du diélectrique. Pour un diélectrique tel que le téréphtalate de polyéthylène ($\varepsilon_r = 3,2$), le niveau de bruit est approximativement de 4 V, en terme de tension équivalente de surface. Le seuil de détection est donc de 8 V de tension équivalente de surface du diélectrique pour un rapport signal sur bruit égal à 2.

Afin de déterminer la dynamique du système de mesure, il faut connaître la variation maximale de charges ou de tension mesurable. Cette valeur peut être limitée par la tension maximum que peut numériser le convertisseur analogique-numérique. Or, cette tension (± 5 V) correspond à une tension équivalente de surface supérieure à 2 000 V. Pour cette tension le champ électrique entre la sonde et le diélectrique, pour une distance sonde-surface typique de 200 µm, est supérieur au champ de claquage dans l'air. C'est donc le champ électrique maximum admissible qui limite la tension équivalente de surface mesurable.

Le potentiel de la surface, localement, change au cours de la mesure. Il est, en effet, modifié par la présence de la sonde au-dessus du diélectrique. Si Q_f est la charge déposée sur une surface S du diélectrique, la tension équivalente de surface, hors mesure, sans électrode au-dessus du diélectrique est donnée par :

$$V_{f_0} = \frac{Q_f}{C_f}.$$

En présence de l'électrode et de son anneau de garde, la tension équivalente de surface devient :

$$V_f = \frac{Q_f}{C_{sf} + C_f}.$$

Comme on a vu dans la section 2.1.2, V_f est égale, en valeur absolue, à la différence de potentiel entre la sonde et la surface, V_{sf}. Si on admet un champ électrique maximum de 30 000 V/cm, la tension V_{sf} maximum est de 600 V, ce qui correspond à un potentiel de surface V_{f_0} égal à 790 V et à une densité superficielle de charges égale à 11 pC/cm². Il est clair que ce sont les risques de claquage entre la sonde et la surface du diélectrique qui limitent la dynamique de la méthode de mesure, du moins tant que le gain G de l'amplificateur reste petit. La dynamique pour $G = 2$ est voisine de 100.

Si la densité superficielle de charges à lire sur la surface est faible et telle que le champ entre la sonde et la surface soit très inférieur au champ disruptif, la mesure nécessite un gain G plus élevé. Dans ce cas, la valeur maximum mesurable est limitée par le convertisseur et la dynamique varie en $1/G$ si l'on suppose que le rapport signal sur bruit est indépendant du gain.

2.4.2 *Sonde à déplacement circulaire.* — Dans ce dispositif, le signal issu de la sonde est amplifié mais n'est pas intégré. Le signal représente donc la variation de charge. Le bruit à la sortie de l'amplificateur, en l'absence de charges à la surface du diélectrique, est de l'ordre de 50 mV, ce qui correspond à une tension équivalente sur la surface de quelques mV.

Afin de calibrer la sensibilité de la sonde, des charges électriques connues sont déposées sur le diélectrique en collectant les ions produits dans l'air par une source de rayons α (^{241}Am). Pour cela, une grille portée à une tension V est placée près de la surface et l'air est ionisé par des rayons α collimatés par une fente de 100 μm de largeur. Les ions sont collectés sur la surface du diélectrique jusqu'à ce que le potentiel équivalent de surface soit égal à V. Après vérification des tensions de surface à l'aide d'un électromètre, la sonde a été calibrée pour différentes valeurs de V, positives et négatives. En utilisant un amplificateur à porte qui permet d'exploiter la périodicité du signal, le résultat obtenu est 12 V par volt de tension équivalente de surface (le diélectrique utilisé est du téréphtalate de polyéthylène de 120 μm d'épaisseur).

Pour un rapport signal sur bruit de 2, le seuil de détection de variation de la tension équivalente de surface est de 8 mV, ce qui correspond à une densité superficielle de charge de l'ordre de 0,2 pC/cm².

Cette sonde dont une des dimensions est de l'ordre de 10 mm est bien adaptée à la lecture de distributions de charges unidimensionnelles comme, par exemple, des traces de particules. En effet, dans ce cas la trace de charges peut être assez étroite et s'étaler sur plusieurs centimètres sans une variation très significative de la densité de charge dans cette direction.

3. Exemples d'applications.

3.1 SONDE À DÉPLACEMENT LINÉAIRE. — Plusieurs applications de cette méthode de lecture de charges ont été envisagées et explorées. L'une d'elles porte sur la révélation d'images radiologiques ou de diffraction (Laue), obtenues par une méthode ionographique sous forme de distributions de charges à la surface d'un diélectrique [9]. La densité superficielle de charge recueillie à la surface du diélectrique est proportionnelle à l'intensité du faisceau de rayons X ou γ qui parvient au capteur. La lecture de la distribution bidimensionnelle de charges permet de reconstituer l'image de l'intensité du faisceau de rayons X ou γ qui caractérise le milieu traversé.

La figure 7 montre une figure de diffraction (présentée en perspective cavalière) d'un cristal de quartz obtenue par cette méthode. L'étalement des pics de diffraction, comparé à la même image obtenue par la méthode classique du film radiologique est dû principalement au capteur

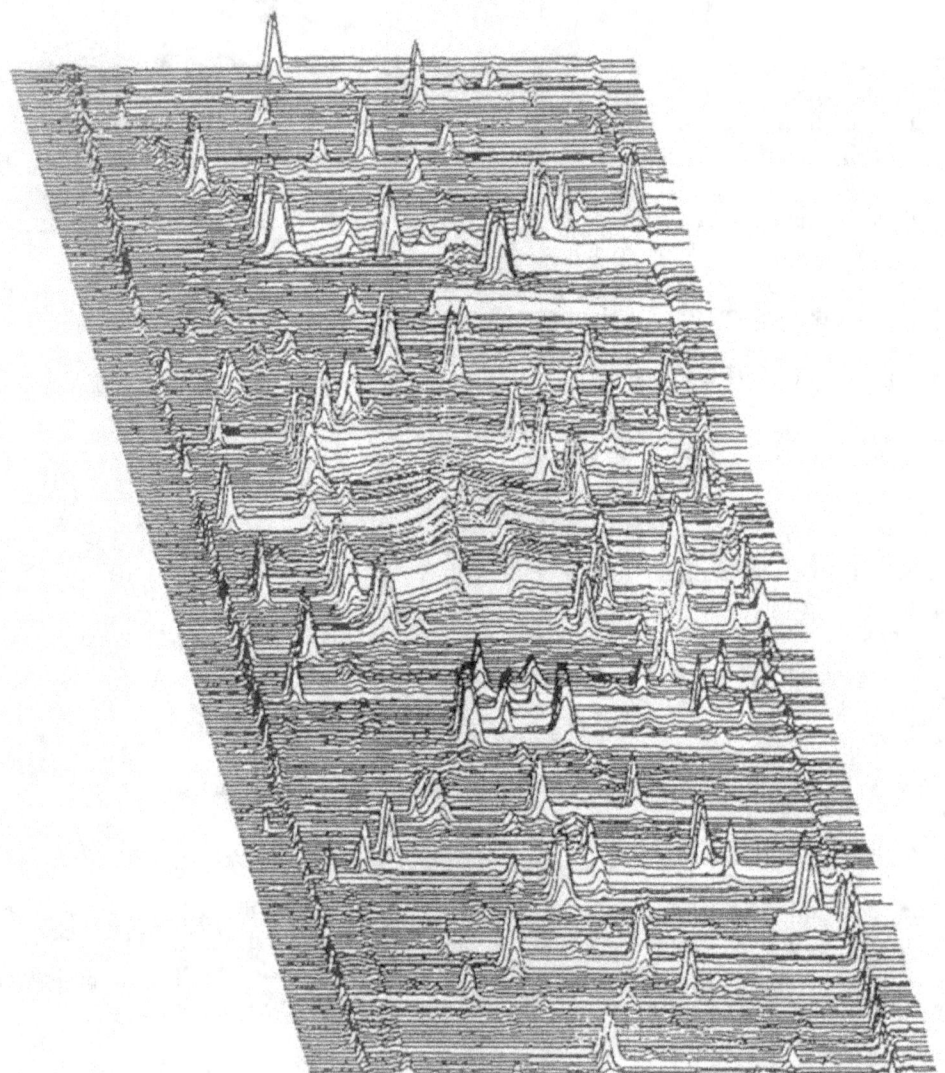

Fig. 7. — Représentation en perspective cavalière d'une figure de diffraction X (Laue en transmission) d'un cristal de quartz, obtenue par une méthode ionographique sous forme d'une distribution de charges superficielles.

[Representation of a X ray diffraction image obtained with a quartz crystal as a surface charge distribution, using an ionographic method.]

utilisé. L'interaction entre les rayons X et le capteur se produit dans un gaz contenant un élément de numéro atomique élevé pour augmenter l'effet photoélectrique. Pour obtenir une assez grande quantité de charges électriques le faisceau de photons doit traverser une épaisseur suffisante de gaz (1 ou 2 mm), ce qui se traduit par un élargissement de la distribution de charges sur le diélectrique.

Dans le but d'une application en contrôle non destructif, des images ionographiques ont été réalisées avec des rayons γ de 660 keV. La figure 8 représente la distribution de charges obtenue sur un diélectrique avec un rayonnement ayant traversé différentes épaisseurs de plomb. Le cache en plomb étant légèrement plus petit que la surface sensible du capteur, on voit apparaître sur les bords du diélectrique, près des bandes de référence de lecture, une augmentation des charges due au rayonnement non absorbé.

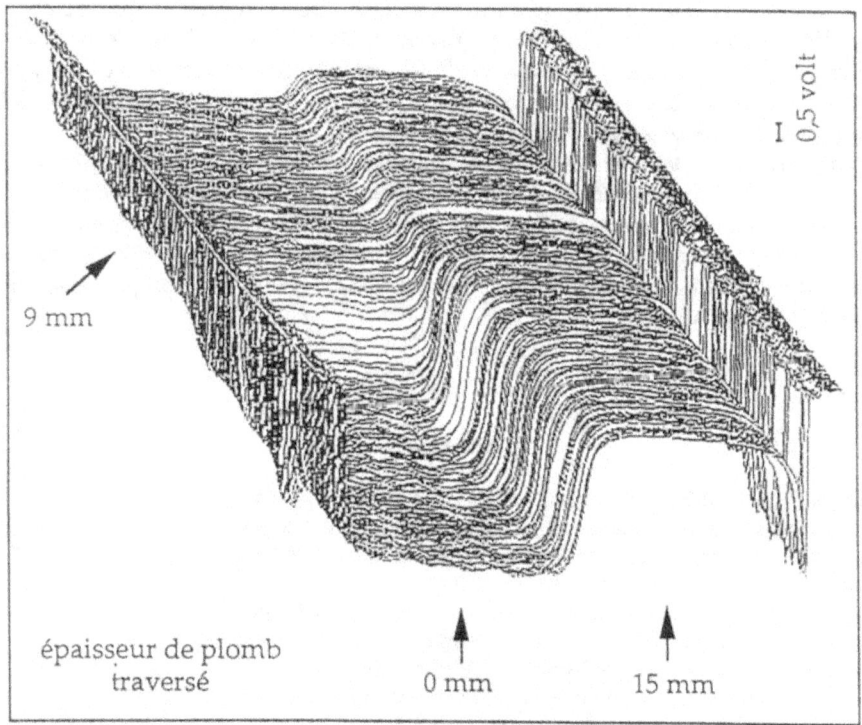

Fig. 8. — Image en rayon γ d'un masque de plomb d'épaisseur variable selon la position, obtenue par une méthode ionographique sous forme d'une distribution de charges superficielles.

[γ ray image, as a surface charge distribution, obtained by a ionographic method, through a lead mask of varying thickness.]

Les charges électriques n'étant ni détruites ni modifiées par notre méthode de lecture, ce dispositif a été également utilisé pour étudier l'écoulement des charges électriques sur des surfaces isolantes [10]. Cette méthode a des applications industrielles dans des domaines où la présence de charges électrostatiques peut provoquer la destruction de composants électroniques. On peut, par exemple, observer ainsi l'influence de différents composants chimiques entrant dans la fabrication des isolants de cartes électroniques.

3.2 SONDE À DÉPLACEMENT CIRCULAIRE. — Cette sonde de grande longueur est bien adaptée à la mesure de répartition unidimensionnelle de charges, telle que celle qui peut être laissée par le passage d'une ou plusieurs particules. Des expériences de détection de particule α ont été réalisées, dans lesquelles le niveau de bruit était suffisamment faible (en particulier, amplitude très faible des charges résiduelles sur le film diélectrique) pour détecter le passage d'une seule particule α, c'est-à-dire une densité de charges déposées de l'ordre de 5×10^{-13} C/cm^2 [11].

4. Conclusion.

Le dispositif de mesure de distribution de charges superficielles qui vient d'être décrit permet l'enregistrement et le stockage sur ordinateur de toutes les informations nécessaires à la reconstruction de l'image de la répartition des charges. Une fois stockée, l'image peut être traitée puis représentée de différentes manières selon les aspects à mettre en évidence. Il est également possible d'enregistrer le signal issu de la sonde, avant intégration, ce qui permet d'envisager le traitement du signal (filtrage...) avant intégration numérique. Certes, la sonde et le dispositif mécanique actuels limitent la résolution spatiale ($\approx 0,5$ mm) qui est sensiblement moins bonne, en radiologie, que celle du film et, la lecture donne une mesure approchée de l'amplitude des charges déposées sur le diélectrique.

Dans son principe, la méthode de lecture qui vient d'être décrite ne modifie, ni l'amplitude, ni la distribution des charges situées sur le diélectrique. Elle permet des lectures successives de la même distribution de charges et donc, par exemple, d'en suivre l'évolution au cours du temps. Cette propriété de la méthode permet d'envisager son développement en milieu industriel, pour l'étude du comportement électrique des matériaux utilisés dans certains domaines de l'électronique.

Bibliographie

[1] Thourson T. L., *IEEE Trans. Electron Devices* **ED-19** (1972) 495-511.
[2] Van Turnhout J., Proc. 1st Intern. Conf. on static electricity, Vienna, May 1970. Publ. in Advances in static electricity, vol. 1 (W. F. de Geest, Auxilia, 1970) pp. 56-81.
[3] Secker P. E., *J. Electrostatics* **1** (1975) 27-36.
[4] Reedyk C. W. et Perlman M. M., *J. Electrochem. Soc.* **115** (1968) 49-51.
[5] Chalmers J. A., Atmospheric Electricity (Pergamon Press, 1967).
[6] Lewiner J. et Charpak G., US Patent n° 4567530 (1986).
[7] Lewiner J., Charpak G. et Pollak E., US Patent n° 4673885 (1987).
[8] Neugebauer H. E. J., *Appl. Opt.* **3** (1964) 385-393.
[9] Cordurié G., Thèse de Doctorat de l'Université Paris VI (1991).
[10] Morisseau D. et Lewiner J., Electrostatics'87, Oxford (IOP Publishing, 1987) pp. 197-201.
[11] Charpak G., Santiard J.-C., Chabot M., Morisseau D. et Cordurié G., CERN Rapport interne EP 87-04 (1987).

Nuclear Instruments and Methods in Physics Research A260 (1987) 365–367
North-Holland, Amsterdam

ELECTROSTATIC IMAGING OF CHARGES LIBERATED IN DIELECTRIC LIQUIDS BY IONIZING RADIATION

G. CHARPAK

CERN-EP, Geneva, Switzerland

D.F. ANDERSON and B.J. KROSS

Particle Detector Group, Fermi National Accelerator Laboratory, Batavia, IL 60510, USA

Received 5 June 1987

The charge liberated by beta particles in a liquid argon–methane mixture has been collected on a mylar film and the image developed with a technique similar to electrophotography. The image of the beta-emitting lines on a chromatographic gel have also been produced in this way.

1. Introduction

Within the framework of a continuing effort to image the trajectories of charged particles in dielectric media [1,2], we have tested the possibility of collecting charges liberated in liquid argon (LAr) on an insulating surface. The image is then developed with a technique similar to electrophotography.

The ultimate goal is to obtain the projected image of the tracks of minimum ionizing particles in a liquid. This requires readout methods with a sensitivity not available at present and thus a research effort is being pursued. It seems of interest to us to verify that charges liberated by ionizing radiation in LAr can be collected on a mylar sheet and imaged with presently available methods, since this technique may also find applications in other fields where the imaging of radioactive sources is of interest.

2. Experimental procedure

A schematic of the experimental setup is shown in fig. 1. A radioactive source is separated from a 125 μm thick mylar sheet by a 2 mm gap filled with a mixture of LAr and methane. The back side of the mylar sheet has a conductive layer of indium tin oxide (ITO) deposited on it. This conductive layer is in contact with the positive high-voltage electrode with the source held at ground potential. Before installation, the mylar was first washed with ethanol to remove any charge from the surface.

We conducted a preliminary study, with beta- and alpha-emitting sources. The image of the collected charges was developed with a liquid toner which is used in electrophotography [3]. Our experiments were conducted with the most sensitive toner that we could find. The toner (CR42A) was produced by Coulter Systems Corp., Bedford, Mas., and kindly put at our disposal by the firm. It is made of a dispersion of particles of a median size of 0.2 μm, which are positively charged, and dispersed in an insulating liquid. With this toner we were able to observe images when the surface potential reached about 25 V.

In order to produce an image of the charge liberated in the LAr, the charge was collected on the mylar sheet for times that varied from a few minutes to several days, depending on the intensity of the source. After comple-

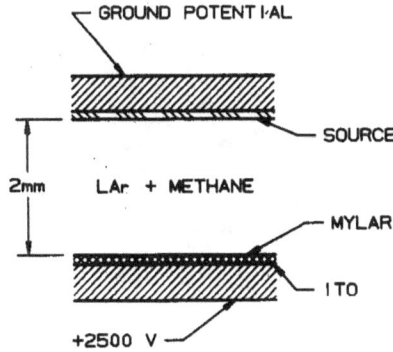

Fig. 1. Schematic of experimental setup.

tion of the charge collection, the LAr/methane was removed and the mylar brought to room temperature to avoid the condensation of moisture. The mylar was then immersed in the toner for a few seconds and then washed by gentle agitation in kerosene. During the warmup period the voltage had to be left on the cell in order to prevent loss of charge.

3. Results

It was found early on that the quality of the image could be improved by the addition of a few percent methane. This was due to a reduction in the diffusion of the electrons. We also found that the results were somewhat improved by an increase in the collection field used, particularly with the LAr/methane mix. Most of our work was done with a collection field of 1250 V/mm.

One object of our effort was to image the beta-emitting lines on a chromatographic gel carrying proteins labeled with [35]S. After an exposure of 12 h, the image in fig. 2a was obtained. Fig. 2b is an image produced by exposing the gel to a special autoradiography film for 6 h. The distance between the three lines to the right of the two darkest lines is about 400 μm, with a width of about 120 μm. The width of the lower strip of gel is about 6.6 mm. We obtained the correct pattern of lines, showing that our experimental procedure does not destroy the images, but the spatial resolution is clearly degraded.

4. Discussion

The densitivity of the toner is far below what is required for the imaging of minimum ionizing particles in LAr. If we assume that the track left by such a particle in LAr had a width of 20 μm after 1 cm of drift, we would have a voltage of about 4 mV at the surface of the 125 μm mylar sheet. For a heavy relativistic ion, the charge density is about 8000 times larger. When recombination in the LAr is taken into account, this heavy ion would yield a surface potential of about 3 V on the surface of the mylar. This is still below the sensitivity of the toner.

However, our observations may prove to be of interest for applications where the acceptable level of sensitivity can be much lower. For instance, the spatial distribution of the beta emitters is of primary importance on a field of research such as biology. A research effort has been invested in studying the possibility of replacing the photographic emulsions used for autoradiography, which suffer several defects: low sensitivity, lack of linearity, and small dynamic range.

Gaseous detectors have also been widely studied. Their major drawback is the large range of the emitted electrons in the gas. This gives them a resolution of 1 mm in the best cases, and more often, several millimeters.

The range of the emitted electrons is about three orders of magnitude smaller in LAr, which could be a great advantage if one could measure the collected charge with sufficient sensitivity and accuracy. At present, experiments ongoing at Ecole de Physique et Chimie in Paris, and at CERN, indicate that sensitivities at least ten times better than the present toner can be obtained and that further substantial improvements in sensitivity can be reasonable expected.

We have shown that LAr can be used as a detection medium for the imaging of ionizing radiation. It would be of interest to perform this research with warm liquids, such as 2,2,4,4-TMP, though the problems of high purity requirements for the warm liquids may make it impractical.

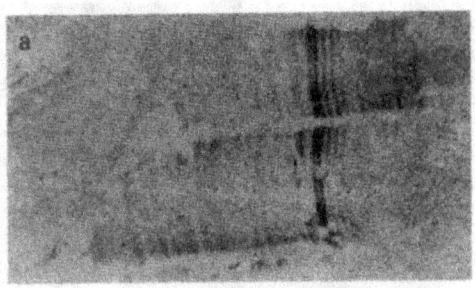

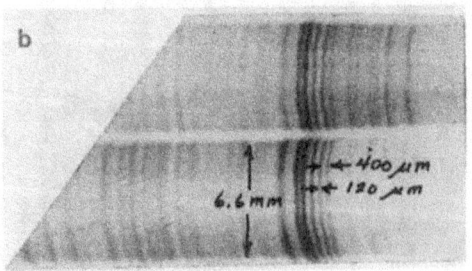

Fig. 2. (a) Image produced on mylar film by beta-emitting lines on a chromatographic gel, and (b) image produced with autoradiography film.

Acknowledgement

The authors would like to thank Mr. H. Howell and Miss Phillis Rorke of Coulter Systems. Corp. for their kind assistance and generous contribution of samples of toner.

References

[1] G. Charpak, R. Bouclier, A. Breskin and R. Chechik, Nucl. Instr. and Meth. 192 (1982) 235.

[2] G. Charpak, H. Czyrkowski, A. Farilla and S. Majewski, Nucl. Instr. and Meth. 196 (1982) 555.

[3] R.M. Schaffer, Electrophotography, 2nd ed. (Focal Press, London, 1975).

Nuclear Instruments and Methods in Physics Research A261 (1987) 445–448
North-Holland, Amsterdam

LIQUID IONIZATION CHAMBERS WITH ELECTRON EXTRACTION AND MULTIPLICATION IN THE GASEOUS PHASE

D.F. ANDERSON [1], G. CHARPAK [2], R.A. HOLROYD [3] and D.C. LAMB [1]

[1] *Particle Detector Group, Fermi National Accelerator Laboratory, Batavia IL 60510, USA*
[2] *CERN-EP, Geneva, Switzerland*
[3] *Department of Chemistry, Brookhaven National Laboratory, Upton NY 11973, USA*

Received 15 June 1987

Electrons have been transferred from liquid to gaseous argon with an estimated efficiency of 100%. Electron extraction has also been accomplished in liquid 2,2,4-TMP with parallel plate amplification in the gas of over 10^3.

1. Introduction

A growing interest has been expressed in the use of large volumes of liquid argon as the detecting medium of time projection chambers [1]. The major applications are for proton decay experiments and neutrino physics.

Even though the energy loss for particle tracks in liquids is much larger than in gases, the absence of charge multiplication leads to the use of expensive low noise amplifiers. There is also a limitation on the spatial accuracy due to limitations on cell segmentation and the electronic noise of such a high capacitance system.

In order to improve the detection ability of liquids, two-phase systems have been studied, where electrons are extracted from liquid argon (LAr) [2] and liquid xenon (LXe) [3] into the gas phase above the liquid. Before these measurements, the possibility of extracting electrons from these materials was not obvious since their conduction band energy (V_0), which is the energy of the excess electrons measured with respect to the vacuum, is negative. Thus, a substantial fraction of an electron volt must be given to the electrons to free them from the liquid into the gas.

There has also been a study where excess electrons have been extracted from liquid n-hexane into the gas phase [4]. Unlike LAr and LXe, this hydrocarbon has a $V_0 > 0$. Thus, it is easy to see how excess electrons can escape the liquid. This result was later verified [5], and in addition, electrons were extracted from isooctane (2,2,4-TMP) with a $V_0 = -0.24$ eV [6].

In our study, we started with a verification that the extraction of excess electrons from LAr held no hidden difficulties. The true object of our work, however, was to explore the possibility of extracting excess electrons from one of the warm liquids that are currently of interest for liquid ionization chambers [7–9]. We have

therefore studied the extraction of electrons from the room temperature liquid 2,2,4-TMP (henceforth referred to as TMP).

2. LAr cell

For our test of the extraction of electrons from LAr, a glass dewar with a 7.5 cm inside diameter was used. A 1 cm-wide strip of the silvering was removed to allow viewing of the liquid level. The dewar was also fitted with a drain so that the level of the LAr could be adjusted.

The cell consisted of a parallel plate ionization chamber, 25 mm in diameter, and with a 4 mm gap. The gas was positioned horizontally so that it could be partially filled with liquid. The lower electrode had an ^{241}Am alpha-particle source plated on it.

3. TMP cell

The glass cell used in our work with TMP is shown in fig. 1. It consisted of a test cell with two electrodes, and a reservoir for the excess liquid. The test cell and the reservoir were connected by a thin glass tube to allow the level of the TMP in the test cell to be adjusted. The electrodes were 15 mm in diameter and separated by a gap of 4 mm, with the lower electrode coated with an ^{241}Am alpha-particle source.

After purification, the TMP was placed in the baked-out glass cell and the cell was sealed off. In the second of two fillings, 80 Torr of argon was also added to the cell. This was to increase the gas pressure above the liquid to improve electron amplification. The electron lifetime in the TMP of this filling was measured to be 7

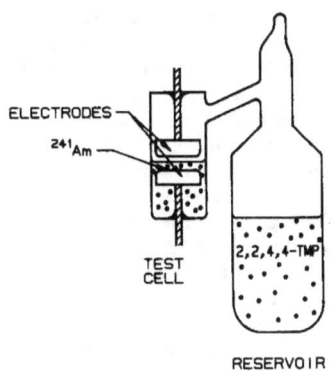

ELECTRODES
^{241}Am

TEST
CELL

2,2,4,4-TMP

RESERVOIR

Fig. 1. Schematic of glass cell.

μs demonstrating that the material's purity was not degraded by the addition of the argon.

4. Experimental results – LAr

The extraction from the LAr of the electrons liberated by the 5.5 MeV alpha particles from the ^{241}Am proved to be very simple. One advantage of LAr is the large amount of charge liberated. At a typical operating voltage for the experiment of 800 V/mm, there were about 2×10^4 electrons liberated [10]. This produces a signal easy to detect.

The ionization gap was first filled with LAr and a pulse height spectrum was taken. At a voltage of 3200 V (800 V/mm) a clean pulse-height spectrum was obtained with an energy resolution of about 25% fwhm. The LAr level was then lowered so that the liquid filled only half of the gap. At the same voltage the pulse height was reduced by 6%, but otherwise had the same shape and energy resolution.

The electric field in the liquid of the partially filled gap was about 643 V/mm. One would anticipate that the pulse height would be about 25% lower than for the liquid-filled gap because of increased recombination at the lower electric field in the liquid [10]. The higher than expected pulse height is attributed to a small amount of amplification in the argon gas above the liquid.

Since the contribution to the signal is proportional to the fraction of the gap across which the electrons drift [10], the pulse height would have been down by a factor of 2 had the electrons not exited the liquid. This, along with the symmetric shape of the pulse-height spectrum, leads us to conclude that all of the electrons escaped the LAr and were collected, even at this low electric field. This result is consistent with earlier work.

The big disadvantage of LAr is that it is very difficult to maintain a constant pressure above the liquid in our small test system. Althouh it is easy to get charge amplification in the gas, the changing pressure made it impossible to maintain a stable gain. This proved not to be the case with TMP.

5. Experimental results – TMP

The major disadvantage of warm liquids for the study of a two-phase detector is the small amount of charge that is liberated. Since we worked with typical electric fields of only about 400 V/mm in the liquid, fewer than 10^3 electrons are liberated by the alpha particles [11]. Thus, the signal for the full-gap measurement was smaller than could be measured with our electronic system. We therefore were not able to study the shape of the alpha peak until the electrons were extracted and amplified in the gas.

Our first tests were made with the cell filled with only TMP and its vapor. Because of TMP's low vapor pressure (about 14 Torr and 40 Torr at 20°C and 40°C, respectively [12]) amplification commenced at relatively low voltages. At room temperature we estimate our maximum charge gain to be about 10^3 at a potential of 1400 V. We increased the temperature of the system in order to increase the vapor pressure of the TMP. At 40°C we achieved a gain of about 7×10^3 at a potential of 2300 V. Unfortunately we had some sparking problems which eventually destroyed the TMP in this filling.

We did estimate that 100% of the electrons escaped the liquid and were collected in the gas. The general results for this test are qualitatively the same as the data discussed below.

The addition of 80 Torr of argon to the second cell filling increased the stability of the system and allowed for higher gain. As expected, as the temperature of the TMP was increased a higher amplification could be achieved, but at a higher voltage. Fig. 2 shows a typical pulse-height spectrum. This spectrum was taken with a charge gain of about 60. From the near symmetry of the peak, we feel that 100% of the electrons liberated by the alpha particles were extracted from the liquid. If there had been an inefficiency in the extraction of the electrons from the liquid, one would expect a low-energy tail on the pulse height distribution. The relatively poor energy resolution we attribute to the fact that, in the small diameter cell used, there was a large meniscus. Since the gap was not uniform, the charge gain would also vary across the cell.

The shoulder on the left in fig. 2 seems real and was present in all of our data, though we are not able to account for its origin. Its pulse height was about 22% of the mean pulse height for the alpha peak. This is about

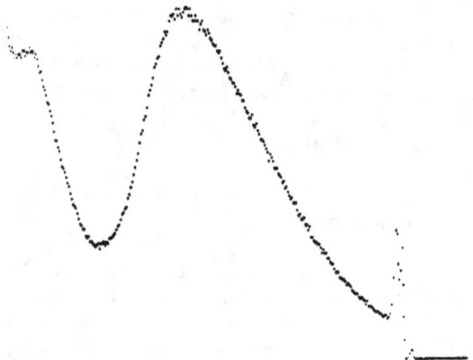

Fig. 2. Typical alpha-particle pulse-height spectrum.

Fig. 3 shows the charge collected per alpha particle as a function of voltage across the gap, for a variety of temperatures. Fig. 3 also shows the amount of charge liberated in the TMP [11]. In all cases, at a charge of a few times 10^6 electrons per alpha particle, the shape of the alpha peak became distorted, consistent with full amplification for a small part of the events and reduced amplification the majority of events. This we believe to be due to the buildup of positive ions in the liquid, since with a charge collection of 3.5×10^6 e$^-$/alpha, and a source intensity of 5700 s, the positive ion density in the liquid would be on the order of 10^7 mm^3. Thus, areas with the greatest source intensity would have lower amplification because of electric field distortions.

a factor of three higher than what would be expected for the gamma rays from ^{241}Am [13]. We placed an intense ^{241}Am source exterior to the cell and were able to verify that the signals from the gamma rays fell approximately where we expected on the pulse-height spectrum.

6. Discussion

Our results confirm that electrons can be conveniently extracted from LAr. In large time projection chambers where the detection medium is LAr, one can envisage a proper mixture of argon and methane, allowing the amplification of electrons in a parallel grid structure with all the inherent advantages of large pulses and greater readout flexibility connected with pulses induced by gaseous amplification. Our results also show that it is as easy to extract electrons from a warm liquid like TMP, widely considered for calorimetry in high-energy physics, with an easy sizeable amplification of the extracted electrons in a parallel plate structure.

Acknowledgements

R.A. Holroyd was supported by the U.S. Department of Energy, Division of Chemical Science, Office of Basic Energy Science under Contract DE-ACO-76CH 00016. D.F. Anderson and D.C. Lamb were supported by the U.S. Department of Energy under Contract DE-ACO2-76CH03000.

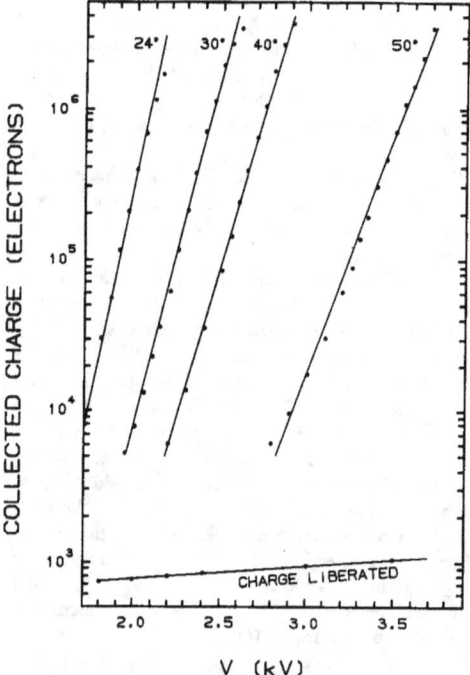

V (kV)

Fig. 3. Charge collected per alpha particle as a function of voltage across the gap for a variety of temperatures. The amount of charge liberated in the TMP as a function of voltage is also shown.

References

[1] ICARUS Collaboration, The ICARUS Proposal, INSN/AE-85/7 (July 1985).

[2] B.A. Dolgoshein et al., Sov. J. Particles Nucl. 4 (1973) 70.

[3] H. Zaklad et al., Liquid xenon multiwire proportional chambers for nuclear medicine applications, submitted to the First World Congress of Nuclear Medicine, Tokyo, Japan (Sept. 30–Oct. 4, 1974) LBL-3000 ABST.

[4] R.M. Minday, L.D. Schmidt and H.T. Davis, J. Chem. Phys. 54 (1971) 3112.

[5] A.A. Balakin, I.S. Boriev and B.S. Yakovlev, Can. J. Chem. 55 (1977) 1985.

[6] R.A. Holroyd, J.M. Press and N. Zevos, J. Chem. Phys. 79 (1983) 483.

[7] J. Engler, H. Keim and B. Wild, Performance test of a TMS calorimeter, KfK 4085 (June 1986) Submitted to Nucl. Instr. and Meth.

[8] J. Engler and H. Keim, Nucl. Instr. and Meth. 223 (1984) 47.

[9] UA1 Collaboration, A proposal to upgrade the UA1 detector in order to extend its physics programme, CERN/SPSC/83-48, SPSC/P92 add. 3 (Aug. 1983).

[10] W.J. Willis and V. Radeka, Nucl. Instr. and Meth. 120 (1974) 221.

[11] R.C. Muñoz, J.B. Cumming and R.A. Holroyd, J. Chem. Phys. 85 (1986) 1104.

[12] CRC Handbook of Chemistry and Physics, 67th ed., R.C. Weast, ed. in chief (CRC Press Inc., Boca Raton, FL, 1986).

[13] R.A. Holroyd and T.K. Sham, J. Phys. Chem. 89 (1985) 2909.

FAST TRACKING DETECTOR USING MULTIDRIFT TUBES

R. BOUCLIER, G. CHARPAK, G.A. ERSKINE, B. GUERARD, J.C. SANTIARD, F. SAULI and N. SOLOMEY

CERN, Geneva, Switzerland

We describe a tracking detector designed as an assembly of multidrift tube modules; because of its fast time-resolution and good position accuracy, coupled with high reliability, the device seems suitable for operation as vertex detector at a collider, and for other high-rate applications. Recent measurements of localization accuracy obtained when operating the modules with a dimethyl ether gas filling are presented.

1. Introduction

The conception of the so-called central tracking or vertex detector around a collider – the most inaccessible part of the experimental apparatus – has to satisfy two major constraints: reliability, and endurance under irradiation. Moreover, the detector has to have the best possible localization accuracy and rate capability. Although solid-state detectors (silicon strips or charge-coupled devices) that suit the above-mentioned requirements are becoming available, their use in large quantities is still restrained by cost and the lack of suitable radiation-resistant large-scale integrated electronics. Most of the present and planned detectors for high-rate colliders include indeed several layers of silicon detectors close to the vacuum pipe, followed by cylindrical multiwire drift chambers of a design optimized for high localization accuracy and multitrack resolution.

Some time ago we proposed [1] to improve the performance and the reliability of a central tracking detector based on multiwire gas chambers, replacing the conventional monolithic construction by a modular assembly of independent elements. For this purpose we have developed a basic detection element, the multidrift module (MDM), capable of making a fast and accurate radial localization of multiple tracks by measuring the drift time on a closely packed array of drift cells. The longitudinal coordinate can be deduced, although with less precision, from a measurement of the ratio of charges detected at the two ends of the resistive anode wires. We have chosen a hexagonal geometry for the drift cell and for the external gas-tight envelope. Fig. 1 shows the basic configuration of a module: in each cell the anode wire is surrounded by six cathode wires, and this arrangement is repeated throughout the module. In the present design, we have adopted a cell radius of 1.45 mm (1.27 mm between wire planes) using as the mechanical support and gas-tight envelope a thin, hexagonal,

carbon-fibre tube 30 mm in diameter, which allows us to mount 70 drift cells in each module. The tube (present thickness 500 μm) is composed of almost equal proportions of carbon and epoxy. So far the length of the modules has had to be limited to 80 cm owing to the unavailability of longer carbon-fibre envelopes. However, detailed studies of the mechanical deformations and electrostatic deflections of the wires have shown that it is feasible to have modules that are several metres long with only a few simple internal supports (see section 4).

Modules can be assembled around an intersection as shown, for example, in fig. 2; in this case, 48 tubes have been used, with a total of about 4000 signal wires. The inner and outer radii of the complete detector are 3 and 12 cm, respectively, but of course these parameters can

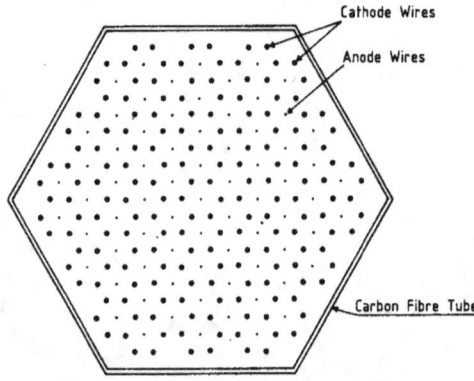

Fig. 1. Schematics of the multidrift module. The basic detection cell consists of an anode wire surrounded by six cathode wires in a hexagonal geometry; the pattern is repeated to fill all the volume contained in a carbon-fibre envelope. In the present design, 70 drift cells are packed within a 30 mm diameter tube.

0168-9002/88/$03.50 © Elsevier Science Publishers B.V.
(North-Holland Physics Publishing Division)

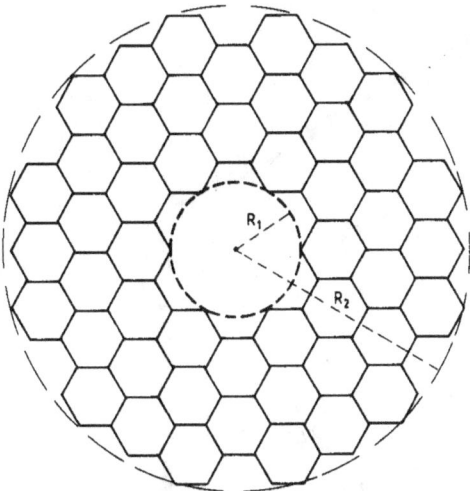

Fig. 2. An example of 48 MDMs assembled around an intersection to constitute a tracking detector. The inner and outer radii of the device are 3 cm and 12 cm, respectively. Tracks are detected and localized both radially (through a drift-time measurement) and longitudinally (through charge division) an average of 30 times.

easily be modified thanks to the modular construction. Each track is sampled an average of 30 times. With the expected (but not yet achieved) radial localization of each wire of around 40 μm rms, a sagitta measurement should therefore be obtained with an accuracy of about 6 μm over 80 mm length (neglecting multiple scattering and systematic errors). Using a resistive anode wire, a longitudinal coordinate localization of 1% or better can easily, be achieved [2]. The total amount of material along the radial direction is about 1.5% of a radiation length, but it is much larger (30% of a radiation length) in the end-plates, including the signal cables.

The present paper describes the recent development work and measurements realized with the multidrift tubes.

2. Construction and operation of the multidrift modules

The method of construction of the MDM has been described in detail elsewhere [3–5] and will be recalled only briefly here. Wires are stretched between two precision-drilled fibre-gas plates and soldered onto miniature brass pins which have V-shaped grooves cut along their axes. The overall radial wire-positioning accuracy thus obtained is better than 20 μm. All cathode pins, and therefore the corresponding wires, are in electrical contact through a layer of copper glued to the external side of the plates, whilst the anode pins, insulated from each other and from the cathodes, allow readout of

individual signals from both sides of the module. The geometry of the fibre-glass end-plate is shown schematically in fig. 3: small circles correspond to the holes receiving the cathode pins, and the double concentric circles represent the position of the anode pins, In order to ensure good insulation, the copper layer covering the external surface of the plate has a larger diameter bore cut around the holes for the anode pins.

The supporting carbon-fibre envelope is glued on the two plates and is electrically insulated from the electrodes; this enables it to be raised to a suitable potential so as to reduce electrostatic deflections of the wires, as will be discussed in section 4.

A close-up of one end of a finished MDM is shown in fig. 4, together with one of the special connectors developed by us for readout. During manufacturing, each end of the module was filled with epoxy, which is used to fasten the carbon-fibre envelope to the end-plates and also to cover all cathode pins completely (leaving just a contact for the high-voltage supply). After the curing process, surfacing on a milling machine exposed the tops of the anode pins, visible in the picture. For the particular model shown here, the electrical contact to the anodes was realized by inserting an anisotropic conducting membrane between the module and the flat printed circuit holding the miniature coaxial cables for the output signals; the contact is established by tightening the two parts together with the help of three threaded pins. This system of connection is cheap and simple to realize; however, because of the rather large variations in the contact serial resistance (a few ohms to several tens of ohms), it is only appropriate for time measurements on the anode wires and not for charge division. For this latter case, we have soldered the miniature cables directly to the anode pins during manufacture. This also helps to reduce the material in the end-plates. Fig. 5 shows several finished modules with various connection systems.

All measurements described here have been realized using 20 μm diameter gold-plated tungsten wires for the anodes, and 70 μm copper–beryllium cathode wires. We have since satisfactorily tested modules with thicker anodes (up to 50 μm) with the aim of increasing their ruggedness and improving the operating characteristics. The signal cables, about 80 cm long and with a characteristic impedance of 50 Ω, fan the high density of anode wire outputs to a system of fast hybrid amplifier–discriminator circuits [6] connected to CAMAC-based time-to-digital and charge-to-digital converters. For the beam runs, the data were written on tape using a small on-line computer and analysed off-line.

3. Experimental results

The MDM can be operated efficiently for the detection of radiation using virtually any gas mixture. How-

II. VERTEX DETECTION

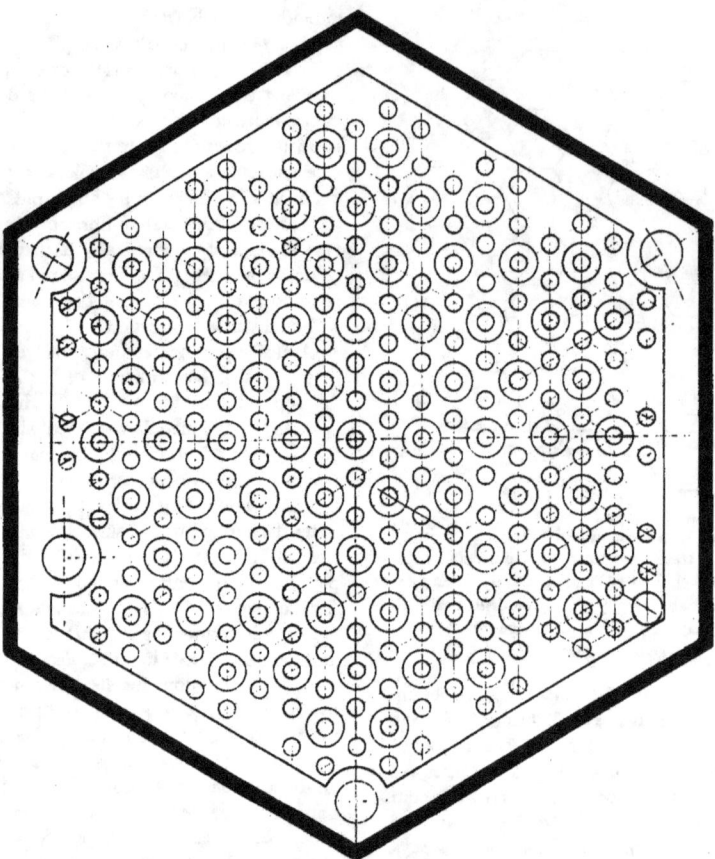

Fig. 3. Schematics of the drilling geometry of the end-plates. Small circles represent the holes for inserting the wire-holding cathode pins; around the anodes, a larger-diameter shallow bore ensures insulation against the copper-covered surface of the plate.

ever, because of the small cell size, the localization accuracy for the detection of fast particles is largely dominated by the primary ionization statistics, i.e. the average distance between primary clusters [$\approx 300 \ \mu$m in an argon (80%) + methane (20%) mixture at atmospheric pressure]. We have therefore tried using pure dimethyl ether [DME: $(CH_3)_2O$] as the gas filling. This is an organic compound with high density at atmospheric pressure (2.4×10^{-3} g/cm³ at 20°C), and which, like heavy hydrocarbons, is expected to have a primary ionization density higher than that of argon (maybe by a factor of 2). Our experimental results, although so far not allowing us to deduce the value of the primary ionization, confirm the substantial improvement in accuracy obtained using DME instead of conventional mixtures [3].

Although originally proposed for use in gaseous detectors because of its low electron diffusion and velocity

at high fields [7], DME has also proved to be impervious to radiation damage due to polymerization. Indeed, proportional chambers operated with this gas have survived prolonged irradiation up to total detected charges above several coulomb per centimetre of wire [8].

The operating characteristics of the MDM filled with DME at atmospheric pressure are illustrated in fig. 6, which shows the detection efficiency for minimum-ionizing particles, and the single wire noise (without radiation) as a function of high voltage. Both measurements have been realized with discriminators set at an equivalent input threshold of about 5×10^{-2} pC. The gas appears to have remarkable quenching properties, and allows us to reach large gains, in excess of 10^6. A detailed analysis of the detected signal shows that even at the higher operating voltage we could not reach the limited streamer regime observed by other authors who

Fig. 4. One end of a finished module. The contacts to the anode pins are visible, as well as the tube for the gas and the pin for the high-voltage supply. The three threaded pins are used for assembly and to hold the signal connector in place, the contact being ensured by an anisotropic conducting membrane, also seen in the picture.

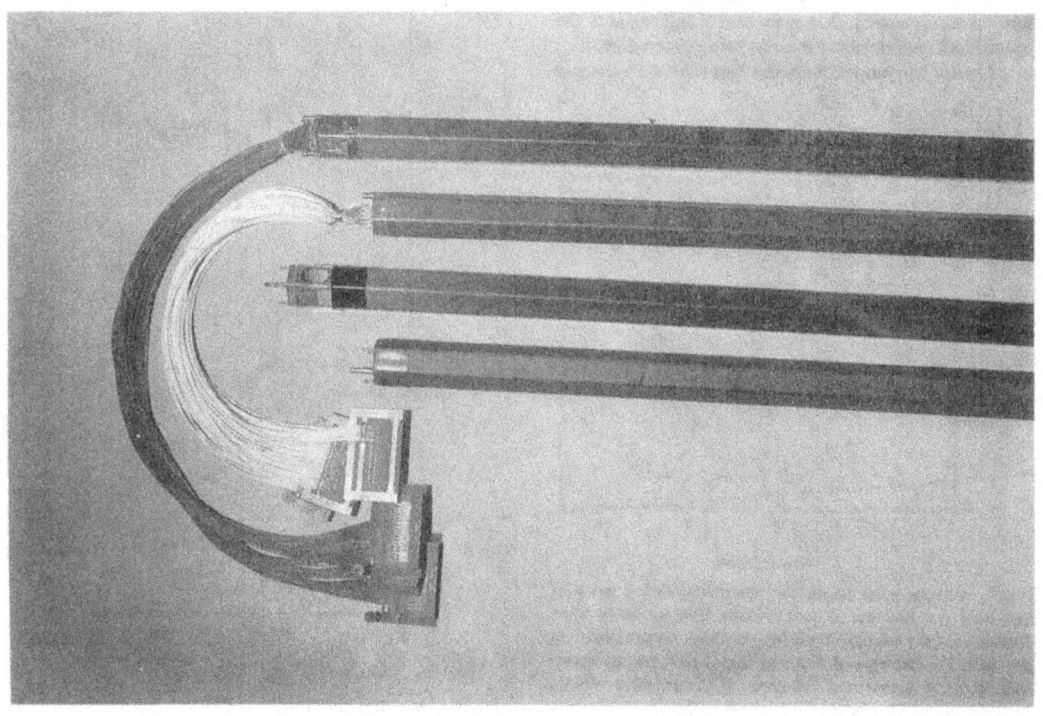

Fig. 5. Several finished modules with different connectors; the modules are 80 cm long.

II. VERTEX DETECTION

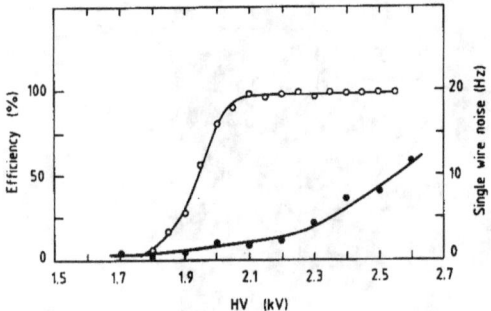

Fig. 6. Efficiency plateau for minimum-ionizing electrons and singles counting rate on one wire without radiation, measured in a fully instrumented module with DME gas filling.

used the same gas but single-wire cylindrical counters [9]. Fig. 7 shows the average pulse height for minimum-ionizing particles recorded as a function of their distance from the anode wire (integrated in 200 μm intervals), for a beam crossing the drift cell perpendicular to one side, and at 2.5 kV. Because of the hexagonal geometry of the cell, the total ionization loss should decrease linearly to zero from a distance corresponding to half the cell radius; the detected charge follows the same trend, meaning that even at the higher gains the operational regime is still close to being proportional.

In order to convert from the recorded drift-time to the real coordinate, we had to use a suitable space–time correlation. In view of the lack of data concerning drift velocities at very high fields in DME, we preferred to measure the correlation directly with a pulsed ultraviolet laser in a special MDM having a small quartz window. The result is shown in fig. 8, measured with a beam traversing the cell parallel to one side. Although these measurements did not enable us to reach the region furthest from the anode (hitting a cathode wire with the laser results in a large swarm of electrons), the curve can be extrapolated sufficiently well for our analysis, making use of existing data at lower fields [10].

Using the space–time correlation, together with a table of reference or zero times (deduced from the data) for each wire, it is possible to compute the drift distance for all wires hit by a track. The reconstruction program then finds the best alignment between the recorded coordinates for each event, employing a standard fitting procedure described in refs. [4] and [5]. Fig. 9 shows a typical reconstructed event in the MDM, and fig. 10 is the distribution of the difference between fitted and

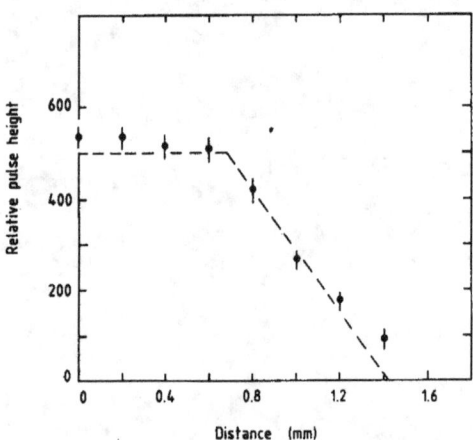

Fig. 7. Average pulse height for minimum-ionizing particles measured as a function of their distance from an anode wire. Tracks cross the hexagonal cell in a direction perpendicular to one side. The decrease of detected charge (full points) corresponds to the geometrical thickness of the sensitive volume (represented by the dashed line), showing the proportionality of the counter (DME at 1 bar, 2.5 kV).

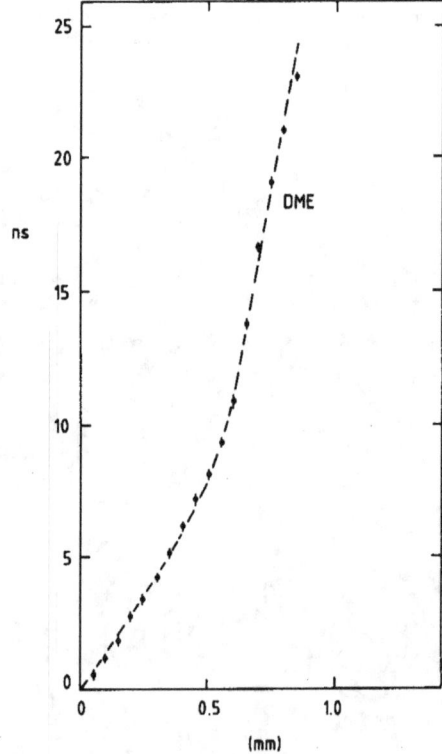

Fig. 8. Space–time correlation in DME in 1 bar, measured in a special drift tube with a pulsed laser beam.

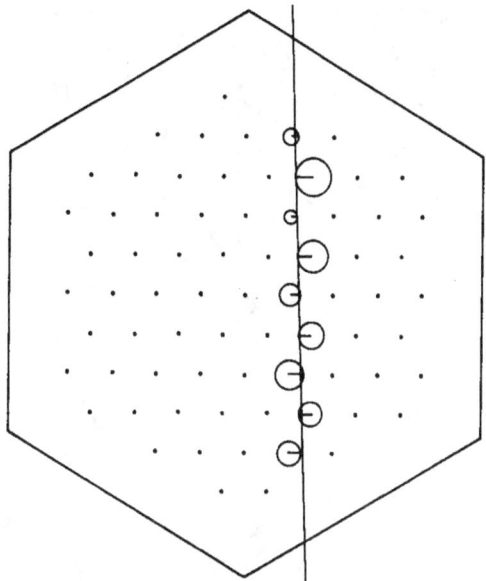

Fig. 9. Example of a single-track reconstructed event. The circles represent the measured drift time on the hit wires.

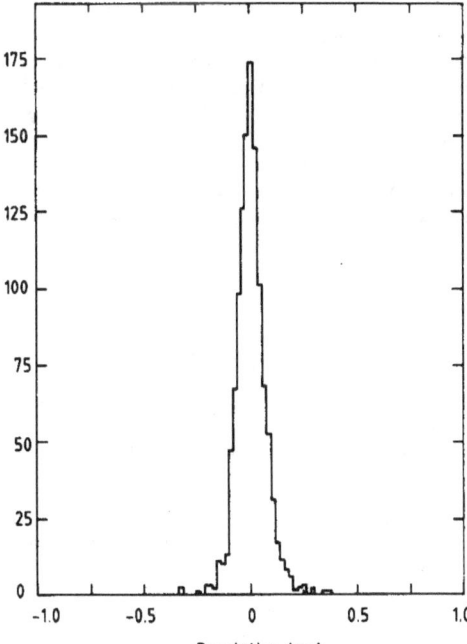

Fig. 10. Distribution of the difference between the fitted and the measured coordinates on a wire for a typical run with minimum-ionizing particles; it has a rms of 70 μm.

measured coordinates on a wire for a typical run; it has a rms of 70 μm.

4. Discussion of the results and future improvements

Various factors contribute to determining the localization accuracy; the already mentioned primary ionization statistics, the electron diffusion, the electronic noise and slewing, the systematic effects such as wire-positioning accuracy and the exact shape of the space–time correlation. A preliminary simulation of the physical processes, which was done with a Monte Carlo program, and some recent results obtained in the same gas with single-wire proportional counters [9], tend to suggest that our main source of dispersion is at present the electronic slewing, i.e. the dependence of the recorded time on the amplitude and shape of the input signal. This is particularly important for tracks close to the wires because of the time structure of the detected signal (with individual clusters reaching the anode over a larger time span) and because of the higher drift velocity in that region. In order to reduce the dispersion, we are currently experimenting with larger anode-wire diameters in the hope of reaching the limited streamer operation that would partly obliterate the slewing effects, and we are developing better and faster electronics.

Another way of improving the localization accuracy of the module is to increase the operating gas pressure. With the present method of construction, the MDM can withstand a pressure above 10 bars without failure, although with a rather substantial deformation of the tube. We had, however, to improve substantially the insulation between electrodes in order to be able to increase the operating voltage, the most critical problem being due to microdischarges between anode and cathode pins on the surface of the epoxy-covered endplates, probably owing to small cracks or to pockets of air trapped during the curing process. One prototype has been successfully operated at 1.75 bars in DME; this required a high-voltage value of around 3.5 kV. This development will be further pursued by installing the chamber and carrying out measurements in a minimum-ionizing particle beam.

In order to be able to build longer modules and to operate them at high voltages, we used a computer program to analyse the electrostatic deflections of the wires in the module as a function of high voltage and length. Because of the cylindrical symmetry of the hexagonal cells in modules, an infinite structure of such cells would be stable at any length and voltage, apart from positioning errors and gravitational sagging effects. In a finite module, however, the symmetry is broken at the boundaries, and this causes a displacement of the outer layers of wires. In normal operating

II. VERTEX DETECTION

conditions (grounded anodes and envelope, cathodes at −2.2 kV), the cathode wires in the outer layer have an outward maximum displacement of 350 μm in the centre of an 80 cm long tube; the outer layer of anode wires move inward by 150 μm under the same conditions. Although the operation of the module seems unaffected by these displacements (in particular, there is no detectable change in the gain of the modified cells), the tracking algorithm would become rather involved. When the outer envelope is kept at an intermediate potential, around −0.5 kV, the maximum displacements are reduced to 200 μm and 30 μm for the outer cathode and anode wires. This is quite acceptable in the present design (see fig. 11). The experimental results described in the previous sections were in fact obtained under these conditions.

The electrostatic displacements can be further reduced by designing the module differently. Inspection of the field map in the module (fig. 12) shows that there is, indeed, an equipotential line waving through the cells close to the cathode wires. Replacement of this line by a physical boundary at the proper potential would leave the symmetry of the infinite structure unperturbed, and no wire would be displaced by electrostatic forces. This geometry can be approximated with a flat surface, such as that of the envelope; this corresponds to a module in which the carbon-fibre tube has been rotated by 30° as compared to the original design. In this case we have

Fig. 12. Computed electric field and equipotentials in the MDM.

computed a maximum displacement of the anode wires of less than 20 μm, always for an 80 cm long module operated at −2.2 kV and with the envelope at −1.8 kV.

For longer modules, or at higher operating voltages, it will be necessary to resort to some kind of regular internal wire support, much as is done in large-size multiwire proportional chambers. Various ways of realizing such supports without spoiling the performance are being investigated.

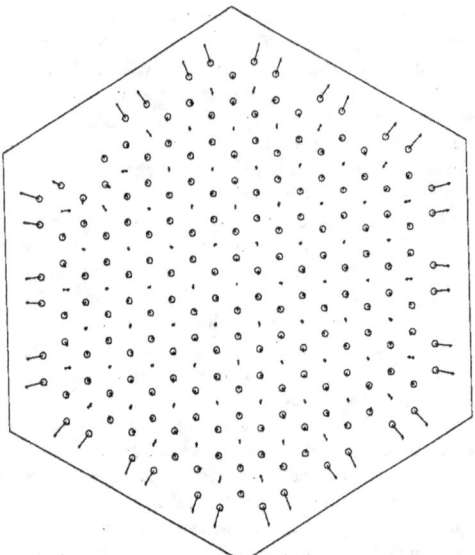

Fig. 11. Computed maximum wire displacements in the centre of a standard MDM, 80 cm long; the scale of the displacement is increased by a factor of 5 in the drawing. Operating conditions: cathodes at −2.2 kV, grounded anodes, outer envelope at −0.5 kV.

References

[1] R. Bouclier, G. Charpak and F. Sauli, CERN EP Internal Report 84–03 (1984).
[2] C.C. Young et al., IEEE Trans. Nucl. Sci. NS-33 (1986) 176.
[3] R. Bouclier et al., IEEE Trans. Nucl. Sci. NS-33 (1986) 169.
[4] R. Bouclier et al., Nucl. Instr. and Meth. A252 (1986) 373.
[5] R. Bouclier et al., preprint CERN-EP/87-14 (1987), to appear in Proc. Workshop on Vertex Detectors, Erice (1986).
[6] J.C. Santiard, CERN EP Internal Report 82–04 (1982).
[7] F. Villa, Nucl. Instr. and Meth. 217 (1983) 273.
[8] S. Majewski, Proc. Workshop on Radiation Damage to Wire Chambers, Berkeley (1986) (LBL-21170, Berkeley, 1986) p. 239.
[9] F. Villa, private communication.
[10] G. Bari et al., Nucl. Instr. and Meth. A251 (1986) 292.

EUROPEAN ORGANIZATION FOR NUCLEAR RESEARCH

EP Internal Report 80-07
21 August 1980

MULTITON GASEOUS DETECTORS

R. Bouclier, G. Charpak, F. Sauli and J.C. Santiard

ABSTRACT

Large volumes of compressed air or any cheap gas used as multiton detectors are being envisaged. The collection of slow positive or negative ions would permit the detection of minimum ionizing particles without any gas amplification, at sufficiently high pressures. Some preliminary measurements of charges liberated in air by α particles show that a good energy resolution can be expected with such gases.

G E N E V A
1980

1. INTRODUCTION

Several physical problems of great interest require heavy detectors, with stringent requirements as regards particle resolution, decay pattern recognition capability, and energy resolution. The two main fields where such detectors are needed are neutrino physics and baryon stability studies.

In a recent approach to this problem, a Moscow group[1] has demonstrated the possibility of a high-pressure argon drift detector where 100 tons of sensitive gaseous volume could be envisaged. Two imperative requirements have to be met for such a detector to work:

- an extremely high purity of argon, to permit the drift of electrons over paths of the order of several metres, at a pressure of several hundred bar, without being captured;

- a complex gas composition, including a small fraction of xenon which appears necessary to permit electron amplification around the wires. The cost of even 1% xenon addition is such that it would appear unrealistic to extrapolate such a detector to sizes 100 times larger, as is required for baryon stability studies which triggered our investigation. We decided to limit this investigation to the possibility of the cheapest gas, namely air, as a detector.

The fact that the electrons are immediately captured means that we have to deal with slow ions and obtain from them all the information we need: position, pattern, energy loss. Since amplification is excluded in any convenient way, we have to investigate the conditions under which the operation in the ionization chamber mode would give us this information.

2. PULSES IN AN AIR CHAMBER

To restrict ourselves to inexpensive electronics we considered that a noise of 10^3 electrons (r.m.s.) could easily be obtained with a simple amplifier. And, indeed, good results were obtained with an amplifier with the following characteristics:

- Noise: 2×10^{-16} C (r.m.s.).

- Sensitivity: 20 V/pC.

- Rise-time: 50 μs. Decay time: 2.5 ms.

- Cost of components: SF 40.

We fixed as our goal the necessity to detect minimum ionizing particles. The energy loss in air is 1.8 MeV/(g/cm^2); this means, at 20 °C, with a pressure of 300 bar, an energy loss of about 0.65 MeV per cm, i.e. about 26,000 ion pairs, far above noise.

The scheme of the detector is given in Fig. 1. The positive ions and the negative ions migrate in opposite directions. They are detected on strips or pads (after traversal of a Frisch grid). The time difference in the induced pulses gives the vertical position; the pulses induced on the strips or pads give the lateral position and the energy loss distribution. In order to find if the available electronics is suitable for the collection of the ions liberated in compressed air by a minimum ionizing particle, we simulated almost the same energy loss, in an air chamber at atmospheric pressure with α particles whose energy loss is close to 1 MeV/cm.

A series of measurements were undertaken in air, with 5.5 MeV α's from ^{241}Am, with the amplifier quoted above. The main obstacle for the detection of ionization pulses appeared to be the production of microphonic noise. This was finally suppressed by suspending the chamber schematized in Fig. 2 from the ceiling, with elastic strings. It then appeared simple to collect the ions, positive or negative, and to obtain a good energy resolution.

An encouraging fact is that for energy losses of the order of several hundred MeV, as generated in proton decay modes, one can expect an energy resolution strictly limited by the systematic errors and dream reasonably of 1%, since there are no sources of ion losses other than columnar recombination -- all the electrons having been captured at the track itself -- and no amplification fluctuations.

3. GASEOUS CALORIMETERS

Table 1 gives some properties of the cheapest gas, air, and of another gas which has interesting properties, namely freon 13 B1, which in a liquid form was one of the filling fluids of Gargamelle.

Unfortunately the radiation length in air is large, 1.1 m at 300 atm at room temperature and 0.5 m at -133 °C. Only large tanks of at least 25 m diameter can be envisaged for the confinement of isotropic showers. The situation is more attractive for physics problems where the showers are in the direction of a beam, since a few metres of lateral dimensions are acceptable and a tube of great length permits a target with a high mass of air.

Freon is clearly more advantageous for electromagnetic showers. Under the manageable conditions of 28 bar at 50 °C, the density is 269 kg/m^3 and the radiation length 60 cm. For some problems in neutrino physics, such as electron scattering, it may be attractive.

4. DRIFT VELOCITIES OF IONS

The mobility of ions varies with their mass M_i and the mass M_g of the gaseous molecules in which they drift. The mobility is roughly proportional to $(1 + M_i/M_g)^{\frac{1}{2}}$.

Table 1

Some properties of air and freon

AIR		
Radiation length 36 g/cm^2, dE/dx = 1.82 MeV/(g/cm^2)		
Pressure (bar)	Density at 17 °C (kg/m^3)	Density at -133 °C (kg/m^3)
100	121.9	573
200	234.7	668
300	326.8	718

CF$_3$Br FREON 13 B1		
Radiation length 16.5 g/cm^2, dE/dx = 1.52 MeV/(g/cm^2) Molar mass 148.9 Critical point 67 °C, 39.8 bar, density 744.8 kg/m^3		
Pressure (in equilibrium with liquid) (bar)	Temperature (°C)	Density (kg/m^3)
12.66	15	102
28.28	50	269

Figure 3 shows the mobility of various ions in nitrogen. We see that a reasonable value for the mobility of ions produced in air is about 3 (cm/s)/(V/cm), while the mobility of freon would be 2 (cm/s)/(V/cm).

High voltages will have to be applied to drift the ions over large distances. Neglecting the technological problems for the time being, let us assume that we can apply 5 kV/cm over the drifting distance. This leads to velocities of 1.5×10^4 cm/s for the air and 0.5×10^4 cm/s for the freon in freon at 1 atm. For 1 m of drift at 300 atm we thus have occupation times of the order of 2 s. The maximum rates that can be tolerated are determined by the granularity of the charge collecting system. This granularity is determined by the noise of the amplifier which controls the minimum length of detector required to detect a minimum ionizing particle, if the sensitivity has to be at this level.

The distance between the shielding grid and the collecting electrodes has to be small compared with the size of the collecting strips or pads.

5. ENERGY RESOLUTION

The test performed with an air gap of 6 cm, where the maximum energy deposited by the α particles of ^{241}Am was 5 MeV, is very encouraging. Figure 4 shows that the pulses are far above noise. With a total energy loss of 1 GeV, averaged over a large number of collecting pads, say 1000, we may expect an improvement of $\sqrt{1000}$ on the energy resolution and aim reasonably at 1%, which would be an enormous asset in a search for rare events.

The problems are however formidable for a detector of 10^4 tons. We could imagine a tube lying on the ground deep in the ocean to give shielding and pressure equilibrium. We would then have a detector with the size of a submarine and maybe a comparable cost to pay for the large quantity of charge-collecting channels with a continuous monitoring of the charge arrival at the anode and cathode.

Owing to the great interest of the physics problems which could be studied with such a detector, we thought it worth presenting these rather unrealistic ideas to the community for comments, suggestions, and criticisms.

* * *

REFERENCE

1) V.K. Chernyatin et al., A pressurized gas detector for high energy neutrinos, Proceedings of the Second IFCA Workshop, Les Diablerets, Switzerland, 4-10 October 1979 (Ed. U. Amaldi) (CERN, Geneva, 1980), p. 320.

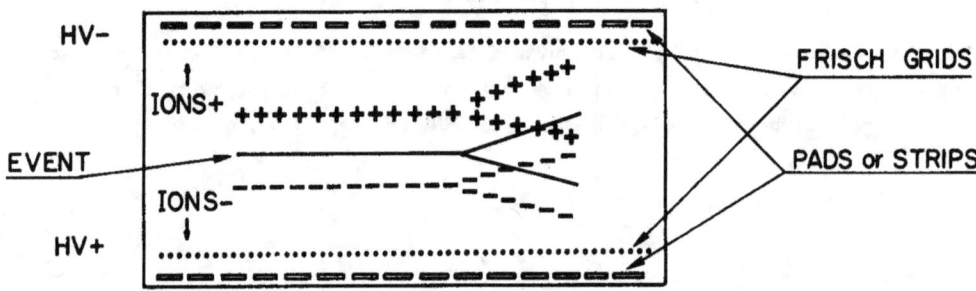

Fig. 1 Schematics of a slow-ion three-dimensional chamber. The filling gas is air or any electronegative gas. The collection of the ions of opposite polarities gives the coordinate along the field lines without requiring a zero time trigger. The charges collected in pads or strips give the orthogonal coordinates and the total energy loss.

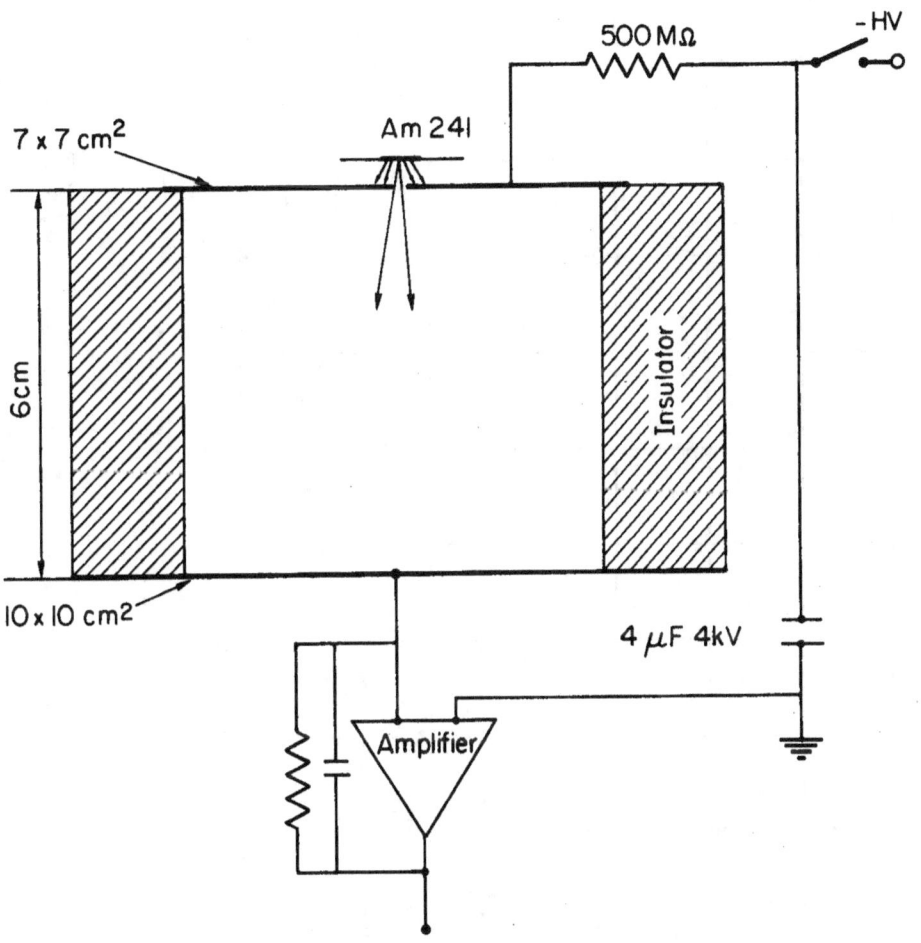

Fig. 2 Test ionization chamber. The 5.5 MeV α particles from [241]Am are
stopped in air at atmospheric pressure and the charges collected and
amplified.

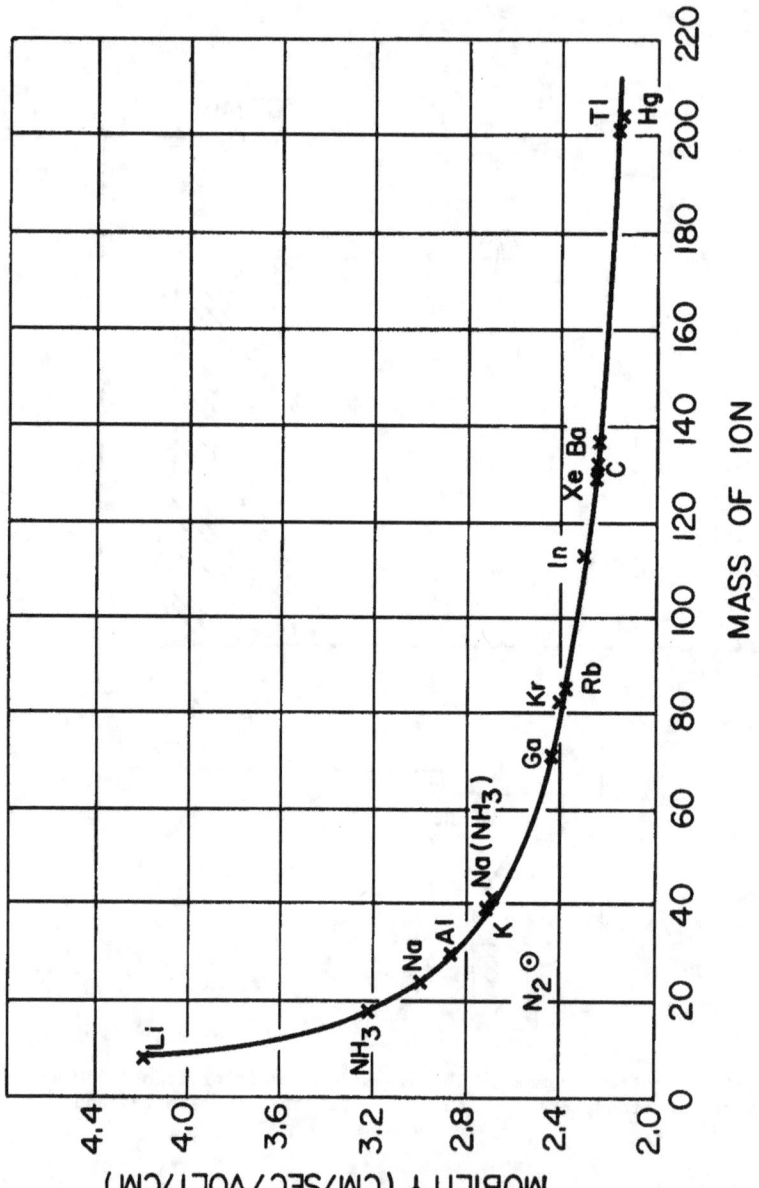

Fig. 3 Mobility in nitrogen of various ions as a function of mass at 1 atm pressure. [From J.H. Mitchell and K.E.W. Ridler, Proc. Royal Soc. (London) A146, 911 (1934).]

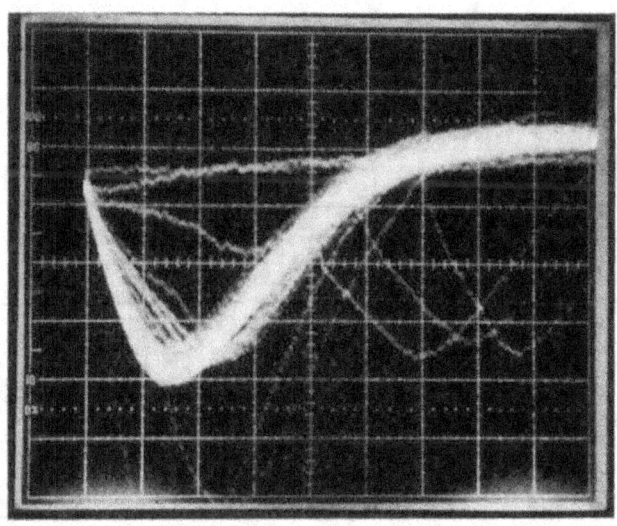

Fig. 4 Pulses from α particles in air. Horizontal scale: 0.5 ms/div. Vertical scale: 50 mV/div. Amplifier sensitivity: 20 V/pC. The α's of an [241]Am source stop in the air in the chamber. The charges are collected by a field of 500 V/cm over a drift length of 6 cm. The noise is only a few per cent of the signal.

Nuclear Instruments and Methods in Physics Research A 346 (1994) 120–126
North-Holland

NUCLEAR
INSTRUMENTS
& METHODS
IN PHYSICS
RESEARCH
Section A

Results of a first beam test of hadron blind trackers

M. Chen [a], D. Luckey [a], M. Smolin [a], K. Sumorok [a], X. Zhang [a], A. Bolozdynya [b], S. Belogurov [b],
D. Churakov [b], A. Koutchenkov [b], A. Kovalenko [b], V. Kuzichev [b], V. Lebedenko [b], V. Sheinkman [b],
G. Smirnov [b], G. Safronov [b], V. Vinogradov [b], Y. Giomataris [c], C. Joseph [c], M. Werlen [c],
G. Charpak [d], B. Blumenfeld [e], A.K. Gougas [c,*], D. Steele [e], M. Akopyan [f]

[a] *Laboratory for Nuclear Science, MIT, Cambridge, MA, USA*
[b] *ITEP, Moscow, Russian Federation*
[c] *IPN, Lausanne University, Lausanne, Switzerland*
[d] *CERN, Geneva, Switzerland*
[e] *Johns Hopkins University, Baltimore, MD, USA*
[f] *Institute for Nuclear Research, Moscow 117312, Russian Federation*

(Received 16 February 1994)

We describe the experimental results of a new type of electron tracker, called Hadron Blind Detector or HBD. An HBD prototype was tested with gas mixtures of CF_4 with He or Ne and a parallel plate avalanche chamber having a CsI photocathode of eight pads. Beam tests confirm the large Cherenkov light bandwidth in the EUV region that can be obtained with such gas mixtures. It results in a large quality factor of about 500 cm^{-1} which allows HBD operation with a much shorter radiator thickness than conventional Cherenkov counters. Full electron efficiency was obtained, while pions were rejected up to momenta of 9 GeV/c. HBD is unique in measuring electron trajectories near the vertex, vetoing Dalitz pairs, and providing trigger on electrons among heavy hadron background. We discuss the use of such detectors for lepton identification and detection in high energy physics experiments and especially in heavy ion colliders.

1. Introduction

In high intensity fixed target, beam dump, or colliding beam experiments using relativistic heavy ions or hadrons it is often required to detect e/γ precisely near the interaction regions in the presence of very large numbers of hadrons produced. For example, in the case of Au ion on Au ion collisions, the number of π's produced is about 10 000 particles/event, and even 2000 at the central region with rapidity $|\eta| < 1$ at RHIC, and about five times larger at the Large Hadron Collider (LHC) energies. In a proton–proton collider, multiplicity is lower but pile-up of many events is expected, due to the high luminosity, and coverage of the forward–backward rapidity region is required.

The main backgrounds for direct electrons are π^0 Dalitz pairs, photon conversions and hadrons, while for direct photons they are mainly π^0/η decays. For muons, they are K decays before the hadron absorbers. A tracker which is sensitive only to electrons near the interaction regions can reconstruct the electron trajectories to obtain the secondary vertex.

The Hadron Blind Detector (HBD), originally proposed by Charpak and Giomataris [1], uses Cherenkov light emitted by relativistic particles to identify electrons and to measure their trajectories. Its application to heavy ion colliders to reject Dalitz pairs and its potential use in a low temperature environment in combination with liquid xenon or krypton calorimeters to increase the pion rejection was introduced in ref. [2].

HBD is useful for experiments emphasizing e/γ physics at heavy ion colliders, such as the PHENIX experiment at RHIC (see ref. [3]) and L3H at the LHC Heavy Ion Collider [4]. In the future proton–proton colliders electron identification provided by the calorimeter is limited by jets faking electrons. One can reject such events using low density material and fast HBD trackers at an early trigger level [5].

HBD uses the same gas for both the radiator and the detection gap. Thus there is no longer a need for windows to separate the radiator region from the amplification/detector region. Such windows limit the transparency in the far UV for all Cherenkov counters. An active research was undertaken to find gas mixtures transparent in the extreme UV region (EUV) which can be used simultaneously as a detector gas filling having a high multiplication gain (10^5–10^6). Laboratory results have shown that CF_4 mixtures with noble gases may serve as an ideal HBD gas [6]. Among the gas quenchers, CF_4 is one of the most

* Corresponding author. Tel. +41 22 707 6111, fax +41 22 707 6555.

0168-9002/94/$07.00 © 1994 – Elsevier Science B.V. All rights reserved
SSDI 0168-9002(94)00328-5

transparent and the radiating Cherenkov light can be extended in the EUV region up to 12 eV.

The Cherenkov light emitted in the radiator is converted into photoelectrons by a thin CsI photocathode [7]. The photoelectrons produced are then amplified and collected by anode wires in a Parallel Plate Avalanche Chamber (PPAC), consisting of an anode wire plane and a plane of cathode pads. The spatial points of the electrons are obtained using the energy weighted average of signals from the anode wires and from the cathode pads. We describe herewith experimental results obtained with an HBD prototype. We present extensive laboratory results, as well as tests in a particle beam.

2. Experimental set-up

The HBD prototype tested is shown in Fig. 1. It consists of the following components: a pressure tank, a front flange with a sapphire UV window at the center, a CsI photocathode, a PPAC and a back flange with gas, HV and signal ports. It permits operation at 10^{-10} < absolute pressure < 5 atm and at temperatures down to $-160°C$. The outgassing rate after baking at 50°C has been measured to be less than 10^{-7} mbar/week.

HBD uses Cherenkov light generated in 40 cm of gas at room temperature. The Cherenkov light is converted into photoelectrons by a thin (0.5 μm) CsI photocathode (40 × 40 mm²), divided into eight 4 mm wide strips with separated readout. The spatial positions of the electrons are obtained using the energy weighted average of signals from the cathode pads. The response of the cathode has been calibrated by a pulsed UV lamp system with 100 ns pulse width. The photoelectrons in turn are amplified and collected by anode wires in a PPAC. The cathode plane of the PPAC is the grounded CsI photocathode plate, and the

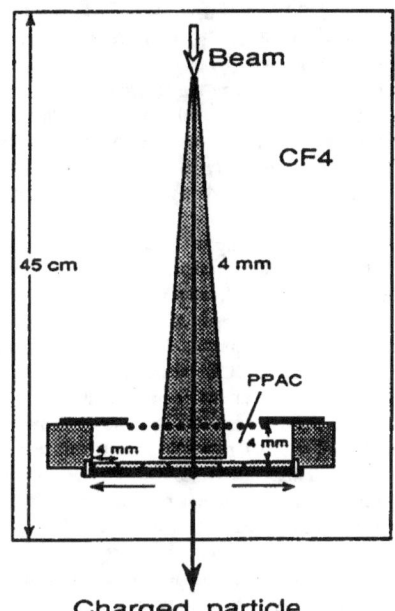

Fig. 1. Schematic of a single cell of a hadron blind tracker. A charged particle (e⁻) produces Cherenkov light in a gas radiator, producing photoelectrons on the photocathode, which are extracted, amplified and collected on the anode grid and readout using both the anode wires and the cathode pads to obtain the electron spatial positions.

positive high voltage was applied on an anode wire mesh in a 4 mm gap.

Cathode strips of palladium–silver are deposited by a sputtering technique on a 0.5 mm ceramic substrate. A thin

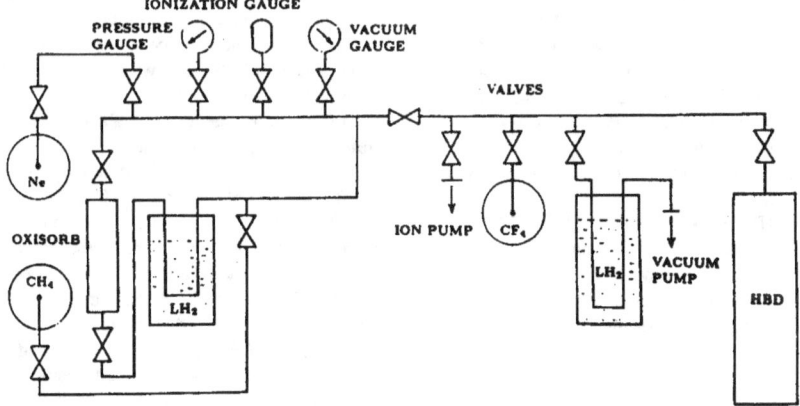

Fig. 2. The HBD gas supply system. 1 – HBD, 2 – valve, 3 – CF₄, 4 – ion pump, 5 – LN₂ trap, 6 – turbo-molecular pump, 7 – filter, 8 – oil pump, 9 – CH₄, 10 – pressure gauge, 11 – thermo-couple vacuum gauge, 12 – vacuum gauge, 13 – Ne, 14 – Oxisorb, 15 – LN₂ pump.

0.5 μm layer of CsI is then deposited by vacuum deposition of CsI powder. Since CsI is hydroscopic and water absorption can deteriorate its quantum efficiency, precautions were taken to minimize the exposure in atmospheric air. We estimate a 10 min total exposure of the coated photocathode in air. The photocathode was mounted inside the HBD volume immediately after the CsI deposition and the HBD vessel was pumped down to the optimum vacuum before gas filling.

Gas purity of a few ppm is required to satisfy the demands for the radiator transparency in the EUV wavelength range. The gas purification system used (see Fig. 2) consists of an Oxisorb cartridge [#1] and a primary turbo pump followed by an ion pump. Once a vacuum of about 10^{-5} Torr is obtained with the turbo system, a higher vacuum of 10^{-7} Torr is achieved with the ion pump. Although such high vacuum conditions are not required for the operation of the HBD, maximal precautions were taken to prevent any transparency losses in the radiator, since there was no on-line transparency monitor. Once a satisfactory vacuum level is reached the gas mixture is introduced through the filter. The gas filling is under control by a set of valves and allows operation at a broad range of pressures, from low pressure to overpressure.

3. Search for HBD gas mixtures

The main goal of laboratory tests is to find gas mixtures which can achieve high and stable gain in a PPAC and which, as Cherenkov radiators, are highly transparent for < 220 nm wavelengths for which a high quantum efficiency of CsI cathode has been measured. The prototype HBD has low outgassing and is vacuum tight, staying a $< 10^{-7}$ Torr after one week. This is important since a gas purity of a few ppm is required to achieve high transparency in the EUV wavelength range.

The detector response was measured using different gas mixtures at various partial pressures by applying a high voltage on the anode plane and measuring the cathode response when the cathode is illuminated by a pulsed or continuous UV lamp. The output signal resulting from the avalanche amplification process in the PPAC gap is fed to a charge preamplifier [8] in the case of pulsed operation mode. In the continuous mode the current was measured directly on a pAmp-meter.

Two types of UV lamps were used: a pulsed $Xe + T_2$ scintillating light source peaked at 172 nm and a continuous Hg lamp with a quartz window. In the pulsed mode the observed current pulses are composed of a sharp peak due to electrons and a long tail due to the slow ion drift. The

[#1] Oxisorb filters are made by Messer Griesheim GmbH, Dusseldorf, FRG

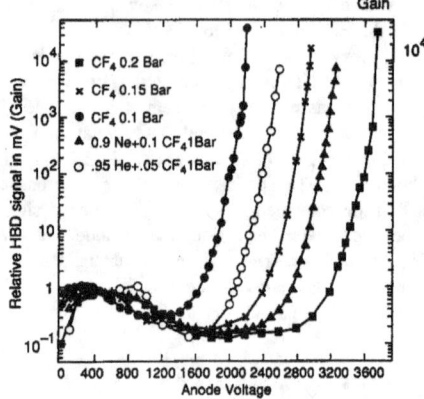

Fig. 3. The measured AC current gain versus high voltage across a 4 mm gap for various gas mixtures in an HBD.

PPAC gain was measured at various electric field strengths by comparing the total collected charge to the one measured at lower voltages below the electron multiplication threshold. The collected charge reaches a plateau at low electric fields and should be flat up to the threshold of the electron multiplication process; it reflects the fact that the total number of photoelectrons created in the CsI photocathode is fully collected on the anode. Above the multiplication threshold an exponential increase of the collected charge is measured as a function of the electric field. When CF_4 mixtures are used, however, a deep fall of the collected charge is observed just below the multiplication threshold. This drop is due to an electron capture phenomenon occurring when photoelectrons reach an energy of about 8 eV [9]. At 8 eV the electron absorption cross-section is of the order of 1 Mb which corresponds to an electron mean free path of the order of several mm (for a typical CF_4 concentration of 10%). This absorption does not affect significantly the gain achieved. Since the chamber operates at high gain, charge multiplication occurs in a mean distance of 300 μm, which is much shorter than the electron mean free path.

Fig. 3 shows the gain obtained with various partial pressures of CF_4 at low pressure as well as gas mixtures of neon and helium with CF_4 at atmospheric pressure. A satisfactory gain of higher than 10^4 was obtained. Table 1 summarizes the maximum gain achieved with various gas mixtures at different partial pressures. Similarly, the multiplication factor is shown in Fig. 4 using a continuous light source. As indicated in Table 1 the best options are either pure CF_4 at various partial pressures or gas mixtures of a noble gas and CF_4. Using CH_4 instead of CF_4 one can achieve satisfactory gains; the transparency, however, is much poorer. The gain limitation observed for some of the gas mixtures containing Ar or Kr as carrier gas is probably due to the photon feedback mechanism since these gases

Table 1
Maximum stable gain and HV reached for various HBD gases and gaps between anode and cathode

Gas mixture	Pressure (bar)	Gap (mm)	Maximum gain	Voltage (kV)
CF_4	0.35	3	10^5	5.15
	0.25	4	10^5	4.7
	0.15	4	5×10^7	3.1
	0.05	4	1.5×10^8	1.45
CH_4	0.5	4	10	5.5
He + 4% CF_4	1.0	4	10^4	3.05
Ar + 4% CF_4	1.0	4	1	4.8
Ar + 1% CF_4	1.0	4	1	3.5
Ar + 0.5% CF_4	2.0	4	1	4.0
	1.0	4	2	3.35
Ar + 1% CH_4	1.0	4	1	3.8
	0.5	4	1	2.5
Ar + 1% CF_4 + 1% CH_4	1.0	4	3	3.9
Ne + 4% CF_4 + 21% He	1.0	4	10	4.7
Kr + 10% CF_4	0.2	4	5×10^3	2.0
Kr + 20% CF_4	0.22	4	10^3	2.3

are scintillating in the bandwidth where CsI is sensitive (120 and 150 nm, respectively).

Since neon and CF_4 are having high indices of refraction $(n - 1) = 80 \times 10^{-6}$ and 400×10^{-6} respectively, the Cherenkov light production is higher compared to a helium gas mixture. We have performed the first beam tests with these two gas fillings: CF_4 or Ne + CF_4.

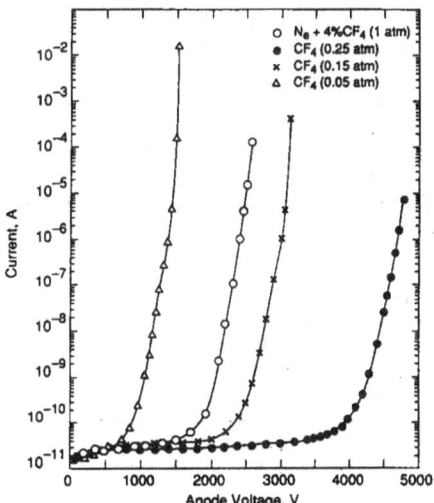

Fig. 4. The measured DC current gain versus high voltage across a 4 mm gap for various gas mixtures in an HBD.

4. Beam tests

A prototype of the detector was tested at a particle beam containing electrons/pions of 6–50 GeV/c. The first anode plane used wires of 50 μm diameter spaced every 800 μm and was found to produce a non-uniform electric field. To improve the field homogeneity we have replaced the anode plane with a wire mesh plane composed of 50 μm wires interwoven every 500 μm.

The triggering system was composed of three plastic scintillators two placed in front of the apparatus and one behind it. The upstream counters had an active area of 3×3 cm^2, while the third one was 1.5 cm^2. A threshold Cherenkov counter was installed upstream of the whole setup to select pions or electrons. The gas filling was He at 3 atm ensuring a pion rejection up to 9 GeV/c. Two multiwire drift chambers placed one in front and one behind the HBD prototype were read out along with the HBD data providing us with track information. Their space resolution is 150 μm. To increase the electron selection efficiency we have installed a BaF$_2$ crystal of 10 X_0 downstream from the HBD. The electron trigger was provided by the coincidence signal of the three hodoscopes in conjunction with the threshold Cherenkov counter and the BaF$_2$ signals. To investigate the HBD pion rejection capabilities we have collected data using a pion trigger consisting of the coincidence of the three hodoscopes vetoed by the BaF$_2$ signal. The threshold Cherenkov was not used since its efficiency was approximately 10%. Fig. 5 shows the schematic arrangement of the apparatus during the tests. We estimate that the pion contamination of the electron sample was approximately 2%. The beam size spot was 15 mm diameter.

The HBD was filled with 250 mbar CF_4 partial pressure. The incident particle momentum was 6 GeV/c, below the pion Cherenkov threshold for both the HBD ($p_{thr} \approx 9$ GeV/c) and the threshold Cherenkov counter. At this gas pressure we expect that the Cherenkov angle θ is 15 mrad and the corresponding spot size at the photodetector plane is 12 mm diameter. Due to the Cherenkov angle of emission we expect that the signal coming from a particle above the Cherenkov threshold will be spread to several adjacent strips in the sensitive detector area. Fig. 6

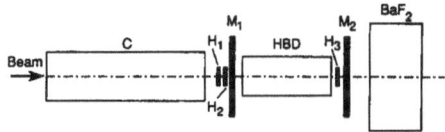

Fig. 5. The schematic drawing of the PS beam setup. The HBD prototype is placed behind a threshold Cherenkov counter (C), two plastic scintillator hodoscopes (H$_1$, H$_2$) and a MWPC (M$_1$). Downstream from the HBD is the H$_3$ counter, a MWPC (M$_2$) and a 20 cm long BaF$_2$ crystal.

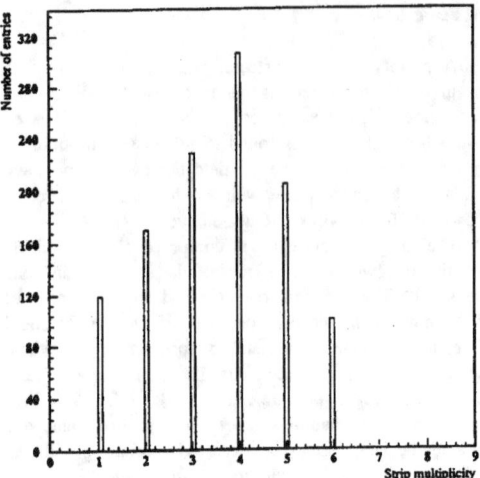

Fig. 6. Distribution of the number of 4 mm wide cathode pads fired by the Cherenkov light of 6 GeV electrons.

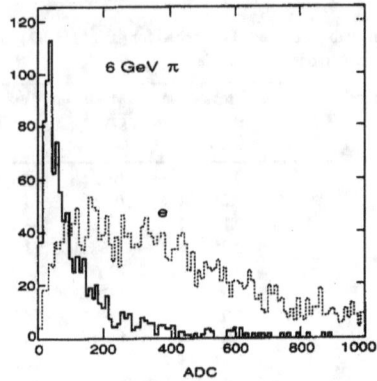

Fig. 8. The measured ADC distributions for 6 GeV electrons and π's, from the test HBD, filled with 0.2 bar of CF_4.

shows the multiplicity of 4 mm wide strips for single particle events above the Cherenkov threshold.

The alignment was performed by varying the focusing magnets provided for the beam extraction. We had the possibility to independently change the horizontal and the vertical beam position. Fig. 7 shows the mean HBD re-

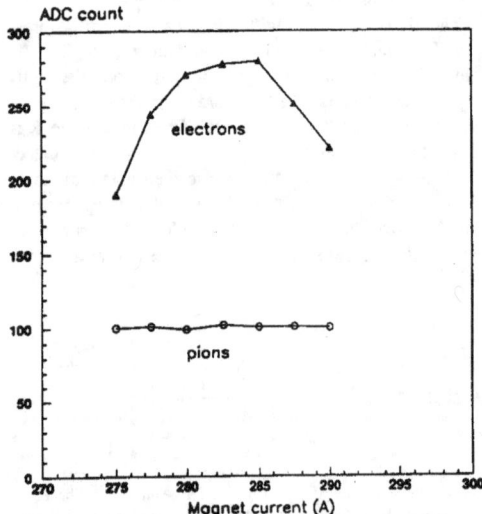

Fig. 7. Mean collected charge vs. the vertical beam position for electrons and pions. Pedestal, about 70 ADC counts, has not been subtracted.

sponse using both electron and pion trigger. As one can see, the mean collected charge for pions remains flat for all values of horizontal and vertical beam displacement, while the corresponding electron curve exhibits a broad peak when the beam is aligned to the prototype. The higher absolute value of the mean collected charge is due to the Cherenkov light produced in the case of electrons, while the peak itself is due to the limited size of our photodetector (30 mm in diameter).

Using the track information from the two drift chambers we have rejected double tracks and plotted the HBD mean collected charge versus the horizontal (or vertical) track position. Again, the electron curve has a mean value that is 4–5 times higher than that of pions and is peaked at the HBD geometrical centre (Fig. 1). The charge distribution, for both electrons and pions, is shown in Fig. 8. Most of the pion collected charge is at the pedestal level confirming the insensitivity of the detector to minimum ionizing particles. The detector inefficiency at the pedestal is about 10%. The number of detected photoelectrons is estimated to be > 4.

The number of photoelectrons N, produced in a photodetector is:

$$N = N_0 L \sin^2\theta_C, \qquad (1)$$

where L is the length of the radiator. θ_C is the Cherenkov emission angle and N_0 is a quality factor depending on the transparency of the radiator and the detector, and on the photodetector quantum efficiency. In most Cherenkov counters a good quality factor is of the order of 100. We have observed a quality factor of 500. This is primarily due to the fact that the wavelength range of sensitivity of our detector is not limited (as usually) by the window transparency cutoff, but extends up to the gas transparency cutoff which is 12 eV for CF_4 [10].

5. Future improvement

As shown in Figs. 8 and 9, although a clear difference between pions and electrons has been observed, several factors should be improved to obtain much better e/π separation:

1) The high tail of the π spectrum is due to contamination of electrons and a better trigger is required.

2) The low tail of electrons is partly the result of non-uniformity in gain due to the woven mesh structure of the anode, which will be improved by using a flat grid or a very fine mesh.

3) Fluctuations due to the electron capture in CF_4 during the electron multiplication process [11]. This process requires more systematic investigation and it may be the cause of time fluctuations observed on the output HBD signal. One way to suppress such fluctuations is the use of a special filter, Nanochem purifier, well known for being able to get rid of electron negative components inside CF_4. Another way is the addition in the binary $Ne + CF_4$ gas mixture of a ternary gas. In a previous work it was demonstrated that the addition of a third component (CO_2 or H_2O) boosts the electron multiplication and has excellent thermalizing properties, resulting in a reduction of the fluctuations and suppression of the electron capture of the CF_4 [12]. In our case we need a far UV transparent third gas mixture. Candidates are nitrogen and hydrogen gases.

4) The expected number of photoelectrons produced by high energy electrons can be more than doubled by using 0.5 bar of CF_4 + 0.5 bar of Ne or 1 bar of Ar + CH_4.

5) The use of fast amplifiers will enhance signals of the photoelectrons produced at the cathode relative to ions produced in the gap.

6. Conclusions and outlook

The results of the first particle beam test of a HBD prototype are encouraging: a large quality factor of 500 was obtained, far exceeding what have been achieved with conventional detectors. It confirms the first goal of the HBD concept: the absence of a window separating the Cherenkov radiator and the photon detector can improve the detected Cherenkov photon yield. To be sensitive to the EUV light, CF_4 based mixtures were used to increase the emitted photon bandwidth. Several mixtures, at low or atmospheric pressure, suitable for the HBD were tested. Admixture of Ne or He carrier gas with CF_4 are very promising. They combine high far UV transparency and stable and high gain for the PPAC detector. The electron capture effect during the avalanche development in CF_4 mixtures is not a serious obstacle. It results in a small loss of about 10–30% primary created photoelectrons, depending on the CF_4 concentration. Pure He or Ne gas mixtures exhibit a photoelectron back-diffusion process, which appears to decrease the quantum efficiency of CsI photocathodes by a factor of 2–3. This loss is recovered by the addition of a few percent of the $CF_4(CH_4)$ quencher.

The second goal of the HBD concept was also demonstrated: the quite low efficiency of the PPAC on the primary ions created in the gas by minimum ionizing particles. Depending on the gas admixture used and the gain of the PPAC, this inefficiency was of the order of 10–30%. Full efficiency was measured for high energy incident electrons. The pion rejection capability of the device was limited by a broad pulse height distribution, larger than the expected statistical fluctuations, observed in the electron sample. This effect, as well as the time instabilities of the output signal observed for some CF_4 gas mixtures, is under study. It could be due to two effects: electric field inhomogeneities near the anode mesh or/and fluctuations during the avalanche process due the CF_4 electron capture effect. Further experimental investigations are needed to improve the photodetector, the electronic chain and the gas mixture. The quantum efficiency of CsI cathodes could be improved by using better polished surfaces, as used in previous investigations [13]. Better electric uniformity can be reached by using very fine meshes. We expect the optimization of such different components should allow one to obtain a fully efficient HBD in the next years.

Acknowledgements

We are grateful to S. Aronson, S. Nagamiya, Y. Akiba; R. Hayano, T. Hemmick, and Nu Xu for their encouragement and support as well as many useful discussions. We are indebted to our colleagues from the L3-SMD collaboration, in particular R. Leiste, T. Coan and B. Zhou, for providing us with the data acquisition system during the

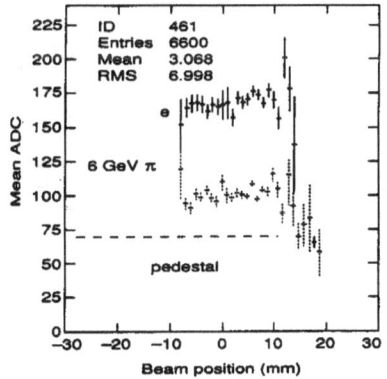

Fig. 9. The measured mean ADC values for 6 GeV electrons and π's, versus the transverse beam position across the test HBD, filled with 0.2 bar of CF_4.

SPS tests. This work is partially supported by DOE Grant # DE-FG02-93ER40790 and by the PHENIX Collaboration.

References

[1] Y. Giomataris and G. Charpak, Nucl. Instr. and Meth. A 310 (1991) 589.

[2] M. Chen et al., Proc. Int. Conf. on Liquid Radiation Detector, Waseda University, April 1992; also PHENIX Workshop, BNL, Aug. 24, 1992.

[3] PHENIX Conceptual Design Report, January 29, 1993.

[4] M. Chen, L3H A Muon–Electron–Photon Detector for LHC Heavy Ion Physics, CERN/LHCC/93–15.

[5] Y. Giomataris and G. Charpak, Proc. Int. Lepton Photon Symp. and Europhys. Conf., Geneva July 1991, 2 (1991) 251.

[6] Y. Giomataris et al., Nucl. Instr. and Meth. A 323 (1992) 431.

[7] J. Seguinot et al., Nucl. Instr. and Meth. A 297 (1990) 133.

[8] J. Fischer et al., Nucl. Instr. and Meth. A 238 (1985) 249.

[9] S.R. Hunter and L.G. Christophorou, J. Chem. Phys. 80 (12) 1984;
L.G. Christophorou et al., J. Appl. Phys. 71 (1992) 15;
J. Va'vra, Nucl. Instr. and Meth. A 323 (1992) 34.

[10] W.R. Harshbarger et al., J. Electron Spectrosc. 1 (1972/73) 319.

[11] S.F. Biagi, Nucl. Instr. and Meth. A 310 (1991) 133.

[12] P.G. Daskos, J.G. Carter and L.G. Christoforou, J. Appl. Phys. 71(1) 1992.

[13] G. Charpak et al., CERN/PPE/92–220, submitted to Nucl. Instr. and Meth.

Nuclear Instruments and Methods in Physics Research A306 (1991) 439–445
North-Holland

A trigger for beauty

G. Charpak [a], Y. Giomataris [b] and L. Lederman [c]

[a] *CERN, Geneva, Switzerland*
[b] *World Lab, Lausanne, Switzerland*
[c] *Fermilab, Batavia, IL, and University of Chicago, Chicago, IL, USA*

Received 21 February 1991

The possibility of B-meson experiments, in a fixed-target high-energy proton machine (Tevatron) is discussed. Compared to a B-meson factory experiment, it can produce 10^5 $B\bar{B}$'s per hour, using 10^8 protons per second, but it suffers from high background and needs high selectivity to cope with the million times higher interaction rate. To overcome these difficulties a technique called the "optical trigger for beauty" is proposed, based on the detection of Cherenkov photons produced in a 2 mm thick LiF crystal, through a fast photodetector. Its virtue is that it is opaque to minimum-bias events originating in a small target, but sensitive to the high impact parameter B-meson decay charged particles from a secondary vertex. Calculations and first simulations results give a good efficiency for B-meson detection. A multistep trigger, combining the "optical trigger" and a tracking detector, allows significant selection and a consequent enrichment of the data sample. Taking into account its fast response (~ 1 ns), the above considerations can be extended to other hadronic machines, especially those with high-rate environments such as the LHC or SSC.

1. Introduction

A major challenge to experimental high-energy physics as carried out with proton accelerators has to do with the fact that protons are complex objects. The standard image of protons, especially as characterized with some justice by advocates of electron machines, resembles a "garbage can", where the banana peels, coffee grounds and egg shells are replaced by quarks and gluons. Gluons themselves are capable of virtual dissociation into pairs of quarks, further increasing the messiness of our picture of the proton as a probe of subnuclear physics.

The collision of two complex objects then has two major disadvantages over the collision of the elegantly simple and point-like (i.e. structureless) electrons. One is the fact that the energy per constituent is decreased over the laboratory energy of the proton; a rule of thumb suggests that this factor is about 5, for an average collision in which the interest is in the quark–quark or the quark–gluon collisions. The other is the presence of a multitude of spectator objects, many of which will acquire some of the energy of the collision and assist in clustering the detector. The saving grace for protons is again twofold in this simplified debate. For one, protons are easier and cheaper to accelerate, and this more than compensates the factor of 5. Also, the cross sections for a quark–quark hard collision to produce a particular interesting object, say W or B particles, is much larger than the equivalent electron–positron cross section. It is not inhibited, as is the e^+e^- collision, by the necessity of exciting only the quantum numbers of the intermediate photon or Z^0.

The key problem then is most dramatically illustrated by (but by no means restricted to) the hadronic production of heavy quarks, i.e. charmed mesons and beauty mesons. Here, the production rates can be a factor of 100 higher than the ones at the equivalent e^+e^- total energy sitting on a resonance like the J/ψ or the ψ', etc. The equivalent hadronic production is, however, about 10^{-3} of σ_t, with Fermilab's 800 GeV protons hitting a target.

The production and study of charmed mesons was the exclusive province of e^+e^- colliders from 1975 until the Fermilab experiment E691, using two new technologies, successfully beat down the backgrounds, actually in a photoproduction experiment. In order to distinguish between an event originating in the target from a decay of a charmed meson near the target, E691 used the then-new silicon microvertex technology [1]. The laboratory lifetimes were such that impact parameters of the order of 50 μm had to be detected. E691 and its follow-up hadroproduction experiments used a very loose trigger and a then-new parallel processing technology (ACP) developed with the specific objective of handling huge amounts of data and doing the event reconstruction rapidly. E761 wrote over 10 000 high-density magnetic tapes in their 1988 run on hadroproduction of charm. The data are being processed with an ACP system.

Whereas it is true that data-recording technology is improving, it seems that a more selective trigger, which provides an enrichment in charm events by a factor of 100 or more, would encourage a more sophisticated on-line analysis, a relatively refined data storage, and perhaps a quicker off-line analysis.

When all these problems are applied to beauty particles, the motivation for trigger development becomes far more forceful. Whereas lifetimes in the laboratory system are far more favourable in the B-meson case, the fraction of the total cross section is now only 10^{-6}. This minuscule cross section (20 nb) is still large compared with the one in e^+e^- colliders, which have again dominated the subject since the discovery of b quarks at Fermilab in 1977. For example, in a beam producing 10^8 interactions per second of 800 GeV protons, some 10^2 $B^0\overline{B}^0$ are produced. Fermilab's duty cycle is such that 2×10^3 $B\overline{B}$'s can be produced per minute or 10^7 per 100 hour week. It is clear that with reasonable attention to acceptances and efficiencies, huge yields can in principle be expected. As of late 1990, not a single B^0 event has been seen in the fixed-target program. The challenge is in the huge rates (10^8 per second is equivalent to a luminosity of 10^{33} cm^{-2} s^{-1} in a collider) and the huge backgrounds.

The motivation for solving these problems is very great. The production and decay of B mesons contains some of the least-known parameters of the standard model. More than that, it is another approach to one of the most crucial problems of our times, the famous CP-symmetry breakdown heretofore only observed in the neutral K system. This process is generally considered to be the origin of matter–antimatter asymmetry in the Universe and is known as the "origin of matter" problem. This has been widely appreciated and has resulted in many proposals throughout the world for the construction of "beauty factories" at a cost of many hundreds of millions of dollars. Thus, the drive to find a way to use the intense fixed-target source is very amply motivated.

2. The virtues of a trigger for b-quarks

Much of the specific examples are derived from a current Fermilab experiment, E789 [2], which hopes to see B-mesons via their presumed two-body decay modes ($\pi^+\pi^-$, $K^+\pi^-$, K^-K^+, $\cdots$). The prevalence of silicon microvertex hardware and possible improvements (e.g. diamond detector [3]) is essential. Suppose one developed a "zero-level" trigger which could select events generated in space near the target and which would thereby enrich the data in B-meson events by a factor of, say, 50. The Fermilab duty cycle has 10^7 buckets per second; with 10 interactions per bucket, the event train following this trigger would have an average of 1 event

per 500 ns. In this time, level-1 on-line event analysis could further filter the events to gain another factor of 50. This in turn permits a very sophisticated on-line analysis (level-2), which can lead to a comfortable $\approx$ 1000 events/min for recording on tape. If the various levels of filtering did not unacceptably reduce the efficiency, some 10 $B\overline{B}$ decays are recorded per minute with a typical 1% acceptance. The very selective E789 spectrometer adds another 10^{-5} in branching ratio for each two-body channel, but even here, some few hundred events would be recorded and fully reconstructed. Thus, the challenge to the instrumentalist is to distinguish between events arising from a target of 200 μm diameter, 1 mm thick, and decays of B's whose laboratory mean path length averages about 1 cm, with a mean impact parameter of 800 μm. If this is to serve as a zero-level trigger, the decision time must be comparable with or less than 20 ns, the radio frequency bucket spacing.

3. The proposal

In the next sections we describe a new detector, based on the Cherenkov photon detection, satisfying the main requirements for a zero-level trigger of a Tevatron fixed-target experiment. This detector, named "optical trigger", is sensitive to the main characteristics of a B-meson event, a high decay-particle impact parameter, high multiplicity, large polar angle with respect to the beam axis. Moreover, its 1 ns (or less) response makes it a candidate for even higher-rate environments, as with the LHC or SSC machines.

3.1. Review of the Cherenkov effect

The emission angle of Cherenkov photons by an ultrarelativistic particle traversing a medium of refractive index n, is given by the relation: $\cos \theta_{Ch} = 1/n$, and the number of emitted photons in an interval of energy ΔE is

$$N = (\alpha/\hbar c) L \, \Delta E \, \sin^2 \theta_{Ch}, \tag{1}$$

where α is the fine structure constant, $\hbar$ Planck's constant, c the velocity of light, and L the thickness of the radiating medium. The limiting angle for total reflection by a plane, for a photon travelling in a medium of index n, is given by the relation: $\sin \theta_{tr} = 1/n$. As a consequence, for particles orthogonal to the limiting surface, if $n = \sqrt{2}$ the emitted Cherenkov photons are exactly at the limit for total reflection, if $n \geq \sqrt{2}$ the photons are trapped, if $n \leq \sqrt{2}$ they escape. If the particles are within a narrow cone around the line perpendicular to the surface, the Cherenkov photons can all be transmitted or reflected, depending on the value and sign of $(n - \sqrt{2})$ [11].

A detector based only on this configuration is sensitive to relatively large polar particle angles, which is the case of the decays of B mesons, but its rejection of minimum-bias events is certainly poor. Another way to control the angular spread of the beam so as to favour total trapping or transmission of the photons is to adjust the curvature of the exit surface limiting the medium.

3.2. The principle of the optical trigger

The beam target is at the centre of the spherical boundary of the medium of index n. A hole follows the target in such a way that non-interacting primaries as well as a substantial fraction of the secondary particles traverse the medium without any interaction.

The refractive index n, close to $\sqrt{2}$, and the curvature are chosen in such a way that Cherenkov photons, produced in the medium by charged particles from the target, escape through the surface. However, a small fraction of the photons is reflected. Since there are over 10^6 times as many tracks from the target as B's, even a small fraction is unacceptable. It is possible to coat the surface in such a way that a substantial part of these photons is transmitted. The principle of antireflecting coating is very efficient for a given wavelength band and for a given angle of incidence. Fortunately all the photons cross the surface at the same angle, within a narrow band determined by the width of the beam and

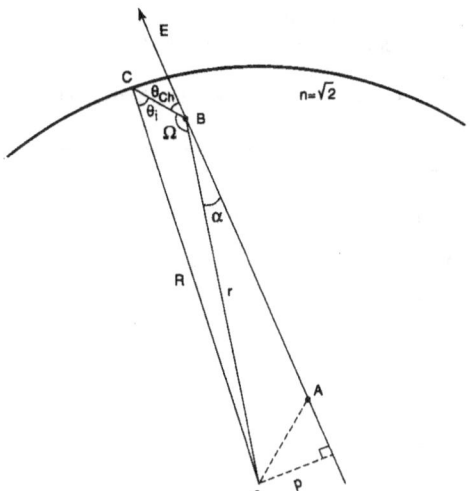

Fig. 1. Schematic of a decay particle passing through a crystal, having a spherical surface, of refractive index $n = \sqrt{2}$. Here OA is the flight of B^0; ABE is the decay particle; BC is the Cherenkov photon; θ_{Ch} is the Cherenkov emission angle in a medium of refraction index n; θ_i is the angle of incidence of the photon with the exit surface.

the target. Although this technique is not excluded, a better one, based on the multireflection suppression of background photons, seems possible. It is discussed below.

The B mesons are produced at some distance from the target and their decays are emitted at angles slightly larger than the average angle of the secondary particles from minimum-bias events, giving rise to a relatively large impact parameter. These particles generate Cherenkov photons, some of which are sufficiently inclined to the exit surface to be totally reflected. They are focused by the surface into a ring or a fraction of a ring whose position depends on the direction of the particle.

An example of the process is illustrated in fig. 1, where a B meson decays at point A, giving a charged particle ABE. An exact calculation has to take into account three-dimensional geometry, since the Cherenkov light can be emitted out of the plane defined by the trajectories of the particle and the B meson. If the Cherenkov photon is emitted in this plane, if p is the impact parameter of the particle ABE, and if $n = \sqrt{2}$, it is straightforward to find the condition for internal total reflection. From triangles OAB and OCB respectively, we get:

$$\sin \alpha = \frac{p}{r}, \tag{2}$$

$$\frac{\sin \theta_i}{r} = \frac{\sin \Omega}{R}. \tag{3}$$

From eq. (3) we find the angle θ_i at which the photon is incident upon the exit surface:

$$\sin \theta_i = \frac{r}{R} \sin(\theta_{Ch} + \alpha). \tag{4}$$

The condition for total internal reflection is then

$$\frac{r}{R} \sin(\theta_{Ch} + \alpha) \geq \sin \theta_{crit} = \frac{1}{n} \cong \frac{1}{\sqrt{2}}, \tag{5}$$

where θ_{crit} is the critical angle; since $\theta_{crit} \cong \theta_{Ch} \cong 45°$,

$$\sin \alpha + \cos \alpha \geq \frac{R}{r}. \tag{6}$$

Since α is generally small, we can use eq. (2) to find:

$$p \geq R - r = \delta r, \tag{7}$$

where R is the radius of the exit surface and r is the flight distance of the decaying particle. In the general three-dimensional case, the exact calculation is very complicated. Using relation (26) from ref. [4] and assuming small particle angles with respect to the beam axis, we can derive a more general relation becoming, to first approximation:

$$\delta r = p \cos \phi,$$

where ϕ is the azimuthal emission angle of the Cherenkov photon. This relation can be satisfied only if $-90°$ ≤ ϕ ≤ $90°$, which excludes half of the Cherenkov photons. The condition indicates also that only a limited part of the particle path is useful. The background photons from particles emerging from the interaction point never satisfy the total-reflection condition since their impact parameter is zero; they are largely transmitted.

3.3. Monte Carlo simulations results

The production of $B\bar{B}$ pairs in proton–proton collisions has been simulated using the PYTHIA [5] program, for 0.9 TeV incident energies on a fixed target. The program assumes gluon–gluon fusion and quark–quark interaction for $B\bar{B}$ production.

Fig. 2 shows the momenta distribution and fig. 3 the polar angular distribution of the produced B mesons. The mean momentum is 160 GeV and the mean polar angle about 20 mrad. The distance of the vertex of B decay particles from the interaction point is of the order of 15 mm, as shown in fig. 4. This distance is certainly a little higher than the B-meson's expected mean path length. However it includes a small part of any D-meson path produced in B-meson decays. This is favoured by our optical trigger, since it raises the mean impact parameter of charged particles produced. As shown in fig. 5 the impact parameter is of the order of 1 mm. The average momentum of B-meson decay particles shown in fig. 6 is 20 GeV. Their polar angle, shown in fig. 7, is quite large (≈ 50 mrad), which also favours our experimental technique.

The main message coming from the above study is that the impact parameter is significant (≈ 1 mm) for Tevatron energies.

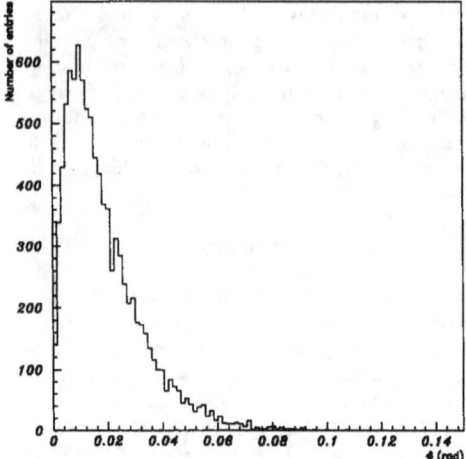

Fig. 3. Laboratory angular distribution of B mesons.

With a 1 mm thick radiator, we can have a first estimate of the number of photoelectrons collected in the photodetector. Using eq. (1), the number of photons produced is

$$N/(eV\,cm) = 370 \sin^2\theta_{Ch} = 185$$

$$[\text{or } N/(eV\,mm) = 18.5].$$

Assuming that only a part ($\sim 0.5 \langle \cos \phi \rangle$) of the photons are lost, because they do not satisfy the total-reflection condition (the exact number should come from a full Monte Carlo), a quantum efficiency of 30%, neglecting losses due to absorption, and taking 1 eV as the photon energy acceptance of the photodetector the

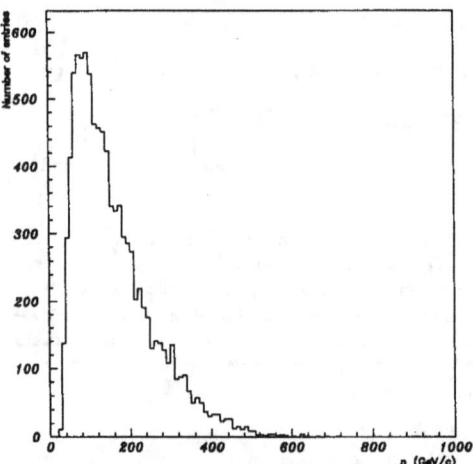

Fig. 2. Laboratory B-meson momenta distribution in a pp interaction, with 900 GeV incident proton energy.

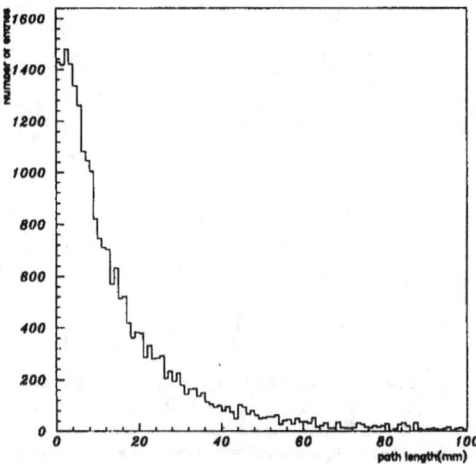

Fig. 4. B-meson path length in millimetres.

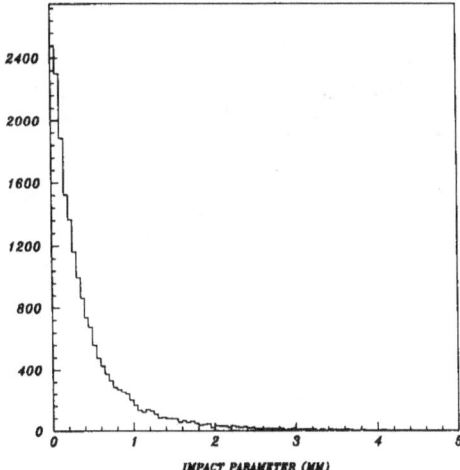

Fig. 5. Impact parameter of B-meson decay particles.

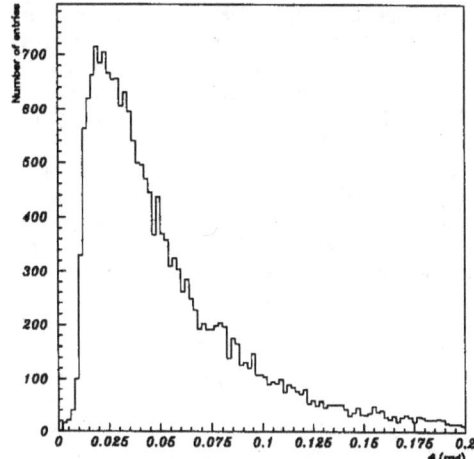

Fig. 7. Laboratory angular distribution of charged particles emerging from B-meson decays.

number of detected photoelectrons per charged decay particle is

$$N/(0.5 \text{ eV} \times 1 \text{ mm}) = 0.5 \times 0.3 \times 18.5 \langle \cos \phi \rangle = 2.8.$$

The number of detected photoelectrons per charged particle should be multiplied by the average charged particle multiplicity in $B\bar{B}$ decays. Assuming an average of 5 particles per B meson, the total number of photoelectrons detected is about 28. This high number can compensate for some possible optimism in our calculations, but gives a reasonable order of magnitude. A reserve factor is the thickness of the radiator, but here

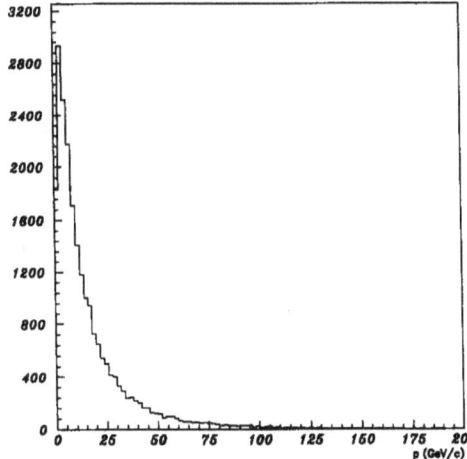

Fig. 6. Momenta distribution of charged particles, emerging from B-meson decays.

we also add nuclear interactions of the target-originated tracks.

3.4. The Cherenkov radiator

A possible, but not unique candidate for the Cherenkov radiator is LiF, chosen for its low Z and absence of scintillating light. For $\lambda = 300$ (500) nm, the index of refraction is 1.405 (1.395), corresponding to an angular dispersion of a few milliradians. This is a very comfortable region since a normal photomultiplier with a Pyrex window can be used. A selection filter is not necessary; the energy bandwidth is automatically defined between the threshold of the photocathode and the cut-off of the window used.

Other crystals having relatively low refractive index are MnF_2 and NaF. The first has about the same index as LiF, it does not seem to scintillate, and it is a stable and reliable material. Some liquids like hydrocarbons (heptane, C_2H_2, $\cdots$) are attractive because of their low dispersion index in the visible, and of their high radiation and interaction lengths, but they need to be enclosed. However, they are not a priori excluded, since at low temperatures their crystals can also be considered. A systematic search may lead to the discovery of other potential candidates for the Cherenkov radiator. The possibility to lower the refractive index of other crystals by doping them, or by changing the temperature or the pressure, can also be envisaged.

3.5. The photodetector

As a photodetector, one can use any photomultiplier having the right photon acceptance, with the highest possible quantum efficiency. Therefore it must be of the reflective type (head-on), requiring an optimum angle between the photocathode plane and the incident photon beam. As was shown in the previous section, the Cherenkov photons are emitted with little dispersion, since the charged particles have an angular distribution of the order of 50 mrad. By choosing the right angle, large quantum efficiencies ($\approx 50\%$) can in principle be obtained. Another important feature of a good photomultiplier is its high gain ($\geq 10^6$), permitting excellent single photoelectron detection spectra. Hence, imposing an on-line threshold on the collected charge should reduce the backgrounds. Finally the fast photomultiplier response, combined with the fast Cherenkov light, permits its operation in a very high rate environment ($> 10^8$ interactions per second).

Silicium photocathodes have very high quantum efficiencies, in particular in the ultra-violet region, but they are not, in general, able to detect single photoelectrons. However, techniques are under development to obtain high gains. A new type of photodiode, presented by Atac [6] at Snowmass, which has high granularity, high quantum efficiency, high gain, should be considered. Although its quantum efficiency is high, 60% between 300 and 600 nm, its response is not very fast. It may be possible to bring the rise-time of this detector to the 10 ns level, with further development.

4. A first design of an optical trigger detector

Only a full Monte Carlo program, simulating not only the detector and the event generation, but also secondary interactions in the crystal, plus possible physical backgrounds, is required for us to reach a final design of our device. However, our preliminary calculations under certain approximations are confirmed by a first Monte Carlo detector simulation program, and a tentative design is presented. Since only the last millimetre or so of the track generates light in the radiator, it is possible to choose a configuration different from that of fig. 1. This consists of a narrow shell, 2 mm thick, between two parallel spherical surfaces, having a radius of curvature of about 50 mm. The thin target sits at the centre of curvature of the crystal. Fig. 8 illustrates the principle: photons satisfying condition (7) are totally reflected and, after multiple reflections (about 30), are collected at the edges and detected. The quality of the crystal's surface must be excellent to ensure that losses be less than 1% per reflection. This is technically possible for small objects. Obviously, after 30 reflections, background photons from target-originated tracks that manage to be reflected should be completely attenuated, and antireflecting coating on the surface is not necessary. The Cherenkov photons are polarized and the reflectivity on the radiator surface is polarization-de-

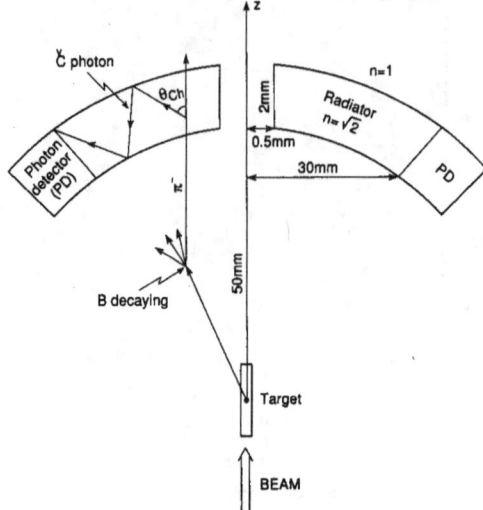

Fig. 8. Illustration of a tentative design of an "optical trigger".

pendent. In our calculation, we have taken these effects into account; they alter our conclusions by only a negligible factor.

Assuming a perfect response of our device, the main limitation will come from hadronic interactions in the crystal producing secondary particles with a high impact parameter. By having only 2×10^{-3} of the interaction length with 2 mm LiF and additional reduction of secondary interactions due to the beam hole, this background should reduce the trigger rate by two or three orders of magnitude. The other limitation will come

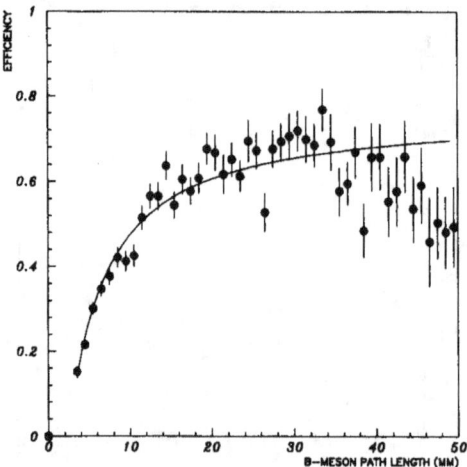

Fig. 9. The detection efficiency of the optical trigger for B mesons, as a function of their path length.

from δ-rays generated in the crystal transversely to the incident particle beam, since their energy can surpass the threshold of about 1 MeV and produce Cherenkov light. This irreducible background has a relatively high probability estimated to be about 1%, in agreement with experimental measurements of the Cherenkov light produced in glass of 2 cm thick [12]. However, it is a background confined in the crystal and therefore a coincidence in a system of two such crystals close to each other should reduce it to a negligible level, with very small expected influence on the signal.

The factor of 50 to several hundred as a zero-level trigger then gives the on-line data acquisition system time to add further requirements to the event, e.g. a fast tracker that identifies the radiator as the source of the trigger, transverse-momentum cuts, and at the level 2 or 3, a vertex processor that confirms the presence of a secondary vertex.

This method can meet the challenge of a high-rate fixed-target experiment. First simulation results show that an efficiency well over 20% can be obtained. Fig. 9 shows the expected B-meson detection efficiency as a function of the B path length. The efficiency is better than 50% for paths greater than 10 mm, containing more than 30% of the produced B mesons. The previous considerations and the first Monte Carlo simulation results are encouraging; further work and tests are necessary to confirm the good functioning of the trigger.

5. Future possible improvements

Improvements of this system will come from the design of photodetectors which will be insensitive to the background emerging from the target. In photomultipliers, the Cherenkov radiation produced in the glass forces heavy shielding or requires that the crystal shell be as far as possible from the target. With proper shielding, scintillation counters surrounding the photon detector may be useful, imposing an additional veto to the optical trigger signal. With solid-state photodiodes, the problem is similar, and even worse because of the direct detection of the charged particles.

Gaseous detectors, with CsI photocathodes of the type recently developed [7], would be an ideal solution, since high quantum efficiency (40%) has been measured in the UV region. Further development is necessary to show if a shift to the visible is possible. However, operation in the UV region of a CsI photocathode with adsorbed TMAE is possible [8] and compatible with our device. This has the advantage of being much less sensitive to direct ionization from charged particles, since the gaseous detectors can be filled with helium gas [9] or with mixtures of gases at very low pressure (≈ 1 Torr) [10].

Improvement could also come from the transfer of the light emerging from the crystal, to a safe distance from the target, with an appropriate optical system made of mirrors and lenses, shaping the end of the crystal.

6. Conclusions

Our study has shown that using the optical properties of the Cherenkov light produced in a radiator of appropriate shape, it is possible to separate photons produced by particles emerging from the target, from those produced by a particle decaying at some distance from the target.

The parameters corresponding to $B\bar{B}$ pairs produced by a 0.9 TeV accelerator seem appropriate for this technique, which appears to be of quite general use, i.e. with accelerators at different energies or with other types of unstable particles.

The optical trigger is best suited for rejecting high-intensity unwanted events, since it avoids the overloading of the triggering counters by drastically reducing their signals.

Acknowledgements

We would like to thank A. Coudert, G. Prost and M.-S. Vascotto, for their competent help in preparing this article for publication. This work was supported in part by the US Department of Energy.

References

[1] J.C. Anjos et al., Phys. Rev. Lett. 60 (1988) 1379.
[2] D.M. Kaplan et al., Fermilab proposal E-789 (1988).
[3] D. Kania and co-workers, Expression of Interest SSC EOI-9 (1990); and
D. Kania et al., submitted to Appl. Phys. Lett. (1990).
[4] T. Ypsilantis, preprint CERN-EP/89-150, talk given at the Workshop on Particle Identification at High Luminosity Hadron Colliders, Fermilab, 1989.
[5] M. Bengtsson et al., The Lund Monte Carlo Programs long write-up, CERN (1989).
[6] R. Atac et al., talk given at DPF Summer Study on High Energy Physics in the 1990s, Snowmass, 1990.
[7] J. Séguinot et al., preprint CERN-EP/90-88 (1990).
[8] D. Anderson et al., preprint Fermilab-Pub-90/182 (1990), submitted to Nucl. Instr. and Meth.
[9] G. Charpak et al., Nucl. Instr. and Meth. 161 (1979) 19; and
Y. Giomataris et al., Nucl. Instr. and Meth. A279 (1989) 322.
[10] A. Breskin et al., Nucl. Instr. and Meth. 220 (1983) 349.
[11] V. Fitch and R. Motley, Phys. Rev. 101 (1956) 496.
[12] C. Biino et al., Nucl. Instr. and Meth. A295 (1990) 102.

8. APPLICATIONS OF DETECTORS TO BIOLOGY OR MEDICINE: 1970–1992

8. APPLICATIONS OF DETECTORS TO BIOLOGY OR MEDICINE: 1970–1992

Nuclear Instruments and Methods 172 (1980) 603–608
© North-Holland Publishing Company

A XENON HIGH-PRESSURE PROPORTIONAL SCINTILLATION-CAMERA FOR X AND γ-RAY IMAGING

Hoan NGUYEN NGOC, Jack JEANJEAN, Hidehiko ITOH * and Georges CHARPAK

Laboratoire de l'Accélérateur Linéaire, Université Paris XI, 91405 Orsay, France

Received 2 October 1979

A γ-ray imaging detector, working at ambient temperature, with good energy resolution, good spatial accuracy and reasonable detection efficiency has been built.

Typical energy resolution is 6 keV at 122 keV, with 3 mm spatial resolution.

Criteria for evaluation of light yield and energy resolution are given.

Sodium iodide crystals have almost monopolized the domain of X-ray and γ-ray imaging in nuclear medicine over the past two decades. It has, however, a drawback for low energy γ-rays: a rather poor energy resolution, typically 10% FWHM at 120 keV [1] leading to a degradation of the spatial resolution.

Putting aside the Si, or Ge based detector, even high-purity germanium, all requiring low-temperature liquid-nitrogen cooling, one research tendency is to look for room-temperature semiconductors like CdTe and HgI_2 [2] allowing a FWHM of 5% at 120 keV. Again, with the present state-of-the art, these devices suffer from a rather small useful area – a few cm^2.

We have been engaged in an applied research-program leading to a large surface gamma-camera, with good spatial and energy-resolution, based on the gas proportional scintillation mechanism. Our previous results [3,4] showed the performances of this type of detector for nuclear medicine imaging with X-rays of up to 50 keV allowing a 10 cm diameter aperture, 5% of energy resolution at 30 keV and 2.7 mm FWHM spatial resolution.

In this study, with a high pressure xenon gas, we prove the feasibility of such a detector with the most commonly used radioisotope 140 keV γ-ray namely ^{99m}Tc, with an energy resolution similar to that of an HgI_2 semiconductor.

This paper reports the detector design and its performance as well as an approach to the light yield evaluation.

The basic structure of the detector is similar to the

* On leave from Saga University, Japan.

one described in a previous paper [3]. Stainless steel shaping field rings were placed every 1 cm in the 10 cm long drift field region. The high electric field light producing gap is 7 mm, and delimited by thin, annealed, stainless-steel mesh (0.05 mm wire on a 0.55 mm spacing).

The optical window made of Tetrasil B, 35 mm thick, with a 50% transmission at 1700 Å, was connected to five two-inch 10-stage venetian blind XP2000-RTC photomultipliers.

Since high-pressure gas requires a high potential difference of up to 15 kV, the linear potentials were brought to the shaping field rings by pressure contact with a ceramic bar covered with a layer of ruthenium

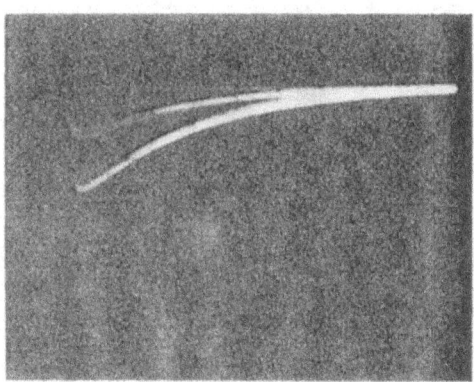

Fig. 1. Signals at the output of the preamplifier, showing the rays at 26, 30 and 60 keV. Xenon pressure was set at 6 atm, electric field at 10 kV/cm, a rather low value. Radiation source: ^{241}Am. Horizontal scale 10 μs/cm.

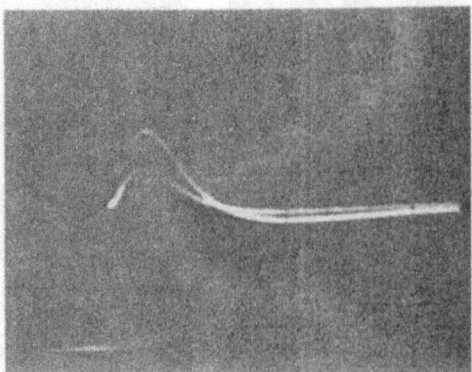

Fig. 2. Same conditions as before, with signals observed at the amplifier output. Integration and decay time both set at 3 μs. Horizontal scale 5 μs/cm.

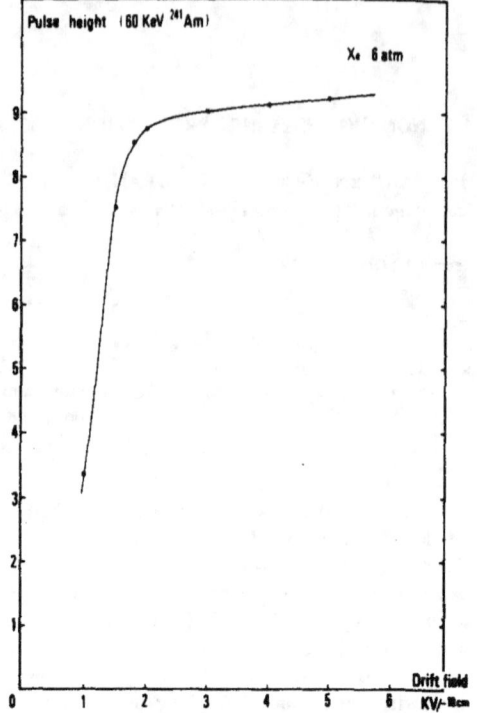

Fig. 4. Effect of the drift-field on the light gain.

oxide and baked at 1000°C as in thin film techniques. The linear resistance was 50 MΩ cm^{-1}. This greatly simplifies the high-voltage connections.

Each photomultiplier signal was fed to a charge preamplifier (30 μs decay time). The outputs were simultaneously directed to Lecroy's peak-sensing ADC, and to a resistor summing-network followed by a spectroscopy amplifier (Canberra 2010) set at

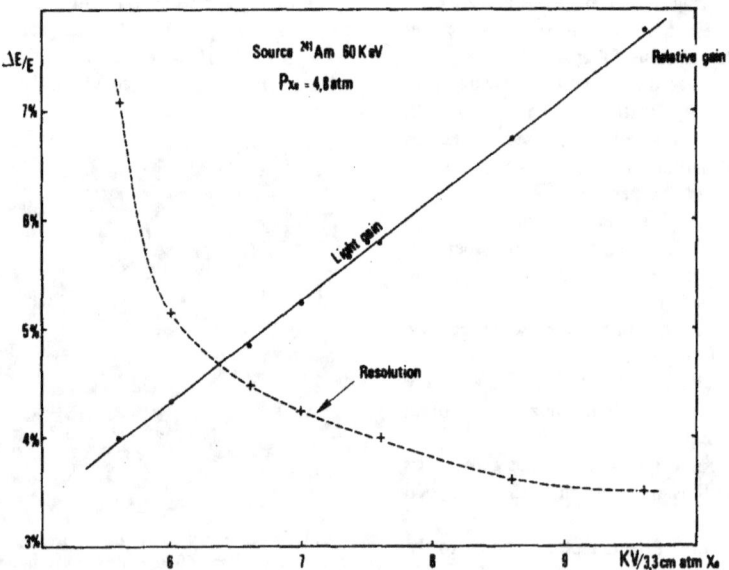

Fig. 3. Typical light gain and energy resolution curves, versus electric-field.

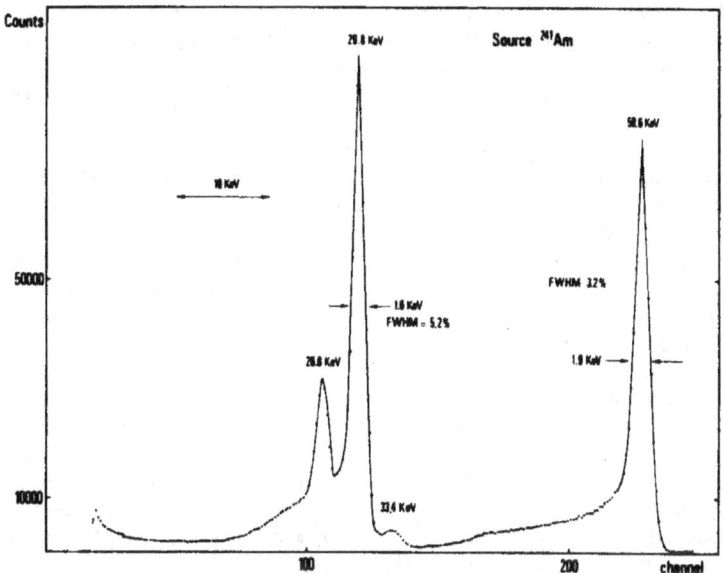

Fig. 5. Energy spectrum of ^{241}Am filtered with 3 cm of xenon above the drifting-field region with an electric field of 13 kV cm^{-1} at 5 atm. Observed peaks correspond to γ-ray 59.60 keV and Xe escapes peaks at 26.0 and 29.8 keV. The small peak at 33.4 keV is due to xenon fluorescence.

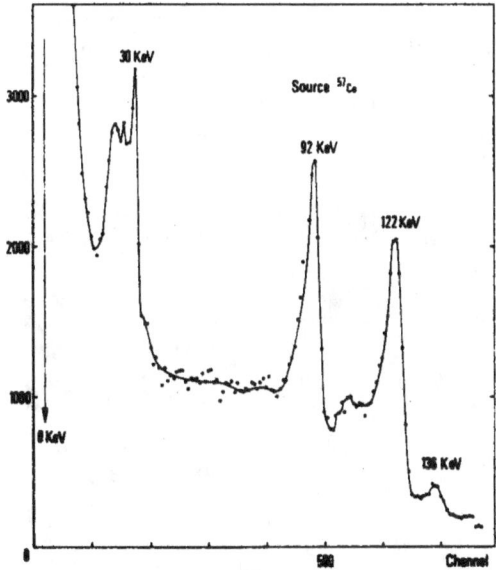

Fig. 6. Energy spectrum of ^{57}Co giving 136 and 122 keV, xenon escape peak at 92 keV and Xe fluorescence around 30 keV were also observed. Electric-field and pressure as in fig. 3.

3–3 μs integration-decay time (electron collection time is ~2 μs). Gain and energy resolution measurements were carried out with an amplifier-multichannel part while spatial resolution was calculated through an ADC + microcomputer.

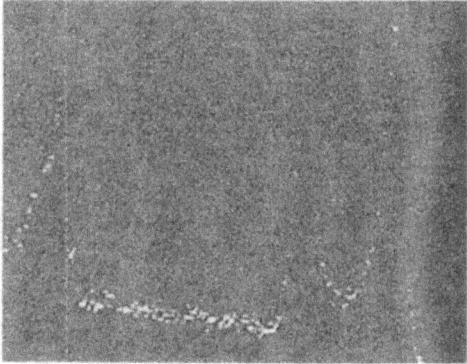

Fig. 7. Energy spectrum of ^{99}Tc source at a field of 11 kV/cm at 6 atm. Observed peaks at 140 keV due to the source, at 110 keV, to the xenon escape peaks and a group around 30 keV corresponding to the fluorescence of the xenon gas in the drifting-field region. Detection efficiency at 140 keV is 25% for 5 cm drift region at 10 atm.

Figures 1 and 2 show the two 60 and 30 keV X-rays at preamplifier and amplifier outputs. Due to the light saturation on the polaroids, the two escape peaks at 26 and 30 keV are not clearly visible on these photographs.

Figures 3 and 4 display the typical behaviour of our gas proportional scintillating detector, gas gain, and energy resolution versus the electric fields of the light-production gap and of the drifting-field region.

Various energy-spectra with monoenergetic γ-rays at different gas pressures were measured. Figs. 5, 6 and 7 show spectra obtained respectively with ^{241}Am(60 keV), ^{57}Co(136, 122 keV), and ^{99m}Tc(140 keV) radioisotopes collimated to 2 mm diameter beam. Figure 8 was obtained with a bare americium source inside the detector to avoid low energy absorption by the window.

Low energy neptunium X-rays were clearly observed. The 5 MeV α particle signal is, however, far beyond the saturation region of the preamplifier. Figure 9 shows the two thorium C, C' at 6.09 and 8.78 MeV α-rays. Energy resolution is 1.7% FWHM at 8.78 MeV. Results on energy linearity and resolution are summarized in fig. 10 for a xenon pressure of 5 atm.

Figure 11 gives an idea about the spatial uniformity of the pulse height as well as of its resolution along one radius.

Since the thin window was not available for the imaging-property tests, a set of equidistant 1 mm

holes every 10 mm was put along a line making 30° with the line passing through three photomultiplier centers so as not to consider the best spatial resolution configuration. Fig. 12 shows the resolution obtained with an ^{241}Am source manually moved across these holes.

For each pressure, a set of relative gain and energy resolution versus the light-production electric field was measured. This was allowed for by plotting the resolution versus the inverse of the gain to separate the contributions from the primary fluctuation and from the amplification stage including detector and photomultipliers.

From the assumed 0.20 quantum efficiency of the photocathode, the overall conversion efficiency amounted to 0.009 photoelectrons at the photocathode per photon in the light-production gap. This enabled us to derive approximate expressions for the light-yield as well as the energy resolution for parallel plane grids, xenon filled, proportional scintillating detector.

Light yield = $4000\,[(E/p) - 1.3)]\,px$ photon keV^{-1} .

Resolution (FWHM) = $\dfrac{14}{\sqrt{\epsilon_{keV}}}\left(1 + \dfrac{9}{px\,[(E/p) - 1.3)]}\right)^{1/2}$

where x, in cm, stands for the light-production-gap distance; p, is xenon pressure in kg cm^{-2}; E, electric field in the light production gap in kV cm^{-1}; ϵ, energy of the X or γ-ray in keV.

The light yield gives the total number of photons centred at 1700 Å per keV deposited in the drifting-field region.

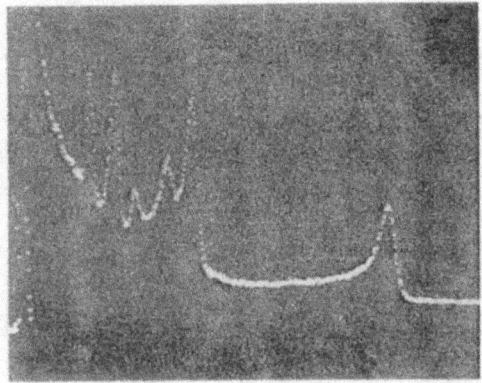

Fig. 8. Energy spectrum of a bare ^{241}Am source set inside of the drifting-field region. Light-producing field was 15 kV/cm at 5 atm. Observed peaks at 14.2, 17.9, 21.1, 26.0, 30.0, 60.0 keV correspond to γ-rays from the source, xenon escape peaks, and the Np X-rays. Data were collected during 4 h, showing the system stability.

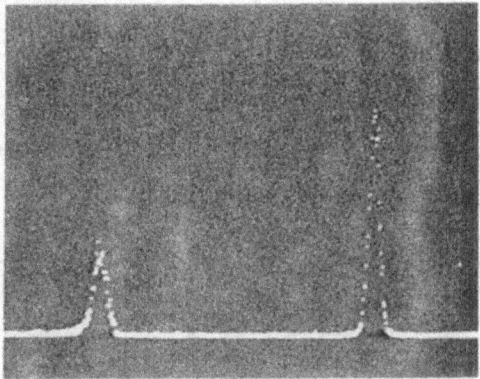

Fig. 9. Energy spectrum of thorium C, C' with α energies at 6.09 and 8.78 MeV. Light producing field was 8 kV/cm at 2 atm.

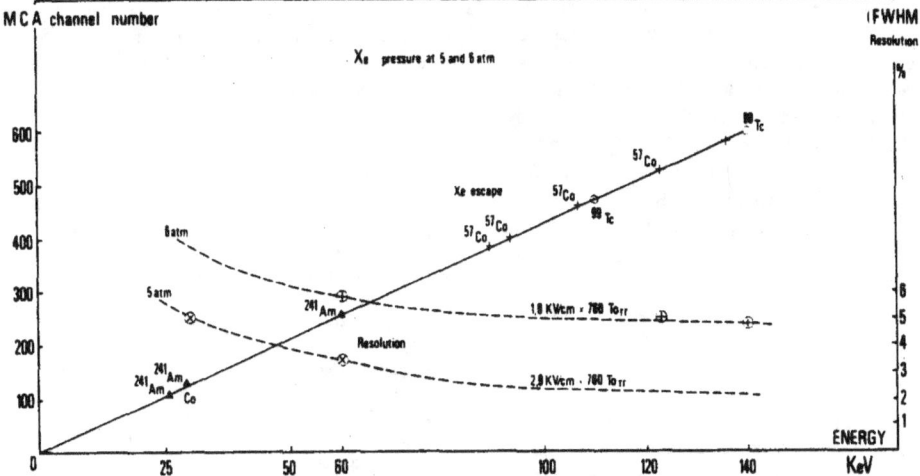

Fig. 10. Shows the energy linearity and typical energy resolution at two reduced fields.

For practical purposes, the reduced field should stay below the limit of 4 kV cm^{-1} atm^{-1} xenon where the charge multiplication occurs, resulting in a poorer energy resolution. In practice, the corona discharge appears before the charge multiplication; with great care in the connection design of the grids, one can easily obtain a reduced field of 3.3 kV cm^{-1} atm^{-1}, in which case the light yield amounts to

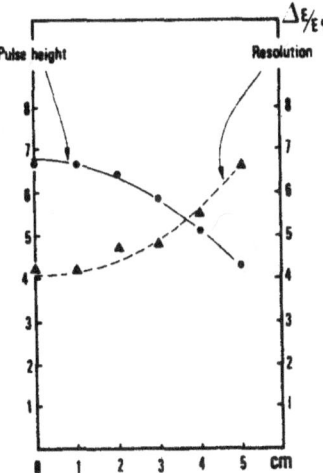

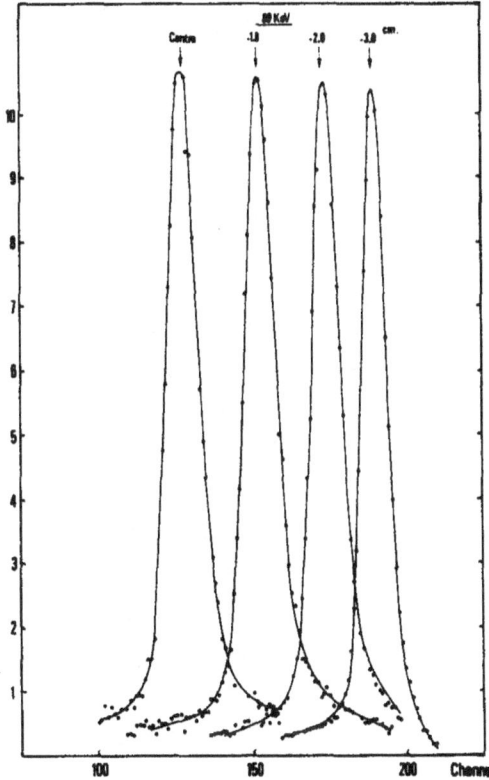

Fig. 11. Spatial uniformity of energy response and energy resolution along one radius. The abscissa origin is the detector center.

Fig. 12. Spatial resolution along one diameter of the detector at $E/p = 11$ kV cm^{-1} at 4 atm. This was taken by moving a 1 mm collimated ^{241}Am source.

$8000px$ photon keV^{-1}. This value is quite compatible – within 20% – with the recent estimate by Anderson et al. [5] and a factor 2 lower than Andresen's [6] rough estimate. Moreover, if one optimistically assumes the INa scintillator-efficiency to be 50 eV per photon, then the light-yield of a GPS detector, with a $px = 2$ cm atm xenon, will be 800 times that of a INa scintillator. This is not in contradiction with Policarpo's value – 100 times that of INa – if one takes into account the fact that the light-collection efficiency in an INa is nearly an order of magnitude greater than that in the GPS detector, and that factor – a hundred – did include the collection efficiency.

This program of research has been sponsored by the Institut National de Physique Nucléaire et de Physique des Particules with financial support from the Délégation Générale de la Recherche Scientifique et Technique.

Thanks are particularly due to Professors J. Yoccoz and J. Perez-y-Jorba for their constant interest and encouragement. Technical designs are due to Courvoisier and N. Greguric.

References

[1] D.E. Persyk and Te Moi, IEEE Trans. Nucl. Sc. NS25 (1978) 615.
[2] Rapport SES-SAI Avancement travaux 1979, Commissariat à l'Energie Atomique.
[3] Nguyen Ngoc Hoan, Nucl. Instr. and Meth. 154 (1978) 597.
[4] Nguyen Ngoc Hoan, S. Majewski, G. Charpak and A.J.P.L. Policarpo, J. Nucl. Med. 20 (1979) 335.
[5] D.F. Anderson, T.T. Hamilton, W.H.M. Ku and K. Novick, Nucl. Instr. and Meth. 163 (1979) 125.
[6] R.D. Andresen, E.A. Leiman and A. Peacock, Nucl. Instr. and Meth. 140 (1977) 371.
[7] A.J.P.L. Policarpo, M.A.F. Alves, M.C.M. Dos Santos and M.J.T. Carvalho, Nucl. Instr. and Meth. 102 (1972) 337.

Nuclear Instruments and Methods in Physics Research A269 (1988) 385–391
North-Holland, Amsterdam

TEST OF A BaF$_2$–TMAE DETECTOR FOR POSITRON-EMISSION TOMOGRAPHY

P. MINÉ, G. CHARPAK, J.-C. SANTIARD and D. SCIGOCKI,

CERN, Geneva, Switzerland

M. SUFFERT

CRN, Strasbourg, France

S. TAVERNIER

IIHE, VUB and ULB, Brussels, Belgium

A detector consisting of BaF$_2$ scintillators and wire chambers has been tested for 511 keV gamma-rays. The wire chamber is filled with photosensitive tetrakis(dimethylamine)ethylene (TMAE) vapour and is operated at a pressure of a few Torr, using different gases at various temperatures. Energy, time, and position resolutions are given for BaF$_2$ crystals with sections of 10×10 mm^2 and 5×5 mm^2. We discuss the potential of this gamma detector for positron-emission tomography (PET) and compare it with other systems.

1. Introduction

Barium fluoride is an inorganic scintillator with a radiation length of 20.5 mm and has two components in its scintillation light. The slow component, which peaks at 310 nm, has a decay-time constant of 620 ns, and the fast component, which peaks at 220 nm, has a decay-time constant of only 0.6 [1]. Hence, it is one of the fastest scintillators known. It is this unique combination of short decay-time constant and ultraviolet (UV) response which is exploited in the solid-state proportional counter (SSPC) [2–6].

In the SSPC the UV scintillation light is detected in a proportional wire chamber. This is achieved by adding a few Torr of tetrakis(dimethylamine)ethylene (TMAE) vapour to the chamber gas. In this way it is possible to combine the good detection efficiency characteristic of dense inorganic scintillators with the good spatial resolution of wire chambers. The SSPC has important potential applications in nuclear physics, in particle physics, and in nuclear medicine. In the present article we are only concerned with the last one.

In nuclear medicine an obvious application of the SSPC is positron emission tomography (PET). In PET one has to detect the 511 keV photons from positron annihilation with the best possible efficiency and time resolution and with a spatial resolution of ≈ 2 mm. The more traditional approach to PET is to use a photomultiplier for each crystal. In this scheme a spatial resolution of a few mm and a large solid-angle coverage leads to a prohibitively expensive device. The commercially available systems use a compromise between these desirable features and the cost. With the SSPC one can envision a PET camera which contains many thousands of small BaF$_2$ crystals, all read by a few wire chambers. Moreover this camera would have an excellent time resolution: to reduce the rate of accidental events it is necessary to have a narrow coincidence window; BaF$_2$ is better than NaI or BGO, owing to the fast decay time of its UV component. A low-pressure wire chamber can take advantage of this speed, owing to its good timing properties and high gain [7].

Another possible solution for PET is the use of a conventional wire chamber with some sort of gamma converter [8–10]. These systems have good spatial resolution but suffer from poor detection efficiency and poor time resolution.

First results using $10 \times 10 \times 50$ mm^3 crystals have been already described elsewhere [11]. Below we present more systematic measurements with $5 \times 5 \times 50$ mm^3 crystals. This results in an improved spatial resolution. We also show that it is possible to operate the system continuously, without interruption for pumping and refilling.

2. Description of the detector

We have built two counters consisting of square matrices of 25 parallelepipedal BaF$_2$ crystals, $10 \times 10 \times 50$ mm^3 for the first one and $5 \times 5 \times 50$ mm^3 for the second one, followed by a multiwire proportional cham-

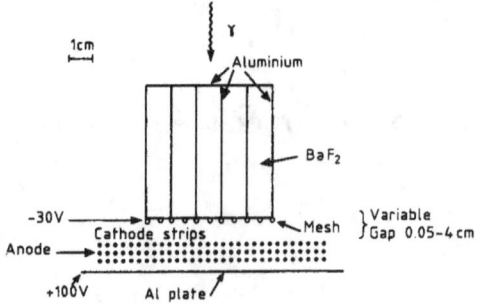

Schematics of the SSPC

Fig. 1. Schematics of the SSPC.

ber (MWPC) (figs. 1 and 2). The crystals are optically separated and covered with reflective aluminium. On the side facing the MWPC they are in contact with a transparent grid, kept at a fixed potential. The 511 keV gamma-rays are absorbed in the BaF_2, with an efficiency of 90%. The MWPC consists of cathode planes of 100 μm diameter wires, grouped into 1 or 0.5 cm strips, in such a way that a strip faces a row of crystals, and an anode plane of 15 μm wires with a gap between planes of 3 mm. The wire spacing is 1 mm in all planes. The distance between the nearest cathode plane of the MWPC and the BaF_2 matrix – hereafter called the collection gap – can be varied from 0.5 mm to 4 cm, and a drift or amplification voltage can be applied, over this gap, in order to operate in the multistep mode. The SSPC is saturated with TMAE vapour at temperatures ranging from 20 to 55°C. A few Torr of another gas are added and the temperature is kept 5°C higher to avoid condensation.

3. Energy resolution and stability

Unlike many other scintillators, the light yield of the fast component of the BaF_2 scintillator is insensitive to temperature [12,13]. Variations in gain are hence entirely due to the wire chamber. We checked with a UV-sensitive photomultiplier that the light output of

Fig. 2. A photograph of the MWPC with the matrix of $10 \times 10 \times 50$ mm³ crystals.

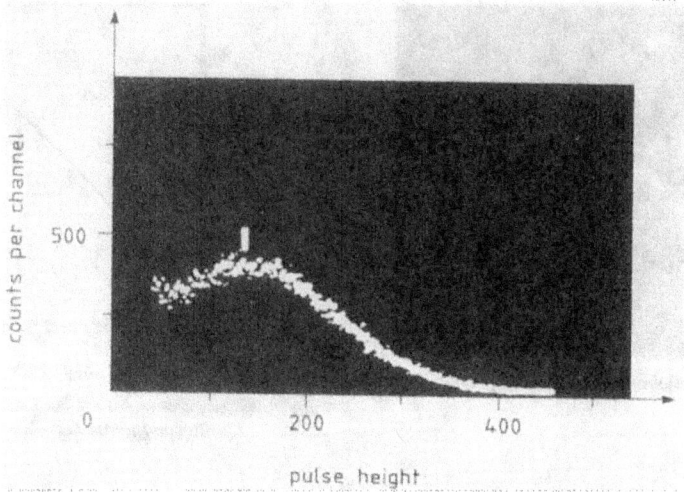

Fig. 3. 661 keV spectrum (^{137}Cs) obtained with pure TMAE at 50 °C and with a 1.5 cm gap collection.

the BaF$_2$ matrix is the same as that of a single crystal of equivalent size. The energy resolution with the SSPC is measured with the anode signal fed into an Ortec 142 preamplifier. The voltage in the drift space is kept at 30 V to avoid avalanche amplification, which could deteriorate the energy resolution. We obtain ^{137}Cs peaks ($E_\gamma = 662$ keV) with isobutane, CH$_4$, and CF$_4$, added to TMAE. Above 40 °C we also observe good working conditions with TMAE vapour alone. Our best signal is obtained with pure TMAE at 50 °C in a gap of 1.5 cm (fig. 3). The energy resolution is poor, owing to the small number of photoelectrons [3]. From the peak width we estimate their number to be of the order of 10. We deduce that the gain of the MWPC is about 2×10^4 without amplification in the collection gap, for a high voltage of 400 V on the anode. This gain is three times higher than what we measure at 20 °C with 3 Torr of isobutane.

The good working conditions we observe at high temperature are due to the rise of the TMAE vapour pressure. Our measurement of this quantity (fig. 4) follows the Clausius–Clapeyron relation, as in ref. [14].

If the chamber is filled and closed we observe a deterioration of the gain in 24 h. This is clearly incompatible with the routine work of a positron camera. The problem is probably caused by the chemical reaction of TMAE with some component or outgassing of the chamber. In an SSPC used in high-energy physics [15] we have tested a continuously flowing system. The TMAE bottle is constantly open on one side of the detector. The other side has a microleak, or a valve drive by a pressure transducer (MKS type 223BHA10), followed by a refrigerated bath to freeze the TMAE,

and by a vacuum pump. With this system we have been able to keep the gain stable within a few percent and to operate the apparatus continuously for three months.

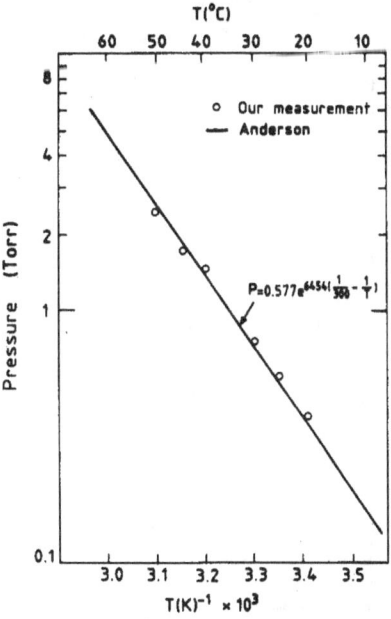

Fig. 4. TMAE vapour pressure as a function of the inverse of the absolute temperature, and the temperature in degrees Celsius, including points from ref. [14].

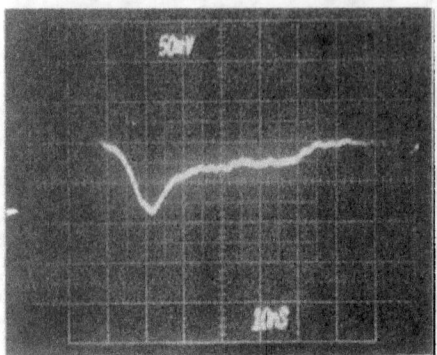

Fig. 5. Signal showing the 10 ns rise-time.

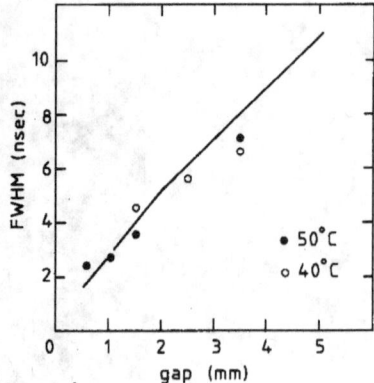

Fig. 7. Time resolution for a ^{22}Na source, as a function of the collection gap (the curve is a simulation).

4. Time resolution

The time resolution is measured with a ^{22}Na source: the positrons emitted by this isotope annihilate in the source, producing two back-to-back photons of 511 keV. The start signal for the time measurement is provided by the detection of one of the photons in a small BaF₂ crystal coupled to a fast UV photomultiplier (XP2020Q). The stop signal is obtained from the anode of the SSPC. For the time-resolution measurements the electric field in the collection gap is increased until

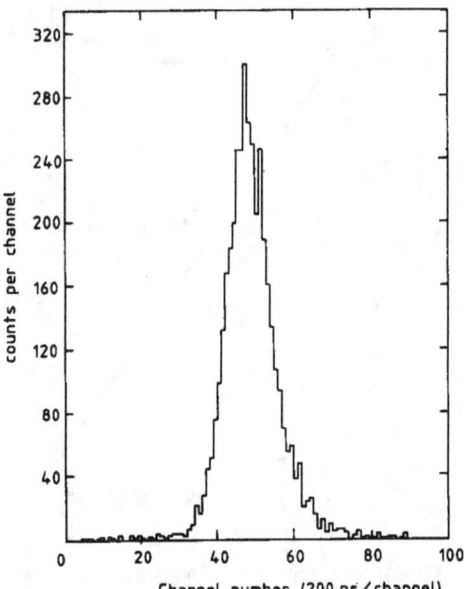

Fig. 6. Time spectrum for a 511 keV ^{22}Na source.

parallel-plate amplification is obtained. Under these conditions the anode signal shows a rise-time of 10 ns (see fig. 5). The drift velocity is measured by recording the delay between the photomultiplier and the chamber signals as a function of the collection gap. We find in pure TMAE a value of 3.8 cm/μs without amplification and 10 cm/μs with maximum amplification.

Results are shown in fig. 6: at 50°C, with a 0.5 mm gap and a 130 V amplification voltage, we obtain 2.4 ns FWHM. The variation of the time resolution with the gap distance is compatible with a simulation assuming 10 UV photons detected by TMAE in an infinite collection gap and 10 cm/μs drift velocity (fig. 7). This simulation includes the Poisson statistical fluctuation in the ion-pair production mechanism by the incoming light. It assumes that the detection of the electron closest to the MWPC determines the timing.

There is clearly a conflict between the requirements of good time and good energy resolution. It seems to us that for PET a good time resolution is more important. A collection gap of 2 mm, corresponding to a time resolution of ≈ 4 ns at 50°C, seems appropriate.

5. Position resolution

One possibility of measuring the position of the incoming gamma-ray is to have a large-diameter crystal and to use the Anger method. This method suffers from the small number of detected photons and cannot reach a value lower than 13 mm [6]. The other possibility, chosen in the present study, is to use a matrix of small closely packed crystals, optically separated, acting as light guides. A few photons are sufficient to identify the crystal which is hit. No attempt is made to localize the impact point inside the crystal.

Our previous measurement [11] of the position reso-

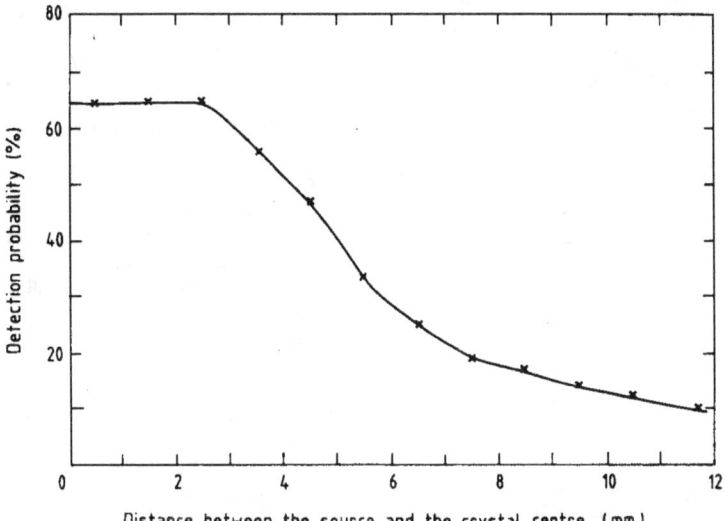

Fig. 8. Probability of hitting a crystal (with a matrix of $10 \times 10 \times 50$ mm³ crystals) when the source is moved away from the centre.

lution was performed with a collimated ^{22}Na source. In the present study we avoid the difficult problem of collimation by using coincidences between two back-to-back 511 keV photons from a ^{22}Na source. The source, 1 mm in diameter, was placed at 12 cm from the crystals. The solid angle was defined by a BaF₂ crystal, 5×5 mm² in section, viewed by a photomultiplier placed at 68 cm from the source, opposite to the SSPC. The uncertainties coming from this experimental procedure are the following (FWHM): 1 mm for the source diameter, 0.9 mm for the geometrical collimation, and 1 mm for the deviation from linearity of the two gamma-rays, as calculated for example by Derenzo and Budinger [16]. We measured the position resolution by moving the source perpendicularly to the axis from the SSPC to the photomultiplier. Each coincidence between the photomultiplier and the matrix was recorded, and no energy threshold was set on individual crystals. The position information was derived from the cathode signals, which were amplified and fed into analog-to-digital converters. Several values were tried for the gap width and the TMAE temperature. The results presented below were obtained with a 2 mm collection gap at 50°C.

If the SSPC detects a 511 keV photon, all the signal is not necessarily confined to one single cathode strip. This is due to: (a) Compton scattering in the BaF₂ such that the energy of the 511 keV photon is deposited in more than one crystal; (b) spreading of the light in the collection gap; and (c) electronic noise and cross talk. If the impact point of the photon is assigned to the strip with the largest signal, one obtains a position resolution of 11.5 mm (fig. 8) and 5.8 mm (fig. 9) for 5×5 mm²

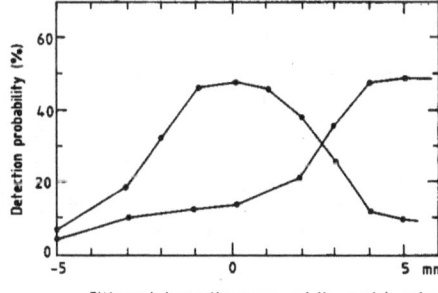

Fig. 9. Probability of hitting a crystal (with a matrix of $5 \times 5 \times 50$ mm³ crystals) when the source is moved away from the centre (two adjacent crystals are represented).

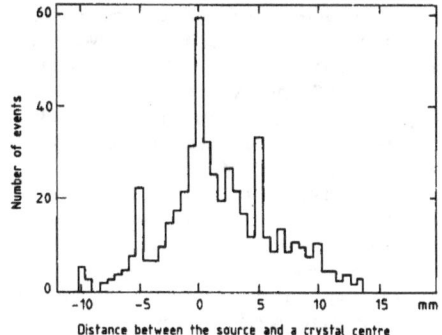

Fig. 10. Continuous distribution obtained by the centre-of-gravity method ($5 \times 5 \times 50$ mm³ crystal).

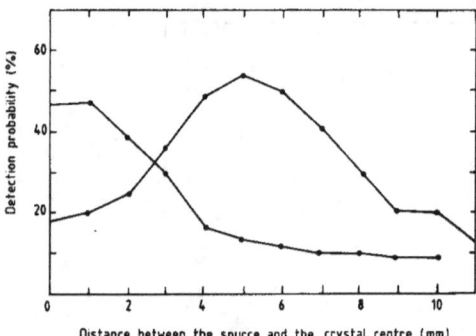

Fig. 11. Probability distribution obtained by the centre-of-gravity method.

and 10×10 mm², respectively. We also tested another method, to see if additional information can improve the performance: we determined the position by the centre-of-gravity method, using the highest signal and the two adjacent ones. This gives a continuous distribution with peaks located at the centre of the crystals (fig. 10). When we assign each measurement to the nearest centre, we find a resolution curve (fig. 11) as before. The FWHM is found to be 6 mm. Thus this method gives a similar but slightly worse result.

At 23°C, with a 1–4 cm collection gap, the resolution was about 22 mm, which demonstrates the importance of keeping a narrow gap. In all the cases, we have checked that the amplification voltage of the PPAC has only a minor influence on the spatial resolution.

6. Discussion

We find a position resolution which is almost equal to the size of the crystals. When we subtract quadratically the contributions due to source diameter, collimation, and nonlinearity, we obtain 5.6 mm FWHM for 5 mm rods. This number, divided by 2, would correspond to a precision of 2.8 mm for the image, not including the degradation due to reconstruction. The resolution can probably be improved further by taking even smaller crystals, e.g. with a section of 3×3 mm².

The performance can certainly be improved if the crystals are separated by lead or tungsten septa, to absorb Compton gamma-rays, at the expense of a decrease in the efficiency. On the other hand, one knows that the accuracy is deteriorated when the photon hits the detector with an angle different from 90°, as occurs for a source distant from the centre of the camera gantry. All these effects have been studied by different authors [17,18]. They deserve a detailed analysis in our

specific case, both experimentally and by simulation, which we plan to carry out in the near future.

7. Possible designs for a new PET camera

Because of the parallax error, a PET camera based on BaF₂ and wire chambers must consist of a cylindrical detector where all crystals point towards the axis of the cylinder, or of a number of small plane detector modules assembled in a ring as illustrated in fig. 12. Each module should be less than 5×10 cm² in section.

Fig. 13 shows a possible detector module. The complete ring has 24 such modules. Each one contains 512 BaF₂ crystals, 3×3 mm² in section. The wire chambers operate at low pressure in pure TMAE at 50°C. The resolution on the time difference in such a system is expected to be between 3 and 6 ns FWHM, slightly better than a BGO ring camera. Below, this system is compared with a high-performance ring camera with 600 BGO crystals 3×10 mm² in section [19]:

- The time resolution for the BaF₂ camera is better by a factor of 3 and, as a result, the rejection factor of the background from accidental coincidences is reduced by the same factor.
- The resolution in the plane of the ring is worse because of Compton scattering and the smaller density of BaF₂; the magnitude of this effect is difficult to evaluate without a detailed study, but it can be at least partly cured by the use of tungsten septa between the crystals.
- The axial resolution is better by a factor of 3.
- The solid angle is about 5 times bigger, and several sections can be imaged; a BGO camera having the same solid angle would be prohibitively expensive. An object the size of a human head can be completely imaged, in one go, as a three-dimensional object.

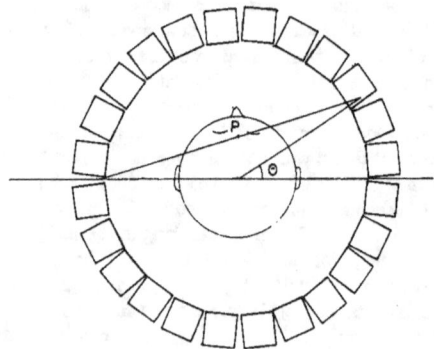

Fig. 12. Possible design of a PET camera with 24 MWPCs, to minimize the parallax error.

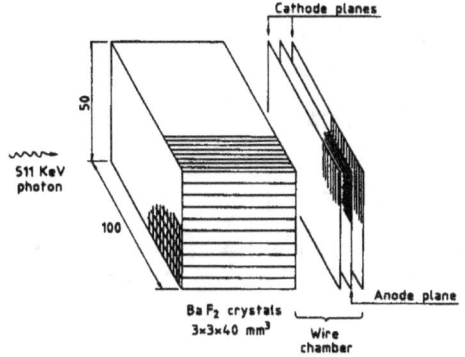

Fig. 13. One of the 24 modules of the camera sketched in fig. 12.

– The BaF$_2$ camera has almost no energy resolution; this will make the rejection of Compton scattered photons less effective; again only a detailed study will allow evaluation of the importance of this effect.

The solution looks attractive. It is simple and probably competitive in price with current commercial ring designs.

8. Conclusion

We have investigated the performances of a SSPC designed for 511 keV gamma-ray imaging. It is possible to operate this device constantly for several months at low pressure, if the TMAE is renewed by a continuously flowing gas system. To ensure good time and space resolution, it is necessary to have a small amplification gap. Consequently the system has to be operated at ≈ 50°C to save the efficiency of the light detection. This temperature and a 2 mm gap collection with amplification are appropriate working parameters.

We have measured a time resolution of 2.4 ns, which is too large for time of flight but gives a significant background reduction compared to NaI or BGO. The spatial precision with a matrix of small crystals is better than with a single large crystal. Despite the small number of photons, crystals with a section of 5 × 5 mm^2 can

be separated by the MWPC. A resolution of 5.6 mm FWHM for the photon detection with a single detector has been measured. This measurement suggests that the spatial resolution could be improved further by taking even smaller crystals. Our present result encourages us to work in more detail on the design of a full-size medical prototype. We expect that a medical PET camera based on an SSPC will have a high-rate capability and will allow a dramatic improvement of the axial precision and of the acceptance.

References

[1] M. Laval et al., Nucl. Instr. and Meth. 206 (1983) 169.
[2] D.F. Anderson, Phys. Lett. B118 (1982) 230.
[3] D.F. Anderson et al., Nucl. Instr. and Meth. 217 (1983) 217.
[4] D.F. Anderson et al., Nucl. Instr. and Meth. 225 (1984) 8.
[5] D.F. Anderson et al., Nucl. Instr. and Meth. 228 (1984) 33.
[6] P. Schotanus et al., Nucl. Instr. and Meth. A252 (1986) 255, and IEEE Trans. Nucl. Sci. NS-34 (1987) 272.
[7] A. Breskin, Nucl. Instr. and Meth. 196 (1982) 11.
[8] D. Townsend et al., IEEE Trans. Nucl. Sci. NS-30 (1983) 594.
[9] M.R. Hawkesworth et al., Nucl. Instr. and Meth. A253 (1986) 145.
[10] R. Bellazini et al., IEEE Trans. Nucl. Sci. NS-31 (1984) 645.
[11] P Miné et al., IEEE Trans. Nucl. Sci. NS-34 (1987) 458.
[12] P. Schotanus et al., Nucl. Instr. and Meth. A328 (1985) 564.
[13] M. Suffert and G. Charpak, CERN EP Internal Report 86-03 (1986).
[14] D.F. Anderson, IEEE Trans. Nucl. Sci. NS-28 (1981) 842.
[15] D. Scigocki, Thesis, Université de Savoie, unpublished, and CERN EP Internal Report 87-03 (1987).
[16] S.E. Derenzo and T.F. Budinger, Advanced instrumentation for positron emission tomography, Berkeley Report LBL 19435 (1985).
[17] E.J. Hoffman et al., IEEE Trans. Nucl. Sci. NS-33 (1986) 420.
[18] Wai-Hoi Wong et al., IEEE Trans. Nucl. Sci. NS-31 (1984) 381.
[19] S.E. Derenzo et al., IEEE Trans. Nucl. Sci. NS-34 (1987) 321.

NUCLEAR INSTRUMENTS AND METHODS 141 (1977) 449-455; © NORTH-HOLLAND PUBLISHING CO.

SOME PROPERTIES OF SPHERICAL DRIFT CHAMBERS

G. CHARPAK, C. DEMIERRE, R. KAHN, J. C. SANTIARD and F. SAULI

CERN, Geneva, Switzerland

Received 22 November 1976

A large-aperture X-ray imaging chamber is described. Angular aperture: 90°. Thickness of spherical absorbing drift space: 10 cm. The photoelectron's position is measured in a proportional chamber of 50×50 cm^2, with an accuracy of 0.8 mm (fwhm) in all coordinates.

1. Introduction

We have investigated the properties of a spherical drift chamber[1]) aimed at measuring the angular distribution of soft X-rays emitted by a point source, crystal or pinhole.

The characteristics of the chamber are the following (fig. 1):

a) *Drift space D.*
- Entrance electrode: spherical shape, aluminium of 40 μm thickness, radius 18 cm.
- Exit electrode: spherical shape, stainless steel grid of 82% transparency, radius of 28 cm.
- Angular acceptance: 90°.

b) *Transfer space T.*
- Minimum thickness, on the central axis: 2.5 cm; maximum thickness, at 45°: 10 cm.

c) *Read-out proportional chamber C.*
- Dimensions: 50 cm × 50 cm.

- Anode wire spacing: 1 mm or 2 mm, of 10 μm diameter gold-plated tungsten; or 2 mm, of 20 μm diameter gold-plated tungsten.
- Cathode wires spacing: 1 mm, of 50 μm diameter bronze beryllium.
- Anode–cathode distance: 6 mm.
- Read-out method: analog computation of avalanche centroid.

2. The proportional chamber

The chamber is made of a plane of anode wires sandwiched between cathode planes and at a distance of 6 mm from each. The cathode planes are made of wires of 50 μm diameter, spaced by 1 mm, parallel to the anode wires in the entrance face and orthogonal to it in the back plane. The wires are connected together by groups of 6. The pulses induced in every strip are mixed together in parallel in an analogic circuit measuring the centre of gravity of the induced pulse;

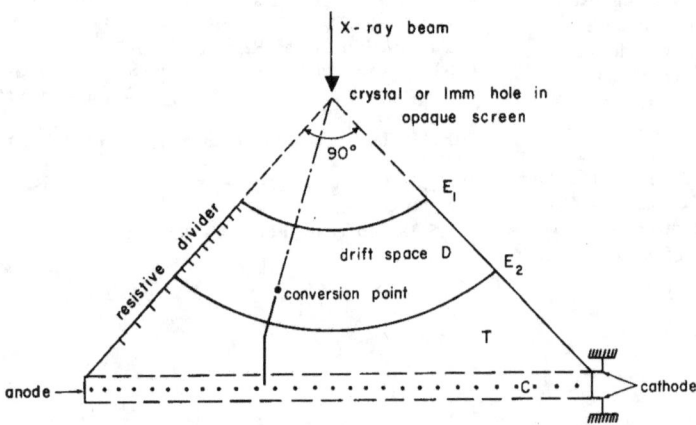

Fig. 1. Spherical drift chamber. Angular aperture: 90°. Drift length: 10 cm. Radii of the entrance and exit electrodes: 18 cm and 28 cm, respectively. MWPC: wire spacing of 1 mm and 2 mm, 50 × 50 cm^2

this device is an extension to 72 channels of a previous one with 26 channels[2]). The total time required for the two-dimensional read-out of the coordinates of an avalanche is 5 μs.

The accuracy is somehow better than 1 mm, which was initially aimed at for this chamber. It first appeared that the limit for the accuracy would be set by the anode wire spacing which was planned to be of 1 mm.

Many inconveniences were connected with such a wire spacing. To overcome the electrostatic instabilities we had to have nylon strings supporting the anode wires every 10 cm. This introduced dead regions which are a nuisance in an instrument of this kind. The operation was also very delicate and wire breaking was a common failure. We finally discovered that because of the diffusion in the drift space, the electron clouds always shared their charge between two wires, even at 2 mm wire spacing, thus permitting an interpolation between wires. We then gave up the 1 mm wire spacing and with 2 mm wire spacing the accuracy is still well below 1 mm, as we will see. This paved the way for chambers of a much simpler construction, with anode wires of 20 μm diameter, spaced at 2 mm. As is well known from high-energy physics practice, chambers of several square metres can be easily built with such parameters; they still give a localization accuracy much better than the wire spacing, if they are connected to a drift space of sufficient thickness to spread the initial electron cloud to dimensions larger than a wire spacing.

While the experiments are far from being completed, many other interesting properties, to some extent surprising, have appeared in the preliminary study which we present here. Whilst the chamber is intended to be used with a mixture rich in xenon, these studies were conducted with a flowing mixture commonly used in multiwire chambers: argon (70%), isobutane (28%), methylal (2%). The use of argon, instead of xenon, is advantageous at this stage since only a fraction of the X-rays is absorbed in the drift space, typically 60% at 8 keV; we can thus compare the transport properties of the different spaces by controlling separately the voltages applied to them. When we studied the variation of the pulse height as a function of the fields in D and T, it appeared that the results were dependent on the voltage of the chamber. The striking phenomenon was that beyond a given voltage in the chamber *the pulses generated by photoelectrons which had transited through the drift space were larger and of better resolution than those from photoelectrons produced in the chamber.*

This is clearly illustrated by the curves of fig. 2. The variation of the average pulse height as a function of voltage is shown in fig. 3. The curve obtained without drifting is approximate since the resolution becomes very poor as the voltage in the chamber increases and the peak is badly defined. We thus see that one advantage of having the photoelectron drifting before reaching the anode wire is that higher amplification can be reached in the linear region than in a thin proportional counter.

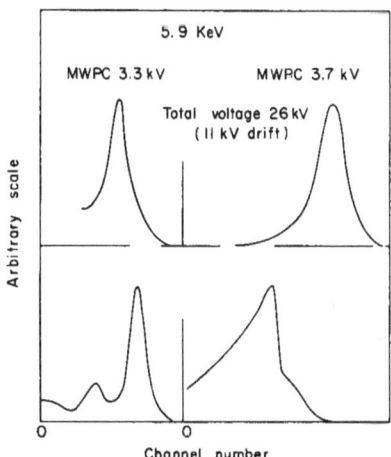

Fig. 2. Space-charge effects. Pulse-height distributions from X-rays of 5.9 keV absorbed directly in the chamber (top curves) or after drifting through a spherical chamber (bottom curves) at two voltages: 3.3 kV and 3.7 kV.

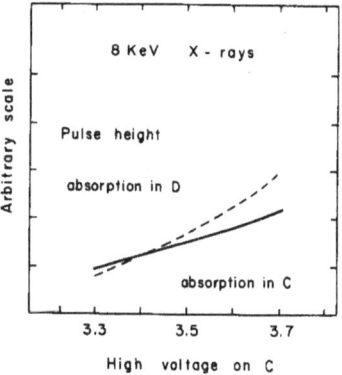

Fig. 3. Space-charge effects. Pulse height variation as a function of voltage on the chamber C for 8 keV photons absorbed in C or in D. 18 kV total voltage.

3. The transfer space

The transfer space is the weakest part of the system since the electrical field varies by a factor of 4 from centre to edge.

By applying a reverse field on the drift space one can study the pulses from X-rays absorbed only in T. Fig. 4 shows the pulse height from 8 keV X-rays, as a function of applied voltage, for different directions of a beam of X-rays; at 0° incidence (on the central axis) we see that the pulse height is decreasing by about 10% from 5 kV to 15 kV. This is a consequence of a change in the electrical transparency for electrons crossing the cathode plane, due to the different ratio of the fields on the two sides.

At 40° inclination of the X-ray beam, the pulse height increases smoothly but steadily and does not saturate at the maximum possible voltage on T, limited by discharges to about 20 kV. The interpretation is not clear. With increasing voltage both the path length and the size of the electron cloud change; because of the electron diffusion the cloud size is slightly reduced as the field increases. This does not, however, explain the increase in pulse height. The change in energy resolution is also interesting; it degrades slightly from 13% to 18% for the central part, thus confirming a loss of electrons as the field in the transfer space increases, and it improves from 42% to 18% at 40°, indicating a higher efficiency of electron collection. From 10 kV to 15 kV we have a region where the resolution is ranging from 15% to 18% for 8 keV X-rays at a given angle, with a pulse-height spread not exceeding 20% from 0 to 40°.

At 3.7 kV we have repeated this study. At 0°, the diffusion in T is insufficient to spread the electron cloud enough to escape space-charge saturation and the energy resolution is so bad that the measurements are too imprecise. At 40°, the resolution is good because of the diffusion introduced by the larger path length.

The pulse height varies exactly as at 3.3 kV, although the absolute value is about 5 times larger, thus confirming that thanks to the long drift space the space-charge effects occurring at 3.7 kV for photoelectrons which have diffused is unimportant. The energy resolution of about 18% to 20% is roughly the same between 10 and 15 kV and somehow better at smaller voltages than with a voltage of 3.3 kV in the chamber.

4. The drift space

We have not studied the variation of the collection efficiency as a function of the field in D, with a constant field in T, but varied the two simultaneously, with a chain of resistances giving the sharing in the total potential between D and T. The sharing has been varied. The response of the chamber to a beam of 8 keV X-rays, collimated to about 1 mm and centred in the chamber, varies with angle and with applied voltage, in a way which we think can be interpreted. Some typical spectra for a total voltage of 26 kV, where 11 kV are applied to the drift space and 15 kV to the transfer space, are seen in fig. 5, with 3.7 kV on chamber C.

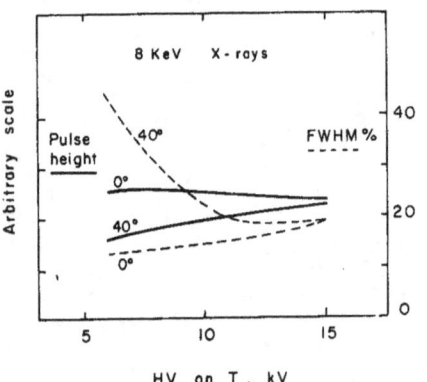

Fig. 4. Absorption and transmission by transfer region T. 3.3 kV on C. 8 keV X-rays at two angles. 0° and 40°.

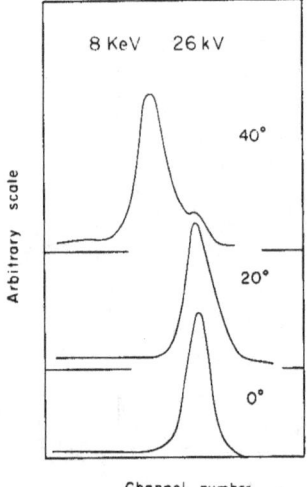

Fig. 5. Response of spherical chamber to 8 keV photons. Hv on C: 3.7 kV. Total voltage: 26 kV (15 kV on T). At 40°, the second peak is due to X-rays absorbed in T. At 0° their pulse height is reduced by space-charge saturation.

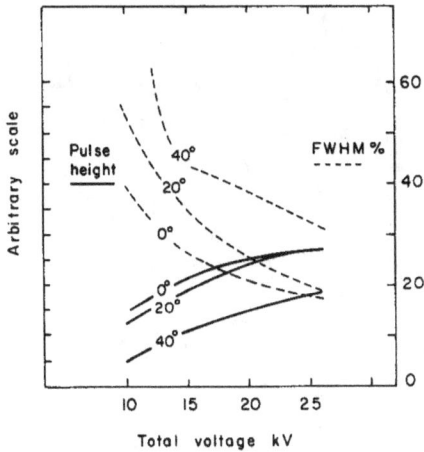

Fig. 6. Response of spherical chamber to 8 keV photons. Pulse height and resolution variation as a function of the total applied voltage.

As a function of total voltage, the pulse height and resolution vary (fig. 6). At a total voltage of 26 kV, the resolution is acceptable at 0° (17%). The peak is symmetric, with no ghost peaks. At 20° the pulse height reaches the same value as at 0°, with a slightly worse resolution, 19% at 26 kV. But as we increase the angle a second peak appears clearly. At 40° we are on a rather steep slope for the pulse height, even at 26 kV total voltage, and we reach only 70% of the value found at 0° and 20°. Since at 15 kV on the transfer voltage (which corresponds to 26 kV total voltage) we reach the same pulse height at 0° and 40°, we may conclude that this loss of 30% occurs when the electrons cross the spherical grid between D and T. The occurrence of the second peak is a function of angle. It is suppressed, at 0°, by the space-charge effect, since it comes from those photons which have been absorbed in the shortest thickness of T or in C.

At 40°, the path length for the photoelectrons absorbed in T is longer and the amplification is higher (figs. 2 and 3). These effects are not important if the X-rays are totally absorbed in D.

5. Spatial resolution

The spatial resolution was studied in the chamber before it was in its final form. The wires were 10 μm thick instead of 20 μm as in the final version and the wire spacing was the same, namely 2 mm. It had nylon strings every 10 cm. It was just the 1 mm spacing chamber with 1 wire out of two removed. When the

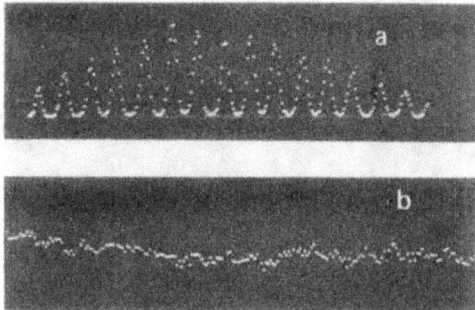

Fig. 7. Localization in the spherical chamber. Source of ^{55}Fe; (a) against C, (b) at the focus S of the spherical chamber.

^{55}Fe source emitting the 5.9 keV photons is placed against the chamber, the localization in the direction orthogonal to the wires shows the pattern of wires distant by 2 mm with a fwhm of about 0.7 mm (fig. 7a). If the source is placed at the focus of the drift chamber, the pattern disappears and the response is continuous (fig. 7b). The question is whether the resolution has been spoiled or whether we are now interpolating between wires; to answer this question we have used a collimator of 0.1 mm diameter on our X-ray generator, which we could direct at all angles through the focal point of the spherical chamber.

Fig. 8 shows the centroid positions in the directions perpendicular to the wires (a and b) and parallel to them (c), and in the direction parallel to the wires when the beam is not centred (d). *The striking feature is that the accuracy is the same in all directions*, within our accuracy measurement of 0.7 mm (fwhm). In other words, the electron cloud, which is initially of an extension well below 100 μm when it is produced, increases by diffusion and reaches a size larger than the wire spacing. This ensures the energy sharing between wires and permits an accuracy much better than the wire spacing.

An image was made with the chamber working as a pinhole camera with a hole of 1 mm diameter in a thin gold foil at the focus. The X-ray source was ^{55}Fe deposited in the grooves of a plastic foil featuring the word "CERN". The image (fig. 9) displayed by the computer was exactly the one obtained by autoradiography except that it was enlarged by a factor of 6. The nylon string supporting the wire is visible.

The chamber was tested with a crystal of CuV_2S_4 of 0.5 mm width (spatial group Fd3m, $a = 9.801$ Å), rotating at the centre. Fig. 10 shows the computer display of the two-dimensional coordinates of photons at a fixed crystal position. The reflections $00\bar{4}$ and $\bar{3}11$ are

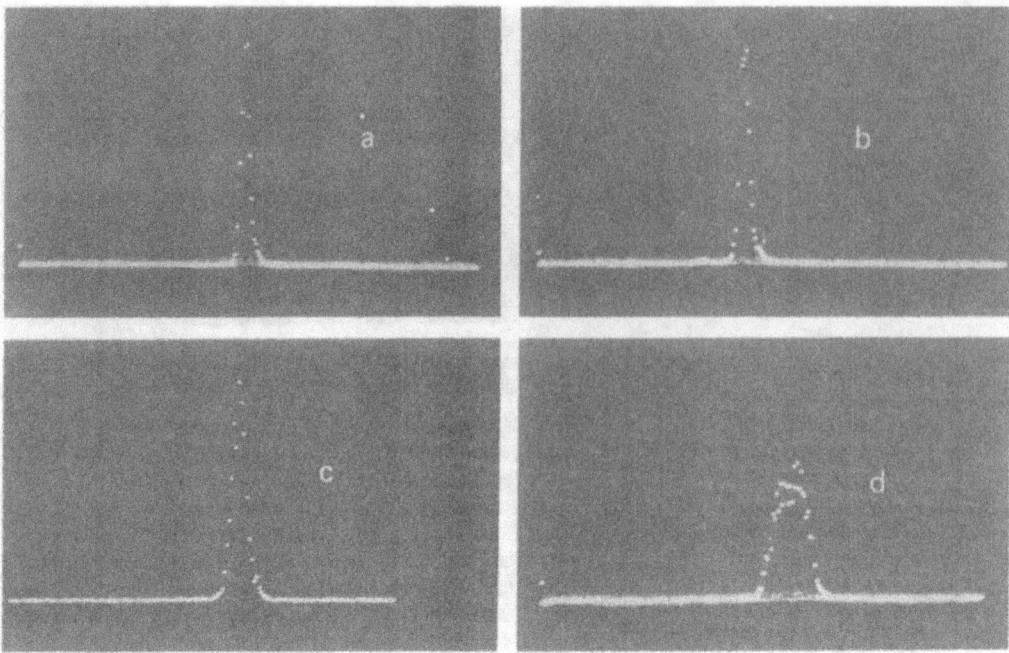

Fig. 8. Accuracy of the spherical chamber. Beam of 8 keV, collimator 0.1 mm. 11 channels = 2 mm. 20 kV total drift. (a) 0°. Coordinates orthogonal to anode wires. (b) Edge of chamber ~42.5°. Coordinates orthogonal to anode wires. (c) Edge of chamber ~42.5°. Coordinates parallel to wires, beam through centre. (d) Edge of chamber ~42.5°. Coordinates parallel to wires, beam out of centre. Fwhm ~0.7 mm in all directions. 0.5 mm accounted for by centroid read-out electronics.

clearly visible. Fig. 11a shows a slice of the distribution in the vertical direction (parallel to the wires) across the reflection 00$\bar{4}$, and fig. 11b that in the nearly horizontal direction (orthogonal to the wires) of the $\bar{3}$11 reflection.

We see that the resolution is around 1 mm (fwhm). The angles of reflection being rather large (36.7° and 30.2°), we see that the quality of large-aperture chambers is encouraging.

Fig. 9. Image of an X-ray source. A solution of ^{55}Fe is deposited on letters grooved in a plastic foil. Vertical letter size: 8 mm. Source at 5 cm from a pinhole of 1 mm diameter in a gold foil. Magnification factor: 6.

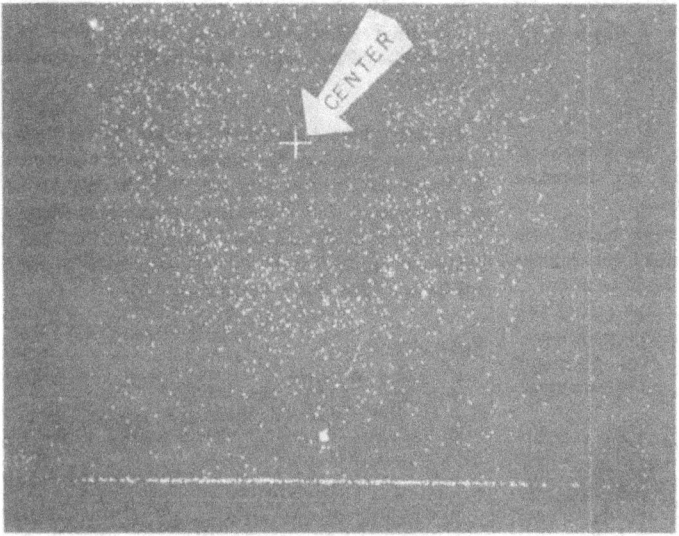

Fig. 10. Diffraction pattern in a CuV$_2$S$_4$ crystal. Drift voltage: 12 kV. Transfer voltage: 12 kV. Hv on the chamber 3.8 kV. Cross at centre. Pattern at a fixed angular position. Bottom reflection 00$\bar{4}$, top left reflection $\bar{3}$11.

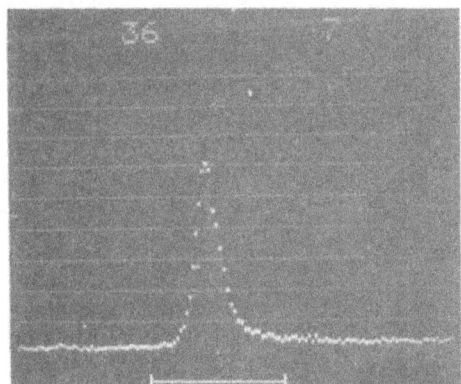

Fig. 11. Diffraction spot width in a CuV$_2$S$_4$ crystal. Drift voltage: 12 kV. Transfer voltage: 12 kV. Hv on chamber: 3.8 kV. Reflection 00$\bar{4}$: $2\theta_B = 36.7°$ (θ_B = Bragg angle). (fuhm = 0.8 mm.)

6. Further developments: the "zoom" X-ray camera

Our results show that a considerable accuracy, much better than the wire spacing, can be achieved if the electrons are given sufficient path length in the drift space to bring the electron cloud produced by a photoelectron to a size of the order of the wire spacing. This renders very easy the construction of large surface chambers. Our method of local measurement of the charge centroid keeps the accuracy to a level almost independent of chamber size. The electronic method which we use for the computation may have to be abandoned in favour of more accurate fast digital calculations[3]). The critical part is the difficulty to make spherical electrodes with large radii and large surfaces.

We have investigated two ways of overcoming this difficulty:

1) The rise-time of the pulse detected in a drift chamber with parallel electrodes is a function of the distance travelled by the electron cluster. We have found[4]) that, with 5.9 keV photoelectrons produced in our gas mixture, an accuracy of around 40% fwhm of the drift length can be achieved, up to 8 cm total drift. This is far from being good enough to compete with spherical drift chambers.

2) In a flat electrode structure, electrical fields of any orientation can be obtained if the limiting electrodes are made of insulators covered with thin conducting lines at proper potentials.

The same lines of force as in cylindrical or spherical structures can then be obtained with flat electrodes. These are of much simpler construction, they are appropriate to large surfaces, and they have the considerable advantage of adjustable focal length. Such structures are under investigation.

7. Conclusions

In drift spaces of large dimensions, the diffusion of electrons spreads the initial cloud of ionization electrons. This presents two advantages:

- The primary electrons can give independent avalanches on the anode wires, and much larger gains can be obtained without space-charge saturation.
- The electron cloud can share its charge among two wires, at least, thus permitting the interpolation of the position between the wires. This facilitates considerably the construction of large-size chambers of high accuracy.
- One disadvantage of the spreading of the electrons may be a reduction of the maximum permissible counting rate, since a larger length of the wire is concerned by every pulse. However, as long as one stays in a linear region it is not clear how it compares with the case of strong local space-charge saturation effects, where the wire is dead over small distances of about 0.2 mm for tens of microseconds.

We are indebted to Dr. A. Breskin for his help in the measurements of the chamber properties, and to R. Bouclier for many contributions to the development of the chamber. The spherical chambers were designed and built by R. Benoît and N. Greguric, who had to overcome many delicate mechanical problems, which will be described in a separate paper.

References

[1] G. Charpak, Z. Hajduk, A. Jeavons, R. Stubbs and R. Kahn, Nucl. Instr. and Meth. **122** (1974) 307.
[2] C. Demierre and G. Vuilleumier, Proc. 2nd Ispra Symp. on *Nuclear electronics*, Stresa, 1975 (EUR 5370e, EURATOM, Luxemburg, 1975) p. 151.
[3] A. Jeavons, N. Ford, B. Lindberg, C. Parkman and Z. Hajduk, IEEE Trans. Nucl. Sci. NS-23, no. 1 (1976) 259.
[4] A. Breskin, G. Charpak and F. Sauli, Nucl. Instr. and Meth. **136** (1976) 497.

Nuclear Instruments and Methods 172 (1980) 337–344
© North-Holland Publishing Company

A FAST X-RAY DIFFRACTOMETER BASED ON A SPHERICAL DRIFT MULTIWIRE PROPORTIONAL CHAMBER

R. KAHN, R. FOURME, B. CAUDRON, R. BOSSHARD

LURE (CNRS–UPS), 91405 Campus d'Orsay, France and Laboratoire de Physicochimie Structurale Université Paris XII, 94010 Créteil, France

R. BENOIT, R. BOUCLIER, G. CHARPAK, J.C. SANTIARD and F. SAULI

CERN, Geneva, Switzerland

An X-ray diffractometer based on a large spherical drift multiwire proportional chamber with a digital position encoder has been designed for operation with rotating anode sources and fully monochromatized synchrotron radiation sources.

Main characteristics are: number of picture elements, 480×240 (480); detective quantum efficiency at 1.54 A: 0.58; operating X-ray band: 1–2.5 A, count rate: 370 kHz with 12% loss; gas mixture: xenon, argon, ethane, carbon dioxide.

Preliminary results are reported, both on the performance of the detector and on the data collection software for protein crystallography.

1. Introduction

High energy storage rings are powerful X-ray sources and, in this context, making area detectors with suitable characteristics is a challenging problem. Each type of experiment has different detector requirements and each design should be optimized to satisfy them. With current techniques, various requirements are frequently contradictory (e.g., high angular resolution and fast read-out) and, in fact, each system will be a somewhat unsatisfactory compromise which is also dictated by financial resources.

Among various experiments, structural studies of macromolecules using X-ray diffraction from single crystals are extremely demanding. An ideal device for this purpose should do the following:

(1) provide angular resolution of reflexions from crystals with long unit-cell axes (say, 50–200 Å) with negligible spatial distortion.

(2) Collect simultaneously all reflexions at a given crystal orientation to a resolution in real space of ~2 Å or better.

(3) Accept wavelengths from ~1 to 2.4 Å which would permit the use of the anomalous scattering of most important heavy atoms.

(4) Detect all quanta entering the detector with high and uniform quantum efficiency.

(5) Accept data at a rate compatible with the intensity of the monochromatized synchrotron radiation.

(6) Give energy resolution for rejection of higher order harmonics, if any.

(7) Provide complete software for crystal alignment, prediction of the spot locations on electronic pictures, on-line data collection and data reduction including background corrections.

(8) Be reasonably simple to operate and reliable for shared experiments.

(9) Be of "reasonable" cost and bulk.

A high ratio of information per photon incident on the specimen — that means *extending crystal lifetime* — depends on the number of separable pixels on the detector area [1].

Conditions (1) and (2) can be simultaneously satisfied by increasing *the number of pixels*. If the number of pixels is not increased, high angular resolution requires a long crystal-to-detector distance, hence poor resolving power; or high resolving power requires a short crystal-to-detector distance, hence poor angular resolution.

In contrast a *high count rate* is not of the utmost importance: any improvement of detector count rate does not extend crystal lifetime but only saves the crystallographer's time [1]. This is not the major problem for normal protein crystals.

In addition, one should not forget that the com-

VII. DETECTORS AND OTHER SUBJECTS

puting time which is required between two consecutive electronic pictures may well become longer than the time which is necessary to record a picture. Any gain in speed will then have a limited effect on the total duration of the experiment. From our experience with the protein crystallography set-up at LURE [2] (oscillation camera and focusing Ge (111) monochromator, storage ring DCI operated at 1.72 GeV, 250 mA), the total background in the 2 Å cone for a protein crystal is about $1-3 \times 10^7$ photons s^{-1}. With a channel-cut Ge (111) monochromator the total background is $0.3-1 \times 10^6$ photons s^{-1}, which is about the same as a conventional high intensity set-up (fine focus 1.6 kW rotating anode tube, graphite monochromator). Such a monochromator provides a very parallel and highly monochromatic beam which is suitable for multiwave length data collection with sufficient flux to get a good balance between exposure time and computing time.

Starting from these considerations and from various technical limitations, the data collection system which is described hereafter is oriented toward accurate data collection with highly monochromatized, tunable radiation, for small and medium-sized proteins. Larger proteins are probably best tackled with the film set-up and larger pass-band, high intensity, focused radiation.

Since the accuracy of data and the angular resolution are, for our purpose, more important than very high counting rates, a multiwire proportional chamber based diffractometer (MWPC) was preferred to a TV detector. The success of planar MWPC as electronic area detectors for recording X-ray diffraction patterns (3) was a strong incentive to try to improve their angular resolution, counting rate capability and the uniformity of quantum detection efficiency across the sensitive area.

Our system is the result of a joint effort between two groups at Geneva and Orsay, and is based on a MWPC with a spherical drift gap. The properties of the spherical drift gap were investigated on a small prototype detector with an opening angle of 50° [7]. A larger Mark II prototype with a useful opening angle of ~81° was then built; the encoding system was based on a read-out of the charge centroid type [7] and the MWPC was filled with a continuous flow of "cheap" gas (argon 70%, isobutane 28%, methylal 2%). Mark II detectors have been tested at LURE [8] and at MIT. The actual Mark III chamber has a similar geometry (except a thicker transfer space); (X, Y) coordinates are calculared by a digital

encoder and a xenon mixture circulating in a closed circuit is used. In the following, the MWPC is described, the electronic read-out and storage system is outlined and some preliminary results obtained with the data collection software are presented.

2. Detector design

The finite thickness of the absorbing mixture introduces an error whenever the trajectories of the quanta are not perpendicular to the detector plane; on the other hand, a sufficient thickness is required for high quantum efficiency.

Three solutions can be proposed to overcome the difficulty: the first one is a pressurized detector [4]. With a mixture containing mostly xenon at a pressure of ~10 atm, 8 keV photons are totally converted in 1 mm, so that the parallax error is kept small. The main advantage of such a flat MWPC is a free choice of angular resolution: the crystal-to-detector distance is not fixed. The design of a large pressurized detector is not trivial with respect to windows, leaks, wire connections and closed gas circuit with purifiers (in the light of our experience with a closed circuit at atmospheric pressure, the latter point is a quite serious one). The second solution is a small flat MWPC which is moved so that the incident X-rays are nearly normal to the detector [5]; the lower ratio of information per incident photon can be compensated for by placing several detectors around the crystal [6]. The third solution is achieved by a spherically symmetric drift gap centered at the crystal in which diffracted photons are absorbed, coupled by means of a transfer drift gap to one planar MWPC; this system eliminates parallax errors provided that all the photons are converted in the drift region. The effect of any penetration of the transfer region can be seen in the "comet" tails of the spots, pointing away from the centre of the detector [7,8]. In the MARK II chamber, the CERN gas mixture absorbs 76% of the flux transmitted through the entrance window at 8 keV by photoelectric conversion so that a large fraction of the photons is converted in the transfer region. In the MARK III chamber, this effect is completely suppressed by the addition of xenon.

In a standard MWPC without drift gap, the coordinate along the direction perpendicular to the anode wire is quantized because the avalanche of secondary electrons is located on one wire. The addition of an extended drift gap provides a means to overcome this

+

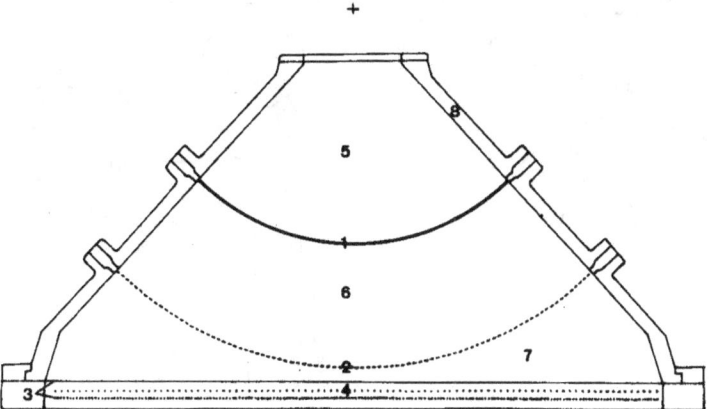

Fig. 1. Spherical drift chamber, schematic. 1, entrance electrode (−18 kV); 2, exit electrode (−6 kV); 3, cathodes (0 V); 4, anode (+4 kV); 5, helium; 6, drift gap; 7, transfer gap; 8, conical edge.

limitation. Diffusion of the primary ionization occurs along the drift path so that the avalanche is expanded and distributed over more than one wire; with drift paths of 12 cm to 17 cm, the response between wires is nearly linear [8]. In this respect, the spherical drift space improves the spatial resolution of the MPWC.

Finally the long drift path is useful for operation with a storage ring filled with one bunch like DCI; even if two photons are scattered by the crystal during one pulse, the pathlengths of the primary photoelectrons will be generally different; since the drift velocity is low (typically ~5 cm μs^{-1}), the time interval between the arrivals at the anode plane may be longer than the encoder dead time, so that the photons are detected as single events. The time structure of the synchrotron radiation is thus partly smoothed. A disadvantage of the spherical drift MWPC is a fixed sample-to-detector distance and thus a fixed angular resolution; the electric field diverges from a fixed point where the sample is located, at the centre of spherical electrodes. Radial fields could be produced by printed electrodes, which would give more flexibility to the design; this is under investigation.

The major features of MARK III detector are illustrated in fig. 1. It is made of non-bakable materials (epoxy, fiberglass, rubber gaskets, . . .). The main characteristics are: entrance electrode: 18 cm radius, hydroformed 40 μm thick aluminium window; exit electrode: 28 cm radius, nickel-plated stainless steel grid; drift gap 10 cm; useful acceptance of the cone

~82°. The flat MWPC is made with three planes of wires spaced by 6 mm. The two cathode planes are made of 480 100 μm wires spaced by 1 mm; the anode plane, made of 240 20 μm wires spaced by 2 mm, is maintained at ~4 kV. The gas used for 8 keV radiation is a mixture of xenon (24%), argon (35%),

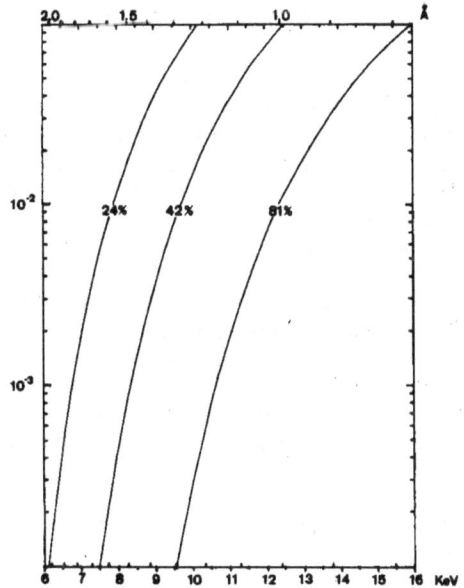

Fig. 2. Fraction of photons not converted in the drift gap versus energy for various fractions of xenon.

VII. DETECTORS AND OTHER SUBJECTS

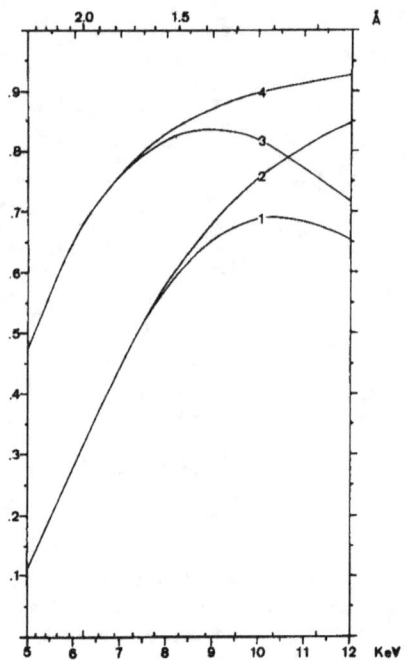

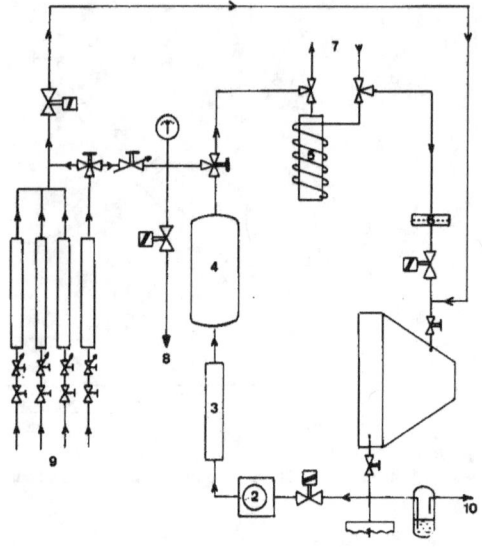

Fig. 4. Gas circuit, schematic. 1, pressostat; 2, pump; 3, calcium oven; 4, mixing tank; 5, purifier; 6, filter; 7, circuit for
regeneration of the purifier (hydrogen + argon); 8, to vacuum
(used prior to filling and regeneration); 9, gas inlet; 10, gas
exhaust.

ethane (35%) and carbon dioxide (6%). The tip of the
conical edge is filled with helium to reduce the
absorption of scattered X-rays by air; this gas is
slightly pressurized so as to protect the thin hemi-
spherical entrance window.

The percentage of flux which reaches the transfer
region has been plotted versus the energy of incoming
photons (fig. 2). It is ~1% at 8 keV with 24% xenon
and ~1% at 12.3 keV ($\lambda = 1$ Å) with 81% xenon.

The quantum counting efficiency of the detector
versus the energy is plotted on fig. 3 for a mixture
with 24% xenon; it is limited at lower energy by the
absorption of the entrance window. (A beryllium
window would be feasible but rather costly). From
fig. 2 and fig. 3, the practical range is ~5–12 keV.

With the continuous flow system of MARK II and
argon, we found no important degradation in the
energy spectrum over long periods. In the case of a
closed circuit, substantial degradation appeared
within about one hour from the introduction of fresh
gases. The main reason is electron attachment by
oxygen, water vapor leaking through the mylar window and various electronegative contaminents outgassing from the detector; this is a serious problem
for such a detector where the pathlength varies from
~3 cm (for a photon converted near the exit grid) to
~17 cm. The gas had to be continuously purified by
flowing over catalytic copper and hot calcium. The
gas circuit is thus rather complex (fig. 4). The mixture of four gases which is now used in the result of
many trials. It is better than with gas flowing without
purification. The addition of a small amount of carbon dioxide suppresses sparks observed after purification. With fresh gases and purification, a good energy
resolution (~12% fwhm) is obtained at 8 keV, with a
minor degradation over a period of 3 weeks. Every
month, the copper is regenerated by hydrogen and
initial performances are restored.

3. (X, Y) Cathode read-out, data storage and com-
puter

To accept data at a rate compatible with a standard rotating anode generator or with fully mono-

chromatized synchrotron radiation, the read-out systems was designed for an average rate of $0.3–0.6 \times 10^6$ detected events, which corresponds at 8 keV to about $0.5–1 \times 10^6$ incident photons. The position encoder must be about one order of magnitude faster to limit loss due to coincident events to ~10%. Such high counting rates have been achieved by a digital encoder with one amplifying circuit per wire linked to a large mass memory. When an event corresponds to the energy discrimination of the anode circuit, the distribution of pulses induced on several wires in each cathode plane is sensed. Let X_1 and X_2 be the farthest channels where a signal higher than a fixed threshold is detected; X_1 and X_2 are found with priority encoders and the midpoint is calculated by a digital processor. The same treatment is simulta-

neously done along the Y direction, giving a pair of coordinates. A similar system has been previously described [9] and another one has been operated on small chambers (up to 128×128 channels) [5]. The crucial point for good spatial resolution is identity of the amplifier–shaper thresholds [5] and each channel includes a potentiometer to make fine adjustments.

The MWPC electronics is mounted partly along the perimeter of the detector and partly in a separate crate. The encoder gives binary coded (X, Y) coordinates which are buffered in a shift register; X, Y are adresses for a storage location (cell) in a mass memory. The corresponding cell is incremented by 1. The mass memory is a 128K, 16 bit word, semiconductor memory [10] plugged in a standard CAMAC crate. Performances of the actual read-out are: maximum count rate 370 kHz with 12% loss; number of pixels 480×240 (to be extended to 480×480) with 64K events per pixel. The maximum count rate is set by the read-increment-write cycle of the mass memory and could be more than double with the faster memories which are now available.

The CAMAC crate is linked to a PDP 11/34 minicomputer by a direct memory access interface so that the read-out and the data storage are treated as a conventional peripheral (fig. 5). An entire image is transferred from the external memory to the computer memory in ~350 ms. The system is now equipped with a computer display terminal which permits monitoring of the mass memory. It is intended to install another display, independent of the computer, based on a silicon target storage tube and a scan converter driving a high resolution monitor.

4. Goniometer and beam monitor

Mechanical spare parts of a commercial Arndt-Wonaçott oscillation camera have been used to assemble a single crystal goniometer. The stepping motor and the solenoid of the beam-shutter are connected to CAMAC drivers. This goniometer has a single rotation axis (Φ-axis). This system is less flexible than a three-circle goniometer but in practice it will not greatly reduce the potential of the multireflexion diffractometer; it has the advantage of simplicity, accuracy (0.002 degree) and easily alignment. For operation with the synchrotron source, it is necessary to monitor the intensity of the monochromatic beam. A small ion chamber [11] will be installed between the collimator and the crystal; the

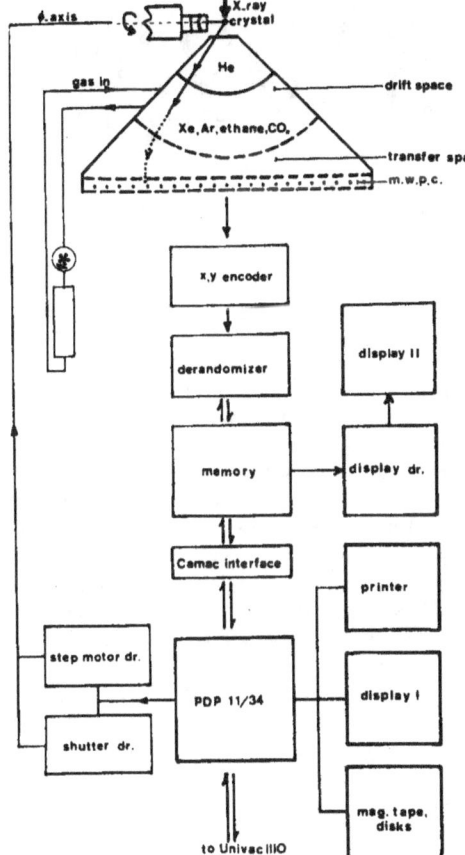

Fig. 5. Schematic of the set-up.

VII. DETECTORS AND OTHER SUBJECTS

Fig. 6. General view of the MWPC-based diffractometer (computer, mass core and peripherals are not shown).

picoammeter is connected to a 10 bit CAMAC digital voltmeter under computer control. These data are used in the data collection software.

The diffractometer is shown in fig. 6.

5. Data collection software and utility routines

Main programs are written in ANSI FORTRAN but all device control routines are written in ASSEMBLER for higher efficiency and speed.

Crystals are approximately oriented using optical observation and electronic pictures. Then a still image is recorded at, say $\Phi = 0°$, and a few strong Bragg reflexions are pointed on the display. The centres of gravity (X, Y, Φ) of those reflexions are calculated from intensity measurements on successive images. The same process is repeated at, say $\Phi = 40$ and $90°$. Reflexions are indexed and a refined orientation matrix is calculated. The (X, Y, Φ) coordinates of all reflexions at a given resolution are then calculated and sorted on a peripheral according to Φ-values, so

that the spot positions on still photographs can be accurately predicted.

The data collection procedure with standard sources is, in essence, the procedure described by Xuong et al. [12]. A series of stationary diffraction patterns are recorded, rotating the crystal by a small fixed angle between each exposure (frame). This is equivalent to the step-scanning technique used in standard diffractometry except that here all simultaneously occurring reflexions are measured. A reflexion will occur in several consecutive frames. Using predicted positions, the computer can read-out the counts accumulated at the reflexion coordinates of a particular frame and store them in a data file associated with this reflexion. Integrated intensities are estimated from the count profile obtained from successive frames. With respect to the standard oscillation technique, the peak-to-background ratio is considerably enhanced and reflexion overlapping is eliminated.

This procudure has to be modified with synchrotron radiation: (i) with highly monochromatic and parallel radiation, the profile of a Bragg reflexion is much sharper than usual [13]; instead of stationary pictures, the crystal is slowly rotated during each exposure. (ii) To compensate for the intensity decay, the computer modifies the rotation speed ω according to the intensity I, monitored by the ion chamber, so that $I\omega^{-1}$ remains constant. In the case of large changes in the intensity, the loss due to coincident events may substantially change and this may also be corrected by the same process. All data are then collected at essentially the same scale.

6. Preliminary results

The equalization of the amplifier—shaper thresholds has not yet been completed for the whole sensitive area. A test circuit for 32 channels, under computer control, will greatly simplify this adjustment. The result of misadjustment is clearly visible on images as uneven distribution of events on various channels *. In spite of this fact, first results have been very encouraging. With beryllium acetate crystals to test the spatial distortion of the MWPC and the automatic indexing routines, it was demonstrated that correct indexing is obtained for unit-cells with cell edges up to at least 157 Å. Using a sealed X-ray tube with the source (10×0.5 mm) set at a take-off angle of 2.5° at a distance of 16 cm from the crystal, a graphite monochromator and a 70 mm long collimator of 0.8 mm diameter, the reflexions cover an area of ~4×4 mm (fig. 7). This will be substantially reduced by a finer collimation and completion of the MWPC adjustment. A preliminary estimation of the maximum resolvable cell edges is given in table 1 for various wavelengths. The closest spacings between spots are obtained when the sphere of reflexion is tangent to a direction parallel to the shortest axis of the reciprocal lattice. We have assumed spacings of 5 mm and 3 mm; the latter is a realistic goal for a finely collimated beam, a small crystal and an optimized set-up.

Finally, a protein single crystal (a crystal of lysozyme in the high temperature form, kindly supplied by Dr J. Berthou) was mounted on the goniometer. Using still electronic pictures, the crystal was oriented with respectively c and a nearly parallel to the oscillation axis and the vertical (fig. 7b). The indexing pass

* Note added in proof: The encoder is now adjusted and these features no longer appear.

Fig. 7. Still picture for a crystal of lysozyme B (orthorhombic-unit cell with $a = 56.3$ A, $b = 73.8$ A and $c = 30.4$ A). The display has no grey level and pixels with an accumulated count higher than an arbitrary threshold are displayed; 7(a) was obtained for a low threshold and (b) with a high threshold; in the latter case, all points belong to Bragg peaks.

and the calculation of the refined orientation matrix were straightforward.

The diffractometer will be extensively tested with the standard source and various protein crystals to investigate spatial distortions and evaluate the uniformity of detective quantum efficiency; for instance, uneven thickness of the entrance window

VII. DETECTORS AND OTHER SUBJECTS

Table 1

λ	0.98 A	1.54 A	1.96 A	2.45 A
Resolution in real space	1.4 A	2.2 A	2.8 A	3.5 A
Max. cell edge 5 mm spacing	59 A	92 A	117 A	147 A
3 mm spacing	98 A	153 A	195 A	245 A

could justify correction of intensity profiles. Then it will be installed on a dedicated beam port of the D2 beam line at DCI; a separate function two-crystal monochromator with fixed exit slit [14] will provide a highly parallel and tunable X-ray beam. The set-up will be used as an image acquisition facility and especially − but not exclusively − for protein crystallography.

References

[1] G.E. Schulz and G. Rosenbaum, Nucl. Instr. and Meth. 152 (1978) 205.

[2] M. Lemonnier, R. Fourme, F. Rousseaux and R. Kahn, Nucl. Instr. and Meth. 152 (1978) 173.

[3] C. Cork, D. Fehr, R. Hamlin, W. Vernon, Ng. h. Xuong and V. Perez Mendez, J. Appl. Cryst. 7 (1973) 319.

[4] S. Sobottka, G. Cornick and R. Kretsinger, SSRL report 78/04 (1978) VIII. 70, SSRL, Stanford, California, USA.

[5] S.E. Baru, G.I. Proviz, G.A. Savinov, V.A. Sidorov, A.G. Khabakhpashev, B.N. Shuvalov and V.A. Yakovlev. Nucl. Instr. and Meth. 152 (1978) 209.

[6] R. Hamlin, C. Cork, C. Nielsen, W. Vernon and Ng. h. Xuong, Eleventh Int. Congr. Crystal. (1978) Warszawa, Poland.

[7] G. Charpak, Z. Hajduk, A.P. Jeavons, R. Kahn and R.J. Stubbs, Nucl. Instr. and Meth. 122 (1974) 1307.

[8] G. Charpak, C. Demierre, R. Kahn, J.C. Santiard and F. Sauli, Nucl. Instr. and Meth. 141 (1977) 141.

[9] C. Parkman, Z. Hajduk, A. Jeavons, N. Ford and B. Lindberg, report DD/75/14 (1975) CERN, Geneva, Switzerland.

[10] R. Klesse and J. Munnier, report 77 MU 18T (1977) Institut Laue−Langevin, Grenoble, France.

[11] Designed according to H. Bartunik (EMBL-Outstation, Hamburg, FRG).

[12] Ng. h. Xuong, S.T. Freer, R. Hamlin, C. Nielsen and W. Vernon, Acta Cryst. A34 (1978) 289.

[13] J.C. Phillips, PhD Thesis (1978), Stanford University, Stanford, California, USA.

[14] Designed in collaboration with J. Goulon and M. Lemonnier at LURE.

Nuclear Instruments and Methods 201 (1982) 193–196
North-Holland Publishing Company

DEVELOPMENT OF A MULTIWIRE PROPORTIONAL CHAMBER AS AN AREA SENSITIVE DETECTOR FOR X-RAY PROTEIN CRYSTALLOGRAPHY

D. BADE, F. PARAK, R.L. MÖSSBAUER, W. HOPPE, N. LEVAI and G. CHARPAK *

Max-Planck-Institut für Biochemie, Abteilung für Strukturforschung I, Am Klopferspitz, D-8033 Martinsried, Fed. Rep. Germany

and

Physikdepartment E15 der Technischen Universität München, James-Franck-Strasse, D-8046 Garching, Fed. Rep. Germany

The prototype of an area sensitive multiwire proportional chamber operated in series with a spherical drift chamber was investigated for its applicability to protein crystallography, employing 14.4 keV ($\lambda = 0.86$ Å) γ-radiation of ^{57}Co Mössbauer sources. The 10 cm long spherical drift field, in this case, provides high efficiency in order to compensate for the small primary intensity of Mössbauer sources. The detector system was tested with different gas mixtures. The improvement of the spatial resolution with increasing xenon content is demonstrated using the $60\bar{3}$ reflection of a myoglobin crystal. The addition of xenon to the counter gas is shown to be necessary for protein crystallography with short wavelengths. The background is sufficiently low for Mössbauer experiments, due to good energy discrimination at the anode (≤ 0.003 counts/(min $\times$ mm^2)).

1. Introduction

The Mössbauer scattering on a ^{57}Fe nucleus offers a new possibility for an experimental solution of the phase problem arising in structure determination of protein crystals [1]. The applicability of this method was demonstrated in the case of a myoglobin single crystal [2]. Such experiments have to be performed with ^{57}Co gamma sources which yield a much lower intensity of the primary beam than X-ray tubes. In order to get reasonable times for the collection of the intensities of all reflections which are necessary for a structure determination, one has to work with a slightly divergent beam. This way, one excites simultaneously a large number of Bragg reflections. This procedure requires a large area sensitive detector. Because of the low intensity scattered into each Bragg reflection (typically 1 count/min) one has to provide for an optimum efficiency of the counter. For the same reason the background has to be as low as possible in an energy window around 14.4 keV ($\lambda = 0.86$ Å). The spatial resolution must be sufficient to allow for a separation of closely spaced protein crystal reflections.

Optimal efficiency together with low background can be achieved by using small Si(Li) detectors for single reflections [2]. However, the

* CERN, Geneva, Switzerland.

extension of this concept to larger Si(Li) counters for many reflections [3] involves the adaption of special detector assemblies to each reciprocal lattice geometry. Flat multiwire proportional chambers (MWPC) might appear to be adequate for the application in protein crystallography, but for usual anode–cathode distances the achievable efficiency is rather low at 14.4 keV. An increase of the thickness of the counter improves the efficiency, but at the same time the parallax problem reduces the spatial resolution at large scattering angles. The spherical drift chamber in combination with a flat MWPC [4] can provide both optimal efficiency for 14.4 keV radiation and sufficient spatial resolution. First experiments with a special prototype chamber designed for protein crystallography with wavelengths below 1 Å are described in the following sections.

2. Detector design

The detector construction was carried out at CERN with a concept similar to two other counters [5,6]. As schematically shown in fig. 1 it consists of three successive chambers made from epoxy and fiberglass. A cylindrical entrance chamber is filled with helium in order to reduce the absorption of the γ-rays. The γ-ray is absorbed in a spherical drift chamber between two spherical

IV. GAS CHAMBER BASED DETECTORS

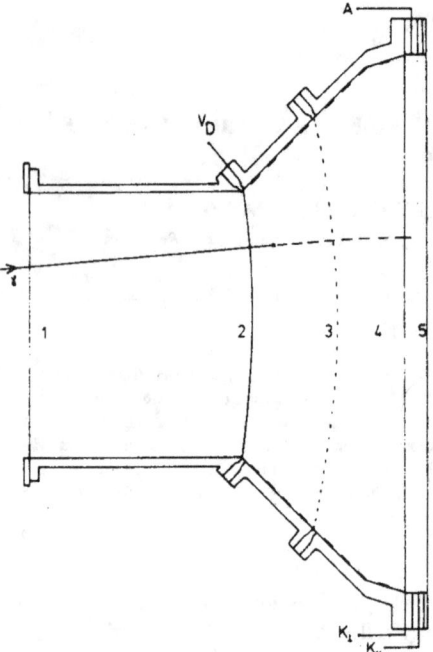

Fig. 1. Scheme of the MWPC with spherical drift chamber. γ-rays penetrate through a hostaphan window (1) into the entrance chamber containing He and cross the spherical entrance electrode (2) of the drift chamber filled with counter gas. After photoelectric absorption the electrons drift through the spherical electric field maintained by the drift voltage V_D (-18 kV) and through the exit electrode (3) and the transfer space (4) into the flat MWPC (5). They cause an avalanche at the anode A ($+3.8$ kV). The position is determined at the cathodes ($K_\parallel$ and $K_\perp$).

electrodes. The crystal under study is placed at the focus of the spheres. All scattered γ-rays will then pass radially through the drift chamber. The electrons produced by absorption of a γ-quantum in the counter gas likewise drift radially. This way, the spatial resolution does no longer depend on the absorption position of a γ-quantum and no parallax problem occurs. The described detector has a spherical entrance electrode of 40 μm Al. The spherical exit electrode consists of a stainless-steel grid. The thickness of the drift chamber is 10 cm.

In contrast to the other detectors mentioned above, the spherical drift chamber of our detector was designed for a radius of 90 cm of the entrance electrode in order to resolve the closely spaced

protein reflections for wavelengths below 1 Å. The size of the entrance window limits the opening angle of the prototype to 18°. At both sides of the drift space the radial boundary conditions of the spherical electric field are matched by cylindrical electrodes which are connected to the drift potential by a resistive divider.

Electrons generated in the drift space pass through the transfer space and enter perpendicularly into the flat MWPC (50×50 cm^2) consisting of three wire planes which are 6 mm apart. Gold-plated tungsten wires with 100 μm diameter and 1 mm separation are used for the rectangular cathode arrays and similar wires with 20 μm diameter and 2 mm separation form the anode plane. Subject to the opening angle of the chamber only the central area of 30×30 cm^2 of the MWPC is presently being used.

3. Readout and data processing

The readout of the MWPC is performed by the analog center-of-gravity readout method [7]. The principle is schematically shown in fig. 2. An avalanche at a single anode wire induces positive pulse distributions on the two rectangular cathode wire planes. The coordinates X and Y of the avalanche can then be determined from the center of gravity which is obtained from a weighted summation of the analog pulses by a resistor network. Diffusion of the primary ionization along the drift path can expand the avalanche over more than one anode wire. Then the analog center-of-gravity method in connection with the spherical drift chamber allows to overcome the spatial quantizing

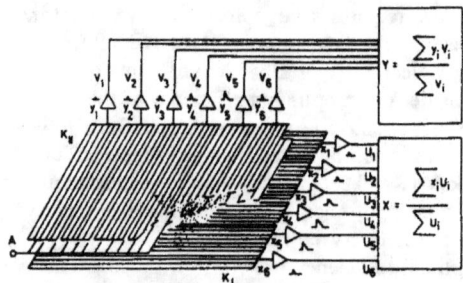

Fig. 2. Scheme of the analog center-of-gravity readout. An avalanche at the anode (A) gives rise to pulse distributions at the cathodes ($K_\parallel$ and $K_\perp$). The position (X, Y) is obtained by analog summation and division.

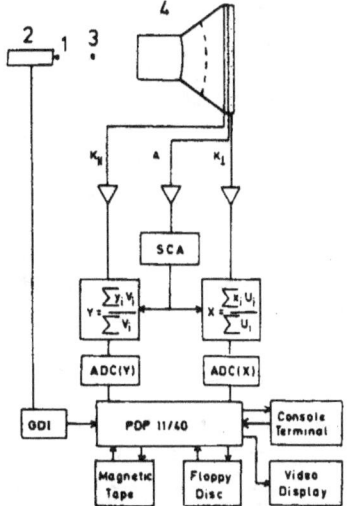

Fig. 3. On-line coupling of the detector to a PDP 11/40. The 14.4 keV γ-radiation from the ^{57}Co source (1) on the Mössbauer drive (2) is scattered by the protein crystal (3) into the detector (4). The Mössbauer drive is synchronized by a general device interface (GDI).

effect of the anode wires, too. For good linearity of the position determination over the whole detector careful and iterative adjustment of the resistor network was found to be essential. In practice this work was done using trimmable helipot resistors in addition to resistors with fixed values. The negative pulses of the anode wires were used for energy discrimination which is necessary to achieve a low background.

The data processing is performed with commercial elements as shown in fig. 3. Whenever an anode pulse appears in the proper energy window and is discriminated by the single channel analyzer, the coordinates X and Y are determined from the cathode signals and become available as analog levels. They can be read into the core memory of the PDP 11/40 computer after analog-to-digital conversion with commercial AR-11 interfaces, which can be started externally. Further data reduction, graphic display or transfer to magnetic tape is performed with special programs residing on a floppy disc. A Mössbauer drive can be synchronized to the readout by a general device interface. In this configuration the maximum counting rate of the detector is limited by the AD-conversion time of the AR-11 units to 30000 counts/s,

which is sufficient for Mössbauer experiments. It should be mentioned, however, that this limitation comes purely from the readout system and is not imposed by the detector itself.

The readout can be speeded up by fast AD-conversion units and transfer to a large ($512K \times 16$ bits) separate semiconductor memory working with microsecond cycle time. In that way the maximum counting rate can be 5×10^5 counts/s.

All interrupt service and data processing routines run under a RT-11 operating system and are written in Assembler for high efficiency and speed. For fast information on the crystal orientation the whole detector area is divided into 96×96 pixels and the data are stored in the computer core memory. In addition to this possibility the interrupt service routine of the two AR-11's is programmed with enough flexibility to restrict the readout to smaller parts of the detector area with a corresponding decrease in pixel size. The smallest pixel size (0.4×0.4 mm^2) is then given by the accuracy of the ADC.

4. Detector operation and results

The detector can be operated with a continuous flow of 80% Ar and 20% C_2H_6. This is however not satisfying for measurements with wavelenghts below 1 Å because the photoelectric absorption amounts only to 26% for 14.4 keV quanta in the 10 cm path within the drift space. A second problem arises from the practical range of the photoelectrons. At low energy ($\lesssim 20$ keV) photoelectrons are ejected preferentially perpendicular to the direction of the absorbed γ-quanta [8]. This effect can spoil the spatial resolution substantially, if the range of the photoelectrons is large. According to ref. 9 the practical range of a 12 keV photoelectron was calculated to be larger than 2 mm in 80% Ar + 20% C_2H_6. This effect can be reduced by the addition of Xe to the counter gas. The high cost and large volume of the detector makes it necessary to circulate and purify the counter gas. In the present assembly the detector chamber cannot be evacuated and therefore helium leak testing cannot be done effectively. The purification system must efficiently remove oxygen and water vapour from the counter gas. For this purpose we have applied a gas cleaning and purification system similar to that in ref. 5. Oxygen and water vapour are

IV. GAS CHAMBER BASED DETECTORS

removed by a copper catalyst (BASF R3-11) and hot calcium, respectively. The circulation of about 100 cm³/min is effected by a small oil-free rotation pump.

Continuous operation up to periods of two weeks was obtained with a gas mixture of 24% Xe, 35% Ar, 35% C_2H_6 and 6% CO_2. This gives 63% efficiency for 14.4 keV γ-rays in the drift space. Better efficiency was obtained once with 41% Xe, 27% Ar, 27% C_2H_6 and 5% CO_2. However, the stability of the working point was still not sufficient at these conditions. The spatial resolution of the prototype detector was tested with an oscillating myoglobin crystal and 17 keV MoK_a radiation ($\lambda = 0.71$ Å). The profile of the $60\bar{3}$ reflection of myoglobin is given in fig. 4 for three different gas mixtures. From comparison with the reflection size on an X-ray film a width of 4 mm is expected because of the crystal dimensions. It can be clearly seen that the counter produces a small additional broadening, particularly when no Xe is added to the counter gas. This effect can be attributed to the practical range of the photoelectrons as discussed above. It should be mentioned, however,

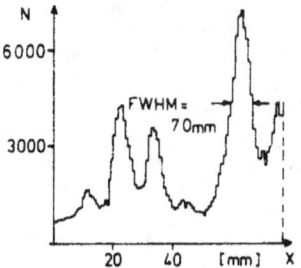

Fig. 5. Profiles of the adjacent reflections $8\bar{2}0$, $7\bar{2}0$, $6\bar{2}0$, $5\bar{2}0$ and $4\bar{2}0$ of the $a^* b^*$ plane of myoglobin.

that this broadening can be tolerated in protein crystallography. Moreover, a further increase of the Xe content of the counter gas would make the broadening of the reflections by the counter negligible.

Fig. 5 gives the profiles of some adjacent reflections of the $a * b *$ plane of myoglobin. Strong and weak neighbouring reflections are well resolved. The well-known intensities of these reflections are accurately reproduced.

The background of the detector was found to be 0.003 counts/(min × mm²) for the gas mixture with 24% Xe. This value is low enough for Mössbauer experiments.

This work was supported by the Bundesministerium für Forschung und Technologie.

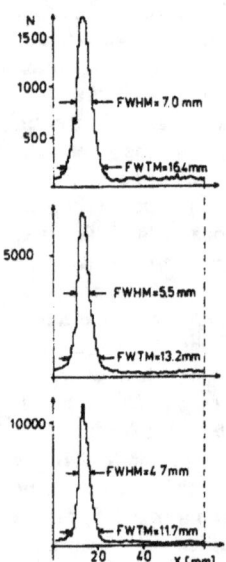

Fig. 4. Profile of the $60\bar{3}$ reflection of myoglobin for different gas mixtures; pixel size = 0.8 mm. Top: 46% Ar + 46% C_2H_6 + 8% CO_2. Center: 24% Xe + 35% Ar + 35% C_2H_6 + 6% CO_2. Bottom: 41% Xe + 27% Ar + 27% C_2H_6 + 5% CO_2.

References

[1] F. Parak, R.L. Mössbauer and W. Hoppe, Ber. Bunsenges. Phys. Chemie 74 (1970) 1207.

[2] F. Parak, R.L. Mössbauer, W. Hoppe, U.F. Thomanek and D. Bade, J. Phys. (Paris) 37, C6 (1976) 703.

[3] U. Biebl and F. Parak, Nucl. Instr. and Meth. 112 (1973) 455.

[4] G. Charpak, Z. Hajduk, A. Jeavons, R. Stubbs and R. Kahn, Nucl. Instr. and Meth. 122 (1974) 307.

[5] R. Kahn, R. Fourme, B. Caudron, R. Bosshard, R. Benoit, R. Bouclier, G. Charpak, J.C. Santiard and F. Sauli, Nucl. Instr. and Meth. 172 (1980) 337.

[6] C. Bolon, M. Deutsch, R. Lanza, G. Quigley and A. Rich, IEEE Trans. Nucl. Sci. NS-26 (1979) 146.

[7] G. Charpak, A. Jeavons, F. Sauli and R. Stubbs, CERN Report No. 73-11 (1973).

[8] E.J. Williams, J.M. Nuttal and J.M. Barlow, Proc. Roy. Soc. (London) A121 (1928) 611.

[9] R.O. Lane and D.J. Zaffarano, Phys. Rev. 94 (1954) 960.

Nuclear Instruments and Methods 195 (1982) 469–473
North-Holland Publishing Company

A MULTISTEP PARALLAX-FREE X-RAY IMAGING COUNTER

A. BRESKIN, G. CHARPAK and J.C. SANTIARD

CERN, Geneva, Switzerland

Received 14 September 1981

With a large preamplifying gap of 2 cm, combined with a localization multiwire chamber, it is shown that measurement of pulse height provides the depth of the atom which absorbs a monoenergetic low-energy X quantum to a great accuracy. The preliminary results show that for 8 keV X-rays, accuracies of a fraction of a millimetre are achieved. No error is introduced by the limited efficiency of the absorbing gap, thus permitting the use of low-cost argon mixture fillings. Large-surface parallax-free gaseous detectors of planar structure can be built for monoenergetic diverging sources.

1. Introduction

The localization of X-ray quanta by means of gaseous detectors involves a basic difficulty. The thickness of the detectors required for reaching a large efficiency introduces localization errors for photons impinging on the entrance window at angles very inclined to the normal and being absorbed at various depth across the gap.

One way to overcome the difficulty is to have the photons absorbed in a wide-gap drift volume with a radial electric field centred at the source [1]. This field is normally provided by spherical electrodes, and several recent experimental results have confirmed the validity of this method [2–5]. However, it applies only to situations where the photon source is at a fixed position, namely at the centre of the entrance window. The construction problems connected with spherical structures make it difficult to store imaging detectors with a great variety of entrance window radii.

The analysis of the properties of multistep avalanche chambers, where amplification occurs in successive amplifying structures [6], has shown that it is possible to build parallel-gap structures that are free of parallax errors [7]. The principle is simple (fig. 1). If monoen-

ergetic photons are absorbed, by the photoelectric effect, in the gas of a parallel-plate amplifying structure, the pulse height is proportional to $e^{\alpha(L-z)}$, where L is the gap width, z is the distance between the entrance window and the photoelectron which is supposed to produce a punctual swarm of ionization electrons, and α is the Townsend coefficient. If this pulse height can be measured, the depth z is obtained with an accuracy that is a function of the pulse-height resolution, the gain, and the gap width. The parallax error due to ignorance of the depth at which an X-ray photon is absorbed in a large gap can then be corrected.

An easy way of measuring this pulse height and of localizing the position of the avalanche is to transfer the electrons of the avalanche from the first stage to a second amplifying gap that consists of a multiwire chamber or another parallel-plate structure. The recent work of our group at CERN [6–8] has shown that this can be done either way, with a rigorous proportionality between the transferred electrons and the initial avalanche size in the first amplifying gap. In this article we report some measurements demonstrating the validity of this concept. The amplifying gap was combined with a multiwire proportional chamber (MWPC). Accuracies as good as 0.5 mm fwhm have been obtained over absorption depths of 2 cm.

2. Accuracy of absorption depth measurement

If a photon of definite energy liberates N_0 ion pairs at a depth z in the gas of a parallel-plate amplifying structure, the total number of electrons collected by the anode is

$$N = N_0 e^{\alpha(L-z)}.$$

In most cases it is rather difficult, without amplifica-

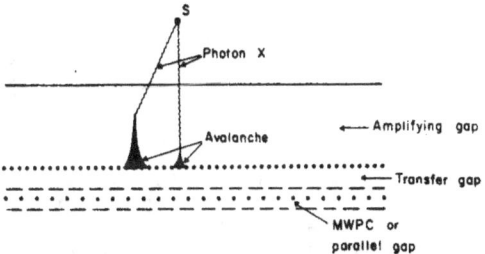

Fig. 1. Principle of the parallax-free multistep counter.

tion, to detect directly the pulse produced by the initial charge N_0 librated by the photoelectrons. If we wish to detect the photons absorbed near the anode, where the gain is close to unity, then a further amplification step is necessary.

It has been shown that a constant fraction of the total number of the electrons produced in the avalanche can be transferred to a following element, for example a drift space followed by a MWPC. This fraction f is very close to the ratio of the electric fields in the drift space and the amplifying gap. It can be conveniently set at values close to 20% or 30%. The electrons transferred to the second amplifying element give rise to a pulse whose resolution is a function of many parameters: initial amplification factor, transfer efficiency, final amplification, electronics noise, and linearity. It has been observed however, that the energy resolution (fwhm) for 6 keV photons can always be kept at values close to $\Delta E/E = 25\%$, which we will take as an average value for our discussion. The final pulse height is proportional to

$$E = Nfm,$$

where m is the gain in the end-step localizing MWPC. Since we may consider that f and m are constant and do not depend on the conversion depth z in the first element, then the resolution on z will be

$$\Delta z = \frac{\Delta E}{E}\frac{L}{\ln G_{max}} = \frac{\Delta E}{E}\frac{l}{\alpha};$$

α is dependent on the maximum gain G_{max} attained across the full first amplifying gap. If we take for $\Delta E/E$ the upper value of 25% and a gap of 2 cm, then for a gain of 10^3, which is easy to achieve in stable conditions, the value of α is 3.5 ion pairs/cm and $|dz| = 0.8$ mm (fwhm), which is very attractive indeed. Fig. 2 shows the z resolution versus G_{max} in the first element.

3. Experimental results

We have verified this concept, for a chamber of 100×100 mm^2, with the following experimental set-up (fig. 1):
— an amplifying gap (L) 20 mm deep;
— a transfer gap 10 mm deep;
— a MWPC made of 20 μm wires, 2.54 mm apart, with a 10 mm gap.
A beam of 8 keV photons collimated to 0.2 mm was sent through a side window, at various depths of the absorption gap. The gas flowing in the counter was a mixture of 98.4% argon and 1.6% propane.

The tests were done at three different gains:

$$e^{\alpha L} = 2.5 \times 10^2, \qquad 4.2 \times 10^3, \qquad 10^5.$$

The gain of the MWPC was adjusted so as to reach the same total maximum gain for the whole counter, i.e. about 5×10^5 in all three cases. The pulse height was measured from the two cathodes of the MWPC gated by

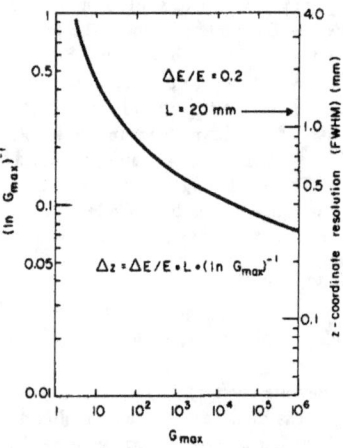

Fig. 2. The z-coordinate resolution function versus the maximum gain of the first amplification element. The right-hand scale gives the resolution for the case $\Delta E/E = 20\%$ (fwhm) and amplification gap $L = 20$ mm.

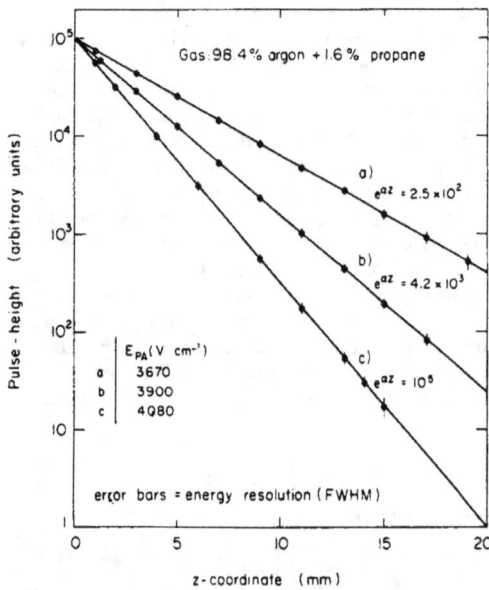

Fig. 3. Pulse height recorded at the MWPC for various gains of the first amplification element, as functions of the photon conversion depth z. The error bars correspond to the energy resolution (fwhm). For all three gains, the mean energy resolution is of the order of 22% fwhm.

the signals of six anode wires. A charge-sensitive pre-amplifier, ORTEC 142, was used, following by a linear amplifier. We were limited by the dynamic range of the preamplifier, so we could not cover the full gap of the first amplifying element. Fig. 3 shows the variations of pulse height with absorption depth and the experimental energy resolution (fwhm) observed at each measuring point. The energy resolutions for the corresponding gains of 2.5×10^2, 4.2×10^3, and 10^5 were 16%–40%, 16.5%–30%, and 17%–50%. In all three cases the mean value over most of the measured range was about 22%, and the large numbers at the end of each range were due to electronics noise.

Fig. 4 shows a 2 mm separation between two collimated beams for the different gains, at different depths across the amplifying gap. The escape peak is visible, and its harmful effect will be discussed later. The position resolution Δz has been calculated from measured $\Delta E/E$ data values and is plotted in fig. 5.

For a gain of 4.2×10^3, which is quite convenient, we see that we have a fwhm of 0.4 mm for most of the chamber depth, as expected.

It is clear that the limit in depth accuracy is such that it can match the lateral accuracy reached by the best read-out methods. We are considering the use of logarithmic amplifiers in order to solve the dynamic range problem.

4. Some advantages and limitations of the method

The big advantage of the method is that flat structures can fully replace spherical structures, without the limitations attached to a fixed radius. The disadvantage is that it applies only to monochromatic radiations. Another advantage is that there is no need for a 100% absorption in the conversion gap. Those photons which are absorbed after the conversion gap give rise to smaller

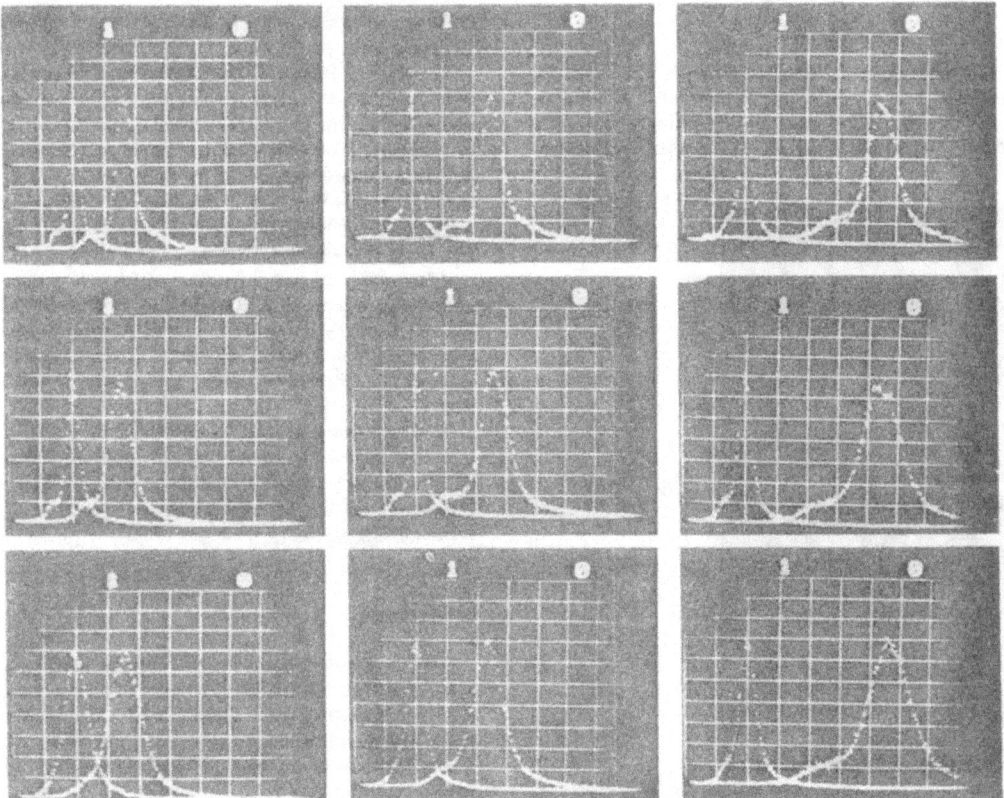

Fig. 4. Energy spectra corresponding to a 2 mm separation between two collimated beams (0.2 mm), at various conversion depths z: *lefthand column* ($G_{max} = 2.5 \times 10^2$) *top*: $z = 3$ mm, 1 mm, *middle*: $z = 10$ mm, 8 mm, *bottom*: $z = 17$ mm, 15 mm; *middle column* ($G_{max} = 4.2 \times 10^3$) *top*: $z = 3$ mm, 1 mm, *middle*: $z = 10$ mm, 8 mm, *bottom*: $z = 17$ mm, 15 mm; *righthand column* ($G_{max} = 10^5$) *top*: $z = 3$ mm, 1 mm, *middle*: $z = 8$ mm, 6 mm, *bottom*: $z = 13$ mm, 11 mm. The energy resolution has deteriorated in the vicinity of the anode owing to electronics noise.

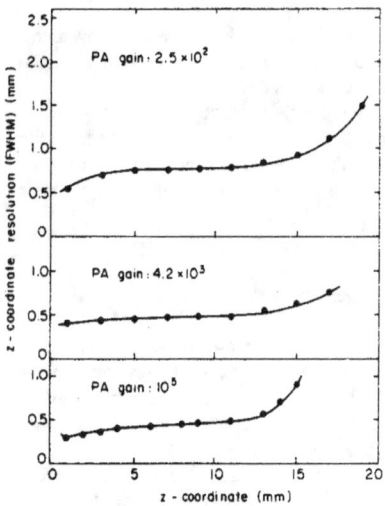

Fig. 5. Experimental depth resolution. The fwhm is computed from the energy resolution values of fig. 3.

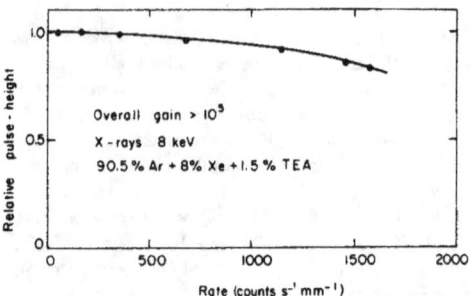

Fig. 7. Pulse-height variation versus the counting rate per millimetre of the MWPC wire. The photon beam was collimated to 0.2 mm. The photon beam penetrated the amplification element in parallel with the electrodes, near to the cathode; this corresponds to the worst case of the higher gain ($>10^5$). The data then correspond to an *upper limit*.

pulses and are easily eliminated by the electronics. In contrast, in a spherical drift chamber it is essential to have full absorption since the traversing photons are detected with full efficiency in the localization chamber.

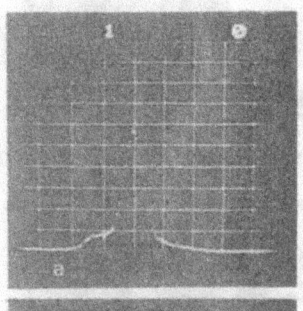

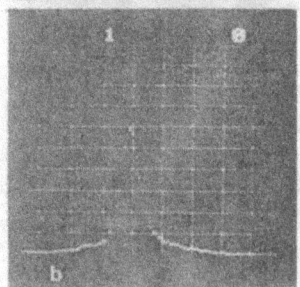

Fig. 6. Energy spectra of 8 keV photons. (a) 98.4% Ar + 1.6% propane; the escape peak is well pronounced. (b) 90.5% Ar + 8% Xe + 1.5% TEA; the escape peak has almost disappeared owing to absorption in Xe.

This leads to the necessity of having xenon as a conversion gas in spherical drift chambers, whilst a cheaper gas such as argon is convenient for the multistep chambers.

However, it should be emphasized that the escape peaks, which are frequently observed satellites of the main photoelectric peaks, may introduce a serious source of background since they make it impossible to correlate, in a non-equivoqual way, the pulse height and the depth. This may also lead to the adoption of xenon in some cases, but without the necessity of reaching full efficiency, thus permitting the use of thin gaps.

Fig. 6 shows an energy spectrum of 5.9 keV photons in the argon–propane mixture and in 90.5% Ar + 8% Xe + 1.5% triethylamine (TEA). It can be seen that the escape peak has almost disappeared; the tails on both sides of the distribution are due to imperfect collimation.

We have tested the rate capability of such a structure, and observed (as shown in fig. 7) that up to rates of $10^3 \, \mathrm{s}^{-1} \, \mathrm{mm}^{-1}$ along a wire of the MWPC, the energy response is not degraded for gains of 10^3–10^4 in the preamplifying gap and a total gain of 10^5.

5. Conclusion

The use of proportional chambers for the imaging of soft X-ray radiation has many attractive features; however, it is limited in many applications by the parallax error introduced when detectors of relatively large thicknesses have to be used to keep the quantum efficiency at a sufficient level.

We have shown that by using a photon absorbing gap with moderate multiplication and transfer to a localizing multiwire chamber, it is possible to achieve accuracies better than 1 mm (fwhm) for a 2 cm deep gap.

The counters are very flexible, and using the same detector structure we can decide, according to the application, whether to use the full amplifying gap with low gains and moderate resolutions (0.7–1 mm fwhm), or a limited depth with higher gain but better resolution (0.3–0.4 mm fwhm) – but also lower efficiency.

Measurements of the spatial accuracy in multistep chambers have shown that the centroid of avalanches can be localized to positions as accurate as 100 μm after two steps of amplification. Parallel-grid amplification has already been tested for surfaces of 0.5 m^2.

Since the necessity to measure the pulse height over a large dynamic range introduces some limitations in the maximum rate capability of such detectors, we may say that for moderate rates our approach should permit the construction of large-surface X-ray detectors with a high spatial accuracy and free of parallax error for X-ray beams emitted by sources within a large range of angles.

We are indebted to Dr. F. Sauli for stimulating remarks.

References

[1] G. Charpak, Z. Hajduk, A.P. Jeavons, R. Kahn and R.J. Stubbs, Nucl. Instr. and Meth. 122 (1974) 1307.

[2] R. Kahn, R. Fourme, B. Caudron, R. Bosshard, R. Benoit, R. Bouclier, G. Charpak, J.-C. Santiard and F. Sauli, Nucl. Instr. and Meth. 172 (1980) 337.

[3] C. Bolon, M. Deutsch and R. Lanza, SSRL Report 78/04 (1978) VIII 84, SSRL Stanford, California.

[4] D. Bade, F. Parak, R.L. Mössbauer, W. Hoffe, N. Levai and G. Charpak, in: ESF–CNRS–EMBL Workshop on X-ray position-sensitive detectors and energy discriminating detectors, Hamburg, 1980, to be published in Nucl. Instr. and Meth.

[5] P.F. Christie, E. Matthieson and D.K. Evans, J. Phys. E: 9 (1978) 673.

[6] A. Breskin, G. Charpak, S. Majewski, G. Melchart, G. Petersen and F. Sauli, Nucl. Instr. and Meth. 161 (1979) 19.

[7] G. Charpak, Preprint CERN-EP/81-07 (1981), in: ESF–CNRS–EMBL Workshop on X-ray position-sensitive detectors and energy discriminating detectors, Hamburg, 1980, to be published in Nucl. Instr. and Meth.

[8] A. Breskin, G. Charpak, S. Majewski, G. Melchart, A. Peisert, F. Sauli, F. Mathy and G. Petersen, Nucl. Instr. and Meth. 178 (1980) 11.

Nuclear Instruments and Methods 176 (1980) 239–244
© North-Holland Publishing Company

THE MULTISTEP AVALANCHE CHAMBER AS A DETECTOR IN RADIOCHROMATOGRAPHY IMAGING

G. PETERSEN

The Niels Bohr Institute, University of Copenhagen, Copenhagen, Denmark

G. CHARPAK, G. MELCHART and F. SAULI

CERN, Geneva, Switzerland

We have investigated the possibility of using the multistep avalanche chamber as an imaging detector in radiochromatography. High position resolution, better than 0.5 mm fwhm, combined with a high sensitivity give very satisfactory results. Measurement on two-dimensional distributions of β-emitting radionuclides such as ^{3}H, ^{14}C and ^{35}S are presented and compared with results obtained from autoradiography.

1. Introduction

Thin layer chromatography using β-ray emitting radionuclides as tracers is a powerful tool in biophysics, biochemistry, pharmacology and clinical medicine, for the analytical separation of chemical compounds. Two-dimensional distributions of the compounds can be created by employing consecutively two different principles of chromatography or electrophoresis combined with chromatography on the same plate [1]. Commonly used tracers are ^{3}H, ^{14}C, ^{32}P, ^{35}S and ^{131}I. The different characteristics of the β-emission spectra of these isotopes may necessitate the use of different techniques in their detection. The classical method consists of the exposure of the chromatogram plate to an X-ray film (autoradiography). Compounds labelled with ^{32}P or ^{131}I often yield satisfactory pictures within a couple of hours, but for low intensity samples, especially those labelled with ^{3}H, the exposure time may extend to several weeks. In the case of ^{3}H the sensitivity of the method is often enhanced by adding a small amount of fluorescing component to the solvent used. Apart from other detection methods liquid scintillation counting and scanning of the chromatogram with one or several Geiger–Müller tubes in a row should be mentioned [2]. Both of these methods can be used in a quantitative determination of the amount of a certain compound on the chromatogram plate, but they are rather time consuming; a scan over a 20×20 cm^2 chromatogram takes typically 24 h. Spark chambers [3], and hybrid spark chambers [4], have been constructed as detectors for radiochromatography. As they use optical readout with registration on film, the above-mentioned quantitative measurements are not easy to perform with these detectors. We have approached the problem in a different way by using the multistep avalance mechanism in a gaseous detector, recently investigated by our group at CERN, as the basis for detection of the β-rays emitted from the surface of the chromatogram plate. We have used a purely electronic readout of the detector with subsequent data handling done in a small computer.

2. The multistep avalanche chamber

2.1. The photon mediated avalanche

It has been shown [5,6], that with a proper choice of gas mixture it is possible to develop an avalanche between parallel grids with the following characteristics: the lateral spread of the avalanche is such that a sizeable fraction of the resulting electron cloud can be extracted into a subsequent transfer space. The transferred fraction is nearly independent of the position of the electron initiating the avalanche. This permits proportionality between the initial ionization and the final transferred charge and avoids modulation of the efficiency as a function of position.

It has been inferred that the emission and absorption of VUV photons play a role in the spread of the avalanche; however other processes, such as diffusion and the Penning effect, may also participate.

239

2.2. The localization effect

The charge multiplication that an electron in the preamplification stage gives rise to varies exponentially with its distance from the second preamplification grid. An electron cloud, obtained by preamplification and transfer of electrons in the ionization trail through the preamplification region, will therefore mainly localize the point where the ionizing particle crossed the first preamplification grid. The localization effect can easily be treated theoretically. Assuming infinitely long ionization tracks emitted uniformly in space from a point on the first preamplification grid we find the following expression for the centre of gravity distribution of preamplified and transferred electron clouds

$$I(z) \simeq 1 \bigg/ \left[1 + \left(\frac{\ln M}{l} z \right)^2 \right] ,$$

where l is the distance between preamplification grids, M is the multiplication that an electron obtains over the distance l, and z is the coordinate in a direction parallel to the grids. The formula is accurate within a few per cent for values of $M > 50$. In the case of no multiplication the expression is modified to

$$I(z) = 1 \bigg/ \left[1 + \left(\frac{2}{l} z \right)^2 \right] .$$

Fig. 1 shows $I(z)$ for $l = 3$ mm and different values of M. In the case of $M = 2000$ we expect a position accuracy of about 1 mm fwhm.

In practice the ionization tracks that a β-emitting isotope gives rise to are neither straight lines nor infinitely long. A large number of the tracks have a very limited range, less than a few hundred μm, in the detector gas thus enhancing the final position accuracy of the detector. This effect is specially pronounced in the case of ^{3}H. For all the isotopes that we have used, a position resolution better than 0.5 mm fwhm was obtained.

3. Experimental set-up

The arrangement of wire planes in the detector is shown in fig. 2. Two electrodes have been mounted on top of a standard multiwire proportional chamber (MWPC) to provide the preamplification and transfer regions. The upper preamplification electrode is a mesh of stainless steel wires, 30 μm in diameter and 400–600 μm apart. The lower electrode is a mesh of Ni wires, 10 μm in diameter and 100 μm apart. The distance between the two electrodes is 3 mm, and the distance between the lower mesh and the MWPC is 5 mm. The sample is mounted inside the detector, as close as possible to the upper preamplification electrode. The MWPC and associated electronics have been described elsewhere [7]. Briefly, the anode plane is made of gold-plated tungsten wires, 10 μm in diameter and 1.27 mm apart. The two orthogonal cathode planes are made of Cu–Be wires, 100 μm in diameter and 1.27 mm apart. The anode–cathode distance is 5 mm, and the active area of the detector is 10×10 cm². The bidimensional readout of the MWPC is done with the centre-of-gravity method where, for each avalanche

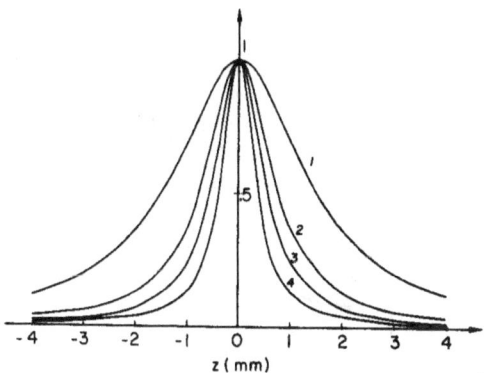

Fig. 1. Theoretical distributions of the centre of gravity of preamplified and transferred electron clouds for different values of the amplification factor M. The curves 1 to 4 correspond to the M values 1, 50, 200, and 2000.

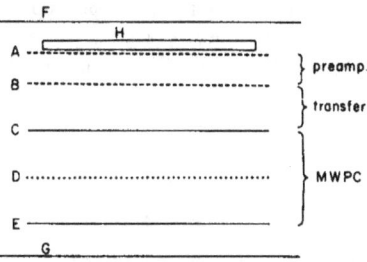

Fig. 2. Schematical view of the detector. F and G are mylar windows; A and B are the preamplification electrodes; C, D, and E constitute a standard MWPC. H is the sample, mounted inside the detector on top of the upper preamplification electrode.

occurring on the anode, we record the positive induced charge distributions on the two cathode planes. The cathode wires are connected together 4 by 4 in strips, and each strip is connected to a charge preamplifier having a sensitivity of about 250 mV/pC, a rise-time of 10 ns, and a decay time of 50 ns *; the amplified signals are then transmitted via coaxial cables to a CAMAC-based 10-bit current integrating ADC **.

All anode wires are connected together; the signal from this plane is recorded in an ADC channel and also provides the gate for the ADC modules after a suitable discrimination and shaping. The gate width is 100 ns. A small on-line computer is used to organize the data acquisition and tranfer to magnetic tape.

As a gas filling we have used a mixture of argon and acetone in the volume concentrations 98−2.

4. Calibration and data handling

A detailed description of the procedure used to calibrate the chains of the preamplifier, coaxial cable, and ADC channel, is given in ref. [7]. The calibration allows us (using software) to correct for nonlinearities in the chains as well as for dispersion between them.

Let us denote by q_i the effective charge signal on strip S_i, as obtained from the raw data using the above-mentioned calibration procedure. We define the centre-of-gravity of a cluster of adjacent induced charges as the quantity:

$$\bar{S} = \sum (q_i - b) S_i \Big/ \sum (q_i - b),$$

where b is a bias level, and the sum extends only to positive values of $q_i - b$. A proper choice of bias level allows reduction of the influence of pick-up and electronics noise in the centre-of-gravity determination. In our case a value of $b = 2\%$ of Σq_i gave satisfying results.

5. Experimental results

5.1. Introduction

We have made measurements on different samples marked with ^{3}H, ^{14}C, and ^{35}S. In the case of ^{3}H and

---

* Developed at CERN by J.C. Santiard and produced with a hybrid thick film technique by CIT-Alcatel, France.
** Lecroy, 12-channel ADC type 2249 A.

---

^{14}C, silica gels on glass plates were used as chromatogram plates; for ^{35}S a polyacryl-amide gel was used. It is important that the chromatogram plates be flat, so as to keep the distance to the upper preamplification electrode as small as possible over the whole surface of the sample. Silica gels on glass plates seem to be perfectly suited for our detector in that respect.

5.2. ^{14}C results

The ^{14}C has an emission spectrum extending to about 150 keV, implying that most of the β particles have a range of many millimetres in the detector gas. The halflife of the isotope is 5600 y. The total activity of our sample was about 90000 disintegrations per minute (DPM). We have measured the counting rate in the detector for different values of anode voltage and preamplification voltage. The result is shown in fig. 3. It is seen that for high values of the preamplification voltage, we obtain the beginning of a plateau. A full plateau cannot be reached in our case, because the preamplification mechanism is not efficient enough to bring the lowest part of the β-emission spectrum over the electronic threshold. The distribution of regions with high activity of the sample is shown in fig. 4. The upper part of the figure is the autoradiogram that was obtained after 48 h of exposure time; the lower part shows the response of our detector after about 20 min of data taking. The scales are 25 mm in both directions. The lower left corners of the pictures cannot be compared directly, because the most intense part of the sample was removed before placing it in the

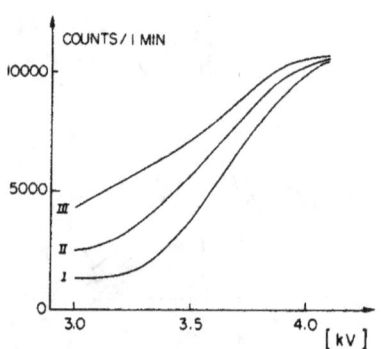

Fig. 3. Counting rate obtained with the ^{14}C sample for different voltages on the chamber. The curves I, II, and III correspond to the anode voltages of 2.25, 2.35 and 2.45 kV, respectively.

V. AVALANCHE CHAMBERS

detector. Fig. 5 is a comparison between a photometer scan of the autoradiogram (top) and the lower part of fig. 4 projected on to the vertical axis. The resolutions of the two methods are comparable.

5.3. ^3H results

The emission spectrum of ^3H extends to about 18 keV, implying ranges of a few hundred μm for a considerable part of the β-rays. The half-life of the isotope is 12.3 y. The sample activity was about 600 000 DPM. After an exposure time of 48 h, the autoradiogram shows only the most intense line,

see fig. 6. The response of our detector after 5 s and 10 min of data taking, respectively, is shown in fig. 7. The scales are 70 mm in both directions. In this case our method is indeed several orders of magnitude more sensitive than the autoradiography.

5.4. ^{35}S results

The emission spectrum of this isotope is similar to that of ^{14}C, but the halflife is much shorter, i.e. 87 d. The upper part of fig. 8 is the autoradiogram after 4 d of exposure time; the lower part is the response of our detector after a few minutes data taking. The

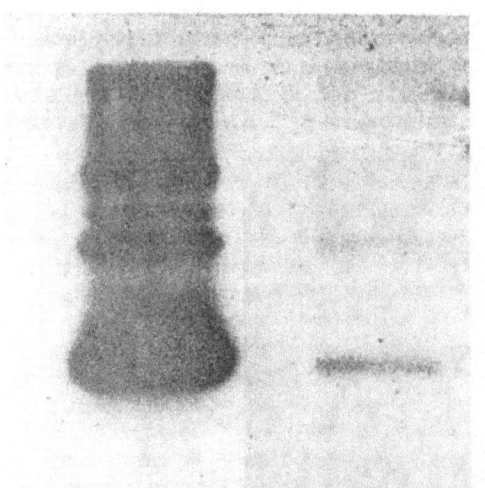

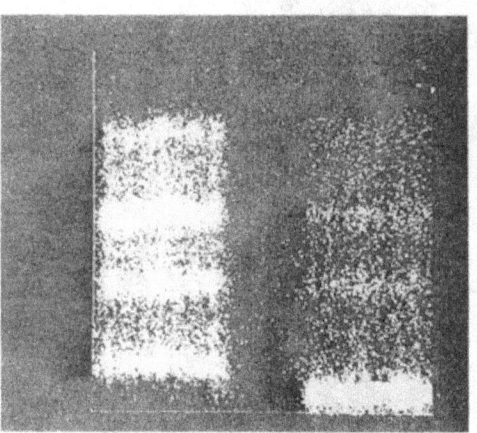

Fig. 4. Comparison between an autoradiogram of the ^{14}C sample and the response of our detector. The real site of the images is 25 × 25 mm^2.

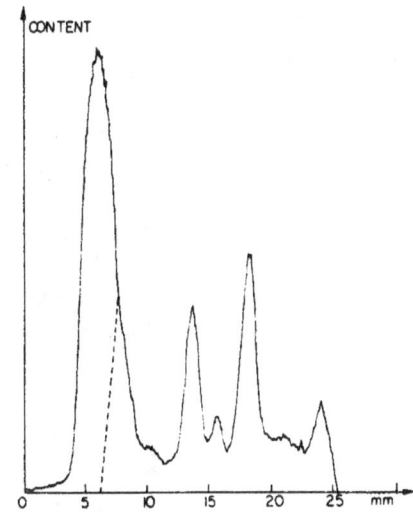

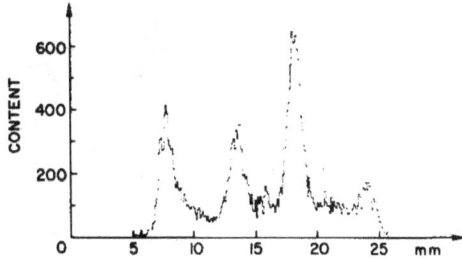

Fig. 5. For ^{14}C: comparison between a photometer scan of the autoradiogram and the response of our detector. The resolutions are comparable, better than 0.5 mm fwhm. The most intense line of the sample was removed before mounting it in the detector. This cuts the high peak as shown on the top curve.

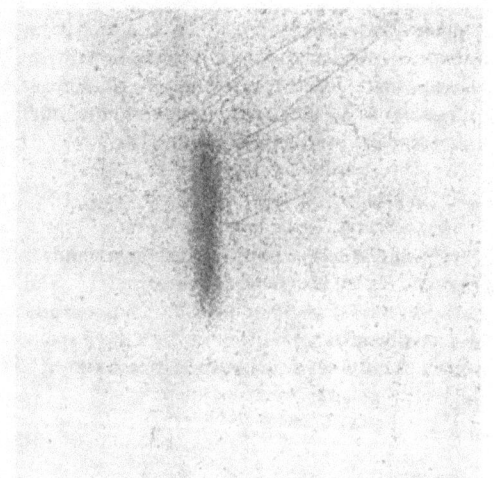

Fig. 6. The autoradiogram of the ³H sample after 48 h of exposure time. Only the most intense line is visible.

scales are 70 mm in both directions. In this case we can only poorly resolve the detailed structure seen in the autoradiogram; after a few minutes more data taking a large part of the picture will be over-exposed. We do not consider this as a drawback of the detector itself. The problem of creating good pictures can easily be solved using a more refined software, and grey-tone or colour display screen.

6. Conclusions

The multistep avalanche chamber seems to be a very useful tool in radiochromatography imaging. Compared with the traditional method of auto-radiography it is extremely sensitive, especially in the case of ³H measurements. In a fast way it provides information about the two-dimensional distributions as well as intensities of the radionuclides. The method

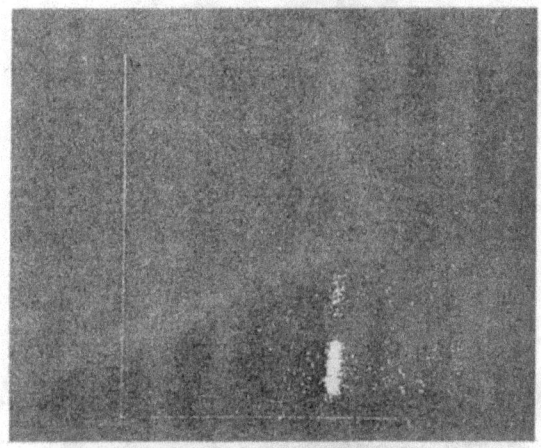

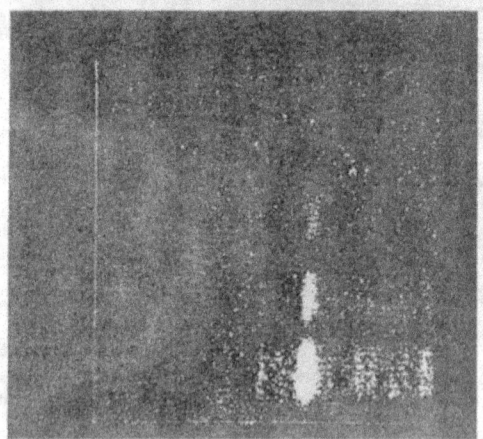

Fig. 7. The response of our detector for the ³H sample for 5 s and 10 min of data taking. The size is 70 × 70 mm².

V. AVALANCHE CHAMBERS

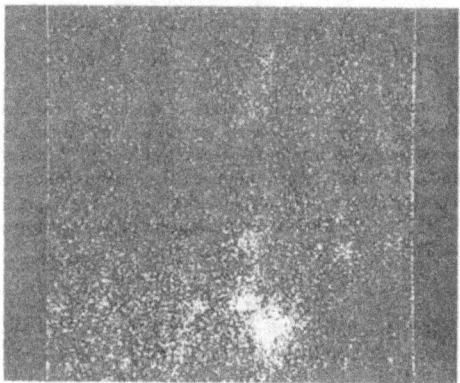

Fig. 8. Distribution of ^{35}S labelled compounds. The upper picture shows the autoradiogram, the lower one shows the response of our detector. The size is 70×70 mm^2.

requires completely flat samples, but this is not considered as a serious problem; it is probably possible in many cases to adhere gels to support plates so as to obtain the desired geometrical tolerances.

We wish to thank Dr. T. Amfred at the Hospital of Aalborg, Denmark, Dr. G. Turnock at the University of Leicester, England, and Dr. P. Prentki at the University of Geneva, Switzerland, for production of test samples and for fruitful discussions.

G. Petersen wishes to thank the Danish Natural Science Council and the Danish Technical Science Council for the support that made it possible for him to join this project.

References

[1] H.Ch. Curtius and M. Roth, eds., Clinical biochemistry, Principles and methods, vol. 1 (W. de Gruyter, Berlin and New York, 1974).
[2] L. Bötter-Jensen, H.J.M. Hansen and P. Theodorsson, Nucl. Instr. and Meth. 144 (1977) 529.
[3] H. Filthuth, Berthold Laboratory Report, Wildbad, Germany (1977).
[4] T. Aoyama and T. Watanabe, Nuc. Instr. and Meth. 150 (1978) 203.
[5] G. Charpak and F. Sauli, Phys. Lett. 78B (1978) 523.
[6] A. Breskin, G. Charpak, S. Majewski, G. Petersen and F. Sauli, Nucl. Instr. and Meth. 161 (1979) 19.
[7] G. Charpak, G. Melchart, G. Petersen and F. Sauli, Nucl. Instr. and Meth. 167 (1979) 455.

Nuclear Instruments and Methods in Physics Research A278 (1989) 779-787
North-Holland, Amsterdam

A GASEOUS DETECTOR FOR HIGH-ACCURACY AUTORADIOGRAPHY OF RADIOACTIVE COMPOUNDS WITH OPTICAL READOUT OF AVALANCHE POSITIONS

W. DOMINIK *, N. ZAGANIDIS, P. ASTIER, G. CHARPAK, J.C. SANTIARD and F. SAULI

CERN, Geneva, Switzerland

E. TRIBOLLET

Centre Médical Universitaire, Geneva, Switzerland

A. GEISSBÜHLER and D. TOWNSEND

Division Médecine Nucléaire, Hôpital Cantonal, Univ. Geneva, Switzerland

Received 9 February 1989

We describe a detector based on the imaging of avalanches in gases initiated by ionization electrons produced by β-particles entering an amplifying gap limited by two parallel electrodes. The avalanches are made visible by the addition of an efficient light-emitting vapour to noble gases such as argon or xenon. Imaging of the avalanches with an image intensifier and a CCD coupled to a computer permits accuracies of the order of 50 μm for the β-rays emitted by tritium. Application of the method to biological tissue slices labelled with tritium shows a gain in time, with respect to the photographic method, by a factor close to 100. Images of gels labelled with ^{32}P or ^{35}S show accuracies better than 0.5 mm (FWHM). The linearity and large dynamic range in intensity measurements are far superior to those of the standard autoradiography method.

1. Introduction

The imaging of β-emitting radionuclides is a routinely used technique in biophysics, biochemistry, pharmacology and clinical medicine. Depending on the methods of separation of chemical compounds one has to deal with one-dimensional or two-dimensional imaging. Commonly used tracers are ^{3}H, ^{14}C, ^{35}S, ^{32}P, ^{125}I.

The differences in the energy spectra of the electrons emitted by these various nuclei result in differences in the detection conditions. The most commonly used method consists of exposing the chromatography plates or the biological slices to photographic films (autoradiography).

Compounds labelled with ^{32}P or ^{125}I may yield satisfactory pictures after a few hours, whilst with ^{3}H it may take several months to obtain an adequate image. In this last case the range of the low-energy electrons of maximum energy 18 keV limits the useful emitting layer to a thickness of the order of a micron.

After an image has been obtained on a film, time-consuming and expensive methods have to be used to obtain a quantitative estimate of the intensity distribution.

Many attempts have been made in the last two decades to design electronic systems that would detect and localize the β-rays. Wire chambers have led to several commercial instruments with an advertised accuracy ranging from 1 to 3 mm (FWHM). Restricting ourselves only to gaseous detectors, one of the main difficulties encountered is the large range of the β-rays in the gases, at atmospheric pressure. One way to overcome this has been the use of the multistep avalanche chamber [1,2]. In this device a wire chamber localizes a cloud of electrons from an avalanche produced in the uniform electric field created between two parallel electrodes. The sample with the radioactive compound is placed against the cathode, either outside the gap or so that its surface is the cathode plane. The anode of the gap is a grid, transparent to electrons, made for instance of a mesh of 50 μm wires spaced by 500 μm. This introduces a significant improvement in the positional accuracy for the following reason: when a long-range β-particle penetrates such a gap, it produces ionization electrons all along the track, but the multiplication of the electrons is not constant. It is proportional to $e^{\alpha x}$, where X is the distance of the electron to the anode and α is the number of ionizing collisions per centimetre, produced in the gas by an electron experiencing ionizing collisions under the influence of an electric field.

---

* On leave of absence from Warsaw University, Poland.

Thus the ionizing electrons produced upon the entrance of the β-particle give rise to the largest amplification. This reduces very much the error due to the finite range of the β-particle, if the position of the maximum of the charge distribution produced by the avalanches is measured.

Accuracies of the order of 0.5 mm (FWHM) had been obtained with β-rays from ^{14}C and ^{35}S [2]. Using a similar design, other groups have obtained two-dimensional images with an accuracy of the same order, with ^{3}H and ^{35}S-labelled compounds [3,4]. In all these attempts, the localization of the avalanches in the last step chamber is obtained by various electronic methods giving the electrical centroid of the avalanches. Another group, using only one step of amplification between parallel electrodes has also illustrated the advantages of parallel electrodes for the amplification [5].

In this article we present a new method [6]. It exploits the advantage introduced by the electron amplification due to avalanches in a uniform electric field in the gas, between parallel electrodes; but the readout method, based on the optical imaging of the atoms excited in an avalanche, leads to a substantial improvement in the accuracy, bringing it to better than 100 μm (FWHM). Not only is this method a very accurate one for measuring the centroid of an avalanche, but the detailed picture it gives of the distribution of the charges in the avalanche may lead to further improvement in the accuracy of the position measurement of the first electron produced in the gap by the entering β-ray.

2. Optical readout of electron avalanches

When avalanches are reproduced by multiplication of electrons in gases under the effect of strong electric fields, a great variety of processes lead to the emission of photons with a spectrum ranging from vacuum ultraviolet (VUV) to visible. The intensity may vary over a considerable range from gas to gas and as a function of the electric field. Several years of research permitted us to find the optimum conditions for the emission of light in a wavelength range convenient for optical imaging with the maximum amount of photons per charge produced in an electron avalanche at the largest charge gains. The initial goal was to visualize the position of charged-particle tracks in gases [7] or to image the VUV photons produced by Cherenkov light. It was concluded that the optimum structures for the maximum amount of light utilize, for the amplification, parallel-grid electrodes and not wires. It was found that addition, to the noble gases, of vapours of triethylamine (TEA) [8] or tetrakis(dimethylamino)ethylene (TMAE) [7,9], results in copious emission of fluorescence photons, spectra peaked at 280 nm and 480 nm, respectively; the second

is obviously more convenient for optical imaging, but TMAE is not easy to handle, because of its low vapour pressure and strong oxidation. Gooch et al. have reached independently the same conclusion [10] on the advantages of TEA for optical imaging of avalanches initiated by Cherenkov VUV photons, and Sauvage et al. have made a detailed study [11] of the light emission of TMAE and TEA in avalanche chambers. It was also found that it is easy to convert the photons of the first spectrum into visible light on appropriate thin plastic wavelength shifters * or by a thin layer of sodium salicilate deposited on the Mylar foil and placed in contact with the anode grid [7]. The average intensity of photons produced in avalanches is close to one per electron. We found it easy to image single electrons produced by the absorption of VUV Cherenkov photons in a multistep structure made of two successive parallel-grid gaps [12,13]. We applied this technique to the imaging of the β-ray-emitting compounds labelled with ^{3}H, ^{32}P, and ^{35}S.

3. The experimental setup

The schematics of the experimental setup is shown in fig. 1, which displays two versions of the detector that we have used for the imaging of β-emitters. The principle of detection is the same in all cases: light emitted by atoms excited during the electron multiplication in the gas is detected by means of an image intensifier coupled to a CCD camera.

It has been observed that the yield of photons that can be obtained, for a given charge gain, from electron avalanches between parallel electrodes [14] is larger than that in the multiwire chamber where multiplication occurs near the thin wires. This fact led us to using crossed-wire meshes for the construction of the charge-amplifying structures. They have the advantage of being transparent to light, and to electrons if multiplication in successive structures is required. A mesh of wires of 50 μm diameter and 500 μm pitch was employed.

The installation of the radioactive sample inside the active volume of the detector is necessary in the case of tritium sources, where the very low energy β-particle (E_{max} = 18.6 keV) would be absorbed by even the thinnest plastic window. We studied two versions of the detector for the ^{3}H-labelled sample tested: sample behind a cathode mesh, in contact with it (fig. 1a), and sample on a glass plate covered by a 200 Å thick layer of gold constituting the cathode of the amplifying gap (fig. 1b). Such a thin layer of gold is transparent for the major fraction of the electrons emitted by tritium and

---

* 100 μm of doped polystyrene, made by M. Bourdinaud, DPhPE, CEN-Saclay, France.

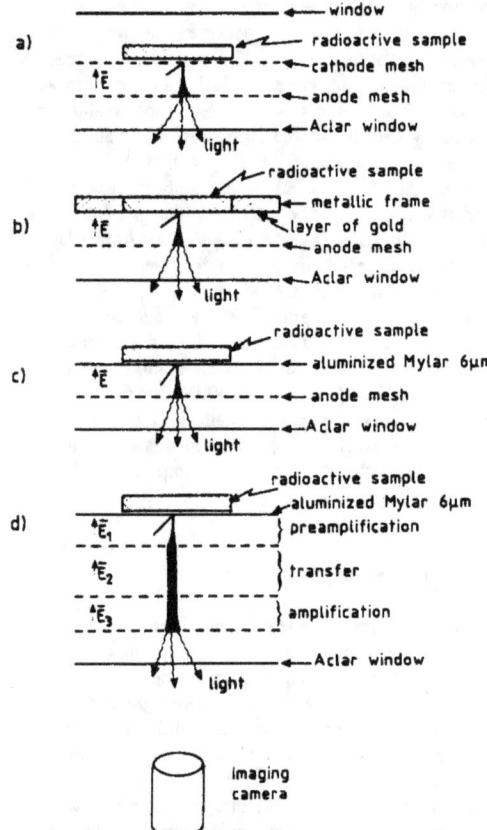

Fig. 1. Charge-amplifying structures and geometrical coupling of radioactive sources. (a) Parallel-grids amplifying gap with the radioactive sample placed in contact with a cathode mesh. (b) The ³H-labelled surface is covered by about 200 Å of gold and is part of the cathode. (c) For emitters of more penetrating β-particles, such as ³²P or ³⁵S, the radioactive samples are separated from the gap by a thin conductive cathode. (d) Multistep parallel-plate structure. The avalanches produced against the anode mesh emit light, which is imaged through Aclar windows that are transparent to UV photons. The light is transmitted through lenses to an image intensifier and a CCD camera, read out by an IBM-PC/AT digitizer.

yet mechanically strong enough to withstand occasional sparking.

For most uses it is, however, more convenient to have the sample separated from the gas volume of the detector; we tested a version with aluminized Mylar of about 6 μm thickness serving as the window and the cathode of the chamber (fig. 1c). Such a configuration can be used only for isotopes emitting β-particles of higher energy such as ^{14}C ($E_{max} = 156$ keV), ^{35}S ($E_{max} = 168$ keV), and ^{32}P ($E_{max} = 1700$ keV). Optimal posi-

tion resolution implies a minimal distance between the source and the amplifying gap.

The use of multistep structures (fig. 1d) permits the gating of the image intensifier and the triggering of the data-acquisition system by the information from the first preamplification stage, since it is detected before the preamplified charge reaches the main amplification gap. The delay of a few hundred nanoseconds due to the drift time in the transfer gap is sufficient for this purpose.

The optical information is detected by the image intensifier coupled to a CCD camera through the Aclar * 22A window, 50 μm thick, which is transparent in the TEA emission region.

Two gas mixtures were used: argon and xenon, with a few percent TEA admixture.

A UV-sensitive photomultiplier was installed close to the camera; light pulses in coincidence with charged signals from the chamber were used for triggering the image digitization and storing system.

4. The optical readout system

A CCD camera equipped with an image intensification system read out the information carried by the photons emitted by the charge avalanches in the chamber.

The CCD (Thomson 7852 CDA-80) camera is a matrix of 144×208 pixels of total area 4.32×5.82 mm², which allows a high granularity of the detector surface readout. The associated electronics produces the chain of electric pulses corresponding to the contents of the pixels for each image detected. This signal is fed to the digitizer card (Data Translation DT 2851) which interfaces with the IBM-PC/AT computer. The digitizer has a precision of 8 bits, which determine the dynamic range of the light intensity measurement in every pixel.

The impact point of the β-particle is reconstructed by calculating the centre of gravity of the detected luminous spot for ³H, where the range of the β is small, and by calculating the centroid of the avalanche within a determined range around the maximum of intensity for ³²P and ³⁵S.

The amount of light emitted by the avalanches is not sufficient to be directly detected by the CCD. Therefore the use of an image intensifier coupled to it is indispensable.

Two different chains of image intensification were used:
(1) Camera A: a proximity-focused image intensifier of 25 mm diameter, with a bialkali photocathode de-

* Polychlorotrifluoroethylene (PCTFE).

posited on the fused silica window and an X3 phosphor screen (100 ns decay time) on the fibre optics plate, served mainly as a wavelength converter because of its small photonic gain. The second-stage multichannel-plate image intensifier, of the same diameter, was optically coupled to the exit window of the first one. The image size was then reduced from 25 mm to 7 mm (CCD diagonal length) by the fibre-optics taper.

(2) Camera B: a multichannel-plate gateable image intensifier, of 18 mm diameter, with an S20 photocathode deposited on the sapphire window and the P36 screen as a first stage (RTC XX 1410/SP 41721.160). The DEP electrostatic magnifying tube reducing the image size from 18 mm to 7 mm was optically coupled to the exit window of the first stage and followed by the CCD.

Both modules were operated at the total gain of about 3000 W/W for the peak wavelength of the incident photons: 280 nm.

With our CCD and the associated electronics, a maximum of 50 images per second could be transferred to the digitizing card in the PC. The rate of data taking in our initial instrument was limited, by the speed of the analysing and storing system, to about 3 images per second. It has now increased to 20 images per second.

An efficient light collection is one of the most important parameters of the optical readout system. We used a reflective-type UV objective 90 mm f/1.1 (Lyman-Alpha II of NYE).

5. Imaging of ^{3}H-labelled compounds

In order to test the accuracy of the method we imaged a structure (produced by Amersham) commonly used in biology laboratories as a calibration standard to analyse autoradiographs quantitatively.

It is made of thin plastic strips divided in rectangular spots labelled with ^{3}H, with variable intensities. We placed the sample behind the cathode grid (fig. 1a). The position accuracy of our method can be demonstrated in a convenient way by the modulation in intensity introduced by the grid of wires (of 50 μm diameter, spaced every 500 μm) that serves as a collimator for the low-energy electrons. The width of the amplifying gap was 4 mm.

With the Ar + TEA (2%) filling gas we tested the performance of our detector using Camera A for light detection. Fig. 2 shows the intensity distributions of the pixels of maximum light intensity in the detected spots and the integrated intensity of these spots. Both distributions have a peaked shape, making it easy to set the criteria for the spot recognition and the analysis by software. The physical size of the single luminous spot was about 1.5 mm total width, and the pixel of maximum intensity contained on average about 10% of the integrated intensity of a spot. A high granularity of the CCD readout results in a good precision of the impact point.

The image of the Amersham calibration standard is displayed in fig. 3. The two parallel arrays of the

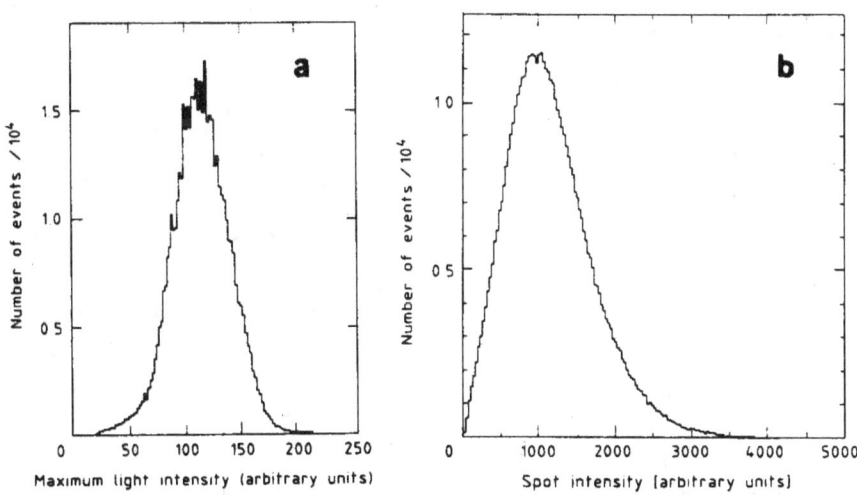

Fig. 2. Pulse-height distributions of the pixels at maximum light intensity in the detected spots (a) and the integrated intensity of spots (b). Tritium-labelled sample detected by parallel-plate chamber filled with Ar + 2% TEA.

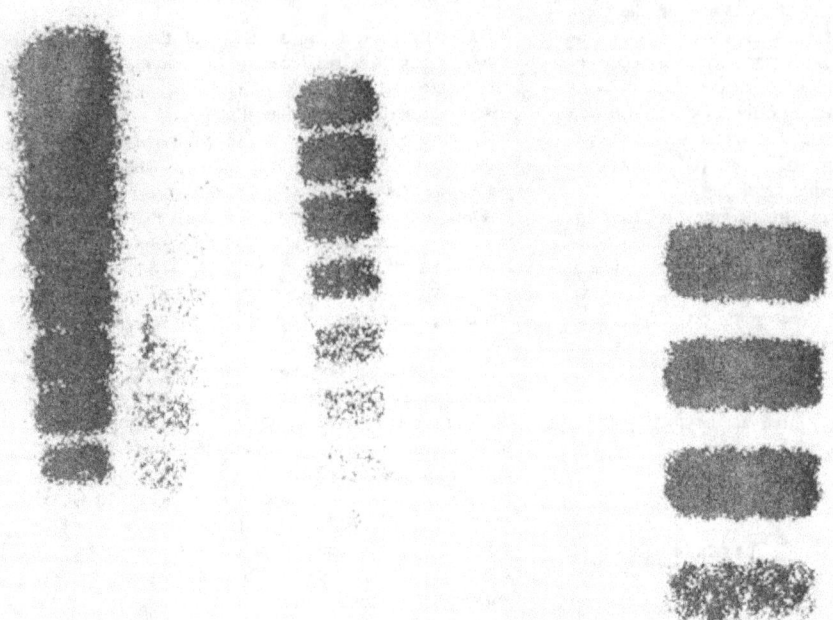

Fig. 3. (Left) Image of ^{3}H-labelled Amersham calibration standard. (Right) Enlarged view of the four most intense bars of the pattern; the bars are 1.4 mm wide and 1 mm apart. Single parallel-plate amplifying gap; in Ar + 2% TEA.

variable-intensity radioactive bars are visible. The ratio of intensity of the strongest bar (top left) to the weakest one (bottom right) is about 1000. The structure appearing between the two arrays comes from contamination by residual radioactivity of the substrate, caused by

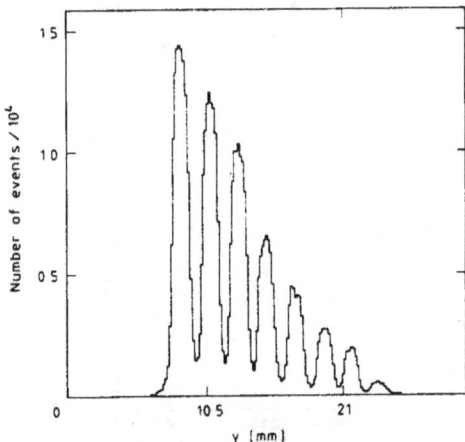

Fig. 4. Intensity distribution along the array of most intense bars of ^{3}H-labelled Amersham calibration standard; same conditions as in fig. 3.

temporary contact with the sample. Unfortunately the dynamic range of our display was smaller than that of the local source rate; this results in the apparent saturation of the most intense part of the image. This region, magnified by a factor of 2, is shown in fig. 3 (right). One can observe the structure of the crossed-wire grid. The accuracy is of the order of 100 μm (FWHM). Fig. 4 shows the distributions of intensity in the most intense array of rectangular spots. The separation is excellent, and the relative intensity of the spots coincides with the one given by the manufacturer.

We also tested slices of biological specimens from a kidney and the brain of a rat, using the chamber of the configuration shown in fig. 1b filled with a gas mixture of Xe (96.5%) + TEA (3.5%). The image of the kidney (fig. 5a) clearly displays the renal ducts, which have a physical width of 30 to 50 μm, as exhibited by the autoradiography, obtained in a 3-month exposure (fig. 5b), of the adjacent slice. The presented image was obtained in 40 h, with only 40% efficiency of the data-acquisition system, caused by the computer deadtime that we have now reduced by a factor of 10. It can nevertheless be analysed for quantitative information. Fig. 5a also shows the intensity distribution along the line crossing the image displayed on the screen of an Apollo DN580, which has a resolution of 1280 × 1024 pixels. The images are displayed as matrices of 256 × 256 pixels, each pixel being represented by 8 bits. Images for which the dynamic range of the data exceeds 8 bits can be displayed with different positions of the 8-bit window, i.e. the 8 bits of the display screen can be mapped onto any 8 bits of the input image. Multiple 8-bit images are therefore used to display the full dynamic range of the data. The single pixel of the CCD corresponded to 180×180 μm^2 area in the detector plane. Such an image can thus be obtained in about 20 h, and this has to be compared with the three months of exposure time for the autoradiography. A gas filling containing mainly xenon is most favourable, because of the small range of the low-energy β-particles emitted by tritium: this is close to 100 μm for the average electrons of 6 keV. Among the other factors determining the position resolution is the statistics of light emitted by a single avalanche and the spread of the avalanche in the image plane, allowing dense sampling of the light spot by the CCD. From this point of view, the situation is

slightly better with argon mixed with TEA, for which higher charge gain is achievable with a lateral charge diffusion larger than with Xe + TEA. All these parameters have to be optimized by choosing the adequate detector geometry, pressure of the gas, and optical readout granularity. With argon at atmospheric pressure, the accuracy is slightly worse than with xenon, because of the larger range of the β-particles.

6. Imaging of ^{32}P- and ^{35}S-labelled compound

The imaging of ^{32}P and ^{35}S tracers gives more freedom in the detector construction, because these offer the possibility of using thin windows between the samples and the detector. On the other hand, the position resolution may be degraded because of the long range of the β-rays. The measurements with these sources were performed in different versions of the chamber

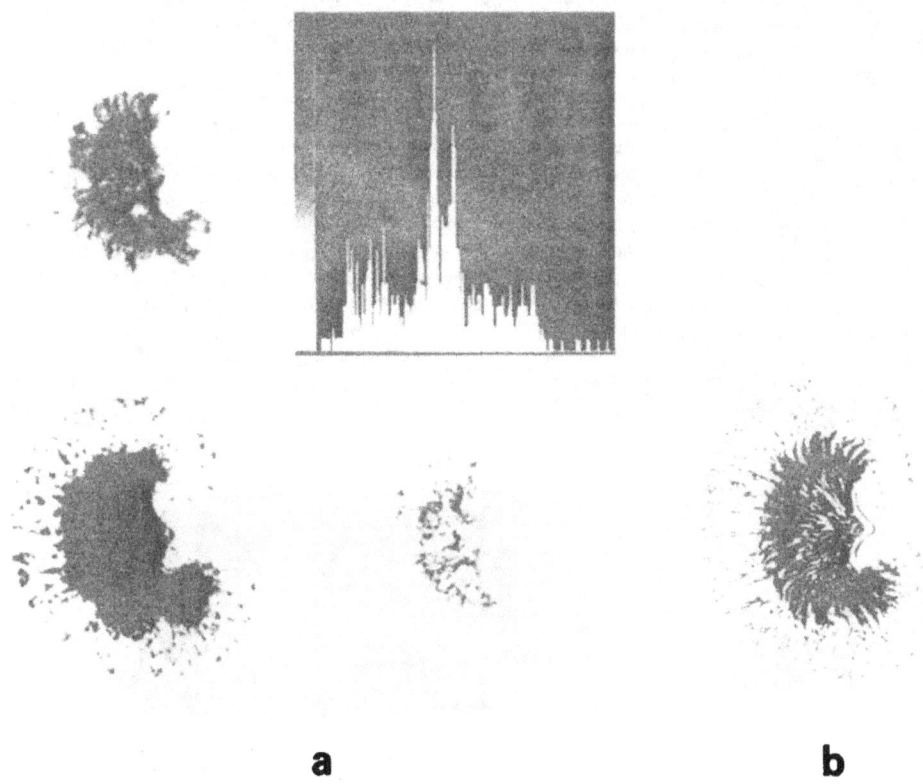

a　　　　　　　**b**

Fig. 5. (a) Images of the tritiated rat-kidney slice obtained with a gaseous detector: 256 grey level window centred at three different levels, in order to display details of very different intensity. Details of the renal ducts of about 50 μm are visible. Gas filling Xe + 3.5% TEA. Top right is the intensity distribution along the oblique line in the top-left picture, with a screen pixel width of 40 μm in the plane of the sample. (b) Autoradiography. Image of the rat kidney slice, adjacent to the preceding one, to which a photographic emulsion was apposed during 3 months. From ref. [6].

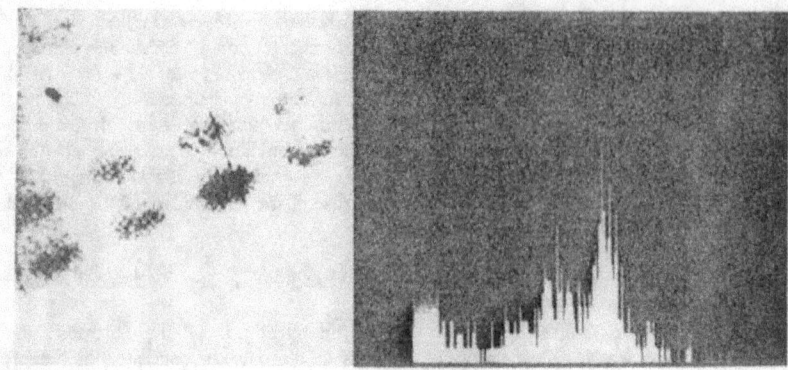

Fig. 6. Image of portion of a chromatography gel, with compounds labelled with ^{32}P, with the intensity distribution along the line. Gas filling Ar + 2.5% TEA. Screen pixels 40 μm. Detector structure of fig. 1a.

filled with an Ar + TEA (2%) gas mixture and by means of the Camera B system.

We have tested a chromatography gel with compounds labelled by ^{32}P placed against the grid (see fig. 1a). The effect of the parallel gap multiplying preferentially the ionization electrons liberated in the gas close to the entrance point is visible on the resulting image (fig. 6); this shows an accuracy of about 0.5 mm whereas the gap thickness was 6 mm and the range of the β-particles emitted by this isotope is much larger. The algorithm used for the reconstruction of the entrance points, from the light spots measurement, was the same as in the ^{3}H experiment, with a small modification, since the structure of the events is very different, the spots being elongated because of the long path of the electrons in the amplifying gap. This illustrates the potential of the parallel-plate detector, with such an imaging device, for the determination of the entrance point of the β-particles. We limited the calculation of

the centroid to 9 × 9 pixels, centred around the pixel of maximum intensity. In principle, a measurement of this pixel alone should allow the optimal precision, but since it is subject to various sources of fluctuations, we found that the accuracy was best when we took the centroid of the pixels around it.

The compound labelled with emitters such as ^{32}P and ^{35}S can be separated from the chamber working volume by a thin plastic foil, transparent to these energetic β-rays. This avoids damaging the sample by the organic compounds in the gas mixture or by accidental sparks; it avoids the chamber pollution by products evaporating from the gel and facilitates the change of the gel under test. The separating foil should serve as the entrance window as well as the cathode plane, in order to exploit the paralle-plate amplification advantage of favouring the detection of the ionization electrons close to the cathode. We have tested this type of chamber configuration (fig. 1c), using aluminized

Fig. 7. Image of an electrophoresis gel with compounds labelled with ^{35}S. Portion of a two-dimensional gel, with the intensity distribution along the oblique line. Gas filling Ar + 2.5% TEA. Detector structure of fig. 1c. From ref. [6].

Mylar of 6 μm thickness as the cathode and a ^{35}S-doped bidimensional gel placed in contact with the foil. The gap thickness was 4.5 mm in this measurement. The resulting image of the pattern is shown in fig. 7. There was no significant degradation of the detection efficiency from the β-ray absorption in the window. The detector worked well for a few days of data taking, under conditions of a high charge gain and high light output, without damage from accidental sparks. Permanent sparking points must be avoided, for example at the edge of the electrode, because they may ultimately lead to chamber damage, through aluminium evaporation from the Mylar window. We have measured the position accuracy of ^{35}S emitters by imaging linear sources of 150 μm width, separated by a distance of 2 mm. We obtained images with a width of 500 μm (FWHM) (fig. 8).

The use of a multistep chamber for radiography offers several advantages. The signal from the first stage may be used for the image intensifier gating and for a short-time increase in the light gain of the second stage by pulsing. In this case, a larger total charge gain and larger intensity of photons can be obtained. However, we expect a small degradation of the position resolution, caused by diffusion of the electron cloud in the transfer gap and its distortion during its passage through the successive meshes separating the regions of strong and weak electric field. We have tested this using the chamber shown schematically in fig. 1d and using the same ^{35}S-labelled pattern as in the measurement described above. Figs. 9a and b show the vertical slices of the central, intense spot from fig. 7, recorded in the single and multistep detector, respectively. The readout

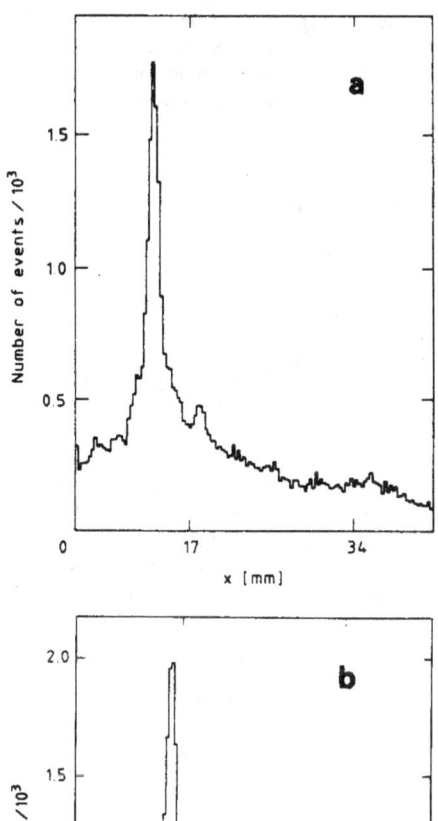

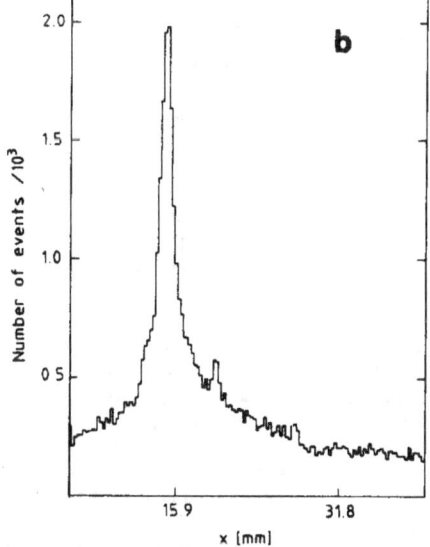

Fig. 9. Position distribution in the vertical cross section of the central intense structure seen in fig. 7, measured with two versions of the detector: (a) single parallel-plate detector of the structure shown in fig. 1c, and (b) multistep parallel-plate chamber (fig. 1d). Gas filling Ar + 2.5% TEA.

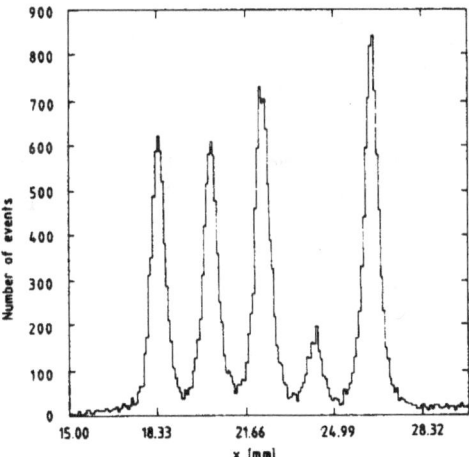

Fig. 8. Localization accuracy with ^{35}S. Image of 5 slits, 150 μm, 2 mm apart, placed on a nonuniform source. Resolution 500 μm (FWHM). Gas filling: Ar + 2.5% TEA. From ref. [6].

granularity from these measurements was 340 μm per pixel and 317 μm per pixel, respectively. The measured width of the spot is 1.71 mm (FWHM) for the single-step detector and 1.99 mm (FWHM) in the multistep case. Although the conditions of operation and the geometry of the multistep detector were not optimized, all significant spots are clearly distinguished on the detected image.

7. Discussion: the limits of accuracy

The description of the results obtained with some radioactive samples routinely used in biological applications shows that the accuracy reached at present should lead to useful practical applications. The gain in time for the imaging of radioactive samples is two orders of magnitude, compared to photographic film. The resolution is still not as good, but is sufficient for many applications, and the results are obtained directly in a digitized form, with a much better linearity.

The width of the avalanches is, physically, of the order of 0.5 mm (FWHM) in the case of a ^{3}H source. The accuracy in the position of the centroid is improved if the image is distributed over many CCD pixels. It is possible to widen the avalanches either by using a single gap with larger electrode spacing, or by using a multistep chamber. In this case one could also obtain a much larger intensity of photons. In order to achieve the best resolution, the charge gain of the first stage should be maximum.

The accuracy is limited by the physical extension of the β-track. This is the reason why, with ^{3}H, the accuracy is better with Xe than with Ar as the main gas component in the chambers. An increase in pressure could lead to improvements. We intend to investigate this further, since any significant progress in the accuracy we have currently reached may lead to new applications.

In the measurements with ^{32}P and ^{35}S we used an Ar + TEA (2%) gas mixture. In argon, the mean distance between ionizing collisions of a minimum ionizing particle is 300 μm, thus limiting the position resolution. It is not easy to find more favourable gas mixtures: they have not only to give a larger primary ionization but also to be compatible with the emission of light. Choosing a reconstruction algorithm better than the centroid of charges around the maximum may lead to some improvement in accuracy, but the one that we obtain

already with ^{35}S sources, 200 μm (rms), is close to the optimum and is convenient for most applications.

One attractive feature of the method is that it allows the user to follow in a continuous way the buildup of the image and thus to take, if necessary, action before any preset time limit. This should also lead to the use of less intense sources for the routine work in biology laboratories.

The linear response and large dynamic range of the detector should permit a better quantitative evaluation in the interpretation of complex images.

References

[1] G. Charpak and F. Sauli, Phys. Lett. B78 (1978) 523.
[2] G. Petersen, G. Charpak, G. Melchart and F. Sauli, Nucl. Instr. and Meth. 176 (1980) 239.
[3] A. Cattai, Nucl. Instr. and Meth. 215 (1983) 3.
[4] J.E. Bateman, R. Stephenson and J.F. Connolly, Nucl. Instr. and Meth. A269 (1988) 415.
[5] F. Angelini, R. Bellazzini, A. Brez, M.M. Massai and M.R. Torquati, Nucl. Instr. and Meth. A269 (1988) 430.
[6] G. Charpak, W. Dominik and N. Zaganidis, CERN-EP/88-165 (25 Nov. 1988), submitted to Proc. Nat. Acad. Sci. USA.
[7] G. Charpak, W. Dominik, J.P. Fabre, J. Gaudaen, F. Sauli and M. Suzuki, Nucl. Instr. and Meth. A269 (1988) 142.
[8] G. Charpak, Proc. Int. Symp. on Lepton and Photon Interactions at High Energy, Kyoto, 1985 (Kyoto University, 1986) p. 514.
[9] G. Charpak, W. Dominik, J.P. Fabre, J. Gaudaen, V. Peskov, F. Sauli and M. Suzuki, Proc. Nuclear Science Symp., San Francisco (1987) [IEEE Trans. Nucl. Sci. NS-35 (1988) 483].
[10] T.K. Gooch et al., Nucl. Instr. and Meth. 241 (1985) 363.
[11] D. Sauvage, A. Breskin and R. Chechik, Weizmann Inst. preprint WIS-88/8/February-PH (1988).
[12] A. Breskin, R. Chechik, Z. Fraenkel, D. Sauvage, V. Steiner, I. Tserruya, G. Charpak, W. Dominik, J.P. Fabre, J. Gaudaen, F. Sauli, M. Suzuki, P. Fischer, P. Glässel, M. Ries, A. Schön and M.J. Specht, Nucl. Instr. and Meth. A273 (1988) 798.
[13] Y. Giomataris, A. Gougas, W. Dominik, G. Charpak, F. Sauli and N. Zaganidis, preprint CERN-EP/88-96, presented at the Conf. on Advanced Technology and Particle Physics, Como (1988).
[14] G. Charpak, W. Dominik, J.C. Santiard, F. Sauli and N. Solomey, Nucl. Instr. and Meth. A274 (1989) 275.

Proc. Natl. Acad. Sci. USA
Vol. 86, pp. 1741–1745, March 1989
Applied Physical Sciences

Optical imaging of the spatial distribution of β-particles emerging from surfaces

(autoradiography/gaseous detectors)

G. CHARPAK, W. DOMINIK, AND N. ZAGANIDIS

EP Division, European Organization for Nuclear Research, 1211 Geneva 23, Switzerland

Contributed by G. Charpak, November 29, 1988

ABSTRACT   The multiplication in gases of ionization electrons, by the effect of the electric fields between parallel electrodes, leads to the emission of light from the molecules excited in the avalanche process. The optical imaging of this light, with intensifiers, on charge-coupled devices permits the localization, in the gaseous volume, of the entrance points of the β-particles emitted by radioactive compounds placed close to or at the cathode electrode. Thin slices of anatomical samples labeled with ³H show detailed structures 30 μm in size. Gels carrying ³²P or ³⁵S are imaged with accuracies of the order of 0.5 mm (full width at half maximum). In comparison with photographic emulsion, the gain in time for data taking is close to a factor of 100, with the advantage of linearity and wider dynamic range in the intensity measurement and a greatly improved signal-to-noise ratio.

Many fields of research use chemicals labeled with electron-emitting radioactive elements. An important step in the measurement is often to determine the intensity distribution of the radioactive compounds localized at various spots on a surface. Photographic emulsion apposed to the surface is the most common technique for the imaging of radioactive elements because of its simplicity, low cost, and excellent spatial resolution. However, several drawbacks hamper this method: the low sensitivity leads sometimes to exposure times of several months; the nonlinearity of the response and saturation effects make it difficult to obtain accurate quantitative measurements of intensities. This has led to various attempts to find a replacement for photographic emulsions.

Multiwire proportional chambers, routinely used in high-energy physics, have been exploited for the detection and localization of electrons emitted by radioactive elements and have given birth to several commercial instruments. They rarely permit accuracies of better than 1–3 mm [full width at half maximum (FWHM)] to be reached because of two main factors: the range of β-particles in gases can be several centimeters at atmospheric pressure, and the anode wire spacing is often of the order of 2 mm for convenience in the operation of the chambers, limiting the accuracy to this value.

Gaseous detectors with a different electrode structure (1) have dealt with these two factors successfully (2). If the detecting region is limited by parallel planes of electrodes, the multiplication of electrons in the gas under the effect of an electric field is an exponential function of the distance x of the electron to the anode plane:

$$M = \exp(\alpha x),$$

where $1/\alpha$ is the mean free path for an electron drifting in the electric field to experience an ionizing collision; α is a function of the nature of the gas filling, of the electric field, and of the pressure.

Under these conditions, electrons nearest to the cathode give rise to the largest multiplication (Fig. 1). A radioactive sample emitting electrons is placed near or against the cathode plane of such a detector. Among all avalanches initiated by ionization electrons along the track of a β-particle, the one that has the highest intensity of charges is that liberated by the electron freed nearest to the entrance point of the ionizing particle. The localization of this maximum will thus, to a great extent, permit the elimination of the error introduced by the large range of the electron and the finite thickness of the detector.

An anode plane can be free of the limitations connected with the anode wire spacing of multiwire chambers and can be made of a continuous conducting surface or of a metallic mesh with very small wire spacing. There exist several electronic methods that permit the localization of the maximum of the charge collected at the anode plane.

The maximum gain M achieved under conditions of proportional amplification in a single gap can be of the order of 10^5 or even higher. It is limited by sparking produced by occasional highly ionizing events or by irregularities in the electrodes. For this reason, it is usually very impractical to work with gains larger than 10^4, which can be sufficient, in some cases, for autoradiography. For instance, with ³H in a gas such as argon at atmospheric pressure, where the average β-particle of 6 keV (1 eV = 1.602×10^{-19} J) will produce about 200 initial electrons, localized within a few hundred micrometers, a gain of 10^3 is quite sufficient for each detection and localization. With radionuclides, such as ³²P, emitting many minimum-ionizing electrons, the number of primary ionizations in argon, for instance, is close to 30 per cm; if an accuracy of a few hundred micrometers is desirable, one has to be able to detect the signal initiated by a small number of electrons, and a larger gain may then be required.

This can easily be achieved by using a succession of gaps separated by transfer spaces, where electron swarms drift without multiplication (1). The electrodes defining the gaps are made of grids permitting the transfer of electrons from gap to gap. This considerably reduces the secondary effects due to photons or ions, which are responsible for sparking. Large gains, $M > 10^6$, are achieved, permitting the detection of single electrons produced, for instance, by vacuum ultraviolet radiation.

In the last few years, several imaging detectors based on such multistep structures have been investigated (2, 3) and accuracies of the order of 500 μm are reported for ³²P and ³⁵S. Similar results are obtained with a single-step multiplying gap (4).

In this work we report on a different approach to the problem of the measurement of the charge distribution of the avalanches produced by ionizing particles in a gap between

Abbreviations: CCD, charge-coupled device; FWHM, full width at half maximum.

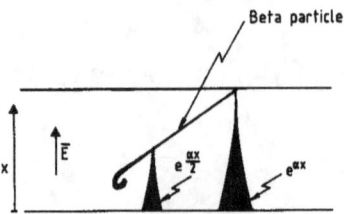

FIG. 1. Localization of the radioactive emitters with avalanches between parallel electrodes. A β-particle entering the gap produces a trail of ionization electrons in the gas filling. If the electric field E is large enough for ionizing collisions of the drifting electrons, the multiplication is an exponential function of the distance x to the anode. Determining the position of the maximum of the charge distribution at the anode gives the coordinates of the electron freed closest to the entrance point of the β-particle.

parallel electrodes. It has allowed us to obtain images of ^{3}H distribution in a biological sample, showing a clear picture of 30-μm anatomical details, and seems most appropriate for the information retrieval from amplifying gaps between parallel electrodes.

EXPERIMENTAL METHOD

Readout of Avalanche Position from the Light Emitted by Excited Atoms. When electrons produce ionization in gases, they also experience collisions resulting in the emission of light. The wavelengths of the emitted photons and their intensity vary over a considerable range, with the nature of the gas, the electric field, and the pressure. It has been found that with an appropriate choice of gases it is possible to obtain an emission of UV or visible photons suitable for the optical readout of the information on the charge distribution of avalanches between parallel electrodes (5). There exists some choice in the nature of the gas mixtures. In the work reported here, the results were obtained with triethylamine added to argon or xenon. The avalanches result in the emission of a broad photon line with the wavelength centered at about 280 nm and an intensity of about one photon per electron in the avalanche.

Several versions of this imaging detector have been studied. For ^{3}H the range of the β-particles is so small that it is impractical to have a window between the chamber and the source. The sample is made conductive by evaporating on it a thin ($\approx$200 Å) layer of gold. The sample holder is embedded in a metallic frame, which constitutes the cathode of the amplifying gap (Fig. 2A).

For ^{32}P and ^{35}S, emitting electrons with a maximum energy of 1.70 MeV and 0.168 MeV, respectively, it is more convenient to have as cathode a thin window with the radioactive sample placed against it (Fig. 2B).

The anode is a mesh made of 50-μm interwoven wires with a pitch of 500 μm, at a distance of 5 mm from the cathode. The gases, mixed with about 2.5% of triethylamine, are flushed through the chamber at atmospheric pressure.

The volume of the chamber is limited on the anode side by a window that is transparent to the light emitted by the excited molecules. This window can be made of quartz or Aclar (polychlorotrifluoroethylene). The anode has a transparency of about 80%. Since the total lateral spread of the avalanches, 0.5 mm (FWHM), is larger than the pitch of the mesh, the images do not exhibit the structure of this mesh, and the centroid of the light distribution is apparently not affected by it, at our level of accuracy. We have studied the images obtained with single gaps, with a succession of two gaps, and also with a thin plastic scintillator converting the UV light into visible light. We will only report here on the

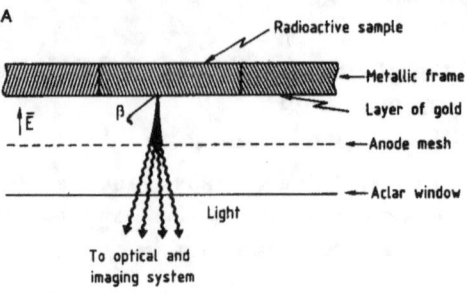

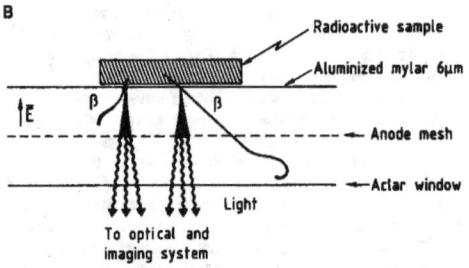

FIG. 2. Geometrical coupling of radioactive surfaces and amplifying gaps. (A) For ^{3}H the surface is covered by about 200 Å of gold and is part of the cathode. (B) For emitters of more penetrating β-particles, such as ^{32}P or ^{35}S, the radioactive samples are separated from the gap by a thin conductive electrode. The gas fillings in the gaps are argon or xenon, mixed with triethylamine vapors, emitting a photon spectrum centered at 280 nm. The light is transmitted through lenses to an image intensifier and CCDs read out by a computer. The avalanches produced against the anode mesh emit light, which is imaged through Aclar windows that are transparent to UV light.

results obtained with single gaps, since they display all the properties that make this approach of interest for most potential users, while the other methods introduce only a small reduction of accuracy with some practical advantages.

A reflective type of UV objective 90 mm f/1.1, placed at 50 cm from the last transparent mesh electrode of the chamber, focuses the light from the avalanches on the photocathode of an image intensifier coupled to a charge-coupled device (CCD) (Thomson 7852 CDA-80), which is a matrix of 144 × 208 pixels with a total area of 4.32 × 5.82 mm^2. Taking into account the optical parameters, each pixel of the CCD corresponds to a pixel on the surface of the sample, of 180 μm × 180 μm. The signal stored in each pixel of the CCD is transmitted to an IBM-PC/AT computer, with an accuracy of 8 bits. This is one limit to the dynamic range for a given amplification in the gas, which can be varied by changing the high voltage applied on the electrodes. With the CCD and our associated electronics, a maximum of 50 images per sec can be transferred to the computer. A photomultiplier placed close to the lens detects a light signal at every event, and the signal from it can be used for gating purposes rejecting, for instance, events producing occasional sparking, which occurs at a rate of 10 per hr.

Imaging of the Spatial Distribution of β-Particle Emitters. *Imaging of ^{3}H.* For every event the pixel distribution is taken into account for the calculation of the position of the electron that has initiated the avalanche. For the β-particles emitted by ^{3}H we have chosen xenon as a filling gas to minimize their range, estimated to be close to 100 μm for the average electrons. With the width of each avalanche being of the order of 0.5 mm (FWHM), the image covers several pixels of the CCD. Taking the center of gravity of the light spots gives an image that displays the quality of the method. Fig. 3*A* shows the results obtained with slices, 20 μm thick, of the kidney of a rat in 40 hr, recording 2.5 events per sec. We can at present, because of the reduction of the dead time caused by the speed of the data-acquisition system, bring this time to 20 hr. Fig. 3*B* is a photograph of an autoradiogram of the adjacent slice of the kidney obtained by apposing a ^{3}H-sensitive film on it during 3 months. It displays details of the distribution of the ^{3}H-labeled compound in the collecting

ducts, which are between 30 μm and 50 μm wide, but contains very little information on the relative intensities because of saturation, which is also responsible for the broadening of the image of the tubules. The image given by our detector can easily be analyzed for quantitative information, and Fig. 3*A* also shows the intensity distribution along the line crossing the image. The accuracy is obviously sufficient for observing details as small as the tubules. It is smaller than the β-particle range, since we measure the position of the centroid of the track. Fig. 3*C* shows a slice of

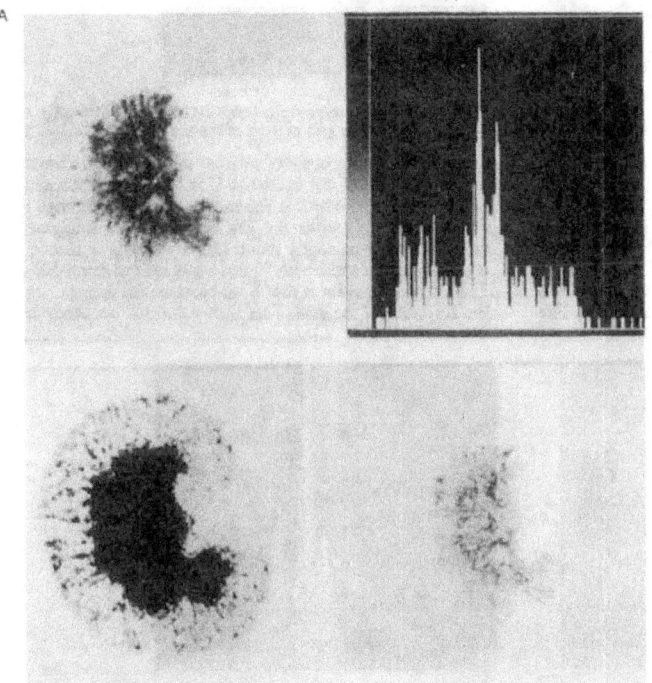

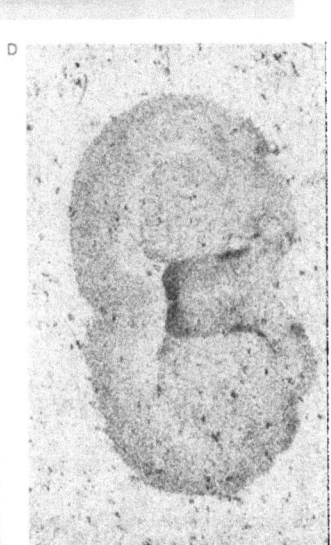

FIG. 3. Images of samples with compounds labeled with ^{3}H. (*A*) Image obtained with the gaseous detector: 256 gray level window, centered at three levels (images at the left and lower right), showing a slice from a ^{3}H-labeled rat kidney. The gas filling was Xe plus 2.5% triethylamine. Details of the renal ducts of about 50 μm are visible. This image was obtained in 20 hr (see text). The intensity distribution along the oblique line in the image at the upper left is shown at the upper right, with a pixel width of 40 μm in the plane of the sample. (*B*) Autoradiography. Image of the rat kidney slice adjacent to the preceding one to which a photographic emulsion was apposed for 3 months. (*C*) Slice from a ^{3}H-labeled rat brain under the same conditions as in *A*, but the activity was 7 times less. (*D*) Autoradiography. Image of the rat brain slice adjacent to the preceding one under the same conditions as in *B*.

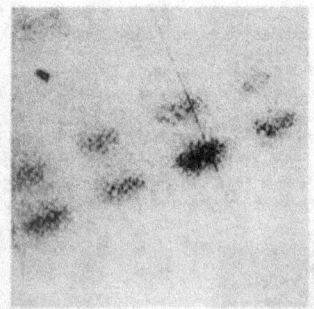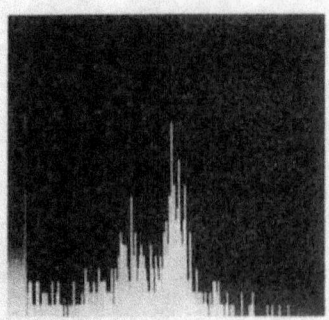

FIG. 4.   Image of chromatography gel with compounds labeled with [32]P. Portion of a chromatography gel with the intensity distribution along a line. The gas filling was Ar plus 2.5% triethylamine; pixels were 40 μm. The structure of the grid in front of the sample is visible (see text).

rat brain with the ^{3}H-labeled compound localized at the septum. In Fig. 3D an autoradiogram of the adjacent slice of brain, obtained in the same conditions as Fig. 3B, is shown for comparison. The same gain in data-taking time is observed as with the kidney, with respect to the use of photographic emulsion.

Imaging of [32]P. We have tested the detector with a chromatography gel with compounds labeled with [32]P. The

distribution of radioactivity consists of lines that are too thick to test the limit of our accuracy (Fig. 4). The results show a width of ≈1 mm, which is the same as the one obtained with autoradiography, while the gap thickness is 5 mm and the range of the β-particles much larger. We have placed the source behind a mesh with 50-μm wires spaced every 500 μm. The interesting point is that if we blow up the images, we see the structure of the grid. This shows that the accuracy is not

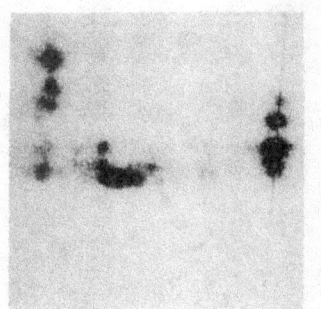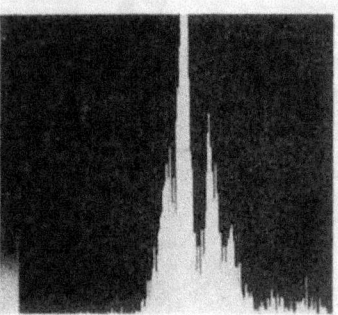

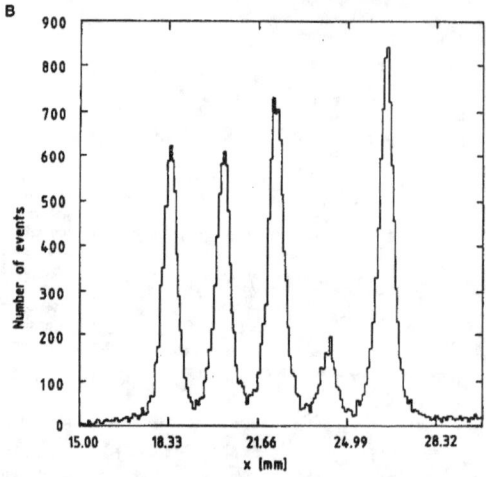

FIG. 5.   (A) Image of an electrophoresis gel with compounds labeled with [35]S. A portion of a two-dimensional gel (*Left*) with the intensity distribution along the oblique line (*Right*) is shown. (B) Localization accuracy with [35]S. Image of five slits, 150 μm, 2 mm apart, placed on a non-uniform ^{35}S source is shown at a resolution of 500 μm (FWHM). The gas filling was Ar plus 2.5% triethylamine.

limited by our readout system. In cases where it is limited by the extension of the radioactive sources, it is possible to image larger surfaces by simply adjusting the optics.

Imaging of ^{35}S. For ^{35}S we used samples of a two-dimensional electrophoresis gel. We have been able to compare our results with an autoradiogram obtained in 15 days. In a period 30 times shorter, we obtained better quantitative data, as is shown in Fig. 5A. We have measured the positioning accuracy of ^{35}S emitters by imaging linear sources of 150-μm width, separated by a distance of 2 mm. We obtained images with a width of 500 μm (FWHM) (Fig. 5B).

DISCUSSION

The optical readout of avalanches in gases initiated by electrons liberated by vacuum ultraviolet Cherenkov photons or charged particles will certainly be the subject of more active research and development in high-energy physics. The discovery that adding vapors to noble gases yields a copious intensity of light (6, *) in avalanches has allowed this approach to get started. It also offers prospects of promising applications in all the domains where autoradiography techniques are used for quantitative measurement of the spatial distribution of radioactive compounds carried by biological samples or gels.

The advantage of using the emission of light instead of the electric pulses induced in avalanches stems from several factors. (*i*) A modern technique widely used in the video industry, the CCD, provides us with hundreds of thousands of pixels at low cost. (*ii*) The gaseous detectors made of parallel planes of electrodes permit, to a large extent, the elimination of the drawback of the large ranges of the

β-particle in the gases. (*iii*) The research on gas mixtures emitting light of a convenient wavelength has resulted in several options permitting the photography of large surfaces on a single CCD, from great distances. Detectors with a surface of 20 × 20 cm², capable of imaging many small samples at the same time, are now feasible. (*iv*) The possibility of having an efficiency of 100% for every electron liberated in the gas volume and the low noise make this instrument particularly suitable for the imaging of low-intensity radioactive samples. The shape of the image of a β-particle is not circular but elliptical and depends on its range: it is almost circular for ^3H and most elongated for ^{32}P, with a peak at the beginning of the track. We have used only the barycenter of the pixels for the localization of the β-particles from ^3H. For ^{32}P and ^{35}S we have calculated the barycenter of a predetermined number of pixels close to the maximum.

We are indebted to Professors B. Roques from the Université René Descartes in Paris and J. J. Dreifuss from the Centre Médical Universitaire in Geneva for inspiring discussions on the problems of autoradiography. We thank Dr. E. Tribollet from the Centre Médical Universitaire in Geneva for her guidance in the choice of biological samples, and Dr. D. Townsend and Mr. A. Geissbühler from the Hôpital Cantonal in Geneva for their help in the image analysis.

1. Charpak, G. & Sauli, F. (1978) *Phys. Lett. B* **78**, 523–528.
2. Petersen, G., Charpak, G., Melchart, G. & Sauli, F. (1980) *Nucl. Instrum. Methods* **176**, 239–244.
3. Bateman, J. E., Stephenson, R. & Connolly, J. F. (1988) *Nucl. Instrum. Methods* **A269**, 415–424.
4. Angelini, F., Bellazini, R., Brez, A., Massai, M. M. & Torquati, M. R. (1988) *Nucl. Instrum. Methods* **A269**, 430–435.
5. Charpak, G., Dominik, W., Fabre, J. P., Gaudaen, J., Sauli, F. & Suzuki, M. (1988) *Nucl. Instrum. Methods* **A269**, 142–148.
6. Gooch, T. K., Gilmore, R. S., Jeffery, D. R. N., Kwan, W. L., Llewellyn, T. J., McArthur, I. C., Malos, J. & Tapper, R. J. (1985) *Nucl. Instrum. Methods* **241**, 363–374.

*Charpak, G., International Symposium on Lepton and Photon Interactions at High Energy, 1985, Kyoto, Japan, p. 514.

Proc. Natl. Acad. Sci. USA
Vol. 88, pp. 1466–1468, February 1991
Applied Physical Sciences

Localization and quantitation of tritiated compounds in tissue sections with a gaseous detector of β particles: Comparison with film autoradiography

(β-particle detector/quantitative autoradiography/³H-labeled ligands/vasopressin receptors)

Eliane Tribollet*, Jean Jacques Dreifuss*, Georges Charpak†, Wojciech Dominik‡,
and Nicolas Zaganidis†

*Department of Physiology, University Medical Centre, 1211 Geneva 4, Switzerland; †European Organization for Nuclear Research, Geneva, Switzerland; and ‡National Centre for Scientific Research, Paris, France

Contributed by Georges Charpak, October 30, 1990

ABSTRACT   Quantitative analysis of tritium polymer standards and of brain sections labeled with tritiated vasopressin was carried out by using a gaseous detector of β particles designed for this purpose. The gaseous detector showed major advantages compared with film autoradiography: the linearity and the large dynamic range of intensity measurements as well as the short time needed for data acquisition.

Radiolabeled molecules can be localized in tissue sections by using film autoradiographic detection. This approach is highly sensitive, offers an excellent anatomical resolution, and is suitable for quantitative analysis provided that appropriate standard samples are exposed along with the sections. However, the method is hampered by several drawbacks. First, when tritiated ligands are used, appropriate exposure times can be as long as several months. Second, the relationship between the optical density of the film and the exposure is not linear (1). Therefore, accurate quantitation in regions containing different amounts of radioisotopes may require the exposure of sections for various lengths of time.

Recently, a gaseous detector was designed that allows imaging of β-particle-emitting isotopes localized in tissue sections or on chromatography gels (2). The principle is to transform the particles emitted into light. The method offers a good anatomical resolution in a short time. For example, analysis of kidney or brain sections labeled with a tritiated compound yielded images with an accuracy of 30–50 μm within hours, while 6–10 weeks was needed to obtain sheet film images from the same sections (2). An additional attractive feature of the method is that data are quantitative and digitized.

To further evaluate the performance of the gaseous detector, we have used it to quantify (*i*) different concentrations of tritium in commercially available tritium polymer standards and (*ii*) [³H][Arg⁸]vasopressin ([³H]AVP) bound to AVP receptors present in the brain of the rat. We compare the results with data obtained with sheet film autoradiography.

MATERIALS AND METHODS

Series of 15-μm-thick coronal sections were cut through the septal area of the brain of adult rats. AVP receptors were labeled *in vitro* with 1.5 nmol of [³H]AVP (68 Ci/mmol; 1 Ci = 37 GBq) (New England Nuclear) as described (3). In some samples nonradioactive AVP or nonradioactive vasopressor agonist [Phe²,Orn⁸]vasotocin ([Phe²,Orn⁸]VT) (4) was used as well as 1.5 nmol of [³H]AVP. A tritium-sensitive film was apposed to labeled sections for 3 months along with tritiated

polymer standard samples (Amersham). After obtainment of autoradiograms, sections and standard samples were analyzed with the gaseous detector as described (2).

Briefly, samples were made conductive by coating them with a 100-Å-thin layer of gold (according to the procedure used in scanning electron microscopy) and then placed in the gas chamber of the detector. Data from standard samples were collected over 2 days, those from brain sections were collected during 24–48 hr. The characteristics of the system are such that each particle generates a light spot of 1 mm diameter, which is read out by a CCD camera. The coordinates of the center of gravity of each light spot are calculated and visualized on the screen of an Apollo DN580. The mean light intensity in a given area is measured and expressed in cps and mm². Measurements on standard samples were made on square surfaces (0.6 × 0.6 mm). For brain sections, they were carried out on 1-mm² areas included within the lateral septal nucleus, and the values obtained were converted into fmol per mg of protein.

Quantitation of the film autoradiograms was performed by means of computer-assisted image analysis (Quantimet 1900; Cambridge Instruments Ltd., Cambridge, U.K.). The mean optical density of autoradiograms of standard samples were measured on 0.75 × 2 mm surfaces. In brain sections, measurements were carried out on the same areas of the lateral septum as were analyzed with the gaseous detector. Values obtained were converted into fmol per mg of protein.

RESULTS AND DISCUSSION

The relationship between the tritium concentration of the standard samples and the optical density of the film autoradiograms that they generated is not linear (Fig. 1 *Lower*). At activities >30–40 nCi/mg, the film progressively saturates. Therefore, to perform accurate quantitation at the highest concentrations of tritium, it would be necessary to decrease the exposure time.

In contrast, the number of particles per unit time and surface obtained by the gaseous detector is a linear function of the tritium concentration (Fig. 1 *Upper*). This characteristic of the method, one of its major advantages, means that, in tissue sections containing heterogeneous densities of radiolabeled molecules, accurate determinations can be obtained in regions of both high and low density with similar times of analysis.

In Fig. 2, results obtained from brain sections are illustrated by photographs taken from film autoradiograms and from the screen where data obtained by the gaseous detector were visualized. The anatomical resolution of the detector is not as good as that of film. However, the images presented

Abbreviations: AVP, [Arg⁸]vasopressin; VT, vasotocin.

1466

Applied Physical Sciences: Tribollet *et al.*

Proc. Natl. Acad. Sci. USA 88 (1991)   1467

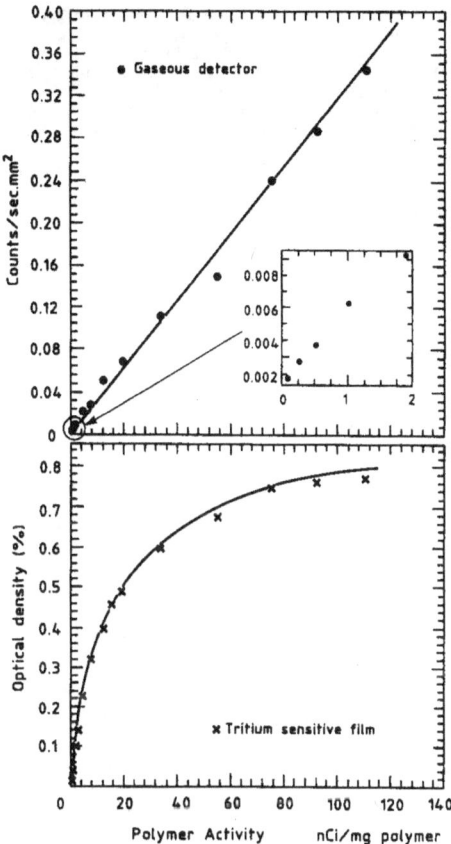

FIG. 1. Analysis of various concentrations of tritium in tritium polymer standards by the gaseous detector (●) and film autoradiography (+). Each value is the average of two measurements. Nonspecific binding was subtracted. Note the linear response of the gaseous detector at all concentrations analyzed, contrasting with the tritium-sensitive film.

here were generated from raw data, without any processing, and could therefore be considerably improved. Nevertheless, one can clearly see, with both methods, a well-delineated

Table 1. Quantitative analysis of [³H]AVP found in brain sections by film autoradiography and with the gaseous detector

Section	fmol per mg of protein	
	Film	Detector
A	275	278
C	200	235
D	160	192
E	115	120

Each value is the average of two measurements obtained as described in the text. Nonspecific binding was subtracted. Data were converted into fmol per mg of protein referring to standard sample specific activity of [³H]AVP and protein concentration (5). The sections analyzed are those shown in Fig. 2. Section A was incubated with 1.5 nmol of [³H]AVP; sections C, D, and E were incubated with the same amount of radioligand and also with 10, 20, or 50 nmol of nonradioactive [Phe², Orn⁸]VT, respectively.

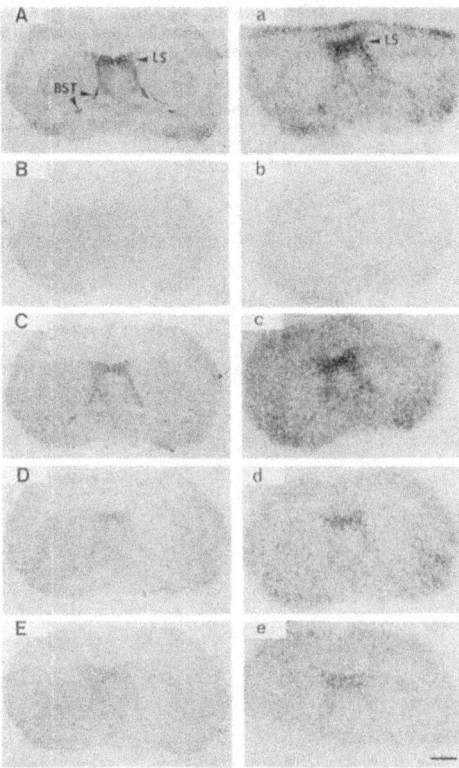

FIG. 2. Distribution of [³H]AVP in brain sections. Photographs of film autoradiograms (*A–E*) are shown alongside those of computer-generated images from the gaseous detector (*a–e*). Sections were incubated with 1.5 nmol of [³H]AVP alone (*A* and *a*), and in the presence of either 2 μmol of nonradioactive AVP (*B* and *b*) or of 10 nmol (*C* and *c*), 50 nmol (*D* and *d*), or 150 nmol (*E* and *e*) of nonradioactive [Phe², Orn⁸]VT. BST, bed nucleus of the stria terminalis; LS, lateral septal nucleus.

septal labeling and a decrease in intensity proportional to the concentration of V_1 agonist added. The amounts of [³H]AVP binding estimated by quantitative analysis of film autoradiograms were similar to those calculated from data recorded by the gaseous detector (Table 1). (The gaseous detector described in this report was built by Biospace Instruments, Paris.) In this particular experiment, the range of radioactivities explored was within the linear range of the film. Therefore, the main advantage of the direct measurement compared with film autoradiography was the considerable gain of time: 2 days as against 3 months.

In summary, a newly designed detector of β rays has proved to be useful for quantitative analysis of tritiated compounds in tissue sections. Advantages of this method compared with sheet film autoradiography include the linearity and large dynamic range in intensity measurements as opposed to the restricted linear range of the film and the considerable gain of time when low concentrations of tritium are to be analyzed, as occurs often in receptor detection studies.

We are grateful to R. Bouclier for his help in the construction of the instrument and to D. Townsend and A. Geissbühler for the analysis of the images at the Cantonal and University Hospital at Geneva. We wish to thank the Lord Michelham of Hellingly Foundation, the Carlos and Elsie de Reuter Foundation, and the Swiss National Foundation for Scientific Research for financial support in the development of the gaseous detector.

1. Unnerstall, J. R., Niehoff, D. L., Kuhar, M. J. & Palacios, J. M. (1982) *J. Neurosci. Methods* **6**, 59–73.
2. Charpak, G., Dominik, W. & Zaganidis, N. (1989) *Proc. Natl. Acad. Sci. USA* **86**, 1741–1745.
3. Tribollet, E., Barberis, C., Jard, S., Dubois-Dauphin, M. & Dreifuss, J. J. (1988) *Brain Res.* **441**, 105–118.
4. Manning, M., Bankowski, K. & Sawyer, W. H. (1987) in *Vasopressin: Principles and Properties*, eds. Gash, D. M. & Boer, G. J. (Plenum, New York), pp. 335–365.
5. Lowry, O. H., Roberts, N. R., Leiner, K. Y., Wu, M. L., Farr, A. L. & Albers, R. W. (1954) *J. Biol. Chem.* **207**, 39–49.